THE OXFORD HANDBOOK OF

NEURONAL ION CHANNELS

THE OXFORD HANDBOOK OF

NEURONAL ION CHANNELS

Edited by

ARIN BHATTACHARJEE

OXFORD
UNIVERSITY PRESS

OXFORD
UNIVERSITY PRESS

Oxford University Press is a department of the University of Oxford. It furthers the University's objective of excellence in research, scholarship, and education by publishing worldwide. Oxford is a registered trade mark of Oxford University Press in the UK and certain other countries.

Published in the United States of America by Oxford University Press
198 Madison Avenue, New York, NY 10016, United States of America.

CIP data is on file at the Library of Congress

ISBN 978-0-19-066916-4

DOI: 10.1093/oxfordhb/9780190669164.001.0001

1 3 5 7 9 8 6 4 2

Printed by Integrated Books International, United States of America

Contents

Preface vii
Editor's Bio ix
List of Contributors xi

SECTION 1 BASIC PRINCIPLES

1. Excitable Membrane Properties of Neurons 3
   Leonard K. Kaczmarek

2. Ion Channel Permeation and Selectivity 33
   Juan J. Nogueira and Ben Corry

3. Gating of Ion Channels 64
   Rene Barro-Soria

SECTION 2 VOLTAGE-GATED CHANNELS

4. The Voltage-Dependent K^+ Channel Family 101
   Hanne B. Rasmussen and James S. Trimmer

5. Potassium Channel Mutations in Epilepsy 144
   Elizabeth E. Palmer

6. The Voltage-Dependent Sodium Channel Family 198
   Mariola Zaleska, Samantha C. Salvage, Andrew J. Thompson, Sivakumar Namadurai, Christopher L.-H. Huang, Trevor Wilkinson, Fiona S. Cusdin, and Antony P. Jackson

7. Specialized Sodium Channels in Pain Transmission 224
   Yucheng Xiao, Zifan Pei, and Theodore R. Cummins

8. Sodium Channelopathies of the Central Nervous System 257
   Paul G. DeCaen, Alfred L. George, Jr., and Christopher H. Thompson

SECTION 3 LIGAND-GATED CHANNELS

9. α-Amino-3-Hydroxy-5-Methyl-4-Isoxazolepropionic Acid and Kainate Receptors 291
G. Brent Dawe, Patricia M. G. E. Brown, and Derek Bowie

10. *N*-Methyl-D-Aspartate Receptors 343
Gary J. Iacobucci and Gabriela K. Popescu

11. Nicotinic Acetylcholine Receptors 374
Roger L. Papke

12. $GABA_A$ Receptor Physiology and Pharmacology 419
Martin Wallner, A. Kerstin Lindemeyer, and Richard W. Olsen

13. P2X Receptors 458
Annette Nicke, Thomas Grutter, and Terrance M. Egan

14. Large Conductance Potassium Channels in the Nervous System 486
Willy Carrasquel-Ursulaez, Yenisleidy Lorenzo, Felipe Echeverria, and Ramon Latorre

15. Hyperpolarization-Activated Cyclic Nucleotide–Gated Channels 545
Alessio Masi, Maria Novella Romanelli, Guido Mannaioni, and Elisabetta Cerbai

SECTION 4 OTHER CHANNELS

16. Tandem Pore Domain Potassium Channels 571
Douglas A. Bayliss

17. TRPC Channels—Insight from the *Drosophila* Light Sensitive Channels 611
Ben Katz, William L. Pak, and Baruch Minke

18. Acid-Sensing Ion Channels 646
Stefan Gründer

19. TMEM16 Ca^{2+} Activated Cl^- Channels and CLC Chloride Channels and Transporters 696
Anna Boccaccio and Michael Pusch

Index 737

Preface

It was in 1939, just before the outbreak of World War II, when Alan Hodkin and Andrew Huxley first recorded an action potential from a squid giant axon. Yet, understanding why there are varied sizes, shapes, and numbers of action potentials elicited by different neurons and how these neurons transition from one mode of firing to another remains an important research pursuit of current-day neuroscience. With the advent of molecular cloning, we now know there are hundreds of ion channel genes particularly expressed in the brains of higher animal systems. The immense diversity of firing properties of neurons within the central and peripheral nervous systems can be in large part attributable to the vast number of ion channel genes. We now appreciate that stable neuronal network responses to diverse environmental stimuli are also likely due to the availability of multiple ion channel genes. With the ever-increasing number of pharmacological tools, including naturally occurring toxins and transgenic animal models, we are starting to understand the contributions different ion channels make to neuronal firing and network responses.

The purpose of this handbook is to provide the reader with an important reference to some of the identified functions of ion channels in neurons. The introductory chapters provide the reader with an overview of the major classes of ion channels and a comprehensive view of the mechanisms of ion channel selectivity and gating. The book is then divided into three categories: voltage-dependent ion channels, ligand-gated channels, and other channels. There are chapters focused on ionotropic neurotransmitter receptors and chapters devoted to understanding the relationship between channel dysfunction and neurological disease states. The plastic nature of neurons should become apparent as examples of channel regulation via phosphorylation, trafficking, and transcriptional changes will be described.

The great number of ion channel genes, many still with unknown function, cannot be encompassed by any one handbook. There are other online resources curating a list of all the identified ion channel genes and gene families, describing tissue distribution, and tabulating basic channel properties. This handbook, on the other hand, provides an in-depth, enriching look into ion channel biology and biophysics, emphasizing how ion channels contribute to neuronal function. It was challenging to bring this collection of chapters together, especially during these uncertain times; nonetheless, the handbook will provide the reader with an important, easy go-to reference on the fundamental roles of ion channels in neurons.

I thank Len Kaczmarek for considering me to spearhead this project. I thank him for his guidance and wisdom as well. I am indebted to all the chapter contributors, for the realization of this book could not have occurred without their immense efforts. I would like to especially thank Ada Brunstein, head of reference at Oxford University Press, who supported the project from the onset and worked hard to bring the production of the handbook together.

Editor's Bio

Arin Bhattacharjee, PhD, obtained his bachelor's degree and pharmacology certificate from the University of Alberta in 1992. He completed doctoral training at the University of South Alabama with Ming Li, PhD, in 1999, and completed postdoctoral training at Yale University with Leonard Kaczmarek, PhD. In 2005, Bhattacharjee was appointed assistant professor at the University at Buffalo in 2005, and currently, he is an associate professor of pharmacology and toxicology. Bhattacharjee investigates how sodium-activated potassium channels contribute to neuronal firing in pain-sensing neurons, during inflammation and nerve injury. He has published 40 papers on ion channel biology and function.

List of Contributors

Rene Barro-Soria
University of Miami, USA

Douglas A. Bayliss
University of Virginia, USA

Arin Bhattacharjee
State University of New York at Buffalo, USA

Anna Boccaccio
Consiglio Nazionale delle Ricerche, Italy

Derek Bowie
McGill University, Canada

Patricia M. G. E. Brown
McGill University, Canada

Willy Carrasquel-Ursulaez
Universidad de Valparaíso, Chile

Elisabetta Cerbai
University of Florence, Italy

Ben Corry
Australian National University, Australia

Theodore R. Cummins
Indiana University, USA

Fiona S. Cusdin
Granta Park, UK

G. Brent Dawe
McGill University, Canada

Paul G. DeCaen
Northwestern University, USA

Felipe Echeverria
Universidad de Valparaíso, Chile

Terrance M. Egan
Saint Louis University School of Medicine, USA

Alfred L. George, Jr
Northwestern University, USA

Stefan Gründer
RWTH Aachen University, Germany

Thomas Grutter
Université de Strasbourg, France; Centre National de la Recherche Scientifique, France

Christopher L.-H. Huang
University of Cambridge, UK

Gary J. Iacobucci
State University of New York at Buffalo, USA

Antony P. Jackson
University of Cambridge, UK

Leonard K. Kaczmarek
Yale University, USA

Ben Katz
Hebrew University, Israel

Ramon Latorre
Universidad de Valparaíso, Chile

A. Kerstin Lindemeyer
University of California, Los Angeles, USA

Yenisleidy Lorenzo
Universidad de Valparaíso, Chile

Guido Mannaioni
University of Florence, Italy

Alessio Masi
University of Florence, Italy

Baruch Minke
Hebrew University, Israel

Sivakumar Namadurai
University of Cambridge, UK

Annette Nicke
Ludwig-Maximilians-Universität München, Germany

Juan J. Nogueira
Australian National University, Australia

Richard W. Olsen
University of California, Los Angeles, USA

William L. Pak
Purdue University, USA

Elizabeth E. Palmer
University of New South Wales, Australia

Roger L. Papke
University of Florida, USA

Zifan Pei
Indiana University-Purdue University, USA

Gabriela K. Popescu
State University of New York at Buffalo, USA

Michael Pusch
Consiglio Nazionale delle Ricerche, Italy

Hanne B. Rasmussen
University of Copenhagen, Denmark

Maria Novella Romanelli
University of Florence, Italy

Samantha C. Salvage
University of Cambridge, UK

Andrew J. Thompson
University of Cambridge, UK

Christopher H. Thompson
Northwestern University, USA

James S. Trimmer
University of California, Davis, USA

Martin Wallner
University of California, Los Angeles, USA

Trevor Wilkinson
Granta Park, UK

Yucheng Xiao
Indiana University-Purdue University, USA

Mariola Zaleska
University of Cambridge, USA

SECTION 1

BASIC PRINCIPLES

CHAPTER 1

EXCITABLE MEMBRANE PROPERTIES OF NEURONS

LEONARD K. KACZMAREK

Introduction

For the first couple of decades after the discovery of the ionic basis of neuronal action potentials using the giant axon of squid (Hodgkin & Huxley, 1952), a very common question among pioneers of cellular neurophysiology was "What is the molecular identity of *the* sodium (or calcium, or potassium) channel?" Clues slowly emerged that there might exist more than one type of each of these channels. For example, the pharmacology of sodium channels was found to vary in different types of neurons (Matsuda, Yoshida, & Yonezawa, 1978). Single-channel recordings revealed the existence of at least three types of unitary calcium channel (Nowycky, Fox, & Tsien, 1985). Rapidly inactivating potassium currents that differ from the traditional Hodgkin-Huxley current, termed the *delayed rectifier*, which repolarizes action potentials in squid axons were discovered in some types of neurons as well as in oocytes, suggesting that there may be two types of potassium channel (Connor & Stevens, 1971; Hagiwara, Kusano, & Saito, 1961). All such considerations were made irrelevant with the cloning of the genes for these classes of ion channels, nine for voltage-dependent sodium channels, 10 for voltage-dependent calcium channels, and over 70 for pore-forming subunits of potassium channels. Moreover, over 30 different genes have been found that encode nonselective cation channels, which are permeable to sodium, potassium, and calcium ions in different ratios. Add to this the array of chloride channels and ionotropic neurotransmitter receptors which are cation or anion channels gated by binding of a neurotransmitter, and the number of possible genetic ways of constructing an excitable cell becomes astronomical.

One potential explanation for the existence of a very large number of channels that apparently serve similar biological functions is that these are required for the diversity of firing patterns encountered in the nervous system. Different neurons have very different electrical properties (Figure 1.1). Some neurons never fire an action potential

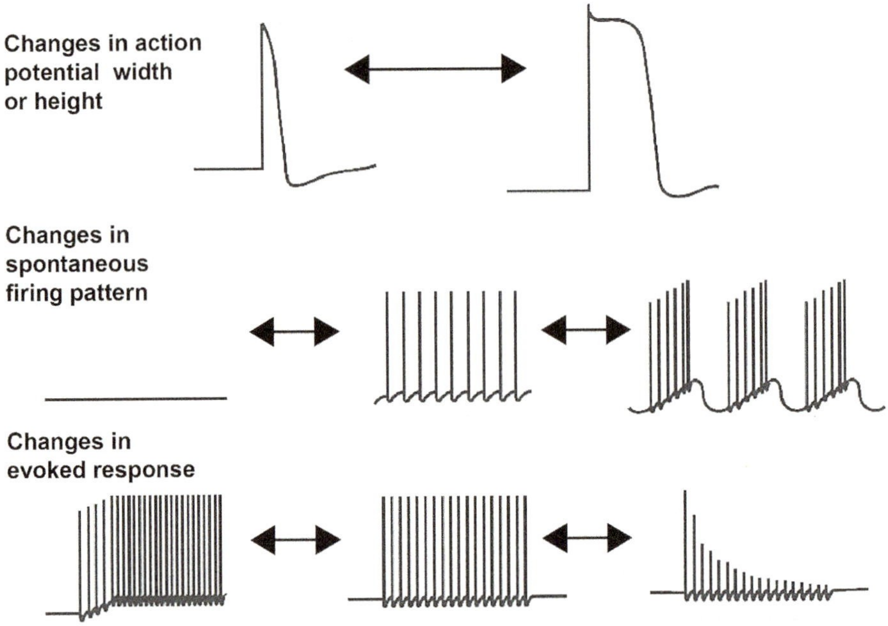

FIGURE 1.1 Schematic diagram illustrating different types of neuronal firing patterns and ways in which they may be altered by modulation of ion channels.

unless activated by a strong excitatory input, while others are capable of firing in a regular pattern or in bursts, even in the absence of any synaptic stimulation. The duration of the action potential in different neurons ranges from hundreds of microseconds to tens of milliseconds. Some neurons fire at high rates for hours in response to maintained stimulation, whereas others cease to fire after only a few seconds of sustained excitatory input. Certain neurons will not fire in response to brief stimulation but will start to fire if their excitatory drive is maintained over a prolonged period. Still other neurons are triggered to fire spontaneously by synaptic stimulation and then continue to fire after the presynaptic input has ceased.

A further rationale for the large number of ion channel genes is that the diverse firing patterns illustrated in Figure 1.1 are not fixed but are known to be modulated by sensory stimuli and by the incoming patterns of synaptic stimuli to which a neuron is exposed (Kaczmarek & Levitan, 1987; Marder, 2012). Experiments first carried out with the nervous system of invertebrates and later confirmed in mammals demonstrated that the height and width of action potentials, the pattern of spontaneous firing, and the patterns evoked by an incoming stimulus can also be altered by synaptic and hormonal events. These changes are clearly linked to changes in animal behaviors. The onset of reproductive and feeding behaviors and simple forms of learning are known to be caused by such rapid changes in the intrinsic excitability of neurons that control these behaviors. Many of these changes can now be attributed to posttranslational modification of ion channels, such as phosphorylation of their regulatory domains. Because there are myriad types of neurons in the nervous system and each type of neuron may respond in a different way to stimulation, this may require a very large set of ion channels, each of which is modulated in a slightly different way by posttranslational mechanisms.

Another potential explanation for the existence of the large assortment of ion channel genes has come from more recent findings that ion channels do more than simply conduct ions. The cytoplasmic and extracellular domains of ion channels bind other cell components such as signaling molecules, molecular motors, cell adhesion molecules, and the cytoskeleton (Kaczmarek, 2006; Lee, Fakler, Kaczmarek, & Isom, 2014). Some of these interactions allow channels to be trafficked to different locations in a cell. They also allow some channels, when activated, to propagate signals to the cell that may alter processes such as cell adhesion, transcription, and RNA translation. These activities have been termed *nonconductive* functions of ion channels.

This chapter will provide a bird's-eye view of the multiple gene families that shape the electrical personality of neurons. Specifically, it will cover sodium, potassium, calcium, and nonselective channels. Of necessity, such an overview can give only a very brief outline of these channels, and other chapters in this volume will provide more specific details. Although ionotropic neurotransmitter receptors are bona fide ion channels, they are not normally considered to shape the intrinsic excitability of neurons; but they will also be covered in other chapters. Moreover, chloride channels can play important roles in shaping neuronal excitability, but their roles are covered elsewhere (Duran, Thompson, Xiao, & Hartzell, 2010; Ha & Cheong, 2017; Rahmati, Hoebeek, Peter, & De Zeeuw, 2018).

Sodium Channels

Voltage-dependent sodium channels shape the firing patterns of neurons in at least three different ways. Their classic role is to provide the upstroke of the action potential in mammalian neurons (Hodgkin & Huxley, 1952). They can also generate a persistent sodium current, sometimes termed I_{NaP}, that produces a persistent depolarization to drive repetitive firing (Crill, 1996). Finally, a type of sodium current termed *resurgent current* is activated during the repolarization of many types of neurons, enhancing their ability to fire at high rates (Lewis & Raman, 2014). There are nine genes for these channels, which are termed $Na_V1.1$–$Na_V1.9$. These encode the large pore-forming α subunits of the channels and have 24 transmembrane domains. These are organized into four tandem domains (domains I–IV), each of which has six transmembrane α-helical segments (termed *S1–S6*) (Catterall, Goldin, & Waxman, 2005) (Figure 1.2A). Segment S4 in each domain has a series of positively charged amino acids at three residue intervals, and it is the movement of these segments in response to changes in membrane potential across the membrane that renders the channels sensitive to transmembrane voltage. A loop between S5 and S6 in each domain provides the lining of the pore that conducts sodium ions.

Of the nine voltage-dependent sodium channels, $Na_V1.1$, $Na_V1.2$, $Na_V1.3$, and $Na_V1.6$ are expressed in neurons of the central nervous system (Table 1.1). Interestingly, the action potentials of peripheral neurons are driven primarily by a different set of channels, $Na_V1.7$, $Na_V1.8$, and $Na_V1.9$. Because many of these peripheral neurons in dorsal root ganglia are nociceptors, they have been a promising target for pharmacological treatments to relieve pain. The neurotoxin tetrodotoxin (TTX), which is found in endosymbiotic bacteria within the gut of puffer fish, is a potent blocker of many of these sodium channels (Bane, Lehane, Dikshit, O'Riordan, & Furey, 2014); and sensitivity to TTX has become a conventional component of the way that individual Na_V subunits, as well as sodium currents in neurons, are characterized (Table 1.1). In addition to the sodium channels listed in Table 1.1, there is a related channel termed *Na_X1*, which is encoded by the *SCN7A* gene. This is insensitive to voltage changes across the plasma membrane but is regulated by changes in sodium concentrations outside the cell (Noda & Hiyama, 2015). It is expressed primarily in glial cells of sensory circumventricular organs.

While many older textbooks described action potentials as "all or none" events, it has become clear that the height and width of action potentials in neurons are subject to modulation by neurotransmitters and hormones. While much of this modulation reflects modifications in calcium or potassium channels, even voltage-dependent sodium channels can be modified by phosphorylation, leading to changes in the shape of action potentials and the ability to sustain repetitive firing (Scheuer, 2011). As one example, the sodium current in pyramidal cells of the prefrontal cortex is suppressed by activation of 5-hydroxytryptamine 2 (5-HT_2) neurotransmitter receptors (Figure 1.2B). This is believed to be mediated by the phosphorylation of serine residues on the $Na_V1.2$ α subunit by protein kinase C (Carr et al., 2003). The partial inactivation of Na^+ current produced by 5-HT develops over many seconds (Figure 1.2C), increasing the threshold at which action potentials are generated in response to depolarization. This increase in threshold during exposure to 5-HT can be detected during repeated depolarizations that normally trigger repetitive firing and eventually prevents sustained firing during depolarization (Figure 1.2D,E) (Carr et al., 2003). Because sustained firing of pyramidal cells in the part of the brain is required for working memory in higher mammals (Constantinidis et al., 2018; Goldman-Rakic, 1995), neurotransmitter modulation of Na^+ channels may influence which sensory events are subject to rapid recall.

Fully functioning sodium channels contain, in addition to their major pore-forming α subunits $Na_V1.1$–$Na_V1.9$, another set of intrinsic subunits, the sodium channel β subunits, which have only a single-membrane-spanning region (Namadurai et al., 2015; O'Malley & Isom, 2015). There are four of these, $\beta1$–$\beta4$, encoded by the genes *Scn1b*–*Scn4b*. All four β subunits are found in different types of neurons, and their binding to the α subunits controls insertion into the plasma membrane as well as the voltage dependence and the kinetics of activation and inactivation of the resultant channel complex. The extracellular region of the β subunits has an immunoglobulin domain, and the $\beta1$ subunit has been shown to function as a cell adhesion molecule that can regulate axonal fasciculation and neurite outgrowth (Figure 1.2F) (O'Malley & Isom, 2015). For example, the $\beta1$ subunits promote the outgrowth of neurites from cerebellar granule neurons, and this requires Na^+ entry through its associated $Na_V1.6$ α subunit (Brackenbury et al., 2010). Loss of the $\beta1$ subunit results in the disorganization of the axons of these neurons in the cerebellum (Figure 1.2G) (Brackenbury et al., 2008).

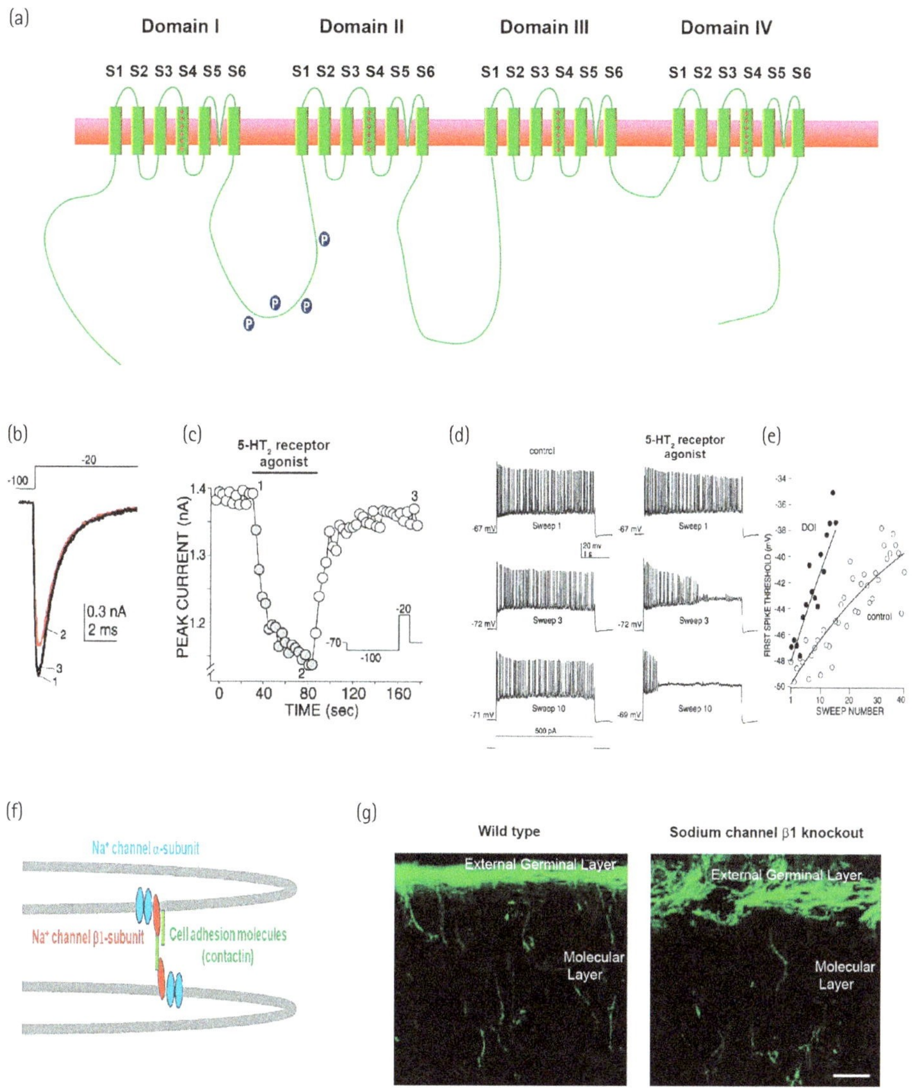

FIGURE 1.2 Voltage-dependent sodium channels. (A) Schematic diagram of the transmembrane topology of a sodium channel α subunit. (B–E) Stimulation of 5-hydroxytryptamine 2 (5-HT_2) receptors reduces Na^+ current in pyramidal cells of mouse prefrontal cortex. (B) An example of Na^+ current measured by stepping the membrane potential from –100 to –20 mV before and after (red trace) application of 2,5-dimethoxy-4-iodoamphetamine (DOI), a 5-HT_2 receptor agonist. (C) Time course of the decrease in current after application of DOI. (D) Examples of the effects of repeated depolarization of pyramidal cells on the firing of action potentials before and after application of DOI. (E) Plots of the membrane threshold for triggering an action potential with repeated stimulation as in (C) before and after DOI application. Data from Carr et al. (2003). (F) Sodium channels contribute to cell adhesion through their β subunits. Schematic diagram of Na^+ channels on the processes of two different neurons. The associated β1 subunits can mediate cell–cell adhesion either directly by β1–β1 binding or indirectly through other associated cell adhesion molecules such as contactin (O'Malley & Isom, 2015). (G) Genetic deletion of the Na^+ channel β1 subunit results in the disorganization of the axons of cerebellar granule neurons in the cerebellum. In this image, the growing axons have been labeled (green) with an antibody to a cell adhesion molecule, Transient Axonal Glycoprotein 1(TAG-1), that is restricted to these neurons (Brackenbury et al., 2008).

Table 1.1 Voltage-Dependent Sodium Channels

Sodium Channel	Gene for α Subunit	Expression	Pharmacology
$Na_V1.1$	*SCN1A*	Central neurons	TTX-sensitive
$Na_V1.2$	*SCN2A*	Central neurons	TTX-sensitive
$Na_V1.3$	*SCN3A*	Central neurons	TTX-sensitive
$Na_V1.4$	*SCN4A*	Skeletal muscle	TTX-sensitive
$Na_V1.5$	*SCN5A*	Cardiac muscle	TTX-resistant
$Na_V1.6$	*SCN8A*	Central neurons	TTX-sensitive
$Na_V1.7$	*SCN9A*	Neurons of dorsal root ganglion	TTX-sensitive
$Na_V1.8$	*SCN10A*	Neurons of dorsal root ganglion, cardiac muscle	TTX-resistant
$Na_V1.9$	*SCN11A*	Neurons, nociceptive neuron in dorsal root ganglion and trigeminal ganglion	TTX-resistant

Note. TTX = tetrodotoxin.

Calcium Channels

The opening of voltage-dependent calcium channels during neuronal firing allows calcium ions to enter the cytoplasm from the extracellular medium. This triggers numerous cellular events, including changes in gene transcription, adenosine triphosphate (ATP) production in mitochondria and neurotransmitter release in presynaptic terminals. In many neurons, calcium currents also contribute to determining the height and width of action potentials and/or to setting the membrane potential and patterns of spontaneous firing. There are 10 genes that encode these channels, which are grouped into three families: $Ca_V1.1$–$Ca_V1.4$, $Ca_V2.1$–$Ca_V2.3$, and $Ca_V3.1$–$Ca_V3.3$ (Table 1.2). As with sodium channels, these are the large pore-forming α subunits of the channels and have 24 transmembrane domains organized into four domains, comparable to those shown in Figure 1.2A for Na_V subunits (Figure 1.3A) (Catterall, 2011; Simms & Zamponi, 2014).

The members of the first family of calcium channels ($Ca_V1.1$–$Ca_V1.4$) are termed *L-type* channels and are characterized by their sensitivity to block by dihydropyridines (Nanou & Catterall, 2018). Because they become activated at positive potentials, such as those encountered during an action potential, they are also described as high-voltage-activated channels. The individual members of the second class, $Ca_V2.1$, $Ca_V2.2$, and $Ca_V2.3$, have been termed *P/Q-*, *N-*, and *R-type* calcium channels, respectively. Like the L-type channels, they activate during action potentials but are insensitive to dihydropyridines. A general distinction between the biological roles of these first two classes is that calcium entry through L-type channels is typically linked to metabolic changes including changes in gene transcription, while neurotransmitter release at

Table 1.2 Voltage-Dependent Calcium Channels

Calcium Channel	Gene for α Subunit	Expression	Characteristics
$Ca_V1.1$	*CACNA1S*	Skeletal muscle	L-type
$Ca_V1.2$	*CACNA1C*	Heart/smooth muscle	L-type
$Ca_V1.3$	*CACNA1D*	Neurons	L-type
$Ca_V1.4$	*CACNA1F*	Photoreceptors	L-type
$Ca_V2.1$	*CACNA1A*	Neurons	P/Q-type
$Ca_V2.2$	*CACNA1B*	Neurons	N-type
$Ca_V2.3$	*CACNA1E*	Neurons	R-type
$Ca_V3.1$	*CACNA1G*	Neurons	T-type
$Ca_V3.2$	*CACNA1H*	Neurons	T-type
$Ca_V3.3$	*CACNA1I*	Neurons	T-type

synaptic endings is typically triggered by P/Q-, N-, or R-type channels. This is not, however, a hard-and-fast rule because the $Ca_V1.3$ and $Ca_V1.4$ channels provide the major source of calcium entry for triggering release at some presynaptic terminals such as those of sensory hair cells in the auditory system and retinal photoreceptors, respectively (Catterall, 2011).

The members of the third class of calcium channels, $Ca_V3.1$–$Ca_V3.3$, are termed *T-type* for "transient" channels and are also described as low-voltage-activated channels. In contrast to the first two classes of calcium channels, they both activate and inactivate at negative membrane potentials close to or below the resting membrane potentials of most neurons. This allows them to shape the endogenous firing pattern of many neurons. For example, repeated bursts of action potentials followed by interburst hyperpolarizations can be shaped by a cycle of activation, inactivation, and then recovery from inactivation of T-type channels (Choi et al., 2010).

As with the voltage-dependent sodium channels, the voltage-activated calcium channels interact with auxiliary subunits that regulate their assembly, trafficking, and biophysical properties. The best characterized of these are the $\alpha 2\delta$, β, and γ subunits that bind the pore-forming voltage-activated channel α subunits (Arikkath & Campbell, 2003; Dolphin, 2013). The Ca_V1 family also interacts with a variety of calcium-binding proteins including calmodulin (CaM) and a family of CaM-like calcium-binding proteins termed CaBPs (Hardie & Lee, 2016).

Voltage-gated calcium channels are potently modulated by a variety of second messenger pathways, including phosphorylation and the direct binding of the $\beta\gamma$ subunits of G proteins (Figure 1.3A) (Catterall, 2011). In the $Ca_V2.1$ and $Ca_V2.2$ channels, which predominate at numerous presynaptic terminals, there exists a sequence, termed the *synprint site*, which allows these channels to interact directly with soluble *N*-ethylmaleimide-sensitive factor activating protein receptor (SNARE) proteins that

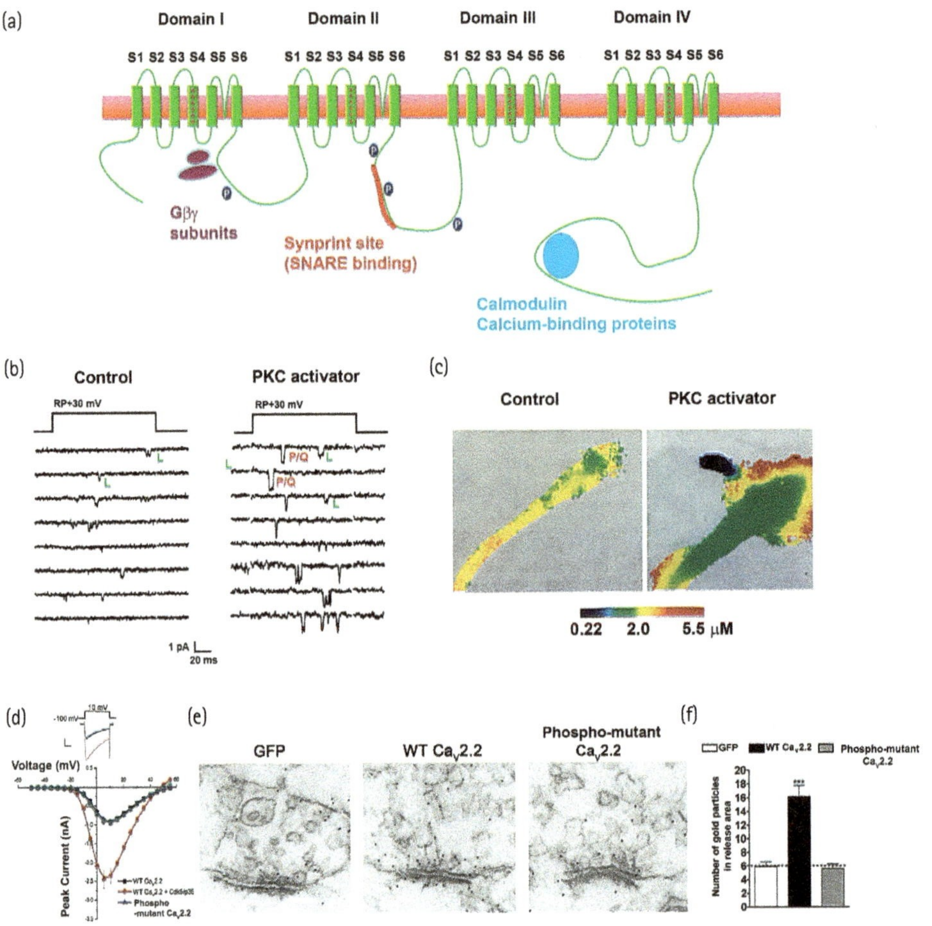

FIGURE 1.3 Modulation of voltage-dependent calcium channels. (A) Schematic diagram of the transmembrane topology of a calcium channel α subunit, indicating sites of interaction with regulatory components in different families of calcium channels. (B) Cell-attached single-channel recordings of channels in neurosecretory bag cell neurons of *Aplysia* (Strong, Fox, Tsien, & Kaczmarek, 1987). Left panels show recordings under control conditions, where only a single species of L-type channels can be detected. Right panels show recordings after application of an activator of protein kinase C (PKC), which causes the insertion of a P/Q channel (White & Kaczmarek, 1997; Zhang et al., 2008). In this condition two sizes of unitary openings can be detected, those of the control L-type channel as well as those of the larger-conductance P/Q channel. (C) Pseudocolor images of levels of intracellular Ca^{2+} in the growth cone of a bag cell neuron after stimulation of a train of action potentials. Images show the same growth cone before and after application of the PKC activator. Insertion of vesicles containing the P/Q channel causes an increase in the size of the membrane and the appearance of new sites of calcium entry (Knox, Quattrocki, Connor, & Kaczmarek, 1992). (D) Voltage clamp traces (upper panel) and current–voltage relation (lower panel) of mammalian $Ca_V2.2$ channels expressed in a cell line (Su et al., 2012). The current of the wild-type channel (WT $Ca_V2.20$ is increased by coexpression with activated Cdk5 (Cdk5/p35), but currents of a $Ca_V2.2$ mutant in which eight Cdk5 phosphorylation sites were mutated are not affected by Cdk5. (E) Immunogold labeling for $Ca_V2.2$ in cultured hippocampal neurons transduced with control green fluorescent protein (GFP) or with the WT or phosphorylation site mutant $Ca_V2.2$ channels. Accumulation of the transduced channel at active zones occurs with the WT channel but not for the mutant. (F) Quantification of colocalization of $Ca_V2.2$ in the three conditions of (E) (Su et al., 2012).

control fusion of synaptic vesicles with the plasma membrane. Thus, modulation of these channels, through mechanisms such as phosphorylation, produces strong direct effects on neurotransmitter release, as well as shaping neuronal excitability (Catterall, 2011). In some cases, activation of a kinase results in the insertion of new calcium channels into the plasma membrane from an intracellular pool. An example of this is provided in Figure 1.3A, which shows single-channel recordings of channels in neurosecretory neurons of *Aplysia*. Under control conditions, only a single species of L-type ($Ca_V1.3$) can be detected, but activation of protein kinase C causes the insertion of a P/Q ($Ca_V2.1$) channel that contributes to increased release of neuropeptides resulting from enhanced calcium influx during action potentials (Figure 1.3B). Another example is shown in Figure 1.3C, which shows that currents of mammalian $Ca_V2.2$ (the N-type channel) are substantially enhanced by activated cyclin-dependent kinase 5 (Cdk5) (Su et al., 2012). Phosphorylation of a serine residue in the distal cytoplasmic C-terminal tail of this channel by Cdk5 not only increases the probability of channel opening with depolarization but also serves to tether the channel to the presynaptic active zone, resulting in enhanced neurotransmitter release (Figure 1.3D,E).

Because Ca^{2+} ions impact a vast number of cellular functions, including neurotransmission, gene expression, and mitochondrial function, the opening of channels that are permeable to Ca^{2+} ions has consequences beyond mere regulation of excitability. Some of the protein–protein interactions of Ca^{2+} channels with cytoplasmic signaling molecules serve to promote these functions (D'Arco & Dolphin, 2012). For example, the close association of $Ca_V1.3$ channels and perhaps other L-type channels with CaM and clusters of CaM kinase II results in the rapid activation of this kinase when the channel opens. This triggers a preferred signaling pathway to the nucleus, resulting in the phosphorylation of nuclear CREB (cyclic adenosine monophosphate (AMP) response element binding protein) and changes in gene transcription (Wheeler et al., 2012). Some of the signaling functions of calcium channel proteins may, however, not require the entry of Ca^{2+} ions. For example, a fragment of the $Ca_V1.2$ channel, termed CCAT (the *calcium channel–associated transcriptional regulator*), can be transcribed independently of the full-length channel itself and acts as a transcription factor in the nucleus (Gomez-Ospina et al., 2013). Perhaps the earliest example of how calcium channels play a key role that does not require calcium flux is found in excitation–contraction coupling in skeletal muscle, where the function of $Ca_V1.1$ is to act as a voltage sensor that is coupled to a different channel, the ryanodine receptor, to release Ca^{2+} ions from internal stores (Bannister & Beam, 2013; Calderon, Bolanos, & Caputo, 2014).

Potassium Channels

The first-described and most traditional role of a potassium channel is to open during depolarization, restoring the membrane potential to a resting negative value after an action potential. It may therefore be surprising that there are over 70 different genes that encode pore-forming α subunits of potassium channels, compared to only nine or 10 for the sodium and calcium channels. Moreover, in contrast to sodium and calcium channel α subunits which are comprised of a single long polypeptide, the pore of most potassium channels is typically comprised of tetramers of α subunits, each of which corresponds to one of the four domains of a sodium channel (Figure 1.4A). A functional potassium channel is either a homomer of four identical α subunits or a heteromer containing different α subunits (Figure 1.4B). This, combined with the fact that there exist multiple splice variants for most of the α subunits, means that the number of potential potassium channel complexes is astronomical.

The potassium channel α subunits can be classified into five groups: (1) voltage-dependent channels of the K_V family, (2) calcium-activated channels (K_{Ca} channels), (3) sodium-activated channels (K_{Na} channels), (4) inwardly rectifying channels (K_{ir} channels), and (5) two pore domain channels (K_{2P} channels). The K_{Ca} channels can be further divided into two types, large-conductance ($K_{Ca}1.1$, also called big-conductance [BK] channels) and small-conductance ($K_{Ca}2$ and $K_{Ca}3$ channels, also called small-conductance [SK] or intermediate-conductance [IK] channels). A diagram illustrating the transmembrane topology of each of these groups is shown in Figure 1.4A.

Voltage-Dependent Potassium Channels

Potassium channels that open selectively when the membrane potential of a cell membrane is depolarized constitute the largest subclass. There are 40 different genes that encode the pore-forming α subunits of these channels (Table 1.3). They all have six α-helical transmembrane segments (S1–S6), with a P-domain between S5 and S6. As with the voltage-dependent sodium and calcium channels, basic amino acids (lysine or arginine) are repeated every three residues in the fourth transmembrane segment (S4), and the movement of these charged residues in response to transmembrane changes in voltage triggers channel opening.

These Kv channels can be divided into 12 groups based simply on homology. Space precludes a detailed description of the channels in each group, but there are some general rules. Kv1 channels are typical delayed rectifiers in the Hodgkin-Huxley sense, and Kv1.1 and Kv1.2 are often clustered at the juxtaparanodes adjacent to the nodes of Ranvier (Rasband & Peles, 2016). Kv2 channels arrange themselves into distinct clusters that are coupled to underlying layers of endoplasmic reticulum (Kirmiz et al., 2018). Kv3 channels are commonly expressed in neurons capable of firing at high rates (Kaczmarek & Zhang, 2017). Kv4 channels inactivate rapidly during maintained depolarization and are expressed in neuronal dendrites (Birnbaum et al., 2004; Chen et al., 2006). Some of the groups, specifically Kv5, Kv6, Kv8, and Kv9 α subunits, are "silent" subunits in that they do not form functional channels by themselves when expressed in heterologous cells. They can, however, coassemble with Kv2.1 channels to generate heteromeric channels with properties that differ from those of Kv2.1 homomeric channels alone (Enyedi & Czirjak, 2010; Hugnot et al., 1996; Richardson & Kaczmarek, 2000).

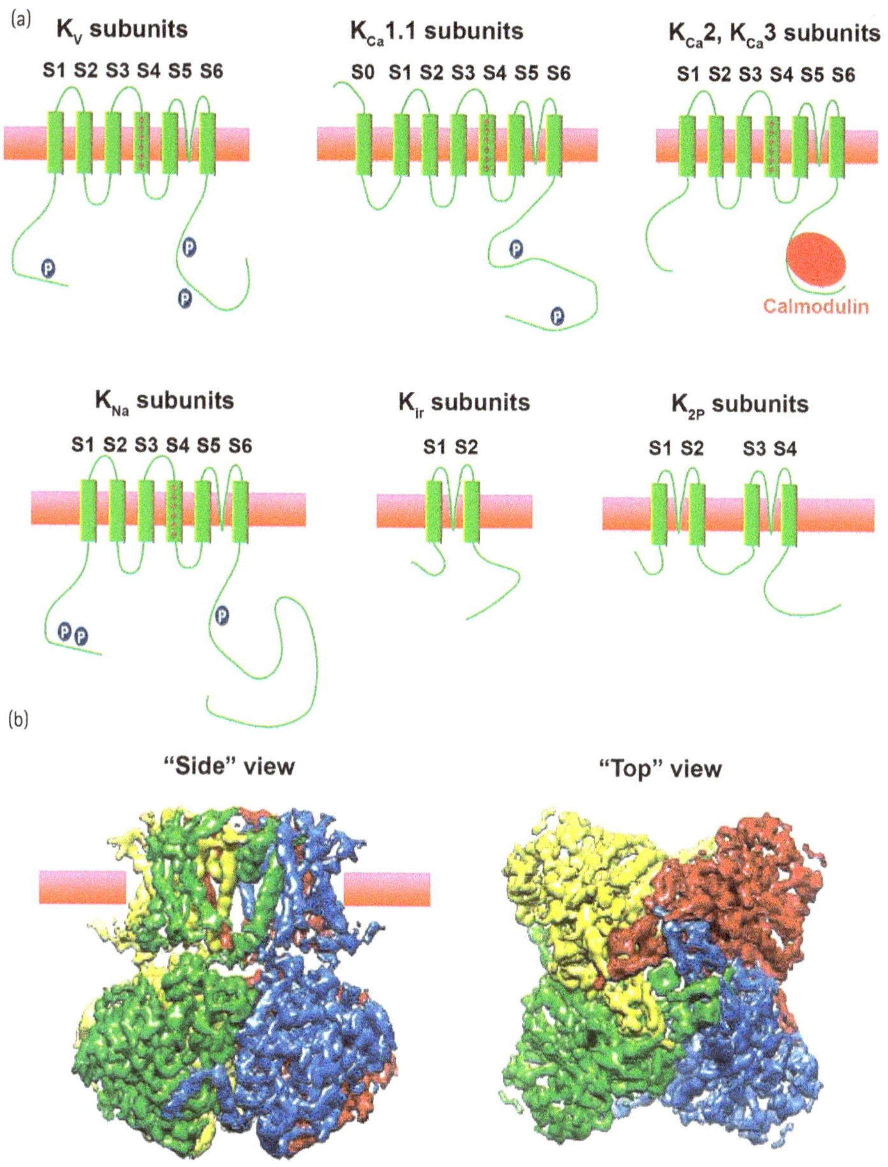

FIGURE 1.4 Potassium channels. (A) Schematic diagrams of the transmembrane topology of each of the five classes of potassium channels. (B) Images of the structure of a $K_{Na}1.1$ potassium channel obtained using cryo-electron microscopy (Hite et al., 2015). Each of the four identical α subunits that make this tetrameric channel is shown in a different color. Side and end-on views are shown.

Table 1.3 Voltage-Dependent Potassium Channels

Voltage-Dependent Potassium Channel	Genes for α Subunit	Characteristics	Alternate Names
K_V1.1, 1.2, 1.3, 1.4, 1.5, 1.6, 1.7. 1.8	*KCNA1, KCNA2, KCNA3, KCNA4, KCNA5, KCNA6, KCNA7, KCNA10*	Contribute to action potential repolarization	Shaker
K_V2.1, 2.2	*KCNB1, KCNB2*	Form clusters at neuronal somata	Shab
K_V3.1, 3.2, 3.3, 3.4	*KCNC1, KCNC2, KCNC3, KCNC4*	Expressed in fast firing neurons	Shaw
K_V4.1, 4.2, 4.3	*KCND1, KCND2, KCND3*	Rapidly inactivating A-current in dendrites	Shab
K_V5.1	*KCNF*	"Silent" modifier subunit	
K_V6.1, 6.2, 6.3, 6.4	*KCNG1, KCNG2, KCNG3, KCNG4*	"Silent" modifier subunits	
K_V7.1, 7.2, 7.3, 7.4, 7.5	*KCNQ1, KCNQ2, KCNQ3, KCNQ4, KCNQ5*	Slowly activating in response to depolarization	LQT ("long QT") channels, "M-current" channels
K_V8.1, 8.2	*KCNV1, KCNV2*	"Silent" modifier subunits	
K_V9.1, 9.2, 9.3	*KCNS1, KCNS2, KCNS3*	"Silent" modifier subunits	
K_V10.1, 10.2	*KCNH1, KCNH5*	Bind calmodulin	EAG channels ("ether-a-gogo")
K_V11.1, 11.2, 11.3	*KCNH2, KCNH6, KCNH7*	Expressed in many tissues including neurons and cardiac cells; loss of Kv11.1 causes cardiac long QT syndrome	HERG channels ("human EAG-related gene")
K_V12.1, 12.2, 12.3	*KCNH8, KCNH3, KCNH4*	Primarily expressed in neurons	ELK channels ("EAG-like")

One example of how regulation of Kv channels influences neuronal firing patterns is provided by the Kv2.1 channel, which contributes the major component of K^+ current in hippocampal neurons, as well as many other types of neurons. The amplitude and voltage dependence of this current are regulated by multiple phosphorylation sites on the Kv2.1 protein. Application of the neurotransmitter glutamate to hippocampal neurons shifts its voltage dependence of activation and inactivation and increases current amplitude (Mohapatra et al., 2009) (Figure 1.5A,B). This reduces neuronal excitability (Figure 1.5C). One of the interesting aspects of the Kv2.1 channel is that many of these channels aggregate to form discrete clusters on the soma and proximal dendrites of the neurons (Figure 1.5D). The channels in these clusters are thought to be inactive as K^+ channels. Instead, the clustered channels serve to link the plasma membrane to layers of endoplasmic reticulum that lie immediately

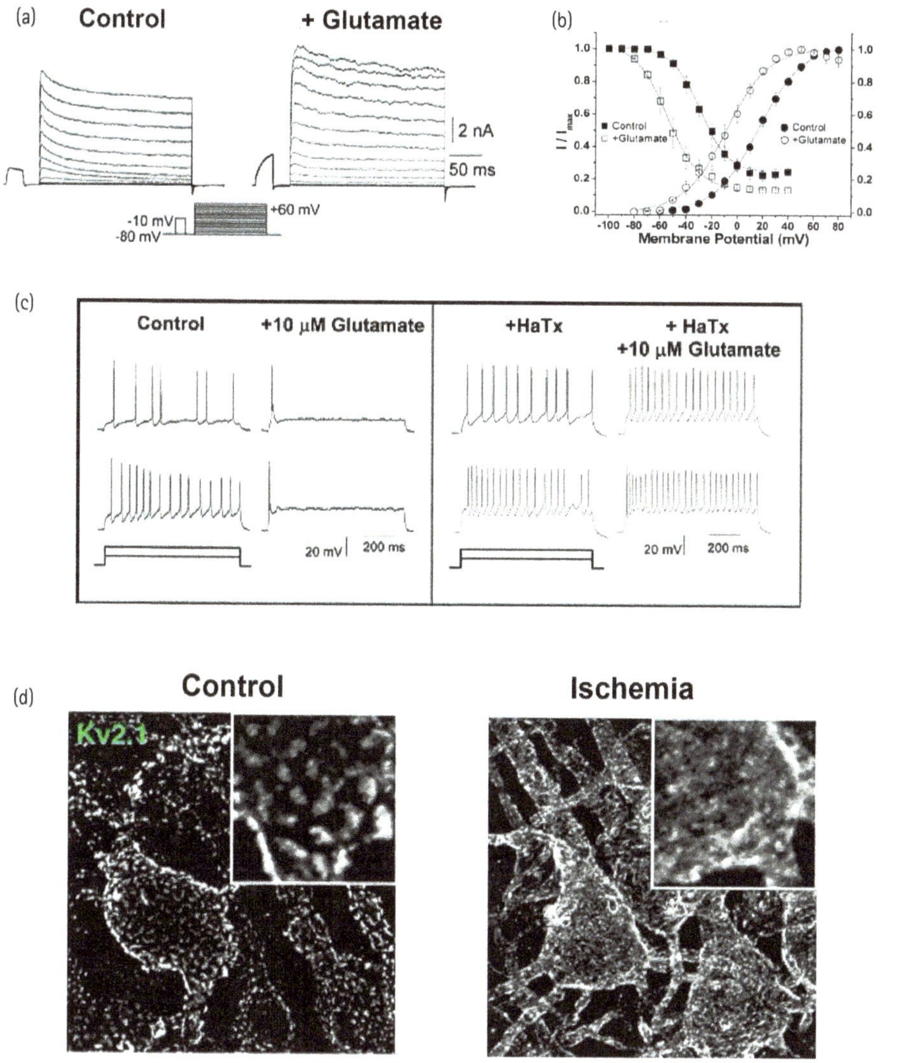

FIGURE 1.5 Modulation of the Kv2.1 channel. (A) Application of a low concentration of glutamate increases K^+ current in cultured hippocampal neurons. (B) Glutamate alters the voltage dependence of activation and inactivation of K^+ current in these neurons. (C) Left panel shows that a low concentration of glutamate suppresses the firing rate of hippocampal neurons in response to two different levels of depolarizing current. Right panel shows that the effects of glutamate on firing are abolished in the presence of hanatoxin (HaTx), an inhibitor of Kv2.1 channels. (A–C) are from Mohapatra et al. (2009). (D) Immunostaining of clusters of Kv2.1 channels on cortical neurons under control conditions and after dissolution of the clusters during acute brain ischemia (Misonou et al., 2008).

under the surface. This is mediated by Kv2.1-binding proteins that couple the two types of membrane (Johnson, Leek, & Tamkun, 2019). The clustering of the channels and their coupling to the endoplasmic reticulum are rapidly modulated in response to many biological stimuli including neurotransmitters such as glutamate as well as by conditions such as seizures or ischemia (Figure 1.5D) (Misonou, Thompson, & Cai, 2008). This is regulated by phosphorylation sites on Kv2.1 that are distinct from those that alter voltage dependence. The full biological significance of the dual biological function of these channels and the independent modulation of each function are active areas of investigation.

Calcium-Activated Potassium Channels

Calcium-activated potassium channels can be divided into two groups with quite distinct properties (Kaczmarek et al., 2017). The first group contains only one member, $K_{Ca}1.1$, which has also been termed the *BK* or *maxi-K* channel because of its large conductance measured in single-channel recordings (Table 1.4). Although it has one more transmembrane segment than K_V subunits (Figure 1.4A), it is essentially a voltage-dependent K_V channel with a large cytoplasmic C-terminal domain that allows it to be activated by Ca^{2+} ions (Latorre et al., 2017). The central action of the binding of Ca^{2+} is to alter the voltage dependence of the channel, increasing open probability at more negative potentials. This channel is very widely expressed in the central nervous system as well as many non-neuronal tissues. These include vascular smooth muscle, where activation of these channels plays a central role in the regulation of blood pressure. Many different splice isoforms can be generated from *KCNMA1*, the gene that encodes $K_{Ca}1.1$, channels, such that the properties and regulation of these channels can differ from cell to cell. One splice isoform specifically targets $K_{Ca}1.1$ channels to the inner membrane of mitochondria (Balderas, Zhang, Stefani, & Toro, 2015).

The second group of Ca^{2+}-activated K^+ channels is comprised of four members ($K_{Ca}2.1$, $K_{Ca}2.2$, $K_{Ca}2.3$, and $K_{Ca}3.1$; Table 1.4) (Adelman, Maylie, & Sah, 2012). The first three in this group are usually termed *SK* channels because of the small conductance relative to that of the BK channels. The fourth channel, $K_{Ca}3.1$, is sometimes termed the

Table 1.4 Calcium-Activated Potassium Channels

Calcium-Dependent Potassium Channels	Genes for α Subunit	Characteristics	Alternate Names
$K_{Ca}1.1$	*KCNMA1*	Voltage- and Ca^{2+}-dependent	BK, Slo1
$K_{Ca}2.1$	*KCNN1*	Voltage-independent, contribute to slow afterhyperpolarization in neurons	SK1
$K_{Ca}2.2$	*KCNN2*	Voltage-independent, contribute to slow afterhyperpolarization in neurons	SK2
$K_{Ca}2.3$	*KCNN3*	Voltage-independent, contribute to slow afterhyperpolarization in neurons	SK3
$K_{Ca}3.1$	*KCNN4*	Primarily in non-neuronal cells; major role in immune cells and red blood cells	SK4, IK, Gardos channel

IK channel because it has an "intermediate" conductance. These channels are not sensitive to changes in voltage. Moreover, in contrast to the BK channels, the pore-forming α subunits do not bind Ca^{2+} ions directly. Instead, the Ca^{2+}-binding protein CaM is an intrinsic component of each channel complex. Because of their insensitivity to transmembrane voltage, they play a very different physiological role from BK channels in neuronal firing. Specifically, they hyperpolarize and slow neuronal firing as long as intracellular Ca^{2+} levels are elevated, as occurs during prolonged neuronal firing. This usually leads to adaptation of the firing rate during maintained stimulation. The activation of SK channels and their trafficking into and out of synaptic spines in neurons of the hippocampus have been shown to regulate the synaptic plasticity that underlies some forms of learning and memory (Adelman et al., 2012).

Sodium-Activated Potassium Channels

Potassium channels activated by elevations in cytoplasmic Na^+ were first described in cardiac cells and were subsequently found to contribute to shaping the excitability of a variety of neurons (Dryer, 1994; Kaczmarek, 2013; Kaczmarek et al., 2017; Kameyama et al., 1984). There are two known pore-forming subunits of these channels, $K_{Na}1.1$ and $K_{Na}1.2$; and these have also been termed *Slack* and *Slick*, respectively (Table 1.5) (Bhattacharjee & Kaczmarek, 2005; Kaczmarek et al., 2017). The first of these, $K_{Na}1.1$, is expressed primarily in the nervous system, while $K_{Na}1.2$ is expressed much more widely. The pore-forming α subunit of $K_{Na}1.1$ binds a variety of other proteins including FMRP (fragile X mental retardation protein), Phactr-1 (phosphatase and actin regulator protein-1), and TMEM16-C (transmembrane protein 16-C) (Brown et al., 2010; Fleming et al., 2016; Huang et al., 2013). Mutations in this subunit result in early-onset epilepsies and very severe intellectual disabilities (Kim & Kaczmarek, 2014).

Inward Rectifier Potassium Channels

Inwardly rectifier potassium channels (K_{ir} channels) are simpler in structure than the channels we have considered thus far in that they are comprised of only two α-helical

Table 1.5 Sodium-Activated Potassium Channels

Sodium-Dependent Potassium Channels	Genes for α Subunit	Characteristics	Alternate Names
$K_{Na}1.1$	*KCNT1*	Primarily expressed in neurons	Slack, Slo2.2
$K_{Na}1.2$	*KCNT2*	Widespread distribution	Slick, Slo2.1

transmembrane segments with an intervening P domain (Hibino et al., 2010). Their name derives from the fact that, in contrast to all of the previous channels discussed, they allow K^+ ions to flow only when the membrane potential is negative (i.e., close to or more negative than the typical resting potential of cells). Depolarization of cells produces a suppression of K^+ ion flow through the channels. This is because at more positive potentials positively charged cytoplasmic polyamines such as spermine, also Mg^{2+} ions, are driven into the pore of the channel, blocking the flow of K^+. This block is sensitive to the concentration of extracellular K^+, and inward rectification is reduced when external K^+ is increased.

There are seven subfamilies of K_{ir} channel α subunits (Table 1.6), and, as with Kv and K_{Ca} channels, a functional channel is a homomer or heteromer of α subunits. Their functional roles are quite varied. In some tissues, such as kidney, their major role is to control bulk movements of potassium ions. In neurons and in cardiac cells, as well as in beta cells of the pancreas, their central role is to control the pattern of firing of action potentials. Particularly important in this regard is the $K_{ir}3$ subfamily of channels. The opening of these channels is controlled by G protein–coupled receptors such M2-muscarinic, A1-adenosine, and metabotropic glutamate receptors, as well as many others. When activated by neurotransmitters, these receptors trigger the release of βγ subunits from G-protein complexes, and these βγ subunits bind directly to the channels to trigger opening. Because of their voltage dependence, K_{ir} channels influence membrane properties primarily at or near the resting potential. Thus, activation of K_{ir} channels generally hyperpolarizes and inhibits neuronal firing and, in cardiac

Table 1.6 Inward Rectifier Potassium Channels

Inwardly Rectifying Potassium Channels	Genes for α Subunit	Characteristics/Location	Alternate Names
$K_{ir}1.1$	*KCNJ1*	K^+ homeostasis in kidney	ROMK
$K_{ir}2.1$, 2.2, 2.3, 2.4	*KCNJ2, KCNJ12, KCNJ4, KCNJ14*	Constitutively active in cardiac cells and neurons	IRK
$K_{ir}3.1$, 3.2, 3.3, 3.4	*KCNJ3, KCNJ6, KCNJ9, KCNJ5*	G protein–coupled channels in neurons and cardiac cells	GIRK1, GIRK2, GIRK3, GIRK4,
$K_{ir}4.1$, 4.2	*KCNJ10, KCNJ15*	Glial cells, kidney stomach, sensitive to pH	
$K_{ir}5.1$	*KCNJ16*	Nonfunctional but forms heteromers with $K_{ir}4.1$ and $K_{ir}2.1$	
$K_{ir}6.1$, 6.2	*KCNJ8, KCNJ11*	Classic ATP-sensitive channel; widespread distribution; form complexes sulfonylurea receptor	K-ATP
$K_{ir}7.1$	*KCNJ13*	Epithelial cells	

Note. ATP = adenosine triphosphate.

cells, slows heart rate. K_{ir} channels, however, have little effect on the height or width of action potentials or of firing patterns in response to stimuli that produce strong depolarization.

Some K_{ir} channels are directly regulated by cytoplasmic levels of ATP and ADP (adenosine diphosphate). Most prominently, each $K_{ir}6.1$ and $K_{ir}6.2$ subunit in a tetrameric complex binds a sulfonylurea receptor (SUR1, SUR2A, or SUR2B) to form an octameric channel complex. This channel complex opens when ATP levels are low and closes with elevations in cytoplasmic ATP levels. In some cells, such as pancreatic beta cells, the closure of the channels following an elevation of ATP levels depolarizes the cells to produce repetitive firing of action potentials, resulting in the exocytosis of secretory granules. In other cells, such as cardiac cells, the channels are closed under normal conditions but are thought to serve a protective role by hyperpolarizing the cells when ATP levels fall. In addition to these classic ATP-sensitive channels, several other K_{ir} channels, such as $K_{ir}1.1$, $K_{ir}4.0$, and $K_{ir}4.1/K_{ir}5.1$ heteromers, are modulated by cytoplasmic ATP levels (Butt & Kalsi, 2006).

Two-Pore Domain K^+ Channel Subunits

In contrast to the other K^+ channel subunits we have covered, which all form tetramers in which each subunit contributes one P domain that lines the ion-conducting pore, the K2P (two-pore domain) subunits form dimers, with each subunit contributing two P domains, each of which is flanked by a membrane-spanning α helix (Enyedi & Czirjak, 2010; Goldstein, Bockenhauer, O'Kelly, & Zilberberg, 2001; Lesage & Lazdunski, 2000). In this respect, a K_{2P} subunit resembles two K_{ir} subunits linked end to end (Figure 1.4A). Functional channels can be either homodimers or heterodimers containing two different K_{2P} subunits.

Under normal conditions, K_{2P} channels are relatively insensitive to the voltage across the plasma membrane. As result, the major function of K_{2P} channels appears to be to contribute to the "leak" or "background" K^+ conductance that sets the resting membrane potential and the input resistance of both excitable and nonexcitable cells. In mammals, there are 15 genes that encode these channels (Table 1.7). These can be further subdivided into six groups (TWIK, TREK, TASK, TALK, THIK, and TRESK) based on homology and functional similarities. The names of these groups give some insights into the initial functional or structural properties first attributed to the member channels.

TWIK reflects **T**andem P domains in a **W**eakly **I**nwardly rectifying **K**$^+$ channel
TREK is short for **T**WIK-**RE**lated **K**$^+$ channel
TASK is **T**wo-pore domain **A**cid-**S**ensitive **K**$^+$ channel
TALK is **T**wo-pore domain **AL**kaline-activated **K**$^+$ channel
THIK is **T**wo-pore domain **H**alothane **I**nhibited **K**$^+$ channel
TRESK is **T**WIK-**RE**lated **S**pinal cord **K**$^+$ channel

Table 1.7 Two Pore Domain Potassium Channels

Two-Pore Domain K^+ Channel Subunits	Genes for α Subunit	Characteristics	Alternate Names
$K_{2P}1$	*KCNK1*	Widespread distribution in neurons and other cell types, modified by SUMOylation	TWIK-1
$K_{2P}2$	*KCNK2*	Widespread distribution, modulated by G_s- and G_q-coupled receptors	TREK-1
$K_{2P}3$	*KCNK3*	Acid-sensitive, blocked by protonation of histidine residue, modulated by external pH	TASK1
$K_{2P}4$	*KCNK4*	Selectively expressed in brain, spinal cord, and retina	TRAAK (member of the TREK subfamily)
$K_{2P}5$	*KCNK5*	Brain and kidney, activated by external alkalinization	TASK-2
$K_{2P}6$	*KCNK6*	Widespread distribution	TWIK-2
$K_{2P}7$	*KCNK7*	Localized to intracellular membranes	Member of the TWIK subfamily
$K_{2P}9$	*KCNK9*	Expressed in brain, decreased by low external pH	TASK-3
$K_{2P}10$	*KCNK10*	Widely expressed, activated by volatile anesthetics	TREK-2
$K_{2P}12$	*KCNK12*	Highly expressed in kidney, "silent" subunit that can form functional channels with $K_{2P}13$	THIK-2
$K_{2P}13$	*KCNK13*	Inhibited by halothane	THIK-1
$K_{2P}15$	*KCNK15*	Uncharacterized, potentially "silent" subunit	TASK-5
$K_{2P}16$	*KCNK16*	Activated by isofluorane, nitric oxide, and increasing external pH	TALK-1
$K_{2P}17$	*KCNK17*	Activated by, nitric oxide, and increasing external pH; inhibited by chloroform	TALK-2
$K_{2P}18$	*KCNK18*	Expressed in brain, activated by cytoplasmic Ca^{2+} and volatile anesthetics, mutations associated with migraine	TRESK

Despite having been assigned by evolution the apparently humdrum task of maintaining the resting membrane potential and K^+ conductance in many cells, K_{2P} channels are subject to regulation by an extraordinarily large range of signaling pathways, with major consequences for the properties of excitable cells such as neurons. The presence of TWIK channels in the plasma membrane is regulated by their trafficking to and from internal membranes. The ability of $K_{2P}1$ channels to even function as channels is controlled by the addition/removal of SUMO (small ubiquitin-like modifier) proteins to the channel itself (Plant, Zuniga, Araki, Marks, & Goldstein, 2012). Other K_{2P} channels are regulated by protein kinases, arachidonic acid and membrane lipids, intracellular or external pH, cytoplasmic Ca^{2+} levels, G protein–coupled receptors, volatile anesthetics, and mechanical stretching of the plasma membrane (Enyedi & Czirjak, 2010). The way that some K_{2P} channels may alter excitability is further enhanced by the finding that, under certain conditions, some K_{2P} channels can alter their selectivity, becoming permeable to Na^+ ions (Ma, Zhang, Zhou, & Chen, 2012; Thomas, Plant, Wilkens, McCrossan, & Goldstein, 2008).

A "nonconducting" function has been described for the $K_{2P}2$ (TREK-1) channel (Lauritzen et al., 2005). Overexpression of this subunit in cultured neurons alters their morphology by inducing the formation of numerous new actin-rich filipodia. The ability of the channels to induce this morphological change does not require ion flow through the channels as it persists in the presence of a channel blocker. It does, however, require phosphorylation of a specific serine residue in the cytoplasmic C-terminal domain of the channel. It has been proposed that the ability of the channel to interact with the actin cytoskeleton may play a role in growth cone extension or synaptogenesis (Lauritzen et al., 2005).

Nonselective Cation Channels

The next category of ion channels that influence the intrinsic excitability of neurons are those that are nonselective between Na^+ and K^+ ions (Table 1.8). In many cases these nonselective cation channels are also permeable to Ca^{2+} ions. Thus, activation of these channels always promotes depolarization. They can be categorized into two general classes, cyclic nucleotide–regulated channels (CNG and HCN channels) and transient receptor potential channels (TRP channels).

Cyclic Nucleotide–Regulated Channels

As their name implies, the activity of cyclic nucleotide–regulated channels depends on the cytoplasmic levels of either cyclic AMP or cyclic guanosine monophosphate (cyclic GMP). Opening of these channels occurs when these nucleotides bind to a cyclic nucleotide–binding domain in the cytoplasmic C terminus of the channel α subunit.

Table 1.8 Non-Selective Cation Channels

Nonselective Cation Channels	Genes for α Subunit	Characteristics	Alternate Names
CNGA1, CNGA2, CNGA3, CNGA4	*CNGA1, CNGA2, CNGA3, CNGA4*	Transduction in photoreceptors and olfactory cells	
HCN1, HCN2, HCN3, HCN4	*HCN1, HCN2, HCN3, HCN4*	Regulation of neuronal firing and synaptic integration	H-current channels
TRPA1	*TRPA1*	Nociception, thiol modification of cytoplasmic residues	Ankyrin family of TRP channels
TRPC1, TRPC3, TRPC4, TRPC5, TRPC6, TRPC7	*TRPC1, TRPC3, TRPC4, TRPC5, TRPC6, TRPC7*	Activated by $G_{q/11}$-coupled receptors, receptor tyrosine kinases	Canonical family of TRP channels
TRPM1, TRPM2, TRPM3, TRPM4, TRPM5, TRPM6, TRPM7, TRPM8	*TRPM1, TRPM2, TRPM3, TRPM4, TRPM5, TRPM6, TRPM7, TRPM8*	Photoreception, olfactory reception, thermosensation	Melastatin family of TRP channels
TRPML1, TRPML2, TRPML3	*MCOLN1, MCOLN2, MCOLN3*	Intracellular vesicles, endosomes	Mucolipin family of TRP channels
TRPP1, TRPP2, TRPP3	*PKD2, PKD2L1, PKD2L2*	Located in primary cilia and on internal membranes	Polycystin family of TRP channels; individual members also termed PKD2, PKD2L1 and PKD2L2, or PC2, PC2L1 and PC2L2, respectively
TRPV1, TRPV2, TRPV3, TRPV4, TRPV5, TRPV6	*TRPV1, TRPV2, TRPV3, TRPV4, TRPV5, TRPV6*	Thermosensation, response to inflammation and noxious heat; calcium transport in non-neuronal cells (TRPV5, TRPV6)	Vanilloid family of TRP channels

Note. CNG = cyclic nucleotide–gated; HCN = hyperpolarization-activated cyclic nucleotide–gated; TRP = transient receptor potential.

These channels can be further subdivided into two classes: the cyclic nucleotide-gated channels (CNG channels) and the hyperpolarization-activated cyclic nucleotide–gated channels (HCN channels). All of these channels have six transmembrane segments and cytoplasmic N and C termini similar to the transmembrane topology of K_V channels (Figure 1.6A,B).

There are four CNG α subunits (CNGA1, CNGA2, CNGA3, and CNGA4), and these are expressed selectively in photoreceptors or olfactory cells, where they play a central

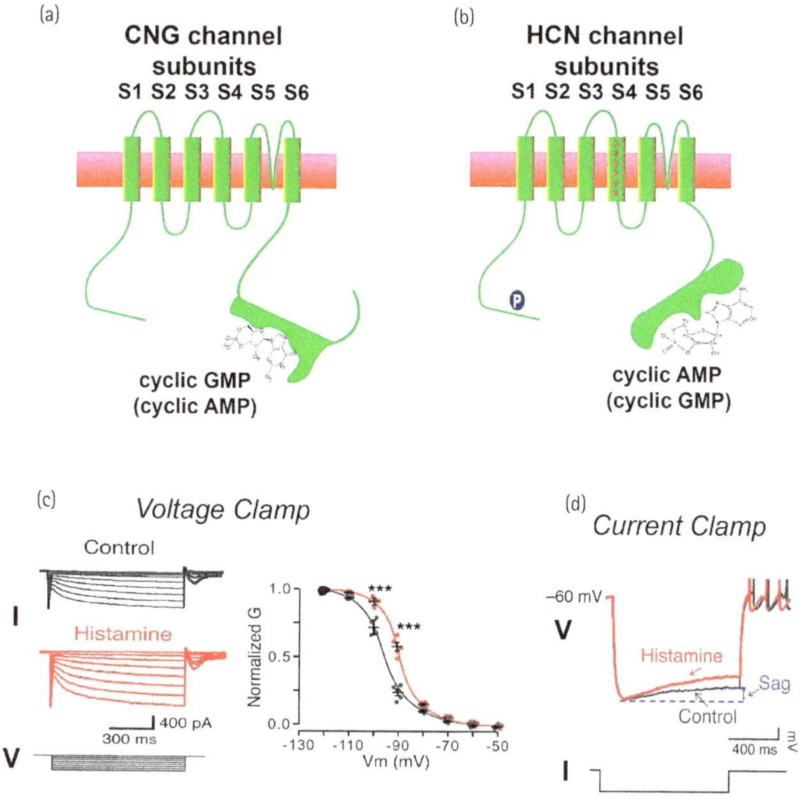

FIGURE 1.6 Cyclic nucleotide–gated (CNG) nonselective cation channels. (A) Transmembrane topology of CNG channels. (B) Transmembrane topology of hyperpolarization-activated cyclic nucleotide–gated (HCN) channels. (C) Voltage clamp traces (left) showing slow activation of HCN channels (h-current) in subthalamic neurons before and after application of histamine. Panel at right shows current–voltage relations before and after histamine. (D) Examples of the sag in membrane potential during a sustained hyperpolarizing current pulse in the same neurons before and after histamine (Zhuang et al., 2018). AMP = adenosine monophosphate; GMP = guanosine monophosphate.

role in transduction of light or odorant stimuli into changes in membrane potential (Kaupp & Seifert, 2002). Like the SK subfamily of K^+ channels, the tetrameric CNG channels are not sensitive to voltage and have no charged residues in their S4 segments (Figure 1.6A). Thus, channel opening depends directly on the cytoplasmic concentration of cyclic AMP or cyclic GMP.

In contrast to CNG channels, the four genes that encode HCN channels (HCN1, HCN2, HCN3, and HCN4) are widely expressed in neurons, while HCN4 also plays a key role in generating cardiac rhythms. HCN channels underlie I_h, or the h current, which regulates the rate of bursting in several types of neurons (He, Chen, Li, & Hu, 2014; Luthi & McCormick, 1998). HCN channels differ from CNG channels in that, in addition to the cyclic nucleotide–binding domains on their cytoplasmic C termini, they have positively charged residues in their S4 domains (Figure 1.6B) (He et al., 2014). In contrast to K_V channels, however, HCN channels close with depolarization and open when the membrane hyperpolarizes. Although, in this respect, tetrameric HCN channels resemble K_{ir} channels, the rate at which HCN channels open and close with changes in voltage is much slower than that of K_{ir} channels and occurs through a very different mechanism (changes in protein conformation following movement of the S4 segment versus channel block for the K_{ir} channels) (Lewis, Estep, & Chetkovich, 2010).

The binding of cyclic nucleotides, primarily cyclic AMP, to HCN channels shifts their voltage dependence to positive potentials in the physiological range. As a result, cyclic AMP binding can alter neuronal firing patterns. As one example, the neurotransmitter histamine elevates cyclic AMP in subthalamic neurons, altering their firing patterns. Figure 1.6C shows the effect of histamine on the amplitude and voltage dependence of inward HCN currents that develop during voltage-clamp pulses to negative potentials (Zhuang et al., 2018).

Activation of the h current has a characteristic effect on the way the membrane potential of a neuron responds to a sustained hyperpolarizing current. After the initial hyperpolarization of the membrane, the potential slowly swings toward more positive potentials, reflecting the slow activation of the HCN channels, as illustrated in Figure 1.6D. This is sometimes called a *sag* in the membrane potential. In neurons that are actively firing, increased HCN channel activity accelerates the rate of depolarization toward the action potential threshold, altering the timing of action potentials. For example, if HCN channels are activated during the slow afterhyperpolarization that often follows a burst of action potentials, this will speed the depolarization of the cells toward a second burst. In addition, modulation of HCN channels can modify other aspects of neuronal excitability, including the integration of synaptic potentials in dendrites (He et al., 2014; Luthi & McCormick, 1998).

Transient Receptor Potential Channels

The first transient receptor potential (TRP) channel to be discovered was found in the genome of the fruit fly *Drosophila*, where it is required for normal vision (Minke, 2010). In common with CNG channels, they have six α-helical transmembrane domains and are not voltage-dependent, lacking the charged residues in S4 that confer voltage sensitivity. There are six families of TRP channels, TRPA, TRPC, TRPM, TRPML, TRPP, and TRPV (Figure 1.7A, Table 1.8). TRP subunits assemble as homo- or heterotetramers to form channels that are permeant to sodium, potassium, and calcium ions (Nilius & Szallasi, 2014; Ramsey, Delling, & Clapham, 2006).

As is the case for CNG channels, the best-established role for TRP channels in the nervous system is in sensory cells, where they are required for the transduction of a variety of changes in the environment (Dhaka, Viswanath, & Patapoutian, 2006). In contrast to most of the other channels discussed here, for which activation is determined primarily by one or two factors such as voltage or Ca^{2+} concentration, many TRP

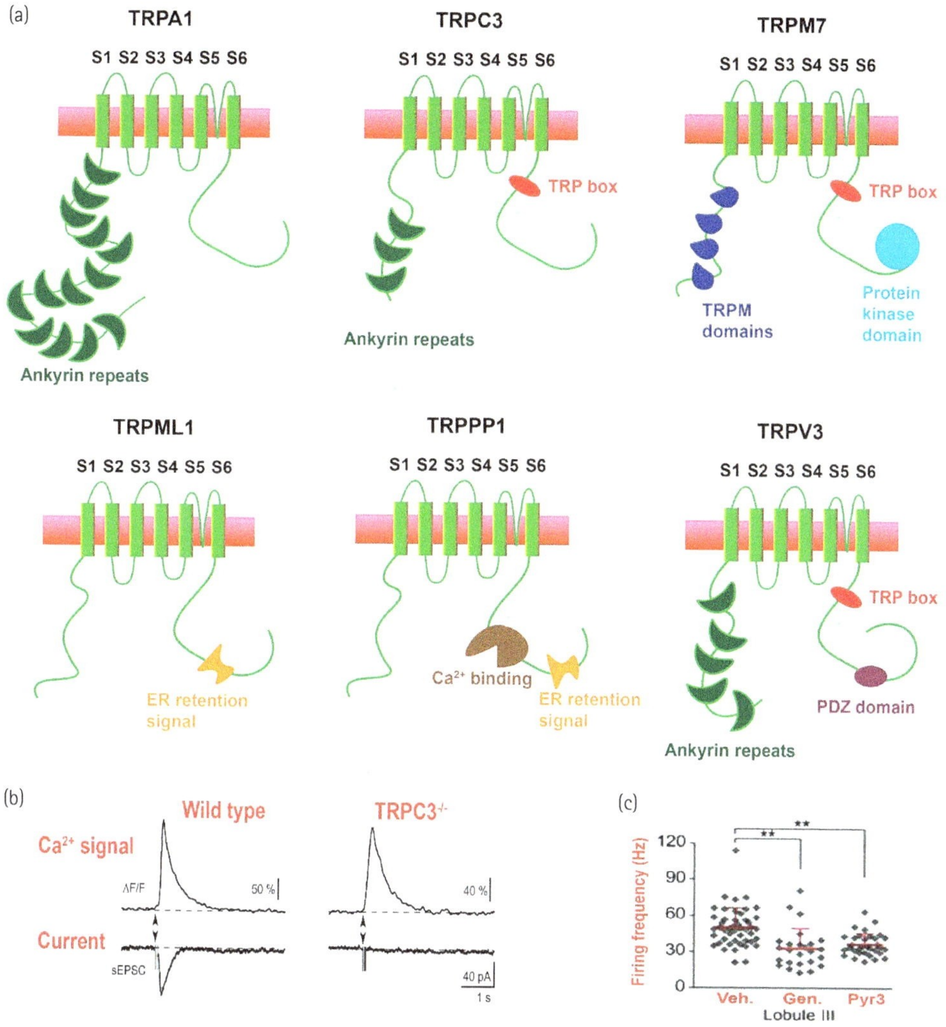

FIGURE 1.7 Transient receptor potential (TRP) channels. (A) Transmembrane topology of TRP channels, with examples of the wide variety of intracellular domains in members of each of the six classes of these channels (modified from Clapham, 2003). Ankyrin domains are mediators of protein–protein interactions. The role of the TRPM domains is not established. The TRP box may control interactions with phosphatidylinositol 4,5-bisphosphate (PIP2) membrane lipids. (B) Genetic elimination of the TRPC3 channel (TRPC3$^{-/-}$) eliminates the slow excitatory current (lower traces) in cerebellar Purkinje cells evoked by stimulation (arrows) of presynaptic parallel fibers. Upper traces show changes in intracellular calcium produced by activation of mGluR1 (metabotropic glutamate receptor 1) receptors, which remain unaffected by loss of TRPC3 (Hartmann et al., 2008). (C) Inhibition of TRPC3 channels using genistein (Gen.) or the more selective agent Pyr3 reduces the firing rate of Purkinje neurons within lobule III of the cerebellum compared to treatment with vehicle (Veh.) (Zhou et al., 2014). ER = endoplasmic reticulum; sEPSC = slow excitatory postsynaptic current.

channels can be activated by a wide range of stimuli. For example, TRPA1 is present in nociceptive neurons of the dorsal root ganglion, which have nerve endings in the skin, as well as in many other cell types such as those in the respiratory system. It contributes to nociception by binding a variety of chemicals such as mustard oil, which becomes covalently linked to the cytoplasmic domains causing sustained channel opening. Numerous other agents, including inflammatory mediators, bind noncovalently to produce more transient pain sensations. Another channel, TRPV1, is also present in nociceptive sensory neurons. TRPV1 is activated by increases in temperature and by direct binding of irritants such as capsaicin, the active ingredient in chili peppers. TRPV2, TRPV3, and TRPV4 are also activated by heat but at different temperatures, allowing a full range of sensations from comfortable warmth to painful heat. Yet another channel, TRPM8, is activated by cold temperatures and binds menthol and other agents that generate sensations of cooling. Thus, TRP channels are responsible for a wide range of responses to changes in the environment, such as itching, coughing, and both pleasurable and painful responses to temperature and chemical agents.

TRP channels also shape the firing patterns of neurons in the central nervous system. As one example, TRPC3 channels are found in Purkinje cells of the cerebellum, where they are required for the slow excitatory postsynaptic current produced on stimulation of the presynaptic parallel fibers (Figure 1.7B) (Hartmann & Konnerth, 2015). Like many of the other TRPC channels, TRPC3 is activated by membrane receptors coupled to G_q-protein signaling, although the full pathway that links mGluR1 (metabotropic glutamate receptor 1) to TRPC3 opening is not established. Pharmacological block of TRPC3 in the Purkinje neurons within certain regions of the cerebellum also reduces their intrinsic firing rate, consistent with the notion that, when activated, these channels provide a sustained depolarizing drive (Figure 1.7C) (Zhou et al., 2014).

There are numerous other examples of how TRP channels are required for different aspects of neuronal plasticity. Several members of the TRPC family are found in the growth cones of developing neurons and can be activated by guidance molecules. Subsequent calcium entry through these channels determines the extension of neurites, the direction of pathfinding, and the collapse of the growth cones (Kaczmarek, Riccio, & Clapham, 2012; Li et al., 2005). In addition to their role in sensory cells and neurons, TRP channels regulate the function of many non-neuronal cells including those in the kidney, endocrine, immune, and cardiovascular systems (Nilius & Szallasi, 2014).

Conclusions

Having the ability to choose to express any combination of α subunits for nine voltage-dependent sodium channels, 10 voltage-dependent calcium channels, 77 potassium channels, and 35 nonselective cation channels confers neurons and other excitable cells with the ability to generate all of the types of firing patterns shown in Figure 1.1, as well as a myriad other forms of electrical activity. If one couples this with the fact that many of the channels come together as multimers of different α subunits with auxiliary subunits and that the RNA transcripts for most α subunit genes can be alternatively spliced to generate different proteins, it appears there exists a limitless number of ways to shape the intrinsic excitability of a neuron. Indeed, it is clear that the same firing pattern can be produced by many different combinations of voltage-dependent or ion-gated channels (Goldman, Golowasch, Marder, & Abbott, 2001). What changes with every different combination of channels, however, is the way the neuron responds to external and internal stimulation and the cellular events that follow activation of the channels during stimulation. Thus, one of the key questions for neuroscientists in this field is to determine how the expression of different combinations of channels is chosen during development and altered in response to changes in the environment.

References

Adelman, J. P., Maylie, J., & Sah, P. (2012). Small-conductance Ca^{2+}-activated K^+ channels: Form and function. *Annual Review of Physiology, 74*, 245–269. doi:10.1146/annurev-physiol-020911-153336

Arikkath, J., & Campbell, K. P. (2003). Auxiliary subunits: Essential components of the voltage-gated calcium channel complex. *Current Opinion in Neurobiology, 13*(3), 298–307.

Balderas, E., Zhang, J., Stefani, E., & Toro, L. (2015). Mitochondrial BKCa channel. *Frontiers in Physiology, 6*, 104. doi:10.3389/fphys.2015.00104

Bane, V., Lehane, M., Dikshit, M., O'Riordan, A., & Furey, A. (2014). Tetrodotoxin: Chemistry, toxicity, source, distribution and detection. *Toxins (Basel), 6*(2), 693–755. doi:10.3390/toxins6020693

Bannister, R. A., & Beam, K. G. (2013). $Ca_V1.1$: The atypical prototypical voltage-gated Ca^{2+} channel. *Biochimica et Biophysica Acta, 1828*(7), 1587–1597. doi:10.1016/j.bbamem.2012.09.007

Bhattacharjee, A., & Kaczmarek, L. K. (2005). For K^+ channels, Na^+ is the new Ca^{2+}. *Trends in Neurosciences, 28*(8), 422–428. doi:10.1016/j.tins.2005.06.003

Birnbaum, S. G., Varga, A. W., Yuan, L. L., Anderson, A. E., Sweatt, J. D., & Schrader, L. A. (2004). Structure and function of Kv4-family transient potassium channels. *Physiological Reviews, 84*(3), 803–833. doi:10.1152/physrev.00039.2003

Brackenbury, W. J., Calhoun, J. D., Chen, C., Miyazaki, H., Nukina, N., Oyama, F., . . . Isom, L. L. (2010). Functional reciprocity between Na^+ channel Nav1.6 and beta1 subunits in the coordinated regulation of excitability and neurite outgrowth. *Proceedings of the National Academy of Sciences of the United States of America, 107*(5), 2283–2288. doi:10.1073/pnas.0909434107

Brackenbury, W. J., Davis, T. H., Chen, C., Slat, E. A., Detrow, M. J., Dickendesher, T. L., . . . Isom, L. L. (2008). Voltage-gated Na^+ channel beta1 subunit–mediated neurite outgrowth requires Fyn kinase and contributes to postnatal CNS development in vivo. *Journal of Neuroscience, 28*(12), 3246–3256. doi:10.1523/JNEUROSCI.5446-07.2008

Brown, M. R., Kronengold, J., Gazula, V. R., Chen, Y., Strumbos, J. G., Sigworth, F. J., . . . Kaczmarek, L. K. (2010). Fragile X mental retardation protein controls gating of the sodium-activated potassium channel Slack. *Nature Neuroscience, 13*(7), 819–821.

Butt, A. M., & Kalsi, A. (2006). Inwardly rectifying potassium channels (Kir) in central nervous system glia: A special role for Kir4.1 in glial functions. *Journal of Cellular and Molecular Medicine, 10*(1), 33–44.

Calderon, J. C., Bolanos, P., & Caputo, C. (2014). The excitation–contraction coupling mechanism in skeletal muscle. *Biophysical Reviews*, *6*(1), 133–160. doi:10.1007/s12551-013-0135-x

Carr, D. B., Day, M., Cantrell, A. R., Held, J., Scheuer, T., Catterall, W. A., & Surmeier, D. J. (2003). Transmitter modulation of slow, activity-dependent alterations in sodium channel availability endows neurons with a novel form of cellular plasticity. *Neuron*, *39*(5), 793–806.

Catterall, W. A. (2011). Voltage-gated calcium channels. *Cold Spring Harbor Perspectives in Biology*, *3*(8), a003947. doi:10.1101/cshperspect.a003947

Catterall, W. A., Goldin, A. L., & Waxman, S. G. (2005). International Union of Pharmacology. XLVII. Nomenclature and structure–function relationships of voltage-gated sodium channels. *Pharmacological Reviews*, *57*(4), 397–409. doi:10.1124/pr.57.4.4

Chen, X., Yuan, L. L., Zhao, C., Birnbaum, S. G., Frick, A., Jung, W. E., . . . Johnston, D. (2006). Deletion of Kv4.2 gene eliminates dendritic A-type K^+ current and enhances induction of long-term potentiation in hippocampal CA1 pyramidal neurons. *Journal of Neuroscience*, *26*(47), 12143–12151. doi:10.1523/JNEUROSCI.2667-06.2006

Choi, S., Yu, E., Kim, D., Urbano, F. J., Makarenko, V., Shin, H. S., & Llinas, R. R. (2010). Subthreshold membrane potential oscillations in inferior olive neurons are dynamically regulated by P/Q- and T-type calcium channels: A study in mutant mice. *Journal of Physiology*, *588*(Pt 16), 3031–3043. doi:10.1113/jphysiol.2009.184705

Clapham, D. E. (2003). TRP channels as cellular sensors. *Nature*, *426*(6966), 517–524. doi:10.1038/nature02196

Connor, J. A., & Stevens, C. F. (1971). Voltage clamp studies of a transient outward membrane current in gastropod neural somata. *Journal of Physiology*, *213*(1), 21–30.

Constantinidis, C., Funahashi, S., Lee, D., Murray, J. D., Qi, X. L., Wang, M., & Arnsten, A. F. T. (2018). Persistent spiking activity underlies working memory. *Journal of Neuroscience*, *38*(32), 7020–7028. doi:10.1523/JNEUROSCI.2486-17.2018

Crill, W. E. (1996). Persistent sodium current in mammalian central neurons. *Annual Review of Physiology*, *58*, 349–362. doi:10.1146/annurev.ph.58.030196.002025

D'Arco, M., & Dolphin, A. C. (2012). L-type calcium channels: On the fast track to nuclear signaling. *Science Signaling*, *5*(237), pe34. doi:10.1126/scisignal.2003355

Dhaka, A., Viswanath, V., & Patapoutian, A. (2006). Trp ion channels and temperature sensation. *Annual Review of Neuroscience*, *29*, 135–161. doi:10.1146/annurev.neuro.29.051605.112958

Dolphin, A. C. (2013). The alpha2delta subunits of voltage-gated calcium channels. *Biochimica et Biophysica Acta*, *1828*(7), 1541–1549. doi:10.1016/j.bbamem.2012.11.019

Dryer, S. E. (1994). Na^+-activated K^+ channels: A new family of large-conductance ion channels. *Trends in Neurosciences*, *17*(4), 155–160.

Duran, C., Thompson, C. H., Xiao, Q., & Hartzell, H. C. (2010). Chloride channels: Often enigmatic, rarely predictable. *Annual Review of Physiology*, *72*, 95–121. doi:10.1146/annurev-physiol-021909-135811

Enyedi, P., & Czirjak, G. (2010). Molecular background of leak K^+ currents: Two-pore domain potassium channels. *Physiological Reviews*, *90*(2), 559–605. doi:10.1152/physrev.00029.2009

Fleming, M. R., Brown, M. R., Kronengold, J., Zhang, Y., Jenkins, D. P., Barcia, G., . . . Kaczmarek, L. K. (2016). Stimulation of slack K^+ channels alters mass at the plasma membrane by triggering dissociation of a phosphatase-regulatory complex. *Cell Reports*, *16*(9), 2281–2288. doi:10.1016/j.celrep.2016.07.024

Goldman, M. S., Golowasch, J., Marder, E., & Abbott, L. F. (2001). Global structure, robustness, and modulation of neuronal models. *Journal of Neuroscience*, *21*(14), 5229–5238.

Goldman-Rakic, P. S. (1995). Cellular basis of working memory. *Neuron*, *14*(3), 477–485.

Goldstein, S. A., Bockenhauer, D., O'Kelly, I., & Zilberberg, N. (2001). Potassium leak channels and the KCNK family of two-P-domain subunits. *Nature Reviews Neuroscience, 2*(3), 175–184. doi:10.1038/35058574

Gomez-Ospina, N., Panagiotakos, G., Portmann, T., Pasca, S. P., Rabah, D., Budzillo, A., ... Dolmetsch, R. E. (2013). A promoter in the coding region of the calcium channel gene CACNA1C generates the transcription factor CCAT. *PLoS One, 8*(4), e60526. doi:10.1371/journal.pone.0060526

Ha, G. E., & Cheong, E. (2017). Calcium-activated chloride channels: A new target to control the spiking pattern of neurons. *BMB Reports, 50*(3), 109–110. doi:10.5483/bmbrep.2017.50.3.033

Hagiwara, S., Kusano, K., & Saito, N. (1961). Membrane changes of *Onchidium* nerve cell in potassium-rich media. *Journal of Physiology, 155*, 470–489.

Hardie, J., & Lee, A. (2016). Decalmodulation of Cav1 channels by CaBPs. *Channels (Austin), 10*(1), 33–37. doi:10.1080/19336950.2015.1051273

Hartmann, J., Dragicevic, E., Adelsberger, H., Henning, H. A., Sumser, M., Abramowitz, J., ... Konnerth, A. (2008). TRPC3 channels are required for synaptic transmission and motor coordination. *Neuron, 59*(3), 392–398. doi:10.1016/j.neuron.2008.06.009

Hartmann, J., & Konnerth, A. (2015). TRPC3-dependent synaptic transmission in central mammalian neurons. *Journa of Molecular Medicine (Berlin, Germany), 93*(9), 983–989. doi:10.1007/s00109-015-1298-7

He, C., Chen, F., Li, B., & Hu, Z. (2014). Neurophysiology of HCN channels: From cellular functions to multiple regulations. *Progress in Neurobiology, 112*, 1–23. doi:10.1016/j.pneurobio.2013.10.001

Hibino, H., Inanobe, A., Furutani, K., Murakami, S., Findlay, I., & Kurachi, Y. (2010). Inwardly rectifying potassium channels: Their structure, function, and physiological roles. *Physiological Reviews, 90*(1), 291–366. doi:10.1152/physrev.00021.2009

Hite, R. K., Yuan, P., Li, Z., Hsuing, Y., Walz, T., & MacKinnon, R. (2015). Cryo-electron microscopy structure of the Slo2.2 Na^+-activated K^+ channel. *Nature, 527*(7577), 198–203. doi:10.1038/nature14958

Hodgkin, A. L., & Huxley, A. F. (1952). A quantitative description of membrane current and its application to conduction and excitation in nerve. *Journal of Physiology, 117*(4), 500–544.

Huang, F., Wang, X., Ostertag, E. M., Nuwal, T., Huang, B., Jan, Y. N., ... Jan, L. Y. (2013). TMEM16C facilitates Na^+-activated K^+ currents in rat sensory neurons and regulates pain processing. *Nature Neuroscience, 16*(9), 1284–1290. doi:10.1038/nn.3468

Hugnot, J. P., Salinas, M., Lesage, F., Guillemare, E., de Weille, J., Heurteaux, C., ... Lazdunski, M. (1996). Kv8.1, a new neuronal potassium channel subunit with specific inhibitory properties towards Shab and Shaw channels. *EMBO Journal, 15*(13), 3322–3331.

Johnson, B., Leek, A. N., & Tamkun, M. M. (2019). Kv2 channels create endoplasmic reticulum/plasma membrane junctions: A brief history of Kv2 channel subcellular localization. *Channels (Austin), 13*(1), 88–101. doi:10.1080/19336950.2019.1568824

Kaczmarek, J. S., Riccio, A., & Clapham, D. E. (2012). Calpain cleaves and activates the TRPC5 channel to participate in semaphorin 3A–induced neuronal growth cone collapse. *Proceedings of the National Academy of Sciences of the United States of America, 109*(20), 7888–7892. doi:10.1073/pnas.1205869109

Kaczmarek, L. K. (2006). Non-conducting functions of voltage-gated ion channels. *Nature Reviews Neuroscience, 7*(10), 761–771.

Kaczmarek, L. K. (2013). Slack, Slick and sodium-activated potassium channels. *ISRN Neuroscience, 2013*(2013). doi:10.1155/2013/354262

Kaczmarek, L. K., Aldrich, R. W., Chandy, K. G., Grissmer, S., Wei, A. D., & Wulff, H. (2017). International Union of Basic and Clinical Pharmacology. C. Nomenclature and properties of calcium-activated and sodium-activated potassium channels. *Pharmacological Reviews*, *69*(1), 1–11. doi:10.1124/pr.116.012864

Kaczmarek, L. K., & Levitan, I. B. (1987). *Neuromodualtion: The biochemical control of neuronal excitability*. New York: Oxford University Press.

Kaczmarek, L. K., & Zhang, Y. (2017). Kv3 channels: Enablers of rapid firing, neurotransmitter release, and neuronal endurance. *Physiological Reviews*, *97*(4), 1431–1468. doi:10.1152/physrev.00002.2017

Kameyama, M., Kakei, M., Sato, R., Shibasaki, T., Matsuda, H., & Irisawa, H. (1984). Intracellular Na^+ activates a K^+ channel in mammalian cardiac cells. *Nature*, *309*(5966), 354–356.

Kaupp, U. B., & Seifert, R. (2002). Cyclic nucleotide-gated ion channels. *Physiological Reviews*, *82*(3), 769–824. doi:10.1152/physrev.00008.2002

Kim, G. E., & Kaczmarek, L. K. (2014). Emerging role of the KCNT1 Slack channel in intellectual disability. *Frontiers in Cellular Neuroscience*, *8*, 209. doi:10.3389/fncel.2014.00209

Kirmiz, M., Palacio, S., Thapa, P., King, A. N., Sack, J. T., & Trimmer, J. S. (2018). Remodeling neuronal ER-PM junctions is a conserved nonconducting function of Kv2 plasma membrane ion channels. *Molecular Biology of the Cell*, *29*(20), 2410–2432. doi:10.1091/mbc.E18-05-0337

Knox, R. J., Quattrocki, E. A., Connor, J. A., & Kaczmarek, L. K. (1992). Recruitment of Ca^{2+} channels by protein kinase C during rapid formation of putative neuropeptide release sites in isolated *Aplysia* neurons. *Neuron*, *8*(5), 883–889.

Latorre, R., Castillo, K., Carrasquel-Ursulaez, W., Sepulveda, R. V., Gonzalez-Nilo, F., Gonzalez, C., & Alvarez, O. (2017). Molecular determinants of BK channel functional diversity and functioning. *Physiological Reviews*, *97*(1), 39–87. doi:10.1152/physrev.00001.2016

Lauritzen, I., Chemin, J., Honore, E., Jodar, M., Guy, N., Lazdunski, M., & Patel, A. J. (2005). Cross-talk between the mechano-gated K2P channel TREK-1 and the actin cytoskeleton. *EMBO Reports*, *6*(7), 642–648. doi:10.1038/sj.embor.7400449

Lee, A., Fakler, B., Kaczmarek, L. K., & Isom, L. L. (2014). More than a pore: Ion channel signaling complexes. *Journal of Neuroscience*, *34*(46), 15159–15169. doi:10.1523/JNEUROSCI.3275-14.2014

Lesage, F., & Lazdunski, M. (2000). Molecular and functional properties of two-pore-domain potassium channels. *American Journal of Physiology Renal Physiology*, *279*(5), F793–F801. doi:10.1152/ajprenal.2000.279.5.F793

Lewis, A. H., & Raman, I. M. (2014). Resurgent current of voltage-gated Na^+ channels. *Journal of Physiology*, *592*(22), 4825–4838. doi:10.1113/jphysiol.2014.277582

Lewis, A. S., Estep, C. M., & Chetkovich, D. M. (2010). The fast and slow ups and downs of HCN channel regulation. *Channels (Austin)*, *4*(3), 215–231. doi:10.4161/chan.4.3.11630

Li, Y., Jia, Y. C., Cui, K., Li, N., Zheng, Z. Y., Wang, Y. Z., & Yuan, X. B. (2005). Essential role of TRPC channels in the guidance of nerve growth cones by brain-derived neurotrophic factor. *Nature*, *434*(7035), 894–898. doi:10.1038/nature03477

Luthi, A., & McCormick, D. A. (1998). H-current: Properties of a neuronal and network pacemaker. *Neuron*, *21*(1), 9–12.

Ma, L., Zhang, X., Zhou, M., & Chen, H. (2012). Acid-sensitive TWIK and TASK two-pore domain potassium channels change ion selectivity and become permeable to sodium in

extracellular acidification. *Journal of Biological Chemistry, 287*(44), 37145–37153. doi:10.1074/jbc.M112.398164
Marder, E. (2012). Neuromodulation of neuronal circuits: Back to the future. *Neuron, 76*(1), 1–11. doi:10.1016/j.neuron.2012.09.010
Matsuda, Y., Yoshida, S., & Yonezawa, T. (1978). Tetrodotoxin sensitivity and Ca component of action potentials of mouse dorsal root ganglion cells cultured in vitro. *Brain Research, 154*(1), 69–82.
Minke, B. (2010). The history of the *Drosophila* TRP channel: The birth of a new channel superfamily. *Journal of Neurogenetics, 24*(4), 216–233. doi:10.3109/01677063.2010.514369
Misonou, H., Thompson, S. M., & Cai, X. (2008). Dynamic regulation of the Kv2.1 voltage-gated potassium channel during brain ischemia through neuroglial interaction. *Journal of Neuroscience, 28*(34), 8529–8538. doi:10.1523/JNEUROSCI.1417-08.2008
Mohapatra, D. P., Misonou, H., Pan, S. J., Held, J. E., Surmeier, D. J., & Trimmer, J. S. (2009). Regulation of intrinsic excitability in hippocampal neurons by activity-dependent modulation of the KV2.1 potassium channel. *Channels (Austin), 3*(1), 46–56.
Namadurai, S., Yereddi, N. R., Cusdin, F. S., Huang, C. L., Chirgadze, D. Y., & Jackson, A. P. (2015). A new look at sodium channel beta subunits. *Open Biology, 5*(1), 140192. doi:10.1098/rsob.140192
Nanou, E., & Catterall, W. A. (2018). Calcium channels, synaptic plasticity, and neuropsychiatric disease. *Neuron, 98*(3), 466–481. doi:10.1016/j.neuron.2018.03.017
Nilius, B., & Szallasi, A. (2014). Transient receptor potential channels as drug targets: From the science of basic research to the art of medicine. *Pharmacological Reviews, 66*(3), 676–814. doi:10.1124/pr.113.008268
Noda, M., & Hiyama, T. Y. (2015). The Na(x) channel: What it is and what it does. *Neuroscientist, 21*(4), 399–412. doi:10.1177/1073858414541009
Nowycky, M. C., Fox, A. P., & Tsien, R. W. (1985). Three types of neuronal calcium channel with different calcium agonist sensitivity. *Nature, 316*(6027), 440–443.
O'Malley, H. A., & Isom, L. L. (2015). Sodium channel beta subunits: Emerging targets in channelopathies. *Annual Review of Physiology, 77*, 481–504. doi:10.1146/annurev-physiol-021014-071846
Plant, L. D., Zuniga, L., Araki, D., Marks, J. D., & Goldstein, S. A. (2012). SUMOylation silences heterodimeric TASK potassium channels containing K2P1 subunits in cerebellar granule neurons. *Science Signaling, 5*(251), ra84. doi:10.1126/scisignal.2003431
Rahmati, N., Hoebeek, F. E., Peter, S., & De Zeeuw, C. I. (2018). Chloride homeostasis in neurons with special emphasis on the olivocerebellar system: Differential roles for transporters and channels. *Frontiers in Cellular Neuroscience, 12*, 101. doi:10.3389/fncel.2018.00101
Ramsey, I. S., Delling, M., & Clapham, D. E. (2006). An introduction to TRP channels. *Annual Review of Physiology, 68*, 619–647. doi:10.1146/annurev.physiol.68.040204.100431
Rasband, M. N., & Peles, E. (2016). The nodes of Ranvier: Molecular assembly and maintenance. *Cold Spring Harbor Perspectives in Biology, 8*, a020495.
Richardson, F. C., & Kaczmarek, L. K. (2000). Modification of delayed rectifier potassium currents by the Kv9.1 potassium channel subunit. *Hearing Research, 147*(1–2), 21–30.
Scheuer, T. (2011). Regulation of sodium channel activity by phosphorylation. *Seminars in Cell & Developmental Biology, 22*(2), 160–165. doi:10.1016/j.semcdb.2010.10.002
Simms, B. A., & Zamponi, G. W. (2014). Neuronal voltage-gated calcium channels: Structure, function, and dysfunction. *Neuron, 82*(1), 24–45. doi:10.1016/j.neuron.2014.03.016

Strong, J. A., Fox, A. P., Tsien, R. W., & Kaczmarek, L. K. (1987). Stimulation of protein kinase C recruits covert calcium channels in *Aplysia* bag cell neurons. *Nature*, *325*(6106), 714–717. doi:10.1038/325714a0

Su, S. C., Seo, J., Pan, J. Q., Samuels, B. A., Rudenko, A., Ericsson, M., . . . Tsai, L. H. (2012). Regulation of N-type voltage-gated calcium channels and presynaptic function by cyclin-dependent kinase 5. *Neuron*, *75*(4), 675–687. doi:10.1016/j.neuron.2012.06.023

Thomas, D., Plant, L. D., Wilkens, C. M., McCrossan, Z. A., & Goldstein, S. A. (2008). Alternative translation initiation in rat brain yields K2P2.1 potassium channels permeable to sodium. *Neuron*, *58*(6), 859–870. doi:10.1016/j.neuron.2008.04.016

Wheeler, D. G., Groth, R. D., Ma, H., Barrett, C. F., Owen, S. F., Safa, P., & Tsien, R. W. (2012). Ca_V1 and Ca_V2 channels engage distinct modes of Ca^{2+} signaling to control CREB-dependent gene expression. *Cell*, *149*(5), 1112–1124. doi:10.1016/j.cell.2012.03.041

White, B. H., & Kaczmarek, L. K. (1997). Identification of a vesicular pool of calcium channels in the bag cell neurons of *Aplysia californica*. *Journal of Neuroscience*, *17*(5), 1582–1595.

Zhang, Y., Helm, J. S., Senatore, A., Spafford, J. D., Kaczmarek, L. K., & Jonas, E. A. (2008). PKC-induced intracellular trafficking of Ca_V2 precedes its rapid recruitment to the plasma membrane. *Journal of Neuroscience*, *28*(10), 2601–2612. doi:10.1523/JNEUROSCI.4314-07

Zhou, H., Lin, Z., Voges, K., Ju, C., Gao, Z., Bosman, L. W., . . . Schonewille, M. (2014). Cerebellar modules operate at different frequencies. *Elife*, *3*, e02536. doi:10.7554/eLife.02536

Zhuang, Q. X., Li, G. Y., Li, B., Zhang, C. Z., Zhang, X. Y., Xi, K., . . . Zhu, J. N. (2018). Regularizing firing patterns of rat subthalamic neurons ameliorates parkinsonian motor deficits. *Journal of Clinical Investigation*, *128*(12), 5413–5427. doi:10.1172/JCI99986

CHAPTER 2

ION CHANNEL PERMEATION AND SELECTIVITY

JUAN J. NOGUEIRA AND BEN CORRY

Introduction

The movement of ions across cell membranes through ion channels underlies a wide range of biological processes such as the transmission of nerve impulses, stimulation of muscle contraction, and regulation of cell volume and blood pressure, to name just a few. Structural alterations of ion channels and subsequent malfunctions may cause several channelopathies (Bagal et al., 2013), including epilepsy, long QT syndrome, deafness, and cerebellar ataxia, among others. While the process of ion transport sounds conceptually simple, it requires the coordinated and rapid movement of specific ions across cell membranes at precise times in response to different stimuli through a very intricate mechanism. The elucidation of the steps involved in this complex mechanism, in which ion channels play a fundamental role, is still an area of intense, ongoing research.

The gating (opening and closing) of ion channels to allow the passage of ions can be induced as a response to chemical stimuli, temperature changes, voltage changes, ligand binding, or mechanical forces. Voltage-gated ion channels are responsible for the generation and propagation of action potentials in neurons and other excitable cells (Catterall, 2012). Under depolarizing conditions (increasing transmembrane potential), voltage-gated sodium (Na_V) channels open and transport Na^+ ions from the extracellular media to the cytosol, causing further depolarization of the membrane. Then, Na_V channels are inactivated, and voltage-gated potassium (K_V) channels open and allow the outward transport of K^+ ions to reach electrochemical equilibrium with the extracellular medium. This repolarizes (decreases the potential of) the membrane and restores the negative resting potential of the cell. Voltage-gated calcium (Ca_V) channels are also able to modify the membrane potential by transporting Ca^{2+} ions from outside to inside the cell. Moreover, this ion transport is also involved in cell signaling processes such as gene transcription (Clapham, 2007).

The delicate process of ion transport requires an exquisite control over the inward and outward permeation of the different ions, whose mechanistic aspects are not yet fully understood. For example, in order to fulfil their role, K_v channels have to prevent the passage of other ions, such as Na^+, that would dissipate the electric gradient. Indeed, K_v channels typically let 100–1000 K^+ ions pass for every Na^+. In addition, K_v channels have to be able to rapidly return the cell to its resting state after an electrical signal and, thus, transport many millions of K^+ ions across the membrane every second. Reconciling how a channel can be exquisitely selective yet let ions pass at such large rates is an unresolved intellectual challenge that has fascinated scientists ever since these channels were proposed more than 60 years ago.

This chapter aims to explain what we know about the passage of ions through channels (permeation) and the mechanisms by which channels distinguish between different ion species (selectivity), based on experimental and theoretical research. We will briefly describe the most common experimental and theoretical approaches used to investigate the transport of ions through ion channels. We then turn our attention to the selection and permeation of ions and the mechanisms by which different channels achieve this, focusing on K_v, Na_v, and Ca_v channels, in which exquisite selection is achieved among very similar ions.

Approaches to Studying Permeation and Selectivity

The knowledge gained on ion permeation and selectivity in the last two decades has considerably increased thanks to the use of advanced experimental and theoretical methodologies, especially electrophysiological measurements, structural analysis and dynamics simulations. These techniques enable the inspection of fundamental aspects of the mechanisms of ion transport through ion channels at macroscopic and microscopic levels of resolution and, in many cases, in a time-resolved picture.

Experimental Approaches

The measurement of currents passing through ion channels remains one of the key tools to understanding the steps involved in permeation and the mechanisms of selectivity, as the rate of permeation through an ion channel can be directly measured through single-channel patch-clamp recordings. The permeability of a channel to different ions can be determined by measuring the current with just one permeant ion type present at a time. The permeability ratio is often defined as the current of one ion type measured in this way compared to the current of another. This, however, should be differentiated from a measure of the selectivity of an ion channel that is performed with multiple ion species present. Selectivity measurements usually proceed under "bi-ionic" conditions, with a different permeant species present on each side of the membrane that compete with each other to pass through the channels. With no applied potential, a nonselective channel will pass zero current. But a selective channel will show a net current carried by the more selected species. The degree of selectivity is then typically determined from the reversal potential of the current–voltage relationship (LeMasurier, Heginbotham, & Miller, 2001), which is defined as the potential at which there is no net flow of a particular ion type.

Because the channel protein has to be attractive to permeating ions to enable rapid transport, it is not uncommon for more than one ion to enter a channel at any one time, something that can also be probed by studying ionic currents. If ions move independently though the pore, then, the ratio of efflux to influx is directly related to the activities of the ion type on each side and the membrane potential (Ussing, 1949). But if ions must interact to permeate the pore, this relationship must be modified to include the so-called flux ratio exponent, with a value greater than one indicating non-independent motion (Hodgkin & Keynes, 1955). While an exponent greater than one yields a clear indication of a multi-ion pore, the exact value can vary anywhere between one and the number of ion binding sites in the pore (Heckmann, 1972; Hille & Schwarz, 1978), and saturation of currents at high concentrations can alter the conclusions (Vora, Corry, & Chung, 2008).

Electrophysiological measurements on mutant ion channels have also provided great mechanistic insight into ion permeation (Naylor et al., 2016; Yue, Navarro, Ren, Ramos, & Clapham, 2002). The alteration of ion conductivity and/or selectivity induced by mutation of a particular residue highlights the relevance of the mutated residue in the transport mechanism. In addition to electrophysiological experiments on both wild-type or mutant channels, the determination of atomic resolution structures of ion channels using X-ray crystallography or cryo-electron microscopy has also contributed to advance our knowledge in ion transport. In the last 20 years, atomic-resolution structures have been published for a number of ion channels (Bagnéris et al., 2013; Brohawn, del Marmol, & MacKinnon, 2012; Doyle et al., 1998; Jiang et al., 2003; Lenaeus et al., 2017; Long, Campbell, & Mackinnon, 2005; Long, Tao, Campbell, & MacKinnon, 2007; McCusker et al., 2012; Morais-Cabral, Zhou, & MacKinnon, 2001; Payandeh, Gamal El-Din, Scheuer, Zheng, & Catterall, 2012; Shaya et al., 2014; Shen et al., 2017; Sula et al., 2017; Tang et al., 2014; Wu et al., 2016; Wu et al., 2015; Yan et al., 2017; Zhang et al., 2012; Zhou, Morais-Cabral, Kaufman, & MacKinnon, 2001), which yielded valuable mechanistic information. For example, the electron-density maps provided by these structural methods enabled researchers to hypothesize the number of ion binding sites and their location in the selectivity filter. The mechanistic information gained from electrophysiological experiments and X-ray and cryo-electron microscopy structures for Na^+, K^+, and Ca^{+2} channels will be discussed in more detail later.

Theoretical Approaches

The microscopic events that take place as ions pass through a channel—e.g., conformational changes in the protein, solvation/desolvation of ions along the journey, or formation/breaking of ion/protein interactions—are almost impossible to directly interrogate with experimental approaches. However, they can be witnessed in advanced dynamic simulations (Chung & Corry, 2005; Corry, 2017; Maffeo, Bhattacharya, Yoo, Wells, & Aksimentiev, 2012; Oakes, Furini, & Domene, 2016), provided that they are carefully validated against experimental data.

The dynamic simulation of ionic permeation events across ion channels requires very long simulation times because ionic permeation usually occurs in the nanosecond to microsecond time scale, or even longer. A drastic but very efficient approach that can be employed to alleviate the computational cost of the simulation is to represent the structure of the entire system, or part of it, by a structureless continuum dielectric material, as represented in Figure 2.1A and 2.1B. This approach is employed in the Poisson-Nernst-Planck (PNP) simulations, in which no atoms are represented explicitly, and in Brownian dynamics (BD) simulations, in which only select atoms (usually the ions) are included explicitly. The theory and the limitations behind these approximations have been extensively discussed (Chung & Corry, 2005; Corry, Kuyucak, & Chung, 2000; Levitt, 1999; Maffeo et al., 2012). An issue with the fully-continuum approach is that the average ion concentration in the pore often cannot adequately represent the forces felt on individual ions as they pass through the narrow pores (Corry et al., 2000). The BD method, while avoiding this issue, also involves many simplifications, including the use of a static protein structure and approximations to account for ion solvation. Both methods allow for the currents through channels to be efficiently determined and have provided important clues about the mechanisms of ion permeation and selectivity, although it remains challenging in these methods to discriminate between very similar ions such as Na^+ and K^+.

A more realistic description of the mechanism of ion transport through protein channels can be obtained when all the atoms of the system are explicitly accounted for in the model (Figure 2.1C). This is the case of the all-atom molecular dynamics (MD) methodology, where the motion of each of the atoms is simulated by integrating numerically the Hamilton equations of motion for many short-time intervals (Adcock & McCammon, 2006). The use of MD to simulate ion channels has gained popularity in the last two decades due to the experimentally determined high-resolution ion-channel structures, which can be employed as initial structures in the dynamics, and improvements in computing power that allow simulations to reach the timescale of ion permeation. The integration of the equations of motion requires the computation of the atomic forces at each time step of the simulation. The vast majority of MD simulations of ion channels makes use of classical molecular mechanics (MM) force fields (Corry, 2017; Oakes et al., 2016) to compute the forces due to the very high computational cost of using quantum mechanical (QM) or hybrid QM/MM techniques. The only exceptions to this are full-QM treatments that have been employed to discuss the selectivity of ion channels by drastically simplifying the system to a reduced model of only a few tens of atoms, which tried to mimic the selectivity filter in a static picture (Dudev, Grauffel, & Lim, 2016; Dudev & Lim, 2014).

Even with classical simulations, the modeling of ionic permeation through ion channels is very computationally demanding since it is a very slow process, which may

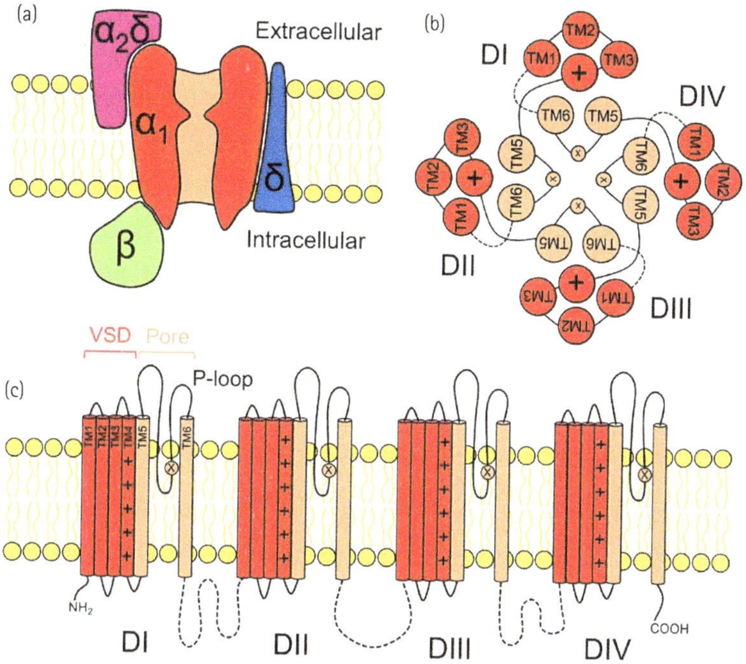

FIGURE 2.1 Theoretical models usually employed in dynamic simulations of ion permeation through ion channels.
A. In Poisson-Nernst-Planck dynamics, the protein, lipid membrane, and water are represented by continuum dielectric materials with specific dielectric constants ε_i and the ions by a continuous charge distribution. **B.** In Brownian dynamics, only the ions are represented explicitly. **C.** All-atom molecular dynamics describe explicitly every atom of the system. The protein, lipids, water, cations, and anions are represented in red, orange, cyan, blue and green, respectively, in panels A, B, and C. **D.** Representation of a possible potential-energy profile for an ion-permeation pathway through the entire ion channel. The dynamics can be accelerated by applying an electric field, which pushes positive ions to travel in the direction of the field, overcoming the energy barriers.

last hundreds of nanoseconds or even microseconds. This means that the simulation of a significant number of events to extract statistically meaningful conclusions would require simulations to reach the millisecond time scale or even more. Consequently, the dynamics are often accelerated by means of several approaches. For example, a large but constant electric field, which helps overcome large potential-energy barriers (Figure 2.1D), can be applied to generate a net flux of ions (Furini & Domene, 2018; Gumbart, Khalili-Araghi, Sotomayor, & Roux, 2012; Ulmschneider et al., 2013). The net flux of ions generated along the protein pore allows the computation of the conductance and the direct comparison with experiment (Ulmschneider et al., 2013). Alternatively, rather than directly simulating ion conduction, a common method to study permeation is to determine the free energy of the conduction process using techniques such as umbrella sampling (Corry & Thomas, 2012; Li, Sun, Liu, & Gong, 2017; Mahdavi, Kuyucak, & Ye,

2015), from which the steps involved in ion conduction and rates of permeation can be inferred.

Potassium Channels

As mentioned in the Introduction, K^+ channels have to rapidly transport K^+ ions while simultaneously being highly selective for K^+ over Na^+ in order to rapidly reset the membrane potential and maintain low resting potentials. The examination of unidirectional fluxes of ions through K^+ channels led Hodgkin and Keynes to postulate a long, narrow pore holding two to three ions, which conducts in a knock-on mechanism (Hodgkin & Keynes, 1955). However, their data could not rule out the possibility of a vacancy mechanism in which vacancies, not just ions, also diffuse inside the filter. A highly conserved signature TXGYG sequence was found among all K^+ selective channels (Heginbotham, Lu, Abramson, & MacKinnon, 1994). This lies in a section of the protein known to interact with a range of toxins, blockers, and permeating ions (MacKinnon & Miller, 1989; MacKinnon & Yellen, 1990). Mutations within this sequence could remove ion selectivity among monovalent ions (Heginbotham et al., 1994), suggesting that this sequence lines a key region of the pore.

More recently, a large number of atomic resolution structures of different K^+ channels have been published (e.g., see Brohawn et al., 2012; Doyle et al., 1998; Jiang et al., 2003; Long et al., 2005; Long et al., 2007; Morais-Cabral et al., 2001; Zhou et al., 2001) showing that common principles underlie ion permeation and selectivity in this diverse class of proteins. Indeed, all eukaryotic and prokaryotic K_v, Na_v, and Ca_v channels share a similar architecture, schematically displayed in Figure 2.2. The structural region in charge of ion permeation is formed by four domains (DI–DIV) symmetrically arranged around the central pore, which are covalently bonded forming a single peptide chain in Ca_v channels and eukaryotic Na_v channels, but is composed of four separate protein chains in K_v channels and in prokaryotic Na_v channels (Grizel, Glukhov, & Sokolova, 2014). In addition to the α subunit, K_v channels also present β and other auxiliary subunits, which regulate the function of the α subunit. Usually, the four domains of the α subunit are identical in K_v channels, although some channels present two or more non-identical domains. Each domain contains six transmembrane helices (TM1–TM6), the first four of which form the voltage-sensing domain, while the pore domain is formed by an outer helix (TM5) and an inner pore lining helix (TM6) joined by a re-entrant P-loop that contains the signature sequence and forms a narrow selectivity filter. Analysis of the crystal structures suggested four ion binding sites within the filter, commonly labelled S1–S4, as well as an external site (S0) and a site within the central cavity (S5) (Morais-Cabral et al., 2001), as seen in Figure 2.3A. In addition, ions would most commonly occupy either sites S1 and S3 or S2 and S4 separated by water molecules, with conduction requiring the entrance of a third ion to generate a knock-on mechanism (Morais-Cabral et al., 2001) in line with the early suggestions of

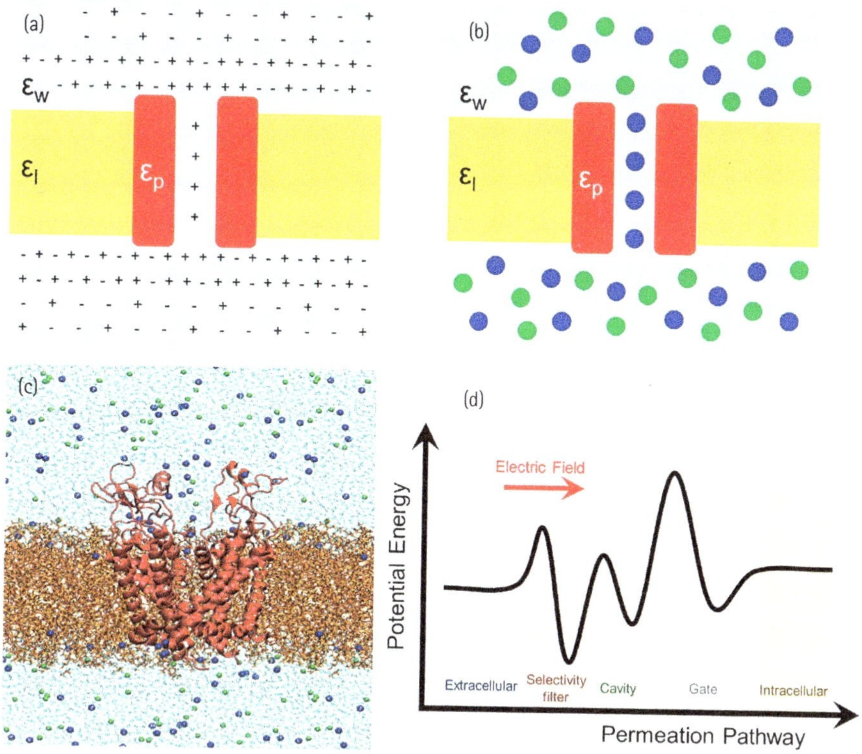

FIGURE 2.2 Schematic representation of the structure of voltage-gated ion channels. **A.** Side view of an ion channel composed by a α subunit (red), which is responsible for ion conduction and is present in all the channels, and additional subunits, which regulate the function of the α subunit and are present only in certain classes of channels. The lipid membrane is represented in yellow. **B.** Top view of the α subunit formed by four domains (DI–DIV), which in turn are composed of six transmembrane helices (TM1–TM6). The TM5 and TM6 helices (light red) form the central pore of the channel along which the ions are transported. These helices are linked by a P-loop that contains the selectivity filter with a signature sequence represented by the letter X. The TM1–TM4 helices form the voltage-sensing domain (VSD) and are located in the external region of the channel. **C.** Side view of the expanded representation of the four domains, which are covalently bonded in Ca_v channels and eukaryotic Na_v channels such that they are formed from one long protein chain but are non-covalently bonded in K_v channels and prokaryotic Na_v channels.

Hodgkin and Keynes (Hodgkin & Keynes, 1955). This was further supported by a range of BD simulations that *indirectly* examined the most likely conduction to suggest that two ions typically occupy the pore with an intervening water molecule, with the entry of a third ion required to generate knock-on conduction of ions and water (Aqvist & Luzhkov, 2000; Berneche & Roux, 2001, 2003; Chung, Allen, Hoyles, & Kuyucak, 1999; Chung, Allen, & Kuyucak, 2002). Improving computer power has since allowed MD simulations to be run for long enough to witness large numbers of permeation events. These also supported the idea that ions and water would move through the selectivity filter together with the most likely conduction process shown in Figure 2.3B (Jensen et al., 2010; Jensen et al., 2012; Jensen, Jogini, Eastwood, & Shaw, 2013; Khalili-Araghi, Tajkhorshid, & Schulten, 2006).

However, other simulation studies have suggested different permeation routes that may either coexist with the traditional steps or replace them altogether (Furini & Domene, 2009; Kopfer et al., 2014). In particular, it has been suggested that ions may pass through the selectivity filter in direct contact with each other, without intervening water molecules, as shown in Figure 2.3C. In this mechanism, three to four ions populate the selectivity filter simultaneously. While it seems unusual for ions to completely dehydrate and be found so close to one another and for vacancies to appear in the filter, the attraction between the carbonyl oxygen atoms and the ions is suggested to be able to stabilize this configuration. Furthermore, the large electrostatic interactions between the ions mean that thermal fluctuations in ion positions maximize the conduction rate. It is suggested that this so-called direct Coulomb knock-on mechanism yields much more rapid conduction than does the traditional alternating-site model (Kopfer et al., 2014).

The new model does, however, conflict with a large amount of experimental data. Flux-ratios suggest only two to three ions in the pore (Hodgkin & Keynes, 1955). The X-ray structures posit alternating ions and water (Doyle et al., 1998; Morais-Cabral et al., 2001), a point most clearly seen in anomalous scattering data in which K^+ was replaced by Tl^+. But it is worth noting that the occupancies seen in these structures cannot be completely accounted for by simple alternating occupancy; for example, the occupancies in S1 and S3 are not exactly equal, nor are those in S2 and S4 (Furini & Domene, 2009; Zhou & MacKinnon, 2004; Zhou & MacKinnon, 2003). A recent re-refinement of the Tl^+ scattering data was found to be consistent with close contacts between permeating ions (Kopfer et al., 2014). However, water has been found to pass through K^+ channels in the absence of K^+ (Hoomann, Jahnke, Horner, Keller, & Pohl, 2013), and streaming potentials suggest at least that one water molecule passes for each K^+ ion (Alcayaga, Cecchi, Alvarez, & Latorre, 1989; Ando, Kuno, Shimizu, Muramatsu, & Oiki, 2005; Iwamoto & Oiki, 2011), which is suggestive of alternating occupancy of ions and water in the pore. Recent ultrafast time-resolved infrared spectroscopy experiments were able to detect specific multi-ion configurations in the pore and also showed water to separate neighboring ions in the channels (Kratochvil et al., 2016). Proponents of the idea of K^+ ions in direct contact, however, suggest that the experimental data from the infrared experiments can be interpreted to be consistent with either model, and that the streaming potential measurements are done in osmotic gradients that favor water permeation (Kopec et al., 2018; Kopfer et al., 2014).

Whether ions pass through the selectivity filter separated by water or not is a contentious issue that needs to be resolved. Specifically, one can ask why some simulations commonly see adjacent ions, but others do not. Is this a consequence of different force fields or simulation conditions? Can new experimental data shed unambiguous light on the issue? What is probably clear is that multiple permeation pathways exist, and it is a question of which is more common rather than which is correct.

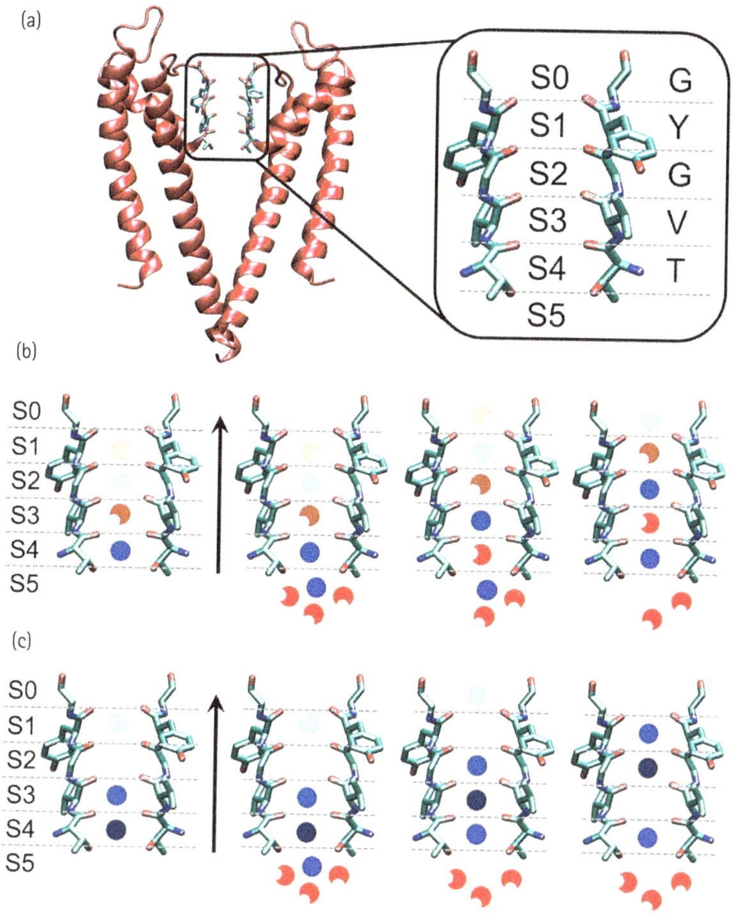

FIGURE 2.3 Permeation mechanism of K^+ ions through a K^+ channel.
A. Structure of the pore of the KcsA channel with half of the pore removed for clarity. Zoom-in view of the selectivity filter with the TVGYG sequence and six binding sites. The hydrogen atoms are not shown. **B.** Traditional knock-on mechanism in which alternating K^+ ions and water molecules permeate through the selectivity filter. **C.** Direct Coulomb knock-on mechanism in which only K^+ ions and vacancies are able to travel along the selectivity filter. The black arrow in B and C indicates the direction of ion motion (efflux transport). C, N, and O atoms are displayed in cyan, blue, and red, respectively. K^+ ions inside the filter are shown in different shades of blue, and the O and H atoms of water in different shades of red and in gray, respectively.

In order to set the negative resting potential of cells and rapidly reset the resting potential after depolarization, K^+ channels have to be extraordinarily selective, with about 100-fold preference for K^+ over Na^+. Tests with other monovalent ions show a selectivity sequence in competitions experiments of $Tl^+ > K^+ > Rb^+ > NH_4^+ \gg Cs^+, Na^+, Li^+$ (Blatz & Magleby, 1984; Block & Jones, 1996; Hagiwara & Takahashi, 1974; Heginbotham & MacKinnon, 1993; Hille, 1973; LeMasurier et al., 2001; Park, 1994; Reuter & Stevens, 1980; Yellen, 1984). Ba^{2+}, which is a similar size to K^+, can enter K^+ channels and block

K^+ currents (Armstrong & Taylor, 1980; Neyton & Miller, 1988). As the ability of K^+ channels to select between K^+ and Na^+ is the most physiologically important feature, we focus on this here.

The selectivity filter of K^+ channels is narrow, so ions must largely dehydrate if they are to enter from bulk. Alkali cations hold their hydrating water very tightly, with smaller ions having a stronger hydration free energy than larger ions. For example, the hydration free energy of Na^+ is about −88 kcal/mol, compared to −70 kcal/mol for K^+ (Marcus, 1994; Schmid, Miah, & Sapunov, 2000). Thus, the energetic cost (dehydration penalty) of Na^+ losing its water and entering a narrow pore is greater than for K^+. This helps us understand the reason why a narrow K^+ channel selects for the larger K^+ over Na^+: a narrow pore has an intrinsic preference for K^+ over Na^+ (provided both can fit inside) (Song & Corry, 2009). However, for ions to rapidly move into the pore, the free energy penalty for dehydration has to be compensated for by a favorable attraction to the protein. Thus, the net selectivity of the pore relates to the balance between the repulsive (dehydration) and attractive (ion–protein) interaction for each ion.

Several possible explanations for K^+ vs. Na^+ selectivity were deduced from electrophysiological and electrochemical experiments prior to the publication of atomic-resolution structures. In his study of glass electrodes, Eisenman argued that ligands of differing field strength could selectively bind differently sized ions, and that this might be applicable to biological systems (Eisenman, 1962). That K^+ over Na^+ selection could be achieved simply by tuning the charge on the ligands (protein) was further supported by pioneering free-energy calculations from MD simulations that highlighted the key role that the electrostatic attraction between ions and their coordinating ligands plays in determining the selectivity of the cyclic antibiotic valinomycin (Aqvist, Alvarez, & Eisenman, 1992). Although this study noted that steric factors that influence the ability of ligands to pack around the ion (and presumably influence the number of ligands contacting the ion) also contribute to the selectivity of the host molecule, this effect has been given less importance than the field strength of the ligands in many discussions of K^+ channels. An alternative was the so-called snug-fit model that pictured a pore that could neatly pack around the larger K^+ ion, but could not adequately coordinate the smaller Na^+ as required to replace the coordinating water molecules (Bezanilla & Armstrong, 1972).

The publication of the first K^+ channel structure was thought to support the concept of the snug-fit model, as the backbone carbonyls were seen to form cavities that beautifully coordinated permeating K^+ ions, but would be too large for Na^+ (Doyle et al., 1998). However, it has since been shown that the inherent flexibility of the protein makes this explanation for selectivity unlikely. In the absence of K^+, for example, the selectivity filter is seen to shrink to a much narrower conformation (Zhou et al., 2001), while even the thermal parameters in the K^+-bound structure suggest significant atomic vibrations (Noskov, Berneche, & Roux, 2004). Simulations of the binding sites in the K^+ channel filter also showed that the protein could retain selectivity even in toy models of the selectivity filter in which the carbonyl ligands were completely free to move so that there was no rigid cavity present (Noskov et al., 2004; Thomas, Jayatilaka, & Corry, 2007).

Numerous alternative proposals have thus been put forward. The dipole moment of the carbonyl ligands could intrinsically yield selectivity, in line with Eisenman's proposal (Noskov et al., 2004). Angstrom-level constraints on the position of the carbonyl atoms can also restrict the number of ligands coordinating the permeating ions, and the presence of eight ligands around each binding site was shown to play an important role in defining selectivity (Bostick, Arora, & Brooks, 2009; Bostick & Brooks, 2007; Thomas et al., 2007; Thomas, Jayatilaka, & Corry, 2011, 2013; Varma & Rempe, 2007). Another possible explanation relates to the concept of strain energy—even if the protein is flexible enough to deform to accommodate Na^+, this may come at an energetic cost for the protein (Yu, Noskov, & Roux, 2010). The snug-fit and liquid-like models of selectivity can therefore be seen as opposite ends of a spectrum. If the binding site is very stiff, then selection arises from a snug-fit mechanism. If the site is completely flexible, then the number and dipole moment of the ligands will dictate selectivity. In between these, a combination of factors, including internal strain on the protein, can act together. Thus, in reality, selectivity probably involves contributions from all these factors.

The mechanisms discussed so far revolve around the thermodynamics of ion binding in the selectivity filter—that is, the relative depth of the energy minima for K^+ and Na^+ in a binding site. But several more recent studies suggest that selectivity may be related to kinetic effects—the different rates to move into or between sites (Liu & Lockless, 2013; Sauer, Zeng, Canty, Lam, & Jiang, 2013; Thompson et al., 2009). Computational and experimental studies have shown that while the cage sites between the planes of carbonyl oxygens selectively bind K^+ as discussed previously, Na^+ can bind in the plane of carbonyl oxygens—a position that favors Na^+ over K^+ (Thompson et al., 2009). Therefore, it is dangerous to examine individual sites on their own when explaining selectivity. The implication of this is that Na^+ may be able to bind in the selectivity filter, but it has difficulty moving through the pore because of large barriers faced when moving Na^+ in a filter that also holds K^+ (Egwolf & Roux, 2010; Nimigean & Allen, 2011; Thompson et al., 2009). Recently, it has been shown using mutations in the non-selective NaK channel selectivity filter that the presence of multiple ion binding sites is essential for selectivity. The NaK channel and the so-called NaK2CNG channel have only three ion binding sites in the selectivity filter and are non-selective for K^+ over Na^+. The NaK2K mutant, however, recovers the fourth ion binding site and has a strong equilibrium preference for K^+, suggesting that the multi-ion nature of ion permeation is essential for selectivity (Derebe et al., 2011; Sauer et al., 2013). To reinforce the kinetic aspects of selectivity, it is seen that at equilibrium, the selectivity filter selectively binds K^+ even in the non-selective channels, suggesting that the selectivity of permeation cannot simply be derived from equilibrium properties (Liu & Lockless, 2013; Sauer, Zeng, Raghunathan, & Jiang, 2011). However, recently an alternative thermodynamic interpretation has been provided for this data. If ions move through the filter completely dehydrated instead of being separated by water molecules (as suggested by the model in which ions permeate in direct contact with one another), then the selectivity is enhanced by maximizing the dehydration energy difference of Na^+ and K^+ (Kopec et al., 2018). Thus, the presence of multiple ion binding sites may simply be required to enforce complete dehydration. This agrees with earlier thermodynamics studies that showed that the presence of a water molecule surrounding one of the binding sites significantly reduces its selectivity (Alam & Jiang, 2009; Fowler, Tai, & Sansom, 2008; Noskov & Roux, 2007; Thomas et al., 2011). While it is fair to say that the principles underlying K^+ channel selectivity are understood, a simple, intuitively clear explanation is lacking—perhaps because it is a complex physical process that cannot be explained by a single principle alone.

Sodium Channels

Na_v channels are responsible for the transport of Na^+ ions in excitable cells from the high-Na^+ concentration extracellular region to the low-Na^+ concentration cytosol (Hille, 2001). The opening of the channel and subsequent transport of Na^+ ions through the protein are induced by the depolarization of the cell membrane. The inward current of ions induces further depolarization, generating an action potential that travels along excitable cells and cell networks (Marban, Yamagishi, & Tomaselli, 1998). The mechanistic knowledge of ion permeation in Na_v channels is not so extensive as that in K_v channels, due to the delay in resolving the structure of Na_v channels with atomic resolution. Specifically, the first crystalized structure, captured in a closed-pore conformation, was reported only in 2011 (Payandeh, Scheuer, Zheng, & Catterall, 2011) from *Arcobacter butzleri* (Na_vAb) with cysteine mutations in the TM6 helices. Since then, a few additional full or pore-only bacterial channel structures have been reported in different gating states (Bagnéris et al., 2013; Lenaeus et al., 2017; McCusker et al., 2012; Payandeh et al., 2012; Shaya et al., 2014; Sula et al., 2017; Zhang et al., 2012). In addition, the structures of eukaryotic Na_v channels from American cockroach (designated Na_vPaS) (Shen et al., 2017), electric eel (EeNa$_v$1.4) (Yan et al., 2017), and humans (Na_v1.4) (Pan et al., 2018) have been obtained by cryo-electron spectroscopy very recently.

The structure of bacterial Na_v channels presents features similar to those of K_v channels. They are composed of a large pore-forming α subunit and smaller auxiliary β subunits (Catterall & Swanson, 2015), as represented in Figure 2.1. The α subunit, responsible for voltage-gated ion permeation, contains four identical domains (DI–DIV), which are not covalently bonded. As for K_v channels, each domain is formed by six transmembrane segments (TM1–TM6), four of which (TM1–TM4) comprise the voltage sensor domain that regulates the channel gating. The other two segments of each domain (TM5–TM6) compose the pore region of the protein. The entrance of ions into the pore domain is regulated by the selectivity filter located in a long P-loop that connects TM5 and TM6. The most common sequence found in the selectivity filter of bacterial Na^+ channels is composed of the TLESWS residues (see Figure 2.4A). Contrary to prokaryotic Na^+ channels, the α subunit of eukaryotic Na^+ channels is formed by four different domains that are covalently linked; i.e., the channel is formed by a single peptide chain. However, the most important structural difference affecting ion conduction is found in the selectivity filter. While the strongest binding site in the selectivity filter of bacterial Na_v channels is formed by one glutamate (E) from each of the four domains, which compose the so-called EEEE ring, the equivalent binding site is formed by the DEKA ring (aspartate [D], glutamate [E], lysine [K], and alanine [A]) in eukaryotic Na_v channels (Favre, Moczydlowski, & Schild, 1996). As explained later, both the EEEE and DEKA motifs play a crucial role in the process of ion conduction and selectivity.

Structural data suggested the existence of three binding sites in the selectivity filter of the Na_vAb channel (Payandeh et al., 2011) composed of the TLESWS sequence (see Figure 2.4A): the high-field-strength anionic site (S_{HFS}) formed by the side chains of glutamate residues of the EEEE ring, and two sites located deeper in the filter and formed by the backbone carbonyls of leucine (site S_{CEN}) and threonine (site S_{IN}), respectively. Free-energy profiles computed by umbrella-sampling classical MD simulations (Corry & Thomas, 2012) and metadynamics and voltage-biased MD simulations (Stock, Delemotte, Carnevale, Treptow, & Klein, 2013) corroborated the existence of the three binding sites in Na_vAb. However, additional umbrella-sampling simulations on the same system found free-energy minima only at the S_{HFS} and S_{CEN} sites (Furini & Domene, 2012), while equilibrium classical MD simulations found significant population of Na^+ only at S_{HFS} and S_{IN}, while the S_{CEN} site was mainly occupied by water molecules (Carnevale, Treptow, & Klein, 2011). Voltage-biased MD simulations on the prokaryotic channel of *Magnetococcus marinus* (Na_vMs) (Ulmschneider et al., 2013) predicted five binding sites in the selectivity filter, which were termed S0, S1, S2, S3, and S4 in analogy to the binding sites of K^+ channels. S1, S3, and S4 corresponded to the already established S_{HFS}, S_{CEN}, and S_{IN} sites for Na_vAb, respectively, while S0 and S5 were identified as new sites located at the vestibule of the selectivity filter, and slightly below S_{HFS}, respectively. However, these two additional ion sites were not found in an experimental study (Naylor et al., 2016), where only S_{HFS}, S_{CEN}, and S_{IN} were clearly identified as ion binding sites. A fourth peak of density was also found between the S_{CEN} and S_{IN} sites, but it was interpreted as a water binding site (Naylor et al., 2016). Contrary to Na_vAb and Na_vMs, whose selectivity filter sequence is TLESWS, only two Na^+ binding sites were identified in the selectivity filter TLSSWE of the marine alphaproteobacterium HIMB114 (Na_vRh) by classical MD simulations (Zhang et al., 2013). Although contradictory results were reported for the same prokaryotic Na_v channels, and several channels present different selectivity filter sequences, in general, one can conclude that Na^+ channels present a smaller number of ion binding sites in the selectivity filter than K^+ channels do.

The mechanism of ion permeation throughout Na^+ channels significantly differs from the mechanism in K^+ channels due to the different size of the selectivity filter. In particular, the selectivity filter of Na_v channels is shorter and wider, a fact that was surprising considering that Na^+ ions are smaller than K^+ ones and therefore should be able to permeate through narrower filters. That the pore is wide enough to accommodate both K^+ and Na^+ eliminates the possibility of a snug-fit model to explain Na^+ selectivity. Two important differences in the ion conduction mechanism between Na_v and K_v channels due to the different pore sizes were highlighted by classical MD simulations on the Na_vAb channel: (i) Na^+ ions were found to be always hydrated to some extent along the permeation pathway, and (ii) Na^+ ions can be located next to each other in the selectivity filter, contrary to the single-file configuration of K^+ ions in K_v channels (Corry & Thomas, 2012; Furini & Domene, 2012). Another theoretical study on the open-pore Na_vMs channel found that water flux may not be correlated with ion flux in Na_v channels since the average dwell time of water molecules in the channel was found to be two orders of magnitude lower than that of Na^+ ions (Ulmschneider et al., 2013). However, this contradicted the results of previous classical MD simulations for the Na_vAb channel in a closed conformation (Carnevale et al., 2011), which predicted a larger dwell time in the channel for the ions than for the water. However, these apparently contradictory results for two channels that present the same sequence in the selectivity filter (TLESWS) could be a consequence of having different gating states in the investigated channels (an open structure for Na_vMs and a closed one for Na_vAb).

An important feature of ion conduction that was evidenced by MD (Boiteux, Vorobyov, & Allen, 2014; Carnevale et al., 2011; Corry & Thomas, 2012; Furini & Domene, 2012; Stock et al., 2013; Ulmschneider et al., 2013) and BD simulations (Vora, Corry, & Chung, 2005; Vora et al., 2008) is that ion transport in Na_v occurs through a multi-ion mechanism, in which the selectivity filter is populated by more than one ion at any time. The following multi-ion permeation mechanism, displayed in Figure 2.4B, through the selectivity filter of Na_vAb was proposed based on umbrella-sampling MD simulations (Corry & Thomas, 2012): (i) A first ion enters the pore and binds to the S_{HFS} site; (ii) a second ion is attracted close to the position of the first one; (iii) the first ion is able to permeate to the inner positions S_{CEN} and S_{IN} while the second ion moves to S_{HFS}; and (iv) the inner ion can permeate to the central cavity of the channel. Along the whole permeation pathway, Na^+ ions move through the channel with their first solvation shell almost complete. The comparison of the free-energy profiles for one-ion and two-ion pathways suggested that the conduction of an ion along the filter is favored by the presence of a second ion. However, the relatively low-energy barriers obtained along the single-ion permeation mechanism suggest that transport of individual ions can also take place. The motion of Na^+ ions in the multi-ion mechanism was found to be less correlated than the motion of K^+ ions through K_v channels; therefore, the Na^+ permeation mechanism through Na_vAb was termed "loosely coupled knock-on conduction." A contrary, strong knock-on mechanism, shown in Figure 2.4C, involving the participation of three ions in the selectivity filter of Na_vAb, was shown by unbiased classical MD simulations (Boiteux et al., 2014). These simulations also predicted the presence of a three-ion pass-way mechanism, where two ions coexist at the EEEE ring before pushing a third ion into the cavity of the protein. All these different results reported for the same ion channel indicate that it is very likely that multiple permeation pathways are operative in the transport of Na^+, and that a simplified picture describing the process is difficult to draw.

Two quite relevant steps in Na^+ conduction, independent of the permeation mechanism, are the attraction of the Na^+ ions from the extracellular media to the S_{HFS} site, composed of the EEEE ring, and the subsequent permeation from S_{HFS} to the inner binding sites. The second process involves overcoming a free-energy barrier, which was found to be ≥2 kcal/mol larger for K^+ and Ca^{2+} than for Na^+ (Corry, 2013; Corry

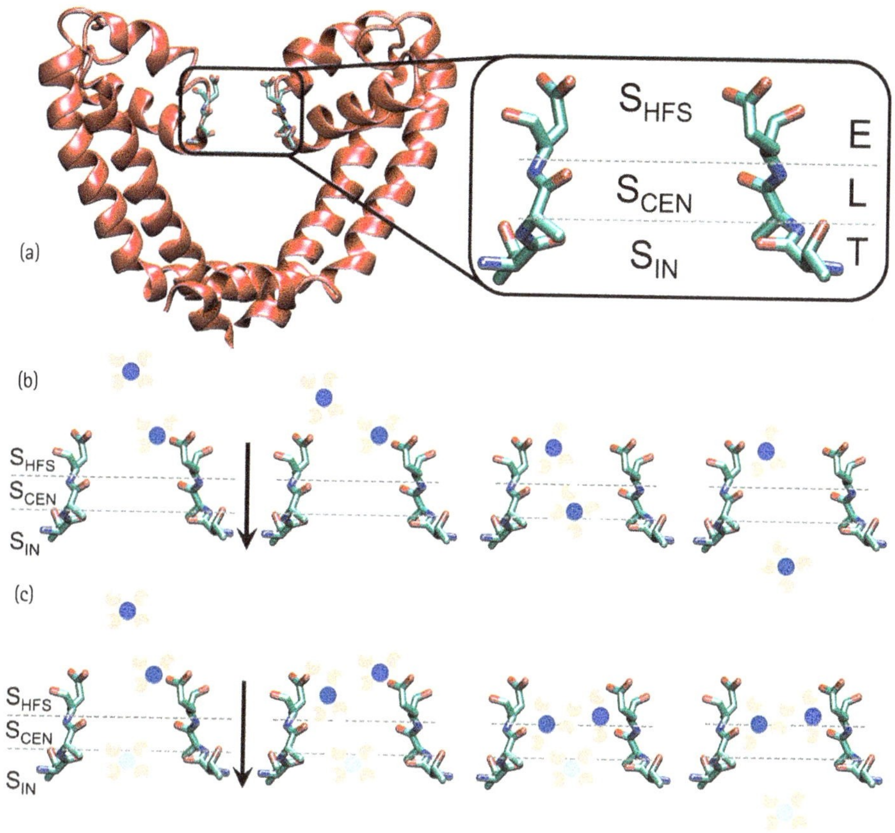

FIGURE 2.4 Permeation mechanism of Na^+ ions through a Na^+ channel.
A. Structure of the pore of the Na_VAb channel with half of the pore removed for clarity. Zoom-in view of the selectivity filter with the TLE sequence involved in the three binding sites. The hydrogen atoms and side chain of leucine are not shown. **B.** Loosely coupled knock-on mechanism in which two solvated Na^+ ions permeate through the selectivity filter. **C.** Strong knock-on mechanism in which three solvated Na^+ ions permeate. The black arrow in B and C indicates the direction of ion motion (influx transport). C, N, and O atoms are displayed in cyan, blue, and red, respectively. Na^+ ions inside the filter are shown in different shades of blue, and the O and H atoms of water in light red and in gray, respectively.

& Thomas, 2012; Furini & Domene, 2012), explaining the preference of Na_VAb for Na^+ over K^+. At the transition point, the solvated ions are located into the plane formed by the four glutamates and, therefore, the available space is limited. Due to its smaller size, the Na^+ ion and its solvating water molecules fit better than the solvated K^+ ion into the EEEE plane. As a consequence, the water molecules solvating Na^+ can adopt an ideal configuration where the intermolecular interactions are more favorable than in the case of solvated K^+ (Corry & Thomas, 2012). Whole-cell patch clamping measurements have also shown that the EEEE ring is crucially involved in ion permeability and selectivity of the Na_VMs channel (Naylor et al., 2016). When the selectivity filter of the channel was mutated from TLESWS to TLDSWS—i.e., the glutamates of the EEEE ring were replaced by aspartates—no electronic density was found at the S_{HFS} binding site of the crystal structure. This means that the ion binding site disappeared upon mutation. As a consequence, the mutant channel presented a lower current density and selectivity for Na^+ than the wild-type channel. The protonation state of the glutamate residues of the EEEE ring is also important for the ion transport mechanism. Umbrella-sampling MD simulations have shown that the side chains of protonated glutamate residues are preferentially oriented towards the protein cavity when compared to unprotonated residues (Furini, Barbini, & Domene, 2014). This induces an increase of the energy barrier for Na^+ translocation and therefore reduces ion conduction. Despite the importance of the S_{HFS} site in the transport of Na^+, additional mutations at different sites of the selectivity filter also strongly affect permeation and selectivity (Shaya et al., 2011; Yue et al., 2002). This result points towards a complex permeation mechanism in which several regions of the selectivity filter are involved.

The conduction mechanism in eukaryotic Na_v channels is much less understood than that in bacterial Na_v channels due to the lack of atomistic structures, which precluded the performance of theoretical simulations up to now. Although a clear mechanistic picture has not emerged yet, several factors are believed to contribute to Na^+ selectivity, including the coordination number of the ion and its partial solvation/desolvation, the electron-donating ability of the residues of the protein pore, and the rigidity and size of the selectivity filter (Dudev & Lim, 2014). The lysine residue present in the DEKA motif is known to be indispensable in Na^+ discrimination. Mutation of the lysine residue in the selectivity filter to another amino acid largely eliminates selectivity for Na^+ ions (Favre et al., 1996; Schlief, Schönherr, Imoto, & Heinemann, 1996; Sun, Favre, Schild, & Moczydlowski, 1997), even when the charge of the filter is preserved. Indeed, not only the presence of lysine is relevant for the conduction mechanism, but also its position in the DEKA ring. Swopping the glutamate and lysine residues from the wild-type DEKA motif to DKEA reduced the Na^+:K^+ permeability ratio by a factor of three (Schlief et al., 1996). This behavior was explained based on quantum-mechanical calculations on simplified models of the DEKA and DKEA motifs (Dudev & Lim, 2014). The interactions between lysine and the other residues in the DEKA ring are stronger than those in the DKEA ring. Therefore, the DEKA configuration makes the pore narrower and more rigid, favoring the permeation of the smaller ion.

Since the selectivity filter of mammalian Na_v channels presents certain similarities with the bacterial Na_vRh channel, a mutant version of the latter has been employed in classical MD simulations to get insight into the conduction mechanisms in eukaryotic channels (Xia, Liu, Li, Yan, & Gong, 2013). Specifically, the first ring of serine residues located at the selectivity filter (TLSSWE) was mutated to DEKA in order to mimic the selectivity filter of mammalian Na_v channels. Three ion binding sites were identified along the simulations, suggesting that a multi-ion mechanism seems to also be operative in eukaryotic channels. Recent voltage-biased MD simulations, where the pore of the DEKA-mutated Na_vRh channel was opened by alignment with the open-conformation of the bacterial Na_vMs channel, investigated the ion permeation mechanism through the full protein (Li et al., 2017). It was found that the multiple ions that can populate the selectivity filter are able to permeate from the filter to the central cavity of the protein through a knock-on mechanism. It was speculated that the coupled ion motion is induced by the presence of the DEKA motif, since ions permeated independently in a non-coupled manner through the wild-type Na_vRh channel. The first theoretical study employing a real structure from a eukaryotic channel (Na_vPaS) was reported only recently (Zhang et al., 2018). Unbiased classical MD simulations showed that Na^+ ions permeate below the DEKA ring of the selectivity filter partially hydrated and in an asymmetrical manner, strongly interacting with the negatively charged aspartate and glutamate residues of the DEKA motif. In contrast, K^+ ions were not able to permeate below the DEKA ring along 300 ns simulations. Despite the relevant mechanistic details obtained from these simulations, the complete absence of K^+ ions, together with the small number of Na^+ ions, inside the selectivity filter during the unbiased simulations might indicate that the results are not fully converged. Thus, better sampling could be needed to obtain more reliable conclusions. The availability of the recently crystalized Na_vPaS (Shen et al., 2017), EeNa_v1.4 (Yan et al., 2017), and Nav1.4 (Pan et al., 2018) eukaryotic channels will encourage further theoretical studies in the coming years, which will provide significant insight into the mechanism of ion conduction through eukaryotic Na_v channels.

CALCIUM CHANNELS

Ca_v channels are one of the main ways that electrical signals are turned into responses in cells. They are extremely discriminating in terms of the ions that they let pass, selecting Ca^{2+} over Na^+ at a ratio of over 1000:1 (Hess, Lansman, & Tsien, 1986). This degree of selectivity is all the more remarkable given that Na^+ is usually many orders of magnitude more numerous than Ca^{2+} in physiological conditions. Ca^{2+} channels must rely on more than just the size of passing ions to discriminate between them, as Na^+ and Ca^{2+} have very similar diameters. Our understanding of ion permeation and selectivity in these channels is that they are intimately linked, as the blocking of one ion species by another has been the prime datum for explaining both properties.

Ca_v channels can be classified according to the degree of membrane depolarization needed to activate them as high-voltage activated channels and low-voltage activated channels. The former ones are protein complexes composed by a pore-forming α1 subunit in addition to ancillary β, α2δ, and γ subunits (see Figure 2.1). Low-voltage activated channels lack the ancillary subunits and possess only the pore region (Catterall, Perez-Reyes, Snutch, & Striessnig, 2005; Simms & Zamponi, 2014). As in Na^+ and K^+ channels, the α1 subunit includes the pore, the voltage sensor, and the gating apparatus of the protein. It comprises four homologous domains composed of six transmembrane fragments (TM1–TM6) covalently linked in a single chain. The selectivity filter is also found in the pore loop between segments TM5 and TM6.

Surprisingly, monovalent ions such as Na^+ conduct through these channels at higher rates than divalent ions when no divalent ions are present (Almers, Mccleskey, & Palade, 1984; Fukushima & Hagiwara, 1985; Hess et al., 1986; Kostyuk, Mironov, & Shuba, 1983; Kuo & Hess, 1993b). However, such monovalent currents are blocked when the Ca^{2+} concentration reaches only 1 μM (Almers et al., 1984; Kostyuk et al., 1983). The ability for some ions to block others provides evidence for the high-affinity binding of ions to the pore—the ions that bind most strongly would block the passage of other ions. Thus, working out which ions block other ions suggests which ions are able to most strongly bind to the pore. A range of experiments suggested that the affinity for the pore follows the following sequence: $La^{3+} > Cd^{2+} > Co^{2+} > Ca^{2+} > Sr^{2+} > Ba^{2+} > Li^+ > Na^+ > K^+ > Cs^+$ (Fenwick, Marty, & Neher, 1982; Hess et al., 1986; Lee & Tsien, 1984; Reuter & Scholz, 1977; Taylor, 1988). However, there are some exceptions to this trend. For example, Na^+ ions can attenuate Ca^{2+} currents in some cases. This can be explained by competition for entry to the pore—i.e., kinetic effects—rather than occupancy of the high-affinity binding site (Corry, Allen, Kuyucak, & Chung, 2001; Polo-Parada & Korn, 1997). Single-channel conductance measurements follow the exact opposite order to that for blocking: $La^{3+} < Cd^{2+} < Co^{2+} < Ca^{2+} < Sr^{2+} < Ba^{2+} < Li^+ < Na^+ < K^+ < Cs^+$ (Hess et al., 1986; Kuo & Hess, 1993a, 1993b). This has been explained by the "sticky pore" hypothesis, in which the ions with the highest affinity pass through the pore more slowly, yielding a lower conductance (Bezanilla & Armstrong, 1972).

A consequence of the sticky pore model is that permeation through Ca^{2+} channels must be a multi-ion process. If the channel binds ions strongly, then it would be hard to generate pico-ampere currents measured with a single-ion pore (Bezanilla & Armstrong, 1972). Furthermore, a number of other strands of evidence support multi-ion permeation. In many cases when two permeating species are mixed, the current through the pore is smaller than for either ion on its own. This so-called anomalous mole fraction effect is seen clearly in mixtures of Ca^{2+}/Ba^{2+} and Ca^{2+}/Na^+ (Almers et al., 1984; Hess & Tsien, 1984). It is most easily explained if the pore holds multiple ions, as in a single-ion pore the current should always lie between the currents found with each ion type on its own. In a multi-ion pore, unfavorable interactions between the ions can slow permeation (although it is not possible to completely exclude a single-ion pore if it undergoes some kind of conformational alteration during permeation). Blockage experiments also support a multi-ion pore. Ca^{2+} blocks Na^+ currents with an affinity of ~1 μM, but Ca^{2+} currents saturate with an affinity of 14 mM. This can be explained if the first affinity represents binding of Ca^{2+} to an empty pore or a pore holding Na^+, but the second affinity represents a second Ca^{2+} binding to a pore already holding one divalent ion.

Site-directed mutagenesis showed that four glutamate residues—one from each P loop of the channel—were responsible for generating the high-affinity Ca^{2+} binding site. Removing any one of these reduces the affinity for Ca^{2+} as well as the conductance and specificity for Ca^{2+} (Ellinor, Yang, Sather, Zhang, & Tsien, 1995; Kim, Morii, Sun, Imoto, & Mori, 1993; Parent & Gopalakrishnan, 1995; Yang, Ellinor, Sather, Zhang, & Tsien, 1993). As there was no evidence for any other high-affinity binding site, this so-called

EEEE locus has been presumed to be the sole cause of high-affinity binding and selectivity, as it is the case in bacterial Na^+ channels. However, reconciling how multi-ion conduction can occur with only a single high-affinity biding site has been a lasting challenge.

The nature of the permeation process makes more sense in the light of recent structural data on Ca_v channels. The first of these structures made use of the similarity (and probable evolutionary connection) of Ca^{2+} channels to Na^+ channels, noting that Ca^{+2} selectivity can be inferred on a Na^+ channel and vice versa, with only a few point mutations (Heinemann, Terlau, Stuhmer, Imoto, & Numa, 1992; Yue et al., 2002). Conferring Ca^{2+} selectivity upon the bacterial Na^+ channel NavAb by mutating its selectivity filter from TLESWSM to TLDDWSD (to generate the CavAb channel) showed a selectivity filter with three potential ion binding sites (Tang et al., 2014), as shown in Figure 2.5 (note that this channel has a DDDD locus rather than EEEE). The site 1 at the entrance of the pore is formed by the carboxyl groups of the ring formed by four aspartate residues. Deeper in the pore, the acidic side chains of another aspartate ring and the carbonyl oxygen atoms of leucine residues form the site 2. Finally, the carbonyls from four threonines form the site 3. It is suggested that Ca^{2+} ions pass through the pore in a knock-off mechanism (see Figure 2.5A) in which the pore cycles between a state with one hydrated Ca^{2+} in the central site and a state with two hydrated ions at the distal sites—similar to the model seen in early BD studies (Corry et al., 2001). This mechanism reconciles the apparently contradictory data that multiple ions are required for rapid permeation, but only a single high-affinity site exits, as only the central site has sufficient affinity to block monovalent currents. A knock-off mechanism was also predicted by classical MD simulations for the TRPV6 channel, but the pore showed a larger ion occupancy cycling between two-ion and three-ion states (Sakipov, Sobolevsky, & Kurnikova, 2018). Thus, it is very likely that several permeation mechanisms simultaneously operate, as in the case of Na^+ and K^+ channels.

Eukaryotic Ca^{2+} channels are made from a single protein chain containing four homologous repeats that surround the pore. The structures of Ca_v1.1 show that this can yield an asymmetrical selectivity filter in which ions bind off-axis (Wu et al., 2016; Wu et al., 2015). While the structure of the selectivity filter is not as well defined in these as for Ca_vAb, it suggests at least two sites at which hydrated Ca^{2+} can bind. It is not clear if this supports the knock-off mechanism, but as for Ca_vAb, only one of the two sites is likely to yield high-affinity binding. It is important to note that the selectivity filters of Ca_vAb and Ca_v1.1 are different. For example, it is remarkable that the EEEE locus is missing in Ca_vAb and, therefore, the permeation and selectivity mechanisms between these two channels might significantly differ.

The multi-ion nature of permeation has been modeled using a number of different approaches. Rate-theory models showed that a pore holding two binding sites can utilize repulsion between the ions to generate high throughput as well as specificity (Almers et al., 1984; Armstrong & Neyton, 1991; Hess & Tsien, 1984; Kuo & Hess, 1993b; Yang et al., 1993). But these models have difficulty in defining the physical causes of the binding sites themselves (Corry & Hool, 2007). BD simulations showed how multiple

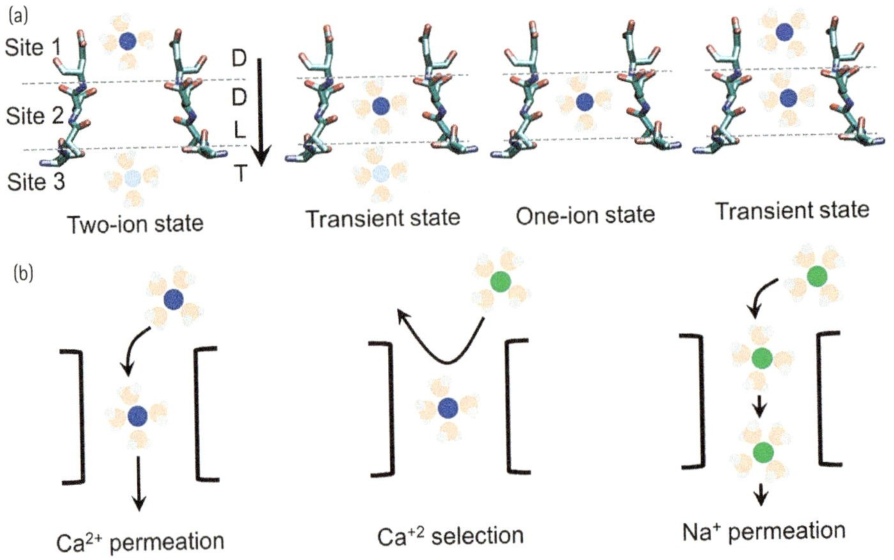

FIGURE 2.5 Permeation and selectivity mechanisms of Ca^{2+} ions through an artificial Ca^{2+} channel (Ca_vAb).
A. Selectivity filter of the Ca_vAb channel with the TLDD sequence involved in the three binding sites. The hydrogen atoms and side chain of leucine are not shown. Ion permeations follow a knock-off mechanism where repulsion between adjacent ions accelerates ion conduction. The black arrow indicates the direction of ion motion (influx transport). C, N, and O atoms are displayed in cyan, blue, and red, respectively. Ca^{2+} ions inside the filter are shown in different shades of blue, and the O and H atoms of water in light red and in gray, respectively. **B.** Schematic representation of conduction events inside a Ca_v channel when different ions are present. Ca^{+2} and Na^+ ions are displayed in blue and green, respectively.

ions can compete to occupy the channel to create rapid transport and selectivity, in a set of studies that closely predicted the inferences from the structural studies (Corry et al., 2001; Corry & Chung, 2006; Corry, Vora, & Chung, 2005). In these, the pore is seen to always hold at least one Ca^{2+} ion that is attracted with high affinity to the ring of glutamate residues. The resident ion can only be displaced by the entrance of another Ca^{2+} ion, with the electrostatic repulsion between the two ions being sufficient to overcome the attraction of the first ion to the protein, as is represented in Figure 2.5B. In contrast, permeation in pure Na^+ solution is seen to involve three ions. Selectivity can be simply explained in electrostatic terms, as once Ca^{2+} occupies the pore it can only be replaced by another Ca^{2+} ion, as the repulsion from an incoming Na^+ is insufficient to eject the resident ion due to the strong attraction of the divalent ion to the negatively charged protein.

An alternative mechanism for generating selectivity was proposed from studies using continuum-based theories that suggested that ions compete to achieve charge neutrality with the protein inside a selectivity filter having limited space (Boda, Busath, Henderson, & Sokolowski, 2000; Nonner, Catacuzzeno, & Eisenberg, 2000). Divalent ions are suggested to easily neutralize the protein while occupying little volume, but not enough Na^+ ions can fit into the filter to achieve neutrality. While conceptually appealing and able to explain selection among a range of different ion types (Gillespie, 2008; Gillespie, Xu, Wang, & Meissner, 2005), this model requires specific restraints on the filter volume that do not seem to correspond to the larger filters seen in the structural studies that can hold multiple hydrated ions. The determination of a high-resolution structure of the selectivity filter of a Ca_V channel may elucidate allow the specific reasons for Ca^{2+}/Na^+ selection.

Summary

Voltage-gated ion channels are involved in many relevant biological processes of prokaryotic and eukaryotic cells. Despite their indisputable biological relevance, the mechanisms by which ion channels are able to transport ions in a selective way across cell membranes are not yet completely understood. Electrophysiological measurements and the analysis of crystalized and cryo-electron microscopy structures have provided significant knowledge. However, current experimental techniques are not able to access many microscopic mechanistic details of ion permeation and the origins of selectivity and can be difficult to interpret. For example, it was inferred based on early structural information that K^+ selectivity in K_V channels arises through a snug-fit mechanism, which appears incorrect in the light of further structures and dynamics simulations. In recent decades, a vast number of theoretical studies based on dynamics simulations has investigated the mechanism of ion transport through ion channels. Dynamics simulations are able to provide a time-dependent picture and great microscopic mechanistic insight. However, simulations should be interpreted carefully and validated against experimental or higher-level theoretical data, because they can provide erroneous or contradictory results. This can be seen in cases in which simulations conducted by different groups have yielded different results, such as ions spaced by water or in direct contact in K_V channels, presumably due to small differences in the force fields and computational implementation. In addition, the most commonly used theoretical models are not able to properly describe some intermolecular interactions that are likely to play a fundamental role in ion conduction, such as polarization, exchange, and charge-transfer interactions.

We have summarized in this chapter the main mechanistic knowledge gained in the last decades in ion transport through K_V, Na_V, and Ca_V channels. Ion permeation most likely occurs with multiple ions via multiple mechanisms in all these channels. Even in the tightly constrained single-file K_V selectivity filter, two permeation pathways probably coexist in which ions are either separated by water or in direct contact. Since Na_V channels present a wider selectivity filter, their mechanisms of permeation are more diverse than for K_V channels. In Na_V channels, ions permeate almost fully hydrated, and two ions can be located next to each other inside the filter. Conduction through bacterial Na_V channels probably occurs via multiple mechanisms, including single-ion conduction, loosely coupled knock-on of two or three ions, or more strongly coupled motions of multiple ions. The conduction mechanism in eukaryotic Na_V channels is less clear than that in bacteria due to there being few computational studies to make use of the recent eukaryotic structures. Ca_V channels also have a wider filter, although it is not clear if ions can bypass each other. Experimental and theoretical work on Ca_V channels points towards a multi-ion knock-off mechanism, in which the system cycles between a state with one hydrated Ca^{2+} bound to the central EEEE binding site and a state with two hydrated ions bound to two distal sites.

How these channels select between different ions is even more difficult to explain. Several reasons behind the selectivity of K_V channels have been suggested, but the common factor in these is that the channels make use of the fact that is easier to dehydrate K^+ than Na^+ as required to enter a narrow pore. However, how the channel is able to balance the dehydration barrier to allow rapid permeation of K^+ but not Na^+ is less clear. Thermodynamic explanations suggest preferential attraction to the filter due to the number and dipole moment of the ligands and the flexibility of the protein. Kinetic effects can also play an important role in selectivity since the energy barriers that K^+ ions have to overcome when moving between binding sites are smaller than the energy barriers for other ions.

Na^+ selectivity in bacterial Na_V channels has both a thermodynamic and a kinetic origin. The binding of Na^+ ions to the EEEE ring is more favorable than the binding of other ions, and the escape from the EEEE site to a different binding site inside the selectivity filter requires overcoming a smaller energy barrier for Na^+ than for other ions. The reasons behind selectivity in eukaryotic Na_V channels are also not clear yet, but it is thought that several factors may be crucial, including the ion solvation/desolvation energy, the coordination number of the ion, the existence of charge-transfer processes between the protein and the solvated ions, and the rigidity of the selectivity filter. In addition, conductivity measurements have clearly shown that the lysine residue of the DEKA ring is crucially involved in selectivity. However, its specific role is unknown.

Selectivity in Ca_V channels seems to be explained by the strong electrostatic interaction between Ca^{2+} and the selectivity filter. Once a Ca^{2+} ion is bound to the EEEE ring, only another Ca^{2+} ion is able to replace it, and the repulsion between the ions favors the permeation process. However, additional theories have emerged. For example, theoretical studies suggest that selectivity arises when ions with a large charge:volume ratio, like divalent ions, are able to neutralize the negative charge of the selectivity filter.

Despite the tremendous progress achieved in the knowledge of ion permeation and selectivity in ion channels, even the most widely studied example, K^+ channels, has many unresolved questions. For example, how many ions are involved in a permeation event? What are the steps in permeation? What is the most important factor in deriving selectivity for K^+? Theoretical simulations will certainly contribute to answering these questions. However, more accurate models, going beyond fixed-charges force fields, able to properly describe all types of interactions present in the system will be required.

Acknowledgements

We thank the Australian Research Council (Grant FT130100781) and the Australian National University for financial support.

References

Adcock, S. A., & McCammon, J. A. (2006). Molecular dynamics: Survey of methods for simulating the activity of proteins. *Chemical Reviews, 106*(5), 1589–1615. doi:10.1021/cr040426m

Alam, A., & Jiang, Y. (2009). Structural analysis of ion selectivity in the Nak channel. *Nature Structural and Molecular Biology, 16*(1), 35–41. doi:10.1038/nsmb.1537

Alcayaga, C., Cecchi, X., Alvarez, O., & Latorre, R. (1989). Streaming potential measurements in $Ca2^+$-activated K^+ channels from skeletal and smooth muscle. Coupling of ion and water fluxes. *Biophysical Journal, 55*(2), 367–371. doi:10.1016/S0006-3495(89)82814-0

Almers, W., McCleskey, E. W., & Palade, P. T. (1984). A non-selective cation conductance in frog-muscle membrane blocked by micromolar external calcium-ions. *Journal of Physiology–London, 353*(Aug), 565–583. doi:10.1113/jphysiol.1984.sp015351

Ando, H., Kuno, M., Shimizu, H., Muramatsu, I., & Oiki, S. (2005). Coupled K^+-water flux through the hERG potassium channel measured by an osmotic pulse method. *Journal of General Physiology, 126*(5), 529–538. doi:10.1085/jgp.200509377

Aqvist, J., Alvarez, O., & Eisenman, G. (1992). Ion-selective properties of a small ionophore in methanol studied by free-energy perturbation simulations. *Journal of Physical Chemistry, 96*(24), 10019–10025.

Aqvist, J., & Luzhkov, V. (2000). Ion permeation mechanism of the potassium channel. *Nature, 404*(6780), 881–884. doi:10.1038/35009114

Armstrong, C. M., & Neyton, J. (1991). Ion permeation through calcium channels—a one-site model. *Annals of the New York Academy of Sciences, 635*, 18–25. doi:10.1111/j.1749-6632.1991.tb36477.x

Armstrong, C. M., & Taylor, S. R. (1980). Interaction of barium ions with potassium channels in squid giant axons. *Biophysical Journal, 30*(3), 473–488. doi:10.1016/S0006-3495(80)85108-3

Bagal, S. K., Brown, A. D., Cox, P. J., Omoto, K., Owen, R. M., Pryde, D. C., . . . Swain, N. A. (2013). Ion channels as therapeutic targets: A drug discovery perspective. [Review]. *Journal of Medicinal Chemistry, 56*(3), 593–624. doi:10.1021/jm3011433

Bagnéris, C., Decaen, P. G., Hall, B. A., Naylor, C. E., Clapham, D. E., Kay, C. W. M., & Wallace, B. A. (2013). Role of the C-terminal domain in the structure and function of tetrameric sodium channels. [Article]. *Nature Communications, 4*, 2465. doi:10.1038/ncomms3465

Berneche, S., & Roux, B. (2001). Energetics of ion conduction through the K^+ channel. *Nature, 414*(6859), 73–77. doi:10.1038/35102067

Berneche, S., & Roux, B. (2003). A microscopic view of ion conduction through the K^+ channel. *Proceedings of the National Academy of Sciences of the United States of America, 100*(15), 8644–8648. doi:10.1073/pnas.1431750100

Bezanilla, F., & Armstrong, C. M. (1972). Negative conductance caused by entry of sodium and cesium ions into potassium channels of squid axons. *Journal of General Physiology, 60*(5), 588-608.

Blatz, A. L., & Magleby, K. L. (1984). Ion conductance and selectivity of single calcium-activated potassium channels in cultured rat muscle. *Journal of General Physiology, 84*(1), 1–23.
Block, B. M., & Jones, S. W. (1996). Ion permeation and block of M-type and delayed rectifier potassium channels. Whole-cell recordings from bullfrog sympathetic neurons. *Journal of General Physiology, 107*(4), 473–488.
Boda, D., Busath, D. D., Henderson, D., & Sokolowski, S. (2000). Monte Carlo simulations of the mechanism for channel selectivity: The competition between volume exclusion and charge neutrality. *Journal of Physical Chemistry B, 104*(37), 8903–8910. doi:10.1021/jp0019658
Boiteux, C., Vorobyov, I., & Allen, T. W. (2014). Ion conduction and conformational flexibility of a bacterial voltage-gated sodium channel. [Article]. *Proceedings of the National Academy of Sciences of the United States of America, 111*(9), 3454–3459. doi:10.1073/pnas.1320907111
Bostick, D. L., Arora, K., & Brooks, C. L. (2009). K^+/Na^+ selectivity in toy cation binding site models is determined by the "host." *Biophysical Journal, 96*(10), 3887–3896. doi:10.1016/j.bpj.2008.12.3963
Bostick, D. L., & Brooks, C. L. (2007). Selectivity in K^+ channels is due to topological control of the permeant ion's coordinated state. *Proceedings of the National Academy of Sciences of the United States of America, 104*(22), 9260–9265. doi:10.1073/pnas.0700554104
Brohawn, S. G., del Marmol, J., & MacKinnon, R. (2012). Crystal structure of the human K2p Traak, a lipid- and mechano-sensitive K^+ ion channel. *Science, 335*(6067), 436–441. doi:10.1126/science.1213808
Carnevale, V., Treptow, W., & Klein, M. L. (2011). Sodium ion binding sites and hydration in the lumen of a bacterial ion channel from molecular dynamics simulations. *Journal of Physical Chemistry Letters, 2*(19), 2504–2508. doi:10.1021/jz2011379
Catterall, W. A. (2012). Voltage-gated sodium channels at 60: Structure, function and pathophysiology. [Review]. *Journal of Physiology, 590*(11), 2577–2589. doi:10.1113/jphysiol.2011.224204
Catterall, W. A., Perez-Reyes, E., Snutch, T. P., & Striessnig, J. (2005). International Union of Pharmacology. Xlviii. Nomenclature and structure-function relationships of voltage-gated calcium channels. [Review]. *Pharmacological Reviews, 57*(4), 411–425. doi:10.1124/pr.57.4.5
Catterall, W. A., & Swanson, T. M. (2015). Structural basis for pharmacology of voltage-gated sodium and calcium channels. [Short survey]. *Molecular Pharmacology, 88*(1), 141–150. doi:10.1124/mol.114.097659
Chung, S. H., Allen, T. W., Hoyles, M., & Kuyucak, S. (1999). Permeation of ions across the potassium channel: Brownian dynamics studies. *Biophysical Journal, 77*(5), 2517–2533. doi:10.1016/S0006-3495(99)77087-6
Chung, S. H., Allen, T. W., & Kuyucak, S. (2002). Conducting-state properties of the Kcsa potassium channel from molecular and Brownian dynamics simulations. *Biophysical Journal, 82*(2), 628–645. doi:10.1016/S0006-3495(02)75427-1
Chung, S. H., & Corry, B. (2005). Three computational methods for studying permeation, selectivity and dynamics in biological ion channels. [Review]. *Soft Matter, 1*(6), 417–427. doi:10.1039/b512455g
Clapham, D. E. (2007). Calcium signaling. [Review]. *Cell, 131*(6), 1047–1058. doi:10.1016/j.cell.2007.11.028
Corry, B. (2013). $Na^+/Ca2^+$ selectivity in the bacterial voltage-gated sodium channel Navab. *PeerJ, 2013*(1), e16. doi:10.7717/peerj.16
Corry, B. (2017). Chapter 6: Computer simulation of ion channels. In *Computational biophysics of membrane proteins* (pp. 161–196). Cambridge The Royal Society of Chemistry.

Corry, B., Allen, T. W., Kuyucak, S., & Chung, S. H. (2001). Mechanisms of permeation and selectivity in calcium channels. *Biophysical Journal, 80*(1), 195–214. doi:10.1016/S0006-3495(01)76007-9

Corry, B., & Chung, S. H. (2006). Mechanisms of valence selectivity in biological ion channels. *Cellular and Molecular Life Sciences, 63*(3), 301–315. doi:10.1007/s00018-005-5405-8

Corry, B., & Hool, L. C. (2007). Calcium channels. In S. H. Chung, O. S. Andersen, & V. Krishnamurthy (Eds.), *Biological membrane ion channels: Dynamics, structure and applications* (pp. 241–299). New York: Springer.

Corry, B., Kuyucak, S., & Chung, S. H. (2000). Tests of continuum theories as models of ion channels. II. Poisson-Nernst-Planck theory versus Brownian dynamics. [Article]. *Biophysical Journal, 78*(5), 2364–2381. doi:10.1016/s0006-3495(00)76781-6

Corry, B., & Thomas, M. (2012). Mechanism of ion permeation and selectivity in a voltage gated sodium channel. [Article]. *Journal of the American Chemical Society, 134*(3), 1840–1846. doi:10.1021/ja210020h

Corry, B., Vora, T., & Chung, S. H. (2005). Electrostatic basis of valence selectivity in cationic channels. *Biochimica et Biophysica Acta–Biomembranes, 1711*(1), 72–86. doi:10.1016/j.bbamem.2005.03.002

Derebe, M. G., Sauer, D. B., Zeng, W., Alam, A., Shi, N., & Jiang, Y. (2011). Tuning the ion selectivity of tetrameric cation channels by changing the number of ion binding sites. *Proceedings of the National Academy of Sciences of the United States of America, 108*(2), 598–602. doi:10.1073/pnas.1013636108

Doyle, D. A., Cabral, J. M., Pfuetzner, R. A., Kuo, A. L., Gulbis, J. M., Cohen, S. L., & MacKinnon, R. (1998). The structure of the potassium channel: Molecular basis of K^+ conduction and selectivity. *Science, 280*(5360), 69–77. doi:10.1126/science.280.5360.69

Dudev, T., Grauffel, C., & Lim, C. (2016). Influence of the selectivity filter properties on proton selectivity in the influenza A M2 channel. *Journal of the American Chemical Society, 138*(39), 13038–13047. doi:10.1021/jacs.6b08041

Dudev, T., & Lim, C. (2014). Ion selectivity strategies of sodium channel selectivity filters. *Accounts of Chemical Research, 47*(12), 3580–3587. doi:10.1021/ar5002878

Egwolf, B., & Roux, B. (2010). Ion selectivity of the Kcsa channel: A perspective from multi-ion free energy landscapes. *Journal of Molecular Biology, 401*(5), 831–842. doi:10.1016/j.jmb.2010.07.006

Eisenman, G. (1962). Cation selective glass electrodes and their mode of operation. *Biophysical Journal, 2*(2 Pt 2), 259–323.

Ellinor, P. T., Yang, J., Sather, W. A., Zhang, J. F., & Tsien, R. W. (1995). $Ca2^+$ Channel selectivity at a single-locus for high-affinity $Ca2^+$ interactions. *Neuron, 15*(5), 1121–1132. doi:10.1016/0896-6273(95)90100-0

Favre, I., Moczydlowski, E., & Schild, L. (1996). On the structural basis for ionic selectivity among Na^+, K^+, and $Ca2^+$in the voltage-gated sodium channel. [Article]. *Biophysical Journal, 71*(6), 3110–3125. doi:10.1016/s0006-3495(96)79505-x

Fenwick, E. M., Marty, A., & Neher, E. (1982). Sodium and calcium channels in bovine chromaffin cells. *Journal of Physiology, 331*, 599–635.

Fowler, P. W., Tai, K., & Sansom, M. S. (2008). The selectivity of K^+ ion channels: Testing the hypotheses. *Biophysical Journal, 95*(11), 5062–5072. doi:10.1529/biophysj.108.132035

Fukushima, Y., & Hagiwara, S. (1985). Currents carried by monovalent cations through calcium channels in mouse neoplastic B lymphocytes. *Journal of Physiology, 358*, 255–284.

Furini, S., Barbini, P., & Domene, C. (2014). Effects of the protonation state of the EEEE motif of a bacterial Na^+-channel on conduction and pore structure. [Article]. *Biophysical Journal, 106*(10), 2175–2183. doi:10.1016/j.bpj.2014.04.005
Furini, S., & Domene, C. (2009). Atypical mechanism of conduction in potassium channels. *Proceedings of the National Academy of Sciences of the United States of America, 106*(38), 16074–16077. doi:10.1073/pnas.0903226106
Furini, S., & Domene, C. (2012). On conduction in a bacterial sodium channel. [Article]. *PLoS Computational Biology, 8*(4), e1002476. doi:10.1371/journal.pcbi.1002476
Furini, S., & Domene, C. (2018). Ion-triggered selectivity in bacterial sodium channels. [Article]. *Proceedings of the National Academy of Sciences of the United States of America, 115*(21), 5450–5455. doi:10.1073/pnas.1722516115
Gillespie, D. (2008). Energetics of divalent selectivity in a calcium channel: The ryanodine receptor case study. *Biophysical Journal, 94*(4), 1169–1184. doi:10.1529/biophysj.107.116798
Gillespie, D., Xu, L., Wang, Y., & Meissner, G. (2005). (De)constructing the ryanodine receptor: Modeling ion permeation and selectivity of the calcium release channel. *Journal of Physical Chemistry B, 109*(32), 15598–15610. doi:10.1021/jp052471j
Grizel, A. V., Glukhov, G. S., & Sokolova, O. S. (2014). Mechanisms of activation of voltage-gated potassium channels. [Review]. *Acta Naturae, 6*(23), 10–26.
Gumbart, J., Khalili-Araghi, F., Sotomayor, M., & Roux, B. (2012). Constant electric field simulations of the membrane potential illustrated with simple systems. [Article]. *Biochimica et Biophysica Acta–Biomembranes, 1818*(2), 294–302. doi:10.1016/j.bbamem.2011.09.030
Hagiwara, S., & Takahashi, K. (1974). The anomalous rectification and cation selectivity of the membrane of a starfish egg cell. *Journal of Membrane Biology, 18*(1), 61–80.
Heckmann, K. (1972). Single file diffusion. *Biomembranes, 3*, 127–153.
Heginbotham, L., Lu, Z., Abramson, T., & MacKinnon, R. (1994). Mutations in the K^+ channel signature sequence. *Biophysical Journal, 66*(4), 1061–1067. doi:10.1016/S0006-3495(94)80887-2
Heginbotham, L., & MacKinnon, R. (1993). Conduction properties of the cloned Shaker K^+ channel. *Biophysical Journal, 65*(5), 2089–2096. doi:10.1016/S0006-3495(93)81244-X
Heinemann, S. H., Terlau, H., Stuhmer, W., Imoto, K., & Numa, S. (1992). Calcium channel characteristics conferred on the sodium channel by single mutations. *Nature, 356*(6368), 441–443. doi:10.1038/356441a0
Hess, P., Lansman, J. B., & Tsien, R. W. (1986). Calcium-channel selectivity for divalent and mono-valent cations—voltage and concentration-dependence of single channel current in ventricular heart-cells. *Journal of General Physiology, 88*(3), 293–319. doi:10.1085/jgp.88.3.293
Hess, P., & Tsien, R. W. (1984). Mechanism of ion permeation through calcium channels. *Nature, 309*(5967), 453–456. doi:10.1038/309453a0
Hille, B. (1973). Potassium channels in myelinated nerve—selective permeability to small cations. *Journal of General Physiology, 61*(6), 669–686. doi:10.1085/jgp.61.6.669
Hille, B. (2001). *Ion channels of excitable membranes* (3rd ed.). Sunderland, MA: Sinauer.
Hille, B., & Schwarz, W. (1978). Potassium channels as multi-ion single-file pores. *Journal of General Physiology, 72*(4), 409–442.
Hodgkin, A. L., & Keynes, R. D. (1955). The potassium permeability of a giant nerve fibre. *Journal of Physiology, 128*(1), 61–88.
Hoomann, T., Jahnke, N., Horner, A., Keller, S., & Pohl, P. (2013). Filter gate closure inhibits ion but not water transport through potassium channels. *Proceedings of the National Academy of Sciences of the United States of America, 110*(26), 10842–10847. doi:10.1073/pnas.1304714110

Iwamoto, M., & Oiki, S. (2011). Counting ion and water molecules in a streaming file through the open-filter structure of the K channel. *Journal of Neuroscience, 31*(34), 12180–12188. doi:10.1523/JNEUROSCI.1377-11.2011

Jensen, M. O., Borhani, D. W., Lindorff-Larsen, K., Maragakis, P., Jogini, V., Eastwood, M. P., . . . Shaw, D. E. (2010). Principles of conduction and hydrophobic gating in K^+ channels. *Proceedings of the National Academy of Sciences of the United States of America, 107*(13), 5833–5838. doi:10.1073/pnas.0911691107

Jensen, M. O., Jogini, V., Borhani, D. W., Leffler, A. E., Dror, R. O., & Shaw, D. E. (2012). Mechanism of voltage gating in potassium channels. *Science, 336*(6078), 229–233. doi:10.1126/science.1216533

Jensen, M. O., Jogini, V., Eastwood, M. P., & Shaw, D. E. (2013). Atomic-level simulation of current–voltage relationships in single-file ion channels. *Journal of General Physiology, 141*(5), 619–632. doi:10.1085/jgp.201210820

Jiang, Y., Lee, A., Chen, J., Ruta, V., Cadene, M., Chait, B. T., & MacKinnon, R. (2003). X-ray structure of a voltage-dependent K^+ channel. *Nature, 423*(6935), 33–41. doi:10.1038/nature01580

Khalili-Araghi, F., Tajkhorshid, E., & Schulten, K. (2006). Dynamics of K^+ ion conduction through Kv1.2. *Biophysical Journal, 91*(6), L2–L4. doi:10.1529/biophysj.106.091926

Kim, M. S., Morii, T., Sun, L. X., Imoto, K., & Mori, Y. (1993). Structural determinants of ion selectivity in brain calcium channel. *FEBS Letters, 318*(2), 145–148.

Kopec, W., Köpfer, D. A., Vickery, O. N., Bondarenko, A. S., Jansen, T. L. C., de Groot, B. L., & Zachariae, U. (2018). Direct knock-on of desolvated ions governs strict ion selectivity in K^+ channels. [Article]. *Nature Chemistry, 10*(8), 813–820. doi:10.1038/s41557-018-0105-9

Kopfer, D. A., Song, C., Gruene, T., Sheldrick, G. M., Zachariae, U., & de Groot, B. L. (2014). Ion permeation in K^+ channels occurs by direct Coulomb knock-on. *Science, 346*(6207), 352–355. doi:10.1126/science.1254840

Kostyuk, P. G., Mironov, S. L., & Shuba, Y. M. (1983). 2 ion-selecting filters in the calcium-channel of the somatic membrane of mollusk neurons. *Journal of Membrane Biology, 76*(1), 83–93. doi:10.1007/Bf01871455

Kratochvil, H. T., Carr, J. K., Matulef, K., Annen, A. W., Li, H., Maj, M., . . . Zanni, M. T. (2016). Instantaneous ion configurations in the K^+ ion channel selectivity filter revealed by 2d IR spectroscopy. *Science, 353*(6303), 1040–1044. doi:10.1126/science.aag1447

Kuo, C. C., & Hess, P. (1993a). Characterization of the high-affinity $Ca2^+$ binding-sites in the L-type $Ca2^+$ channel pore in rat pheochromocytoma cells. *Journal of Physiology-London, 466*, 657–682.

Kuo, C. C., & Hess, P. (1993b). Ion permeation through the L-type $Ca2^+$ channel in rat pheochromocytoma cells—2 sets of ion-binding sites in the pore. *Journal of Physiology-London, 466*, 629–655.

Lee, K. S., & Tsien, R. W. (1984). High selectivity of calcium channels in single dialysed heart cells of the guinea-pig. *Journal of Physiology, 354*, 253–272.

LeMasurier, M., Heginbotham, L., & Miller, C. (2001). Kcsa: It's a potassium channel. *Journal of General Physiology, 118*(3), 303–313. doi:10.1085/jgp.118.3.303

Lenaeus, M. J., Gamal El-Din, T. M., Ing, C., Ramanadane, K., Pomès, R., Zheng, N., & Catterall, W. A. (2017). Structures of closed and open states of a voltage-gated sodium channel. [Article]. *Proceedings of the National Academy of Sciences of the United States of America, 114*(15), E3051–E3060. doi:10.1073/pnas.1700761114

Levitt, D. G. (1999). Modeling of ion channels. [Review]. *Journal of General Physiology, 113*(6), 789–794. doi:10.1085/jgp.113.6.789

Li, Y., Sun, R., Liu, H., & Gong, H. (2017). Molecular dynamics study of ion transport through an open model of voltage-gated sodium channel. [Article]. *Biochimica et Biophysica Acta–Biomembranes, 1859*(5), 879–887. doi:10.1016/j.bbamem.2017.02.003

Liu, S. A., & Lockless, S. W. (2013). Equilibrium selectivity alone does not create K^+-selective ion conduction in K^+ channels. *Nature Communications, 4*, 2746. doi:10.1038/ncomms3746

Long, S. B., Campbell, E. B., & Mackinnon, R. (2005). Crystal structure of a mammalian voltage-dependent Shaker family K^+ channel. *Science, 309*(5736), 897–903. doi:10.1126/science.1116269

Long, S. B., Tao, X., Campbell, E. B., & MacKinnon, R. (2007). Atomic structure of a voltage-dependent K^+ channel in a lipid membrane-like environment. *Nature, 450*(7168), 376–382. doi:10.1038/nature06265

MacKinnon, R., & Miller, C. (1989). Mutant potassium channels with altered binding of charybdotoxin, a pore-blocking peptide inhibitor. *Science, 245*(4924), 1382–1385.

MacKinnon, R., & Yellen, G. (1990). Mutations affecting tea blockade and ion permeation in voltage-activated K^+ channels. *Science, 250*(4978), 276–279.

Maffeo, C., Bhattacharya, S., Yoo, J., Wells, D., & Aksimentiev, A. (2012). Modeling and simulation of ion channels. *Chemical Reviews, 112*(12), 6250–6284. doi:10.1021/cr3002609

Mahdavi, S., Kuyucak, S., & Ye, S. (2015). Mechanism of ion permeation in mammalian voltage-gated sodium channels. *PLoS ONE, 10*(8). e0133000 doi:10.1371/journal.pone.0133000

Marban, E., Yamagishi, T., & Tomaselli, G. F. (1998). Structure and function of voltage-gated sodium channels. *Journal of Physiology, 508*(Pt 3), 647–657. doi:10.1111/j.1469-7793.1998.647bp.x

Marcus, Y. (1994). A Simple empirical-model describing the thermodynamics of hydration of ions of widely varying charges, sizes, and shapes. *Biophysical Chemistry, 51*(2–3), 111–127. doi:10.1016/0301-4622(94)00051-4

McCusker, E. C., Bagnéris, C., Naylor, C. E., Cole, A. R., D'Avanzo, N., Nichols, C. G., & Wallace, B. A. (2012). Structure of a bacterial voltage-gated sodium channel pore reveals mechanisms of opening and closing. [Article]. *Nature Communications, 3*, 1102. doi:10.1038/ncomms2077

Morais-Cabral, J. H., Zhou, Y. F., & MacKinnon, R. (2001). Energetic optimization of ion conduction rate by the K^+ selectivity filter. *Nature, 414*(6859), 37–42. doi:10.1038/35102000

Naylor, C. E., Bagnéris, C., Decaen, P. G., Sula, A., Scaglione, A., Clapham, D. E., & Wallace, B. (2016). Molecular basis of ion permeability in a voltage-gated sodium channel. [Article]. *EMBO Journal, 35*(8), 820–830. doi:10.15252/embj.201593285

Neyton, J., & Miller, C. (1988). Discrete $Ba2^+$ block as a probe of ion occupancy and pore structure in the high-conductance $Ca2^+$-activated K^+ channel. *Journal of General Physiology, 92*(5), 569–586. doi:10.1085/jgp.92.5.569

Nimigean, C. M., & Allen, T. W. (2011). Origins of ion selectivity in potassium channels from the perspective of channel block. *Journal of General Physiology, 137*(5), 405–413. doi:10.1085/jgp.201010551

Nonner, W., Catacuzzeno, L., & Eisenberg, B. (2000). Binding and selectivity in L-type calcium channels: A mean spherical approximation. *Biophysical Journal, 79*(4), 1976–1992. doi:10.1016/S0006-3495(00)76446-0

Noskov, S. Y., Berneche, S., & Roux, B. (2004). Control of ion selectivity in potassium channels by electrostatic and dynamic properties of carbonyl ligands. *Nature, 431*(7010), 830–834. doi:10.1038/nature02943

Noskov, S. Y., & Roux, B. (2007). Importance of hydration and dynamics on the selectivity of the Kcsa and Nak channels. *Journal of General Physiology, 129*(2), 135–143. doi:10.1085/jgp.200609633

Oakes, V., Furini, S., & Domene, C. (2016) Voltage-gated sodium channels: Mechanistic insights from atomistic molecular dynamics simulations. *Current Topics in Membranes, 78*, 183–214.

Pan, X., Li, Z., Zhou, Q., Shen, H., Wu, K., Huang, X., … Yan, N. (2018). Structure of the human voltage-gated sodium channel Na_V 1.4 in complex with B1. [Article]. *Science, 362*(6412). doi:10.1126/science.aau2486

Parent, L., & Gopalakrishnan, M. (1995). Glutamate substitution in repeat IV alters divalent and monovalent cation permeation in the heart $Ca2^+$ channel. *Biophysical Journal, 69*(5), 1801–1813. doi:10.1016/S0006-3495(95)80050-0

Park, Y. B. (1994). Ion selectivity and gating of small conductance $Ca(2^+)$-activated K^+ channels in cultured rat adrenal chromaffin cells. *Journal of Physiology, 481*(Pt 3), 555–570.

Payandeh, J., Gamal El-Din, T. M., Scheuer, T., Zheng, N., & Catterall, W. A. (2012). Crystal structure of a voltage-gated sodium channel in two potentially inactivated states. [Article]. *Nature, 486*(7401), 135–139. doi:10.1038/nature11077

Payandeh, J., Scheuer, T., Zheng, N., & Catterall, W. A. (2011). The crystal structure of a voltage-gated sodium channel. [Article]. *Nature, 475*(7356), 353–359. doi:10.1038/nature10238

Polo-Parada, L., & Korn, S. J. (1997). Block of N-type calcium channels in chick sensory neurons by external sodium. *Journal of General Physiology, 109*(6), 693–702.

Reuter, H., & Scholz, H. (1977). A study of the ion selectivity and the kinetic properties of the calcium dependent slow inward current in mammalian cardiac muscle. *Journal of Physiology, 264*(1), 17–47.

Reuter, H., & Stevens, C. F. (1980). Ion conductance and ion selectivity of potassium channels in snail neurons. *Journal of Membrane Biology, 57*(2), 103–118. doi:10.1007/Bf01868997

Sakipov, S., Sobolevsky, A. I., & Kurnikova, M. G. (2018). Ion permeation mechanism in epithelial calcium channel Trvp6. [Article]. *Scientific Reports, 8*(1), 5715. doi:10.1038/s41598-018-23972-5

Sauer, D. B., Zeng, W., Canty, J., Lam, Y., & Jiang, Y. (2013). Sodium and potassium competition in potassium-selective and non-selective channels. *Nature Communications, 4*, 2721. doi:10.1038/ncomms3721

Sauer, D. B., Zeng, W., Raghunathan, S., & Jiang, Y. (2011). Protein interactions central to stabilizing the K^+ channel selectivity filter in a four-sited configuration for selective K^+ permeation. *Proceedings of the National Academy of Sciences of the United States of America, 108*(40), 16634–16639. doi:10.1073/pnas.1111688108

Schlief, T., Schönherr, R., Imoto, K., & Heinemann, S. H. (1996). Pore properties of rat brain II sodium channels mutated in the selectivity filter domain. [Article]. *European Biophysics Journal, 25*(2), 75–91. doi:10.1007/s002490050020

Schmid, R., Miah, A. M., & Sapunov, V. N. (2000). A new table of the thermodynamic quantities of ionic hydration: Values and some applications (enthalpy-entropy compensation and born radii). *Physical Chemistry Chemical Physics, 2*(1), 97–102. doi:10.1039/a907160a

Shaya, D., Findeisen, F., Abderemane-Ali, F., Arrigoni, C., Wong, S., Nurva, S. R., … Minor Jr, D. L. (2014). Structure of a prokaryotic sodium channel pore reveals essential gating elements and an outer ion binding site common to eukaryotic channels. [Article]. *Journal of Molecular Biology, 426*(2), 467–483. doi:10.1016/j.jmb.2013.10.010

Shaya, D., Kreir, M., Robbins, R. A., Wong, S., Hammon, J., Brüggemann, A., & Minor Jr, D. L. (2011). Voltage-gated sodium channel (Na V) protein dissection creates a set of functional pore-only proteins. [Article]. *Proceedings of the National Academy of Sciences of the United States of America, 108*(30), 12313–12318. doi:10.1073/pnas.1106811108

Shen, H., Zhou, Q., Pan, X., Li, Z., Wu, J., & Yan, N. (2017). Structure of a eukaryotic voltage-gated sodium channel at near-atomic resolution. *Science, 355*(6328), eaal4326. doi:10.1126/science.aal4326

Simms, B. A., & Zamponi, G. W. (2014). Neuronal voltage-gated calcium channels: Structure, function, and dysfunction. *Neuron, 82*(1), 24–45. doi:10.1016/j.neuron.2014.03.016

Song, C., & Corry, B. (2009). Intrinsic ion selectivity of narrow hydrophobic pores. *Journal of Physical Chemistry B, 113*(21), 7642–7649. doi:10.1021/jp810102u

Stock, L., Delemotte, L., Carnevale, V., Treptow, W., & Klein, M. L. (2013). Conduction in a biological sodium selective channel. *Journal of Physical Chemistry B, 117*(14), 3782–3789. doi:10.1021/jp401403b

Sula, A., Booker, J., Ng, L. C. T., Naylor, C. E., Decaen, P. G., & Wallace, B. A. (2017). The complete structure of an activated open sodium channel. [Article]. *Nature Communications, 8*, 14205. doi:10.1038/ncomms14205

Sun, Y. M., Favre, I., Schild, L., & Moczydlowski, E. (1997). On the structural basis for size-selective permeation of organic cations through the voltage-gated sodium channel: Effect of alanine mutations at the Deka locus on selectivity, inhibition by $Ca2^{+}$and H^{+}, and molecular sieving. [Article]. *Journal of General Physiology, 110*(6), 693–715. doi:10.1085/jgp.110.6.693

Tang, L., Gamal El-Din, T. M., Payandeh, J., Martinez, G. Q., Heard, T. M., Scheuer, T., . . . Catterall, W. A. (2014). Structural basis for $Ca2^{+}$ selectivity of a voltage-gated calcium channel. *Nature, 505*(7481), 56–61. doi:10.1038/nature12775

Taylor, W. R. (1988). Permeation of barium and cadmium through slowly inactivating calcium channels in cat sensory neurones. *Journal of Physiology, 407*, 433–452.

Thomas, M., Jayatilaka, D., & Corry, B. (2007). The predominant role of coordination number in potassium channel selectivity. *Biophysical Journal, 93*(8), 2635–2643. doi:10.1529/biophysj.107.108167

Thomas, M., Jayatilaka, D., & Corry, B. (2011). Mapping the importance of four factors in creating monovalent ion selectivity in biological molecules. *Biophysical Journal, 100*(1), 60–69. doi:10.1016/j.bpj.2010.11.022

Thomas, M., Jayatilaka, D., & Corry, B. (2013). how does overcoordination create ion selectivity? *Biophysical Chemistry, 172*, 37–42. doi:10.1016/j.bpc.2012.11.005

Thompson, A. N., Kim, I., Panosian, T. D., Iverson, T. M., Allen, T. W., & Nimigean, C. M. (2009). Mechanism of potassium-channel selectivity revealed by Na^{+} and Li^{+} binding sites within the KcsA pore. *Nature Structural and Molecular Biology, 16*(12), 1317–1324. doi:10.1038/nsmb.1703

Ulmschneider, M. B., Bagnéris, C., McCusker, E. C., DeCaen, P. G., Delling, M., Clapham, D. E., . . . Wallace, B. A. (2013). Molecular dynamics of ion transport through the open conformation of a bacterial voltage-gated sodium channel. [Article]. *Proceedings of the National Academy of Sciences of the United States of America, 110*(16), 6364–6369. doi:10.1073/pnas.1214667110

Ussing, H. H. (1949). The distinction by means of tracers between active transport and diffusion—the transfer of iodide across the isolated frog skin. *Acta Physiologica Scandinavica, 19*(1), 43–56. doi:10.1111/j.1748-1716.1949.tb00633.x

Varma, S., & Rempe, S. B. (2007). Tuning ion coordination architectures to enable selective partitioning. *Biophysical Journal, 93*(4), 1093–1099. doi:10.1529/biophysj.107.107482

Vora, T., Corry, B., & Chung, S. H. (2005). A model of sodium channels. [Article]. *Biochimica et Biophysica Acta—Biomembranes, 1668*(1), 106–116. doi:10.1016/j.bbamem.2004.11.011

Vora, T., Corry, B., & Chung, S. H. (2008). Brownian dynamics study of flux ratios in sodium channels. *European Biophysics Journal with Biophysics Letters, 38*(1), 45–52. doi:10.1007/s00249-008-0353-5

Wu, J., Yan, Z., Li, Z., Qian, X., Lu, S., Dong, M., . . . Yan, N. (2016). Structure of the voltage-gated calcium channel Ca(V)1.1 at 3.6 a resolution. *Nature, 537*(7619), 191–196. doi:10.1038/nature19321

Wu, J., Yan, Z., Li, Z., Yan, C., Lu, S., Dong, M., & Yan, N. (2015). Structure of the voltage-gated calcium channel Cav1.1 complex. *Science, 350*(6267), aad2395. doi:10.1126/science.aad2395

Xia, M., Liu, H., Li, Y., Yan, N., & Gong, H. (2013). The mechanism of Na^+/K^+ selectivity in mammalian voltage-gated sodium channels based on molecular dynamics simulation. [Article]. *Biophysical Journal, 104*(11), 2401–2409. doi:10.1016/j.bpj.2013.04.035

Yan, Z., Zhou, Q., Wang, L., Wu, J., Zhao, Y., Huang, G., . . . Yan, N. (2017). Structure of the Nav1.4-B1 complex from electric eel. *Cell, 170*(3), 470–482. e411. doi:10.1016/j.cell.2017.06.039

Yang, J., Ellinor, P. T., Sather, W. A., Zhang, J. F., & Tsien, R. W. (1993). Molecular determinants of $Ca2^+$ selectivity and ion permeation in L-Type $Ca2^+$ Channels. *Nature, 366*(6451), 158–161. doi:10.1038/366158a0

Yellen, G. (1984). Ionic permeation and blockade in $Ca2^+$-activated K^+ channels of bovine chromaffin cells. *Journal of General Physiology, 84*(2), 157–186.

Yu, H., Noskov, S. Y., & Roux, B. (2010). Two mechanisms of ion selectivity in protein binding sites. *Proceedings of the National Academy of Sciences of the United States of America, 107*(47), 20329–20334. doi:10.1073/pnas.1007150107

Yue, L., Navarro, B., Ren, D., Ramos, A., & Clapham, D. E. (2002). The cation selectivity filter of the bacterial sodium channel, Nachbac. [Article]. *Journal of General Physiology, 120*(6), 845–853. doi:10.1085/jgp.20028699

Zhang, J., Mao, W., Ren, Y., Sun, R. N., Yan, N., & Gong, H. (2018). Simulating the ion permeation and ion selection for a eukaryotic voltage-gated sodium channel Navpas. [Letter]. *Protein and Cell, 9*(6), 580–585. doi:10.1007/s13238-018-0522-y

Zhang, X., Ren, W., DeCaen, P., Yan, C., Tao, X., Tang, L., . . . Yan, N. (2012). Crystal structure of an orthologue of the nachbac voltage-gated sodium channel. *Nature, 486*, 130. doi:10.1038/nature11054

Zhang, X., Xia, M., Li, Y., Liu, H., Jiang, X., Ren, W., . . . Gong, H. (2013). Analysis of the selectivity filter of the voltage-gated sodium channel Na V Rh. [Article]. *Cell Research, 23*(3), 409–422. doi:10.1038/cr.2012.173

Zhou, M., & MacKinnon, R. (2004). A mutant Kcsa K^+ channel with altered conduction properties and selectivity filter ion distribution. *Journal of Molecular Biology, 338*(4), 839–846. doi:10.1016/j.jmb.2004.03.020

Zhou, Y., Morais-Cabral, J. H., Kaufman, A., & MacKinnon, R. (2001). Chemistry of ion coordination and hydration revealed by a K^+ channel-Fab complex at 2.0 a resolution. *Nature, 414*(6859), 43–48. doi:10.1038/35102009

Zhou, Y. F., & MacKinnon, R. (2003). The occupancy of ions in the K^+ selectivity filter: Charge balance and coupling of ion binding to a protein conformational change underlie high conduction rates. *Journal of Molecular Biology, 333*(5), 965–975. doi:10.1016/j.jmb.2003.09.022

CHAPTER 3

GATING OF ION CHANNELS

RENE BARRO-SORIA

Introduction

Excitable cells use electricity as a fundamental mechanism for signaling across membranes. Examples include the electrical signals underlying the complex neuronal circuitry in the brain, the secretion of hormones and neurotransmitters, and skeletal and cardiac muscle contraction. For these electrical events to occur, cells move ions across membranes using a wide variety of membrane proteins, including transporters and ion channels, that respond to specific stimuli. These signaling proteins respond to changes in temperature, chemicals, physical forces, and voltage across the membrane by opening and closing ion conductance pathways (Hille, 2001).

Ion channels are specialized pore-forming proteins in cell membranes. Their role in facilitating ion conduction down its electrochemical gradient in a rapid and selective manner across membranes constitutes the basis of membrane excitability (Hille, 2001). To trigger and shape the signals and responses underlying excitation, ion channels have evolved a wide variety of structures called *sensors* that function as transducers of the stimuli. This facilitates the opening and closing of the channel by altering the conformation of the channel pore in a process referred to as *channel gating*. In a manner analogous to traditional gates that regulate movement between two compartments, the gates of membrane ion channels form a barrier between the extracellular and intracellular milieux that control the flow of ions between these compartments. Understanding how these ion channels gate open or closed and how this gating is regulated remains a major goal of modern biophysics. Crystallographic and cryo-electron microscopic (EM) high-resolution structural studies combined with structural dynamic techniques such as electron paramagnetic resonance (EPR), electrophysiological, and computational approaches have provided unique experimental structure–function models to study these families of proteins and have contributed to the rapid progress in our understanding of ion channel gating (reviewed in Domene, Haider, & Sansom, 2003). Nonetheless, there are still numerous gaps in our understanding of the complex functionality of these channels. The continuous refinement of approaches and technologies to gain deeper insights into the biophysical properties governing ion channel functioning will not only impact our understanding of these proteins, but also provide a framework for the development of therapeutic strategies to treat disorders associated with these channels.

This introductory chapter will provide an overview of our understanding of ion channel gating with special emphasis on voltage-gated potassium channels, the largest family of ion channels and by far the most extensively investigated.

Ion Channel Architecture

For years, ion channel biophysicists have developed molecular gating models with remarkable accuracy, even when structures at atomic-level resolution were not available. The rapid advancement in X-ray crystallographic and cryo-EM technologies has provided real structural data about the architecture of ion channels. These studies combined with molecular modeling have shown that these signaling molecules adopt different conformational states that in turn facilitated electrophysiological structure–function analysis. The combination of structural and computational analysis with electrophysiological approaches has drastically expanded and refined our knowledge of the mechanisms underlying ion channel gating of the pore.

Ion channels are composed of two main modules: the ion-conducting pore, containing the selectivity filter and the gate that opens and closes the pore, and, for most channels, the sensor that regulates the gate (Figure 3.1A,D). In this section, I will summarize the main features of these structural modules. Although I will focus my discussion mainly on voltage-gated K^+ (Kv) channels, the concepts are relevant for understanding how ion channels function in general.

Historical View

Classical studies from the squid giant axon by Hodgkin and Huxley were the first to show and effectively to model that changes in Na^+ and K^+ permeability were facilitated by the movement of charged "particles" or dipoles in the membrane due to changes in the membrane potential (Hodgkin & Huxley, 1952c). In the following decades, multiple different methodologies and approaches, including pharmacological and electrophysiological, were developed to identify the membrane proteins responsible for voltage-activated Na^+ and K^+ conductance (Armstrong & Bezanilla, 1973; Hille, 1968; Hladky & Haydon, 1970; White & Miller, 1979).

Early works on the squid giant axon and on skeletal muscle experimentally demonstrated the occurrence of charge movements within the membrane that conferred voltage sensitivity to ion permeation through what are now known as

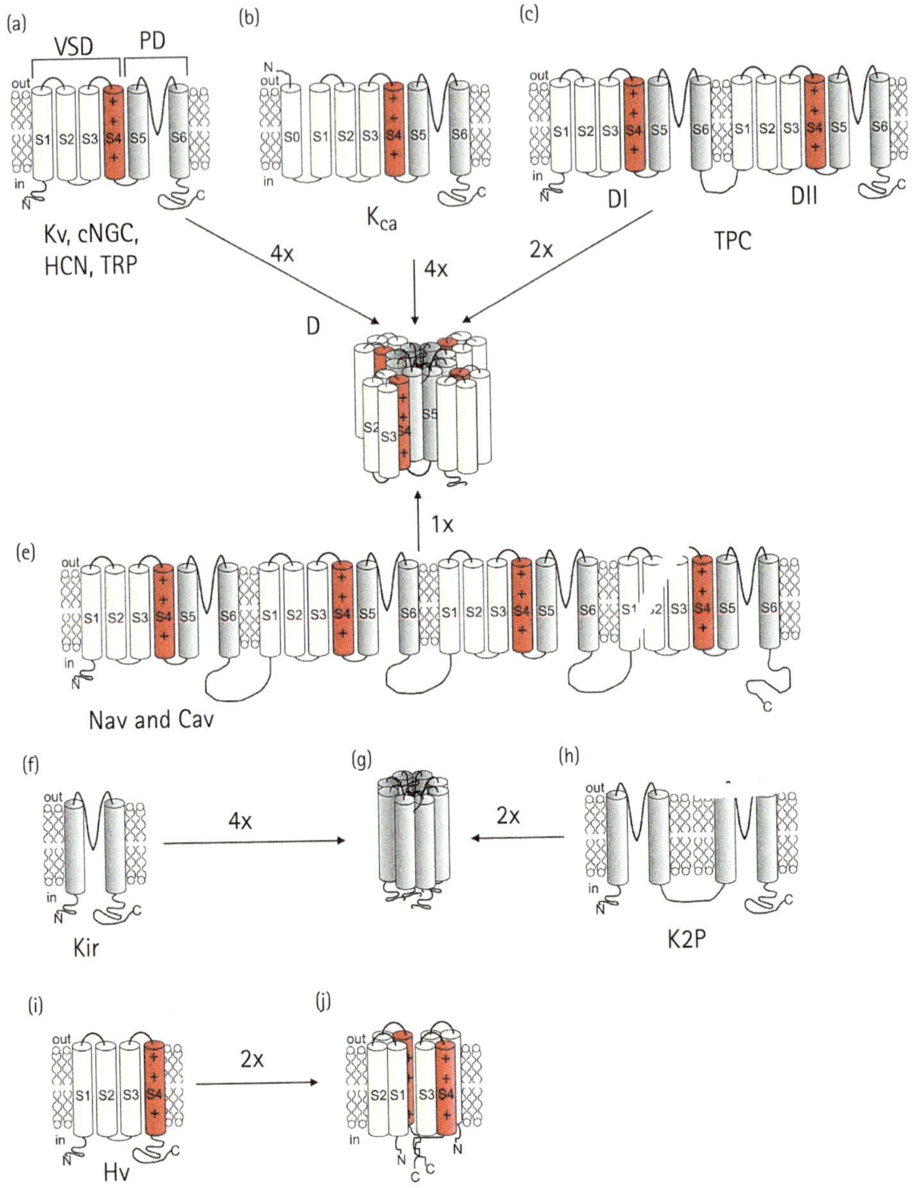

FIGURE 3.1 Membrane topology of voltage-gated cation channels. Four (A,B,F), two (C,H), and one (E) alpha subunits assemble to form tetrameric channels (D,G). The S1–S4 (yellow) and S5–S6 (gray) transmembrane segments form the voltage-sensing domain (VSD) and the conducting pore domain (PD), respectively. The fourth transmembrane segment, S4 (red), has several positively charged residues. Voltage-gated human proton (Hv) channels contain only a VSD (I), which also functions as the ion conduction pore in a dimeric arrangement (J). Cav = voltage-gated Ca^{2+} channel; cNGC = cyclic nucleotide–gated ion channel; DI/DII = domain I/domain II; HCN = hyperpolarization-activated cyclic nucleotide-gated; Hv = voltage-gated proton (H^+) channel; Kca = Ca^{2+}-activated K^+ channels; Kir = inwardly rectifying K^+ channel; K2P = two-pore domain K^+ channel; Kv = voltage-gated K^+ channel; Nav = voltage-gated Na^+ channel; TPC = two-pore channel; TRP = transient receptor potential.

voltage-gated ion channels (Armstrong & Bezanilla, 1973, 1974; Keynes & Rojas, 1974; Schneider & Chandler, 1976). The realization that the total charge moved in response to depolarizations correlated with estimates of Na^+ channel density, together with the observation that the fast time course of charge movement correlated with that of the channel opening, supported the idea that ion permeability likely occurred as a consequence of the movement of charged particles through the membrane (Armstrong & Bezanilla, 1973, 1974). Later structure–function studies inferred by the cloning and sequence analysis of the electric organ of the eel *Electrophorus electricus* Na^+ channel supported the notion that these voltage-sensing charges were indeed positively charged amino acids within the fourth transmembrane segment (S4) (Noda et al., 1984). Subsequent electrophysiological and site-directed mutagenesis studies showed that neutralization of these charged residues effectively decreased the voltage sensitivity of the channels, further supporting the idea that S4 acts as the voltage sensor during channel activation (Papazian, Timpe, Jan, & Jan, 1991; Stuhmer et al., 1989).

One area that puzzled scientists for many years was the structural determinants of ion selectivity and the location of the gate that allows ions to flow through the channel pore. The pioneering work of Hille (1973) and Bezanilla and Armstrong (1972) has helped to shape our understanding of the size of the pore, the thermodynamic requirements for ion permeations, and the location and putative motion of the gate. These studies, performed on myelinated nerve fibers from *Rana pipiens* (Hille, 1973) and on the squid giant axon (Bezanilla & Armstrong, 1972), suggested that the ion movement through the "potassium channel tunnel" (selectivity filter) required the coordination of four carbonyl oxygens from the tunnel backbone. The tunnel backbone surrounds and stabilizes K^+ ions in a similar manner as oxygen dipoles from water molecules solvate K^+ ions in solution (Bezanilla & Armstrong, 1972). This energetically costly transfer of a hydrated K^+ ion in solution to its dehydrated state inside the pore helped explain how K^+ ions quickly translocate through the pore of potassium channels while concomitantly selecting against impermeable ions.

The groundbreaking work of Armstrong used quaternary ammonium (QA) derivatives to block the intracellular side of the channel pore of squid axons and kinetic experiments to elucidate the location and dynamics of the gate (Armstrong, 1966, 1971, 1974). These kinetics studies revealed that, even when the QA derivatives were applied before activation of the K^+ channel currents, the inhibitory effect of QA only occurred after the channels were opened, as if the blockers could only bind the channel when the "gate" was in the open conformation. Furthermore, closure of the channel required the removal of QA blockers from its binding site, in a manner that suggested that the gate could not close with the blocker site occupied (the "foot-in-the-door" effect). These experiments helped shape our understanding that the activation gating requires a dynamic accessibility to the aqueous inner vestibule of the pore.

Consistent with these models, subsequent cysteine scanning experiments by the Yellen laboratory in the Shaker K^+ channel showed state-dependent cysteine modification at the C-terminal end of the S6 segment, as if the motion of a putative gate in or around this region was needed to open and close the channel (del Camino, Holmgren,

Liu, & Yellen, 2000; Liu, Holmgren, Jurman, & Yellen, 1997). They compared the effects of different open channel blockers on methanethiosulfonate (MTS) reagent modification of cysteines introduced in the S6 of the Shaker K^+ channel (del Camino et al., 2000; del Camino & Yellen, 2001; Liu et al., 1997). They found that below certain residues within the intracellular half of S6, all introduced cysteines were modified regardless of the size of the blocker used, as if the pore would become pretty broad. These results led them to propose that during activation gating a bend of the S6 segment occurs near the two prolines at positions 473 and 475, thereby producing a larger water vestibule below the site of bundle crossing (del Camino et al., 2000). These elegant studies helped solidify the notion that the transmembrane helices forming the gate coordinately move to open and close the channel (Armstrong, 1971; del Camino, Kanevsky, & Yellen, 2005; Yellen, 1998a, 1998b).

Structural Evidence for the Pore and the Gate of K^+ Channels

The simplest structural rearrangement of an ion conduction pathway is probably exemplified by the structures of the prokaryotic K^+ channel of *Streptomyces* A (KcsA), the K^+ channel from *Methanobacterium thermoautotrophicum* (MthK), the Kv channel from *Aeropyrum pernix* (KvAP), and the inwardly rectifying K^+ channel from bacteria (KirBac) (Doyle et al., 1998; Jiang et al., 2002a, 2003; Kuo et al., 2003) (Figure 3.2). X-ray crystallographic structural data on these pore-forming proteins showed a structure consisting of four subunits, each made up of two membrane-spanning segments termed *TM1* and *TM2*. The three-dimensional structural data from X-ray crystallography of the KcsA channels showed that the second helix from all four subunits formed a bundle that occludes the ion conduction pathway in an "inverted teepee" arrangement around a central pore (Doyle et al., 1998) (Figure 3.2A,B). The extracellular portion of the pore is formed by the membrane-re-entrant pore (P) loop containing the TVGYG sequence motif that forms the selectivity filter. This region forms a narrow path delineated by the oxygen atoms from the carboxy groups of the TVGYG amino acid sequence that permits rapid movements of permeant ions between binding sites (Figure 3.2, in black). In the intracellular portion of the pore—the region between the selectivity filter and the intracellular bundle crossing the second helices—resides a large aqueous cavity with a hydrophobic lining at the intracellular end in a closed conformation (Doyle et al., 1998). In MthK (Figure 3.2C,D) and KvAP channels captured in the open conformation, the TM2 helices are bent at a conserved glycine, allowing the intracellular solution to access the pore (Jiang et al., 2002a, 2003).These crystal structures revealed that the gate can adopt either a closed or an open conformations (Figure 3.2B,D).

These studies provided graphic hints into the possible movement of the gate during channel opening. Moreover, they shed light on the gating mechanism of K^+ channels, placing the conserved glycine on the TM2 helix as a putative gate hinge that facilitates

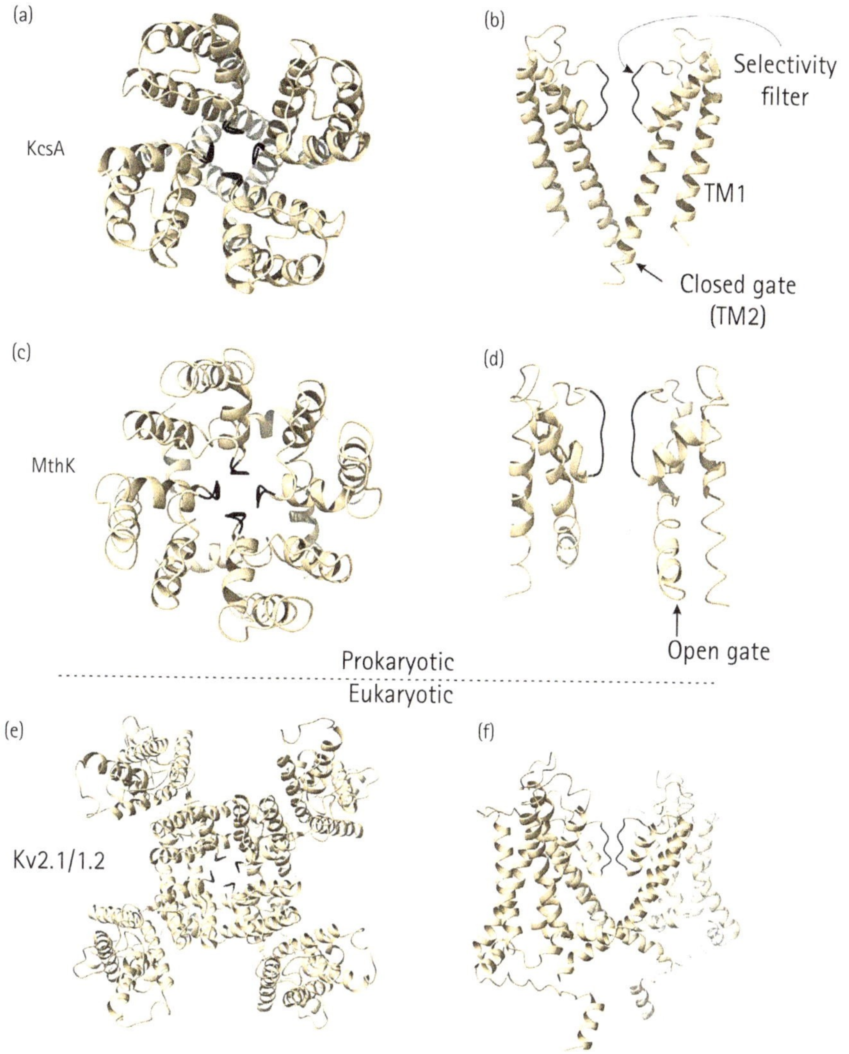

FIGURE 3.2 Cryogenic-electron microscopy (Cryo-EM) structure of prokaryotic and eukaryotic K^+ channels. Top (A,C,E) and side (B,D,F) views of K^+ channel of *Streptomyces* A (KcsA), K+ channel from *Methanobacterium thermoautotrophicum* (MthK), and voltage-gated K^+ (Kv) 2.1/1.2 paddle chimera channels shown as tan ribbons. The selectivity filter is depicted in black. Protein database accession code (1J95 for KcsA, 1LNQ for MthK, and 6EBM for Kv2.1/1.2 paddle chimera). Structures were generated using UCSF ChimeraX, version 1.1 (2020). TM = transmembrane domain.

the opening of the intracellular mouth of the pore (Jiang et al., 2002b). These structural studies were complemented with subsequent functional studies that validated their findings. The pioneering EPR measurements of Perozo and colleagues (Liu, Sompornpisut, & Perozo, 2001; Perozo, Cortes, & Cuello, 1998), combined with early simulation studies (Shrivastava, Capener, Forrest, & Sansom, 2000; Shrivastava &

Sansom, 2000), provided direct physical evidence for motion of the second helices of the gate. Further functional studies using cysteine accessibility experiments helped to clarify how the extracellular region stays rigid during gating while the intracellular portion undergoes significant conformational changes, which collectively supports the gating mechanism of KcsA at this conserved glycine hinge (Kelly & Gross, 2003).

Comparison of the amino acid sequence and the tertiary structure of the prokaryotic potassium KscA channel shows that it is similar to the eukaryotic Shaker K^+ channel in this pore-forming region (MacKinnon, Cohen, Kuo, Lee, & Chait, 1998). Initially, high-resolution EM structural data at first glance suggested that the architecture of the pore region is relatively similar across most functionally diverse ion channels (MacKinnon et al., 1998) (Figure 3.2). Although the structure of the bundle-crossing gate of Kv and KcsA channels showed similarities, the early functional work by Yellen and colleagues revealed important differences in the manner in which the activation gate operates (del Camino & Yellen, 2001; Liu et al., 1997). Thus, unlike KcsA in which the glycine hinge controls a large opening and closing of the gate, the hinge from most Kv channels is composed of the highly conserved amino acid sequence PxP or PxG in the S6 helix, with "x" being typically a hydrophobic residue (Long, Campbell, & Mackinnon, 2005a; Long, Tao, Campbell, & MacKinnon, 2007). These initial static snapshots of prokaryotic ion channel architecture in the different conformational conducting states together with the early functional work on the gate (del Camino et al., 2000; del Camino & Yellen, 2001; Liu et al., 1997) and the more recent structural studies from eukaryotic potassium channels (Cuello, Jogini, Cortes, Pan, et al., 2010; Jiang et al., 2002b, 2003; Long et al., 2005a, 2007) have formed the basis of our current understanding of the gating mechanisms.

As shown in Figure 3.1 in gray, helices S5 and S6 of four subunits (analogous to prokaryotic M1 and M2 TM), the P-helix loop connecting the helices, and the C-terminal end of the S6 segments form the conducting pore and gate of most ion channels (del Camino & Yellen, 2001; Ho et al., 1993; Kubo, Baldwin, Jan, & Jan, 1993; Long et al., 2005a, 2007). The addition of other domains, such as the voltage-sensing domain (VSD), calcium binding domains, and cyclic nucleotide–binding domains—to name just a few of the most studied sensor domains—to the pore assembly provides the functional diversity across ion channels (Figure 3.1). In Kv channels, the pore is flanked by four domains, each composed of S1–S4 TM segments that together form the VSD (Figure 3.1A,D). The fourth TM segment of each VSD, called S4, has several positively charged amino acids that function as the voltage sensor (Aggarwal & MacKinnon, 1996; Larsson, Baker, Dhillon, & Isacoff, 1996; Mannuzzu, Moronne, & Isacoff, 1996; Seoh, Sigg, Papazian, & Bezanilla, 1996) (Figure 3.1, in red). Other families of ion channels like eukaryotic two-pore channels, big potassium (BK), and voltage-gated sodium (Nav) and calcium (Cav) channels are structurally similar to Kv channels (Hofmann, Lacinova, & Klugbauer, 1999; Yu & Catterall, 2004), with the difference that Nav and Cav channels have four Kv-like homologous domains covalently connected into a single polypeptide (Figure 3.1). Other ion channels have also been extensively studied and have been reviewed elsewhere (Csanady, Vergani, & Gadsby, 2019; Duran, Thompson, Xiao, & Hartzell, 2010; T. J. Jentsch & Pusch, 2018; Yu & Catterall, 2004).

Different Sensors Interfacing the Gating Domain and/or Pore Region

Understanding how cells adapt and cope with multiple environmental challenges has been a central subject of modern biology. To survive, cells have evolved a vast set of membrane signaling proteins equipped with a wide variety of sensors. These sensors facilitate communication between the extracellular and intracellular milieux. How these signaling molecules, including ion channels, sense changes in temperature, chemical exposures, physical forces, and voltage change has intrigued researchers for decades.

Physical and Chemical Sensors

Thermal and mechanosensation at the molecular level were mysteries for science until the discovery of the transient receptor potential (TRP) channel family. TRP channels are found in different cell types throughout the body (Ramsey, Delling, & Clapham, 2006). Overall, the activation of TRP channels by diverse chemical and physical stimuli, including endogenous lipids such as diacylglycerol (Palazzo, Rossi, de Novellis, & Maione, 2013) and phosphoinositide-4,5-bisphosphate (Ong & Ambudkar, 2017), temperature (Islas, 2017; Naziroglu & Braidy, 2017), and mechanical stretch (Yamaguchi, Iribe, Nishida, & Naruse, 2017), leads to membrane depolarization and increased intracellular Ca^{2+} concentration. Thus, high temperature–activated TRP channels include members of the TRPV1-4 and TRPM2 families, whereas TRPA1 and TRPM8 are activated by cold temperatures (Naziroglu & Braidy, 2017). In addition to temperature, TRPV and TRPM8 function as ligand-gated channels that can be activated by capsaicin and low intracellular Ca^{2+} levels (Caterina & Julius, 2001). Two members of the TRPC subfamily (TRPC3 and TRPC6) have been implicated in stretch-induced force responses in the heart (Yamaguchi et al., 2017). These wide spectra of physiological stimuli combined with the effect of intracellular Ca^{2+} as a powerful secondary messenger signal allow this family of ion channels to play crucial roles in health and disease. This is indeed reflected by the growing number of studies reporting novel functions associated with TRP channels every year (Islas, 2017). Not surprisingly, mutations in these channels have been associated with a wide variety of pathological conditions, including certain cancers, chronic pain, deafness, inflammation, vascular smooth muscle dysfunction, Alzheimer disease, Parkinson disease, and ischemic stroke (Julius, 2013; Martin-Bornez, Galeano-Otero, Del Toro, & Smani, 2020; Wang, Tu, Zhang, & Shao, 2020). Given this diversity of function, there is an ongoing need to understand both the genetics and the molecular determinants of TRP channel activation to facilitate the development of effective therapeutic approaches.

Like Kv channels, TRP channels form tetramers (Hoenderop et al., 2003) (Figure 3.1A,D). Each subunit consists of six transmembrane segments (S1–S6), where S5–S6 from each of the four subunits forms the centrally located selectivity filter and ion-conducting pore. In general, TRP channels are nonselective for cations and have similar preferences for both monovalent (e.g., Na^+) and divalent (e.g., Ca^{2+}) cations (Venkatachalam & Montell, 2007). Cryo-EM structures have shown that in TRP channels the S1–S4 domain of one subunit is closely aligned with the S5–S6 pore domain (PD) of the adjacent subunit, thereby making a domain-swapped configuration that resembles that seen in Kv channels (Long et al., 2005a; Long, Campbell, & Mackinnon, 2005b; Long et al., 2007). However, unlike classical Kv channels, TRP channels are not primarily gated by changes in voltage since their voltage sensors have fewer—and more poorly conserved—positively charged amino acids than most voltage-gated ion channels (Jardin et al., 2017; Ramsey et al., 2006). Although the gating mechanism in thermo-TRP channels is not complete understood, several studies suggest that large changes in state enthalpy and entropy are required to open the S6 gate. It has been suggested that the voltage and temperature activating mechanisms may be intertwined to indistinguishably contribute to conformational changes that lead to channel gating (reviewed in Islas, 2017). Using whole-cell patch-clamp and inside-out patch experiments on cells expressing TRPV1 and TRPM8 channels, Nilius and colleagues showed that depolarizing voltages beyond the physiological voltage range (+100 mV) are able to activate these channels in a voltage-dependent manner (Voets et al., 2004). In contrast, other studies suggest that the VSD may serve as a temperature-independent anchoring scaffold that allosterically bridges the ligand binding module with the PD to open the channel (Islas, 2017). Whole-cell and outside-out patch-clamp recordings of TRPV1 channels show that heat evoked a significant TRPV1 current even at strong negative voltages (<−140 mV), as if S4 movement were not required to gate the channel. More strikingly, the same study shows that temperature-independent activation of TRPV1 through treatment with capsaicin was also possible at even stronger negative voltages (<−140 mV). This further suggests that the S4 charge movement may not play a significant role in TRPV1 channel gating (Yao, Liu, & Qin, 2010). Supporting this hypothesis, another group found that cold temperatures cause both a leftward shift of the G(V) relation and an increase in the maximum open probability of TRPM8 channels (Brauchi, Orio, & Latorre, 2004). These results suggest that while temperature changes seem to independently affect voltage-dependent and voltage-independent mechanisms, their effects may allosterically contribute to enhance pore opening (Brauchi et al., 2004; Latorre, Brauchi, Orta, Zaelzer, & Vargas, 2007). Together, these findings support the notion that independent gating sensors in TRP channels may coexist such that the S4 movement is separated from that of temperature sensors, favoring allosteric coupling models of gating. Whether voltage sensors are directly or allosterically coupled to TRP-channel gating is currently a subject of intense investigation and is outside the scope of this introductory chapter. We here only aim to highlight some of the best-studied gating mechanisms with an emphasis on the need for a thorough molecular understanding of these physiologically important sensors.

Voltage Sensing

The molecular mechanisms by which proteins sense voltage changes have been the most intensely investigated physiological processes in ion channels. This is probably due to the importance of understanding the mechanisms underlying cellular excitability.

The introduction of the cut-open oocyte voltage-clamp technique by Stefani and Bezanilla provided a refined methodology to effectively measure the movement of gating charges across the membrane (Stefani & Bezanilla, 1998; Taglialatela, Toro, & Stefani, 1992). Using a nonconducting Shaker Kv channel to avoid contamination with ionic currents, they directly measured rapid and transient outward (on-gating) and inward (off-gating) currents upon depolarization and hyperpolarization, respectively. The amount of charge moved during channel activation was effectively calculated by integrating the on- and off-gating currents experimentally recorded (Stefani & Bezanilla, 1998). It was then established for the Shaker K^+ channel that about 3 to 3.5 charges (per subunit) move across the membrane electric field upon channel activation (Aggarwal & MacKinnon, 1996; Schoppa, McCormack, Tanouye, & Sigworth, 1992; Seoh et al., 1996). Estimates of how much charge moves during channel activation have been successfully assessed for several other voltage-gated ion channels (Bean & Rios, 1989; Carmona et al., 2018; De La Rosa & Ramsey, 2018; Hirschberg, Rovner, Lieberman, & Patlak, 1995; Miceli, Cilio, Taglialatela, & Bezanilla, 2009; Noceti et al., 1996; Shenkel & Bezanilla, 1991; Wang, Dou, Goodchild, Es-Salah-Lamoureux, & Fedida, 2013). With the refinement of methodologies and models to measure the effective charged moved during channel activation, one fundamental question remained: How would the energy used in moving charges across the membrane be transferred or coupled to the pore region and translated into channel gating?

Although this chapter does not aim to cover all approaches and methodologies used to study voltage-dependent ion channel gating, I will attempt in the following section to offer an overview of the wealth of techniques and tools used to understand how voltage-activated ion channels open and close the pore in response to changes in the transmembrane voltage.

Biophysical and Chemical Tools to Study Structure and Function of Voltage-Gated Ion Channel

The variety of approaches and techniques developed to study the structure and function of membrane proteins have been recently reviewed elsewhere (Braun, Sheikh, & Pless, 2020; Cowgill & Chanda, 2019; Paoletti, Ellis-Davies, & Mourot, 2019). Here, I will only cover the basic aspects of three of the most used techniques: Cysteine Accessibility, Voltage-Clamp, and Patch-Clamp Fluorometry.

Cysteine-Scanning Mutagenesis

The combination of site-directed mutagenesis and chemical modification has been a powerful tool to study conformational changes during ion channel gating. Thus, the mutation of one cysteine (Cys) at a time in a channel protein facilitated the determination of whether these substituted Cys react with small, charged, sulfhydryl-specific reagents such as MTS. The placement of the Cys residues at different positions within the voltage sensor (or the gate) allowed for the mapping of sensor (or gate) movement. Structural insights into channel function were originally inferred by state-dependent Cys accessibility approaches of MTS derivatives, first in the nicotinic acetylcholine (ACh) receptor channel (Akabas, Stauffer, Xu, & Karlin, 1992) and later in voltage-gated ion channels (Larsson et al., 1996; Yang, George, & Horn, 1996; Yusaf, Wray, & Sivaprasadarao, 1996). This simple methodology involving site-directed mutagenesis assumed that upon covalent modification of substituted Cys in a residue of interest, a functional change in channel gating occurs (Figure 3.3).

Studies in the Nav from human skeletal muscle (Yang et al., 1996) and in the Shaker K^+ channel (Larsson et al., 1996; Yusaf et al., 1996) probed the state-dependent intracellular and extracellular solvent exposure of S4 movement using the Cys modifying membrane-impermeable thiol reagent (2-[trimethylammonium]ethyl methanethiosulfonate [MTSET]) (Larsson et al., 1996; Yang et al., 1996) or by p-chloromercuribenzene sulfonate (Yusaf et al., 1996). These pioneering studies provided the first experimental snapshot of the voltage-dependent distribution of S4 residues in the resting and activated conformations and inferred which positively charged residues within S4 contributed most to gating charge. Since these groundbreaking studies, site-directed mutagenesis and thiol-specific modifications have been successfully applied to study the voltage- and ligand-activated mechanisms of other ion channels. Thus, the state-dependent Cys modifications in the human proton channel Hv1 showed that the fourth transmembrane segment S4 moves and acts as a voltage sensor to allow proton permeation (Gonzalez, Koch, Drum, & Larsson, 2010). Cys accessibility of externally applied MTSET on the S4 of the human slowly activating delayed rectifier potassium (I_{Ks}) channel revealed that S4 moves in two consecutive steps during channel opening (Barro-Soria et al., 2014). More recently, Cys-scanning mutagenesis has also established the pore-lining residues of the acid-activated chloride channel TMEM206 (Ullrich et al., 2019), as well as uncovering the agonist binding site and ionic permeation pathways of the P2X ATP-gated cation channels (Allsopp, El Ajouz, Schmid, & Evans, 2011) and nicotinic ACh receptors (New, Del Villar, Mazzaferro, Alcaino, & Bermudez, 2018; Sullivan, Chiara, & Cohen, 2002).

In a series of seminal works published between 1998 and 2000, the Yellen group expanded these methodologies to gain insights into the, until then, elusive pore structure of voltage-gated ion channels and thus to understand the gating process itself. Using the unique chemistry of the thiol side chain of substituted Cys in the S6 gate to create metal binding sites for Cd^{2+} ions, they inferred the underlying molecular mechanisms of the activation gating in the Shaker K^+ channel (Holmgren, Shin, & Yellen, 1998). Since Cd^{2+}

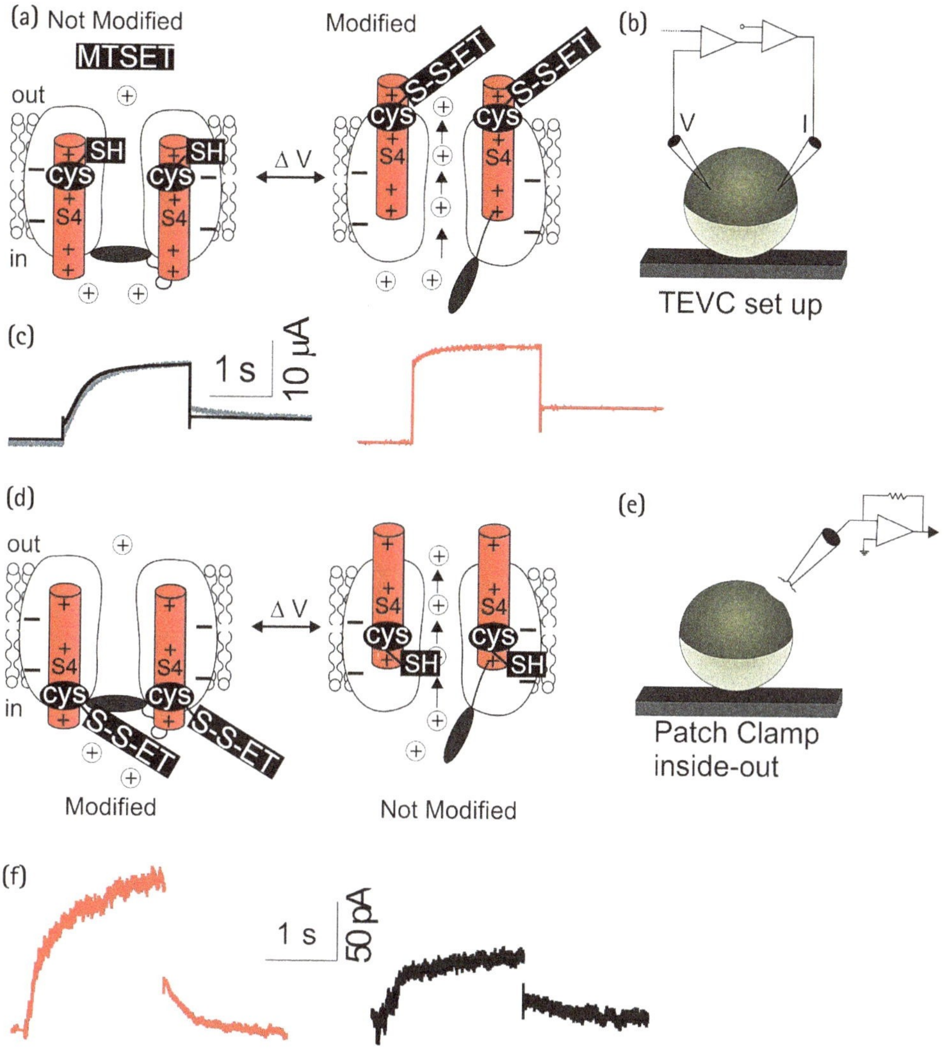

FIGURE 3.3 Cysteine accessibility with 2-(trimethylammonium)ethyl methanethiosulfonate (MTSET). Cartoon of extracellular (A) and intracellular (D) MTSET modification of substituted Cys. Currents from voltage-gated K^+ (Kv) 7.2 (C) and Kv7.2/7.3 (F) channels before (black) and after (gray and red) extracellular (A-C) and intracellular (D-F) MTSET application in the closed (left) and open (right) states of a cys-substituted residue in the S4 region, respectively. The currents are recorded using two-electrode voltabe clamp (TEVC; B) and patch clamp in the inside-out configuration (E).

trapped the gate in the open state by forming metal bridges between residue 476C in S6 (above the bundle crossing) of one subunit and H486 (below the bundle crossing) in another subunit, they concluded that, during gate opening, there is a rearrangement of S6 in the form of a tilting or sharp bending, facilitating the physical intersubunit interaction. This constraining of S6 in the open state fit well with experimental evidence showing a wider pore below the bundle crossing. By comparing their functional results with the structure of the bacterial KcsA channel (Doyle et al., 1998), they realized that the bend of S6 happened at the bundle crossing (Holmgren et al., 1998) of the conserved Pro-X-Pro sequence of the inner S6 gate of Kv channels (del Camino et al., 2000). Based on these observations, they correctly predicted that the open state structure of the gate in voltage-gated (Shaker) K^+ channels, and consequently the activation gating, differed from that of non-voltage-gated bacterial (KcsA) channels (del Camino et al., 2000). These results were later corroborated with the publication of the first eukaryotic Kv structure chimera Kv1.2/2.1 (Long et al., 2005a).

Although site-directed mutagenesis and chemical modification approaches did not offer a dynamic view of the voltage sensor (S4) and gate movements, they provided invaluable insights into the structural changes of S4 and S6 in the two energetically different conformation statuses (resting and activated). To complement the lack of kinetics information usually lacking from Cys accessibility approaches, voltage-clamp fluorometry (VCF) and patch-clamp fluorometry (PCF) were developed (Mannuzzu et al., 1996; Zheng & Zagotta, 2000).

Voltage-Clamp Fluorometry VCF has become a powerful and versatile tool for biophysicists to understand conformational rearrangements during activation gating in voltage-gated ion channels as it allows simultaneous measurements of S4 movement (by fluorescence) and gate opening (by ionic current) (Mannuzzu et al., 1996). The basic principle of VCF relies on the distinct physicochemical properties of fluorescent probes in different environments and on their short half-life once excited (Lakowicz, 2006). Fluorophores tethered to thiol reactive moieties are covalently attached to substituted Cys residues located on the extracellular side and generally near the voltage sensor (S4). The voltage-dependent fluorescence changes associated with the relative movement of the S4-fluorescent probe between hydrophobic and aqueous interphases alongside quenching groups such as tryptophan, can report on structural changes (Chanda & Bezanilla, 2002; Cowgill & Chanda, 2019; Mannuzzu et al., 1996; Pantazis & Olcese, 2012) (Figure 3.4). With this technique, it was then possible to determine the kinetics and voltage dependence of the S4 movement and directly compare these parameters to gate opening. In 1996, Mannuzzu et al. were the first to successfully record S4 motion in the Shaker K^+ channel using VCF. They combined tetramethylrhodamine maleimide fluorescent labeling on the extracellular end of the S4 segment with whole-cell ionic and gating currents under two-electrode voltage clamp. Using this approach, they showed in real time that changes in fluorescence tracked the gating charge displacement, which in turn directly correlated to channel activation. This seminal study provided a direct physical demonstration that the fourth transmembrane segment (S4) in Kv channels acted as the voltage sensor during activation gating.

Since this groundbreaking study, VCF has been extensively used to study the kinetics on conformational changes, leading to activation gating in a variety of other channels and transporters. VCF experiments have been key in the development of kinetics models of several transporters, including the glucose transporter SGLT1

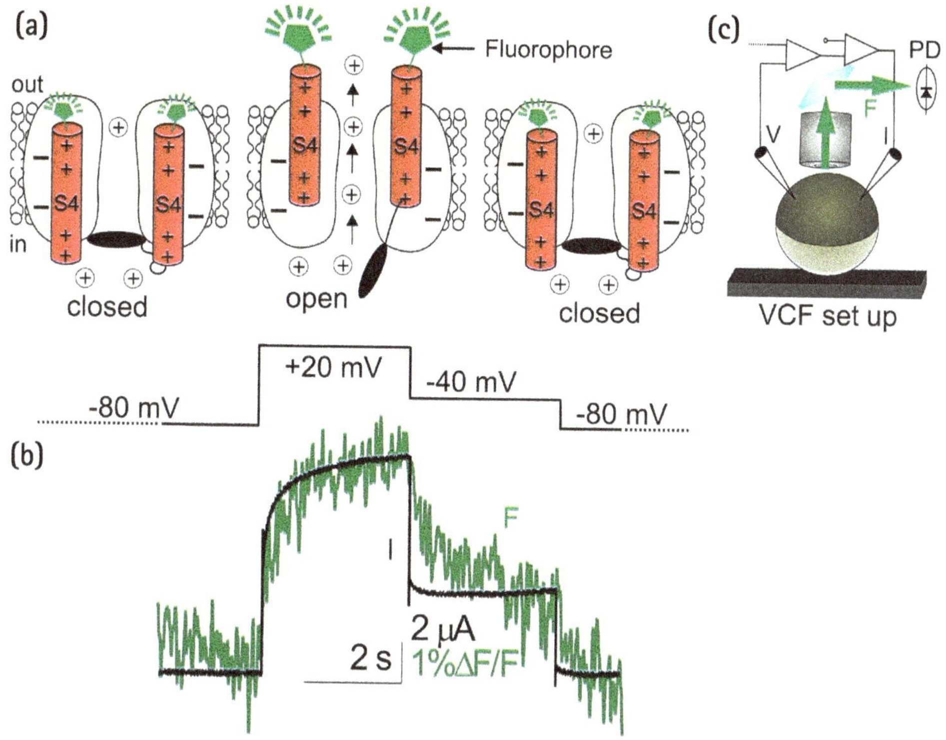

FIGURE 3.4 Voltage-clamp fluorometry (VCF). (A) Cysteines are introduced close to the voltage sensor (S4) and labeled with a fluorophore tethered to a maleimide group. As in Figure 3.3, only two subunits are shown. Upon voltage changes, the S4s move out and drag along the fluorophore attached. The environmental change around the fluorophore alters its fluorescence intensity. (B) Representative current (black) and fluorescence (green) traces from rat voltage-gated K^+ 7.2 channel for the indicated voltage protocol. (C) Schematic illustration of a VCF setup where the current and fluorescence are recorded simultaneously. I = current; F = fluorescence; V = voltage; PD = Photodiode.

(Loo et al., 1998), the GABA transporter GAT1 (Li, Farley, & Lester, 2000), the glutamate transporter EAAT3 (Larsson, Tzingounis, Koch, & Kavanaugh, 2004; Larsson et al., 2010), the serotonin transporter SERT (Li & Lester, 2002; Soderhielm, Andersen, Munro, Nielsen, & Kristensen, 2015), the Na^+/K^+- and H^+/K^+-ATPases (Geibel, Kaplan, Bamberg, & Friedrich, 2003; Geibel et al., 2003), and the type IIb Na^+-coupled phosphate cotransporter SLC34 (Virkki, Murer, & Forster, 2006). VCF experiments on the human proton (Hv) channel heterologously expressed in *Xenopus* oocytes directly demonstrated that S4 moves and acts as the voltage sensor (Gonzalez et al., 2010). Moreover, labeling Hv1 mutants from *Ciona intestinalis* with the thiol-reactive fluorophore Alexa488-maleimide on S4 showed that channel opening was preceded by voltage-dependent conformational changes that are independent of S4 charge movements from both subunits and that a second cooperative rearrangement of S4s was needed to open the channel (Qiu, Rebolledo, Gonzalez, & Larsson, 2013). Similarly,

VCF studies on the cardiac rapidly activating delayed rectifier potassium (I_{Kr}) channel encoded by the human ether-a-go-go-related gene (hERG) (Es-Salah-Lamoureux, Fougere, Xiong, Robertson, & Fedida, 2010; Smith & Yellen, 2002) and the I_{Ks} channel (encoded by KCNQ1 and KCNE1 genes) (Barro-Soria et al., 2014; Osteen et al., 2010; Zaydman et al., 2014) showed that upon voltage changes two types of S4 movement seem to be required to open the channels: a fast component at negative voltages that tracked the kinetics and voltage dependence of inactivation in the case of hERG or generated the main gating charge movement in the I_{Ks} channel and a slow component at more positive voltages that correlates with both the kinetics and voltage dependence of activation gating. Additional studies that combined optical and electrophysiological approaches have been performed on several ion channels, including the mammalian Shaker-related KCNA5 channel (Vaid, Claydon, Rezazadeh, & Fedida, 2008), the ether-à-go-go EAG channel (Bannister, Chanda, Bezanilla, & Papazian, 2005), the BK channel (Savalli, Kondratiev, de Quintana, Toro, & Olcese, 2007; Savalli, Kondratiev, Toro, & Olcese, 2006), Kv7.1 (Hou et al., 2017; Osteen et al., 2012), and Kv7.3 (Barro-Soria, 2019; Kim, Pless, & Kurata, 2017), and the hyperpolarization-activated cyclic nucleotide-gated (HCN) channel (Bruening-Wright, Elinder, & Larsson, 2007; Bruening-Wright & Larsson, 2007). These studies were instrumental in building comprehensive kinetics models of the different conformational changes of S4 occurring prior to and during channel gating.

Despite the ultrafast absorption and emission kinetic nature of most fluorophores, VCF under the two-electrode voltage-clamp (TEVC) technique was unable to reliably allow measurements of the time course of S4 movement on fast-gating channels like Nav channels due to the slow voltage clamp in TEVC of *Xenopus* oocytes. Cha and Bezanilla (1997) tackled this issue by coupling VCF to the cut-open oocyte technique (Taglialatela et al., 1992), thereby simultaneously measuring fluorescence signals, gating charge displacements, and ionic currents under a controlled voltage clamp setting. Subsequently, VCF experiments by Chanda and Bezanilla on the skeletal muscle sodium channel unambiguously showed that channel activation correlated with the fluorescence from the S4 segment of domains DI–DIII, which in turn followed the kinetics of the fast component of gating currents, whereas the fluorescence from the S4 segment from domain DIV correlated more with that of the inactivation gate (Capes, Goldschen-Ohm, Arcisio-Miranda, Bezanilla, & Chanda, 2013; Chanda & Bezanilla, 2002). Additional VCF studies further assessed how different β subunits (Zhu et al., 2017) affected the voltage dependence and kinetics of individual voltage sensors in the four domains of human Nav1.5 (Varga et al., 2015). These fluorescence measurements coupled to gating or ionic currents have shaped our understanding of the underlying structural changes occurring during activation of voltage-gated sodium channel with and without regulatory β subunits (reviewed in Barro-Soria, Liin, & Larsson, 2017).

VCF has also been used to more accurately determine the free-energy coupling between interacting residues. A strong coupling energy (>1.5 kT) extracted from a thermodynamic mutant cycle analysis typically indicates functional interaction between residues (Ranganathan, Lewis, & MacKinnon, 1996). Theoretical analysis of protein thermodynamics suggests that the free-energy differences between the resting and activated states are best estimated from the gating charge/voltage curve, Q(V) (reflecting gating charge movement), rather than the G(V) curves (reflecting channel opening) (Chowdhury & Chanda, 2013). However, because measuring gating current has been technically challenging, especially for slowly activating channels, the use of steady-state fluorescence/voltage curves, F(V), through VCF represents a more accurate alternative than G(V) curves to estimate the energetics associated with interactions. Recently, as an alternative to fluorescence resonance energy transfer (FRET)–based assays to estimate atomic distances on the human BK channel, Pantazis, Westerberg, Althoff, Abramson, and Olcese (2018) developed an elegant VCF-modified methodology termed *distance-encoding photoinduced electron transfer*. This methodology uses the electron-transfer properties of moieties such as Trp to quench the excited state of organic fluorophores in a distance-dependent manner (Doose, Neuweiler, & Sauer, 2009; Jones Brunette & Farrens, 2014). A comprehensive review on the use of VCF to study channels and transporters can be found in Cowgill and Chanda (2019).

Patch-Clamp Fluorometry

Alternative labeling approaches to VCF to study structure–function relationships of membrane proteins have been developed in part because of the lack of structural and dynamic information of the resting and intermediate states of ion channels together with the fact that key regions of the gating machinery, such as the activation gate, mainly face the intracellular side of channels. The use of PCF, a variation of VCF, circumvented some of the intrinsic limitations of VCF, which can only label and track extracellularly accessible regions, thereby allowing simultaneous site-specific fluorescence recording of inside-facing regions and inside-out patch-clamp current recording. PCF was first used to study conformational changes in the C-linker region upon ligand binding that led to activation gating of cyclic nucleotide-gated (CNG) channels (Zheng & Zagotta, 2000). Subsequent studies where PCF was paired with FRET facilitated the elucidation of the dynamic rearrangements around the gating rings upon Ca^{2+} and allosteric ligand binding during activation gating in BK (Miranda et al., 2013) and CNG (Taraska & Zagotta, 2007) channels, respectively. Using PCF in giant inside-out patches from fluorescently labeled Hv1 channels, the Isacoff group showed that besides the canonical outward S4 movement necessary to open the channel, a concomitant allosteric rearrangement of the S1 segment occurs (Mony, Berger, & Isacoff, 2015).

The introduction of genetically encoded unnatural amino acids (UAAs) into proteins (Noren, Anthony-Cahill, Griffith, & Schultz, 1989), including the fluorescent noncanonical amino acid L-Anap pioneered by the Schultz lab (Chatterjee, Guo, Lee, & Schultz, 2013), represented a turning point in the efforts to probe the structure and dynamics of membrane proteins in general, and ligand-gated and voltage-gated ion channels in particular (Nowak et al., 1995; Pless, Galpin, Frankel, & Ahern, 2011; Pless, Galpin, Niciforovic, & Ahern, 2011). Because UAA mutagenesis is a genetically encoded technology, it allows for the labeling of single sites in any part of the channel, including the cytosolic side. Fluorescent UAAs can be incorporated into proteins expressed in

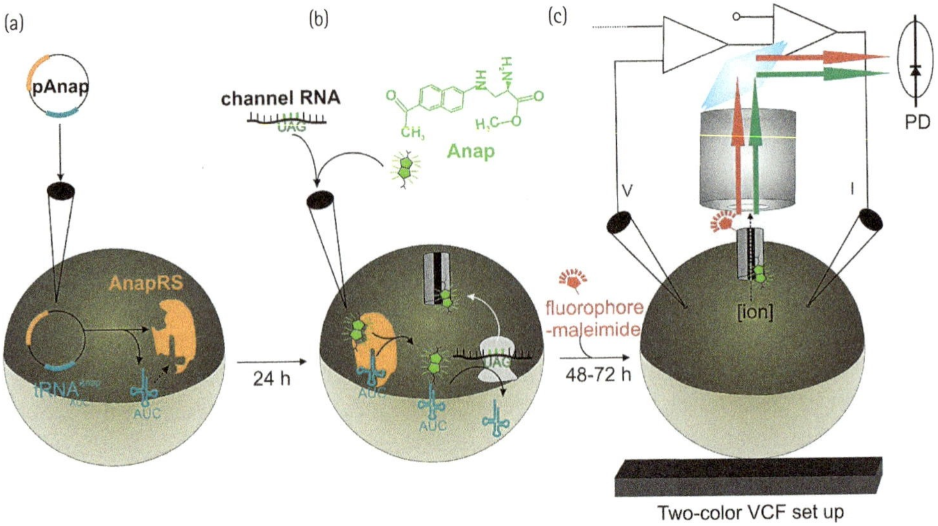

FIGURE 3.5 Unnatural amino acid (UAA) mutagenesis. Cartoon showing incorporation of the fluorescent UAA 3-[(6-acetyl-2-naphthalenyl)amino]-L-alanine, monotrifluoroacetate (L-ANAP) in *Xenopus leavis* oocytes. (A) The plasmid encoding the AnapRS/transfer RNA pair is injected into the nucleus. (B) The L-Anap and the channel RNA are injected into the cytosol. (C) Schematic illustration of a two-color voltage-clamp fluorometry (VCF) setup where the current and fluorescence changes from two different channel regions are recorded simultaneously. PD = Photodiode.

mammalian cells using amber suppression. This approach requires either in vitro acylation of transfer RNAs (tRNAs) to encode the UAA or expression of an orthogonal aminoacyl tRNA synthetase–tRNA pair specific for the UAA along with the protein of interest containing an amber stop codon at the desired amino acid residue (Pless & Ahern, 2013, 2015; Pless, Leung, Galpin, & Ahern, 2011) (Figure 3.5). Taking advantage of the small size of L-Anap, which is only slightly bigger than tryptophan but with a short linker to the protein backbone and its spectral properties, the Blunck laboratory was able to encode L-Anap into the Shaker channel using an orthogonal tRNA–synthetase pair and combined it with two-color VCF to study S4 movement and activation gating (Kalstrup & Blunck, 2013). Using this approach, they were able to simultaneously measure gating current and two independently introduced optic reporters, allowing them to assess different regions of the channel at the same time. Through these experiments, they elegantly described the entire process of S4 movement from different closed states to activation gating. This technology allowed them to uncover the conformational changes underlying inside-facing regions like the S1 segment, which was generally assumed to play a static role during gating but was revealed to have a more dynamic role during channel gating (Kalstrup & Blunck, 2013, 2018).

Incorporation of ANAP has also been used in transition metal ion FRET (tmFRET)-based PCF studies to understand, for instance, the underlying voltage-dependent gating mechanisms of HCN channels. For example, Dai, Aman, DiMaio, and Zagotta (2019) used tmFRET between Anap (as a donor) and transition metal or 1-(2-(pyridin-2-yldisulfanyl)ethyl)-1,4,7,10-tetraazacyclododecane–Cu^{2+} (TETAC, as acceptors) pairs. They showed that a kink in the most C-terminal part of the S4 segment occurred after a large S4 downward movement of around 10 Å during hyperpolarization, providing important information on both the hyperpolarized state and the dynamics of the VSD during HCN channel activation (Dai et al., 2019). Recently, the Pless group introduced modifications to the PCF technique to probe fluorescence protein dynamics with a considerably higher signal-to-noise ratio compared to conventional VCF (Wulf & Pless, 2018).

Recent Findings: Nonclassical Gating

Even with the high-resolution structural models of many of these channels available, one should not at first glance assume that the gate resides and operates only at the innermost part of the helix (TM2 or S6) containing the gating hinge at the mouth of the pore in all ion channels. This is the case, as summarized next, in the bacterial KcsA channel (Cuello, Jogini, Cortes, Pan, et al., 2010; Cuello, Jogini, Cortes, & Perozo, 2010) and the eukaryotic outwardly rectifying background "leak" conductance two-PD potassium (K2P) channels (Schewe et al., 2016).

The Selectivity Filter as the Activating Gate

In KcsA and K2P channels, which lack the VSD (Figures 3.1G,H and 3.2A,B), the canonical gating mechanism at the intracellular entrance of the S6 gate of the channel coexists with a C-type inactivation gating mechanism at the selectivity filter toward the extracellular side of the channel (reviewed in Douguet & Honore, 2019; Kim & Nimigean, 2016; Mathie, Al-Moubarak, & Veale, 2010). Using giant patch recordings in an inside-out configuration and molecular dynamic simulations, the Baukrowitz group showed that the intrinsic movement of three to four ions into the electric field of the inactive selectivity filter is what confers voltage sensitivity in K2P channels (Schewe et al., 2016). Mutations of the highly conserved threonine residues (T157C and T266C) in the K^+-selective motif (TIGFG) in the first (P1) or second (P2) pore loop of TREK-1 removed the voltage-dependent gating behavior, suggesting that these residues in the selectivity filter were essential for the voltage-gating mechanism. Further experiments in which intracellular K^+ was replaced by Rb^+ (and not by other cations) showed that the voltage-dependent activation of TREK-1 largely increased, suggesting that the direction of ion flux is critical for K2P channel gating. This study suggests that, at least in K2P channels, the selectivity filter works as a one-way valve whose opening and closing is dictated by the outward and inward movement of potassium ions, respectively (Schewe et al., 2016).

Other K^+ channels, such as the Ca^{2+}-activated K^+ channels (K_{Ca}), might also gate at the selectivity filter even in the absence of calcium ions. In a manner that resembles the C-type inactivation mechanism, the selectivity filter of these channels may transiently experience conformational changes that reduce the open probability gating, thereby hampering permeant ion flow (Kurata & Fedida, 2006; McCoy & Nimigean, 2012). Interestingly, in the human voltage-gated ether-à-go-go–related gene (hERG) channels, besides the classical voltage-dependent motion of the lower gate, a rapid inactivation mechanism can also occur through a partial collapse of the selectivity filter (Smith, Baukrowitz, & Yellen, 1996; Vandenberg, Perozo, & Allen, 2017), which also acts as a channel gate.

Variations on a Theme: VSD-to-PD Coupling

Classic work by Hodgkin and Huxley on the squid giant axon potassium conductance predicted that channel activation was preceded by multiple voltage-dependent transitions of independent "voltage-sensing particles" (the four S4 segments in Kv channels) (Hodgkin & Huxley, 1952a, 1952b, 1952c; Hodgkin, Huxley, & Katz, 1952). This sequential electromechanical gating model, which has since been used for many other Kv channels, assumes a physical interaction of the inside-facing S4–S5 linker with the S6 gate upon voltage-dependent rearrangement of S4 (Bezanilla, 2008; Bezanilla & Perozo, 2003; Chowdhury & Chanda, 2012; Long et al., 2005b). In 2015, Lörinczi et al published a surprising result that sparked new ideas in the way we look at ion channel gating. They showed that voltage-dependent gating in KCNH channels is retained even if the VSD is not linked to the S5–S6 PD by cutting the S4–S5 linker (Lorinczi et al., 2015). Subsequently, cryo-EM structural studies on the EAG (KCNH1) channel (Whicher & MacKinnon, 2016) revealed that, unlike what is seen in most Kv channels, the S4–S5 linker architecture formed a short loop that lies adjacent to its own S5–S6 PD. This non-domain-swapped arrangement of the VSD and PD of KCNH1 channels provided a structural framework to explain how the VSD may transduce transmembrane electric field changes to the channel pore using a short S4–S5 linker even in the absence of a physical linkage between these domains. Similar to the Hv1 proton channels (Figure 3.1I,J) or the voltage-sensing phosphatase that lacks the S5–S6 PD, this hitherto unanticipated activation gating mechanism may reveal how the VSD evolved independently from the PD to confer functional diversity. While significant progress has been made in our understanding of S4 movement coupled to pore opening in classical ion channels, it has become apparent that not all ion channels use the same mechanisms. Other nonclassical activating gating mechanisms have been recently proposed for other channels, including the HCN (Dai et al., 2019) and KAT1 (Clark, Contreras, Shen, & Perozo, 2020) channels.

Another example of the diversity of mechanisms used for activation gating is the Kv7 channel family, which has recently received a lot of attention due to its relevance in physiological and pathophysiological conditions. Kv7 channels mediate one of the major potassium currents in excitable cells. Kv7 channels are encoded by the KCNQ genes (1–5) and comprise five members, Kv7.1–Kv7.5 (Abbott & Pitt, 2014; Jentsch, 2000; Jespersen, Grunnet, & Olesen, 2005). Kv7.1 channels are expressed in tissues as diverse as the heart, gut, and inner ear, where they contribute to the termination of the cardiac action potential and regulation of K^+ homeostasis needed for electrolyte transport in epithelial cells (Liin, Barro-Soria, & Larsson, 2015). Different heterotetrameric combinations of Kv7.2, Kv7.3, and Kv7.5 form the I_{KM} channel, which mediates the muscarine-regulated M potassium currents in neurons (Schroeder, Hechenberger, Weinreich, Kubisch, & Jentsch, 2000; Wang et al., 1998). The I_{KM} channel activates at subthreshold voltages and during action potential initiation. This channel contributes to maintaining the resting membrane potential and impedes repetitive neuronal firing, therefore playing a crucial role in controlling neuronal excitability (Brown & Adams, 1980; Halliwell & Adams, 1982). Kv7.4 channels, cloned from the human retina cDNA library, are expressed in the outer hair cells of the inner ear and in the structures of the brainstem, where they control potassium homeostasis (Jentsch, 2000; Jentsch, Schroeder, Kubisch, Friedrich, & Stein, 2000; Kubisch et al., 1999). Mutations in Kv7 channels result in a wide variety of debilitating and life-threatening conditions including cardiac arrhythmias, congenital deafness, neuropathic pain, and abnormal neuronal activity that drive networks of neurons into epileptic seizures (reviewed in Jentsch, 2000; Jentsch et al., 2000; Maljevic & Lerche, 2014).

Since Kv7 channels are central to physiological and pathophysiological events, the mechanisms of voltage-dependent gating have been the subject of extensive study. Kv7 channels share a similar domain-swap tetrameric architecture as that seen in most Kv channels (Li et al., 2021; Sun & MacKinnon, 2017) (Figure 3.6). As in Kv channels, the positively charged S4 TM segment functions as the voltage sensor that responds to depolarization, leading to movement in the outward direction and hence triggering channel opening (Figure 3.6, in red). Voltage sensor activation of Kv7 channels has been demonstrated by 1) gating currents (to date, only resolved in Kv7.1 [Ruscic et al., 2013] and Kv7.4 [Miceli, Vargas, Bezanilla, & Taglialatela, 2012] channels), 2) VCF in Kv7.1 (Osteen et al., 2010) and Kv7.3 (Kim et al., 2017) channels, and 3) Cys accessibility approaches in Kv7.1 (Nakajo & Kubo, 2007; Rocheleau & Kobertz, 2008; Wu et al., 2010), Kv7.2 (Gourgy-Hacohen et al., 2014), and Kv7.3 (Gourgy-Hacohen et al., 2014). VCF measurements in Kv7.1 (Osteen et al., 2010; Ruscic et al., 2013), Kv7.3 (Barro-Soria, 2019; Kim et al., 2017), and Kv7.2 channels showed that the kinetics and voltage dependence of S4 movement corresponded with that of channel opening (Figure 3.4B); and in the case of Kv7.1 these biophysical properties were similar to those seen for the gating charge movements, suggesting that S4 moves during activation gating (Ruscic et al., 2013). S4 movement assessed by MTSET accessibility in Kv7.1 (Rocheleau & Kobertz, 2008; Wu et al., 2010) and in Kv7.2 (Figure 3.3A–C) showed that residues located in the N-terminal

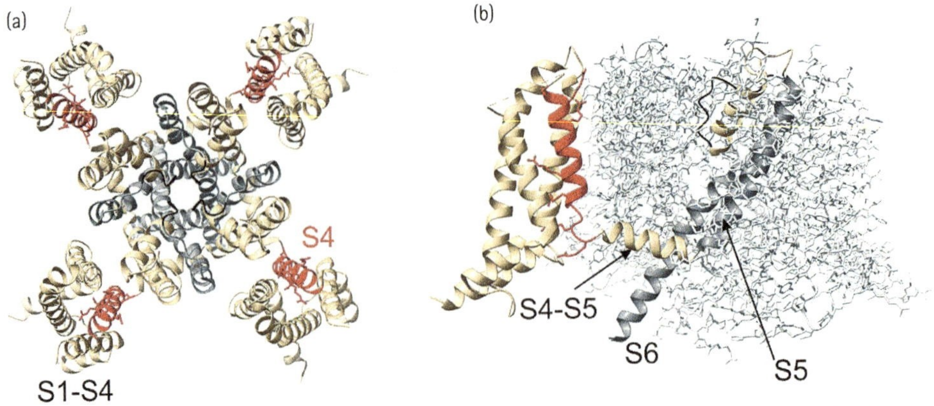

FIGURE 3.6 Cryogenic-electron microscopy (Cryo-EM) structure of voltage-gated K^+ (Kv) 7.2 channel. Top (A) and side (B) views of homotetrameric Kv7.2 channel in the apo state as in Li et al. (2021). In (A) the four subunits are shown as ribbons, and in (B) three subunits are shown as wire frame and the S1–S6 transmembrane helices of one subunit as ribbons. The gate is formed by the S6 segment, and the selectivity filter is shown in black. The voltage sensing domains, formed by helices S1–S4, lie close to the S5–S6 pore domain of the adjacent subunit in a domain-swapped configuration. The voltage sensor (S4), shown as a red ribbon, has several positively charged residues (red sticks). Protein database accession code (7cro). Structures were generated using UCSF ChimeraX, version 1.1 (2020).

portion of S4 are modified much faster at depolarized voltages than at hyperpolarized voltages, suggesting that these residues are buried in the membrane in the closed state and move out during voltage-dependent activation. In contrast, residues located toward the most C-terminal portion of S4 seem to get modified faster at hyperpolarized voltages than at depolarized voltages, suggesting that these residues are facing the cytosolic side in the closed state but could move inside the membrane during voltage-dependent activation (Figure 3.3D–F). Moreover, state-dependent electrostatic interactions between S4 arginine residues and the conserved negative charge residue in the S2 segment of Kv7.1 channels further support the outward movement of S4 during channel gating (Wu et al., 2010). These and other studies not only suggest that the canonical interaction between the S6 and the S4–S5 linker play a critical role for the coupling of the VSD and PD domains but also support a role for conformational changes from regions more distal than the S6 gate such as interactions with the S1, S1–S2 loop, and the outer vestibule of the pore (Gourgy-Hacohen et al., 2014; Lee, Banerjee, & MacKinnon, 2009; Wu et al., 2010).

In contrast to most Kv channels, Kv7.1 has a PXG motif, instead of the more conserved PXP sequence, in the S6 gate and lacks two important positively charged residues in S4. These structural fingerprints have been suggested to account for the flexible activation gating and/or the unusual VSD–PD coupling observed in Kv7.1 (reviewed in Cui, 2016; Liin et al., 2015). Unlike classical Hodgkin and Huxley gating models, Kv7.1 has the unusual property of opening before all S4s in the tetramer have been activated

(Osteen et al., 2012). Using VCF on linked concatemers of Kv7.1, the Larsson and Kass groups found that the four S4s move relatively independently in Kv7.1 channels. They also showed that, compared to Shaker channels, the S4 movement in Kv7.1 altered the open probability <50-fold (compared to $>10^7$ fold in Shaker). This implies a much less rigid coupling between S4 and the gate (Osteen et al., 2012). Another study showed that S4 can move between the resting and activated states even when the pore was locked in the open state (Zaydman et al., 2013). The Cui group showed that charge reversal mutations in the S2 and S4 segments can trap the voltage sensor in Kv7.1 in intermediate and activated states (Wu et al., 2010). VCF recordings uncovered a biphasic F(V) curve, suggesting that S4 moves from a resting-closed state at negative voltages, through an intermediate state at intermediate depolarized voltages, and finally to a fully activated state at more depolarized voltages. Both the intermediate and fully activated states of S4 allow for channel opening (Hou et al., 2017, 2020). These states, which can also be pharmacologically dissected, have different ion selectivity and differences in VSD–PD coupling. Conformational changes in the whole VSD during activation gating, and not merely in the C-terminal portion of S4, seem to influence the nature of PD operation through VSD–PD interactions (Hou et al., 2020; Taylor et al., 2020; Zaydman et al., 2014). Disruption of the state-dependent interactions of these energetically different intermediate and activated states, and not the direct activation of the VSD and PD, may alternatively explain the underlying mechanism by which β subunits modulate the Kv7.1 channel (Cui, 2016; Zaydman et al., 2014). These findings have been the foundation of an allosteric VSD–PD coupling model that has helped to explain Kv7.1 channel gating flexibility in response to disease-causing mutations, modulators, and tissue-specific regulatory β subunits (Liin et al., 2015). The unexpected interdependency of VSD activation and gate opening dissected by intermediate and activated states of S4 in the Kv7.1 channel provides a framework to understand voltage-dependent gating not only of neuronal Kv7.2–Kv7.5 channels, but also of voltage-gated ion channels in general.

Outlook

A combination of methodologies using the power of site-directed fluorescent labeling, gating currents, mathematical simulations, and high-resolution structural information has facilitated the uncovering of gating mechanisms of ion channels and transporters with unprecedented details. The structure–function versatility of these membrane signaling molecules underlined by the ability to respond to a wide variety of physical and chemical stimuli form the physiological basis that allows for the integration and processing of information required for survival. Despite the wealth of knowledge already gained about ion channel function derived from these multidisciplinary scientific discoveries, many challenging questions remain to be answered to fully understand the role of these molecules in cellular signaling in health and disease.

Acknowledgments

I thank Michaela A. Edmond for the assemble of Figures 3.3 and 3.5 and helpful comments. I also thank Drs. Derek Dyxhoorn and H. Peter Larsson for helpful comments on the manuscript. Work in the Barro-Soria Lab is supported by the National Institutes of Health grant R01NS110847.

References

Abbott, G. W., & Pitt, G. S. (2014). Ion channels under the sun. *FASEB Journal*, *28*(5), 1957–1962. doi:10.1096/fj.14-0501ufm

Aggarwal, S. K., & MacKinnon, R. (1996). Contribution of the S4 segment to gating charge in the Shaker K^+ channel. *Neuron*, *16*(6), 1169–1177. Retrieved from http://www.ncbi.nlm.nih.gov/pubmed/8663993

Akabas, M. H., Stauffer, D. A., Xu, M., & Karlin, A. (1992). Acetylcholine receptor channel structure probed in cysteine-substitution mutants. *Science*, *258*(5080), 307–310. doi:10.1126/science.1384130

Allsopp, R. C., El Ajouz, S., Schmid, R., & Evans, R. J. (2011). Cysteine scanning mutagenesis (residues Glu52-Gly96) of the human P2X1 receptor for ATP: Mapping agonist binding and channel gating. *Journal of Biological Chemistry*, *286*(33), 29207–29217. doi:10.1074/jbc.M111.260364

Armstrong, C. M. (1971). Interaction of tetraethylammonium ion derivatives with the potassium channels of giant axons. *Journal of General Physiology*, *58*(4), 413–437. doi:10.1085/jgp.58.4.413

Armstrong, C. M. (1974). Ionic pores, gates, and gating currents. *Quarterly Reviews of Biophysics*, *7*(2), 179–210. doi:10.1017/s0033583500001402

Armstrong, C. M. (1966). Time course of TEA^+-induced anomalous rectification in squid giant axons. *Journal of General Physiology*, *50*(2), 491–503. doi:10.1085/jgp.50.2.491

Armstrong, C. M., & Bezanilla, F. (1974). Charge movement associated with the opening and closing of the activation gates of the Na channels. *Journal of General Physiology*, *63*(5), 533–552. doi:10.1085/jgp.63.5.533

Armstrong, C. M., & Bezanilla, F. (1973). Currents related to movement of the gating particles of the sodium channels. *Nature*, *242*(5398), 459–461. Retrieved from http://www.ncbi.nlm.nih.gov/entrez/query.fcgi?cmd=Retrieve&db=PubMed&dopt=Citation&list_uids=4700900

Bannister, J. P., Chanda, B., Bezanilla, F., & Papazian, D. M. (2005). Optical detection of rate-determining ion-modulated conformational changes of the ether-à-go-go K^+ channel voltage sensor. *Proceedings of the National Academy of Sciences of the United States of America*, *102*(51), 18718–18723. doi:10.1073/pnas.0505766102

Barro-Soria, R. (2019). Epilepsy-associated mutations in the voltage sensor of KCNQ3 affect voltage dependence of channel opening. *Journal of General Physiology*, *151*(2), 247–257. doi:10.1085/jgp.201812221

Barro-Soria, R., Liin, S. I., & Larsson, H. P. (2017). Using fluorescence to understand β subunit–Na_V channel interactions. *Journal of General Physiology*, *149*(8), 757–762. doi:10.1085/jgp.201711843

Barro-Soria, R., Rebolledo, S., Liin, S. I., Perez, M. E., Sampson, K. J., Kass, R. S., & Larsson, H. P. (2014). KCNE1 divides the voltage sensor movement in KCNQ1/KCNE1 channels into two steps. *Nature Communications*, *5*, 3750. doi:10.1038/ncomms4750

Bean, B. P., & Rios, E. (1989). Nonlinear charge movement in mammalian cardiac ventricular cells. Components from Na and Ca channel gating. *Journal of General Physiology*, *94*, 65–93.

Bezanilla, F. (2008). How membrane proteins sense voltage. *Nature Reviews Molecular Cell Biology*, *9*(4), 323–332. doi:10.1038/nrm2376

Bezanilla, F., & Armstrong, C. M. (1972). Negative conductance caused by entry of sodium and cesium ions into the potassium channels of squid axons. *Journal of General Physiology*, *60*(5), 588–608. doi:10.1085/jgp.60.5.588

Bezanilla, F., & Perozo, E. (2003). The voltage sensor and the gate in ion channels. *Advances in Protein Chemistry*, *63*, 211–241. Retrieved from https://www.ncbi.nlm.nih.gov/pubmed/12629972

Brauchi, S., Orio, P., & Latorre, R. (2004). Clues to understanding cold sensation: Thermodynamics and electrophysiological analysis of the cold receptor TRPM8. *Proceedings of the National Academy of Sciences of the United States of America*, *101*(43), 15494–15499. doi:10.1073/pnas.0406773101

Braun, N., Sheikh, Z. P., & Pless, S. A. (2020). The current chemical biology tool box for studying ion channels. *Journal of Physiology*, *598*(20), 4455–4471. doi:10.1113/JP276695

Brown, D. A., & Adams, P. R. (1980). Muscarinic suppression of a novel voltage-sensitive K^+ current in a vertebrate neurone. *Nature*, *283*(5748), 673–676. Retrieved from http://www.ncbi.nlm.nih.gov/pubmed/6965523

Bruening-Wright, A., Elinder, F., & Larsson, H. P. (2007). Kinetic relationship between the voltage sensor and the activation gate in spHCN channels. *Journal of General Physiology*, *130*(1), 71–81. Retrieved from http://www.ncbi.nlm.nih.gov/entrez/query.fcgi?cmd=Retrieve&db=PubMed&dopt=Citation&list_uids=17591986

Bruening-Wright, A., & Larsson, H. P. (2007). Slow conformational changes of the voltage sensor during the mode shift in hyperpolarization-activated cyclic-nucleotide-gated channels. *Journal of Neuroscience*, *27*(2), 270–278. Retrieved from http://www.ncbi.nlm.nih.gov/entrez/query.fcgi?cmd=Retrieve&db=PubMed&dopt=Citation&list_uids=17215386

Capes, D. L., Goldschen-Ohm, M. P., Arcisio-Miranda, M., Bezanilla, F., & Chanda, B. (2013). Domain IV voltage-sensor movement is both sufficient and rate limiting for fast inactivation in sodium channels. *Journal of General Physiology*, *142*(2), 101–112. doi:10.1085/jgp.201310998

Carmona, E. M., Larsson, H. P., Neely, A., Alvarez, O., Latorre, R., & Gonzalez, C. (2018). Gating charge displacement in a monomeric voltage-gated proton (Hv1) channel. *Proceedings of the National Academy of Sciences of the United States of America*, *115*(37), 9240–9245. doi:10.1073/pnas.1809705115

Caterina, M. J., & Julius, D. (2001). The vanilloid receptor: A molecular gateway to the pain pathway. *Annual Reviews of Neuroscience*, *24*, 487–517. doi:10.1146/annurev.neuro.24.1.487

Cha, A., & Bezanilla, F. (1997). Characterizing voltage-dependent conformational changes in the Shaker K^+ channel with fluorescence. *Neuron*, *19*(5), 1127–1140. Retrieved from http://www.ncbi.nlm.nih.gov/pubmed/9390525

Chanda, B., & Bezanilla, F. (2002). Tracking voltage-dependent conformational changes in skeletal muscle sodium channel during activation. *Journal of General Physiology*, *120*(5), 629–645. doi:10.1085/jgp.20028679

Chatterjee, A., Guo, J., Lee, H. S., & Schultz, P. G. (2013). A genetically encoded fluorescent probe in mammalian cells. *Journal of the American Chemical Society, 135*(34), 12540–12543. doi:10.1021/ja4059553

Chowdhury, S., & Chanda, B. (2013). Free-energy relationships in ion channels activated by voltage and ligand. *Journal of General Physiology, 141*(1), 11–28. doi:10.1085/jgp.201210860

Chowdhury, S., & Chanda, B. (2012). Perspectives on: Conformational coupling in ion channels: Thermodynamics of electromechanical coupling in voltage-gated ion channels. *Journal of General Physiology, 140*(6), 613–623. doi:10.1085/jgp.201210840

Clark, M. D., Contreras, G. F., Shen, R., & Perozo, E. (2020). Electromechanical coupling in the hyperpolarization-activated K^+ channel KAT1. *Nature, 583*(7814), 145–149. doi:10.1038/s41586-020-2335-4

Cowgill, J., & Chanda, B. (2019). The contribution of voltage clamp fluorometry to the understanding of channel and transporter mechanisms. *Journal of General Physiology, 151*(10), 1163–1172. doi:10.1085/jgp.201912372

Csanady, L., Vergani, P., & Gadsby, D. C. (2019). Structure, gating, and regulation of the CFTR anion channel. *Physiological Reviews, 99*(1), 707–738. doi:10.1152/physrev.00007.2018

Cuello, L. G., Jogini, V., Cortes, D. M., Pan, A. C., Gagnon, D. G., Dalmas, O., . . . Perozo, E. (2010). Structural basis for the coupling between activation and inactivation gates in K^+ channels. *Nature, 466*(7303), 272–275. doi:10.1038/nature09136

Cuello, L. G., Jogini, V., Cortes, D. M., & Perozo, E. (2010). Structural mechanism of C-type inactivation in K^+ channels. *Nature, 466*(7303), 203–208. doi:10.1038/nature09153

Cui, J. (2016). Voltage-dependent gating: Novel insights from KCNQ1 channels. *Biophysical Journal, 110*(1), 14–25. doi:10.1016/j.bpj.2015.11.023

Dai, G., Aman, T. K., DiMaio, F., & Zagotta, W. N. (2019). The HCN channel voltage sensor undergoes a large downward motion during hyperpolarization. *Nature Structural & Molecular Biology, 26*(8), 686–694. doi:10.1038/s41594-019-0259-1

De La Rosa, V., & Ramsey, I. S. (2018). Gating currents in the Hv1 proton channel. *Biophysical Journal, 114*(12), 2844–2854. doi:10.1016/j.bpj.2018.04.049

del Camino, D., Holmgren, M., Liu, Y., & Yellen, G. (2000). Blocker protection in the pore of a voltage-gated K^+ channel and its structural implications. *Nature, 403*(6767), 321–325.

del Camino, D., Kanevsky, M., & Yellen, G. (2005). Status of the intracellular gate in the activated-not-open state of shaker K^+ channels. *Journal of General Physiology, 126*(5), 419–428. Retrieved from http://www.ncbi.nlm.nih.gov/entrez/query.fcgi?cmd=Retrieve&db=PubMed&dopt=Citation&list_uids=16260836

del Camino, D., & Yellen, G. (2001). Tight steric closure at the intracellular activation gate of a voltage-gated K^+ channel. *Neuron, 32*(4), 649–656. Retrieved from http://www.ncbi.nlm.nih.gov/11719205/

Domene, C., Haider, S., & Sansom, M. S. (2003). Ion channel structures: A review of recent progress. *Current Opinion in Drug Discovery & Development, 6*(5), 611–619. Retrieved from https://www.ncbi.nlm.nih.gov/pubmed/14579510

Doose, S., Neuweiler, H., & Sauer, M. (2009). Fluorescence quenching by photoinduced electron transfer: A reporter for conformational dynamics of macromolecules. *Chemphyschem, 10*(9–10), 1389–1398. doi:10.1002/cphc.200900238

Douguet, D., & Honore, E. (2019). Mammalian mechanoelectrical transduction: Structure and function of force-gated ion channels. *Cell, 179*(2), 340–354. doi:10.1016/j.cell.2019.08.049

Doyle, D. A., Morais Cabral, J., Pfuetzner, R. A., Kuo, A., Gulbis, J. M., Cohen, S. L., . . . MacKinnon, R. (1998). The structure of the potassium channel: Molecular basis of K^+

conduction and selectivity. *Science*, *280*(5360), 69–77. Retrieved from http://www.ncbi.nlm.nih.gov/pubmed/9525859

Duran, C., Thompson, C. H., Xiao, Q., & Hartzell, H. C. (2010). Chloride channels: Often enigmatic, rarely predictable. *Annual Reviews of Physiology*, *72*, 95–121. doi:10.1146/annurev-physiol-021909-135811

Es-Salah-Lamoureux, Z., Fougere, R., Xiong, P. Y., Robertson, G. A., & Fedida, D. (2010). Fluorescence-tracking of activation gating in human ERG channels reveals rapid S4 movement and slow pore opening. *PLoS One*, *5*(5), e10876. doi:10.1371/journal.pone.0010876

Geibel, S., Kaplan, J. H., Bamberg, E., & Friedrich, T. (2003). Conformational dynamics of the Na^+/K^+-ATPase probed by voltage clamp fluorometry. *Proceedings of the National Academy of Sciences of the United States of America*, *100*(3), 964–969. doi:10.1073/pnas.0337336100

Geibel, S., Zimmermann, D., Zifarelli, G., Becker, A., Koenderink, J. B., Hu, Y. K., ... Bamberg, E. (2003). Conformational dynamics of Na^+/K^+- and H^+/K^+-ATPase probed by voltage clamp fluorometry. *Annals of the New York Academy of Sciences*, *986*, 31–38. doi:10.1111/j.1749-6632.2003.tb07136.x

Gonzalez, C., Koch, H. P., Drum, B. M., & Larsson, H. P. (2010). Strong cooperativity between subunits in voltage-gated proton channels. *Nature Structural & Molecular Biology*, *17*(1), 51–56. Retrieved from http://www.ncbi.nlm.nih.gov/entrez/query.fcgi?cmd=Retrieve&db=PubMed&dopt=Citation&list_uids=20023639

Gourgy-Hacohen, O., Kornilov, P., Pittel, I., Peretz, A., Attali, B., & Paas, Y. (2014). Capturing distinct KCNQ2 channel resting states by metal ion bridges in the voltage-sensor domain. *Journal of General Physiology*, *144*(6), 513–527. doi:10.1085/jgp.201411221

Halliwell, J. V., & Adams, P. R. (1982). Voltage-clamp analysis of muscarinic excitation in hippocampal neurons. *Brain Research*, *250*(1), 71–92. Retrieved from http://www.ncbi.nlm.nih.gov/entrez/query.fcgi?cmd=Retrieve&db=PubMed&dopt=Citation&list_uids=6128061

Hille, B. (2001). *Ion channels of excitable membranes* (3rd ed.). Sunderland, MA: Sinauer Associates.

Hille, B. (1968). Pharmacological modifications of the sodium channels of frog nerve. *Journal of General Physiology*, *51*(2), 199–219. doi:10.1085/jgp.51.2.199

Hille, B. (1973). Potassium channels in myelinated nerve. Selective permeability to small cations. *Journal of General Physiology*, *61*(6), 669–686. doi:10.1085/jgp.61.6.669

Hirschberg, B., Rovner, A., Lieberman, M., & Patlak, J. (1995). Transfer of twelve charges is needed to open skeletal muscle Na^+ channels. *Journal of General Physiology*, *106*(6), 1053–1068. Retrieved from http://www.ncbi.nlm.nih.gov/entrez/query.fcgi?cmd=Retrieve&db=PubMed&dopt=Citation&list_uids=8786350

Hladky, S. B., & Haydon, D. A. (1970). Discreteness of conductance change in bimolecular lipid membranes in the presence of certain antibiotics. *Nature*, *225*(5231), 451–453. doi:10.1038/225451a0

Ho, K., Nichols, C. G., Lederer, W. J., Lytton, J., Vassilev, P. M., Kanazirska, M. V., & Hebert, S. C. (1993). Cloning and expression of an inwardly rectifying ATP-regulated potassium channel. *Nature*, *362*(6415), 31–38. doi:10.1038/362031a0

Hodgkin, A. L., & Huxley, A. F. (1952a). The components of membrane conductance in the giant axon of *Loligo*. *Journal of Physiology*, *116*(4), 473–496. Retrieved from http://www.ncbi.nlm.nih.gov/entrez/query.fcgi?cmd=Retrieve&db=PubMed&dopt=Citation&list_uids=14946714

Hodgkin, A. L., & Huxley, A. F. (1952b). The dual effect of membrane potential on sodium conductance in the giant axon of *Loligo*. *Journal of Physiology, 116*(4), 497–506. Retrieved from http://www.ncbi.nlm.nih.gov/entrez/query.fcgi?cmd=Retrieve&db=PubMed&dopt=Citation&list_uids=14946715

Hodgkin, A. L., & Huxley, A. F. (1952c). A quantitative description of membrane current and its application to conduction and excitation in nerve. *Journal of Physiology, 117*(4), 500–544. Retrieved from http://www.ncbi.nlm.nih.gov/entrez/query.fcgi?cmd=Retrieve&db=PubMed&dopt=Citation&list_uids=12991237

Hodgkin, A. L., Huxley, A. F., & Katz, B. (1952). Measurement of current–voltage relations in the membrane of the giant axon of *Loligo*. *Journal of Physiology, 116*(4), 424–448. Retrieved from http://www.ncbi.nlm.nih.gov/entrez/query.fcgi?cmd=Retrieve&db=PubMed&dopt=Citation&list_uids=14946712

Hoenderop, J. G., Voets, T., Hoefs, S., Weidema, F., Prenen, J., Nilius, B., & Bindels, R. J. (2003). Homo- and heterotetrameric architecture of the epithelial Ca^{2+}+ channels TRPV5 and TRPV6. *EMBO Journal, 22*(4), 776–785. doi:10.1093/emboj/cdg080

Hofmann, F., Lacinova, L., & Klugbauer, N. (1999). Voltage-dependent calcium channels: From structure to function. *Reviews of Physiology, Biochemistry and Pharmacology, 139*, 33–87. doi:10.1007/BFb0033648

Holmgren, M., Shin, K. S., & Yellen, G. (1998). The activation gate of a voltage-gated K^+ channel can be trapped in the open state by an intersubunit metal bridge. *Neuron, 21*(3), 617–621.

Hou, P., Eldstrom, J., Shi, J., Zhong, L., McFarland, K., Gao, Y., . . . Cui, J. (2017). Inactivation of KCNQ1 potassium channels reveals dynamic coupling between voltage sensing and pore opening. *Nature Communications, 8*(1), 1730. doi:10.1038/s41467-017-01911-8

Hou, P., Kang, P. W., Kongmeneck, A. D., Yang, N. D., Liu, Y., Shi, J., . . . Cui, J. (2020). Two-stage electro-mechanical coupling of a KV channel in voltage-dependent activation. *Nature Communications, 11*(1), 676. doi:10.1038/s41467-020-14406-w

Islas, L. D. (2017). Molecular mechanisms of temperature gating in TRP channels. In T. L. R. Emir (Ed.), *Neurobiology of TRP channels* (pp. 11–25). Boca Raton, FL: CRC Press.

Jardin, I., Lopez, J. J., Diez, R., Sanchez-Collado, J., Cantonero, C., Albarran, L., . . . Rosado, J. A. (2017). TRPs in pain sensation. *Frontiers in Physiology, 8*, 392. doi:10.3389/fphys.2017.00392

Jentsch, T. J. (2000). Neuronal KCNQ potassium channels: Physiology and role in disease. *Nature Reviews Neuroscience, 1*(1), 21–30.

Jentsch, T. J., & Pusch, M. (2018). CLC chloride channels and transporters: Structure, function, physiology, and disease. *Physiological Revies, 98*(3), 1493–1590. doi:10.1152/physrev.00047.2017

Jentsch, T. J., Schroeder, B. C., Kubisch, C., Friedrich, T., & Stein, V. (2000). Pathophysiology of KCNQ channels: Neonatal epilepsy and progressive deafness. *Epilepsia, 41*(8), 1068–1069.

Jespersen, T., Grunnet, M., & Olesen, S. P. (2005). The KCNQ1 potassium channel: From gene to physiological function. *Physiology (Bethesda), 20*, 408–416. doi:10.1152/physiol.00031.2005

Jiang, Y., Lee, A., Chen, J., Cadene, M., Chait, B. T., & MacKinnon, R. (2002a). Crystal structure and mechanism of a calcium-gated potassium channel. *Nature, 417*(6888), 515–522. Retrieved from http://www.ncbi.nlm.nih.gov/entrez/query.fcgi?cmd=Retrieve&db=PubMed&dopt=Citation&list_uids=12037559

Jiang, Y., Lee, A., Chen, J., Cadene, M., Chait, B. T., & MacKinnon, R. (2002b). The open pore conformation of potassium channels. *Nature, 417*(6888), 523–526. Retrieved from http://www.ncbi.nlm.nih.gov/entrez/query.fcgi?cmd=Retrieve&db=PubMed&dopt=Citation&list_uids=12037560

Jiang, Y., Lee, A., Chen, J., Ruta, V., Cadene, M., Chait, B. T., & MacKinnon, R. (2003). X-ray structure of a voltage-dependent K^+ channel. *Nature, 423*(6935), 33–41. Retrieved from http://www.ncbi.nlm.nih.gov/entrez/query.fcgi?cmd=Retrieve&db=PubMed&dopt=Citation&list_uids=12721618

Jones Brunette, A. M., & Farrens, D. L. (2014). Distance mapping in proteins using fluorescence spectroscopy: Tyrosine, like tryptophan, quenches bimane fluorescence in a distance-dependent manner. *Biochemistry, 53*(40), 6290–6301. doi:10.1021/bi500493r

Julius, D. (2013). TRP channels and pain. *Annual Reviews of Cell and Developmental Biology, 29*, 355–384. doi:10.1146/annurev-cellbio-101011-155833

Kalstrup, T., & Blunck, R. (2013). Dynamics of internal pore opening in K_V channels probed by a fluorescent unnatural amino acid. *Proceedings of the National Academy of Sciences of the United States of America, 110*(20), 8272–8277. doi:10.1073/pnas.1220398110

Kalstrup, T., & Blunck, R. (2018). S4–S5 linker movement during activation and inactivation in voltage-gated K^+ channels. *Proceedings of the National Academy of Sciences of the United States of America, 115*(29), E6751–E6759. doi:10.1073/pnas.1719105115

Kelly, B. L., & Gross, A. (2003). Potassium channel gating observed with site-directed mass tagging. *Nature Structural Biology, 10*(4), 280–284. doi:10.1038/nsb908

Keynes, R. D., & Rojas, E. (1974). Kinetics and steady-state properties of the charged system controlling sodium conductance in the squid giant axon. *Journal of Physiology, 239*(2), 393–434. doi:10.1113/jphysiol.1974.sp010575

Kim, D. M., & Nimigean, C. M. (2016). Voltage-gated potassium channels: A structural examination of selectivity and gating. *Cold Spring Harbor Perspectives in Biology, 8*(5), a029231. doi:10.1101/cshperspect.a029231

Kim, R. Y., Pless, S. A., & Kurata, H. T. (2017). PIP2 mediates functional coupling and pharmacology of neuronal KCNQ channels. *Proceedings of the National Academy of Sciences of the United States of America, 114*(45), E9702–E9711. doi:10.1073/pnas.1705802114

Kubisch, C., Schroeder, B. C., Friedrich, T., Lutjohann, B., El Amraoui, A., Marlin, S., . . . Jentsch, T. J. (1999). KCNQ4, a novel potassium channel expressed in sensory outer hair cells, is mutated in dominant deafness. *Cell, 96*(3), 437–446.

Kubo, Y., Baldwin, T. J., Jan, Y. N., & Jan, L. Y. (1993). Primary structure and functional expression of a mouse inward rectifier potassium channel. *Nature, 362*(6416), 127–133. doi:10.1038/362127a0

Kuo, A., Gulbis, J. M., Antcliff, J. F., Rahman, T., Lowe, E. D., Zimmer, J., . . . Doyle, D. A. (2003). Crystal structure of the potassium channel KirBac1.1 in the closed state. *Science, 300*(5627), 1922–1926.

Kurata, H. T., & Fedida, D. (2006). A structural interpretation of voltage-gated potassium channel inactivation. *Progress in Biophysics and Molecular Biology, 92*(2), 185–208. doi:10.1016/j.pbiomolbio.2005.10.001

Lakowicz, J. R. (2006). *Principles of fluorescence spectroscopy*. New York, NY: Springer.

Larsson, H. P., Baker, O. S., Dhillon, D. S., & Isacoff, E. Y. (1996). Transmembrane movement of the Shaker K^+ channel S4. *Neuron, 16*(2), 387–397. Retrieved from http://www.ncbi.nlm.nih.gov/pubmed/8789953

Larsson, H. P., Tzingounis, A. V., Koch, H. P., & Kavanaugh, M. P. (2004). Fluorometric measurements of conformational changes in glutamate transporters. *Proceedings of the National Academy of Sciences of the United States of America, 101*(11), 3951–3956. doi:10.1073/pnas.0306737101

Larsson, H. P., Wang, X., Lev, B., Baconguis, I., Caplan, D. A., Vyleta, N. P., . . . Noskov, S. Y. (2010). Evidence for a third sodium-binding site in glutamate transporters suggests an ion/substrate coupling model. *Proceedings of the National Academy of Sciences of the United States of America, 107*(31), 13912–13917. Retrieved from http://www.ncbi.nlm.nih.gov/entrez/query.fcgi?cmd=Retrieve&db=PubMed&dopt=Citation&list_uids=20634426

Latorre, R., Brauchi, S., Orta, G., Zaelzer, C., & Vargas, G. (2007). ThermoTRP channels as modular proteins with allosteric gating. *Cell Calcium, 42*(4–5), 427–438. doi:10.1016/j.ceca.2007.04.004

Lee, S. Y., Banerjee, A., & MacKinnon, R. (2009). Two separate interfaces between the voltage sensor and pore are required for the function of voltage-dependent K^+ channels. *PLoS Biology, 7*(3), e47. Retrieved from http://www.ncbi.nlm.nih.gov/entrez/query.fcgi?cmd=Retrieve&db=PubMed&dopt=Citation&list_uids=19260762

Li, M., Farley, R. A., & Lester, H. A. (2000). An intermediate state of the gamma-aminobutyric acid transporter GAT1 revealed by simultaneous voltage clamp and fluorescence. *Journal of General Physiology, 115*(4), 491–508. doi:10.1085/jgp.115.4.491

Li, M., & Lester, H. A. (2002). Early fluorescence signals detect transitions at mammalian serotonin transporters. *Biophysical Journal, 83*(1), 206–218. doi:10.1016/S0006-3495(02)75162-X

Li, X., Zhang, Q., Guo, P., Fu, J., Mei, L., Lv, D., . . . Guo, J. (2021). Molecular basis for ligand activation of the human KCNQ2 channel. *Cell Research, 31*(1), 52–61. doi:10.1038/s41422-020-00410-8

Liin, S. I., Barro-Soria, R., & Larsson, H. P. (2015). The KCNQ1 channel—Remarkable flexibility in gating allows for functional versatility. *Journal of Physiology, 593*(12), 2605–2615. doi:10.1113/jphysiol.2014.287607

Liu, Y., Holmgren, M., Jurman, M. E., & Yellen, G. (1997). Gated access to the pore of a voltage-dependent K^+ channel. *Neuron, 19*(1), 175–184.

Liu, Y. S., Sompornpisut, P., & Perozo, E. (2001). Structure of the KcsA channel intracellular gate in the open state. *Nature Structural Biology, 8*(10), 883–887. Retrieved from http://www.ncbi.nlm.nih.gov/entrez/query.fcgi?cmd=Retrieve&db=PubMed&dopt=Citation&list_uids=11573095

Long, S. B., Campbell, E. B., & Mackinnon, R. (2005a). Crystal structure of a mammalian voltage-dependent Shaker family K^+ channel. *Science, 309*(5736), 897–903. Retrieved from http://www.ncbi.nlm.nih.gov/entrez/query.fcgi?cmd=Retrieve&db=PubMed&dopt=Citation&list_uids=16002581

Long, S. B., Campbell, E. B., & Mackinnon, R. (2005b). Voltage sensor of Kv1.2: Structural basis of electromechanical coupling. *Science, 309*(5736), 903–908. Retrieved from http://www.ncbi.nlm.nih.gov/entrez/query.fcgi?cmd=Retrieve&db=PubMed&dopt=Citation&list_uids=16002579

Long, S. B., Tao, X., Campbell, E. B., & MacKinnon, R. (2007). Atomic structure of a voltage-dependent K^+ channel in a lipid membrane-like environment. *Nature, 450*(7168), 376–382. Retrieved from http://www.ncbi.nlm.nih.gov/entrez/query.fcgi?cmd=Retrieve&db=PubMed&dopt=Citation&list_uids=18004376

Loo, D. D., Hirayama, B. A., Gallardo, E. M., Lam, J. T., Turk, E., & Wright, E. M. (1998). Conformational changes couple Na^+ and glucose transport. *Proceedings of the National Academy of Sciences of the United States of America, 95*(13), 7789–7794. doi:10.1073/pnas.95.13.7789

Lorinczi, E., Gomez-Posada, J. C., de la Pena, P., Tomczak, A. P., Fernandez-Trillo, J., Leipscher, U., ... Pardo, L. A. (2015). Voltage-dependent gating of KCNH potassium channels lacking a covalent link between voltage-sensing and pore domains. *Nature Communications*, *6*, 6672. doi:10.1038/ncomms7672

MacKinnon, R., Cohen, S. L., Kuo, A., Lee, A., & Chait, B. T. (1998). Structural conservation in prokaryotic and eukaryotic potassium channels. *Science*, *280*(5360), 106–109. Retrieved from https://pubmed.ncbi.nlm.nih.gov/9525854/

Maljevic, S., & Lerche, H. (2014). Potassium channel genes and benign familial neonatal epilepsy. *Progress in Brain Research*, *213*, 17–53. doi:10.1016/B978-0-444-63326-2.00002-8

Mannuzzu, L. M., Moronne, M. M., & Isacoff, E. Y. (1996). Direct physical measure of conformational rearrangement underlying potassium channel gating. *Science*, *271*(5246), 213–216. Retrieved from http://www.ncbi.nlm.nih.gov/entrez/query.fcgi?cmd=Retrieve&db=PubMed&dopt=Citation&list_uids=8539623

Martin-Bornez, M., Galeano-Otero, I., Del Toro, R., & Smani, T. (2020). TRPC and TRPV channels' role in vascular remodeling and disease. *International Journal of Molecular Sciences*, *21*(17), 6125. doi:10.3390/ijms21176125

Mathie, A., Al-Moubarak, E., & Veale, E. L. (2010). Gating of two pore domain potassium channels. *Journal of Physiology*, *588*(Pt 17), 3149–3156. doi:10.1113/jphysiol.2010.192344

McCoy, J. G., & Nimigean, C. M. (2012). Structural correlates of selectivity and inactivation in potassium channels. *Biochimica et Biophysica Acta*, *1818*(2), 272–285. doi:10.1016/j.bbamem.2011.09.007

Miceli, F., Cilio, M. R., Taglialatela, M., & Bezanilla, F. (2009). Gating currents from neuronal $K_V7.4$ channels: General features and correlation with the ionic conductance. *Channels (Austin)*, *3*(4), 274–283.

Miceli, F., Vargas, E., Bezanilla, F., & Taglialatela, M. (2012). Gating currents from Kv7 channels carrying neuronal hyperexcitability mutations in the voltage-sensing domain. *Biophysical Journal*, *102*(6), 1372–1382. doi:10.1016/j.bpj.2012.02.004

Miranda, P., Contreras, J. E., Plested, A. J., Sigworth, F. J., Holmgren, M., & Giraldez, T. (2013). State-dependent FRET reports calcium- and voltage-dependent gating-ring motions in BK channels. *Proceedings of the National Academy of Sciences of the United States of America*, *110*(13), 5217–5222. doi:10.1073/pnas.1219611110

Mony, L., Berger, T. K., & Isacoff, E. Y. (2015). A specialized molecular motion opens the Hv1 voltage-gated proton channel. *Nature Structural & Molecular Biology*, *22*(4), 283–290. doi:10.1038/nsmb.2978

Nakajo, K., & Kubo, Y. (2007). KCNE1 and KCNE3 stabilize and/or slow voltage sensing S4 segment of KCNQ1 channel. *Journal of General Physiology*, *130*(3), 269–281. doi:10.1085/jgp.200709805

Naziroglu, M., & Braidy, N. (2017). Thermo-sensitive TRP channels: Novel targets for treating chemotherapy-induced peripheral pain. *Frontiers in Physiology*, *8*, 1040. doi:10.3389/fphys.2017.01040

New, K., Del Villar, S. G., Mazzaferro, S., Alcaino, C., & Bermudez, I. (2018). The fifth subunit of the $(\alpha4\beta2)_2$ β2 nicotinic ACh receptor modulates maximal ACh responses. *British Journal of Pharmacology*, *175*(11), 1822–1837. doi:10.1111/bph.13905

Noceti, F., Baldelli, P., Wei, X., Qin, N., Toro, L., Birnbaumer, L., & Stefani, E. (1996). Effective gating charges per channel in voltage-dependent K^+ and Ca^{2+} channels. *Journal of General Physiology*, *108*(3), 143–155. doi:10.1085/jgp.108.3.143

Noda, M., Shimizu, S., Tanabe, T., Takai, T., Kayano, T., Ikeda, T., . . . Numa, S. (1984). Primary structure of *Electrophorus electricus* sodium channel deduced from cDNA sequence. *Nature*, *312*(5990), 121–127.

Noren, C. J., Anthony-Cahill, S. J., Griffith, M. C., & Schultz, P. G. (1989). A general method for site-specific incorporation of unnatural amino acids into proteins. *Science, 244*(4901), 182–188. doi:10.1126/science.2649980

Nowak, M. W., Kearney, P. C., Sampson, J. R., Saks, M. E., Labarca, C. G., Silverman, S. K., . . . Lester, H. A. (1995). Nicotinic receptor binding site probed with unnatural amino acid incorporation in intact cells. *Science, 268*(5209), 439–442. doi:10.1126/science.7716551

Ong, H. L., & Ambudkar, I. S. (2017). STIM-TRP pathways and microdomain organization: Contribution of TRPC1 in store-operated Ca^{2+} entry: Impact on Ca^{2+} signaling and cell function. *Advances in Experimental Medicine and Biology, 993*, 159–188. doi:10.1007/978-3-319-57732-6_9

Osteen, J. D., Barro-Soria, R., Robey, S., Sampson, K. J., Kass, R. S., & Larsson, H. P. (2012). Allosteric gating mechanism underlies the flexible gating of KCNQ1 potassium channels. *Proceedings of the National Academy of Sciences of the United States of America, 109*(18), 7103–7108. doi:10.1073/pnas.1201582109

Osteen, J. D., Gonzalez, C., Sampson, K. J., Iyer, V., Rebolledo, S., Larsson, H. P., & Kass, R. S. (2010). KCNE1 alters the voltage sensor movements necessary to open the KCNQ1 channel gate. *Proceedings of the National Academy of Sciences of the United States of America, 107*(52), 22710–22715. doi:10.1073/pnas.1016300108

Palazzo, E., Rossi, F., de Novellis, V., & Maione, S. (2013). Endogenous modulators of TRP channels. *Current Topics in Medicinal Chemistry, 13*(3), 398–407. doi:10.2174/1568026611313030014

Pantazis, A., & Olcese, R. (2012). Relative transmembrane segment rearrangements during BK channel activation resolved by structurally assigned fluorophore-quencher pairing. *Journal of General Physiology, 140*(2), 207–218. doi:10.1085/jgp.201210807

Pantazis, A., Westerberg, K., Althoff, T., Abramson, J., & Olcese, R. (2018). Harnessing photoinduced electron transfer to optically determine protein sub-nanoscale atomic distances. *Nature Communications, 9*(1), 4738. doi:10.1038/s41467-018-07218-6

Paoletti, P., Ellis-Davies, G. C. R., & Mourot, A. (2019). Optical control of neuronal ion channels and receptors. *Nature Reviews Neuroscience, 20*(9), 514–532. doi:10.1038/s41583-019-0197-2

Papazian, D. M., Timpe, L. C., Jan, Y. N., & Jan, L. Y. (1991). Alteration of voltage-dependence of Shaker potassium channel by mutations in the S4 sequence. *Nature, 349*(6307), 305–310. doi:10.1038/349305a0

Perozo, E., Cortes, D. M., & Cuello, L. G. (1998). Three-dimensional architecture and gating mechanism of a K^+ channel studied by EPR spectroscopy. *Nature Structural Biology, 5*(6), 459–469. doi:10.1038/nsb0698-459

Pless, S. A., & Ahern, C. A. (2015). Introduction. In C. Ahern & S. Pless (Eds.), *Advances in experimental medicine and biology: Vol. 869. Novel chemical tools to study ion channel biology* (pp. 1–4). New York, NY: Springer.

Pless, S. A., & Ahern, C. A. (2013). Unnatural amino acids as probes of ligand–receptor interactions and their conformational consequences. *Annual Review of Pharmacology and Toxicology, 53*, 211–229. doi:10.1146/annurev-pharmtox-011112-140343

Pless, S. A., Galpin, J. D., Frankel, A., & Ahern, C. A. (2011). Molecular basis for class Ib anti-arrhythmic inhibition of cardiac sodium channels. *Nature Communications, 2*, 351. doi:10.1038/ncomms1351

Pless, S. A., Galpin, J. D., Niciforovic, A. P., & Ahern, C. A. (2011). Contributions of counter-charge in a potassium channel voltage-sensor domain. *Nature Chemical Biology, 7*(9), 617–623. doi:10.1038/nchembio.622

Pless, S. A., Leung, A. W., Galpin, J. D., & Ahern, C. A. (2011). Contributions of conserved residues at the gating interface of glycine receptors. *Journal of Biological Chemistry, 286*(40), 35129–35136. doi:10.1074/jbc.M111.269027

Qiu, F., Rebolledo, S., Gonzalez, C., & Larsson, H. P. (2013). Subunit interactions during cooperative opening of voltage-gated proton channels. *Neuron, 77*(2), 288–298. doi:10.1016/j.neuron.2012.12.021

Ramsey, I. S., Delling, M., & Clapham, D. E. (2006). An introduction to TRP channels. *Annual Review of Physiology, 68*, 619–647. doi:10.1146/annurev.physiol.68.040204.100431

Ranganathan, R., Lewis, J. H., & MacKinnon, R. (1996). Spatial localization of the K^+ channel selectivity filter by mutant cycle-based structure analysis. *Neuron, 16*(1), 131–139. doi:10.1016/s0896-6273(00)80030-6

Rocheleau, J. M., & Kobertz, W. R. (2008). KCNE peptides differently affect voltage sensor equilibrium and equilibration rates in KCNQ1 K^+ channels. *Journal of General Physiology, 131*(1), 59–68. doi:10.1085/jgp.200709816

Ruscic, K. J., Miceli, F., Villalba-Galea, C. A., Dai, H., Mishina, Y., Bezanilla, F., & Goldstein, S. A. (2013). IKs channels open slowly because KCNE1 accessory subunits slow the movement of S4 voltage sensors in KCNQ1 pore-forming subunits. *Proceedings of the National Academy of Sciences of the United States of America, 110*(7), E559–E566. doi:10.1073/pnas.1222616110

Savalli, N., Kondratiev, A., de Quintana, S. B., Toro, L., & Olcese, R. (2007). Modes of operation of the BKCa channel beta2 subunit. *Journal of General Physiology, 130*(1), 117–131. doi:10.1085/jgp.200709803

Savalli, N., Kondratiev, A., Toro, L., & Olcese, R. (2006). Voltage-dependent conformational changes in human Ca^{2+}- and voltage-activated K^+ channel, revealed by voltage-clamp fluorometry. *Proceedings of the National Academy of Sciences of the United States of America, 103*(33), 12619–12624. doi:10.1073/pnas.0601176103

Schewe, M., Nematian-Ardestani, E., Sun, H., Musinszki, M., Cordeiro, S., Bucci, G., . . . Baukrowitz, T. (2016). A non-canonical voltage-sensing mechanism controls gating in K2P K^+ channels. *Cell, 164*(5), 937–949. doi:10.1016/j.cell.2016.02.002

Schneider, M. F., & Chandler, W. K. (1976). Effects of membrane potential on the capacitance of skeletal muscle fibers. *Journal of General Physiology, 67*(2), 125–163. doi:10.1085/jgp.67.2.125

Schoppa, N. E., McCormack, K., Tanouye, M. A., & Sigworth, F. J. (1992). The size of gating charge in wild-type and mutant Shaker potassium channels. *Science, 255*(5052), 1712–1715.

Schroeder, B. C., Hechenberger, M., Weinreich, F., Kubisch, C., & Jentsch, T. J. (2000). KCNQ5, a novel potassium channel broadly expressed in brain, mediates M-type currents. *Journal of Biological Chemistry, 275*(31), 24089–24095.

Seoh, S. A., Sigg, D., Papazian, D. M., & Bezanilla, F. (1996). Voltage-sensing residues in the S2 and S4 segments of the Shaker K^+ channel. *Neuron, 16*(6), 1159–1167. Retrieved from http://www.ncbi.nlm.nih.gov/pubmed/8663992

Shenkel, S., & Bezanilla, F. (1991). Patch recordings from the electrocytes of electrophorus. Na channel gating currents. *Journal of General Physiology, 98*(3), 465–478. doi:10.1085/jgp.98.3.465

Shrivastava, I. H., Capener, C. E., Forrest, L. R., & Sansom, M. S. (2000). Structure and dynamics of K channel pore-lining helices: A comparative simulation study. *Biophysical Journal, 78*(1), 79–92. doi:10.1016/S0006-3495(00)76574-X

Shrivastava, I. H., & Sansom, M. S. (2000). Simulations of ion permeation through a potassium channel: Molecular dynamics of KcsA in a phospholipid bilayer. *Biophysical Journal, 78*(2), 557–570. doi:10.1016/S0006-3495(00)76616-1

Smith, P. L., Baukrowitz, T., & Yellen, G. (1996). The inward rectification mechanism of the HERG cardiac potassium channel [see comments]. *Nature, 379*(6568), 833–836.

Smith, P. L., & Yellen, G. (2002). Fast and slow voltage sensor movements in HERG potassium channels. *Journal of General Physiology, 119*(3), 275–293. doi:10.1085/jgp.20028534

Soderhielm, P. C., Andersen, J., Munro, L., Nielsen, A. T., & Kristensen, A. S. (2015). Substrate and inhibitor-specific conformational changes in the human serotonin transporter revealed by voltage-clamp fluorometry. *Molecular Pharmacology, 88*(4), 676–688. doi:10.1124/mol.115.099911

Stefani, E., & Bezanilla, F. (1998). Cut-open oocyte voltage-clamp technique. *Methods in Enzymology, 293*, 300–318.

Stuhmer, W., Conti, F., Suzuki, H., Wang, X. D., Noda, M., Yahagi, N., . . . Numa, S. (1989). Structural parts involved in activation and inactivation of the sodium channel. *Nature, 339*(6226), 597–603. Retrieved from http://www.ncbi.nlm.nih.gov/entrez/query.fcgi?cmd=Retrieve&db=PubMed&dopt=Citation&list_uids=2543931

Sullivan, D., Chiara, D. C., & Cohen, J. B. (2002). Mapping the agonist binding site of the nicotinic acetylcholine receptor by cysteine scanning mutagenesis: Antagonist footprint and secondary structure prediction. *Molecular Pharmacology, 61*(2), 463–472. doi:10.1124/mol.61.2.463

Sun, J., & MacKinnon, R. (2017). Cryo-EM structure of a KCNQ1/CaM complex reveals insights into congenital long QT syndrome. *Cell, 169*(6), 1042–1050.e9. doi:10.1016/j.cell.2017.05.019

Taglialatela, M., Toro, L., & Stefani, E. (1992). Novel voltage clamp to record small, fast currents from ion channels expressed in *Xenopus* oocytes. *Biophysical Journal, 61*(1), 78–82. doi:10.1016/S0006-3495(92)81817-9

Taraska, J. W., & Zagotta, W. N. (2007). Structural dynamics in the gating ring of cyclic nucleotide–gated ion channels. *Nature Structural & Molecular Biology, 14*(9), 854–860. doi:10.1038/nsmb1281

Taylor, K. C., Kang, P. W., Hou, P., Yang, N. D., Kuenze, G., Smith, J. A., . . . Sanders, C. R. (2020). Structure and physiological function of the human KCNQ1 channel voltage sensor intermediate state. *Elife, 9*, e53901. doi:10.7554/eLife.53901

Ullrich, F., Blin, S., Lazarow, K., Daubitz, T., von Kries, J. P., & Jentsch, T. J. (2019). Identification of TMEM206 proteins as pore of PAORAC/ASOR acid-sensitive chloride channels. *Elife, 8*, e49187. doi:10.7554/eLife.49187

Vaid, M., Claydon, T. W., Rezazadeh, S., & Fedida, D. (2008). Voltage clamp fluorimetry reveals a novel outer pore instability in a mammalian voltage-gated potassium channel. *Journal of General Physiology, 132*(2), 209–222. doi:10.1085/jgp.200809978

Vandenberg, J. I., Perozo, E., & Allen, T. W. (2017). Towards a structural view of drug binding to hERG K^+ channels. *Trends in Pharmacological Sciences, 38*(10), 899–907. doi:10.1016/j.tips.2017.06.004

Varga, Z., Zhu, W., Schubert, A. R., Pardieck, J. L., Krumholz, A., Hsu, E. J., . . . Silva, J. R. (2015). Direct measurement of cardiac Na^+ channel conformations reveals molecular pathologies of inherited mutations. *Circulation Arrhythmia and Electrophysiology, 8*(5), 1228–1239. doi:10.1161/CIRCEP.115.003155

Venkatachalam, K., & Montell, C. (2007). TRP channels. *Annual Review of Biochemistry, 76*, 387–417. doi:10.1146/annurev.biochem.75.103004.142819

Virkki, L. V., Murer, H., & Forster, I. C. (2006). Voltage clamp fluorometric measurements on a type II Na^+-coupled Pi cotransporter: Shedding light on substrate binding order. *Journal of General Physiology, 127*(5), 539–555. doi:10.1085/jgp.200609496

Voets, T., Droogmans, G., Wissenbach, U., Janssens, A., Flockerzi, V., & Nilius, B. (2004). The principle of temperature-dependent gating in cold- and heat-sensitive TRP channels. *Nature, 430*(7001), 748–754. doi:10.1038/nature02732

Wang, H. S., Pan, Z., Shi, W., Brown, B. S., Wymore, R. S., Cohen, I. S., . . . McKinnon, D. (1998). KCNQ2 and KCNQ3 potassium channel subunits: Molecular correlates of the M-channel. *Science, 282*(5395), 1890–1893.

Wang, R., Tu, S., Zhang, J., & Shao, A. (2020). Roles of TRP channels in neurological diseases. *Oxidative Medicine and Cellular Longevity, 2020*, 7289194. doi:10.1155/2020/7289194

Wang, Z., Dou, Y., Goodchild, S. J., Es-Salah-Lamoureux, Z., & Fedida, D. (2013). Components of gating charge movement and S4 voltage-sensor exposure during activation of hERG channels. *Journal of General Physiology, 141*(4), 431–443. doi:10.1085/jgp.201210942

Whicher, J. R., & MacKinnon, R. (2016). Structure of the voltage-gated K^+ channel Eag1 reveals an alternative voltage sensing mechanism. *Science, 353*(6300), 664–669. doi:10.1126/science.aaf8070

White, M. M., & Miller, C. (1979). A voltage-gated anion channel from the electric organ of *Torpedo californica*. *Journal of Biological Chemistry, 254*(20), 10161–10166. Retrieved from https://www.ncbi.nlm.nih.gov/pubmed/489590

Wu, D., Delaloye, K., Zaydman, M. A., Nekouzadeh, A., Rudy, Y., & Cui, J. (2010). State-dependent electrostatic interactions of S4 arginines with E1 in S2 during Kv7.1 activation. *Journal of General Physiology, 135*(6), 595–606. Retrieved from http://www.ncbi.nlm.nih.gov/entrez/query.fcgi?cmd=Retrieve&db=PubMed&dopt=Citation&list_uids=20479111

Wulf, M., & Pless, S. A. (2018). High-sensitivity fluorometry to resolve ion channel conformational dynamics. *Cell Reports, 22*(6), 1615–1626. doi:10.1016/j.celrep.2018.01.029

Yamaguchi, Y., Iribe, G., Nishida, M., & Naruse, K. (2017). Role of TRPC3 and TRPC6 channels in the myocardial response to stretch: Linking physiology and pathophysiology. *Progress in Biophysics and Molecular Biology, 130*(Pt B), 264–272. doi:10.1016/j.pbiomolbio.2017.06.010

Yang, N., George, A. L., Jr., & Horn, R. (1996). Molecular basis of charge movement in voltage-gated sodium channels. *Neuron, 16*(1), 113–122.

Yao, J., Liu, B., & Qin, F. (2010). Kinetic and energetic analysis of thermally activated TRPV1 channels. *Biophysical Journal, 99*(6), 1743–1753. doi:10.1016/j.bpj.2010.07.022

Yellen, G. (1998a). The moving parts of voltage-gated ion channels. *Quarterly Reviews of Biophysics, 31*(3), 239–295.

Yellen, G. (1998b). Premonitions of ion channel gating [see comment]. *Nature Structural Biology, 5*(6), 421. Retrieved from PM:0009628476

Yu, F. H., & Catterall, W. A. (2004). The VGL-chanome: A protein superfamily specialized for electrical signaling and ionic homeostasis. *Science's STKE, 2004*(253), re15. doi:10.1126/stke.2532004re15

Yusaf, S. P., Wray, D., & Sivaprasadarao, A. (1996). Measurement of the movement of the S4 segment during the activation of a voltage-gated potassium channel. *Pflugers Archic, 433*(1–2), 91–97. doi:10.1007/s004240050253

Zaydman, M. A., Kasimova, M. A., McFarland, K., Beller, Z., Hou, P., Kinser, H. E., . . . Cui, J. (2014). Domain–domain interactions determine the gating, permeation, pharmacology, and subunit modulation of the IKs ion channel. *Elife, 3*, e03606. doi:10.7554/eLife.03606

Zaydman, M. A., Silva, J. R., Delaloye, K., Li, Y., Liang, H., Larsson, H. P., . . . Cui, J. (2013). Kv7.1 ion channels require a lipid to couple voltage sensing to pore opening. *Proceedings of the National Academy of Sciences of the United States of America, 110*(32), 13180–13185. doi:10.1073/pnas.1305167110

Zheng, J., & Zagotta, W. N. (2000). Gating rearrangements in cyclic nucleotide-gated channels revealed by patch-clamp fluorometry. *Neuron*, *28*(2), 369–374. doi:10.1016/s0896-6273(00)00117-3

Zhu, W., Voelker, T. L., Varga, Z., Schubert, A. R., Nerbonne, J. M., & Silva, J. R. (2017). Mechanisms of noncovalent beta subunit regulation of NaV channel gating. *Journal of General Physiology*, *149*(8), 813–831. doi:10.1085/jgp.201711802

SECTION 2

VOLTAGE-GATED CHANNELS

CHAPTER 4

THE VOLTAGE-DEPENDENT K^+ CHANNEL FAMILY

HANNE B. RASMUSSEN AND JAMES S. TRIMMER

Introduction

Ionic currents originating from voltage-dependent K^+ (Kv) channels were first described by Hodgkin and Huxley in 1952 (Hodgkin & Huxley, 1952). In these classical experiments, they deduced the ionic currents underlying the neuronal action potential by performing voltage-clamp experiments on the squid giant axon. This included the identification of the major current responsible for the repolarization of the action potential, an outward potassium (K^+) current activated by membrane depolarization. At that time, the size and complexity of the Kv channel gene family that we now know is responsible for voltage-dependent K^+ currents could not have been imagined. A total of 40 genes encoding Kv channel α subunits have been identified in the human genome, and thus far, 10 of these have been directly associated with human diseases. The number of distinct Kv channels is further increased by alternative splicing and editing of the mRNA products of these Kv channel α subunit genes, by the ability of the α subunit polypeptides to combine in different homo- and hetero-tetrameric subunit configurations to form the functional Kv channels, and by the association of Kv channels with accessory subunits and other interacting proteins in native channel protein complexes. This results in a huge number of structurally distinct Kv channel complexes with unique biophysical and pharmacological properties, subcellular localization, and sensitivity to dynamic modulation by cellular signaling pathways. This diversity allows Kv channels to uniquely impact diverse aspects of electrical information processing. They shape neuronal firing properties, modulate neurotransmitter release, and influence dendritic signal processing. Moreover, the robust dynamic modulation of Kv channels by signaling pathways can impact their expression level, subcellular localization, and functional properties, which further enhances the diversity of this already complex family of ion channels. Kv channels therefore represent an inherently diverse toolbox for neurons to fine-tune their electrical properties, and the function of these cells and their circuits, and they also provide a rich substrate for dynamic modulation. In this chapter, we will review our current fundamental knowledge of this important, large, and diverse ion channel family.

Nomenclature of Kv Channels

Many of the 40 Kv channel α subunits, when originally identified, were assigned multiple names, making it somewhat difficult to navigate the early literature. However, based on a suggestion by Chandy, a systematic nomenclature for Kv channel α subunits was adopted by the International Union of Pharmacology (IUPHAR) (Chandy, 1991; Gutman et al., 2005), which was subsequently employed for other ion channels. The IUPHAR system has grouped the 40 human α subunits into 12 major subfamilies, Kv1–Kv12, based on sequence homology, the name "Kv" referring to a potassium channel (K) whose activation is voltage-dependent (v). In parallel, official names have been assigned to the α subunit genes by the HUGO Gene Nomenclature Committee (HGNC, http://www.genenames.org/cgi-bin/genefamilies/set/274). The genes are named KCN for K^+ (K) channel (CN), followed by a letter characteristic of one, or in a few cases, several related subfamilies. In certain, but not all cases, the nomenclature of auxiliary subunits follows these conventions. Today most publications utilize the IUPHAR nomenclature when referring to the α and auxiliary subunit polypeptides, and the HGNC nomenclature for gene names. This systematic approach, which is summarized in Table 4.1, has greatly facilitated the unambiguous navigation of the scientific literature, albeit in the context of the somewhat intimidating genomic and proteomic complexity of the channels formed by this large gene family.

Cloning of Kv Channels

The first Kv channel α subunit was cloned by positional cloning in the *Drosophila melanogaster Shaker* mutant (Kamb, Iverson, & Tanouye, 1987; Papazian, Schwarz, Tempel, Jan, & Jan, 1987; Tempel, Papazian, Schwarz, Jan, & Jan, 1987). Within a year, the cDNA clone for the first mammalian Kv channel α subunit orthologue, appropriately named Kv1.1, was isolated from a mouse cDNA library by homology screening with *Shaker* cDNA probe (Tempel, Jan, & Jan, 1988). Over the next 12 years, the diversity of genes encoding Kv channel α subunits was uncovered, primarily through homology-based screening methods based on the high degree of sequence similarity within the core region comprising transmembrane segments S1–S6 (see the next section in this chapter), and employing low stringency hybridization techniques, reverse transcription-polymerase chain reaction (RT-PCR)based screening with degenerate primers or Basic

Table 4.1 Overview of Kv Channels: Naming, Subcellular Localization in Neurons, Basic Current Features, and Disease Associations

IUPHAR	HGNC	Predominant subcellular localization	Current characteristics	Associated human disease
Kv1 family				
Kv1.1	*KCNA1*	AIS, juxtaparanodes, preterminal segment of axon	Fast, low activation threshold delayed rectifier	Episodic ataxia type 1
Kv1.2	*KCNA2*	AIS, juxtaparanodes, preterminal segment of axon	Fast, low activation threshold delayed rectifier	Early infantile epileptic encephalopathy-32
Kv1.3	*KCNA3*	Presynaptic (Calyx of Held)	Fast, low activation threshold delayed rectifier	–
Kv1.4	*KCNA4*	Preterminal segment of axon	Fast, low activation threshold A-current	–
Kv1.5	*KCNA5*	Mostly non-neuronal in brain	Fast, low activation threshold delayed rectifier	–
Kv1.6	*KCNA6*	Somatodendritic	Fast, low activation threshold delayed rectifier	–
Kv1.7	*KCNA7*	–	Fast, low activation threshold delayed rectifier	–
Kv1.8	*KCNA10*	–	Fast, low activation threshold delayed rectifier	–
Kv2 family				
Kv2.1	*KCNB1*	Large clusters on soma, proximal dendrites, and AIS	Slow, high activation threshold delayed rectifier	–
Kv2.2	*KCNB2*	Large clusters on soma, proximal dendrites, and AIS	Slow, high activation threshold delayed rectifier	–
Kv3 family				
Kv3.1	*KCNC1*	Presynaptic terminals, nodes of Ranvier	Fast, high activation threshold delayed rectifier	–

(continued)

Table 4.1 Continued

IUPHAR	HGNC	Predominant subcellular localization	Current characteristics	Associated human disease
Kv3.2	*KCNC2*	Presynaptic terminals	Fast, high activation threshold delayed rectifier	–
Kv3.3	*KCNC3*	Presynaptic terminals	Fast, high activation threshold A-type current	Spinocerebellar ataxia type 13
Kv3.4	*KCNC4*	Presynaptic terminals	Fast, high activation threshold A-type current	–
Kv4 family				
Kv4.1	*KCND1*	Somatodendritic	Fast, low activation threshold A-type current	–
Kv4.2	*KCND2*	Somatodendritic	Fast, low activation threshold A-type current	–
Kv4.3	*KCND3*	Somatodendritic	Fast, low activation threshold A-type current	–
Kv5 family				
Kv5.1	*KCNF1*	–	Modifier of Kv2 currents	–
Kv6 family				
Kv6.1	*KCNG1*	–	Modifier of Kv2 currents	–
Kv6.2	*KCNG2*	–	Modifier of Kv2 currents	–
Kv6.3	*KCNG3*	–	Modifier of Kv2 currents	–
Kv6.4	*KCNG4*	–	Modifier of Kv2 currents	–
Kv7 family				
Kv7.1	*KCNQ1*	–	Slow, low activation threshold delayed rectifier	Long QT syndrome, Jervell Lange-Nielsen syndrome
Kv7.2	*KCNQ2*	AIS, nodes of Ranvier, presynapses	Slow, low activation threshold delayed rectifier	Benign familial neonatal epilepsy
Kv7.3	*KCNQ3*	AIS, nodes of Ranvier, presynapses	Slow, low activation threshold delayed rectifier	Benign familial neonatal epilepsy
Kv7.4	*KCNQ4*	–	Slow, low activation threshold delayed rectifier	Deafness, non-syndromic autosomal dominant 2
Kv7.5	*KCNQ5*	Possibly presynaptic	Slow, low activation threshold delayed rectifier	

Table 4.1 Continued

IUPHAR	HGNC	Predominant subcellular localization	Current characteristics	Associated human disease
Kv8 family				
Kv8.1	*KCNV1*	–	Modifier of Kv2 currents	–
Kv8.2	*KCNV2*	Inner segment of cone and rod photoreceptors of the retina	Modifier of Kv2 currents	cone-dystrophy with supernormal rod electroretinogram
Kv9 family				
Kv9.1	*KCNS1*	–	Modifier of Kv2 currents	–
Kv9.2	*KCNS2*	–	Modifier of Kv2 currents	–
Kv9.3	*KCNS3*	–	Modifier of Kv2 currents	–
Kv10 family				
Kv10.1	*KCNH1*	Presynaptic	Slow, low activation threshold delayed rectifier	Zimmermann-Laband and Temple-Baraitser syndromes
Kv10.2	*KCNH5*	–	Slow, low activation threshold delayed rectifier	–
Kv11 family				
Kv11.1	*KCNH2*	–	Slow, low activation threshold delayed rectifier	Long QT syndrome
Kv11.2	*KCNH6*	–	Slow, low activation threshold delayed rectifier	–
Kv11.3	*KCNH7*	–	Slow, low activation threshold delayed rectifier	–
Kv12 family				
Kv12.1	*KCNH8*	–	Slow, low activation threshold delayed rectifier	–
Kv12.2	*KCNH3*	–	Slow, low activation threshold delayed rectifier	–
Kv13.3	*KCNH4*	–	Slow, low activation threshold delayed rectifier	–

Faded channel subunits are not reported to have significant expression in neurons.

AIS: axon initial segment.

Local Alignment Search Tool (BLAST) searches in the expressed sequence tag or genomic databases (reviewed in Vacher, Mohapatra, & Trimmer, 2008). Exceptions were the discoveries of the Kv2 and Kv7 subfamilies of Kv channel α subunits. A cDNA encoding Kv2.1 was isolated by functionally screening pools and then sub-pools of mRNAs generated from a rat brain cDNA library in *Xenopus* oocytes for Kv currents activated by membrane depolarization, ultimately leading to the isolation of a single clone, Kv2.1 (Frech, VanDongen, Schuster, Brown, & Joho, 1989). The human Kv7.1 α subunit gene was isolated by positional cloning and mutational analysis in an affected family due to its association with the inherited cardiac arrhythmia, long QT syndrome (Q. Wang et al., 1996).

Structure of Kv Channels

The polypeptides encoded by the 40 human Kv channel α subunit genes share several common structural characteristics. They all have a core domain of six transmembrane-spanning helices (S1–S6), a membrane-embedded P-loop connecting the S5 and S6 transmembrane segments, and cytoplasmic N- and C-terminal domains (Figure 4.1). A functional Kv channel is formed upon the assembly of four Kv channel α subunits (Long, Campbell, & MacKinnon, 2005; MacKinnon, 1991). The initial cloning of multiple isoforms of the *Shaker* α subunit suggested that Kv channel α subunits might assemble into both homo- and hetero-multimeric channel complexes, which turned out to be the case (see, e.g., Isacoff, Jan, & Jan, 1990; Ruppersberg et al., 1990). Indeed, the ability of Kv channel subunits to combine in different tetrameric subunit configurations greatly contributes to the variety of functional Kv channels expressed in neurons.

The fundamental attributes of a Kv channel, to create a voltage-regulated passageway for K^+ ions across the lipid bilayer, are encoded within the α subunit core domain comprising the S1–S6 transmembrane segments, which contains all of the transmembrane and extracellular domains of the α subunit as well as cytoplasmic linkers (Figure 4.1). The conductance passageway or pore is created by the assembly of the pore-forming domain, created by the S5 and S6 transmembrane segments and connecting P-loop present on each of the four α subunits, assembled in four-fold symmetry around the central pore (Figure 4.1). The resulting structure of the Kv channel pore domain (Long et al., 2005) is highly similar to the structure of KcsA, a K^+ channel α subunit from *Streptomyces lividans* that has a simpler structure with only the two transmembrane segments and P-loop that forms the pore, and lacking the voltage-sensing domain present on Kv channels, and was the first K^+ channel whose structure was solved at the atomic level (Doyle et al., 1998). The pore of these channels contains an inner water-filled cavity followed by a narrower selectivity filter (Figure 4.1). The selectivity filter is formed by the P-loop and contains a conserved signature sequence, TVGY(F)G, whose structure favors the coordination of dehydrated K^+ ions since it mimics the water oxygens that surround hydrated K^+ ions in solution. The open-channel pore

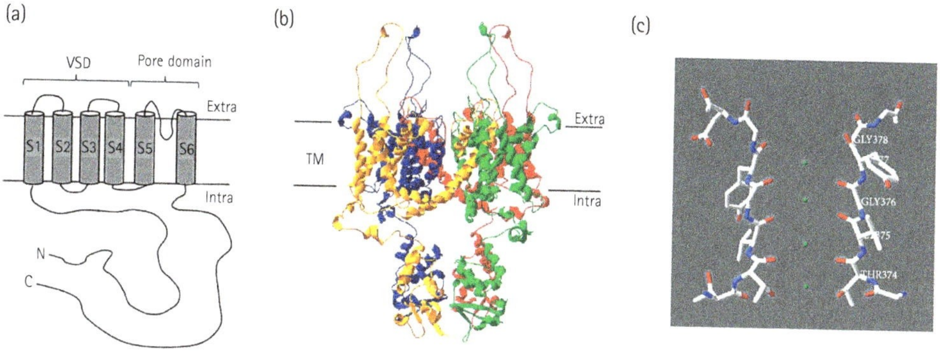

The structure of Kv channels.

FIGURE 4.1 **A.** Schematic overview of the domain structure of a Kv channel α subunit. Each α subunit contains six transmembrane helical segments (S1–S6). S1–S4 make up the voltage-sensing domain (VSD), while S5–S6 and the connecting P-loop constitute the pore domain. The N- and C-terminal domains are cytoplasmic. "Extra" and "intra" refer to the extracellular and intracellular side of the PM, respectively. **B.** Ribbon representation of the crystal structure of Kv α subunits, in this case rat Kv1.2, illustrating the tetrameric assembly of Kv channels. Each subunit is labeled in a different color. "TM" refers to the transmembrane portion of the channel. **C.** The crystal structure of the Kv1.2 selectivity filter with the residues of the signature sequence (TVGYG) indicated on one side. For clarity, only two opposing α subunits are shown. The green dots represent the K^+ ions in the four positions in the filter. Only two positions are in reality occupied at one time, with water molecules occupying the remaining two sites (2,4 and 1,3-configurations). Images **B** and **C** were created using Swiss-PDB Viewer and PDB ID: 3LUT.

creates an energetically favorable passageway for K^+ ions that is extremely efficient as it allows conduction at near diffusion-limited rates (10^8 ions s^{-1}) in a strongly K^+ selective manner, such that in certain K^+ channels, K^+ ions are more than 10,000 times more permeant than sodium ions (Doyle et al., 1998; Morais-Cabral, Zhou, & MacKinnon, 2001). The voltage-sensing function of Kv channels primarily resides in the S1–S4 transmembrane segments, also known as the voltage-sensing domain (VSD), present in each α subunit and that in the Kv channel structure lie peripheral to the central pore (Figure 4.1). Importantly, the otherwise hydrophobic S4 transmembrane segment contains a positively charged amino acid residue at every third position, such that the S4 segment can sense and move in response to changes in membrane potential. The movements of the VSD are transferred to the pore domain either within the subunit (Whicher & MacKinnon, 2016) or between the VSD domain of one subunit and the pore-forming domain of the adjacent subunit, which is referred to as "domain-swapping" (Long et al., 2005). In either case, the transfer of movement between the VSD and the pore domain allows for voltage-dependent control of pore opening.

In contrast to the highly conserved S1–S6 core domain, the N-terminal cytoplasmic domain that precedes it and the C-terminal cytoplasmic domain that follows it differ considerably between the different Kv channel α subunits, and in particular between subunits from different Kv channel subfamilies. These cytoplasmic domains play diverse roles in dictating subfamily-specific subunit co-assembly, interaction with accessory subunits, and through their interaction with one another, they impact the voltage-dependent gating of the channel. Further, they serve as platforms for interaction with regulatory proteins that impact the function and subcellular localization of the Kv channel complex.

The Diversity of Kv Channels

Although the different members of the Kv channel family share many structural features, the ionic currents that arise from Kv channels formed by the co-assembly of different α subunits are remarkably different (Figure 4.2). Kv channels formed from α subunits from certain subfamilies respond to small membrane depolarizations (Kv1, Kv4, Kv7, Kv10–12), while those formed from others require the strong depolarization of the type obtained during action potentials to be activated (Kv2, Kv3). The rate with which this activation happens can also differ substantially between channels formed by the different α subunits. Kv channels formed from Kv1, Kv3, and Kv4 α subunits display fast activation kinetics, while those formed from Kv2, Kv7, and Kv10–12 α subunits activate more slowly. Finally, certain types of Kv channels inactivate while others conduct sustained currents, although this can be further impacted by co-assembly with certain

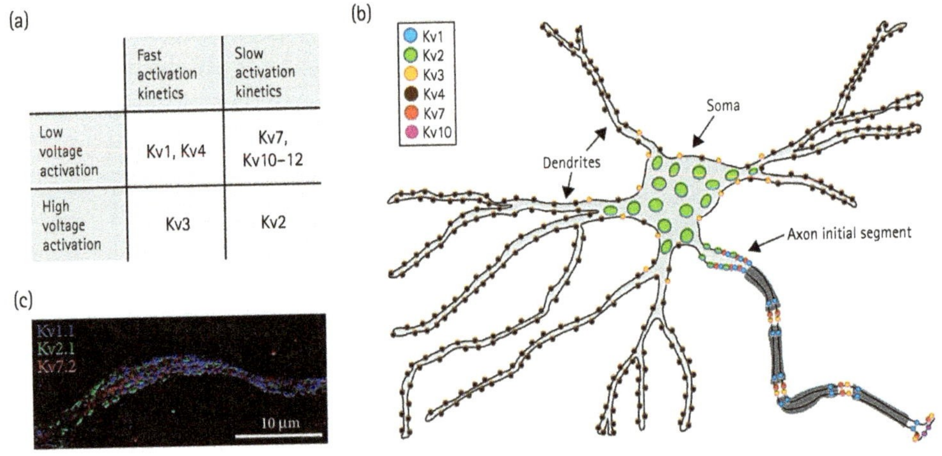

The diversity of Kv channels.

FIGURE 4.2 A. A simplified overview of the biophysical properties of different Kv channels formed upon expression of the individual α subunits as indicated, and in which the various channels are arranged according to their voltage-dependent activation properties. **B.** Illustration of the primary subcellular localization(s) of different Kv channel α subunits. **C.** Confocal image of the axon initial segment (AIS) of a 17 days in vitro (DIV) cultured rat hippocampal neuron immunolabeled for the Kv1.1, Kv2.1, and Kv7.2 α subunits, illustrating their non-overlapping localizations within this specialized neuronal compartment.

auxiliary subunits. Four of the Kv α subunit subfamilies, namely Kv5, Kv6, Kv8, and Kv9, fail to produce currents when expressed alone in heterologous expression systems. These subfamilies are considered modulatory subunits for the Kv2 subfamily.

In general, the different subfamilies of Kv channel α subunits preferentially localize to different neuronal sub-compartments (Figure 4.2). The Kv1 and Kv7 subfamilies are predominantly axonal channels, while Kv4 channels are exclusively localized to the somatodendritic region. Kv3 channels can be localized in either the axonal or the somatodendritic compartment, depending on the Kv3 subtype and the neuronal cell type. Kv2 channels are selectively found in the soma and proximal portions of both the axon and dendrites, but are noticeably absent in the distal regions of these processes. Even within these neuronal compartments, different Kv channel α subunits display further compartmentalization. For instance, α subunits from the Kv1, Kv2, and Kv7 subfamilies are all present at the axon initial segment (AIS), the site of action potential initiation (Figure 4.2). However, each Kv subfamily member has its own unique localization within this compartment, resulting in an impressive mosaic of Kv channel localization patterns when the members of these three subfamilies are visualized in the same cell (Figure 4.2).

The 40 different Kv channel α subunits encoded in mammalian genomes are thus endowed with distinct functional characteristics as well as distinct cellular and subcellular expression patterns. This allows them to regulate different aspects of neuronal information processing and also to be subjected to modulation by distinct local signaling events. The cellular expression patterns of distinct Kv channel α subunits in different neuronal cell types yield distinct repertoires of homo- and hetero-tetrameric Kv channel complexes localized at specific subcellular sites, and a diversity of expression, function, and regulation that greatly contributes to defining the specific firing properties and shaping of other functions of individual neurons. While the Kv1, Kv2, Kv3, Kv4, and Kv7 subfamilies have been extensively studied, and relatively more is known in terms of their expression, subcellular localization, and contributions to neuronal functions, relatively little is known when it comes to the electrically silent Kv channels as well as the Kv10–Kv12 subfamilies. These latter Kv channels comprise a substantial component of what is currently regarded as the under-studied druggable genome (https://commonfund.nih.gov/idg).

The Kv1 Subfamily

The members of the Kv1 channel subfamily are orthologues of the *Drosophila Shaker* channel and are sometimes referred to as the *Shaker* subfamily. They constitute the largest single Kv channel α subunit subfamily, with eight members named Kv1.1–1.8 (Table 4.1). Of these, Kv1.1–Kv1.6 are expressed in brain, with Kv1.1, Kv1.2, and Kv1.4 being the most prominently expressed subunits (reviewed in Vacher et al., 2008). Kv1 channel α subunits contain a large cytoplasmic N-terminus and a relatively smaller C-terminal domain (Figure 4.3). Upon expression in heterologous expression systems, Kv1 α subunits give rise to low-voltage activated K^+ currents that activate with fast activation kinetics (Figure 4.3). The exact activation threshold and activation kinetics differ among homo- and hetero-tetrameric channels made from different Kv1 α subunits, with homo-tetrameric Kv1.1 channels displaying the lowest activation threshold and the fastest activation kinetics (reviewed in Ovsepian et al., 2016). Homo-tetrameric channels formed from most Kv1 subfamily members display little or no inactivation, resulting in sustained delayed rectifier-type K^+ currents, with the clear exception of Kv1.4, which produces transient A-type currents due to a pronounced N-type inactivation (Figure 4.3). Although homo-tetrameric Kv1 channels are produced in heterologous expression systems and are also found to some extent in brain neurons, most neuronal Kv1 channels are hetero-tetramers of two or more distinct α subunit *in vivo* (Coleman, Newcombe, Pryke, & Dolly, 1999). The subfamily-specific assembly of Kv1 channel α subunits into homo- or hetero-tetramers depends on an N-terminal T1 tetramerization domain that allows Kv1 subunits to assemble with other members of the Kv1 subfamily, but not with those of other Kv subfamilies (Figure 4.3; Li, Jan, & Jan, 1992). Hetero-tetrameric Kv1 channels display biophysical and pharmacological properties, as well as trafficking properties (Manganas & Trimmer, 2000) that are shaped by the contributing channel α subunits, thereby creating further diversity in neuronal Kv1 currents (Coleman et al., 1999).

Kv1 Accessory Proteins

The N-terminal T1 domain also serves as a docking site for the regulatory Kvβ auxiliary subunits (Figure 4.3). These are cytoplasmic oxidoreductase proteins that co-assemble with and regulate the biophysical properties and intracellular trafficking of Kv1 channels. The critical role of Kvβ subunits in Kv1 channel function has been revealed by loss-of-function mutations in Kvβ subunits that phenocopy mutants lacking Kv1 expression (Chouinard, Wilson, Schlimgen, & Ganetzky, 1995; Heilstedt et al., 2001). The binding of Kvβ subunits to the T1 domain has been elegantly demonstrated by crystal structures of both the isolated T1 domain as well as full-length Kv1.2 in association with Kvβs. The crystal structures have revealed that the T1 tetramer is separated from the intracellular pore opening as a "hanging gondola," with the Kvβ subunits attached to its lower side (Figure 4.3; Kobertz, Williams, & Miller, 2000; Long et al., 2005). In mammals, three Kvβ subunit genes have been identified—Kvβ1–3—and transcripts from each are subject to alternative splicing. The Kvβ subunits exert distinct effects on the biophysical properties of Kv1 channels (reviewed in Pongs & Schwarz, 2010). The most well-described effect is that of Kvβ1.1, which confers rapid N-type inactivation to Kv1 channels that would otherwise lack it due to the presence of an N-terminal inactivation peptide on Kvβ1.1 that blocks the open channel pore (Figure 4.3; Pongs & Schwarz, 2010). A similar mechanism causes the rapid inactivation in Kv1.4-containing channels, as the extended N-terminus of the Kv1.4 α subunit contains a similar inactivation peptide (Figure 4.3). The physiological role of the

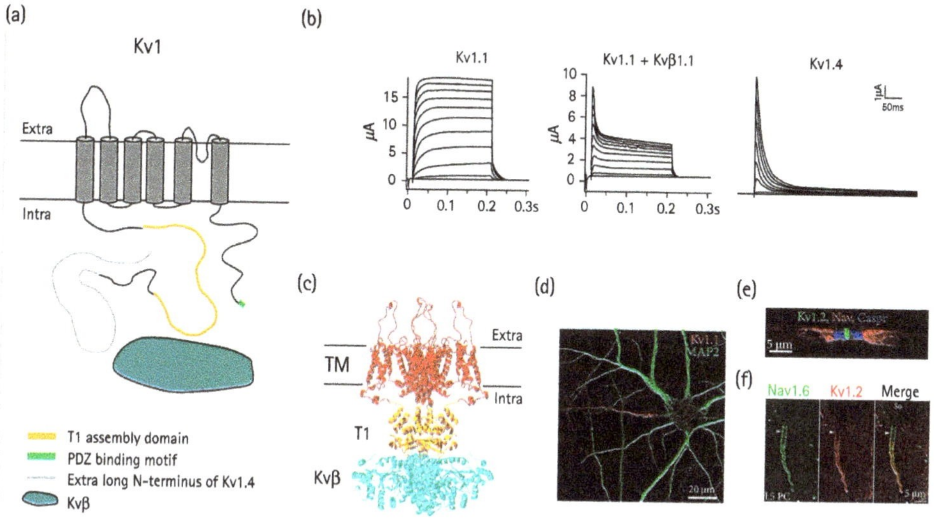

The Kv1 subfamily.

FIGURE 4.3 **A.** Schematic overview of the domain structure of a Kv1 α subunit. The associated cytoplasmic Kvβ auxiliary subunit is included. **B.** *Left side*: Representative current trances obtained in *Xenopus laevis* oocytes expressing Kv1.1, Kv1.1 in combination with Kvβ1.1 or Kv1.4. Current traces were recorded by applying depolarizing potentials from −60 to +50 mV with a 10 mV increment for 200 ms from a holding potential at −80 mV. *Right side*: Representative whole-cell current traces for Kv1.4 transiently expressed in CHO-K1 cells. Currents were evoked by step depolarization to test potentials between −90 and +70 mV for 500 ms in 20 mV increments. (Adapted with permission from Fan, Bi, Jin, & Qi, 2010; Jow, Zhang, Kopsco, Carroll, & Wang, 2004.) **C.** Ribbon representation of the crystal structure of the rat Kv1.2 α subunit in complex with rat Kvβ2 (PDB ID: 3LUT). Kv1.2 is shown in red except for the T1 domain, which is orange. Kvβ2 is represented in cyan. (Image created with Swiss-PDB Viewer.) **D.** Confocal image of a 17 DIV cultured rat hippocampal neuron immunolabeled for the Kv1.1 α subunit and the somatodendritic marker microtubule associated protein 2 (MAP2), illustrating primarily axonal localization of this α subunit. **E.** Kv1.2 localization in paranodes in sciatic nerve axons. Kv1.2 is expressed in the juxtaparanodal regions that surround the node of Ranvier (labeled for voltage-gated sodium channels in green) and paranodes (labeled for Caspr in blue). (Adapted with permission from Horresh et al., 2008; Jensen, Rasmussen, & Misonou, 2011.) **F.** Kv1.2 localization at the axon initial segment (AIS) of rat layer 5 pyramidal cells (L5PC). The localization of the AIS was visualized by immunolabeling for Nav1.6. "So" indicates the location of the soma. (Adapted with permission from Lorincz & Nusser, 2008.)

oxidoreductase activity of the Kvβ subunits remains unclear, though there appears to be a coupling between the oxidoreductase activity and the Kv1 current modulation by Kvβ subunits (Pongs & Schwarz, 2010).

Kv1 channels have also been found associated with the leucine-rich glioma-inactivated 1 (LGI1) protein, a secreted neuronal protein that is linked to autosomal-dominant lateral temporal lobe epilepsy (Schulte et al., 2006). LGI1 regulates both inactivation kinetics and expression levels of Kv1 channels, resulting in increased current levels (Schulte et al., 2006; Seagar et al., 2017). While the mechanism by which LGI1 regulates Kv1 channels remains obscure, LGI1 is a ligand for the catalytically inactive metalloprotease ADAM22, another component of the Kv1 complex (Fukata et al., 2006; Ogawa et al., 2010; Schulte et al., 2006; Yamagata et al., 2018). Intriguingly, individually knocking out *Lgi1, Adam22, Kv1.1, Kv1.2*, and *Kvβ2* leads to an epileptic phenotype in mice, potentially supportive of a functional link between these proteins present in native brain Kv1 channel complexes (Brew et al., 2007; Connor et al., 2005; Fukata et al., 2006; Sagane et al., 2005; Smart et al., 1998).

Kv1 Subcellular Localization

Kv1 channels are predominantly axonal channels, with the exception of those containing Kv1.6 that are somatodendritic (reviewed in Vacher et al., 2008). The axonal Kv1 channels localize to the AIS, the juxtaparanode, and the preterminal segment of the axon (Figure 4.3). The localization appears to involve Kvβ subunits that promote axonal localization of Kv1 channels upon heterologous expression in cultured hippocampal neurons (Campomanes et al., 2002; Gu, Jan, & Jan, 2003). However, the localization of the Kv1.1 and Kv1.2 α subunits is unaffected in *Kvβ2* and *Kvβ1.1/Kvβ2* double knockout mice, suggesting that the Kvβ subunits that are abundantly associated with these Kv1 α subunits in brain (Rhodes et al., 1997) are not absolutely required for their axonal localization, a conundrum that remains to be solved (Connor et al., 2005; McCormack et al., 2002). In their extreme C-terminus, Kv1 α subunits contain a post synaptic density protein (PSD95), Drosophila disc large tumor suppressor (Dlg1), and zonula occludens-1 protein (zo-1) (PDZ)-binding motif that interacts with members of the membrane-associated guanylate kinase (MAGUK) family (Figure 4.3; Kim, Niethammer, Rothschild, Jan, & Sheng, 1995), and MAGUKs co-purify with Kv1.1 from rat brain (Schulte et al., 2006). However, the role of MAGUKs in the localization of Kv1 channels remains obscure as axonal Kv1 channels are normally clustered in both PSD-93 and PSD-95 knockout mice (Ogawa et al., 2010; Rasband et al., 2002).

Kv1 Function in Neurons

Due to a combination of their fast, low-threshold activation and their subcellular localization, Kv1 channels significantly impact the axonal action potential by regulating its threshold, waveform, and frequency (see, e.g., Johnston, Forsythe, & Kopp-Scheinpflug, 2010; Kole, Letzkus, & Stuart, 2007). They further regulate the release of neurotransmitter from presynaptic terminals (see Yang et al., 2014). Here, inactivating Kv1 currents display cumulative inactivation during trains of action potentials, which leads to a broadening of the presynaptic action potential, increased calcium influx, and potentiation of neurotransmitter release (reviewed in Dodson & Forsythe, 2004). Mutations in the Kv1.1 gene *KCNA1* are associated with inherited episodic ataxia type 1 (EA1; Rajakulendran, Schorge, Kullmann, & Hanna, 2007), and de novo mutations in the Kv1.2 gene *KCNA2* with early infantile epileptic encephalopathy-32 (EIEE32; Pena & Coimbra, 2015).

The Kv2 Subfamily

The Kv2 subfamily consists of two members named Kv2.1 and Kv2.2, which are orthologues of the *Drosophila* Shab channel (Table 4.1). The Kv2 α subunits are broadly and robustly expressed in the nervous system, with Kv2.1 having the broadest expression (Trimmer, 2015). They are equipped with a long N-terminal domain and an extremely long C-terminus (Figure 4.4). When expressed in heterologous expression systems, the channels form sustained delayed rectifier-type currents that activate with slowactivation kinetics (Figure 4.4). Though the formation of Kv2.1/Kv2.2 heteromers is possible (Kihira, Hermanstyne, & Misonou, 2010), the overall extent to which this happens in vivo is not clear (Bishop et al., 2015). Kv2 α subunits can co-assemble with members of the electrically silent Kv5, Kv6, Kv8, and Kv9 subfamilies, which leads to altered biophysical properties of Kv2-mediated K^+ currents (see section on KvS subunits, this chapter). Again, the extent to which this heteromerization takes place in vivo is unclear, as most studies are based on heterologous expression (Bocksteins, 2016). As in Kv1 channels, the specific assembly of α subunits within the Kv2 subfamily, and with Kv5, Kv6, Kv8, and Kv9 α subunits, is regulated by an N-terminal T1 tetramerization domain (Figure 4.4). Electron microscopy and single-particle image analysis have revealed that the Kv2.1 T1 domain is almost completely surrounded by the long Kv2.1 C-terminus that wraps around it and occupies the space below it (Figure 4.4; Adair et al., 2008; Grizel et al., 2014).

Kv2 Accessory Proteins

Kv2 channels interact and co-localize extensively with amphoterin-induced gene and ORF 1 (AMIGO-1), a single-pass transmembrane cell adhesion molecule that can also act to promote neurite outgrowth (Figure 4.4; Bishop et al., 2018; Mandikian et al., 2014; Peltola, Kuja-Panula, Lauri, Taira, & Rauvala, 2011). AMIGO-1 appears to be an integral component of both Kv2.1- and Kv2.2-containing channel complexes in adult brain neurons, as it exhibits an almost complete overlap in cellular and subcellular localization with Kv2 channels, and it depends on Kv2 α subunits for its proper expression and subcellular localization (Bishop et al., 2018; Mandikian et al.,

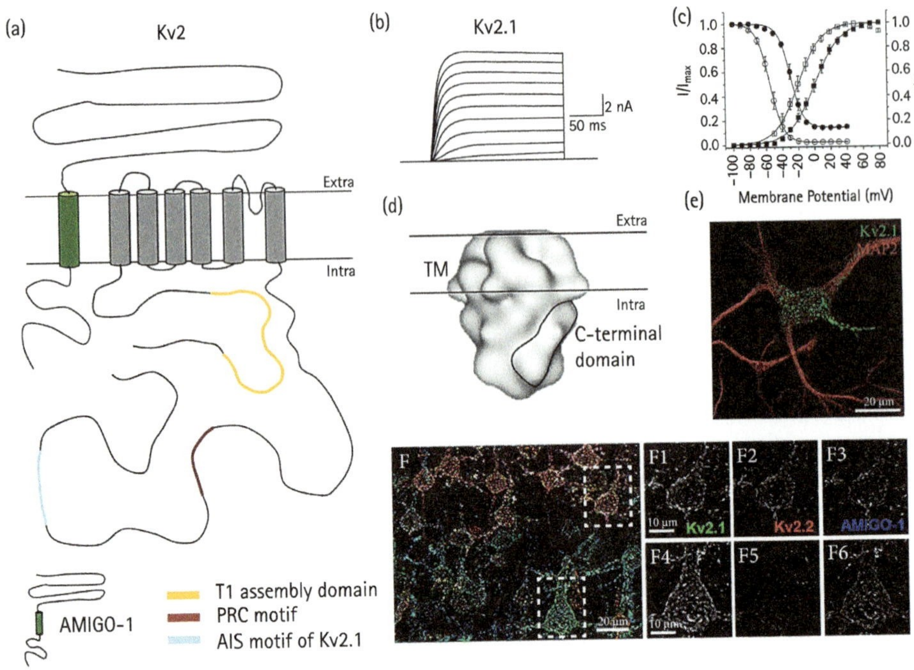

The Kv2 subfamily.

FIGURE 4.4 **A.** Schematic overview of the domain structure of a Kv2 α subunit. The associated single-pass transmembrane cell adhesion molecule AMIGO-1 is included. AIS: axon initial segment, PRC: proximal restriction and clustering. **B.** Representative whole-cell current traces from HEK293 expressing the Kv2.1 α subunit. The cells were held at –100 mV and step depolarized to + 80 mV for 200 ms in 10 mV increments. **C.** Voltage-dependent activation (*squares*) and steady-state inactivation (*circles*) relationships of Kv2.1 currents in HEK293 cells before (*filled symbols*) and after (*open symbols*) dephosphorylation induced by intracellular dialysis of alkaline phosphatase for 30 minutes, illustrating the hyperpolarized shift in the dephosphorylated channels. Error bars represent mean ± standard error of the mean (SEM). (**B** and **C** were adapted from Mohapatra & Trimmer, 2006.) **D.** Surface-shaded 3D density map of human Kv2.1 at ≈25 Å resolution. The location of the C-terminus is indicated by the black line. "TM" refers to the transmembrane portion of the channel. (Adapted with permission from Adair et al., 2008.) **E.** Confocal image of a 17 DIV cultured rat hippocampal neuron immunolabeled for the Kv2.1 α subunit and MAP2, illustrating the clustered localization of Kv2.1 on the soma and proximal dendrites and AIS. **F.** Localization of Kv2.1 (green), Kv2.2 (red), and the accessory protein AMIGO-1 (blue) in mouse somatosensory cortical neurons, illustrating the highly clustered Kv2 localization on soma and proximal neurites. Different pyramidal cells of the cortex express different levels of Kv2.1 and Kv2.2, as demonstrated in the inserts to the right (F1–F6). (Adapted from Bishop et al., 2018.)

2014; Peltola et al., 2011). It is tempting to speculate that, given its established role as a cell adhesion molecule, AMIGO-1 could be involved in the localization of neuronal Kv2 clusters at contact sites for astrocytic processes (see section on Kv2 subcellular localization, next).

Kv2 Subcellular Localization

The subcellular localization of Kv2 channels beautifully illustrates that, while ion channels are often localized to pre-established membrane domains, they can also impact and even generate the molecular and structural organization of the domains at which they are found. Kv2 channels exhibit a unique, highly restricted subcellular localization in neurons, where they are localized at high density in large clusters restricted to the soma, the proximal segment of the dendrites, and the AIS (Figure 4.4). Kv2 clusters are located at endoplasmic reticulum (ER)–plasma membrane (PM) or ER-PM junctions, termed "subsurface cisternae" in neurons, and are juxtaposed to astrocytic processes (Bishop et al., 2015; Du, Tao-Cheng, Zerfas, & McBain, 1998; Fox et al., 2015; Misonou, Thompson, & Cai, 2008). This specific localization is due to a Kv2-specific motif localized in the extended C-terminal domain, the proximal restriction and clustering (PRC) motif that is present in both the Kv2.1 and Kv2.2 α subunits (Figure 4.4; Bishop et al., 2015; Lim, Antonucci, Scannevin, & Trimmer, 2000). The motif interacts with ER-resident vesicle-associated membrane-associated proteins isoform A and B (VAPA and VAPB) and this interaction probably drives the clustering of Kv2.1 and Kv2.2 as well as organizes the ER-PM junctions themselves (Johnson et al., 2018; Kirmiz, Palacio, et al., 2018; Kirmiz, Vierra, Palacio, & Trimmer, 2018). The remodeling of ER-PM junctions *via* interaction with VAP proteins is retained in Kv2.1 and Kv2.2 mutants lacking ionic conductance, showing that it is a bona fide non-conducting function of Kv2 channels (Kirmiz, Palacio, et al., 2018). This is especially intriguing as the bulk of exogenously expressed Kv2.1 channels are in a non-conducting state (Benndorf, Koopmann, Lorra, & Pongs, 1994; Fox, Loftus, & Tamkun, 2013; K. M. S. O'Connell, Loftus, & Tamkun, 2010). While the Kv2-specific PRC motif supports localization to the AIS and also clusters VAPA/VAPB at this location, Kv2.1 has an additional motif in the C-terminus, the AIS motif, which also contributes to its clustered localization in this subcellular compartment by acting as a motif for directing its non-canonical trafficking to the AIS (Figure 4.4, Jensen et al., 2017). How this motif mediates Kv2 channel clustering remains unknown, but it could be though binding to another ER resident protein specific to the AIS. Intriguingly, the AIS motif is not well conserved in Kv2.2, which suggests that its localization on the AIS could have a distinct underlying mechanism. The clustered localization of Kv2 channels is dynamically regulated such that increasing neuronal activity causes dispersion of Kv2 clusters and retraction of the subsurface cisternae from the PM, most likely caused by a disruption of the interaction between Kv2s and VAPs (Fox et al., 2015; Kirmiz, Vierra, et al., 2018; Misonou et al., 2004). The dispersion is associated with and probably driven by dephosphorylation of Kv2.1, and is associated with a hyperpolarizing shift in the voltage-dependence of activation (Figure 4.4; Misonou et al., 2004; Murakoshi, Shi, Scannevin, & Trimmer, 1997). Kv2.1 is more responsive to these activity-dependent changes in subcellular localization and voltage activation than is Kv2.2 (Bishop et al., 2015).

Kv2 Function in Neurons

In accordance with their high expression levels, Kv2 channels constitute the major part of somatic delayed rectifier-type K^+ current in many neuronal cell types (Du, Haak, Phillips-Tansey, Russell, & McBain, 2000; Guan, Tkatch, Surmeier, Armstrong, & Foehring, 2007; Liu & Bean, 2014; Malin & Nerbonne, 2002; Murakoshi & Trimmer, 1999; Palacio et al., 2017). As Kv2 channels activate with slow kinetics, they have their most significant impact during periods of repetitive firing, where they can facilitate high frequency-firing, but also limit dendritic calcium transients (Du et al., 2000; Guan, Armstrong, & Foehring, 2013; J. Johnston et al., 2008; Liu & Bean, 2014). Further, when neuronal activity levels are high, the resulting hyperpolarized activation of Kv2.1 results in a suppression of excitability (Mohapatra et al., 2009). Due to their subcellular localization, high expression level, and activity-dependent regulation, Kv2 channels are thus key regulators of intrinsic excitability. In support of this, Kv2.1 knockout mice display neuronal and behavioral hyperexcitability (Speca et al., 2014). A number of recent studies have found de novo mutations in the *KCNB1* gene encoding Kv2.1 associated with pediatric encephalopathic epilepsy patients (Saitsu et al., 2015; Thiffault et al., 2015; Torkamani et al., 2014).

The Kv3 Subfamily

Kv3 α subunits are mammalian orthologues of the *Drosophila Shaw* gene products. The Kv3 α subunit subfamily includes four members, Kv3.1–Kv3.4, all of which are expressed in the nervous system (Table 4.1). These α subunits contain a large N-terminus and a C-terminal domain of variable length, with Kv3.3 having the longest C-terminus (Figure 4.5). Due to extensive alternative mRNA splicing, Kv3 subfamily members are present in multiple isoforms that differ in the length and sequence of the C-terminal domain (for review, see Kaczmarek & Zhang, 2017). When expressed in heterologous cells, Kv3 α subunits can generate non-inactivating delayed rectifier-type currents (Kv3.1 and Kv3.2) or A-type currents (Kv3.3 and Kv3.4) (Figure 4.5). Kv3-mediated currents are characterized by a high activation threshold and very fast activation and deactivation kinetics. The inactivation observed in Kv3.3 and Kv3.4 α subunits is due to the presence of an N-terminal inactivation peptide in these α subunits (Figure 4.5) similar to the mechanism observed in Kv1.4. Kv3 α subunits can form hetero-tetrameric channels, a process regulated by the N-terminal T1 tetramerization domain (Figure 4.5).

Kv3 Accessory Proteins

The C-terminus of Kv3.3 binds the cell survival protein Hax-1, which activates Arp2/3-dependent formation of a cortical actin filament network just below the PM (Zhang et al., 2016). The presence of the cortical actin cytoskeleton slows channel inactivation

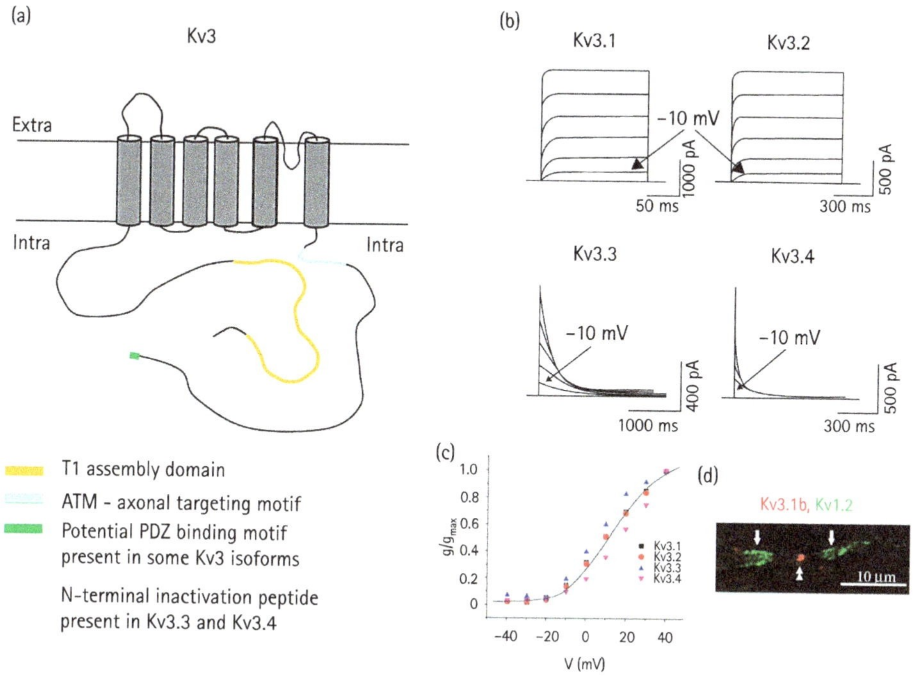

The Kv3 subfamily.

FIGURE 4.5 **A.** Schematic overview of the domain structure of a Kv3 α subunit. **B.** Representative Kv3 currents in transfected Chinese hamster ovary (CHO) cells, illustrating the different kinetics of inactivation for the different isoforms. The cells were subjected to depolarizing pulses from −40 mV to +40 mV (in 10 mV increments, from a holding potential of −80 mV). **C.** Normalized conductance–voltage relationship of Kv3 currents expressed in CHO cells. The conductance at the indicated voltage [g= IV/(V-VK)], divided by the maximum conductance (gmax) for the currents shown in **A**, is plotted as a function of membrane potential. (**A** and **B** were adapted with permission from Rudy & McBain, 2001.) **D.** Immunolocalization of the Kv3.1b α subunit in nodes of Ranvier of the spinal cord. The juxtaparanodes were immunolabeled with a Kv1.2 antibody, demonstrating the compartmentalized localization of the two Kv channels in this axonal subdomain. (Adapted with permission from J. Devaux et al., 2003.)

and is most likely the explanation for the different rates of inactivation observed between Kv3.3 and Kv3.4. The modulation of the submembranous actin organization exerted by Kv3.3 through its interaction partner Hax-1 is another beautiful demonstration of the impact ion channels can have on the structure and molecular organization of the domains in which they are embedded. We will most likely encounter many more of these alternative actions of ion channels in the future.

Kv3 Subcellular Localization

Kv3 channels can be localized to both the axonal and somatodendritic domains. Most often, they are localized to the soma, proximal dendrites, nodes of Ranvier, and presynapses (Figure 4.5). The localization varies with the neuronal cell type, even for the same Kv3 isoform (reviewed in Vacher et al., 2008). Axonal targeting appears to involve conditional binding to the scaffold protein ankyrin-G (ankG) through an axonal targeting motif present in the proximal C-terminal tail of all Kv3 subfamily members (Figure 4.5; Xu, Cao, Xiao, Zhu, & Gu, 2007). Kv3 targeting to distal dendrites appears to involve a potential PDZ-binding motif present at the extreme C-terminus of some Kv3 isoforms (Figure 4.5; Deng et al., 2005).

Kv3 Function in Neurons

Since Kv3 channels are characterized by a high activation threshold, they are only activated during action potentials, and due to their fast activation kinetics, they contribute substantially to membrane repolarization. In Kv3-expressing neurons, loss of Kv3 current broadens the action potential and eliminates the afterhyperpolarization, which impairs the ability of neurons to fire at high frequencies due to accumulated inactivation of Nav channels (reviewed in Kaczmarek & Zhang, 2017). Accordingly, Kv3 α subunits are robustly expressed in fast-spiking neurons such as auditory brain stem neurons, parvalbumin-positive GABAergic interneurons, and Purkinje cells of the cerebellum (Trimmer, 2015). In addition to their critical role in permitting high-frequency firing, Kv3 channels present at presynaptic terminals are important regulators of synaptic efficacy, with inhibition of Kv3 current resulting in a broadening of the action potential, increased presynaptic calcium influx, and enhanced neurotransmitter release (reviewed in Kaczmarek & Zhang, 2017). Certain inherited mutations in the *KCNC3* gene that encodes Kv3.3 cause neurodegenerative spinocerebellar ataxia type 13 (SCA13; Waters et al., 2006).

The Kv4 Subfamily

The Kv4 subfamily consists of α subunits that are orthologues of the *Drosophila Shal* gene product. The Kv4 subfamily contains three members, Kv4.1–4.3, all of which are expressed in the nervous system with Kv4.2 and Kv4.3 being the most predominantly expressed α subunits (Table 4.1). Kv4 α subunits contain large N- and C-terminal domains (Figure 4.6). Kv4 channels generate low threshold, fast activating and inactivating A-type currents upon heterologous expression (Figure 4.6). The fast inactivation is of the N-type and can be mediated by the N-terminus of the α subunit itself, or by the N-terminus of the accessory dipeptidyl peptidase-like (DPPL) α subunits (Figure 4.6; Jerng & Pfaffinger, 2014). Kv4 currents are furthermore unique in that they are typically inactivated at resting membrane potentials and require a prior hyperpolarization to activate. This is due to a closed state inactivation that requires hyperpolarized membrane potentials to recover (Jerng & Pfaffinger, 2014). Heteromeric assembly between

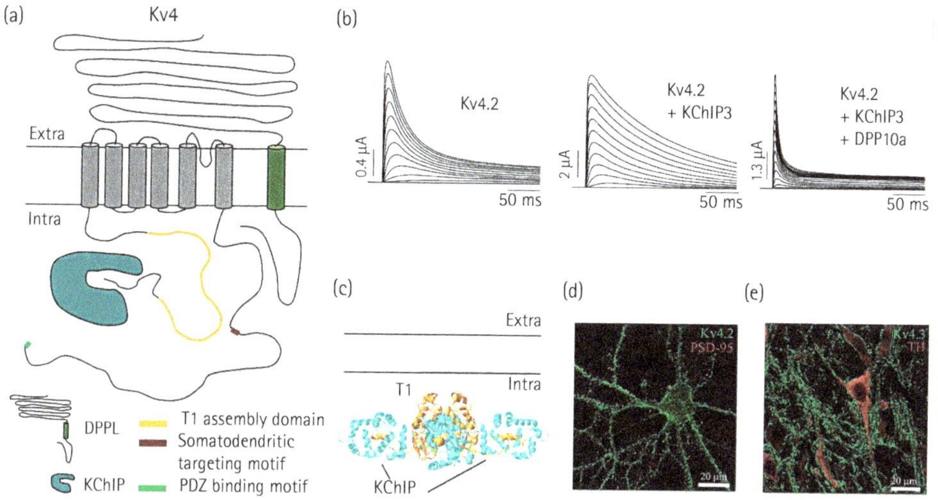

The Kv4 subfamily.

FIGURE 4.6 A. Schematic overview of the domain structure of a Kv4 α subunit. The cytoplasmic Kv channel-interacting protein (KChIP) as well as the single-pass transmembrane dipeptidyl peptidase-like (DPPL) auxiliary subunits are included. **B.** Representative current trances obtained in *Xenopus laevis* oocytes expressing Kv4.2 alone or in combination with KChIP3 and DPP10a as indicated. Currents were elicited by 1-second-long depolarizing pulses from −100 to +60 mV in 10 mV increments from a holding potential of −100 mV. Only the first 250 ms are shown. (Adapted with permission from Jerng, Lauver, & Pfaffinger, 2007.) **C.** Ribbon representation of the crystal structure of the isolated N-terminus of the human Kv4.3 α subunit in complex with KChIP1. The Kv4.3 N-terminus is represented in yellow and KChIP in cyan. The image was created with Swiss-PDB Viewer and PDB ID: 2NZ0. **D.** Cultured rat hippocampal neuron immunolabeled for the Kv4.2 α subunit, illustrating its expression on the soma and dendrites. The localization of dendritic spines that can also contain Kv4.2 was visualized by immunolabeling for PSD-95. (Figure adapted with permission from C. S. Jensen et al., 2011.) **E.** Confocal image from rat substantia nigra illustrating the somatodendritic localization of Kv4.3 α subunits in dopaminergic neurons. "TH" is tyrosine hydroxylase, a marker of dopaminergic neurons.

different Kv4 channel α subunits is possible and is dictated by an N-terminal T1 tetramerization domain (Figure 4.6).

Kv4 Accessory Proteins

Kv4 members interact with and are modulated by the cytoplasmic, calcium-binding proteins known as Kv channel-interacting proteins (Figure 4.6, KChIPs; An et al., 2000). Due to a number of start sites and alternative splicing, the four members of this subfamily, KChIP1–4, exist in an impressive 17 isoforms with distinct expression patterns in brain (reviewed in Jerng & Pfaffinger, 2014). KChIPs interact with Kv4 channels through a hydrophobic sequence located in the extreme N-terminus of one channel α subunit and the T1 domain of the adjacent α subunit, thereby clamping them (Pioletti, Findeisen, Hura, & Minor, 2006; Scannevin et al., 2004; H. Wang et al., 2007). Crystal structures of the isolated Kv4.3 N-terminus in association with KChIP1 reveal that the KChIP subunits are located lateral to the T1 domain, in contrast to the Kvβ subunits that are placed below the T1 domain of Kv1 α subunits (Figure 4.6). In general, KChIPs promote surface-expression of Kv4 channels (Shibata et al., 2003), slow N-type inactivation, and accelerate recovery from inactivation (Figure 4.6; reviewed in Jerng & Pfaffinger, 2014). That one of the primary functions of KChIPs is to regulate Kv4 function is supported by the observation that KChIP protein levels are drastically reduced in Kv4.2 knockout mice (Menegola & Trimmer, 2006), probably due to destabilization of the KChIPs in the absence of their Kv4 binding partners (Foeger, Marionneau, & Nerbonne, 2010).

Two DPPLs named DPP6 and DPP10 also interact with and critically regulate Kv4 channels (Figure 4.6; Nadal et al., 2003; Zagha et al., 2005). Like KChIPs, they exist in various isoforms. DPPLs are single-pass transmembrane proteins that profoundly impact the biophysical properties of Kv4 channels by shifting both activation and inactivation in the hyperpolarized direction, accelerating inactivation and altering channel conductance (Figure 4.6; reviewed in Jerng & Pfaffinger, 2014). DPPLs interact with the Kv4 VSD, which possibly explains the strong impact of these subunits on channel activation (Dougherty, Tu, Deutsch, & Covarrubias, 2009). Like KChIPs, DPPLs promote surface expression of Kv4 channels (Foeger, Norris, Wren, & Nerbonne, 2012; Seikel & Trimmer, 2009). DDP6 knockout mice show significant reductions in both Kv4 and KChIP expression, suggesting that DPPLs are critical regulators of the overall stability of Kv4 complexes (Sun et al., 2011). While DPP6 is generally expressed in neurons that also express Kv4.2, DPP10 shows significant co-expression with Kv4.3 (Clark et al., 2008; Nadal, Amarillo, Vega-Saenz de Miera, & Rudy, 2006; Zagha et al., 2005). That selective co-expression patterns also hold for KChIPs (Rhodes et al., 2004) that can form a ternary complex with Kv4 α subunits and DPPLs (Jerng, Kunjilwar, & Pfaffinger, 2005) suggests a preferential association between certain Kv4 α and auxiliary subunits to impact the characteristics of Kv4 channels.

Kv4 Subcellular Localization

Kv4 channels are exclusively expressed in the somatodendritic compartment (Figure 4.6; Kerti, Lorincz, & Nusser, 2012; Rhodes et al., 2004), and functional Kv4 channels are found with an increasing gradient of expression towards the distal dendrites (Hoffman, Magee, Colbert, & Johnston, 1997). The somatodendritic gradient is lost in DPP6 knockout mice, demonstrating a critical role of this accessory subunit in establishing the gradient (Sun et al., 2011). The somatodendritic localization furthermore requires a dileucine motif present in the cytoplasmic C-terminal tail (Figure 4.6; Rivera, Ahmad, Quick, Liman, & Arnold, 2003). How this motif mechanistically confers somatodendritic localization remains unclear, but its mutation in Kv4.2 impacts the post-Golgi vesicular trafficking of the channel, suggesting a possible role in vesicular sorting (Jensen et al., 2014). Kv4 channels also contain a C-terminal PDZ-binding motif, and while the motif can mediate interaction with PSD-95, its exact function is currently not clear (Wong, Newell, Jugloff, Jones, & Schlichter, 2002).

Kv4 function in Neurons

Kv4 channels in complex with KChIP and DPPL subunits underlie the subthreshold A-type transient current (I_{SA}) observed in the somatodendritic compartment of most neurons (Jerng & Pfaffinger, 2014). The most striking feature of this current is its impact on dendritic signal processing. Due to its localization and low threshold, rapid activation, the I_{SA} current plays an important role in dampening the back-propagation of action potentials into dendrites (Hoffman et al., 1997). However, since Kv4 channels require hyperpolarization to recover from inactivation, Kv4 currents are selectively suppressed in dendrites that have experienced recent depolarizing activity. These dendrites experience larger back-propagating action potentials and increased calcium influx, which contribute to the compartmentalized regulation of long-term potentiation of excitatory synaptic transmission (LTP) (Johnston et al., 2003). In accordance, hippocampal CA1 neurons of Kv4.2 knockout mice display an increase in the amplitude of back-propagating action potentials, associated elevations in dendritic calcium influx, and altered LTP induction (Chen et al., 2006).

The Kv7 Subfamily

The Kv7 α subunits were initially referred to by their gene name KCNQ, and even today some publications utilize this name. The subfamily contains five members named Kv7.1–7.5, and all except Kv7.1 display neuronal expression (Table 4.1). The most prominent neuronal α subunits are Kv7.2, Kv7.3, and Kv7.5, as Kv7.4 displays a very restricted expression: only being present in certain auditory nuclei (Kharkovets et al., 2000). While the N-terminus in Kv7 channels is shorter than in most other Kv subfamilies, they contain extremely long C-terminal domains that make up around half the protein (Figure 4.7). In heterologous expression systems, channels formed upon expression of Kv7 α subunits generate low-threshold, slowly activating delayed rectifier-type currents that display little or no inactivation (Figure 4.7). The channels require binding of the phospholipid phosphatidylinositol-4,5-bisphosphate (PIP_2) for channel opening, and its depletion results in the inhibition of the current, most likely due to an uncoupling between the VSD and the pore domain (Suh & Hille, 2002; Sun & MacKinnon, 2017; Zaydman & Cui, 2014). Hetero-tetramerization within this subfamily is extensive, with the most commonly observed neuronal Kv7 channels most likely being Kv7.2/Kv7.3 heteromers (Hadley et al., 2003). The Kv7 α subunits do not contain the conserved N-terminal T1 assembly domain

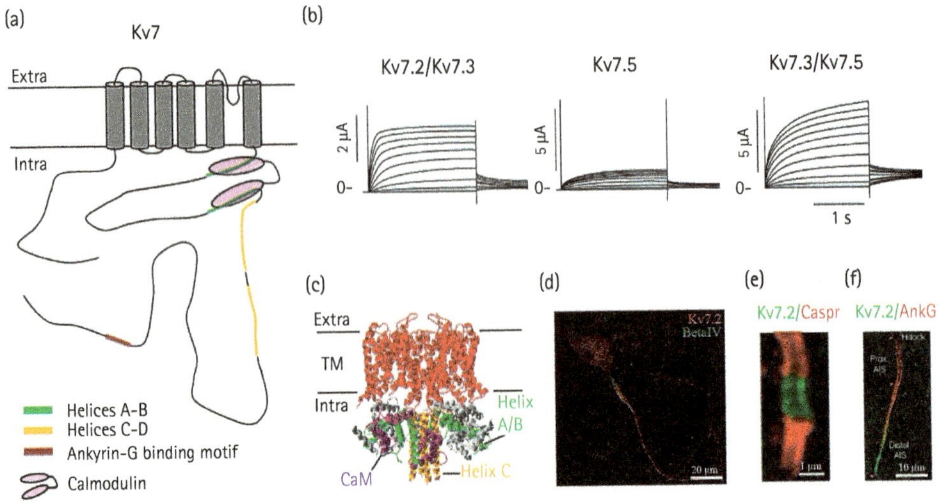

The Kv7 subfamily.

FIGURE 4.7 **A.** Schematic overview of the domain structure of a Kv7 α subunit. The associated regulatory calmodulin (CaM) subunit is included. **B.** Representative current trances obtained in *Xenopus laevis* oocytes expressing Kv7.5 alone, Kv7.2 and Kv7.3, or Kv7.3 and Kv7.5, as indicated. Currents were activated by voltage-steps from −80 mV to +40 mV in 10 mV increments. (Adapted from Gilling et al., 2013.) **C.** Ribbon representation of the cryoEM structure of *Xenopus* Kv7.1 in complex with CaM. Kv7.1 is shown in red except for helices A and B, which are shown in green, and helix C, which is represented in orange. One CaM molecule is shown in purple. For better visualization, the other three CaM molecules are colored gray. The image was created with Swiss-PDB Viewer and PDB ID: 5VMS. **D.** Confocal image of a 17 DIV cultured rat hippocampal neuron immunolabeled for the Kv7.2 α subunit and betaIV-spectrin (a marker of the AIS), illustrating the primarily axonal localization of Kv7.2. **E.** High-magnification image of a larger node of Ranvier within the L5 somatosensory cortex. $K_v7.2$ antibodies immunolabel the nodal membrane, which is flanked by Caspr at the paranodes. **F.** Maximal projection image of a large L5 somatosensory neuron co-labeled for the $K_v7.2$ α subunit and the AIS marker ankG, illustrating the localization of this α subunit in the distal AIS. (E and F were adapted with permission from Battefeld, Tran, Gavrilis, Cooper, & Kole, 2014.)

observed in most other Kv channel α subunits. Instead, tetramerization is regulated by two C-terminal coiled-coil regions, helices C and D (Figure 4.7; Schwake et al., 2006). While helix C is required for the tetramerization, helix D appears to regulate the subtype-specific heteromerization.

Kv7 Accessory Proteins

The proximal Kv7 C-terminal tail contains two other helical structures, helices A and B, which bind the calcium-binding protein calmodulin (CaM; Figure 4.7; Wen & Levitan, 2002; Yus-Najera, Santana-Castro, & Villarroel, 2002). CaM appears to be an integral part of the Kv7 channel complex, as Kv7 channels interact with both the calcium-free Apo/CaM as well as the calcium-loaded CaM form. Crystal structures of helices A and B in complex with CaM reveal that CaM clamps the proximal C-terminal tail in the absence of calcium, a clamp that is released upon calcium binding, leading to inhibition of voltage-dependent activation of the neuronal Kv7 channels (Figure 4.7; Chang et al., 2018; Gamper & Shapiro, 2003). Interestingly, Apo/CaM binding also appears to be essential for efficient cell surface expression of Kv7 channels, most likely by promoting exit from the ER (Cavaretta et al., 2014; Chang et al., 2018; Etxeberria et al., 2008). This suggests that only the Kv7 channels that are equipped with the machinery for the proper calcium-dependent regulation of their activity are allowed to traffic to the cell surface. The presence of CaM on ER-localized Kv7 channels raises the possibility that their exit from the ER can be modulated by intracellular calcium levels.

Kv7 Subcellular Localization

The Kv7 channels are predominantly axonal channels with localizations to the AIS, nodes of Ranvier, and presynapses (Figure 4.7; Devaux, Kleopa, Cooper, & Scherer, 2004; Vacher et al., 2008). Localization to the AIS requires a sequence in the distal C-terminus that mediates interaction with the scaffolding protein ankG (Figure 4.7; Chung, Jan, & Jan, 2006; Pan et al., 2006; Rasmussen et al., 2007). This interaction stabilizes Kv7 channels at the AIS (Benned-Jensen et al., 2016). While the ankG binding sequence is present in several of the neuronal Kv7 subfamily members, the binding motif of Kv7.3 appears to be the primary anchor at the AIS (Rasmussen et al., 2007). The reason for the differential importance of the ankG binding motifs remains elusive, but it could be explained by differential phosphorylation of the ankG binding motifs present in diverse Kv7 α subunits, as the binding to ankG can be regulated by phosphorylation (Xu & Cooper, 2015).

Kv7 function in Neurons

Kv7 channels underlie the neuronal "M current" observed in most brain neurons (Brown & Adams, 1980; Wang et al., 1998). The name stems from the fact that the M current is inhibited upon activation of muscarinic acetylcholine receptors, resulting in increased neuronal excitability. Due to their low-threshold activation, Kv7 channels can regulate the resting membrane potential and the action potential threshold at the AIS (reviewed in Brown & Passmore, 2009). However, as Kv7 channels display very slow activation kinetics, they do not contribute significantly to the action potential repolarization. Instead, they activate upon repetitive firing, on which they can exert a strong dampening effect. In agreement with their critical role in inhibiting repetitive firing, certain mutations in the genes encoding Kv7.2 and

Kv7.3 cause benign familiar neonatal epilepsy (BFNE, formerly known as BFNC), a specific form of epilepsy that affects newborns (Biervert et al., 1998; Charlier et al., 1998; Singh et al., 1998). Kv7.4 mutations can cause DFNA2, a form of non-syndromic hearing loss, consistent with expression of this α subunit in the inner ear (Kubisch et al., 1999). Further, mutations in the non-neuronal member of this subfamily, Kv7.1, are among the primary causes of the inherited cardiac arrhythmia long QT syndrome, and the recessive Jervell and Lange-Nielsen syndrome, which in addition to cardiac arrhythmias features progressive hearing loss (Neyroud et al., 1997; Q. Wang et al., 1996).

The KvS (Kv5, Kv6, Kv8, and Kv9) Subfamily

Members of the Kv5, Kv6, Kv8, and Kv9 subfamilies of Kv channel α subunits are atypical in that they fail to produce currents in heterologous expression systems and are therefore referred to as the electrically silent Kv channels (KvS). Together these subfamilies comprise ten members, and all except Kv6.2 have a reported expression in brain (Table 4.1; reviewed in Bocksteins, 2016). The KvS α subunits have the same overall structure as the functional Kv α subunits and are equipped with a long N-terminus and a relatively short C-terminus (Figure 4.8). The lack of functional expression in these subfamilies appears, at least partly, to stem from the inability of KvS α subunits to assemble into homotetramers, with the unassembled channel α subunits being retained in the ER (Ottschytsch, Raes, Van Hoorick, & Snyders, 2002). While the KvS α subunits contain an N-terminal T1 tetramerization domain similar to Kv1–4 (Figure 4.8), these N-termini fail to interact with one another, which suggests that their homo-tetrameric assembly is not supported by their T1 domains (Post, Kirsch, & Brown, 1996). The KvS α subunits can, however, assemble with members of the Kv2 subfamily, an assembly that promotes KvS surface expression and results in functional Kv2/KvS heteromers in co-expressing heterologous cells (Ottschytsch et al., 2002). The biophysical profiles of Kv2/KvS heteromers differ from homo-tetrameric Kv2 channels, and KvS are therefore considered modulatory subunits of the Kv2 channels (Bocksteins & Snyders, 2012). In general, KvS co-expression reduces Kv2 current densities, except for Kv9.3, which enhances it. KvS α subunits can furthermore shift the voltage dependence of both activation and inactivation, as well as modify the kinetics of activation, inactivation, and/or deactivation of co-expressed Kv2 channels (Figure 4.8; reviewed in Bocksteins & Snyders, 2012). The degree to which Kv2/KvS complexes exist in vivo remains unclear, as most studies on these channels have been performed in heterologous expression systems. However, knockdown of certain KvS α subunits impacts neuronal excitability (e.g., Tsantoulas et al., 2012), supporting an in vivo role in ion channel function.

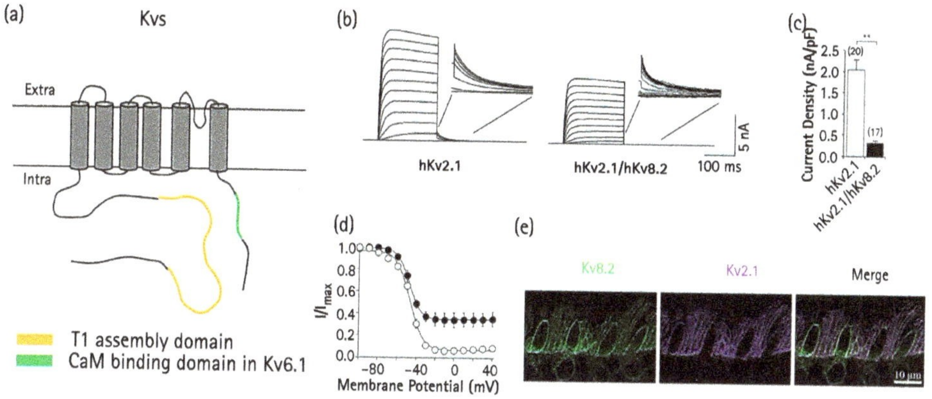

The KvS subfamily.

FIGURE 4.8 **A.** Schematic overview of the domain structure of a KvS α subunit. **B.** Example of Kv2 modulation by a KvS subunit. Shown are Kv2.1 and hKv2.1/hKv8.2 currents recorded from HEK293 cells expressing Kv2.1 alone or in combination with Kv8.2. Currents were elicited by 200 ms voltage steps from −80 to +80 mV (10 mV increments) followed by a step to −40 mV. **C.** Bar diagram illustrating the reduction in current density observed upon co-expression of Kv2.1 and Kv8.2 compared to homomeric Kv2.1. Currents were recorded at +30 mV in HEK293 cells. **D.** Graph illustrating the prominent effect of Kv8.2 on the Kv2.1 voltage dependence of steady-state inactivation. Channels were inactivated for 20 s (Kv2.1, open circles) or 30 s (Kv2.1/Kv8.2, closed circles) at prepulse potentials ranging from −100 to +40 mV, followed by a test pulse to +60 mV to activate residual non-inactivated channels. The normalized amplitudes were plotted against the prepulse potentials and fitted with a Boltzmann function ($n = 7$–9). (**B–D** were adapted with permission from Smith et al., 2012.) **E.** Co-immunolabeling for the Kv8.2 and Kv2.1 α subunits in human retinal inner segments. Kv8.2 is present in both rod and cone inner segment membranes, where it is co-localized with Kv2.1. (Adapted from Gayet-Primo et al., 2018.)

KvS Accessory Proteins

Kv6.1 interacts with CaM in a calcium-dependent manner (O'Connell et al., 2010). The interaction is mediated through a C-terminal binding motif that is exclusively found in Kv6.1 and not in other KvS members (Figure 4.8). The possible binding of CaM to Kv2/Kv6.1 heteromers, and its potential impact, has not been determined.

KvS Subcellular Localization

The specific subcellular localization of KvS α subunits in neurons remains largely unknown. Kv8.2 α subunits have been localized in the inner segment of retinal photoreceptors, where they co-localize with Kv2 channels (Figure 4.8; Gayet-Primo, Yaeger, Khanjian, & Puthussery, 2018). However, in general, studies of the subcellular localization of the different KvS α subunits are lacking, including whether KvS subunits co-localize with Kv2 channels in the characteristic clusters observed in brain neurons. Such studies, as well as determining whether KvS α subunits can influence the subcellular localization of Kv2 channels in neurons, would add important knowledge to our current understanding of these intriguing electrically silent subunits.

KvS function in Neurons

Profiling of mRNA levels suggests that KvS α subunits are far more restricted in their cellular expression patterns than the broadly expressed Kv2 channels, making it possible that any in vivo role for KvS α subunits in modulating Kv2 currents would occur in a restricted, cell-specific manner. The neuronal functions of KvS α subunits are, however, quite unexplored, due to the lack of specific pharmacological agents that would distinguish KvS-containing channels, or antibodies to label endogenous KvS α subunits in tissue samples. As most studies on KvS channels have been performed in heterologous expression systems, whether regulation of Kv2 channels is the primary neuronal function of KvS subfamily members remains to be determined. Studies in retinal photoreceptor cells nevertheless support that Kv2.1 and Kv8.2 associate and co-localize in the inner segment of both rod and cone photoreceptors and together underlie the I_{Kx} current that stabilizes the dark resting potential and accelerates the voltage response to dim light (Beech & Barnes, 1989; Gayet-Primo et al., 2018). In support of this, certain mutations in the *KCNV2* gene that encodes Kv8.2 cause the retinal disorder "cone-dystrophy with supernormal rod electroretinogram," that leads to life-long visual loss (Wu et al., 2006).

THE KV10–12 SUBFAMILIES

The members of the Kv10–Kv12 subfamilies are all orthologues of the *Drosophila* ether-à-go-go (EAG) channel and are often referred to as the "EAG superfamily." They are also known as eag (Kv10), eag-related gene (erg) (Kv11), and eag-like (elk, Kv12) K^+ channels. In total, the superfamily counts eight members that all exhibit neuronal expression at the mRNA level (Table 4.1). These α subunits contain very large N- and C-terminal cytoplasmic domains (Figure 4.9). The N-terminal is characterized by the presence of the EAG domain that consists of a cap sequence and a Per-Arnt-Sim (PAS) domain (Figure 4.9). The C-terminus contains a cyclic nucleotide-binding homology domain (CNBHD). However, this domain does not bind cyclic nucleotides, as the binding domain is occupied by the distal part of the CNBHD itself (reviewed in Bauer & Schwarz, 2018). Although all members of the Kv10–Kv12 subfamilies generate delayed rectifier-type channels when expressed in heterologous expression systems, the ionic currents generated by individual members of the three subfamilies have their own characteristic functional properties. Kv channels formed by Kv10 α subunits conduct delayed rectifier-type currents that activate with slow-activation kinetics (Figure 4.9). Characteristically, Kv10-mediated currents display activation kinetics that depend on the prepulse potential with accelerated activation kinetics at depolarized potentials (Ludwig et al., 1994). In contrast, Kv11-mediated currents display slow activation kinetics, but inactivate rapidly (Figure 4.9). As their recovery from inactivation is fast and their deactivation kinetics slow, Kv11 currents are largest upon repolarization (Figure 4.9). Kv12-mediated currents activate at more hyperpolarized potentials than do those mediated by Kv10 α subunits. They display some (Kv12.2) or no (Kv12.1, Kv12.3) inactivation. In contrast to Kv10 channels, their activation kinetics do not depend on the prepulse potential. The differences in gating characteristics between Kv channels formed by Kv10–Kv12 α subunits can at least partly be explained by sequence variations in the EAG domain that impacts channel gating (reviewed in Bauer & Schwarz, 2018). Kv channels from these three subfamilies are also distinctly regulated, with Kv10 channels regulated by intracellular calcium, Kv11 by extracellular K^+, and Kv12 by external pH (reviewed in Bauer & Schwarz, 2018). Tetramerization of the component α subunits requires a coiled-coil domain in their distal C-terminus (the carboxyl assembly domain, CAD) and hetero-tetramerization is possible, but restricted to α subunits within their own subfamily (Figure 4.9; Jenke et al., 2003; Ludwig, Owen, & Pongs, 1997; Wimmers, Wulfsen, Bauer, & Schwarz, 2001). The subfamily-specific association is regulated by the C-terminal coiled-coil as well as unidentified regions in the N-terminus (Jenke et al., 2003; Lin et al., 2014).

Kv10–12 Accessory Proteins

Kv10.1 α subunits bind CaM in a calcium-dependent manner through three distinct binding motifs located in the N- and C-terminal domains (Figure 4.9; Schönherr, Löber, & Heinemann, 2000; Ziechner et al., 2006). CaM binding at calcium levels above 100 nM results in a potent inhibition of channels containing Kv10 α subunits (Schönherr et al., 2000). The crystal structure of the Kv10.1 α subunit in complex with CaM suggest that CaM binds the N- and C-termini to clamp the two domains, resulting in pore closing (Figure 4.9; Whicher & MacKinnon, 2016). The CaM binding domains are conserved in the related Kv10.2 α subunit but are absent in Kv11 and Kv12 subfamily members, making the CaM-dependent regulation unique to the Kv10 subfamily.

Kv10–12 Subcellular Localization

The subcellular localization of Kv10–Kv12 α subunits remains largely unknown. Experiments on Kv10.1 knockout mice demonstrate a presynaptic function of this subunit in granule cells of the cerebellum, where it was also detected by electron microscopy (Mortensen et al., 2015). Whether Kv10 α subunits also localize to other neuronal compartments has not been determined. The primary subcellular localization of Kv11 and Kv12 α subunits has not yet been defined.

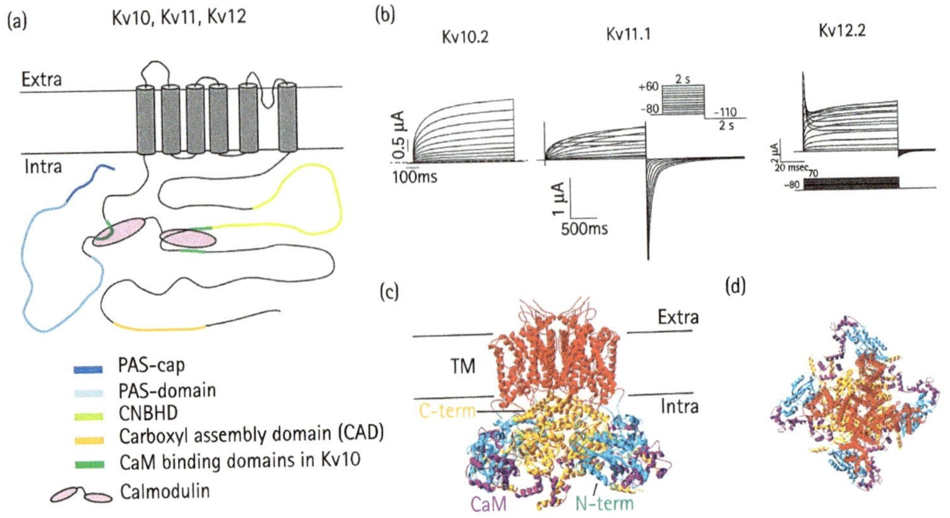

The Kv10–Kv12 subfamilies.

FIGURE 4.9 **A.** Schematic overview of the domain structure of a Kv10–Kv12 α subunit. The Kv10-associated regulatory CaM subunit is included. PAS: Per-Arnt-Sim; CNBHD: cyclic nucleotide-binding homology domain. **B.** Representative current trances obtained in *Xenopus laevis* oocytes expressing Kv10.2, Kv11.1, or Kv12.2 α subunits, as indicated. For Kv11.1 and Kv12.2, the applied voltage protocol is shown. For Kv10.2, currents were activated by voltage steps from −90 mV to +40 mV in 10 mV increments, and tail currents recorded at −80 mV. (Adapted with permission from Saganich et al., 1999; Trudeau, Titus, Branchaw, Ganetzky, & Robertson, 1999; Van Slyke et al., 2010.) **C.** Ribbon representation of the cryoEM structure of the rat Kv10.1 α subunit in complex with CaM. The transmembrane and extracellular domains are colored red, while the N- and C-termini are blue and yellow, respectively. CaM is shown in purple. The image was created with Swiss-PDB Viewer and PDB ID: 5K7L. **D.** Illustration of the structure in C as seen from the extracellular side to visualize how the same CaM molecule simultaneously binds the N- and C-terminal domains. Color coding is as in C.

Kv10–12 function in Neurons

The lack of specific pharmacological inhibitors for Kv10 and Kv12 channels has hampered the investigation of their native currents. Therefore, the current knowledge on these two subfamilies is based on studies in knockout animals. Most data suggest a presynaptic function of Kv10.1 α subunits, where they limit the action potential–induced calcium influx into the presynaptic terminals. Since channels formed from Kv10 α subunits display relatively slow activation kinetics, they do not impact presynaptic calcium levels significantly during single action potentials. However, as channel activation accelerates with depolarized potentials, Kv10.1-mediated currents activate faster during repeated action potentials, thereby impacting presynaptic calcium influx in a frequency- and pulse number-dependent manner (Mortensen et al., 2015). Specific blockers are available for Kv11-containing channels, and Kv11 currents can stabilize the resting membrane potential and cause frequency adaptation due to the progressive build-up of current during successive firings (reviewed in Bauer & Schwarz, 2018). However, studies on Kv11 currents are also sparse, since native Kv11 currents are reportedly relatively small and difficult to isolate. A role similar to Kv11 subfamily members' has been reported for Kv12.2 α subunit, based on knockout studies (Zhang et al., 2010).

De novo mutations in *KCNH1* that encode the Kv10.1 α subunit are linked to two neurological diseases, the Zimmermann-Laband and Temple-Baraitser syndromes (Kortüm et al., 2015; Simons et al., 2015). These syndromes have neurological features such as intellectual disability and epileptic seizures, but they are also characterized by extraneuronal characteristics, including facial dysmorphism and hypoplasia or aplasia of nails and terminal phalanges. Effects outside of the nervous system could be caused by the impact of Kv10.1-containing channels on cell cycle control and proliferation, and the α subunit is ectopically expressed in 70 percent of human cancers (Urrego, Tomczak, Zahed, Stühmer, & Pardo, 2014). Interestingly, all reported mutations cause gain-of-function effects in Kv10.1, perhaps consistent with the observation that Kv10.1 knockout mice develop normally (Ufartes et al., 2013). It suggests that, whereas absence of Kv10.1 expression can be tolerated, augmented activity of the channel is not. Mutations in the *KCNH2* gene that encodes Kv11.1 are one of the common causes of the inherited cardiac arrhythmia long QT syndrome (Curran et al., 1995).

Conclusion

Much progress in understanding the structure, function, and regulation of Kv channels has been made since Kv currents were first described in the squid giant axon. Aided by the cloning of Kv channel α and auxiliary subunits, we have learned much about the biophysical and pharmacological properties of Kv channels formed by distinct subunit combinations. The development of reliable, subunit-specific antibodies has provided critical insights into their specific cellular and subcellular localization patterns, and the extent of their association with one another in native channel complexes. Classical biochemical experiments, and more recently, proteomics have begun to reveal the nature and identity of specific post-translational modifications and interacting partners. The integrated knowledge from these studies has allowed for a reliable correlation of specific Kv channel α and auxiliary subunit combinations with native neuronal currents. Molecular cloning studies revealed that the mammalian Kv channel family is not only large, but also very diverse. Although all Kv channels exhibit voltage activation and conduct K^+ ions, they each have their individual biophysical and pharmacological characteristics, and display a variety of subcellular localization patterns, even within highly restricted sub-compartments. Their diversity in structure, function, expression, localization, and dynamic modulation provides neurons with a wide palette of Kv channels that can be utilized to fine-tune specific aspects of neuronal signaling. That the diversity is important is demonstrated by the large number of Kv channels associated with human disease, further underscoring that, while they exhibit many similarities, their functions are in many cases non-redundant.

Future Directions

We have learned a lot about the basic characteristics of different Kv channels since the first channel was cloned in 1987. However, not all Kv subfamilies are well described. The KvS and Kv10–Kv12 subfamilies still remain somewhat mysterious in regard to their neuronal functions. This is mainly due to the lack of the proper tools, such as specific pharmacological blockers and antibodies for these α subunits, such that they remain the "dark matter" of the ion channel universe. As many of these relatively unknown α subunits show strong neuronal expression and some are linked to neurological diseases, it will be important to develop those tools. Knockout studies of the different subunits should help clarify their neuronal functions.

Further, even with regard to the well-described Kv channel α and auxiliary subunits, we still have much to learn. It has become increasingly clear that Kv channel expression and function are not static features. Neuronal signaling pathways can dynamically impact Kv channel expression levels, localization, and function. For instance, many Kv channels display altered gene expression in response to prolonged changes in overall neuronal activity (Lee et al., 2015). There is a whole avenue of research ahead of us, with the aim to understand the regulation of Kv channel function and localization and its physiological impact. Further, we need to obtain the broader picture of how the regulation is mastered. This will require knowledge of the signaling networks that the individual Kv channels operate within. As indicated by the mosaic localization of Kv channels at the AIS (Figure 4.2), the local signaling networks at each of these sites will most likely differ. Proteomic characterization of different Kv channel complexes should help us delineate the associated signaling networks and their role in regulating function within the distinct subcellular compartments.

Finally, the studies on members of the Kv2 subfamily serve as a beautiful example that Kv channel function is not limited to functioning as a voltage-dependent pathway for the flux of K^+ ions across the PM. In fact, the ability of Kv2 to remodel ER-PM junctions in a variety of cell types, including neurons, does not appear to require its K^+ conductance (Kirmiz, Palacio, et al., 2018). This provides yet another prominent example of the non-conducting functions of Kv channels (Kaczmarek, 2006). The elucidation of the protein interaction partners of Kv channels should be instrumental in further defining the interplay between the diverse family of Kv channels and cellular function. Further, as also demonstrated in the Kv2 example, high-resolution imaging of the nanodomain subcellular structures associated with Kv channels could also provide important information in this context. Clarifying both the signaling networks of different Kv channels as well as the whole range of Kv functions, including those that do not involve K^+ flux, should create a basis for us to understand why mutations in some Kv channel α subunits are disease-causing, and those of others are not.

Acknowledgements

Work in the Rasmussen laboratory is supported by the Novo Nordisk Foundation grant no. NNF17OC0028930 and Lundbeckfonden grant no. R273-2017-2988. Related work in the Trimmer lab is supported by National Institutes of Health grants R01 HL144071 and R21 NS101648.

References

Adair, B., Nunn, R., Lewis, S., Dukes, I., Philipson, L., & Yeager, M. (2008). Single particle image reconstruction of the human recombinant Kv2.1 channel. *Biophysical Journal, 94*(6), 2106–2114. https://doi.org/10.1529/biophysj.107.118562

An, W. F., Bowlby, M. R., Betty, M., Cao, J., Ling, H. P., Mendoza, G., . . . Rhodes, K. J. (2000). Modulation of A-type potassium channels by a family of calcium sensors. *Nature, 403*(6769), 553–556. https://doi.org/10.1038/35000592

Battefeld, A., Tran, B. T., Gavrilis, J., Cooper, E. C., & Kole, M. H. P. (2014). Heteromeric Kv7.2/7.3 channels differentially regulate action potential initiation and conduction in neocortical myelinated axons. *Journal of Neuroscience, 34*(10), 3719–3732. https://doi.org/10.1523/JNEUROSCI.4206-13.2014.

Bauer, C. K., & Schwarz, J. R. (2018). Ether-à-go-go K^+ channels: Effective modulators of neuronal excitability. *Journal of Physiology, 596*(5), 769–783. https://doi.org/10.1113/JP275477

Beech, D. J., & Barnes, S. (1989). Characterization of a voltage-gated K^+ channel that accelerates the rod response to dim light. *Neuron, 3*(5), 573–581.

Benndorf, K., Koopmann, R., Lorra, C., & Pongs, O. (1994). Gating and conductance properties of a human delayed rectifier K^+ channel expressed in frog oocytes. *Journal of Physiology, 477*(Pt 1), 1–14.

Benned-Jensen, T., Christensen, R. K., Denti, F., Perrier, J.-F., Rasmussen, H. B., & Olesen, S.-P. (2016). Live imaging of Kv7.2/7.3 cell surface dynamics at the axon initial segment: High steady-state stability and calpain-dependent excitotoxic downregulation revealed. *Journal of Neuroscience, 36*(7), 2261–2266. https://doi.org/10.1523/JNEUROSCI.2631-15.2016

Biervert, C., Schroeder, B. C., Kubisch, C., Berkovic, S. F., Propping, P., Jentsch, T. J., & Steinlein, O. K. (1998). A potassium channel mutation in neonatal human epilepsy. *Science, 279*(5349), 403–406.

Bishop, H. I., Cobb, M. M., Kirmiz, M., Parajuli, L. K., Mandikian, D., Philp, A. M., . . . Trimmer, J. S. (2018). Kv2 ion channels determine the expression and localization of the associated AMIGO-1 cell adhesion molecule in adult brain neurons. *Frontiers in Molecular Neuroscience, 11*, 1. https://doi.org/10.3389/fnmol.2018.00001

Bishop, H. I., Guan, D., Bocksteins, E., Parajuli, L. K., Murray, K. D., Cobb, M. M., . . . Trimmer, J. S. (2015). Distinct cell- and layer-specific expression patterns and independent regulation of Kv2 channel subtypes in cortical pyramidal neurons. *Journal of Neuroscience, 35*(44), 14922–14942. https://doi.org/10.1523/JNEUROSCI.1897-15.2015

Bocksteins, E. (2016). Kv5, Kv6, Kv8, and Kv9 subunits: No simple silent bystanders. *The Journal of General Physiology*, *147*(2), 105–125. https://doi.org/10.1085/jgp.201511507

Bocksteins, E., & Snyders, D. J. (2012). Electrically silent Kv subunits: Their molecular and functional characteristics. *Physiology (Bethesda, Md.)*, *27*(2), 73–84. https://doi.org/10.1152/physiol.00023.2011

Brew, H. M., Gittelman, J. X., Silverstein, R. S., Hanks, T. D., Demas, V. P., Robinson, L. C., . . . Tempel, B. L. (2007). Seizures and reduced life span in mice lacking the potassium channel subunit Kv1.2, but hypoexcitability and enlarged Kv1 currents in auditory neurons. *Journal of Neurophysiology*, *98*(3), 1501–1525. https://doi.org/10.1152/jn.00640.2006

Brown, D. A., & Adams, P. R. (1980). Muscarinic suppression of a novel voltage-sensitive K^+ current in a vertebrate neurone. *Nature*, *283*(5748), 673–676.

Brown, D. A., & Passmore, G. M. (2009). Neural KCNQ (Kv7) channels. *British Journal of Pharmacology*, *156*(8), 1185–1195. https://doi.org/10.1111/j.1476-5381.2009.00111.x

Campomanes, C. R., Carroll, K. I., Manganas, L. N., Hershberger, M. E., Gong, B., Antonucci, D. E., . . . Trimmer, J. S. (2002). Kv beta subunit oxidoreductase activity and Kv1 potassium channel trafficking. *Journal of Biological Chemistry*, *277*(10), 8298–8305. https://doi.org/10.1074/jbc.M110276200

Cavaretta, J. P., Sherer, K. R., Lee, K. Y., Kim, E. H., Issema, R. S., & Chung, H. J. (2014). Polarized axonal surface expression of neuronal KCNQ potassium channels is regulated by calmodulin interaction with KCNQ2 subunit. *PLoS One*, *9*(7), e103655. https://doi.org/10.1371/journal.pone.0103655

Chandy, K. G. (1991). Simplified gene nomenclature. *Nature*, *352*(6330), 26. https://doi.org/10.1038/352026b0

Chang, A., Abderemane-Ali, F., Hura, G. L., Rossen, N. D., Gate, R. E., & Minor, D. L. (2018). A calmodulin C-lobe $Ca2^+$-dependent switch governs Kv7 channel function. *Neuron*, *97*(4), 836–852.e6. https://doi.org/10.1016/j.neuron.2018.01.035

Charlier, C., Singh, N. A., Ryan, S. G., Lewis, T. B., Reus, B. E., Leach, R. J., & Leppert, M. (1998). A pore mutation in a novel KQT-like potassium channel gene in an idiopathic epilepsy family. *Nature Genetics*, *18*(1), 53–55. https://doi.org/10.1038/ng0198-53

Chen, X., Yuan, L.-L., Zhao, C., Birnbaum, S. G., Frick, A., Jung, W. E., . . . Johnston, D. (2006). Deletion of Kv4.2 gene eliminates dendritic A-type K^+ current and enhances induction of long-term potentiation in hippocampal CA1 pyramidal neurons. *Journal of Neuroscience*, *26*(47), 12143–12151. https://doi.org/10.1523/JNEUROSCI.2667-06.2006

Chouinard, S. W., Wilson, G. F., Schlimgen, A. K., & Ganetzky, B. (1995). A potassium channel beta subunit related to the aldo-keto reductase superfamily is encoded by the Drosophila hyperkinetic locus. *Proceedings of the National Academy of Sciences of the United States of America*, *92*(15), 6763–6767.

Chung, H. J., Jan, Y. N., & Jan, L. Y. (2006). Polarized axonal surface expression of neuronal KCNQ channels is mediated by multiple signals in the KCNQ2 and KCNQ3 C-terminal domains. *Proceedings of the National Academy of Sciences of the United States of America*, *103*(23), 8870–8875. https://doi.org/10.1073/pnas.0603376103

Clark, B. D., Kwon, E., Maffie, J., Jeong, H.-Y., Nadal, M., Strop, P., & Rudy, B. (2008). DPP6 localization in brain supports function as a Kv4 channel associated protein. *Frontiers in Molecular Neuroscience*, *1*, 8. https://doi.org/10.3389/neuro.02.008.2008

Coleman, S. K., Newcombe, J., Pryke, J., & Dolly, J. O. (1999). Subunit composition of Kv1 channels in human CNS. *Journal of Neurochemistry*, *73*(2), 849–858.

Connor, J. X., McCormack, K., Pletsch, A., Gaeta, S., Ganetzky, B., Chiu, S.-Y., & Messing, A. (2005). Genetic modifiers of the Kv beta2-null phenotype in mice. *Genes, Brain, and Behavior*, *4*(2), 77–88. https://doi.org/10.1111/j.1601-183X.2004.00094.x

Curran, M. E., Splawski, I., Timothy, K. W., Vincent, G. M., Green, E. D., & Keating, M. T. (1995). A molecular basis for cardiac arrhythmia: HERG mutations cause long QT syndrome. *Cell*, *80*(5), 795–803.

Deng, Q., Rashid, A. J., Fernandez, F. R., Turner, R. W., Maler, L., & Dunn, R. J. (2005). A C-terminal domain directs Kv3.3 channels to dendrites. *Journal of Neuroscience*, *25*(50), 11531–11541. https://doi.org/10.1523/JNEUROSCI.3672-05.2005

Devaux, J., Alcaraz, G., Grinspan, J., Bennett, V., Joho, R., Crest, M., & Scherer, S. S. (2003). Kv3.1b is a novel component of CNS nodes. *Journal of Neuroscience*, *23*(11), 4509–4518.

Devaux, J. J., Kleopa, K. A., Cooper, E. C., & Scherer, S. S. (2004). KCNQ2 is a nodal K^+ channel. *Journal of Neuroscience*, *24*(5), 1236–1244. https://doi.org/10.1523/JNEUROSCI.4512-03.2004

Dodson, P. D., & Forsythe, I. D. (2004). Presynaptic K^+ channels: Electrifying regulators of synaptic terminal excitability. *Trends in Neurosciences*, *27*(4), 210–217. https://doi.org/10.1016/j.tins.2004.02.012

Dougherty, K., Tu, L., Deutsch, C., & Covarrubias, M. (2009). The dipeptidyl-aminopeptidase-like protein 6 is an integral voltage sensor-interacting beta-subunit of neuronal K(V)4.2 channels. *Channels*, *3*(2), 122–128.

Doyle, D. A., Morais Cabral, J., Pfuetzner, R. A., Kuo, A., Gulbis, J. M., Cohen, S. L., . . . MacKinnon, R. (1998). The structure of the potassium channel: Molecular basis of K^+ conduction and selectivity. *Science*, *280*(5360), 69–77.

Du, J., Haak, L. L., Phillips-Tansey, E., Russell, J. T., & McBain, C. J. (2000). Frequency-dependent regulation of rat hippocampal somatodendritic excitability by the K^+ channel subunit Kv2.1. *Journal of Physiology*, *522*(Pt 1), 19–31.

Du, J., Tao-Cheng, J. H., Zerfas, P., & McBain, C. J. (1998). The K^+ channel, Kv2.1, is apposed to astrocytic processes and is associated with inhibitory postsynaptic membranes in hippocampal and cortical principal neurons and inhibitory interneurons. *Neuroscience*, *84*(1), 37–48.

Etxeberria, A., Aivar, P., Rodriguez-Alfaro, J. A., Alaimo, A., Villacé, P., Gómez-Posada, J. C., . . . Villarroel, A. (2008). Calmodulin regulates the trafficking of KCNQ2 potassium channels. *The FASEB Journal*, *22*(4), 1135–1143. https://doi.org/10.1096/fj.07-9712com

Fan, Z., Bi, L., Jin, G., & Qi, Z. (2010). Electrostatic interaction in the NH2-terminus accelerates inactivation of the Kv1.4 channel. *Biochimica et Biophysica Acta (BBA)—Biomembranes*, *1798*(11), 2076–2083. https://doi.org/10.1016/j.bbamem.2010.07.029

Foeger, N. C., Marionneau, C., & Nerbonne, J. M. (2010). Co-assembly of Kv4 {alpha} subunits with K^+ channel-interacting protein 2 stabilizes protein expression and promotes surface retention of channel complexes. *Journal of Biological Chemistry*, *285*(43), 33413–33422. https://doi.org/10.1074/jbc.M110.145185

Foeger, N. C., Norris, A. J., Wren, L. M., & Nerbonne, J. M. (2012). Augmentation of Kv4.2-encoded currents by accessory dipeptidyl peptidase 6 and 10 subunits reflects selective cell surface Kv4.2 protein stabilization. *Journal of Biological Chemistry*, *287*(12), 9640–9650. https://doi.org/10.1074/jbc.M111.324574

Fox, P. D., Haberkorn, C. J., Akin, E. J., Seel, P. J., Krapf, D., & Tamkun, M. M. (2015). Induction of stable ER-plasma-membrane junctions by Kv2.1 potassium channels. *Journal of Cell Science*, *128*(11), 2096–2105. https://doi.org/10.1242/jcs.166009

Fox, P. D., Loftus, R. J., & Tamkun, M. M. (2013). Regulation of Kv2.1 K(+) conductance by cell surface channel density. *Journal of Neuroscience, 33*(3), 1259–1270. https://doi.org/10.1523/JNEUROSCI.3008-12.2013

Frech, G. C., VanDongen, A. M., Schuster, G., Brown, A. M., & Joho, R. H. (1989). A novel potassium channel with delayed rectifier properties isolated from rat brain by expression cloning. *Nature, 340*(6235), 642–645. https://doi.org/10.1038/340642a0

Fukata, Y., Adesnik, H., Iwanaga, T., Bredt, D. S., Nicoll, R. A., & Fukata, M. (2006). Epilepsy-related ligand/receptor complex LGI1 and ADAM22 regulate synaptic transmission. *Science, 313*(5794), 1792–1795. https://doi.org/10.1126/science.1129947

Gamper, N., & Shapiro, M. S. (2003). Calmodulin mediates Ca2+-dependent modulation of M-type K+ channels. *Journal of General Physiology, 122*(1), 17–31. https://doi.org/10.1085/jgp.200208783

Gayet-Primo, J., Yaeger, D. B., Khanjian, R. A., & Puthussery, T. (2018). Heteromeric KV2/KV8.2 channels mediate delayed rectifier potassium currents in primate photoreceptors. *Journal of Neuroscience, 38*(14), 3414–3427. https://doi.org/10.1523/JNEUROSCI.2440-17.2018

Gilling, M., Rasmussen, H. B., Calloe, K., Sequeira, A. F., Baretto, M., Oliveira, G., . . . Tommerup, N. (2013). Dysfunction of the heteromeric KV7.3/KV7.5 potassium channel is associated with autism spectrum disorders. *Frontiers in Genetics, 4*, 54. https://doi.org/10.3389/fgene.2013.00054

Grizel, A., Popinako, A., Kasimova, M. A., Stevens, L., Karlova, M., Moisenovich, M. M., & Sokolova, O. S. (2014). Domain structure and conformational changes in rat KV2.1 ion channel. *Journal of Neuroimmune Pharmacology, 9*(5), 727–739. https://doi.org/10.1007/s11481-014-9565-x

Gu, C., Jan, Y. N., & Jan, L. Y. (2003). A conserved domain in axonal targeting of Kv1 (*Shaker*) voltage-gated potassium channels. *Science, 301*(5633), 646–649. https://doi.org/10.1126/science.1086998

Guan, D., Armstrong, W. E., & Foehring, R. C. (2013). Kv2 channels regulate firing rate in pyramidal neurons from rat sensorimotor cortex. *Journal of Physiology, 591*(19), 4807–4825. https://doi.org/10.1113/jphysiol.2013.257253

Guan, D., Tkatch, T., Surmeier, D. J., Armstrong, W. E., & Foehring, R. C. (2007). Kv2 subunits underlie slowly inactivating potassium current in rat neocortical pyramidal neurons. *Journal of Physiology, 581*(Pt 3), 941–960. https://doi.org/10.1113/jphysiol.2007.128454

Gutman, G. A., Chandy, K. G., Grissmer, S., Lazdunski, M., McKinnon, D., Pardo, L. A., . . . Wang, X. (2005). International Union of Pharmacology. LIII. Nomenclature and molecular relationships of voltage-gated potassium channels. *Pharmacological Reviews, 57*(4), 473–508. https://doi.org/10.1124/pr.57.4.10

Hadley, J. K., Passmore, G. M., Tatulian, L., Al-Qatari, M., Ye, F., Wickenden, A. D., & Brown, D. A. (2003). Stoichiometry of expressed KCNQ2/KCNQ3 potassium channels and subunit composition of native ganglionic M channels deduced from block by tetraethylammonium. *Journal of Neuroscience, 23*(12), 5012–5019.

Heilstedt, H. A., Burgess, D. L., Anderson, A. E., Chedrawi, A., Tharp, B., Lee, O., . . . Shapira, S. K. (2001). Loss of the potassium channel beta-subunit gene, KCNAB2, is associated with epilepsy in patients with 1p36 deletion syndrome. *Epilepsia, 42*(9), 1103–1111.

Hodgkin, A. L., & Huxley, A. F. (1952). Currents carried by sodium and potassium ions through the membrane of the giant axon of Loligo. *Journal of Physiology, 116*(4), 449–472.

Hoffman, D. A., Magee, J. C., Colbert, C. M., & Johnston, D. (1997). K^+ channel regulation of signal propagation in dendrites of hippocampal pyramidal neurons. *Nature, 387*(6636), 869–875. https://doi.org/10.1038/43119

Horresh, I., Poliak, S., Grant, S., Bredt, D., Rasband, M. N., & Peles, E. (2008). Multiple molecular interactions determine the clustering of Caspr2 and Kv1 channels in myelinated axons. *Journal of Neuroscience, 28*(52), 14213–14222. https://doi.org/10.1523/JNEUROSCI.3398-08.2008

Isacoff, E. Y., Jan, Y. N., & Jan, L. Y. (1990). Evidence for the formation of heteromultimeric potassium channels in *Xenopus* oocytes. *Nature, 345*(6275), 530–534. https://doi.org/10.1038/345530a0

Jenke, M., Sánchez, A., Monje, F., Stühmer, W., Weseloh, R. M., & Pardo, L. A. (2003). C-terminal domains implicated in the functional surface expression of potassium channels. *EMBO Journal, 22*(3), 395–403. https://doi.org/10.1093/emboj/cdg035

Jensen, C. S., Rasmussen, H. B., & Misonou, H. (2011). Neuronal trafficking of voltage-gated potassium channels. *Molecular and Cellular Neurosciences, 48*(4), 288–297. https://doi.org/10.1016/j.mcn.2011.05.007

Jensen, C. S., Watanabe, S., Rasmussen, H. B., Schmitt, N., Olesen, S.-P., Frost, N. A., . . . Misonou, H. (2014). Specific sorting and post-Golgi trafficking of dendritic potassium channels in living neurons. *Journal of Biological Chemistry, 289*(15), 10566–10581. https://doi.org/10.1074/jbc.M113.534495

Jensen, C. S., Watanabe, S., Stas, J. I., Klaphaak, J., Yamane, A., Schmitt, N., . . . Misonou, H. (2017). Trafficking of Kv2.1 channels to the axon initial segment by a novel nonconventional secretory pathway. *Journal of Neuroscience, 37*(48), 11523–11536. https://doi.org/10.1523/JNEUROSCI.3510-16.2017

Jerng, H. H., Kunjilwar, K., & Pfaffinger, P. J. (2005). Multiprotein assembly of Kv4.2, KChIP3 and DPP10 produces ternary channel complexes with ISA-like properties. *Journal of Physiology, 568*(Pt 3), 767–788. https://doi.org/10.1113/jphysiol.2005.087858

Jerng, H. H., Lauver, A. D., & Pfaffinger, P. J. (2007). DPP10 splice variants are localized in distinct neuronal populations and act to differentially regulate the inactivation properties of Kv4-based ion channels. *Molecular and Cellular Neurosciences, 35*(4), 604–624. https://doi.org/10.1016/j.mcn.2007.03.008

Jerng, H. H., & Pfaffinger, P. J. (2014). Modulatory mechanisms and multiple functions of somatodendritic A-type K(+) channel auxiliary subunits. *Frontiers in Cellular Neuroscience, 8*, 82. https://doi.org/10.3389/fncel.2014.00082

Johnson, B., Leek, A. N., Solé, L., Maverick, E. E., Levine, T. P., & Tamkun, M. M. (2018). Kv2 potassium channels form endoplasmic reticulum/plasma membrane junctions via interaction with VAPA and VAPB. *Proceedings of the National Academy of Sciences of the United States of America, 115*(31), E7331–E7340. https://doi.org/10.1073/pnas.1805757115

Johnston, D., Christie, B. R., Frick, A., Gray, R., Hoffman, D. A., Schexnayder, L. K., . . . Yuan, L.-L. (2003). Active dendrites, potassium channels and synaptic plasticity. *Philosophical Transactions of the Royal Society of London. Series B, Biological Sciences, 358*(1432), 667–674. https://doi.org/10.1098/rstb.2002.1248

Johnston, J., Forsythe, I. D., & Kopp-Scheinpflug, C. (2010). Going native: Voltage-gated potassium channels controlling neuronal excitability. *Journal of Physiology, 588*(Pt 17), 3187–3200. https://doi.org/10.1113/jphysiol.2010.191973

Johnston, J., Griffin, S. J., Baker, C., Skrzypiec, A., Chernova, T., & Forsythe, I. D. (2008). Initial segment Kv2.2 channels mediate a slow delayed rectifier and maintain high frequency action potential firing in medial nucleus of the trapezoid body neurons. *Journal of Physiology, 586*(14), 3493–3509. https://doi.org/10.1113/jphysiol.2008.153734

Jow, F., Zhang, Z.-H., Kopsco, D. C., Carroll, K. C., & Wang, K. (2004). Functional coupling of intracellular calcium and inactivation of voltage-gated Kv1.1/Kvbeta1.1 A-type K^+ channels.

Proceedings of the National Academy of Sciences of the United States of America, 101(43), 15535–15540. https://doi.org/10.1073/pnas.0402081101

Kaczmarek, L. K. (2006). Non-conducting functions of voltage-gated ion channels. *Nature Reviews Neuroscience, 7*(10), 761–771. https://doi.org/10.1038/nrn1988

Kaczmarek, L. K., & Zhang, Y. (2017). Kv3 channels: Enablers of rapid firing, neurotransmitter release, and neuronal endurance. *Physiological Reviews, 97*(4), 1431–1468. https://doi.org/10.1152/physrev.00002.2017

Kamb, A., Iverson, L. E., & Tanouye, M. A. (1987). Molecular characterization of *Shaker*, a Drosophila gene that encodes a potassium channel. *Cell, 50*(3), 405–413.

Kerti, K., Lorincz, A., & Nusser, Z. (2012). Unique somatodendritic distribution pattern of Kv4.2 channels on hippocampal CA1 pyramidal cells. *European Journal of Neuroscience, 35*(1), 66–75. https://doi.org/10.1111/j.1460-9568.2011.07907.x

Kharkovets, T., Hardelin, J. P., Safieddine, S., Schweizer, M., El-Amraoui, A., Petit, C., & Jentsch, T. J. (2000). KCNQ4, a K^+ channel mutated in a form of dominant deafness, is expressed in the inner ear and the central auditory pathway. *Proceedings of the National Academy of Sciences of the United States of America, 97*(8), 4333–4338.

Kihira, Y., Hermanstyne, T. O., & Misonou, H. (2010). Formation of heteromeric Kv2 channels in mammalian brain neurons. *Journal of Biological Chemistry, 285*(20), 15048–15055. https://doi.org/10.1074/jbc.M109.074260

Kim, E., Niethammer, M., Rothschild, A., Jan, Y. N., & Sheng, M. (1995). Clustering of *Shaker*-type K^+ channels by interaction with a family of membrane-associated guanylate kinases. *Nature, 378*(6552), 85–88. https://doi.org/10.1038/378085a0

Kirmiz, M., Palacio, S., Thapa, P., King, A. N., Sack, J. T., & Trimmer, J. S. (2018). Remodeling neuronal ER-PM junctions is a conserved nonconducting function of Kv2 plasma membrane ion channels. *Molecular Biology of the Cell, 29*(20), 2410–2432. https://doi.org/10.1091/mbc.E18-05-0337

Kirmiz, M., Vierra, N. C., Palacio, S., & Trimmer, J. S. (2018). Identification of VAPA and VAPB as Kv2 channel-interacting proteins defining endoplasmic reticulum-plasma membrane junctions in mammalian brain neurons. *Journal of Neuroscience, 38*(35), 7562–7584. https://doi.org/10.1523/JNEUROSCI.0893-18.2018

Kobertz, W. R., Williams, C., & Miller, C. (2000). Hanging gondola structure of the T1 domain in a voltage-gated $K(^+)$ channel. *Biochemistry, 39*(34), 10347–10352.

Kole, M. H. P., Letzkus, J. J., & Stuart, G. J. (2007). Axon initial segment Kv1 channels control axonal action potential waveform and synaptic efficacy. *Neuron, 55*(4), 633–647. https://doi.org/10.1016/j.neuron.2007.07.031

Kortüm, F., Caputo, V., Bauer, C. K., Stella, L., Ciolfi, A., Alawi, M., . . . Kutsche, K. (2015). Mutations in KCNH1 and ATP6V1B2 cause Zimmermann-Laband syndrome. *Nature Genetics, 47*(6), 661–667. https://doi.org/10.1038/ng.3282

Kubisch, C., Schroeder, B. C., Friedrich, T., Lütjohann, B., El-Amraoui, A., Marlin, S., . . . Jentsch, T. J. (1999). KCNQ4, a novel potassium channel expressed in sensory outer hair cells, is mutated in dominant deafness. *Cell, 96*(3), 437–446.

Lee, K. Y., Royston, S. E., Vest, M. O., Ley, D. J., Lee, S., Bolton, E. C., & Chung, H. J. (2015). N-methyl-D-aspartate receptors mediate activity-dependent down-regulation of potassium channel genes during the expression of homeostatic intrinsic plasticity. *Molecular Brain, 8*, 4. https://doi.org/10.1186/s13041-015-0094-1

Li, M., Jan, Y. N., & Jan, L. Y. (1992). Specification of subunit assembly by the hydrophilic amino-terminal domain of the *Shaker* potassium channel. *Science, 257*(5074), 1225–1230.

Lim, S. T., Antonucci, D. E., Scannevin, R. H., & Trimmer, J. S. (2000). A novel targeting signal for proximal clustering of the Kv2.1 K^+ channel in hippocampal neurons. *Neuron, 25*(2), 385–397.
Lin, T.-F., Lin, I.-W., Chen, S.-C., Wu, H.-H., Yang, C.-S., Fang, H.-Y., ... Jeng, C.-J. (2014). The subfamily-specific assembly of Eag and Erg K^+ channels is determined by both the amino and the carboxyl recognition domains. *Journal of Biological Chemistry, 289*(33), 22815–22834. https://doi.org/10.1074/jbc.M114.574814
Liu, P. W., & Bean, B. P. (2014). Kv2 channel regulation of action potential repolarization and firing patterns in superior cervical ganglion neurons and hippocampal CA1 pyramidal neurons. *Journal of Neuroscience, 34*(14), 4991–5002. https://doi.org/10.1523/JNEUROSCI.1925-13.2014
Long, S. B., Campbell, E. B., & MacKinnon, R. (2005). Crystal structure of a mammalian voltage-dependent *Shaker* family K^+ channel. *Science, 309*(5736), 897–903. https://doi.org/10.1126/science.1116269
Lorincz, A., & Nusser, Z. (2008). Cell-type-dependent molecular composition of the axon initial segment. *Journal of Neuroscience, 28*(53), 14329–14340. https://doi.org/10.1523/JNEUROSCI.4833-08.2008
Ludwig, J., Owen, D., & Pongs, O. (1997). Carboxy-terminal domain mediates assembly of the voltage-gated rat ether-à-go-go potassium channel. *EMBO Journal, 16*(21), 6337–6345. https://doi.org/10.1093/emboj/16.21.6337
Ludwig, J., Terlau, H., Wunder, F., Brüggemann, A., Pardo, L. A., Marquardt, A., ... Pongs, O. (1994). Functional expression of a rat homologue of the voltage gated ether á go-go potassium channel reveals differences in selectivity and activation kinetics between the Drosophila channel and its mammalian counterpart. *EMBO Journal, 13*(19), 4451–4458.
MacKinnon, R. (1991). Determination of the subunit stoichiometry of a voltage-activated potassium channel. *Nature, 350*(6315), 232–235. https://doi.org/10.1038/350232a0
Malin, S. A., & Nerbonne, J. M. (2002). Delayed rectifier K^+ currents, IK, are encoded by Kv2 alpha-subunits and regulate tonic firing in mammalian sympathetic neurons. *Journal of Neuroscience, 22*(23), 10094–10105.
Mandikian, D., Bocksteins, E., Parajuli, L. K., Bishop, H. I., Cerda, O., Shigemoto, R., & Trimmer, J. S. (2014). Cell type-specific spatial and functional coupling between mammalian brain Kv2.1 K^+ channels and ryanodine receptors. *Journal of Comparative Neurology, 522*(15), 3555–3574. https://doi.org/10.1002/cne.23641
Manganas, L. N., & Trimmer, J. S. (2000). Subunit composition determines Kv1 potassium channel surface expression. *Journal of Biological Chemistry, 275*(38), 29685–29693. https://doi.org/10.1074/jbc.M005010200
McCormack, K., Connor, J. X., Zhou, L., Ho, L. L., Ganetzky, B., Chiu, S.-Y., & Messing, A. (2002). Genetic analysis of the mammalian K^+ channel beta subunit Kvbeta 2 (Kcnab2). *Journal of Biological Chemistry, 277*(15), 13219–13228. https://doi.org/10.1074/jbc.M111465200
Menegola, M., & Trimmer, J. S. (2006). Unanticipated region- and cell-specific downregulation of individual KChIP auxiliary subunit isotypes in Kv4.2 knock-out mouse brain. *Journal, 26*(47), 12137–12142. https://doi.org/10.1523/JNEUROSCI.2783-06.2006
Misonou, H., Mohapatra, D. P., Park, E. W., Leung, V., Zhen, D., Misonou, K., ... Trimmer, J. S. (2004). Regulation of ion channel localization and phosphorylation by neuronal activity. *Nature Neuroscience, 7*(7), 711–718. https://doi.org/10.1038/nn1260
Misonou, H., Thompson, S. M., & Cai, X. (2008). Dynamic regulation of the Kv2.1 voltage-gated potassium channel during brain ischemia through neuroglial interaction. *Journal of Neuroscience, 28*(34), 8529–8538. https://doi.org/10.1523/JNEUROSCI.1417-08.2008

Mohapatra, D. P., Misonou, H., Pan, S.-J., Held, J. E., Surmeier, D. J., & Trimmer, J. S. (2009). Regulation of intrinsic excitability in hippocampal neurons by activity-dependent modulation of the KV2.1 potassium channel. *Channels, 3*(1), 46–56.

Mohapatra, D. P., & Trimmer, J. S. (2006). The Kv2.1 C terminus can autonomously transfer Kv2.1-like phosphorylation-dependent localization, voltage-dependent gating, and muscarinic modulation to diverse Kv channels. *Journal of Neuroscience, 26*(2), 685–695. https://doi.org/10.1523/JNEUROSCI.4620-05.2006

Morais-Cabral, J. H., Zhou, Y., & MacKinnon, R. (2001). Energetic optimization of ion conduction rate by the K^+ selectivity filter. *Nature, 414*(6859), 37–42. https://doi.org/10.1038/35102000

Mortensen, L. S., Schmidt, H., Farsi, Z., Barrantes-Freer, A., Rubio, M. E., Ufartes, R., ... Pardo, L. A. (2015). KV10.1 opposes activity-dependent increase in Ca^{2+} influx into the presynaptic terminal of the parallel fibre-Purkinje cell synapse. *Journal of Physiology, 593*(1), 181–196. https://doi.org/10.1113/jphysiol.2014.281600

Murakoshi, H., Shi, G., Scannevin, R. H., & Trimmer, J. S. (1997). Phosphorylation of the Kv2.1 K^+ channel alters voltage-dependent activation. *Molecular Pharmacology, 52*(5), 821–828.

Murakoshi, H., & Trimmer, J. S. (1999). Identification of the Kv2.1 K^+ channel as a major component of the delayed rectifier K^+ current in rat hippocampal neurons. *Journal of Neuroscience, 19*(5), 1728–1735.

Nadal, M. S., Amarillo, Y., Vega-Saenz de Miera, E., & Rudy, B. (2006). Differential characterization of three alternative spliced isoforms of DPPX. *Brain Research, 1094*(1), 1–12. https://doi.org/10.1016/j.brainres.2006.03.106

Nadal, M. S., Ozaita, A., Amarillo, Y., Vega-Saenz de Miera, E., Ma, Y., Mo, W., ... Rudy, B. (2003). The CD26-related dipeptidyl aminopeptidase-like protein DPPX is a critical component of neuronal A-type K^+ channels. *Neuron, 37*(3), 449–461.

Neyroud, N., Tesson, F., Denjoy, I., Leibovici, M., Donger, C., Barhanin, J., ... Guicheney, P. (1997). A novel mutation in the potassium channel gene KVLQT1 causes the Jervell and Lange-Nielsen cardioauditory syndrome. *Nature Genetics, 15*(2), 186–189. https://doi.org/10.1038/ng0297-186

O'Connell, D. J., Bauer, M. C., O'Brien, J., Johnson, W. M., Divizio, C. A., O'Kane, S. L., ... Cahill, D. J. (2010). Integrated protein array screening and high throughput validation of 70 novel neural calmodulin-binding proteins. *Molecular & Cellular Proteomics9*(6), 1118–1132. https://doi.org/10.1074/mcp.M900324-MCP200

O'Connell, K. M. S., Loftus, R., & Tamkun, M. M. (2010). Localization-dependent activity of the Kv2.1 delayed-rectifier K^+ channel. *Proceedings of the National Academy of Sciences of the United States of America, 107*(27), 12351–12356. https://doi.org/10.1073/pnas.1003028107

Ogawa, Y., Oses-Prieto, J., Kim, M. Y., Horresh, I., Peles, E., Burlingame, A. L., ... Rasband, M. N. (2010). ADAM22, a Kv1 channel-interacting protein, recruits membrane-associated guanylate kinases to juxtaparanodes of myelinated axons. *Journal of Neuroscience, 30*(3), 1038–1048. https://doi.org/10.1523/JNEUROSCI.4661-09.2010

Ottschytsch, N., Raes, A., Van Hoorick, D., & Snyders, D. J. (2002). Obligatory heterotetramerization of three previously uncharacterized Kv channel alpha-subunits identified in the human genome. *Proceedings of the National Academy of Sciences of the United States of America, 99*(12), 7986–7991. https://doi.org/10.1073/pnas.122617999

Ovsepian, S. V., LeBerre, M., Steuber, V., O'Leary, V. B., Leibold, C., & Oliver Dolly, J. (2016). Distinctive role of KV1.1 subunit in the biology and functions of low threshold $K^{(+)}$ channels with implications for neurological disease. *Pharmacology & Therapeutics*, *159*, 93–101. https://doi.org/10.1016/j.pharmthera.2016.01.005

Palacio, S., Chevaleyre, V., Brann, D. H., Murray, K. D., Piskorowski, R. A., & Trimmer, J. S. (2017). Heterogeneity in Kv2 channel expression shapes action potential characteristics and firing patterns in CA1 versus CA2 hippocampal pyramidal neurons. *ENeuro*, *4*(4). https://doi.org/10.1523/ENEURO.0267-17.2017

Pan, Z., Kao, T., Horvath, Z., Lemos, J., Sul, J.-Y., Cranstoun, S. D., . . . Cooper, E. C. (2006). A common ankyrin-G-based mechanism retains KCNQ and NaV channels at electrically active domains of the axon. *Journal of Neuroscience*, *26*(10), 2599–2613. https://doi.org/10.1523/JNEUROSCI.4314-05.2006

Papazian, D. M., Schwarz, T. L., Tempel, B. L., Jan, Y. N., & Jan, L. Y. (1987). Cloning of genomic and complementary DNA from *Shaker*, a putative potassium channel gene from Drosophila. *Science*, *237*(4816), 749–753.

Peltola, M. A., Kuja-Panula, J., Lauri, S. E., Taira, T., & Rauvala, H. (2011). AMIGO is an auxiliary subunit of the Kv2.1 potassium channel. *EMBO Reports*, *12*(12), 1293–1299. https://doi.org/10.1038/embor.2011.204

Pena, S. D. J., & Coimbra, R. L. M. (2015). Ataxia and myoclonic epilepsy due to a heterozygous new mutation in KCNA2: Proposal for a new channelopathy. *Clinical Genetics*, *87*(2), e1–e3. https://doi.org/10.1111/cge.12542

Pioletti, M., Findeisen, F., Hura, G. L., & Minor, D. L. (2006). Three-dimensional structure of the KChIP1-Kv4.3 T1 complex reveals a cross-shaped octamer. *Nature Structural & Molecular Biology*, *13*(11), 987–995. https://doi.org/10.1038/nsmb1164

Pongs, O., & Schwarz, J. R. (2010). Ancillary subunits associated with voltage-dependent K⁺ channels. *Physiological Reviews*, *90*(2), 755–796. https://doi.org/10.1152/physrev.00020.2009

Post, M. A., Kirsch, G. E., & Brown, A. M. (1996). Kv2.1 and electrically silent Kv6.1 potassium channel subunits combine and express a novel current. *FEBS Letters*, *399*(1–2), 177–182.

Rajakulendran, S., Schorge, S., Kullmann, D. M., & Hanna, M. G. (2007). Episodic ataxia type 1: A neuronal potassium channelopathy. *Neurotherapeuticss*, *4*(2), 258–266. https://doi.org/10.1016/j.nurt.2007.01.010

Rasband, M. N., Park, E. W., Zhen, D., Arbuckle, M. I., Poliak, S., Peles, E., . . . Trimmer, J. S. (2002). Clustering of neuronal potassium channels is independent of their interaction with PSD-95. *Journal of Cell Biology*, *159*(4), 663–672. https://doi.org/10.1083/jcb.200206024

Rasmussen, H. B., Frøkjaer-Jensen, C., Jensen, C. S., Jensen, H. S., Jørgensen, N. K., Misonou, H., . . . Schmitt, N. (2007). Requirement of subunit co-assembly and ankyrin-G for M-channel localization at the axon initial segment. *Journal of Cell Science*, *120*(Pt 6), 953–963. https://doi.org/10.1242/jcs.03396

Rhodes, K. J., Carroll, K. I., Sung, M. A., Doliveira, L. C., Monaghan, M. M., Burke, S. L., . . . Trimmer, J. S. (2004). KChIPs and Kv4 alpha subunits as integral components of A-type potassium channels in mammalian brain. *Journal of Neuroscience*, *24*(36), 7903–7915. https://doi.org/10.1523/JNEUROSCI.0776-04.2004

Rhodes, K. J., Strassle, B. W., Monaghan, M. M., Bekele-Arcuri, Z., Matos, M. F., & Trimmer, J. S. (1997). Association and colocalization of the Kvbeta1 and Kvbeta2 beta-subunits with Kv1 alpha-subunits in mammalian brain K⁺ channel complexes. *Journal of Neuroscience*, *17*(21), 8246–8258.

Rivera, J. F., Ahmad, S., Quick, M. W., Liman, E. R., & Arnold, D. B. (2003). An evolutionarily conserved dileucine motif in Shal K⁺ channels mediates dendritic targeting. *Nature Neuroscience*, *6*(3), 243–250. https://doi.org/10.1038/nn1020

Rudy, B., & McBain, C. J. (2001). Kv3 channels: Voltage-gated K^+ channels designed for high-frequency repetitive firing. *Trends in Neurosciences, 24*(9), 517–526.

Ruppersberg, J. P., Schröter, K. H., Sakmann, B., Stocker, M., Sewing, S., & Pongs, O. (1990). Heteromultimeric channels formed by rat brain potassium-channel proteins. *Nature, 345*(6275), 535–537. https://doi.org/10.1038/345535a0

Sagane, K., Hayakawa, K., Kai, J., Hirohashi, T., Takahashi, E., Miyamoto, N., . . . Nagasu, T. (2005). Ataxia and peripheral nerve hypomyelination in ADAM22-deficient mice. *BMC Neuroscience, 6*, 33. https://doi.org/10.1186/1471-2202-6-33

Saganich, M. J., Vega-Saenz de Miera, E., Nadal, M. S., Baker, H., Coetzee, W. A., & Rudy, B. (1999). Cloning of components of a novel subthreshold-activating $K(^+)$ channel with a unique pattern of expression in the cerebral cortex. *Journal of Neuroscience, 19*(24), 10789–10802.

Saitsu, H., Akita, T., Tohyama, J., Goldberg-Stern, H., Kobayashi, Y., Cohen, R., . . . Matsumoto, N. (2015). De novo KCNB1 mutations in infantile epilepsy inhibit repetitive neuronal firing. *Scientific Reports, 5*, 15199. https://doi.org/10.1038/srep15199

Scannevin, R. H., Wang, K., Jow, F., Megules, J., Kopsco, D. C., Edris, W., . . . Rhodes, K. J. (2004). Two N-terminal domains of Kv4 $K(^+)$ channels regulate binding to and modulation by KChIP1. *Neuron, 41*(4), 587–598.

Schönherr, R., Löber, K., & Heinemann, S. H. (2000). Inhibition of human ether à go-go potassium channels by $Ca(2^+)$/calmodulin. *EMBO Journal, 19*(13), 3263–3271. https://doi.org/10.1093/emboj/19.13.3263

Schulte, U., Thumfart, J.-O., Klöcker, N., Sailer, C. A., Bildl, W., Biniossek, M., . . . Fakler, B. (2006). The epilepsy-linked Lgi1 protein assembles into presynaptic Kv1 channels and inhibits inactivation by Kvbeta1. *Neuron, 49*(5), 697–706. https://doi.org/10.1016/j.neuron.2006.01.033

Schwake, M., Athanasiadu, D., Beimgraben, C., Blanz, J., Beck, C., Jentsch, T. J., . . . Friedrich, T. (2006). Structural determinants of M-type KCNQ (Kv7) K^+ channel assembly. *Journal of Neuroscience, 26*(14), 3757–3766. https://doi.org/10.1523/JNEUROSCI.5017-05.2006

Seagar, M., Russier, M., Caillard, O., Maulet, Y., Fronzaroli-Molinieres, L., De San Feliciano, M., . . . El Far, O. (2017). LGI1 tunes intrinsic excitability by regulating the density of axonal Kv1 channels. *Proceedings of the National Academy of Sciences of the United States of America, 114*(29), 7719–7724. https://doi.org/10.1073/pnas.1618656114

Seikel, E., & Trimmer, J. S. (2009). Convergent modulation of Kv4.2 channel alpha subunits by structurally distinct DPPX and KChIP auxiliary subunits. *Biochemistry, 48*(24), 5721–5730. https://doi.org/10.1021/bi802316m

Shibata, R., Misonou, H., Campomanes, C. R., Anderson, A. E., Schrader, L. A., Doliveira, L. C., . . . Trimmer, J. S. (2003). A fundamental role for KChIPs in determining the molecular properties and trafficking of Kv4.2 potassium channels. *Journal of Biological Chemistry, 278*(38), 36445–36454. https://doi.org/10.1074/jbc.M306142200

Simons, C., Rash, L. D., Crawford, J., Ma, L., Cristofori-Armstrong, B., Miller, D., . . . Taft, R. J. (2015). Mutations in the voltage-gated potassium channel gene KCNH1 cause Temple-Baraitser syndrome and epilepsy. *Nature Genetics, 47*(1), 73–77. https://doi.org/10.1038/ng.3153

Singh, N. A., Charlier, C., Stauffer, D., DuPont, B. R., Leach, R. J., Melis, R., . . . Leppert, M. (1998). A novel potassium channel gene, KCNQ2, is mutated in an inherited epilepsy of newborns. *Nature Genetics, 18*(1), 25–29. https://doi.org/10.1038/ng0198-25

Smart, S. L., Lopantsev, V., Zhang, C. L., Robbins, C. A., Wang, H., Chiu, S. Y., . . . Tempel, B. L. (1998). Deletion of the K(V)1.1 potassium channel causes epilepsy in mice. *Neuron, 20*(4), 809–819.

Smith, K. E., Wilkie, S. E., Tebbs-Warner, J. T., Jarvis, B. J., Gallasch, L., Stocker, M., & Hunt, D. M. (2012). Functional analysis of missense mutations in Kv8.2 causing cone dystrophy with supernormal rod electroretinogram. *Journal of Biological Chemistry, 287*(52), 43972–43983. https://doi.org/10.1074/jbc.M112.388033

Speca, D. J., Ogata, G., Mandikian, D., Bishop, H. I., Wiler, S. W., Eum, K., . . . Trimmer, J. S. (2014). Deletion of the Kv2.1 delayed rectifier potassium channel leads to neuronal and behavioral hyperexcitability. *Genes, Brain, and Behavior, 13*(4), 394–408. https://doi.org/10.1111/gbb.12120

Suh, B.-C., & Hille, B. (2002). Recovery from muscarinic modulation of M current channels requires phosphatidylinositol 4,5-bisphosphate synthesis. *Neuron, 35*(3), 507–520.

Sun, J., & MacKinnon, R. (2017). Cryo-EM structure of a KCNQ1/CaM complex reveals insights into congenital long QT syndrome. *Cell, 169*(6), 1042–1050.e9. https://doi.org/10.1016/j.cell.2017.05.019

Sun, W., Maffie, J. K., Lin, L., Petralia, R. S., Rudy, B., & Hoffman, D. A. (2011). DPP6 establishes the A-type K(+) current gradient critical for the regulation of dendritic excitability in CA1 hippocampal neurons. *Neuron, 71*(6), 1102–1115. https://doi.org/10.1016/j.neuron.2011.08.008

Tempel, B. L., Jan, Y. N., & Jan, L. Y. (1988). Cloning of a probable potassium channel gene from mouse brain. *Nature, 332*(6167), 837–839. https://doi.org/10.1038/332837a0

Tempel, B. L., Papazian, D. M., Schwarz, T. L., Jan, Y. N., & Jan, L. Y. (1987). Sequence of a probable potassium channel component encoded at *Shaker* locus of Drosophila. *Science, 237*(4816), 770–775.

Thiffault, I., Speca, D. J., Austin, D. C., Cobb, M. M., Eum, K. S., Safina, N. P., . . . Sack, J. T. (2015). A novel epileptic encephalopathy mutation in KCNB1 disrupts Kv2.1 ion selectivity, expression, and localization. *Journal of General Physiology, 146*(5), 399–410. https://doi.org/10.1085/jgp.201511444

Torkamani, A., Bersell, K., Jorge, B. S., Bjork, R. L., Friedman, J. R., Bloss, C. S., . . . Kearney, J. A. (2014). De novo KCNB1 mutations in epileptic encephalopathy. *Annals of Neurology, 76*(4), 529–540. https://doi.org/10.1002/ana.24263

Trimmer, J. S. (2015). Subcellular localization of K^+ channels in mammalian brain neurons: Remarkable precision in the midst of extraordinary complexity. *Neuron, 85*(2), 238–256. https://doi.org/10.1016/j.neuron.2014.12.042

Trudeau, M. C., Titus, S. A., Branchaw, J. L., Ganetzky, B., & Robertson, G. A. (1999). Functional analysis of a mouse brain Elk-type K^+ channel. *Journal of Neuroscience, 19*(8), 2906–2918.

Tsantoulas, C., Zhu, L., Shaifta, Y., Grist, J., Ward, J. P. T., Raouf, R., . . . McMahon, S. B. (2012). Sensory neuron downregulation of the Kv9.1 potassium channel subunit mediates neuropathic pain following nerve injury. *Journal of Neuroscience, 32*(48), 17502–17513. https://doi.org/10.1523/JNEUROSCI.3561-12.2012

Ufartes, R., Schneider, T., Mortensen, L. S., de Juan Romero, C., Hentrich, K., Knoetgen, H., . . . Stuehmer, W. (2013). Behavioural and functional characterization of Kv10.1 (Eag1) knockout mice. *Human Molecular Genetics, 22*(11), 2247–2262. https://doi.org/10.1093/hmg/ddt076

Urrego, D., Tomczak, A. P., Zahed, F., Stühmer, W., & Pardo, L. A. (2014). Potassium channels in cell cycle and cell proliferation. *Philosophical Transactions of the Royal Society of London. Series B, Biological Sciences, 369*(1638), 20130094. https://doi.org/10.1098/rstb.2013.0094

Vacher, H., Mohapatra, D. P., & Trimmer, J. S. (2008). Localization and targeting of voltage-dependent ion channels in mammalian central neurons. *Physiological Reviews, 88*(4), 1407–1447. https://doi.org/10.1152/physrev.00002.2008

Van Slyke, A. C., Rezazadeh, S., Snopkowski, M., Shi, P., Allard, C. R., & Claydon, T. W. (2010). Mutations within the S4–S5 linker alter voltage sensor constraints in hERG K^+ channels. *Biophysical Journal*, *99*(9), 2841–2852. https://doi.org/10.1016/j.bpj.2010.08.030

Wang, H. S., Pan, Z., Shi, W., Brown, B. S., Wymore, R. S., Cohen, I. S., . . . McKinnon, D. (1998). KCNQ2 and KCNQ3 potassium channel subunits: Molecular correlates of the M-channel. *Science*, *282*(5395), 1890–1893.

Wang, H., Yan, Y., Liu, Q., Huang, Y., Shen, Y., Chen, L., . . . Chai, J. (2007). Structural basis for modulation of Kv4 K^+ channels by auxiliary KChIP subunits. *Nature Neuroscience*, *10*(1), 32–39. https://doi.org/10.1038/nn1822

Wang, Q., Curran, M. E., Splawski, I., Burn, T. C., Millholland, J. M., VanRaay, T. J., . . . Keating, M. T. (1996). Positional cloning of a novel potassium channel gene: KVLQT1 mutations cause cardiac arrhythmias. *Nature Genetics*, *12*(1), 17–23. https://doi.org/10.1038/ng0196-17

Waters, M. F., Minassian, N. A., Stevanin, G., Figueroa, K. P., Bannister, J. P. A., Nolte, D., . . . Pulst, S. M. (2006). Mutations in voltage-gated potassium channel KCNC3 cause degenerative and developmental central nervous system phenotypes. *Nature Genetics*, *38*(4), 447–451. https://doi.org/10.1038/ng1758

Wen, H., & Levitan, I. B. (2002). Calmodulin is an auxiliary subunit of KCNQ2/3 potassium channels. *Journal of Neuroscience*, *22*(18), 7991–8001.

Whicher, J. R., & MacKinnon, R. (2016). Structure of the voltage-gated K^+ channel Eag1 reveals an alternative voltage sensing mechanism. *Science*, *353*(6300), 664–669. https://doi.org/10.1126/science.aaf8070

Wimmers, S., Wulfsen, I., Bauer, C. K., & Schwarz, J. R. (2001). Erg1, erg2 and erg3 K channel subunits are able to form heteromultimers. *Pflügers Archiv: European Journal of Physiology*, *441*(4), 450–455.

Wong, W., Newell, E. W., Jugloff, D. G. M., Jones, O. T., & Schlichter, L. C. (2002). Cell surface targeting and clustering interactions between heterologously expressed PSD-95 and the Shal voltage-gated potassium channel, Kv4.2. *Journal of Biological Chemistry*, *277*(23), 20423–20430. https://doi.org/10.1074/jbc.M109412200

Wu, H., Cowing, J. A., Michaelides, M., Wilkie, S. E., Jeffery, G., Jenkins, S. A., . . . Webster, A. R. (2006). Mutations in the gene KCNV2 encoding a voltage-gated potassium channel subunit cause "cone dystrophy with supernormal rod electroretinogram" in humans. *American Journal of Human Genetics*, *79*(3), 574–579. https://doi.org/10.1086/507568

Xu, M., Cao, R., Xiao, R., Zhu, M. X., & Gu, C. (2007). The axon-dendrite targeting of Kv3 (Shaw) channels is determined by a targeting motif that associates with the T1 domain and ankyrin G. *Journal of Neuroscience*, *27*(51), 14158–14170. https://doi.org/10.1523/JNEUROSCI.3675-07.2007

Xu, M., & Cooper, E. C. (2015). An ankyrin-G N-terminal gate and protein kinase CK2 dually regulate binding of voltage-gated sodium and KCNQ2/3 potassium channels. *Journal of Biological Chemistry*, *290*(27), 16619–16632. https://doi.org/10.1074/jbc.M115.638932

Yamagata, A., Miyazaki, Y., Yokoi, N., Shigematsu, H., Sato, Y., Goto-Ito, S., . . . Fukai, S. (2018). Structural basis of epilepsy-related ligand-receptor complex LGI1-ADAM22. *Nature Communications*, *9*(1), 1546. https://doi.org/10.1038/s41467-018-03947-w

Yang, Y.-M., Wang, W., Fedchyshyn, M. J., Zhou, Z., Ding, J., & Wang, L.-Y. (2014). Enhancing the fidelity of neurotransmission by activity-dependent facilitation of presynaptic potassium currents. *Nature Communications*, *5*, 4564. https://doi.org/10.1038/ncomms5564

Yus-Najera, E., Santana-Castro, I., & Villarroel, A. (2002). The identification and characterization of a noncontinuous calmodulin-binding site in noninactivating voltage-dependent

KCNQ potassium channels. *Journal of Biological Chemistry*, *277*(32), 28545–28553. https://doi.org/10.1074/jbc.M204130200

Zagha, E., Ozaita, A., Chang, S. Y., Nadal, M. S., Lin, U., Saganich, M. J., . . . Rudy, B. (2005). DPP10 modulates Kv4-mediated A-type potassium channels. *Journal of Biological Chemistry*, *280*(19), 18853–18861. https://doi.org/10.1074/jbc.M410613200

Zaydman, M. A., & Cui, J. (2014). PIP2 regulation of KCNQ channels: Biophysical and molecular mechanisms for lipid modulation of voltage-dependent gating. *Frontiers in Physiology*, *5*, 195. https://doi.org/10.3389/fphys.2014.00195

Zhang, X., Bertaso, F., Yoo, J. W., Baumgärtel, K., Clancy, S. M., Lee, V., . . . Jegla, T. (2010). Deletion of the potassium channel Kv12.2 causes hippocampal hyperexcitability and epilepsy. *Nature Neuroscience*, *13*(9), 1056–1058. https://doi.org/10.1038/nn.2610

Zhang, Y., Zhang, X.-F., Fleming, M. R., Amiri, A., El-Hassar, L., Surguchev, A. A., . . . Kaczmarek, L. K. (2016). Kv3.3 channels bind Hax-1 and Arp2/3 to assemble a stable local actin network that regulates channel gating. *Cell*, *165*(2), 434–448. https://doi.org/10.1016/j.cell.2016.02.009

Ziechner, U., Schönherr, R., Born, A.-K., Gavrilova-Ruch, O., Glaser, R. W., Malesevic, M., . . . Heinemann, S. H. (2006). Inhibition of human ether à go-go potassium channels by Ca2$^+$/calmodulin binding to the cytosolic N- and C-termini. *FEBS Journal*, *273*(5), 1074–1086. https://doi.org/10.1111/j.1742-4658.2006.05134.x

CHAPTER 5

POTASSIUM CHANNEL MUTATIONS IN EPILEPSY

ELIZABETH E. PALMER

Introduction

Epilepsy, characterized pathophysiologically by abnormal cerebral electrical activity and clinically by recurrent seizures, is a common neurological disorder with a worldwide prevalence of approximately 1% (de Boer, Mula, & Sander, 2008). Most epilepsy syndromes are treatable with antiepileptic drugs (AEDs); however, up to 30% of seizure disorders are pharmacoresistant (Kong, Ho, Ho, & Lim, 2014), and a sobering recent study showed that despite the introduction of many new AEDs with differing mechanisms of action, outcomes in newly diagnosed epilepsy have not improved over the last two decades (Chen, Brodie, Liew, & Kwan, 2018). Treatment-resistant seizures are associated with poor quality of life, physical and psychological comorbidities, and increased risk of sudden unexplained death in epilepsy (SUDEP) (Jacoby & Baker, 2008). A particularly severe type of early childhood-onset drug-resistant epilepsy associated with developmental delay/intellectual disability is developmental and epileptic encephalopathy (DEE). Patients with DEE have a high burden of care for families and the health service, a significant risk of comorbidities, and an increased risk of SUDEP, as high as 60% for some subtypes (Khan & Al Baradie, 2012).

Thus, better understanding of the underlying pathophysiology of epilepsy is required in order to more appropriately develop and target AEDs for children and adults.

The etiology of epilepsy is heterogeneous and includes acquired and genetic causes. The subset of epilepsy where a genetic diagnosis is currently most likely to be found is DEE. A recent prospective study performed over a 3-year period in Scotland has suggested a minimal incidence of early childhood-onset genetic epilepsy of 1 in 2,000 live births (Symonds et al., 2019). This would place genetic epilepsy as one of the more common types of childhood genetic disorders, for example, compared to an incidence of approximately 1 in 3,650 births for cystic fibrosis (Ahern et al., 2017) and 1 in 11,000 for spinal muscular atrophy (Sugarman et al., 2012).

Over 200 monogenic causes for epilepsy are reported, and there is significant phenotypic overlap. The ability to find a genetic cause in individual patients has been facilitated by the advent of massively parallel sequencing (MPS), which enables the interrogation of a large number of genes (MPS panel), the coding regions of the genome (exome sequencing), or the entire genome (whole-genome sequencing) (Biesecker & Green, 2014; Bowdin et al., 2016). Currently, a molecular diagnosis can be found in some subsets of epilepsy in up to 50% of individuals (Palmer et al., 2018; Symonds et al., 2019).

Many genetic epilepsies are due to rare variants in genes encoding ion channel subunits. This chapter focuses on the current literature linking variants in genes encoding the α subunits of potassium channels to epilepsy: recent excellent articles summarizing more general causes of genetic epilepsy are available (e.g., Helbig, Heinzen, Mefford, & Commission, 2018; McTague, Howell, Cross, Kurian, & Scheffer, 2016).

Determining the underlying molecular cause of epilepsy can be important for several reasons. Firstly, a genetic diagnosis allows genetic counseling for an individual affected by a genetic epilepsy and that individual's family, for example, an individual with a heterozygous variant in a gene would have a 1 in 2 chance of passing on the same genetic variant and predisposition to epilepsy to any future children. Knowledge of the inheritance pattern of a genetic epilepsy can open family planning options such as preimplantation genetic screening.

Secondly, identification of the cause of an epilepsy disorder can provide "closure" for individuals or their parents, who may have unwarranted guilt arising from misconceptions regarding etiology (Berkovic, 2015). Providing genetic diagnostic information to families has been shown to empower the parents of affected individuals by reducing feelings of isolation and improving quality of life, and there are now many individual genetic epilepsy support groups and foundations that can help link families with the same diagnosis globally (Vears, Dunn, Wake, & Scheffer, 2015; Walters, Wells-Kilpatrick, & Pandeleos, 2014).

Thirdly, and a focus of this chapter, is that knowing the molecular cause might directly influence clinical management. Knowledge of the natural history of certain genetic epilepsies can guide overall management approaches. For example, *KCNQ2*-related seizures typically abate in mid-childhood to early adolescence, allowing a trial of weaning off AEDs. Appropriate health surveillance can also be directed by knowledge of expected comorbidities with specific genetic epilepsies; the identification of variants in ion channel genes, including those of the potassium channels (Goldman, 2015), may result in aggressive treatment of nocturnal generalized seizures, and direct parental education aimed at reducing the chance of SUDEP, a complication particularly linked to certain ion channelopathies, and identification of a variant in TSC1 or TSC2 should prompt surveillance for cerebral tumors and renal, cardiac, and pulmonary complications that can be associated with these genetic conditions. For a, albeit currently relatively limited, number of genetic epilepsies, understanding the molecular cause influences the choice of AED (Symonds et al., 2019). For example, sodium channel blockers are recommended to be avoided for *SCN1A*-related epilepsies as they have been demonstrated to worsen seizure control (Guerrini et al., 1998; Perucca et al., 2017) but should be trialed for patients with *SCN8A* gain-of-function mutations (Møller & Johannesen, 2016).

However, the quality of clinical evidence linking specific genetic causes of epilepsy to targeted AEDs is often suboptimal due to many factors, including incomplete understanding of the underlying pathophysiology for each specific genetic variant, limited availability of targeted therapies, difficulties in monitoring clinical outcomes, small patient cohorts, and lack of funding opportunities and infrastructure for clinical trials. An example of challenges moving from a molecular diagnosis to good evidence for targeted therapies is discussed in more detail in this chapter with regard to the evidence for and against quinidine for *KCNT1* gain-of-function variants. A sure molecular diagnosis will be required to allow epilepsy patients to enter future clinical trials, which may include not only channel modifiers but genetic therapies, such as antisense oligonucleotides aimed at reducing the expression of mutant channels (Petrou et al., 2018). Truty et al. (2019) reported a cohort of over 9,000 patients referred for epilepsy gene panel testing and reported that 33% of the 1,502 patients with a molecular diagnosis had a clinically actionable variant and 25% were eligible to enroll in a clinical trial.

Thus, there has never been a more pressing need for combined basic and translational research into understanding how variants in potassium and other ion channel genes result in epilepsy and the optimal targeted therapies for each molecular cause so that a true precision medicine approach can be applied to epilepsy management. Such a precision medicine approach should result in more cost-effective therapies that improve health and developmental outcomes.

Overview of Potassium Channels Related to Epilepsy

Potassium channels are ancient and thought likely to exist in all biological cells. Their roles are diverse, including direct control of neuronal excitability as well as roles in ion regulation and metabolism (Kohling & Wolfart, 2016).

Potassium channels consist of four α subunits, which cluster together within a cell membrane to form a small water-filled hole, the pore, through which potassium ions can flow between the inside (where K^+ concentration is high) and the outside (where K^+ concentration is low) of the cell. The α subunits are subdivided into protein families. Each α subunit includes at least two transmembrane (TM) helices and a short loop between them called the pore forming (P) loop (Choe, 2002; Kuang, Purhonen, & Hebert, 2015).

The different α-subunit families differ in the number of TM and P domains. The largest is the six transmembrane (6TM) family, which includes α subunits with six TM domains and one P loop between the two most carboxy-terminal helices (Figure 5.1A). This family includes the voltage-gated channels (α subunits designated K_v) and the calcium and sodium gated channels (α subunits designated K_{Ca} or K_{Na}). K_v channels are activated by depolarization and deactivated by repolarization, relatively quickly. They are understood to be mainly located in axons and play important roles in the regulation of the shape and delay of axonal action potentials (Kohling & Wolfart, 2016). K_{Ca} channels have roles in mediating different parts of the action potential after hyperpolarization. K_{Na} channels are regulated by both sodium and membrane potential and have roles in the control of cellular excitability as well as possibly more fundamental roles in brain development and function.

The second family is the four transmembrane (4TM) family (Figure 5.1B), which consists of the inward rectifier potassium channels (K_{ir}) that conduct inward current better than outward current and, under physiological conditions, contribute to the

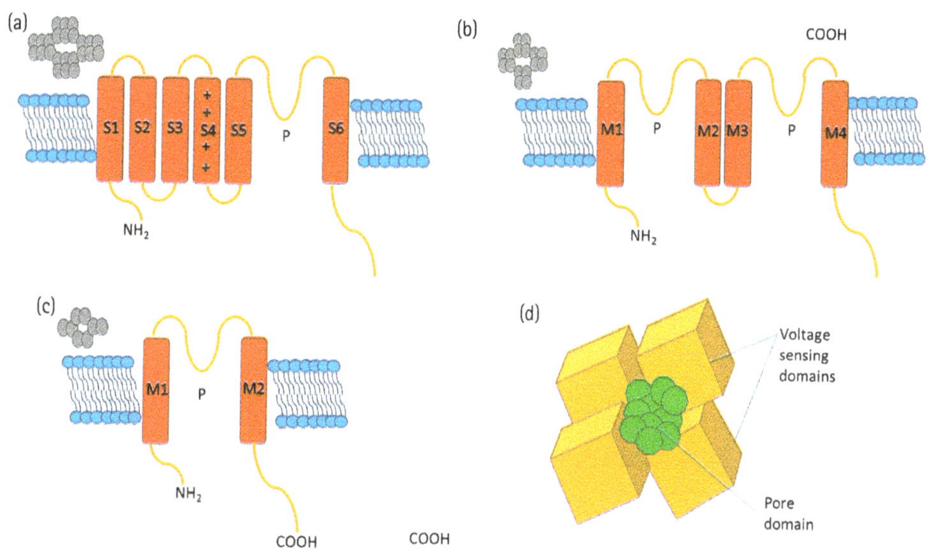

FIGURE 5.1 Structure of the four main classes of potassium channels. (A) 6TM/P channels consist of six transmembrane (TM) helices (S1–S6) with a pore forming (P) loop between S5 and S6. These channels include voltage-gated K^+ channels and calcium- and sodium-gated K^+ channels. Voltage-gated K^+ channel α subunits include four TM helices (S1–S4) preceding the 2TM/P structure to allow the channel to sense and respond to changes in membrane potential (designated by the plus signs in S4). (B) 4TM/2P channels consist of four TM helices (M1–M4), with two P loops between M1/M2 and M3/M4. (C) 2TM/P channels consist of two TM helices (M1 and M2) with a P loop between them, exemplified by inwardly rectifying K^+ channels. Voltage-sensitive potassium α subunits form a tetrameric complex with the voltage-sensitive domains around a central pore region. The aqueous pore is surrounded by the pore loop, which includes a signature sequence (amino acids [(T/SxxTxGxG]) and the most carboxy-terminal TM helix (S6). These regions thus carry the structural determinants of the high potassium selectivity of the channel (Choe, 2002).

basic potassium current that maintains the resting membrane potential (Kohling & Wolfart, 2016). $K_{ir}4.1$ channels are found exclusively in glial cells. $K_{ir}6$ channels are coupled to the intracellular energy support: $K_{ir}6$ channels open when adenosine triphosphate (ATP) levels are low, for example, after prolonged action potential firing (Higashi et al., 2001).

The third α-subunit family is the two transmembrane (2TM) family (Figure 5.1C). Members of this family are also understood to contribute to current leak, important in maintaining the resting membrane potential (Feliciangeli, Chatelain, Bichet, & Lesage, 2015).

Potassium channels can be made up of different combinations of subunits within families (homo- or heterotetramers), which can be associated with a range of β subunits. Different combinations of subunits vary in how they turn on (gate) or turn off (inactivate) in response to external signals or according to in-built mechanisms (Choe, 2002; Kohling & Wolfart, 2016; Kuang et al., 2015). This huge variety of possible channel structures and functions and differences in their expression across different subcellular compartments of the neuron, brain regions, and circuits allow potassium channels to fine-tune electrical activity of the brain in myriad ways. For example, potassium channels can regulate the resting membrane potential, control the repolarization rate of action potentials, control spike frequency adaptation, and regulate membrane resistance (Jan & Jan, 2012). Moreover, potassium channels can have more indirect roles in regulation of ion milieu or metabolism.

This chapter focuses on variants described within the 20 genes currently known to encode for α potassium channel subunits which, at the time of publication, have been associated with a neurological phenotype with an epilepsy component. It is important to recognize that new associations between potassium channel encoding genes and epilepsy continue to be made, and it is anticipated that many more "potassium channelopathies" will be described in the coming years. Table 5.1 shows that there are at least 23 brain-expressed potassium subunit encoding genes not yet ascribed to a human neurological function: variants in these genes may well be described linked to epilepsy phenotypes in the future.

Potassium Channelopathies in Epilepsy

This section reviews the current literature linking variants in α potassium channel subunits to epilepsy, highlighting what is known regarding the genotypic and phenotypic spectrum of each channelopathy, our understanding of the pathophysiology underlying these genetic conditions, the clinical implications of each diagnosis, and the evidence for targeted therapeutics.

Table 5.1 Genes Encoding α Potassium Channel Subunits, Expression in Humans, and Information on Any Known Phenotype in Humans with Online Mendelian Inheritance in Man (https://omim.org) MIM Number or PubMed IDENTIFICATION NUMBER (PMID) if no MIM Reference

K^+ Channel Family	Gene Encoding Subunit	Protein Subunit	Expression in Humans	Human Epilepsy Phenotype Described by 2020? (OMIM or PMID reference)
Voltage-gated K channels	*KCNA1*	$K_v1.1$	Brain	Yes (MIM 160120)
	KCNA2	$K_v1.2$	Brain	Yes (MIM 616366)
	KCNA3	$K_v1.3$	Brain	No
	KCNA4	$K_v1.4$	Brain	Neurological phenotype (MIM 618284)
	KCNA5	$K_v1.5$	Heart	
	KCNA6	$K_v1.6$	Colon, soft tissue	
	KCNA7	$K_v1.7$	Skeletal muscle	Yes (MIM 616056)
	KCNA8	$K_v1.8$	Unknown	Yes (MIM 616187)
	KCNB1	$K_v2.1$	Brain	Neurological phenotype (MIM 605259)
	KCNB2	$K_v2.2$	Brain	
	KCNC1	$K_v3.1$	Brain	
	KCNC2	$K_v3.2$	Brain	Possible; PMID: 24501278
	KCNC3	$K_v3.3$	Brain	
	KCNC4	$K_v3.4$	Brain, muscle	Neurological phenotype (MIM 607346)
	KCND1	$K_v4.1$	Brain	
	KCND2	$K_v4.2$	Brain, heart	
	KCND3	$K_v4.3$	Brain, heart	Yes (MIM 613720; 121200)
	KCNF1	$K_v5.1$	Brain	
	KCNG1	$K_v6.1$	Brain, multiorgan	Yes (MIM 121201)
	KCNG2	$K_v6.2$	Brain, heart	Yes (MIM 617601)
	KCNG3	$K_v6.3$	Brain	Yes (MIM 611816; 135500)
	KCNG4	$K_v6.4$	Testis	
	KCNQ1	$K_v7.1$	Multiorgan	Possible epilepsy (MIM 605716)
	KCNQ2	$K_v7.2$	Brain	
	KCNQ3	$K_v7.3$	Heart	
	KCNQ4	$K_v7.4$	Brain, ear, heart	
	KCNQ5	$K_v7.5$	Brain, muscle	
	KCNV1	$K_v8.1$	Brain	
	KCNV2	$K_v8.2$	Pancreas	
	KCNS1	$K_v9.1$	Brain	
	KCNS2	$K_v9.2$	Brain	
	KCNS3	$K_v9.3$	Heart	
	KCNH1	$K_v10.1$	Brain	
	KCNH5	$K_v10.2$	Brain	
	KCNH2	$K_v10.2$	Heart, brain	
	KCNH6	$K_v11.1$	Brain (low), glands	
	KCNH7	$K_v11.2$	Brain	
	KCNH3	$K_v12.2$	Brain	
	KCNH8	$K_v12.3$	Brain	
	KCNH4	$K_v12.4$	Brain	

(*continued*)

Table 5.1 Continued

K^+ Channel Family	Gene Encoding Subunit	Protein Subunit	Expression in Humans	Human Epilepsy Phenotype Described by 2020? (OMIM or PMID reference)
Ca- and Na-activated potassium channels	*KCNMA1*	$K_{Ca}1.1$	Brain	Yes (MIM 617643; 609446; 618596)
	KCNN1	$K_{Ca}2.1$	Brain	
	KCNN2	$K_{Ca}2.2$	Brain, glands	Yes (MIM 618658)
	KCNN3	$K_{Ca}2.3$	Breast, renal, brain	Yes (MIM 615005; 614959)
	KCNN4	$K_{Ca}3.1$	Blood, muscle	
	KCNT1	$K_{Na}1.1$	Brain	Yes (MIM 617771)
	KCNT2	$K_{Na}1.2$	Brain	
	KCNU1	$K_{Ca}5.1$	Testis only	
Inwardly rectifying potassium channels	*KCNJ1*	$K_{ir}1.1$	Heart	Yes (MIM 618381)
	KCNJ2	$K_{ir}2.1$	Heart	Yes (MIM 606176)
	KCNJ3	$K_{ir}3.1$	Brain, heart	
	KCNJ4	$K_{ir}2.3$	Heart	
	KCNJ5	$K_{ir}3.4$	Brain, inner ear, heart	
	KCNJ6	$K_{ir}3.2$		
	KCNJ8	$K_{ir}6.1$	Brain	
	KCNJ9	$K_{ir}3.3$	Heart	
	KCNJ10	$K_{ir}4.1$	Heart	
	KCNJ11	$K_{ir}6.2$		
	KCNJ12	$K_{ir}2.2$		
	KCNJ13	$K_{ir}7.1$		
	KCNJ14	$K_{ir}2.4$		
	KCNJ15	$K_{ir}4.2$		
	KCNJ16	$K_{ir}5.1$		
Two P domain potassium channels	*KCNK1*	$K_{2P}1.1$	Brain	Yes (MIM 618381)
	KCNK2	$K_{2P}2.1$		
	KCNK3	$K_{2P}3.1$		
	KCNK4	$K_{2P}4.1$		
	KCNK5	$K_{2P}5.1$		
	KCNK6	$K_{2P}6.1$		
	KCNK7	$K_{2P}7.1$		
	KCNK9	$K_{2P}9.1$		
	KCNK10	$K_{2P}10.1$		
	KCNK12	$K_{2P}12.1$		
	KCNK13	$K_{2P}13.1$		
	KCNK15	$K_{2P}15.1$		
	KCNK16	$K_{2P}16.1$		
	KCNK17	$K_{2P}17.1$		
	KCNK18	$K_{2P}18.1$		

Sources: Protein nomenclature from Aldrich et al. (2019). Gene expression and OMIM data from MARRVEL (http://marrvel.org/), accessed January 2020 (Wang et al., 2017).

Six Transmembrane Domain Family: Voltage-Gated Potassium Channels

KCNA1 Encoding Kv1.1

Expression and Function in Human Brain

$K_v1.1$ channels are widely expressed including in the central and peripheral nervous systems and skeletal muscles. They have roles in controlling the excitability of neurons within the cerebellum, hippocampus, cortex, and peripheral nervous system (D'Adamo et al., 2014). The $K_v1.1$ homomeric channel consists of four identical subunits encoded by *KCNA1*, but $K_v1.1$ subunits can also join with α subunits of other members of the K_v1 family, most commonly $K_v1.2$ subunits, to form heterotetrameric channels. $K_v1.1$:$K_v1.2$ channels are expressed in the cerebellar basket cell terminals and at the juxtaparanodal region of motor axons (Hasan & D'Adamo, 2010). The electrophysiological properties of $K_v1.2$ containing channels can be further varied by association with diverse β subunits.

Reported Human Neurological Conditions, Associated Variants, and Insights into Pathophysiology

A wide range of missense variants in KCNA1 affecting the amino acid structure in diverse functional domains of the Kv1.1 subunit (Figure 5.2), as well as a smaller number of nonsense variants and small deletions, have been reported in affected individuals with the phenotype of episodic ataxia type 1 (EA1) (MIM 160120)[1] with our without epilepsy (Hasan & D'Adamo, 2010). EA1 is characterized clinically by dramatic episodes of spastic contractions of the skeletal muscles of the head and limbs, with loss of motor coordination and balance, muscle cramps, and stiffness and myokymia (muscle twitching with a rippling appearance). Symptoms of EA1 can start in childhood or adolescence, and often there is a positive family history with an affected parent from whom the *KCNA1* variant is inherited. Affected individuals have a higher chance of epilepsy but may also remain seizure-free. EA1 is associated with variable specific learning difficulties or intellectual disability and other movement disorders including choreoathetosis and carpal spasm. Electrophysiological assessments can be diagnostic. The severity and frequency of EA1 attacks and comorbidities can vary even between affected individuals in the same family.

A distinctive phenotype including epilepsy, infantile contractures, postural abnormalities, and skeletal deformities is reported in association with the p.(Thr226Arg) variant, affecting the amino acid structure in the S2 helix (Zuberi et al., 1999) (Figure 5.2).

A more severe epilepsy phenotype, associated with severe intellectual disability and a variable episodic ataxia and/or tremor, has been described in association with de novo *KCNA1* pathogenic variants, particularly those altering the amino acid structure in the proline–valine–proline (PVP) motif at the base of S6 (Rogers et al., 2018) (Figure 5.2).

In general, most studies have shown that EA1-related variants impair channel function and reduce the outward flux of potassium through the channel, with highly

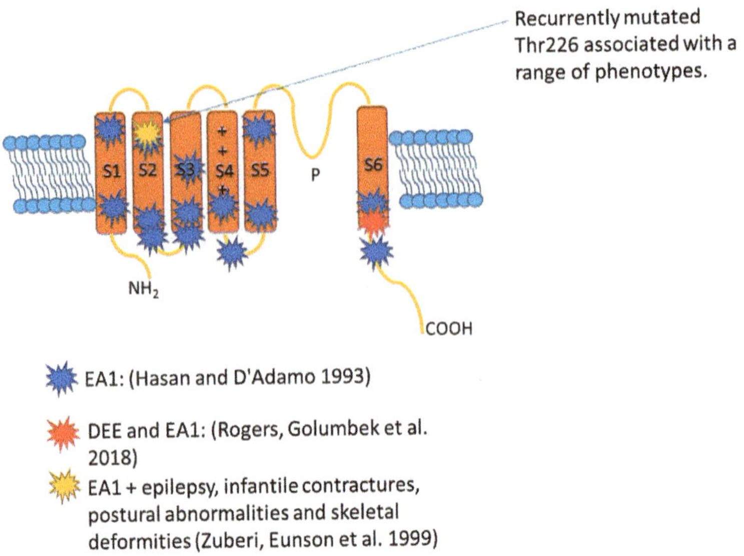

FIGURE 5.2 Position of affected amino acid residues altered by rare *KCNA1* missense variants associated with neurological phenotypes. The red star indicates an altered residue linked to DEE and EA1, blue stars indicate altered residues linked to EA1, and the yellow star is the altered residue described by Zuberi and colleagues (1999) linked to a more complex systemic condition.

variable effects on aspects of channel expression and gating (D'Adamo et al., 2014; Hasan & D'Adamo, 2010; Zuberi et al., 1999). As variants are heterozygous, channels may be formed which include both mutated and wild-type subunits, and coexpression systems have shown that some mutated subunits exert dominant negative effects on wild-type subunits, reducing surface expression and current amplitude, whereas other subunits do not seem to affect expression. Moreover, pathogenic *KCNA1* variants can disrupt the function of other closely related proteins, due to effects on the ability of K_v1.1 to associate with other K_v1 α subunits and diverse β subunits (D'Adamo, Imbrici, Sponcichetti, & Pessia, 1999). For example, D'Adamo and colleagues (1999) suggest that abnormal potassium channels including mutant K_v1.1 and wild-type K_v1.2 α subunits in the presynaptic basket cells in the cerebellum may increase membrane excitability, leading to prolonged action potential duration and enhanced calcium ion influx. This could lead to an increased γ-aminobutyric acid (GABA) release from basket cell terminals, reducing the inhibitory outputs of the relevant Purkinje cells and thus disrupting the complex cerebellum circuitry. This work highlights how mutations in certain α subunits can have profound functional impacts on diverse potassium channels, altering the balance of excitability and neurotransmission in the brain in complex manners.

Clinical Management Implications of Channelopathy

The carbonic anhydrase inhibitor acetazolamide has been shown to reduce the frequency and severity of episodic ataxia in some, but not all, affected individuals and to reduce seizure frequency in some patients with a DEE phenotype (Hasan & D'Adamo, 2010; Rogers et al., 2018). A range of AEDs have been reported to have positive effects in some affected individuals, including phenytoin, carbamazepine, and lamotrigine. In addition, certain behavioral modifications, aimed at reducing stress, abrupt movements, loud noises, and the use of caffeine, have been shown to reduce symptoms. Females often have exacerbation of symptoms in association with pregnancy. Caution should be taken with anesthesia as severe myokymia has been reported with anesthesia induction (Hasan & D'Adamo, 2010; Rogers et al., 2018).

KCNA2 Encoding $K_v1.2$

Expression and Function in Human Brain

KCNA2 is widely expressed in the human brain in a broad range of both excitatory and inhibitory neurons, and $K_v1.2$ containing channels play an important role in enabling efficient neuronal repolarization after an action potential (Lorincz & Nusser, 2008).

Reported Human Neurological Conditions, Associated Variants, and Insights into Pathophysiology

Rare de novo missense variants in *KCNA2* were reported in association with a severe developmental and epileptic encephalopathy (MIM 616366) (Syrbe et al., 2015) (Figure 5.3). Some variants resulted in a constitutionally open channel (gain of function) and other reduced current amplifications (loss of function) (Figure 5.3). The variants causing a loss of function were associated with infantile or early childhood seizure onset and frequent febrile and afebrile focal motor and dyscognitive seizures. Seizures resolved by the age of 15 years, but a mild to moderate intellectual disability persisted, in association with variable ataxia and myoclonus.

The variants causing dominant gain-of-function effects were associated with more severe clinical syndromes: with moderate intellectual disability, more severe ataxia, and persistent seizures with generalized epileptic discharges on electroencephalography (EEG) (Syrbe et al., 2015).

Corbett and colleagues (2016) also described a family with an inherited rare sequence variant resulting in a three–amino acid deletion within the S3 domain of $K_v1.2$ associated with variable neurocognitive phenotypes including episodic ataxia, self-limited seizures, generalized epilepsy, focal seizures, or a more severe DEE phenotype in the youngest family member. In a *Xenopus* model this deletion resulted in reduced current amplitudes consistent with a loss-of-function effect.

Thus, de novo and inherited amino acid altering variants in *KCNA2* have been reported in association with variable epilepsy and intellectual disability, frequently associated with episodic ataxia, showing some similarity to the phenotypic range seen in *KCNA1*-related encephalopathies. It is particularly striking that recurrent de novo missense variants in both *KCNA1* and *KCNA2* that alter the structure of the highly conserved PVP motif at the C terminal end of the S6 region, thought to link the gate to the voltage sensor (Long, Campbell, & Mackinnon, 2005), are reported in association

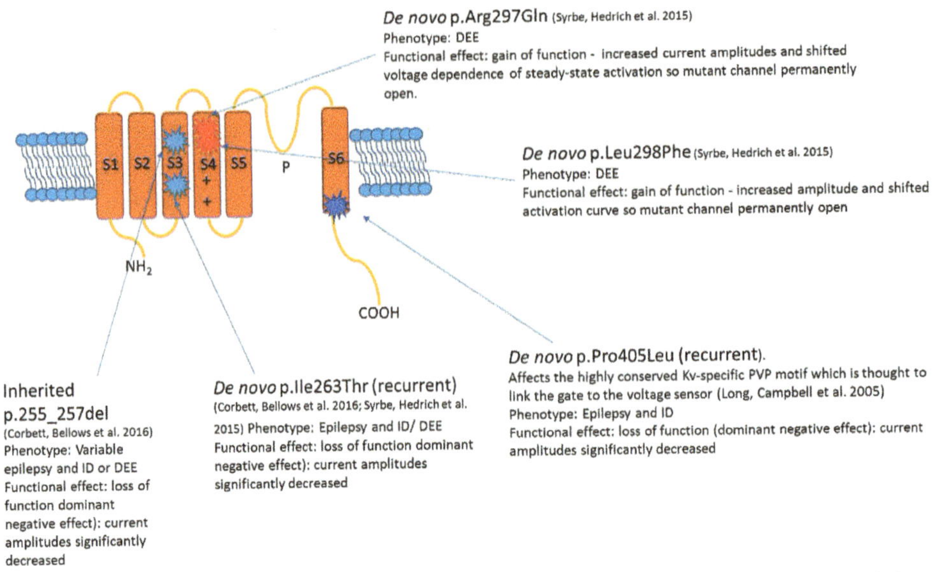

FIGURE 5.3 Range of rare *KCNA2* variants associated with neurological phenotypes and their functional effect in cellular models.

with overlapping intellectual disability/epilepsy and episodic ataxia phenotype (Rogers et al., 2018; Syrbe et al., 2015) (Figure 5.3).

Clinical Management Implications of Channelopathy

To date, no specific therapeutic agents have been recommended for *KCNA2*-related encephalopathy. As for all channelopathies, care should be taken to note the pathophysiological effect of individual patient variants as likely different agents would be most efficacious in variants associated with dominant negative loss of function or gain of function.

KCNA4 Encoding Kv1.4

Expression and Function in Human Brain

$K_v1.4$ likely contributes to the fast repolarizing phase of action potentials and has been found in striatal neurons. It also has high expression in a variety of brain regions and the outer retina.

Reported Human Neurological Conditions, Associated Variants, and Insights into Pathophysiology

A homozygous missense variant with the NH2-terminal domain of $K_v1.4$ (NM_ 002233.4: c.266G>C; p.[Arg89Gln]) segregated with a novel autosomal recessive neurological and eye condition (MIM 618284) syndrome in one family. Clinical features include congenital cataract, intellectual disability, and attention deficit disorder. On magnetic resonance imaging (MRI), affected individuals had striatal thinning. None of

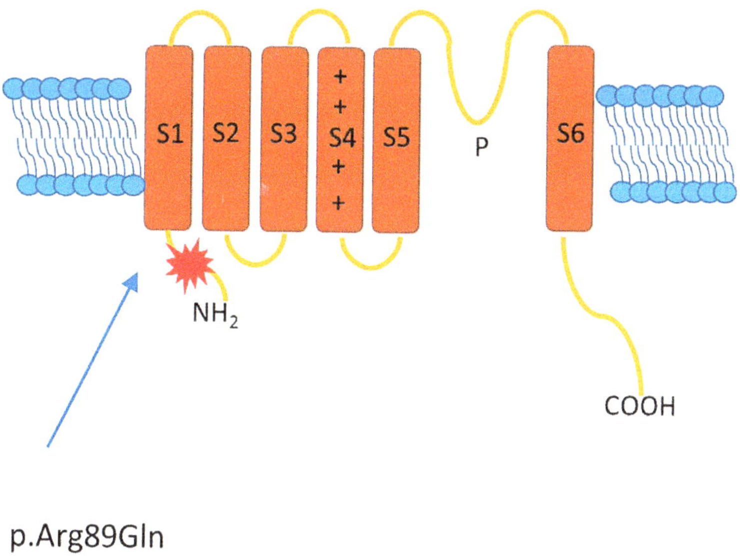

FIGURE 5.4 Location of the substituted amino acid in the N-terminal region of $K_v1.4$, which may affect the ability of the channel to localize correctly in the membrane.

the affected individuals had seizures. The affected residue lay within the NH2-terminal domain, which is thought to be responsible for binding of the $K_v1.4$ subunit to actinin, which may result in decreased efficiency of the channel's activity (Figure 5.4). The variant was shown to have a functional effect, reducing current amplitude in injected *Xenopus* oocytes (Kaya et al., 2016).

KCNB1 Encoding $K_v2.1$

Expression and Function in Human Brain

$K_v2.1$ is variably expressed in cortical layers and pyramidal cells in the brain, in a pattern distinct from $K_v2.2$ (Bishop et al., 2015). The protein has four known domains. The T1 domain is involved in combining with other $K_v2.1$ proteins or other voltage-gated potassium channel proteins (such as $K_v3.1$, K_v6, and K_v9 members) to form homotetramers or heterotetramers. The TM domain contains six elements: S1–S4 form the voltage sensor domain, and S5 and S6 form the pore domain. The extracellular loop between S5 and S6 contains the potassium selectivity filter.

$K_v2.1$ containing channels conduct delayed rectifier potassium current, which modulates membrane repolarization of electrically excitable cells (Murakoshi & Trimmer, 1999).

Reported Human Neurological Conditions, Associated Variants, and Insights into Pathophysiology (MIM 616056)

De novo missense variants in *KCNB1* were first reported by Torkamani and colleagues (2014) in association with early-onset epileptic encephalopathy. These variants resulted in functional studies in a cellular model with loss of ion selectivity and gain of depolarizing inward cation conductance. Since then over 26 patients have been described with (likely) pathogenic de novo variants in the gene, summarized by de Kovel and colleagues (2017) and Kang and colleagues (2019). Types of variants include missense and putative loss of function (nonsense or frameshift variants which lead to a premature termination codon). All missense variants (see Figure 5.5) lie within the 6TM regions and their extracellular and intracellular connecting loops—with a concentration in the voltage sensor S4 and the selectivity filter between S5 and S6. Loss-of-function variants are more spread throughout the gene, including in DNA encoding the C-terminal domain.

Kang and colleagues (2019) developed a high-throughput functional assay which allowed them to look at the biophysical properties and cell-surface expression of variant $K_v2.1$ channels expressed in heterologous cells. They reported that all variants (missense and two loss-of-function variants) resulted in channel hypofunction, with molecular mechanisms underlying the functional phenotype including 1) decreased potassium conductance, 2) altered voltage dependence, 3) reduced protein expression, or 4) altered cell-surface expression. The functional impact of some was rescued by coexpression with wild-type subunit, whereas others variants resulted in a worsening of functional effect in this situation, suggesting that they may have a dominant negative effect.

De Kovel and colleagues looked at any correlations between genotype and phenotype. All reported patients with a de novo missense or loss-of-function variant had developmental delay (typically severe), and intractable epilepsy was very common, affecting 84% of the patient cohort. The only individuals without epilepsy had loss-of-function variants in the C-terminal region of the gene or (one patient) a missense variant in the S1–S2 linker loop. In general, patients with more severe epilepsy had a missense variant in the voltage sensor (S4), particularly affecting the arginine residues and the P domains (S5, P loop, and S6). However, both they and Kang and colleagues note that epilepsy severity can be very variable, even for patients with the same recurrent variant, so factors other than *KCNB2* genotype are likely to influence overall clinical presentation.

For those with epilepsy, seizures typically started around 12 months of age, with an age range between 3 and 18 months. Epileptic spasms were a common first epilepsy presentation in 36%. Most patients went on to develop a range of seizure types, including tonic, focal-clonic, myoclonic, and atypical absences. EEGs were typically highly abnormal. Examples of EEG presentations include hypsarrhythmia, or general slowing of background activity with multifocal spikes and polyspikes. A quarter had a very distinctive EEG pattern of continuous spike and wave during slow-wave sleep or electrical status epilepticus during slow-wave sleep.

Autism was a common comorbidity, reported in 50%. Nearly half had hypotonia, and ataxia was reported in 17%. MRI brain scans, when conducted, were typically normal or showed nonspecific features such as supratentorial or cerebellar volume loss and white matter abnormalities.

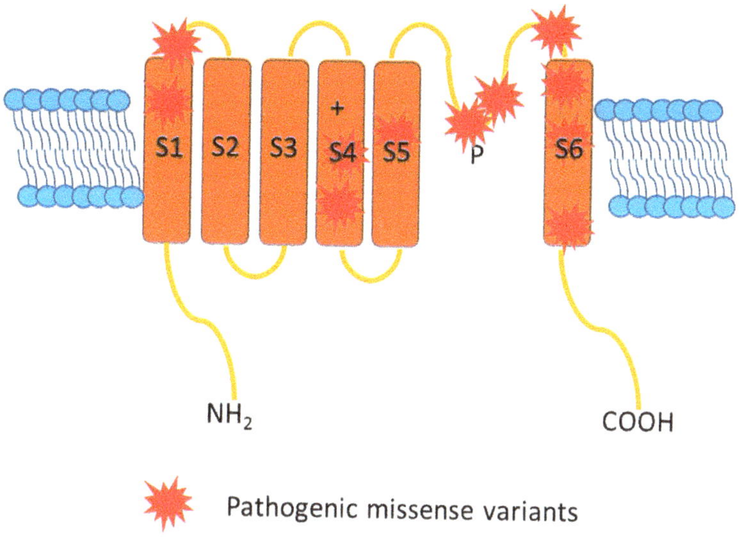

FIGURE 5.5 Location of (likely) pathogenic missense variants in *KCNB1*.

Clinical Management Implications of Channelopathy

Unfortunately, most patients with a *KCNB1*-related encephalopathy have a severe epilepsy phenotype refractory to multiple antiepileptic medications. de Kovel and colleagues (2017) note, anecdotally, a partial efficacy of the ketogenic diet; however, this was not effective in all patients. Kang and colleagues (2019) note that for those affected individuals whose variant is associated with a reduced cell-surface expression of *KCNB1*, molecular chaperones which could increase cell-surface expression might be a good therapeutic option. Conversely, for those whose variant results in normal expression but reduced function of the channel, subtype-selective activators of $K_v2.1$ might be a viable future therapeutic option. Therapeutic approaches that result in increased expression of wild-type channel subunits might also hold promise.

KCNC1 Encoding $K_v3.1$

Expression and Function in Human Brain

KCNC1 encodes $K_V3.1$, which functions as a highly conserved potassium ion channel α subunit of the K_V3 subfamily of voltage-gated tetrameric potassium ion channels. K_V3 channel subunits consist of six membrane-spanning segments (S1–S6) and can form heterotetramers with other members of the family (Aggarwal & MacKinnon, 1996; Seoh, Sigg, Papazian, & Bezanilla, 1996). $K_v3.1$ expression is limited to the central nervous system and a subpopulation of T lymphocytes. In the brain it is preferentially expressed in subsets of fast-spiking neurons, especially in inhibitory GABAergic interneurons (Gan & Kaczmarek, 1998).

K_V3 channels have more positively shifted voltage-dependent activation and faster activation and deactivation rates than other K_v channels and are thought to regulate high-frequency firing in several types of neurons within the brain and neurotransmitter release (Rudy & McBain, 2001).

Reported Human Neurological Conditions, Associated Variants, and Insights into Pathophysiology

A range of human phenotypes have been described in association with variants in *KCNC1*. Not all of these have been reported to include epilepsy.

1. *Myoclonus epilepsy and ataxia due to potassium channel mutation (MEAK) (MIM 616187)*. The first pathogenic *KCNC1* variant reported in humans was the recurrent variant p.(Arg320His) causal of a progressive myoclonus phenotype. The phenotype was characterized by myoclonus, often with tremor, typically starting in a previously normally developing child, between the ages of 6 and 14 years. The myoclonus worsened progressively, and many children became wheelchair-dependent for mobility by adolescence. Patients typically had infrequent tonic-clonic seizures and progressive but relatively mild cognitive decline (Muona et al., 2015). The variant affects the critical arginine residues in the S4 TM domain, which acts as a voltage sensor. In a *Xenopus laevis* oocyte expression system, p.(Arg320His) was shown to result in channel hypofunction, with a reduction in current amplitude and a hyperpolarizing shift in the activation curve. The effect was dominant negative.

2. *Developmental and epileptic encephalopathy*. Subsequently, Cameron and colleagues (2019) identified a separate phenotype, characterized by infantile-onset seizures (frequently myoclonus), with moderate to severe intellectual disability and variable ataxia and dysmorphism. All patients had a recurrent de novo missense variant p.(Ala421Val), which affects a residue in the P S6 domain. Seizure control was very variable: some achieved seizure control with polypharmacy, while others remained refractory to treatments. This variant was also reported by Park and colleagues (2019) with the same phenotype. The functional effect of this variant differed between the two groups, with Cameron and colleagues finding no dominant negative effect but Park and colleagues reporting a dominant negative effect.

3. *Intellectual disability without epilepsy*. Cameron and colleagues (2019) described three patients with mild to moderate intellectual disability and no seizures: one had a de novo missense variant p.(Arg317His), affecting the S4 voltage sensor domain, whereas a mother–son pair had an inherited putative loss-of-function variant (p.[Gln492*]). Poirier and colleagues (2017) also described a loss-of-function variant (p.[Arg339*]) segregating with intellectual disability in a family.

4. *Nonprogressive myoclonus without epilepsy or intellectual disability*. This phenotype was associated with a de novo missense variant in the S1 domain (p.[Cys208Tyr]) (Park et al., 2019).

Clinical Management Implications of Channelopathy

For MEAK, it was noted that several patients had transient clinical improvement with febrile illnesses, and in vitro experiments supported an increased availability of wild-type channels with fever (Oliver et al., 2017). It is hard to extrapolate this phenomenon to a safe therapeutic option. The most effective AED combinations for MEAK

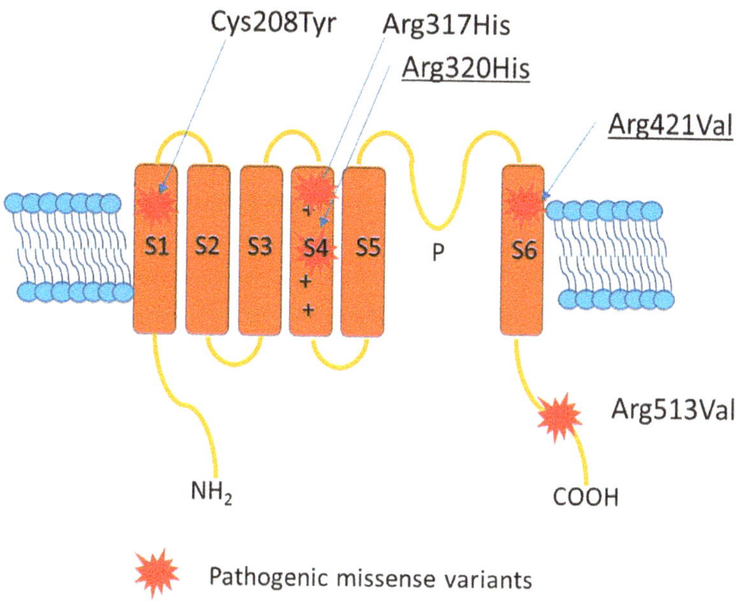

FIGURE 5.6 Location of amino acid residues altered in $K_V3.1$ by pathogenic missense variants in *KCNC1*. The recurrent Arg320His associated with MEAK and the recurrent Arg421Val associated with a DEE are underlined.

included valproate, zonisamide, clonazepam, and levetiracetam, whereas lamotrigine appeared to exacerbate the myoclonus (Oliver et al., 2017). Recently, small molecular K_v3 modulators, such as REO1, have been shown to enhance the open probability of channels with the recurrent MEAK variant p.(Arg320His) (Figure 5.6). Such modulators may represent a therapeutic option for *KCNC1*-related epilepsy where there is a loss of function of the channel (Munch et al., 2018).

KCNC3 Encoding $K_v3.3$

KCNC3 is expressed at high levels in cerebellar Purkinje cells, auditory brainstem nuclei, and other neurons capable of firing at high rates. Different missense variants in *KCNC3*, as well as C-terminal proline deletions, have been reported in association with the autosomal dominant progressive neurological condition spinocerebellar ataxia-13 (MIM 605259) that, to date, does not include seizures (Zhang & Kaczmarek, 2016).

KCND2 Encoding $K_v4.2$ Is a Candidate Gene for an Epilepsy Phenotype

Expression and Function in Human Brain

KCND2 is expressed in the human brain, with particularly high expression in the cerebellum (Serodio & Rudy, 1998). It is thought to mediate an A-type/transient current and limits low-frequency firing and backpropagation of the action potential.

Reported Human Neurological Conditions, Associated Variants, and Insights into Pathophysiology

One family has been reported in which a pair of monozygotic twin girls affected by infantile-onset refractory severe epilepsy (hundreds of myoclonic seizures per day), intellectual disability, and autism were found to have a de novo variant in *KCND2* NM_012281.2:c.1210G>A; p.(Val404Met). The affected residue was at a highly conserved residue in the C-terminal end of the S6 helix (Figure 5.7). This residue is predicted to be part of the ion permeation pathway. In vitro functional expression studies in *Xenopus* oocytes showed that the p.(Val404Met) mutant protein reached peak amplitude significantly later than wild type, and the decay of the current was significantly slower and less complete, owing to impaired closed-state inactivation of the potassium channel. Thus, this variant may result in a gain of function of the channel.

A frameshift variant (N587fs*1) had also been reported in association with a temporal lobe epilepsy (TLE) in one individual. The deleted region would include three critical extracellular signal–regulated kinase phosphorylation sites (Singh et al., 2006). Although functional studies suggest a loss of function, the variant was also present in the patient's asymptomatic father and cannot be confirmed as a definite pathogenic variant, especially as various more proximal nonsense and frameshift variants are now reported in healthy individuals in the gnomAD database who do not have a neurological phenotype.[2] This variant may increase susceptibility to TLE: larger association studies will be required to confirm this.

KCND3 Encoding $K_v4.3$

Expression and Function in Human Brain

KCND3 encodes the α subunit $K_v4.3$, which has similar functions to other members of the K_v4 family in that $K_v4.3$ containing channels rapidly activate and inactivate in response to membrane depolarization, contributing to the neuronal subthreshold A-type potassium currents and controlling the action potential repolarization and frequency,

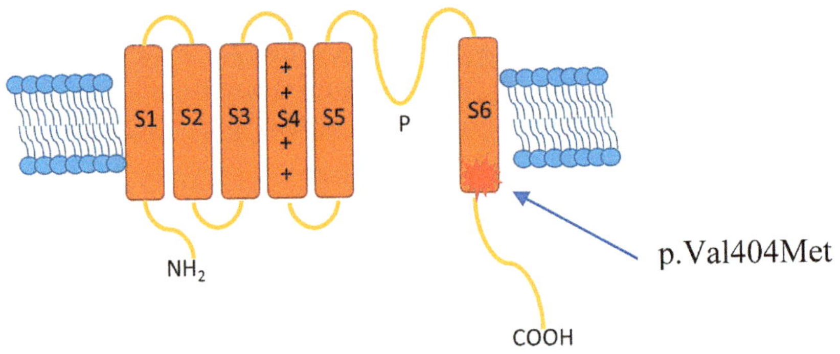

FIGURE 5.7 Position of the amino acid residue altered by the de novo *KCND2* variant in identical twins with a severe epilepsy, intellectual disability, and autism phenotype.

and thus neuronal excitability (Tsaur, Chou, Shih, & Wang, 1997). The channel is widely expressed in the brain, including in the cortex, basal ganglia, hypothalamus, and cerebellum. The channel characteristics of Kv4.3, including protein trafficking and channel expression and activity, can be modified by interactions with the K_v channel-interacting protein 2 (Wang et al., 2007). The channel is also expressed in the heart and has been associated with cardiac phenotypes.

Reported Human Neurological Conditions, Associated Variants, and Insights into Pathophysiology

Several inherited variants, including an in-frame amino acid deletion (p.[Phe227del]) and a missense variant (p.[Thr352Pro]) located within the third extracellular loop of the K_v4.3 channel, have been reported in families with an autosomal dominant spinocerebellar atrophy (MIM 607346) associated with mild cerebellar ataxia with onset between 10 and 55 years and cerebellar atrophy with variable intellectual disability but without an epilepsy phenotype (Duarri et al., 2012; Lee et al., 2012). Both *KCND3* variants appeared to result in reduced membrane expression of the mutant channel in cellular models (i.e., may cause loss of channel function).

Smets and colleagues (2015) subsequently reported a de novo *KCND3* mutation (c.877_885dupCGCGTCTTC; p.[Arg293_Phe295dup]) in a child with a very early-onset cerebellar ataxia (symptoms noted at the age of 3 years), a mild intellectual disability, and a generalized epilepsy that was controlled with monotherapy with valproate. The patient also had attention deficit hyperactivity disorder, strabismus, oral apraxia associated with continuous hypersalivation, and joint hyperlaxity. The variant resulted in a duplication of the arginine–valine–phenylalanine motif in the S4 segment of K_v4.3. Functional studies showed that it added an extra positive charge to K_v4.3 in the voltage-sensor domain, causing a severe shift of the voltage-dependent gating to more depolarized voltages, that is, a reduction in channel function.

A more severe developmental and epileptic encephalopathy phenotype, encompassing refractory epilepsy, psychomotor regression, attention deficit, and visual decline, has more recently been reported in a patient with a de novo missense variant, c.1174G > A, p.(Val392Ile), altering an amino acid residue in the S6 segment (Wang et al., 2019). This child had seizures commencing at 18 months of age, typically one or two times a month, frequently triggered by a fever. EEG showed multifocal spikes, spike and wave discharges, sharp waves, and sharp and slow waves primarily in the right temporal lobe. Visual decline was noted by the age of 4 years. His development was normal prior to seizure onset, but he showed decline in cognition and memory after onset of seizures. This rare variant has previously been reported in an individual with no prior neurological symptoms who was investigated for sudden unexplained death aged 20 (Giudicessi et al., 2012). Thus, it remains uncertain if this variant is causal of the child's DEE, although no satisfactory additional genetic diagnosis could be made by exome sequencing. The functional effect of this variant was examined and reported to be associated with complex changes in channel function: increasing peak current density and slowing recovery from inactivation (Giudicessi et al., 2012).

Clinical Management Implications of Channelopathy

Missense variants resulting in a gain of channel function have been associated with Brugada syndrome and sudden unexplained death (Giudicessi et al., 2011). To date, no specific therapies have been reported to be beneficial for loss-of-function variants associated with neurocognitive phenotypes.

KCNQ2 Encoding $K_v7.2$

Expression and Function of KCNQ2 in Human Brain

KCNQ2 encoded subunits combine with other K_v7 subunits to form heteromultimeric channels in the central nervous system. Channels formed by $K_v7.2$ and $K_v7.3$ (encoded by *KCNQ3*) are believed to play a major role in mediating the "M current," which is a very slowly activating and slowly inactivating potassium current that opposes sustained depolarizations and repetitive action potential firing. M channels can powerfully control the number of action potentials fired by a neuron receiving a strong excitatory input. Thus, they have at least two different roles in the brain. Firstly, they can serve as a restraint on repetitive neuronal discharges, overall reducing excitability. Secondly, they can act locally to mediate transient increases in excitability resulting from the release of modulatory neurotransmitters such as acetylcholine. Therefore, both reduction and increase in M current may result in abnormalities in excitability and a predisposition to epilepsy (Cooper & Jan, 2003). These different roles of M channels are important when considering why variants demonstrated to cause loss of function as well as variants demonstrated to result in gain of function of $K_v7.2$ or $K_v7.3$ containing channels have been associated with human epilepsy phenotypes.

Reported Human Neurological Conditions, Associated Variants, and Insights into Pathophysiology

Variants in *KCNQ2* have been reported in a range of epilepsy phenotypes, from KCNQ2-self-limited seizures (also known as benign familial neonatal epilepsy [KCNQ2-BFNE]) to KCNQ2-related neonatal epileptic encephalopathy (KCNQ2-NEE).

1. KCNQ2-BFNE (MIM 121200) is associated with a wide range of typically brief seizure types including tonic/apneic episodes, focal clonic episodes, or autonomic changes. Seizures start in the neonatal period, typically at 2 to 8 days of life. Babies are otherwise healthy and develop typically. Seizures characteristically cease by 1 year of life, with the EEG completely normalizing by 2 years. However, up to 30% of affected children may have recurrence of seizures in later life, including febrile seizures.

2. KCNQ2-NEE (MIM 613720) is associated with a severe epilepsy phenotype, starting in the first week of life. Seizures are frequently tonic, with associated focal motor and autonomic features like in KCNQ2-BFNE but more frequent (typically multiple in a day), and may be associated with a burst suppression pattern or multifocal epileptiform activity on EEG. The affected individual has typically moderate to severe developmental impairment which persists even after the seizure frequency reduces, central hypotonia, and progressive spasticity. Seizures typically cease between 9 months and 4 years of age.

Most variants causing KCNQ2-BFNE are inherited from a parent with a similar history of neonatal-onset seizures and are variants predicted to reduce expression (i.e., cause haploinsufficiency) of the channel subunit, for example, due to alteration in a start or stop codon, insertions or deletions changing the reading frame, or premature stop codon variants. However, certain missense variants are also associated with BFNE, particularly variants localizing in the intracellular domain between S2 and S3. Most tested missense variants result in a mild to moderate reduction in the M current (reduction in channel function).

In comparison, most variants causing KCNQ2-NEE are de novo and result in an alteration (change or loss) of an amino acid in one of the structurally important regions of the $K_v7.2$ structure, as shown in Figure 5.8. However, this distinction is not absolute, and missense variants in these regions have sometimes been reported in association with a KCNQ2-BFNE phenotype.

When tested, most missense or small in-frame deletion variants associated with KCNQ2-NEE result in a more severe reduction of the M current than those missense variants associated with KCNQ2-BFNE (Allen, Weckhuysen, Gorman, King, & Lerche, 2020). However, a small number of patients with a DEE phenotype have also been identified whose missense variants have been shown in cellular functional studies to result in a gain of channel function (Miceli, Soldovieri, et al., 2015). These variants tend to cluster in the voltage sensing domain (S2, S3, and S4 TM helices) and result in stabilization of the activated (open) state of the channel. It is suggested that such a gain of function would result in complex changes in network interactions within the brain. To date, such gain-of–channel function variants are associated with a very severe clinical course. A better understanding of the specific pathophysiological effects of different variants

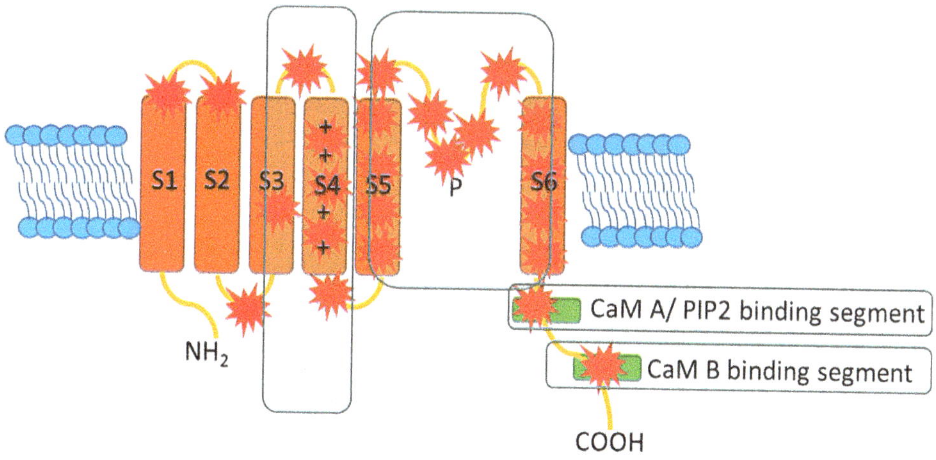

FIGURE 5.8 Diagram of the *KCNQ2* subunits where missense (substituted amino acid) or indel (deleted amino acid) variants have been reported in association with *KCNQ2*-NEE (red stars). Pathogenic variants can be seen to cluster in the regions highlighted by gray boxes: namely, the S4 voltage sensor, the pore (P), and two binding domains in the C-terminal domain (COOH): a proximal region which binds calmodulin (CaM A) and phosphatidylinositol-4,5-bisphosphanate (PIP2) and a more distal calmodulin binding region (CaM B)

in the developing human brain will be very important in guiding the development of targeted therapies.

Clinical Management Implications of Channelopathy

Most children with KCNQ2-BFNE respond to monotherapy with phenobarbitone, which decreases neuronal excitability through enhancement of GABA-related hyperpolarization and blockade of glutamate signaling. There is limited evidence that sodium channel blockers (e.g., carbamazepine, oxycarbazepine, phenytoin, lamotrigine) are effective for the limited time period of seizures and may be associated with less drowsiness and hence possible feeding difficulties compared to phenobarbitone. Due to the natural history of this condition, AEDs can be rationally reduced by 2 years of age.

Due to a number of cohort studies showing more rapid seizure control with early introduction of the sodium blocker carbamazepine or phenytoin, these agents are now recommended as first-tier therapies for proven or suspected KCNQ2-NEE (Miceli et al., 2010; Pisano et al., 2015). Again, due to the known natural history, a trial in reduction of AEDs can be considered in mid-childhood.

Although there is some suggestion that early effective treatment that reduces seizure frequency will improve final diagnostic outcome, this is yet to be systematically proven (Miceli et al., 2010; Pisano et al., 2015).

Ezogabine (also known as retigabine) is a pharmacological agent which acts directly on $K_v7.2$ channels, increasing their opening, and has been suggested as a targeted treatment for *KCNQ2*-encephalopathy related to a variant that would be predicted to reduce channel activity. Some preliminary evidence for the utility of ezogabine against refractory seizures for KCNQ2-BFNE when started early has been obtained in a small retrospective study (Millichap et al., 2016). However, this drug has been withdrawn in some countries as it has been associated with blueish discoloration of the nails and skin, and some individuals have reported changes in vision and pigment changes in the retina. There are current preclinical trials aiming to identify novel compounds which more potently and selectively modulate $K_v7.2$ and $K_v7.3$ channel function, allowing early control of epilepsy and hopefully preservation of neurodevelopment (Allen et al., 2020). Again, knowing the exact effect of an individual's genetic variant on the function of the channel in individual cells and brain circuits will be required to target the best therapies.

KCNQ3 Encoding $K_v7.3$

Expression and Function in Human Brain

$K_v7.3$ is also important in the formation of the M current.

Reported Human Neurological Conditions, Associated Variants, and Insights into Pathophysiology

Three related phenotypes have been associated to date with rare variants in *KCNQ3*, namely BFNE, benign familial infantile epilepsy (BFIE), and *KCNQ3*-related developmental disability (Miceli et al., 2014).

1. *KCNQ3*-BFNE (MIM 121201) is characterized by brief seizures starting between days 2 and 8 of life. The seizure types include tonic and apneic seizures, focal-clonic, and autonomic changes. Seizure are self-limiting, tending to cease before 1 year of age. Typically, development is normal; but some individuals do develop an intellectual disability, and the risk of permanent developmental disability varies even within a family (Miceli, Striano, et al., 2015).
2. *KCNQ3*-BFIE is characterized by brief seizures, often in clusters, starting within the first year of life but after the neonatal period. The seizure types are most commonly focal but can be generalized. Seizure are self-limiting, tending to cease before 2 years of age; and development is typically normal. Variants causing *KCNQ3*-BFNE or BFIE are typically inherited from a parent who had similar symptoms in infancy. Occasionally, parents may carry a variant and be asymptomatic due to incomplete penetrance or possibly because the seizures were so brief that they were not recognized.
3. *KCNQ3*-related developmental disability refers to intellectual disability in association with rare de novo variants in KCNQ3 without a prominent epilepsy phenotype. One variant in the S4 region (p.[Arg230Cys]) has been reported recurrently de novo in children with developmental disability, both with and without epilepsy; however, detailed phenotypic information is lacking for these children (Bosch et al., 2016; Epi4K Consortium et al., 2013; Miceli, Soldovieri, et al., 2015; Rauch et al., 2012).

As shown in Figure 5.9, rare *KCNQ3* variants associated with neurological phenotypes are more common in the S4, S5, and S5–S6 pore regions and in the S6 and carboxy-terminal regions. Most pathogenic variants reported to date are missense variants, altering the amino acid structure in these domains. Indeed, heterozygous loss-of-function variants may be tolerated as they are reported in individuals reportedly without a neurological phenotype in the gnomAD database.

Inherited missense variants associated with BFIE and BFNE that have been investigated in cellular models typically reduce current by 20–60%—with no obvious correlation demonstrated between the extent of the functional effect and phenotypic severity (Miceli, Striano, et al., 2015; Soldovieri et al., 2014). In contrast, the de novo p.(Arg230Cys) variant in the S4 domain, which has been recurrently associated with intellectual disability, has been shown to stabilize the channel's open state, similar to the effect of some rare variants reported in *KCNQ2* that alter residues in the S4 domain of $K_v7.2$ (Miceli, Soldovieri, et al., 2015).

Clinical Management Implications of Channelopathy

Seizures associated with inherited *KCNQ3* missense variants can generally be controlled by carbamazepine, phenobarbitone, and phenytoin; and AEDs can be withdrawn by age 3–6 months for BNIE or by age 1–3 years for BFNE.

There is emerging evidence that early treatment with carbamazepine is particularly associated with a more rapid resolution of seizures in infants with neonatal-onset seizures and a family history of similar seizures, with inherited variants in *KCNQ3* (Sands et al., 2016). The demonstration that carbamazepine is safe and effective is reassuring as it means that phenobarbitone's side effects of sedation and hypotonia,

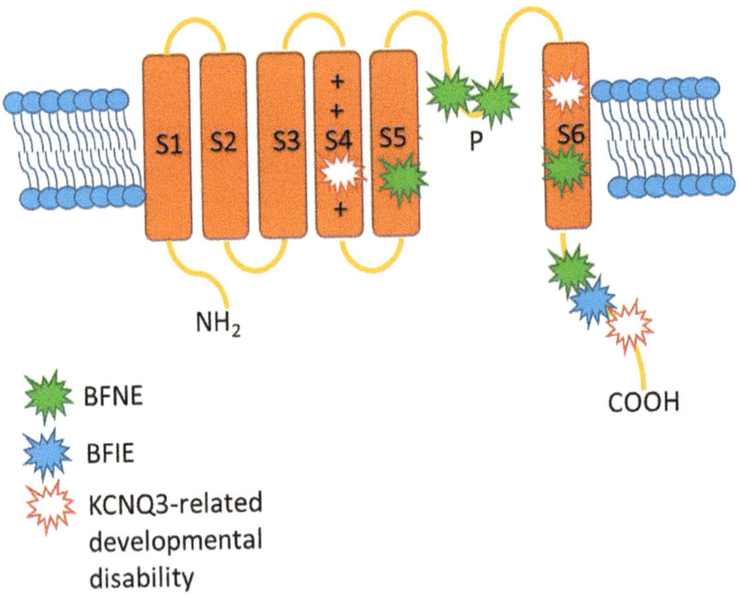

FIGURE 5.9 Location in the $K_v7.3$ subunit of amino acid substitutions reported in association with a neurological phenotype (*KCNQ3*-BFNE, *KCNQ3*-BFIE, or *KCNQ3*-related developmental disability).

which can interfere with the establishment of feeds and delay discharge from intensive care, may be avoidable.

In contrast, it is likely that alternative therapeutic approaches will be required for the *KCNQ3*-encephalopathy associated with the recurrent de novo p.(Arg230Cys) variant as this variant appears to have completely different effects on channel dynamics. Preclinical trials aiming to more specifically correct the different pathophysiological effect of variants in $K_v7.2$ and 3 are in process (Allen et al., 2020).

KCNQ5 Encoding $K_v7.5$

Expression and Function in Human Brain

KCNQ5 is widely expressed in the human brain. $K_v7.5$ containing channels are also important in regulating the slow activation and deactivating M current.

Reported Human Neurological Conditions, Associated Variants, and Insights into Pathophysiology

Lehman and colleagues (2017) reported four rare de novo missense variants in *KCNQ5* with varyingly severe neurological phenotypes (MIM 617601). Two variants, p.(Leu341Ile) and p.(Val145Gly), were associated with mild intellectual disability without seizures; another two were associated with more severe intellectual disability and treatment-resistant epilepsy. These variants, when tested in a cellular model, resulted in reduced channel expression and electrophysiological changes consistent with a loss

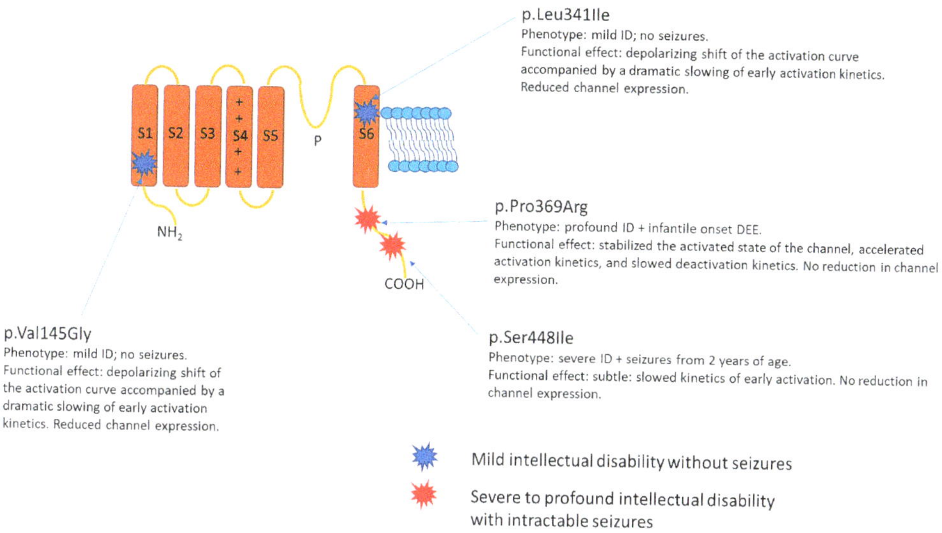

FIGURE 5.10 Position of mutated amino acid residues in the $K_v7.5$ subunit in affected individuals with a neurological phenotype.

of channel function (Lehman et al., 2017). In comparison, the variant p.(Pro369Arg) was associated with a profound DEE. This variant resulted in a clear gain of channel function in the absence of alteration of channel expression in a cellular model. Lehman and colleagues suggest that this variant should increase the function of the M current. Increased M current, for example, secondary to certain gain-of-function mutations in *KCNQ2* and *KCNQ3*, might increase neuronal excitability due to overactivation of hyperpolarization-activated nonselective cation channels (Miceli, Soldovieri, et al., 2015; Miceli, Striano, et al., 2015) (Figure 5.10).

The fourth variant reported by Lehman et al. is not yet convincingly demonstrated as causal. This is the variant p.(Ser448Ile), which resulted only in a subtle loss of channel function in a cellular model.

Subsequently, Rosti and colleagues (2019) reported an individual with mild intellectual disability and adolescent-onset, well-controlled absence epilepsy who had a de novo intragenic duplication in *KCNQ5*. This variant would be predicted to result in gene haploinsufficiency due to aberrant splicing and introduction of a premature stop codon. The authors speculate that reduced function of $K_v7.5$ channels could result in a lowered seizure threshold due to decreased neuronal repolarization reserve mediated by the M current. Thus, it appears that variants resulting in either loss or gain of channel function might be associated with epilepsy. More studies and a larger patient cohort will be required to clarify the phenotypic–genotypic spectrum for *KCNQ5*-encephalopathy.

Clinical Management Implications of Channelopathy

No specific therapies have yet been reported to be helpful for *KCNQ5*-related seizures. It is noted that ezogabine/retigabine has a similar functional effect to the p.(Pro369Arg) mutation and, thus, could worsen seizures further for this particular variant and should not be trialed (Lehman et al., 2017).

KCNH1 Encoding Kv10.1

Expression and Function in Human Brain

KCNH1 encodes the potassium channel α subunit K_v10.1, which has high expression in the brain and is thought to regulate neurotransmitter release and synaptic transmission. It is homologous to the *Drosophila* ether-à-go-go-related gene potassium channel.

Reported Human Neurological Conditions, Associated Variants, and Insights into Pathophysiology

The first de novo missense variants in *KCNH1* in humans were described in the syndromic conditions Temple-Baraister syndrome (TBS; MIM 611816) by Simons and colleagues (2015) and Zimmermann-Laband syndrome (ZLS; MIM 135500) by Kortüm and colleagues (2015). These two conditions were described before the genetic cause was identified, based on their distinctive constellations of clinical features. TBS was reported to be particularly associated with absent or hypoplastic nails, thumbs, and great toes; and ZLS was reported to be associated with more coarsened facial features, hypertricosis, hepatosplenomegaly, and hypoplasia of the terminal phalanges and nails. However, the two conditions shared many overlapping characteristics including severe to profound intellectual disability, dysmorphic facial features (most commonly hypertelorism, a broad nasal tip, and a wide mouth), and nail aplasia or hypoplasia. Epilepsy is a very common core clinical feature for both TBS and ZLS.

Subsequently, overlap between genetic causes, in that some de novo missense variants in *KCNH1* have been described in individuals with a clinical diagnosis of either TBS or ZLS, and broadening of the phenotype—more recently, individuals have been reported with de novo missense variants in *KCNH1*, with a prominent epilepsy and intellectual disability phenotype but only very mild systemic features (Mastrangelo et al., 2016)—means that KCNH1-*encephalopathy* might be a more appropriate term to describe this group of conditions associated with variants in *KCNH1*. Seizure onset in *KCNH1*-encephalopathy is typically in the first year of life. First seizures may be generalized or focal motor, evolving to generalized tonic-clonic seizures. Convulsive status epilepticus is common. Seizures are often refractory to polytherapy, although some patients have achieved control with a variety of AED combinations. Brain MRI findings, if present, are nonspecific and can include increased extra-axial spaces, ventricular asymmetry, and corpus callosal hypoplasia (Mastrangelo et al., 2016).

Pathogenic missense variants mainly affect residues involved in the voltage sensor S4 or the P sections S5 and S6 (Figure 5.11). Functional studies are supportive of most tested variants resulting in a gain of channel function, for example, Kortüm and colleagues (2015) found that the four tested ZLS-associated *KCNH1* variants led to the channel open state being preferred, a lower conductance, or a change in the closed–open transition

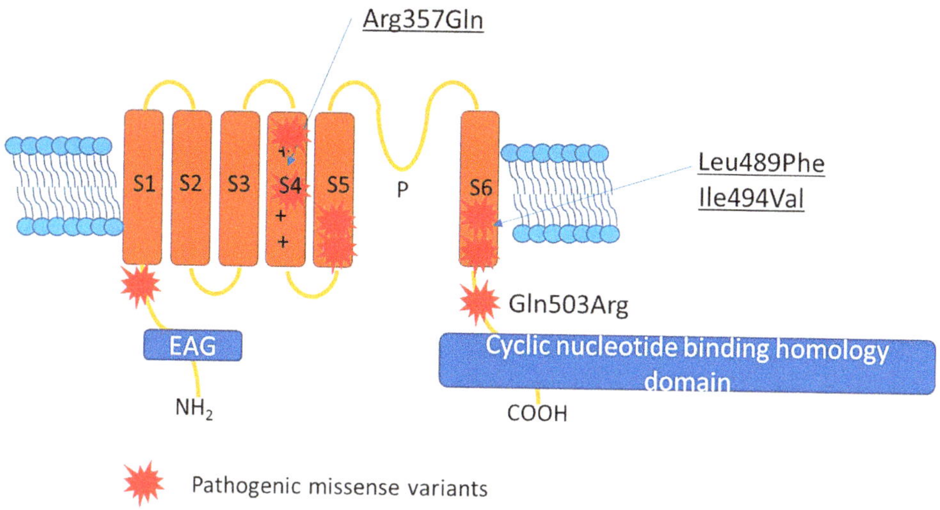

FIGURE 5.11 Localization of amino acid residues affected by pathogenic missense variants reported in association with TBS, ZLS, or nonspecific *KCNH1*-related encephalopathy. Recurrent variants in the S4 and S6 segments are underlined.

of the voltage-gated potassium channel. Simons and colleagues (2015) found that certain *KCNH1* variants lowered the activation threshold of the mutant K_v10.1 containing channels and delayed their deactivation.

Clinical Implications of Channelopathy Diagnosis

To date, no clear AED or combination of AEDs has been found to be particularly helpful for *KCNH1*-related epilepsy (Mastrangelo et al., 2016). There is a lack of specific inhibitors of the "eag" potassium channel subfamily, and nonspecific blockers that might reduce *KCNH2* function in cardiac tissue could lead to serious side effects including long QT syndrome and fatal arrhythmia. Screening to identify highly specific compounds that would selectively block *KCNH1* and gene therapy/oligonucleotide therapy to reduce expression of abnormal gene products could be helpful therapeutic strategies in the future.

KCNH5 Encoding Kv10.2

Expression and Function in Human Brain

K_v10.2 is expressed in the human nervous system, but its functions are not well understood (Yang et al., 2013).

Reported Human Neurological Conditions, Associated Variants, and Insights into Pathophysiology

To date, one de novo variant in *KCNH5* has been identified in a child with autism and epileptic encephalopathy (Veeramah et al., 2013). This variant results in an amino acid substitution in the S4 helix (p.[Arg327His]). Structural modeling showed that this amino acid substitution would reduce the ionic interactions between this residue and negatively charged residues in the S1–S3 helices in the closed and early activation states, thus favoring channel opening (a gain in channel function) (Yang et al., 2013).

Clinical Management Implications of Channelopathy Diagnosis

As is the case for *KCNH1*-related encephalopathy, the development of subfamily-specific blockers of K_v10.2 containing channels would be potentially useful for affected individuals, with variants resulting in gain of function of the channel

Six Transmembrane Domain Family: Ca- and Na-Activated Potassium Channels

KCNMA1 Encoding K_{Ca}1.1

Expression and Function in Human Brain

KCNMA1 encodes K_{Ca}1.1, the pore forming α subunit of "Big K^+" (BK) large-conductance calcium and voltage-activated potassium channels.

K_{Ca}1.1 consists of an extracellular N terminus, seven TM domains (S0–S6), a P loop between S5 and S6, and a large intracellular C-terminal domain that includes two calcium binding domains, RCK (regulator of K^+ conductance) 1 and 2, which comprise the intracellular gating ring (Figure 5.12). S1–S4 comprise the voltage sensing domain, and S5 and S6 form the pore through which K can pass, along with the P loop. Four K_{Ca}1.1 subunits assemble to form a homotetrameric channel, which can combine with different accessory β and γ units, which bind on the extracellular N terminus and S0 (Yuan, Leonetti, Pico, Hsiung, & MacKinnon, 2010).

BK channel activation requires both membrane depolarization and an increase in intracellular calcium. These channels are widely expressed in both excitable and nonexcitable cells, but expression is highest in brain and muscle, where these channels regulate neuronal excitability and muscle contractility (Bailey, Moldenhauer, Park, Keros, & Meredith, 2019). In the brain, BK channels activate to allow potassium current out of the cell when there is membrane depolarization or increased intracellular calcium, leading to hyperpolarization of the membrane and decreased excitability. Typically, activation of BK channels reduces neuronal firing rates and presynaptic neurotransmitter release and, in muscle, reduces muscle contraction. This suggests that variants which alter BK activity would alter the balance of excitation and inhibition in the brain, as well as muscle contractibility and tone (Bailey et al., 2019).

Reported Human Neurological Conditions, Associated Variants, and Insights into Pathophysiology

Several rare missense variants in *KCNMA1* have been reported in association with human neurological phenotypes. These variants mainly affect residues in the P loop S6

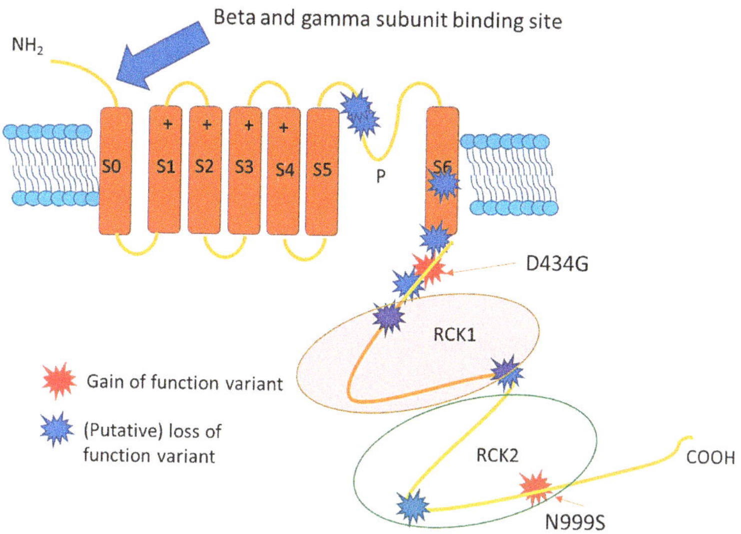

FIGURE 5.12 Structure of the $K_{Ca}1.1$ subunit with residues affected by missense variants associated with neurocognitive conditions.

and in the intracellular carboxy-terminal domains, which include the RCK1 and RCK2 domains, important for calcium binding and regulation of channel gating (Figure 5.12).

The main symptoms reported in affected individuals with rare, predicted pathogenic variants in *KCNMA1* are seizures and movement disorders.

1. *Paroxysmal nonkinesigneic dyskinesia, with or without generalized epilepsy (MIM 609446) or generalized epilepsy (MIM 618596).* One phenotype is of generalized epilepsy and/or paroxysmal dyskinesia (MIM 609446). This phenotype was reported to segregate with the inherited *KCNMA1* variant D434G in 13 members of a large family (Du et al., 2005). The variant affects the RCK1 domain in the carboxy-terminal region of the subunit. In a heterologous cell line model, this variant conferred a gain of channel function. Another variant resulting in gain of function in a heterologous model is the recurrent p.(Asn999Ser) variant, located in the RCK2 domain. This variant has been reported de novo in seven unrelated patients with combinations of generalized epilepsy, paroxysmal dyskinesia, and developmental delay/intellectual disability (Bailey et al., 2019).

2. *Liang-Wang syndrome (MIM 618729).* Liang and colleagues (2019) reported a recurrent de novo variant, p.(Gly375Arg), which results in an amino acid substitution within the S6 helix. This variant is reported to result in a loss of channel function, and affected individuals heterozygous for this variant have a distinctive multisystemic phenotype consisting of facial dysmorphism, aortic root dilatation and hernia or megacystis, severe intellectual disability, feeding difficulties and intestinal mobility difficulties, and axial hypotonia. Two of the reported patients had generalized epilepsy (absence seizures), and one patient had mild cerebellar atrophy. None had paroxysmal dyskinesia, ataxia, or dystonia.

3. *Cerebellar atrophy, developmental delay, and seizures (MIM 617643).* In comparison to the heterozygous variants, a homozygous frameshift duplication variant,

KCNMA1 (p.[Tyr676Leufs*7]), was reported in two affected siblings within a consanguineous family (Tabarki, AlMajhad, AlHashem, Shaheen, & Alkuraya, 2016). This homozygous variant should result in nonsense mediated decay of both *KCNMA1* alleles, resulting in a complete loss of channel function. Both affected children had severe developmental delay and diffuse hypotonia with marked nonprogressive cerebellar atrophy on MRI. Neither had paroxysmal dyskinesia. They had myoclonic seizures, starting in early infancy. In one child the seizures were well controlled with sodium valproate; but in the second the seizures evolved into tonic seizures, and the EEG was consistent with Lennox-Gestaut syndrome. Reasonable control was achieved with the addition of levetiracetam. Both heterozygous carrier parents were asymptomatic, without movement disorder or epilepsy. The authors note that the phenotype in the homozygous children is similar to that in mouse *Kcnma1* homozygous knockouts that have a severe cerebellar ataxia (Sausbier et al., 2004). Symptoms included epilepsy, developmental delay, and severe cerebellar atrophy but an absence of paroxysmal dyskinesia.

Clinical Management Implications of Channelopathy

A true targeted therapeutic approach for *KCNMA1*-linked channelopathy is limited by the lack of clinically available and approved specific BK channel pharmacological modulators and limited understanding of how the underlying *KCNMA1* variants lead to the variety of neurological symptoms seen in patients. Therefore, an array of standard antiepileptic medications has been trialed in patients with *KCNMA1*-linked channelopathy with a variety of responses noted. Some studies report a positive effect of sodium valproate and/or lamotrigine and/or levetiracetam. A couple of reports suggest a worsening of seizures with ethosuxinamide (Bailey et al., 2019). For paroxysmal nonkinesigenic dyskinesia, several reports suggest the utility of clonazepam and of acetazolamide. Acetazolamide is an interesting therapeutic choice as it is a carbonic anhydrase inhibitor and has been shown in rat models to have a direct agonic action on BK channels (although this has not been proven in human neurons) and thus might be particularly beneficial in individuals whose variant results in a reduction in channel function (Tricarico, Barbieri, Mele, Carbonara, & Camerino, 2004). Of note, some AEDs also have a carbonic anhydrase action (topiramate, zonisamide, and sulthiame), and it would be helpful to systematically review if any of these medications have been more successful in seizure control for patients with variants in *KCNMA1* resulting in reduced channel function.

Bailey and colleagues (2019) present an interesting discussion on potential future therapeutic targets for *KCNMA1*-related channelopathies and stress the importance of understanding the pathophysiological consequences of an affected individual's variant, that is, how it affects channel function under different physiological conditions and in the intact human brain. They point to several possible therapeutic avenues to evaluate, including the neuromodulator docosahexaenoic acid, an omega 3 fatty acid found in oily fish and some AEDs (acetazolamide and zonisamide), which has been shown to increase

BK currents and therefore may have some efficacy for enhancing hypofunctioning channels. Another possible avenue they report is to normalize BK channel activity associated with gain-of–channel function mutations using calcium channel inhibitors, which could reduce calcium-dependent activation of BK channels. They highlight that the calcium channel blocker verapamil has been previously used in refractory epilepsy and hyperkinetic movement disorder. However, care should be taken to monitor for potential cardiovascular side effects

KCNN3 Encoding $K_{Ca}2.2$

$K_{Ca}2.2$ is one α subunit of the small-conductance calcium-activated potassium channel subfamily, also known as the SK channel subfamily. SK channels are widely expressed in the nervous system and important for the regulation of cellular excitability. They are part of large multiprotein complexes which include the P channel subunits, the constitutively bound calcium sensor calmodulin, protein kinase CK2, and protein phosphatase 2A. Binding of calcium to calmodulin results in opening of the channel, and protein kinase CK2 and protein phosphatase 2A binding further modulate the calcium sensitivity of the channels by phosphorylating or dephosphorylating bound calmodulin (Bauer et al., 2019).

Reported Human Neurological Conditions, Associated Variants, and Insights into Pathophysiology

Bauer and colleagues (2019) recently reported three unrelated individuals with de novo missense variants in *KCNN3* who had a clinical diagnosis of ZLS (MIM 618658), based on their clinical features of dysmorphic facial features, nail hypoplasia, and mild to moderate intellectual disability. None of these individuals had developed seizures at the time of reporting. As discussed, de novo missense variants in *KCNH1* causing gain of function in $K_v10.1$ containing channels have previously been linked to a ZLS phenotype. The three variants affected residues in different domains of the channel subunit, as shown in Figure 5.13. The variants were predicted to lie in functional domains important for the regulation of channel gating by the binding of calcium to calmodulin. Patch-clamp whole-cell studies demonstrated that all three variants resulted in increased calcium sensitivity, leading to faster and more complete activation of the mutant channels, in effect a gain of channel function.

Clinical Management Implications of Channelopathy

To date, all three individuals with de novo variants in *KCNN3* have not developed seizures. However, it is likely that other individuals will be described with variants in this gene and a neurocognitive phenotype, and it remains to be seen whether epilepsy will become a prominent clinical feature for this new channelopathy. If so, specific blockers or openers of the channel may be helpful, depending on the pathophysiological impact of each specific variant.

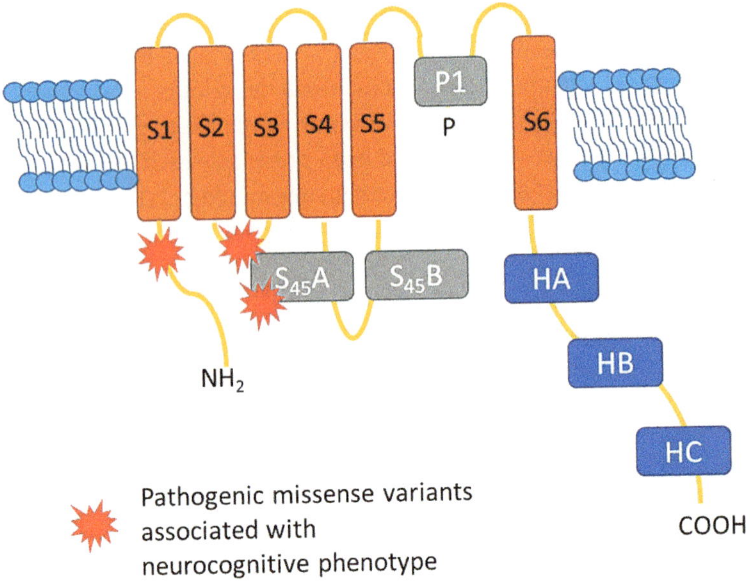

FIGURE 5.13 Rare de novo variants resulting in amino acid substitutions in $K_{Ca}2.2$.

KCNT1 Encoding $K_{Na}1.1$

Expression and Function in Human Brain

KCNT1 encodes one of the two known sodium-activated potassium channel subunits and is also known as the SLACK (sequence like a calcium-activated K^+) channel. The gene is expressed throughout the brain, with particularly high expression in the cerebellum, olfactory bulb, and brainstem nuclei (Bhattacharjee, Gan, & Kaczmarek, 2002), as well as in dorsal root ganglia and renal and cardiac tissues. $K_{Na}1.1$ is a large subunit. It includes six TM domains and a P domain between S5 and S6, like voltage potassium channels, as well as a very large carboxy-terminal domain which includes two RCK domains and an oxidized nicotinamide adenine dinucleotide (NAD^+) binding domain (Figure 5.14). The RCK domains form a ring cytoplasmically, and, in the presence of sodium, they expose the channel pore. They are important in functional tetramer formation. The NAD^+ binding domain reduces the sodium requirement of the channel in the presence of increased NAD^+ concentrations (Tamsett, Picchione, & Bhattacharjee, 2009). The carboxy-terminal domain also interacts with the *FMR1* encoded Fragile X mental retardation protein (FMRP), affecting channel opening (Brown et al., 2008). The association with FMRP may be important in overall control of brain development and function.

Reported Human Neurological Conditions, Associated Variants, and Insights into Pathophysiology

Variants in *KCNT1* have been reported in two main clinical epilepsy phenotypes: epilepsy of infancy with migrating focal seizures (EIMFS) and autosomal dominant nocturnal

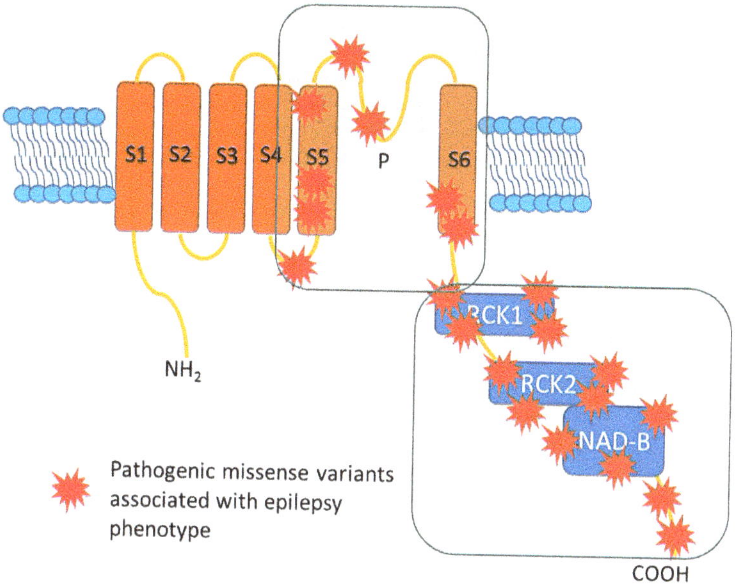

FIGURE 5.14 Rare de novo or inherited missense variants resulting in amino acid substitutions in $K_{Na}1.1$ in patients with epilepsy phenotypes.

frontal lobe epilepsy (ADNFLE). *KCNT1* variants can also be associated with less specific epilepsy and intellectual disability phenotypes.

1. *EIMFS (MIM 614959).* This is a DEE characterized by drug-resistant seizures that start in the first 6 months of life, with median age at onset being 3.5 weeks (McTague et al., 2018). Seizure types are primary focal motor but may secondarily generalize. Multiple other seizure types, including tonic, clonic, tonic-clonic, myoclonic, and epileptic spasms, have also been reported. A characteristic EEG finding is of focal ictal discharges that migrate across adjacent cortical regions and arise independently at multiple locations. Affected individuals are centrally hypotonic, and some develop peripheral spasticity with age. A movement disorder, including choreoathetosis, dyskinesia, and dystonias, is commonly also present. Development typically stalls or regresses with seizure onset, and children have severe to profound intellectual disability. Progressive microcephaly is common. Brain MRI may show frontal cerebral atrophy, delayed myelination, cerebellar atrophy, and/or hippocampal volume loss. It is yet unclear if *KCNT1*-EIMFS variants also increase the risk of cardiac arrhythmias.

2. *ADNFLE (MIM 615005).* Affected individuals have clusters of nocturnal motor seizures and often have an affected first degree relative with similar symptoms, although this is not always the case as penetrance is less than 100%. Cognition may be normal or relatively mildly affected, and seizures typically start in late childhood or adolescence. Psychiatric and behavioral comorbidities are common.

Figure 5.14 summarizes the clustering of missense variants reported to date in individuals with a *KCNT1*-related encephalopathy. Variants particularly cluster in the S5, P loop, and S6 region that form the central pore, as well as in the C-terminal domain.

Functional studies to date indicate that most tested missense variants result in a gain of channel function, with increases in peak current amplitudes of *KCNT1*-containing channels of 3- to 12-fold (Barcia et al., 2012; Heron et al., 2012; Martin et al., 2014; McTague et al., 2018). An exception is a de novo p.(Phe932Ile) variant, which affects a residue in the carboxy-terminal domain, distal to the NAD-B functional domain, which has been identified in a child with a severe intractable epilepsy and abnormal myelination. Mutant channels displayed a reduction in channel function (Vanderver et al., 2014).

Quraishi and colleagues (2019) recently investigated how variants in KCNT2 might contribute to hyperexcitability by generating and evaluating a human induced pluripotent stem cell–derived neuronal line, which was engineered to harbor the p.(Phe924Leu) variant in both KCNT1 alleles, using nuclease-mediated engineering. The electrophysiological properties of the neuronal cells derived from these engineered cells were evaluated using voltage-clamp recordings as well as multielectrode arrays and current-clamp recordings. This allowed characterization of firing patterns and action potential properties. They found that the p.(Phe924Leu) variant did not alter the expression of $K_{Na}1.1$ protein, and there were no compensatory changes in the other channels thought to interact with $K_{Na}1.1$ containing channels: BK channels or Nav1.1 channels. They found, however, that p.(Phe924Leu) expressing neurons had increased K_{Na} currents, that action potentials were of shortened duration, and that there was increased amplitude of the afterhyperpolarization following each action potential. In networks of spontaneous active neuronal cells, these electrophysiological changes resulted in increases in mean firing rate, frequency of rapid bursts of action potentials, and intensity of firing within a burst. This was consistent with this variant, which in single-cell models had previously been shown to cause gain of channel function, resulting in hyperexcitability in both isolated neuronal cells and neural networks. This study is an exciting example of how newer technologies can be applied to more closely model the effect of potassium channel variants on the human brain; further studies will be helpful to evaluate the effect of a heterozygous mutation. The authors also point out that $K_{Na}1.1$ variants result in severe neuropsychiatric and neurodevelopmental consequences that are likely not just due to seizures, and evaluation of the role of variants on the function of known interactors including FMRP and Phactr-1 will be also required (Quraishi et al., 2019).

McTague and colleagues (2018) looked at any evidence of genotype–phenotype correlation, but there is no clear difference in the location of variants or electrophysiological effect in functional models between more severe DEE phenotypes and ADNFLE.

Clinical Implications of Channelopathy

KCNT1-epilepsy is frequently refractory to multiple AEDs, and there is therefore an urgent need to develop better targeted therapeutics. Bearden and colleagues (2014) reported the response of a 2-year-old girl with a severe EIMF phenotype associated with a *KCNT1* p.(Arg428Gln) missense variant to the *KCNT1* inhibitor quinidine. The patient's variant had been previously reported in *KCNT1*-related encephalopathy and shown in functional studies to result in gain of channel function. Quinidine was cautiously added to an existing multiple AED and ketogenic diet regime in hospital, and there was an improvement, but not complete control, of her seizures, with several adjustments in dose required. Her development showed slight improvement. The authors note that quinidine is a weak *KCNT1* inhibitor and only partially crosses the blood–brain barrier. Moreover, it can result in prolongation of the QT interval, and close electrocardiographic monitoring is required. Quinidine can also inhibit the metabolism of many other AEDs. Thus, care is recommended with consideration of quinidine for *KCNT1*-related encephalopathy, especially as subsequent studies did not universally replicate the original findings. For example, Mullen and colleagues (2018) conducted a small, randomized, placebo-controlled, crossover clinical trial of *KCNT1*-related ADNFLE in adolescent and adult patients which was negative. Indeed, it has been suggested that multiple considerations may affect response to quinidine, including the age of the patient. Abdelnour and colleagues (2018) reviewed eight clinical patients with *KCNT1*-related epilepsy trialed on quinidine and found that only children aged under 4 years at the start of the quinidine trial had a positive response. More studies are required to clarify which patients, if any, should be trialed on quinidine.

Other AEDs which have had positive effects in some patients include stiripentol, in combination with a benzodiazepine, and levetiracetam, as well as the ketogenic diet. Vagal nerve stimulation has not been shown to be helpful.

KCNT2 Encoding $K_{Na}1.2$

Expression and Function in Human Brain

$K_{Na}1.2$ or SLICK (sequence like an intermediate conductance K^+) containing channels are expressed in the brain, particularly in the subplate of the cerebral cortex in utero and the hippocampus and cortex in the adult, suggesting that the channel plays an ongoing role in cerebral function. $K_{Na}1.2$ subunits are postulated to heterotetramerize with $K_{Na}1.1$ subunits in at least some brain regions (Chen et al., 2009; Gururaj et al., 2017). $K_{Na}1.2$ containing channels have fast activation kinetics and are sensitive to intracellular chloride levels and cell volume variations. They also have limited sodium permeability.

Reported Human Neurological Conditions, Associated Variants, and Insights into Pathophysiology

We described a rare de novo missense variant in a male child with a severe developmental and epileptic encephalopathy phenotype, with generalized seizures from 3 months of age, evolving to epileptic spasms (Gururaj et al., 2017). His seizures have continued into childhood and are intractable to multiple antiepileptic mediations. Predominant seizure types are prolonged tonic seizures, myoclonic jerk, and atypical absences (MIM 617771). EEG is persistently abnormal with disorganized background, multifocal epileptogenic activity, or hypsarrhythmia. On brain MRI, there is white matter thinning, especially in the corpus callosum. Developmental impact is profound; he is nonverbal and nonambulant.

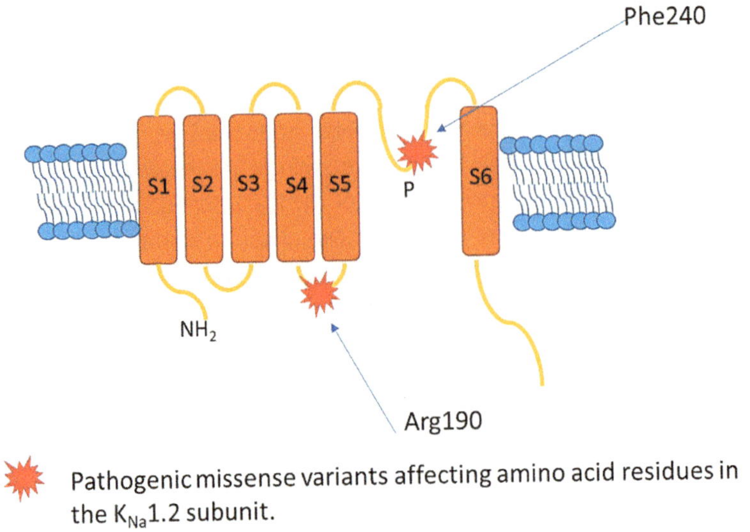

FIGURE 5.15 Pathogenic missense variants in *KCNT2*-related encephalopathy.

The *KCNT2* variant resulted in an amino acid substitution, p.(Phe240Leu). The affected residue is situated in the channel pore helix between TM domains S5 and S6, a residue which had previously been demonstrated to be critical to normal K_{Na}1.2 channel gating (Suzuki, Hansen, & Sanguinetti, 2016) (Figure 5.15). Electrophysiological modeling of the mutated channel showed that the mutation resulted in a dramatic change of function, essentially converting the wild-type potassium channel upregulated by chloride to a mutant channel which was no longer selective to potassium but likely permeable to sodium and downregulated by chloride. This is likely to result in an increased inward I_{Na} disrupting the balance between sodium and potassium exchange, which is required for normal neuronal excitability.

The mutation also resulted in reduced membrane expression in *Xenopus laevis* oocytes, consistent with reduced membrane trafficking or more rapid protein degradation. Expression of the mutant channel in embryonic rat dorsal root ganglion neurons resulted in severe neuronal toxicity, and in the few surviving neurons there was evidence of membrane hyperexcitability, demonstrating a plausible pathophysiological link to the epilepsy phenotype in the proband.

Subsequently, Ambrosino and colleagues (2018) reported two females with separate de novo missense variants in *KCNT2* and severe epilepsy phenotypes: one had West syndrome which evolved to Lennox-Gastaut syndrome and the other had a DEE with migrating focal seizures. Both children's epilepsy was persistent despite multiple AEDs and trial of a ketogenic diet. Both children had general developmental delay and moderate to severe intellectual disability. Both could walk unassisted; one was nonverbal, and the other could only talk in short sentences. The two girls were noted to have subtle distinctive features including busy eyebrows and long eyelashes with broad mouths.

Both children had a rare missense variant affecting the same amino acid residue, Arg190 (p.[Arg190His] and p.[Arg190Pro]). This amino acid residue lies in the intracellular S4–S5 linker region. Functional studies in HEK-293 cells showed that both variants resulted in larger outwardly rectifying currents with a hyperpolarizing shift in half-maximal voltage, consistent with a gain of–channel function effect.

Mao and colleagues (2020) recently reported two patients with de novo truncating variants in *KCNT2*: a frameshift variant (p.[Leu48Glnfs43]) in a patient with early-onset epileptic encephalopathy and some features consistent with EIMFS and a nonsense variant (p.[Lys564*]) in a patient with classical EIMFS phenotype. Both variants, when tested in a Chinese hamster ovary cell model, reduced the global current density of heteromeric channels. This publication reinforces the role of *KCNT2* in early-onset epilepsies.

Clinical Management Implications of Channelopathy

Ambrosino and colleagues (2018), in the light of recent literature describing the response of some *KCNT1*-encephalopathy patients to quinidine, investigated and showed an inhibitory effect of quinidine on both wild-type and mutant $KCNT2^{R190H}$ and $KCNT2^{R190P}$ channels. Quinidine was trialed in the first child with an initial reduction in seizure frequency (which could not be sustained) and some improvement in alertness, according to her parents. However, doses needed to be reduced due to significant prolongation of the QT interval (Ambrosino et al., 2018).

Quinidine was not trialed in the child with the Phe240Leu variant as this did not result in a clear gain of function but rather resulted in a more complex change in function, which quinidine was unlikely to help. A ketogenic diet has resulted in some improvement in seizure frequency and severity in this child. Targeted therapies for *KCNT2*-related encephalopathy are urgently required in view of the poor developmental progress and severe epilepsy in the affected children reported to date.

Two Transmembrane Domain Family, Including the Inwardly Rectifying Potassium Channels

KCNJ10 Encoding $K_{ir}4.1$

Expression and Function in Human Brain

KCNJ10 encodes the inwardly rectifying potassium channel α subunit $K_{ir}4.1$. It is expressed in the brain, ear, eye, and renal tubular systems. $K_{ir}4.1$ subunits form homotetramers or heterotetramers with, for example, $K_{ir}5.1$ subunits (Higashi et al., 2001). Studies in mice show that *KCNJ10* is expressed in brain astroglia, predominantly in the cerebral and cerebellar cortices and in the caudate nucleus and putamen (Bockenhauer et al., 2009). Mice with *KCNJ10* deletion develop seizures and ataxia and die shortly after birth, with impaired oligodendrocyte development and myelination (Neusch, Rozengurt, Jacobs, Lester, & Kofuji, 2001). Mice with conditional glial cell Kcnj10 knockout die at approximately 3 weeks of age, with evidence of glial membrane depolarization, inhibition of potassium and glutamate uptake, and enhanced short-term synaptic potentiation (Djukic, Casper, Philpot, Chin, & McCarthy, 2007). It is proposed that $K_{ir}4.1$ containing channels provide a protective "potassium sink," taking up extruded potassium from repetitively firing neuronal cells, which, if left around neurons, would otherwise decrease the membrane potential, resulting in hyperexcitability and seizures (Bockenhauer et al., 2009).

In the mouse ear, *KCNJ10* is expressed in the stria vasculitis of the inner ear, where it has a role in the generation of the endocochlear potential required for hearing. $K_{ir}4.1$ containing channels are also important for spiral ganglion neuron excitation. In the kidney, $K_{ir}4.1$ and $K_{ir}5.1$ containing channels are required for maintenance of electrolyte and fluid homeostasis through the control of resting membrane potentials and transepithelial voltages. They are required for potassium recycling in the distal convoluted tubules and cortical collecting ducts.

Reported Human Neurological Conditions, Associated Variants, and Insights into Pathophysiology

Bockenhauer and colleagues (2009) first linked homozygous rare missense variants in *KCNJ10* to the autosomal recessive syndromic epilepsy condition EAST syndrome by linkage analysis. EAST syndrome is characterized by **e**pilepsy, **a**taxia, **s**ensorineural deafness, and **t**ubulopathy, a renal salt-losing tubulopathy with normotensive hypokalemic metabolic acidosis. The missense variants altered amino acids in the first transmembrane domain of the α subunit. Studying the effect of the mutations in *Xenopus* oocytes showed that the mutations resulted in a significant reduction in the inwardly rectifying potassium current. Subsequently, over 16 different pathogenic variants have been reported in association with EAST syndrome, which typically cluster in the transmembrane regions and the C-terminal region (Figure 5.16). Most variants are missense and, when functionally tested, result in reduction or loss of channel function. A small number of nonsense or frameshift variants are also reported.

Independently Scholl and colleagues (2009) reported an overlapping clinical condition, SeSAME (seizures, sensorineural deafness, ataxia, mental retardation, and electrolyte imbalance), to be also associated with missense or nonsense variants in *KCNJ10*.

The neurological features are the most disabling for EAST/SeSAME syndrome. Seizures are almost universal, typically with infantile onset between 3 and 9 months of age. Seizures are generalized (commonly generalized tonic-clonic) or secondarily generalized. They are commonly brief and frequently can be controlled with a single antiepileptic medication. Around half of affected individuals are seizure-free by adolescence; others develop focal seizures or have ongoing generalized seizures, especially with intercurrent illness. However, occasional status epilepticus has been reported with this condition. EEGs are usually normal or include nonspecific findings (Cross et al., 2013; Mir et al., 2019). Ataxia is noted from when affected children try to walk and can be very disabling. Ambulation is always delayed, and some children do not achieve

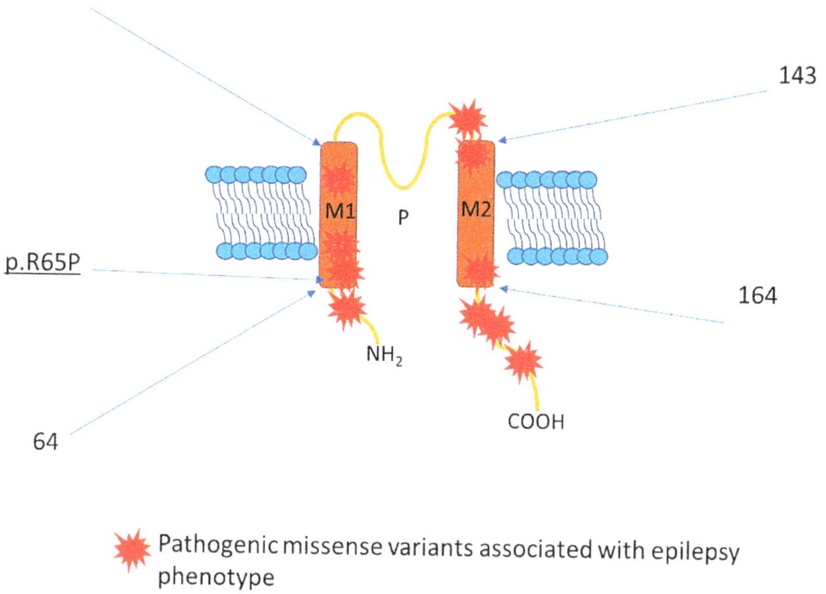

FIGURE 5.16 Position of amino acids affected by pathogenic variants in the $K_{ir}4.1$ channel subunit in affected individuals with EAST/SeSAME syndrome.

independent ambulation. Those who can walk have a broad-based gait. Other cerebellar signs include intention tremor, scanning speech, and dysmetria. Neurological symptoms are generally reported to be nonprogressive, although some children have been reported to have increased tone (Celmina et al., 2019).

Sensorineural hearing loss is nearly universal and may progress (Celmina et al., 2019). Hearing loss can be linked to delayed speech development, and speech therapy and hearing aides are recommended.

Renal involvement is variable, and the degree of electrolyte imbalance can vary from mild to severe, even within a family. Patients should be monitored for hypokalemic metabolic alkalosis, hypomagnesemia, hyponatremia, or hypochloremia. Electrolytes will need to be corrected when aberrant.

There is some discussion in the literature about whether intellectual disability is a core feature of EAST syndrome or whether academic difficulties reflect the severe ataxia, hearing impairment, and seizures. Intellectual disability, when reported, is often mild (Mir et al., 2019).

Sicca and colleagues (2011) reported two unrelated families with a maternally inherited rare heterozygous *KCNJ10* missense variant present in children affected by autism and epilepsy. In one family the mother had a history of tic disorder, obsessive–compulsive symptoms, and motor clumsiness; in the other family the mother was reported as healthy. One variant was in the N-terminal domain, the other in the first TM domain. Functional studies in a cellular model suggested a pathological gain-of-function effect. However, the lack of segregation with epilepsy in the family means it remains uncertain if heterozygous variants are truly causative of the epilepsy phenotype in these two families, and larger clinical cohorts will be required to confirm this.

Clinical Management Implications of Channelopathy

The seizure management in EAST syndrome has been reviewed by Cross and colleagues (2013) and Mir and colleagues (2019). Both groups reported that monotherapy with an AED was often effective. However, recurrence of seizures with AED weaning, including with status epilepticus, has been reported in some cases, even after a prolonged period of being seizure-free, so it is recommended that if weaning is being trialed, this should be accompanied by advice regarding the possibility of recurrence and the prescription of a rescue medication such as buccal midazolam.

Mir and colleagues recommended the use of low-dose carbamazepine as a first-line AED, with lamotrigine as another good option. They point to evidence that carbamazepine may act on ATP-sensitive potassium channels. They recommended avoidance of valproate as it may cause tremors as a side effect, which could worsen the tremor and unsteadiness that often accompany this condition, and because valproate can be linked to hepatic deterioration when started early in childhood. They also suggest avoiding topiramate as a first-line drug if possible due to possible side effects of cognitive impairment and impact on verbal fluency.

KCNJ11 Encoding $K_{ir}6.2$

Expression and Function in Human Brain

KCNJ11 is widely expressed, including in the brain and the pancreas. It encodes the $K_{ir}6.2$ subunit of the ATP-dependent potassium (K_{ATP}) channel. The K_{ATP} channel is formed by the combination of four $K_{ir}6.2$ subunits with four sulfonylurea receptors (SUR1 or SUR2). Opening of the channel depends on intracellular ATP/adenosine diphosphate levels, and thus the channel allows the metabolic state of a cell to be directly coupled to its electrical activity (Liss & Roeper, 2001). Increased glucose in the cell leads to increased ATP, which inhibits the K_{ATP} channel, causing channel closure. Channel closure leads to membrane depolarization. In the pancreas, K_{ATP} channel closure and membrane depolarization lead to activation of voltage-dependent calcium channels, which triggers an increase in intracellular calcium, which then results in insulin secretion.

In the brain, *KCNJ11* is expressed widely, with particularly high levels of expression in the cerebellum. The precise role of $K_{ir}6.2$ containing channels in the brain is not fully understood, but they likely play a role in glucose sensing and adapting neuronal activity to metabolic demands, as well as seizure propagation (Liss & Roeper, 2001).

Reported Human Neurological Conditions, Associated Variants, and Insights into Pathophysiology

The first clinical symptom associated with activating variants in the *KCNJ11* gene was that of permanent neonatal diabetes mellitus (PNDM; MIM 606176), which can be, in >90% of affected individuals, appropriately treated with oral sulfonylurea medication, avoiding the need for insulin injections (Pearson et al., 2006). This is because activating *KCNJ11* variant cause a reduced response of the K_{ATP} channel to ATP (i.e., it does not tend to close with high intracellular glucose, so insulin secretion is not provoked). However, sulfonylurea medication can bypass this malfunction, causing the K_{ATP} channel to close, resulting in insulin secretion (Gloyn et al., 2004; Pearson et al., 2006).

It rapidly became clear that children and adults with PNDM due to *KCNJ11* variants also had a range of neuromuscular and neurocognitive symptoms. In some individuals these symptoms were fairly subtle, including specific learning difficulties (abnormalities in attention, visuospatial coordination, and executive function) and mental health difficulties (anxiety disorder, psychiatric comorbidities, and sleep disturbances). In others a more overt intellectual disability and epilepsy were present, often associated with muscle weakness, and such individuals may be described as having DEND syndrome (developmental delay, epilepsy and neonatal diabetes) (Gloyn et al., 2004). A small proportion of children have a very severe type of DEND, with marked developmental delay, muscle weakness, and severe infantile-onset generalized epilepsy starting before the first year of life, associated with abnormal EEG that could show bilateral epileptiform activity or hypsarrhythmia.

There is emerging evidence of genotype–phenotype correlation. Certain *KCNJ11* variants are more associated with DEND and others with only mild neuropsychiatric manifestations. Residues associated with neonatal diabetes alone (transient or permanent neonatal diabetes) tend to lie within the ATP binding site in the N terminal region or are located at the interfaces between $K_{ir}6.2$ subunits. In contrast, variants associated with DEND cluster at residues some distance from the ATP binding site, for example, residues associated with functional coupling to the SUR1 subunit, or affect residues implicated in the regulation of channel gating (Hattersley & Ashcroft, 2005). Variants associated with neonatal diabetes alone tend to impair ATP-dependent channel inhibition without affecting significantly the open probability of the channel in the absence of ATP, whereas variants associated with prominent neurological features markedly bias the channel to the open state. This results indirectly in reduction of the ability of ATP to block the channel because ATP stabilizes the long-closed state of the channel, which is now less frequent (Hattersley & Ashcroft, 2005).

Clinical Management Implications of Channelopathy

Finding the underlying genetic cause is recommended for neonatal-onset diabetes mellitus as the finding of a *KCNJ11* variant should prompt close neurodevelopmental follow-up and early treatment of any emerging seizures. Sulfonylurea medication will typically be started on genetic diagnosis, allowing discontinuation of insulin. However, variants which affect the open probability of the channel (i.e., those which are most associated with neurological symptoms) are less sensitive to sulfonylureas and may require higher drug doses. The extent to which sulfonylureas will improve neurological features independently of diabetes control is uncertain, but it is recommended that less specific sulfonylureas may be preferred over SUR1-selective sulfonylureas (such as glicazide and tolbutamide) as they will bind to both the SUR1 receptors in the pancreas and the SUR2 receptors associated with muscle and brain K_{ATP} channels (Hattersley & Ashcroft, 2005).

Four Transmembrane Domain Family, Also Known as the Two P Domain Potassium Channels

KCNK4 Encoding $K_{2P}4.1$

Expression and Function in Human Brain

KCNK4 encodes a subunit of a two P domain potassium channel which is sensitive to lipid and mechanical stimulation: specifically, $K_{2P}4.1$ containing channels are opened by arachidonic acid and polyunsaturated fatty acids, membrane stretch, and cell swelling and closed by hyperosmolarity. The channel is also known as TRAAK (TWIK-related arachidonic acid-stimulated K^+ channel) and is dimeric, with each subunit contributing two pore domains (Figure 5.17). Each subunit also has four transmembrane domains and an extracellular cap covering the pore. Different models of gating have been proposed. One hypothesis suggests that bending of M4 converts the channel between the open and closed states: when M4 is straight, lipid moieties can insert in the lateral fenestration and potassium flux is blocked (closed), whereas bending of M4 closes the fenestrations, prevents lipid insertions, and leads to a conducting (open) state of the channel. In other models, channel activation involves conformational transitions of the selectivity filter, formed by the P domains (Bauer et al., 2018).

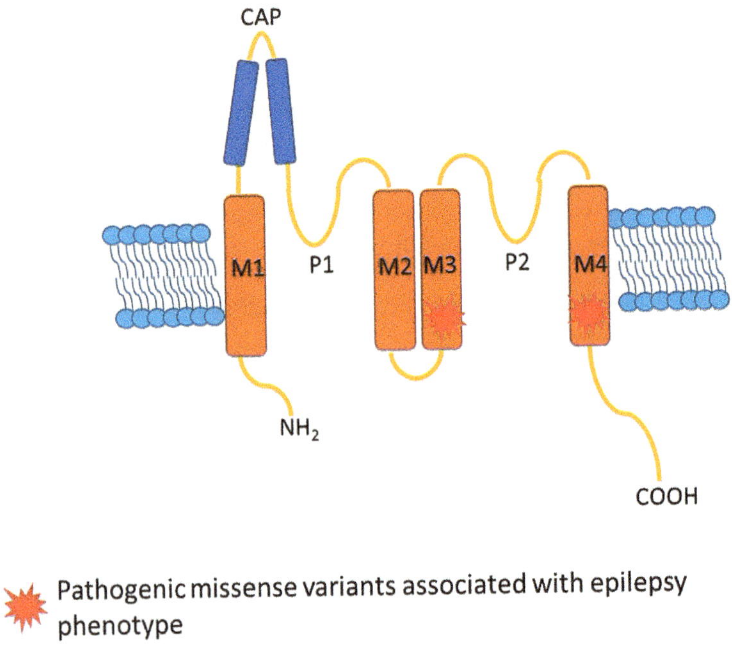

FIGURE 5.17 Position of pathogenic missense variants in the $K_{2P}4.1$ channel.

$K_{2P}4.1$ containing channels allow passive potassium flux out of cells. These channels are involved in cell volume regulation. In the central nervous system, they are understood to maintain the resting membrane potential and to be important targets for modulation of neuronal activity. They are expressed in the human and rodent brains. In the mouse brain they localize to neurons but not astrocytes and are particularly highly expressed in the hippocampus. Their exact role in brain function is incompletely understood, knockout in mice resulted in relative protection from ischemic stroke (Laigle, Confort-Gouny, Le Fur, Cozzone, & Viola, 2012).

Reported Human Neurological Conditions, Associated Variants, and Insights into Pathophysiology

Bauer and colleagues (2018) reported two de novo missense variants, p.(Ala172Glu and p.(Ala244Pro) (Figure 5.17), in *KCNK4* in three unrelated individuals resulting in a recognizable multisystemic condition which they called FHEIG (facial dysmorphism, hypertrichosis, epilepsy, intellectual disability/developmental delay, and gingival overgrowth; MIM 618381). Facial features include bitemporal narrowing, short philtrum, bushy eyebrows, and long eyelashes with gingival overgrowth.

Two of the three individuals had epilepsy: one had severe generalized seizures (tonic-clonic and absence seizures) starting in the second year of life, the other had right focal clonic seizures with secondary generalization starting at 10 months of age, and the third patient had EEG abnormalities in the absence of overt seizures. Seizure control was achieved with a combination of AEDs in the first patient (levetiracetam, oxcarbamazepine, carbamazepine, clonazepam, and pyridoxine) and with carbamazepine therapy alone in the second patient. Intellectual disability was severe in the two patients with the p.(Ala172Glu) variant and milder in the patient with the p.(Ala244Pro) variant. The variants lay in the M3 and M4 domains, and functional studies showed that both variants resulted in a significant gain of function basally and impaired sensitivity to mechanical stimulation and arachidonic acid. Molecular dynamics simulations indicated that both variants resulted in sealing of the lateral intramembrane fenestration that has been proposed to negatively control K^+ flow by allowing lipid access to the central cavity of the channel, thus supporting the first gating hypothesis described.

Clinical Management Implications of Channelopathy

Reporting on more patients is required to make clear statements about which AEDs may be helpful. The efficacy of carbamazepine as a single agent in one of the patients reported by Bauer and colleagues (2018) is noted.

Conclusions and Future Directions

Table 5.1 summarizes the potassium channel families and which subunits are known to be expressed in brain and, to date, reported to be associated with a neurological phenotype, especially if that includes epilepsy. As new associations between rare variants in K channel subunits and epilepsy phenotypes continue to be made, due to advances in the ability to offer unbiased genomic testing, combined with international collaborations to rapidly delineate clinical cohorts, it is likely that the list of "K channelopathies" in epilepsy will continue to grow over the coming years, further improving our ability to diagnose the underlying cause of epilepsy.

This review also highlights that missense variants reported in association with severe epilepsy phenotypes typically alter the three-dimensional amino acid structure of certain critical parts of potassium channels. Such mutation "hot spots" include the amino acid residues that line the pore, making up the selectivity filter for the channel, and residues that are involved in the regulation of channel gating or inactivation. This is highlighted by the channel subunit diagrams for individual genes (Figures 5.2–5.17). Such missense variants may result in a range of differences in channel function, such as an increase in the likelihood or duration of opening (often labeled a "gain" of channel function) or an increase in duration or likelihood of closing (often labeled a "loss" of channel function). Some variants, however, clearly result in much more complex changes in channel function, such as an alteration of the permeability of the channel to potassium or other ions or the chance of channel opening or closing under different physiological or disease states (which should perhaps more accurately be labeled a "change" in channel function). For a good example of this complexity, see the range of functional effects of different variants in channels including subunits encoded by *KCNT1* and *KCNT2*. Yet other variants may affect the ability of the channel to be expressed on the membrane surface or aggregate with other subunits or modifying proteins.

Some channelopathies are linked to genetic variants described as nonsense or frameshift variants which will result in reduced expression of the channel. In genetic terms these are called *loss-of-function variants*. It is notable that often, although not always, such loss-of-function variants in genes encoding channel subunits are linked to a milder neurological phenotype, for example, without prominent epilepsy features or with self-limiting or easily treated seizures. For example, see the discussion regarding variants in *KCNQ2*.

It is also clear that genotype–phenotype correlations are often not at all straightforward. The same variant may be associated with widely differing clinical features, even when present in affected members of the same family. Divisions of channelopathies into different clinical groups (e.g., *KCNJ10* SeSAME/EAST syndrome, EIMFS/ADNFLE, *KCNH1*-related TBS/ZLS, *KCNQ2*-BFNE/NEE) can often be confusing; for example, the same variant sometimes results in presentations consistent with a range of clinical groups.

However, understanding not only the involved gene but the actual variant, how the variant in the gene affects gene expression and channel function, and the predominant clinical symptoms is critical for efforts to move toward targeted therapies. A more accurate (albeit longer) description of channelopathies may therefore be helpful, one that summarizes these features. An example could be along these lines: *KCNQ5*-encephalopathy [gene], due to de novo p.(Pro369Arg) variant [variant]; a variant which results in gain of channel function in a single-cell model [functional effect] and a DEE phenotype [phenotype]. Such a description may help clinicians in decision-making about therapies, for example, thinking about channel inhibitors only for patients whose variant in the gene results in a gain of channel function, not all patients with a variant (causing gain, loss, or a more complex change of channel function).

As summarized, many potassium channelopathies are associated with severe epilepsies, neurodevelopmental delays or regression, alterations in tone or movement disorders, or extra neurological features. Despite improvements in understanding of the cause of these channelopathies, in terms of identification of a (likely) pathogenic gene variant, very few conditions have a targeted treatment, supported by evidence from clinical trials. There is, however, an urgent need to move from diagnosis to improved diagnosis.

To enable this move to targeted therapies, it would be ideal to gain a precise understanding of how each individual variant affects the expression and function of that channel subunit and the range of complex channels it forms a part of, not only in individual human neuronal or glial cells but also in complex brain circuits under a range of physiological and pathophysiological conditions. This requires the ability to move from heterologous single-cell electrophysiological studies to more complex models better representing the human brain, such as three-dimensional brain organoid models (Lee, Bendriem, Wu, & Shen, 2017). However, as the clinical features can vary widely among individuals with the same variant, possibly due to the impact of different genetic or environmental backgrounds, more personalized models may be required, for example, along the lines of organoid studies tailored to different genetic backgrounds which are being trialed in cystic fibrosis (Noordhoek, Gulmans, van der Ent, & Beekman, 2016). The development of such personalized models of individual variants may allow rational drug screening and return of a personalized medication prescription for all affected individuals.

A complementary approach is to better document the natural history and treatment responses of individual channelopathies. This is likely to require international consented patient registries and optimal ways to accurately record seizure types, severity and frequency, and neurodevelopmental outcomes. As an example of how useful this approach can be, some recommendations for therapies have emerged from surveys of patients with different genetic epilepsies which record patient responses to different drug regimes (Sadleir et al., 2020).

When different therapeutic approaches are considered, approaches that attempt to upregulate wild-type copies of the affected gene, downregulate mutations of the gene, or correct the precise mutation in the gene by gene editing may become viable options in the near future due to advances in technologies such as crispr/cas9 gene editing and targeted oligonucleotide approaches (Lee et al., 2017; Liu & Jaenisch, 2019; Petrou et al., 2018). Such approaches may impact on a range of clinical symptoms, not just those related to epilepsy. A critical question is how early such therapies could be commenced, especially if neurodevelopment is hoped to be preserved. As discussed, most severe potassium channelopathies are due to de novo variants in potassium α-subunit genes in the absence of a family history. Therefore, genetic diagnoses are typically only made after commencement of symptoms, most often seizures, in early infancy or childhood. Is this, however, too late to consider that therapies might correct the neurodevelopmental trajectory? May in utero therapies be feasible to achieve this aim? This would require prenatal genetic diagnosis, which is not outside the realm of possibility, although it requires careful ethical consideration (Adam, 2018).

In summary, the last 10 years have seen an explosion of genetic discoveries linking rare variants in genes encoding the α subunits of potassium channels. It is likely that genetic discoveries will continue to be made and improve our understanding of genotype–phenotype correlations. We may make inroads into the role of these channels into later-onset and milder seizures. However, it is critical that these genetic and clinical discoveries are coupled with basic and translational science aimed at improving our knowledge of the pathophysiological impact of these variants in each affected individual and how best to target therapies. This concerted effort is required in order to improve not only seizure control for affected patients but their overall quality of life.

Notes

1. MIM refers to the indexed phenotype in the Online Inheritance in Man (OMIM) database: https://www.omim.org/
2. gnomAD v2.11 non neuro: https://gnomad.broadinstitute.org/gene/ENSG00000184408?dataset=gnomad_r2_1_non_neuro

References

Abdelnour, E., Gallentine, W., McDonald, M., Sachdev, M., Jiang, Y. H., & Mikati, M. A. (2018). Does age affect response to quinidine in patients with KCNT1 mutations? Report of three new cases and review of the literature. *Seizure, 55*, 1–3. doi:10.1016/j.seizure.2017.11.017

Adam, M. P. (2018). Is prenatal genomic testing ready for prime time? *Genetics in Medicine, 20*(7), 695–696. doi:10.1038/gim.2018.5

Aggarwal, S. K., & MacKinnon, R. (1996). Contribution of the S4 segment to gating charge in the Shaker K^+ channel. *Neuron, 16*(6), 1169–1177. doi:10.1016/s0896-6273(00)80143-9

Ahern, S., Sims, G., Tacey, M., Esler, M., Oldroyd, J., Dean, J., & Bell, S. (2017). *The Australian Cystic Fibrosis Data Registry Annual Report, 2015* (Report No. 18). Retrieved from https://research.monash.edu/en/publications/australian-cystic-fibrosis-data-registry-annual-report-2015

Aldrich R, Chandy KG, Grissmer S, Gutman GA, Kaczmarek LK, Wei AD, & Wulff H. (2019). Calcium- and sodium-activated potassium channels (version 2019.4) in the IUPHAR/BPS Guide to Pharmacology Database. IUPHAR/BPS Guide to Pharmacology CITE. Retrieved from https://doi.org/10.2218/gtopdb/F69/2019.4.

Allen, N. M., Weckhuysen, S., Gorman, K., King, M. D., & Lerche, H. (2020). Genetic potassium channel–associated epilepsies: Clinical review of the K_v family. *European Journal of Paediatric Neurology, 24*, 105–116. doi:10.1016/j.ejpn.2019.12.002

Ambrosino, P., Soldovieri, M. V., Bast, T., Turnpenny, P. D., Uhrig, S., Biskup, S., . . . Lemke, J. R. (2018). De novo gain-of-function variants in KCNT2 as a novel cause of developmental and epileptic encephalopathy. *Annals of Neurology*, *83*(6), 1198–1204. doi:10.1002/ana.25248

Bailey, C. S., Moldenhauer, H. J., Park, S. M., Keros, S., & Meredith, A. L. (2019). KCNMA1-linked channelopathy. *Journal of General Physiology*, *151*(10), 1173–1189. doi:10.1085/jgp.201912457

Barcia, G., Fleming, M. R., Deligniere, A., Gazula, V. R., Brown, M. R., Langouet, M., . . . Nabbout, R. (2012). De novo gain-of-function KCNT1 channel mutations cause malignant migrating partial seizures of infancy. *Nature Genetics*, *44*(11), 1255–1259. doi:10.1038/ng.2441

Bauer, C. K., Calligari, P., Radio, F. C., Caputo, V., Dentici, M. L., Falah, N., . . . Tartaglia, M. (2018). Mutations in KCNK4 that affect gating cause a recognizable neurodevelopmental syndrome. *American Journal of Human Genetics*, *103*(4), 621–630. doi:10.1016/j.ajhg.2018.09.001

Bauer, C. K., Schneeberger, P. E., Kortüm, F., Altmüller, J., Santos-Simarro, F., Baker, L., . . . Kutsche, K. (2019). Gain-of-function mutations in KCNN3 encoding the small-conductance Ca^{2+}-activated K^+ channel SK3 cause Zimmermann-Laband syndrome. *American Journal of Human Genetics*, *104*(6), 1139–1157. doi:10.1016/j.ajhg.2019.04.012

Bearden, D., Strong, A., Ehnot, J., DiGiovine, M., Dlugos, D., & Goldberg, E. M. (2014). Targeted treatment of migrating partial seizures of infancy with quinidine. *Annals of Neurology*, *76*(3), 457–461. doi:10.1002/ana.24229

Berkovic, S. F. (2015). Genetics of epilepsy in clinical practice. *Epilepsy Currents*, *15*(4), 192–196. doi:10.5698/1535-7511-15.4.192

Bhattacharjee, A., Gan, L., & Kaczmarek, L. K. (2002). Localization of the Slack potassium channel in the rat central nervous system. *Journal of Comparative Neurology*, *454*(3), 241–254. doi:10.1002/cne.10439

Biesecker, L. G., & Green, R. C. (2014). Diagnostic clinical genome and exome sequencing. *New England Journal of Medicine*, *370*(25), 2418–2425. doi:10.1056/NEJMra1312543

Bishop, H. I., Guan, D., Bocksteins, E., Parajuli, L. K., Murray, K. D., Cobb, M. M., . . . Trimmer, J. S. (2015). Distinct cell- and layer-specific expression patterns and independent regulation of Kv2 channel subtypes in cortical pyramidal neurons. *Journal of Neuroscience*, *35*(44), 14922–14942. doi:10.1523/JNEUROSCI.1897-15.2015

Bockenhauer, D., Feather, S., Stanescu, H. C., Bandulik, S., Zdebik, A. A., Reichold, M., . . . Kleta, R. (2009). Epilepsy, ataxia, sensorineural deafness, tubulopathy, and KCNJ10 mutations. *New England Journal of Medicine*, *360*(19), 1960–1970. doi:10.1056/NEJMoa0810276

Bosch, D. G., Boonstra, F. N., de Leeuw, N., Pfundt, R., Nillesen, W. M., de Ligt, J., . . . de Vries, B. B. (2016). Novel genetic causes for cerebral visual impairment. *European Journal of Human Genetics*, *24*(5), 660–665. doi:10.1038/ejhg.2015.186

Bowdin, S., Gilbert, A., Bedoukian, E., Carew, C., Adam, M. P., Belmont, J., . . . Krantz, I. D. (2016). Recommendations for the integration of genomics into clinical practice. *Genetics in Medicine*, *18*(11), 1075–1084. doi:10.1038/gim.2016.17

Brown, M. R., Kronengold, J., Gazula, V. R., Spilianakis, C. G., Flavell, R. A., von Hehn, C. A., . . . Kaczmarek, L. K. (2008). Amino-termini isoforms of the Slack K^+ channel, regulated by alternative promoters, differentially modulate rhythmic firing and adaptation. *Journal of Physiology*, *586*(21), 5161–5179. doi:10.1113/jphysiol.2008.160861

Cameron, J. M., Maljevic, S., Nair, U., Aung, Y. H., Cogne, B., Bezieau, S., . . . Berkovic, S. F. (2019). Encephalopathies with KCNC1 variants: Genotype–phenotype–functional correlations. *Annals of Clinical and Translational Neurology, 6*(7), 1263–1272. doi:10.1002/acn3.50822

Celmina, M., Micule, I., Inashkina, I., Audere, M., Kuske, S., Pereca, J., . . . Strautmanis, J. (2019). EAST/SeSAME syndrome: Review of the literature and introduction of four new Latvian patients. *Clinical Genetics, 95*(1), 63–78. doi:10.1111/cge.13374

Chen, H., Kronengold, J., Yan, Y., Gazula, V. R., Brown, M. R., Ma, L., . . . Kaczmarek, L. K. (2009). The N-terminal domain of Slack determines the formation and trafficking of Slick/Slack heteromeric sodium-activated potassium channels. *Journal of Neuroscience, 29*(17), 5654–5665. doi:10.1523/JNEUROSCI.5978-08.2009

Chen, Z., Brodie, M. J., Liew, D., & Kwan, P. (2018). Treatment outcomes in patients with newly diagnosed epilepsy treated with established and new antiepileptic drugs: A 30-year longitudinal cohort study. *JAMA Neurology, 75*(3), 279–286. doi:10.1001/jamaneurol.2017.3949

Choe, S. (2002). Potassium channel structures. *Nature Reviews Neuroscience, 3*(2), 115–121. doi:10.1038/nrn727

Cooper, E. C., & Jan, L. Y. (2003). M-channels: Neurological diseases, neuromodulation, and drug development. *Archives of Neurology, 60*(4), 496–500. doi:10.1001/archneur.60.4.496

Corbett, M. A., Bellows, S. T., Li, M., Carroll, R., Micallef, S., Carvill, G. L., . . . Gecz, J. (2016). Dominant KCNA2 mutation causes episodic ataxia and pharmacoresponsive epilepsy. *Neurology, 87*(19), 1975–1984. doi:10.1212/WNL.0000000000003309

Cross, J. H., Arora, R., Heckemann, R. A., Gunny, R., Chong, K., Carr, L., . . . Bockenhauer, D. (2013). Neurological features of epilepsy, ataxia, sensorineural deafness, tubulopathy syndrome. *Developmental Medicine and Child Neurology, 55*(9), 846–856. doi:10.1111/dmcn.12171

D'Adamo, M. C., Gallenmuller, C., Servettini, I., Hartl, E., Tucker, S. J., Arning, L., . . . Klopstock, T. (2014). Novel phenotype associated with a mutation in the KCNA1(Kv1.1) gene. *Frontiers in Physiology, 5*, 525. doi:10.3389/fphys.2014.00525

D'Adamo, M. C., Imbrici, P., Sponcichetti, F., & Pessia, M. (1999). Mutations in the KCNA1 gene associated with episodic ataxia type-1 syndrome impair heteromeric voltage-gated K^+ channel function. *FASEB Journal, 13*(11), 1335–1345. doi:10.1096/fasebj.13.11.1335

de Boer, H. M., Mula, M., & Sander, J. W. (2008). The global burden and stigma of epilepsy. *Epilepsy & Behavior, 12*(4), 540–546. doi:10.1016/j.yebeh.2007.12.019

de Kovel, C. G. F., Syrbe, S., Brilstra, E. H., Verbeek, N., Kerr, B., Dubbs, H., . . . Koeleman, B. P. C. (2017). Neurodevelopmental disorders caused by de novo variants in KCNB1 genotypes and phenotypes. *JAMA Neurology, 74*(10), 1228–1236. doi:10.1001/jamaneurol.2017.1714

Djukic, B., Casper, K. B., Philpot, B. D., Chin, L. S., & McCarthy, K. D. (2007). Conditional knock-out of Kir4.1 leads to glial membrane depolarization, inhibition of potassium and glutamate uptake, and enhanced short-term synaptic potentiation. *Journal of Neuroscience, 27*(42), 11354–11365. doi:10.1523/JNEUROSCI.0723-07.2007

Du, W., Bautista, J. F., Yang, H., Diez-Sampedro, A., You, S. A., Wang, L., . . . Wang, Q. K. (2005). Calcium-sensitive potassium channelopathy in human epilepsy and paroxysmal movement disorder. *Nature Genetics, 37*(7), 733–738. doi:10.1038/ng1585

Duarri, A., Jezierska, J., Fokkens, M., Meijer, M., Schelhaas, H. J., den Dunnen, W. F., . . . Verbeek, D. S. (2012). Mutations in potassium channel kcnd3 cause spinocerebellar ataxia type 19. *Annals of Neurology, 72*(6), 870–880. doi:10.1002/ana.23700

Epi4K Consortium, Epilepsy Phenome/Genome Project, Allen, A. S., Berkovic, S. F., Cossette, P., Delanty, N., . . . Winawer, M. R. (2013). De novo mutations in epileptic encephalopathies. *Nature, 501*(7466), 217–221. doi:10.1038/nature12439

Feliciangeli, S., Chatelain, F. C., Bichet, D., & Lesage, F. (2015). The family of K2P channels: Salient structural and functional properties. *Journal of Physiology*, *593*(12), 2587–2603. doi:10.1113/jphysiol.2014.287268

Gan, L., & Kaczmarek, L. K. (1998). When, where, and how much? Expression of the Kv3.1 potassium channel in high-frequency firing neurons. *Journal of Neurobiology*, *37*(1), 69–79. doi:10.1002/(sici)1097-4695(199810)37:1<69::aid-neu6>3.0.co;2-6

Giudicessi, J. R., Ye, D., Kritzberger, C. J., Nesterenko, V. V., Tester, D. J., Antzelevitch, C., & Ackerman, M. J. (2012). Novel mutations in the KCND3-encoded Kv4.3 K^+ channel associated with autopsy-negative sudden unexplained death. *Human Mutation*, *33*(6), 989–997. doi:10.1002/humu.22058

Giudicessi, J. R., Ye, D., Tester, D. J., Crotti, L., Mugione, A., Nesterenko, V. V., . . . Ackerman, M. J. (2011). Transient outward current (I(to)) gain-of-function mutations in the KCND3-encoded Kv4.3 potassium channel and Brugada syndrome. *Heart Rhythm*, *8*(7), 1024–1032. doi:10.1016/j.hrthm.2011.02.021

Gloyn, A. L., Pearson, E. R., Antcliff, J. F., Proks, P., Bruining, G. J., Slingerland, A. S., . . . Hattersley, A. T. (2004). Activating mutations in the gene encoding the ATP-sensitive potassium-channel subunit Kir6.2 and permanent neonatal diabetes. *New England Journal of Medicine*, *350*(18), 1838–1849. doi:10.1056/NEJMoa032922

Goldman, A. M. (2015). Mechanisms of sudden unexplained death in epilepsy. *Current Opinion in Neurology*, *28*(2), 166–174. doi:10.1097/WCO.0000000000000184

Guerrini, R., Dravet, C., Genton, P., Belmonte, A., Kaminska, A., & Dulac, O. (1998). Lamotrigine and seizure aggravation in severe myoclonic epilepsy. *Epilepsia*, *39*(5), 508–512.

Gururaj, S., Palmer, E. E., Sheehan, G. D., Kandula, T., Macintosh, R., Ying, K., . . . Bhattacharjee, A. (2017). A de novo mutation in the sodium-activated potassium channel KCNT2 alters ion selectivity and causes epileptic encephalopathy. *Cell Reports*, *21*(4), 926–933. doi:10.1016/j.celrep.2017.09.088

Hasan, S. M., & D'Adamo, M. C. (2010). Episodic ataxia type 1. In M. P. Adam, H. H. Ardinger, R. A. Pagon, S. E. Wallace, L. J. H. Bean, K. Stephens, & A. Amemiya (Eds.), *GeneReviews*. Seattle, WA: University of Washington.

Hattersley, A. T., & Ashcroft, F. M. (2005). Activating mutations in Kir6.2 and neonatal diabetes: New clinical syndromes, new scientific insights, and new therapy. *Diabetes*, *54*(9), 2503–2513. doi:10.2337/diabetes.54.9.2503

Helbig, I., Heinzen, E. L., Mefford, H. C., International League Against Epilepsy Genetics Commission. (2018). Genetic literacy series: Primer part 2-Paradigm shifts in epilepsy genetics. *Epilepsia*, *59*(6), 1138–1147. doi:10.1111/epi.14193

Heron, S. E., Smith, K. R., Bahlo, M., Nobili, L., Kahana, E., Licchetta, L., . . . Dibbens, L. M. (2012). Missense mutations in the sodium-gated potassium channel gene KCNT1 cause severe autosomal dominant nocturnal frontal lobe epilepsy. *Nature Genetics*, *44*(11), 1188–1190. doi:10.1038/ng.2440

Higashi, K., Fujita, A., Inanobe, A., Tanemoto, M., Doi, K., Kubo, T., & Kurachi, Y. (2001). An inwardly rectifying K^+ channel, Kir4.1, expressed in astrocytes surrounds synapses and blood vessels in brain. *American Journal of Physiology Cell Physiology*, *281*(3), C922-C931. doi:10.1152/ajpcell.2001.281.3.C922

Jacoby, A., & Baker, G. A. (2008). Quality-of-life trajectories in epilepsy: A review of the literature. *Epilepsy & Behavior*, *12*(4), 557–571. doi:10.1016/j.yebeh.2007.11.013

Jan, L. Y., & Jan, Y. N. (2012). Voltage-gated potassium channels and the diversity of electrical signalling. *Journal of Physiology*, *590*(11), 2591–2599. doi:10.1113/jphysiol.2011.224212

Kang, S. K., Vanoye, C. G., Misra, S. N., Echevarria, D. M., Calhoun, J. D., O'Connor, J. B., . . . Kearney, J. A. (2019). Spectrum of $K_V2.1$ dysfunction in *KCNB1*-associated neurodevelopmental disorders. *Annals of Neurology*, *86*(6), 899–912. doi:10.1002/ana.25607

Kaya, N., Alsagob, M., D'Adamo, M. C., Al-Bakheet, A., Hasan, S., Muccioli, M., . . . Al-Owain, M. (2016). *KCNA4* deficiency leads to a syndrome of abnormal striatum, congenital cataract and intellectual disability. *Journal of Medical Genetics*, *53*(11), 786–792. doi:10.1136/jmedgenet-2015-103637

Khan, S., & Al Baradie, R. (2012). Epileptic encephalopathies: An overview. *Epilepsy Research and Treatment*, *2012*, 403592. doi:10.1155/2012/403592

Kohling, R., & Wolfart, J. (2016). Potassium channels in epilepsy. *Cold Spring Harbor Perspectives in Medicine*, *6*(5), a022871. doi:10.1101/cshperspect.a022871

Kong, S. T., Ho, C. S., Ho, P. C., & Lim, S. H. (2014). Prevalence of drug resistant epilepsy in adults with epilepsy attending a neurology clinic of a tertiary referral hospital in Singapore. *Epilepsy Research*, *108*(7), 1253–1262. doi:10.1016/j.eplepsyres.2014.05.005

Kortüm, F., Caputo, V., Bauer, C. K., Stella, L., Ciolfi, A., Alawi, M., . . . Kutsche, K. (2015). Mutations in KCNH1 and ATP6V1B2 cause Zimmermann-Laband syndrome. *Nature Genetics*, *47*(6), 661–667. doi:10.1038/ng.3282

Kuang, Q., Purhonen, P., & Hebert, H. (2015). Structure of potassium channels. *Cellular and Molecular Life Sciences*, *72*(19), 3677–3693. doi:10.1007/s00018-015-1948-5

Laigle, C., Confort-Gouny, S., Le Fur, Y., Cozzone, P. J., & Viola, A. (2012). Deletion of TRAAK potassium channel affects brain metabolism and protects against ischemia. *PLoS One*, *7*(12), e53266. doi:10.1371/journal.pone.0053266

Lee, C. T., Bendriem, R. M., Wu, W. W., & Shen, R. F. (2017). 3D brain organoids derived from pluripotent stem cells: Promising experimental models for brain development and neurodegenerative disorders. *Journal of Biomedical Science*, *24*(1), 59. doi:10.1186/s12929-017-0362-8

Lee, Y. C., Durr, A., Majczenko, K., Huang, Y. H., Liu, Y. C., Lien, C. C., . . . Soong, B. W. (2012). Mutations in KCND3 cause spinocerebellar ataxia type 22. *Annals of Neurology*, *72*(6), 859–869. doi:10.1002/ana.23701

Lehman, A., Thouta, S., Mancini, G. M. S., Naidu, S., van Slegtenhorst, M., McWalter, K., . . . Claydon, T. (2017). Loss-of-function and gain-of-function mutations in KCNQ5 cause intellectual disability or epileptic encephalopathy. *American Journal of Human Genetics*, *101*(1), 65–74. doi:10.1016/j.ajhg.2017.05.016

Liang, L., Li, X., Moutton, S., Schrier Vergano, S. A., Cogné, B., Saint-Martin, A., . . . Wang, Q. K. (2019). De novo loss-of-function KCNMA1 variants are associated with a new multiple malformation syndrome and a broad spectrum of developmental and neurological phenotypes. *Human Molecular Genetics*, *28*(17), 2937–2951. doi:10.1093/hmg/ddz117

Liss, B., & Roeper, J. (2001). Molecular physiology of neuronal K-ATP channels (review). *Molecular Membrane Biology*, *18*(2), 117–127.

Liu, X. S., & Jaenisch, R. (2019). Editing the epigenome to tackle brain disorders. *Trends in Neurosciences*, *42*(12), 861–870. doi:10.1016/j.tins.2019.10.003

Long, S. B., Campbell, E. B., & Mackinnon, R. (2005). Voltage sensor of Kv1.2: Structural basis of electromechanical coupling. *Science*, *309*(5736), 903–908. doi:10.1126/science.1116270

Lorincz, A., & Nusser, Z. (2008). Cell-type-dependent molecular composition of the axon initial segment. *Journal of Neuroscience*, *28*(53), 14329–14340. doi:10.1523/JNEUROSCI.4833-08.2008

Mao, X., Bruneau, N., Gao, Q., Becq, H., Jia, Z., Xi, H., . . . Aniksztejn, L. (2020). The epilepsy of infancy with migrating focal seizures: Identification of de novo mutations of the KCNT2 gene that exert inhibitory effects on the corresponding heteromeric $K_{Na}1.1/K_{Na}1.2$ potassium channel. *Frontiers in Cell Neuroscience, 14*, 1. doi:10.3389/fncel.2020.00001

Martin, H. C., Kim, G. E., Pagnamenta, A. T., Murakami, Y., Carvill, G. L., Meyer, E., ... Fleming, M. R. (2014). Clinical whole-genome sequencing in severe early-onset epilepsy reveals new genes and improves molecular diagnosis. *Human Molecular Genetics, 23*(12), 3200–3211.

Mastrangelo, M., Scheffer, I. E., Bramswig, N. C., Nair, L. D., Myers, C. T., Dentici, M. L., ... Leuzzi, V. (2016). Epilepsy in KCNH1-related syndromes. *Epileptic Disorders, 18*(2), 123–136. doi:10.1684/epd.2016.0830

McTague, A., Nair, U., Malhotra, S., Meyer, E., Trump, N., Gazina, E. V., ... Kurian, M. A. (2018). Clinical and molecular characterization of KCNT1-related severe early-onset epilepsy. *Neurology, 90*(1), e55–e66. doi:10.1212/WNL.0000000000004762

McTague, A., Howell, K. B., Cross, J. H., Kurian, M. A., Scheffer, I. E. (2016). The genetic landscape of the epileptic encephalopathies of infancy and childhood. *Lancet Neurology, 15*(3), 304–316. doi:10.1016/S1474-4422(15)00250-1

Miceli, F., Soldovieri, M. V., Ambrosino, P., De Maria, M., Migliore, M., Migliore, R., & Taglialatela, M. (2015). Early-onset epileptic encephalopathy caused by gain-of-function mutations in the voltage sensor of Kv7.2 and Kv7.3 potassium channel subunits. *Journal of Neuroscience, 35*(9), 3782–3793. doi:10.1523/JNEUROSCI.4423-14.2015

Miceli, F., Soldovieri, M. V., Joshi, N., Weckhuysen, S., Cooper, E., & Taglialatela, M. (2010). *KCNQ2*-related disorders. In M. P. Adam, H. H. Ardinger, R. A. Pagon, S. E. Wallace, L. J. H. Bean, K. Stephens, & A. Amemiya (Eds.), *GeneReviews*. Seattle, WA: University of Washington.

Miceli, F., Soldovieri, M. V., Joshi, N., Weckhuysen, S., Cooper, E. C., & Taglialatela, M. (2014). *KCNQ3*-related disorders. In M. P. Adam, H. H. Ardinger, R. A. Pagon, S. E. Wallace, L. J. H. Bean, K. Stephens, & A. Amemiya (Eds.), *GeneReviews*. Seattle, WA: University of Washington.

Miceli, F., Striano, P., Soldovieri, M. V., Fontana, A., Nardello, R., Robbiano, A., ... Mangano, S. (2015). A novel KCNQ3 mutation in familial epilepsy with focal seizures and intellectual disability. *Epilepsia, 56*(2), e15–e20. doi:10.1111/epi.12887

Millichap, J. J., Park, K. L., Tsuchida, T., Ben-Zeev, B., Carmant, L., Flamini, R., ... Cooper, E. C. (2016). KCNQ2 encephalopathy: Features, mutational hot spots, and ezogabine treatment of 11 patients. *Neurology Genetics, 2*(5), e96. doi:10.1212/NXG.0000000000000096

Mir, A., Chaudhary, M., Alkhaldi, H., Alhazmi, R., Albaradie, R., & Housawi, Y. (2019). Epilepsy in patients with EAST syndrome caused by mutation in the KCNJ10. *Brain & Development, 41*(8), 706–715. doi:10.1016/j.braindev.2019.03.009

Møller, R. S., Johannesen, K. M. (2016). Precision medicine: SCN8A encephalopathy treated with sodium channel blockers. *Neurotherapeutics, 13*(1), 190–191. doi:10.1007/s13311-015-0403-5

Mullen, S. A., Carney, P. W., Roten, A., Ching, M., Lightfoot, P. A., Churilov, L., ... Scheffer, I. E. (2018). Precision therapy for epilepsy due to KCNT1 mutations: A randomized trial of oral quinidine. *Neurology, 90*(1), e67–e72. doi:10.1212/WNL.0000000000004769

Munch, A. S., Saljic, A., Boddum, K., Grunnet, M., Hougaard, C., & Jespersen, T. (2018). Pharmacological rescue of mutated K. *European Journal of Pharmacology, 833*, 255–262. doi:10.1016/j.ejphar.2018.06.015

Muona, M., Berkovic, S. F., Dibbens, L. M., Oliver, K. L., Maljevic, S., Bayly, M. A., ... Lehesjoki, A. E. (2015). A recurrent de novo mutation in KCNC1 causes progressive myoclonus epilepsy. *Nature Genetics, 47*(1), 39–46. doi:10.1038/ng.3144

Murakoshi, H., & Trimmer, J. S. (1999). Identification of the Kv2.1 K^+ channel as a major component of the delayed rectifier K^+ current in rat hippocampal neurons. *Journal of Neuroscience*, *19*(5), 1728–1735.

Neusch, C., Rozengurt, N., Jacobs, R. E., Lester, H. A., & Kofuji, P. (2001). Kir4.1 potassium channel subunit is crucial for oligodendrocyte development and in vivo myelination. *Journal of Neuroscience*, *21*(15), 5429–5438.

Noordhoek, J., Gulmans, V., van der Ent, K., & Beekman, J. M. (2016). Intestinal organoids and personalized medicine in cystic fibrosis: A successful patient-oriented research collaboration. *Current Opinion in Pulmonary Medicine*, *22*(6), 610–616. doi:10.1097/MCP.0000000000000315

Oliver, K. L., Franceschetti, S., Milligan, C. J., Muona, M., Mandelstam, S. A., Canafoglia, L., . . . Berkovic, S. F. (2017). Myoclonus epilepsy and ataxia due to KCNC1 mutation: Analysis of 20 cases and K. *Annals of Neurology*, *81*(5), 677–689. doi:10.1002/ana.24929

Palmer, E. E., Schofield, D., Shrestha, R., Kandula, T., Macintosh, R., Lawson, J. A., . . . Sachdev, R. K. (2018). Integrating exome sequencing into a diagnostic pathway for epileptic encephalopathy: Evidence of clinical utility and cost effectiveness. *Molecular Genetics & Genomic Medicine*, *6*(2), 186–199. doi:10.1002/mgg3.355

Park, J., Koko, M., Hedrich, U. B. S., Hermann, A., Cremer, K., Haberlandt, E., . . . Haack, T. B. (2019). KCNC1-related disorders: New de novo variants expand the phenotypic spectrum. *Annals of Clinical and Translational Neurology*, *6*(7), 1319–1326. doi:10.1002/acn3.50799

Pearson, E. R., Flechtner, I., Njolstad, P. R., Malecki, M. T., Flanagan, S. E., Larkin, B., . . . Neonatal Diabetes International Collaborative Group. (2006). Switching from insulin to oral sulfonylureas in patients with diabetes due to Kir6.2 mutations. *New England Journal of Medicine*, *355*(5), 467–477. doi:10.1056/NEJMoa061759

Perucca, P., Scheffer, I. E., Harvey, A. S., James, P. A., Lunke, S., Thorne, N., . . . Kwan, P. (2017). Real-world utility of whole exome sequencing with targeted gene analysis for focal epilepsy. *Epilepsy Research*, *131*, 1–8. doi:10.1016/j.eplepsyres.2017.02.001

Petrou, S., Li, M., Jancovsk, N., Jafar-najad, P., Burbano, L., Nemiroff, A., . . . Rigo, F. (2018). *Antisense oligonucleotide therapy for SCN2A gain-of-function epilepsies.* Paper presented at the American Epilepsy Society's Annual Meeting, New Orleans, LA.

Pisano, T., Numis, A. L., Heavin, S. B., Weckhuysen, S., Angriman, M., Suls, A., . . . Cilio, M. R. (2015). Early and effective treatment of KCNQ2 encephalopathy. *Epilepsia*, *56*(5), 685–691. doi:10.1111/epi.12984

Poirier, K., Viot, G., Lombardi, L., Jauny, C., Billuart, P., & Bienvenu, T. (2017). Loss of function of KCNC1 is associated with intellectual disability without seizures. *European Journal of Human Genetics*, *25*(5), 560–564. doi:10.1038/ejhg.2017.3

Quraishi, I. H., Stern, S., Mangan, K. P., Zhang, Y., Ali, S. R., Mercier, M. R., . . . Kaczmarek, L. K. (2019). An epilepsy-associated KCNT1 mutation enhances excitability of human iPSC-Derived neurons by increasing Slack K_{Na} currents. *Journal of Neuroscience*, *39*(37), 7438–7449. doi:10.1523/JNEUROSCI.1628-18.2019

Rauch, A., Wieczorek, D., Graf, E., Wieland, T., Endele, S., Schwarzmayr, T., . . . Strom, T. M. (2012). Range of genetic mutations associated with severe non-syndromic sporadic intellectual disability: An exome sequencing study. *Lancet*, *380*(9854), 1674–1682. doi:10.1016/S0140-6736(12)61480-9

Rogers, A., Golumbek, P., Cellini, E., Doccini, V., Guerrini, R., Wallgren-Pettersson, C., . . . Gurnett, C. A. (2018). De novo KCNA1 variants in the PVP motif cause infantile epileptic encephalopathy and cognitive impairment similar to recurrent KCNA2 variants. *American Journal of Medical Genetics A, 176*(8), 1748–1752. doi:10.1002/ajmg.a.38840

Rosti, G., Tassano, E., Bossi, S., Divizia, M. T., Ronchetto, P., Servetti, M., . . . Puliti, A. (2019). Intragenic duplication of KCNQ5 gene results in aberrant splicing leading to a premature termination codon in a patient with intellectual disability. *European Journal of Human Genetics, 62*(9), 103555. doi:10.1016/j.ejmg.2018.10.007

Rudy, B., & McBain, C. J. (2001). Kv3 channels: Voltage-gated K^+ channels designed for high-frequency repetitive firing. *Trends in Neurosciences, 24*(9), 517–526. doi:10.1016/s0166-2236(00)01892-0

Sadleir, L. G., Kolc, K. L., King, C., Mefford, H. C., Dale, R. C., Gecz, J., & Scheffer, I. E. (2020). Levetiracetam efficacy in PCDH19 girls clustering epilepsy. *European Journal of Paediatric Neurology, 24*, 142–147. doi:10.1016/j.ejpn.2019.12.020

Sands, T. T., Balestri, M., Bellini, G., Mulkey, S. B., Danhaive, O., Bakken, E. H., . . . Cilio, M. R. (2016). Rapid and safe response to low-dose carbamazepine in neonatal epilepsy. *Epilepsia, 57*(12), 2019–2030. doi:10.1111/epi.13596

Sausbier, M., Hu, H., Arntz, C., Feil, S., Kamm, S., Adelsberger, H., . . . Ruth, P. (2004). Cerebellar ataxia and Purkinje cell dysfunction caused by Ca^{2+}-activated K^+ channel deficiency. *Proceedings of the National Academy of Sciences USA, 101*(25), 9474–9478. doi:10.1073/pnas.0401702101

Scholl, U. I., Choi, M., Liu, T., Ramaekers, V. T., Hausler, M. G., Grimmer, J., . . . Lifton, R. P. (2009). Seizures, sensorineural deafness, ataxia, mental retardation, and electrolyte imbalance (SeSAME syndrome) caused by mutations in KCNJ10. *Proceedings of the National Academy of Sciences USA, 106*(14), 5842–5847. doi:10.1073/pnas.0901749106

Seoh, S. A., Sigg, D., Papazian, D. M., & Bezanilla, F. (1996). Voltage-sensing residues in the S2 and S4 segments of the Shaker K^+ channel. *Neuron, 16*(6), 1159–1167. doi:10.1016/s0896-6273(00)80142-7

Serodio, P., & Rudy, B. (1998). Differential expression of Kv4 K^+ channel subunits mediating subthreshold transient K^+ (A-type) currents in rat brain. *Journal of Neurophysiology, 79*(2), 1081–1091. doi:10.1152/jn.1998.79.2.1081

Sicca, F., Imbrici, P., D'Adamo, M. C., Moro, F., Bonatti, F., Brovedani, P., . . . Pessia, M. (2011). Autism with seizures and intellectual disability: Possible causative role of gain-of-function of the inwardly-rectifying K^+ channel Kir4.1. *Neurobiology of Disease, 43*(1), 239–247. doi:10.1016/j.nbd.2011.03.016

Simons, C., Rash, L. D., Crawford, J., Ma, L., Cristofori-Armstrong, B., Miller, D., . . . Taft, R. J. (2015). Mutations in the voltage-gated potassium channel gene KCNH1 cause Temple-Baraitser syndrome and epilepsy. *Nature Genetics, 47*(1), 73–77. doi:10.1038/ng.3153

Singh, B., Ogiwara, I., Kaneda, M., Tokonami, N., Mazaki, E., Baba, K., . . . Yamakawa, K. (2006). A Kv4.2 truncation mutation in a patient with temporal lobe epilepsy. *Neurobiology of Disease, 24*(2), 245–253. doi:10.1016/j.nbd.2006.07.001

Smets, K., Duarri, A., Deconinck, T., Ceulemans, B., van de Warrenburg, B. P., Züchner, S., . . . Baets, J. (2015). First de novo KCND3 mutation causes severe Kv4.3 channel dysfunction leading to early onset cerebellar ataxia, intellectual disability, oral apraxia and epilepsy. *BMC Medical Genetics, 16*, 51. doi:10.1186/s12881-015-0200-3

Soldovieri, M. V., Boutry-Kryza, N., Milh, M., Doummar, D., Heron, B., Bourel, E., . . . Lesca, G. (2014). Novel KCNQ2 and KCNQ3 mutations in a large cohort of families with benign neonatal epilepsy: First evidence for an altered channel regulation by syntaxin-1A. *Human Mutatation, 35*(3), 356–367. doi:10.1002/humu.22500

Sugarman, E. A., Nagan, N., Zhu, H., Akmaev, V. R., Zhou, Z., Rohlfs, E. M., . . . Allitto, B. A. (2012). Pan-ethnic carrier screening and prenatal diagnosis for spinal muscular atrophy: clinical laboratory analysis of >72,400 specimens. *European Journal of Human Genetics, 20*(1), 27–32. doi:10.1038/ejhg.2011.134

Suzuki, T., Hansen, A., & Sanguinetti, M. C. (2016). Hydrophobic interactions between the S5 segment and the pore helix stabilizes the closed state of Slo2.1 potassium channels. *Biochimica et Biophysica Acta, 1858*(4), 783–792. doi:10.1016/j.bbamem.2015.12.024

Symonds, J. D., Zuberi, S. M., Stewart, K., McLellan, A., O'Regan, M., MacLeod, S., . . . Wilson, M. (2019). Incidence and phenotypes of childhood-onset genetic epilepsies: A prospective population-based national cohort. *Brain, 142*(8), 2303–2318. doi:10.1093/brain/awz195

Syrbe, S., Hedrich, U. B. S., Riesch, E., Djemie, T., Muller, S., Moller, R. S., . . . Lemke, J. R. (2015). De novo loss- or gain-of-function mutations in KCNA2 cause epileptic encephalopathy. *Nature Genetics, 47*(4), 393–399. doi:10.1038/ng.3239

Tabarki, B., AlMajhad, N., AlHashem, A., Shaheen, R., & Alkuraya, F. S. (2016). Homozygous KCNMA1 mutation as a cause of cerebellar atrophy, developmental delay and seizures. *Human Genetics, 135*(11), 1295–1298. doi:10.1007/s00439-016-1726-y

Tamsett, T. J., Picchione, K. E., & Bhattacharjee, A. (2009). NAD^+ activates KNa channels in dorsal root ganglion neurons. *Journal of Neuroscience, 29*(16), 5127–5134. doi:10.1523/JNEUROSCI.0859-09.2009

Torkamani, A., Bersell, K., Jorge, B. S., Bjork, R. L., Friedman, J. R., Bloss, C. S., . . . Kearney, J. A. (2014). De novo KCNB1 mutations in epileptic encephalopathy. *Annals of Neurology, 76*(4), 529–540. doi:10.1002/ana.24263

Tricarico, D., Barbieri, M., Mele, A., Carbonara, G., & Camerino, D. C. (2004). Carbonic anhydrase inhibitors are specific openers of skeletal muscle BK channel of K^+-deficient rats. *FASEB Journal, 18*(6), 760–761. doi:10.1096/fj.03-0722fje

Truty, R., Patil, N., Sankar, R., Sullivan, J., Millichap, J., Carvill, G., . . . Aradhya, S. (2019). Possible precision medicine implications from genetic testing using combined detection of sequence and intragenic copy number variants in a large cohort with childhood epilepsy. *Epilepsia Open, 4*(3), 397–408. doi:10.1002/epi4.12348

Tsaur, M. L., Chou, C. C., Shih, Y. H., & Wang, H. L. (1997). Cloning, expression and CNS distribution of Kv4.3, an A-type K^+ channel alpha subunit. *FEBS Letters, 400*(2), 215–220. doi:10.1016/s0014-5793(96)01388-9

Vanderver, A., Simons, C., Schmidt, J. L., Pearl, P. L., Bloom, M., Lavenstein, B., . . . Taft, R. J. (2014). Identification of a novel de novo p.Phe932Ile KCNT1 mutation in a patient with leukoencephalopathy and severe epilepsy. *Pediatric Neurology, 50*(1), 112–114. doi:10.1016/j.pediatrneurol.2013.06.024

Vears, D. F., Dunn, K. L., Wake, S. A., & Scheffer, I. E. (2015). "It's good to know": Experiences of gene identification and result disclosure in familial epilepsies. *Epilepsy Research, 112*, 64–71. doi:10.1016/j.eplepsyres.2015.02.011

Veeramah, K. R., Johnstone, L., Karafet, T. M., Wolf, D., Sprissler, R., Salogiannis, J., . . . Hammer, M. F. (2013). Exome sequencing reveals new causal mutations in children with epileptic encephalopathies. *Epilepsia, 54*(7), 1270–1281. doi:10.1111/epi.12201

Walters, J., Wells-Kilpatrick, K., & Pandeleos, T. (2014). My epilepsy story—PCDH19 alliance. *Epilepsia, 55*(7), 968–969. doi:10.1111/epi.12555

Wang, H., Yan, Y., Liu, Q., Huang, Y., Shen, Y., Chen, L., . . . Chai, J. (2007). Structural basis for modulation of Kv4 K^+ channels by auxiliary KChIP subunits. *Nature Neuroscience, 10*(1), 32–39. doi:10.1038/nn1822

Wang, J., Al-Ouran, R., Hu, Y., Kim, S. Y., Wan, Y. W., Wangler, M. F., . . . Bellen, H. J. (2017). MARRVEL: Integration of human and model organism genetic resources to facilitate functional annotation of the human genome. *American Journal of Human Genetics, 100*(6), 843–853. doi:10.1016/j.ajhg.2017.04.010

Wang, J., Wen, Y., Zhang, Q., Yu, S., Chen, Y., Wu, X., . . . Bao, X. (2019). Gene mutational analysis in a cohort of Chinese children with unexplained epilepsy: Identification of a new KCND3 phenotype and novel genes causing Dravet syndrome. *Seizure, 66*, 26–30. doi:10.1016/j.seizure.2019.01.025

Yang, Y., Vasylyev, D. V., Dib-Hajj, F., Veeramah, K. R., Hammer, M. F., Dib-Hajj, S. D., & Waxman, S. G. (2013). Multistate structural modeling and voltage-clamp analysis of epilepsy/autism mutation Kv10.2-R327H demonstrate the role of this residue in stabilizing the channel closed state. *Journal of Neuroscience, 33*(42), 16586–16593. doi:10.1523/JNEUROSCI.2307-13.2013

Yuan, P., Leonetti, M. D., Pico, A. R., Hsiung, Y., & MacKinnon, R. (2010). Structure of the human BK channel Ca^{2+}-activation apparatus at 3.0 A resolution. *Science, 329*(5988), 182–186. doi:10.1126/science.1190414

Zhang, Y., & Kaczmarek, L. K. (2016). Kv3.3 potassium channels and spinocerebellar ataxia. *Journal of Physiology, 594*(16), 4677–4684. doi:10.1113/JP271343

Zuberi, S. M., Eunson, L. H., Spauschus, A., De Silva, R., Tolmie, J., Wood, N. W., . . . Hanna, M. G. (1999). A novel mutation in the human voltage-gated potassium channel gene (Kv1.1) associates with episodic ataxia type 1 and sometimes with partial epilepsy. *Brain, 122*(Pt 5), 817–825. doi:10.1093/brain/122.5.817

CHAPTER 6

THE VOLTAGE-DEPENDENT SODIUM CHANNEL FAMILY

MARIOLA ZALESKA, SAMANTHA C. SALVAGE, ANDREW J. THOMPSON, SIVAKUMAR NAMADURAI, CHRISTOPHER L.-H. HUANG, TREVOR WILKINSON, FIONA S. CUSDIN, AND ANTONY P. JACKSON

ACTIVATION of voltage-dependent sodium (Na_v) channels is responsible for the initial membrane depolarization phase of the action potential. Within a few milliseconds of opening, the Na_v channels typically enter an inactive state, during which they are functionally refractory to any further membrane depolarization signals. Restoration of the membrane potential by opening voltage-dependent potassium (K_v) channels permits Na_v channel recovery from inactivation back to their resting state, resetting the channel and permitting further activation (Vandenberg & Waxman, 2012). Na_v channels are of major research interest in neurobiology, pharmacology, biophysics, and structural biology. Over a thousand mutations have been identified in different Na_v channel isoforms that are related to a variety of inherited diseases, including epilepsy, cardiopathologies, myotonias, and chronic pain syndromes (Huang, Liu, Yan, & Yan, 2017; Kruger & Isom, 2016). Consequently, the development of drugs that target Na_v channels is of major pharmacological interest.

The minimum functional component of the eukaryotic Na_v channel is a single 250 kDa α-subunit that contains the ion-selective pore. Nine mammalian α-subunits, designated Na_V1.1–1.9, have been identified, as well as an atypical channel Na_{vx}—the product of distinct genes *SCN1A–10A* (Catterall, 2017; de Lera Ruiz & Kraus, 2015). Different isoforms vary in their gating behavior that reflect their physiological roles, and many are expressed in complex tissue-specific and developmentally regulated patterns. Further structural diversity is generated by alternative mRNA splicing and post-translational modifications, including N-linked glycosylation, phosphorylation, ubiquitination, arginine methylation, palmitylation, sulphation, and S-nitrosylation (Onwuli & Beltran-Alvarez, 2016).

The eukaryotic α-subunit is formed by a single polypeptide chain (approximately 2,000 amino acid residues long) containing four homologous-but non-identical-domains, designated DI–DIV (Figure 6.1A). Each domain contains six transmembrane α-helical segments, designated S1–S6, which are connected through short or moderate-length extracellular and intracellular loops. Helices S1–S4 from each domain form a voltage-sensing module (VSM). The pore module (PM) contains helices S5 and S6, connected to each other through extracellular loop regions and the re-entrant P-loop (pore loop) helices (Figure 6.1B). The domains create a pseudotetrameric unit in which the PMs from each domain line the central ion-conducting pore. Within each domain, the VSMs lie on the perimeter, and the VSM of one domain makes close contact with the clockwise PM from the adjacent domain, as viewed from above (Figure 6.2). This interleaved arrangement is characteristic of all known eukaryotic voltage-dependent ion channels and probably underlies and facilitates the coupling of VSM movement with pore opening (see further, this article). Both VSMs and PMs of Na_V channels can be expressed as functionally isolated modules (McCusker, D'Avanzo, Nichols, & Wallace, 2011; Paramonov et al., 2017). This suggests an independent evolutionary origin of the PMs and VSMs. Indeed, the subunits of some tetrameric prokaryotic ion channels consist of PMs only (Anderson & Greenberg, 2001). Furthermore, some VSM homologues occur in otherwise functionally unrelated molecules. A particularly striking example is the voltage-sensitive phosphatase whose membrane-embedded VSM controls its phosphoinositide phosphatase activity in response to changes in membrane potential (Murata, Iwasaki, Sasaki, Inaba, & Okamura, 2005; Piao, Rajakumar, Kang, Kim, & Baker, 2015).

The Na_V channels belong to a large superfamily of voltage-gated ion channels, including voltage-gated potassium (K_V) and calcium (Ca_V) channels. Phylogenetic analysis suggests that the genes encoding Na_V and Ca_V channels evolved by two separate gene duplication events from an ancestral K_V-like channel containing a single domain. Subsequent duplication and divergence led to the separate evolution of Na_V channel and Ca_V channel gene families (Anderson & Greenberg, 2001; Moran, Barzilai, Liebeskind, & Zakon, 2015). Sodium-selective channels are also widespread in prokaryotes. Unlike their eukaryotic equivalents, the prokaryotic Na_V channels are tetramers of four identical subunits, with each subunit corresponding to an individual eukaryotic Na_V channel domain (Koishi et al., 2004). Their relative simplicity, and the availability of several high-resolution atomic structures, has made prokaryotic channels popular models to investigate the molecular mechanisms of gating behavior (Catterall & Zheng, 2015). However, it should be noted that detailed phylogenetic analysis strongly suggests that sodium-selectivity arose independently in prokaryotic and eukaryotic Na_V channel families (Liebeskind, Hillis, & Zakon, 2013). This is important to bear in mind when interpreting structural experiments, especially when applied to the mechanism of ion-selectivity and inactivation.

Although structures for related molecules such as a mammalian voltage-dependent calcium channels have been solved (Wu et al., 2016), at the time of writing, the structure of only one eukaryotic Na_V channel (from the American cockroach, *Periplaneta*

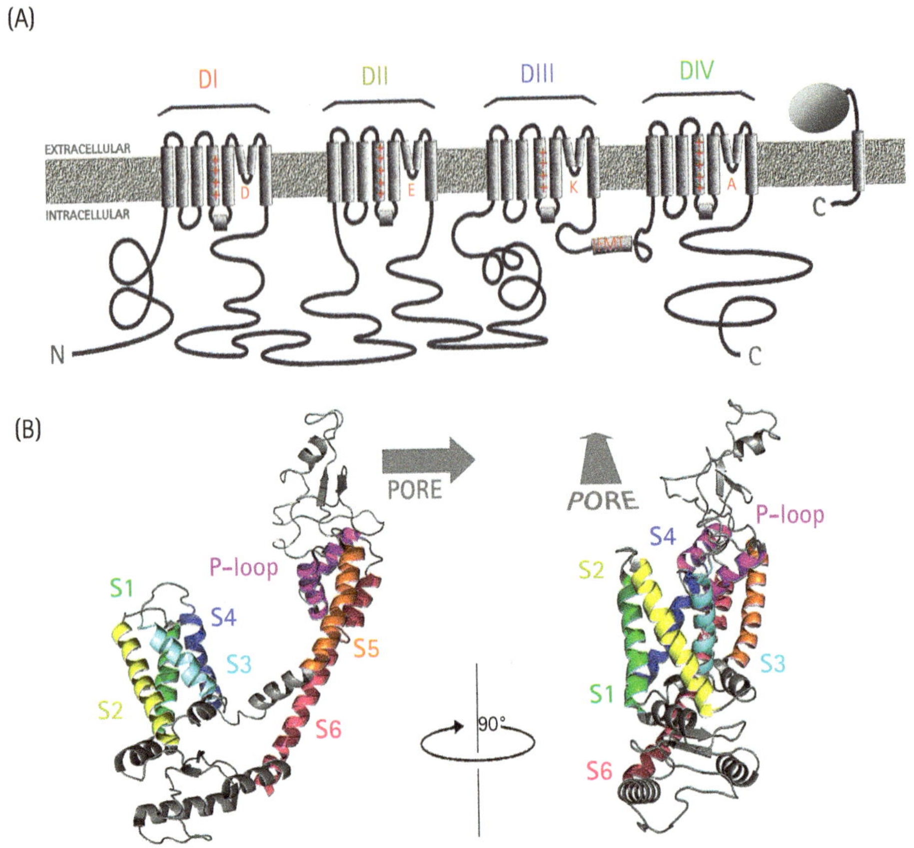

FIGURE 6.1 A cartoon of the voltage-gated sodium channel (Na_v) and associated β-subunit. *A*. The primary structure of Na_v consists of four domains (DI–DIV), each of which contains six transmembrane α-helices (S1–S6) and two smaller P-loop α-helices. Ion-selectivity is governed by a ring of amino acids (DEKA, red text) that converge from each of the P-loop regions of all four domains. An α-helical inactivation gate between DIII and DIV contains a cluster of hydrophobic residues (IFMT) that can occlude the pore. Charged residues that act as a voltage sensor are found in S4 of each domain (+, red text; also see Figure 6.3). The β-subunit consists of a single transmembrane α-helix joined to an extracellular immunoglobulin domain. *B*. A single domain from the crystal structure of the cockroach Na_v channel (PDBID: 5X0M) showing the arrangement of segments S1–S6 and the P-loop. The right-hand side of the panel is rotated by 90°, and viewed from outside of the channel as if looking towards the center of the pore. The α-helices are represented as cylinders and the adjoining polypeptide chains as black lines.

americana) has been solved to near-atomic resolution (Shen et al., 2017) (Figures 6.1B, 6.2). Rapid technical developments in structural biology, especially in cryo-electron microscopy, should see structures for several mammalian Na_v channels becoming available in the near future. This will undoubtedly have a major impact on the field of Na_v channel research, and will greatly extend our understanding of these important molecules.

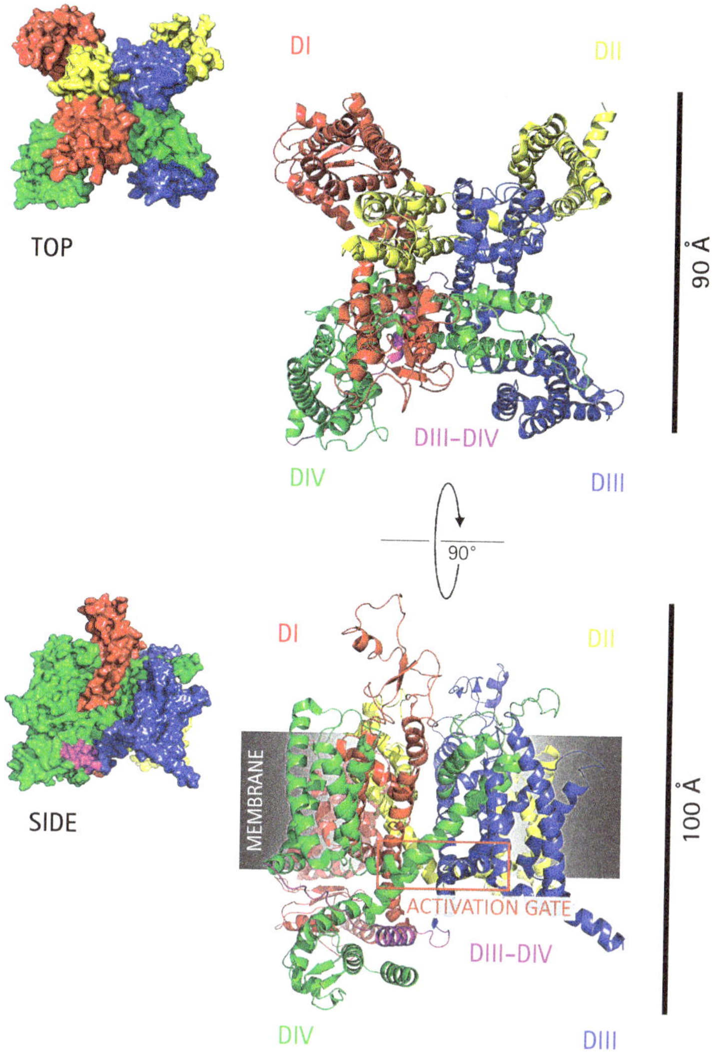

FIGURE 6.2 The crystal structure of the cockroach voltage-gated sodium channel (PDBID: 5XOM) from above and from the side, showing the central ion-selective pore and the four domains that surround it. Each domain is colored differently, and the position of the cell membrane is shown as a grey box. The activation gate, mentioned in the text, is shown in the red-outlined box.

This review does not aim to provide a comprehensive analysis of current experimental approaches to Na_v channel biology. For reviews with such emphasis, we recommend, for example, Ahern, Payandeh, Bosmans, and Chanda (2016). Rather, we provide a general overview of Na_v channels for the non-specialist reader, place them in their broader physiological context, and note the pathological effects of some Na_v channel mutants. The potential for protein and peptide-based pharmacological tools to modulate Na_v channel behavior is then discussed in the light of this background.

The Na$_V$ Channel Structure and Gating Mechanism

The S5 and S6 helices of each PM form the central pore cavity wall. At the intracellular face, the S6 helices from each domain draw together to form an intracellular cavity containing hydrophobic amino acids. This forms the activation gate and is constricted when the channel is closed (Figure 6.2) (Clairfeuille, Xu, Koth, & Payandeh, 2016). An extracellular linker connects helix S5 to a membrane-descending P-helix (P1), followed by an ascending P-helix (P2) and a further short extracellular loop connecting to helix S6 (Figures 6.1, 6.2). The extracellular loops from each domain create a turret-like structure for the outer mouth, which extends above the pore. They are glycosylated and form a pre-selection vestibule filter. In the cockroach structure, the extracellular loops contain disulphide bonds. Sequence comparison of the cockroach Na$_v$ channel with mammalian Na$_v$ channels shows that the cysteines (and thus, most likely, the disulphide bonds) are fully conserved, implying an important role in stabilizing the ion channel preselection filter (Shen et al., 2017). The narrowest point in the vestibule occurs where the P1 and P2 helices reverse direction within the membrane (Figures 6.1A, 6.3). Here charged residues provide a high field strength (HFS) site at the constriction point (Stephens, Guan, Zhorov, & Spafford, 2015). In eukaryotic Na$_v$ channels, the residues constituting this site are aspartate (DI), glutamate (DII), lysine (DIII), and alanine (DIV). This creates an asymmetrical selectivity filter that is largely responsible for favoring sodium ions over other positively charged ions such as potassium or calcium (Heinemann, Terlau, Stuhmer, Imoto, & Numa, 1992).

To open the pore, the VSM must detect changes in membrane potential and transmit this information electromechanically to the PM by inducing an allosteric rearrangement

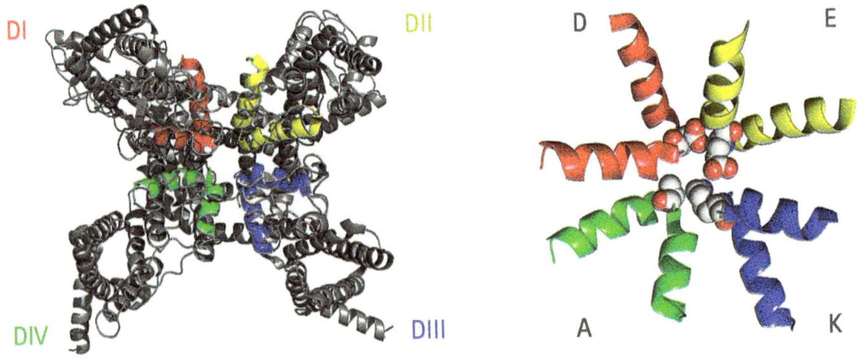

FIGURE 6.3 The ion-selectivity filter of Na$_v$ (P-loop). Top view of the cockroach Na$_v$ channel (PDBID: 5XoM). The selectivity filter of the Na$_v$ is formed by two short α-helices that extend towards the pore from each domain. These are located between S5 and S6 and converge to creating a narrow, sodium-permeable constriction that is surrounded by the amino acids DEKA from each of DI–DIV, respectively.

of the S5 and S6 helices. Transmembrane helix S4 of the VSM contains four to six positively charged arginine and lysine residues that serve as the gating charges. They occur every three residues, so that the positive charges approximately lie along one face of the S4 helix (Noda et al., 1984) (Figure 6.4). For this arrangement to be thermodynamically feasible, the S4 positive charges within the membrane must be neutralized by forming ion-pairs with corresponding acidic groups from residues within the surrounding S1–S3 helices. According to the "sliding helix" (Catterall, 1986) and "helical screw" (Guy & Seetharamulu, 1986) models, the negative internal membrane potential provides a "pulling" force to keep these charges facing inward in the resting state. Following depolarization, this force is transiently weakened, enabling the S4 helix to move outward, probably in a spiral path, so that ion-pairs exchange partners. There is a wealth of evidence that supports the general outline of the model. This includes chemical modification, fluorescent labelling experiments, and mutagenesis studies guided by atomic-resolution structures of prokaryotic Na_v channels (Catterall, 2010; Clairfeuille et al., 2016; DeCaen, Yarov-Yarovoy, Zhao, Scheuer, & Catterall, 2008; Zhang et al., 2012). Nevertheless, some questions remain. For example, there is some debate as to whether the inner part of the S4 helix may transiently adopt a 3_{10} helix during this movement (Ahern et al., 2016; Villalba-Galea, Sandtner, Starace, & Bezanilla, 2008). If so, this would enable the inner region of the S4 helix to stretch during the activation process. A subtle point is that the transition from α-helix to high-energy 3_{10} helix, driven by the electrical field, would provide a means of capturing electrostatic potential energy, which can subsequently be used to drive the rearrangements needed to open the pore (Yarov-Yarovoy et al., 2012). Structures of prokaryotic Na_v channels trapped in distinct conformational states suggest that following activation, the VSM rotates in the membrane plane around the PM, thus exerting a torque on the S4–S5 linker (Catterall,

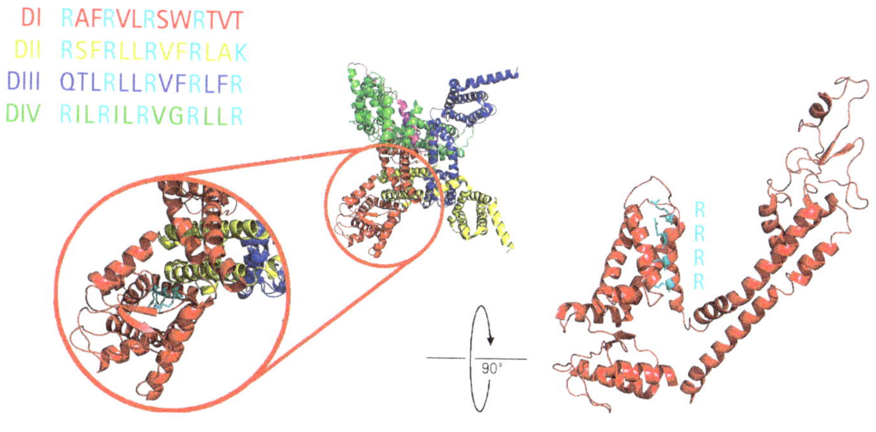

FIGURE 6.4 The voltage-sensor: In each domain, charged residues in S4 detect changes in the cell membrane potential. Here the Na_v channel is viewed from below and DI is highlighted in more detail. Charged residues in S4 are shown in the enlarged image (cyan sidechains) with an accompanying alignment of S4 residues from each of the four domains. The image is of the cockroach Na_v channel crystal structure (PDBID: 5XoM).

Wisedchaisri, & Zheng, 2017; Clairfeuille et al., 2016). This pulls the lower end of the S5 helix outward and shifts the positions of the PM helices, with the S6 helix twisting in a counterclockwise manner (as seen from the intracellular face), and thereby opening the pore. The characteristic feature of Na_V channels, whereby each VSM is most closely associated with the PM of its neighbor (Figure 6.2), can now be rationalized, as this arrangement will facilitate gating by enforcing a concerted opening.

In mammalian and probably other eukaryotic Na_V channels, the four VSMs activate with differing kinetics. Movement of the DI–DIII VSMs is the most rapid and is sufficient to begin ion flow (Chanda & Bezanilla, 2002). Interestingly, in the cockroach Na_V channel structure, two of the four S4 helices adopt a 3_{10} conformation and are in different relative positions (Shen et al., 2017), suggesting that this structure may have captured some of the presumed heterogeneity in eukaryotic VSM activation. The DIV VSM activates with the slowest kinetics (Bosmans, Martin-Eauclaire, & Swartz, 2008). Its movement frees an intracellular linker called the *inactivation gate* that connects DIII helix S6 to DIV helix S1 (Figure 6.1A) (Capes, Goldschen-Ohm, Arcisio-Miranda, Bezanilla, & Chanda, 2013) The inactivation gate contains a cluster of hydrophobic residues containing the amino acid sequence IFMT (the IFMT motif) that can now bind to a corresponding inactivation particle receptor lying within the S4–S5 linkers of DII, DIII, and DIV (Popa, Alekov, Bail, Lehmann-Horn, & Lerche, 2004). As a result, the inactivation gate occludes the pore and inactivates the channel within a few milliseconds of opening. This is the molecular basis of the fast inactivation pathway and ensures that the action potential can only be propagated in the forward direction.

The Na_V Channel Auxiliary Subunits

The Na_V channel α-subunit *in vivo,* typically exists in association with auxiliary subunits and other proteins within larger macromolecular assemblies localized to discrete regions of the plasma membrane (Abriel, 2010; Abriel, Rougier, & Jalife, 2015; Heine, Ciuraszkiewicz, Voigt, Heck, & Bikbaev, 2016; Lee, Fakler, Kaczmarek, & Isom, 2014; Meadows & Isom, 2005). The best-characterized auxiliary proteins are the Na_V β-subunits for which four genes (*Scn1b, Scn2b, Scn3b,* and *Scn4b*) encode the proteins β1, β2, β3, and β4. In addition, alternative splicing of the *Scn1b* gene can generate a secreted β1 Ig domain, lacking a transmembrane domain (Qin et al., 2003). All β-subunits have a type I membrane topology containing one extracellular amino-terminal single V-type immunoglobulin (Ig) domain connected through a short neck to a transmembrane alpha-helical domain and a small intracellular carboxy-terminal region. The β-subunits modulate intracellular Na_V channel traffic, surface expression, and protein stability; they generally enhance rates of channel activation, inactivation, and recovery from inactivation, and they modulate the voltage-dependencies of these parameters. Some β-subunits are also involved in transmediated cell adhesion (Brackenbury & Isom, 2011; Cusdin, Clare, & Jackson, 2008; Namadurai et al., 2015).

A characteristic feature of the β-subunits is their ability to shift the half-maximal voltages ($V_{1/2}$) for activation and inactivation, usually in a hyperpolarizing direction (i.e., the voltage where half the channels activate or inactivate is shifted to more negative values compared to the values shown by the α-subunit alone). As a result, the action potential threshold is lowered, leading to an increased probability of firing (Namadurai et al., 2015). For the case of β1, these shifts can be abolished under conditions that inhibit the addition of sialic acid to N-linked sugar residues (Johnson, Montpetit, Stocker, & Bennett, 2004). All four of the β-subunit Ig domains are heterogeneously glycosylated in vivo. Hence, we have suggested that the β-subunit Ig domains might be positioned on the α-subunit in such a way that they can present negative charges from N-linked sugars close enough to one or more of the VSMs to influence its local field (Namadurai et al., 2015). Topological considerations indicate that the simplest way this can be achieved is for the Ig domain to bind a site or sites on the large extracellular S5–S6 pore loops and the β2-binding site has indeed been mapped to this loop on DII (Das, Gilchrist, Bosmans, & Van Petegem, 2016). Because of the close sequence and structural similarity between the β2 and β4 Ig domains (Das et al., 2016; Gilchrist, Das, Van Petegem, & Bosmans, 2013), it is likely that these two β-subunits bind to the same region on the α-subunit. In both cases, the β2 and β4 Ig domains bind covalently to the α-subunit through a disulphide bond. In contrast, the β1 and β3 Ig domains bind non-covalently to a different site. There is some evidence that the β1 Ig domain contacts sites on the S5–S6 extracellular loop from DI and DIV (Makita, Bennett, & George, 1996), but the binding site for the β3-subunit Ig domain has not yet been identified.

A striking feature of the β3-subunit is its ability to form homo-dimers and trimers in vivo,that can cross-link Na_v α-subunits (Namadurai et al., 2014). There are literature reports suggesting that even in the absence of β-subunits, the heart-specific channel Nav1.5 α-subunits can associate together on the plasma membrane (Clatot et al., 2012). So, the β3-subunit may promote and further stabilize this natural tendency for oligomerization (Namadurai et al., 2014). Furthermore, several β-subunits, including β3, can bind in a *cis* configuration to cell-adhesion molecules NrCAM, neurofascin, and contactins (Kazarinova-Noyes et al., 2001; Ratcliffe, Westenbroek, Curtis, & Catterall, 2001). Since these molecules contain multiple Ig-like binding sites, they could in principle bind more than one Na_v channel and thus further enhance and extend local Na_v channel clustering. There is some evidence that the β1-subunit can also cross-link Na_v channel α-subunits in vivo. For example, a mutant Na_v 1.5 channel is retained in the endoplasmic reticulum (ER) and thus leads to a form of Brugada syndrome. Remarkably, the wild-type Na_v1.5 channel is also trapped in the ER when co-expressed with both the mutant and in the presence of the β1-subunit (Mercier et al., 2012; Namadurai et al., 2015). This suggests that β1, like β3, may promote the formation of Na_v channel oligomers. The β1-subunit can also bind to the K_v4.2 potassium channel, a major regulator of neuronal excitability (Marionneau et al., 2012), and proteomic analysis suggest that the heart-specific Na_v1.5 channel can also stably interact with potassium channels;although whether this is a direct association or mediated by additional proteins is not clear (Willis, Ponce-Balbuena, & Jalife, 2015). The existence of such cross-linked Na_v channel α-subunits raises the question of whether they become functionally coupled under these conditions. This is a controversial idea (McCormick, Shu, & Yu, 2007), but functional coupling between different α-subunits could contribute to rapid action potential initiation and gating. It may also help explain why many inherited sodium channel pathologies exhibit dominant-negative phenotypes (Hoshi et al., 2014; Keller et al., 2005; Poelzing et al., 2006; Sottas & Abriel, 2016). Dominant negative behavior is a common feature when a mutant subunit is incorporated into a multicomponent assembly and blocks the activity of all subunits in the complex (Veitia, 2007).

The β1 and β3 subunits also enhance the rates of inactivation and recovery from inactivation of the channel. The DIII–DIV linker region containing the inactivation gate also contains a separate binding site for the carboxy-terminus of the α-subunit, and binding between these two regions stabilizes the inactivated state (Kass, 2006). The likely importance of this interaction is illustrated by the existence of an epilepsy-inducing mutation in the carboxy-terminus of $Na_v1.1$ that disrupts the interaction and slows inactivation (Spampanato et al., 2004). The intracellular carboxy-termini of β1 and β3 also bind to the α-subunit carboxy-terminus. Thus, the β1 and β3-subunits may facilitate fast inactivation by enhancing the binding of the α-subunit carboxy-terminus and the inactivation gate. Furthermore, the α-subunit carboxy-terminal domain contains two EF hands and a calmodulin-binding IQ motif, both structural features involved in calcium sensing (Miloushev et al., 2009). In $Na_v1.4$, $Na_v1.5$, and $Na_v1.6$, calmodulin binds to both the α-subunit carboxy-terminal domain and the inactivation gate, and confers calcium-sensitivity on the inactivation properties of the channel (Gabelli et al., 2014; Sarhan, Tung, Van Petegem, & Ahern, 2012). The importance of this property is shown by several distinct arrhythmogenic mutations in the IQ motif of the $Na_v1.5$ carboxy-terminus (Rook et al., 1999).

As with almost all intrinsic membrane proteins, Na_v channels first fold and assemble in the lumen of the ER. It is not uncommon for large intrinsic membrane proteins to fold with relatively low efficiency, leading to accumulation in the ER. Under these circumstances, chaperones typically enhance folding efficiency (Araki & Nagata, 2011). One role of the β-subunits may be to act as a chaperone, since they increase Na_v channel trafficking out of the ER and to the plasma membrane (Cusdin et al., 2008). Na_v channel trafficking can also be disrupted by mutations that prevent the normal anchoring of the channel into the plasma membrane. For example, the membrane-bound cytoskeletal protein ankyrin-G enhances the clustering of $Na_v1.2$ and $Na_v1.6$ into the nodes of Ranvier and axon initial segment (Cusdin et al., 2008). In $Na_v1.6$, mutations in a cytoplasmic linker sequence between DII helix S6 and DIII helix S1 prevent the binding to ankyrin-G and inhibit channel association with the axon initial segment (Gasser et al., 2012).

"Non-classical" Roles for Na_V Channels

Na_v channels are also expressed in cells not usually thought to be electrically excitable (Black & Waxman, 2013). A notable example is the expression of the cardiac channel

$Na_V1.5$ in phagosomes and late endosomes of activated macrophages. Both selective siRNA knockdown and tetrodotoxin treatment inhibit phagocytosis in these cells, suggesting that the Na_V channel has an important function in these processes. A likely role for $Na_V1.5$ is to allow sodium ion efflux from the endolysosomes, to provide charge counterbalance for the protons pumped into the organelles. Since the $Na_V1.5$ channel is inhibited by low pH, this could act as a feedback inhibitor of excessive acidification (Carrithers et al., 2007).

The $Na_V1.5$ channel is also expressed in astrocytes under pathological conditions leading to astrogliosis (Black, Newcombe, & Waxman, 2010). In murine models of multiple sclerosis, the relative abundance of $Na_V1.5$ in astrocytes is correlated with disease severity (Pappalardo, Liu, Black, & Waxman, 2014). An alternatively spliced neonatal form of $Na_V1.5$ was expressed in astrocytomas and its level correlated with increasing astrocytoma grade (Xing et al., 2014). Additionally, in the human U251 astrocytoma model cell-line, siRNA knockdown of $Na_V1.5$ expression conferred a loss or reduction in the proliferative, invasive, and migratory properties of these cells, and enhanced their apoptosis (Xing et al., 2014). The enhanced invasive properties that $Na_V1.5$ confers in this context has also been observed in human breast cancer cells (Nelson, Yang, Millican-Slater, & Brackenbury, 2015). The causal mechanisms connecting $Na_V1.5$ expression to enhanced invasiveness is not clear, but one attractive hypothesis suggests that the $Na_V1.5$ channels co-localize in lipid rafts with the Na^+/H^+ exchanger (NHE-1). Sodium influx activates the exchanger leading to enhanced proton extrusion and the subsequent activation of acid-dependent cell-surface proteases (Brisson et al., 2011). Similarly, an enhanced expression of $Na_V1.7$ has been observed in several prostate cancer cell-lines, where the expression correlates with invasive potential (Brackenbury & Djamgoz, 2007). Collectively, these findings suggest a key role for Na_V channels in many pathologies, including those that regulate neuroinflammatory processes and enhance the aggressive nature of cancer cells, with roles that are distinct from that of its classic function in cell excitability.

How Different Na_V Channel Isoforms Combine in Physiological Context: Examples from Cardiac Cells and Peripheral Pain Neurones

Most electrically excitable cells express multiple Na_V channel isoforms, each with different gating behaviors and distinct opening and inactivation kinetics. In such cases, the combined functional interactions between these distinct Na_V channel isoforms, acting together, can generate complex and emergent behavior. To illustrate this concept, two examples are discussed: cardiac cells and dorsal root ganglia (DRG) sensory neurones.

Na_V Channel Behavior in Cardiac Cells

The heart expresses not only cardiac $Na_v1.5$, but also the neuronal, $Na_v1.1$, $Na_v1.3$, and $Na_v1.6$ channels (Maier et al., 2003). The occurrence and abundance of these different Na_v channel isoforms vary between different cardiac tissue types. Regenerative voltage-dependent Na_v channel opening triggers the large, rapid, inward depolarizing current that initiates the action potential upstroke. This is fundamental to *propagation* of the cardiac electrical activity through successive cardiomyocytes, triggering the heartbeat. In addition, Na_v channel activation thresholds determine the onset of the upstroke, relative to the background pacing ion channel activity. This determines the *frequency* of pacemaker excitation (Huang, Lei, Matthews, Zhang, & Lei, 2012). Both processes underlie function in the cardiac sino-atrial node (SAN), the source of the heart's natural pacemaker activity, in which co-expression of and interaction between Na_v channel α-subunit isoforms play central roles.

The SAN pacemaker comprises central and peripheral regions. The structure is in turn surrounded by the atrial tissue to which it conducts the resulting rhythmic excitation. Pacemaker activity begins from a repolarization of the preceding action potential. The extent of this depends upon the magnitude of the outward current mediated by the rapid K^+ channel (Verheijck, van Ginneken, Bourier, & Bouman, 1995). The resulting hyperpolarization activates an inward depolarizing, so-called *funny current*, carried primarily by the HCN4 hyperpolarization-activated cyclic nucleotide-gated (HCN) channel. Later phases of diastolic depolarization involve contributions from the depolarizing late, L-type, calcium current and transient, T-type, Ca^{2+} channel current (Sanders, Rakovic, Lowe, Mattick, & Terrar, 2006). The resulting membrane potential depolarization activates the Na_v channels that in turn trigger the action potential. However, SAN function not only involves a pacing process, but also requires conduction of the resulting excitation between its successive component cells from the center to the periphery of the SAN and between the outermost cells of the SAN and its coupled atrial cells. Here, different SAN regions and distinct Na_v isoforms play distinct roles. Whereas $Na_v1.1$ occurs throughout the SAN, occurrence of the $Na_v1.5$ channel is restricted to peripheral as opposed to central SAN cells (Lei et al., 2004). Cells in the central SAN pacemaker region are smaller in size and therefore also in total membrane capacitance. Their consequently larger input impedance, together with the rapid kinetics of activation shown by their $Na_v1.1$ channels, enhances excitability and therefore their primarily pacemaker role within the SAN. This contrasts with the larger size, larger total membrane capacitance and consequent lower input impedance of the peripheral region cells that express $Na_v1.5$. Consequently, their function consequently appears primarily to involve conduction of the resulting action potential from the central pacemaker region to the atrial cells that surround the SAN. Both conduction and activation processes vary with the relationship between active and passive electrophysiological properties between coupled cells within the SAN. They would also be affected by current-load matching properties between peripheral SAN cells and the atrial myocytes to which they are directly coupled.

Such distinct pacing and conducting roles shown by $Na_v1.1$ and $Na_v1.5$ were demonstrated through experiments separating these contributions through the greater tetrodotoxin (TTX)-sensitivity of $Na_v1.1$ compared to $Na_v1.5$. Firstly, challenge by TTX at nM-concentrations that would selectively inhibit $Na_v1.1$ reduced pacemaker rates by 65%, 22%, and 15% in intact mouse hearts (Maier et al., 2002), isolated SA nodes, and isolated SAN pacemaker cells, respectively (Lei et al., 2004). Secondly, action potential clamp studies demonstrated that such TTX-sensitive Na_v currents are activated within voltage ranges traversed by the pacemaker potential, consistent with its additional participation in action potential conduction. Thirdly, block of both the TTX-sensitive and TTX-resistant Na^+ current by TTX at μM concentrations, but not selective block of TTX-sensitive Na^+ current by 10 or 100 nM TTX, increased SAN conduction times from the leading pacemaker site in the center of the SAN through the periphery to surrounding atrial muscle. Finally, modifications in either $Na_v1.1$ or $Na_v1.5$ influenced the emergent heart rates. Thus, $Na_v1.5$ haploinsufficient *Scn5a*$^{+/-}$ mice replicated the depressed heart rates and sino-atrial block clinically observed in sinus node disorder patients (Asseman et al., 1983). Their isolated hearts showed sinus bradycardia, with both slowed, and episodes of blocked, sino-atrial conduction. Isolated SAN and atrial tissue preparations from these *Scn5a*$^{+/-}$ mice similarly exhibited both slowed and blocked sino-atrial conduction. SAN cells from the *Scn5a*$^{+/-}$ mice demonstrated about a 30% reduction in maximum Na^+ currents compared to wild-type cells (Lei et al., 2005). These findings may also form the basis for human SAN syndromes associated with genetic defects in $Na_v1.5$, whose features are of significant clinical importance (Lei et al., 2004; Maier et al., 2003). Thus, sinus node dysfunction affects ~1 in 600 cardiac patients aged over 65 years and constitutes the clinical indication for ~50% of the million permanent pacemaker implants per year worldwide (Dobrzynski, Boyett, & Anderson, 2007).

Peripheral Pain-Sensing Neurones

Sensory neurones in the DRG detect painful stimuli that are transmitted to the spinal cord (see the chapter by Xiao et al., this volume). These neurones express a number of distinct Na_v channel isoforms, in particular, $Na_v1.7$, $Na_v1.8$, and $Na_v1.9$, with smaller amounts of $Na_v1.3$ (Rogers, Tang, Madge, & Stevens, 2006). Associated β-subunits, particularly β3, whose expression correlates closely with that of $Na_v1.7$, may fine-tune the gating behavior of the channels (Shah et al., 2000). Acting together with voltage-dependent calcium and potassium channels, these molecules set the resting potential, action potential threshold, and neuronal firing rate (Waxman, 2012). To add further complexity, the relative expression of these channel isoforms is dynamic, and it can change dramatically following peripheral nerve damage (Chahine & O'Leary, 2014).

The existence of rare individuals congenitally insensitive to pain (CIP) led to the identification of $Na_v1.7$ as an Na_v channel with a critical role in pain perception (Cox et al., 2006). Most patients with congenital pain insensitivity possess $Na_v1.7$-deletion mutations that prevent functional expression of the protein. However, some missense $Na_V1.7$ mutations precipitate syndromes with the opposite pathology;a variety of extreme and often chronic pain conditions such as inherited erythromelalgia (IEM) and paroxysmal extreme pain disorder (PEPD) (Habib, Wood, & Cox, 2015). The IEM and PEPD mutant channels are hyperexcitable, but act in different ways. In IEM, the $Na_V1.7$ mutation leads to a pronounced hyperpolarizing shift in the voltage-dependenceof activation, causing a lowered action potential threshold (Cummins, Dib-Hajj, & Waxman, 2004). In contrast, PEPD mutations are more associated with depolarizing shifts in steady-state fast inactivation, which leads to a higher persistent current (Fertleman et al., 2006; Lampert, O'Reilly, Reeh, & Leffler, 2010). In both cases, the mutant channels become more easily activated by the initiating impulses. Figure 6.5 maps the locations of these mutations onto the cockroach Na_V channel (Shen et al., 2017). A striking feature of this analysis is how many of the mutations map to known functionally critical regions of the channels. Some 80% are found in the VSMs and PMs, with no clear distinction between different domains. In some cases, the position of a mutation immediately suggests an explanation for the functional impairment shown by the channel. For example, several of the IEM and PEPD mutants map to the S4 helix of one or more VSMs, and to the S4–S5 cytosolic linker connecting the VSM to the PMs. Other PEPD mutants map to the IFMT inactivation gate (Figures 6.1, 6.2, 6.5), most likely compromising fast inactivation and leading to persistent currents. On the other hand, it is not clear how some of the mutations affect activity. The rare cases of missense CIP mutants are good examples. All three of these mutations occur within the S5–S6 extracellular pore-loops where they are not obviously close to any of the recognized and important functional regions of the channel (Figure 6.5). Hence, the full atomic-resolution structure of $Na_V1.7$ may be required to further understand their pathology.

The $Na_V1.7$ channels are mainly expressed in the terminals of sensory neurones. They activate and inactivate rapidly, but recover from inactivation relatively slowly. Their normal role is to activate in response to small depolarizations close to the resting potential. In doing so, they respond to and amplify ramp stimuli, and so bring the neurone closer to its action potential threshold (Herzog, Cummins, Ghassemi, Dib-Hajj, & Waxman, 2003). However, the $Na_V1.8$ isoform, also present in DRG neurones, activates at more depolarized potentials and has a fast recovery from inactivation profile. It will therefore respond to the initial $Na_V1.7$-driven depolarizations by repetitive firing (Renganathan, Cummins, & Waxman, 2001). The $Na_V1.3$ isoform, normally associated with the central nervous system, is up-regulated in DRGs in a variety of chronic pain conditions (Shah et al., 2000). This isoform shows a fast recovery from inactivation and a significant persistent current (Lampert, Hains, & Waxman, 2006). Both properties will probably contribute to hyperexcitability following nerve injury.

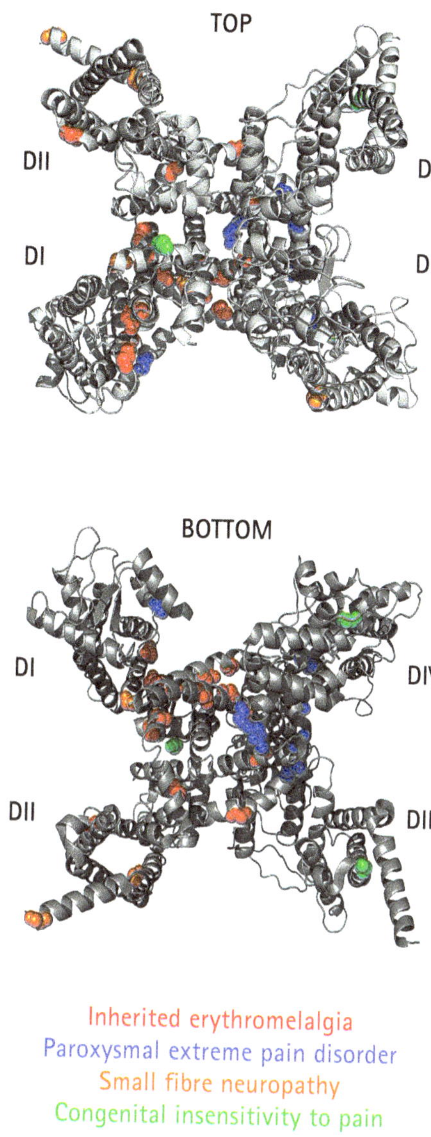

FIGURE 6.5 Pathological mutations of $Na_V1.7$. Here we see a selection of residues that have been identified by associated with inherited erythromelalgia (red), paroxysmal extreme pain disorder (blue), small fiber neuropathy (orange), and congenital insensitivity to pain (green). In the top panel we see the Na_V structure as viewed from above, and in the bottom panel we see it from below. The structure shown here is a cartoon representation of cockroach Na_V (PDBID: 5XoM), and the corresponding backbones of the mutated residues (shown as spheres) have been highlighted following ClustalW alignment of cockroach Na_V and human $Na_V1.7$ sequences. These mutations can be found in W. Huang et al., 2017.

Towards the Specific Targeting of Voltage-Gated Sodium Channels: Toxins and Antibodies

Na_V channels are major drug targets, and inhibitors have been clinically exploited to provide therapeutics acting as antiarrthythmics, anticonvulsants, and local anesthetics. Agents such as flecainide and lidocaine are key examples (Salvage et al., 2017; Sheets, Fozzard, Lipkind, & Hanck, 2010). Unfortunately, the lack of Na_V channel isoform selectivity of these inhibitors can limit their therapeutic use. Given the prominent role of Na_V channels in such a wide range of diseases, there is significant interest in developing *isoform-selective* inhibitors. This is particularly important in the field of pain disorders (Kwong & Carr, 2015). The compelling genetic evidence for the role of $Na_V1.7$ in pain perception has made this Na_V channel isoform a major pharmacological target (Vetter et al., 2017). It should also be noted that CIP patients who lack functional $Na_V1.7$ show no other serious developmental defects—although they may be anosmic (Weiss et al., 2011). This indicates that, unlike some Na_V channel isoforms, $Na_V1.7$ is functionally specialized, making it a highly attractive pharmacological target. Recent reports have described a number of promising and selective small-molecule inhibitors that have advanced to the stage of clinical trials (Kwong & Carr, 2015), including aryl sulphonamides ICA-12143, PF-04856264, and GX-936, which target the S1–S4 voltage sensor domain of DIV (Ahuja et al., 2015; McCormack et al., 2013). The structural basis for binding GX-936 was elucidated by co-crystallizing the compound with a chimeric protein composed of the VSM from a bacterial Na_V channel engineered to contain portions of the DIV VSM of human Na_V 1.7 (Ahuja et al., 2015). This approach, combining traditional small-molecule screening methods, but guided by structural insights into the target protein, offers a powerful general method for new drug discovery.

Another area of research in the discovery of subtype selective inhibitors is the exploitation of natural toxins that target Na_V channels: in particular, peptide-based toxins isolated from the venoms of sea anemones, spiders, snails, scorpions, and centipedes (de Lera Ruiz & Kraus, 2015). This is a rich and still largely untapped resource, and it has been estimated that there may be millions of spider-venom peptides (Escoubas, Sollod, & King, 2006). Given their diversity, the study of these molecules has identified interesting therapeutic leads that are often more potent than small molecules. Despite their variety, these toxins act in a limited number of ways, usually by binding to the pore region and thus blocking sodium entry, or by binding to and inhibiting movement of the VSM (Gilchrist, Olivera, & Bosmans, 2014). Two promising examples are protoxin-II (ProTx-II) and huwentoxin-IV (HwTx-IV), which exhibit a degree of Na_V subtype selectivity. Both target the domain II voltage sensor of Na_V 1.7 (Klint et al., 2012). In the case of huwentoxin-IV, the toxin inhibits Na_V 1.7 with an IC_{50} of 26nM but has an IC_{50} of $> 10\mu M$ for Na_V 1.5, demonstrating the pharmacological selectivity of the toxin (Revell et al., 2013). In other cases, the isoform specificity of the toxin may not be sufficient for immediate clinical use. Here, a variety of approaches might be adopted to engineer the proteins to have improved potency and selectivity. These methods include generation of peptide libraries, directed evolution, saturation mutagenesis, and chemical modification (alone or in combination), and analysis of structure–activity relationships to identify improved variants (Flinspach et al., 2017). As an example, a structure–function study of HwTx-IV was performed in which chemically synthesized and oxidatively folded peptide analogues were made and studied by automated electrophysiology. Using this approach, several peptides with enhanced potency for $Na_V1.7$ (up to 45-fold compared to the wild-type peptide) were produced (Revell et al., 2013). As a second example, native tarantula ceratotoxin-1 (CcoTx1) binds to a number of different Na_V channel isoforms. Through a combination of directed evolution and structure-based mutagenesis, a variant was generated that was highly potent for $Na_V1.7$ compared to other Na_V channels (Shcherbatko et al., 2016).

Antibody therapeutics represent a further alternative approach (Reichert, 2012). A major part of their attraction is the ability to achieve exceptionally high specificity for the target. Protein engineering techniques can also be applied during lead optimization to enhance the affinity (in the nanomolar to subpicomolar range) and potency, as well as the effector function and pharmacokinetic properties to improve cellular cytotoxicity and plasma half-life (Wilkinson, Gardener, & Williams, 2015).

The potential for monoclonal antibody–based blocking of Na_V channels has been appreciated for some time (Meiri et al., 1986). However, for clinical use, monoclonal antibodies would have to be specific for a particular Na_V channel isoform. There are now several examples of functional antibodies that have been raised to peptides representing either the first or second loops of the DI or DII voltage sensors of Na_V channels or the pore loops (Finney et al., 2016; J. H. Lee et al., 2014; Macdonald, Murphy, Papadopoulos, Stahl, & Alessandri-Haber, 2014; Ulrichts et al., 2015; Xu et al., 2005). A number of antibodies bound to the pore loop with a K_D in the picomolar to nanomolar range, an affinity level that is requiredappropriate for therapeutic use. An example of a VSM antagonist is a monoclonal antibody (SVmAb1) raised against the second extracellular loop of the DII S3–S4 voltage sensor. This is the same site targeted by HwTx-IV and ProTx-II. Application of SVmAb1 stabilized the closed state, thus inhibiting activation, and was claimed to be efficacious in mouse models of neuropathic and inflammatory pain as well in suppressing acute and chronic itch. SVmAb1 was shown to inhibit the function of $Na_V1.7$ with high subtype specificity and potency in a state-dependent manner (J. H. Lee et al., 2014). Unfortunately, these claims were not confirmed by a subsequent study (Liu et al., 2016). Indeed, Liu et al. could not even demonstrate binding to $Na_V1.7$ of a recombinant antibody made to the same sequence as SVmAb1. The striking disparity between the results of these two groups is not resolved, but it highlights the challenging task of screening for anti-(Na_V channel) functional antibodies using electrophysiology techniques that can be extremely sensitive to many variables such as buffer components and pH.

Antibodies generally show a limited permeation through the blood–brain barrier. This this can be an advantage if the aim is to target only peripheral subtypes (such as the pain-associated $Na_v1.7$) and in doing so, to avoid antagonizing subtypes in the central nervous system, which may result in undesirable side-effects, but therelatively large size of antibodies can restrict penetration into tissues more generally. Recent developments have exploited camelid antibodies that lack the two immunoglobulin light-chains and the first immunoglobulin heavy-chain constant domain. These antibodies are thus smaller than conventional antibodies and allow penetration into more inaccessible epitopes (Nguyen, Desmyter, & Muyldermans, 2001).

Alternative biological approaches have been investigated that utilize micro RNAs/short hairpin RNAs (shRNAs) to knockdown expression of the protein at the mRNA level (Muroi et al., 2011; Shao et al., 2016; Spencer, 2016). Providing suitable delivery methods are developed in the future, this type of drug could be more effective and more specific than a small molecule, antibody, or peptide antagonist.

Summary

In the 66 years since the Hodgkin Huxley model was first described (Hodgkin & Huxley, 1952), voltage-dependent Na_v channels have moved from a necessary mathematical abstraction to purified proteins, and now increasingly to atomic-resolution understanding. During this time, work on Na_v channels has inspired and sometimes driven experimental innovations in protein chemistry, electrophysiology, and pharmacology. The imminent arrival over the next few years of atomic-resolution structures for mammalian Na_v channels will undoubtedly inspire further hypotheses to provide better molecular insights into channel behavior, both normal and pathological. This in turn should encourage new approaches to rational drug development.

But there are still major unresolved questions. For example, the broader physiological and cell-biological context of Na_v channels, as they exist on the plasma membrane of neurones and muscle cells, is far from clear. This includes the roles of additional and auxiliary subunits and other Na_v channel interactors that modify channel behavior in vivo. To address such questions will probably require the application of high-resolution imaging techniques such as cryo-electron tomography, and of analytical techniques such as quantitative proteomics. This will represent a second revolution in understanding Na_v channel biology.

Acknowledgements

MZ was supported by postdoctoral funding from Medimmune. SCS was supported by a British Heart Foundation postdoctoral project grant PG/14/79/31102 (to APJ and CLH). AJT was funded by the British Heart Foundation PG/13/39/3029. CLH acknowledges additional grant support from the Medical Research Council, MR/M001288/1 and the Wellcome Trust, 105727/Z/14/Z.

References

Abriel, H. (2010). Cardiac sodium channel Na(v)1.5 and interacting proteins: Physiology and pathophysiology. *Journal of Molecular and Cellular Cardiology*, *48*(1), 2–11. doi:10.1016/j.yjmcc.2009.08.025

Abriel, H., Rougier, J. S., & Jalife, J. (2015). Ion channel macromolecular complexes in cardiomyocytes: Roles in sudden cardiac death. *Circulation Research*, *116*(12), 1971–1988. doi:10.1161/CIRCRESAHA.116.305017

Ahern, C. A., Payandeh, J., Bosmans, F., & Chanda, B. (2016). The hitchhiker's guide to the voltage-gated sodium channel galaxy. *Journal of General Physiology*, *147*(1), 1–24. doi:10.1085/jgp.201511492

Ahuja, S., Mukund, S., Deng, L., Khakh, K., Chang, E., Ho, H., . . . Payandeh, J. (2015). Structural basis of Nav1.7 inhibition by an isoform-selective small-molecule antagonist. *Science*, *350*(6267), aac5464. doi:10.1126/science.aac5464

Anderson, P. A., & Greenberg, R. M. (2001). Phylogeny of ion channels: Clues to structure and function. *Comparative Biochemistry and Physiology. Part B, Biochemistry and Molecular Biology*, *129*(1), 17–28.

Araki, K., & Nagata, K. (2011). Protein folding and quality control in the ER. *Cold Spring Harbor Perspectives in Biology*, *3*(11), a007526. doi:10.1101/cshperspect.a007526

Asseman, P., Berzin, B., Desry, D., Vilarem, D., Durand, P., Delmotte, C., . . . Thery, C. (1983). Persistent sinus nodal electrograms during abnormally prolonged postpacing atrial pauses in sick sinus syndrome in humans: Sinoatrial block vs overdrive suppression. *Circulation*, *68*(1), 33–41.

Black, J. A., Newcombe, J., & Waxman, S. G. (2010). Astrocytes within multiple sclerosis lesions upregulate sodium channel Nav1.5. *Brain*, *133*(Pt 3), 835–846. doi:10.1093/brain/awq003

Black, J. A., & Waxman, S. G. (2013). Noncanonical roles of voltage-gated sodium channels. *Neuron*, *80*(2), 280–291. doi:10.1016/j.neuron.2013.09.012

Bosmans, F., Martin-Eauclaire, M. F., & Swartz, K. J. (2008). Deconstructing voltage sensor function and pharmacology in sodium channels. *Nature*, *456*(7219), 202–208. doi:10.1038/nature07473

Brackenbury, W. J., & Djamgoz, M. B. (2007). Nerve growth factor enhances voltage-gated Na+ channel activity and Transwell migration in Mat-LyLu rat prostate cancer cell line. *Journal of Cellular Physiology*, *210*(3), 602–608. doi:10.1002/jcp.20846

Brackenbury, W. J., & Isom, L. L. (2011). Na channel beta subunits: Overachievers of the ion channel family. *Frontiers in Pharmacology*, *2*, 53. doi:10.3389/fphar.2011.00053

Brisson, L., Gillet, L., Calaghan, S., Besson, P., Le Guennec, J. Y., Roger, S., & Gore, J. (2011). Na(V)1.5 enhances breast cancer cell invasiveness by increasing NHE1-dependent H(+) efflux in caveolae. *Oncogene*, *30*(17), 2070–2076. doi:10.1038/onc.2010.574

Capes, D. L., Goldschen-Ohm, M. P., Arcisio-Miranda, M., Bezanilla, F., & Chanda, B. (2013). Domain IV voltage-sensor movement is both sufficient and rate limiting for fast inactivation in sodium channels. *Journal of General Physiology*, *142*(2), 101–112. doi:10.1085/jgp.201310998

Carrithers, M. D., Dib-Hajj, S., Carrithers, L. M., Tokmoulina, G., Pypaert, M., Jonas, E. A., & Waxman, S. G. (2007). Expression of the voltage-gated sodium channel NaV1.5 in the macrophage late endosome regulates endosomal acidification. *Journal of Immunology*, *178*(12), 7822–7832.

Catterall, W. A. (1986). Molecular properties of voltage-sensitive sodium channels. *Annual Review of Biochemistry*, *55*, 953–985. doi:10.1146/annurev.bi.55.070186.004513

Catterall, W. A. (2010). Ion channel voltage sensors: Structure, function, and pathophysiology. *Neuron, 67*(6), 915–928. doi:10.1016/j.neuron.2010.08.021

Catterall, W. A. (2017). Forty years of sodium channels: Structure, function, pharmacology, and epilepsy. *Neurochemical Research, 42*(9), 2495–2504. doi:10.1007/s11064-017-2314-9

Catterall, W. A., Wisedchaisri, G., & Zheng, N. (2017). The chemical basis for electrical signaling. *Nature Chemical Biology, 13*(5), 455–463. doi:10.1038/nchembio.2353

Catterall, W. A., & Zheng, N. (2015). Deciphering voltage-gated Na(+) and Ca(2+) channels by studying prokaryotic ancestors. *Trends in Biochemical Sciences, 40*(9), 526–534. doi:10.1016/j.tibs.2015.07.002

Chahine, M., & O'Leary, M. E. (2014). Regulation/modulation of sensory neuron sodium channels. *Handbook of Experimental Pharmacology, 221*, 111–135. doi:10.1007/978-3-642-41588-3_6

Chanda, B., & Bezanilla, F. (2002). Tracking voltage-dependent conformational changes in skeletal muscle sodium channel during activation. *Journal of General Physiology, 120*(5), 629–645.

Clairfeuille, T., Xu, H., Koth, C. M., & Payandeh, J. (2016). Voltage-gated sodium channels viewed through a structural biology lens. *Current Opinion in Structural Biology, 45*, 74–84. doi:10.1016/j.sbi.2016.11.022

Clatot, J., Ziyadeh-Isleem, A., Maugenre, S., Denjoy, I., Liu, H., Dilanian, G., . . . Neyroud, N. (2012). Dominant-negative effect of SCN5A N-terminal mutations through the interaction of Na(v)1.5 alpha-subunits. *Cardiovascular Research, 96*(1), 53–63. doi:10.1093/cvr/cvs211

Cox, J. J., Reimann, F., Nicholas, A. K., Thornton, G., Roberts, E., Springell, K., . . . Woods, C. G. (2006). An SCN9A channelopathy causes congenital inability to experience pain. *Nature, 444*(7121), 894–898. doi:10.1038/nature05413

Cummins, T. R., Dib-Hajj, S. D., & Waxman, S. G. (2004). Electrophysiological properties of mutant Nav1.7 sodium channels in a painful inherited neuropathy. *Journal of Neuroscience, 24*(38), 8232–8236. doi:10.1523/JNEUROSCI.2695-04.2004

Cusdin, F. S., Clare, J. J., & Jackson, A. P. (2008). Trafficking and cellular distribution of voltage-gated sodium channels. *Traffic, 9*(1), 17–26. doi:10.1111/j.1600-0854.2007.00673.x

Das, S., Gilchrist, J., Bosmans, F., & Van Petegem, F. (2016). Binary architecture of the Nav1.2-beta2 signaling complex. *Elife, 5*, e10960. doi:10.7554/eLife.10960

de Lera Ruiz, M., & Kraus, R. L. (2015). Voltage-gated sodium channels: Structure, function, pharmacology, and clinical indications. *Journal of Medicinal Chemistry, 58*(18), 7093–7118. doi:10.1021/jm501981g

DeCaen, P. G., Yarov-Yarovoy, V., Zhao, Y., Scheuer, T., & Catterall, W. A. (2008). Disulfide locking a sodium channel voltage sensor reveals ion pair formation during activation. *Proceedings of the National Academy of Sciences of the United States of America, 105*(39), 15142–15147. doi:10.1073/pnas.0806486105

Dobrzynski, H., Boyett, M. R., & Anderson, R. H. (2007). New insights into pacemaker activity: Promoting understanding of sick sinus syndrome. *Circulation, 115*(14), 1921–1932. doi:10.1161/CIRCULATIONAHA.106.616011

Escoubas, P., Sollod, B., & King, G. F. (2006). Venom landscapes: Mining the complexity of spider venoms via a combined cDNA and mass spectrometric approach. *Toxicon, 47*(6), 650–663. doi:10.1016/j.toxicon.2006.01.018

Fertleman, C. R., Baker, M. D., Parker, K. A., Moffatt, S., Elmslie, F. V., Abrahamsen, B., . . . Rees, M. (2006). SCN9A mutations in paroxysmal extreme pain disorder: Allelic variants underlie distinct channel defects and phenotypes. *Neuron, 52*(5), 767–774. doi:10.1016/j.neuron.2006.10.006

Finney, H. M., Baker, T. S., Lawson, A. D. G., Miller, K. M., de Ryck, M. R., Wolff, C. G. J. (2016). U.S. Patent No. US9, 266, 953 B2. Washington, DC: U.S. Patent and Trademark Office.

Flinspach, M., Xu, Q., Piekarz, A. D., Fellows, R., Hagan, R., Gibbs, A., . . . Wickenden, A. D. (2017). Insensitivity to pain induced by a potent selective closed-state Nav1.7 inhibitor. *Scientific Reports, 7*, 39662. doi:10.1038/srep39662

Gabelli, S. B., Boto, A., Kuhns, V. H., Bianchet, M. A., Farinelli, F., Aripirala, S., . . . Amzel, L. M. (2014). Regulation of the NaV1.5 cytoplasmic domain by calmodulin. *Nature Communications, 5*, 5126. doi:10.1038/ncomms6126

Gasser, A., Ho, T. S., Cheng, X., Chang, K. J., Waxman, S. G., Rasband, M. N., & Dib-Hajj, S. D. (2012). An ankyrinG-binding motif is necessary and sufficient for targeting Nav1.6 sodium channels to axon initial segments and nodes of Ranvier. *Journal of Neuroscience, 32*(21), 7232–7243. doi:10.1523/JNEUROSCI.5434-11.2012

Gilchrist, J., Das, S., Van Petegem, F., & Bosmans, F. (2013). Crystallographic insights into sodium-channel modulation by the beta4 subunit. *Proceedings of the National Academy of Sciences of the United States of America, 110*(51), E5016–5024. doi:10.1073/pnas.1314557110

Gilchrist, J., Olivera, B. M., & Bosmans, F. (2014). Animal toxins influence voltage-gated sodium channel function. *Handbook of Experimental Pharmacology, 221*, 203–229. doi:10.1007/978-3-642-41588-3_10

Guy, H. R., & Seetharamulu, P. (1986). Molecular model of the action potential sodium channel. *Proceedings of the National Academy of Sciences of the United States of America, 83*(2), 508–512.

Habib, A. M., Wood, J. N., & Cox, J. J. (2015). Sodium channels and pain. *Handbook of Experimental Pharmacology, 227*, 39–56. doi:10.1007/978-3-662-46450-2_3

Heine, M., Ciuraszkiewicz, A., Voigt, A., Heck, J., & Bikbaev, A. (2016). Surface dynamics of voltage-gated ion channels. *Channels (Austin), 10*(4), 267–281. doi:10.1080/19336950.2016.1153210

Heinemann, S. H., Terlau, H., Stuhmer, W., Imoto, K., & Numa, S. (1992). Calcium channel characteristics conferred on the sodium channel by single mutations. *Nature, 356*(6368), 441–443. doi:10.1038/356441a0

Herzog, R. I., Cummins, T. R., Ghassemi, F., Dib-Hajj, S. D., & Waxman, S. G. (2003). Distinct repriming and closed-state inactivation kinetics of Nav1.6 and Nav1.7 sodium channels in mouse spinal sensory neurons. *Journal of Physiology, 551*(Pt 3), 741–750. doi:10.1113/jphysiol.2003.047357

Hodgkin, A. L., & Huxley, A. F. (1952). A quantitative description of membrane current and its application to conduction and excitation in nerve. *Journal of Physiology, 117*(4), 500–544.

Hoshi, M., Du, X. X., Shinlapawittayatorn, K., Liu, H., Chai, S., Wan, X., . . . Deschenes, I. (2014). Brugada syndrome disease phenotype explained in apparently benign sodium channel mutations. *Circulation: Cardiovascular Genetics, 7*(2), 123–131. doi:10.1161/CIRCGENETICS.113.000292

Huang, C. L., Lei, L., Matthews, G. D., Zhang, Y., & Lei, M. (2012). Pathophysiological mechanisms of sino-atrial dysfunction and ventricular conduction disease associated with SCN5A deficiency: Insights from mouse models. *Frontiers in Physiology, 3*, 234. doi:10.3389/fphys.2012.00234

Huang, W., Liu, M., Yan, S. F., & Yan, N. (2017). Structure-based assessment of disease-related mutations in human voltage-gated sodium channels. *Protein and Cell*, *8*(6), 401–438. doi:10.1007/s13238-017-0372-z

Johnson, D., Montpetit, M. L., Stocker, P. J., & Bennett, E. S. (2004). The sialic acid component of the beta1 subunit modulates voltage-gated sodium channel function. *Journal of Biological Chemistry*, *279*(43), 44303–44310. doi:10.1074/jbc.M408900200

Kass, R. S. (2006). Sodium channel inactivation in heart: a novel role of the carboxy-terminal domain. *Journal of Cardiovascular Electrophysiology*, *17*(Suppl 1), S21–S25. doi:10.1111/j.1540-8167.2006.00381.x

Kazarinova-Noyes, K., Malhotra, J. D., McEwen, D. P., Mattei, L. N., Berglund, E. O., Ranscht, B., . . . Xiao, Z. C. (2001). Contactin associates with Na+ channels and increases their functional expression. *Journal of Neuroscience*, *21*(19), 7517–7525.

Keller, D. I., Rougier, J. S., Kucera, J. P., Benammar, N., Fressart, V., Guicheney, P., . . . Abriel, H. (2005). Brugada syndrome and fever: Genetic and molecular characterization of patients carrying SCN5A mutations. *Cardiovascular Research*, *67*(3), 510–519. doi:10.1016/j.cardiores.2005.03.024

Klint, J. K., Senff, S., Rupasinghe, D. B., Er, S. Y., Herzig, V., Nicholson, G. M., & King, G. F. (2012). Spider-venom peptides that target voltage-gated sodium channels: Pharmacological tools and potential therapeutic leads. *Toxicon*, *60*(4), 478–491. doi:10.1016/j.toxicon.2012.04.337

Koishi, R., Xu, H., Ren, D., Navarro, B., Spiller, B. W., Shi, Q., & Clapham, D. E. (2004). A superfamily of voltage-gated sodium channels in bacteria. *Journal of Biological Chemistry*, *279*(10), 9532–9538. doi:10.1074/jbc.M313100200

Kruger, L. C., & Isom, L. L. (2016). Voltage-gated Na+ channels: Not just for conduction. *Cold Spring Harbor Perspectives in Biology*, *8*(6). doi:10.1101/cshperspect.a029264

Kwong, K., & Carr, M. J. (2015). Voltage-gated sodium channels. *Current Opinion in Pharmacology*, *22*, 131–139. doi:10.1016/j.coph.2015.04.007

Lampert, A., Hains, B. C., & Waxman, S. G. (2006). Upregulation of persistent and ramp sodium current in dorsal horn neurons after spinal cord injury. *Experimental Brain Research*, *174*(4), 660–666. doi:10.1007/s00221-006-0511-x

Lampert, A., O'Reilly, A. O., Reeh, P., & Leffler, A. (2010). Sodium channelopathies and pain. *Pflugers Archiv. European Journal of Physiology*, *460*(2), 249–263. doi:10.1007/s00424-009-0779-3

Lee, A., Fakler, B., Kaczmarek, L. K., & Isom, L. L. (2014). More than a pore: Ion channel signaling complexes. *Journal of Neuroscience*, *34*(46), 15159–15169. doi:10.1523/JNEUROSCI.3275-14.2014

Lee, J. H., Park, C. K., Chen, G., Han, Q., Xie, R. G., Liu, T., . . . Lee, S. Y. (2014). A monoclonal antibody that targets a NaV1.7 channel voltage sensor for pain and itch relief. *Cell*, *157*(6), 1393–1404. doi:10.1016/j.cell.2014.03.064

Lei, M., Goddard, C., Liu, J., Leoni, A. L., Royer, A., Fung, S. S., . . . Huang, C. L. (2005). Sinus node dysfunction following targeted disruption of the murine cardiac sodium channel gene Scn5a. *Journal of Physiology*, *567*(Pt 2), 387–400. doi:10.1113/jphysiol.2005.083188

Lei, M., Jones, S. A., Liu, J., Lancaster, M. K., Fung, S. S., Dobrzynski, H., . . . Boyett, M. R. (2004). Requirement of neuronal- and cardiac-type sodium channels for murine sinoatrial node pacemaking. *Journal of Physiology*, *559*(Pt 3), 835–848. doi:10.1113/jphysiol.2004.068643

Liebeskind, B. J., Hillis, D. M., & Zakon, H. H. (2013). Independent acquisition of sodium selectivity in bacterial and animal sodium channels. *Current Biology*, *23*(21), R948–R949. doi:10.1016/j.cub.2013.09.025

Liu, D., Tseng, M., Epstein, L. F., Green, L., Chan, B., Soriano, B., . . . Moyer, B. D. (2016). Evaluation of recombinant monoclonal antibody SVmab1 binding to Na V1.7 target sequences and block of human Na V1.7 currents. *F1000Research*, 5, 2764. doi:10.12688/f1000research.9918.1

Macdonald, L., Murphy, A. J., Papadopoulos, N. J., Stahl, N., & Alesssandri-Haber, N. (2014). U.S. Patent No. WO 2014/159595 A2. (n.d.). Washington, DC: U.S. Patent and Trademark Office.

Maier, S. K., Westenbroek, R. E., Schenkman, K. A., Feigl, E. O., Scheuer, T., & Catterall, W. A. (2002). An unexpected role for brain-type sodium channels in coupling of cell surface depolarization to contraction in the heart. *Proceedings of the National Academy of Sciences of the United States of America*, *99*(6), 4073–4078. doi:10.1073/pnas.261705699

Maier, S. K., Westenbroek, R. E., Yamanushi, T. T., Dobrzynski, H., Boyett, M. R., Catterall, W. A., & Scheuer, T. (2003). An unexpected requirement for brain-type sodium channels for control of heart rate in the mouse sinoatrial node. *Proceedings of the National Academy of Sciences of the United States of America*, *100*(6), 3507–3512. doi:10.1073/pnas.2627986100

Makita, N., Bennett, P. B., & George, A. L., Jr. (1996). Molecular determinants of beta 1 subunit-induced gating modulation in voltage-dependent Na+ channels. *Journal of Neuroscience*, *16*(22), 7117–7127.

Marionneau, C., Carrasquillo, Y., Norris, A. J., Townsend, R. R., Isom, L. L., Link, A. J., & Nerbonne, J. M. (2012). The sodium channel accessory subunit Navbeta1 regulates neuronal excitability through modulation of repolarizing voltage-gated K(+) channels. *Journal of Neuroscience*, *32*(17), 5716–5727. doi:10.1523/JNEUROSCI.6450-11.2012

McCormack, K., Santos, S., Chapman, M. L., Krafte, D. S., Marron, B. E., West, C. W., . . . Castle, N. A. (2013). Voltage sensor interaction site for selective small molecule inhibitors of voltage-gated sodium channels. *Proceedings of the National Academy of Sciences of the United States of America*, *110*(29), E2724–E2732. doi:10.1073/pnas.1220844110

McCormick, D. A., Shu, Y., & Yu, Y. (2007). Neurophysiology: Hodgkin and Huxley model—still standing? *Nature*, *445*(7123), E1–E2; discussion E2–E3. doi:10.1038/nature05523

McCusker, E. C., D'Avanzo, N., Nichols, C. G., & Wallace, B. A. (2011). Simplified bacterial "pore" channel provides insight into the assembly, stability, and structure of sodium channels. *Journal of Biological Chemistry*, *286*(18), 16386–16391. doi:10.1074/jbc.C111.228122

Meadows, L. S., & Isom, L. L. (2005). Sodium channels as macromolecular complexes: Implications for inherited arrhythmia syndromes. *Cardiovascular Research*, *67*(3), 448–458. doi:10.1016/j.cardiores.2005.04.003

Meiri, H., Goren, E., Bergmann, H., Zeitoun, I., Rosenthal, Y., & Palti, Y. (1986). Specific modulation of sodium channels in mammalian nerve by monoclonal antibodies. *Proceedings of the National Academy of Sciences of the United States of America*, *83*(21), 8385–8389.

Mercier, A., Clement, R., Harnois, T., Bourmeyster, N., Faivre, J. F., Findlay, I., . . . Chatelier, A. (2012). The beta1-subunit of Na(v)1.5 cardiac sodium channel is required for a dominant negative effect through alpha-alpha interaction. *PLoS One*, *7*(11), e48690. doi:10.1371/journal.pone.0048690

Miloushev, V. Z., Levine, J. A., Arbing, M. A., Hunt, J. F., Pitt, G. S., & Palmer, A. G., 3rd. (2009). Solution structure of the NaV1.2 C-terminal EF-hand domain. *Journal of Biological Chemistry*, *284*(10), 6446–6454. doi:10.1074/jbc.M807401200

Moran, Y., Barzilai, M. G., Liebeskind, B. J., & Zakon, H. H. (2015). Evolution of voltage-gated ion channels at the emergence of Metazoa. *Journal of Experimental Biology*, *218*(Pt 4), 515–525. doi:10.1242/jeb.110270

Murata, Y., Iwasaki, H., Sasaki, M., Inaba, K., & Okamura, Y. (2005). Phosphoinositide phosphatase activity coupled to an intrinsic voltage sensor. *Nature, 435*(7046), 1239–1243. doi:10.1038/nature03650

Muroi, Y., Ru, F., Kollarik, M., Canning, B. J., Hughes, S. A., Walsh, S., ... Undem, B. J. (2011). Selective silencing of Na(V)1.7 decreases excitability and conduction in vagal sensory neurons. *Journal of Physiology, 589*(Pt 23), 5663–5676. doi:10.1113/jphysiol.2011.215384

Namadurai, S., Balasuriya, D., Rajappa, R., Wiemhofer, M., Stott, K., Klingauf, J., ... Jackson, A. P. (2014). Crystal structure and molecular imaging of the Nav channel beta3 subunit indicates a trimeric assembly. *Journal of Biological Chemistry, 289*(15), 10797–10811. doi:10.1074/jbc.M113.527994

Namadurai, S., Yereddi, N. R., Cusdin, F. S., Huang, C. L., Chirgadze, D. Y., & Jackson, A. P. (2015). A new look at sodium channel beta subunits. *Open Biology, 5*(1), 140192. doi:10.1098/rsob.140192

Nelson, M., Yang, M., Millican-Slater, R., & Brackenbury, W. J. (2015). Nav1.5 regulates breast tumor growth and metastatic dissemination in vivo. *Oncotarget, 6*(32), 32914–32929. doi:10.18632/oncotarget.5441

Nguyen, V. K., Desmyter, A., & Muyldermans, S. (2001). Functional heavy-chain antibodies in Camelidae. *Advances in Immunology, 79*, 261–296.

Noda, M., Shimizu, S., Tanabe, T., Takai, T., Kayano, T., Ikeda, T., ... et al. (1984). Primary structure of *Electrophorus electricus* sodium channel deduced from cDNA sequence. *Nature, 312*(5990), 121–127.

Onwuli, D. O., & Beltran-Alvarez, P. (2016). An update on transcriptional and post-translational regulation of brain voltage-gated sodium channels. *Amino Acids, 48*(3), 641–651. doi:10.1007/s00726-015-2122-y

Pappalardo, L. W., Liu, S., Black, J. A., & Waxman, S. G. (2014). Dynamics of sodium channel Nav1.5 expression in astrocytes in mouse models of multiple sclerosis. *Neuroreport, 25*(15), 1208–1215. doi:10.1097/WNR.0000000000000249

Paramonov, A. S., Lyukmanova, E. N., Myshkin, M. Y., Shulepko, M. A., Kulbatskii, D. S., Petrosian, N. S., ... Shenkarev, Z. O. (2017). NMR investigation of the isolated second voltage-sensing domain of human Nav1.4 channel. *Biochimica et Biophysica Acta, 1859*(3), 493–506. doi:10.1016/j.bbamem.2017.01.004

Piao, H. H., Rajakumar, D., Kang, B. E., Kim, E. H., & Baker, B. J. (2015). Combinatorial mutagenesis of the voltage-sensing domain enables the optical resolution of action potentials firing at 60 Hz by a genetically encoded fluorescent sensor of membrane potential. *Journal of Neuroscience, 35*(1), 372–385. doi:10.1523/JNEUROSCI.3008-14.2015

Poelzing, S., Forleo, C., Samodell, M., Dudash, L., Sorrentino, S., Anaclerio, M., ... Deschenes, I. (2006). SCN5A polymorphism restores trafficking of a Brugada syndrome mutation on a separate gene. *Circulation, 114*(5), 368–376. doi:10.1161/CIRCULATIONAHA.105.601294

Popa, M. O., Alekov, A. K., Bail, S., Lehmann-Horn, F., & Lerche, H. (2004). Cooperative effect of S4-S5 loops in domains D3 and D4 on fast inactivation of the Na+ channel. *Journal of Physiology, 561*(Pt 1), 39–51. doi:10.1113/jphysiol.2004.065912

Qin, N., D'Andrea, M. R., Lubin, M. L., Shafaee, N., Codd, E. E., & Correa, A. M. (2003). Molecular cloning and functional expression of the human sodium channel beta1B subunit, a novel splicing variant of the beta1 subunit. *European Journal of Biochemistry, 270*(23), 4762–4770.

Ratcliffe, C. F., Westenbroek, R. E., Curtis, R., & Catterall, W. A. (2001). Sodium channel beta1 and beta3 subunits associate with neurofascin through their extracellular immunoglobulin-like domain. *Journal of Cell Biology, 154*(2), 427–434.

Reichert, J. M. (2012). Marketed therapeutic antibodies compendium. *mAbs, 4*(3), 413–415. doi:10.4161/mabs.19931

Renganathan, M., Cummins, T. R., & Waxman, S. G. (2001). Contribution of Na(v)1.8 sodium channels to action potential electrogenesis in DRG neurons. *Journal of Neurophysiology, 86*(2), 629–640.

Revell, J. D., Lund, P. E., Linley, J. E., Metcalfe, J., Burmeister, N., Sridharan, S., . . . Bednarek, M. A. (2013). Potency optimization of Huwentoxin-IV on hNav1.7: A neurotoxin TTX-S sodium-channel antagonist from the venom of the Chinese bird-eating spider *Selenocosmia huwena. Peptides, 44*, 40–46. doi:10.1016/j.peptides.2013.03.011

Rogers, M., Tang, L., Madge, D. J., & Stevens, E. B. (2006). The role of sodium channels in neuropathic pain. *Seminars in Cell and Developmental Biology, 17*(5), 571–581. doi:10.1016/j.semcdb.2006.10.009

Rook, M. B., Bezzina Alshinawi, C., Groenewegen, W. A., van Gelder, I. C., van Ginneken, A. C., Jongsma, H. J., . . . Wilde, A. A. (1999). Human SCN5A gene mutations alter cardiac sodium channel kinetics and are associated with the Brugada syndrome. *Cardiovascular Research, 44*(3), 507–517.

Salvage, S. C., Chandrasekharan, K. H., Jeevaratnam, K., Dulhunty, A. F., Thompson, A. J., Jackson, A. P., & Huang, C. L. (2017). Multiple targets for flecainide action: Implications for cardiac arrhythmogenesis. *British Journal of Pharmacology*. doi:10.1111/bph.13807

Sanders, L., Rakovic, S., Lowe, M., Mattick, P. A., & Terrar, D. A. (2006). Fundamental importance of Na+-Ca2+ exchange for the pacemaking mechanism in guinea-pig sino-atrial node. *Journal of Physiology, 571*(Pt 3), 639–649. doi:10.1113/jphysiol.2005.100305

Sarhan, M. F., Tung, C. C., Van Petegem, F., & Ahern, C. A. (2012). Crystallographic basis for calcium regulation of sodium channels. *Proceedings of the National Academy of Sciences of the United States of America, 109*(9), 3558–3563. doi:10.1073/pnas.1114748109

Shah, B. S., Stevens, E. B., Gonzalez, M. I., Bramwell, S., Pinnock, R. D., Lee, K., & Dixon, A. K. (2000). Beta3, a novel auxiliary subunit for the voltage-gated sodium channel, is expressed preferentially in sensory neurons and is upregulated in the chronic constriction injury model of neuropathic pain. *European Journal of Neuroscience, 12*(11), 3985–3990.

Shao, J., Cao, J., Wang, J., Ren, X., Su, S., Li, M., . . . Zang, W. (2016). MicroRNA-30b regulates expression of the sodium channel Nav1.7 in nerve injury-induced neuropathic pain in the rat. *Molecular Pain, 12*. doi:10.1177/1744806916671523

Shcherbatko, A., Rossi, A., Foletti, D., Zhu, G., Bogin, O., Galindo Casas, M., . . . Strop, P. (2016). Engineering highly potent and selective microproteins against Nav1.7 sodium channel for treatment of pain. *Journal of Biological Chemistry, 291*(27), 13974–13986. doi:10.1074/jbc.M116.725978

Sheets, M. F., Fozzard, H. A., Lipkind, G. M., & Hanck, D. A. (2010). Sodium channel molecular conformations and antiarrhythmic drug affinity. *Trends in Cardiovascular Medicine, 20*(1), 16–21. doi:10.1016/j.tcm.2010.03.002

Shen, H., Zhou, Q., Pan, X., Li, Z., Wu, J., & Yan, N. (2017). Structure of a eukaryotic voltage-gated sodium channel at near-atomic resolution. *Science, 355*(6328). doi:10.1126/science.aal4326

Sottas, V., & Abriel, H. (2016). Negative-dominance phenomenon with genetic variants of the cardiac sodium channel Nav1.5. *Biochimica et Biophysica Acta, 1863*(7 Pt B), 1791–1798. doi:10.1016/j.bbamcr.2016.02.013

Spampanato, J., Kearney, J. A., de Haan, G., McEwen, D. P., Escayg, A., Aradi, I., . . . Meisler, M. H. (2004). A novel epilepsy mutation in the sodium channel SCN1A identifies a cytoplasmic domain for beta subunit interaction. *Journal of Neuroscience, 24*(44), 10022–10034. doi:10.1523/JNEUROSCI.2034-04.2004

Spencer, N. J. (2016). Switching off pain at the source: Is this the end for opioid pain relief? *Pain Management, 6*(1), 39–47. doi:10.2217/pmt.15.52

Stephens, R. F., Guan, W., Zhorov, B. S., & Spafford, J. D. (2015). Selectivity filters and cysteine-rich extracellular loops in voltage-gated sodium, calcium, and NALCN channels. *Frontiers in Physiology, 6*, 153. doi:10.3389/fphys.2015.00153

Ulrichts, P., Van der Woning, S., De Boeck, G., Hofman, E., Blanchetot, C., Saunders, M., & De Haard, J. J. W. (2015). U.S. Patent No. WO 2015/032916 A1. (n.d.). Washington, DC: U.S. Patent and Trademark Office.

Vandenberg, J. I., & Waxman, S. G. (2012). Hodgkin and Huxley and the basis for electrical signaling: A remarkable legacy still going strong. *Journal of Physiology, 590*(11), 2569–2570. doi:10.1113/jphysiol.2012.233411

Veitia, R. A. (2007). Exploring the molecular etiology of dominant-negative mutations. *Plant Cell, 19*(12), 3843–3851. doi:10.1105/tpc.107.055053

Verheijck, E. E., van Ginneken, A. C., Bourier, J., & Bouman, L. N. (1995). Effects of delayed rectifier current blockade by E-4031 on impulse generation in single sinoatrial nodal myocytes of the rabbit. *Circulation Research, 76*(4), 607–615.

Vetter, I., Deuis, J. R., Mueller, A., Israel, M. R., Starobova, H., Zhang, A., . . . Mobli, M. (2017). NaV1.7 as a pain target—From gene to pharmacology. *Pharmacology and Therapeutics, 172*, 73–100. doi:10.1016/j.pharmthera.2016.11.015

Villalba-Galea, C. A., Sandtner, W., Starace, D. M., & Bezanilla, F. (2008). S4-based voltage sensors have three major conformations. *Proceedings of the National Academy of Sciences of the United States of America, 105*(46), 17600–17607. doi:10.1073/pnas.0807387105

Waxman, S. G. (2012). Sodium channels, the electrogenisome and the electrogenistat: Lessons and questions from the clinic. *Journal of Physiology, 590*(11), 2601–2612. doi:10.1113/jphysiol.2012.228460

Weiss, J., Pyrski, M., Jacobi, E., Bufe, B., Willnecker, V., Schick, B., . . . Zufall, F. (2011). Loss-of-function mutations in sodium channel Nav1.7 cause anosmia. *Nature, 472*(7342), 186–190. doi:10.1038/nature09975

Wilkinson, T. C., Gardener, M. J., & Williams, W. A. (2015). Discovery of functional antibodies targeting ion channels. *Journal of Biomolecular Screening, 20*(4), 454–467. doi:10.1177/1087057114560698

Willis, B. C., Ponce-Balbuena, D., & Jalife, J. (2015). Protein assemblies of sodium and inward rectifier potassium channels control cardiac excitability and arrhythmogenesis. *American Journal of Physiology—Heart and Circulatory Physiology, 308*(12), H1463–1473. doi:10.1152/ajpheart.00176.2015

Wu, J., Yan, Z., Li, Z., Qian, X., Lu, S., Dong, M., . . . Yan, N. (2016). Structure of the voltage-gated calcium channel Ca(v)1.1 at 3.6 A resolution. *Nature, 537*(7619), 191–196. doi:10.1038/nature19321

Xing, D., Wang, J., Ou, S., Wang, Y., Qiu, B., Ding, D., ... Gao, Q. (2014). Expression of neonatal Nav1.5 in human brain astrocytoma and its effect on proliferation, invasion and apoptosis of astrocytoma cells. *Oncology Reports*, *31*(6), 2692–2700. doi:10.3892/or.2014.3143

Xu, S. Z., Zeng, F., Lei, M., Li, J., Gao, B., Xiong, C., ... Beech, D. J. (2005). Generation of functional ion-channel tools by E3 targeting. *Nature Biotechnology*, *23*(10), 1289–1293. doi:10.1038/nbt1148

Yarov-Yarovoy, V., DeCaen, P. G., Westenbroek, R. E., Pan, C. Y., Scheuer, T., Baker, D., & Catterall, W. A. (2012). Structural basis for gating charge movement in the voltage sensor of a sodium channel. *Proceedings of the National Academy of Sciences of the United States of America*, *109*(2), E93–E102. doi:10.1073/pnas.1118434109

Zhang, X., Ren, W., DeCaen, P., Yan, C., Tao, X., Tang, L., ... Yan, N. (2012). Crystal structure of an orthologue of the NaChBac voltage-gated sodium channel. *Nature*, *486*(7401), 130–134. doi:10.1038/nature11054

CHAPTER 7

SPECIALIZED SODIUM CHANNELS IN PAIN TRANSMISSION

YUCHENG XIAO, ZIFAN PEI, AND THEODORE R. CUMMINS

INTRODUCTION

PAIN is an unpleasant sensory and emotional experience that severely affects an individual's quality of life (Millan, 1999). However, pain sensations are essential for survival. They can warn the individual of tissue damage and to seek appropriate actions to ameliorate the painful condition. Pain sensations typically originate in peripheral sensory neurons called *nociceptors*. The cell bodies of these nociceptive sensory neurons with a single bifurcating axon are located in the dorsal root ganglia (DRG) and trigeminal ganglia (Devor, 1999). The axon branch that projects to peripheral tissue is responsible for transduction of noxious stimuli into nociceptive impulses, and the branch that projects to the dorsal horn of the spinal cord relays these impulses to the central nerve system. The transmission of nerve impulses and overall excitability of nociceptive neurons is determined mostly by an array of voltage-gated ion channels, including sodium channels, potassium channels, and calcium channels. Modification of ion channel function can result in changes in excitability of neurons. Changes that induce over-excitability of nociceptive neurons can lead to increased pain sensation. Voltage-gated sodium channels (VGCS, or Na_V) are especially interesting, as the opening of VGSCs is fundamental to generation of action potentials in almost all excitable tissues (Catterall, Goldin, & Waxman, 2005). The potential power of targeting nociceptor excitability and VGSCs to reduce pain sensations is strongly supported by pharmacology studies in cells and animal models, as well as by genetic channelopathy studies in rare inherited mutations linked to abnormal painful neuropathy (Dib-Hajj, Cummins, Black, &

Waxman, 2010). In particular, as will be discussed later, loss-of-function mutations of $Na_V1.7$ or gain-of-function mutations of $Na_V1.9$ have been shown to be linked to the congenital inability to experience pain. These remarkable findings clearly indicate not only that VGSCs play important roles in pain sensation, but also that some VGSC isoforms are likely to be ideal targets for the development of novel agents for treating pain. In this chapter, we will review the roles of the VGSC isoforms specialized in pain transmission.

Overview of Voltage-Gated Sodium Channels

The opening of VGSCs can selectively allow permeation of sodium ions across the cell membrane, underlying the initiation and propagation of action potentials (Goldin, 2001). VGSCs are complex transmembrane proteins (Figure 7.1A) composed of a functional pore-forming α-subunit associated with one or two auxiliary β-subunits (β1–β4) associated via covalent and/or non-covalent interactions to a single α-subunit (O'Malley & Isom, 2015). Although expression of the α-subunit alone is sufficient to produce a functional sodium channel, β-subunits can regulate expression level and gating properties of α-subunits. The α-subunit is a large polypeptide (220–260 kD) consisting of ~2000 amino acid residues (Noda & Numa, 1987). The folded α-subunit protein is assembled into four homologous domains (DI–DIV), each consisting of six transmembrane α-helices (S1–S6). While the S5–S6 segments of each domain coordinate to form a highly conserved Na+-selective central pore, the S1–S4 segments serve as voltage sensing modules. The S4 segment (also often called the *voltage sensor*) contains positively charged residues (Arg/Lys) at every third position that can sense the change in membrane potential to regulate channel gating (Yarov-Yarovoy et al., 2012). While the S4 segments in DI–DIII are responsible for channel activation, the S4 segment in DIV is believed to play a vital role in channel inactivation (although DIII may have a role in inactivation, too). VGSCs typically undergo transition among three different basic states (closed, open, and fast-inactivated) in response to changes in membrane potential during action-potential firing (Figure 7.1B). At resting membrane potentials, the majority of sodium channels are in the closed (or resting) configuration, and current through channel pore is prohibited. Upon depolarization, the S4 segments of DI–DIII are positioned in an outward configuration so that the channel pore opens ("activates"), allowing influx of sodium ions (Xiao, Blumenthal, & Cummins, 2014). In general, the majority of VGSCs remain open only within a millisecond or so, after which sodium current can be terminated by a process known as *ast-inactivationf* (Figure 7.1C). Fast-inactivation is thought to be coupled to outward movement of DIV S4 segment, which can induce a short polypeptide (an isoleucine-phenylalanine-methionine-threonine, or IFMT, motif) on the cytoplasmic loop connecting DIII and DIV to bind within or near the inner pore region, therefore occluding the inflow of sodium ions (West et al., 1992).

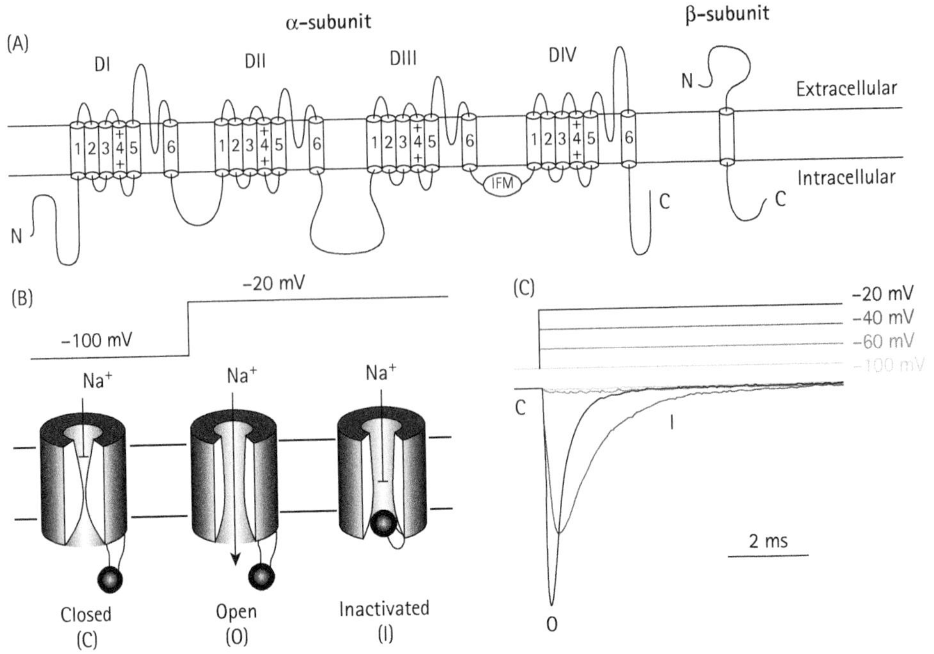

FIGURE 7.1 **(A)**, Schematic diagram of sodium channel α- and β-subunits. The voltage sensor (segment 4) of each domain is marked with "++." A short three-residue peptide "IFM" (Ile-Phe-Met) represents inactivation particle. **(B)**, A simplified cartoon indicating the typical activity of sodium channels transitions among three different states (closed, open, and fast-inactivated). **(C)**, Currents fluxing through sodium channels elicited by depolarizing steps to various potentials (−60, −40, −20 mV) from a holding potential of −100 mV.

VGSCs can recover from inactivation when the cell membrane is repolarized to negative (resting) membrane potentials. Recovery from inactivation allows VSGCs to go back to the closed state so that they are able to open again in response to subsequent membrane depolarization. The recovery rate from fast-inactivation can play a vital role in determining the firing frequency of excitable cells.

There are ten α-subunit isoforms (Na_V1.1–1.9 and Nax) that have been cloned and functionally characterized from mammals (Catterall et al., 2005). Most of them are predominantly expressed in central or peripheral neurons (the major exceptions being Na_V1.4 and Na_V1.5, which are predominantly expressed on skeletal muscle and cardiac myocytes, respectively).

VGSC isoforms can show up to 75% sequence similarity to each other. However, even subtle sequence differences can determine important differences in their functional and pharmacological profiles. Tetrodotoxin (TTX) is a well-known small molecule VSGC blocker initially isolated from the liver of the puffer fish (Ritchie, 1975). It occludes the ion-permeation pathway by docking at the outer vestibule of the channel pore. However, not all VGSCs are highly sensitive to TTX (Rogart & Regan, 1985). A crucial pore residue (Y362 in Na_V1.7) has been identified as the major determinant for the sensitivity of VGSC isoforms to TTX (found to be either Tyr or Phe in sensitive isoforms and either Cys or Ser in resistant isoforms) (Leffler, Herzog, Dib-Hajj, Waxman, & Cummins, 2005). Accordingly, VSGC isoforms are often classified as tetrodotoxin (TTX)-sensitive ($Na_V1.1$–$Na_V1.4$, $Na_V1.6$ and $Na_V1.7$) and TTX-resistant ($Na_V1.5$, $Na_V1.8$, $Na_V1.9$, and, most likely, Nax).

Adult DRG neurons express multiple VSGC isoforms, including $Na_V1.1$, $Na_V1.6$, $Na_V1.7$, $Na_V1.8$, $Na_V1.9$, and Nax (S. Dib-Hajj, Black, Cummins, & Waxman, 2002; Felts, Yokoyama, Dib-Hajj, Black, & Waxman, 1997). $Na_V1.3$ can also be found in nociceptive neurons, but expression of $Na_V1.3$ strongly depends on the developmental stage of sensory neurons (Waxman, Kocsis, & Black, 1994). Three isoforms ($Na_V1.7$, $Na_V1.8$, and $Na_V1.9$) play specialized roles in nociceptive sensory neurons, although $Na_V1.1$, $Na_V1.3$, and $Na_V1.6$ have also been implicated in pain sensations.

$Na_V1.7$

$Na_V1.7$ (also called hNE9 or PN1 in the past) is encoded by the gene *SCN9A* located on chromosome 2q224.3. $Na_V1.7$ protein is not significantly expressed in CNS, spinal cord, muscle or heart tissue (Klugbauer, Lacinova, Flockerzi, & Hofmann, 1995; Toledo-Aral et al., 1997), but is highly expressed in most peripheral nociceptive DRG neurons as well as sympathetic ganglia neurons (Black et al., 1996). However, it may play a role in some CNS tissues (Branco et al., 2016). $Na_V1.7$ channel distributes over the multiple compartments of nociceptive neurons with expression extending from the free peripheral terminals in the peripheral axon branch to the terminals in the dorsal horn. $Na_V1.7$-specific antibodies helped determine that in DRG neurons, $Na_V1.7$ protein has more intense expression levels in C-type neurons than in A-type neurons (Djouhri, Newton, et al., 2003; Gould et al., 2000). Moreover, $Na_V1.7$ is the major TTX-sensitive VGSC isoform in C-type neurons, although these small diameter neurons also can express a second TTX-sensitive VGSC isoform ($Na_V1.6$ that is predominantly expressed in A-type neurons) (Wilson et al., 2011). Expression of $Na_V1.7$ can be upregulated by multiple factors including the sodium channel β2-subunit, nerve growth factor (NGF), tumor necrosis factor-α (TNF-α) as well as protein kinase C (PKC) and p38 (Chattopadhyay, Mata, & Fink, 2008) (Gould et al., 2000; Tamura et al., 2014). There is also compelling evidence of an increase in the mRNA and protein expression with this channel in rat DRG tissue following injection of complete Freund's adjuvant (CFA) or carrageenan into the hind paw (Black, Liu, Tanaka, Cummins, & Waxman, 2004; Nassar et al., 2004). Importantly, many of these factors are inflammatory mediators or, like CFA, cause tissue inflammation.

Compared with $Na_V1.6$ and other VGSCs found in the central nervous system (CNS), $Na_V1.7$ channels exhibit several unique gating properties which may substantially affect DRG neuronal excitability. Closed-state fast-inactivation (inactivation that occurs at negative potentials where the vast majority of channels do not open) can modulate channel availability. This process reflects the direct transition from the closed ("resting") configuration to a fast-inactivated state. $Na_V1.7$ shows sevenfold slower closed-state fast-inactivation at negative potentials close to the resting membrane potential than $Na_V1.6$. Slow closed-state fast-inactivation has been proposed to underlie the generation of ramp currents in response to small, slow, subthreshold depolarizations (Cummins, Howe, & Waxman, 1998). Accordingly, $Na_V1.7$ is found to generate significantly larger ramp current in DRG neurons (Cummins et al., 1998; Herzog, Cummins, Ghassemi, Dib-Hajj, & Waxman, 2003). Because $Na_V1.7$ ramp currents occur near resting membrane potential, and this channel is the major TTX-sensitive isoform in small DRG neurons, $Na_V1.7$ is thought to set the threshold of action-potential generation in nociceptive neurons (Cummins et al., 1998). On the other hand, $Na_V1.7$ can exhibit significantly slower repriming (the rate of recovery from inactivation) kinetics. The recovery from inactivation, reflecting primarily the rate of the channel's transitioning between the closed-inactivated and closed-available states, is approximately fivefold slower near the resting membrane potential of neurons for $Na_V1.7$ currents than for $Na_V1.6$ currents. This suggests that $Na_V1.7$ channels take longer to return to the closed configuration, such that $Na_V1.7$ channels exhibit a delay before they are available for reopening and able to facilitate the next action-potential firing. This slow recovery from inactivation may underlie the mechanism for the lower frequency of action-potential firing observed with C-type DRG neurons that predominantly express $Na_V1.7$ than with A-type DRG neurons that predominantly express $Na_V1.6$ (Djouhri & Lawson, 2004; Djouhri, Newton, et al., 2003). These unique biophysical properties are observed in small-diameter DRG neurons and also with cloned human $Na_V1.7$ channels heterogeneously expressed in *HEK293* cells. There is substantial evidence supporting the hypothesis that $Na_V1.7$ serves as a crucial contributor to pain sensation. The most compelling evidence initially came from studies on patients from Northern Pakistan with a congenital inability to experience pain (CIP). Neurological examinations by Cox and colleagues (Cox et al., 2006) indicated that these patients apparently had never suffered from any pain sensations. However, they appeared normal on other physiological functions that were observed, including proprioception, touch, temperature perception, vision, hearing, and intelligence. No motor or, surprisingly, autonomic dysfunction was detected, either. Ahmad et al. (Ahmad et al., 2007) studied a Canadian family from Newfoundland with members who also displayed CIP. Like the patients from Northern Pakistan, these patients showed normal neurological examination. In addition, the patients had normal sweating. All of the patients studied displayed no pain behavior in childhood, although several initially affected individuals claimed to start to feel pain during adolescence, which could be a learned behavior. Subsequent analysis indicated that CIP patients also exhibit deficits in their ability to smell, and are anosmic. These independent studies all linked CIP phenotype to loss-of-function mutations in $Na_V1.7$. Sequence analysis of the *SCN9A* gene indicated several distinct homozygous nonsense mutations in affected individuals from the families that were studied. These nonsense mutations were identified as deletion, insertion, or substitution of a single nucleotide base, which led to a frameshift and a truncated $Na_V1.7$ protein. The phenomenon was also observed in another study, focusing on nine families from seven different nationalities in which affected individuals met the diagnostic criteria for CIP (Goldberg et al., 2007). In the absence of $Na_V1.7$, humans have an intact axon reflex arc and no discernable autonomic disturbances, suggesting again that $Na_V1.7$ in humans does not play an essential role in mediating autonomic effects. This lack of obvious autonomic dysfunction is surprising, given the strong expression of $Na_V1.7$ in sympathetic neurons, but it may reflect sodium channel redundancy or compensation in these neurons that is lacking in nociceptive neurons. The truncated $Na_V1.7$ is nonfunctional because no current is observed when recombinant $hNa_V1.7$ containing these CIP mutation sites were transfected into *HEK293* or *ND7/23* cells (Figure 7.2A,B). Interestingly, a recent study further linked reduced activity (not complete elimination) of $Na_V1.7$ channels to CIP. In contrast to CIP-related mutations identified previously, the two new mutations (*W1775R* and *L1831X*) still endow $Na_V1.7$ protein with some function, but decrease current density, negatively shifting steady-state inactivation and positively shifting activation (Emery et al., 2015). However, family members of CIP patients with only one nonfunctional $Na_V1.7$ allele seem to experience normal pain sensations, raising questions about how much $Na_V1.7$ would need to be inhibited in the majority of patients with two functional alleles in order to obtain substantial reductions in pain sensations.

The phenotype of CIP occurring in humans could be recapitulated on global $Na_V1.7$ knockout mice (Gingras et al., 2014), although $Na_V1.7$ expression was previously demonstrated to show species-specific profiles in central endocrine brain regions and the adrenal medulla (Ahmad et al., 2007). Like individuals with loss-of-function $Na_V1.7$, knockout mice are completely insensitive to pain in response to tactile, thermal, and chemical stimuli. Deficits in olfaction similar to those observed in some individuals with CIP are also detected in knockout mice. Conversely, the knockouts showed no defects in mechanical sensitivity or overall movement. Therefore, mice can be considered as reasonably suitable models for developing better $Na_V1.7$-targeted analgesics. One caveat is that sequence differences between rodent and human $Na_V1.7$ channels may impact channel pharmacological properties.

In contrast to loss-of-function mutations in $Na_V1.7$, gain-of-function mutations that alter voltage-dependent gating properties of $Na_V1.7$ increase pain sensations (but again, somewhat surprisingly, have not been reported to have pronounced effects on autonomic function). Gain-of-function mutations in $Na_V1.7$ have been widely identified in families with a painful inherited neuropathy, inherited erythromelalgia (IEM) or paroxysmal extreme pain disorder (PEPD). IEM is the first human pain syndrome linked to naturally occurring point mutation in voltage-gated sodium channels (Yang et al., 2004). It is characterized by severe chronic burning pain sensations in the hands and feet. More than 15 point mutations in $Na_V1.7$ have been identified in patients with IEM syndrome to date (Dib-Hajj, Yang, Black, & Waxman, 2013). These mutations are located predominantly in the first half of the protein sequence, but they can be found within N-terminal, transmembrane segments or intracellular loops of the $Na_V1.7$ protein. They are found to substantially enhance the excitability of nociceptor neurons by lowering the current threshold for generation of action potentials and increasing repetitive firing (Dib-Hajj et al., 2005). The overwhelming majority of IEM mutations cause a hyperpolarizing shift

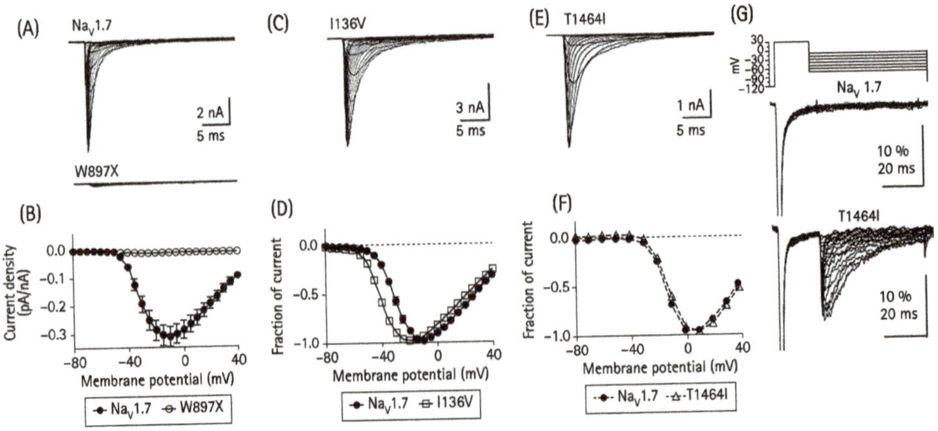

FIGURE 7.2 Typical current traces from the wild type hNa$_V$1.7 channels (A) and from hNa$_V$1.7 channels containing a CIP mutation (W897X) (A, *below*), an IEM mutation (I136V) (C), and a PEPD mutation (T1464I) (E). (B) shows the effects of W897X on current density. (D) and (F) show the effects of I136V and T1464I on current–voltage relationships of hNa$_V$1.7 channels, respectively. Families of sodium currents were induced by 50-ms depolarizing steps to various potentials ranging from −80 to +40 mV in 5-mV increments. Cells were held at −120 mV. (G) The T1464I PEPD mutation enhances Na$_V$1.7 resurgent currents.

in the voltage-dependent activation (Figure 7.2C,D) (Cummins, Dib-Hajj, & Waxman, 2004; Dib-Hajj et al., 2013). The shift accordingly induces the channels to generate ramp currents at more negative potentials. Most of the IEM mutations slow channel deactivation and accelerate recovery rate from inactivation, allowing DRG neurons to fire at a higher frequency (Dib-Hajj et al., 2005). IEM mutations generate larger ramp current, but some also shift steady-state inactivation to more positive potentials, or impair slow inactivation. However, some of these additional effects are not consistently observed, and they can even be contradictory among IEM mutations (Cheng et al., 2011; Cummins et al., 2004). This suggests that these additional effects might not be the crucial factors responsible for IEM phenotype, but rather modulate the severity of pain symptoms caused by specific IEM mutations.

Paroxysmal extreme pain disorder (PEPD), characterized by severe, burning pain in the submandibular, ocular, and rectal areas, as well as flushing, is a second autosomal-dominant chronic pain disorder linked to naturally occurring mutations in the Na$_V$1.7 channel (Fertleman et al., 2006; Fertleman et al., 2007). Among ten PEPD mutations identified, two are located within the DIV voltage sensor (S4), while others are within the inactivation particle (IFMT) or intracellular loops that form the docking site for inactivation particle (Fertleman et al., 2006). Dib-Hajj et al. (Dib-Hajj et al., 2008) demonstrated that transient transfection of PEPD mutant (*M1267K*) Na$_V$1.7 channels endowed DRG neurons with hyperexcitability. The mutant channels not only significantly lower the current threshold for generation of action potentials, but also substantially increase the firing frequency (Fischer & Waxman, 2010). PEPD mutations modify gating properties of Na$_V$1.7 through a mechanism different from IEM mutations described before. PEPD mutations fail to significantly affect channel activation and deactivation (Dib-Hajj et al., 2008; Fertleman et al., 2006; Jarecki, Sheets, Jackson, & Cummins, 2008). One of the most striking features for PEPD mutations is that they impair channel inactivation (Figure 7.2E,F). They slow the rate of fast-inactivation, shift steady-state inactivation to more positive potentials, and accelerate recovery from inactivation. An increase in persistent current is also observed, which partially results from the augmented window currents due to the significant positive shift in steady-state inactivation. The other striking feature is to augment β4-mediated resurgent currents of $Na_V1.7$ channels (Jarecki, Piekarz, Jackson, & Cummins, 2010), an atypical sodium current typically evoked during the repolarizing phase of action potentials. Resurgent currents, identified widely in many neuronal populations, including in specific neurons in the cerebellum, brainstem, trigeminal ganglia, and DRG (Afshari et al., 2004; Enomoto, Han, Hsiao, Wu, & Chandler, 2006; Kim, Kushmerick, & von Gersdorff, 2010), are believed to facilitate generation of higher frequency action potentials (Raman & Bean, 1997; Xie et al., 2016). $Na_V1.6$ is the major carrier of resurgent current in CNS and DRG neurons, but $Na_V1.7$ channels also exhibit an intrinsic ability to generate resurgent currents (Grieco & Raman, 2004; Jarecki et al., 2010; Patel, Barbosa, Xiao, & Cummins, 2015). In general, however, under control conditions, only a small population (~23.8%) of DRG neurons expressing wild type $Na_V1.7$ generated $Na_V1.7$ resurgent currents, with an average amplitude of 1% of classical peak transient current (Jarecki et al., 2010). The IEM mutations failed to influence the generation of resurgent currents, but most of the PEPD mutations substantially upregulated the percentage of nociceptive neurons generating $Na_V1.7$ resurgent currents and also significantly increased $Na_V1.7$ resurgent current amplitude (Figure 7.2G) (Theile, Jarecki, Piekarz, & Cummins, 2011). Furthermore, computer simulations indicated that resurgent currents associated with the *I1461T* mutation could induce high-frequency action-potential firing in nociceptive neurons, which is consistent with the observation in DRG neurons transiently transfected by *M1627K*, another PEPD mutant (Dib-Hajj et al., 2008; Jarecki et al., 2008). The conclusion that PEPD mutations increase pain sensations at least in part by enhancing resurgent currents is also supported by the pharmacological studies of scorpion α-toxins that alter gating properties of $Na_V1.7$ channels through a mechanism similar to the PEPD mutations. In nociceptive neurons, scorpion α-toxins can selectively impair fast-inactivation of $Na_V1.7$ without evidently affecting TTX-resistant sodium channels (Rowe et al., 2011). They can lower threshold, enhance resurgent currents, increase firing frequency, and broaden duration of action potentials, thus resulting in an increase in pain sensations (Abbas et al., 2013; Schiavon et al., 2010). Interestingly, there is a lack of striking evidence from either IEM or PEPD patients that the sense of smell is affected by the gain of function $Na_V1.7$ mutations that cause pain.

These fantastic findings, together with the fact that loss of $Na_V1.7$ would not increase the expression level of TTX-resistant VGSCs in DRG neurons (Nassar et al., 2004), suggest that $Na_V1.7$ should be an extremely promising target for developing novel analgesics without significant side effects. Local anesthetics (such as lidocaine) are small molecules used in a wide range of clinical situations to treat pain, and are thought to act in part by inhibiting nociceptor sodium channels, including $Na_V1.7$. However, they also can induce several severe side effects (such as cardiac-toxicity and motor block) when applied systemically, because they are non-selective towards VGSC isoforms due to their binding sites in the highly conserved inner pore of VGSC isoforms (Fozzard, Sheets, & Hanck, 2011; Mulroy, 2002; Yanagidate & Strichartz, 2007). Recently, many peptide toxins from the venoms of spiders, scorpions, and sea cone snails have been shown to amazingly exhibit much higher isoform-specific selectivity and divergent pharmacological functions towards VSGCs than local anesthetics and TTX (Billen, Bosmans, & Tytgat, 2008; Catterall, Trainer, & Baden, 1992). Unlike TTX and local anesthetics, most of these toxins function as VGSC modifiers by docking on voltage sensors (Xiao et al., 2014). HWTX-IV and ProTx-II are two short peptide toxins isolated from tarantula venoms. While having no effect on $Na_V1.4$, $Na_V1.5$, or neuronal TTX-resistant VGSCs, HWTX-IV preferentially inhibits $Na_V1.7$ by trapping the DII voltage sensor in the closed configuration (Xiao et al., 2008). ProTx-II inhibits VGSCs through a mechanism similar to that of HWTX-IV, and shows ~70-fold higher binding affinity for $Na_V1.7$ than for other VGSC isoforms (Xiao, Blumenthal, Jackson, Liang, & Cummins, 2010). While peptide toxins have greatly contributed to our knowledge of the physiological roles and structure-function of VGSCs, it is hoped that some of them may even provide promising candidates for developing novel analgesics without significant side effects. A $Na_V1.7$-specific antibody that targets the exact same region of the DII voltage sensor targeted by HWTX-IV was reported to effectively suppress pain sensation in mice (Lee et al., 2014; Liu et al., 2014; Xiao et al., 2008). This $Na_V1.7$-specific antibody also indicated that $Na_V1.7$ might be involved in the pathway of itch sensations. While data from patients with CIP seems to indicate that $Na_V1.7$ inhibitors might also reduce olfaction, it is likely that patients with severe pain could tolerate this side effect. In contrast, an argument could be made that autonomic effects could be also encountered with $Na_V1.7$ inhibitors in some patient populations, and this could be important to monitor in clinical trials.

$Na_V1.8$

$Na_V1.8$ channels, encoded by *SCN10A* (Akopian, Sivilotti, & Wood, 1996), are highly expressed in nociceptive neurons and display unique biophysical properties that are believe to underlie their specialized role in nociceptive signaling. This TTX-resistant channel is predominately expressed in small-diameter sensory neurons, and is also called the *sensory neuron-specific* (SNS) sodium channel (Akopian et al., 1996; Arbuckle & Docherty, 1995; Caffrey, Eng, Black, Waxman, & Kocsis, 1992). Many studies have revealed the importance of $Na_V1.8$ in action-potential generation in nociceptive neurons. Studies of DRG neurons from $Na_V1.8$ knockout mice suggests that C-type DRG neurons can generate TTX-resistant sodium-dependent action potentials, and $Na_V1.8$ contributes 80–90% of the inward current during the rapid upstroke phase of the action potential in these neurons (Renganathan, Cummins, & Waxman, 2001). Studies also show that an increased $Na_V1.8$ expression level is associated with broadened action-potential durations and larger action-potential overshoots in DRG neurons (Djouhri, Fang, et al., 2003). Interestingly, while $Na_V1.8$ is considered a hallmark of nociceptive neurons, aberrant expression is observed in cerebellar neurons in animal models of multiple sclerosis (Craner et al., 2003). The heterologous expression of $Na_V1.8$ in Purkinje cells substantially modifies action-potential properties of Purkinje cells and generated pacemaker-like firing activities upon depolarization (Renganathan, Gelderblom, Black, & Waxman, 2003). Overall, $Na_V1.8$ plays a critical role in modulating neuronal excitability and largely plays a specialized role in nociceptive cell signaling.

$Na_V1.8$ exhibits unique biophysical properties that can be easily distinguished from TTX-sensitive sodium currents. These differences in gating properties of $Na_V1.8$ contribute to its specific role in sensory neuronal excitability (Akopian et al., 1999). $Na_V1.8$ shows a significant slower rate of activation (Akopian et al., 1996). And the voltage-dependence of activation is shifted towards more depolarized membrane potentials compared with other sodium currents, indicating that more depolarizing force is required for $Na_V1.8$ channel opening. Therefore, $Na_V1.8$ is most likely activated after other types of sodium channels. Moreover, the fast-inactivation of $Na_V1.8$ is also significantly slower, and the voltage-dependence of inactivation is shifted to more depolarized potentials (Cummins & Waxman, 1997). The slower rate of fast-inactivation implies an increased sodium conductance during action-potential repolarization and is potentially a major contributor to prolonged action-potential duration in excitable cells with high level of $Na_V1.8$ expression. Indeed, small-diameter DRG sensory neurons that express substantial $Na_V1.8$ protein generate action potentials with much larger durations than the vast majority of neurons that do not express $Na_V1.8$. The depolarized steady-state inactivation enables $Na_V1.8$ to activate from depolarized membrane potentials where other sodium channels tend to be fully inactivated (Han, Huang, & Waxman, 2016). This may allow nociceptive neurons to continue to fire action potentials under compromised conditions associated with tissue damage. $Na_V1.8$ recovery from fast-inactivation, on the other hand, is substantially faster than observed with typical TTX-sensitive channels expressed in nociceptive neurons. The enhanced recovery rate suggests that the channels undergo shorter refractory periods and can be reactivated quickly while other channels still remain inactivated, thus potentially contributing to the repetitive firing of nociceptors (Liu & Wood, 2011; Rush, Cummins, & Waxman, 2007). However, Blair and Bean (Blair & Bean, 2003) found that $Na_V1.8$ channels can also exhibit pronounced slow inactivation (a gating process that is distinct from fast-inactivation and independent of the IFMT motif) and that this slow inactivation can induce adaptation that limits the duration of repetitive firing for nociceptive neurons in response to sustained stimulation. The role of slow inactivation in modulating neuronal excitability is complex. Unlike other VGSCs (including $Na_V1.7$), the voltage dependence of slow inactivation of $Na_V1.8$ is reportedly completely resistant to cold temperature, and it has been hypothesized that this makes $Na_V1.8$ critical for pain signaling at low temperature (Zimmermann et al., 2007).

Pain perception generally is driven by activity from peripheral nociceptive neurons, where the information from external stimuli is transmitted to the CNS. The electrical excitability in peripheral sensory neurons is essential in determining most pain sensations. Increased pain sensation can be induced by hyperexcitable nociceptors, which can occur when the membrane potential falls into the range that is depolarized enough to reach the activation threshold of $Na_V1.8$, but not enough to fully inactivate these channels. On the contrary, the absence of $Na_V1.8$ can cause the hypoexcitability of the neurons, since other sodium channel isoforms are largely inactivated with prolonged depolarized membrane potentials (Han, Huang, et al., 2016). Many animal models have demonstrated the importance of $Na_V1.8$ in regulating neuronal excitability and pain sensation (Akopian et al., 1999). $Na_V1.8$-null mutant mice, with a lowered activation threshold due to the exclusive expression of TTX-sensitive sodium channels, display a decreased sensitivity to noxious stimuli and delayed development of inflammatory hyperalgesia (Akopian et al., 1999). Selective knockdown of $Na_V1.8$ expression diminished the neuropathic pain caused by spinal nerve injury (Lai et al., 2002). Another study, using a $Na_V1.8$-cre mouse model, revealed that optogenetic silencing of nociceptive afferents reduces inflammatory pain (Daou et al., 2016). On the other hand, increased expression of $Na_V1.8$ and its functional currents were observed in the rat model of bone cancer pain and with scorpion sting–induced pain (Liu et al., 2014; Ye et al., 2016). Upregulated $Na_V1.8$ expression was also found in rats with chronic peripheral inflammation (Belkouch et al., 2014). These studies suggest that $Na_V1.8$ is essential in nociception and inflammatory pain. A recent study also reveals that an unusually slow TTX-resistant resurgent current is probably carried by $Na_V1.8$ (Tan, Piekarz, et al., 2014). The enhancement of TTX-resistant resurgent currents by inflammatory mediators may be an important mechanism of sensory neuron hyperexcitability induced by inflammatory pain. These slow resurgent currents may prolong the nociceptor action potential or contribute to repetitive firing in nociceptor neurons.

Pain is generally considered an important adaptive function to signal damage in response to external stimuli or pathological conditions. However, counter-adaptive evolution, even though very rare, may also occur to reduce pain sensation under special occasions. An interesting study has revealed that counter-adaptive evolution is associated with amino acid changes in $Na_V1.8$. Bark scorpions produce powerful toxins in their venom that can induce an intense pain sensation in humans and mice (Rowe & Rowe, 2008). Grasshopper mice, however, use these scorpions as a food source and seem immune to the venom (Figure 7.3). Normally $Na_V1.8$ is insensitive to this scorpion venom, but $Na_V1.8$ channels from grasshopper mice have sequence modifications that facilitate channel interaction with venom toxins. The interaction of $Na_V1.8$ with venom blocks $Na_V1.8$ currents, substantially attenuating action-potential firing and probably inhibiting pain signaling (Rowe, Xiao, Rowe, Cummins, & Zakon, 2013). It is important to note that this adaptation does not reduce overall pain sensations but rather targets a specific noxious stimulus, allowing the mice to take advantage of a local nutrient source.

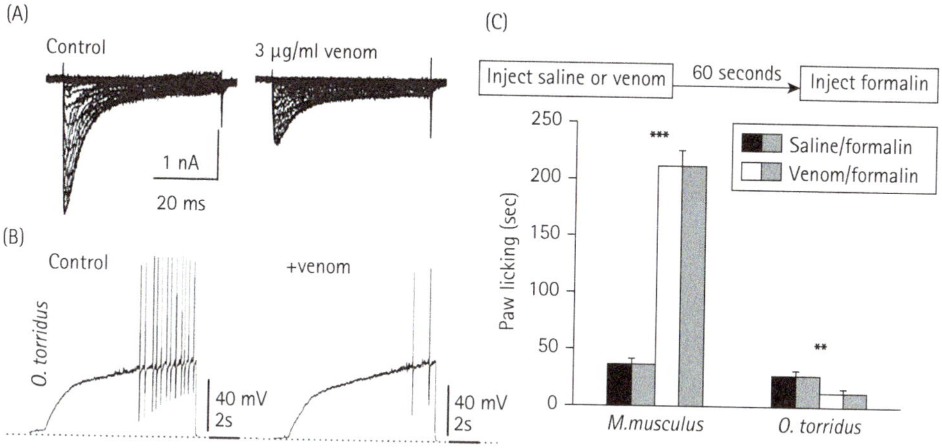

FIGURE 7.3 Adaptations to painful stimuli. $Na_V1.8$ currents from grasshopper mice (*O. torridus*) DRG neurons are inhibited by bark scorpion venom (**A**), whereas lab mouse (*M. musculus*) $Na_V1.8$ currents are insensitive (data not shown). Venom greatly reduces action-potential firing in grasshopper mouse neurons (**B**) but enhances firing in lab mouse neurons (data not shown). Venom injections greatly increase pain responses in lab mice but reduces the response to formalin in grasshopper mice (**C**).

Adapted from Rowe, Xiao, Rowe, Cummins, and Zakon, 2013.

Apart from the studies on different animal models, there is also evidence for direct association of $Na_V1.8$ with human pain perception. A reduced expression of $Na_V1.8$ in sensory cell bodies was found after spinal cord injury in humans (Coward et al., 2000). A significant increase in the number and intensity of $Na_V1.8$ immunoreactive nerve terminals was found in a patient's finger that developed symptoms of causalgia (intractable burning pain and hypersensitivity) (Shembalkar et al., 2001). $Na_V1.8$ gain-of-function mutations that either enhance activation or impair inactivation are associated with small-fiber neuropathy and other painful syndromes (Faber et al., 2012; Han, Huang, et al., 2016; Han et al., 2014). These observations are consistent with what we have concluded from animal models.

Due to the involvement of $Na_V1.8$ in inflammatory and neuropathic pain, it has been recognized as a promising target for novel analgesics. Many previously discovered sodium channel blockers show poor isoform selectivity and may induce serious side effects. Over years of study, VGSC blockers with $Na_V1.8$ preference have been reported. For example, one study discovered sodium channel blockers that preferentially block $Na_V1.8$ currents and successfully reduced pain symptoms in animal models of chronic neuropathic and inflammatory pain (Gaida, Klinder, Arndt, & Weiser, 2005). A-803467 is another sodium channel blocker that has established $Na_V1.8$ selectivity. This compound can effectively block $Na_V1.8$ currents at an IC_{50} of 140nM, ablate action-potential generation in DRG neurons, and alleviate neuropathic and inflammatory pain in a dose-dependent manner (Jarvis et al., 2007; McGaraughty et al., 2008). However, most existing preferential $Na_V1.8$ blockers have shown poor oral bioavailability. Later studies have targeted developing novel $Na_V1.8$ modulators with better oral pharmacokinetics and demonstrated their efficacy in both rodent models (Payne et al., 2015) and preclinical studies of neuropathic and inflammatory pain (Bagal et al., 2015).

$Na_V1.9$

$Na_V1.9$ (also known as SNS2/NaN) is another sodium channel subtype that is highly resistant to TTX ($IC_{50} \sim 200\ \mu M$). It is encoded by the gene *SCN11A*, co-localized in human chromosome 3 (3p21–24) along with *SCN5A* and *SCN10A* (the genes encoding $Na_V1.5$ and $Na_V1.8$, respectively). $Na_V1.9$ has been shown to be expressed in neurons of DRG and trigeminal ganglia and in intrinsic myenteric neurons (Coste, Crest, & Delmas, 2007; Dib-Hajj, Tyrrell, et al., 1999; Rugiero et al., 2003). Immunological and electrophysiological studies indicated that in the DRG, $Na_V1.9$ is selectively expressed in small- and medium-sized nociceptive neurons (Coste et al., 2007; Fang et al., 2002). $Na_V1.9$ has been localized within the soma of DRG neurons and free nerve terminals, as well as central terminals within the outer layers of the substantia gelatinosa in spinal cord (Amaya et al., 2006; Dib-Hajj, Black, & Waxman, 2015). Interestingly, the majority of magnocellular neurosecretory cells of the supraoptic nucleus also seem to express $Na_V1.9$ with normal function; therefore, the channel has been proposed to be involved in osmoregulation (Black, Vasylyev, Dib-Hajj, & Waxman, 2014). Expression of $Na_V1.9$ can increase in rat DRG neurons with age. While this begins at embryonic day 15, the expression can reach adult levels by postnatal day 7 (Benn, Costigan, Tate, Fitzgerald, & Woolf, 2001). Expression of $Na_V1.9$ can be altered following tissue injury or under disease conditions. For example, an infraorbital nerve–chronic constriction injury has been showed to downregulate $Na_V1.9$ at mRNA and protein levels in rat trigeminal ganglion (Xu, Zhang, Wang, Wang, & Wang, 2016). A similar downregulation is observed in neurons within the submucosal and myenteric plexus in Hirschsprung's disease, too (O'Donnell, Coyle, & Puri, 2016).

$Na_V1.9$ current was originally identified and characterized in DRG neurons from $Na_V1.8$-null mice (Cummins et al., 1999). Electrophysiological analysis showed that $Na_V1.9$ exhibited gating properties only grossly similar to other sodium channel isoforms in that the channels can transition between closed, open, and inactivated states in a voltage-dependent fashion. However, $Na_V1.9$ activates at much more negative potentials, especially when compared to $Na_V1.8$. It activates at potentials that are close to resting membrane potential (−60 to −70 mV), and $Na_V1.9$ currents display extremely slow activation and slow fast-inactivation kinetics. $Na_V1.9$ also produces large, persistent TTX-resistant sodium currents in small DRG neurons (Figure 7.4A). TTX-resistant persistent currents are not measurable in $Na_V1.9$-null DRG neurons at potentials more negative than −30 mV (Priest et al., 2005). These unique characteristics are consistent with the findings on cloned $Na_V1.9$ heterologously expressed in *ND7/23* or *HEK293* cells (Vanoye, Kunic, Ehring, & George, 2013) (Lin et al., 2016). In sharp contrast to the

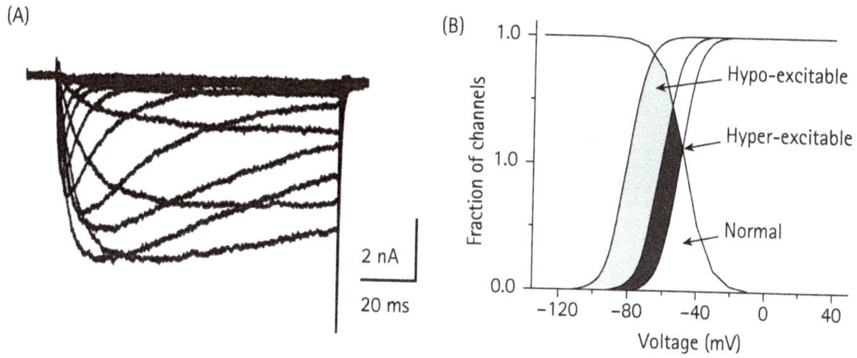

FIGURE 7.4 (A) Representative $Na_V1.9$ persistent currents recorded from DRG neuron. (B) Overlap between voltage-dependence of activation and inactivation for $Na_V1.9$ currents. White overlap region indicates overlap for wild-type channels contributing to normal nociception. The black region indicates enhanced overlap associated with gain-of-function mutations that cause increased pain. The gray region indicates extreme overlap associated with mutations that dramatically shift activation, leading to large window currents and probable depolarization block and lack of pain sensations.

commonly accepted paradigm where VGSCs are mainly responsible for the upstroke of the action potential, $Na_V1.9$ channels are thought to help set resting membrane potential and contribute to subthreshold electrogenesis in nociceptive neurons, but not make a major contribution to the rapid upstroke of the action potential (Cummins et al., 1999; Herzog, Cummins, & Waxman, 2001). This hypothesis is supported by the findings that human disease-related mutations that enhance $Na_V1.9$ activation can depolarize resting membrane potential (Leipold et al., 2013). However, strangely, knockout of $Na_V1.9$ did not significantly change resting membrane potential in DRG neurons (Amaya et al., 2006; Priest et al., 2005). This might be due to altered expression of other proteins, induced by $Na_V1.9$ knockout that compensates the impact of $Na_V1.9$ loss on resting membrane potential. Indeed, in $Na_V1.9$-null DRG neurons, mRNA levels of $Na_V1.1$, $Na_V1.7$, and $Na_V1.8$, as well as the β3 subunit, are all substantially upregulated (Priest et al., 2005).

Studies on $Na_V1.9$-null mice indicate that $Na_V1.9$ knockout causes a reduction in inflammatory pain behavior in response to peripheral administration of prostaglandin E2 (PGE2), bradykinin, interleukin-1, capsaicin, guanosine triphosphate (GTP), and nonhydrolyzable GTP analogs (GTPγS), and P2X3 and P2Y receptor agonists, without disrupting basal or neuropathic pain (Priest et al., 2005), suggesting that $Na_V1.9$ is a major effector of peripheral inflammatory pain hypersensitivity. These inflammatory mediators can substantially increase $Na_V1.9$ persistent current via a pathway involving G-proteins (Maingret et al., 2008; Ostman, Nassar, Wood, & Baker, 2008; Rush & Waxman, 2004; Vanoye et al., 2013). PGE2 selectively acts through $G_{i/o}$ but not G_s to cause a twofold increase in current amplitude accompanied by a 6–8 mV hyperpolarizing shift of voltage dependence of activation. GTPγS, the nonselective activator of G-protein signaling, causes a more pronounced increase in current amplitude. Exposure to GTPγS also alters gating properties of $Na_V1.9$, including increasing channel open probability and positively shifting the voltage dependence of inactivation (~10 mV) (Maingret et al., 2008; Vanoye et al., 2013).

Naturally occurring mutations in human $Na_V1.9$ have been linked to pain disorders. The first mutation (*L811P*) in human $Na_V1.9$ was identified in individuals from a German family diagnosed with an apparent congenital insensitivity to pain (Leipold et al., 2013). Examination indicated that the affected individuals presented with normal intelligence, muscle biopsy, electromyography, and sensory axonal distribution. However, they also showed self-mutilations, slow-healing wounds, mild muscular weakness, delayed motor development, and gastrointestinal dysfunction. The phenotypes could be replicated when the same heterozygous mutation was introduced into mice. Electrophysiological analysis showed that the mutation had multiple effects on gating properties of $Na_V1.9$ channels, including shifting channel activation to more negative potentials, slowing down the kinetics of channel deactivation, and shifting steady-state inactivation to more negative potentials. The most striking effect is that the *L811P* mutation substantially increased $Na_V1.9$ current at voltages near resting membrane potentials in DRG neurons. In contrast to wild-type (WT) small DRG neurons, knockin neurons displayed a +6.7 mV shift of the resting membrane potential. It is unknown yet how this mutation affects the firing frequency of action potentials, but it evidently reduces action-potential duration in nociceptor neurons and substantially decreases the frequency of miniature excitatory postsynaptic currents in the spinal cord. It has been supposed that this positive shift of resting membrane potential may induce other action-potential contributors, including $Na_V1.7$, $Na_V1.8$ and calcium channels, to undergo progressive inactivation, leading to impaired neurotransmitter release at presynaptic nerve terminals that is necessary to transmit pain signals to the spinal cord (Leipold et al., 2013).

In contrast to the mutation *L811P*, seven other missense mutations (*R222H/S, R225C, A808G, G699R, I183T,* and *L1158P*) of $Na_V1.9$ identified more recently have been reported to be associated with painful peripheral neuropathy (Han et al., 2015; Han, Yang, et al., 2016; Huang et al., 2014; Zhang et al., 2013). Most of them are found to also shift the voltage dependence of activation of $Na_V1.9$ in a hyperpolarizing direction. Some of them also can slow down channel deactivation or positively shift steady-state inactivation. Interestingly, expression of each mutant channel slightly depolarizes resting membrane potentials of DRG neurons by +3.5 ~ +5.5 mV (Han et al., 2015; Huang et al., 2014). They can significantly enhance DRG excitability by increasing firing frequency and lowering the current threshold for firing. These observations are not entirely consistent with the mechanism proposed for the pain insensitivity with the *L811P* mutation that also depolarized resting membrane potential (Leipold et al., 2013). The +6.7 mV shift observed with *L811P* is marginally larger, and since the *L811P* mutation causes a more pronounced depolarization in resting membrane potentials, it may be reasonably expected that expression of the seven other mutant channels would induce a smaller fraction of $Na_V1.7$ and $Na_V1.8$ to undergo progressive inactivation than with the *L811P* mutant channel. Another $Na_V1.9$ mutation, *L1320F*, induces a 27 mV negative shift in $Na_V1.9$ activation (similar to that observed with *L811P*) and also causes loss of pain sensitivity (Huang et al., 2017). Therefore, one explanation might be that, unlike *L811P*, the fraction of $Na_V1.7$ and $Na_V1.8$ inactivated by the other gain-of-function mutations is not enough to depress the excitability of nociceptive neurons. On the contrary, slight depolarization can lower the current threshold for action-potential generation, leading to neuronal hyperexcitability. Indeed, experiments by Huang et al. indicate that a U-shaped relationship exists between the DRG resting membrane potential and neuronal excitability that helps dictate whether a given mutation induces hyper- or hypo-excitability. Thus, mutations that induce a small negative shift in the voltage-dependence of activation enhance pain, but mutations that are associated with large negative shifts in activation lead to a loss of pain sensations (Figure 7.4B). We also raise the alternative possibility that the *L811P* and/or *L1320F* mutations in $Na_V1.9$ channels might selectively induce compensatory changes that would not happen for other $Na_V1.9$ gain-of-function mutations. Peptide toxins or small molecules that selectively enhance $Na_V1.9$ activation might be helpful to clarify the divergent roles of gain-of-function mutations of $Na_V1.9$ in pain sensation.

$Na_V1.9$ also plays a crucial role in the perception of pain triggered by noxious cold. Low ambient temperatures may facilitate $Na_V1.9$ activation in the population of cold-sensitive and heat-sensitive DRG neurons (Lolignier et al., 2015). Disruption of $Na_V1.9$ expression not only markedly increased the cold-induced pain threshold in healthy rodents but also significantly alleviated the pathological hypersensitivity to cold induced by oxaliplatin, a platinum-based antineoplastic agent widely used in cancer chemotherapy. Interestingly, oxaliplatin-induced pain is not reduced in transgenic animals that lack $Na_V1.7$ in nociceptors, while pain is reduced in $Na_V1.7$ knockouts for many other pain models (Minett et al., 2014). Gain-of-function mutations (*I381T* and *V1184A*) that enhance the activities of $Na_V1.9$ channels can induce burning pain in response to cold temperature (Huang et al., 2014; Lolignier et al., 2015).

$Na_V1.9$ channels might be a good target for specific blockers or potent agonists to effectively treat pain. Because $Na_V1.9$ is less conserved with CNS and cardiac sodium channels than other isoforms, there is some hope that it will be easier to develop $Na_V1.9$-selective modulators than $Na_V1.7$-selective modulators. This has been hampered, however, by difficulties in expressing $Na_V1.9$ in heterologous systems. One concern is that previous studies have shown that sustained Na+ influx through sodium channels can injure small-diameter central axons (Stys, Sontheimer, Ransom, & Waxman, 1993) and that some gain-of-function mutations of sodium channels can influence the integrity of DRG neuron axons (Persson et al., 2013). However, Leipold et al. (Leipold et al., 2013) showed that the *L811P* mutation of $Na_V1.9$ failed to impact the normal density of small-diameter and large-diameter sensory neuron axons or small nerve fibers. Together with the finding that the *L811P* mutation did not impair intellectual ability, this could suggest that $Na_V1.9$ hyperactivation is a possible treatment strategy. It is still possible that $Na_V1.9$ activators may reduce pain in some patients due to pronounced depolarization block, but, if not strong enough due to genetic or other modifiers, they may induce pain in others. In addition, since insensitivity to pain induced by the *L811P* mutation in $Na_V1.9$ is also accompanied by multiple side effects, including slow healing, muscular weakness, delayed motor development, and gastrointestinal dysfunction, that are not detected in patients with truncated $Na_V1.7$ channels, $Na_V1.9$

modulators might have disadvantages with systemic application in clinic pain treatment compared to $Na_V1.7$ blockers. In particular, because $Na_V1.9$ channels are expressed in enteric neurons, there is concern that pharmacologically targeting $Na_V1.9$ for pain therapy could induce diarrhea or other gastrointestinal side effects.

$Na_V1.6$

The $Na_V1.6$ (also known as PN4 or sodium channel protein VIII) α subunit encoded by the *SCN8A* gene, is another TTX-sensitive sodium channel isoform broadly expressed in unmyelinated and myelinated neurons in CNS and the peripheral nervous system (PNS). It can be detected in multiple subcellular compartments of neurons, including axons, nodes, dendrites, cell bodies, and synapses. Immunofluorescence evidence indicates that $Na_V1.6$ is the major VGSC isoform highly concentrated at nodes of Ranvier of myelinated central and peripheral sensory neurons (Caldwell, Schaller, Lasher, Peles, & Levinson, 2000). Ankyrin-G plays a crucial role in this clustering of $Na_V1.6$ at nodes of Ranvier and axon initial segments (Gasser et al., 2012). Although $Na_V1.6$ is the major isoform at nodes of Ranvier in peripheral sensory neurons, it is not entirely clear whether these neurons have a structure akin to the axon initial segment of CNS neurons. $Na_V1.6$ is expressed at high levels in medium- to large-diameter DRG neurons (neurons that generally give rise to faster conducting, myelinated axons), while only low levels of mRNA for this isoform are detected in small-diameter DRG neurons (associated with slow-conducting C-fibers or Aδ fibers) (Black et al., 1996). In DRG neurons, $Na_V1.6$ channels display rapid activation and fast-inactivation kinetics similar to $Na_V1.7$ channels, but they also have two different properties that greatly affect the excitability of peripheral sensory neurons. First, $Na_V1.6$ channels exhibit a fivefold faster recovery rate from inactivation near resting membrane potential than $Na_V1.7$ channels (Herzog et al., 2003), which may allow neurons with high $Na_V1.6$ expression to fire action potentials at a high frequency. This is consistent with the observation that A-type DRG neurons display higher firing frequency of action potentials than C-type DRG neurons do. As discussed previously, C-type DRG neurons predominantly express $Na_V1.7$ TTX-sensitive channels. Rush and colleagues (Rush et al., 2006) showed that the recovery rate of $Na_V1.6$ can be substantially slowed by FHF2A, one member of the fibroblast growth factor homologous factor subfamily. This suggests that the excitability of A-type DRG neurons may be influenced through modulation of $Na_V1.6$. Second, $Na_V1.6$ channels are able to generate TTX-sensitive $Na_V\beta4$-mediated resurgent currents (Barbosa et al., 2015; Cummins, Dib-Hajj, Herzog, & Waxman, 2005). This property is also confirmed in heterologous expression systems (Patel et al., 2015). $Na_V1.6$ channels are the major carriers of TTX-sensitive resurgent currents in DRG neurons, although other VGSC isoforms are also found to produce resurgent currents (Cummins et al., 2005; Jarecki et al., 2010). The relative amplitude of resurgent currents produced by $Na_V1.6$ is more than twofold larger than that produced by other VGSC isoforms, such as $Na_V1.4$, $Na_V1.5$, and $Na_V1.7$. As resurgent currents are typically elicited during the repolarization phase of action potentials, $Na_V1.6$-mediated resurgent currents are predicted to contribute to higher firing frequency in DRG neurons. Bant and Raman (Bant & Raman, 2010) showed that reduction of resurgent currents by knockdown of the $Na_V\beta4$ subunit caused a significant decrease in repetitive firing frequency in cerebellar granule cells. As with $Na_V1.8$ resurgent currents, TTX-sensitive resurgent currents in DRG neurons are substantially upregulated by inflammatory mediators (Tan, Piekarz, et al., 2014) and *in vivo* inflammation of the DRG (Xie et al., 2016).

There are more than 31 genetic mutations identified in $Na_V1.6$ since the first was reported in 2012 (Veeramah et al., 2012). Some of them (gain-of-function) may increase activities of $Na_V1.6$ channels such as causing a hyperpolarizing shift in the voltage dependence of activation, slowing fast-inactivation, or elevating persistent currents. Some of them (loss-of-function) may reduce activities of $Na_V1.6$ channels, often by protein truncation (Blanchard et al., 2015; Meisler et al., 2016). However, all mutations except *M136V* are reported to be associated with epilepsy, intellectual disability, ataxia, infantile spasms, or Dravet syndrome. *M136V* is the only mutation reported so far to be associated with pain; specifically, idiopathic trigeminal neuralgia, a debilitating pain disorder characterized by brief, electric-shock-like pains (Tanaka et al., 2016). The mutation does not significantly alter the gating properties of $Na_V1.6$ channels, but it increases current density and resurgent currents in trigeminal ganglia (TRG) neurons. The *M136V* mutation accordingly upregulates the excitability of TRG neurons by reducing the current threshold and increasing the firing frequency. It is unknown yet how this mutation affects the excitability of DRG neurons. Interestingly, oxaliplatin, like the *M136V* mutation, is found to increase $Na_V1.6$-mediated resurgent currents and persistent currents in large-sized DRG neurons (Sittl et al., 2012). Oxaliplatin can induce cold allodynia in humans and animal models, and the induced cold allodynia is inhibited by μ-conotoxin GIIIA, a preferential blocker of $Na_V1.6$ in neurons (Deuis et al., 2013). Furthermore, the studies by Xie et al. (Xie, Strong, Ye, Mao, & Zhang, 2013; Xie, Strong, & Zhang, 2015) showed that local knockdown of $Na_V1.6$ by siRNA reduced mechanical pain behaviors in a rat spinal nerve ligation model of neuropathic pain and in a rat DRG local inflammation pain model. Therefore, this genetic, pharmacological, and biochemical evidence positively suggests an important role of $Na_V1.6$ in peripheral pain pathways. However, $Na_V1.6$ plays crucial roles in CNS function, so targeting $Na_V1.6$ without impacting other neurological activity may be difficult. Drugs that do not cross the blood–brain barrier may help, but impaired motor neuron activity could still be a problem. As resurgent currents may not be important at the nodes of Ranvier, targeting $Na_V1.6$ resurgent currents with drugs that do not penetrate the CNS may be an effective strategy to treat some forms of pain, such as oxaliplatin-induced pain.

$Na_V1.3$

$Na_V1.3$ coded by the *SCN3A* gene is a TTX-sensitive VGSC isoform abundantly expressed in CNS and PNS neurons during embryonic and neonatal stages of development. $Na_V1.3$ is also expressed in adult CNS neurons, but is normally absent or present at low levels in adult PNS neurons (Beckh, Noda, Lubbert, & Numa, 1989). $Na_V1.3$ displays current properties with rapid activation and inactivation kinetics similar to other TTX-sensitive VGSC isoforms, but it also shows two interesting and unique gating properties that may contribute to hyperexcitability of DRG neurons. First, compared with $Na_V1.7$, $Na_V1.3$ exhibits three times faster recovery from inactivation (Cummins et al., 2001; Cummins et al., 1998). The study by Cummins and Waxman (Cummins & Waxman, 1997) showed that TTX-sensitive sodium currents in axotomized DRG neurons recover from inactivation four times faster than in normal DRG neurons. The increase in recovery rate can be diminished by intrathecal delivery of NGF or glial cell–derived neurotrophic factor (GDNF), two factors reversing the increase in $Na_V1.3$ expression in the axotomized DRG, suggesting that $Na_V1.3$ is responsible for the increased recovery rate in axotomized DRG neurons (Leffler et al., 2002). Second, $Na_V1.3$ can generate a relatively large ramp current with two components in response to slow ramp depolarization (Cummins et al., 2001; Estacion & Waxman, 2013). While the more hyperpolarized component results from window current, the more depolarized component is caused by persistent current. Because ramp currents are able to reduce the threshold for action-potential generation and because faster recovery rate allows neurons to sustain firing at higher frequency, upregulation of $Na_V1.3$ expression is believed to increase the excitability of DRG neurons.

In general, $Na_V1.3$ is likely to have a limited role in pain sensation in normal adult animal models due to its absence, or presence at low level, in mature sensory neurons. As expected, global knockout of $Na_V1.3$ changes neither acute pain nor neuropathic pain behavior in mice (Nassar et al., 2006). However, $Na_V1.3$ expression can be significantly upregulated in PNS neurons following periphery nerve injury and inflammation (Black et al., 1999; Dib-Hajj, Fjell, et al., 1999; Kim, Oh, Chung, & Chung, 2002; Waxman et al., 1994). In rat models of neuropathic pain induced by spared nerve injury, Samad and colleagues (Samad et al., 2013) demonstrated that pain sensation was attenuated by knockdown of $Na_V1.3$ in DRG using virus-mediated shRNA. Knockdown of $Na_V1.3$ channels also alleviates tactile allodynia in DRG of streptozotocin-induced diabetic rats (Tan, Samad, Dib-Hajj, & Waxman, 2015). However, it is surprising, as these findings are in contrast with previous observations (Lindia, Kohler, Martin, & Abbadie, 2005; Nassar et al., 2006). Nassar and colleagues reported that $Na_V1.3$ knockout did not alter inflammatory pain behavior in knockout mice. Lindia and colleagues (2005) reported that $Na_V1.3$ antisense oligodeoxynucleotides could strongly inhibit upregulation of $Na_V1.3$ channels induced by inflammation and spared nerve injury, but the $Na_V1.3$ antisense oligodeoxynucleotides failed to influence mechanical or cold allodynia associated with nerve injury in rats. To clarify the controversial roles of $Na_V1.3$ channels in pain signaling pathways, more pharmacological evidence will be required. Because there is a lack of genetic evidence implicating $Na_V1.3$ in human pain, identifying $Na_V1.3$-specific inhibitors has not been a major priority.

$Na_V1.1$

$Na_V1.1$ encoded by the *SCN1A* gene is a TTX-sensitive VGSC isoform widely expressed in both CNS and PNS neurons. $Na_V1.1$ currents display gating properties with kinetics of fast activation and inactivation grossly similar to other TTX-sensitive VGSC isoforms. *In situ* hybridization histochemistry showed that $Na_V1.1$ mRNA was expressed primarily by large- and medium-sized myelinated neurons within DRGs (Osteen et al., 2016), and that small-diameter neurons expressed low levels of $Na_V1.1$ transcript (Black et al., 1996), suggesting that $Na_V1.1$ might have a limited role in pain sensation mediated by C-type nociceptive neurons. Indeed, although more than 100 different mutations have been identified in $Na_V1.1$, the majority of them are linked to genetic epilepsy syndromes. Only a handful have been reported to be associated with familial hemiplegic migraine (Kahlig et al., 2008; Toldo et al., 2011). Recently, pharmacological experiments showed that purported activation of $Na_V1.1$ by Hm1a, a spider toxin capable of slowing inactivation of $Na_V1.1$ current, elicited acute pain and mechanical allodynia (Osteen et al., 2016), consistent with a previous finding that myelinated Aδ fibers are involved in mechanonociception (Eilers & Schumacher, 2005). However, it remains unknown whether blockade of $Na_V1.1$ channels affects pain sensation, and it is not clear if Hm1a activity on other channels contributed to some of the pain behaviors observed in the 2016 study. Since Kahlig and colleagues (Kahlig et al., 2008) reported that familial hemiplegic migraine could be induced by both gain-of-function and loss-of-function mutations in $Na_V1.1$, it seems likely that $Na_V1.1$-selective inhibitors might not be effective in reducing pain in some patients. However, mutations that reduce $Na_V1.1$ channel expression are associated with the Dravet epilepsy syndrome (Steinlein, 2014), and this raises concerns that inhibiting $Na_V1.1$ could increase seizure susceptibility in some patients.

Nax

Nax, also known as NaG, SCL11, or Na_V2, is encoded by the *SCN7A* gene located on chromosome 2q21–23. It is believed to be highly resistant to TTX (although expression studies are lacking) and broadly distributed in PNS, CNS, muscle, and heart. Nax shows only approximately 50% amino acid sequence identity with Na_V1 isoforms. Especially, Nax possesses fewer positively charged residues in voltage sensor S4 segment than Na_V1 isoforms, and evidence suggests that Nax fails to sense or respond to changes in membrane potential. Nax can reportedly be activated by high extracellular Na+ concentration (Hiyama et al., 2002). There is strong evidence showing that Nax plays an important role in maintaining sodium homeostasis in CNS and in epithelial tissues (Watanabe, Hiyama, Kodama, & Noda, 2002; Xu et al., 2015). However, there is no evidence indicating that Nax plays a special role in pain sensation except that Nax exhibits high expression in DRG neurons (Akopian, Souslova, Sivilotti, & Wood, 1997). Its function in sensory neurons, if any, is unclear.

Conclusions

Sensory neurons involved in pain sensations are intriguing because they express an array of voltage-gated sodium channels. In the early age of voltage-clamp electrophysiology, it was postulated that VGSCs lacked significant diversity and that the primary function of these channels is to give rise to the regenerative upstroke of action potentials (Hille, 1984). This may be true for muscle cells and many neurons, but it clearly is not the case for nociceptive neurons. A given neuron may express four or more isoforms. Rates of activation, inactivation, and recovery from inactivation can differ by well over an order of magnitude. The voltage-dependence of activation and inactivation can differ by tens of millivolts for the different isoforms. Yet a 5 mV negative shift in activation of $Na_V1.7$ is believed to be able to induce severe neuropathic pain in erythromelalgia patients, and subtle changes in the rate of inactivation of either $Na_V1.7$ or $Na_V1.8$ channels may underlie paroxysmal extreme pain disorder and small-fiber neuropathies. These observations suggest that VGSCs in sensory neurons play specialized roles and are finely tuned to aid the reproducible transmission of nociceptive information to the CNS, even under compromised conditions such as those associated with tissue injury. VGSCs in sensory neurons are promising targets for pharmacological interventions, in part because of their diversity and complexity, but also because several isoforms (i.e., $Na_V1.7$, $Na_V1.8$, and $Na_V1.9$) are predominantly involved in nociceptive signaling but have limited roles in other physiological activities.

Future Directions

Although much has been learned about VGSCs in sensory neurons, there are many questions that have not been fully answered. Sensory neuron VGSCs can be finely tuned, but their properties can also be dynamically regulated. The extent to which these proteins and their activity are regulated by post-translational modifications is unclear. Phosphorylation has been actively investigated (Hudmon et al., 2008; Tan, Priest, et al., 2014), but recent evidence indicates that VGSCs can also be dynamically regulated by palmitoylation (Pei, Xiao, Meng, Hudmon, & Cummins, 2016) and SUMOylation (Plant, Marks, & Goldstein, 2016). An important question is to what extent sensory neuronal sodium channels are modulated by methylation, palmitoylation, SUMOylation, and other post-translational modifications. VGSC activity can also be modulated by an expanding array of accessory proteins (Barbosa et al., 2017; Dustrude et al., 2016). It will be interesting to see the extent to which interactions with accessory proteins are altered by different pain states. Importantly, while much is known about the gating properties of sensory neuronal VGSCs, little is known about their subcellular distribution. Is there a specialized density of sodium channels in the distal axon similar to the axon initial segment in CNS neurons? Are the different isoforms expressed in unmyelinated nociceptive neurons differentially trafficked and distributed in the various neuronal compartments? Finally, although progress is being made in the development of isoform-specific blockers, it remains to be determined if isoform-specific blockers of VGSCs will be efficacious in common pain syndromes. This remains a vibrant area of research.

REFERENCES

Abbas, N., Gaudioso-Tyzra, C., Bonnet, C., Gabriac, M., Amsalem, M., Lonigro, A., ... Delmas, P. (2013). The scorpion toxin Amm VIII induces pain hypersensitivity through gain-of-function of TTX-sensitive Na(+) channels. *Pain, 154*(8), 1204–1215. doi:10.1016/j.pain.2013.03.037

Afshari, F. S., Ptak, K., Khaliq, Z. M., Grieco, T. M., Slater, N. T., McCrimmon, D. R., & Raman, I. M. (2004). Resurgent Na currents in four classes of neurons of the cerebellum. *Journal of Neurophysiology, 92*(5), 2831–2843. doi:10.1152/jn.00261.2004

Ahmad, S., Dahllund, L., Eriksson, A. B., Hellgren, D., Karlsson, U., Lund, P. E., ... Krupp, J. J. (2007). A stop codon mutation in SCN9A causes lack of pain sensation. *Human Molecular Genetics, 16*(17), 2114–2121. doi:ddm160

Akopian, A. N., Sivilotti, L., & Wood, J. N. (1996). A tetrodotoxin-resistant voltage-gated sodium channel expressed by sensory neurons. *Nature, 379*(6562), 257–262.

Akopian, A. N., Souslova, V., England, S., Okuse, K., Ogata, N., Ure, J., ... Wood, J. N. (1999). The tetrodotoxin-resistant sodium channel SNS has a specialized function in pain pathways. *Nature Neuroscience, 2*(6), 541–548.

Akopian, A. N., Souslova, V., Sivilotti, L., & Wood, J. N. (1997). Structure and distribution of a broadly expressed atypical sodium channel. *FEBS Letters, 400*(2), 183–187.

Amaya, F., Wang, H., Costigan, M., Allchorne, A. J., Hatcher, J. P., Egerton, J., ... Woolf, C. J. (2006). The voltage-gated sodium channel Na(v)1.9 is an effector of peripheral inflammatory pain hypersensitivity. *Journal of Neuroscience, 26*(50), 12852–12860.

Arbuckle, J. B., & Docherty, R. J. (1995). Expression of tetrodotoxin-resistant sodium channels in capsaicin-sensitive dorsal root ganglion neurons of adult rats. *Neuroscience Letters, 185*(1), 70–73.

Bagal, S. K., Bungay, P. J., Denton, S. M., Gibson, K. R., Glossop, M. S., Hay, T. L., ... Thompson, L. R. (2015). Discovery and optimization of selective Nav1.8 modulator series that demonstrate efficacy in preclinical models of pain. *ACS Medicinal Chemistry Letters, 6*(6):650–654. doi:10.1021/acsmedchemlett.5b00059

Bant, J. S., & Raman, I. M. (2010). Control of transient, resurgent, and persistent current by open-channel block by Na channel beta4 in cultured cerebellar granule neurons. *Proceedings of the National Academy of Sciences of the United States of America, 107*(27), 12357–12362. doi:10.1073/pnas.1005633107

Barbosa, C., Tan, Z. Y., Wang, R., Xie, W., Strong, J. A., Patel, R. R., ... Cummins, T. R. (2015). Navbeta4 regulates fast resurgent sodium currents and excitability in sensory neurons. *Molecular Pain, 11*, 60. doi:10.1186/s12990-015-0063-9

Barbosa, C., Xiao, Y., Johnson, A. J., Xie, W., Strong, J. A., Zhang, J. M., & Cummins, T. R. (2017). FHF2 isoforms differentially regulate Nav1.6-mediated resurgent sodium currents in dorsal root ganglion neurons. *Pflugers Archiv*, *469*(2):195–212. doi:10.1007/s00424-016-1911-9

Beckh, S., Noda, M., Lubbert, H., & Numa, S. (1989). Differential regulation of three sodium channel messenger RNAs in the rat central nervous system during development. *EMBO Journal*, *8*(12), 3611–3616.

Belkouch, M., Dansereau, M. A., Tetreault, P., Biet, M., Beaudet, N., Dumaine, R., ... Sarret, P. (2014). Functional up-regulation of Nav1.8 sodium channel in Abeta afferent fibers subjected to chronic peripheral inflammation. *Journal of Neuroinflammation*, *11*, 45. doi:10.1186/1742-2094-11-45

Benn, S. C., Costigan, M., Tate, S., Fitzgerald, M., & Woolf, C. J. (2001). Developmental expression of the TTX-resistant voltage-gated sodium channels Nav1.8 (SNS) and Nav1.9 (SNS2) in primary sensory neurons. *Journal of Neuroscience*, *21*(16), 6077–6085. doi:21/16/6077

Billen, B., Bosmans, F., & Tytgat, J. (2008). Animal peptides targeting voltage-activated sodium channels. *Current Pharmaceutical Design*, *14*(24), 2492–2502.

Black, J. A., Cummins, T. R., Plumpton, C., Chen, Y. H., Hormuzdiar, W., Clare, J. J., & Waxman, S. G. (1999). Upregulation of a silent sodium channel after peripheral, but not central, nerve injury in DRG neurons. *Journal of Neurophysiology*, *82*(5), 2776–2785.

Black, J. A., Dib-Hajj, S., McNabola, K., Jeste, S., Rizzo, M. A., Kocsis, J. D., & Waxman, S. G. (1996). Spinal sensory neurons express multiple sodium channel alpha-subunit mRNAs. *Brain Research: Molecular Brain Research*, *43*(1–2), 117–131.

Black, J. A., Liu, S., Tanaka, M., Cummins, T. R., & Waxman, S. G. (2004). Changes in the expression of tetrodotoxin-sensitive sodium channels within dorsal root ganglia neurons in inflammatory pain. *Pain*, *108*(3), 237–247.

Black, J. A., Vasylyev, D., Dib-Hajj, S. D., & Waxman, S. G. (2014). Nav1.9 expression in magnocellular neurosecretory cells of supraoptic nucleus. *Experimental Neurology*, *253*, 174–179. doi:10.1016/j.expneurol.2014.01.004

Blair, N. T., & Bean, B. P. (2003). Role of tetrodotoxin-resistant Na+ current slow inactivation in adaptation of action potential firing in small-diameter dorsal root ganglion neurons. *Journal of Neuroscience*, *23*(32), 10338–10350.

Blanchard, M. G., Willemsen, M. H., Walker, J. B., Dib-Hajj, S. D., Waxman, S. G., Jongmans, M. C., ... Kamsteeg, E. J. (2015). De novo gain-of-function and loss-of-function mutations of SCN8A in patients with intellectual disabilities and epilepsy. *Journal of Medical Genetics*, *52*(5), 330–337. doi:10.1136/jmedgenet-2014-102813

Branco, T., Tozer, A., Magnus, C. J., Sugino, K., Tanaka, S., Lee, A. K., ... Sternson, S. M. (2016). Near-perfect synaptic integration by Nav1.7 in hypothalamic neurons regulates body weight. *Cell*, *165*(7), 1749–1761. doi:10.1016/j.cell.2016.05.019

Caffrey, J. M., Eng, D. L., Black, J. A., Waxman, S. G., & Kocsis, J. D. (1992). Three types of sodium channels in adult rat dorsal root ganglion neurons. *Brain Research*, *592*(1–2), 283–297.

Caldwell, J. H., Schaller, K. L., Lasher, R. S., Peles, E., & Levinson, S. R. (2000). Sodium channel Na(v)1.6 is localized at nodes of Ranvier, dendrites, and synapses. *Proceedings of the National Academy of Sciences of the United States of America*, *97*(10), 5616–5620. doi:10.1073/pnas.090034797

Catterall, W. A., Goldin, A. L., & Waxman, S. G. (2005). International Union of Pharmacology. XLVII. Nomenclature and structure–function relationships of voltage-gated sodium channels. *Pharmacology Review*, *57*(4), 397–409.

Catterall, W. A., Trainer, V., & Baden, D. G. (1992). Molecular properties of the sodium channel: A receptor for multiple neurotoxins. *Bulletin de la Societé de Pathologie Exotique, 85*(5 Pt 2), 481–485.

Chattopadhyay, M., Mata, M., & Fink, D. J. (2008). Continuous delta-opioid receptor activation reduces neuronal voltage-gated sodium channel (NaV1.7) levels through activation of protein kinase C in painful diabetic neuropathy. *Journal of Neuroscience, 28*(26), 6652–6658. doi:10.1523/JNEUROSCI.5530-07.2008

Cheng, X., Dib-Hajj, S. D., Tyrrell, L., Te Morsche, R. H., Drenth, J. P., & Waxman, S. G. (2011). Deletion mutation of sodium channel Na(V)1.7 in inherited erythromelalgia: Enhanced slow inactivation modulates dorsal root ganglion neuron hyperexcitability. *Brain, 134*(Pt 7), 1972–1986. doi:10.1093/brain/awr143

Coste, B., Crest, M., & Delmas, P. (2007). Pharmacological dissection and distribution of NaN/Nav1.9, T-type Ca2+ currents, and mechanically activated cation currents in different populations of DRG neurons. *Journal of General Physiology, 129*(1), 57–77. doi:jgp.200609665

Coward, K., Plumpton, C., Facer, P., Birch, R., Carlstedt, T., Tate, S., . . . Anand, P. (2000). Immunolocalization of SNS/PN3 and NaN/SNS2 sodium channels in human pain states. *Pain, 85*(1–2), 41–50.

Cox, J. J., Reimann, F., Nicholas, A. K., Thornton, G., Roberts, E., Springell, K., . . . Woods, C. G. (2006). An SCN9A channelopathy causes congenital inability to experience pain. *Nature, 444*(7121), 894–898.

Craner, M. J., Kataoka, Y., Lo, A. C., Black, J. A., Baker, D., & Waxman, S. G. (2003). Temporal course of upregulation of Na(v)1.8 in Purkinje neurons parallels the progression of clinical deficit in experimental allergic encephalomyelitis. *Journal of Neuropathology and Experimental Neurology, 62*(9), 968–975.

Cummins, T. R., Aglieco, F., Renganathan, M., Herzog, R. I., Dib-Hajj, S. D., & Waxman, S. G. (2001). Nav1.3 sodium channels: Rapid repriming and slow closed-state inactivation display quantitative differences after expression in a mammalian cell line and in spinal sensory neurons. *Journal of Neuroscience, 21*(16), 5952–5961.

Cummins, T. R., Dib-Hajj, S. D., Black, J. A., Akopian, A. N., Wood, J. N., & Waxman, S. G. (1999). A novel persistent tetrodotoxin-resistant sodium current in SNS-null and wild-type small primary sensory neurons. *Journal of Neuroscience, 19*(24), RC43.

Cummins, T. R., Dib-Hajj, S. D., Herzog, R. I., & Waxman, S. G. (2005). Nav1.6 channels generate resurgent sodium currents in spinal sensory neurons. *FEBS Letters, 579*(10), 2166–2170.

Cummins, T. R., Dib-Hajj, S. D., & Waxman, S. G. (2004). Electrophysiological properties of mutant Nav1.7 sodium channels in a painful inherited neuropathy. *Journal of Neuroscience, 24*(38), 8232–8236.

Cummins, T. R., Howe, J. R., & Waxman, S. G. (1998). Slow closed-state inactivation: A novel mechanism underlying ramp currents in cells expressing the hNE/PN1 sodium channel. *Journal of Neuroscience, 18*(23), 9607–9619.

Cummins, T. R., & Waxman, S. G. (1997). Downregulation of tetrodotoxin-resistant sodium currents and upregulation of a rapidly repriming tetrodotoxin-sensitive sodium current in small spinal sensory neurons after nerve injury. *Journal of Neuroscience, 17*(10), 3503–3514.

Daou, I., Beaudry, H., Ase, A. R., Wieskopf, J. S., Ribeiro-da-Silva, A., Mogil, J. S., & Séguéla, P. (2016). Optogenetic silencing of Nav1.8-positive afferents alleviates inflammatory and neuropathic pain. *eNeuro, 3*(1). doi:10.1523/ENEURO.0140-15.2016

Deuis, J. R., Zimmermann, K., Romanovsky, A. A., Possani, L. D., Cabot, P. J., Lewis, R. J., & Vetter, I. (2013). An animal model of oxaliplatin-induced cold allodynia reveals a crucial role for Nav1.6 in peripheral pain pathways. *Pain, 154*(9), 1749–1757. doi:10.1016/j.pain.2013.05.032

Devor, M. (1999). Unexplained peculiarities of the dorsal root ganglion. *Pain*, Suppl 6, S27–S35.

Dib-Hajj, S., Black, J. A., Cummins, T. R., & Waxman, S. G. (2002). NaN/Nav1.9: A sodium channel with unique properties. *Trends in Neuroscience, 25*(5), 253–259.

Dib-Hajj, S. D., Black, J. A., & Waxman, S. G. (2015). NaV1.9: A sodium channel linked to human pain. *Nature Reviews Neuroscience, 16*(9), 511–519. doi:10.1038/nrn3977

Dib-Hajj, S. D., Cummins, T. R., Black, J. A., & Waxman, S. G. (2010). Sodium channels in normal and pathological pain. *Annual Review of Neuroscience, 33*, 325–347. doi:10.1146/annurev-neuro-060909-153234

Dib-Hajj, S. D., Estacion, M., Jarecki, B. W., Tyrrell, L., Fischer, T. Z., Lawden, M., . . . Waxman, S. G. (2008). Paroxysmal extreme pain disorder M1627K mutation in human Nav1.7 renders DRG neurons hyperexcitable. *Molecular Pain, 4*, 37.

Dib-Hajj, S. D., Fjell, J., Cummins, T. R., Zheng, Z., Fried, K., LaMotte, R., . . . Waxman, S. G. (1999). Plasticity of sodium channel expression in DRG neurons in the chronic constriction injury model of neuropathic pain. *Pain, 83*(3), 591–600.

Dib-Hajj, S. D., Rush, A. M., Cummins, T. R., Hisama, F. M., Novella, S., Tyrrell, L., . . . Waxman, S. G. (2005). Gain-of-function mutation in Nav1.7 in familial erythromelalgia induces bursting of sensory neurons. *Brain, 128*(Pt 8), 1847–1854. doi:awh514

Dib-Hajj, S. D., Tyrrell, L., Escayg, A., Wood, P. M., Meisler, M. H., & Waxman, S. G. (1999). Coding sequence, genomic organization, and conserved chromosomal localization of the mouse gene Scn11a encoding the sodium channel NaN. *Genomics, 59*(3), 309–318. doi:10.1006/geno.1999.5890

Dib-Hajj, S. D., Yang, Y., Black, J. A., & Waxman, S. G. (2013). The Na(V)1.7 sodium channel: From molecule to man. *Nature Reviews Neuroscience, 14*(1), 49–62. doi:10.1038/nrn3404

Djouhri, L., Fang, X., Okuse, K., Wood, J. N., Berry, C. M., & Lawson, S. N. (2003). The TTX-resistant sodium channel Nav1.8 (SNS/PN3): Expression and correlation with membrane properties in rat nociceptive primary afferent neurons. *Journal of Physiology, 550*(Pt 3), 739–752.

Djouhri, L., & Lawson, S. N. (2004). Abeta-fiber nociceptive primary afferent neurons: A review of incidence and properties in relation to other afferent A-fiber neurons in mammals. *Brain Research: Brain Research Reviews, 46*(2), 131–145. doi:10.1016/j.brainresrev.2004.07.015

Djouhri, L., Newton, R., Levinson, S. R., Berry, C. M., Carruthers, B., & Lawson, S. N. (2003). Sensory and electrophysiological properties of guinea-pig sensory neurones expressing Nav 1.7 (PN1) Na+ channel alpha subunit protein. *Journal of Physiology, 546*(Pt 2), 565–576.

Dustrude, E. T., Moutal, A., Yang, X., Wang, Y., Khanna, M., & Khanna, R. (2016). Hierarchical CRMP2 posttranslational modifications control NaV1.7 function. *Proceedings of the National Academy of Sciences of the United States of America, 113*(52), E8443–E8452. doi:10.1073/pnas.1610531113

Eilers, H., & Schumacher, M. A. (2005). Mechanosensitivity of primary afferent nociceptors in the pain pathway. In: Kamkin A, Kiseleva I, editors. *Mechanosensitivity in Cells and Tissues.* Moscow: Academia. doi:NBK7514 [bookaccession]

Emery, E. C., Habib, A. M., Cox, J. J., Nicholas, A. K., Gribble, F. M., Woods, C. G., & Reimann, F. (2015). Novel SCN9A mutations underlying extreme pain phenotypes: Unexpected electrophysiological and clinical phenotype correlations. *Journal of Neuroscience, 35*(20), 7674–7681. doi:10.1523/JNEUROSCI.3935-14.2015

Enomoto, A., Han, J. M., Hsiao, C. F., Wu, N., & Chandler, S. H. (2006). Participation of sodium currents in burst generation and control of membrane excitability in mesencephalic trigeminal neurons. *Journal of Neuroscience, 26*(13), 3412–3422. doi:26/13/3412

Estacion, M., & Waxman, S. G. (2013). The response of Na(V)1.3 sodium channels to ramp stimuli: Multiple components and mechanisms. *Journal of Neurophysiology, 109*(2), 306–314. doi:10.1152/jn.00438.2012

Faber, C. G., Lauria, G., Merkies, I. S., Cheng, X., Han, C., Ahn, H. S., ... Waxman, S. G. (2012). Gain-of-function Nav1.8 mutations in painful neuropathy. *Proceedings of the National Academy of Sciences of the United States of America, 109*(47), 19444–19449. doi:10.1073/pnas.1216080109

Fang, X., Djouhri, L., Black, J. A., Dib-Hajj, S. D., Waxman, S. G., & Lawson, S. N. (2002). The presence and role of the tetrodotoxin-resistant sodium channel Na(v)1.9 (NaN) in nociceptive primary afferent neurons. *Journal of Neuroscience, 22*(17), 7425–7433.

Felts, P. A., Yokoyama, S., Dib-Hajj, S., Black, J. A., & Waxman, S. G. (1997). Sodium channel alpha-subunit mRNAs I, II, III, NaG, Na6 and hNE (PN1): Different expression patterns in developing rat nervous system. *Brain Research: Molecular Brain Research, 45*(1), 71–82.

Fertleman, C. R., Baker, M. D., Parker, K. A., Moffatt, S., Elmslie, F. V., Abrahamsen, B., ... Rees, M. (2006). SCN9A mutations in paroxysmal extreme pain disorder: Allelic variants underlie distinct channel defects and phenotypes. *Neuron, 52*(5), 767–774.

Fertleman, C. R., Ferrie, C. D., Aicardi, J., Bednarek, N. A., Eeg-Olofsson, O., Elmslie, F. V., ... Stephenson, J. B. (2007). Paroxysmal extreme pain disorder (previously familial rectal pain syndrome). *Neurology, 69*(6), 586–595.

Fischer, T. Z., & Waxman, S. G. (2010). Familial pain syndromes from mutations of the Nav1.7 sodium channel. *Annals of the New York Academy of Sciences, 1184*, 196–207. doi:10.1111/j.1749-6632.2009.05110.x

Fozzard, H. A., Sheets, M. F., & Hanck, D. A. (2011). The sodium channel as a target for local anesthetic drugs. *Frontiers in Pharmacology, 2*, 68. doi:10.3389/fphar.2011.00068

Gaida, W., Klinder, K., Arndt, K., & Weiser, T. (2005). Ambroxol, a Nav1.8-preferring Na(+) channel blocker, effectively suppresses pain symptoms in animal models of chronic, neuropathic and inflammatory pain. *Neuropharmacology, 49*(8), 1220–1227. doi:10.1016/j.neuropharm.2005.08.004

Gasser, A., Ho, T. S., Cheng, X., Chang, K. J., Waxman, S. G., Rasband, M. N., & Dib-Hajj, S. D. (2012). An ankyrinG-binding motif is necessary and sufficient for targeting Nav1.6 sodium channels to axon initial segments and nodes of Ranvier. *Journal of Neuroscience, 32*(21), 7232–7243. doi:10.1523/JNEUROSCI.5434-11.2012

Gingras, J., Smith, S., Matson, D. J., Johnson, D., Nye, K., Couture, L., ... McDonough, S. I. (2014). Global Nav1.7 knockout mice recapitulate the phenotype of human congenital indifference to pain. *PLoS One, 9*(9), e105895. doi:10.1371/journal.pone.0105895

Goldberg, Y. P., MacFarlane, J., MacDonald, M. L., Thompson, J., Dube, M. P., Mattice, M., ... Hayden, M. R. (2007). Loss-of-function mutations in the Nav1.7 gene underlie congenital indifference to pain in multiple human populations. *Clinical Genetics, 71*(4), 311–319.

Goldin, A. L. (2001). Resurgence of sodium channel research. *Annual Review of Physiology, 63*, 871–894. doi:10.1146/annurev.physiol.63.1.871

Gould, H. J., 3rd, Gould, T. N., England, J. D., Paul, D., Liu, Z. P., & Levinson, S. R. (2000). A possible role for nerve growth factor in the augmentation of sodium channels in models of chronic pain. *Brain Research*, *854*(1–2), 19–29. doi:S0006-8993(99)02216-7

Grieco, T. M., & Raman, I. M. (2004). Production of resurgent current in NaV1.6-null Purkinje neurons by slowing sodium channel inactivation with beta-pompilidotoxin. *Journal of Neuroscience*, *24*(1), 35–42. doi:10.1523/JNEUROSCI.3807-03.2004

Han, C., Huang, J., & Waxman, S. G. (2016). Sodium channel Nav1.8: Emerging links to human disease. *Neurology*, *86*(5), 473–483. doi:10.1212/wnl.0000000000002333

Han, C., Vasylyev, D., Macala, L. J., Gerrits, M. M., Hoeijmakers, J. G., Bekelaar, K. J., . . . Waxman, S. G. (2014). The G1662S NaV1.8 mutation in small fibre neuropathy: Impaired inactivation underlying DRG neuron hyperexcitability. *Journal of Neurology, Neurosurgery, and Psychiatry*, *85*(5), 499–505. doi:10.1136/jnnp-2013-306095

Han, C., Yang, Y., de Greef, B. T., Hoeijmakers, J. G., Gerrits, M. M., Verhamme, C., . . . Waxman, S. G. (2015). The Domain II S4–S5 linker in Nav1.9: A missense mutation enhances activation, impairs fast inactivation, and produces human painful neuropathy. *Neuromolecular Medicine*, *17*(2), 158–169. doi:10.1007/s12017-015-8347-9

Han, C., Yang, Y., Te Morsche, R. H., Drenth, J. P., Politei, J. M., Waxman, S. G., & Dib-Hajj, S. D. (2016). Familial gain-of-function Nav1.9 mutation in a painful channelopathy. *Journal of Neurology, Neurosurgery, and Psychiatry*. doi:jnnp-2016-313804

Herzog, R. I., Cummins, T. R., Ghassemi, F., Dib-Hajj, S. D., & Waxman, S. G. (2003). Distinct repriming and closed-state inactivation kinetics of Nav1.6 and Nav1.7 sodium channels in mouse spinal sensory neurons. *Journal of Physiology*, *551*(Pt 3), 741–750. doi:10.1113/jphysiol.2003.047357

Herzog, R. I., Cummins, T. R., & Waxman, S. G. (2001). Persistent TTX-resistant Na+ current affects resting potential and response to depolarization in simulated spinal sensory neurons. *Journal of Neurophysiology*, *86*(3), 1351–1364.

Hille, B. (1984). *Ionic channels of excitable membranes*. Sunderland, MA: Sinauer Associates.

Hiyama, T. Y., Watanabe, E., Ono, K., Inenaga, K., Tamkun, M. M., Yoshida, S., & Noda, M. (2002). Na(x) channel involved in CNS sodium-level sensing. *Nature Neuroscience*, *5*(6), 511–512. doi:10.1038/nn856

Huang, J., Han, C., Estacion, M., Vasylyev, D., Hoeijmakers, J. G., Gerrits, M. M., . . . Group, P. S. (2014). Gain-of-function mutations in sodium channel Na(v)1.9 in painful neuropathy. *Brain*, *137*(Pt 6), 1627–1642. doi:10.1093/brain/awu079

Huang, J., Vanoye, C. G., Cutts, A., Goldberg, Y. P., Dib-Hajj, S. D., Cohen, C. J., . . . George, A. L., Jr. (2017). Sodium channel NaV1.9 mutations associated with insensitivity to pain dampen neuronal excitability. *Journal of Clinical Investigation*. doi:10.1172/JCI92373

Hudmon, A., Choi, J. S., Tyrrell, L., Black, J. A., Rush, A. M., Waxman, S. G., & Dib-Hajj, S. D. (2008). Phosphorylation of sodium channel Na(v)1.8 by p38 mitogen-activated protein kinase increases current density in dorsal root ganglion neurons. *Journal of Neuroscience*, *28*(12), 3190–3201. doi:10.1523/JNEUROSCI.4403-07.2008

Jarecki, B. W., Piekarz, A. D., Jackson, J. O., 2nd, & Cummins, T. R. (2010). Human voltage-gated sodium channel mutations that cause inherited neuronal and muscle channelopathies increase resurgent sodium currents. *Journal of Clinical Investigation*, *120*(1), 369–378. doi:10.1172/JCI40801

Jarecki, B. W., Sheets, P. L., Jackson, J. O., 2nd, & Cummins, T. R. (2008). Paroxysmal extreme pain disorder mutations within the D3/S4-S5 linker of Nav1.7 cause moderate destabilization of fast inactivation. *Journal of Physiology*, *586*(Pt 17), 4137–4153.

Jarvis, M. F., Honore, P., Shieh, C. C., Chapman, M., Joshi, S., Zhang, X. F., . . . Krafte, D. S. (2007). A-803467, a potent and selective Nav1.8 sodium channel blocker, attenuates neuropathic and inflammatory pain in the rat. *Proceedings of the National Academy of Sciences of the United States of America, 104*(20), 8520–8525.

Kahlig, K. M., Rhodes, T. H., Pusch, M., Freilinger, T., Pereira-Monteiro, J. M., Ferrari, M. D., . . . George, A. L., Jr. (2008). Divergent sodium channel defects in familial hemiplegic migraine. *Proceedings of the National Academy of Sciences of the United States of America, 105*(28), 9799–9804. doi:10.1073/pnas.0711717105

Kim, C. H., Oh, Y., Chung, J. M., & Chung, K. (2002). Changes in three subtypes of tetrodotoxin sensitive sodium channel expression in the axotomized dorsal root ganglion in the rat. *Neuroscience Letters, 323*(2), 125–128.

Kim, J. H., Kushmerick, C., & von Gersdorff, H. (2010). Presynaptic resurgent Na+ currents sculpt the action potential waveform and increase firing reliability at a CNS nerve terminal. *Journal of Neuroscience, 30*(46), 15479–15490. doi:10.1523/JNEUROSCI.3982-10.2010

Klugbauer, N., Lacinova, L., Flockerzi, V., & Hofmann, F. (1995). Structure and functional expression of a new member of the tetrodotoxin-sensitive voltage-activated sodium channel family from human neuroendocrine cells. *EMBO Journal, 14*(6), 1084–1090.

Lai, J., Gold, M. S., Kim, C. S., Bian, D., Ossipov, M. H., Hunter, J. C., & Porreca, F. (2002). Inhibition of neuropathic pain by decreased expression of the tetrodotoxin-resistant sodium channel, NaV1.8. *Pain, 95*(1–2), 143–152.

Lee, J. H., Park, C. K., Chen, G., Han, Q., Xie, R. G., Liu, T., . . . Lee, S. Y. (2014). A monoclonal antibody that targets a NaV1.7 channel voltage sensor for pain and itch relief. *Cell, 157*(6), 1393–1404. doi:10.1016/j.cell.2014.03.064

Leffler, A., Cummins, T. R., Dib-Hajj, S. D., Hormuzdiar, W. N., Black, J. A., & Waxman, S. G. (2002). GDNF and NGF reverse changes in repriming of TTX-sensitive Na(+) currents following axotomy of dorsal root ganglion neurons. *Journal of Neurophysiology, 88*(2), 650–658.

Leffler, A., Herzog, R. I., Dib-Hajj, S. D., Waxman, S. G., & Cummins, T. R. (2005). Pharmacological properties of neuronal TTX-resistant sodium channels and the role of a critical serine pore residue. *Pflugers Archiv, 451*(3), 454–463.

Leipold, E., Liebmann, L., Korenke, G. C., Heinrich, T., Giesselmann, S., Baets, J., . . . Kurth, I. (2013). A de novo gain-of-function mutation in SCN11A causes loss of pain perception. *Nature Genetics, 45*(11), 1399–1404. doi:10.1038/ng.2767

Lin, Z., Santos, S., Padilla, K., Printzenhoff, D., & Castle, N.A. (2016) Biophysical and Pharmacological Characterization of Nav1.9 Voltage Dependent Sodium Channels Stably Expressed in HEK-293 Cells. *PLoS One, 11*(8): e0161450. doi: 10.1371/journal.pone.0161450

Lindia, J. A., Kohler, M. G., Martin, W. J., & Abbadie, C. (2005). Relationship between sodium channel NaV1.3 expression and neuropathic pain behavior in rats. *Pain, 117*(1–2), 145–153. doi:10.1016/j.pain.2005.05.027

Liu, M., & Wood, J. N. (2011). The roles of sodium channels in nociception: Implications for mechanisms of neuropathic pain. *Pain Medicine, 12*(Suppl 3), S93–S99. doi:10.1111/j.1526-4637.2011.01158.x

Liu, X. D., Yang, J. J., Fang, D., Cai, J., Wan, Y., & Xing, G. G. (2014). Functional upregulation of Nav1.8 sodium channels on the membrane of dorsal root ganglia neurons contributes to the development of cancer-induced bone pain. *PLoS One, 9*(12), e114623. doi:10.1371/journal.pone.0114623

Liu, Y., Wu, Z., Tang, D., Xun, X., Liu, L., Li, X., ... Yi, J. (2014). Analgesic effects of Huwentoxin-IV on animal models of inflammatory and neuropathic pain. *Protein and Peptide Letters, 21*(2), 153–158. doi:PPL-EPUB-57214

Lolignier, S., Bonnet, C., Gaudioso, C., Noel, J., Ruel, J., Amsalem, M., ... Busserolles, J. (2015). The Nav1.9 channel is a key determinant of cold pain sensation and cold allodynia. *Cell Reports, 11*(7), 1067–1078. doi:10.1016/j.celrep.2015.04.027

Maingret, F., Coste, B., Padilla, F., Clerc, N., Crest, M., Korogod, S. M., & Delmas, P. (2008). Inflammatory mediators increase Nav1.9 current and excitability in nociceptors through a coincident detection mechanism. *Journal of General Physiology, 131*(3), 211–225. doi:10.1085/jgp.200709935

McGaraughty, S., Chu, K. L., Scanio, M. J. C., Kort, M. E., Faltynek, C. R., & Jarvis, M. F. (2008). A selective Nav1.8 sodium channel blocker, A-803467 [5-(4-chlorophenyl-N-(3,5-dimethoxyphenyl)furan-2-carboxamide], attenuates spinal neuronal activity in neuropathic rats. *Journal of Pharmacology and Experimental Therapeutics, 324*(3), 1204–1211. doi:10.1124/jpet.107.134148

Meisler, M. H., Helman, G., Hammer, M. F., Fureman, B. E., Gaillard, W. D., Goldin, A. L., ... Scheffer, I. E. (2016). SCN8A encephalopathy: Research progress and prospects. *Epilepsia, 57*(7), 1027–1035. doi:10.1111/epi.13422

Millan, M. J. (1999). The induction of pain: An integrative review. *Progress in Neurobiology, 57*(1), 1–164.

Minett, M. S., Falk, S., Santana-Varela, S., Bogdanov, Y. D., Nassar, M. A., Heegaard, A. M., & Wood, J. N. (2014). Pain without nociceptors? Nav1.7-independent pain mechanisms. *Cell Reports, 6*(2), 301–312. doi:10.1016/j.celrep.2013.12.033

Mulroy, M. F. (2002). Systemic toxicity and cardiotoxicity from local anesthetics: Incidence and preventive measures. *Regional Anesthesia Pain Medicine, 27*(6), 556–561. doi:S1098733902000779

Nassar, M. A., Baker, M. D., Levato, A., Ingram, R., Mallucci, G., McMahon, S. B., & Wood, J. N. (2006). Nerve injury induces robust allodynia and ectopic discharges in Nav1.3 null mutant mice. *Molecular Pain*, 2, 33.

Nassar, M. A., Stirling, L. C., Forlani, G., Baker, M. D., Matthews, E. A., Dickenson, A. H., & Wood, J. N. (2004). Nociceptor-specific gene deletion reveals a major role for Nav1.7 (PN1) in acute and inflammatory pain. *Proceedings of the National Academy of Sciences of the United States of America, 101*(34), 12706–12711. doi:10.1073/pnas.0404915101

Noda, M., & Numa, S. (1987). Structure and function of sodium channel. *Journal of Receptor Research, 7*(1–4), 467–497.

O'Donnell, A. M., Coyle, D., & Puri, P. (2016). Decreased Nav1.9 channel expression in Hirschsprung's disease. *Journal of Pediatric Surgery, 51*(9), 1458–1461. doi:10.1016/j.jpedsurg.2016.05.007

O'Malley, H. A., & Isom, L. L. (2015). Sodium channel beta subunits: Emerging targets in channelopathies. *Annual Review of Physiology, 77*, 481–504. doi:10.1146/annurev-physiol-021014-071846

Osteen, J. D., Herzig, V., Gilchrist, J., Emrick, J. J., Zhang, C., Wang, X., ... Julius, D. (2016). Selective spider toxins reveal a role for the Nav1.1 channel in mechanical pain. *Nature, 534*(7608), 494–499. doi:10.1038/nature17976

Ostman, J. A., Nassar, M. A., Wood, J. N., & Baker, M. D. (2008). GTP up-regulated persistent Na+ current and enhanced nociceptor excitability require NaV1.9. *Journal of Physiology, 586*(4), 1077–1087.

Patel, R. R., Barbosa, C., Xiao, Y., & Cummins, T. R. (2015). Human Nav1.6 channels generate larger resurgent currents than human Nav1.1 channels, but the Navbeta4 peptide does not protect either isoform from use-dependent reduction. *PLoS One*, *10*(7), e0133485. doi:10.1371/journal.pone.0133485

Payne, C. E., Brown, A. R., Theile, J. W., Loucif, A. J., Alexandrou, A. J., Fuller, M. D., ... Stevens, E. B. (2015). A novel selective and orally bioavailable Nav 1.8 channel blocker, PF-01247324, attenuates nociception and sensory neuron excitability. *British Journal of Pharmacology*, *172*(10), 2654–2670. doi:10.1111/bph.13092

Pei, Z., Xiao, Y., Meng, J., Hudmon, A., & Cummins, T. R. (2016). Cardiac sodium channel palmitoylation regulates channel availability and myocyte excitability with implications for arrhythmia generation. *Nature Communications*, *7*, 12035. doi:10.1038/ncomms12035

Persson, A. K., Liu, S., Faber, C. G., Merkies, I. S., Black, J. A., & Waxman, S. G. (2013). Neuropathy-associated Nav1.7 variant I228M impairs integrity of dorsal root ganglion neuron axons. *Annals of Neurology*, *73*(1), 140–145. doi:10.1002/ana.23725

Plant, L. D., Marks, J. D., & Goldstein, S. A. (2016). SUMOylation of NaV1.2 channels mediates the early response to acute hypoxia in central neurons. *Elife*, *5*, 1–23. doi:10.7554/eLife.20054

Priest, B. T., Murphy, B. A., Lindia, J. A., Diaz, C., Abbadie, C., Ritter, A. M., ... Martin, W. J. (2005). Contribution of the tetrodotoxin-resistant voltage-gated sodium channel NaV1.9 to sensory transmission and nociceptive behavior. *Proceedings of the National Academy of Sciences of the United States of America*, *102*(26), 9382–9387.

Raman, I. M., & Bean, B. P. (1997). Resurgent sodium current and action potential formation in dissociated cerebellar Purkinje neurons. *Journal of Neuroscience*, *17*(12), 4517–4526.

Renganathan, M., Cummins, T. R., & Waxman, S. G. (2001). Contribution of Na(v)1.8 sodium channels to action potential electrogenesis in DRG neurons. *Journal of Neurophysiology*, *86*(2), 629–640.

Renganathan, M., Gelderblom, M., Black, J. A., & Waxman, S. G. (2003). Expression of Nav1.8 sodium channels perturbs the firing patterns of cerebellar Purkinje cells. *Brain Research*, *959*(2), 235–242.

Ritchie, J. M. (1975). Binding of tetrodotoxin and saxitoxin to sodium channels. *Philosophical Transactions of the Royal Society of London, B: Biological Sciences*, *270*(908), 319–336.

Rogart, R. B., & Regan, L. J. (1985). Two subtypes of sodium channel with tetrodotoxin sensitivity and insensitivity detected in denervated mammalian skeletal muscle. *Brain Research*, *329*(1–2), 314–318.

Rowe, A. H., & Rowe, M. P. (2008). Physiological resistance of grasshopper mice (*Onychomys spp.*) to Arizona bark scorpion (*Centruroides exilicauda*) venom. *Toxicon*, *52*(5), 597–605. doi:10.1016/j.toxicon.2008.07.004

Rowe, A. H., Xiao, Y., Rowe, M. P., Cummins, T. R., & Zakon, H. H. (2013). Voltage-gated sodium channel in grasshopper mice defends against bark scorpion toxin. *Science*, *342*(6157), 441–446. doi:10.1126/science.1236451

Rowe, A. H., Xiao, Y., Scales, J., Linse, K. D., Rowe, M. P., Cummins, T. R., & Zakon, H. H. (2011). Isolation and characterization of CvIV4: A pain inducing alpha-scorpion toxin. *PLoS One*, *6*(8), e23520. doi:10.1371/journal.pone.0023520

Rugiero, F., Mistry, M., Sage, D., Black, J. A., Waxman, S. G., Crest, M., ... Gola, M. (2003). Selective expression of a persistent tetrodotoxin-resistant Na+ current and NaV1.9 subunit in myenteric sensory neurons. *Journal of Neuroscience*, *23*(7), 2715–2725. doi:23/7/2715

Rush, A. M., Cummins, T. R., & Waxman, S. G. (2007). Multiple sodium channels and their roles in electrogenesis within dorsal root ganglion neurons. *Journal of Physiology*, *579*(Pt 1), 1–14. doi:10.1113/jphysiol.2006.121483

Rush, A. M., & Waxman, S. G. (2004). PGE2 increases the tetrodotoxin-resistant Nav1.9 sodium current in mouse DRG neurons via G-proteins. *Brain Research*, *1023*(2), 264–271.

Rush, A. M., Wittmack, E. K., Tyrrell, L., Black, J. A., Dib-Hajj, S. D., & Waxman, S. G. (2006). Differential modulation of sodium channel Na(v)1.6 by two members of the fibroblast growth factor homologous factor 2 subfamily. *European Journal of Neuroscience*, *23*(10), 2551–2562.

Samad, O. A., Tan, A. M., Cheng, X., Foster, E., Dib-Hajj, S. D., & Waxman, S. G. (2013). Virus-mediated shRNA knockdown of Na(v)1.3 in rat dorsal root ganglion attenuates nerve injury-induced neuropathic pain. *Molecular Therapy*, *21*(1), 49–56. doi:10.1038/mt.2012.169

Schiavon, E., Stevens, M., Zaharenko, A. J., Konno, K., Tytgat, J., & Wanke, E. (2010). Voltage-gated sodium channel isoform-specific effects of pompilidotoxins. *The FEBS Journal*, *277*(4), 918–930. doi:10.1111/j.1742-4658.2009.07533.x

Shembalkar, P. K., Till, S., Boettger, M. K., Terenghi, G., Tate, S., Bountra, C., & Anand, P. (2001). Increased sodium channel SNS/PN3 immunoreactivity in a causalgic finger. *European Journal of Pain*, *5*(3), 319–323. doi:10.1053/eujp.2001.0251

Sittl, R., Lampert, A., Huth, T., Schuy, E. T., Link, A. S., Fleckenstein, J., . . . Carr, R. W. (2012). Anticancer drug oxaliplatin induces acute cooling-aggravated neuropathy via sodium channel subtype Na(V)1.6-resurgent and persistent current. *Proceedings of the National Academy of Sciences of the United States of America*, *109*(17), 6704–6709. doi:10.1073/pnas.1118058109

Steinlein, O. K. (2014). Mechanisms underlying epilepsies associated with sodium channel mutations. *Progress in Brain Research*, *213*, 97–111. doi:10.1016/B978-0-444-63326-2.00005-3

Stys, P. K., Sontheimer, H., Ransom, B. R., & Waxman, S. G. (1993). Noninactivating, tetrodotoxin-sensitive Na+ conductance in rat optic nerve axons. *Proceedings of the National Academy of Sciences of the United States of America*, *90*(15), 6976–6980.

Tamura, R., Nemoto, T., Maruta, T., Onizuka, S., Yanagita, T., Wada, A., . . . Tsuneyoshi, I. (2014). Up-regulation of NaV1.7 sodium channels expression by tumor necrosis factor-alpha in cultured bovine adrenal chromaffin cells and rat dorsal root ganglion neurons. *Anesthesia and Analgesia*, *118*(2), 318–324. doi:10.1213/ANE.0000000000000085

Tan, A. M., Samad, O. A., Dib-Hajj, S. D., & Waxman, S. G. (2015). Virus-mediated knockdown of Nav1.3 in dorsal root ganglia of STZ-induced diabetic rats alleviates tactile allodynia. *Molecular Medicine*. doi:10.2119/molmed.2015.00063

Tan, Z. Y., Piekarz, A. D., Priest, B. T., Knopp, K. L., Krajewski, J. L., McDermott, J. S., . . . Cummins, T. R. (2014). Tetrodotoxin-resistant sodium channels in sensory neurons generate slow resurgent currents that are enhanced by inflammatory mediators. *Journal of Neuroscience*, *34*(21), 7190–7197. doi:10.1523/JNEUROSCI.5011-13.2014

Tan, Z. Y., Priest, B. T., Krajewski, J. L., Knopp, K. L., Nisenbaum, E. S., & Cummins, T. R. (2014). Protein kinase C enhances human sodium channel hNav1.7 resurgent currents via a serine residue in the domain III–IV linker. *FEBS Letters*, *588*(21), 3964–3969. doi:10.1016/j.febslet.2014.09.011

Tanaka, B. S., Zhao, P., Dib-Hajj, F. B., Morisset, V., Tate, S., Waxman, S. G., & Dib-Hajj, S. D. (2016). A gain-of-function mutation in Nav1.6 in a case of trigeminal neuralgia. *Molecular Medicine*, *22*, 338–348. doi:10.2119/molmed.2016.00131

Theile, J. W., Jarecki, B. W., Piekarz, A. D., & Cummins, T. R. (2011). Nav1.7 mutations associated with paroxysmal extreme pain disorder, but not erythromelalgia, enhance Navbeta4 peptide-mediated resurgent sodium currents. *Journal of Physiology, 589*(Pt 3), 597–608. doi:jphysiol.2010.200915

Toldo, I., Bruson, A., Casarin, A., Salviati, L., Boniver, C., Sartori, S., . . . Clementi, M. (2011). Polymorphisms of the SCN1A gene in children and adolescents with primary headache and idiopathic or cryptogenic epilepsy: Is there a linkage? *Journal of Headache Pain, 12*(4), 435–441. doi:10.1007/s10194-011-0359-8

Toledo-Aral, J. J., Moss, B. L., He, Z. J., Koszowski, A. G., Whisenand, T., Levinson, S. R., . . . Mandel, G. (1997). Identification of PN1, a predominant voltage-dependent sodium channel expressed principally in peripheral neurons. *Proceedings of the National Academy of Sciences of the United States of America, 94*(4), 1527–1532.

Vanoye, C. G., Kunic, J. D., Ehring, G. R., & George, A. L., Jr. (2013). Mechanism of sodium channel NaV1.9 potentiation by G-protein signaling. *Journal of General Physiology, 141*(2), 193–202. doi:10.1085/jgp.201210919

Veeramah, K. R., O'Brien, J. E., Meisler, M. H., Cheng, X., Dib-Hajj, S. D., Waxman, S. G., . . . Hammer, M. F. (2012). De novo pathogenic SCN8A mutation identified by whole-genome sequencing of a family quartet affected by infantile epileptic encephalopathy and SUDEP. *American Journal of Human Genetics, 90*(3), 502–510. doi:10.1016/j.ajhg.2012.01.006

Watanabe, E., Hiyama, T. Y., Kodama, R., & Noda, M. (2002). NaX sodium channel is expressed in non-myelinating Schwann cells and alveolar type II cells in mice. *Neuroscience Letters, 330*(1), 109–113.

Waxman, S. G., Kocsis, J. D., & Black, J. A. (1994). Type III sodium channel mRNA is expressed in embryonic but not adult spinal sensory neurons, and is reexpressed following axotomy. *Journal of Neurophysiology, 72*(1), 466–470.

West, J. W., Patton, D. E., Scheuer, T., Wang, Y., Goldin, A. L., & Catterall, W. A. (1992). A cluster of hydrophobic amino acid residues required for fast Na(+)-channel inactivation. *Proceedings of the National Academy of Sciences of the United States of America, 89*(22), 10910–10914.

Wilson, M. J., Yoshikami, D., Azam, L., Gajewiak, J., Olivera, B. M., Bulaj, G., & Zhang, M. M. (2011). mu-Conotoxins that differentially block sodium channels NaV1.1 through 1.8 identify those responsible for action potentials in sciatic nerve. *Proceedings of the National Academy of Sciences of the United States of America, 108*(25), 10302–10307. doi:10.1073/pnas.1107027108

Xiao, Y., Bingham, J. P., Zhu, W., Moczydlowski, E., Liang, S., & Cummins, T. R. (2008). Tarantula huwentoxin-IV inhibits neuronal sodium channels by binding to receptor site 4 and trapping the domain ii voltage sensor in the closed configuration. *Journal of Biological Chemistry, 283*(40), 27300–27313. doi:10.1074/jbc.M708447200

Xiao, Y., Blumenthal, K., & Cummins, T. R. (2014). Gating-pore currents demonstrate selective and specific modulation of individual sodium channel voltage-sensors by biological toxins. *Molecular Pharmacology, 86*(2), 159–167. doi:10.1124/mol.114.092338

Xiao, Y., Blumenthal, K., Jackson, J. O., 2nd, Liang, S., & Cummins, T. R. (2010). The tarantula toxins ProTx-II and huwentoxin-IV differentially interact with human Nav1.7 voltage sensors to inhibit channel activation and inactivation. *Molecular Pharmacology, 78*(6), 1124–1134. doi:10.1124/mol.110.066332

Xie, W., Strong, J. A., Ye, L., Mao, J. X., & Zhang, J. M. (2013). Knockdown of sodium channel NaV1.6 blocks mechanical pain and abnormal bursting activity of afferent neurons in inflamed sensory ganglia. *Pain, 154*(8), 1170–1180. doi:10.1016/j.pain.2013.02.027

Xie, W., Strong, J. A., & Zhang, J. M. (2015). Local knockdown of the NaV1.6 sodium channel reduces pain behaviors, sensory neuron excitability, and sympathetic sprouting in rat models of neuropathic pain. *Neuroscience, 291*, 317–330. doi:10.1016/j.neuroscience.2015.02.010

Xie, W., Tan, Z. Y., Barbosa, C., Strong, J. A., Cummins, T. R., & Zhang, J. M. (2016). Upregulation of the sodium channel NaVbeta4 subunit and its contributions to mechanical hypersensitivity and neuronal hyperexcitability in a rat model of radicular pain induced by local dorsal root ganglion inflammation. *Pain, 157*(4), 879–891. doi:10.1097/j.pain.0000000000000453

Xu, W., Hong, S. J., Zhong, A., Xie, P., Jia, S., Xie, Z., . . . Mustoe, T. A. (2015). Sodium channel Nax is a regulator in epithelial sodium homeostasis. *Science Translational Medicine, 7*(312), 312ra177. doi:10.1126/scitranslmed.aad0286

Xu, W., Zhang, J., Wang, Y., Wang, L., & Wang, X. (2016). Changes in the expression of voltage-gated sodium channels Nav1.3, Nav1.7, Nav1.8, and Nav1.9 in rat trigeminal ganglia following chronic constriction injury. *Neuroreport, 27*(12), 929–934. doi:10.1097/WNR.0000000000000632

Yanagidate, F., & Strichartz, G. R. (2007). Local anesthetics. *Handbook of Experimental Pharmacology, 177*, 95–127.

Yang, Y., Wang, Y., Li, S., Xu, Z., Li, H., Ma, L., . . . Shen, Y. (2004). Mutations in SCN9A, encoding a sodium channel alpha subunit, in patients with primary erythermalgia. *Journal of Medical Genetics, 41*(3), 171–174.

Yarov-Yarovoy, V., DeCaen, P. G., Westenbroek, R. E., Pan, C. Y., Scheuer, T., Baker, D., & Catterall, W. A. (2012). Structural basis for gating charge movement in the voltage sensor of a sodium channel. *Proceedings of the National Academy of Sciences of the United States of America, 109*(2), E93–E102. doi:10.1073/pnas.1118434109

Ye, P., Hua, L., Jiao, Y., Li, Z., Qin, S., Fu, J., . . . Ji, Y. (2016). Functional up-regulation of Nav1.8 sodium channel on dorsal root ganglia neurons contributes to the induction of scorpion sting pain. *Acta Biochimica et Biophysica Sinica (Shanghai), 48*(2), 132–144. doi:10.1093/abbs/gmv123

Zhang, X. Y., Wen, J., Yang, W., Wang, C., Gao, L., Zheng, L. H., . . . Liu, J. Y. (2013). Gain-of-function mutations in SCN11A cause familial episodic pain. *American Journal of Human Genetics, 93*(5), 957–966. doi:10.1016/j.ajhg.2013.09.016

Zimmermann, K., Leffler, A., Babes, A., Cendan, C. M., Carr, R. W., Kobayashi, J., . . . Reeh, P. W. (2007). Sensory neuron sodium channel Nav1.8 is essential for pain at low temperatures. *Nature, 447*(7146), 855–858.

CHAPTER 8

SODIUM CHANNELOPATHIES OF THE CENTRAL NERVOUS SYSTEM

PAUL G. DECAEN, ALFRED L. GEORGE, JR., AND CHRISTOPHER H. THOMPSON

The Structure of Voltage-Gated Sodium Channels

Evolutionary Origins of Voltage-Gated Sodium Channels

The earliest known life forms on Earth are putative fossilized microorganisms found in hydrothermal vent precipitates. In the genomes of these halophilic archaea is evidence that the first ion channels appeared 3 billion years ago as an evolutionary adaptation to this extreme osmotic condition (Anderson & Greenberg, 2001; Strong, Chandy, & Gutman, 1993). First, a singular domain channel of unknown ion selectivity probably evolved from a common ancestor shared by voltage-gated sodium (Na_V) and potassium (K_V) channels (Strong et al., 1993). Based on phylogenetic analysis, the lineage of prokaryotic Na_V probably diverged from a common progenitor with eukaryotic Na_V channels (Liebeskind, Hillis, & Zakon, 2013; Zakon, 2012). At that time, the prokaryotic progenitor gene duplicated, giving rise to a two domain channel topology found in contemporary two-pore channels (TPC), which reside in endosomal and lysosomal compartments (Galione et al., 2009; Liebeskind, Hillis, & Zakon, 2012). Then a second round of gene duplication occurred, creating four domain channels, which provided the topological blueprint for eukaryotic Na_V and Ca_V channels (Cai, 2012). In the first section of this chapter, we will compare the recently resolved high-resolution structures of prokaryotic and eukaryotic Na_V channels, and how their intramolecular interactions control their function in the cell membrane.

Assembly of Archetypal Prokaryotic Na_V Channel Structures

Although prokaryotic Na_V channel structures have garnered considerable interest, the principal value is their relationship to more complex human homologues. These simple Na_V archetypes have provided a simple protein platform to investigate the structural basis for Na^+ selectivity and the mechanism of channel activation. The simplicity of prokaryotic Na_V channels provides a more tractable target than their mammalian counterparts for structural analysis using X-ray crystallography, and thus structures from this family were the first to be solved.

Each subunit of prokaryotic Na_V channels has six transmembrane-spanning segments (S1–6) that contain a voltage-sensing module (VSM, S1–4) and pore-forming module (PM, S5–6) (Payandeh, Scheuer, Zheng, & Catterall, 2011; Ren et al., 2001). Each subunit or monomer of prokaryotic Na_V channels is viewed as analogous to the four domains of eukaryotic Na_V channels (Figure 8.1A) (Scheuer, 2014; Tang et al., 2014). Four prokaryotic Na_V subunits oligomerize to form a homotetramer, where the PM and cytosolic C-terminal coiled-coil motifs form homotypical contacts around a central ion-conducting pathway (Figure 8.1A, B). Three nearly full-length (Na_VCt, Na_VAb, and Na_VRh) and two complete prokaryotic Na_V structures (Na_VMs and Na_VAb) have been determined (Payandeh et al., 2011; Tsai et al., 2013; Zhang et al., 2012; Lenaeus et al., 2017; Sula et al., 2017). These structures have been captured in unique conformations, which have provided much needed spatial context in which state-dependent interactions may occur during Na_V function.

Relationship of Structure to Function

Voltage-gated sodium channels turn on the flow of ionic current in response to changes in membrane potential. Figure 8.1A illustrates the "pre-open" state of Na_VAb (3.2 Å) and the open state of Na_VM channels (2.4 Å), which represent the conformational changes undertaken when the electrical potential of the cell membrane is shifted from negative to zero millivolts (Figure 8.1B) (Lenaeus et al., 2017; Sula et al., 2017). As discussed in other sections of this book (see Zaleska et al., this volume), and reiterated here, the opening and closing of the pore is mechanically linked to outward movement of the S4 segment that contains a series of positively charged residues (R1–R4) called "gating charges." Transfer of the S4 gating charges has been measured at 3–4e– per voltage sensor module, which precedes the opening and flow of the ionic current in eukaryotic and prokaryotic Na_V channels (Armstrong & Bezanilla, 1974; Keynes, Rojas, & Rudy, 1974; Kuzmenkin, Bezanilla, & Correa, 2004). The movement of gating charges is stabilized by conserved negatively charged resides. The outward trajectory of the S4 has been tracked in a series of disulfide trapping experiments, and the estimated displacement is ≈ 8 Å (DeCaen, Yarov-Yarovoy, Scheuer, & Catterall, 2011; DeCaen, Yarov-Yarovoy, Sharp, Scheuer, & Catterall, 2009; DeCaen, Yarov-Yarovoy,

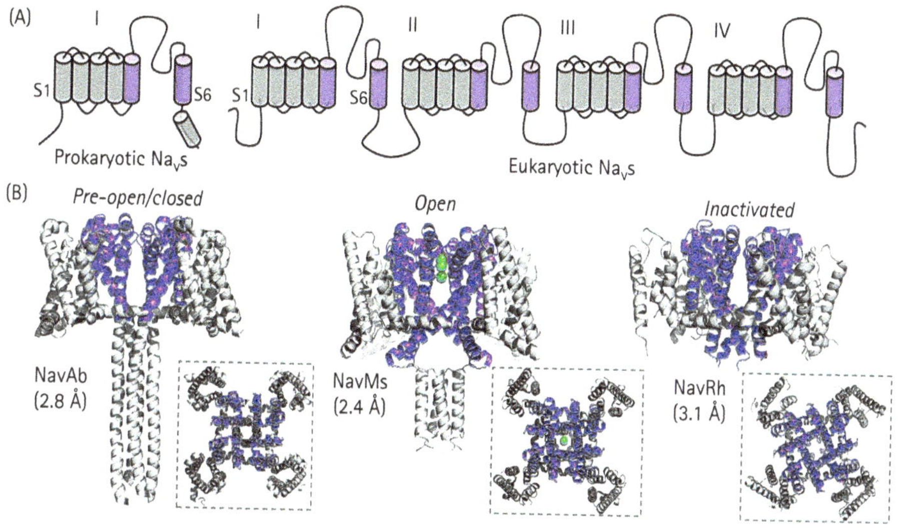

FIGURE 8.1 Structures of prokaryotic Na$_V$ channels.
(A) Topology of prokaryotic and eukaryotic sodium channels. Transmembrane pore-forming helices are colored purple. (B) Transmembrane views of the Na$_V$Ab, Na$_V$Ms, and Na$_V$Rh prokaryotic structures depicting three putative states visited during membrane depolarization. Pore-forming helices S5 and S6 are colored purple. *Insets*, extracellular views of each channel structure.

Zhao, Scheuer, & Catterall, 2008; Yarov-Yarovoy et al., 2012). The onset of the trapping effect has led to the proposal that the VSM has at least two activated states (Yarov-Yarovoy et al., 2012). These measurements predicted that the S4 secondary structure must adopt a full or partial 3_{10}-helix during activation, an unprecedented feature that was confirmed in each of the prokaryotic (and eukaryotic) Na$_V$ structures. After the membrane is depolarized and the VSM has activated, the pore is opened by displacement of the S4–S5 linker that physically interacts with the cytosolic S6 lower gate through a network of hydrogen bonds and charge–counter-charge interactions (Sula et al., 2017). This movement in turn causes each of the four S6 to splay 12° from the center axis, widening the conduction pathway to accommodate the passaging of partially hydrated Na^+ ions (Figure 8.1B).

The mechanism of ion selectivity has become clearer from recent structural revelations. The extracellular mouth of the pore contains the ion selectivity filter. In prokaryotic Na$_V$ channels, the filter contains a symmetrical ring of four aspartate residues and two offset rings of backbone carbonyls (Payandeh et al., 2011; Sula et al., 2017). These atoms coordinate a row of three partially hydrated Na^+ ions by replacing a square array of oxygen atoms found in first hydration shell. Based on functional analysis, partially hydrated Na^+ and the smaller Li^+ pass, whereas larger monovalent (e.g., K^+) and divalent cations do not (DeCaen, Takahashi, Krulwich, Ito, & Clapham, 2014; Naylor et al., 2016; Zhang et al., 2013). Other cations are proposed to be excluded by size and permeation, and this is dictated by the fit of the filter to the incoming ion hydration shells. After the channel has opened, all Na_V channels exhibit slow inactivation, a process wherein the channel is shifted into a non-conductive state that is distinct from the closed state.

Both eukaryotic and prokaryotic Na_V channels exhibit slow inactivation that operates with a time constant of tens to hundreds of milliseconds (Ren et al., 2001; Ulmschneider et al., 2013), but eukaryotic Na_V channels also exhibit fast inactivation, which is structurally and kinetically distinct from slow inactivation (see Structural Elements of Eukaryotic Na_V Channels). Structurally, the slow-inactivated state has been captured as distortions of the selectivity filter and pore domain, which in turn causes occlusion or collapse of the ion coordination sites in the inactivated Na_VRh, Na_VCt, and Na_VAb channel structures (Figure 8.1B) (Payandeh, Gamal El-Din, Scheuer, Zheng, & Catterall, 2012; Tsai et al., 2013; Zhang et al., 2012). Ironically, the most enigmatic state of the activation cycle of Na_V channels is the closed state, which has not been captured in crystal structures because the membrane potential is insufficiently hyperpolarized during the crystallization process. Thus, a complete set of structures that capture prokaryotic Na_V channels in the major conformational states occurring during membrane depolarization is not available.

Structural Elements of Eukaryotic Na_V Channels

In the human genome, there are ten Na_V channel pore-forming subunit genes (*SCN1A* to *SCN11A*) that encode nine functional channels (designated Na_V1.1 to Na_V1 .9) and one atypical homologue (Na_V2.1 or Na_X). These large (260 kDa) and complex (multiple subunits with heteromeric stoichiometry) transmembrane proteins have posed a major challenge for structural biologists. The structure of the eukaryotic Na_V channels is best described as a four-domain pseudoheterotetramer. Each of the four domains (DI–IV) contains six transmembrane segments, which contribute to its total 24-transmembrane-helix assembly (Figure 8.1A). The first structural view of a eukaryotic Na_V channel structure was determined by single-particle cryo-electron microscopy (cryo-EM) applied to the channel purified from the electric organ of *Electrophorus electricus* (Sato et al., 2001). Although the structure was resolved to only 19 Å, some coarse features of the folded protein were apparent, including a bell-shaped outer surface and several inner cavities. Due to their size and structural complexity, solving the atomic structures of eukaryotic Na_V channels has proven to be challenging using X-ray crystallography. However, over the past eight years, the technology to acquire and analyze single-particle cryo-EM data sets has improved immensely, allowing unprecedented structural analysis of macromolecules at atomic resolution. Using this approach, the first two high-resolution structures of a eukaryotic Na_V channels were determined (Figure 8.2).

The first solved eukaryotic Na_V channel structure was that of a channel cloned from the cockroach *Periplaneta americana* at 3.8-Å resolution (Shen et al., 2017). The second structure was of the electric eel channel, EeNa_V1.4 at 4.0-Å, which was solved in complex with an accessory β1 subunit (more on these heteromeric interactions in Eukaryotic Na_Vβ Subunit Complex) (Z. Yan et al., 2017). These channels share 36–66% amino acid

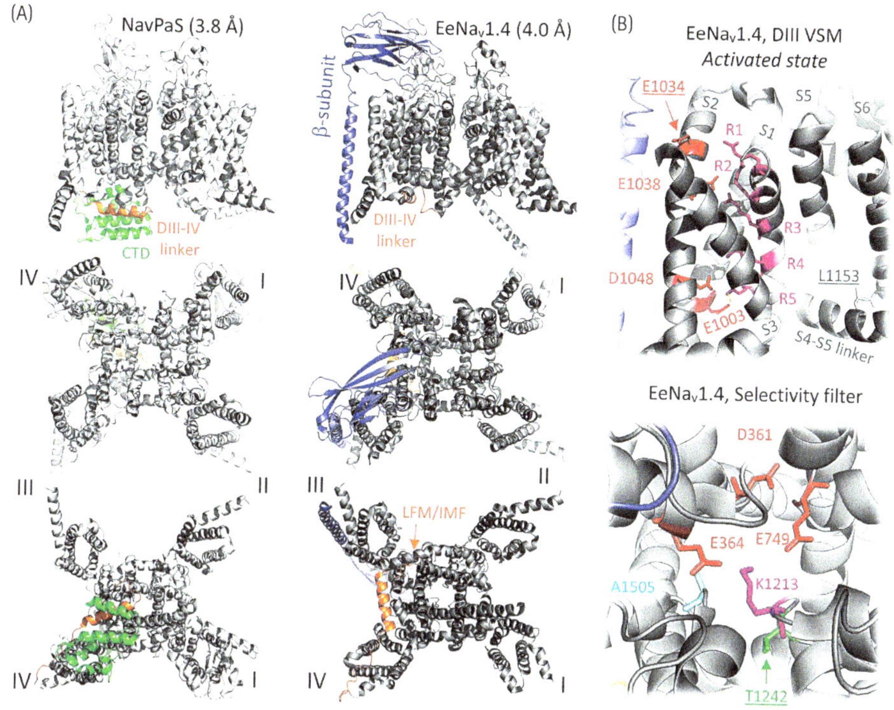

FIGURE 8.2 Structures of eukaryotic Na$_v$ channels.
(A) Transmembrane (*top*), extracellular (*middle*), and intracellular (*bottom*) views of the Na$_v$PaS and EeNa$_v$1.4-β1 complex. (B) The EeNa$_v$1.4 DIII voltage sensor module (*top*) highlighting gating charge (R1–R5) interactions. The selectivity filter (*bottom*) highlighting the putative side chains involved in Na ion selectivity. The sites of disease-causing variants (E1034, L1153, and T1242) are underlined.

sequence identity with human Na$_V$ channels and thus provide a structural template to understand disease-causing variants (Huang, Liu, Yan, & Yan, 2017). As expected, each of the Na$_V$PaS and EeNa$_V$1.4 channels exhibits pseudosymmetry, resembling each subunit from prokaryotic Na$_V$ channels. However, the functional states of the published Na$_V$PaS and EeNa$_V$1.4 structures are ambiguous. The VSMs from each domain appear to represent a mixture of activated and deactivated conformations. Additionally, the C-terminal gate in Na$_V$PaS is narrow, prohibiting the passage of Na^+ ions, whereas this site is more dilated in the EeNa$_V$1.4 structure (Figure 8.2). Thus, the state of the channel gate was tentatively assigned as "closed" or "open" for Na$_V$PaS or EeNa$_V$1.4, respectively. Many of sequential gating charge interactions sampled by each of the VSMs in the eukaryotic structures were captured in the previously discussed structure-function studies of prokaryotic channels (Figure 8.2B, *top*). The moieties proposed to determine Na^+ selectivity in Na$_V$PaS and EeNa$_V$1.4 are shared by human Na$_V$ channels, including an asymmetrical ring of four amino acids (Asp, Glu, Lys, and Ala) contributed by each of the four domains (Figure 8.2B, *bottom*) (Sun, Favre, Schild, & Moczydlowski, 1997). Although it is postulated that eukaryotic channels use a similar mechanism to select for and conduct Na^+, it is not clear how or which of the selectivity filter residues are involved because putative ions and protein–ion contacts in the filter were not resolved in these structures. Nonetheless, the structures of Na_V selectivity filters have provided a basis to further examine parts of the filter never before considered, such as the role of the intervening pore loops and the impact of the pore-directed negative dipole generated by the eight pore helices.

As mentioned previously, eukaryotic Na_V have evolved a fast inactivation mechanism that rapidly (1–2 ms) turns off the flow of ions. Na_V channels recover from fast inactivation within a few hundred microseconds after membrane repolarization, which facilitates the rapid frequency of action potentials conducted by nerve and muscle cells (Hodgkin & Huxley, 1952b). Early functional studies demonstrated that fast inactivation can be altered by mutating sites within the DIII–IV linker (e.g., Isoleucine-phenylalanine-methionine (IFM) or leucine-methionine-phenylalanine (LMF) motif) and residues within the C-terminal domain (CTD), and that activation of the VSM domain III is rate-limiting (Capes, Goldschen-Ohm, Arcisio-Miranda, Bezanilla, & Chanda, 2013; Motoike et al., 2004; Stuhmer et al., 1989). However, because these sites are located at seemingly different locations within the channel, the molecular basis for the asynchronous gating model of fast inactivation has been an enigma. Nonetheless, insights into the mechanism of fast inactivation can be gleaned from comparing the Na_VPaS and EeNa_V1.4 structures. In the Na_VPaS structure, the III–IV linker rests on the cytoplasmic side the DIV VSM and physically links this domain with S6 of DIII pore module. This feature is also present in the EeNa_V1.4 channel, but here the LMF motif of the III–IV linker engages the corner interface of DIII VSD and DIV S6. Based on these structures, channel inactivation is proposed to be initiated by the LMF–linker association, which in turn causes the DIV S6 to move right-handedly and collapse the ion-conducting pathway at the lower gate. Importantly, the CTD and DIII–DIV linker are hotspots of disease variants, which implicates impairment of fast-inactivation in hyperexcitability of affected cell membranes as a common disease mechanism (e.g., forms of epilepsy) (Huang et al., 2017).

Eukaryotic $Na_V\beta$ Subunit Complex

In situ, Na_V channels form complexes with a small family of accessory proteins called β subunits, which are single-pass transmembrane proteins belonging to the immunoglobulin (Ig) domain superfamily of cell-adhesion molecules. Humans have four genes (*SCN1B–SCN4B*) encoding β1–β4 subunits. Although heterologous expression of any of the mammalian pore-forming subunit genes is sufficient to generate measurable Na^+ current, co-expression with β subunits enhances current density and shifts the voltage dependence of activation and inactivation to more negative membrane potentials (Makita, Bennett, & George, 1996; Patino & Isom, 2010; Qu et al., 1999), and as discussed in other sections of this book, may perform non-canonical roles as well as interact with other families of voltage-gated ion channels (see Zaleska et al., this volume). Prior to their structural determination, biochemical analysis of the β subunits determined that β2 and β4 form disulfide bonds with the pore-forming subunits (Yereddi et al., 2013). However, little else was understood regarding the sites of their association and the molecular regulation of Na_V function. In the aforementioned $EeNa_V1.4$-β1 structure, the β1 transmembrane helix interacts with the DIII S2 segment through hydrophobic interactions, while the β1 Ig domain interacts with the extracellular S1–S2 loop of DIII through five intermolecular charge–counter-charge interactions (Figure 8.2B). Thus, it seems that β1 regulates the voltage dependence of activation of $Na_V1.4$ by interactions with the DIII VSM. Future work should validate these putative intersubunit interactions and determine whether the transmembrane portion and/or the Ig motif are responsible for regulating channel function.

In isolation, human β3 and β4 subunit Ig domains have been crystalized at atomic resolution as trimers, which suggests that β subunits may coordinate multimers of Na_V channel complexes on the cell membrane (Gilchrist, Das, Van Petegem, & Bosmans, 2013; Namadurai et al., 2014). Thus, it possible that β subunits may be responsible for coordinating the membrane distribution of the Na_V alpha subunit in excitable cell membranes, as observed for $Na_V1.2$-β4 and $Na_V1.6$-β1 at the neuronal nodes of Ranvier; and $Na_V1.5$-β2 in the cardiac T-tubule system (Buffington & Rasband, 2013). As the interactome of Na_V channels is elucidated, one can envision a time when more protein complexes involving Na_V channels will emerge. Recently resolved macromolecular complexes such as the octameric exocyst (>750 kDa) (Mei et al., 2018) and ribosomal Sec-rRNA complexes (>900 kDa) (Frauenfeld et al., 2011) demonstrate the seemingly limitless ability of cryo-EM to undertake a more complete structural view of protein regulation. Thus, it seems that the future of Na_V structural biology resides in determining how these proteins associate with endogenous modifiers and how these associations impact electrical and chemical signaling within the cell.

Sodium Channels in the Central Nervous System

In excitable cells, action potential generation and propagation is governed by members of the Na_V channel family. Of the ten members of this ion channel family, four ($Na_V1.1$, $Na_V1.2$, $Na_V1.3$, and $Na_V1.6$) reside primarily within the central nervous system (Whitaker et al., 2000; Whitaker et al., 2001). As noted from the structural analysis discussed previously, these proteins have evolved to be highly sensitive to changes in membrane potential, and developed activation and inactivation processes that enable the high-frequency action potential firing associated with normal neuronal function (Catterall, 1992, 2000). Currents mediated by these channels are activated very rapidly, resulting in a large influx of sodium underlying the rising phase of the action potential. Under normal circumstances, channel activation is transient due to fast inactivation, which occurs on the sub-millisecond time scale. Inactivated channels cannot reopen until they recover from inactivation, and the duration of recovery is intimately linked to the refractory period, which limits the maximal firing rate of neuronal cells (Hodgkin & Huxley, 1952a). There is also a persistent component of sodium current that is essential for enhancing synaptic currents, but, as discussed later in this chapter, enhanced persistent current can have pathophysiological consequences (Crill, 1996). As discussed in other chapters of this book, voltage-gated sodium channels may also perform non-canonical functions, and neuronal isoforms are not expressed strictly within the central nervous system (see Zaleska et al., this volume).

Despite the fact that all neuronal Na_V channels perform similar functions, they do not share the same developmental expression patterns, cellular localizations, or interacting partners. While this information has been difficult to glean from human tissue, experiments from rodent brain have been informative. Experiments from rodents suggest that $Na_V1.3$ is the predominant Na_V channel in embryonic and early postnatal brain, with mRNA expression falling to nearly undetectable levels by postnatal day 30 (P30) (Felts, Yokoyama, Dib-Hajj, Black, & Waxman, 1997). However, $Na_V1.3$ expression has been detected in layer III pyramidal cells in the adult brain (Vacher, Mohapatra, & Trimmer, 2008). The drop in $Na_V1.3$ expression is coordinated with increasing expression of $Na_V1.1$ and $Na_V1.6$, both of which reach steady-state values between P15 and P39. $Na_V1.2$ maintains relatively high expression throughout development, but expression modestly rises from P0 to P9 (Gazina et al., 2010).

In addition to changes in overall expression levels during development, all neuronal Na_V channels undergo developmentally regulated alternative mRNA splicing, which leads to the incorporation of an alternate exon-encoding part of S3 and S4 of domain I (Copley, 2004; Kasai et al., 2001). Functionally, alternatively spliced Na_V channel proteoforms show differences in channel properties, with subtle differences in voltage dependence of activation or inactivation. For example, the neonatal $Na_V1.2$ proteoform exhibits faster inactivation compared to the adult proteoform, which is predicted to limit the excitability of immature neurons (Gazina et al., 2015; Gazina et al., 2010). Additionally, while rodent *Scn1a* does not show developmentally regulated alternative mRNA splicing, the human *SCN1A* gene undergoes alterative splicing at this exon that results in proteoforms with divergent sensitivity to commonly prescribed antiepileptic drugs such as phenytoin and lamotrigine (Thompson, Kahlig, & George, 2011).

As stated before, while these channels perform similar functions, they do not necessarily have overlapping cellular expression. $Na_V1.1$ is strongly localized to the axon initial segment (AIS) of parvalbumin and somatostatin-positive interneurons throughout the neocortex and hippocampus (Ogiwara et al., 2007; Westenbroek, Merrick, & Catterall, 1989). This channel is also strongly expressed in cerebellar Purkinje neurons (Kalume, Yu, Westenbroek, Scheuer, & Catterall, 2007). $Na_V1.2$ is expressed at the AIS and nodes of Ranvier in excitatory cells throughout the neocortex (Hu et al., 2009; Tian, Wang, Ke, Guo, & Shu, 2014). Early in development, $Na_V1.2$ may be found along the entire AIS, but it later becomes restricted to the proximal AIS and is replaced by $Na_V1.6$ at the remaining portions of the AIS and in the nodes of Ranvier (Liao, Deprez, et al., 2010). In addition, some evidence suggests that $Na_V1.2$ is expressed in unmyelinated mossy fibers of hippocampal granule cells and cerebellar granule cells along with $Na_V1.6$, as well as subpopulations of gamma-aminobutyric-acid releasing (GABAergic) interneurons (Gong, Rhodes, Bekele-Arcuri, & Trimmer, 1999; Hu et al., 2009; Miyazaki et al., 2014; Westenbroek et al., 1989; Yamagata, Ogiwara, Mazaki, Yanagawa, & Yamakawa, 2017). $Na_V1.3$ expression appears to be largely somatodendritic in both pyramidal cells and interneuronal populations (Vacher et al., 2008). Thus, mutations in these channels may selectively alter excitatory or inhibitory tone, as discussed later in this chapter.

In addition to the effects of β subunits, which modify the function and trafficking of Na_V channels, other proteins may modulate Na_V function or are necessary for subcellular localization. As stated, $Na_V1.1$, $Na_V1.2$, and $Na_V1.6$ have been shown to accumulate at the AIS. Guidance to the AIS is promoted by the cytoskeletal proteins ankyrin-G and βIV spectrin as well as FGF14 (Hund et al., 2010; Jenkins & Bennett, 2001; Pablo, Wang, Presby, & Pitt, 2016). In addition to promoting proper subcellular localization, FGF14 also modulates the biophysical properties of $Na_V1.2$ and 1.6, inducing a depolarized shift in voltage dependence of inactivation, which may enhance neuronal excitability (Laezza et al., 2009). FGF12 also promotes a depolarized shift in voltage dependence of inactivation for $Na_V1.2$, but not $Na_V1.1$ (Wang, Wang, Hoch, & Pitt, 2011). Recently, $Na_V1.2$ has been shown to be regulated by CaMKII. Phosphorylation of $Na_V1.2$ by CaMKII induces a depolarized shift in voltage dependence of inactivation and promotes a large increase in the amplitude of persistent sodium current, resulting in greater excitability of hippocampal pyramidal neurons (Thompson, Hawkins, Kearney, & George, 2017). Interestingly, βIV spectrin is important for correct localization of CaMKII to the AIS (Hund et al., 2010). Inactivation parameters of $Na_V1.2$ and $Na_V1.6$ channels are also independently modulated by calmodulin, presumably through interaction with the C-terminal IQ domain or the DIII–DIV linker (Pitt & Lee, 2016; Yan, Wang, Marx, & Pitt, 2017). This interaction suppresses persistent current mediated by these channels. These data suggest that intracellular calcium may regulate sodium channel activity at the AIS. Our knowledge of Na_V interacting partners continues to grow, and this will be a critical area of research for interrogation of Na_V channel modulation and potential non-canonical functions.

Sodium Channel Diseases of the Central Nervous System

While Na_V channels are essential for normal neuronal physiology, the genes encoding these channels have been implicated in many neurological diseases. Most often, mutations in these genes lead to disorders such as epilepsy, autism, and migraine, which

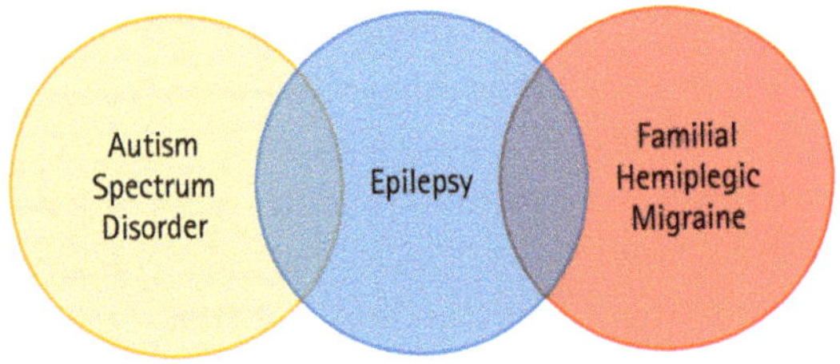

FIGURE 8.3 Sodium channelopathies of the central nervous system.
Venn diagram depicting comorbidity of sodium channelopathies of the central nervous system.

have overlapping clinical features (Figure 8.3). In this section, we will discuss the various diseases affecting the central nervous system that have been associated with genetic variants in Na_V channel genes. We will discuss genetic evidence and what we have learned regarding the cellular and network mechanisms underlying these diseases.

Much of the information about pathogenic functions of mutant Na_V channels has been gleaned from studies performed in heterologous systems, and it has become customary to assign classifications such as gain of function or loss of function based on these experiments. Therefore, it is important that we define these classifications and how these conclusions are made, based on changes in specific functional properties of Na_V channels. As one might expect, "gain of function" and "loss of function" are meant to describe changes in Na_V channel properties predicted to either accentuate or dampen Na^+ flux, some of which are depicted in Figure 8.4. Examples of functional effects consistent with gain of function include hyperpolarized voltage dependence of activation, depolarized voltage dependence of inactivation, slowed inactivation kinetics, faster recovery from inactivation, or enhanced persistent sodium current. In contrast, smaller whole-cell current density, depolarized voltage dependence of activation, hyperpolarized voltage dependence of inactivation, faster inactivation kinetics, and slowed recovery from inactivation would be consistent with loss of function. It is important to emphasize that these are only predictions of channel behavior, and the impact of these changes on cell autonomous excitability and on network activity depends upon many factors, including the neuron types expressing the mutant Na_V channels.

Epilepsy

Epilepsy is a neurological disorder affecting approximately 1% of the population. The condition is diagnosed when two or more unprovoked seizures of unknown cause occur (Hauser, Annegers, & Kurland, 1993; Helbig, Scheffer, Mulley, & Berkovic, 2008). The etiology of most epilepsy cases is without an identifiable cause and is thought to have an underlying genetic basis. Over the past decade, mutations in voltage-gated ion channel genes have emerged as a major cause of monogenic epilepsy syndromes, with variants in each predominantly brain-expressed Na_V channel genes having associations with one

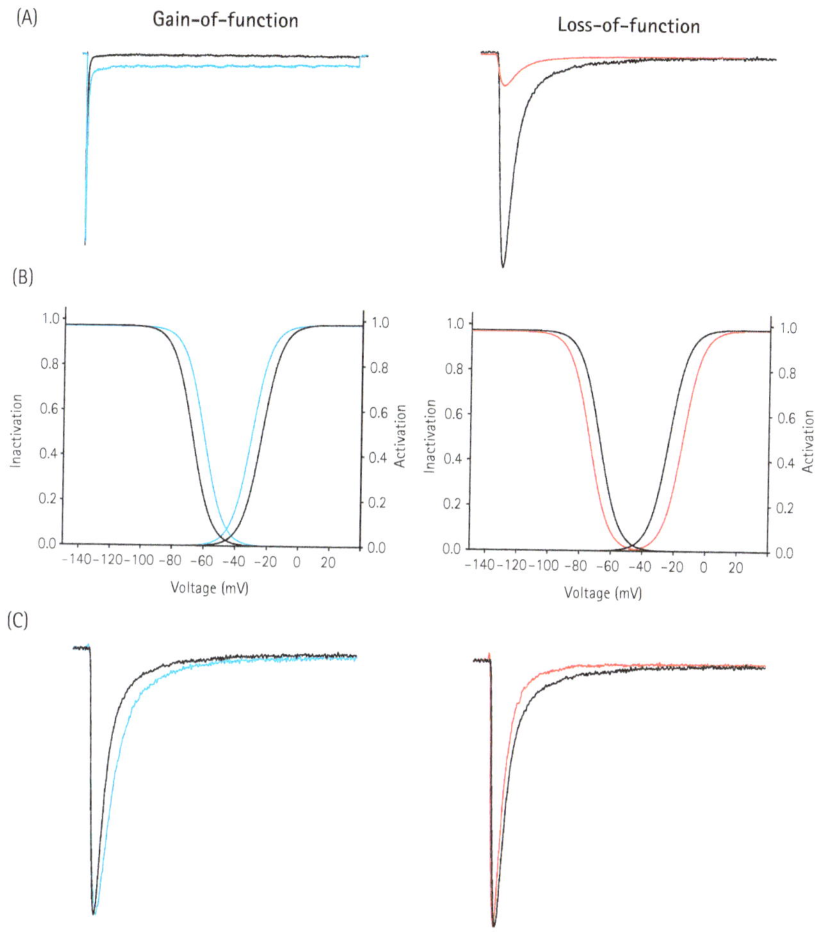

FIGURE 8.4 Biophysical properties of mutant Na_V channels.
Representative gain- and loss-of-function effects associated with (A) abnormal whole-cell current (enhanced persistent current or reduced current density); (B) altered voltage-dependent gating properties; and (C) altered channel inactivation kinetics.

or more syndromes. Recent reports have suggested that variants in *SCN1A*, *SCN2A*, and *SCN8A* account for approximately 40% of pathogenic cases (Butler, da Silva, Alexander, Hegde, & Escayg, 2017; Lindy et al., 2018; Mercimek-Mahmutoglu et al., 2015; Parrini et al., 2017). Sodium channel–related epilepsies range in severity from severe (e.g., Dravet syndrome, Ohtahara syndrome) to relatively benign (e.g., genetic epilepsy with febrile seizures plus benign familial neonatal seizures) (George, 2005; Kaplan, Isom, & Petrou, 2016). In the next section of this chapter, we will discuss various forms of Na_V channel–related epilepsies, with an emphasis on disease etiology, underlying functional and cellular mechanisms, and knowledge gleaned from advanced animal and cellular models.

Epileptic Encephalopathy

Dravet syndrome, first described in 1978, is an epileptic encephalopathy with onset during infancy and a poor prognosis (Dravet, 2011). Patients present with generalized tonic-clonic or hemiclonic seizures, often precipitated by fever, within the first year of life. Beginning in the second year of life, patients develop afebrile seizures and show significant developmental delays resulting in permanent cognitive impairment, ataxia, and motor dysfunction. Approximately 80% of Dravet syndrome patients have de novo heterozygous missense or truncating mutations in *SCN1A*, with currently more than 1400 variants identified (Dravet, 2011; Parihar & Ganesh, 2013; Steel, Symonds, Zuberi, & Brunklaus, 2017). Dravet syndrome is generally not responsive to treatment with most commonly prescribed antiepileptic drugs, and Na_V channel blocking drugs are widely considered contraindicated. However, seizure control has been achieved in some patients with the ketogenic diet (Brunklaus, Ellis, Reavey, Forbes, & Zuberi, 2012; Guerrini et al., 1998; Nabbout et al., 2013; Steel et al., 2017) and certain antiepileptic drug combinations (e.g., clobazam and stiripentol). Important advances in understanding the pathogenesis of Dravet syndrome using many experimental platforms have inspired new therapeutic strategies.

Whereas truncating variants are generally predicted to be complete loss-of-function, most missense variants in *SCN1A* have uncertain functional consequences. Utilizing heterologous expression, a large number of Dravet syndrome–associated $Na_V1.1$ missense mutations have been functionally evaluated. Despite the fact that Dravet syndrome mutations do not show a tendency to cluster within specific functional domains of the channel, most mutations exhibit severe loss of function compared to wild-type $Na_V1.1$, as depicted in Figure 8.5 (Claes et al., 2003; Claes et al., 2001; Ohmori, Kahlig, Rhodes, Wang, & George, 2006; Rhodes, Lossin, Vanoye, Wang, & George, 2004; Thompson, Porter, Kahlig, Daniels, & George, 2012). Additionally, many mutations exhibit impaired trafficking to the plasma membrane, suggesting that the mutations may evoke protein misfolding (Thompson et al., 2012).

Demonstrating loss of function for many mutations was an early clue that haploinsufficiency may be the primary mechanism responsible for Dravet syndrome, but this notion raised the question of how a loss of Na_V channel function results in hyperexcitable network activity and epilepsy. An answer was provided in 2006 when two groups independently showed that heterozygous loss of mouse *Scn1a* recapitulated many of the disease phenotypes associated with Dravet syndrome, including spontaneous generalized seizures, hyperthermia-induced seizures, ataxia, and premature death (Ogiwara et al., 2007; Yu et al., 2006). Whole-cell voltage clamp recording of acutely isolated hippocampal neurons showed lower sodium current density, specifically in GABAergic interneurons (Ogiwara et al., 2007; Yu et al., 2006). This loss of sodium current density resulted in blunted interneuron excitability, suggesting that Dravet syndrome arises from disinhibition of neuronal networks. Cell-type-specific deletion of *Scn1a* in parvalbumin- or somatostatin-positive interneurons was sufficient

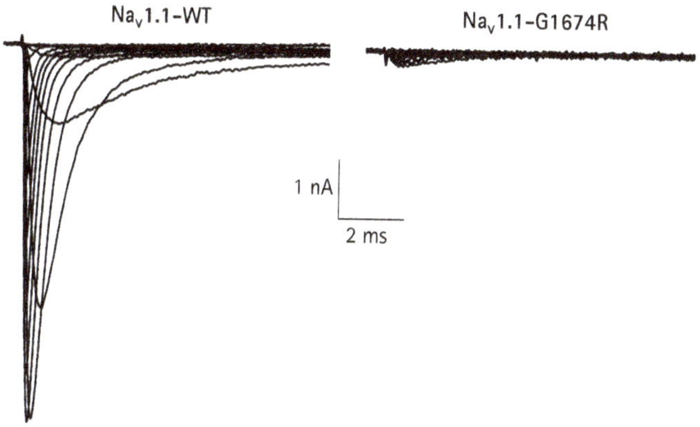

FIGURE 8.5 Loss of function of *SCN1A* leads to Dravet syndrome.
Whole-cell sodium currents from WT-Na$_V$1.1 and Dravet syndrome-associated mutations G1674R.

to recapitulate many of the behavioral and electrophysiological features identified in the global *Scn1a*$^{+/-}$ knockout (Cheah et al., 2012; Tai, Abe, Westenbroek, Scheuer, & Catterall, 2014). Interestingly, cerebellar Purkinje neurons from *Scn1a*$^{+/-}$ mice also exhibit smaller sodium currents and impaired excitability compared to neurons from wild-type animals, a likely explanation for ataxia observed in Dravet syndrome (Kalume et al., 2007). However, it is likely that defects in the basal ganglia also contribute to motor defects of these patients (Gataullina & Dulac, 2017). Other investigations of *Scn1a*$^{+/-}$ mice suggest that Dravet syndrome may not arise solely from impaired interneuron excitability. Importantly, lower interneuron sodium current density is observable at an age when mice are not observed to have seizures. In addition to finding lower interneuron sodium current density, these is evidence for higher sodium current density and greater excitability of hippocampal pyramidal cells from *Scn1a*$^{+/-}$ mice, potentially resulting from upregulation of Na$_V$1.6 (Anderson, Hawkins, Thompson, Kearney, & George, 2017; Mistry et al., 2014). This greater sodium current density in excitatory neurons exhibits a strong age dependence, with no difference observed in neurons from P14 mice, and a nearly twofold increase in current density compared to wild-type mice at age P21, which corresponds to the age when seizure frequency rises sharply. Unlike other forms of epilepsy, Dravet syndrome does not remit with age; therefore, an age-dependent increase in excitatory neuron excitability may be a critical contributor to the progression of Dravet syndrome.

Patient-derived induced pluripotent stem cells (iPSCs) have been used recently to interrogate the cellular and network mechanisms of Dravet syndrome. Early experiments examining excitability of glutamatergic neurons derived from *SCN1A* mutant Dravet syndrome iPSC lines showed greater excitability of these cells compared with neurons from control lines, consistent with the observation of increased excitatory neuron excitability (Liu et al., 2013). More recent work demonstrated reduced excitability of

GABAergic interneurons, but it has failed to recapitulate hyperexcitability of excitatory neurons (Jiao et al., 2013; Kim et al., 2018; Liu et al., 2016; Sun et al., 2016). Use of emerging technologies such as brain organoid cultures may offer opportunities to investigate additional behaviors of neural networks in Dravet syndrome.

Mutations in *SCN2A* are another major cause of epileptic encephalopathies, including Ohtahara syndrome, West syndrome, and epilepsy of infancy with migrating focal seizures. Patients with these syndromes have multiple seizure types, including focal seizures, tonic seizures, generalized tonic-clonic seizures, and infantile spasms. Seizures typically present in the first days or weeks of life, with some reports of seizures occurring on the day of birth (Howell et al., 2015; Nakamura et al., 2013; Wolff et al., 2017). Indeed, Ohtahara syndrome has the earliest age of onset among infantile epileptic encephalopathy associated with Na_V channel mutations. In addition to seizures, many *SCN2A*-related epileptic encephalopathy patients also have significant neurodevelopmental and intellectual disabilities.

Investigations of the electrophysiological properties of *SCN2A* mutations associated with early onset epilepsy have suggested gain of function as a common pathophysiological mechanism. For example, E1211K (see Fig. 1), which is associated with intractable childhood epilepsy, has been shown to evoke a strongly hyperpolarized voltage dependence of activation (>10 mV) and a depolarized shift in inactivation (<20 mV) (Ogiwara et al., 2009). Computational modeling of this and one other *SCN2A* mutation (L1473M) demonstrated hyperexcitability of both immature neurons, where $Na_V1.2$ is the sole axonal Na_V channel, and mature neurons, where $Na_V1.2$ is restricted to the proximal AIS (Ben-Shalom et al., 2017). Another mutation, A263V, has been shown to have enhanced persistent current, slower inactivation kinetics, and a depolarized shift in the voltage dependence of inactivation, all of which are consistent with promoting a hyperexcitable state (Liao, Anttonen, et al., 2010).

A knock-in mouse model bearing the A263V variant in the *Scn2a* gene is the only mouse model mimicking a human *SCN2A* epileptic encephalopathy. Interestingly, while heterozygous mice do not show a severe seizure phenotype, they do have altered neuronal excitability. Because $Na_V1.2$ is largely expressed in excitatory neurons, one can predict that gain-of-function defects in $Na_V1.2$ will cause increased excitability in these cells. Indeed, whole-cell current clamp recording in brain slices from *Scn2a*$^{A263V+/-}$ mice show hyperexcitability of CA1 pyramidal neurons (Schattling et al., 2016). Interestingly, when A263V mice are bred to homozygosity, the resulting mice have frequent spontaneous seizures and premature death. Unlike Dravet syndrome, many patients with *SCN2A*-related epileptic encephalopathies respond well to sodium channel blocking antiepileptic drugs, including phenytoin, lamotrigine, and oxcarbazepine, as would be expected from heighted activity of excitatory neurons. However, some evidence suggests that Na_V channel blocking drugs are more effective in early onset epilepsy patients, whereas later onset epilepsies exhibit resistance to these drugs. This may reflect diminished expression of $Na_V1.2$ in excitatory neurons as neuronal development progresses, with later onset epilepsy being driven by loss of $Na_V1.2$ function in interneurons (Wolff et al., 2017).

A third severe form of epilepsy termed "early infantile epileptic encephalopathy type 13" (EIEE13) is associated with mutations in *SCN8A* encoding $Na_V1.6$ (Wagnon & Meisler, 2015). Most patients with EIEE13 have generalized seizures with onset prior to 18 months of age. Following seizure onset, developmental regression occurs, with mild to severe intellectual disability. EIEE13 also typically presents with movement disorders such as mild ataxia, choreoathetosis, and occasionally quadriplegia (Lopez-Santiago et al., 2017; McNally et al., 2016; Meisler et al., 2016; Wagnon et al., 2016; Wagnon & Meisler, 2015). Importantly, most patients with EIEE13 have treatment-refractory seizures, but approximately 40% of patients respond well to antiepileptic drugs that target Na_V channels. While the number of variants associated with EIEE13 is growing, functional data are available for only a few of these variants (Barker et al., 2016; de Kovel et al., 2014; Estacion et al., 2014; Patel, Barbosa, Brustovetsky, Brustovetsky, & Cummins, 2016). Due to overlapping expression with $Na_V1.2$, one can predict that epilepsy in patients carrying $Na_V1.6$ mutations may result from overactive excitatory neurons. Functional analysis has revealed that $Na_V1.6$ variants associated with EIEE13 show primarily gain-of-function effects, including enhanced persistent current and slower entry into inactivated states. A mouse model of a prototypical $Na_V1.6$ mutation, N1768D, exhibits infrequent spontaneous seizures and premature death (Lopez-Santiago et al., 2017; Wagnon et al., 2015). However, unlike in humans, seizures do not have early onset in the mice. Brain slice recording performed at P20 shows evidence of increased neuronal excitability, including a greater number of action potentials per degree of electrical stimulation and the occurrence of early afterdepolarizations, which are thought to be derived from the large persistent currents associated with this variant. Like the *Scn2a*^{A263V} model, *Scn8a*^{N1768D} mice show a much more severe phenotype when bred to homozygosity, with 100% mortality by three weeks of age.

Finally, de novo *SCN3A* variants were identified in a small cohort of patients diagnosed with an early onset epileptic encephalopathy. Mutations in $Na_V1.3$ have previously been reported in less severe forms of epilepsy. However, these patients show early onset seizures within the first 18 months of life, with focal, tonic, generalized tonic-clonic, or myoclonic seizures. All patients also show significant developmental delay (Zaman et al., 2018). Functional analysis revealed predominantly gain-of-function phenotypes for these variants, including hyperpolarized voltage dependence of inactivation and large, persistent sodium currents (Zaman et al., 2018). While it is too early to draw definitive conclusions regarding the mechanism of *SCN3A*-related epileptogenesis, one can speculate that *SCN3A*-related epileptic encephalopathies arise from abnormal excitability during embryonic development.

While each epileptic encephalopathy stems from mutations in different Na_V channels, and the mechanisms underlying epileptogenesis are likely to be different, some common conclusions may be drawn. First, variants that dramatically alter channel function drastically alter neuronal excitability, resulting in either heightened excitability of excitatory neurons or dampened excitability of inhibitory neurons; either mechanism capable of promoting hyperexcitability of neuronal networks. Secondly, this altered network excitability early in development induces long-term impairment of neuronal development that results in severe cognitive disability. This long-term impairment of cognitive development, or in some cases cognitive decline, distinguishes severe epileptic encephalopathies from more benign syndromes that are discussed in the next section.

Other Monogenic Epilepsy Syndromes

In addition to severe epileptic encephalopathy, mutations in *SCN1A*, *SCN2A*, and *SCN3A* are also associated with more benign forms of epilepsy, including genetic epilepsy with febrile seizures plus (GEFS+; *SCN1A*), benign familial neonatal infantile seizures (BFNIS; *SCN2A*) and cryptogenic focal epilepsy (*SCN3A*). Unlike epileptic encephalopathies, patients with these type of epilepsies do not generally have persistent cognitive deficits and in some cases show seizure remission with age. Also, these disorders tend to exhibit genetic transmission and classic Mendelian inheritance (typically autosomal-dominant), as opposed to de novo mutations found in most cases of epileptic encephalopathy.

GEFS+ is an autosomal-dominant familial epilepsy syndrome with febrile seizures presenting in early childhood that often persist beyond six years of age. There is strong phenotypical variability, with clinical phenotypes ranging from only mild febrile seizures to afebrile focal or generalized seizures, including atonic, myoclonic, or absence seizure types (Berkovic & Scheffer, 2001; Scheffer & Berkovic, 1997). This phenotypical variability is further demonstrated within families, wherein family members carrying the same mutations may be asymptomatic, have febrile seizures, or present with febrile seizures and generalized epilepsy (Escayg et al., 2000). GEFS+ typically arises from missense mutations in *SCN1A* (Brunklaus, Ellis, Reavey, Semsarian, & Zuberi, 2014). Unlike Dravet syndrome, which primarily results from loss of function of $Na_V1.1$, GEFS+ mutations typically show more subtle biophysical defects when measured in heterologous cells. These include enhanced persistent sodium current (R1648H), depolarized shifts in voltage dependence of activation (R859C, I1656M, R1657C), or hyperpolarized shifts in voltage dependence of inactivation (R1916G) (Barela et al., 2006; Lossin et al., 2003; Rusconi et al., 2009; Vanoye, Lossin, Rhodes, & George, 2006). Some mutations also show either moderate or complete loss of whole-cell sodium currents, including V1353L, A1685V, and R1916G (Vanoye et al., 2006; Barela et al., 2006; Bechi et al., 2015; Lossin et al., 2003; Rhodes et al., 2005; Volkers et al., 2011). Computational modeling of the R859C mutation predicted reduced neuronal excitability (Barela et al., 2006). This suggests that GEF+ may arise by cellular mechanisms similar to Dravet syndrome, but with less devastating progression. While it appears that both GEFS+ and Dravet syndrome may arise from similar, but not identical, network mechanisms, one can speculate that complete loss of one functional $Na_V1.1$ allele in Dravet syndrome may result in a more severe phenotype.

A knock-in mouse model of GEFS+, bearing a mutation analogous to human R1648H (*Scn1a*$^{R1648H+/-}$), exhibits infrequent spontaneous seizures and reduced seizure threshold to hyperthermia or chemoconvulsant exposure (Martin et al., 2010). Acutely dissociated bipolar neurons from animals homozygous for the mutation show reduced excitability compared to both wild-type and heterozygous animals (Martin et al., 2010). In addition, inhibitory neurons within the thalamic nucleus reticularis (nRt), fast-spiking cortical layer IV interneurons, and CA1 hippocampal *stratum oriens* interneurons showed impaired excitability, while excitatory neurons were unaffected (Hedrich et al., 2014). Additionally, these animals exhibited disrupted GABA-mediated neurotransmission. Interestingly, if animals were exposed to a prolonged febrile event prior to slice recording, CA3 pyramidal neurons showed greater excitability compared to wild-type or *Scn1a*$^{R1648H+/-}$ animals that had not been exposed to a febrile event (Dutton et al., 2017). This "priming" with a febrile seizure has also been shown to increase spontaneous seizure frequency in *Scn1a*$^{+/-}$ mice, and this experimental strategy may represent a more clinically relevant paradigm for studying these animals (Hawkins et al., 2017). Oddly, voltage clamp recording of nucleated patches of neurons from *Scn1a*$^{R1648H+/-}$ mice did not exhibit enhanced persistent current compared to wild-type neurons and instead showed currents nearly indistinguishable from wild-type neurons, with only a slight slowing of recovery from inactivation. (Hedrich et al., 2014; Martin et al., 2010). This may be due to two possibilities: (1) rundown of persistent current during measurements, or (2) the same variant in mouse and human $Na_V1.1$ behaves differently. The second possibility is intriguing, and not without precedent. The most common mutation in *CFTR* , the gene associated with cystic fibrosis in the United States (ΔF508) produces a protein that does not properly traffic to the plasma membrane in humans, but traffics and functions properly in rodents (Ostedgaard et al., 2007).

Variants in both *SCN2A* and *SCN3A* are also associated with less severe epilepsy phenotypes, including BFNIS (*SCN2A*), and cryptogenic focal epilepsy (*SCN3A*) (Wolff et al., 2017; Vanoye, Gurnett, Holland, George, & Kearney, 2014). Functional evaluation of BFNIS-associated $Na_V1.2$ variants has shown a mix of gain- and loss-of-function effects, including enhanced persistent current (M252V), depolarized voltage dependence of inactivation (L1563V), or impaired cell surface expression (L1330M) (Liao, Deprez, et al., 2010; Misra, Kahlig, & George, 2008; Scalmani et al., 2006; Schwarz et al., 2016; Xu et al., 2007). Interestingly, both M252V and L1563V showed biophysical defects only when expressed in the context of the neonatal $Na_V1.2$ alternatively spliced proteoform (Liao, Deprez, et al., 2010; Xu et al., 2007). This suggests that developmentally regulated mRNA splicing of $Na_V1.2$, as well as a developmental switch from $Na_V1.2$ to $Na_V1.6$ in the AIS and nodes of Ranvier, may critically contribute to seizure remission in some patients. While no mouse models have been developed that mimic BFNIS, computational modeling of neurons with L1330M (L1153 in Figure 8.2) shows modestly increased neuronal excitability that normalizes as the neuron matures (Ben-Shalom et al., 2017). Thus far, it is difficult to make conclusions regarding the mechanisms underlying benign epilepsy associated with variants in *SCN3A*. Electrophysiological data show both gain and loss of function, though enhanced persistent sodium current appears to be a common feature (Estacion, Gasser, Dib-Hajj, & Waxman, 2010; Vanoye et al., 2014). Transfection of mouse hippocampal neurons with a gain-of-function *SCN3A* variant, K354Q, increases neuronal excitability (Estacion et al., 2010). However, mice deficient in *Scn3a* show greater susceptibility to electroconvulsive and chemiconvulsive-induced seizures than wild-type littermates (Lamar et al., 2017).

Although neuronal Na_V channels perform similar functions, they do so in specific cell populations. Thus mutations in different Na_V channels give rise to epilepsy phenotypes that vary in age of onset, seizure types, and therapeutic outcome. Epilepsy is a major source of disease burden in the United States, and continued research is needed to understand the nature of epileptogenesis so that improvements can be made in currently available treatments.

Autism Spectrum Disorder

Autism spectrum disorders (ASD) are a group of lifelong neurodevelopmental disorders with childhood onset. These disorders are associated with impaired social and communication skills and are characterized by ritualistic, repetitive, and rigid behavioral patterns. ASD has a population prevalence of 1 in 68 individuals in the United States with a male preponderance of 4 to 1 (Srivastava & Sahin, 2017; Weiss et al., 2003; Woodbury-Smith & Scherer, 2018). This represents a stunning rise in prevalence from historical rates of 4/10,000, although it is unknown if this is due to greater awareness, improvements in diagnostic criteria, or increases in environmental factors associated with ASD. ASD is typically diagnosed within the first three years of life, which is a critical period of neurodevelopment when neurite outgrowth and synaptogenesis occur. Post-mortem analysis of ASD brains has demonstrated neuroanatomical abnormalities and cellular-level differences in dendritic branching and spine density (Lin, Frei, Kilander, Shen, & Blatt, 2016).

It is widely accepted that the etiology of ASD has a very strong genetic component, with some estimates of heritability exceeding 90% (Weiss et al., 2003; Woodbury-Smith & Scherer, 2018). While ASD is generally considered to be a complex genetic trait, recent population-based gene discovery efforts have identified several de novo variants associated with the disorder (Codina-Sola et al., 2015; D'Gama et al., 2015; O'Roak et al., 2011; Sanders et al., 2012; Wang et al., 2016). Among the genes identified, *SCN2A* has been strongly associated with ASD. Interestingly, a significant percentage of ASD patients, 9–22% by some estimates, has comorbid epilepsy (Strasser, Downes, Kung, Cross, & De Haan, 2018).

Voltage-gated sodium channels were first identified as ASD risk genes in 2003 in a study that found missense or in-frame deletion variants in *SCN1A*, *SCN2A*, and *SCN3A* (Weiss et al., 2003). Interestingly, one variant in *SCN2A* (R1902C) affects a region of the $Na_V1.2$ protein important for calmodulin binding. Analysis of calmodulin-binding affinity to a synthetic peptide encoding either the wildtype (WT) or variant protein sequence showed that the variant peptide exhibits twofold lower affinity for calmodulin. Interestingly, in the presence of 10 μM Ca^{2+}, R1902C showed a hyperpolarized shift in voltage dependence of activation and a depolarized shift of voltage-dependence of inactivation, while WT channel function was not modulated by Ca^{2+} (Wang et al., 2014). Functional analysis of a number of other $Na_V1.2$ variants, including T1420M (equivalent to T1242 in the $Na_V1.4$ structure), depicted in Figure 8.2, demonstrated smaller whole-cell sodium currents when measured in heterologous systems (Ben-Shalom et al., 2017). This loss of function was explained by either a loss of cell surface expression or a functionally defective channel. Computational modeling of neurons carrying these variants showed that in simulated immature cortical excitatory neurons, mutant $Na_V1.2$ channels caused impaired neuronal excitability. This is in contrast to epilepsy-associated mutations in $Na_V1.2$, which showed greater neuronal excitability in this model neuron. Importantly, in mature neurons, where $Na_V1.6$ replaces $Na_V1.2$ as the major Na_V channel in the proximal AIS, ASD-associated $Na_V1.2$ variants have no impact on neuronal excitability (Ben-Shalom et al., 2017). This led to the hypothesis that loss of $Na_V1.2$ function early in development, associated with impaired neuronal excitability, may be a critical pathophysiological mechanism in ASD. Interestingly, hetero- or homozygous knockout of mouse *Scn2a* does not produce an epilepsy phenotype (Planells-Cases et al., 2000). Thus, a critical test of this hypothesis will be to determine if *Scn2a*-deficient mice recapitulate any of the behavioral features associated with ASD.

As discussed, ASD is a commonly observed comorbidity of various epilepsy syndromes, with estimates of comorbid ASD in Dravet syndrome ranging 24–47% (Strasser et al., 2018). One proposed mechanism for development of ASD is an increased excitation:inhibition ratio stemming from a reduction in key inhibitory pathways in the neocortex (Chao et al., 2010; Coghlan et al., 2012; Rubenstein & Merzenich, 2003). Consistent with this hypothesis, *Scn1a*$^{+/-}$ mice show many autism spectrum behaviors, in addition to the well-documented seizure/premature death phenotype. Specifically, *Scn1a*$^{+/-}$ mice exhibit hyperactivity, anxiety, and deficits in social interactions compared to wild-type mice (Han et al., 2012; Ito et al., 2012; Tatsukawa, Ogiwara, Mazaki, Shimohata, & Yamakawa, 2018). Conditional deletion of *Scn1a* in forebrain parvalbumin-positive interneurons recapitulated the autistic-like traits of the global *Scn1a*$^{+/-}$ model, suggesting that parvalbumin-positive interneurons may be key contributors to the pathogenesis of ASD in this model (Tatsukawa et al., 2018). Furthermore, ASD phenotypes in *Scn1a*$^{+/-}$ mice could be rescued either by enhancement of GABAergic neurotransmission by treatment with low-dose clonazepam, a positive allosteric modulator of $GABA_A$-receptors, or by treatment with cannabidiol (Han et al., 2012). Thus, ASD associated with mutations in Na_V channels may result from a disrupted excitation/inhibition balance, but further work dissecting specific mechanisms of *SCN1A*- and *SCN2A*-related ASD will provide valuable insight into ASD disease progression and inform therapeutic decision making.

Familial Hemiplegic Migraine

Migraine is a complex, paroxysmal neurovascular disorder that is characterized by recurring severe headaches lasting one to three days. Migraine has a lifetime incidence of 14–18% worldwide, with more than 30 million migraine suffers in the United States (Lipton et al., 2007; Sutherland & Griffiths, 2017). Approximately 35% of all migraine suffers have aura preceding the headache. The transient neurological symptoms associated with aura correspond to cortical spreading depression, a wave of neuronal depolarization that slowly spreads across the cerebral cortex, generating transient periods of intense neuronal activity followed by long-lasting suppression (Mantegazza & Cestele, 2018). Although migraine is most likely a complex genetic trait, there are monogenic forms. Variants in *SCN1A* have been identified in patients diagnosed with familial hemiplegic migraine type 3 (FHM3), a rare autosomal-dominant form of migraine with aura characterized by transient motor weakness, visual aura, and speech disturbances.

To date, only 10 *SCN1A* mutations have been identified in FHM3 families. Like $Na_V1.1$ mutations associated with mild forms of epilepsy, functional evaluation of FHM3 variants has yielded a constellation of biophysical phenotypes, ranging from reduced cell surface expression (Q1489K) to alterations in channel activation or inactivation properties. Some variants, such as L263V and Q1489K, have altered, albeit divergent, inactivation properties compared to wild-type $Na_V1.1$ (Cestele et al., 2008; Kahlig et al., 2008). While L263V shows slower inactivation rate, depolarized voltage dependence of activation, and reduced use-dependent channel rundown, Q1489K exhibits faster inactivation and enhanced use-dependent channel rundown (Kahlig et al., 2008). Still a third variant, T1174S, exhibits depolarized voltage dependence of activation, while both L263V and Q1489K have hyperpolarized activation compared to wild-type channels (Cestele, Labate, et al., 2013; Kahlig et al., 2008). Enhanced persistent sodium current is shared among many of the FHM3-associated mutations. Current clamp recording of neurons transfected with Q1489K, an FHM3 mutation with increased persistent sodium current and faster inactivation kinetics, showed a unique excitability profile. These experiments showed self-limiting hyperexcitability, whereby neurons initially showed increased excitability with weak stimulation, but as stimulation strength increased, excitability converged with wild-type levels (Cestele et al., 2008). Mutant L1649Q also promoted neuronal hyperexcitability (Cestele, Schiavon, Rusconi, Franceschetti, & Mantegazza, 2013). Thus, while epilepsy-associated variants in *SCN1A* would generally be predicted to reduced interneuron excitability, this variant showed an initial increase in excitability. Work on FHM1 and FHM2, which stem from mutations in *CACNA1A* encoding the $Ca_V2.1$ voltage-gated calcium channel and *ATP1A2* encoding a glial-expressed isoform of the Na/K-ATPase, respectively, have suggested that increased intracellular calcium and neurotransmitter release promote a lower threshold for cortical spreading depression (Kahlig et al., 2008; Sutherland & Griffiths, 2017). Future work should be devoted to interrogating the functional and molecular effects of these mutations in order to determine how mutations in the same gene, *SCN1A*, can result in two different disorders such as migraine and epilepsy.

Summary

Over the past few decades, research has provided critical insights into the structure of Na_V channels and their function within the central nervous system. Even though we have gained much information into the structural states of the channel, researchers have yet to resolve the closed state of the channel, which may be a critical factor for structure-based drug design. Also, future research should be directed towards understanding the dynamics of structural rearrangements in response to physiological stimuli. It will also be important to devote research efforts to more fully understanding the interactions between Na_V channels and their many interacting partners, and determining how these interactions modulate channel function and localization. Sodium channel diseases within the central nervous system significantly overlap. It is clear that variants in neuronal sodium channel that can lead to epilepsy, autism, and migraine may arise from either gain- or loss-of-function effects, and that disruption of the excitatory/inhibitory balance at key stages in development is a critical factor in pathogenesis. However, we do not fully understand how variants with similar functional defects may result in more or less severe forms of epilepsy, for example. Much of the work devoted to cataloging functional defects in sodium channels has been performed in heterologous cells, and future work needs to be devoted to understanding channel dysfunction in native neuronal cells. Future efforts will be needed to determine what factors guide neuronal development down specific pathophysiological pathways. We need to more fully understand which, and when, specific networks are affected in these disorders. Only then can effective treatment strategies be devised.

REFERENCES

Anderson, L. L., Hawkins, N. A., Thompson, C. H., Kearney, J. A., & George, A. L., Jr. (2017). Unexpected efficacy of a novel sodium channel modulator in Dravet syndrome. *Scientific Reports*, *7*(1), 1682. doi:10.1038/s41598-017-01851-9

Anderson, P. A., & Greenberg, R. M. (2001). Phylogeny of ion channels: clues to structure and function. *Comparative Biochemistry and Physiology Part B: Biochemistry and Molecular Biology*, *129*(1), 17–28.

Armstrong, C. M., & Bezanilla, F. (1974). Charge movement associated with the opening and closing of the activation gates of the Na channels. *Journal of General Physiology*, *63*(5), 533–552.

Barela, A. J., Waddy, S. P., Lickfett, J. G., Hunter, J., Anido, A., Helmers, S. L., . . . Escayg, A. (2006). An epilepsy mutation in the sodium channel *SCN1A* that decreases channel excitability. *Journal of Neuroscience*, *26*(10), 2714–2723.

Barker, B. S., Ottolini, M., Wagnon, J. L., Hollander, R. M., Meisler, M. H., & Patel, M. K. (2016). The *SCN8A* encephalopathy mutation p.Ile1327Val displays elevated sensitivity to the anticonvulsant phenytoin. *Epilepsia*, *57*(9), 1458–1466. doi:10.1111/epi.13461

Bechi, G., Rusconi, R., Cestele, S., Striano, P., Franceschetti, S., & Mantegazza, M. (2015). Rescuable folding defective NaV1.1 (*SCN1A*) mutants in epilepsy: properties, occurrence, and novel rescuing strategy with peptides targeted to the endoplasmic reticulum. *Neurobiology of Disease*, *75*, 100–114. doi:10.1016/j.nbd.2014.12.028

Ben-Shalom, R., Keeshen, C. M., Berrios, K. N., An, J. Y., Sanders, S. J., & Bender, K. J. (2017). Opposing effects on NaV1.2 function underlie differences between *SCN2A* variants observed in individuals with autism spectrum disorder or infantile seizures. *Biological Psychiatry*, *82*(3), 224–232. doi:10.1016/j.biopsych.2017.01.009

Berkovic, S. F., & Scheffer, I. E. (2001). Genetics of the epilepsies. *Epilepsia*, *42*(Suppl 5), 16–23.

Brunklaus, A., Ellis, R., Reavey, E., Forbes, G. H., & Zuberi, S. M. (2012). Prognostic, clinical and demographic features in *SCN1A* mutation-positive Dravet syndrome. *Brain*, *135*(Pt 8), 2329–2336. doi:10.1093/brain/aws151

Brunklaus, A., Ellis, R., Reavey, E., Semsarian, C., & Zuberi, S. M. (2014). Genotype phenotype associations across the voltage-gated sodium channel family. *Journal of Medical Genetics*, *51*(10), 650–658. doi:10.1136/jmedgenet-2014-102608

Buffington, S. A., & Rasband, M. N. (2013). Na+ channel-dependent recruitment of Navbeta4 to axon initial segments and nodes of Ranvier. *Journal of Neuroscience*, *33*(14), 6191–6202. doi:10.1523/JNEUROSCI.4051-12.2013

Butler, K. M., da Silva, C., Alexander, J. J., Hegde, M., & Escayg, A. (2017). Diagnostic yield from 339 epilepsy patients screened on a clinical gene panel. *Pediatric Neurology*, *77*, 61–66. doi:10.1016/j.pediatrneurol.2017.09.003

Cai, X. (2012). Ancient origin of four-domain voltage-gated Na+ channels predates the divergence of animals and fungi. *Journal of Membrane Biology*, *245*(2), 117–123. doi:10.1007/s00232-012-9415-9

Capes, D. L., Goldschen-Ohm, M. P., Arcisio-Miranda, M., Bezanilla, F., & Chanda, B. (2013). Domain IV voltage-sensor movement is both sufficient and rate limiting for fast inactivation in sodium channels. *Journal of General Physiology*, *142*(2), 101–112. doi:10.1085/jgp.201310998

Catterall, W. A. (1992). Cellular and molecular biology of voltage-gated sodium channels. *Physiological Review*, *72*, S15–S48.

Catterall, W. A. (2000). From ionic currents to molecular mechanisms: the structure and function of voltage-gated sodium channels. *Neuron*, *26*(1), 13–25.

Cestele, S., Labate, A., Rusconi, R., Tarantino, P., Mumoli, L., Franceschetti, S., . . . Gambardella, A. (2013). Divergent effects of the T1174S *SCN1A* mutation associated with seizures and hemiplegic migraine. *Epilepsia*, *54*(5), 927–935. doi:10.1111/epi.12123

Cestele, S., Scalmani, P., Rusconi, R., Terragni, B., Franceschetti, S., & Mantegazza, M. (2008). Self-limited hyperexcitability: functional effect of a familial hemiplegic migraine mutation of the Nav1.1 (*SCN1A*) Na+ channel. *Journal of Neuroscience*, *28*(29), 7273–7283.

Cestele, S., Schiavon, E., Rusconi, R., Franceschetti, S., & Mantegazza, M. (2013). Nonfunctional NaV1.1 familial hemiplegic migraine mutant transformed into gain of function by partial rescue of folding defects. *Proceedings of the National Academy of Sciences of the United States of America*, *110*(43), 17546–17551. doi:10.1073/pnas.1309827110

Chao, H. T., Chen, H., Samaco, R. C., Xue, M., Chahrour, M., Yoo, J., . . . Zoghbi, H. Y. (2010). Dysfunction in GABA signalling mediates autism-like stereotypies and Rett syndrome phenotypes. *Nature*, *468*(7321), 263–269. doi:10.1038/nature09582

Cheah, C. S., Yu, F. H., Westenbroek, R. E., Kalume, F. K., Oakley, J. C., Potter, G. B., . . . Catterall, W. A. (2012). Specific deletion of NaV1.1 sodium channels in inhibitory interneurons causes seizures and premature death in a mouse model of Dravet syndrome. *Proceedings of the National Academy of Sciences of the United States of America*, *109*(36), 14646–14651. doi:10.1073/pnas.1211591109

Claes, L., Ceulemans, B., Audenaert, D., Smets, K., Lofgren, A., Del Favero, J., . . . De Jonghe, P. (2003). De novo *SCN1A* mutations are a major cause of severe myoclonic epilepsy of infancy. *Human Mutation*, *21*(6), 615–621.

Claes, L., Del Favero, J., Ceulemans, B., Lagae, L., Van Broeckhoven, C., & De Jonghe, P. (2001). De novo mutations in the sodium-channel gene *SCN1A* cause severe myoclonic epilepsy of infancy. *American Journal of Human Genetics*, *68*(6), 1327–1332.

Codina-Sola, M., Rodriguez-Santiago, B., Homs, A., Santoyo, J., Rigau, M., Aznar-Lain, G., ... Cusco, I. (2015). Integrated analysis of whole-exome sequencing and transcriptome profiling in males with autism spectrum disorders. *Molecular Autism*, *6*, 21. doi:10.1186/s13229-015-0017-0

Coghlan, S., Horder, J., Inkster, B., Mendez, M. A., Murphy, D. G., & Nutt, D. J. (2012). GABA system dysfunction in autism and related disorders: from synapse to symptoms. *Neuroscience & Biobehavioral Reviews*, *36*(9), 2044–2055. doi:10.1016/j.neubiorev.2012.07.005

Copley, R. R. (2004). Evolutionary convergence of alternative splicing in ion channels. *Trends in Genetics*, *20*(4), 171–176.

Crill, W. E. (1996). Persistent sodium current in mammalian central neurons. *Annual Review of Physiology*, *58*, 349–362. doi:10.1146/annurev.ph.58.030196.002025

D'Gama, A. M., Pochareddy, S., Li, M., Jamuar, S. S., Reiff, R. E., Lam, A. N., ... Walsh, C. A. (2015). Targeted DNA sequencing from autism spectrum disorder brains implicates multiple genetic mechanisms. *Neuron*, *88*(5), 910–917. doi:10.1016/j.neuron.2015.11.009

de Kovel, C. G., Meisler, M. H., Brilstra, E. H., van Berkestijn, F. M., van 't Slot, R., van Lieshout, S., ... Koeleman, B. P. (2014). Characterization of a de novo *SCN8A* mutation in a patient with epileptic encephalopathy. *Epilepsy Research*, *108* (9), 1511–1518. doi:10.1016/j.eplepsyres.2014.08.020

DeCaen, P. G., Takahashi, Y., Krulwich, T. A., Ito, M., & Clapham, D. E. (2014). Ionic selectivity and thermal adaptations within the voltage-gated sodium channel family of alkaliphilic Bacillus. *Elife*, *3*. doi:10.7554/eLife.04387

DeCaen, P. G., Yarov-Yarovoy, V., Scheuer, T., & Catterall, W. A. (2011). Gating charge interactions with the S1 segment during activation of a Na+ channel voltage sensor. *Proceedings of the National Academy of Sciences of the United States of America*, *108*(46), 18825–18830. doi:10.1073/pnas.1116449108

DeCaen, P. G., Yarov-Yarovoy, V., Sharp, E. M., Scheuer, T., & Catterall, W. A. (2009). Sequential formation of ion pairs during activation of a sodium channel voltage sensor. *Proceedings of the National Academy of Sciences of the United States of America*, *106*(52), 22498–22503. doi:10.1073/pnas.0912307106

DeCaen, P. G., Yarov-Yarovoy, V., Zhao, Y., Scheuer, T., & Catterall, W. A. (2008). Disulfide locking a sodium channel voltage sensor reveals ion pair formation during activation. *Proceedings of the National Academy of Sciences of the United States of America*, *105*(39), 15142–15147. doi:10.1073/pnas.0806486105

Dravet, C. (2011). The core Dravet syndrome phenotype. *Epilepsia*, *52*(Suppl 2), 3–9. doi:10.1111/j.1528-1167.2011.02994.x

Dutton, S. B. B., Dutt, K., Papale, L. A., Helmers, S., Goldin, A. L., & Escayg, A. (2017). Early-life febrile seizures worsen adult phenotypes in *Scn1a* mutants. *Experimental Neurology*, *293*, 159–171. doi:10.1016/j.expneurol.2017.03.026

Escayg, A., MacDonald, B. T., Meisler, M. H., Baulac, S., Huberfeld, G., An-Gourfinkel, I., ... Malafosse, A. (2000). Mutations of *SCN1A*, encoding a neuronal sodium channel, in two families with GEFS+2. *Nature Genetics*, *24*(4), 343–345.

Estacion, M., Gasser, A., Dib-Hajj, S. D., & Waxman, S. G. (2010). A sodium channel mutation linked to epilepsy increases ramp and persistent current of Nav1.3 and induces hyperexcitability in hippocampal neurons. *Experimental Neurology*, *224*(2), 362–368. doi:10.1016/j.expneurol.2010.04.012

Estacion, M., O'Brien, J. E., Conravey, A., Hammer, M. F., Waxman, S. G., Dib-Hajj, S. D., & Meisler, M. H. (2014). A novel de novo mutation of *SCN8A* (Nav1.6) with enhanced channel activation in a child with epileptic encephalopathy. *Neurobiology of Disease*, *69*, 117–123. doi:10.1016/j.nbd.2014.05.017

Felts, P. A., Yokoyama, S., Dib-Hajj, S., Black, J. A., & Waxman, S. G. (1997). Sodium channel alpha-subunit mRNAs I, II, III, NaG, Na6 and hNE (PN1): different expression patterns in developing rat nervous system. *Brain Research. Molecular Brain Research*, *45*(1), 71–82.

Frauenfeld, J., Gumbart, J., Sluis, E. O., Funes, S., Gartmann, M., Beatrix, B., . . . Beckmann, R. (2011). Cryo-EM structure of the ribosome-SecYE complex in the membrane environment. *Nature Structural & Molecular Biology*, *18*(5), 614–621. doi:10.1038/nsmb.2026

Galione, A., Evans, A. M., Ma, J., Parrington, J., Arredouani, A., Cheng, X., & Zhu, M. X. (2009). The acid test: the discovery of two-pore channels (TPCs) as NAADP-gated endolysosomal Ca(2+) release channels. *Pflügers Archiv*, *458*(5), 869–876. doi:10.1007/s00424-009-0682-y

Gataullina, S., & Dulac, O. (2017). From genotype to phenotype in Dravet disease. *Seizure*, *44*, 58–64. doi:10.1016/j.seizure.2016.10.014

Gazina, E. V., Leaw, B. T., Richards, K. L., Wimmer, V. C., Kim, T. H., Aumann, T. D., . . . Petrou, S. (2015). "Neonatal" Nav1.2 reduces neuronal excitability and affects seizure susceptibility and behaviour. *Human Molecular Genetics*, *24*(5), 1457–1468. doi:10.1093/hmg/ddu562

Gazina, E. V., Richards, K. L., Mokhtar, M. B., Thomas, E. A., Reid, C. A., & Petrou, S. (2010). Differential expression of exon 5 splice variants of sodium channel alpha subunit mRNAs in the developing mouse brain. *Neuroscience*, *166*(1), 195–200. doi:10.1016/j.neuroscience.2009.12.011

George, A. L., Jr. (2005). Inherited disorders of voltage-gated sodium channels. *Journal of Clinical Investigation*, *115*, 1990–1999.

Gilchrist, J., Das, S., Van Petegem, F., & Bosmans, F. (2013). Crystallographic insights into sodium-channel modulation by the beta4 subunit. *Proceedings of the National Academy of Sciences of the United States of America*, *110*(51), E5016–E5024. doi:10.1073/pnas.1314557110

Gong, B., Rhodes, K. J., Bekele-Arcuri, Z., & Trimmer, J. S. (1999). Type I and type II Na(+) channel alpha-subunit polypeptides exhibit distinct spatial and temporal patterning, and association with auxiliary subunits in rat brain. *Journal of Comparative Neurology*, *412*(2), 342–352.

Guerrini, R., Dravet, C., Genton, P., Belmonte, A., Kaminska, A., & Dulac, O. (1998). Lamotrigine and seizure aggravation in severe myoclonic epilepsy. *Epilepsia*, *39*(5), 508–512.

Han, S., Tai, C., Westenbroek, R. E., Yu, F. H., Cheah, C. S., Potter, G. B., . . . Catterall, W. A. (2012). Autistic-like behaviour in *Scn1a*+/− mice and rescue by enhanced GABA-mediated neurotransmission. *Nature*, *489*(7416), 385–390.

Hauser, W. A., Annegers, J. F., & Kurland, L. T. (1993). Incidence of epilepsy and unprovoked seizures in Rochester, Minnesota: 1935–1984. *Epilepsia*, *34*(3), 453–468.

Hawkins, N. A., Anderson, L. L., Gertler, T. S., Laux, L., George, A. L., Jr., & Kearney, J. A. (2017). Screening of conventional anticonvulsants in a genetic mouse model of epilepsy. *Annals of Clinical and Translational Neurology*, *4*(5), 326–339. doi:10.1002/acn3.413

Hedrich, U. B., Liautard, C., Kirschenbaum, D., Pofahl, M., Lavigne, J., Liu, Y., . . . Lerche, H. (2014). Impaired action potential initiation in GABAergic interneurons causes hyperexcitable networks in an epileptic mouse model carrying a human Na(V)1.1 mutation. *Journal of Neuroscience*, *34*(45), 14874–14889. doi:10.1523/JNEUROSCI.0721-14.2014

Helbig, I., Scheffer, I. E., Mulley, J. C., & Berkovic, S. F. (2008). Navigating the channels and beyond: unraveling the genetics of the epilepsies. *Lancet Neurology*, *7*, 231–245.

Hodgkin, A. L., & Huxley, A. F. (1952*a*). Currents carried by sodium and potassium ion through the membrane of the giant axon of *Loligo*. *Journal of Physiology (London)*, *116*, 449–472.

Hodgkin, A. L., & Huxley, A. F. (*1952b*). The dual effect of membrane potential on sodium conductance in the giant axon of Loligo. *Journal of Physiology (London)*, *116*(4), 497–506.

Howell, K. B., McMahon, J. M., Carvill, G. L., Tambunan, D., Mackay, M. T., Rodriguez-Casero, V., . . . Scheffer, I. E. (2015). *SCN2A* encephalopathy: A major cause of epilepsy of infancy with migrating focal seizures. *Neurology*, *85*(11), 958–966. doi:10.1212/WNL.0000000000001926

Hu, W., Tian, C., Li, T., Yang, M., Hou, H., & Shu, Y. (2009). Distinct contributions of Na(v)1.6 and Na(v)1.2 in action potential initiation and backpropagation. *Nature Neuroscience*, *12*(8), 996–1002. doi:10.1038/nn.2359

Huang, W., Liu, M., Yan, S. F., & Yan, N. (2017). Structure-based assessment of disease-related mutations in human voltage-gated sodium channels. *Protein Cell*, *8*(6), 401–438. doi:10.1007/s13238-017-0372-z

Hund, T. J., Koval, O. M., Li, J., Wright, P. J., Qian, L., Snyder, J. S., . . . Mohler, P. J. (2010). A beta(IV)-spectrin/CaMKII signaling complex is essential for membrane excitability in mice. *Journal of Clinical Investigation*, *120*(10), 3508–3519. doi:10.1172/JCI43621

Ito, S., Ogiwara, I., Yamada, K., Miyamoto, H., Hensch, T. K., Osawa, M., & Yamakawa, K. (2012). Mouse with Na(v)1.1 haploinsufficiency, a model for Dravet syndrome, exhibits lowered sociability and learning impairment. *Neurobiology of Disease*, *49C*, 29–40.

Jenkins, S. M., & Bennett, V. (2001). Ankyrin-G coordinates assembly of the spectrin-based membrane skeleton, voltage-gated sodium channels, and L1 CAMs at Purkinje neuron initial segments. *Journal of Cell Biology*, *155*(5), 739–746. doi:10.1083/jcb.200109026

Jiao, J., Yang, Y., Shi, Y., Chen, J., Gao, R., Fan, Y., . . . Gao, S. (2013). Modeling Dravet syndrome using induced pluripotent stem cells (iPSCs) and directly converted neurons. *Human Molecular Genetics*, *22*(21), 4241–4252. doi:10.1093/hmg/ddt275

Kahlig, K. M., Rhodes, T. H., Pusch, M., Freilinger, T., Pereira-Monteiro, J. M., Ferrari, M. D., . . . George, A. L., Jr. (2008). Divergent sodium channel defects in familial hemiplegic migraine. *Proceedings of the National Academy of Sciences of the United States of America*, *105*(28), 9799–9804.

Kalume, F., Yu, F. H., Westenbroek, R. E., Scheuer, T., & Catterall, W. A. (2007). Reduced sodium current in Purkinje neurons from NaV1.1 mutant mice: implications for ataxia in severe myoclonic epilepsy in infancy. *Journal of Neuroscience*, *27*(41), 11065–11074.

Kaplan, D. I., Isom, L. L., & Petrou, S. (2016). Role of sodium channels in epilepsy. *Cold Spring Harbor Perspectives in Medicine*, *6*(6). doi:10.1101/cshperspect.a022814

Kasai, N., Fukushima, K., Ueki, Y., Prasad, S., Nosakowski, J., Sugata, K., . . . Smith, R. J. (2001). Genomic structures of *SCN2A* and *SCN3A*—candidate genes for deafness at the DFNA16 locus. *Gene*, *264*(1), 113–122.

Keynes, R. D., Rojas, E., & Rudy, B. (1974). Proceedings: Demonstration of a first-order voltage-dependent transition of the sodium activation gates. *Journal of Physiology*, *239*(2), 100P–101P.

Kim, H. W., Quan, Z., Kim, Y. B., Cheong, E., Kim, H. D., Cho, M., . . . Kang, H. C. (2018). Differential effects on sodium current impairments by distinct *SCN1A* mutations in GABAergic neurons derived from Dravet syndrome patients. *Brain Development*, *40*(4), 287–298. doi:10.1016/j.braindev.2017.12.002

Kuzmenkin, A., Bezanilla, F., & Correa, A. M. (2004). Gating of the bacterial sodium channel, NaChBac: voltage-dependent charge movement and gating currents. *Journal of General Physiology*, *124*(4), 349–356. doi:10.1085/jgp.200409139

Laezza, F., Lampert, A., Kozel, M. A., Gerber, B. R., Rush, A. M., Nerbonne, J. M., . . . Ornitz, D. M. (2009). FGF14 N-terminal splice variants differentially modulate Nav1.2 and Nav1.6-encoded sodium channels. *Molecular & Cellular Neuroscience*, *42*, 90–101.

Lamar, T., Vanoye, C. G., Calhoun, J., Wong, J. C., Dutton, S. B. B., Jorge, B. S., . . . Kearney, J. A. (2017). *SCN3A* deficiency associated with increased seizure susceptibility. *Neurobiology of Disease*, *102*, 38–48. doi:10.1016/j.nbd.2017.02.006

Lenaeus, M. J., Gamal El-Din, T. M., Ing, C., Ramanadane, K., Pomes, R., Zheng, N., & Catterall, W. A. (2017). Structures of closed and open states of a voltage-gated sodium channel. *Proceedings of the National Academy of Sciences of the United States of America*, *114*(15), E3051–E3060. doi:10.1073/pnas.1700761114

Liao, Y., Anttonen, A. K., Liukkonen, E., Gaily, E., Maljevic, S., Schubert, S., . . . Lehesjoki, A. E. (2010). *SCN2A* mutation associated with neonatal epilepsy, late-onset episodic ataxia, myoclonus, and pain. *Neurology*, *75*(16), 1454–1458. doi:10.1212/WNL.0b013e3181f8812e

Liao, Y., Deprez, L., Maljevic, S., Pitsch, J., Claes, L., Hristova, D., . . . Lerche, H. (2010). Molecular correlates of age-dependent seizures in an inherited neonatal-infantile epilepsy. *Brain*, *133*(Pt 5), 1403–1414. doi:10.1093/brain/awq057

Liebeskind, B. J., Hillis, D. M., & Zakon, H. H. (2012). Phylogeny unites animal sodium leak channels with fungal calcium channels in an ancient, voltage-insensitive clade. *Molecular Biology and Evolution*, *29*(12), 3613–3616. doi:10.1093/molbev/mss182

Liebeskind, B. J., Hillis, D. M., & Zakon, H. H. (2013). Independent acquisition of sodium selectivity in bacterial and animal sodium channels. *Current Biology*, *23*(21), R948–R949. doi:10.1016/j.cub.2013.09.025

Lin, Y. C., Frei, J. A., Kilander, M. B., Shen, W., & Blatt, G. J. (2016). A subset of autism-associated genes regulate the structural stability of neurons. *Frontiers in Cellular Neuroscience*, *10*, 263. doi:10.3389/fncel.2016.00263

Lindy, A. S., Stosser, M. B., Butler, E., Downtain-Pickersgill, C., Shanmugham, A., Retterer, K., . . . McKnight, D. A. (2018). Diagnostic outcomes for genetic testing of 70 genes in 8565 patients with epilepsy and neurodevelopmental disorders. *Epilepsia*, *59*(5), 1062–1071. doi:10.1111/epi.14074

Lipton, R. B., Bigal, M. E., Diamond, M., Freitag, F., Reed, M. L., & Stewart, W. F. (2007). Migraine prevalence, disease burden, and the need for preventive therapy. *Neurology*, *68*(5), 343–349.

Liu, J., Gao, C., Chen, W., Ma, W., Li, X., Shi, Y., . . . Li, Z. (2016). CRISPR/Cas9 facilitates investigation of neural circuit disease using human iPSCs: mechanism of epilepsy caused by an *SCN1A* loss-of-function mutation. *Translational Psychiatry*, *6*, e703. doi:10.1038/tp.2015.203

Liu, Y., Lopez-Santiago, L. F., Yuan, Y., Jones, J. M., Zhang, H., O'Malley, H. A., . . . Parent, J. M. (2013). Dravet syndrome patient-derived neurons suggest a novel epilepsy mechanism. *Annals of Neurology*, *74*(1), 128–139. doi:10.1002/ana.23897

Lopez-Santiago, L. F., Yuan, Y., Wagnon, J. L., Hull, J. M., Frasier, C. R., O'Malley, H. A., . . . Isom, L. L. (2017). Neuronal hyperexcitability in a mouse model of *SCN8A* epileptic encephalopathy. *Proceedings of the National Academy of Sciences of the United States of America*, *114*(9), 2383–2388. doi:10.1073/pnas.1616821114

Lossin, C., Rhodes, T. H., Desai, R. R., Vanoye, C. G., Wang, D., Carniciu, S., . . . George, A. L., Jr. (2003). Epilepsy-associated dysfunction in the voltage-gated neuronal sodium channel *SCN1A*. *Journal of Neuroscience*, *23*(36), 11289–11295.

Makita, N., Bennett, P. B., & George, A. L., Jr. (1996). Molecular determinants of beta 1 subunit-induced gating modulation in voltage-dependent Na+ channels. *Journal of Neuroscience*, *16*(22), 7117–7127.

Mantegazza, M., & Cestele, S. (2018). Pathophysiological mechanisms of migraine and epilepsy: Similarities and differences. *Neuroscience Letters, 667*, 92–102. doi:10.1016/j.neulet.2017.11.025

Martin, M. S., Dutt, K., Papale, L. A., Dube, C. M., Dutton, S. B., de, H. G., ... Escayg, A. (2010). Altered function of the *SCN1A* voltage-gated sodium channel leads to gamma-aminobutyric acid-ergic (GABAergic) interneuron abnormalities. *Journal of Biological Chemistry, 285*(13), 9823–9834.

McNally, M. A., Johnson, J., Huisman, T. A., Poretti, A., Baranano, K. W., Baschat, A. A., & Stafstrom, C. E. (2016). *SCN8A* epileptic encephalopathy: Detection of fetal seizures guides multidisciplinary approach to diagnosis and treatment. *Pediatric Neurology, 64*, 87–91. doi:10.1016/j.pediatrneurol.2016.08.003

Mei, K., Li, Y., Wang, S., Shao, G., Wang, J., Ding, Y., ... Guo, W. (2018). Cryo-EM structure of the exocyst complex. *Nature Structural & Molecular Biology, 25*(2), 139–146. doi:10.1038/s41594-017-0016-2

Meisler, M. H., Helman, G., Hammer, M. F., Fureman, B. E., Gaillard, W. D., Goldin, A. L., ... Scheffer, I. E. (2016). *SCN8A* encephalopathy: Research progress and prospects. *Epilepsia, 57*(7), 1027–1035. doi:10.1111/epi.13422

Mercimek-Mahmutoglu, S., Patel, J., Cordeiro, D., Hewson, S., Callen, D., Donner, E. J., ... Snead, O. C., 3rd. (2015). Diagnostic yield of genetic testing in epileptic encephalopathy in childhood. *Epilepsia, 56*(5), 707–716. doi:10.1111/epi.12954

Misra, S. N., Kahlig, K. M., & George, A. L., Jr. (2008). Impaired NaV1.2 function and reduced cell surface expression in benign familial neonatal-infantile seizures. *Epilepsia, 49*(9), 1535–1545.

Mistry, A. M., Thompson, C. H., Miller, A. R., Vanoye, C. G., George, A. L., Jr., & Kearney, J. A. (2014). Strain- and age-dependent hippocampal neuron sodium currents correlate with epilepsy severity in Dravet syndrome mice. *Neurobiology of Disease, 65*, 1–11. doi:10.1016/j.nbd.2014.01.006

Miyazaki, H., Oyama, F., Inoue, R., Aosaki, T., Abe, T., Kiyonari, H., ... Nukina, N. (2014). Singular localization of sodium channel beta4 subunit in unmyelinated fibres and its role in the striatum. *Nature Communications, 5*, 5525. doi:10.1038/ncomms6525

Motoike, H. K., Liu, H., Glaaser, I. W., Yang, A. S., Tateyama, M., & Kass, R. S. (2004). The Na+ channel inactivation gate is a molecular complex: a novel role of the COOH-terminal domain. *Journal of General Physiology, 123*(2), 155–165. doi:10.1085/jgp.200308929

Nabbout, R., Chemaly, N., Chipaux, M., Barcia, G., Bouis, C., Dubouch, C., ... Chiron, C. (2013). Encephalopathy in children with Dravet syndrome is not a pure consequence of epilepsy. *Orphanet Journal of Rare Diseases, 8*, 176. doi:10.1186/1750-1172-8-176

Nakamura, K., Kato, M., Osaka, H., Yamashita, S., Nakagawa, E., Haginoya, K., ... Saitsu, H. (2013). Clinical spectrum of *SCN2A* mutations expanding to Ohtahara syndrome. *Neurology, 81*(11), 992–998. doi:10.1212/WNL.0b013e3182a43e57

Namadurai, S., Balasuriya, D., Rajappa, R., Wiemhofer, M., Stott, K., Klingauf, J., ... Jackson, A. P. (2014). Crystal structure and molecular imaging of the Nav channel beta3 subunit indicates a trimeric assembly. *Journal of Biological Chemistry, 289*(15), 10797–10811. doi:10.1074/jbc.M113.527994

Naylor, C. E., Bagneris, C., DeCaen, P. G., Sula, A., Scaglione, A., Clapham, D. E., & Wallace, B. A. (2016). Molecular basis of ion permeability in a voltage-gated sodium channel. *EMBO Journal, 35* (8), 820–830. doi:10.15252/embj.201593285

O'Roak, B. J., Deriziotis, P., Lee, C., Vives, L., Schwartz, J. J., Girirajan, S., . . . Eichler, E. E. (2011). Exome sequencing in sporadic autism spectrum disorders identifies severe de novo mutations. *Nature Genetics*, *43*(6), 585–589.

Ogiwara, I., Ito, K., Sawaishi, Y., Osaka, H., Mazaki, E., Inoue, I., . . . Yamakawa, K. (2009). De novo mutations of voltage-gated sodium channel alpha II gene *SCN2A* in intractable epilepsies. *Neurology*, *73*(13), 1046–1053.

Ogiwara, I., Miyamoto, H., Morita, N., Atapour, N., Mazaki, E., Inoue, I., . . . Yamakawa, K. (2007). NaV1.1 localizes to axons of parvalbumin-positive inhibitory interneurons: a circuit basis for epileptic seizures in mice carrying an *Scn1a* gene mutation. *Journal of Neuroscience*, *27*(22), 5903–5914.

Ohmori, I., Kahlig, K. M., Rhodes, T. H., Wang, D. W., & George, A. L., Jr. (2006). Nonfunctional *SCN1A* is common in severe myoclonic epilepsy of infancy. *Epilepsia*, *47*(10), 1636–1642.

Ostedgaard, L. S., Rogers, C. S., Dong, Q., Randak, C. O., Vermeer, D. W., Rokhlina, T., . . . Welsh, M. J. (2007). Processing and function of CFTR-DeltaF508 are species-dependent. *Proceedings of the National Academy of Sciences of the United States of America*, *104*(39), 15370–15375. doi:10.1073/pnas.0706974104

Pablo, J. L., Wang, C., Presby, M. M., & Pitt, G. S. (2016). Polarized localization of voltage-gated Na+ channels is regulated by concerted FGF13 and FGF14 action. *Proceedings of the National Academy of Sciences of the United States of America*, *113*(19), E2665–E2674. doi:10.1073/pnas.1521194113

Parihar, R., & Ganesh, S. (2013). The *SCN1A* gene variants and epileptic encephalopathies. *Journal of Human Genetics*, *58*(9), 573–580. doi:10.1038/jhg.2013.77

Parrini, E., Marini, C., Mei, D., Galuppi, A., Cellini, E., Pucatti, D., . . . Guerrini, R. (2017). Diagnostic targeted resequencing in 349 patients with drug-resistant pediatric epilepsies identifies causative mutations in 30 different genes. *Human Mutation*, *38*(2), 216–225. doi:10.1002/humu.23149

Patel, R. R., Barbosa, C., Brustovetsky, T., Brustovetsky, N., & Cummins, T. R. (2016). Aberrant epilepsy-associated mutant Nav1.6 sodium channel activity can be targeted with cannabidiol. *Brain*, *139*(Pt 8), 2164–2181. doi:10.1093/brain/aww129

Patino, G. A., & Isom, L. L. (2010). Electrophysiology and beyond: multiple roles of Na+ channel beta subunits in development and disease. *Neuroscience Letters*, *486*(2), 53–59. doi:10.1016/j.neulet.2010.06.050

Payandeh, J., Gamal El-Din, T. M., Scheuer, T., Zheng, N., & Catterall, W. A. (2012). Crystal structure of a voltage-gated sodium channel in two potentially inactivated states. *Nature*, *486*(7401), 135–139. doi:10.1038/nature11077

Payandeh, J., Scheuer, T., Zheng, N., & Catterall, W. A. (2011). The crystal structure of a voltage-gated sodium channel. *Nature*, *475*(7356), 353–358. doi:10.1038/nature10238

Pitt, G. S., & Lee, S. Y. (2016). Current view on regulation of voltage-gated sodium channels by calcium and auxiliary proteins. *Protein Science*, *25*(9), 1573–1584. doi:10.1002/pro.2960

Planells-Cases, R., Caprini, M., Zhang, J., Rockenstein, E. M., Rivera, R. R., Murre, C., . . . Montal, M. (2000). Neuronal death and perinatal lethality in voltage-gated sodium channel alpha(II)-deficient mice. *Biophysical Journal*, *78*(6), 2878–2891. doi:10.1016/S0006-3495(00)76829-9

Qu, Y., Rogers, J. C., Chen, S. F., McCormick, K. A., Scheuer, T., & Catterall, W. A. (1999). Functional roles of the extracellular segments of the sodium channel alpha subunit in voltage-dependent gating and modulation by beta1 subunits. *Journal of Biological Chemistry*, *274*(46), 32647–32654.

Ren, D., Navarro, B., Xu, H., Yue, L., Shi, Q., & Clapham, D. E. (2001). A prokaryotic voltage-gated sodium channel. *Science, 294*(5550), 2372–2375. doi:10.1126/science.1065635

Rhodes, T. H., Lossin, C., Vanoye, C. G., Wang, D. W., & George, A. L., Jr. (2004). Noninactivating voltage-gated sodium channels in severe myoclonic epilepsy of infancy. *Proceedings of the National Academy of Sciences of the United States of America, 101*, 11147–11152.

Rhodes, T. H., Vanoye, C. G., Ohmori, I., Ogiwara, I., Yamakawa, K., & George, A. L., Jr. (2005). Sodium channel dysfunction in intractable childhood epilepsy with generalized tonic-clonic seizures. *Journal of Physiology, 569*, 433–445.

Rubenstein, J. L., & Merzenich, M. M. (2003). Model of autism: increased ratio of excitation/inhibition in key neural systems. *Genes, Brain & Behavior, 2*(5), 255–267.

Rusconi, R., Combi, R., Cestele, S., Grioni, D., Franceschetti, S., Dalpra, L., & Mantegazza, M. (2009). A rescuable folding defective Nav1.1 (*SCN1A*) sodium channel mutant causes GEFS+: common mechanism in Nav1.1 related epilepsies? *Human Mutation, 30*(7), E747–E760. doi:10.1002/humu.21041

Sanders, S. J., Murtha, M. T., Gupta, A. R., Murdoch, J. D., Raubeson, M. J., Willsey, A. J., . . . State, M. W. (2012). De novo mutations revealed by whole-exome sequencing are strongly associated with autism. *Nature, 485*(7397), 237–241.

Sato, C., Ueno, Y., Asai, K., Takahashi, K., Sato, M., Engel, A., & Fujiyoshi, Y. (2001). The voltage-sensitive sodium channel is a bell-shaped molecule with several cavities. *Nature, 409*(6823), 1047–1051. doi:10.1038/35059098

Scalmani, P., Rusconi, R., Armatura, E., Zara, F., Avanzini, G., Franceschetti, S., & Mantegazza, M. (2006). Effects in neocortical neurons of mutations of the Na(v)1.2 Na+ channel causing benign familial neonatal-infantile seizures. *Journal of Neuroscience, 26*(40), 10100–10109.

Schattling, B., Fazeli, W., Engeland, B., Liu, Y., Lerche, H., Isbrandt, D., & Friese, M. A. (2016). Activity of NaV1.2 promotes neurodegeneration in an animal model of multiple sclerosis. *JCI Insight, 1*(19), e89810. doi:10.1172/jci.insight.89810

Scheffer, I. E., & Berkovic, S. F. (1997). Generalized epilepsy with febrile seizures plus. A genetic disorder with heterogeneous clinical phenotypes. *Brain, 120*(Pt 3), 479–490.

Scheuer, T. (2014). Bacterial sodium channels: models for eukaryotic sodium and calcium channels. *Handbook of Experimental Pharmacology, 221*, 269–291. doi:10.1007/978-3-642-41588-3_13

Schwarz, N., Hahn, A., Bast, T., Muller, S., Loffler, H., Maljevic, S., . . . Hedrich, U. B. S. (2016). Mutations in the sodium channel gene *SCN2A* cause neonatal epilepsy with late-onset episodic ataxia. *Journal of Neurology, 263*(2), 334–343. doi:10.1007/s00415-015-7984-0

Shen, H., Zhou, Q., Pan, X., Li, Z., Wu, J., & Yan, N. (2017). Structure of a eukaryotic voltage-gated sodium channel at near-atomic resolution. *Science, 355*(6328). doi:10.1126/science.aal4326

Srivastava, S., & Sahin, M. (2017). Autism spectrum disorder and epileptic encephalopathy: common causes, many questions. *Journal of Neurodevelopmental Disorders, 9*, 23. doi:10.1186/s11689-017-9202-0

Steel, D., Symonds, J. D., Zuberi, S. M., & Brunklaus, A. (2017). Dravet syndrome and its mimics: Beyond *SCN1A*. *Epilepsia, 58*(11), 1807–1816. doi:10.1111/epi.13889

Strasser, L., Downes, M., Kung, J., Cross, J. H., & De Haan, M. (2018). Prevalence and risk factors for autism spectrum disorder in epilepsy: a systematic review and meta-analysis. *Developmental Medicine and Child Neurology, 60*(1), 19–29. doi:10.1111/dmcn.13598

Strong, M., Chandy, K. G., & Gutman, G. A. (1993). Molecular evolution of voltage-sensitive ion channel genes: on the origins of electrical excitability. *Molecular Biology and Evolution, 10*(1), 221–242. doi:10.1093/oxfordjournals.molbev.a039986

Stuhmer, W., Conti, F., Suzuki, H., Wang, X. D., Noda, M., Yahagi, N., . . . Numa, S. (1989). Structural parts involved in activation and inactivation of the sodium channel. *Nature, 339*(6226), 597–603. doi:10.1038/339597a0

Sula, A., Booker, J., Ng, L. C., Naylor, C. E., DeCaen, P. G., & Wallace, B. A. (2017). The complete structure of an activated open sodium channel. *Nature Communications, 8*, 14205. doi:10.1038/ncomms14205

Sun, Y., Pasca, S. P., Portmann, T., Goold, C., Worringer, K. A., Guan, W., . . . Dolmetsch, R. E. (2016). A deleterious Nav1.1 mutation selectively impairs telencephalic inhibitory neurons derived from Dravet syndrome patients. *Elife, 5*. doi:10.7554/eLife.13073

Sun, Y. M., Favre, I., Schild, L., & Moczydlowski, E. (1997). On the structural basis for size-selective permeation of organic cations through the voltage-gated sodium channel. Effect of alanine mutations at the DEKA locus on selectivity, inhibition by Ca2+ and H+, and molecular sieving. *Journal of General Physiology, 110*(6), 693–715.

Sutherland, H. G., & Griffiths, L. R. (2017). Genetics of migraine: Insights into the molecular basis of migraine disorders. *Headache, 57*(4), 537–569. doi:10.1111/head.13053

Tai, C., Abe, Y., Westenbroek, R. E., Scheuer, T., & Catterall, W. A. (2014). Impaired excitability of somatostatin- and parvalbumin-expressing cortical interneurons in a mouse model of Dravet syndrome. *Proceedings of the National Academy of Sciences of the United States of America, 111*(30), E3139–E3148. doi:10.1073/pnas.1411131111

Tang, L., Gamal El-Din, T. M., Payandeh, J., Martinez, G. Q., Heard, T. M., Scheuer, T., . . . Catterall, W. A. (2014). Structural basis for Ca2+ selectivity of a voltage-gated calcium channel. *Nature, 505*(7481), 56–61. doi:10.1038/nature12775

Tatsukawa, T., Ogiwara, I., Mazaki, E., Shimohata, A., & Yamakawa, K. (2018). Impairments in social novelty recognition and spatial memory in mice with conditional deletion of *Scn1a* in parvalbumin-expressing cells. *Neurobiology of Disease, 112*, 24–34. doi:10.1016/j.nbd.2018.01.009

Thompson, C. H., Hawkins, N. A., Kearney, J. A., & George, A. L., Jr. (2017). CaMKII modulates sodium current in neurons from epileptic *Scn2a* mutant mice. *Proceedings of the National Academy of Sciences of the United States of America, 114*(7), 1696–1701. doi:10.1073/pnas.1615774114

Thompson, C. H., Kahlig, K. M., & George, A. L., Jr. (2011). *SCN1A* splice variants exhibit divergent sensitivity to commonly used antiepileptic drugs. *Epilepsia, 52*(5), 1000–1009.

Thompson, C. H., Porter, J. C., Kahlig, K. M., Daniels, M. A., & George, A. L., Jr. (2012). Nontruncating *SCN1A* mutations associated with severe myoclonic epilepsy of infancy impair cell surface expression. *Journal of Biological Chemistry, 287*(50), 42001–42008.

Tian, C., Wang, K., Ke, W., Guo, H., & Shu, Y. (2014). Molecular identity of axonal sodium channels in human cortical pyramidal cells. *Frontiers in Cellular Neuroscience, 8*, 297. doi:10.3389/fncel.2014.00297

Tsai, C. J., Tani, K., Irie, K., Hiroaki, Y., Shimomura, T., McMillan, D. G., . . . Li, X. D. (2013). Two alternative conformations of a voltage-gated sodium channel. *Journal of Molecular Biology, 425*(22), 4074–4088. doi:10.1016/j.jmb.2013.06.036

Ulmschneider, M. B., Bagneris, C., McCusker, E. C., Decaen, P. G., Delling, M., Clapham, D. E., . . . Wallace, B. A. (2013). Molecular dynamics of ion transport through the open conformation of

a bacterial voltage-gated sodium channel. *Proceedings of the National Academy of Sciences of the United States of America, 110*(16), 6364–6369. doi:10.1073/pnas.1214667110
Vacher, H., Mohapatra, D. P., & Trimmer, J. S. (2008). Localization and targeting of voltage-dependent ion channels in mammalian central neurons. *Physiological Reviews, 88*(4), 1407–1447.
Vanoye, C. G., Gurnett, C. A., Holland, K. D., George, A. L., & Kearney, J. A. (2014). Novel *SCN3A* variants associated with focal epilepsy in children. *Neurobiology of Disease, 62*, 313–322. doi:10.1016/j.nbd.2013.10.015
Vanoye, C. G., Lossin, C., Rhodes, T. H., & George, A. L., Jr. (2006). Single-channel properties of human NaV1.1 and mechanism of channel dysfunction in *SCN1A*-associated epilepsy. *Journal of General Physiology, 127*(1), 1–14.
Volkers, L., Kahlig, K. M., Verbeek, N. E., Das, J. H., van Kempen, M. J., Stroink, H., ... Rook, M. B. (2011). Nav 1.1 dysfunction in genetic epilepsy with febrile seizures-plus or Dravet syndrome. *European Journal of Neuroscience, 34*(8), 1268–1275.
Wagnon, J. L., Barker, B. S., Hounshell, J. A., Haaxma, C. A., Shealy, A., Moss, T., ... Meisler, M. H. (2016). Pathogenic mechanism of recurrent mutations of *SCN8A* in epileptic encephalopathy. *Annals of Clinical and Translational Neurology, 3*(2), 114–123. doi:10.1002/acn3.276
Wagnon, J. L., Korn, M. J., Parent, R., Tarpey, T. A., Jones, J. M., Hammer, M. F., ... Meisler, M. H. (2015). Convulsive seizures and SUDEP in a mouse model of *SCN8A* epileptic encephalopathy. *Human Molecular Genetics, 24*(2), 506–515. doi:10.1093/hmg/ddu470
Wagnon, J. L., & Meisler, M. H. (2015). Recurrent and non-recurrent mutations of *SCN8A* in epileptic encephalopathy. *Frontiers in Neurology, 6*, 104. doi:10.3389/fneur.2015.00104
Wang, C., Chung, B. C., Yan, H., Wang, H. G., Lee, S. Y., & Pitt, G. S. (2014). Structural analyses of Ca(2)(+)/CaM interaction with NaV channel C-termini reveal mechanisms of calcium-dependent regulation. *Nature Communications, 5*, 4896. doi:10.1038/ncomms5896
Wang, C., Wang, C., Hoch, E. G., & Pitt, G. S. (2011). Identification of novel interaction sites that determine specificity between fibroblast growth factor homologous factors and voltage-gated sodium channels. *Journal of Biological Chemistry, 286*(27), 24253–24263.
Wang, T., Guo, H., Xiong, B., Stessman, H. A., Wu, H., Coe, B. P., ... Eichler, E. E. (2016). De novo genic mutations among a Chinese autism spectrum disorder cohort. *Nature Communications, 7*, 13316. doi:10.1038/ncomms13316
Weiss, L. A., Escayg, A., Kearney, J. A., Trudeau, M., MacDonald, B. T., Mori, M., ... Meisler, M. H. (2003). Sodium channels *SCN1A, SCN2A* and *SCN3A* in familial autism. *Molecular Psychiatry, 8*(2), 186–194.
Westenbroek, R. E., Merrick, D. K., & Catterall, W. A. (1989). Differential subcellular localization of the RI and RII Na+ channel subtypes in central neurons. *Neuron, 3*(6), 695–704.
Whitaker, W. R., Clare, J. J., Powell, A. J., Chen, Y. H., Faull, R. L., & Emson, P. C. (2000). Distribution of voltage-gated sodium channel a-subunit and b-subunit mRNAs in human hippocampal formation, cortex, and cerebellum. *Journal of Comparative Neurology, 422*(1), 123–139.
Whitaker, W. R., Faull, R. L., Waldvogel, H. J., Plumpton, C. J., Emson, P. C., & Clare, J. J. (2001). Comparative distribution of voltage-gated sodium channel proteins in human brain. *Brain Research. Molecular Brain Research, 88*(1–2), 37–53.
Wolff, M., Johannesen, K. M., Hedrich, U. B. S., Masnada, S., Rubboli, G., Gardella, E., ... Moller, R. S. (2017). Genetic and phenotypic heterogeneity suggest therapeutic implications in *SCN2A*-related disorders. *Brain, 140*(5), 1316–1336. doi:10.1093/brain/awx054

Woodbury-Smith, M., & Scherer, S. W. (2018). Progress in the genetics of autism spectrum disorder. *Developmental Medicine and Child Neurology, 60*(5), 445–451. doi:10.1111/dmcn.13717

Xu, R., Thomas, E. A., Jenkins, M., Gazina, E. V., Chiu, C., Heron, S. E., . . . Petrou, S. (2007). A childhood epilepsy mutation reveals a role for developmentally regulated splicing of a sodium channel. *Molecular & Cellular Neuroscience, 35(2), 292–301* .

Yamagata, T., Ogiwara, I., Mazaki, E., Yanagawa, Y., & Yamakawa, K. (2017). Nav1.2 is expressed in caudal ganglionic eminence-derived disinhibitory interneurons: Mutually exclusive distributions of Nav1.1 and Nav1.2. *Biochemical and Biophysical Research Communications, 491*(4), 1070–1076. doi:10.1016/j.bbrc.2017.08.013

Yan, H., Wang, C., Marx, S. O., & Pitt, G. S. (2017). Calmodulin limits pathogenic Na+ channel persistent current. *Journal of General Physiology, 149*(2), 277–293. doi:10.1085/jgp.201611721

Yan, Z., Zhou, Q., Wang, L., Wu, J., Zhao, Y., Huang, G., . . . Yan, N. (2017). Structure of the Nav1.4-beta1 complex from electric eel. *Cell, 170*(3), 470–482 e411. doi:10.1016/j.cell.2017.06.039

Yarov-Yarovoy, V., DeCaen, P. G., Westenbroek, R. E., Pan, C. Y., Scheuer, T., Baker, D., & Catterall, W. A. (2012). Structural basis for gating charge movement in the voltage sensor of a sodium channel. *Proceedings of the National Academy of Sciences of the United States of America, 109*(2), E93–E102. doi:10.1073/pnas.1118434109

Yereddi, N. R., Cusdin, F. S., Namadurai, S., Packman, L. C., Monie, T. P., Slavny, P., . . . Jackson, A. P. (2013). The immunoglobulin domain of the sodium channel beta3 subunit contains a surface-localized disulfide bond that is required for homophilic binding. *FASEB Journal, 27*(2), 568–580. doi:10.1096/fj.12-209445

Yu, F. H., Mantegazza, M., Westenbroek, R. E., Robbins, C. A., Kalume, F., Burton, K. A., . . . Catterall, W. A. (2006). Reduced sodium current in GABAergic interneurons in a mouse model of severe myoclonic epilepsy in infancy. *Nature Neuroscience, 9*(9), 1142–1149.

Zakon, H. H. (2012). Adaptive evolution of voltage-gated sodium channels: the first 800 million years. *Proceedings of the National Academy of Sciences of the United States of America, 109*(Suppl 1), 10619–10625. doi:10.1073/pnas.1201884109

Zaman, T., Helbig, I., Bozovic, I. B., DeBrosse, S. D., Bergqvist, A. C., Wallis, K., . . . Goldberg, E. M. (2018). Mutations in *SCN3A* cause early infantile epileptic encephalopathy. *Annals of Neurology, 83*(4), 703–717. doi:10.1002/ana.25188

Zhang, X., Ren, W., DeCaen, P., Yan, C., Tao, X., Tang, L., . . . Yan, N. (2012). Crystal structure of an orthologue of the NaChBac voltage-gated sodium channel. *Nature, 486*(7401), 130–134. doi:10.1038/nature11054

Zhang, X., Xia, M., Li, Y., Liu, H., Jiang, X., Ren, W., . . . Gong, H. (2013). Analysis of the selectivity filter of the voltage-gated sodium channel Na(v)Rh. *Cell Research, 23*(3), 409–422. doi:10.1038/cr.2012.173

SECTION 3

LIGAND-GATED CHANNELS

CHAPTER 9

α-AMINO-3-HYDROXY-5-METHYL-4-ISOXAZOLEPROPIONIC ACID AND KAINATE RECEPTORS

G. BRENT DAWE, PATRICIA M. G. E. BROWN, AND DEREK BOWIE

INTRODUCTION TO THE IONOTROPIC GLUTAMATE RECEPTOR FAMILY

Ionotropic glutamate receptors (iGluRs) are tetrameric proteins that mediate almost all fast-excitatory neurotransmission in the central nervous system. The arrival of an action potential in the presynaptic neuron of glutamatergic synapses triggers the vesicular release of the neurotransmitter L-glutamate, which binds to and activates iGluRs located on the postsynaptic neuron. The binding of the neurotransmitter triggers iGluRs to enter an activated state with the opening of a transmembrane ion channel pore that permits the rapid transport of mono- and divalent cations. Cation influx into the postsynaptic neuron causes membrane depolarization and, depending on its magnitude, may trigger action potential firing in the postsynaptic neuron (Figure 9.1A). Glutamatergic synaptic transmission occurs on a millisecond timescale, endowing neuronal circuits with the capability of responding rapidly to incoming signals, while giving rise to complex cognitive functions and behaviors, such as sensory perception, thought, movement, and memory.

There are 18 mammalian iGluR subunits, which are grouped into four classes largely based on the names of agonists that selectively activate them (Figure 9.1B). Accordingly, iGluR subgroups sensitive to α-amino-3-hydroxy-5-methyl-4-isoxazolepropionic acid (AMPA), kainic acid (KA, or kainate), and *N*-methyl-D-aspartic acid (NMDA) are termed AMPA receptors (AMPARs), KA receptors (KARs), and NMDA receptors

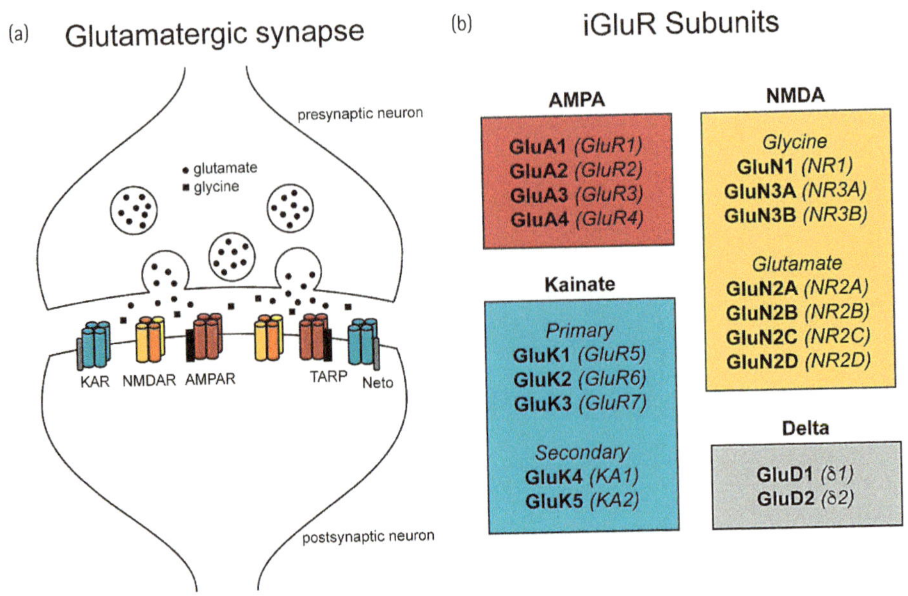

FIGURE 9.1 The iGluR family. (A) Depiction of a glutamatergic synapse, including AMPARs and KARs and their principal auxiliary subunits, as well as NMDARs, which require glutamate and glycine to activate. (B) List of iGluR subunits, according to the subfamily, with older nomenclature shown in brackets. Note that some NMDAR subunits bind glycine, rather than glutamate. Also, KARs are divided into primary subunits, capable of forming functional homomers, and secondary or high-affinity subunits, which must heteromerize with a primary subunit to form functional channels.

(NMDARs). Furthermore, there are two related, "nonconducting" subunits, the orphan/delta subunits (GluD1-2), which apparently fail to be activated in response to known iGluR agonists when expressed alone or with other iGluR subunits (Traynelis et al., 2010). The orphan class receptors appear to have other, non-ionotropic roles that are important in synapse signaling (Yuzaki & Aricescu, 2017). The current nomenclature was implemented beginning in 2009 (Collingridge, Olsen, Peters, & Spedding, 2009); previously, AMPARs were known as GluR1–4; KARs as GluR5–7, KA-1, and KA-2; NMDARs as NR1, NR2A-D, and NR3A-B; and the orphan/delta receptors as δ1–2 (Dingledine, Borges, Bowie, & Traynelis, 1999) (Figure 9.1B).

AMPARs were originally identified as being selectively activated by the synthetic agonist AMPA (Krogsgaard-Larsen, Honore, Hansen, Curtis, & Lodge, 1980), which does not elicit responses at NMDARs though it does activate some homo- and heteromeric KARs (Swanson, Gereau, Green, & Heinemann, 1997). The AMPAR subunits can heteromerize interchangeably, though they also retain function as homomeric channels (Boulter et al., 1990; Keinänen et al., 1990). Despite the many possible combinations, it is generally agreed that the most abundant AMPARs in the mammalian brain are heteromers composed of either GluA1/GluA2 or GluA2/GluA3 subunits (Bowie, 2018). AMPAR subunits cannot assemble with KAR subunits to form functional channels (Bettler et al., 1990; Sommer et al., 1992), rendering the two iGluR subfamilies entirely separable at the molecular level.

The KAR subunits are discriminated based on their relatively high affinity for the agonist KA, first isolated from algae off the coast of Japan (Murakami, Takemoto, & Shimizu, 1953). These subunits only assemble with each other, rather than AMPAR or NMDAR subunits (Wenthold, Trumpy, Zhu, & Petralia, 1994). However, the functional distinction between AMPARs and KARs is blurred by the fact that most AMPAR subunits are responsive to KA, while some KAR subunits (e.g., GluK1) are responsive to AMPA (e.g., Sommer et al., 1992). KAR subunits are further divided into "primary" subunits (GluK1–3), capable of forming functional channels when expressed alone, and "secondary" subunits (GluK4–5), which require co-assembly with at least one primary subunit to exhibit functionality (reviewed in Contractor, Mulle, & Swanson, 2011). The secondary or "high-affinity" KAR subunits notably bind KA with 10-fold higher affinity than the primary subunits, possessing dissociation constant (K_d) values in the 10 nM range (Herb et al., 1992; Werner, Voigt, Keinanen, Wisden, & Seeburg, 1991).

In terms of antagonism, AMPARs and KARs are functionally separable from NMDARs based on their sensitivity to quinoxalinedione antagonists, such as 6-cyano-7-nitroquinoxaline-2,3-dione (CNQX) and 6,7-dinitroquinoxaline-2,3-dione (DNQX) (Honore et al., 1988). These compounds do not inhibit the NMDAR-mediated response at glutamatergic synapses, where all three subfamilies are present. Likewise, the classic NMDAR competitive antagonist 2-amino-5-phosphonopentanoic acid (Davies & Watkins, 1982) does not affect AMPAR or KAR responses. Consistent with the similarity in their pharmacological profiles, the amino acid sequence identity between AMPAR and KAR subunits is roughly 40% but drops to 25% when either receptor subtype is compared to NMDARs (Hollmann & Heinemann, 1994). It is therefore unsurprising that NMDARs are distinguished from other iGluRs by a variety of functional properties, notably slower gating kinetics and distinct sites of allosteric regulation (Paoletti, Bellone, & Zhou, 2013).

An Overview of iGluR Topology

All iGluR subunits possess a similar overall topology, including a lengthy (>500 amino acid) N-terminal extracellular domain, four transmembrane (TM1–4) regions, and an intracellular, largely disordered C-terminal tail (CTD) of variable length (Figure 9.2A). The extracellular region is further divided into two globular, self-interacting domains known as the amino-terminal domain (ATD; alternatively termed "NTD" in other publication) and the agonist or ligand-binding domain (LBD). Interestingly, TM2 partially enters the membrane and turns back into the intracellular space (Hollmann, Maron, & Heinemann, 1994). The LBD is formed of two discontinuous extracellular segments that are located between the ATD and TM1 (S1) as well as between TM3 and TM4 (S2) (Stern-Bach et al., 1994).

Situated distal from the channel pore, the ATD contains approximately 400 amino acids, about half of the entire protein for AMPARs and KARs. Structurally, the ATD is divided into the upper R1 and lower R2 lobes, and the cleft in between plays an important role in the allosteric regulation of NMDARs, notably by inhibitory zinc ions (Hansen, Furukawa, & Traynelis, 2010). Despite its large size, the ATD is required for

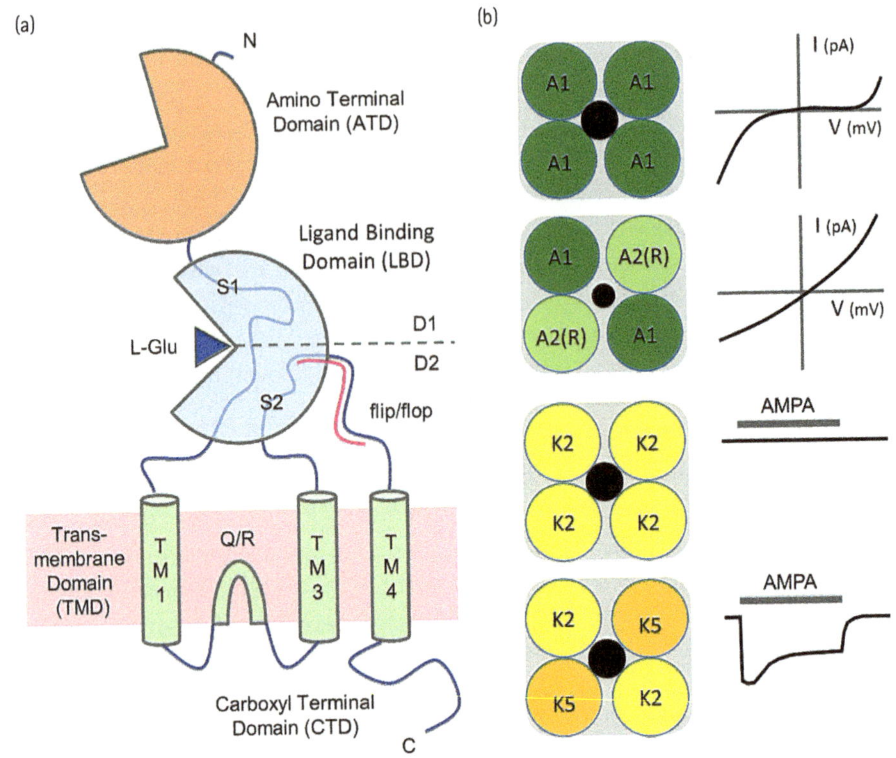

FIGURE 9.2 iGluR topology and subunit assembly. (A) Topological illustration of an iGluR subunit, including the four principal domains: ATD, LBD, TMD, and CTD. The upper and lower lobes of the LBD are referred to as *D1* and *D2*, while the discontinuous amino acid segments that form the LBD are *S1* and *S2*. Note the flip alternate splicing region is only found in AMPARs. (B) AMPAR subunits (green) and KAR subunits (yellow) can form into homomeric and heteromeric complexes. Ion channels formed by GluA2-containing AMPARs have lower conductance but are unaffected by polyamine block at positive membrane potentials, resulting in more linear current–voltage (*I–V*) plots. While GluK2 heteromers are unresponsive to AMPA, GluK2/GluK5 heteromers are activated by the same agonist.

neither receptor assembly nor channel gating since the different iGluR subfamilies retain channel function if their respective ATDs have been deleted (e.g., Fayyazuddin, Villarroel, Le Goff, Lerma, & Neyton, 2000; Pasternack et al., 2002). At the same time, several studies suggest that the ATD facilitates the preferential assembly of specific combinations of subunits (Mayer, 2011), often resulting in 2:2 heteromeric arrangements (Figure 9.2B). In a physiological context, the ATD also appears to mediate iGluR association with other pre- and postsynaptic proteins (Garcia-Nafria, Herguedas, Watson, & Greger, 2016).

Separated by a short linker from the ATD, the 300–amino acid LBD orchestrates the transduction of agonist binding into channel opening. The LBD is formed from both S1 and S2, which together define agonist selectivity (Stern-Bach et al., 1994). When considered in isolation, the LBD structure is divided into domains 1 and 2 (D1 and D2), representing the upper and lower lobes of the clamshell-shaped agonist-binding cleft (Armstrong, Sun, Chen, & Gouaux, 1998). The isolated LBD has, in fact, proven quite amenable to structural characterization, and a number of atomic resolution structures of this domain, particularly from GluA2 AMPARs, have contributed to the refinement of molecular mechanisms for iGluR channel gating (Pohlsgaard, Frydenvang, Madsen, & Kastrup, 2011).

The LBD is also an important site for both genetic and allosteric regulation in AMPARs and KARs. For example, alternative splicing of an exon encoding the S2–TM4 linker region produces either "flip" or "flop" isoforms for all AMPAR subunits (Sommer et al., 1990) (Figure 9.3). GluA2–4 flop isoform receptors are notable for desensitizing much more rapidly than their flip counterparts (e.g., Mosbacher et al., 1994). Furthermore, the flip/flop region is immediately preceded by an R/G RNA

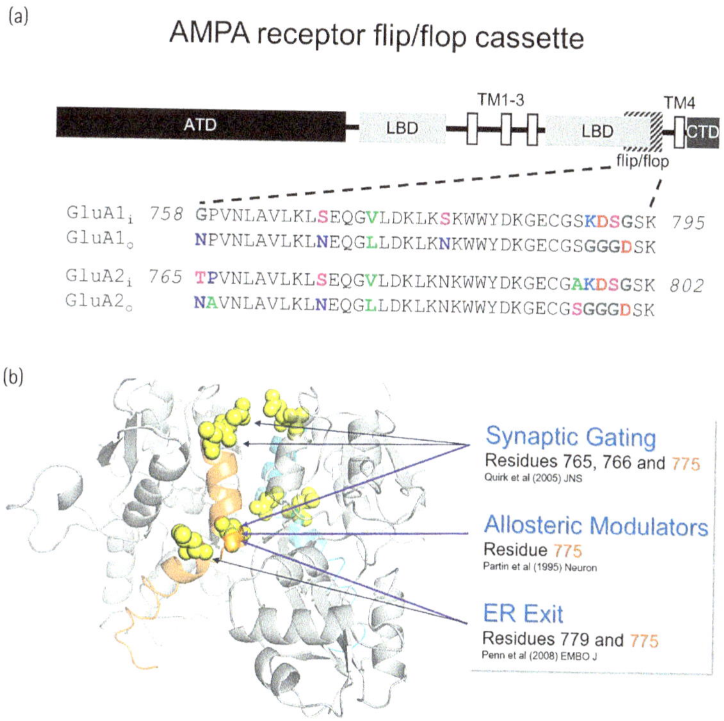

FIGURE 9.3 Importance of the flip/flop cassette in AMPAR gating and allosteric modulation. (A) Amino acid sequence alignment of the alternately spliced flip (i) and flop (o) cassettes of GluA1 and GluA2 AMPARs. Notable amino acid substitutions are highlighted. (B) Structural depiction of amino acid residues impacted by flip/flop alternative splicing. Situated at the apex of the LBD are residues that affect gating kinetics, whereas the 775 position influences sensitivity to cyclothiazide and allosteric anions. Positions lower in the subunit interface have been reported to affect subunit assembly and trafficking.

Adapted from G. B. Dawe et al. Nanoscale mobility of the apo state and TARP stoichiometry dictate the gating behavior of alternatively spliced AMPA receptors. *Neuron*, *102*(5), 976–992. © 2019, with permission from Elsevier.

editing site (Lomeli et al., 1994). The conversion of arginine to glycine in flip and flop variants of GluA2–4 produces a modest slowing of desensitization and acceleration of recovery from desensitization (Lomeli et al., 1994). In contrast, a number of positive allosteric modulators, such as cyclothiazide, bind to the AMPAR LBD and greatly attenuate desensitization (Partin, Bowie, & Mayer, 1995; Sun et al., 2002). Recent work has shown that the flip/flop cassette regulates the nanoscale mobility of the apo state of the AMPAR, which in turn acts as a master switch to control channel gating, allosteric regulation, and regulation by auxiliary subunits (Dawe et al., 2019). For the KAR subfamily, allosteric ions acting at the LBD have a significant influence on desensitization and other gating properties of specific subunits (Bowie, 2002). Whether these effects of ions also work by regulating the mobility of the apo state remains to be investigated.

The TM domain (TMD) is comprised of discontinuous tracts from TM1 to TM4 and is generally permeable to monovalent cations of varying diameter but less permeable to divalent cations (Bowie, 2018; Huettner, 2015). More specifically, TM1 and TM4 flank the outside of the pore, whereas the TM2 re-entrant loop forms the pore itself, and TM3 lines the upper segment of the permeation pathway (Sobolevsky, Rosconi, & Gouaux, 2009). The selectivity filter is thought to be located where TM2 bends back toward the intracellular face of the membrane (Kuner, Beck, Sakmann, & Seeburg, 2001).

A second locus of RNA editing found in TM2, known as the Q/R site, is critical for ion permeation in both AMPARs and KARs. Channels containing GluA2 subunits, in which the glutamine in the pore region is almost always substituted by an arginine, have a linear current–voltage (*I–V*) relationship, whereas channels comprised of unedited GluA1, GluA3, and GluA4 subunits yield whole-cell currents that inwardly rectify, unless forming heteromers with GluA2 (Verdoorn, Burnashev, Monyer, Seeburg, & Sakmann, 1991) (Figure 9.2B). This rectification is due to channel block by intracellular polyamines at more positive holding potentials (Bowie, 2018; Bowie & Mayer, 1995). Moreover, the presence of an arginine at this position also reduces AMPAR calcium permeability (Burnashev, Monyer, Seeburg, & Sakmann, 1992; Hume, Dingledine, & Heinemann, 1991), leading to the classification of GluA2-lacking AMPAR complexes as "calcium-permeable." Meanwhile in KARs, the Q/R position is edited to an intermediate extent (Sommer, Kohler, Sprengel, & Seeburg, 1991).

Below the plasma membrane, the iGluR CTD represents an important site of interaction with both kinases (e.g., Banke et al., 2000) and scaffolding proteins, which specify localization (Tomita, Nicoll, & Bredt, 2001). However, little structural information exists for this region, except within the NMDAR subfamily (Ataman, Gakhar, Sorensen, Hell, & Shea, 2007; Choi, Xiao, Wollmuth, & Bowen, 2011). On a functional level, AMPAR and KAR subtypes with a deleted C-terminal tail retain gating capabilities (Salussolia et al., 2011; S. Yan et al., 2004). As such, an in-depth discussion of the CTD is not provided in this chapter.

An Overview of iGluR Stoichiometry

To form functional channel complexes, iGluR subunits must first assemble as tetramers, whether homomeric or heteromeric in nature (Figure 9.2B). This stoichiometry was first derived indirectly using targeted mutations that altered agonist affinity (e.g., Laube, Kuhse, & Betz, 1998), as well as by analysis of single-channel conductance (Rosenmund, Stern-Bach, & Stevens, 1998). Nevertheless, more definitive proof came from the tetrameric subunit arrangement of the first intact iGluR structure to be obtained at atomic resolution (Sobolevsky et al., 2009). The *Y*-shaped GluA2 AMPAR structure revealed many insights into the modular organization of iGluR complexes, and intact NMDAR (Karakas & Furukawa, 2014; Lee et al., 2014) and KAR (Meyerson et al., 2016, 2014) structures have since been elucidated.

Adding to the complexity of iGluR signaling is the fact that synaptic AMPARs and KARs are often associated with auxiliary or regulatory proteins in the brain. Although there is an expanding list of iGluR auxiliary protein families (especially for AMPARs), they share the common ability to enhance membrane trafficking and/or positively modulate channel activity (Jackson & Nicoll, 2011). The precise stoichiometry between native iGluRs and different auxiliary subunits remains largely unresolved, especially because it is unclear how different classes of auxiliary proteins might occlude each other from association with pore-forming subunits. For the well-studied transmembrane AMPAR regulatory protein (TARP) family of AMPAR auxiliary subunits, there are emerging structural mechanisms that describe how they associate with and modulate pore-forming subunits. As discussed in the section *AMPAR-TARP Assembly & Stoichiometry*, recent work reports that the apparent TARP stoichiometry of native AMPARs varies in a cell-specific manner (Dawe et al., 2019).

Gating and Pharmacology of iGluRs: Theory of Ion Channel Gating

Much of the terminology used to describe iGluR gating originated from pioneering studies of voltage-gated ion channels (VGICs) and nicotinic acetylcholine receptors in the 1950s, long before the molecular identity of receptor–channel complexes could be elucidated. The term *gating* itself describes processes that open and close channels, while *permeation* reflects the transport of ions through an open channel (Horn, 1990). For ligand-gated ion channels (LGICs), the energy required for opening the channel pore (activation) is derived from the binding of one or more agonist molecules (Andersen & Koeppe, 1992). Meanwhile, there are two fundamental gating processes that result in the closure of ion channels. The first, deactivation, is observed upon removal of the agonist. The second, desensitization, can be described as a "progressive reduction in ionic flux in the prolonged presence of agonist" (Keramidas & Lynch, 2013). In other words, receptors become less sensitive over time to their chemical stimulus, favoring channel closure while agonist molecules are still bound.

It is also possible for iGluR desensitization to manifest in several other forms apart from decaying current responses. For instance, the high apparent affinity of desensitized receptors for certain agonists means that desensitization can persist long after an initial agonist application, preventing channel opening in response to subsequent agonist exposures. In this context, the extent of desensitization is reflected in the fractional current response following a second agonist pulse, relative to the maximal response after an initial pulse. Furthermore, iGluRs can become "pre-desensitized" by equilibration with low (inert) concentrations of an agonist, such that a rapid escalation of its concentration to saturating levels still cannot elicit any channel activity. Due to its various manifestations, it can be useful to define desensitization at the molecular level as the transition of an agonist-bound receptor into a nonconducting or (in some cases) poorly conducting state. Similarly, deactivation represents the transition(s) from an open to a closed, agonist-unbound state (Hinard et al., 2016).

The conceptualization of LGIC gating processes as transitions/reactions between states was inspired by earlier work on enzymes, which exist in multiple conformational states that are differentially occupied following the binding of a chemical substrate (Andersen & Koeppe, 1992). In this context, the binding of an agonist does not directly open the channel but rather reduces the energetic barrier for another state transition that corresponds to opening—indeed, in exceptionally rare instances, LGIC currents have been observed without any agonist present (Auerbach, 2015). Although definitions vary, one can reasonably say that affinity is determined by the initial agonist-binding reaction, whereas efficacy is the sum of all other gating transitions (Colquhoun, 1998). Accordingly, partial agonists are chemical compounds that display reduced efficacy at saturating concentrations compared to some maximally efficacious compound (i.e., L-glutamate), while competitive antagonists possess no efficacy (Stephenson, 1956). An additional category of ligands known as *noncompetitive antagonists* reduce receptor responsiveness by interacting somewhere distinct from the orthosteric agonist-binding site and are thus said to act in an "allosteric" manner (Colquhoun, 1998).

Contributions of iGluR Subunits to Channel Gating

The binding of two agonist molecules is necessary for AMPAR and KAR channels to open (Clements, Feltz, Sahara, & Westbrook, 1998). This was proposed by examining the activation kinetics of non-NMDARs in cultured hippocampal neurons and comparing them with gating models that had one, two, or three binding sites (Clements et al., 1998). Though this study could not determine the exact number of binding sites, Rosenmund and colleagues (1998) further developed this idea and suggested a tetrameric stoichiometry. In this case, they engineered a nondesensitizing GluA3–GluK2 chimeric channel and studied single-channel responses. To observe the rapid transitions between the three observed subconductance states, the agonist binding rate was slowed by imposing a much slower step, namely the unbinding of the high-affinity antagonist 2,3-dioxo-6-nitro-7-sulfamoyl-benzo(F)quinoxaline (NBQX) (Rosenmund et al., 1998). The transition into the smallest conducting state was shaped by two time constants, consistent with the idea that two molecules are necessary for activation. Accordingly, the simplest explanation for the three-conductance observation was that receptors are tetramers with two additional agonist-binding steps accounting for the higher conductance levels (Rosenmund et al., 1998).

In AMPARs, the relative contribution of conductance states is agonist-dependent, with higher conductance levels being more abundant in higher agonist concentrations (Rosenmund et al., 1998; Smith & Howe, 2000). However, though the distribution of conductance levels in KARs does not appear to be agonist concentration–dependent, the results remain unclear due to the lack of pharmacological tools, such as cyclothiazide for AMPARs, to block KAR desensitization (though see Dawe et al., 2013).

Gating Behavior of Native AMPARs and KARs

AMPARs and KARs exhibit rapid gating behavior, particularly in comparison to the slower current rise times and decay properties of NMDARs. Generally, AMPAR state transitions occur on a scale of 10 ms or less (Baranovic & Plested, 2016), and this can be considered a good benchmark to divide fast and slow for most gating processes. The earliest observations of AMPAR and KAR gating behavior came from neurons as improvements in the voltage-clamp recording technique and the discovery of subfamily-selective agonists and antagonists made it possible to isolate these receptor populations (Mayer & Westbrook, 1987). At the same time, the accurate measurement of AMPAR and KAR currents also requires the delivery of a fixed agonist concentration over a precise time interval. Various "concentration-clamp" systems can be implemented (e.g., Franke, Hatt, & Dudel, 1987; Krishtal, Marchenko, & Pidoplichko, 1983) so that cells or membrane patches are quickly exposed to agonist-containing solutions on a submillisecond timescale. Without such systems, diffuse agonist release will obscure the visualization membrane current responses because many AMPARs and KARs will rapidly desensitize before nearby receptors can themselves activate to generate a concerted peak response.

Neuronal recordings using the concentration-clamp technique demonstrate that non-NMDAR responses to glutamate almost completely desensitize within tens of milliseconds, with maximal response amplitudes occurring at very high (i.e., >10 mM) glutamate concentrations (Kiskin, Krishtal, & Tsyndrenko, 1986). In contrast, recovery from desensitization typically occurs over hundreds of milliseconds (Bowie & Lange, 2002; Trussell & Fischbach, 1989). Yet for many years, mixed populations of native AMPARs and KARs were unable to be studied in isolation, a problem recognized after KA was shown to activate recombinant AMPAR subunits (i.e., Keinänen et al., 1990). Fortunately, 2,3-benzodiazepine (GYKI) compounds are now used as noncompetitive antagonists, selective for AMPARs over KARs (Clarke et al., 1997; Wilding & Huettner, 1995), facilitating the isolation of KAR-mediated synaptic responses. Often, the residual agonist-evoked currents remaining after GYKI application are quite small (<20%) compared to those mediated by AMPAR populations (e.g., Wilding & Huettner, 1997).

Gating Behavior of Recombinant AMPARs and KARs

The most precise measurements of AMPAR and KAR gating have arguably come from recombinantly expressed GluA2 and GluK2 (Q/R unedited) receptors because these two subunits form homomeric channels with excellent exogenous expression in widely used cell lines (i.e., HEK 293 cells). Gating time constants of recombinant GluA2 receptors largely reflect those of neuronal AMPAR populations, whereas recombinant GluK2 properties often differ from native KAR behavior due to heteromerization with secondary and auxiliary subunits. A summary of GluA2 and GluK2 gating properties (see Table 9.1) entails rapid activation (~200 μs), deactivation (0.5–3 ms), and desensitization (5–10 ms), as well as low glutamate potency (~1 mM). The gating behavior of other iGluR subunits can be gleaned from more extensive reviews (e.g., Dingledine et al., 1999; Traynelis et al., 2010).

Single-channel activity of Q/R unedited AMPARs and KARs can be directly observed because unitary conductance (>5 pS) is greater than that of edited channels (0.2–0.5 pS) (Swanson, Kamboj, & Cull-Candy, 1997), where such values must be

Table 9.1 Biophysical Properties of AMPAR and KAR Gating

	GluA2(Q), flip	GluK2(Q)
Gating Process	time constant (ms)	
Desensitization	5–10	5–8
Deactivation	0.5–1	2–3
Recovery from desensitization	20–25	2000–3000
Rise time (10–90%)	0.2–0.3	0.2
Equilibrium current (% of peak)	~1	0.3–0.4
Agonist potency	EC_{50} (μM)	
Peak current	1,000–2,000	~500
Equilibrium response	300–500 (in CTZ)	30

Sources. The following papers were referenced to develop this table: GluA2 (Carbone & Plested, 2012; Dawe et al., 2016; Horning & Mayer, 2004; Koike, Tsukada, Tsuzuki, Kijima, & Ozawa, 2000; MacLean et al., 2014; Robert, Armstrong, Gouaux, & Howe, 2005; Salazar, Eibl, Chebli, & Plested, 2017; Sun et al., 2002; Yu et al., 2016) and GluK2 (Bowie, 2002; Bowie, Garcia, Marshall, Traynelis, & Lange, 2003; Bowie & Lange, 2002; Carbone & Plested, 2012; Heckmann, Bufler, Franke, & Dudel, 1996). CTZ indicates cyclothiazide.

inferred from noise analysis (see Traynelis & Jaramillo, 1998). During sustained agonist applications, these receptors typically display a single burst of channel openings over a few milliseconds, prior to the rapid onset of desensitization, after which additional openings are extremely rare (e.g., Dawe et al., 2013). An interesting property of these bursts is that they exhibit fast transitions between multiple subconductance levels of approximately 8, 16, and 24 pS, with the mean open time at each level ranging between 0.3 and 0.9 ms (W. Zhang, Cho, Lolis, & Howe, 2008; W. Zhang et al., 2009). The occurrence of subconductance levels has been thought to correlate with the number of bound agonist molecules—two, three, or four (Rosenmund et al., 1998). A caveat to this interpretation is that GluA2 unitary activity in saturating glutamate still exhibits frequently occurring low and intermediate subconductance levels (Prieto & Wollmuth, 2010; W. Zhang et al., 2008), suggesting that other factors besides agonist occupancy dictate conductance level. Furthermore, a fourth, larger conductance level >30 pS is also occasionally visited, though it is more common if auxiliary subunits are present (Howe, 2015).

The Unique Pharmacology of AMPARs and KARs

From a functional sense, AMPARs are unique from KARs because they respond differently to certain agonists, excluding the indiscriminate neurotransmitter glutamate. Within the KAR subfamily, there are few other agonists that elicit maximal (>80% of glutamate) GluK2 peak current responses. Notably, the anthelmintic plant extract quisqualate (Biscoe, Evans, Headley, Martin, & Watkins, 1975), which the AMPAR subfamily was originally named after, and the synthetic molecule 2S,4R-4-methylglutamate (or SYM2081) fit this criterion (Fay, Corbeil, Brown, Moitessier, & Bowie, 2009). Somewhat less efficacious are KA as well as the related algal neurotoxin domoate, which produce GluK2 current responses that, respectively, comprise 50% and 15% of the peak amplitude evoked by glutamate (Fay et al., 2009). Interestingly, domoate responses are slow to desensitize, though deactivation following domoate removal is extremely sluggish, occurring over many seconds (Swanson et al., 1997), and consistent with high-affinity binding, common among toxins.

Further complicating the pharmacology of the KAR subfamily, there is an unusual divide between GluK1 and the other primary subunits (Table 9.2). Specifically, GluK1 is responsive to AMPA, while GluK2 and GluK3 are not, unless co-expressed with high-affinity KAR subunits (Herb et al., 1992; Schiffer, Swanson, & Heinemann, 1997). The same relationship holds for the AMPAR agonist (S)-5-iodowillardiine (Swanson, Green, & Heinemann, 1998), a modified relative of the amino acid willardiine found in plant seeds. A similar pattern also exists for the AMPA analogue (RS)-2-amino-3-(3-hydroxy-5-tert-butylisoxazol-4-yl)propanoic acid (ATPA), though this compound notably has greater potency at GluK1 than GluA1–4 subunits (Stensbol et al., 1999), enabling selective GluK1 KAR activation in neuronal preparations (Clarke et al., 1997). An added benefit of these agonists is that they have provided great templates from which to engineer multiple classes of GluK1-selective antagonists. In some cases, such antagonists (e.g., LY294486 and LY382884) are highly selective for GluK1 over AMPAR subunits, in addition to GluK2 and GluK3 (Bortolotto et al., 1999; Clarke et al., 1997). In other cases, antagonists like UBP310 inhibit GluK1 and GluK3 current responses more potently than those of GluK2 receptors (Mayer, Ghosal, Dolman, & Jane, 2006; Perrais, Pinheiro, Jane, & Mulle, 2009).

Among AMPARs, there is a more consistent pharmacological profile between subunits (Table 9.2). AMPA and quisqualate are considered full agonists at recombinant receptors, and in comparison to glutamate they induce slower deactivation and produce left-shifted dose–response curves, suggestive of higher affinity (W. Zhang, Robert, Vogensen, & Howe, 2006). KA and domoate can also elicit AMPAR activation (Boulter et al., 1990), but KA has a much lower potency than glutamate (Armstrong, Mayer, & Gouaux, 2003) and yields minimal current responses (<1% of glutamate) that do not decay over time (e.g., Dawe et al., 2016). Moreover, the willardiine agonist series consists of moderate to weak partial agonists, with larger halogen substituents reducing efficacy (Jin, Banke, Mayer, Traynelis, & Gouaux, 2003). At present, there are no agonists with remarkable selectivity for a particular subset of AMPAR subunits because the amino acid residues forming the agonist-binding pocket are almost entirely conserved among GluA1–4. The compound 2-Bn-Tet-AMPA, which exhibits a half maximal effective concentration (EC_{50}) that is 10-fold lower at GluA4 versus GluA1–3 receptors (Jensen et al., 2007), is unique in this regard.

Structure and Function of the iGluR LBD

The molecular mechanism of agonist efficacy at iGluRs has been unraveled using a combination of pharmacological interrogation and high-resolution protein structures. The first of these structures to be resolved was a KA-bound GluA2 (flop) LBD, crystallized using an S1-/S2-linked construct (Armstrong et al., 1998). The individual LBD is "kidney-shaped," containing upper and lower domain 1 (D1) and domain 2 (D2) lobes that both form contacts with the agonist molecule (Armstrong et al., 1998). In many cases, the LBD also crystallizes as a dimer, where the openings to the agonist-binding cleft are directed away from the central dimer interface (Armstrong & Gouaux, 2000). Interestingly, when full agonists like AMPA and glutamate bind to the isolated LBD they induce 20-degree closure of the agonist-binding cleft relative to the unliganded (apo) state, based on the angle between the D1 and D2 domains (Figure 9.4A). Meanwhile, the partial agonist KA and the competitive antagonist DNQX induce only 12 degrees and 5 degrees of additional cleft closure, respectively (Armstrong & Gouaux, 2000) (Figure 9.4B). Based on this spectrum of cleft closure, a structural model of agonist efficacy has developed, whereby more efficacious agonists are thought to induce greater closure,

Table 9.2 Agonists and Antagonists of AMPARs and KARs

	GluA1–4	GluK1	GluK2	GluK3	GluK1/GluK5	GluK2/GluK5	GluK3/Gluk5	NMDAR
Agonist								
L-Glutamate	+	+	+	+	+	+	+	+
Quisqualate	+(1–3)		+					+
AMPA	+	+	–	–	+	+	+	–
ATPA	+	+	–		+	+		
KA	+	+	+	+	+	+	+	–
Domoate	+(1,3)	+	+	–	+	+	–	
(S)-5-Iodowillardiine	+(1,2)	+	–	–	+	+	+	
Antagonist								
CNQX (10 μM)	+	+	+	+				–
NBQX	+	+	+					–
GYKI 52466 (100 μM)	+(1)		–			–		
LY 294486 (10 μM)	–(1–4)	+	–					
UBP 310 (30 μM)	–(2)	+	–	+				

Sources. The following papers were referenced to develop this table, though they do not represent the entirety of studies that examine AMPAR and KAR pharmacology: AMPAR agonists (Boulter et al., 1990; Jin et al., 2003; Keinänen et al., 1990; Kizelsztein, Eisenstein, Strutz, Hollmann, & Teichberg, 2000; Robert et al., 2005; Stensbol et al., 1999), KAR agonists (Alt et al., 2004; Herb et al., 1992; Schiffer et al., 1997; Swanson et al., 1997, 1998), NMDARs (Moriyoshi et al., 1991; Nakanishi, Shneider, & Axel, 1990), CNQX (Bettler et al., 1992; Egebjerg, Bettler, Hermans-Borgmeyer, & Heinemann, 1991; Sommer et al., 1992), NBQX (Bleakman et al., 1996), GYKI 52466 (Clarke et al., 1997; Stein, Cox, Seeburg, & Verdoorn, 1992), LY 294486 (Clarke et al., 1997), and UBP 310 (Mayer et al., 2006; Perrais et al., 2009).

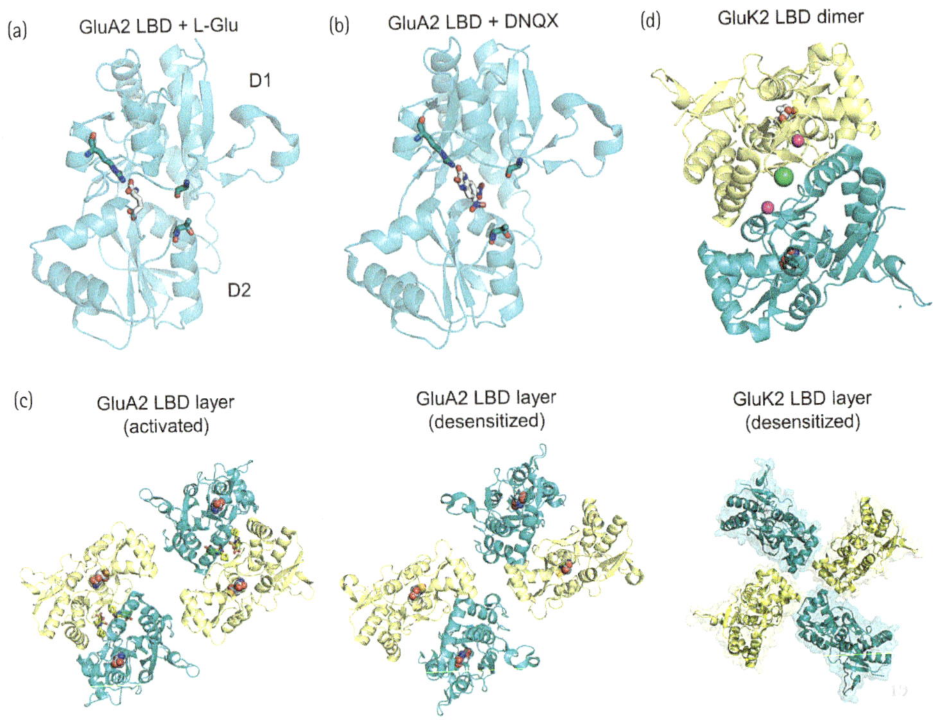

FIGURE 9.4 Dynamics of the AMPAR/KAR LBD underlie agonist efficacy and desensitization. (A, B) The isolated GluA2 LBD adopts a bilobed, clamshell-like arrangement with an agonist-binding cleft in between the D1 and D2 lobes. Binding of the full agonist glutamate (L-Glu; gray stick) induces closure of the cleft (PDB: 1FTJ), whereas binding of the competitive antagonist DNQX retains separation of residues on opposing faces of the cleft (PDB: 1FTL) (Armstrong & Gouaux, 2000). The same three residues are highlighted (cyan sticks) in each structure to provide perspective on the degree of cleft closure. (C) Top view of the GluA2 LBD layer from intact structures. In the presence of glutamate and cyclothiazide (yellow sticks), which attenuates desensitization, the subunits adopt an activated conformation, marked by closely held dimer pairs (left; PDB: 5WEO) (Twomey, Yelshanskaya, Grassucci, Frank, & Sobolevsky, 2017a. In a desensitized conformation, elicited by the agonist quisqualate, there is modest separation between subunits in each dimer pair (right, PBD: 5VHZ) (Meyerson et al., 2016; Twomey, Yelshanskaya, Grassucci, Frank, & Sobolevsky, 2017b). (D) Top view of the GluK2 LBD dimer (top), illustrating the binding pockets of allosteric sodium and chloride ions (PBD: 3G3F) (Chaudhry, Weston, Schuck, Rosenmund, & Mayer, 2009), as well as a similar view of complete GluK2 LBD layer (bottom) from an intact, desensitized structure (PDB: 4UQQ) (Meyerson et al., 2014). Note the extreme, >100-degree rotation of the B/D subunits (cyan) relative to those of the activated GluA2 structure.

which in turn facilitates gating of the channel pore (Armstrong & Gouaux, 2000). Further validation of this model came from the willardiine agonist series, for which increasingly bulky substituent groups reduced cleft closure in crystal structures, as well as relative efficacy of steady-state activation in electrophysiological assays (e.g., Jin et al., 2003). In addition, single-molecule fluorescence-based measurements have reported a higher probability of cleft closure for agonist-bound versus apo-state AMPAR LBDs (Landes, Rambhadran, Taylor, Salatan, & Jayaraman, 2011).

Despite the one-dimensional nature of the cleft closure paradigm, the measurement strongly correlates with agonist efficacy (Pohlsgaard et al., 2011) and has provided a starting point for thinking about how iGluR structure regulates channel gating. However, among the various LBD structures published since 2000 it is clear that certain bound agonists have not induced a degree of cleft closure commensurate with their agonist activity (e.g., Venskutonyte et al., 2012). Indeed, it has even been postulated that efficacy may be governed by a twisting motion of the LBD, rather than cleft closure (Birdsey-Benson, Gill, Henderson, & Madden, 2010). As more complete or intact receptor complexes are solved in different states, it will be appropriate to refine explanations of agonist efficacy to complement the observed changes in quaternary structure. For example, intersubunit crosslinking experiments indicate that the LBD layer explores more conformations, reducing its overall stability, when bound by partial, rather than full, agonists (Baranovic et al., 2016).

Allosteric Modulators of AMPARs

Several positive allosteric modulators of AMPARs have been identified from their effects on non-NMDAR responses in neuronal recordings. For example, the cognition-enhancing drug aniracetam potentiates AMPAR responses (Ito, Tanabe, Kohda, & Sugiyama, 1990) by slowing desensitization and excitatory postsynaptic current (EPSC) decay (Vyklicky, Patneau, & Mayer, 1991). Likewise, cyclothiazide, a benzothiadiazide compound originally developed as a diuretic, also enhances AMPAR responses (Yamada & Tang, 1993) but almost completely blocks agonist-induced desensitization, with minimal effect on deactivation kinetics (Patneau, Vyklicky, & Mayer, 1993; Yamada & Tang, 1993). In recombinant expression systems, cyclothiazide acts exclusively on AMPARs but not KARs (Partin, Patneau, Winters, Mayer, & Buonanno, 1993). Cyclothiazide exerts a greater modulation of flip AMPAR isoforms than flop versions, owing to the amino acid residue at position 775 (in GluA2), which is a serine in flip and asparagine in flop receptors (Partin et al., 1995; Partin, Fleck, & Mayer, 1996). As noted, this amino acid is positioned at the elbow point of an alpha-helical region on the interface between subunits of the dimer pairs in the LBD layer (Armstrong & Gouaux, 2000; Sobolevsky et al., 2009) (Figure 9.3B).

The LBD Dimer Interface and Receptor Desensitization

Slightly higher in the LBD dimer interface from the cyclothiazide binding site, residue 504 (in GluA2) can produce dramatic reductions in AMPAR desensitization when mutated from leucine to other aromatic-containing amino acids. In particular, the introduction of a tyrosine (known as the L/Y mutant) prevents desensitization almost entirely (Stern-Bach, Russo, Neuman, & Rosenmund, 1998). Because ultracentrifugation studies found that cyclothiazide-bound or L/Y mutant GluA2 LBDs exhibit greater protein dimerization, it was proposed that rupturing of the dimer interface is critical for desensitization to proceed for intact receptor complexes (Sun et al., 2002). In fact, a large number of residues along the AMPAR and KAR dimer interfaces appear to influence desensitization kinetics, based on the functional properties of receptors harboring mutations at these sites (e.g., Horning & Mayer, 2004; Y. Zhang, Nayeem, Nanao, & Green, 2006). Recent atomic resolution structures of intact GluA2 and GluK2 tetramers show that dimers within the LBD layer exhibit various degrees of subunit separation among proteins thought to be captured in desensitized states (e.g., Durr et al., 2014; Meyerson et al., 2014) (Figure 9.4C).

Within the KAR subfamily, for which there are no modulatory compounds able to disrupt desensitization to the extent of cyclothiazide at AMPARs, other approaches have been used to circumvent this gating process. Notably, the introduction of disulfide crosslinks across the LBD dimer interface imparts the GluK2 receptor with nondecaying responses to glutamate (Priel, Selak, Lerma, & Stern-Bach, 2006; Weston, Schuck, Ghosal, Rosenmund, & Mayer, 2006). However, subsequent analysis of the Y512C/L783C mutant has shown single-channel openings to be sporadic, despite their occurring with equal probability during saturating agonist applications (Daniels, Andrews, Aurousseau, Accardi, & Bowie, 2013). A separate point mutation in GluK2 (D776K) also disrupts macroscopic current decay associated with desensitization (Nayeem, Zhang, Schweppe, Madden, & Green, 2009). From a structural perspective, the mutant lysine residue points across the apex of the LBD dimer interface, tethering into an electronegative pocket on the opposing subunit (Nayeem, Mayans, & Green, 2011). Accordingly, the D776K mutation acts in a similar manner to constrain the dimer interface as Y521C/L783C, though single-channel recordings demonstrate its open probability to be much higher in glutamate and thus truly nondesensitizing (Dawe et al., 2013).

The Requirement of External Ions for KAR Channel Gating

The gating properties of native and recombinant KARs are regulated by external anions and cations (reviewed in Bowie, 2010). Specifically, lowering the external ionic strength results in faster deactivation and desensitization rates of GluK2 receptors, while increasing ionic strength slows current decay kinetics (Bowie, 2002; Bowie & Lange, 2002). The cation and anion species composing the extracellular solution also influence gating. For GluK2 KARs, sodium produces the largest peak current and slowest desensitization, whereas substitution with progressively larger cations reduces peak amplitudes and accelerates desensitization (Bowie, 2002). Likewise, among anions, chloride and bromide yield maximal peak currents, but substitution with fluoride or iodide lowers peak amplitudes and accelerates desensitization (Bowie, 2002). Similar trends are also exhibited by GluK1 and GluK3 KARs (Plested & Mayer, 2007). Consistent with their vital role in channel gating, removal of all external monovalent ions eliminates detectable current responses from GluK2 KARs (even outward currents at positive holding potentials) but not of GluA1 or GluA2 AMPARs (Dawe et al., 2016; Wong, Fay, & Bowie, 2006), demonstrating that allosteric sodium and chloride ions are, in fact, required for GluK2 gating.

The site for allosteric ion binding at KARs is contained within the LBD dimer interface, where two sodium ions reside in electronegative pockets atop opposing subunits and a single chloride is located in between, at the middle of the interface (Plested & Mayer, 2007; Plested, Vijayan, Biggin, & Mayer, 2008) (Figure 9.4D). Mutations at residues surrounding the sodium pocket, notably M770K, limit the variability in GluK2 behavior between different ionic conditions (MacLean, Wong, Fay, & Bowie, 2011; Paternain, Cohen, Stern-Bach, & Lerma, 2003; Wong, MacLean, & Bowie, 2007). The nondesensitizing mutant D776K also mimics sodium binding by introducing a stably tethered charge into the sodium pocket (Nayeem et al., 2011), causing cation sensitivity to be lost (Dawe et al., 2013; Nayeem et al., 2009). Thus, it has been proposed that the role of ions at GluK2 KARs is to maintain the LBD dimer organization, preventing the separation of subunits that accompanies desensitization (Dawe et al., 2013). In this model, ion unbinding is a key molecular event that triggers the onset of receptor desensitization (Dawe, Aurousseau, Daniels, & Bowie, 2015).

Surprisingly, heteromeric KARs are less sensitive to changes in external anion species than their homomeric counterparts (Paternain et al., 2003; Plested & Mayer, 2007). Assuming the LBD dimer pairs are composed of different subunits (e.g., GluK2 and GluK5), as in heteromeric AMPAR structures (Herguedas et al., 2016), the architecture of the dimer interface may account for the reduced sensitivity. In keeping with this, functional evidence shows that external lithium ions uniquely slow GluK2/GluK5 desensitization when substituted for sodium (Paramo, Brown, Musgaard, Bowie, & Biggin, 2017), an observation that is absent from homomeric channels. Interestingly, AMPARs also harbor the structural hallmarks of an allosteric cation-binding pocket as there are conserved electronegative residues in the equivalent location where sodium binds to KARs. Though ionic strength and external cation species generally do not influence AMPAR gating (Bowie, 2002; Bowie & Lange, 2002), an exception again occurs for lithium, which has been resolved in multiple GluA2 LBD structures (Assaf et al., 2013; Harms, Benveniste, MacLean, Partin, & Jamieson, 2013). Binding in a similar manner to sodium at KARs, lithium slows GluA2 desensitization, though without editing at the R/G position, as in GluA1, this effect is not observed (Dawe et al., 2016). Recent work has also shown that GluA2 flip receptors are exquisitely sensitive to external halide anions through a different binding site found at the LBD dimer interface, near position 775. Notably, larger halide species elicit faster entry into desensitization, though the effect is attenuated in GluA2 flop receptors (Dawe et al., 2019).

Structure and Function of the iGluR Amino-Terminal Domain

The ATD is not necessary for iGluR assembly and channel function, even though it encompasses roughly half of the entire protein (Kumar, Schuck, & Mayer, 2011). NMDARs (Fayyazuddin et al., 2000; Meddows et al., 2001), AMPARs (Pasternack et al., 2002), and KARs (Plested & Mayer, 2007) lacking their respective ATDs all remain capable of yielding current responses. Despite this apparent redundancy, it has long been known that the ATD is a regulatory site for gating, particularly in NMDARs (Hansen et al., 2010). For instance, variability within the ATD region of GluN2 subunits accounts for slower desensitization and deactivation in GluN2D versus GluN2A receptors (Gielen, Siegler Retchless, Mony, Johnson, & Paoletti, 2009; Monyer, Burnashev, Laurie, Sakmann, & Seeburg, 1994; Yuan, Hansen, Vance, Ogden, & Traynelis, 2009). The NMDAR ATD also contains binding sites for several allosteric modulators, including the divalent ion zinc (Peters, Koh, & Choi, 1987; Westbrook & Mayer, 1987) and the anti-ischemic drug ifenprodil (Carter et al., 1988), both of which antagonize responses. Among AMPARs, removal of the ATD slows desensitization roughly 2-fold and accelerates recovery from desensitization by a similar factor (Moykkynen, Coleman, Semenov, & Keinanen, 2014), though GluK2 KAR desensitization and deactivation appear to be unaffected without the N-terminal region (Plested & Mayer, 2007).

The ATD in Receptor Assembly

The structures of isolated AMPAR (GluA2), KAR (GluK2), and NMDAR (GluN2B) ATDs all possess a similar architecture (Jin et al., 2009; Karakas, Simorowski, & Furukawa, 2009; Kumar, Schuck, Jin, & Mayer, 2009). The AMPAR and KAR ATDs crystallize as dimers, with each subunit displaying a bilobed, or clamshell-shaped, organization (Kumar et al., 2009). Though the ATD of iGluRs shows some homology with the binding domain of metabotropic glutamate receptors and bacterial amino acid–binding proteins (O'Hara et al., 1993), there is little structural indication of conserved ligand recognition. This is due to poor sequence conservation at key amino acid–binding residues, as well as several structural features hindering domain closure (Jin et al., 2009; Kumar et al., 2009). Consequently, the most interesting property of the ATD is arguably its propensity for dimerization. Consistent with earlier analytical ultracentrifugation experiments showing that the GluA4 ATD and LBD formed dimers and monomers, respectively (Kuusinen, Abele, Madden, & Keinanen, 1999), analysis of the GluA2 and GluK2 ATDs indicates that their monomer–dimer dissociation constants are orders of magnitude lower than the LBDs of the same subunits (Jin et al., 2009; Kumar et al., 2009). It has therefore been hypothesized that the ATD might facilitate the initial dimerization step during assembly of the tetrameric receptor complex (Gan, Salussolia,

& Wollmuth, 2015). That being said, the structural template for tetramerization, at least for AMPARs, may be elsewhere, such as the TMD (Gan, Dai, Zhou, & Wollmuth, 2016)), since the AMPAR ATD cannot fully assemble on its own (H. Zhao et al., 2012). The ATD of GluK2/GluK5 KARs, however, is able to crystallize as a heterotetramer (Kumar et al., 2011), suggesting that some of the details of AMPAR and KAR assembly may be different.

A more refined interpretation of the ATD is that it biases AMPAR assembly in favor of specific subunit combinations, explaining the predominance of GluA1/GluA2 and/or GluA2/GluA3 heteromers at synapses (Henley & Wilkinson, 2016). This interpretation originated from investigation of AMPAR/KAR chimeras, from which it was concluded that a mismatched ATD region can prevent co-assembly of otherwise similar subunits (Ayalon & Stern-Bach, 2001; Leuschner & Hoch, 1999). Sedimentation velocity analysis of isolated AMPAR ATDs has revealed that the K_d values of ATD dimerization differ considerably between GluA1 (~100 nM), GluA2 (<10 nM), and GluA3 (>1 μM) (Rossmann et al., 2011; H. Zhao et al., 2012). However, the K_d for heterodimerization is reduced to around 1 nM for both GluA1/GluA2 and GluA2/GluA3 heteromers (Rossmann et al., 2011), implying that neither GluA1 nor GluA3 would be likely to assemble as a homomer in the presence of GluA2 subunits. Perhaps not coincidentally, it has been argued that most, if not all, AMPARs at the CA1 hippocampal synapse are GluA1/GluA2 (80%) or GluA2/GluA3 (15%) heteromers (Lu et al., 2009). Similarly, the GluK2/GluK5 heteromer has been described as the most common KAR complex in the brain (Petralia, Wang, & Wenthold, 1994). Not surprisingly then, the ATD heterodimer comprised of GluK2 and GluK5 has a lower K_d value than homodimers of either subunit (Kumar et al., 2011).

Protein Interactions Mediated by the ATD

The ATD is the principal site where the iGluRs are glycosylated as they are trafficked through the endoplasmic reticulum (ER) and Golgi (Everts, Villmann, & Hollmann, 1997). Though N-linked glycosylation notably facilitates subunit assembly and surface expression, it can also influence gating behavior as some of the N-linked oligosaccharides are association sites for extracellular modulators like lectins (carbohydrate-binding proteins). For example, the plant lectin concanavalin-A modifies current responses in KARs to a greater extent than AMPARs (e.g., Everts et al., 1997; Partin et al., 1993), an effect that can be disrupted by ablation of the glycosylation sites (Everts et al., 1999; Fay & Bowie, 2006). Other exogenous lectins, like agglutinin (Yue, MacDonald, Pekhletski, & Hampson, 1995), as well as marine and vertebrate galectins, also modulate AMPAR and KAR function in a subunit-dependent manner (Copits, Vernon, Sakai, & Swanson, 2014; Ueda et al., 2013). More specifically, galectins slow AMPAR and KAR desensitization (Copits et al., 2014) to an extent that is proportional to the number of N-glycosylation sites (Garcia-Nafria et al., 2016). It has also been proposed that the iGluR ATD influences assembly of multiprotein complexes at synapses. In one case, the ATD of GluA2 has been shown to bind N-cadherin, an interaction which promotes spine formation in cultured hippocampal neurons (Saglietti et al., 2007). Likewise, neuronal pentraxin, a lectin protein expressed on axons, colocalizes with and clusters neuronal GluA4 AMPARs but not if the ATD is deleted (Sia et al., 2007).

Effect of Heteromerization on Channel Gating

iGluR heteromerization has profound effects on channel gating as the pharmacological and kinetic properties of heteromeric receptors often differ quite substantially from those of homomeric tetramers of their constituent subunits (summarized in Tables 9.1 and 9.2). That being said, the differing incorporation of the flip or flop cassettes into AMPAR complexes represents another type of heteromerization that can profoundly affect the gating properties of native receptors (Dawe et al., 2019). Interestingly, by comparing the kinetics and anion sensitivity of native AMPARs of the cerebellum with recombinant receptors, the authors were able to propose that cerebellar stellate and Purkinje cells express heteromers that contain both flip and flop isoforms (Dawe et al., 2019).

Heteromerization can also affect ion permeation through the channel pore region. As mentioned, the presence of the Q/R edited GluA2 subunit in heteromeric AMPAR complexes reduces calcium permeability and polyamine block (Burnashev et al., 1992; Geiger et al., 1995; Washburn, Numberger, Zhang, & Dingledine, 1997). While transcripts of the GluA2 subunit are entirely edited at the Q/R site (and other AMPAR subunits are not edited at all), editing of the KAR subunits is less extensive and less well understood (Puchalski et al., 1994; Schmitt, Dux, Gissel, & Paschen, 1996; Sommer et al., 1991). For GluK1 and GluK2 subunits, the extent of editing appears to depend on brain region and developmental regulation (Bernard et al., 1999; Sommer et al., 1991). Additional RNA editing at other residues in the TM1 segment of GluK2 can also determine divalent permeability and polyamine sensitivity, and these sites are also regulated to different extents (Köhler, Burnashev, Sakmann, & Seeburg, 1993).

Among KARs, the assembly of primary (GluK1–3) and secondary (GluK4–5) KAR subunits is a critical determinant of their functional properties. For example, KARs containing GluK4 or GluK5 are responsive to AMPA (see Figure 9.2B) and exhibit slower deactivation kinetics (Barberis, Sachidhanandam, & Mulle, 2008; Herb et al., 1992; Mott, Rojas, Fisher, Dingledine, & Benveniste, 2010). The secondary KAR subunits also have a higher affinity for glutamate than the primary subunits as the dose–response curve for GluK2/GluK5 channels is left-shifted compared to that of homomeric GluK2 (Barberis et al., 2008; Fisher & Mott, 2011). Within heteromeric KARs, the primary subunits are thought to drive channel desensitization, while activation of only the GluK4–5 subunits sustains channel activation because current responses to agonists selective for heteromers, as well as low concentrations of nonselective agonists, exhibit much weaker desensitization (Fisher & Mott, 2011; Mott et al., 2010).

Regulation of the AMPAR and KAR Channel Pore

The TMD is comprised of four TM regions, of which TM1 and TM4 reside on the outside of the pore, the TM2 re-entrant loop forms the pore, and TM3 lines the upper segment of the permeation pathway (Bowie, 2018; Sobolevsky et al., 2009; Twomey, Yelshanskaya, Vassilevski, & Sobolevsky, 2018). The precise location of the selectivity filter is the segment immediately following the Q/R site, from which TM2 bends back toward the intracellular face of the membrane (Kuner et al., 2001). Mutations in this region greatly reduce AMPAR channel permeability to large organic cations (Kuner et al., 2001). Interestingly, iGluRs are permeable to a range of different-sized monovalent cations from lithium to cesium, and non-NMDARs can even be permeated by some organic cations like tris(hydroxymethyl)aminomethane, resulting in estimates of minimal pore diameter of 5.5 Å for NMDARs (Villarroel, Burnashev, & Sakmann, 1995) and 7.5 to 8.0 Å for AMPARs and KARs (Burnashev, Villarroel, & Sakmann, 1996). At the same time, NMDAR permeability to divalent calcium ions is about 3-fold greater (~10% fractional current) than for AMPARs, even when the Q/R site is unedited (Burnashev, Zhou, Neher, & Sakmann, 1995).

Polyamine Block

Polyamines are organic nonprotein cations containing two or more charged amine groups. In mammals, the naturally occurring polyamines are putrescine, spermidine, and spermine, with the most abundant being the latter two (Pegg, 2009; Pegg & McCann, 1982). Polyamines are found in high concentrations in mammalian cells, with estimates ranging from 10 to 100 μM (Bowie & Mayer, 1995; Watanabe, Kusama-Eguchi, Kobayashi, & Igarashi, 1991); but importantly, they are involved in a large number of cellular processes (Pegg, 2009). Given their cationic nature, they interact with negatively charged domains of biomolecules (Tabor & Tabor, 1984), including the pore regions of cation-selective VGICs and LGICs, where they bind with micromolar affinity and hinder ion flow (e.g., Bowie & Mayer, 1995; Haghighi & Cooper, 1998; Lopatin, Makhina, & Nichols, 1994). In this capacity, cytoplasmic polyamines are recognized as important determinants of neuronal signaling by regulating action potential firing rates (Fleidervish, Libman, Katz, & Gutnick, 2008) as well as the strength of neurotransmission (Aizenman, Munoz-Elias, & Cline, 2002; Rozov & Burnashev, 1999).

The inward rectification of *I–V* relationships obtained during AMPAR and KAR whole-cell recordings (e.g., Verdoorn et al., 1991) is due to channel block by intracellular polyamines (Bowie & Mayer, 1995; Donevan & Rogawski, 1995; Kamboj, Swanson, & Cull-Candy, 1995; Koh, Burnashev, & Jonas, 1995). As cations, polyamines are attracted into the channel pore, where, due to their larger cross-sectional diameter and slower permeation rates, they hinder the passage of other smaller cations such as sodium and calcium (Bowie, 2018). For example, the conductance of Q/R unedited GluK2 receptors at +50 mV is a mere 2% of the conductance at −100 mV, where virtually no polyamine block is detectable (Bowie & Mayer, 1995). At membrane potentials greater than +50 mV, the membrane electric field and cation permeation make conditions less favorable for polyamine binding (Bowie, Lange, & Mayer, 1998) such that the polyamines pass all the way through the pore (Bahring, Bowie, Benveniste, & Mayer, 1997). The ability of polyamines to both block and permeate unedited AMPARs and KARs defines them, in pharmacological terms, as permeant channel blockers, unlike the synthetic compounds MK-801 and phencyclidine, which block NMDARs in an effectively irreversible manner (Bowie, 2018). An added complication is that AMPAR and KAR pores are able to accommodate polyamines while in a closed state (Bowie et al., 1998). Interestingly, however, repetitive receptor activation can relieve much of this closed-channel block (Bowie et al., 1998; Rozov, Zilberter, Wollmuth, & Burnashev, 1998), which is thought to represent a novel mechanism of short-term plasticity in the mammalian brain (Rozov & Burnashev, 1999).

Given the prevalence of polyamines in the cytoplasm of almost all cells, most native AMPARs and KARs have evolved distinct mechanisms to prevent the occurrence of polyamine channel block. At AMPARs, two distinct mechanisms prevail that include (1) the formation of an electrostatic repulsion site at the apex of the pore, called the *Q/R site*, and (2) the co-assembly of AMPAR subunits with auxiliary proteins, such as TARPs and cornichon homologs (CNIHs; see below, Other AMPAR Auxiliary Proteins) (Bowie, 2018). Most native AMPARs are assembled with the GluA2 subunit, which contains a positively charged arginine residue at the Q/R site that repels polyamine from entering the pore. The *I–V* relationship of cells expressing GluA2-containing AMPARs is linear. In contrast, native AMPARs composed of GluA1, A3, and/or A4 subunits contain glutamine residues at the Q/R site, which favors polyamine block giving rise to inward rectification at positive membrane potentials. This difference in *I–V* relationships between GluA2-containing AMPARs, which exhibit a linear *I–V* relationship, and GluA2-lacking AMPARs, which exhibit a rectifying *I–V* relationship, is frequently used by synaptic physiologists as a marker to distinguish them in brain tissue. KARs have evolved three distinct mechanisms to attenuate polyamine block: (1) electrostatic repulsion at the Q/R site, (2) structural instability of pore helices via proline residues, and (3) the modulatory effect of neuropilin and tolloid-like (NETO) proteins (Bowie, 2018). Only GluK4 and GluK5 KAR subunits possess a proline residue in the pore helix, which means that native receptors assembled with either subunit lack cytoplasmic polyamine block. However, KARs assembled from GluK1, GluK2, and/or GluK3 all exhibit a high affinity for block by cytoplasmic polyamines.

Although polyamines are able to permeate AMPAR and KARs pores at extreme (>+50 mV) positive membrane potentials (Bahring et al., 1997), this has not been considered to be particularly significant in terms of cellular physiology. However, recent work has shown that the relief of polyamine block observed following auxiliary protein co-assembly with both AMPARs (e.g., TARP γ2 and CNIH3) and KARs (Neto1 and Neto2) (Soto, Coombs, Kelly, Farrant, & Cull-Candy, 2007) is achieved by facilitating polyamine permeation through the pore (Brown, Aurousseau, Musgaard, Biggin, & Bowie, 2016; Brown, McGuire, & Bowie, 2018). Surprisingly, polyamine flux can contribute to a significant conductance through the channel pore, particularly at positive membrane potentials. As mentioned, the lack of polyamine block of most native KARs containing either the GluK4 or GluK5 subunit (Barberis et al., 2008) is due to a single proline residue, conserved among GluK4 and GluK5 subunits, which is proposed to alter pore geometry around the selectivity filter (Brown et al., 2016).

Modulation by Fatty Acids

As with intracellular polyamines, externally applied fatty acids have also been shown to inhibit neuronal AMPARs and KARs, depending on their Q/R editing status (Huettner, 2015). Notably, arachidonic acid, a constituent of cell membranes, can attenuate neuronal AMPAR and KAR responses following several minutes of application (Kovalchuk, Miller, Sarantis, & Attwell, 1994; Wilding, Chai, & Huettner, 1998). Given that KAR inhibition occurs in a voltage-independent manner, it is noteworthy that increased susceptibility to inhibition occurs for Q/R edited receptors (Wilding, Zhou, & Huettner, 2005), the opposite of polyamine block. It remains unresolved whether fatty acids act as channel blockers or integrate into the membrane, altering the lipid environment around the TMD (Huettner, 2015).

Intact AMPAR and KAR Structures: Organization of Tetrameric Complexes

The first "intact" iGluR structure, resolved at atomic resolution (3.6 Å), was published in 2009, revealing the antagonist-bound GluA2 AMPAR to be a tall (180 Å), *Y*-shaped tetramer (Sobolevsky et al., 2009) (Figure 9.5A). Though earlier single-particle cryo-EM images had illustrated a "dimer of dimers" arrangement within the extracellular domains (Safferling et al., 2001; Tichelaar, Safferling, Keinanen, Stark, & Madden, 2004), the 2009 structure provided a wealth of new information regarding the arrangement of subunits, while offering hints at the structural basis of activation. Notably, the A/B and

C/D subunit pairs form ATD dimers, but the B and D subunits "cross over" to form closely packed pairs of LBD dimers comprised of A/D and B/C subunits. To achieve radial symmetry at the pore, the LBD–TMD linker orientations differ considerably between subunits, especially in the TM3–S2 linker, which is extended to reach the distal B and D subunit LBDs but compressed to connect with the more proximal A and C subunit LBDs (Sobolevsky et al., 2009).

The first high-resolution cryo-EM structures of intact, recombinant GluA2/GluA3 and GluA1/GluA2 heteromers have exhibited a 2:2 subunit stoichiometry,

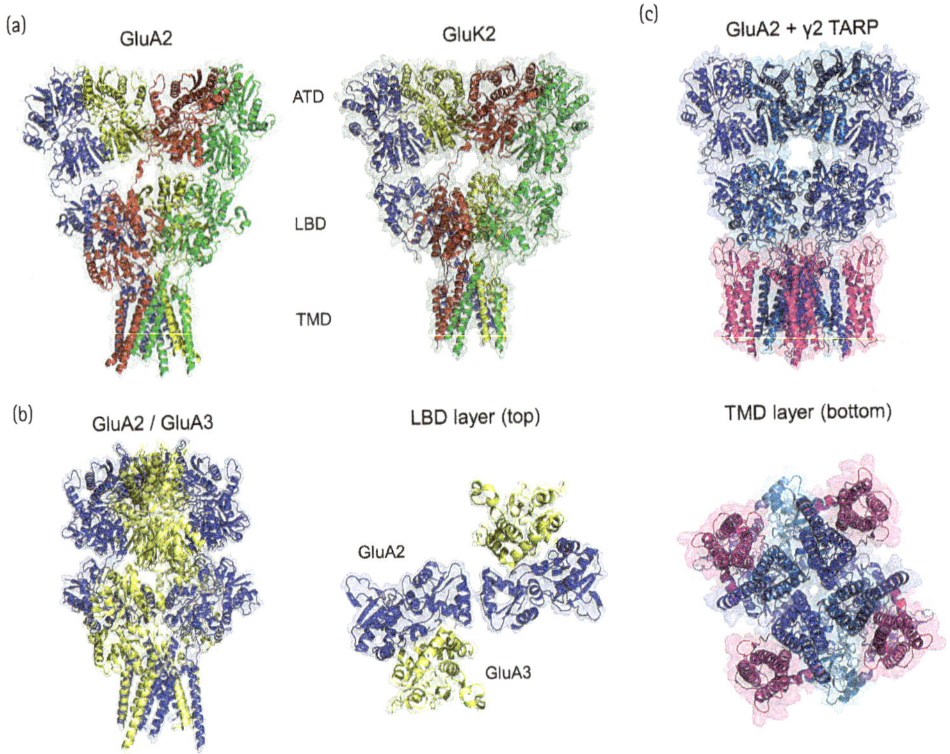

FIGURE 9.5 Arrangement of subunits in intact AMPAR and KAR complexes. (A) Side view of an intact tetrameric GluA2 AMPAR (left, PDB: 3KG2) in an antagonist-bound state (Sobolevsky et al., 2009) and a similarly arranged GluK2 KAR (right, PDB: 5KUF) in a desensitized state (Meyerson et al., 2016). Both structures are colored by subunit, as in the original 2009 paper: A, green; B, red; C, blue; D, yellow. Note the B and D subunits cross over between the ATD and LBD layers, forming dimer pairs with different opposing subunits in each layer. (B) Structure of an intact heterotetrameric AMPAR (left, PDB: 5IDE) formed by GluA2 (blue) and GluA3 (yellow) subunits (Herguedas et al., 2016). The arrangement of the subunits means that each LBD (and ATD) dimer comprises one GluA2 and one GluA3 subunit (right). (C) Structure of an intact AMPAR-TARP complex (top, PDB: 5WEO) formed by GluA2 (blue) and γ2 (magenta) subunits (Twomey et al., 2017a). Looking through the receptor from underneath the TMD layer (bottom), the channel pore can be resolved.

within which one subunit occupied the A/C positions, while the other subunit occupied the B/D positions (Figure 9.5B). Consequently, the ATD and LBD layers were both comprised of heterodimers, and the pore region, which could not be resolved, was expected to be surrounded in a 2:3:2:3 or 1:2:1:2 configuration, placing like subunits opposite to one another, rather than adjacent (Herguedas et al., 2016, 2019). Recent data suggest a potentially more complicated situation with native AMPAR heteromers, which may also assemble with a 1:3 stoichiometry, and with the apparent positional preference of specific subunits, such as GluA2, not necessarily fixed (Y. Zhao, Chen, Swensen, Qian, & Gouaux, 2019). There are not yet any intact structures of heteromeric KARs containing GluK4 or GluK5 subunits that would provide information regarding subunit position and its influence over their emergent gating properties. For NMDARs, which are obligate heteromers that are unable to form functional receptors from GluN1 or GluN2 subunits alone (Monyer et al., 1992), the intact structure reveals that the GluN1 subunits occupy the A/C positions, with the GluN2B subunits occupying the B/D positions (Karakas & Furukawa, 2014; Lee et al., 2014). Although representing an important advance, our understanding of the subunit arrangement within NMDA tetramers is likely to be revised as we learn more about the importance and abundance of triheteromeric NMDARs expressed in native brain tissue (Stroebel, Casado, & Paoletti, 2018; Yi, Bhattacharya, Thompson, Traynelis, & Hansen, 2019).

The GluA2/GluA3 apo state structures are also notable for their *O*-shaped conformation (Herguedas et al., 2016), featuring separated LBD dimers reminiscent of earlier EM reconstructions (i.e., Midgett, Gill, & Madden, 2012), rather than the *Y*-shaped crystal form. It should be noted that more recent structures where the tetramer is not constrained by stabilizing mutations do not exhibit this feature (Herguedas et al., 2019). Together, these structures suggest that the AMPAR apo state is capable of greater conformational flexibility than indicated by crystal structures. Fitting with this idea, EM images of native AMPAR complexes in unliganded and glutamate-bound conformations have illustrated that the compact organization of the ATD is lost during desensitization (Nakagawa, Cheng, Ramm, Sheng, & Walz, 2005; Nakagawa, Cheng, Sheng, & Walz, 2006). More recently, direct measurement of the conformational change induced by agonist binding and receptor desensitization was measured by atomic force microscopy (AFM) to represent almost 10% of the overall height of the receptor (Dawe et al., 2019). Another unexpected insight from AFM measurements has been the observation that nanoscale movement of the ATD in the apo state is regulated by the flip/flop cassette in the LBD (Dawe et al., 2019), which provides a mechanistic understanding of why the overall ATD architecture adopts several different conformations when the receptor is desensitized (Durr et al., 2014; Meyerson et al., 2014). This latter finding of bottom-up (LBD to ATD) conformational regulation of AMPARs differs from NMDARs, where it has been argued that top-down movements (ATD to LBD) dictate agonist open-channel probability and the degree of allosteric regulation (Gielen et al., 2008; Gielen et al., 2009; Yuan et al., 2009).

Mechanism of Pore Opening

A detailed, molecular model of iGluR activation has only recently come together, owing to the difficulty of capturing receptors in an open-channel state. Indeed, the brief, submillisecond open times measured during single-channel recordings of AMPARs and KARs (i.e., Swanson et al., 1997; W. Zhang et al., 2008) suggest that open states are inherently unstable. To better visualize AMPARs in an activated form, agonist-bound protein structures have been obtained with positive allosteric modulators also present (Durr et al., 2014; Meyerson et al., 2014). In such cases, AMPARs exhibit greater closure of the agonist-binding cleft, along with 5–20 Å of vertical compression in the ATD and LBD layers, when compared to apo and antagonist-bound states. These structures also featured an outward expansion of the LBD–TMD linkers, proposed to generate the mechanical force that pulls open the channel pore (i.e., Dong & Zhou, 2011; Sobolevsky et al., 2009). Nevertheless, the first intact structure with the resolution of an open channel pore was obtained relatively recently by imaging GluA2–TARP complexes with cyclothiazide bound (Twomey et al., 2017a) (Figure 9.5C). Based on the architecture of the pore, it has been hypothesized that upward pulling by the LBD–TMD linkers (especially TM3–S2) flips apart pore-lining TM2 segments near the selectivity filter, along with TM3 residues that form an "upper gate," allowing ion permeation (Twomey et al., 2017a). Consistent with this framework, the insertion of amino acid residues into TM3–S2 linkers reduces NMDAR open probability, presumably counteracting the tension that pulls open the pore during gating (Kazi, Dai, Sweeney, Zhou, & Wollmuth, 2014).

LBD Layer Rotation During Desensitization

Several intact, agonist-bound AMPAR and KAR structures resolved in the absence of positive modulators have been presumed to represent desensitized conformations, on account of the high probability that these receptors are desensitized in the continued presence of the agonist. For GluK2 KARs, cryo-EM mapping of its resting and desensitized states indicates that the LBD layer undergoes dynamic rearrangements during the desensitization process (Schauder et al., 2013) (Figure 9.4C). Specifically, the resting state LBD is formed by two closely situated dimer pairs that separate into isolated domains as desensitization proceeds. Remarkably, this separation involves an extreme 125-degree rotation of the distal B/D subunits relative to the resting state (Meyerson et al., 2016, 2014). Rearrangements on a similar scale have not been observed for AMPARs, perhaps because their desensitized states are less stable and/or more short-lived (e.g., Bowie & Lange, 2002). In one agonist-bound GluA2 AMPAR structure, a 105-degree rotation was seen in one subunit within the LBD layer, though the other dimer was largely intact (Durr et al., 2014). Furthermore, a quisqualate-bound AMPAR complexed with the auxiliary protein GSG1L (which slows recovery from desensitization) displayed 14-degree rotation of the A/C subunits, relative to their position in another, presumably nondesensitized structure (Twomey et al., 2017a) (Figure 9.4C). These results suggest that structural arrangements during AMPAR desensitization might be complex and variable between agonists or with auxiliary protein association. Interestingly, despite some rotation of the overall ATD layer, ATD dimers remain intact in desensitized AMPAR structures (Meyerson et al., 2016; Twomey et al.,2017b).

Auxiliary Subunits

Though several proteins are considered auxiliary subunits of AMPARs and KARs, numerous other proteins may interact with or regulate these receptor families without being classified as auxiliary to pore-forming subunits. A useful framework defines auxiliary subunits as being unable of forming ion channels alone (1) but able to interact directly with a pore-forming subunit (2) and modulate its trafficking and/or gating properties in heterologous cells (3), as well as have some effect in vivo (4) (D. Yan & Tomita, 2012). Extensive investigation into the modulation of AMPARs by TARPs (and other auxiliary proteins), as well as KARs by NETO1 and NETO2, has pointed to a fundamental role for auxiliary proteins in the physiology of glutamatergic synapses (Jackson & Nicoll, 2011).

TARPs as AMPAR Chaperones

The first known AMPAR auxiliary protein (TARP γ2) was originally referred to as *stargazin* because it is encoded by the gene disrupted in the stargazer mutant mouse, noted for its absence seizures (Letts et al., 1998; Noebels, Qiao, Bronson, Spencer, & Davisson, 1990). Interestingly, synaptic AMPAR-mediated responses are almost entirely absent in stargazer mice in cerebellar granule cells (Hashimoto et al., 1999). This observation led to the discovery that AMPAR surface expression is enhanced in the presence of stargazin and that their association is required for synaptic localization of AMPARs, which depends on interactions between stargazin and the scaffolding protein PSD-95 (L. Chen et al., 2000; Schnell et al., 2002). Further to this point, interactions between AMPAR and γ2 subunits have been detected in the plasma membrane and ER (Bedoukian, Weeks, & Partin, 2006), suggesting that TARP augmentation of receptor trafficking might stem from early intervention in protein folding or subunit assembly (Figure 9.6A).

The cloning of additional TARP subunits has led to the grouping of γ2, γ3, γ4, and γ8 (type I) based on a conserved TTPV amino acid motif at the intracellular C terminus, while two smaller subgroups of γ5 and γ7 (type II), as well as γ1 and γ6, also exist (Burgess, Gefrides, Foreman, & Noebels, 2001). Protein expression of these type I TARP subunits occurs differentially throughout the brain, with a predominance of γ2 in the cerebellum, γ3 in the cerebral cortex, and γ8 in the hippocampus (Tomita et al.,

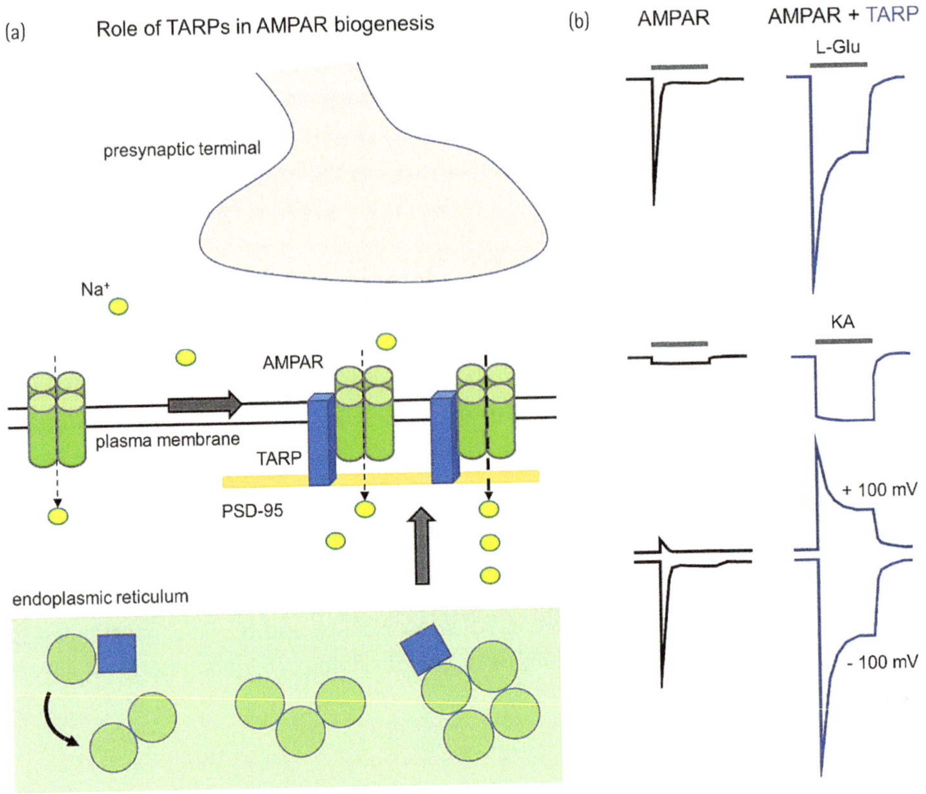

FIGURE 9.6 Role of auxiliary subunits in AMPAR expression and gating. (A) Cartoon illustration of the various roles that TARPs (and other auxiliary proteins) are thought to play in AMPAR biogenesis. TARPs associate with AMPARs in the ER and promote forward trafficking to the plasma membrane. At the membrane, TARPs help AMPARs to anchor in the postsynaptic density through binding to PSD-95. Finally, TARPs also enhance AMPAR gating. (B) TARPs modulate AMPAR function in numerous ways, including slowing agonist-induced desensitization (top), increasing the efficacy of KA (middle), and relieving polyamine block at positive membrane potentials (bottom).

2003). On the whole, there is minor variability in the functional phenotype imparted onto AMPARs by the type I TARP subunits (Kott, Werner, Korber, & Hollmann, 2007; Milstein, Zhou, Karimzadegan, Bredt, & Nicoll, 2007), and they all generally modulate surface trafficking and the duration of channel gating in a positive manner (summarized in Jackson & Nicoll, 2011). For the type II subunits γ5 and γ7, which possess a shorter intracellular C terminus, less modulation of AMPAR gating occurs, despite the fact that both can immunoprecipitate with AMPAR subunits from brain tissue (Kato et al., 2007). The γ5 subunit actually modestly accelerates the deactivation and desensitization of recombinant GluA2 receptors (with no effect on GluA1), despite simultaneously increasing their current response (Kato, Siuda, Nisenbaum, & Bredt, 2008). This current understanding is likely to be revised as we learn more about the regional and cell-type expression of TARPs in the mammalian brain and their effect on recombinant receptors.

Functional Modulation of AMPARs by TARPs

Numerous gating and permeation properties of recombinant AMPARs are positively modulated by stargazin and other type I TARPs. Notably, deactivation and desensitization are slower, the relative efficacy of KA versus glutamate is increased, the glutamate dose–response curve is leftward-shifted, and single-channel properties such as channel conductance and burst length are increased (Priel et al., 2005; Tomita et al., 2005; Turetsky, Garringer, & Patneau, 2005) (Figure 9.6B). In keeping with their ability to favor and stabilize the open state, TARP binding to AMPARs has also been shown to convert quinoxalinedione antagonists (i.e., CNQX and DNQX) into weak partial agonists (MacLean & Bowie, 2011; Menuz, Stroud, Nicoll, & Hays, 2007), promote "resensitization" of the channel during long agonist pulses (Kato et al., 2007), reduce the ability of cyclothiazide to potentiate AMPARs (Cho, St-Gelais, Zhang, Tomita, & Howe, 2007), trigger the occurrence of modal gating behavior (W. Zhang, Devi, Tomita, & Howe, 2014), and uncover enhanced recovery from desensitization, or "superactivation," where the test response amplitude surpasses the initial/conditioning response (Carbone & Plested, 2016). Other than their effects on channel gating, TARPs also affect the channel pore by attenuating channel block by cytoplasmic polyamines (Soto et al., 2007) by enhancing polyamine permeation (Brown et al., 2018). Importantly, the appreciation of TARP-mediated effects has helped explain the biophysical properties of native AMPARs that differed from findings on recombinant receptors. As a result, it is generally assumed that almost all native AMPARs expressed in the mammalian brain are associated either fully or partially with TARP subunits (reviewed in Kato, Gill, Yu, Nisenbaum, & Bredt, 2010; see also Dawe et al., 2019).

AMPAR–TARP Assembly and Stoichiometry

Because AMPAR–TARP fusion proteins retain the same altered biophysical properties as observed during co-expression (Morimoto-Tomita et al., 2009; Shi, Lu, Milstein, & Nicoll, 2009), it was initially assumed that one TARP subunit associated with every pore-forming AMPAR subunit. Without the constraint of protein fusion, experimental approaches quantifying the molecular weight of AMPAR complexes or counting fluorescent-tagged TARP subunits have suggested a variable stoichiometry of between one and four TARPs per channel (Hastie et al., 2013; Kim, Yan, & Tomita, 2010). Such estimates have also been supported by cryo-EM studies of GluA2–γ2 complexes, consisting of one, two, or four TARPs per receptor (Twomey, Yelshanskaya, Grassucci, Frank, & Sobolevsky, 2016; Y. Zhao, Chen, Yoshioka, Baconguis, & Gouaux, 2016) (Figure 9.7). A recent study exploring AMPAR–TARP stoichiometry in the cerebellum has shown that some cells (i.e., Purkinje cells) behave as though having a full TARP contingent, whereas other cells, such as inhibitory stellate cells, are only partially TARP-modified (Dawe et al., 2019). Whether these observations can be extended to other brain

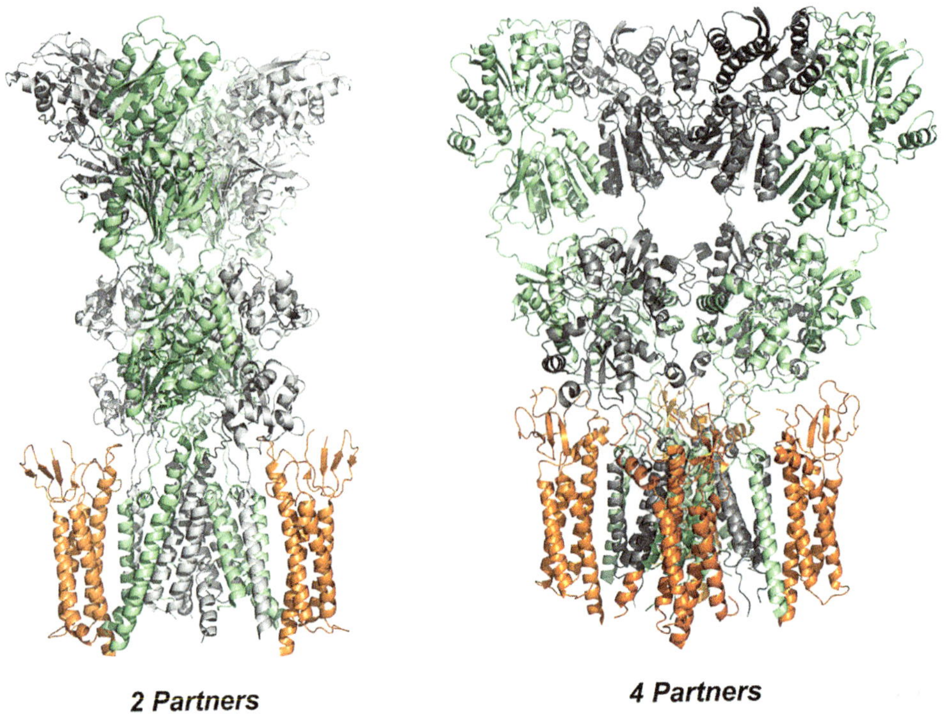

FIGURE 9.7 Variable stoichiometry of AMPAR auxiliary proteins. Synaptic AMPARs display current responses consistent with intermediate and full TARP association, depending on the cell type recorded from. Variable auxiliary protein stoichiometry represents an additional layer to the regulation of glutamatergic signaling.

regions, such as the hippocampus, where γ8 rather than γ2 expression predominates, remains to be investigated.

The structural basis for TARP enhancement of AMPAR gating likely involves multiple interaction sites within the LBD, TMD, and CTD since TARPs have four TM segments with relatively small exterior regions. Indeed, deletion of the ATD still permits functional modulation by γ2 (Cais et al., 2014; Tomita, Shenoy, Fukata, Nicoll, & Bredt, 2007). Yet closer to the membrane, γ2 induces domain closure of the AMPAR agonist-binding cleft, presumably through some extracellular contact point, accounting for the increased efficacy of partial agonists (MacLean, Ramaswamy, Du, Howe, & Jayaraman, 2014). Likewise, the mutation of positively charged LBD residues, namely the KGK motif, predicted to interact with the extracellular loop of TARPs, greatly attenuates the effect of γ2 on GluA2 gating (Dawe et al., 2016). Full polyamine block is not recovered by this mutation, suggesting that other regions of the TARP structure are responsible for its effects on the channel pore (Dawe et al., 2016). On this note, evidence for the TMD being the principal region of TARP association has come from the study of GluK2–GluA3 chimeras. In this case, substitution of the AMPAR TM segments and CTD into the KAR backbone confers TARP association and modulation of some gating properties (Ben-Yaacov et al., 2017).

Whether AMPAR–TARP contact is maintained over time or more transient in nature has been debated, following initial reports that TARPs can dissociate from the AMPAR during agonist binding/activation in a process termed *autoinactivation* (Morimoto-Tomita et al., 2009). Given insight from recent work, the dissociation may simply represent disengagement from the KGK motif (Dawe et al., 2016), although this possibility would need to be formally investigated. For synaptic AMPARs, tracked at the single-molecule level, glutamate exposure appears to induce greater mobility, and the effect can be prevented by fusion to TARPs, suggesting that TARP uncoupling may occur (Constals et al., 2015). However, various indicators of TARP modulation (i.e., enhanced KA-evoked currents, reduced polyamine block) remain intact following long, desensitizing agonist applications, supporting the idea that dissociation does not occur (Coombs, MacLean, Jayaraman, Farrant, & Cull-Candy, 2017). Clearly, more work is needed if these apparent differences are to be resolved.

Other AMPAR Auxiliary Proteins

Cornichon homologs 2 and 3 (CNIH-2 and CNIH-3) have also been identified as AMPAR auxiliary proteins (Schwenk et al., 2009). Although initial studies suggested that these proteins are likely to have three TM helices (Schwenk et al., 2009), a recent cryo-EM structure has shown that they possess four TM regions with little or no extracellular domain (Nakagawa, 2019). In agreement with the recent full-length structure (Nakagawa, 2019), CNIHs have been shown to associate with the TMD of AMPARs through membrane-proximal residues of the extracellular and intracellular domains (Shanks et al., 2014). Like TARPs, heterologously expressed CNIH-2 and CNIH-3 enhance AMPAR surface expression but have an even more profound impact on slowing current decay kinetics (Brown et al., 2018; Coombs et al., 2012; Schwenk et al., 2009). In addition, they increase single-channel conductance and attenuate voltage-dependent polyamine block of unedited AMPARs (Coombs et al., 2012) through a mechanism that promotes polyamine permeation (Brown et al., 2018). CNIH proteins are important in the brain, where conditional knockout of both CNIH-2 and CNIH-3 can greatly reduce synaptic AMPAR-mediated currents (Herring et al., 2013). Interestingly, CNIH-2, but not CNIH-3, is found in high relative abundance throughout the rodent brain (Schwenk et al., 2014).

Germ cell-specific gene 1-like (GSG1L) is another four-pass TM protein found in native AMPAR complexes (Schwenk et al., 2012; Shanks et al., 2012) and shown to associate with multiple AMPAR subunits in vitro (Shanks et al., 2012). When co-expressed with GluA2, GSG1L slows desensitization and recovery from desensitization (Shanks et al., 2012), though interestingly, unlike TARP and CNIH subunits, it reduces single-channel conductance and modestly enhances polyamine block (McGee, Bats, Farrant, & Cull-Candy, 2015; though see Bowie, 2018). In neurons, GSG1L impairs membrane trafficking to reduce the amplitude of AMPAR-mediated EPSCs (Gu et al., 2016; McGee et al., 2015). The full-length AMPAR–GSG1L cryo-EM structure suggests that GSG1L modulates AMPAR gating via an extracellular loop in a manner similar to γ2 TARPs, which is surprising given that GSG1L dramatically slows recovery from desensitization, whereas stargazin speeds it up.

Another class of AMPAR auxiliary proteins is known as cystine-knot AMPAR-modulating proteins (CKAMPs) or Shisa proteins (Haering, Tapken, Pahl, & Hollmann, 2014). CKAMP members are thought to possess a single-pass TM topology with an intracellular PDZ domain-binding motif (Farrow et al., 2015). Two members (CKAMPs 44 and 52 or Shisas 9 and 6, respectively) have been specifically identified as co-localizing with AMPARs at excitatory synapses (Klaassen et al., 2016; von Engelhardt et al., 2010) and are thought to stabilize AMPARs in the postsynaptic density through PDZ interactions (Khodosevich et al., 2014; Klaassen et al., 2016). Biophysical investigation in recombinant systems has revealed variable effects on current amplitudes and gating properties, depending on the AMPAR and CKAMP subunits being expressed (Farrow et al., 2015; Khodosevich et al., 2014; Klaassen et al., 2016), though the most striking effect is to slow recovery from desensitization much like GSG1L. Since AMPAR behavior in CKAMP44 and CKAMP52 knockout mice is not particularly perturbed (Khodosevich et al., 2014; Klaassen et al., 2016; von Engelhardt et al., 2010), it was initially questioned whether CKAMPs play a fundamental role in AMPAR physiology. However, more recent work on knockout mice has demonstrated the importance of CKAMP44 in modulating synaptic short-term depression and input integration of the visual pathway in the lateral geniculate nucleus (X. Chen, Aslam, Gollisch, Allen, & von Engelhardt, 2018).

Other putative AMPAR auxiliary subunits have also been identified, including SynDIG1 (Kalashnikova et al., 2010), Porcupine (Erlenhardt et al., 2016), and SOL-1,2 in *Caenorhabditis elegans* (Wang et al., 2012; Zheng et al., 2006); but less is known about their role in AMPAR function at synapses. For a review of the modulatory effects of some of these auxiliary proteins, see Haering et al. (2014).

KAR Auxiliary Proteins

Neto1 and Neto2 are the two principal KAR auxiliary subunits (Tang et al., 2011; W. Zhang et al., 2009). They are predicted to be single-pass TM proteins containing two extracellular complement C1r/C1s, Uegf, Bmp1 domains and a low-density lipoprotein receptor domain A domain (Stöhr, Berger, Froehlich, & Weber, 2002). Both Neto1 and Neto2 interact with the GluK1 and GluK2 KAR subunits, enhancing trafficking to synaptic membranes (Sheng, Shi, Lomash, Roche, & Nicoll, 2015; Tang et al., 2011) as well as modulating functional properties (Copits, Robbins, Frausto, & Swanson, 2011; W. Zhang et al., 2009) in a subunit-dependent manner (Fisher, 2015). At present, the stoichiometry of Neto1 and Neto2 with homomeric and heteromeric KARs remains unknown.

Generally speaking, Neto1 and Neto2 slow the desensitization and deactivation kinetics of recombinant and synaptic KARs (Straub et al., 2011; W. Zhang et al., 2009). For Neto2, this effect results from an increase in the open probability and burst length of GluK2 single-channel currents (W. Zhang et al., 2009). Structurally, interactions of Neto2 with the GluK2 M3–S2 linker, as well as with the D1–D1 dimer interface, are critical for its modulatory effects on gating (Griffith & Swanson, 2015). Like TARPs, the Neto proteins also attenuate voltage-dependent polyamine block of KARs (Fisher & Mott, 2012) by facilitating polyamine permeation through the channel pore (Brown et al., 2016) in much the same way that TARPs and CNIHs attenuate channel block of AMPARs.

Conclusion

The vast majority of iGluR subunits were initially cloned in the early 1990s. Since that time, the first decade of research uncovered the complex gating behavior and pharmacology of recombinantly expressed receptors and rules governing heteromeric assembly. The second decade heralded the first atomic resolution structures of individual iGluR domains, as well as the discovery of auxiliary proteins that modulate AMPAR and KAR activity at the synapse. Finally, the years leading up to 2020 have been accompanied by structures of intact receptor complexes and models that relate protein movements to gating processes. Looking ahead, several fundamental questions connecting the biophysical properties of iGluRs with their physiological functions remain to be answered. Specifically, what combinations of auxiliary proteins accompany iGluRs in different cell types, and is their stoichiometry important in distinguishing responses between synapses? Moreover, how are iGluR–auxiliary protein interactions temporally regulated, and can such interactions contribute to synaptic plasticity? On a separate note, can the now extensive contingent of iGluR-regulating proteins provide novel routes for pharmacological regulation of glutamatergic signaling? An even more fundamental question involves the variability in AMPAR, KAR, and TARP subunit expression between brain regions. What is the need for local enrichment of specific subunits that generally exhibit the same gating properties as other subunits? And does the utility of flip/flop alternative splicing extend beyond the fine-tuning of desensitization, affecting native receptor assembly and function in other ways? Even after so many breakthrough advances in our understanding of AMPAR and KAR biology, it is still true, both literally and metaphorically, that exciting times remain ahead.

References

Aizenman, C. D., Munoz-Elias, G., & Cline, H. T. (2002). Visually driven modulation of glutamatergic synaptic transmission is mediated by the regulation of intracellular polyamines. *Neuron*, *34*(4), 623–634.

Alt, A., Weiss, B., Ogden, A. M., Knauss, J. L., Oler, J., Ho, K., . . . Bleakman, D. (2004). Pharmacological characterization of glutamatergic agonists and antagonists at recombinant human homomeric and heteromeric kainate receptors in vitro. *Neuropharmacology*, *46*(6), 793–806. Retrieved from http://www.ncbi.nlm.nih.gov/entrez/query.fcgi?cmd=Retrieve&db=PubMed&dopt=Citation&list_uids=15033339

Andersen, O. S., & Koeppe, R. E., 2nd. (1992). Molecular determinants of channel function. *Physiological Reviews*, *72*(4, Suppl.), S89–S158.

Armstrong, N., & Gouaux, E. (2000). Mechanisms for activation and antagonism of an AMPA-sensitive glutamate receptor: Crystal structures of the GluR2 ligand binding core. *Neuron*, *28*(1), 165–181.

Armstrong, N., Mayer, M., & Gouaux, E. (2003). Tuning activation of the AMPA-sensitive GluR2 ion channel by genetic adjustment of agonist-induced conformational changes. *Proceedings of the National Academy of Sciences of the United States of America*, *100*(10), 5736–5741. doi:10.1073/pnas.1037393100

Armstrong, N., Sun, Y., Chen, G. Q., & Gouaux, E. (1998). Structure of a glutamate-receptor ligand-binding core in complex with kainate. *Nature*, *395*(6705), 913–917.

Assaf, Z., Larsen, A. P., Venskutonyte, R., Han, L., Abrahamsen, B., Nielsen, B., . . . Bunch, L. (2013). Chemoenzymatic synthesis of new 2,4-syn-functionalized (S)-glutamate analogues and structure–activity relationship studies at ionotropic glutamate receptors and excitatory amino acid transporters. *Journal of Medicinal Chemistry*, *56*(4), 1614–1628. doi:10.1021/jm301433m

Ataman, Z. A., Gakhar, L., Sorensen, B. R., Hell, J. W., & Shea, M. A. (2007). The NMDA receptor NR1 C1 region bound to calmodulin: Structural insights into functional differences between homologous domains. *Structure*, *15*(12), 1603–1617.

Auerbach, A. (2015). Agonist activation of a nicotinic acetylcholine receptor. *Neuropharmacology*, *96*(Pt. B), 150–156. doi:10.1016/j.neuropharm.2014.10.004

Ayalon, G., & Stern-Bach, Y. (2001). Functional assembly of AMPA and kainate receptors is mediated by several discrete protein-protein interactions. *Neuron*, *31*(1), 103–113.

Bahring, R., Bowie, D., Benveniste, M., & Mayer, M. L. (1997). Permeation and block of rat GluR6 glutamate receptor channels by internal and external polyamines. *Journal of Physiology*, *502*(Pt. 3), 575–589.

Banke, T. G., Bowie, D., Lee, H., Huganir, R. L., Schousboe, A., & Traynelis, S. F. (2000). Control of GluR1 AMPA receptor function by cAMP-dependent protein kinase. *Journal of Neuroscience*, *20*(1), 89–102.

Baranovic, J., Chebli, M., Salazar, H., Carbone, A. L., Faelber, K., Lau, A. Y., . . . Plested, A. J. (2016). Dynamics of the ligand binding domain layer during AMPA receptor activation. *Biophysical Journal*, *110*(4), 896–911. doi:10.1016/j.bpj.2015.12.033

Baranovic, J., & Plested, A. J. (2016). How to build the fastest receptor on earth. *Biological Chemistry*, *397*(3), 195–205. doi:10.1515/hsz-2015-0182

Barberis, A., Sachidhanandam, S., & Mulle, C. (2008). GluR6/KA2 kainate receptors mediate slow-deactivating currents. *Journal of Neuroscience*, *28*(25), 6402–6406.

Bedoukian, M. A., Weeks, A. M., & Partin, K. M. (2006). Different domains of the AMPA receptor direct stargazin-mediated trafficking and stargazin-mediated modulation of kinetics. *Journal of Biological Chemistry*, *281*(33), 23908–23921. doi:10.1074/jbc.M600679200

Ben-Yaacov, A., Gillor, M., Haham, T., Parsai, A., Qneibi, M., & Stern-Bach, Y. (2017). Molecular mechanism of AMPA receptor modulation by TARP/stargazin. *Neuron*, *93*(5), 1126–1137. doi:10.1016/j.neuron.2017.01.032

Bernard, A., Ferhat, L., Dessi, F., Charton, G., Represa, A., Ben-Ari, Y., & Khrestchatisky, M. (1999). Q/R editing of the rat GluR5 and GluR6 kainate receptors in vivo and in vitro: Evidence for independent developmental, pathological and cellular regulation. *European Journal of Neuroscience*, *11*(2), 604–616.

Bettler, B., Boulter, J., Hermans-Borgmeyer, I., O'Shea-Greenfield, A., Deneris, E. S., Moll, C., . . . Heinemann, S. (1990). Cloning of a novel glutamate receptor subunit, GluR5: Expression in the nervous system during development. *Neuron, 5*(5), 583–595.

Bettler, B., Egebjerg, J., Sharma, G., Pecht, G., Hermans-Borgmeyer, I., Moll, C., . . . Heinemann, S. (1992). Cloning of a putative glutamate receptor: A low affinity kainate-binding subunit. *Neuron, 8*(2), 257–265.

Birdsey-Benson, A., Gill, A., Henderson, L. P., & Madden, D. R. (2010). Enhanced efficacy without further cleft closure: Reevaluating twist as a source of agonist efficacy in AMPA receptors. *Journal of Neuroscience, 30*(4), 1463–1470. doi:10.1523/JNEUROSCI.4558-09.2010

Biscoe, T. J., Evans, R. H., Headley, P. M., Martin, M., & Watkins, J. C. (1975). Domoic and quisqualic acids as potent amino acid excitants of frog and rat spinal neurones. *Nature, 255*(5504), 166–167.

Bleakman, D., Ballyk, B. A., Schoepp, D. D., Palmer, A. J., Bath, C. P., Sharpe, E. F., . . . Lodge, D. (1996). Activity of 2,3-benzodiazepines at native rat and recombinant human glutamate receptors in vitro: Stereospecificity and selectivity profiles. *Neuropharmacology, 35*(12), 1689–1702.

Bortolotto, Z. A., Clarke, V. R., Delany, C. M., Parry, M. C., Smolders, I., Vignes, M., . . . Collingridge, G. L. (1999). Kainate receptors are involved in synaptic plasticity. *Nature, 402*(6759), 297–301.

Boulter, J., Hollmann, M., O'Shea-Greenfield, A., Hartley, M., Deneris, E., Maron, C., & Heinemann, S. (1990). Molecular cloning and functional expression of glutamate receptor subunit genes. *Science, 249*(4972), 1033–1037.

Bowie, D. (2002). External anions and cations distinguish between AMPA and kainate receptor gating mechanisms. *Journal of Physiology*, 539(Pt. 3), 725–733.

Bowie, D. (2010). Ion-dependent gating of kainate receptors. *Journal of Physiology, 588*, 67–81.

Bowie, D. (2018). Polyamine-mediated channel block of ionotropic glutamate receptors and its regulation by auxiliary proteins. *Journal of Biological Chemistry, 293*(48), 18789–18802. doi:10.1074/jbc.TM118.003794

Bowie, D., Garcia, E. P., Marshall, J., Traynelis, S. F., & Lange, G. D. (2003). Allosteric regulation and spatial distribution of kainate receptors bound to ancillary proteins. *Journal of Physiology, 547*(Pt. 2), 373–385.

Bowie, D., & Lange, G. D. (2002). Functional stoichiometry of glutamate receptor desensitization. *Journal of Neuroscience, 22*(9), 3392–3403.

Bowie, D., Lange, G. D., & Mayer, M. L. (1998). Activity-dependent modulation of glutamate receptors by polyamines. *Journal of Neuroscience, 18*(20), 8175–8185.

Bowie, D., & Mayer, M. L. (1995). Inward rectification of both AMPA and kainate subtype glutamate receptors generated by polyamine-mediated ion channel block. *Neuron, 15*(2), 453–462.

Brown, P. M. G. E., Aurousseau, M. R. P., Musgaard, M., Biggin, P. C., & Bowie, D. (2016). Kainate receptor pore-forming and auxiliary subunits regulate channel block by a novel mechanism. *Journal of Physiology, 594*(7), 1821–1840.

Brown, P., McGuire, H., & Bowie, D. (2018). Stargazin and cornichon-3 relieve polyamine block of AMPA receptors by enhancing blocker permeation. *Journal of General Physiology, 150*(1), 67–82. doi:10.1085/jgp.201711895

Burgess, D. L., Gefrides, L. A., Foreman, P. J., & Noebels, J. L. (2001). A cluster of three novel Ca^{2+} channel gamma subunit genes on chromosome 19q13.4: Evolution and expression profile of the gamma subunit gene family. *Genomics, 71*(3), 339–350. doi:10.1006/geno.2000.6440

Burnashev, N., Monyer, H., Seeburg, P. H., & Sakmann, B. (1992). Divalent ion permeability of AMPA receptor channels is dominated by the edited form of a single subunit. *Neuron, 8*, 189–198.

Burnashev, N., Villarroel, A., & Sakmann, B. (1996). Dimensions and ion selectivity of recombinant AMPA and kainate receptor channels and their dependence on Q/R site residues. *Journal of Physiology, 496*(Pt. 1), 165–173.

Burnashev, N., Zhou, Z., Neher, E., & Sakmann, B. (1995). Fractional calcium currents through recombinant GluR channels of the NMDA, AMPA and kainate receptor subtypes. *Journal of Physiology, 485*(Pt. 2), 403–418.

Cais, O., Herguedas, B., Krol, K., Cull-Candy, S. G., Farrant, M., & Greger, I. H. (2014). Mapping the interaction sites between AMPA receptors and TARPs reveals a role for the receptor N-terminal domain in channel gating. *Cell Reports, 9*(2), 728–740. doi:10.1016/j.celrep.2014.09.029

Carbone, A. L., & Plested, A. J. (2012). Coupled control of desensitization and gating by the ligand binding domain of glutamate receptors. *Neuron, 74*(5), 845–857. doi:10.1016/j.neuron.2012.04.020

Carbone, A. L., & Plested, A. J. (2016). Superactivation of AMPA receptors by auxiliary proteins. *Nature Communications, 7*, 10178. doi:10.1038/ncomms10178

Carter, C., Benavides, J., Legendre, P., Vincent, J. D., Noel, F., Thuret, F., . . . MacKenzie E. T. (1988). Ifenprodil and SL 82.0715 as cerebral anti-ischemic agents. II. Evidence for N-methyl-D-aspartate receptor antagonist properties. *Journal of Pharmacology and Experimental Therapeutics, 247*(3), 1222–1232.

Chaudhry, C., Weston, M. C., Schuck, P., Rosenmund, C., & Mayer, M. L. (2009). Stability of ligand-binding domain dimer assembly controls kainate receptor desensitization. *EMBO Journal, 28*(10), 1518–1530.

Chen, L., Chetkovich, D. M., Petralia, R. S., Sweeney, N. T., Kawasaki, Y., Wenthold, R. J., . . . Nicoll, R. A. (2000). Stargazin regulates synaptic targeting of AMPA receptors by two distinct mechanisms. *Nature, 408*(6815), 936–943. doi:10.1038/35050030

Chen, X., Aslam, M., Gollisch, T., Allen, K., & von Engelhardt, J. (2018). CKAMP44 modulates integration of visual inputs in the lateral geniculate nucleus. *Nature Communications, 9*(1), 261. doi:10.1038/s41467-017-02415-1

Cho, C. H., St-Gelais, F., Zhang, W., Tomita, S., & Howe, J. R. (2007). Two families of TARP isoforms that have distinct effects on the kinetic properties of AMPA receptors and synaptic currents. *Neuron, 55*(6), 890–904. doi:10.1016/j.neuron.2007.08.024

Choi, U. B., Xiao, S., Wollmuth, L. P., & Bowen, M. E. (2011). Effect of Src kinase phosphorylation on disordered C-terminal domain of N-methyl-D-aspartic acid (NMDA) receptor subunit GluN2B protein. *Journal of Biological Chemistry, 286*(34), 29904–29912. doi:10.1074/jbc.M111.258897

Clarke, V. R., Ballyk, B. A., Hoo, K. H., Mandelzys, A., Pellizzari, A., Bath, C. P., . . . Bleakman, D. (1997). A hippocampal GluR5 kainate receptor regulating inhibitory synaptic transmission. *Nature, 389*(6651), 599–603. doi:10.1038/39315

Clements, J. D., Feltz, A., Sahara, Y., & Westbrook, G. L. (1998). Activation kinetics of AMPA receptor channels reveal the number of functional agonist binding sites. *Journal of Neuroscience, 18*(1), 119–127.

Collingridge, G. L., Olsen, R. W., Peters, J., & Spedding, M. (2009). A nomenclature for ligand-gated ion channels. *Neuropharmacology, 56*(1), 2–5. doi:10.1016/j.neuropharm.2008.06.063

Colquhoun, D. (1998). Binding, gating, affinity and efficacy: The interpretation of structure–activity relationships for agonists and of the effects of mutating receptors. *British Journal of Pharmacology, 125*(5), 924–947.

Constals, A., Penn, A. C., Compans, B., Toulme, E., Phillipat, A., Marais, S., . . . Choquet, D. (2015). Glutamate-induced AMPA receptor desensitization increases their mobility and modulates short-term plasticity through unbinding from stargazin. *Neuron, 85*(4), 787–803. doi:10.1016/j.neuron.2015.01.012

Contractor, A., Mulle, C., & Swanson, G. T. (2011). Kainate receptors coming of age: Milestones of two decades of research. *Trends in Neurosciences, 34*(3), 154–163. https://doi.org/10.1016/j.tins.2010.12.002

Coombs, I. D., MacLean, D. M., Jayaraman, V., Farrant, M., & Cull-Candy, S. G. (2017). Dual effects of TARP γ-2 on glutamate efficacy can account for AMPA receptor autoinactivation. *Cell Reports, 20*(5), 1123–1135. doi:10.1016/j.celrep.2017.07.014

Coombs, I. D., Soto, D., Zonouzi, M., Renzi, M., Shelley, C., Farrant, M., & Cull-Candy, S. G. (2012). Cornichons modify channel properties of recombinant and glial AMPA receptors. *Journal of Neuroscience, 32*(29), 9796–9804. doi:10.1523/JNEUROSCI.0345-12.2012

Copits, B. A., Robbins, J. S., Frausto, S., & Swanson, G. T. (2011). Synaptic targeting and functional modulation of GluK1 kainate receptors by the auxiliary neuropilin and tolloid-like (NETO) proteins. *Journal of Neuroscience, 31*(20), 7334–7340.

Copits, B. A., Vernon, C. G., Sakai, R., & Swanson, G. T. (2014). Modulation of ionotropic glutamate receptor function by vertebrate galectins. *Journal of Physiology, 592*(10), 2079–2096. doi:10.1113/jphysiol.2013.269597

Daniels, B. A., Andrews, E. D., Aurousseau, M. R., Accardi, M. V., & Bowie, D. (2013). Crosslinking the ligand-binding domain dimer interface locks kainate receptors out of the main open state. *Journal of Physiology, 591*(Pt. 16), 3873–3885. doi:10.1113/jphysiol.2013.253666

Davies, J., & Watkins, J. C. (1982). Actions of D and L forms of 2-amino-5-phosphonovalerate and 2-amino-4-phosphonobutyrate in the cat spinal cord. *Brain Research, 235*(2), 378–386.

Dawe, G. B., Aurousseau, M. R., Daniels, B. A., & Bowie, D. (2015). Retour aux sources: Defining the structural basis of glutamate receptor activation. *Journal of Physiology, 593*(1), 97–110. doi:10.1113/jphysiol.2014.277921

Dawe, G. B., Kadir, M. F., Venskutonyte, R., Perozzo, A. M., Yan, Y., Alexander, R. P. D., . . . Bowie, D. (2019). Nanoscale mobility of the apo state and TARP stoichiometry dictate the gating behavior of alternatively spliced AMPA receptors. *Neuron, 102*(5), 976–992. doi:10.1016/j.neuron.2019.03.046

Dawe, G. B., Musgaard, M., Andrews, E. D., Daniels, B. A., Aurousseau, M. R. P., Biggin, P. C., & Bowie, D. (2013). Defining the structural relationship between kainate-receptor deactivation and desensitization. *Nature Structural & Molecular Biology, 20*(9), 1054–1061.

Dawe, G. B., Musgaard, M., Aurousseau, M. R. P., Nayeem, N., Green, T., Biggin, P. C., & Bowie, D. (2016). Distinct structural pathways coordinate the activation of AMPA receptor-auxiliary subunit complexes. *Neuron, 89*(6), 1264–1276.

Dingledine, R., Borges, K., Bowie, D., & Traynelis, S. F. (1999). The glutamate receptor ion channels. *Pharmacological Reviews, 51*(1), 7–61. Retrieved from http://www.ncbi.nlm.nih.gov/entrez/query.fcgi?cmd=Retrieve&db=PubMed&dopt=Citation&list_uids=10049997

Donevan, S. D., & Rogawski, M. A. (1995). Intracellular polyamines mediate inward rectification of Ca^{2+}-permeable alpha-amino-3-hydroxy-5-methyl-4-isoxazolepropionic acid receptors. *Proceedings of the National Academy of Sciences of the United States of America, 92*(20), 9298–9302.

Dong, H., & Zhou, H. X. (2011). Atomistic mechanism for the activation and desensitization of an AMPA-subtype glutamate receptor. *Nature Communications, 2*, 354. doi:10.1038/ncomms1362

Durr, K. L., Chen, L., Stein, R. A., De Zorzi, R., Folea, I. M., Walz, T., . . . Gouaux, E. (2014). Structure and dynamics of AMPA receptor GluA2 in resting, pre-open, and desensitized states. *Cell, 158*(4), 778–792. doi:10.1016/j.cell.2014.07.023

Egebjerg, J., Bettler, B., Hermans-Borgmeyer, I., & Heinemann, S. (1991). Cloning of a cDNA for a glutamate receptor subunit activated by kainate but not AMPA. *Nature, 351*(6329), 745–748. doi:10.1038/351745a0

Erlenhardt, N., Yu, H., Abiraman, K., Yamasaki, T., Wadiche, J. I., Tomita, S., & Bredt, D. S. (2016). Porcupine controls hippocampal AMPAR levels, composition, and synaptic transmission. *Cell Reports, 14*(4), 782–794.

Everts, I., Petroski, R., Kizelsztein, P., Teichberg, V. I., Heinemann, S. F., & Hollmann, M. (1999). Lectin-induced inhibition of desensitization of the kainate receptor GluR6 depends on the activation state and can be mediated by a single native or ectopic N-linked carbohydrate side chain. *Journal of Neuroscience, 19*(3), 916–927.

Everts, I., Villmann, C., & Hollmann, M. (1997). N-Glycosylation is not a prerequisite for glutamate receptor function but is essential for lectin modulation. *Molecular Pharmacology, 52*(5), 861–873.

Farrow, P., Khodosevich, K., Sapir, Y., Schulmann, A., Aslam, M., Stern-Bach, Y., . . . von Engelhardt, J. (2015). Auxiliary subunits of the CKAMP family differentially modulate AMPA receptor properties. *Elife, 4*, e09693. doi:10.7554/eLife.09693

Fay, A.-M. L., & Bowie, D. (2006). Concanavalin-A reports agonist-induced conformational changes in the intact GluR6 kainate receptor. *Journal of Physiology, 572*(Pt. 1), 201–213.

Fay, A. M., Corbeil, C. R., Brown, P., Moitessier, N., & Bowie, D. (2009). Functional characterization and in silico docking of full and partial GluK2 kainate receptor agonists. *Molecular Pharmacology, 75*(5), 1096–1107. doi:10.1124/mol.108.054254

Fayyazuddin, A., Villarroel, A., Le Goff, A., Lerma, J., & Neyton, J. (2000). Four residues of the extracellular N-terminal domain of the NR2A subunit control high-affinity Zn^{2+} binding to NMDA receptors. *Neuron, 25*(3), 683–694.

Fisher, J. L. (2015). The auxiliary subunits Neto1 and Neto2 have distinct, subunit-dependent effects at recombinant GluK1- and GluK2-containing kainate receptors. *Neuropharmacology, 99*, 471–480.

Fisher, J. L., & Mott, D. D. (2011). Distinct functional roles of subunits within the heteromeric kainate receptor. *Journal of Neuroscience, 31*(47), 17113–17122. doi:10.1523/JNEUROSCI.3685-11.2011

Fisher, J. L., & Mott, D. D. (2012). The auxiliary subunits Neto1 and Neto2 reduce voltage-dependent inhibition of recombinant kainate receptors. *Journal of Neuroscience, 32*(37), 12928–12933. doi:10.1523/JNEUROSCI.2211-12.2012

Fleidervish, I. A., Libman, L., Katz, E., & Gutnick, M. J. (2008). Endogenous polyamines regulate cortical neuronal excitability by blocking voltage-gated Na^+ channels. *Proceedings of the National Academy of Sciences of the United States of America*, *105*(48), 18994–18999. doi:10.1073/pnas.0803464105

Franke, C., Hatt, H., & Dudel, J. (1987). Liquid filament switch for ultra-fast exchanges of solutions at excised patches of synaptic membrane of crayfish muscle. *Neuroscience Letters*, *77*(2), 199–204.

Gan, Q., Dai, J., Zhou, H. X., & Wollmuth, L. P. (2016). The transmembrane domain mediates tetramerization of alpha-amino-3-hydroxy-5-methyl-4-isoxazolepropionic acid (AMPA) receptors. *Journal of Biological Chemistry*, *291*(12), 6595–6606. doi:10.1074/jbc.M115.686246

Gan, Q., Salussolia, C. L., & Wollmuth, L. P. (2015). Assembly of AMPA receptors: Mechanisms and regulation. *Journal of Physiology*, *593*(1), 39–48. doi:10.1113/jphysiol.2014.273755

Garcia-Nafria, J., Herguedas, B., Watson, J. F., & Greger, I. H. (2016). The dynamic AMPA receptor extracellular region: A platform for synaptic protein interactions. *Journal of Physiology*, *594*(19), 5449–5458. doi:10.1113/JP271844

Geiger, J. R., Melcher, T., Koh, D. S., Sakmann, B., Seeburg, P. H., Jonas, P., & Monyer, H. (1995). Relative abundance of subunit mRNAs determines gating and Ca^{2+} permeability of AMPA receptors in principal neurons and interneurons in rat CNS. *Neuron*, *15*(1), 193–204.

Gielen, M., Le Goff, A., Stroebel, D., Johnson, J. W., Neyton, J., & Paoletti, P. (2008). Structural rearrangements of NR1/NR2A NMDA receptors during allosteric inhibition. *Neuron*, *57*(1), 80–93. doi:10.1016/j.neuron.2007.11.021

Gielen, M., Siegler Retchless, B., Mony, L., Johnson, J. W., & Paoletti, P. (2009). Mechanism of differential control of NMDA receptor activity by NR2 subunits. *Nature*, *459*(7247), 703–707. doi:10.1038/nature07993

Griffith, T. N., & Swanson, G. T. (2015). Identification of critical functional determinants of kainate receptor modulation by auxiliary protein Neto2. *Journal of Physiology*, *593*(22), 4815–4833. doi:10.1113/JP271103

Gu, X., Mao, X., Lussier, M. P., Hutchison, M. A., Zhou, L., Hamra, F. K., . . . Lu, W. (2016). GSG1L suppresses AMPA receptor–mediated synaptic transmission and uniquely modulates AMPA receptor kinetics in hippocampal neurons. *Nature Communications*, *7*, 10873. doi:10.1038/ncomms10873

Haering, S. C., Tapken, D., Pahl, S., & Hollmann, M. (2014). Auxiliary subunits: Shepherding AMPA receptors to the plasma membrane. *Membranes (Basel)*, *4*(3), 469–490. doi:10.3390/membranes4030469

Haghighi, A. P., & Cooper, E. (1998). Neuronal nicotinic acetylcholine receptors are blocked by intracellular spermine in a voltage-dependent manner. *Journal of Neuroscience*, *18*(11), 4050–4062.

Hansen, K. B., Furukawa, H., & Traynelis, S. F. (2010). Control of assembly and function of glutamate receptors by the amino-terminal domain. *Molecular Pharmacology*, *78*(4), 535–549. doi:10.1124/mol.110.067157

Harms, J. E., Benveniste, M., MacLean, J. K., Partin, K. M., & Jamieson, C. (2013). Functional analysis of a novel positive allosteric modulator of AMPA receptors derived from a structure-based drug design strategy. *Neuropharmacology*, *64*, 45–52. doi:10.1016/j.neuropharm.2012.06.008

Hashimoto, K., Fukaya, M., Qiao, X., Sakimura, K., Watanabe, M., & Kano, M. (1999). Impairment of AMPA receptor function in cerebellar granule cells of ataxic mutant mouse stargazer. *Journal of Neuroscience*, *19*(14), 6027–6036.

Hastie, P., Ulbrich, M. H., Wang, H. L., Arant, R. J., Lau, A. G., Zhang, Z., ... Chen, L. (2013). AMPA receptor/TARP stoichiometry visualized by single-molecule subunit counting. *Proceedings of the National Academy of Sciences of the United States of America, 110*(13), 5163–5168. doi:10.1073/pnas.1218765110

Heckmann, M., Bufler, J., Franke, C., & Dudel, J. (1996). Kinetics of homomeric GluR6 glutamate receptor channels. *Biophysical Journal, 71*(4), 1743–1750. doi:10.1016/S0006-3495(96)79375-X

Henley, J. M., & Wilkinson, K. A. (2016). Synaptic AMPA receptor composition in development, plasticity and disease. *Nature Reviews Neuroscience, 17*(6), 337–350. doi:10.1038/nrn.2016.37

Herb, A., Burnashev, N., Werner, P., Sakmann, B., Wisden, W., & Seeburg, P. H. (1992). The KA-2 subunit of excitatory amino acid receptors shows widespread expression in brain and forms ion channels with distantly related subunits. *Neuron, 8*(4), 775–785.

Herguedas, B., Garcia-Nafria, J., Cais, O., Fernandez-Leiro, R., Krieger, J., Ho, H., & Greger, I. H. (2016). Structure and organization of heteromeric AMPA-type glutamate receptors. *Science, 352*(6285), aad3873. doi:10.1126/science.aad3873

Herguedas, B., Watson, J. F., Ho, H., Cais, O., Garcia-Nafria, J., & Greger, I. H. (2019). Architecture of the heteromeric GluA1/2 AMPA receptor in complex with the auxiliary subunit TARP gamma8. *Science, 364*(6438), eaav9011. doi:10.1126/science.aav9011

Herring, B. E., Shi, Y., Suh, Y. H., Zheng, C. Y., Blankenship, S. M., Roche, K. W., & Nicoll, R. A. (2013). Cornichon proteins determine the subunit composition of synaptic AMPA receptors. *Neuron, 77*(6), 1083–1096. doi:10.1016/j.neuron.2013.01.017

Hinard, V., Britan, A., Rougier, J. S., Bairoch, A., Abriel, H., & Gaudet, P. (2016). ICEPO: The ion channel electrophysiology ontology. *Database (Oxford), 2016*, baw017. doi:10.1093/database/baw017

Hollmann, M., & Heinemann, S. (1994). Cloned glutamate receptors. *Annual Review of Neuroscience, 17*, 31–108. doi:10.1146/annurev.ne.17.030194.000335

Hollmann, M., Maron, C., & Heinemann, S. (1994). N-Glycosylation site tagging suggests a three transmembrane domain topology for the glutamate receptor GluR1. *Neuron, 13*(6), 1331–1343.

Honore, T., Davies, S. N., Drejer, J., Fletcher, E. J., Jacobsen, P., Lodge, D., & Nielsen, F. E. (1988). Quinoxalinediones: Potent competitive non-NMDA glutamate receptor antagonists. *Science, 241*(4866), 701–703.

Horn, R. (1990). A primer of permeation and gating. In A. Borsellino, L. Cervetto, & V. Torre (Eds.), *Sensory Transduction: Vol. 194. NATO ASI Series (Series A: Life Sciences)* (pp. 3–16). Boston, MA: Springer.

Horning, M. S., & Mayer, M. L. (2004). Regulation of AMPA receptor gating by ligand binding core dimers. *Neuron, 41*(3), 379–388.

Howe, J. R. (2015). Modulation of non-NMDA receptor gating by auxiliary subunits. *Journal of Physiology, 593*(1), 61–72. doi:10.1113/jphysiol.2014.273904

Huettner, J. E. (2015). Glutamate receptor pores. *Journal of Physiology, 593*(1), 49–59. doi:10.1113/jphysiol.2014.272724

Hume, R. I., Dingledine, R., & Heinemann, S. F. (1991). Identification of a site in glutamate receptor subunits that controls calcium permeability. *Science, 253*(5023), 1028–1031.

Ito, I., Tanabe, S., Kohda, A., & Sugiyama, H. (1990). Allosteric potentiation of quisqualate receptors by a nootropic drug aniracetam. *Journal of Physiology, 424*, 533–543.

Jackson, A. C., & Nicoll, R. A. (2011). The expanding social network of ionotropic glutamate receptors: TARPs and other transmembrane auxiliary subunits. *Neuron, 70*(2), 178–199. doi:10.1016/j.neuron.2011.04.007

Jensen, A. A., Christesen, T., Bolcho, U., Greenwood, J. R., Postorino, G., Vogensen, S. B., ... Clausen, R. P. (2007). Functional characterization of Tet-AMPA [tetrazolyl-2-amino-3-(3-hydroxy-5-methyl- 4-isoxazolyl)propionic acid] analogues at ionotropic glutamate receptors GluR1–GluR4. The molecular basis for the functional selectivity profile of 2-Bn-Tet-AMPA. *Journal of Medicinal Chemistry*, *50*(17), 4177–4185. doi:10.1021/jm070532r

Jin, R., Banke, T. G., Mayer, M. L., Traynelis, S. F., & Gouaux, E. (2003). Structural basis for partial agonist action at ionotropic glutamate receptors. *Nature Neuroscience*, *6*(8), 803–810. doi:10.1038/nn1091

Jin, R., Singh, S. K., Gu, S., Furukawa, H., Sobolevsky, A. I., Zhou, J., ... Gouaux, E. (2009). Crystal structure and association behaviour of the GluR2 amino-terminal domain. *EMBO Journal*, *28*(12), 1812–1823. doi:10.1038/emboj.2009.140

Kalashnikova, E., Lorca, R. A., Kaur, I., Barisone, G. A., Li, B., Ishimaru, T., ... Diaz, E. (2010). SynDIG1: An activity-regulated, AMPA-receptor-interacting transmembrane protein that regulates excitatory synapse development. *Neuron*, *65*(1), 80–93. doi:10.1016/j.neuron.2009.12.021

Kamboj, S. K., Swanson, G. T., & Cull-Candy, S. G. (1995). Intracellular spermine confers rectification on rat calcium-permeable AMPA and kainate receptors. *Journal of Physiology*, *486*(Pt 2), 297–303.

Karakas, E., & Furukawa, H. (2014). Crystal structure of a heterotetrameric NMDA receptor ion channel. *Science*, *344*(6187), 992–997. doi:10.1126/science.1251915

Karakas, E., Simorowski, N., & Furukawa, H. (2009). Structure of the zinc-bound amino-terminal domain of the NMDA receptor NR2B subunit. *EMBO Journal*, *28*(24), 3910–3920. doi:10.1038/emboj.2009.338

Kato, A. S., Gill, M. B., Yu, H., Nisenbaum, E. S., & Bredt, D. S. (2010). TARPs differentially decorate AMPA receptors to specify neuropharmacology. *Trends in Neurosciences*, *33*(5), 241–248. doi:10.1016/j.tins.2010.02.004

Kato, A. S., Siuda, E. R., Nisenbaum, E. S., & Bredt, D. S. (2008). AMPA receptor subunit-specific regulation by a distinct family of type II TARPs. *Neuron*, *59*(6), 986–996. doi:10.1016/j.neuron.2008.07.034

Kato, A. S., Zhou, W., Milstein, A. D., Knierman, M. D., Siuda, E. R., Dotzlaf, J. E., ... Bredt, D. S. (2007). New transmembrane AMPA receptor regulatory protein isoform, gamma-7, differentially regulates AMPA receptors. *Journal of Neuroscience*, *27*(18), 4969–4977. doi:10.1523/JNEUROSCI.5561-06.2007

Kazi, R., Dai, J., Sweeney, C., Zhou, H. X., & Wollmuth, L. P. (2014). Mechanical coupling maintains the fidelity of NMDA receptor–mediated currents. *Nature Neuroscience*, *17*(7), 914–922. doi:10.1038/nn.3724

Keinänen, K., Wisden, W., Sommer, B., Werner, P., Herb, A., Verdoorn, T. A., ... Seeburg, P. H. (1990). A family of AMPA-selective glutamate receptors. *Science*, *249*, 556–560.

Keramidas, A., & Lynch, J. W. (2013). An outline of desensitization in pentameric ligand-gated ion channel receptors. *Cellular and Molecular Life Sciences*, *70*(7), 1241–1253. doi:10.1007/s00018-012-1133-z

Khodosevich, K., Jacobi, E., Farrow, P., Schulmann, A., Rusu, A., Zhang, L., ... von Engelhardt, J. (2014). Coexpressed auxiliary subunits exhibit distinct modulatory profiles on AMPA receptor function. *Neuron*, *83*(3), 601–615. doi:10.1016/j.neuron.2014.07.004

Kim, K. S., Yan, D., & Tomita, S. (2010). Assembly and stoichiometry of the AMPA receptor and transmembrane AMPA receptor regulatory protein complex. *Journal of Neuroscience*, *30*(3), 1064–1072. doi:10.1523/JNEUROSCI.3909-09.2010

Kiskin, N. I., Krishtal, O. A., & Tsyndrenko, A. (1986). Excitatory amino acid receptors in hippocampal neurons: Kainate fails to desensitize them. *Neuroscience Letters, 63*(3), 225–230.

Kizelsztein, P., Eisenstein, M., Strutz, N., Hollmann, M., & Teichberg, V. I. (2000). Mutant cycle analysis of the active and desensitized states of an AMPA receptor induced by willardiines. *Biochemistry, 39*(42), 12819–12827. doi:10.1021/bi000962i

Klaassen, R. V., Stroeder, J., Coussen, F., Hafner, A. S., Petersen, J. D., Renancio, C., . . . Smit, A. B. (2016). Shisa6 traps AMPA receptors at postsynaptic sites and prevents their desensitization during synaptic activity. *Nature Communications, 7*, 10682. doi:10.1038/ncomms10682

Koh, D. S., Burnashev, N., & Jonas, P. (1995). Block of native Ca^{2+}-permeable AMPA receptors in rat brain by intracellular polyamines generates double rectification. *Journal of Physiology, 486*(Pt. 2), 305–312.

Köhler, M., Burnashev, N., Sakmann, B., & Seeburg, P. H. (1993). Determinants of Ca^{2+} permeability in both TM1 and TM2 of high affinity kainate receptor channels: Diversity by RNA editing. *Neuron, 10*(3), 491–500.

Koike, M., Tsukada, S., Tsuzuki, K., Kijima, H., & Ozawa, S. (2000). Regulation of kinetic properties of GluR2 AMPA receptor channels by alternative splicing. *Journal of Neuroscience, 20*(6), 2166–2174.

Kott, S., Werner, M., Korber, C., & Hollmann, M. (2007). Electrophysiological properties of AMPA receptors are differentially modulated depending on the associated member of the TARP family. *Journal of Neuroscience, 27*(14), 3780–3789. doi:10.1523/JNEUROSCI.4185-06.2007

Kovalchuk, Y., Miller, B., Sarantis, M., & Attwell, D. (1994). Arachidonic acid depresses non-NMDA receptor currents. *Brain Research, 643*(1–2), 287–295.

Krishtal, O. A., Marchenko, S. M., & Pidoplichko, V. I. (1983). Receptor for ATP in the membrane of mammalian sensory neurones. *Neuroscience Letters, 35*(1), 41–45.

Krogsgaard-Larsen, P., Honore, T., Hansen, J. J., Curtis, D. R., & Lodge, D. (1980). New class of glutamate agonist structurally related to ibotenic acid. *Nature, 284*(5751), 64–66.

Kumar, J., Schuck, P., Jin, R., & Mayer, M. L. (2009). The amino terminal domain of GluR6-subtype glutamate receptor ion channels. *Nature Structural and Molecular Biology, 16*(6), 631–638. doi:10.1038/nsmb.1613

Kumar, J., Schuck, P., & Mayer, M. L. (2011). Structure and assembly mechanism for heteromeric kainate receptors. *Neuron, 71*(2), 319–331. doi:10.1016/j.neuron.2011.05.038

Kuner, T., Beck, C., Sakmann, B., & Seeburg, P. H. (2001). Channel-lining residues of the AMPA receptor M2 segment: Structural environment of the Q/R site and identification of the selectivity filter. *Journal of Neuroscience, 21*(12), 4162–4172.

Kuusinen, A., Abele, R., Madden, D. R., & Keinanen, K. (1999). Oligomerization and ligand-binding properties of the ectodomain of the alpha-amino-3-hydroxy-5-methyl-4-isoxazole propionic acid receptor subunit GluRD. *Journal of Biological Chemistry, 274*(41), 28937–28943.

Landes, C. F., Rambhadran, A., Taylor, J. N., Salatan, F., & Jayaraman, V. (2011). Structural landscape of the isolated agonist-binding domain of single AMPA receptors. *Nature Chemical Biology, 7*(3), 168–173. doi:10.1038/nchembio.523

Laube, B., Kuhse, J., & Betz, H. (1998). Evidence for a tetrameric structure of recombinant NMDA receptors. *Journal of Neuroscience, 18*(8), 2954–2961.

Lee, C. H., Lu, W., Michel, J. C., Goehring, A., Du, J., Song, X., & Gouaux, E. (2014). NMDA receptor structures reveal subunit arrangement and pore architecture. *Nature, 511*(7508), 191–197. doi:10.1038/nature13548

Letts, V. A., Felix, R., Biddlecome, G. H., Arikkath, J., Mahaffey, C. L., Valenzuela, A., . . . Frankel, W. N. (1998). The mouse stargazer gene encodes a neuronal Ca^{2+}-channel gamma subunit. *Nature Genetics, 19*(4), 340–347. doi:10.1038/1228

Leuschner, W. D., & Hoch, W. (1999). Subtype-specific assembly of alpha-amino-3-hydroxy-5-methyl-4-isoxazole propionic acid receptor subunits is mediated by their N-terminal domains. *Journal of Biological Chemistry, 274*(24), 16907–16916.

Lomeli, H., Mosbacher, J., Melcher, T., Hoger, T., Geiger, J. R., Kuner, T., . . . Seeburg, P. H. (1994). Control of kinetic properties of AMPA receptor channels by nuclear RNA editing. *Science, 266*(5191), 1709–1713.

Lopatin, A. N., Makhina, E. N., & Nichols, C. G. (1994). Potassium channel block by cytoplasmic polyamines as the mechanism of intrinsic rectification. *Nature, 372*(6504), 366–369. doi:10.1038/372366a0

Lu, W., Shi, Y., Jackson, A. C., Bjorgan, K., During, M. J., Sprengel, R., . . . Nicoll, R. A. (2009). Subunit composition of synaptic AMPA receptors revealed by a single-cell genetic approach. *Neuron, 62*(2), 254–268. doi:10.1016/j.neuron.2009.02.027

MacLean, D. M., & Bowie, D. (2011). TARP regulation of AMPA receptor antagonism: A problem of interpretation. *Journal of Physiology*, 589(Pt. 22), 5383–5390. doi:10.1113/jphysiol.2011.219485

MacLean, D. M., Ramaswamy, S. S., Du, M., Howe, J. R., & Jayaraman, V. (2014). Stargazin promotes closure of the AMPA receptor ligand-binding domain. *Journal of General Physiology, 144*(6), 503–512. doi:10.1085/jgp.201411287

MacLean, D. M., Wong, A. Y., Fay, A. M., & Bowie, D. (2011). Cations but not anions regulate the responsiveness of kainate receptors. *Journal of Neuroscience, 31*(6), 2136–2144. doi:10.1523/JNEUROSCI.4314-10.2011

Mayer, M. L. (2011). Emerging models of glutamate receptor ion channel structure and function. *Structure, 19*(10), 1370–1380. doi:10.1016/j.str.2011.08.009

Mayer, M. L., Ghosal, A., Dolman, N. P., & Jane, D. E. (2006). Crystal structures of the kainate receptor GluR5 ligand binding core dimer with novel GluR5-selective antagonists. *Journal of Neuroscience, 26*(11), 2852–2861.

Mayer, M. L., & Westbrook, G. L. (1987). The physiology of excitatory amino acids in the vertebrate central nervous system. *Progress in Neurobiology, 28*(3), 197–276.

McGee, T. P., Bats, C., Farrant, M., & Cull-Candy, S. G. (2015). Auxiliary subunit GSG1L acts to suppress calcium-permeable AMPA receptor function. *Journal of Neuroscience, 35*(49), 16171–16179. doi:10.1523/JNEUROSCI.2152-15.2015

Meddows, E., Le Bourdelles, B., Grimwood, S., Wafford, K., Sandhu, S., Whiting, P., & McIlhinney, R. A. (2001). Identification of molecular determinants that are important in the assembly of N-methyl-D-aspartate receptors. *Journal of Biological Chemistry, 276*(22), 18795–18803. doi:10.1074/jbc.M101382200

Menuz, K., Stroud, R. M., Nicoll, R. A., & Hays, F. A. (2007). TARP auxiliary subunits switch AMPA receptor antagonists into partial agonists. *Science, 318*(5851), 815–817. doi:10.1126/science.1146317

Meyerson, J. R., Chittori, S., Merk, A., Rao, P., Han, T. H., Serpe, M., . . . Subramaniam, S. (2016). Structural basis of kainate subtype glutamate receptor desensitization. *Nature, 537*(7621), 567–571. doi:10.1038/nature19352

Meyerson, J. R., Kumar, J., Chittori, S., Rao, P., Pierson, J., Bartesaghi, A., . . . Subramaniam, S. (2014). Structural mechanism of glutamate receptor activation and desensitization. *Nature, 514*(7522), 328–334. doi:10.1038/nature13603

Midgett, C. R., Gill, A., & Madden, D. R. (2012). Domain architecture of a calcium-permeable AMPA receptor in a ligand-free conformation. *Frontiers in Molecular Neuroscience, 4*, 56. doi:10.3389/fnmol.2011.00056

Milstein, A. D., Zhou, W., Karimzadegan, S., Bredt, D. S., & Nicoll, R. A. (2007). TARP subtypes differentially and dose-dependently control synaptic AMPA receptor gating. *Neuron, 55*(6), 905–918. doi:10.1016/j.neuron.2007.08.022

Monyer, H., Burnashev, N., Laurie, D. J., Sakmann, B., & Seeburg, P. H. (1994). Developmental and regional expression in the rat brain and functional properties of four NMDA receptors. *Neuron, 12*(3), 529–540.

Monyer, H., Sprengel, R., Schoepfer, R., Herb, A., Higuchi, M., Lomeli, H., . . . Seeburg, P. H. (1992). Heteromeric NMDA receptors: Molecular and functional distinction of subtypes. *Science, 256*(5060), 1217–1221.

Morimoto-Tomita, M., Zhang, W., Straub, C., Cho, C. H., Kim, K. S., Howe, J. R., & Tomita, S. (2009). Autoinactivation of neuronal AMPA receptors via glutamate-regulated TARP interaction. *Neuron, 61*(1), 101–112. doi:10.1016/j.neuron.2008.11.009

Moriyoshi, K., Masu, M., Ishii, T., Shigemoto, R., Mizuno, N., & Nakanishi, S. (1991). Molecular cloning and characterization of the rat NMDA receptor. *Nature, 354*(6348), 31–37. doi:10.1038/354031a0

Mosbacher, J., Schoepfer, R., Monyer, H., Burnashev, N., Seeburg, P. H., & Ruppersberg, J. P. (1994). A molecular determinant for submillisecond desensitization in glutamate receptors. *Science, 266*(5187), 1059–1062.

Mott, D. D., Rojas, A., Fisher, J. L., Dingledine, R. J., & Benveniste, M. (2010). Subunit-specific desensitization of heteromeric kainate receptors. *Journal of Physiology, 588*(Pt. 4), 683–700.

Moykkynen, T., Coleman, S. K., Semenov, A., & Keinanen, K. (2014). The N-terminal domain modulates alpha-amino-3-hydroxy-5-methyl-4-isoxazolepropionic acid (AMPA) receptor desensitization. *Journal of Biological Chemistry, 289*(19), 13197–13205. doi:10.1074/jbc.M113.526301

Murakami, S., Takemoto, T., & Shimizu, Z. (1953). Studies on the effective principles of Digenea simplex Aq. 1. separation of the effective fraction by liquid chromatography. *Yakugaku Zasshi, 73*(9), 1026–1028.

Nakagawa, T. (2019). Structures of the AMPA receptor in complex with its auxiliary subunit cornichon. *Science, 366*(6470), 1259–1263. doi:10.1126/science.aay2783

Nakagawa, T., Cheng, Y., Ramm, E., Sheng, M., & Walz, T. (2005). Structure and different conformational states of native AMPA receptor complexes. *Nature, 433*(7025), 545–549. doi:10.1038/nature03328

Nakagawa, T., Cheng, Y., Sheng, M., & Walz, T. (2006). Three-dimensional structure of an AMPA receptor without associated stargazin/TARP proteins. *Biological Chemistry, 387*(2), 179–187. doi:10.1515/BC.2006.024

Nakanishi, N., Shneider, N. A., & Axel, R. (1990). A family of glutamate receptor genes: Evidence for the formation of heteromultimeric receptors with distinct channel properties. *Neuron, 5*(5), 569–581.

Nayeem, N., Mayans, O., & Green, T. (2011). Conformational flexibility of the ligand-binding domain dimer in kainate receptor gating and desensitization. *Journal of Neuroscience, 31*(8), 2916–2924. doi:10.1523/JNEUROSCI.4771-10.2011

Nayeem, N., Zhang, Y., Schweppe, D. K., Madden, D. R., & Green, T. (2009). A nondesensitizing kainate receptor point mutant. *Molecular Pharmacology, 76*(3), 534–542. doi:10.1124/mol.109.056598

Noebels, J. L., Qiao, X., Bronson, R. T., Spencer, C., & Davisson, M. T. (1990). Stargazer: A new neurological mutant on chromosome 15 in the mouse with prolonged cortical seizures. *Epilepsy Research*, *7*(2), 129–135.

O'Hara, P. J., Sheppard, P. O., Thogersen, H., Venezia, D., Haldeman, B. A., McGrane, V., . . . Mulvihill, E. R. (1993). The ligand-binding domain in metabotropic glutamate receptors is related to bacterial periplasmic binding proteins. *Neuron*, *11*(1), 41–52.

Paoletti, P., Bellone, C., & Zhou, Q. (2013). NMDA receptor subunit diversity: Impact on receptor properties, synaptic plasticity and disease. *Nature Reviews Neuroscience*, *14*(6), 383–400. doi:10.1038/nrn3504

Paramo, T., Brown, P., Musgaard, M., Bowie, D., & Biggin, P. C. (2017). Functional validation of heteromeric kainate receptor models. *Biophysical Journal*, *113*(10), 2173–2177. doi:10.1016/j.bpj.2017.08.047

Partin, K. M., Bowie, D., & Mayer, M. L. (1995). Structural determinants of allosteric regulation in alternatively spliced AMPA receptors. *Neuron*, *14*(4), 833–843.

Partin, K. M., Fleck, M. W., & Mayer, M. L. (1996). AMPA receptor flip/flop mutants affecting deactivation, desensitization, and modulation by cyclothiazide, aniracetam, and thiocyanate. *Journal of Neuroscience*, *16*(21), 6634–6647.

Partin, K. M., Patneau, D. K., Winters, C. A., Mayer, M. L., & Buonanno, A. (1993). Selective modulation of desensitization at AMPA versus kainate receptors by cyclothiazide and concanavalin A. *Neuron*, *11*(6), 1069–1082.

Pasternack, A., Coleman, S. K., Jouppila, A., Mottershead, D. G., Lindfors, M., Pasternack, M., & Keinanen, K. (2002). Alpha-amino-3-hydroxy-5-methyl-4-isoxazolepropionic acid (AMPA) receptor channels lacking the N-terminal domain. *Journal of Biological Chemistry*, *277*(51), 49662–49667. doi:10.1074/jbc.M208349200

Paternain, A. V., Cohen, A., Stern-Bach, Y., & Lerma, J. (2003). A role for extracellular Na^+ in the channel gating of native and recombinant kainate receptors. *Journal of Neuroscience*, *23*(25), 8641–8648.

Patneau, D. K., Vyklicky, L., Jr., & Mayer, M. L. (1993). Hippocampal neurons exhibit cyclothiazide-sensitive rapidly desensitizing responses to kainate. *Journal of Neuroscience*, *13*(8), 3496–3509.

Pegg, A. E. (2009). Mammalian polyamine metabolism and function. *IUBMB Life*, *61*(9), 880–894. doi:10.1002/iub.230

Pegg, A. E., & McCann, P. P. (1982). Polyamine metabolism and function. *American Journal of Physiology*, *243*, C212–C221.

Perrais, D., Pinheiro, P. S., Jane, D. E., & Mulle, C. (2009). Antagonism of recombinant and native GluK3-containing kainate receptors. *Neuropharmacology*, *56*(1), 131–140.

Peters, S., Koh, J., & Choi, D. W. (1987). Zinc selectively blocks the action of N-methyl-D-aspartate on cortical neurons. *Science*, *236*(4801), 589–593.

Petralia, R. S., Wang, Y. X., & Wenthold, R. J. (1994). Histological and ultrastructural localization of the kainate receptor subunits, KA2 and GluR6/7, in the rat nervous system using selective antipeptide antibodies. *Journal of Comparative Neurology*, *349*(1), 85–110. doi:10.1002/cne.903490107

Plested, A. J., & Mayer, M. L. (2007). Structure and mechanism of kainate receptor modulation by anions. *Neuron*, *53*(6), 829–841. doi:10.1016/j.neuron.2007.02.025

Plested, A. J., Vijayan, R., Biggin, P. C., & Mayer, M. L. (2008). Molecular basis of kainate receptor modulation by sodium. *Neuron*, *58*(5), 720–735. doi:10.1016/j.neuron.2008.04.001

Pohlsgaard, J., Frydenvang, K., Madsen, U., & Kastrup, J. S. (2011). Lessons from more than 80 structures of the GluA2 ligand-binding domain in complex with agonists, antagonists and allosteric modulators. *Neuropharmacology*, *60*(1), 135–150. doi:10.1016/j.neuropharm.2010.08.004

Priel, A., Kolleker, A., Ayalon, G., Gillor, M., Osten, P., & Stern-Bach, Y. (2005). Stargazin reduces desensitization and slows deactivation of the AMPA-type glutamate receptors. *Journal of Neuroscience*, *25*(10), 2682–2686. doi:10.1523/JNEUROSCI.4834-04.2005

Priel, A., Selak, S., Lerma, J., & Stern-Bach, Y. (2006). Block of kainate receptor desensitization uncovers a key trafficking checkpoint. *Neuron*, *52*(6), 1037–1046. doi:10.1016/j.neuron.2006.12.006

Prieto, M. L., & Wollmuth, L. P. (2010). Gating modes in AMPA receptors. *Journal of Neuroscience*, *30*(12), 4449–4459. doi:10.1523/JNEUROSCI.5613-09.2010

Puchalski, R. B., Louis, J. C., Brose, N., Traynelis, S. F., Egebjerg, J., Kukekov, V., . . . Moran, T. (1994). Selective RNA editing and subunit assembly of native glutamate receptors. *Neuron*, *13*(1), 131–147.

Robert, A., Armstrong, N., Gouaux, J. E., & Howe, J. R. (2005). AMPA receptor binding cleft mutations that alter affinity, efficacy, and recovery from desensitization. *Journal of Neuroscience*, *25*(15), 3752–3762. doi:10.1523/JNEUROSCI.0188-05.2005

Rosenmund, C., Stern-Bach, Y., & Stevens, C. F. (1998). The tetrameric structure of a glutamate receptor channel. *Science*, *280*(5369), 1596–1599.

Rossmann, M., Sukumaran, M., Penn, A. C., Veprintsev, D. B., Babu, M. M., & Greger, I. H. (2011). Subunit-selective N-terminal domain associations organize the formation of AMPA receptor heteromers. *EMBO Journal*, *30*(5), 959–971. doi:10.1038/emboj.2011.16

Rozov, A., & Burnashev, N. (1999). Polyamine-dependent facilitation of postsynaptic AMPA receptors counteracts paired-pulse depression. *Nature*, *401*(6753), 594–598. doi:10.1038/44151

Rozov, A., Zilberter, Y., Wollmuth, L. P., & Burnashev, N. (1998). Facilitation of currents through rat Ca^{2+}-permeable AMPA receptor channels by activity-dependent relief from polyamine block. *Journal of Physiology*, *511*(Pt. 2), 361–377.

Safferling, M., Tichelaar, W., Kummerle, G., Jouppila, A., Kuusinen, A., Keinanen, K., & Madden, D. R. (2001). First images of a glutamate receptor ion channel: Oligomeric state and molecular dimensions of GluRB homomers. *Biochemistry*, *40*(46), 13948–13953.

Saglietti, L., Dequidt, C., Kamieniarz, K., Rousset, M. C., Valnegri, P., Thoumine, O., . . . Passafaro, M. (2007). Extracellular interactions between GluR2 and N-cadherin in spine regulation. *Neuron*, *54*(3), 461–477. doi:10.1016/j.neuron.2007.04.012

Salazar, H., Eibl, C., Chebli, M., & Plested, A. (2017). Mechanism of partial agonism in AMPA-type glutamate receptors. *Nature Communications*, *8*, 14327. doi:10.1038/ncomms14327

Salussolia, C. L., Corrales, A., Talukder, I., Kazi, R., Akgul, G., Bowen, M., & Wollmuth, L. P. (2011). Interaction of the M4 segment with other transmembrane segments is required for surface expression of mammalian alpha-amino-3-hydroxy-5-methyl-4-isoxazolepropionic acid (AMPA) receptors. *Journal of Biological Chemistry*, *286*(46), 40205–40218. doi:10.1074/jbc.M111.268839

Schauder, D. M., Kuybeda, O., Zhang, J., Klymko, K., Bartesaghi, A., Borgnia, M. J., . . . Subramaniam, S. (2013). Glutamate receptor desensitization is mediated by changes in quaternary structure of the ligand binding domain. *Proceedings of the National Academy of Sciences of the United States of America*, *110*(15), 5921–5926. doi:10.1073/pnas.1217549110

Schiffer, H. H., Swanson, G. T., & Heinemann, S. F. (1997). Rat GluR7 and a carboxy-terminal splice variant, GluR7b, are functional kainate receptor subunits with a low sensitivity to glutamate. *Neuron, 19*(5), 1141–1146.

Schmitt, J., Dux, E., Gissel, C., & Paschen, W. (1996). Regional analysis of developmental changes in the extent of GluR6 mRNA editing in rat brain. *Brain Research Developmental Brain Research, 91*(1), 153–157.

Schnell, E., Sizemore, M., Karimzadegan, S., Chen, L., Bredt, D. S., & Nicoll, R. A. (2002). Direct interactions between PSD-95 and stargazin control synaptic AMPA receptor number. *Proceedings of the National Academy of Sciences of the United States of America, 99*(21), 13902–13907. doi:10.1073/pnas.172511199

Schwenk, J., Baehrens, D., Haupt, A., Bildl, W., Boudkkazi, S., Roeper, J., . . . Schulte, U. (2014). Regional diversity and developmental dynamics of the AMPA-receptor proteome in the mammalian brain. *Neuron, 84*(1), 41–54. doi:10.1016/j.neuron.2014.08.044

Schwenk, J., Harmel, N., Brechet, A., Zolles, G., Berkefeld, H., Muller, C. S., . . . Fakler, B. (2012). High-resolution proteomics unravel architecture and molecular diversity of native AMPA receptor complexes. *Neuron, 74*(4), 621–633. doi:10.1016/j.neuron.2012.03.034

Schwenk, J., Harmel, N., Zolles, G., Bildl, W., Kulik, A., Heimrich, B., . . . Klocker, N. (2009). Functional proteomics identify cornichon proteins as auxiliary subunits of AMPA receptors. *Science, 323*(5919), 1313–1319. doi:10.1126/science.1167852

Shanks, N. F., Cais, O., Maruo, T., Savas, J. N., Zaika, E. I., Azumaya, C. M., . . . Nakagawa, T. (2014). Molecular dissection of the interaction between the AMPA receptor and cornichon homolog-3. *Journal of Neuroscience, 34*(36), 12104–12120. doi:10.1523/JNEUROSCI.0595-14.2014

Shanks, N. F., Savas, J. N., Maruo, T., Cais, O., Hirao, A., Oe, S., . . . Nakagawa, T. (2012). Differences in AMPA and kainate receptor interactomes facilitate identification of AMPA receptor auxiliary subunit GSG1L. *Cell Reports, 1*(6), 590–598. doi:10.1016/j.celrep.2012.05.004

Sheng, N., Shi, Y. S., Lomash, R. M., Roche, K. W., & Nicoll, R. A. (2015). Neto auxiliary proteins control both the trafficking and biophysical properties of the kainate receptor GluK1. *Elife, 4*, e11682.

Shi, Y., Lu, W., Milstein, A. D., & Nicoll, R. A. (2009). The stoichiometry of AMPA receptors and TARPs varies by neuronal cell type. *Neuron, 62*(5), 633–640. doi:10.1016/j.neuron.2009.05.016

Sia, G. M., Beique, J. C., Rumbaugh, G., Cho, R., Worley, P. F., & Huganir, R. L. (2007). Interaction of the N-terminal domain of the AMPA receptor GluR4 subunit with the neuronal pentraxin NP1 mediates GluR4 synaptic recruitment. *Neuron, 55*(1), 87–102. doi:10.1016/j.neuron.2007.06.020

Smith, T. C., & Howe, J. R. (2000). Concentration-dependent substate behavior of native AMPA receptors. *Nature Neuroscience, 3*(10), 992–997. doi:10.1038/79931

Sobolevsky, A. I., Rosconi, M. P., & Gouaux, E. (2009). X-ray structure, symmetry and mechanism of an AMPA-subtype glutamate receptor. *Nature, 462*(7274), 745–756. doi:10.1038/nature08624

Sommer, B., Burnashev, N., Verdoorn, T. A., Keinanen, K., Sakmann, B., & Seeburg, P. H. (1992). A glutamate receptor channel with high affinity for domoate and kainate. *EMBO Journal, 11*(4), 1651–1656.

Sommer, B., Keinanen, K., Verdoorn, T. A., Wisden, W., Burnashev, N., Herb, A., . . . Seeburg, P. H. (1990). Flip and flop: A cell-specific functional switch in glutamate-operated channels of the CNS. *Science, 249*(4976), 1580–1585.

Sommer, B., Kohler, M., Sprengel, R., & Seeburg, P. H. (1991). RNA editing in brain controls a determinant of ion flow in glutamate-gated channels. *Cell*, *67*(1), 11–19. https://doi.org/10.1016/0092-8674(91)90568-J

Soto, D., Coombs, I. D., Kelly, L., Farrant, M., & Cull-Candy, S. G. (2007). Stargazin attenuates intracellular polyamine block of calcium-permeable AMPA receptors. *Nature Neuroscience*, *10*(10), 1260–1267. doi:10.1038/nn1966

Stein, E., Cox, J. A., Seeburg, P. H., & Verdoorn, T. A. (1992). Complex pharmacological properties of recombinant alpha-amino-3-hydroxy-5-methyl-4-isoxazole propionate receptor subtypes. *Molecular Pharmacology*, *42*(5), 864–871.

Stensbol, T. B., Borre, L., Johansen, T. N., Egebjerg, J., Madsen, U., Ebert, B., & Krogsgaard-Larsen, P. (1999). Resolution, absolute stereochemistry and molecular pharmacology of the enantiomers of ATPA. *European Journal of Pharmacology*, *380*(2–3), 153–162. doi:10.1016/s0014-2999(99)00512-9

Stephenson, R. P. (1956). A modification of receptor theory. *British Journal of Pharmacology and Chemotherapy*, *11*(4), 379–393.

Stern-Bach, Y., Bettler, B., Hartley, M., Sheppard, P. O., O'Hara, P. J., & Heinemann, S. F. (1994). Agonist selectivity of glutamate receptors is specified by two domains structurally related to bacterial amino acid-binding proteins. *Neuron*, *13*(6), 1345–1357.

Stern-Bach, Y., Russo, S., Neuman, M., & Rosenmund, C. (1998). A point mutation in the glutamate binding site blocks desensitization of AMPA receptors. *Neuron*, *21*(4), 907–918.

Stöhr, H., Berger, C., Froehlich, S., & Weber, B. H. F. (2002). A novel gene encoding a putative transmembrane protein with two extracellular CUB domains and a low-density lipoprotein class A module: Isolation of alternatively spliced isoforms in retina and brain. *Gene*, *286*(2), 223–231.

Straub, C., Hunt, D. L., Yamasaki, M., Kim, K. S., Watanabe, M., Castillo, P. E., & Tomita, S. (2011). Distinct functions of kainate receptors in the brain are determined by the auxiliary subunit Neto1. *Nature Neuroscience*, *14*(7), 866–873. doi:10.1038/nn.2837

Stroebel, D., Casado, M., & Paoletti, P. (2018). Triheteromeric NMDA receptors: From structure to synaptic physiology. *Current Opinion in Physiology*, 2, 1–12. doi:10.1016/j.cophys.2017.12.004

Sun, Y., Olson, R., Horning, M., Armstrong, N., Mayer, M., & Gouaux, E. (2002). Mechanism of glutamate receptor desensitization. *Nature*, *417*(6886), 245–253. doi:10.1038/417245a

Swanson, G. T., Gereau, R. W., Green, T., & Heinemann, S. F. (1997). Identification of amino acid residues that control functional behavior in GluR5 and GluR6 kainate receptors. *Neuron*, *19*(4), 913–926. Retrieved from http://www.ncbi.nlm.nih.gov/pubmed/9354337

Swanson, G. T., Green, T., & Heinemann, S. F. (1998). Kainate receptors exhibit differential sensitivities to (S)-5-iodowillardiine. *Molecular Pharmacology*, *53*(5), 942–949.

Swanson, G. T., Kamboj, S. K., & Cull-Candy, S. G. (1997). Single-channel properties of recombinant AMPA receptors depend on RNA editing, splice variation, and subunit composition. *Journal of Neuroscience*, *17*(1), 58–69.

Tabor, C. W., & Tabor, H. (1984). Polyamines. *Annual Review of Biochemistry*, *53*, 749–790. doi:10.1146/annurev.bi.53.070184.003533

Tang, M., Pelkey, K. A., Ng, D., Ivakine, E., McBain, C. J., Salter, M. W., & McInnes, R. R. (2011). Neto1 is an auxiliary subunit of native synaptic kainate receptors. *Journal of Neuroscience*, *31*(27), 10009–10018.

Tichelaar, W., Safferling, M., Keinanen, K., Stark, H., & Madden, D. R. (2004). The three-dimensional structure of an ionotropic glutamate receptor reveals a dimer-of-dimers assembly. *Journal of Molecular Biology, 344*(2), 435–442. doi:10.1016/j.jmb.2004.09.048

Tomita, S., Adesnik, H., Sekiguchi, M., Zhang, W., Wada, K., Howe, J. R., . . . Bredt, D. S. (2005). Stargazin modulates AMPA receptor gating and trafficking by distinct domains. *Nature, 435*(7045), 1052–1058. doi:10.1038/nature03624

Tomita, S., Chen, L., Kawasaki, Y., Petralia, R. S., Wenthold, R. J., Nicoll, R. A., & Bredt, D. S. (2003). Functional studies and distribution define a family of transmembrane AMPA receptor regulatory proteins. *Journal of Cell Biology, 161*(4), 805–816. doi:10.1083/jcb.200212116

Tomita, S., Nicoll, R. A., & Bredt, D. S. (2001). PDZ protein interactions regulating glutamate receptor function and plasticity. *Journal of Cell Biology, 153*(5), F19–F24. doi:10.1083/jcb.153.5.f19

Tomita, S., Shenoy, A., Fukata, Y., Nicoll, R. A., & Bredt, D. S. (2007). Stargazin interacts functionally with the AMPA receptor glutamate-binding module. *Neuropharmacology, 52*(1), 87–91. doi:10.1016/j.neuropharm.2006.07.012

Traynelis, S. F., & Jaramillo, F. (1998). Getting the most out of noise in the central nervous system. *Trends in Neurosciences, 21*(4), 137–145.

Traynelis, S. F., Wollmuth, L. P., McBain, C. J., Menniti, F. S., Vance, K. M., Ogden, K. K., . . . Dingledine, R. (2010). Glutamate receptor ion channels: structure, regulation, and function. *Pharmacological Reviews, 62*(3), 405–496. doi:10.1124/pr.109.002451

Trussell, L. O., & Fischbach, G. D. (1989). Glutamate receptor desensitization and its role in synaptic transmission. *Neuron, 3*(2), 209–218.

Turetsky, D., Garringer, E., & Patneau, D. K. (2005). Stargazin modulates native AMPA receptor functional properties by two distinct mechanisms. *Journal of Neuroscience, 25*(32), 7438–7448. doi:10.1523/JNEUROSCI.1108-05.2005

Twomey, E. C., Yelshanskaya, M. V., Grassucci, R. A., Frank, J., & Sobolevsky, A. I. (2016). Elucidation of AMPA receptor–stargazin complexes by cryo-electron microscopy. *Science, 353*(6294), 83–86. doi:10.1126/science.aaf8411

Twomey, E. C., Yelshanskaya, M. V., Grassucci, R. A., Frank, J., & Sobolevsky, A. I. (2017a). Channel opening and gating mechanism in AMPA-subtype glutamate receptors. *Nature, 549*(7670), 60–65. doi:10.1038/nature23479

Twomey, E. C., Yelshanskaya, M. V., Grassucci, R. A., Frank, J., & Sobolevsky, A. I. (2017b). Structural bases of desensitization in AMPA receptor–auxiliary subunit complexes. *Neuron, 94*(3), 569–580. doi:10.1016/j.neuron.2017.04.025

Twomey, E. C., Yelshanskaya, M. V., Vassilevski, A. A., & Sobolevsky, A. I. (2018). Mechanisms of channel block in calcium-permeable AMPA receptors. *Neuron, 99*(5), 956–968. doi:10.1016/j.neuron.2018.07.027

Ueda, T., Nakamura, Y., Smith, C. M., Copits, B. A., Inoue, A., Ojima, T., . . . Sakai, R. (2013). Isolation of novel prototype galectins from the marine ball sponge *Cinachyrella* sp. guided by their modulatory activity on mammalian glutamate-gated ion channels. *Glycobiology, 23*(4), 412–425. doi:10.1093/glycob/cws165

Venskutonyte, R., Frydenvang, K., Hald, H., Rabassa, A. C., Gajhede, M., Ahring, P. K., & Kastrup, J. S. (2012). Kainate induces various domain closures in AMPA and kainate receptors. *Neurochemistry International, 61*(4), 536–545. doi:10.1016/j.neuint.2012.02.016

Verdoorn, T. A., Burnashev, N., Monyer, H., Seeburg, P. H., & Sakmann, B. (1991). Structural determinants of ion flow through recombinant glutamate receptor channels. *Science, 252*(5013), 1715–1718.

Villarroel, A., Burnashev, N., & Sakmann, B. (1995). Dimensions of the narrow portion of a recombinant NMDA receptor channel. *Biophysical Journal, 68*(3), 866–875. doi:10.1016/S0006-3495(95)80263-8

von Engelhardt, J., Mack, V., Sprengel, R., Kavenstock, N., Li, K. W., Stern-Bach, Y., ... Monyer, H. (2010). CKAMP44: A brain-specific protein attenuating short-term synaptic plasticity in the dentate gyrus. *Science, 327*(5972), 1518–1522. doi:10.1126/science.1184178

Vyklicky, L., Jr., Patneau, D. K., & Mayer, M. L. (1991). Modulation of excitatory synaptic transmission by drugs that reduce desensitization at AMPA/kainate receptors. *Neuron, 7*(6), 971–984.

Wang, R., Mellem, J. E., Jensen, M., Brockie, P. J., Walker, C. S., Hoerndli, F. J., ... Maricq, A. V. (2012). The SOL-2/Neto auxiliary protein modulates the function of AMPA-subtype ionotropic glutamate receptors. *Neuron, 75*(5), 838–850.

Washburn, M. S., Numberger, M., Zhang, S., & Dingledine, R. (1997). Differential dependence on GluR2 expression of three characteristic features of AMPA receptors. *Journal of Neuroscience, 17*(24), 9393–9406.

Watanabe, S., Kusama-Eguchi, K., Kobayashi, H., & Igarashi, K. (1991). Estimation of polyamine binding to macromolecules and ATP in bovine lymphocytes and rat liver. *Journal of Biological Chemistry, 266*(31), 20803–20809.

Wenthold, R. J., Trumpy, V. A., Zhu, W. S., & Petralia, R. S. (1994). Biochemical and assembly properties of GluR6 and KA2, two members of the kainate receptor family, determined with subunit-specific antibodies. *Journal of Biological Chemistry, 269*(2), 1332–1339.

Werner, P., Voigt, M., Keinanen, K., Wisden, W., & Seeburg, P. H. (1991). Cloning of a putative high-affinity kainate receptor expressed predominantly in hippocampal CA3 cells. *Nature, 351*(6329), 742–744. doi:10.1038/351742a0

Westbrook, G. L., & Mayer, M. L. (1987). Micromolar concentrations of Zn^{2+} antagonize NMDA and GABA responses of hippocampal neurons. *Nature, 328*(6131), 640–643. doi:10.1038/328640a0

Weston, M. C., Schuck, P., Ghosal, A., Rosenmund, C., & Mayer, M. L. (2006). Conformational restriction blocks glutamate receptor desensitization. *Nature Structural & Molecular Biology, 13*(12), 1120–1127. doi:10.1038/nsmb1178

Wilding, T. J., Chai, Y. H., & Huettner, J. E. (1998). Inhibition of rat neuronal kainate receptors by cis-unsaturated fatty acids. *Journal of Physiology, 513*(Pt 2), 331–339.

Wilding, T. J., & Huettner, J. E. (1995). Differential antagonism of alpha-amino-3-hydroxy-5-methyl-4-isoxazolepropionic acid–preferring and kainate-preferring receptors by 2,3-benzodiazepines. *Molecular Pharmacology, 47*(3), 582–587.

Wilding, T. J., & Huettner, J. E. (1997). Activation and desensitization of hippocampal kainate receptors. *Journal of Neuroscience, 17*(8), 2713–2721.

Wilding, T. J., Zhou, Y., & Huettner, J. E. (2005). Q/R site editing controls kainate receptor inhibition by membrane fatty acids. *Journal of Neuroscience, 25*(41), 9470–9478. doi:10.1523/JNEUROSCI.2826-05.2005

Wong, A. Y., Fay, A. M., & Bowie, D. (2006). External ions are coactivators of kainate receptors. *Journal of Neuroscience, 26*(21), 5750–5755. doi:10.1523/JNEUROSCI.0301-06.2006

Wong, A. Y., MacLean, D. M., & Bowie, D. (2007). Na^+/Cl^- dipole couples agonist binding to kainate receptor activation. *Journal of Neuroscience, 27*(25), 6800–6809. doi:10.1523/JNEUROSCI.0284-07.2007

Yamada, K. A., & Tang, C. M. (1993). Benzothiadiazides inhibit rapid glutamate receptor desensitization and enhance glutamatergic synaptic currents. *Journal of Neuroscience, 13*(9), 3904–3915.

Yan, D., & Tomita, S. (2012). Defined criteria for auxiliary subunits of glutamate receptors. *Journal of Physiology, 590*(1), 21–31. doi:10.1113/jphysiol.2011.213868

Yan, S., Sanders, J. M., Xu, J., Zhu, Y., Contractor, A., & Swanson, G. T. (2004). A C-terminal determinant of GluR6 kainate receptor trafficking. *Journal of Neuroscience, 24*(3), 679–691. doi:10.1523/JNEUROSCI.4985-03.2004

Yi, F., Bhattacharya, S., Thompson, C. M., Traynelis, S. F., & Hansen, K. B. (2019). Functional and pharmacological properties of triheteromeric GluN1/2B/2D NMDA receptors. *Journal of Physiology, 597*(22), 5495–5514. doi:10.1113/JP278168

Yu, A., Alberstein, R., Thomas, A., Zimmet, A., Grey, R., Mayer, M. L., & Lau, A. Y. (2016). Molecular lock regulates binding of glycine to a primitive NMDA receptor. *Proceedings of the National Academy of Sciences of the United States of America, 113*(44), E6786–E6795. doi:10.1073/pnas.1607010113

Yuan, H., Hansen, K. B., Vance, K. M., Ogden, K. K., & Traynelis, S. F. (2009). Control of NMDA receptor function by the NR2 subunit amino-terminal domain. *Journal of Neuroscience, 29*(39), 12045–12058. doi:10.1523/JNEUROSCI.1365-09.2009

Yue, K. T., MacDonald, J. F., Pekhletski, R., & Hampson, D. R. (1995). Differential effects of lectins on recombinant glutamate receptors. *European Journal of Pharmacology, 291*(3), 229–235.

Yuzaki, M., & Aricescu, A. R. (2017). A GluD coming-of-age story. *Trends in Neurosciences, 40*(3), 138–150. doi:10.1016/j.tins.2016.12.004

Zhang, W., Cho, Y., Lolis, E., & Howe, J. R. (2008). Structural and single-channel results indicate that the rates of ligand binding domain closing and opening directly impact AMPA receptor gating. *Journal of Neuroscience, 28*(4), 932–943. doi:10.1523/JNEUROSCI.3309-07.2008

Zhang, W., Devi, S. P., Tomita, S., & Howe, J. R. (2014). Auxiliary proteins promote modal gating of AMPA- and kainate-type glutamate receptors. *European Journal of Neuroscience, 39*(7), 1138–1147. doi:10.1111/ejn.12519

Zhang, W., Robert, A., Vogensen, S. B., & Howe, J. R. (2006). The relationship between agonist potency and AMPA receptor kinetics. *Biophysical Journal, 91*(4), 1336–1346. doi:10.1529/biophysj.106.084426

Zhang, W., St-Gelais, F., Grabner, C. P., Trinidad, J. C., Sumioka, A., Morimoto-Tomita, M., ... Tomita, S. (2009). A transmembrane accessory subunit that modulates kainate-type glutamate receptors. *Neuron, 61*(3), 385–396. doi:10.1016/j.neuron.2008.12.014

Zhang, Y., Nayeem, N., Nanao, M. H., & Green, T. (2006). Interface interactions modulating desensitization of the kainate-selective ionotropic glutamate receptor subunit GluR6. *Journal of Neuroscience, 26*(39), 10033–10042. doi:10.1523/JNEUROSCI.2750-06.2006

Zhao, H., Berger, A. J., Brown, P. H., Kumar, J., Balbo, A., May, C. A., ... Schuck, P. (2012). Analysis of high-affinity assembly for AMPA receptor amino-terminal domains. *Journal of General Physiology, 139*(5), 371–388. doi:10.1085/jgp.201210770

Zhao, Y., Chen, S., Swensen, A. C., Qian, W. J., & Gouaux, E. (2019). Architecture and subunit arrangement of native AMPA receptors elucidated by cryo-EM. *Science, 364*(6438), 355–362. doi:10.1126/science.aaw8250

Zhao, Y., Chen, S., Yoshioka, C., Baconguis, I., & Gouaux, E. (2016). Architecture of fully occupied GluA2 AMPA receptor–TARP complex elucidated by cryo-EM. *Nature*, *536*, 108–111. doi:10.1038/nature18961

Zheng, Y., Brockie, P. J., Mellem, J. E., Madsen, D. M., Walker, C. S., Francis, M. M., & Maricq, A. V. (2006). SOL-1 is an auxiliary subunit that modulates the gating of GLR-1 glutamate receptors in *Caenorhabditis elegans*. *Proceedings of the National Academy of Sciences of the United States of America*, *103*(4), 1100–1105.

CHAPTER 10

N-METHYL-D-ASPARTATE RECEPTORS

GARY J. IACOBUCCI AND GABRIELA K. POPESCU

Introduction and Historical Perspective

N-Methyl-D-aspartate (NMDA) receptors are glutamate- and glycine-gated, cation-permeable channels with primary expression in the central nervous system (CNS), where they mediate the development, transmission, and plasticity of excitatory synapses. Along with α-amino-3-hydroxyl-5-methyl-4-isoxazole-propionate (AMPA), kainate, and delta receptors, they form the superfamily of ionotropic glutamate receptors (iGluRs), a class of mainly postsynaptic proteins that generate the majority of excitatory synaptic transmission in the brain and spinal cord. It will surprise most of today's students just how controversial this fact has been over more than a century of neurophysiology research (Meldrum, 2000).

The first hypothesis for how neurotransmission occurs in the vertebrate brain held electricity to be the main information carrier. This idea was based on experiments pioneered by Fritsch and Hitzig (1870), who showed that electrical stimulation of cortical brain regions in vivo can elicit involuntary motor movement of specific body parts. This electrical hypothesis remained firmly ingrained for the next half-century (Fulton, 1940) despite Cajal's exquisite neuroanatomical drawings, which depicted synapses as inter-neuronal junctions that contained a physical gap (Ramon y Cajal, 1906). Evidence favoring a diffusible chemical as the carrier of information across this gap between two neurons accumulated slowly and gained universal acceptance only after winning over deeply entrenched and domineering critics who fought bitterly what has become known as "the war of soups and sparks" (Eccles, Fatt, & Koketsu, 1954).

The quest for the nature of the endogenous chemical responsible for excitation across central synapses was similarly sinuous and polarizing (Paton, 1959). Many scientists believed firmly that the substance serving as a neurotransmitter in the brain must be highly specialized for this function and must play no other physiological role. Therefore, glutamate was dismissed by many as an implausible candidate, given its prominent roles as a universal building block of proteins and principal actor in basic metabolic processes. Yet, when injected directly into primate brains, glutamate produced immediate and violent convulsions, an incontrovertible proof of its strong and direct excitatory action (Hayashi, 1954). Speculations flourished to explain away this obvious result, and it took more than 20 years of increasingly precise electrophysiological measurements to establish glutamate as the principal endogenous excitatory neurotransmitter at the central vertebrate synapse (Watkins, 2000).

Equally elusive was the molecular nature of the excitatory proteins that respond to glutamate. Using newly developed spinal cell preparations and extracellular recording technologies, Curtis and Watkins first identified, and then synthesized, numerous chemical analogues of glutamate, including NMDA and 2-amino-5-phosphonopentanoate (AP5), which either had intrinsic excitatory activity or interfered specifically with the glutamate-elicited excitation, respectively (Curtis & Watkins, 1960; Watkins, Curtis, & Brand, 1977). By examining the geometric arrangement of their functional moieties, the authors concluded that, whether acting as agonists or antagonists, the excitatory chemicals fell into three structural classes. Watkins and Evans inferred correctly that this fact likely reflected intrinsic differences in the chemical and geometrical features of the agonist-binding sites of three distinct classes of glutamate-gated excitatory channels, which they named AMPA, kainate, and NMDA receptors according to their most effective synthetic agonists (Watkins, 2000). The cloning revolution of the 1990s certified this influential insight by identifying a large family of homologous yet molecularly distinct proteins, whose pharmacology largely mirrored their sequence homology (Hollmann & Heinemann, 1994). Even if cumbersome, this nomenclature remains in effect today for this historical reason and because the pharmacologic approach continues to be the primary means of identifying the types of receptors responsible for excitation across the myriad synapses of the mammalian CNS (Collingridge, Olsen, Peters, & Spedding, 2009; Lodge, 2009).

Family-specific pharmacology was also instrumental in delineating differential biological roles for the three glutamate receptor classes. Specifically, these approaches have helped to galvanize massive interest in NMDA receptors by identifying their critical roles in synaptic plasticity, a form of cellular memory (Bliss & Gardner-Medwin, 1973), and in excitotoxicity, a type of glutamate-induced neuronal death (Choi, 1985; Lucas & Newhouse, 1957; Olney, 1969). Here, we briefly summarize the current knowledge of the properties and biological functions of NMDA receptors. Despite these substantial advances, given their key roles as mediators of fundamental processes in the CNS, the basis of many higher brain functions, and as important actors in neuropsychiatric conditions, NMDA receptors remain the target of assiduous basic and clinical research and no doubt still hold many surprises.

Molecular Identity, Evolutionary Origins, and Protein Expression

Interest in the structure, function, and biological roles of NMDA receptors originates with the critical and powerful signals they generate in the human CNS (Traynelis et al., 2010). Likely, the fundamental roles in the development and physiology of excitatory synapses in humans mirror those they play in other mammals, such as rats and mice, which have provided the majority of the current experimental evidence. As for many other signaling proteins, the distribution of NMDA receptors, and indeed of all iGluR proteins in neuronal membranes, is relatively sparse. In consequence, their molecular identification had to await the development of technologies that allowed functional cloning, including patch clamp electrophysiology (Hamill, Marty, Neher, Sakmann, & Sigworth, 1981) and the exogenous expression of mammalian genetic material (Sumikawa, Houghton, Emtage, Richards, & Barnard, 1981). Functional cloning and subsequently homology cloning led to the identification of a family of 18 homologous iGluR subunits in mammals (Hollmann & Heinemann, 1994) and an abundance of orthologs across virtually all species (De Bortoli, Teardo, Szabo, Morosinotto, & Alboresi, 2016).

These observations placed iGluR research on a firm molecular foundation but also produced significant surprises. First, they demonstrated that glutamate receptors were unrelated to the only family of neurotransmitter receptors known at the time, the pentameric Cys-loop receptors, thus revealing unsuspected, at the time, diversity among synaptic excitatory channels. Second, iGluR subunits appeared to have wide-ranging nonsynaptic expression and, therefore, broad biological effects apart from their conspicuous participation in neurotransmission. Last, by identifying GluD subunits, which form a class of iGluRs that appear impermeable to ions, it was suggested that proteins in the iGluR superfamily may also have non-ionotropic signaling functions (Dore, Aow, & Malinow, 2015).

With sequences in hand for all mammalian iGluR-encoded peptides, homology analyses provided definitive proof that iGluRs represented a new class of synaptic proteins, separate from other neurotransmitter receptors, and identified partial homology domains with other known proteins. These early bioinformatics analyses and subsequent biochemical evidence for a unique membrane topology led to the influential insight that iGluR subunits are modular in structure (Wo & Oswald, 1995; Wood, VanDongen, & VanDongen, 1995). This seminal observation had two immediate and powerful implications. On the one hand, it ushered in new hypotheses that iGluRs may have evolved by repeated gene duplication and fusion events; and on the other, it was suggested that individual modules may retain three-dimensional structure and perhaps functionality, even when separated from the entire protein. Presently, definitive evidence backs both of these hypotheses.

Molecular Diversity and Modular Structure

Mammalian iGluR subunits fall into four homology groups: GluA, GluK, GluN, and GluD (Figure 10.1A). Within each group, functional proteins consist of tetrameric assemblies. Further, of these, only GluA, GluK, and GluN tetramers form glutamate-gated pores. The presence of GluD tetramers at synapses is necessary for normal synaptic functions, but glutamate does not agonize an ionotropic function from these proteins. Therefore, synaptic GluD proteins may be orphan ligand-gated channels, or they may have purely non-ionotropic roles (Yuzaki & Aricescu, 2017). Of the iGluRs with demonstrated ionotropic function, AMPA and kainate receptors, often referred to as non-NMDA receptors, assemble as homo- or heterotetramers of GluA (1–4) and GluK (1–5) subunits, respectively. In contrast, NMDA receptors are necessarily heterotetramers that contain at least one GluN1 subunit and at least two GluN2 and/or GluN3 subunits.

Of the 18 mammalian iGluR genes, seven encode NMDA receptor subunits (Figure 10.1A, B). A single gene encodes the obligatory GluN1 subunit, which binds the required co-agonist glycine. Its transcript is subject to alternative pre-mRNA splicing of exons 5, 21, and/or 23, resulting in eight molecularly distinct variants (Zukin & Bennett, 1995). Four separately encoded GluN2 subunits, A–D, bind the neurotransmitter glutamate. The remaining two genes encode GluN3 subunits, A and B, which bind glycine or d-serine (Ciabarra et al., 1995; Nishi, Hinds, Lu, Kawata, & Hayashi, 2001; Sucher et al., 1995) (Figure 10.1B).

The NMDA receptor subunits have membrane topologies similar to all iGluR subunits, including external N-terminal (NTD) and ligand-binding (LBD) domains, which are distal and proximal to the membrane, respectively, connected to a transmembrane domain (TMD) formed by three transmembrane helices and a re-entrant loop, which extends into the cytoplasm a C-terminal domain (CTD). Homologous modules interact across subunits to form extracellular tetrameric NTD and LBD layers, respectively (Figure 10.1B). Within each subunit, the two external domains organize as two lobes connected by a flexible hinge, which allows relative movement of lobes and variable interlobe aperture. The LBD lobe connects through three short flexible linkers to the three transmembrane helices within the TMD, which contain the channel gate and the cation-selective pore. The cytoplasmic domain is intrinsically disordered and harbors many post-translational modifications and protein–protein interaction sites. As with most modular proteins, each domain has its own unique evolutionary origins (Anderson & Greenberg, 2001). Thus, examining the evolutionary history of each domain offers important clues into the NMDA receptors' ancestors.

Evolutionary Origin

Bioinformatics approaches made possible by a rapidly expanding database of genomic and protein sequences from an increasing variety of organisms provided initial clues

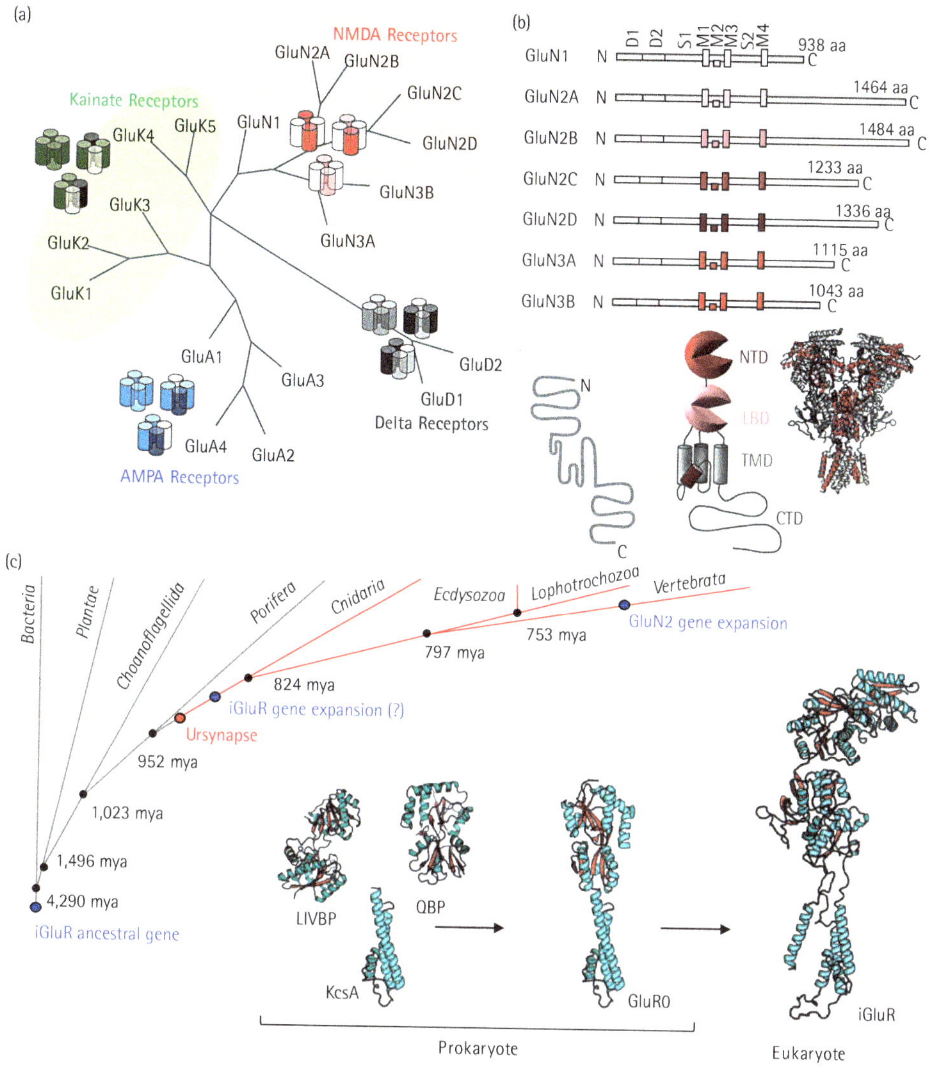

FIGURE 10.1 Diversity, structure, and molecular evolution of NMDA receptors. (A) Eighteen mammalian iGluR subunits segregate into four homology classes. Tetrameric proteins produce AMPA, kainate, and NMDA-gated excitatory channels and orphan GluD channels/receptors. (B) Seven mammalian NMDA receptor subunits share membrane topology and overall three-dimensional architecture. (C) Phylogeny of taxons expressing iGluRs with divergence dates (Timetree, in million years). Red lines, taxons with functional synapses. (Adapted from Ryan and Grant, 2009). Structure of hypothesized evolutionary bacterial precursors: K^+ channels (KcsA, PBD: 5J9P), bacterial periplasmic binding proteins (QBP, PDB: 1WDN; and LIVBP, PDB: 2LIV), and Glu-gated K^+ channels (GluR0, hypothetical structure based on PDB ID 1IIT and 5J9P) and modern mammalian eukaryotic iGluR (GluN2B, PDB: 4PE5).

for the evolutionary origin of iGluR subunits (Figure 10.1C). In prokaryotes, a single gene, encoding GluRo, has been identified to date, strongly suggesting that the four mammalian iGluR subtypes (A, K, D, and N) likely arose within the eukaryotic lineage. Functionally, the GluRo subunit forms a homotetrameric K^+-selective channel that lacks the mammalian NTDs, is gated by glutamate, but is insensitive to AMPA, kainate, or NMDA (Chen, Cui, Mayer, & Gouaux, 1999). More specifically, the differentiation of mammalian subtypes likely occurred after the evolution of plant species because the plant iGluR-like genes, identified in *Arabidopsis*, form a separate homology clade (Wudick, Michard, Oliveira Nunes, & Feijo, 2018). Among eukaryotes, genetic evidence identifies all four mammalian iGluR subtypes (A, K, D, and N) as far back as invertebrates, in *Drosophila* and *Caenorhabditis*, which diverged from the *Homo* lineage about 797 million years ago (mya). Functional results from testing AMPA, kainate, and NMDA agonism on *Hydra vulgaris* suggest the presence of different subtypes in cnidarians, which diverged from *Homo sapiens* ~824 mya (Kumar, Stecher, Suleski, & Hedges, 2017). Thus, pharmacologic diversity, whether or not supported by genetic diversity, appeared within the eukaryotic lineage after the divergence from plants but before the divergence from invertebrates and *Cnidaria*. Rotifers, such as *Adineta vaga*, which appear to have a single iGluR gene, respond functionally to all three mammalian agonists, with preference decreasing from kainate to AMPA and to NMDA (Janovjak, Sandoz, & Isacoff, 2011). Possibly, GluK subunits are the most ancient of the specialized mammalian iGluRs. Alternatively, the ancestral iGluR subunit recognized all three mammalian agonists prior to their genetic divergence.

Given their widespread roles in inter-neuronal communication, the function of iGluRs in unicellular organisms and in species that predate the evolution of synapses is unclear. Genomic analyses have identified iGluR genes in unicellular eukaryotic organisms, such as *Choanoflagellides* (Burkhardt, 2015). Evolutionarily, these organisms existed before the appearance of the first ancestral synapse, the ursynapse, an assembly hypothesized to foreshadow the collection of synaptic proteins identified in the genomes of early marine animals such as cnidarians (Figure 10.1C). In plants such as *Arabidopsis*, iGluRs may serve as amino acid sensors to alert the plant to cell damage from potential threats like insects or fungi (Toyota et al., 2018). Understanding the function of these early precursors may provide clues for possible roles of non-neuronal NMDA receptors in mammals and presumably in humans.

The strong sequence homology within the NMDA receptor subfamily of genes suggests that they originate from a common eukaryotic GluN ancestor, although it remains unclear when the separation happened. Likely, the GluN2 subfamily arose by gene duplication. Invertebrates, such as insects and nematodes, have a single GluN2 ortholog (*Nmdar2* in *Drosophila*, *nmr-2* in *Caenorhabditis*). High similarity in exon structure of GluA2 genes and the analysis of paralogous chromosomal regions surrounding the GluN2 genes (Teng et al., 2010) support the hypothesis that the GluN2 subfamily arose from two rounds of gene duplication, which occurred ~550 mya, at the point of divergence between arthropods and vertebrates (McLysaght, Hokamp, & Wolfe, 2002). The first round of duplication resulted in GluN2A/B and GluN2C/D ancestors, and the second round led to the four individual genes that currently exist in vertebrates (Figure 10.1C).

Partial sequence homology between iGluRs and voltage-gated channels introduced the now widely accepted hypothesis that the two families share the pore of a common ancestor (Beck, Wollmuth, Seeburg, Sakmann, & Kuner, 1999; Kuner, Wollmuth, Karlin, Seeburg, & Sakmann, 1996; Panchenko, Glasser, & Mayer, 2001; Wood et al., 1995). Further, structural and functional evidence indicates that, like potassium-selective channels, functional iGluRs assemble as tetrameric proteins (Karakas & Furukawa, 2014; Laube, Kuhse, & Betz, 1998; Lee et al., 2014; Mano & Teichberg, 1998; Rosenmund, Stern-Bach, & Stevens, 1998; Sobolevsky, Rosconi, & Gouaux, 2009). Finally, the discovery of prokaryotic glutamate-gated K^+-selective channels such as GluRo provides a potential timeline for the evolutionary development of eukaryotic iGluRs (Figure 10.1C).

The iGluRs also share partial sequence homology with the large family of structurally characterized periplasmic bacterial proteins (PBPs) (Quiocho, Phillips, Parsons, & Hogg, 1974). This observation invited the hypothesis that the extracellular portion of each iGluR subunit may consist of two adjacent modules each related to leucine–isoleucine–valine-binding protein (LIVBP) and to glutamine-binding protein (QBP) (Moriyoshi et al., 1991; Stern-Bach et al., 1994; Wo & Oswald, 1994). The idea that the extracellular domains of iGluR subunits may exist as quasi-independent domains energized efforts to isolate and produce these domains, leading to the consequential observation that the separated peptides not only folded correctly but also retained some of the pharmacological properties characteristic of the intact receptor (Lampinen, Pentikainen, Johnson, & Keinanen, 1998). Indeed, structural determination of iGluR LBDs and later of NTDs confirmed this hypothesis and led to more detailed views of the structures and activation mechanisms of iGluRs.

The intracellular domain of iGluRs is the most variable portion of the protein. Conspicuously, vertebrates have a much longer (up to 647 residues in humans) cytoplasmic domain relative to their invertebrate counterparts. In addition, the cytoplasmic domain is the least conserved region of iGluRs, with only 29% sequence identity across the mouse GluN2 paralogs (Ryan, Emes, Grant, & Komiyama, 2008). This domain is unrelated in sequence to known proteins and is intrinsically disordered (Ryan et al., 2008). Clear distinctions of CTD functions have been observed in NMDA receptors even for the closest relatives. For example, truncation of the GluN2B CTD is lethal, similar to complete GluN2B knockout (Kutsuwada et al., 1996; Mori et al., 1998; Sprengel et al., 1998). In contrast, GluN2A knockouts and animals with GluN2A subunits that lack cytoplasmic tails are viable (Sakimura et al., 1995; Sprengel et al., 1998). Consistent with distinct roles of CTDs across NMDA receptor subtypes, genetically swapping the CTDs between GluN2A and GluN2B subunits, produced separate behavioral phenotypes across domains of learning, emotion/motivation, and motor skills (Ryan et al., 2013). Interestingly, electrophysiological recordings demonstrate wild-type properties for these mutant channels (Maki, Aman, Amico-Ruvio, Kussius, & Popescu, 2012). Therefore, the evolutionary divergence of C-termini may have afforded adaptive advantage as a means for bidirectional communication with cell-specific factors. Consistent with this hypothesis, invertebrate GluN orthologs, which have shorter C-termini, lack many of the known protein-binding motifs present on the vertebrate subunits (Ryan et al., 2008).

Expression Across the Life Span, Cell Types, and Specialized Membrane Segments

The seven GluN subunits display unique and highly regulated expression patterns across development, in specific cell types, and across subcellular locations (Paoletti, Bellone, & Zhou, 2013). Controlled expression of NMDA receptor subunits represents a major mechanism regulating neuronal excitability and the physiology of excitatory synapses. Several GluN subunits co-localize at synaptic and nonsynaptic locations, and the observed glutamate-elicited response varies with the identity and amount of the specific subunits expressed.

The GluN1 subunit occurs ubiquitously throughout the embryonic and adult mammalian brain and spinal cord, reflecting its critical role in the assembly and surface trafficking of functional NMDA receptors. Differential splicing generates eight molecularly distinct GluN1 subunits, and additional cellular mechanisms regulate their differential regional and subcellular distribution. Two types of splice variants, GluN1-a and -b, differ in the sequence of the external NTD and produce receptors with distinct kinetic and pharmacological properties. The expression of GluN1-a/b variants responds to physiologic cues, such as patterns of activity, and to pathologic cues, such as following experimental spinal cord injury. Four GluN1 variants (1–4) differ in the molecular structure of the intracellular CTD. GluN1-2 is enriched in cortical and hippocampal regions, whereas GluN1-4 has a complementary profile. Within a given neuron, GluN1-1 localizes more to synaptic regions relative to GluN1-2, which is found mostly in nonsynaptic regions. This may reflect differences in cytoplasmic domain sequences that are critical for synaptic targeting such as binding to neurofilament-L and PDZ proteins. The functional significance of these distributions is unknown as no differences in electrophysiological properties have been identified so far.

GluN2 and GluN3 subunits have differential and often complementary expression patterns across the life span and cell types. GluN2B, GluN2D, and GluN3A subunits predominate early in development. GluN2B displays a broad expression level shortly after birth and becomes largely restricted to the forebrain as the animal develops. GluN2D expression declines during transition to adulthood and becomes restricted in the diencephalon and mesencephalon. GluN3A has a characteristic peak in expression after birth and drops to low but constant levels in adults. In contrast, the expression levels of GluN2A increase gradually throughout development, and this subunit becomes predominant in adult tissue. GluN2C expression levels are relatively constant throughout the life span, dominating only in the cerebellum and the olfactory bulb.

Accumulating evidence suggests that in aged animals NMDA receptor expression levels decrease, and this correlates functionally with reduced NMDA receptor–mediated signals and behaviorally with cognitive deficits. As individuals age, memory is among the earliest cognitive functions to decline (Gallagher & Nicolle, 1993). Specifically, aging associates with deficits in spatial, short-term, and long-term memory in humans, non-human primates, and rodents. Consistent with the pervasive role of NMDA receptors in memory formation, storage, and retrieval, advanced aging correlates strongly with reduced NMDA receptor function.

Evidence from radiolabeled binding studies in mammals indicates that aging-related receptor hypofunction relates primarily to decreased receptor density in cortical and hippocampal tissue and to decreased performance in long-term and spatial memory tasks in rodents. Consistent with these observations, aged animals display reduced long-term potentiation. Specifically, the expression of GluN1 and GluN2B subunits declines with age in rodents and primate models (Gazzaley, Siegel, Kordower, Mufson, & Morrison, 1996; Magnusson, Nelson, & Young, 2002; Sheng, Cummings, Roldan, Jan, & Jan, 1994).

Unique Functional Properties

Relative to other members of the iGluR family, NMDA receptors possess a variety of properties which distinguish them functionally and physiologically. Given the overall similarity in quaternary structure, these significant divergences in function likely reflect subtle structural differences between iGluR subtypes. A thorough understanding of the breadth of NMDA receptor functional properties provides insight into the role of these receptors in physiological and pathophysiological contexts.

Kinetic Properties of the NMDA Receptor EPSC

Most excitatory synapses express several types of iGluRs; therefore, the excitatory postsynaptic current (EPSC) reflects the combined response of all the receptors present. Presently, methods to ascertain the molecular identity and amount of iGluR types at a given synapse are imprecise; moreover, the composition of synaptic receptors is dynamic, changing with the physiological state of the synapse. To date, the majority of the information regarding the types of functional iGluRs at mammalian synapses relies largely on pharmacologic methods of isolating and characterizing the component ionic fluxes of the observed overall current.

The NMDA receptor–mediated synaptic current was revealed experimentally when synaptic activity was recorded in the absence of Mg^{2+}, which in physiologic conditions blocks the channel, and in the presence of exogenous glycine, which is an obligatory co-agonist. Thus enhanced, the NMDA receptor–mediated EPSC was isolated from the co-localized non-NMDA receptors, using AMPA receptor–specific inhibitors such as cyanquixaline. These and other pharmacologic manipulations revealed that the EPSC rise phase closely followed the kinetics of the AMPA-sensitive component, whereas the decay phase followed the decay of the NMDA-responsive component (50–500 ms). Further, because the NMDA component lasted longer than the lifetime of synaptically released glutamate (~1.2 ms), it became clear that the EPSC decay time, which is critical to synaptic processing, integration, and plasticity, depended primarily on properties intrinsic to the postsynaptic NMDA receptors present, especially their gating kinetics (Lester, Clements, Westbrook, & Jahr, 1990). For this reason, over the past two decades, the gating kinetics of NMDA receptors has represented an important area of research. The mechanisms that control the activation of NMDA receptors are most accurately investigated by examining the behaviors of single molecules with statistical methods.

As for most ion channels, currents recorded from a single NMDA receptor display complex patterns of activity. These reflect the stochastic transitions between closed- and open-channel conformations and are governed primarily by thermodynamic constraints (Figure 10.2A) (Iacobucci & Popescu, 2017). Experimental conditions that enhance the signal amplitude and simplify its kinetics allowed the accumulation of sufficient one-channel current records, which when processed with statistical methods exposed a multistate kinetic model for NMDA receptor gating. This model has been extensively validated across the four diheteromeric GluN1/GluN2(A–D) receptor types (Figure 10.2A, inset). It proposes that the rising phase of the macroscopic current reflects the rate with which individual agonist-bound receptors transition across at least three kinetically distinct closed states (families of conformations) to slowly reach open states which allow transmembrane ionic flux. Conversely, the macroscopic current decay phase reflects the rate with which individual open-channel receptors, whose occupancy is maximal during the peak of the macroscopic current, transition back into closed-channel conformations, whose structure restricts the passage of current but allows agonist dissociation to prevent subsequent openings. The model also incorporates off-path desensitized states, whose time-dependent increase in occupancy corresponds to the slow (1–2 s) fade of the macroscopic response recorded during prolonged exposures to agonists, as is likely the case with extra synaptic receptors.

This high-resolution understanding of the NMDA receptor gating process represents a springboard from which to address two important and yet unanswered questions. First, combining macroscopic current recordings and statistical analyses of one-channel currents may reveal how pharmacologic and endogenous ligands, as well as intracellular signaling, impact NMDA receptor gating kinetics and, therefore, the time course of the EPSC (Figure 10.2A). Second, combining electrophysiological approaches with molecular dynamics simulations of transitions between structural conformations may expose the sequence of intramolecular rearrangements that represent receptor activation and therefore the structural identity and lifetime of each of the functional states assumed in the kinetic model (Figure 10.2B).

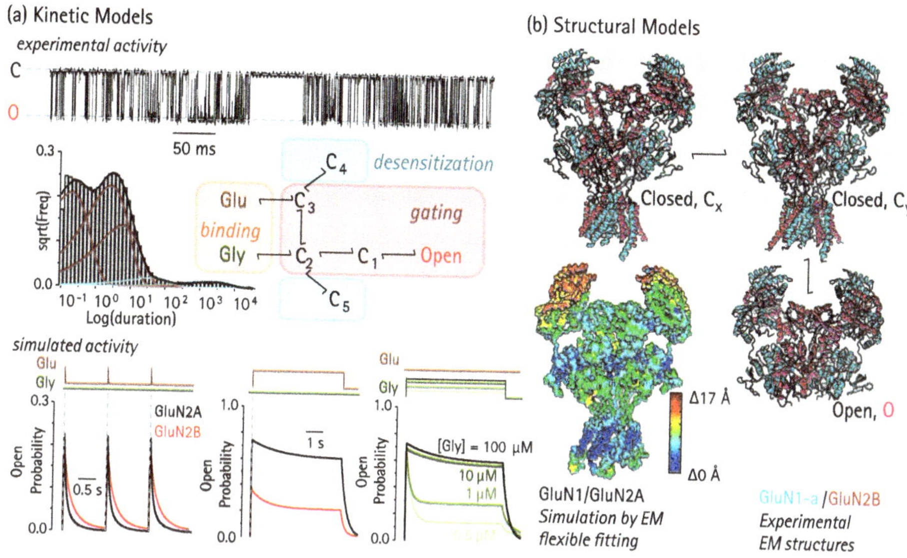

FIGURE 10.2 NMDA receptor gating: kinetic and structural models. (**A**) *Top*, Current trace recorded from a channel exposed to high concentrations of agonists (Glu, Gly) in the absence of divalent cations (Na^+ currents) illustrates stochastic oscillations between 0 and 10 pA (70 pS) current levels, indicative of thermodynamically controlled transitions between conformations with closed (C) and open (O) pores. *Middle*, Histogram illustrates the compound distribution of closed event durations inferred from experimentally recorded one-channel currents; black line illustrates the distribution predicted by the multistate model at right. *Bottom*, Macroscopic responses predicted by the model under brief repetitive stimulation with Glu and under prolonged exposure to agonists reproduce well experimentally recorded NMDA receptor currents (Iacobucci & Popescu 2017). (**B**) Highlighted in gray, three conformations of GluN1/GluN2B receptors observed with cryo-electron microscopy (EM) likely represent closed (C_x, C_y) and open (O) receptors (Tajima et al. 2016; PDB 5FXH, 5FXI, and 5FXG); they are arranged in a temporal sequence that may underlie receptor gating. Molecular modeling and dynamic simulations of the GluN1/GluN2A receptor suggest a temporal sequence of intramolecular conformational change by which closed conformations can transition into open conformations. Heat map illustrates the magnitude of displacement in alpha-carbon positions between closed and open states and identifies hot spots of conformational change (Zheng et al. 2017).

Reaction Mechanism

Historically, the NMDA receptor activation mechanism has been modeled intuitively as a simple two-step binding/gating reaction, where *binding* referred collectively to the four required association/dissociation equilibria (two for glutamate and two for glycine) and *gating* referred globally to any host of conformational changes that transformed the closed (impermeable) receptor into an open (permeable) receptor. However, it became quickly apparent that in most experimental conditions, NMDA receptor current, even when recorded in the continuous presence of agonists, faded in time and that this loss of activity could be reversed with rest. To account for this behavior, the binding/gating scheme was amended with a desensitization step, imagined to represent transitions into a family of closed receptor conformations (states) that required (agonist) binding but did not allow (channel) gating (Lester & Jahr, 1992).

Since the advent of single-channel electrophysiological recordings, NMDA receptor activity has been probed in exquisite detail over several time domains (milliseconds to tens of minutes). In certain experimental conditions, where single NMDA receptors produce uniformly large currents, it is possible to measure the duration of all periods when channels are closed (C) or open (O) as they occur during the normal operation of the channel. The observed distributions of closed and open event durations always include multiple kinetic components that can be queried with statistical methods to develop models of receptor activation, which include all the observed kinetic states (Figure 10.2A) (Iacobucci & Popescu, 2017; Popescu, 2012). Although this method simplifies the underlying reality of myriad conformations, which intermorph during receptor activity, into a bare minimum of kinetically and functionally distinct states (thermodynamically equivalent conformations), it has the power to first approximate transition rates between states based on the single-molecule record and then predict macroscopic current responses to countless stimulation protocols and conditions (Figure 10.2A). This approach has produced a general understanding of the functional changes that a receptor experiences during periods of activity, their relative timing, and how these are modified by mutations and modulators (Popescu, 2005).

A complementary line of investigation seeks to define with atomic resolution the structural arrangement of all functionally relevant conformations. The present knowledge of structure–function relationships in iGluR proteins originates with the influential observation that NTDs and LBDs share homology with bacterial amino-acid binding proteins. This observation motivated efforts to produce the individual domains as water-soluble proteins and thus facilitated structure determination by nuclear magnetic resonance (McFeeters & Oswald, 2002) and later X-ray crystallography (Furukawa & Gouaux, 2003; Inanobe, Furukawa, & Gouaux, 2005; Karakas, Regan, & Furukawa, 2015). Results showed that the external portion of the NMDA receptor can be envisioned as a collection of semi-independent globular domains, which like their ancestors (QBP and LIVBP) bind agonist at the interface between two lobes and can move around a flexible linker. Agonist binding changes the relative position of the two lobes, and in intact receptors, this movement can be communicated as mechanical force to covalently connected sequences (linkers) or to adjacent domains. Given the interleaved structure of the functional tetramers, one can imagine that the initially small local movement produced by agonist binding can ripple onto adjacent domains and produce a chain of back-and-forth conformational changes. The sequence of conformational change likely mirrors the sequence of state transitions inferred with kinetic approaches and explains the probability of populating open receptor conformations and therefore the time course of the NMDA receptor current.

Such a view of receptor activity has been substantiated recently by a series of cryo-electron microscopic structures of intact GluN1/GluN2B receptors that appear to represent closed and open receptor conformations. Comparing these structures revealed two positions for each heterodimer pair within the LBD that differ by a 13.5° rotation. This difference likely places tension on the short linkers between the LBD and TMD, as envisioned for a gating movement (Tajima et al., 2016) (Figure 10.2B). Based on these observed structures, molecular modeling predicts a substantial wave of spatially and temporally organized motions as the physical origin for NMDA receptor gating (Figure 10.2B) (Zheng, Wen, Iacobucci, & Popescu, 2017). Importantly, areas predicted to be most dynamic during gating overlap with disease-related sites identified in human patients (Hu, Chen, Myers, Yuan, & Traynelis, 2016) (Figure 10.3). Present efforts seek to match functionally defined states with structurally defined conformations in a way that would integrate structural, thermodynamic, and kinetic considerations to produce a satisfying understanding of the physical basis of the NMDA receptor signal.

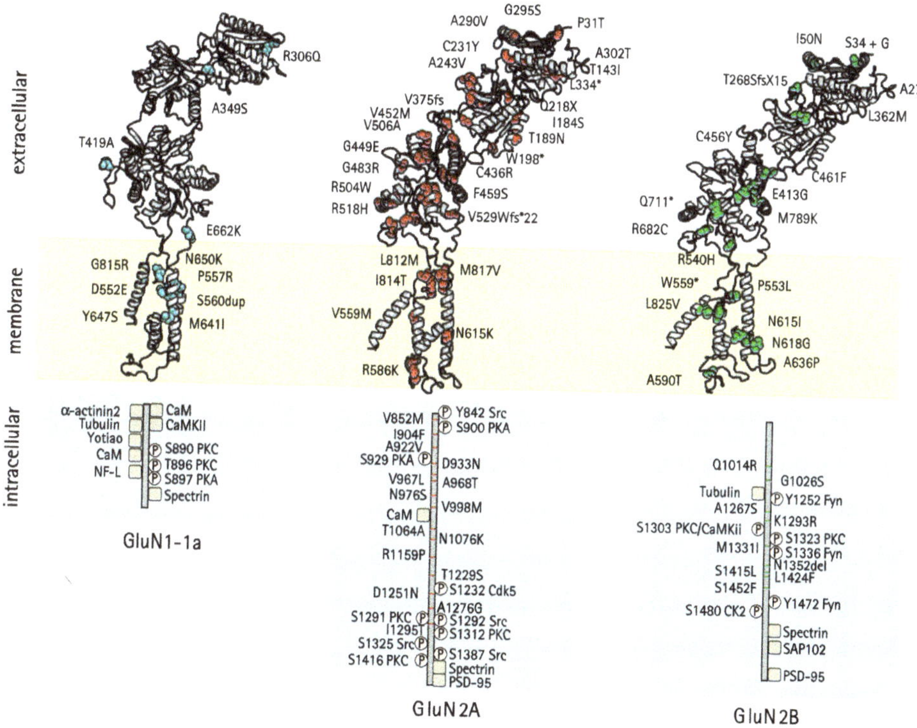

FIGURE 10.3 Disease-associated mutations mapped onto GluN1-a (*left*, PDB ID 4PE5), GluN2A (*middle*, homology model), and GluN2B (*right*, PDB ID 4PE5) subunits. C-terminal structures are not resolved and are displayed in schematic format here. In all subunits, disease-associated mutations span all major receptor domains: extracellular (*top*), transmembrane (*middle*), and intracellular (*bottom*).

Permeation Properties of NMDA Receptors

Aside from their unique gating kinetics, NMDA receptors also have characteristic permeation properties, including relatively large conductance and high Ca^{2+} permeability. In physiologic conditions, the currents gated by NMDA receptors consist primarily of inward Na^+ and Ca^{2+} fluxes (Jahr & Stevens, 1993; Maki & Popescu, 2014). Unitary conductance can reach ~70 pS for GluN1/GluN2(A-B) receptors and ~50 pS for GluN1/GluN2(C-D) (Glasgow, Siegler, Retchless, & Johnson, 2015). Notably, up to 11% of this current may be carried by Ca^{2+}.

Together, long activations and large unitary conductance allow NMDA receptors to produce substantial increases in spine and dendritic Na^+. Ratiometric Na^+ imaging showed that a transient NMDA receptor activation can elevate dendritic spine Na^+ concentrations to 30–40 mM, and during trains of activity produced by high-frequency stimulation, this value can reach 100 mM (Rose & Konnerth, 2001), as was observed in neural networks under epileptic activity (Karus, Mondragao, Ziemens, & Rose, 2015). These substantial alterations in the Na^+ electrochemical gradient can significantly influence excitatory synaptic current reversal potentials and influence global neuronal activity by membrane depolarization, affecting the activity of Na^+ transporters (Blaustein & Lederer, 1999; Mondragao et al., 2016). Surprisingly, increases in intracellular Na^+ can itself alter NMDA receptor function. Single-channel current recordings revealed that Na^+ influx through neighboring glutamate receptors boosts the receptor's open probability. Pharmacological manipulations revealed that this sensitivity to modulation by intracellular Na^+ is determined by Src kinases (Yu & Salter, 1998).

Simultaneously, their high unitary Ca^{2+} flux and prolonged activations result in substantial increases in dendritic Ca^{2+}, whose bulk concentration can reach 10 μM (Sabatini, Oertner, & Svoboda, 2002). Due to these unique properties, NMDA receptors are the predominant source of Ca^{2+} in dendritic spines in a variety of brain regions (Higley & Sabatini, 2012). The intracellular Ca^{2+} flux generated by NMDA receptor activation follows the same kinetics as the macroscopic current and is governed by the receptor gating kinetics (Kovalchuk, Eilers, Lisman, & Konnerth, 2000; Murthy, Sejnowski, & Stevens, 2000). Several endogenous mechanisms increase specifically the amplitude of the Ca^{2+} current, by changing the fractional content of Ca^{2+} in the total current. The metabolic status of the neuron, explicitly through protein kinase A, modulates Ca^{2+} permeability of the NMDA receptor by direct phosphorylation of intracellular residues of the GluN1 subunit (Aman, Maki, Ruffino, Kasperek, & Popescu, 2014; Skeberdis et al., 2006). Separately, protein kinase C may enhance the NMDA component of synaptic transmission by relieving Mg^{2+} block and increasing the lifetime of open receptor states (Chen & Huang, 1992). Importantly, pharmacological inhibition of these kinases interferes with the induction of NMDA receptor–dependent synaptic plasticity.

NMDA Receptors as Nonionic Signaling Hubs

In addition to their critical ionotropic function in shaping the EPSCs, NMDA receptors can execute signal transduction that is independent of their ionic current. This function depends critically on the large intracellular domains of GluN subunits, which include numerous protein–protein interaction sites (Figure 10.3).

Repeated glutamate applications onto NMDA receptors induce dephosphorylation of tyrosine residues on intracellular GluN domains and correlate with a use-dependent decrease in current. Surprisingly, this effect persists when currents are blocked with high levels of extracellular Mg^{2+} (Vissel, Krupp, Heinemann, & Westbrook, 2001) and can result in dendritic spine shrinkage (Stein, Gray, & Zito, 2015). Conversely, glycine applications prime NMDA receptors for subsequent activity-dependent internalization, a form of NMDA receptor plasticity (Han, Campanucci, Cooke, & Salter, 2013; Nong et al., 2003).

Consistent with a biologically important non-ionotropic function of NMDA receptors, transgenic mice engineered to express GluN2 subunits lacking cytoplasmic domains have altered synaptic plasticity, although this manipulation leaves unchanged the NMDA receptor permeation properties (Kohr et al., 2003). Additionally, animals that express GluN2A and GluN2B subunits whose C-termini were swapped have severe behavioral deficits but normal synaptic transmission (Ryan et al., 2013). In vitro, exchanging the C-termini of GluN2A and GluN2B subunits produced NMDA receptors with wild type–like conductance and gating (Maki et al., 2012). More recently, fluorescence resonance energy transfer studies revealed that sustained NMDA applications onto neurons produced conformational shifts in the CTDs of the NMDA receptor. Importantly, competitive antagonists (AP5) prevented the agonist-induced conformational change, whereas pore blockers (MK801 and 7CK) did not prevent these structural rearrangements (Dore et al., 2015). Together these observations support an important biological role for NMDA receptors in signal transduction, aside from their depolarizing and Ca^{2+}-fluxing functions.

The cytoplasmic domains of each NMDA receptor subunit are responsible for extensive protein interactions. The GluN1 subunit is a converging point of interaction between actinin (Wyszynski et al., 1998), calmodulin (Ehlers, Zhang, Bernhadt, & Huganir, 1996), yotiao (Lin et al., 1998), neurofilament-L (Ehlers, Fung, O'Brien, & Huganir, 1998), spectrin (Wechsler & Teichberg, 1998), tubulin (van Rossum, Kuhse, & Betz, 1999), and membrane-associated guanylate kinases (Standley, Roche, McCallum, Sans, & Wenthold, 2000). The GluN2 subunits collectively interact with Ca^{2+}/calmodulin-dependent protein kinase II (CaMKII), actinin (Wyszynski et al., 1998), postsynaptic density (PSD) protein-93, , PSD-95, synapse-associated protein-102 (Sheng & Sala, 2001), synaptic scaffolding molecule (Hirao et al., 1998), channel-interacting PDZ protein (Kurschner, Mermelstein, Holden, & Surmeier, 1998), Src kinase (Yu, Askalan, Keil, & Salter, 1997), spectrin (Wechsler & Teichberg, 1998), phospholipase C (Gurd &

Bissoon, 1997), and tubulin (van Rossum et al., 1999). The GluN3A subunit associates with protein phosphatase 2A (Ma & Sucher, 2004), microtubule-associated protein 1B (Eriksson et al., 2010), cell cycle and apoptosis regulatory protein 1 (Jiang et al., 2010), G protein pathway suppressor 2 (Eriksson et al., 2007), and Rheb (Sucher et al., 2010). Therefore, in addition to the ionotropic roles of channel function, the extensive protein interaction network and metabotropic functions make the NMDA receptor a veritable hub of intracellular signaling.

Physiological Roles

The canonical role of NMDA receptors has been their function as postsynaptic responders to fluctuations in extracellular glutamate concentrations. This role is central to normal synaptic function and pathological missense variants in NMDA receptors have been causally associated with disease through perturbations in this process. However, since their original discovery as excitatory postsynaptic glutamate receptors, research has provided additional insights into their physiological roles in both presynaptic neuronal compartments and non-neuronal cell types.

Neuronal Postsynaptic Signal

The critical role of intracellular Ca^{2+} in the induction of synaptic plasticity was recognized by observations that Ca^{2+} chelators introduced in postsynaptic hippocampal neurons prevented activity-dependent changes in synaptic strength (Lynch, Larson, Kelso, Barrionuevo, & Schottler, 1983). Their high Ca^{2+} permeability over other non-NMDA receptors coupled with their higher sensitivity to synaptically released glutamate made NMDA receptors primary candidates as mediators of the intracellular Ca^{2+} signals that trigger synaptic changes (Jahr & Stevens, 1987; MacDermott, Mayer, Westbrook, Smith, & Barker, 1986; Mayer & Westbrook, 1987).

Further, the ability of NMDA receptors to respond to a diverse array of cellular stimuli allows them to integrate a variety of inputs and thus to function as crucial signaling nodes in the neuron. The sensitivity of NMDA receptors to membrane voltage is imparted by the constitutive block by Mg^{2+} in the pore which is released during depolarization (Nowak, Bregestovski, Ascher, Herbet, & Prochiantz, 1984). During an EPSC event initiated by vesicular glutamate release, the initial depolarization produced by AMPA receptor activation and the resultant Na^{+} influx alleviates the blocking effect of Mg^{2+} to allow Ca^{2+} entry. NMDA receptor–mediated Ca^{2+} flux activates Ca^{2+}-dependent signaling cascades that control the surface expression of postsynaptic AMPA receptors to modify the EPSC amplitude and in effect produce long-term potentiation (LTP) or depression (LTD). Alternatively, the postsynaptic depolarization necessary to relieve NMDA receptor blockade can occur due to the back-propagation of an action potential. The ability of NMDA receptors to respond to postsynaptic action potentials provides the basis of spike timing–dependent plasticity. The sensitivity to synaptic inputs tends to increase when the synaptic input immediately precedes the output action potential. Conversely, the synaptic strength is weakened when the EPSC immediately follows the output action potential (Feldman, 2012). The slow gating kinetics of NMDA receptors, which prolongs the time course of the synaptic signal, facilitates the temporal summation of excitatory signals generated by neighboring spines. This allows for spatial integration of numerous synaptic inputs to regulate the action potential probability of the neuron. Thus, NMDA receptors have an important role in the final neuronal firing pattern, which represents the interaction of all excitatory and inhibitory components in the neuron (Ma, Kelly, & Wu, 2002; Wu, Ma, & Kelly, 2004).

In addition to activating intracellular cascades directly involved in synaptic plasticity, the large intracellular Ca^{2+} signals generated by NMDA receptors modulate the activity of surrounding Ca^{2+}-activated K^+ channels (SK channels) (Ngo-Anh et al., 2005). SK channels mediate K^+ efflux to repolarize the postsynaptic membrane. This repolarization forms a negative feedback loop on NMDA receptors by reinstating the Mg^{2+} block. Furthermore, the inhibition of SK channels by the coordinated activation of type I mGluRs is required for LTP induction in spike timing-dependent plasticity (Tigaret, Olivo, Sadowski, Ashby, & Mellor, 2016).

During development, the establishment and maintenance of mature neuronal networks depend on the activity-dependent pruning of synaptic circuits. The coordination between excitatory and inhibitory inputs dictates the extent of structural plasticity of dendritic spines (Holtmaat & Svoboda, 2009). NMDA receptor–dependent processes involved in synaptic plasticity, LTP and LTD, are associated with synaptic spine development. LTP and LTD are associated with dendritic spine enlargement and shrinkage, respectively (Fortin et al., 2010; Wiegert & Oertner, 2013; Zhou, Homma, & Poo, 2004).

The composite model for synaptic strengthening involves NMDA receptors responding to high-amplitude, brief glutamate concentration released at high frequencies. The resultant activation of NMDA receptors initiates mitogen-activated protein kinase signaling to induce LTP in hippocampal synapses (Banko et al., 2005; Bateup, Takasaki, Saulnier, Denefrio, & Sabatini, 2011; Kelleher, Govindarajan, & Tonegawa, 2004; Malenka, 1994). The Ca^{2+} influx into the dendritic spine through NMDA receptors activated from high-frequency stimuli recruits CaMKII. This activation of CaMKII results in the phosphorylation of AMPA receptors to increase their unitary conductance (Benke, Luthi, Isaac, & Collingridge, 1998) and facilitates trafficking and insertion to the membrane (Ehlers, 2000). In contrast, lower-frequency synaptic stimuli are associated with LTD. The association between LTD and dendritic spine shrinkage led to the hypothesis that the magnitude of Ca^{2+} influx induced by low-frequency stimuli activates protein phosphatases (calcineurin and protein phosphatase 1) in the dendritic spine that reduce AMPA expression and induce spine shrinkage (Mulkey, Endo, Shenolikar, & Malenka, 1994). In addition to this model of synaptic weakening, low-amplitude, steady glutamate concentration (as from spillover from neighboring spines) selectively activate type I mGluRs, which trigger protein phosphatase 2A to initiate AMPA receptor removal (Gross et al., 2015; Huber, Gallagher, Warren, & Bear, 2002; Niere, Wilkerson, & Huber, 2012). This model of NMDA receptor involvement in synaptic plasticity accounts for synaptic excitatory/inhibitory coordination among a region of proximal dendritic spines.

However, the recent evidence that agonist binding alone was sufficient to induce LTP independent of ionic flux led to the hypothesis that ionic flux may not be involved in spine shrinkage (Dore et al., 2015; Nabavi et al., 2013; Stein et al., 2015). Consistent with this hypothesis, glutamate uncaging at LTD-inducing frequencies in hippocampal slices in the presence of the NMDA channel blocker MK801 did not interfere with dendritic spine shrinkage (Stein et al., 2015). Thus, structural rearrangements in the cytoplasmic domain of the NMDA receptor channel are sufficient to alter intracellular biochemical signals to initiate both functional and structural changes in spines.

Neuronal Presynaptic Signal

NMDA receptors can be expressed at presynaptic sites as well (Bouvier, Bidoret, Casado, & Paoletti, 2015). This hypothesis was ushered in by the observation that NMDA receptor antagonists decrease glutamate and aspartate release in CA1 hippocampal slices. Thus, in addition to their postsynaptic roles in controlling the direction and magnitude of long-term plasticity, NMDA receptors can control short-term synaptic plasticity by influencing the probability of neurotransmitter release. Presynaptic modulation of neurotransmitter may occur by several mechanisms: 1) Ca^{2+} influx via NMDA receptors directly activating vesicle fusion machinery, 2) membrane depolarization activating voltage-gated Ca^{2+} channels coupled to vesicle fusion machinery, or 3) modulation of downstream signaling pathways.

The spontaneous release of neurotransmitter can be either Ca^{2+}-dependent or independent. Given the high permeability of NMDA receptors for Ca^{2+}, it is likely that presynaptic regulation by NMDA receptors is largely Ca^{2+}-driven. However, emergent research has indicated that, at specific subsets of synapses, presynaptic NMDA receptors exhibit a tonic enhancement of transmitter release (Brasier & Feldman, 2008; Corlew, Wang, Ghermazien, Erisir, & Philpot, 2007; Crabtree, Lodge, Bashir, & Isaac, 2013; Duguid & Smart, 2004; Sjostrom, Turrigiano, & Nelson, 2003). This tonic enhancement likely involves NMDA receptor activation by ambient glutamate. In addition, the incorporation of receptor subunits with reduced Mg^{2+} sensitivity may further facilitate tonic activation (Banerjee et al., 2009; Larsen et al., 2011; Mameli, Carta, Partridge, & Valenzuela, 2005).

Unlike spontaneous transmitter release which is independent of a stimulus, the role of NMDA receptors in evoked neurotransmitter release is optimized to facilitate glutamate release at specific physiological stimuli frequencies in a given brain region. NMDA receptors in the visual cortex enhance glutamate release at 10 Hz (Larsen et al., 2014; Sjostrom et al., 2003), while receptors in the cerebellar fiber-to-Purkinje cell synapses work best between 40 Hz and 1 kHz (Bidoret, Ayon, Barbour, & Casado, 2009). In hippocampal CA3-to-CA1 synapses, presynaptic NMDA receptors facilitate glutamate release at theta-like, 5 Hz, frequencies (McGuinness et al., 2010). Thus, the sensitivity of presynaptic NMDA receptors to specific frequencies likely reflects brain region–specific subunit expression patterns and regulatory elements similar to their postsynaptic counterparts. The presynaptic composition of NMDA receptors also undergoes developmental changes.

Less understood is how presynaptic NMDA receptors modulate long-term synaptic plasticity. At L4-to-L2/3 synapses in the somatosensory cortex, spike timing–dependent LTD required the activation of presynaptic NMDA receptors, while LTP required only postsynaptic NMDA receptors (Bender, Bender, Brasier, & Feldman, 2006). In addition, this LTD required activation of voltage-gated Ca^{2+} channels, inositol 1,4,5-trisphosphate receptors, and metabotropic glutamate receptors and subsequent engagement of downstream endocannabinoid signaling. Work in L4-to-L2/3 synapses of the barrel cortex showed the same endocannabinoid signaling requirement (Banerjee et al., 2009; Hardingham, Wright, Dachtler, & Fox, 2008; Rodriguez-Moreno et al., 2013). Therefore, the role of presynaptic NMDA receptors in modulating long-term plasticity likely depends on synapse-specific factors that remain to be identified.

Glial Cells

Since their discovery, the majority of research efforts have focused on the eminent role of NMDA receptors in neuron physiology and plasticity. However, emerging research is providing evidence of ionotropic glutamate receptor expression in a much wider array of cell types than previously appreciated. Besides neurons, the prominent cell class within the CNS are glial cells. These cells maintain homeostasis in the CNS. Glia achieve this by regulating K^+ levels, acidity, neurotransmitter levels, cerebral blood flow, antioxidant formation, and water transport. These regulatory roles imply a sensory mechanism for changes in the controlled variable. Expression of iGluRs on glial cells allows these cells to detect changes in glutamate concentration at and around synapses.

The first evidence of glial NMDA receptors came from mRNA work which detected expression of GluN2B and GluN2C in astrocytes from adult rats. However, no transcripts were found for GluN1, the obligatory subunit of NMDA receptors (Akazawa, Shigemoto, Bessho, Nakanishi, & Mizuno, 1994; Conti, Minelli, Molnar, & Brecha, 1994; Luque & Richards, 1995). Subsequent work in young and newborn animals showed that GluN1 expression decreases with culture time, suggesting that the methods of glial cell cultivation may affect gene expression. Human astrocytes have been shown to possess all known NMDA receptor subunit mRNA (Lee et al., 2010).

In addition to RNA expression, electron microscopy of immunolabeled subunits provided evidence for GluN1 and GluN2A/B subunit expression in astrocytes from the visual cortex (Aoki, Venkatesan, Go, Mong, & Dawson, 1994; Conti, DeBiasi, Minelli, & Melone, 1996) as well as in the amygdala, stria terminalis, and nucleus locus coeruleus (Farb, Aoki, & Ledoux, 1995; Gracy & Pickel, 1995). However, no immunoreactivity was detected in CA1 of the adult hippocampus (Gottlieb & Matute, 1997; Krebs, Fernandes, Sheldon, Raymond, & Baimbridge, 2003). Thus, the expression of functional NMDA receptors on astrocytes may be brain region–specific.

In addition to astrocytes, transcripts of GluN1 and GluN2D were identified in the CG-4 oligodendroglial cell line (Yoshioka, Ikegaki, Williams, & Pleasure, 1996). The expression of NMDA receptor subunits in oligodendrocytes was definitively confirmed in mouse oligodendrocytes where transcripts for GluN1, GluN2A-D, and GluN3A were detected, with GluN1, GluN2C, and GluN3A being most abundant (Salter & Fern, 2005). Electrophysiological studies provided direct evidence that these transcripts yield functional receptors in white matter oligodendrocytes in the cerebellum, corpus callosum, and rat optic nerve (Karadottir, Cavelier, Bergersen, & Attwell, 2005; Kolodziejczyk, Hamilton, Wade, Karadottir, & Attwell, 2009; Micu et al., 2006). Interestingly, these NMDA-evoked responses were largely restricted to the cell processes rather than the soma.

Another pioneering study suggested the expression of NMDA receptor subunits on polydendrocytes using neonatal rat oligodendrocyte precursors (Wang et al., 1996). This initial finding was later confirmed with the detection of GluN1 on O4-positive polydendrocytes in both human and postnatal rat white matter (Manning et al., 2008). In addition, GluN2A and GluN2C subunits were found on somas and processes of polydendrocytes in the transgenic NG2-dsRed mouse (Hamilton, Vayro, Wigley, & Butt, 2010).

Microglia also express NMDA receptors. Injection of NMDA into the somatosensory cortex of rats led to the transient activation of microglia as determined histologically (Acarin, Gonzalez, Castellano, & Castro, 1996). This finding was substantiated by detection of GluN1 and GluN2A-D transcripts (Murugan, Sivakumar, Lu, Ling, & Kaur, 2011).

Other Cell Types

Less well understood but beginning to come to light is the role of NMDA receptors in non-neuronal cell types. Recently, functional NMDA receptors have been found on erythrocytes, specifically erythroid precursor cells and immature red blood cells, reticulocytes. Their activation leads to a rapid increase in intracellular Ca^{2+} (Makhro et al., 2010). The function of NMDA receptors on erythrocytes is unclear. However, channel activation modulates intracellular pH, which is critical for hemoglobin oxygen affinity. Thus, NMDA receptors may play a role in tuning the availability of oxygen to peripheral tissues (Makhro et al., 2013). In addition to red blood cells, neuroepithelial cells, which form part of the blood–brain barrier, express GluN1 and GluN2A/B subunit transcripts and protein. Whether these subunits constitute functional receptors remains to be elucidated (Sharp et al., 2003).

The endogenous ligand for NMDA receptor activation in these peripheral tissues is less clear. A recent report shows that mechanical forces produced by cellular deformation can gate NMDA receptor currents in the absence of agonist (Maneshi et al., 2017).

Such alternative modalities of NMDA receptor activation may provide clues to the function of these receptors in non-neuronal tissues.

References

Acarin, L., Gonzalez, B., Castellano, B., & Castro, A. J. (1996). Microglial response to *N*-methyl-D-aspartate-mediated excitotoxicity in the immature rat brain. *Journal of Comparative Neurology, 367*(3), 361–374. doi:10.1002/(SICI)1096-9861(19960408)367:3<361::AID-CNE4>3.0.CO;2-3

Akazawa, C., Shigemoto, R., Bessho, Y., Nakanishi, S., & Mizuno, N. (1994). Differential expression of five *N*-methyl-D-aspartate receptor subunit mRNAs in the cerebellum of developing and adult rats. *Journal of Comparative Neurology, 347*(1), 150–160. doi:10.1002/cne.903470112

Aman, T. K., Maki, B. A., Ruffino, T. J., Kasperek, E. M., & Popescu, G. K. (2014). Separate intramolecular targets for protein kinase A control *N*-methyl-D-aspartate receptor gating and Ca^{2+} permeability. *Journal of Biological Chemistry, 289*(27), 18805–18817. doi:10.1074/jbc.M113.537282

Anderson, P. A., & Greenberg, R. M. (2001). Phylogeny of ion channels: Clues to structure and function. *Comparative Biochemistry and Physiology Part B Biochemistry & Molecular Biology, 129*(1), 17–28.

Aoki, C., Venkatesan, C., Go, C. G., Mong, J. A., & Dawson, T. M. (1994). Cellular and subcellular localization of NMDA-R1 subunit immunoreactivity in the visual cortex of adult and neonatal rats. *Journal of Neuroscience, 14*(9), 5202–5222.

Banerjee, A., Meredith, R. M., Rodriguez-Moreno, A., Mierau, S. B., Auberson, Y. P., & Paulsen, O. (2009). Double dissociation of spike timing-dependent potentiation and depression by subunit-preferring NMDA receptor antagonists in mouse barrel cortex. *Cerebral Cortex, 19*(12), 2959–2969. doi:10.1093/cercor/bhp067

Banko, J. L., Poulin, F., Hou, L., DeMaria, C. T., Sonenberg, N., & Klann, E. (2005). The translation repressor 4E-BP2 is critical for eIF4F complex formation, synaptic plasticity, and memory in the hippocampus. *Journal of Neuroscience, 25*(42), 9581–9590. doi:10.1523/JNEUROSCI.2423-05.2005

Bateup, H. S., Takasaki, K. T., Saulnier, J. L., Denefrio, C. L., & Sabatini, B. L. (2011). Loss of Tsc1 in vivo impairs hippocampal mGluR-LTD and increases excitatory synaptic function. *Journal of Neuroscience, 31*(24), 8862–8869. doi:10.1523/JNEUROSCI.1617-11.2011

Beck, C., Wollmuth, L. P., Seeburg, P. H., Sakmann, B., & Kuner, T. (1999). NMDAR channel segments forming the extracellular vestibule inferred from the accessibility of substituted cysteines. *Neuron, 22*(3), 559–570. doi:10.1016/S0896-6273(00)80710-2

Bender, V. A., Bender, K. J., Brasier, D. J., & Feldman, D. E. (2006). Two coincidence detectors for spike timing-dependent plasticity in somatosensory cortex. *Journal of Neuroscience, 26*(16), 4166–4177. doi:10.1523/JNEUROSCI.0176-06.2006

Benke, T. A., Luthi, A., Isaac, J. T., & Collingridge, G. L. (1998). Modulation of AMPA receptor unitary conductance by synaptic activity. *Nature, 393*(6687), 793–797. doi:10.1038/31709

Bidoret, C., Ayon, A., Barbour, B., & Casado, M. (2009). Presynaptic NR2A-containing NMDA receptors implement a high-pass filter synaptic plasticity rule. *Proceedings of the National Academy of Sciences of the United States of America, 106*(33), 14126–14131. doi:10.1073/pnas.0904284106

Blaustein, M. P., & Lederer, W. J. (1999). Sodium/calcium exchange: Its physiological implications. *Physiological Reviews, 79*(3), 763–854. doi:10.1152/physrev.1999.79.3.763

Bliss, T. V., & Gardner-Medwin, A. R. (1973). Long-lasting potentiation of synaptic transmission in the dentate area of the unanaesthetized rabbit following stimulation of the perforant path. *Journal of Physiology*, *232*(2), 357–374.

Bouvier, G., Bidoret, C., Casado, M., & Paoletti, P. (2015). Presynaptic NMDA receptors: Roles and rules. *Neuroscience*, *311*, 322–340. doi:10.1016/j.neuroscience.2015.10.033

Brasier, D. J., & Feldman, D. E. (2008). Synapse-specific expression of functional presynaptic NMDA receptors in rat somatosensory cortex. *Journal of Neuroscience*, *28*(9), 2199–2211. doi:10.1523/JNEUROSCI.3915-07.2008

Burkhardt, P. (2015). The origin and evolution of synaptic proteins—Choanoflagellates lead the way. *Journal of Experimental Biology*, *218*(Pt 4), 506–514. doi:10.1242/jeb.110247

Chen, G. Q., Cui, C., Mayer, M. L., & Gouaux, E. (1999). Functional characterization of a potassium-selective prokaryotic glutamate receptor. *Nature*, *402*(6763), 817–821. doi:10.1038/45568

Chen, L., & Huang, L. Y. (1992). Protein kinase C reduces Mg^{2+} block of NMDA-receptor channels as a mechanism of modulation. *Nature*, *356*(6369), 521–523. doi:10.1038/356521a0

Choi, D. W. (1985). Glutamate neurotoxicity in cortical cell culture is calcium dependent. *Neuroscience Letters*, *58*(3), 293–297.

Ciabarra, A. M., Sullivan, J. M., Gahn, L. G., Pecht, G., Heinemann, S., & Sevarino, K. A. (1995). Cloning and characterization of chi-1: A developmentally regulated member of a novel class of the ionotropic glutamate receptor family. *Journal of Neuroscience*, *15*(10), 6498–6508.

Collingridge, G. L., Olsen, R. W., Peters, J., & Spedding, M. (2009). A nomenclature for ligand-gated ion channels. *Neuropharmacology*, *56*(1), 2–5. doi:10.1016/j.neuropharm.2008.06.063

Conti, F., DeBiasi, S., Minelli, A., & Melone, M. (1996). Expression of NR1 and NR2A/B subunits of the NMDA receptor in cortical astrocytes. *Glia*, *17*(3), 254–258. doi:10.1002/(SICI)1098-1136(199607)17:3<254::AID-GLIA7>3.0.CO;2-0

Conti, F., Minelli, A., Molnar, M., & Brecha, N. C. (1994). Cellular localization and laminar distribution of NMDAR1 mRNA in the rat cerebral cortex. *Journal of Comparative Neurology*, *343*(4), 554–565. doi:10.1002/cne.903430406

Corlew, R., Wang, Y., Ghermazien, H., Erisir, A., & Philpot, B. D. (2007). Developmental switch in the contribution of presynaptic and postsynaptic NMDA receptors to long-term depression. *Journal of Neuroscience*, *27*(37), 9835–9845. doi:10.1523/JNEUROSCI.5494-06.2007

Crabtree, J. W., Lodge, D., Bashir, Z. I., & Isaac, J. T. (2013). GABAA, NMDA and mGlu2 receptors tonically regulate inhibition and excitation in the thalamic reticular nucleus. *European Journal of Neuroscience*, *37*(6), 850–859. doi:10.1111/ejn.12098

Curtis, D. R., & Watkins, J. C. (1960). The excitation and depression of spinal neurones by structurally related amino acids. *Journal of Neurochemistry*, *6*, 117–141.

De Bortoli, S., Teardo, E., Szabo, I., Morosinotto, T., & Alboresi, A. (2016). Evolutionary insight into the ionotropic glutamate receptor superfamily of photosynthetic organisms. *Biophysical Chemistry*, *218*, 14–26. doi:10.1016/j.bpc.2016.07.004

Dore, K., Aow, J., & Malinow, R. (2015). Agonist binding to the NMDA receptor drives movement of its cytoplasmic domain without ion flow. *Proceedings of the National Academy of Sciences of the United States of America*, *112*(47), 14705–14710. doi:10.1073/pnas.1520023112

Duguid, I. C., & Smart, T. G. (2004). Retrograde activation of presynaptic NMDA receptors enhances GABA release at cerebellar interneuron–Purkinje cell synapses. *Nature Neuroscience*, *7*(5), 525–533. doi:10.1038/nn1227

Eccles, J. C., Fatt, P., & Koketsu, K. (1954). Cholinergic and inhibitory synapses in a pathway from motor-axon collaterals to motoneurones. *Journal of Physiology*, *126*(3), 524–562.

Ehlers, M. D. (2000). Reinsertion or degradation of AMPA receptors determined by activity-dependent endocytic sorting. *Neuron, 28*(2), 511–525.
Ehlers, M. D., Fung, E. T., O'Brien, R. J., & Huganir, R. L. (1998). Splice variant-specific interaction of the NMDA receptor subunit NR1 with neuronal intermediate filaments. *Journal of Neuroscience, 18*(2), 720–730.
Ehlers, M. D., Zhang, S., Bernhadt, J. P., & Huganir, R. L. (1996). Inactivation of NMDA receptors by direct interaction of calmodulin with the NR1 subunit. *Cell, 84*(5), 745–755.
Eriksson, M., Nilsson, A., Samuelsson, H., Samuelsson, E. B., Mo, L., Akesson, E., . . . Sundstrom, E. (2007). On the role of NR3A in human NMDA receptors. *Physiology & Behavior, 92*(1–2), 54–59. doi:10.1016/j.physbeh.2007.05.026
Eriksson, M., Samuelsson, H., Bjorklund, S., Tortosa, E., Avila, J., Samuelsson, E. B., . . . Sundstrom, E. (2010). MAP1B binds to the NMDA receptor subunit NR3A and affects NR3A protein concentrations. *Neuroscience Letters, 475*(1), 33–37. doi:10.1016/j.neulet.2010.03.039
Farb, C. R., Aoki, C., & Ledoux, J. E. (1995). Differential localization of NMDA and AMPA receptor subunits in the lateral and basal nuclei of the amygdala: A light and electron microscopic study. *Journal of Comparative Neurology, 362*(1), 86–108. doi:10.1002/cne.903620106
Feldman, D. E. (2012). The spike-timing dependence of plasticity. *Neuron, 75*(4), 556–571. doi:10.1016/j.neuron.2012.08.001
Fortin, D. A., Davare, M. A., Srivastava, T., Brady, J. D., Nygaard, S., Derkach, V. A., & Soderling, T. R. (2010). Long-term potentiation–dependent spine enlargement requires synaptic Ca^{2+} -permeable AMPA receptors recruited by CaM-kinase I. *Journal of Neuroscience, 30*(35), 11565–11575. doi:10.1523/JNEUROSCI.1746-10.2010
Fritsch, G., & Hitzig, E. (1870). Über die elektrische Erregbarkeit des Grosshirns. *Archiv für Anatomie, Physiologie und wissenschaftliche Medicin, 3*, 300–332.
Fulton, J. F. (1940). The central nervous system. *Annual Review of Physiology, 2*(1), 243–262.
Furukawa, H., & Gouaux, E. (2003). Mechanisms of activation, inhibition and specificity: Crystal structures of the NMDA receptor NR1 ligand-binding core. *EMBO Journal, 22*(12), 2873–2885.
Gallagher, M., & Nicolle, M. M. (1993). Animal models of normal aging: Relationship between cognitive decline and markers in hippocampal circuitry. *Behavioural Brain Research, 57*(2), 155–162. doi:10.1016/0166-4328(93)90131-9
Gazzaley, A. H., Siegel, S. J., Kordower, J. H., Mufson, E. J., & Morrison, J. H. (1996). Circuit-specific alterations of *N*-methyl-D-aspartate receptor subunit 1 in the dentate gyrus of aged monkeys. *Proceedings of the National Academy of Sciences of the United States of America, 93*(7), 3121–3125.
Glasgow, N. G., Siegler Retchless, B., & Johnson, J. W. (2015). Molecular bases of NMDA receptor subtype-dependent properties. *Journal of Physiology, 593*(1), 83–95. doi:10.1113/jphysiol.2014.273763
Gottlieb, M., & Matute, C. (1997). Expression of ionotropic glutamate receptor subunits in glial cells of the hippocampal CA1 area following transient forebrain ischemia. *Journal of Cerebral Blood Flow & Metabolism, 17*(3), 290–300. doi:10.1097/00004647-199703000-00006
Gracy, K. N., & Pickel, V. M. (1995). Comparative ultrastructural localization of the NMDAR1 glutamate receptor in the rat basolateral amygdala and bed nucleus of the stria terminalis. *Journal of Comparative Neurology, 362*(1), 71–85. doi:10.1002/cne.903620105
Gross, C., Chang, C. W., Kelly, S. M., Bhattacharya, A., McBride, S. M., Danielson, S. W., . . . Bassell, G. J. (2015). Increased expression of the PI3K enhancer PIKE mediates

deficits in synaptic plasticity and behavior in fragile X syndrome. *Cell Reports, 11*(5), 727–736. doi:10.1016/j.celrep.2015.03.060

Gurd, J. W., & Bissoon, N. (1997). The *N*-methyl-D-aspartate receptor subunits NR2A and NR2B bind to the SH2 domains of phospholipase C-gamma. *Journal of Neurochemistry, 69*(2), 623–630.

Hamill, O. P., Marty, A., Neher, E., Sakmann, B., & Sigworth, F. J. (1981). Improved patch-clamp techniques for high-resolution current recording from cells and cell-free membrane patches. *Pflugers Archiv, 391*(2), 85–100.

Hamilton, N., Vayro, S., Wigley, R., & Butt, A. M. (2010). Axons and astrocytes release ATP and glutamate to evoke calcium signals in NG2-glia. *Glia, 58*(1), 66–79. doi:10.1002/glia.20902

Han, L., Campanucci, V. A., Cooke, J., & Salter, M. W. (2013). Identification of a single amino acid in GluN1 that is critical for glycine-primed internalization of NMDA receptors. *Molecular Brain, 6*, 36. doi:10.1186/1756-6606-6-36

Hardingham, N., Wright, N., Dachtler, J., & Fox, K. (2008). Sensory deprivation unmasks a PKA-dependent synaptic plasticity mechanism that operates in parallel with CaMKII. *Neuron, 60*(5), 861–874. doi:10.1016/j.neuron.2008.10.018

Hayashi, T. (1954). Effects of sodium glutamate on the nervous system. *Keio Journal of Medicine, 3*, 192–193.

Higley, M. J., & Sabatini, B. L. (2012). Calcium signaling in dendritic spines. *Cold Spring Harbor Perspectives in Biology, 4*(4), a005686. doi:10.1101/cshperspect.a005686

Hirao, K., Hata, Y., Ide, N., Takeuchi, M., Irie, M., Yao, I., . . . Takai, Y. (1998). A novel multiple PDZ domain-containing molecule interacting with *N*-methyl-D-aspartate receptors and neuronal cell adhesion proteins. *Journal of Biological Chemistry, 273*(33), 21105–21110.

Hollmann, M., & Heinemann, S. (1994). Cloned glutamate receptors. *Annual Review of Neuroscience, 17*, 31–108. doi:10.1146/annurev.ne.17.030194.000335

Holtmaat, A., & Svoboda, K. (2009). Experience-dependent structural synaptic plasticity in the mammalian brain. *Nature Reviews Neuroscience, 10*(9), 647–658. doi:10.1038/nrn2699

Hu, C., Chen, W., Myers, S. J., Yuan, H., & Traynelis, S. F. (2016). Human GRIN2B variants in neurodevelopmental disorders. *Journal of Pharmacological Sciences, 132*(2), 115–121. doi:10.1016/j.jphs.2016.10.002

Huber, K. M., Gallagher, S. M., Warren, S. T., & Bear, M. F. (2002). Altered synaptic plasticity in a mouse model of fragile X mental retardation. *Proceedings of the National Academy of Sciences of the United States of America, 99*(11), 7746–7750. doi:10.1073/pnas.122205699

Iacobucci, G. J., & Popescu, G. K. (2017). NMDA receptors: Linking physiological output to biophysical operation. *Nature Reviews Neuroscience, 18*(4), 236–249. doi:10.1038/nrn.2017.24

Inanobe, A., Furukawa, H., & Gouaux, E. (2005). Mechanism of partial agonist action at the NR1 subunit of NMDA receptors. *Neuron, 47*(1), 71–84.

Jahr, C. E., & Stevens, C. F. (1993). Calcium permeability of the *N*-methyl-D-aspartate receptor channel in hippocampal neurons in culture. *Proceedings of the National Academy of Sciences of the United States of America, 90*(24), 11573–11577.

Jahr, C. E., & Stevens, C. F. (1987). Glutamate activates multiple single channel conductances in hippocampal neurons. *Nature, 325*(6104), 522–525.

Janovjak, H., Sandoz, G., & Isacoff, E. Y. (2011). A modern ionotropic glutamate receptor with a K^+ selectivity signature sequence. *Nature Communications, 2*, 232. doi:10.1038/ncomms1231

Jiang, Y., Puliyappadamba, V. T., Zhang, L., Wu, W., Wali, A., Yaffe, M. B., . . . Rishi, A. K. (2010). A novel mechanism of cell growth regulation by cell cycle and apoptosis regulatory protein (CARP)-1. *Journal of Molecular Signaling, 5*, 7. doi:10.1186/1750-2187-5-7

Karadottir, R., Cavelier, P., Bergersen, L. H., & Attwell, D. (2005). NMDA receptors are expressed in oligodendrocytes and activated in ischaemia. *Nature, 438*(7071), 1162–1166. doi:10.1038/nature04302

Karakas, E., & Furukawa, H. (2014). Crystal structure of a heterotetrameric NMDA receptor ion channel. *Science, 344*(6187), 992–997. doi:10.1126/science.1251915

Karakas, E., Regan, M. C., & Furukawa, H. (2015). Emerging structural insights into the function of ionotropic glutamate receptors. *Trends in Biochemical Sciences, 40*(6), 328–337. doi:10.1016/j.tibs.2015.04.002

Karus, C., Mondragao, M. A., Ziemens, D., & Rose, C. R. (2015). Astrocytes restrict discharge duration and neuronal sodium loads during recurrent network activity. *Glia, 63*(6), 936–957. doi:10.1002/glia.22793

Kelleher, R. J., 3rd, Govindarajan, A., & Tonegawa, S. (2004). Translational regulatory mechanisms in persistent forms of synaptic plasticity. *Neuron, 44*(1), 59–73. doi:10.1016/j.neuron.2004.09.013

Kohr, G., Jensen, V., Koester, H. J., Mihaljevic, A. L., Utvik, J. K., Kvello, A., . . . Hvalby, O. (2003). Intracellular domains of NMDA receptor subtypes are determinants for long-term potentiation induction. *Journal of Neuroscience, 23*(34), 10791–10799.

Kolodziejczyk, K., Hamilton, N. B., Wade, A., Karadottir, R., & Attwell, D. (2009). The effect of *N*-acetyl-aspartyl-glutamate and *N*-acetyl-aspartate on white matter oligodendrocytes. *Brain, 132*(Pt 6), 1496–1508. doi:10.1093/brain/awp087

Kovalchuk, Y., Eilers, J., Lisman, J., & Konnerth, A. (2000). NMDA receptor–mediated subthreshold Ca^{2+} signals in spines of hippocampal neurons. *Journal of Neuroscience, 20*(5), 1791–1799.

Krebs, C., Fernandes, H. B., Sheldon, C., Raymond, L. A., & Baimbridge, K. G. (2003). Functional NMDA receptor subtype 2B is expressed in astrocytes after ischemia in vivo and anoxia in vitro. *Journal of Neuroscience, 23*(8), 3364–3372.

Kumar, S., Stecher, G., Suleski, M., & Hedges, S. B. (2017). TimeTree: A resource for timelines, timetrees, and divergence times. *Molecular Biology and Evolution, 34*(7), 1812–1819. doi:10.1093/molbev/msx116

Kuner, T., Wollmuth, L. P., Karlin, A., Seeburg, P. H., & Sakmann, B. (1996). Structure of the NMDA receptor channel M2 segment inferred from the accessibility of substituted cysteines. *Neuron, 17*(2), 343–352. doi:S0896-6273(00)80165-8

Kurschner, C., Mermelstein, P. G., Holden, W. T., & Surmeier, D. J. (1998). CIPP, a novel multivalent PDZ domain protein, selectively interacts with Kir4.0 family members, NMDA receptor subunits, neurexins, and neuroligins. *Molecular and Cellular Neuroscience, 11*(3), 161–172. doi:10.1006/mcne.1998.0679

Kutsuwada, T., Sakimura, K., Manabe, T., Takayama, C., Katakura, N., Kushiya, E., . . . Mishina, M. (1996). Impairment of suckling response, trigeminal neuronal pattern formation, and hippocampal LTD in NMDA receptor epsilon 2 subunit mutant mice. *Neuron, 16*(2), 333–344.

Lampinen, M., Pentikainen, O., Johnson, M. S., & Keinanen, K. (1998). AMPA receptors and bacterial periplasmic amino acid–binding proteins share the ionic mechanism of ligand recognition. *EMBO Journal, 17*(16), 4704–4711. doi:10.1093/emboj/17.16.4704

Larsen, R. S., Corlew, R. J., Henson, M. A., Roberts, A. C., Mishina, M., Watanabe, M., . . . Philpot, B. D. (2011). NR3A-containing NMDARs promote neurotransmitter release and spike timing-dependent plasticity. *Nature Neuroscience, 14*(3), 338–344. doi:10.1038/nn.2750

Larsen, R. S., Smith, I. T., Miriyala, J., Han, J. E., Corlew, R. J., Smith, S. L., & Philpot, B. D. (2014). Synapse-specific control of experience-dependent plasticity by presynaptic NMDA receptors. *Neuron, 83*(4), 879–893. doi:10.1016/j.neuron.2014.07.039

Laube, B., Kuhse, J., & Betz, H. (1998). Evidence for a tetrameric structure of recombinant NMDA receptors. *Journal of Neuroscience, 18*(8), 2954–2961.

Lee, C. H., Lu, W., Michel, J. C., Goehring, A., Du, J., Song, X., & Gouaux, E. (2014). NMDA receptor structures reveal subunit arrangement and pore architecture. *Nature, 511*(7508), 191–197. doi:10.1038/nature13548

Lee, M. C., Ting, K. K., Adams, S., Brew, B. J., Chung, R., & Guillemin, G. J. (2010). Characterisation of the expression of NMDA receptors in human astrocytes. *PLoS One, 5*(11), e14123. doi:10.1371/journal.pone.0014123

Lester, R. A., Clements, J. D., Westbrook, G. L., & Jahr, C. E. (1990). Channel kinetics determine the time course of NMDA receptor–mediated synaptic currents. *Nature, 346*(6284), 565–567.

Lester, R. A., & Jahr, C. E. (1992). NMDA channel behavior depends on agonist affinity. *Journal of Neuroscience, 12*(2), 635–643. Retrieved from http://www.jneurosci.org/cgi/reprint/12/2/635

Lin, J. W., Wyszynski, M., Madhavan, R., Sealock, R., Kim, J. U., & Sheng, M. (1998). Yotiao, a novel protein of neuromuscular junction and brain that interacts with specific splice variants of NMDA receptor subunit NR1. *Journal of Neuroscience, 18*(6), 2017–2027.

Lodge, D. (2009). The history of the pharmacology and cloning of ionotropic glutamate receptors and the development of idiosyncratic nomenclature. *Neuropharmacology, 56*(1), 6–21.

Lucas, D. R., & Newhouse, J. P. (1957). The toxic effect of sodium L-glutamate on the inner layers of the retina. *AMA Archives of Ophthalmology, 58*(2), 193–201.

Luque, J. M., & Richards, J. G. (1995). Expression of NMDA 2B receptor subunit mRNA in Bergmann glia. *Glia, 13*(3), 228–232. doi:10.1002/glia.440130309

Lynch, G., Larson, J., Kelso, S., Barrionuevo, G., & Schottler, F. (1983). Intracellular injections of EGTA block induction of hippocampal long-term potentiation. *Nature, 305*(5936), 719–721.

Ma, C. L., Kelly, J. B., & Wu, S. H. (2002). AMPA and NMDA receptors mediate synaptic excitation in the rat's inferior colliculus. *Hearing Research, 168*(1–2), 25–34.

Ma, O. K., & Sucher, N. J. (2004). Molecular interaction of NMDA receptor subunit NR3A with protein phosphatase 2A. *Neuroreport, 15*(9), 1447–1450. doi:10.1097/01.wnr.0000132773.41720.2d

MacDermott, A. B., Mayer, M. L., Westbrook, G. L., Smith, S. J., & Barker, J. L. (1986). NMDA-receptor activation increases cytoplasmic calcium concentration in cultured spinal cord neurones. *Nature, 321*(6069), 519–522.

Magnusson, K. R., Nelson, S. E., & Young, A. B. (2002). Age-related changes in the protein expression of subunits of the NMDA receptor. *Brain Research Molecular Brain Research, 99*(1), 40–45.

Makhro, A., Hanggi, P., Goede, J. S., Wang, J., Bruggemann, A., Gassmann, M., . . . Bogdanova, A. (2013). *N*-Methyl-D-aspartate receptors in human erythroid precursor cells and in circulating red blood cells contribute to the intracellular calcium regulation. *American Journal of Physiology Cell Physiology, 305*(11), C1123–C1138. doi:10.1152/ajpcell.00031.2013

Makhro, A., Wang, J., Vogel, J., Boldyrev, A. A., Gassmann, M., Kaestner, L., & Bogdanova, A. (2010). Functional NMDA receptors in rat erythrocytes. *American Journal of Physiology Cell Physiology, 298*(6), C1315–C1325. doi:10.1152/ajpcell.00407.2009

Maki, B. A., Aman, T. K., Amico-Ruvio, S. A., Kussius, C. L., & Popescu, G. K. (2012). C-terminal domains of *N*-methyl-D-aspartic acid receptor modulate unitary channel conductance and gating. *Journal of Biological Chemistry, 287*(43), 36071–36080. doi:10.1074/jbc.M112.390013

Maki, B. A., & Popescu, G. K. (2014). Extracellular Ca^{2+} ions reduce NMDA receptor conductance and gating. *Journal of General Physiology, 144*(5), 379–392. doi:10.1085/jgp.201411244

Malenka, R. C. (1994). Synaptic plasticity in the hippocampus: LTP and LTD. *Cell, 78*(4), 535–538.

Mameli, M., Carta, M., Partridge, L. D., & Valenzuela, C. F. (2005). Neurosteroid-induced plasticity of immature synapses via retrograde modulation of presynaptic NMDA receptors. *Journal of Neuroscience, 25*(9), 2285–2294. doi:10.1523/JNEUROSCI.3877-04.2005

Maneshi, M. M., Maki, B., Gnanasambandam, R., Belin, S., Popescu, G. K., Sachs, F., & Hua, S. Z. (2017). Mechanical stress activates NMDA receptors in the absence of agonists. *Scientific Reports, 7*, 39610. doi:10.1038/srep39610

Manning, S. M., Talos, D. M., Zhou, C., Selip, D. B., Park, H. K., Park, C. J., . . . Jensen, F. E. (2008). NMDA receptor blockade with memantine attenuates white matter injury in a rat model of periventricular leukomalacia. *Journal of Neuroscience, 28*(26), 6670–6678. doi:10.1523/JNEUROSCI.1702-08.2008

Mano, I., & Teichberg, V. I. (1998). A tetrameric subunit stoichiometry for a glutamate receptor-channel complex. *Neuroreport, 9*(2), 327–331.

Mayer, M. L., & Westbrook, G. L. (1987). Permeation and block of *N*-methyl-D-aspartic acid receptor channels by divalent cations in mouse cultured central neurones. *Journal of Physiology, 394*, 501–527.

McFeeters, R. L., & Oswald, R. E. (2002). Structural mobility of the extracellular ligand-binding core of an ionotropic glutamate receptor. Analysis of NMR relaxation dynamics. *Biochemistry, 41*(33), 10472–10481.

McGuinness, L., Taylor, C., Taylor, R. D., Yau, C., Langenhan, T., Hart, M. L., . . . Emptage, N. J. (2010). Presynaptic NMDARs in the hippocampus facilitate transmitter release at theta frequency. *Neuron, 68*(6), 1109–1127. doi:10.1016/j.neuron.2010.11.023

McLysaght, A., Hokamp, K., & Wolfe, K. H. (2002). Extensive genomic duplication during early chordate evolution. *Nature Genetics, 31*(2), 200–204. doi:10.1038/ng884

Meldrum, B. S. (2000). Glutamate as a neurotransmitter in the brain: Review of physiology and pathology. *Journal of Nutrition, 130*(4S Suppl.), 1007S–1015S.

Micu, I., Jiang, Q., Coderre, E., Ridsdale, A., Zhang, L., Woulfe, J., . . . Stys, P. K. (2006). NMDA receptors mediate calcium accumulation in myelin during chemical ischaemia. *Nature, 439*(7079), 988–992. doi:10.1038/nature04474

Mondragao, M. A., Schmidt, H., Kleinhans, C., Langer, J., Kafitz, K. W., & Rose, C. R. (2016). Extrusion versus diffusion: Mechanisms for recovery from sodium loads in mouse CA1 pyramidal neurons. *Journal of Physiology, 594*(19), 5507–5527. doi:10.1113/JP272431

Mori, H., Manabe, T., Watanabe, M., Satoh, Y., Suzuki, N., Toki, S., . . . Mishina, M. (1998). Role of the carboxy-terminal region of the GluR epsilon2 subunit in synaptic localization of the NMDA receptor channel. *Neuron, 21*(3), 571–580.

Moriyoshi, K., Masu, M., Ishii, T., Shigemoto, R., Mizuno, N., & Nakanishi, S. (1991). Molecular cloning and characterization of the rat NMDA receptor. *Nature, 354*(6348), 31–37.

Mulkey, R. M., Endo, S., Shenolikar, S., & Malenka, R. C. (1994). Involvement of a calcineurin/inhibitor-1 phosphatase cascade in hippocampal long-term depression. *Nature, 369*(6480), 486–488. doi:10.1038/369486a0

Murthy, V. N., Sejnowski, T. J., & Stevens, C. F. (2000). Dynamics of dendritic calcium transients evoked by quantal release at excitatory hippocampal synapses. *Proceedings of the National Academy of Sciences of the United States of America, 97*(2), 901–906.

Murugan, M., Sivakumar, V., Lu, J., Ling, E. A., & Kaur, C. (2011). Expression of *N*-methyl D-aspartate receptor subunits in amoeboid microglia mediates production of nitric oxide via NF-kappaB signaling pathway and oligodendrocyte cell death in hypoxic postnatal rats. *Glia, 59*(4), 521–539. doi:10.1002/glia.21121

Nabavi, S., Kessels, H. W., Alfonso, S., Aow, J., Fox, R., & Malinow, R. (2013). Metabotropic NMDA receptor function is required for NMDA receptor–dependent long-term depression. *Proceedings of the National Academy of Sciences of the United States of America, 110*(10), 4027–4032. doi:10.1073/pnas.1219454110

Ngo-Anh, T. J., Bloodgood, B. L., Lin, M., Sabatini, B. L., Maylie, J., & Adelman, J. P. (2005). SK channels and NMDA receptors form a Ca^{2+}-mediated feedback loop in dendritic spines. *Nature Neuroscience, 8*(5), 642–649.

Niere, F., Wilkerson, J. R., & Huber, K. M. (2012). Evidence for a fragile X mental retardation protein-mediated translational switch in metabotropic glutamate receptor-triggered Arc translation and long-term depression. *Journal of Neuroscience, 32*(17), 5924–5936. doi:10.1523/JNEUROSCI.4650-11.2012

Nishi, M., Hinds, H., Lu, H. P., Kawata, M., & Hayashi, Y. (2001). Motoneuron-specific expression of NR3B, a novel NMDA-type glutamate receptor subunit that works in a dominant-negative manner. *Journal of Neuroscience, 21*(23), RC185.

Nong, Y., Huang, Y. Q., Ju, W., Kalia, L. V., Ahmadian, G., Wang, Y. T., & Salter, M. W. (2003). Glycine binding primes NMDA receptor internalization. *Nature, 422*(6929), 302–307.

Nowak, L., Bregestovski, P., Ascher, P., Herbet, A., & Prochiantz, A. (1984). Magnesium gates glutamate-activated channels in mouse central neurones *Nature, 307*(5950), 462–465.

Olney, J. W. (1969). Brain lesions, obesity, and other disturbances in mice treated with monosodium glutamate. *Science, 164*(3880), 719–721.

Panchenko, V. A., Glasser, C. R., & Mayer, M. L. (2001). Structural similarities between glutamate receptor channels and K^+ channels examined by scanning mutagenesis. *Journal of General Physiology, 117*(4), 345–360.

Paoletti, P., Bellone, C., & Zhou, Q. (2013). NMDA receptor subunit diversity: Impact on receptor properties, synaptic plasticity and disease. *Nature Reviews Neuroscience, 14*(6), 383–400. doi:10.1038/nrn3504

Paton, W. D. (1959). Mechanisms of transmission in the central nervous system. *Anaesthesia, 14*(1), 3–27.

Popescu, G. (2005). Principles of *N*-methyl-D-aspartate receptor allosteric modulation. *Molecular Pharmacology, 68*(4), 1148–1155. doi:10.1124/mol.105.013896

Popescu, G. K. (2012). Modes of glutamate receptor gating. *Journal of Physiology, 590*(1), 73–91. doi:10.1113/jphysiol.2011.223750

Quiocho, F. A., Phillips, G. N., Jr., Parsons, R. G., & Hogg, R. W. (1974). Letter: Crystallographic data of an L-arabinose-binding protein from *Escherichia coli*. *Journal of Molecular Biology, 86*(2), 491–493.

Ramon y Cajal, S. (1906). *The structures and connections of neurons*. Amsterdam, the Netherlands: Elsevier. (Reprinted 1967)

Rodriguez-Moreno, A., Gonzalez-Rueda, A., Banerjee, A., Upton, A. L., Craig, M. T., & Paulsen, O. (2013). Presynaptic self-depression at developing neocortical synapses. *Neuron, 77*(1), 35–42. doi:10.1016/j.neuron.2012.10.035

Rose, C. R., & Konnerth, A. (2001). NMDA receptor-mediated Na^+ signals in spines and dendrites. *Journal of Neuroscience, 21*(12), 4207–4214.
Rosenmund, C., Stern-Bach, Y., & Stevens, C. F. (1998). The tetrameric structure of a glutamate receptor channel. *Science, 280*(5369), 1596–1599.
Ryan, T. J., Emes, R. D., Grant, S. G., & Komiyama, N. H. (2008). Evolution of NMDA receptor cytoplasmic interaction domains: Implications for organisation of synaptic signalling complexes. *BMC Neuroscience, 9*, 6. doi:10.1186/1471-2202-9-6
Ryan, T. J. & Grant, S. G. (2009). The origin and evolution of synapses. *Nature Reviews Neuroscience, 10*(10), 701–712. doi:10.1038/nrn2717
Ryan, T. J., Kopanitsa, M. V., Indersmitten, T., Nithianantharajah, J., Afinowi, N. O., Pettit, C., . . . Komiyama, N. H. (2013). Evolution of GluN2A/B cytoplasmic domains diversified vertebrate synaptic plasticity and behavior. *Nature Neuroscience, 16*(1), 25–32. doi:10.1038/nn.3277
Sabatini, B. L., Oertner, T. G., & Svoboda, K. (2002). The life cycle of Ca^{2+} ions in dendritic spines. *Neuron, 33*(3), 439–452.
Sakimura, K., Kutsuwada, T., Ito, I., Manabe, T., Takayama, C., Kushiya, E., . . . Mishina, M. (1995). Reduced hippocampal LTP and spatial learning in mice lacking NMDA receptor epsilon 1 subunit. *Nature, 373*(6510), 151–155. doi:10.1038/373151a0
Salter, M. G., & Fern, R. (2005). NMDA receptors are expressed in developing oligodendrocyte processes and mediate injury. *Nature, 438*(7071), 1167–1171. doi:10.1038/nature04301
Sharp, C. D., Fowler, M., Jackson, T. H., Houghton, J., Warren, A., Nanda, A., . . . Alexander, J. S. (2003). Human neuroepithelial cells express NMDA receptors. *BMC Neuroscience, 4*, 28. doi:10.1186/1471-2202-4-28
Sheng, M., Cummings, J., Roldan, L. A., Jan, Y. N., & Jan, L. Y. (1994). Changing subunit composition of heteromeric NMDA receptors during development of rat cortex. *Nature, 368*(6467), 144–147. doi:10.1038/368144a0
Sheng, M., & Sala, C. (2001). PDZ domains and the organization of supramolecular complexes. *Annual Review of Neuroscience, 24*, 1–29. doi:10.1146/annurev.neuro.24.1.1
Sjostrom, P. J., Turrigiano, G. G., & Nelson, S. B. (2003). Neocortical LTD via coincident activation of presynaptic NMDA and cannabinoid receptors. *Neuron, 39*(4), 641–654.
Skeberdis, V. A., Chevaleyre, V., Lau, C. G., Goldberg, J. H., Pettit, D. L., Suadicani, S. O., . . . Zukin, R. S. (2006). Protein kinase A regulates calcium permeability of NMDA receptors. *Nature Neuroscience, 9*(4), 501–510. doi:10.1038/nn1664
Sobolevsky, A. I., Rosconi, M. P., & Gouaux, E. (2009). X-ray structure, symmetry and mechanism of an AMPA-subtype glutamate receptor. *Nature, 462*(7274), 745–756. doi:10.1038/nature08624
Sprengel, R., Suchanek, B., Amico, C., Brusa, R., Burnashev, N., Rozov, A., . . . Seeburg, P. H. (1998). Importance of the intracellular domain of NR2 subunits for NMDA receptor function in vivo. *Cell, 92*(2), 279–289.
Standley, S., Roche, K. W., McCallum, J., Sans, N., & Wenthold, R. J. (2000). PDZ domain suppression of an ER retention signal in NMDA receptor NR1 splice variants. *Neuron, 28*(3), 887–898.
Stein, I. S., Gray, J. A., & Zito, K. (2015). Non-ionotropic NMDA receptor signaling drives activity-induced dendritic spine shrinkage. *Journal of Neuroscience, 35*(35), 12303–12308. doi:10.1523/JNEUROSCI.4289-14.2015
Stern-Bach, Y., Bettler, B., Hartley, M., Sheppard, P. O., O'Hara, P. J., & Heinemann, S. F. (1994). Agonist selectivity of glutamate receptors is specified by two domains structurally related to bacterial amino acid-binding proteins. *Neuron, 13*(6), 1345–1357. doi:0896-6273(94)90420-0

Sucher, N. J., Akbarian, S., Chi, C. L., Leclerc, C. L., Awobuluyi, M., Deitcher, D. L., . . . Lipton, S. A. (1995). Developmental and regional expression pattern of a novel NMDA receptor-like subunit (NMDAR-L) in the rodent brain. *Journal of Neuroscience, 15*(10), 6509–6520.

Sucher, N. J., Yu, E., Chan, S. F., Miri, M., Lee, B. J., Xiao, B., . . . Jensen, F. E. (2010). Association of the small GTPase Rheb with the NMDA receptor subunit NR3A. *Neurosignals, 18*(4), 203–209. doi:10.1159/000322206

Sumikawa, K., Houghton, M., Emtage, J. S., Richards, B. M., & Barnard, E. A. (1981). Active multi-subunit ACh receptor assembled by translation of heterologous mRNA in *Xenopus* oocytes. *Nature, 292*(5826), 862–864.

Tajima, N., Karakas, E., Grant, T., Simorowski, N., Diaz-Avalos, R., Grigorieff, N., & Furukawa, H. (2016). Activation of NMDA receptors and the mechanism of inhibition by ifenprodil. *Nature, 534*(7605), 63–68. doi:10.1038/nature17679

Teng, H., Cai, W., Zhou, L., Zhang, J., Liu, Q., Wang, Y., . . . Sun, Z. (2010). Evolutionary mode and functional divergence of vertebrate NMDA receptor subunit 2 genes. *PLoS One, 5*(10), e13342. doi:10.1371/journal.pone.0013342

Tigaret, C. M., Olivo, V., Sadowski, J. H., Ashby, M. C., & Mellor, J. R. (2016). Coordinated activation of distinct Ca^{2+} sources and metabotropic glutamate receptors encodes Hebbian synaptic plasticity. *Nature Communications, 7*, 10289. doi:10.1038/ncomms10289

Toyota, M., Spencer, D., Sawai-Toyota, S., Jiaqi, W., Zhang, T., Koo, A. J., . . . Gilroy, S. (2018). Glutamate triggers long-distance, calcium-based plant defense signaling. *Science, 361*(6407), 1112–1115. doi:10.1126/science.aat7744

Traynelis, S. F., Wollmuth, L. P., McBain, C. J., Menniti, F. S., Vance, K. M., Ogden, K. K., . . . Dingledine, R. (2010). Glutamate receptor ion channels: Structure, regulation, and function. *Pharmacological Reviews, 62*(3), 405–496. doi:10.1124/pr.109.002451

van Rossum, D., Kuhse, J., & Betz, H. (1999). Dynamic interaction between soluble tubulin and C-terminal domains of *N*-methyl-D-aspartate receptor subunits. *Journal of Neurochemistry, 72*(3), 962–973.

Vissel, B., Krupp, J. J., Heinemann, S. F., & Westbrook, G. L. (2001). A use-dependent tyrosine dephosphorylation of NMDA receptors is independent of ion flux. *Nature Neuroscience, 4*(6), 587–596.

Wang, C., Pralong, W. F., Schulz, M. F., Rougon, G., Aubry, J. M., Pagliusi, S., . . . Kiss, J. Z. (1996). Functional *N*-methyl-D-aspartate receptors in O-2A glial precursor cells: A critical role in regulating polysialic acid–neural cell adhesion molecule expression and cell migration. *Journal of Cell Biology, 135*(6 Pt. 1), 1565–1581.

Watkins, J. C. (2000). l-Glutamate as a central neurotransmitter: Looking back. *Biochemical Society Transactions, 28*(4), 297–309.

Watkins, J. C., Curtis, D. R., & Brand, S. S. (1977). Phosphonic analogues as antagonists of amino acid excitants. *Journal of Pharmacy and Pharmacology, 29*(5), 324.

Wechsler, A., & Teichberg, V. I. (1998). Brain spectrin binding to the NMDA receptor is regulated by phosphorylation, calcium and calmodulin. *EMBO Journal, 17*(14), 3931–3939. doi:10.1093/emboj/17.14.3931

Wiegert, J. S., & Oertner, T. G. (2013). Long-term depression triggers the selective elimination of weakly integrated synapses. *Proceedings of the National Academy of Sciences of the United States of America, 110*(47), E4510–E4519. doi:10.1073/pnas.1315926110

Wo, Z. G., & Oswald, R. E. (1994). Transmembrane topology of two kainate receptor subunits revealed by *N*-glycosylation. *Proceedings of the National Academy of Sciences of the United States of America, 91*(15), 7154–7158. doi:10.1073/pnas.91.15.7154

Wo, Z. G., & Oswald, R. E. (1995). Unraveling the modular design of glutamate-gated ion channels. *Trends in Neurosciences, 18*(4), 161–168.

Wood, M. W., VanDongen, H. M., & VanDongen, A. M. (1995). Structural conservation of ion conduction pathways in K channels and glutamate receptors. *Proceedings of the National Academy of Sciences of the United States of America, 92*(11), 4882–4886.

Wu, S. H., Ma, C. L., & Kelly, J. B. (2004). Contribution of AMPA, NMDA, and GABA(A) receptors to temporal pattern of postsynaptic responses in the inferior colliculus of the rat. *Journal of Neuroscience, 24*(19), 4625–4634. doi:10.1523/JNEUROSCI.0318-04.2004

Wudick, M. M., Michard, E., Oliveira Nunes, C., & Feijo, J. A. (2018). Comparing plant and animal glutamate receptors: Common traits but different fates? *Journal of Experimental Botany, 69*(17), 4151–4163. doi:10.1093/jxb/ery153

Wyszynski, M., Kharazia, V., Shanghvi, R., Rao, A., Beggs, A. H., Craig, A. M., . . . Sheng, M. (1998). Differential regional expression and ultrastructural localization of alpha-actinin-2, a putative NMDA receptor-anchoring protein, in rat brain. *Journal of Neuroscience, 18*(4), 1383–1392.

Yoshioka, A., Ikegaki, N., Williams, M., & Pleasure, D. (1996). Expression of *N*-methyl-D-aspartate (NMDA) and non-NMDA glutamate receptor genes in neuroblastoma, medulloblastoma, and other cells lines. *Journal of Neuroscience Research, 46*(2), 164–178. doi:10.1002/(SICI)1097-4547(19961015)46:2<164::AID-JNR4>3.0.CO;2-F

Yu, X. M., Askalan, R., Keil, G. J., 2nd, & Salter, M. W. (1997). NMDA channel regulation by channel-associated protein tyrosine kinase Src. *Science, 275*(5300), 674–678.

Yu, X. M., & Salter, M. W. (1998). Gain control of NMDA-receptor currents by intracellular sodium. *Nature, 396*(6710), 469–474.

Yuzaki, M., & Aricescu, A. R. (2017). A GluD coming-of-age story. *Trends in Neurosciences, 40*(3), 138–150. doi:10.1016/j.tins.2016.12.004

Zheng, W., Wen, H., Iacobucci, G. J., & Popescu, G. K. (2017). Probing the structural dynamics of the NMDA receptor activation by coarse-grained modeling. *Biophysical Journal, 112*(12), 2589–2601. doi:10.1016/j.bpj.2017.04.043

Zhou, Q., Homma, K. J., & Poo, M. M. (2004). Shrinkage of dendritic spines associated with long-term depression of hippocampal synapses. *Neuron, 44*(5), 749–757. doi:10.1016/j.neuron.2004.11.011

Zukin, R. S., & Bennett, M. V. (1995). Alternatively spliced isoforms of the NMDARI receptor subunit. *Trends in Neurosciences, 18*(7), 306–313.

CHAPTER 11

NICOTINIC ACETYLCHOLINE RECEPTORS

ROGER L. PAPKE

The Discovery and Initial Characterization of Nicotinic Acetylcholine Receptors

Virtually all of the integrated information of the central nervous system (CNS) that comes out through the somatic and autonomic nervous system relies on acetylcholine as a chemical neurotransmitter. Although the connection between the "animal electricity" of a nerve fiber and the contraction of frog muscle tissue was demonstrated by Luigi Galvani as far back as the 1600s, the existence of a chemical intermediate was not conclusively demonstrated until 1921, by the work of Otto Loewi (Loewi, 1957). It was known that in a dissected preparation of vagus nerve and frog heart, electrical stimulation of the vagus would slow the rate of heart contractions, but it was not known whether this was a direct effect of the electrical stimulation, as was supposed to be the case with the earlier experiments of Galvani, or whether there was a chemical intermediate. Loewi's experiment, which came to him in a dream, and he conducted immediately upon wakening, was to drip the perfusate from the electrically stimulated heart onto a second, electrically isolated heart to see if the second heart would also slow down. This experiment usually does not work, and in the words of Loewi (taken from his Nobel prize acceptance speech):

> On mature consideration, in the cold light of the morning, I would not have done it. After all, it was an unlikely enough assumption that the vagus should secrete an inhibitory substance; it was still more unlikely that a chemical substance that was supposed to be effective at very close range between nerve terminal and muscle be

secreted in such large amounts that it would spill over and, after being diluted by the perfusion fluid, still be able to inhibit another heart

(Loewi, 1921).

Acetylcholine (ACh), first identified by Sir Henry Dale in 1914 as a potential neurotransmitter, was later confirmed to be Otto Loewi's "*Vagusstoff*." ACh is synthesized in cholinergic neurons by choline acetyl-transferase (ChAT), taking acetate from acetyl coenzyme A produced by mitochondria and combining it with the ubiquitous membrane phospholipid precursor, choline. The hydrolysis of ACh back into its precursors is rapidly catalyzed by acetylcholine esterase (AChE), an enzyme that is present throughout the brain and especially concentrated at cholinergic synapses in the peripheral nervous system. It was a lucky accident that AChE expression was unusually low in the frogs used by Loewi at the time of year when he did his first experiments.

In Loewi's experiment, the primary target for the ACh released by the vagus was not the heart muscle itself, but a type of ACh receptor on the cell bodies of the parasympathetic neurons in ganglia proximal to the heart. These neurons also make ACh, and that ACh is released diffusely through the pacemaker region of the frog sinus venous, where it acts on a different type of ACh receptor. The two types of ACh receptors in this system have come to be identified as either nicotinic or muscarinic, based on their respective sensitivity to the alkaloids nicotine or muscarine. The nicotinic ACh receptors (nAChR), which are the focus of this chapter, are ligand-gated ion channels. In the heart tissue itself, ACh binds to muscarinic receptors (mAChR), which are seven-transmembrane G-Protein-coupled metabotropic receptors, the activation of which initiates a signal transduction cascade that, among other things, ultimately increases the activity of potassium channels, thus slowing the heart. It is likely that the ACh that spilled over in Loewi's experiment was not actually the "stuff of the vagus" but rather the ACh released by the ganglionic neurons.

Prior to Loewi's work, the British physiologist John Langley postulated the existence of receptive substance in tissues (Langley, 1905), anticipating the discovery of the molecular entities that would respond differentially and in tissue-specific manners to both acetylcholine and either nicotine or muscarine.

Several key elements are present in the structure of ACh (Figure 11.1) that can define the pharmacophore (drug-carrying activity) for either nAChR or mAChR. Essential for the activation of both receptor subtypes is the positively charged nitrogen, a quaternary amine in the case of ACh and muscarine, and a tertiary amine likely to be charged at physiological pH, in the case of nicotine. A second element common to ACh and muscarine, but lacking in nicotine, is the ester oxygen with similar spacing from the charged nitrogen. A second key element in the nicotine pharmacophore, first identified by Beers and Reich (Beers & Reich, 1970), is a hydrogen bond acceptor group that has similar distance also from the charged nitrogen: the carbonyl oxygen in ACh, and the aromatic nitrogen in nicotine.

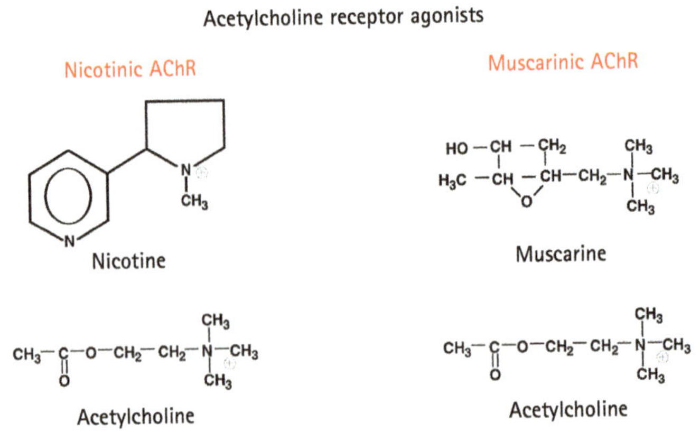

FIGURE 11.1 Cholinergic drugs. Acetylcholine and the prototypical drugs that selectively stimulate the two classes of acetylcholine receptors.

Nicotine was one of several important pharmacological tools for the early studies of nAChR. While nicotine is only a weak agonist for the nAChR of the neuromuscular junction, it is a far more effective agonist for the receptors of autonomic ganglia. A key insight into the physiological functioning of nAChR was that the duration of nicotine applications determined their effects on ganglionic function. Brief applications produced stimulation similar to the activation of preganglionic fibers, but prolonged application blocked transmission through the ganglia through a process we now know to be receptor desensitization (Katz & Thesleff, 1957) (see further in the chapter). Two other important pharmacological tools for early studies of the nicotinic receptor of the neuromuscular junction (NMJ) were antagonists: curare, a reversible poison used by South American natives to poison arrows; and the snake toxin α-bungarotoxin (α-BTX), which binds essentially irreversibly to muscle-type nAChR.

It was shown that curare would block transmission at NMJ without diminishing the nerve action potential or preventing the muscle from responding to direct electrical stimulation. The tight binding of α-BTX to muscle receptors ultimately proved useful for the biochemical characterization and the calculation of density of nAChR at the NMJ.

Electrophysiology of nAChR

Before the age of molecular cloning, patch-clamp, and brain slice recording, the frog neuromuscular junction was a favorite site for ferreting out the physiological and pharmacological properties of nAChR. In 1951, Fatt and Katz published their seminal study of endplate potentials recorded with intracellular glass electrodes (Fatt & Katz, 1951), which began the quest for the elementary signals mediated by single nAChR. The change in muscle membrane potential following nerve stimulation began with the nAChR-mediated end-plate potential, which if sufficiently strong could produce enough depolarization to produce a muscle action potential. With the manipulation of extracellular calcium and magnesium concentrations, Katz and colleagues discovered that they could reduce the probability of ACh release until they could, with some defined likelihood, observe small unitary responses, miniature endplate potentials (mEPPs), which they deduced represented responses to single quanta of ACh (Katz & Miledi, 1972). This quantal analysis anticipated the later discovery of vesicular release of ACh and the electron micrographic observation of transmitter storage vesicles (Salpeter, 1987).

The quantal events observed by Katz, whether evoked by nerve stimulation or observed as spontaneous events, represented the change in membrane voltage produced by the relatively synchronized activation of many thousands of nAChR. In order to better understand the behavior of the nAChR, the technique of voltage-clamp was applied. Voltage clamping involves measuring membrane voltage and using a feedback amplifier to inject current to keep the membrane potential constant under conditions when ion channels have been activated. The currents required to keep the membrane potential fixed (clamped) are equal and opposite to the channel-mediated currents. Voltage-clamp methods were applied with great effect by Hodgkin and Huxley in the early 1950s, using squid giant axons to obtain an understanding standing of the ionic currents of nerve action potentials (Hodgkin & Huxley, 1952).

Under voltage-clamp conditions, the receptor-mediated activity at the muscle endplate could be studied in a quantitative manner, allowing the role of AChE activity to be evaluated. When the AChE activity of the neuromuscular junction was blocked with organo-phosphate poisoning, the miniature end-plate currents (mEPCs) increased slightly in amplitude and more significantly in duration. Elegant studies by the laboratory of Miriam Salpeter combined quantitative electron-microscopic analysis of radiolabeled α-BTX binding with mEPC measurements, ultrastructural data, and diffusion modeling of the NMJ to generate what they referred to as the "saturated disk model" (Bartol, Land, Salpeter, & Salpeter, 1991). The model proposed that a single quantum of ACh, containing an average of about 10,000 molecules of ACh, would be released to diffuse through the synaptic gap and be exposed to AChE bound to the fibrous network known as the *basal lamina*, prior to reaching the post-synaptic surface, where nAChR density was approximately 9,000 receptors per square micron. Although analysis of the effects of the organo-phosphate poisoning suggested that about 10–15% of the ACh is metabolized when it first diffuses across the cleft, the remaining ACh would reach a concentration of about 1 mM over a surface of a couple square microns, sufficient to saturate the ACh binding sites of about 18,000 receptors. They estimated that about 80% of the bound receptors opened synchronously within a millisecond of the synaptic release to generate the mEPC. When the ACh was released from the receptors after the first bout of activation, nearly all was metabolized before it was able to bind to and activate a receptor a second time.

Studies of the neuromuscular junction yielded many insights that have remained fundamental to concepts related to nAChR as mediators of synaptic function, and many of these insights have been found to have validity for glutamate receptor-mediated synaptic currents in the CNS. The essential mode for synaptic receptor function is to respond to a rapid rise in the concentration of the activating ligand. While receptors respond well to rapid changes in agonist concentration, they respond relatively poorly to steady-state increases in agonist concentration. This is because, following a brief period when receptors can be activated, they convert to non-responsive "desensitized" states (Katz & Thesleff, 1957), accounting for the block of ganglionic transmission observed by Langley (Langley & Anderson, 1892).

Although ineffective at generating large currents, prolonged applications of relatively low concentrations of agonist were observed to produce an increase in the "noise" of the electrical records, noise that was hypothesized to arise from the random opening and closing of single channels transitioning between desensitized and activatible states. This hypothesis lead to the hopeful development of "noise analysis" (Neher & Sakmann, 1976a), the goal of which was to extract from the noise estimates of the amplitude and frequency of single-channel events. While in pursuit of such data, scientists in the group led by Erwin Neher and Bert Sakmann at the Max Planck Institute in Goettingen discovered a phenomenon that was soon to transform much of electrophysiology and ultimately win Neher and Sakmann the Nobel prize in 1991. Rather than use traditional sharp electrodes, they approached the membranes of denervated frog muscle fibers with blunt fire-polished electrodes filled with solution containing nicotinic agonists and discovered that such electrodes would form a tight seal to the membranes so that they could observe currents corresponding to the opening and closing of single nAChR channels (Neher & Sakmann, 1976b). The technique became known as "patch-clamping", and with further refinement, it led to the development of an entirely new field of single-channel analysis. Amplifiers and computers of the era were challenged to provide sufficiently high-resolution low-noise signals and acquire data with a speed that matched the rate at which the single channels opened and closed. The best computers that an academic laboratory might be expected to have in the early 1980s might be the size of a telephone booth with less computational power than a modern smart phone. In order to slow the channels down and better resolve currents, some laboratories did their recordings at 4°C. In the laboratory where the author did his dissertation work in the mid-1980s, data were recorded onto an FM tape-recorder so that it could be played back and digitized at a 16-fold slower rate than it was originally acquired.

From the early days of electrophysiology, macroscopic currents were interpreted as the convolution of multiple first-order processes, and assigned activation and inactivation or desensitization rate constants that would fit the time course of the currents observed. These rates were described as variously dependent on membrane voltage or ligand concentration. The basic equation to describe an ionic current (I), is $I = (E_m - E_{rev})NP_o\gamma$, where: $(E_m - E_{rev})$ is the difference between the membrane potential and the potential where the net driving force for current through the channel is zero (i.e., the reversal potential), N is the total number of channels that could contribute to the current, P_o is the probability of a given channel being open (a function of factors such as voltage, ligand binding, and time after a change in voltage or ligand binding), and γ is the conductance of a single open channel. For the analysis of typical macroscopic currents, it can be difficult deconvolute N and P_o, since a large number of channels with a low P_o will produce a similar current to that of a small number of high P_o channels (Papke, 2010). The first single-channel studies of voltage-dependent sodium channels actually led to a reevaluation of the classic studies of sodium channels done by Hodgkin and Huxley, since the single-channel studies suggested that sodium currents arose from many channels with low P_o, rather than the reverse, which was the solution offered by the original analysis (Aldrich, Corey, & Stevens, 1983).

Studies of single-channel currents, when N is by definition equal to one (or at least one at a time), required new analytical approaches and statistical methods, which were largely developed by David Colquhoun in collaboration with Bert Sakmann (Colquhoun & Hawkes, 1983), with additional refinements by Fred Sigworth (Sigworth & Sine, 1987) and others. The observation of unitary and nearly instantaneous conversion between closed and open states of the channels suggested that individual channels had a small number of discreet conformational states and that conversions among these states could be described by Markov models (Horn & Vandenberg, 1984). Simple examples of such models are illustrated in Figure 11.2, along with simulated single-channel data derived from those models (data were generated in the channel simulation module (SIMU) of Quantify Unknown Biophysics (QUB), https://qub.mandelics.com). Transitions between states are assigned first-order rate constants, and, although of random durations, the event duration distributions can be described by exponential functions with time constants equal to the inverse of the rate constants. An alternative approach for an equivalent display of such models is based on Eyring rate theory (Dani, 1986), in which the energy barriers (height of the drawn hill in the energy landscape) to interstate conversions are shown scaled in proportion to the log of the rate constants in the Markov models (Figure 11.2, upper right insets).

In Figure 11.2A, the rate constants for conversions between the closed (C) and open (O) states are equal (100 s^{-1}), so dwells in both of these states are on average 10 ms. The data shown in 2B were generated with the channel closing rate set four-fold higher than the opening rate, so closed times are on average 10 ms and open times average only 2.5 ms. The data in Figure 11.2C were generated using a model with a second closed state (C_2) connected to the open state. Channels in the open state may close to either C_2 or the long-lived closed state (C_1). The closing rate to C_2 is twice the closing rate to C_1, so there are likely to be twice as many dwell times in C_2 as in C_1. However, the rate at which channels in C_2 return to O is rather high (100 s^{-1}), so dwells in C_2 are brief, on average only 1 ms. Note that because there are two paths out of the open state in model C, the dwell times in the open state will be briefer than those in model B; on average only 830 ns. However, openings in this model occur often as bursts of transitions between O and C_2. These bursts will often contain multiple dwells in O and C_2, although the total time in the O state within a burst will be the same as the average open time in model B. A complete analysis of the dwell times for this model would show open times fit by a single exponential (τ_O = 830 ns) and the closed time distribution fit by the sum of two exponentials (τ_{C1} = 10 ms, and τ_{C2} = 1 ms), with the number of events fit to the fast component twice that of the slower component. Evaluation of data like those discussed here

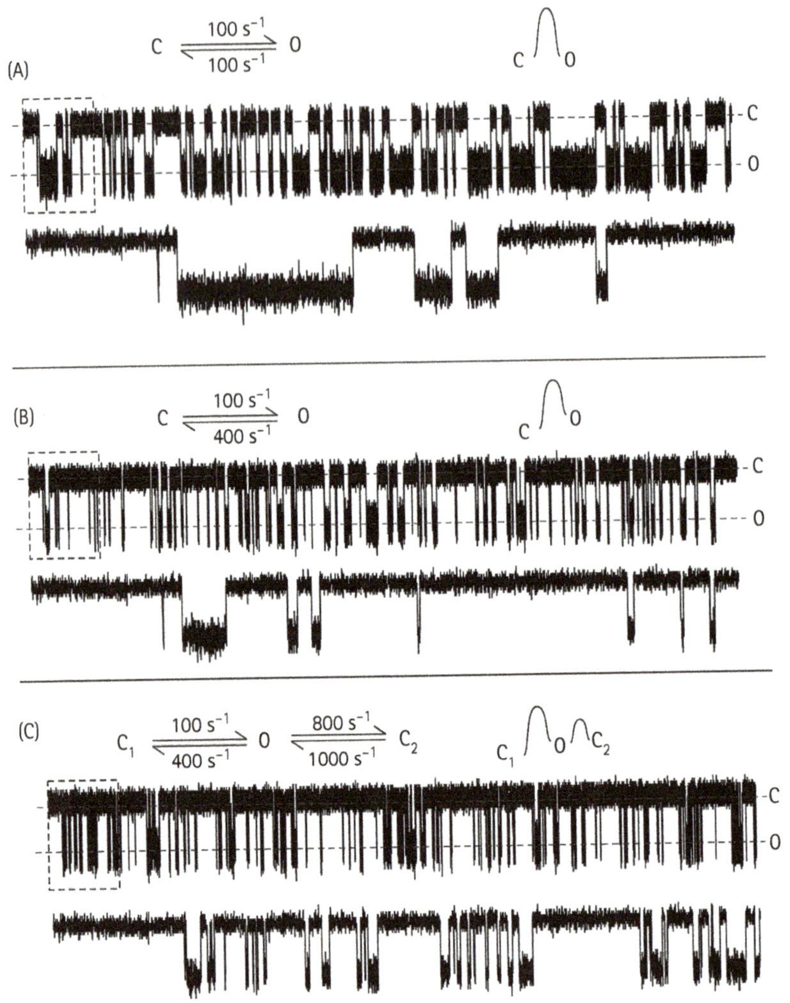

FIGURE 11.2 Kinetic models, energy landscapes, and simulated single channel currents. (A) A simple two-state model in which channels have an equal probability of being in either the open or the closed state. The lower trace is the first segment (in dashed box) at an expanded time scale. (B) A simple two-state model in which channels have a fourfold greater probability of being in the closed rather than the open state. The lower trace is the first segment (in dashed box) at an expanded time scale. (C) A three-state model in which open channels can close to either a relatively long-lived closed state or to a brief intraburst closed state. The lower trace is the first segment (in dashed box) at an expanded time scale.

for model C is referred to as "burst analysis," and the interpretation of such analyses has been instrumental for understanding fundamental processes such as activation (Horn & Vandenberg, 1984), desensitization (Sine & Steinbach, 1987), and the block of open channels by local anesthetics (Neher & Steinbach, 1978) or other drugs, including ACh (Sine & Steinbach, 1984b).

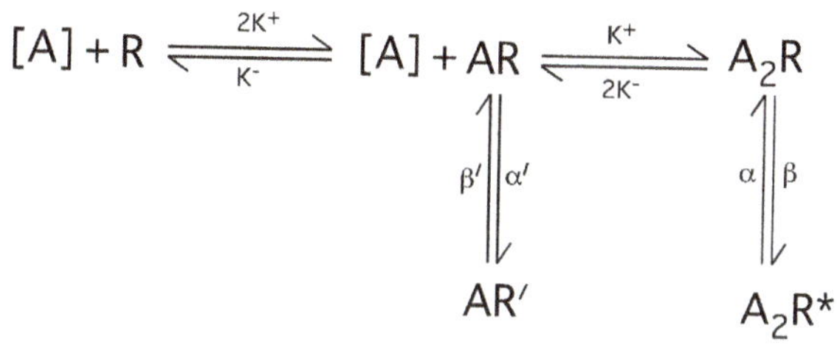

FIGURE 11.3 Kinetic model for acetylcholine binding and channel activation used for describing early single-channel recordings of muscle-type nAChR: [A], agonist concentration; R. resting closed state; AR, agonist bound still in closed state; AR′, brief open state; A_2R, receptor with two agonists bound but still in the closed state; A_2R^*, long-lived open state; K^+, agonist forward binding rate as a product of a microscopic rate and the agonist concentration; K^-, agonist dissociation rate; α, channel opening rate: β, channel closing rate.

Figure 11.3 shows the model for frog muscle nAChR published by Colquhoun and Sakmann in 1985 (Colquhoun & Sakmann, 1985), which has been applied extensively since that time. This model is consistent with an early study of the muscle-type nAChR expressed in the BC3H-1 cell line (Sine & Taylor, 1980) that suggested that AChR channel activation required the binding of agonist on each of two agonist binding sites per receptor and anticipated the later studies of purified receptors described later. Although this model has proven useful for explaining the bursting behavior and the concentration dependence of activation, it is clearly incomplete because it does not account for receptor desensitization and would predict steady-state currents that are not observed. The model, however, remains useful for explaining the behavior of the open channels that appear in the single-channel recordings. The model includes two open states, identified as O′ and O*, a feature that has been consistently observed in single-channel records from all types of heteromeric nAChR (Colquhoun & Sakmann, 1985; Papke, 1993; Papke, Boulter, Patrick, & Heinemann, 1989; Papke, Millhauser, Lieberman, & Oswald, 1988; Sine & Steinbach, 1984a). Dwells in the O′ state are brief and most often occur as single openings. The O* state is more long-lived, and often the O* openings occur in bursts. Data from the work of Colquhoun and other groups (Colquhoun & Sakmann, 1985; Papke et al., 1988) reported that the relative frequency of brief opening was greatest at low concentrations of agonist, while at higher ACh concentrations, most openings were to the O* state. This observation led to the hypothesis that the O′ state arose preferentially from receptors that had agonist bound to just one of the two orthosteric activation sites. Connecting the two levels of agonist occupancy with the two identifiable open states made an attractive hypothesis, but it was challenged by the observation that fast events continued to occur at relatively high concentrations of agonist, when the predicted number of singly liganded channels would have been very low. The hypothesis that receptors with a single ligand bound would open only into the O′ state remained untested until recently, when Williams et al. (Williams, Stokes, Horenstein, & Papke, 2011) used site-directed mutants to selectively knock out one of the two agonist binding sites in muscle-type nAChR. When both agonist binding sites are available, singly bound receptors exist only transiently, since at low agonist concentrations, the receptors are likely to revert to the unbound state, while at higher agonist concentrations, they are likely to acquire a second ligand. By blocking the accessibly of one of the binding sites, one can expose receptors to high agonist concentration and retain the agonist at the single open site. Under these conditions, while the singly bound receptors were observed to preferentially enter the brief open state, they were also able to enter a long-lived open state with bursting activity similar to that described for the O*.

These observations led to the proposed energy landscapes shown in Figure 11.4 that depend on the level of agonist occupancy. The relative vertical positions of the states in these models represent equilibrium free energy of the states, while the lines connecting the states represent the energy barriers for conversions among the states (proportionate to the conversion rate constants, as in Figure 11.2). These models also include a desensitized state, which is ultimately the most stable state in the prolonged presence of agonist and may therefore be useful to describe the nonstationary behavior of receptors following a rapid elevation of ACh concentration, as occurs at the NMJ. Before the release of ACh, receptors are exclusively in the resting closed state (R in Figure 11.3 or C in Figure 11.4). This defines the initial condition at the instant when receptor binding sites are rapidly saturated. From this initial condition, channels are at first most likely to follow the path over the lowest connected energy barrier, accounting for an initial phase of synchronous activation. Ultimately, channels seek out the most stable state and are predominantly desensitized until the agonist concentration decreases and the energy landscape reverts to that of the unbound receptors.

The nAChR Proteins; Biochemical Isolation and Molecular Cloning

The NMJ proved invaluable for revealing the nature of nAChR as mediators of electrochemical transduction, and it was an unusual analog of muscle-type receptors that defined a path toward the first biochemical isolation of nAChR and ultimately led to the cloning of a large family of related proteins that mediate nicotinic cholinergic function in both vertebrates and invertebrates. In 1942, Feldberg and Fessard published evidence that the large electrical currents generated by the electric organ of the ray, *Torpedo marmorata,* were mediated by a receptive substance similar to that identified by Dale and colleagues at the NMJ (Feldberg & Fessard, 1942). The organ, which has the appearance of an oversized NMJ but lacking the muscle's contractile apparatus, generated electrical responses to ACh that could be blocked by curare and were increased by AChE inhibitors. By the early 1970s, efforts were ongoing in several laboratories (Karlin, Weill, McNamee, & Valderrama, 1976; Reynolds & Karlin, 1978) to isolate and purify the nAChR proteins from *Torpedo*. Additionally, beginning in the 1980s, *Torpedo*

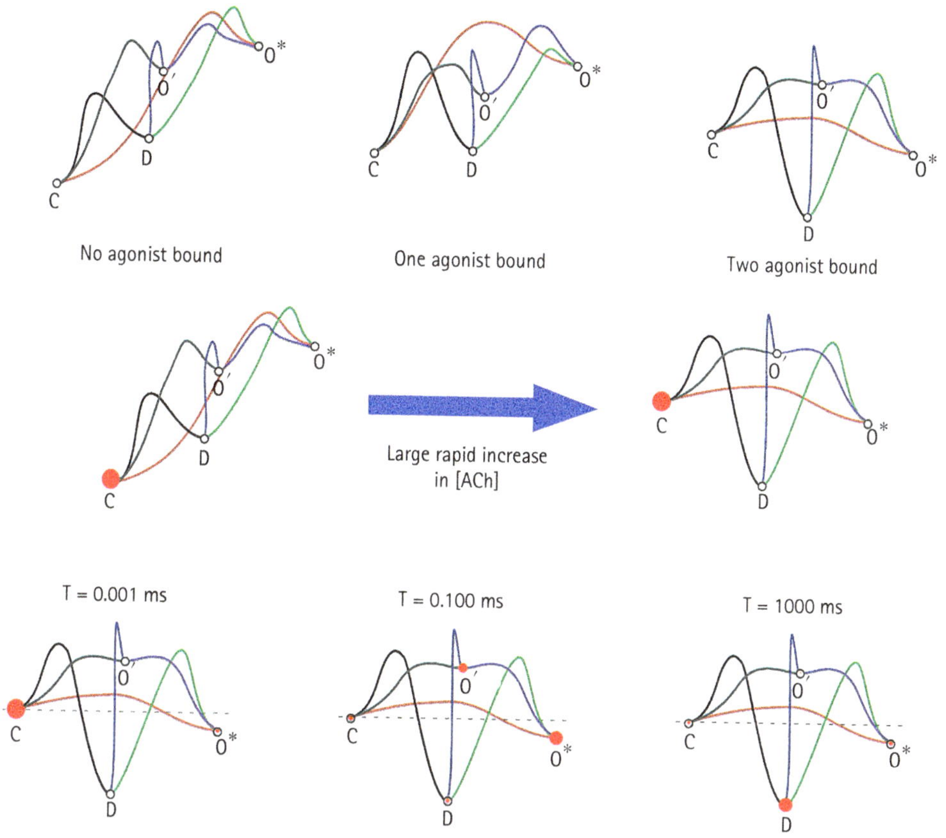

FIGURE 11.4 Energy landscapes describing the probability for transitions among the conformational states described for heteromeric nAChR: C, the resting closed state; O′, the brief open state; O*, the long-lived open state; and D, the desensitized state. The upper third shows transitions among these states differing greatly with different levels of agonist occupancy. The middle shows how, immediately after a rapid elevation in ACh concentration that saturates the binding sites, the channels all switch from one condition to another. The size of the red dot represents the fraction of channels in each state, and immediately after the jump, they are all in the C state. The lower third shows occupancy of each state at times (T) indicated. The size of the red dots represents the changing fractions of channels in each state. Channels immediately tend to cross the lowest energy barrier into the open state, but over time, most channels become stuck in the low-energy desensitized state.

membranes were used to provide progressively higher resolution electron micrographic images of the *Torpedo* nAChR (Unwin, Toyoshima, & Kubalek, 1988). In top views, receptors appeared as rosettes with five subunits, and in cross section, the images showed each having a large extracellular domain with a vestibule extending into a channel surrounded by transmembrane domains (Unwin, 1993, 2005).

Cobra (*Naja naja*) toxin was used to affinity-purify *Torpedo* nAChR, which were shown to contain four different types of protein subunits designated, α, β, γ, and δ based on size (α being the smallest) (Hucho, 1979, 1986; Weber & Changeux, 1974). The protein gels used to separate the subunits indicated that, although the α subunits had the smallest molecular weight, their abundance in the purified receptor preparations suggested that there were two α subunits per receptor subunit complex. It was also shown that the irreversible antagonist α-BTX bound exclusively to the α subunits (Oswald & Changeux, 1982). The *Torpedo* receptors could be reconstituted into membrane vesicles, which permitted pharmacological studies with ion flux methods (Cash, Aoshima, Pasquale, & Hess, 1985; Eldefrawi, Aronstam, Bakry, Eldefrawi, & Albuquerque, 1980). Based on all of these data, it was concluded that muscle/*Torpedo* nAChR were pentamers with the stoichiometry of two α subunits and one each of β, γ, and δ subunits and that the principle elements of the ACh binding sites were on the α subunits. However, it was also shown that the two binding sites were not equivalent, suggesting potential roles for other subunits in the configuration of the binding sites (Dunn & Raftery, 1982).

The isolation and partial amino acid sequencing of the *Torpedo* receptor proteins led to the cDNA cloning of first the *Torpedo* (Ballivet, Patrick, Lee, & Heinemann, 1982; Noda et al., 1982) and then the mammalian muscle nAChR subunits (Kubo et al., 1985; Noda et al., 1983).

With the cloning of the nAChR, techniques were needed to study their electrophysiological properties. In 1982, Ricardo Miledi and his co-workers reported endogenous ACh-activated current in voltage-clamped oocytes obtained from *Xenopus laevis* frogs. These currents were insensitive to curare or α-BTX but could be blocked by the muscarinic antagonist atropine (Kusano, Miledi, & Stinnakre, 1982) and were identified as chloride channel currents, which were secondary to the activation of muscarinic AChR. Nicotinic AChR, however, could be expressed in the oocytes if the cells were first injected with mRNA isolated from cat muscle (Miledi, Parker, & Sumikawa, 1982a). They also showed that if chick brain mRNA was injected, oocytes would form functional receptors that responded to the inhibitory neurotransmitter gamma-aminobutyric acid (GABA) (Miledi, Parker, & Sumikawa, 1982b). In the years that followed, and to the present day, *Xenopus* oocytes have proven to be a useful and flexible heterologous system for the study of cloned receptors (Papke & Smith-Maxwell, 2009). Although primarily used for the study of macroscopic currents, oocytes can also be used for single-channel studies (Methfessel et al., 1986).

In 1985, Takai and colleagues reported the cloning of a novel subunit from bovine muscle (Takai et al., 1985), which began an explosive expansion in our understanding of the nicotinic gene family. The newly discovered subunit was designated *epsilon* (ε) and was preferentially expressed in adult muscle instead of γ, which was found in fetal muscle. A comparison of single-channel currents in excised patches from oocytes expressing αβγδ or αβεδ subunits, from either fetal or adult cow muscle, confirmed the role of specific subunits on single-channel properties. Adult muscle and αβεδ receptors had larger single-channel conductance but briefer open times than fetal αβγδ receptors (Mishina et al., 1986). While αβγδ and αβεδ receptors express over the entire surface of the oocyte, this is not the case for the αβεδ receptors on muscle cells *in vivo*. A developmental switch is stimulated by the interaction between motor nerve and muscle that, among other things, leads to the switch from γ to ε expression and to the clustering of receptors with intracellular scaffolding of the developing NMJ (Froehner, 1991). If the nerve fiber to a muscle cell is cut, some of the developmental switches are set back to the fetal position, γ expression returns, and the nAChR diffuse over the entire surface of the cell, a phenomenon known as *denervation supersensitivity* (Merlie, Isenberg, Russell, & Sanes, 1984). It is interesting to note that, in the *Torpedo* electric organ, although the αβγδ subunit composition and diffuse expression over the entire surface of the electroplaques are common to fetal nAChR, the single-channel properties of *Torpedo* receptors are similar to those of mammalian adult αβεδ receptors (Ortiz-Miranda, Lasalde, Pappone, & McNamee, 1997), suggesting that the evolution of the organ is an example of neoteny, the retention of juvenile features in the adult.

While the studies of muscle-type nAChR and the function of the NMJ remain a tremendous boon to our knowledge of synaptic physiology, by the mid-1980s, interest shifted to the study of the nicotinic receptors of the brain. At that time, there was a strong awareness that nicotine addiction was at the center of an international health care crisis, with cigarette smoking being the single most preventable cause of cancer. *Ex vivo* (Wonnacott, 1986) and *in situ* (Clarke, Schwartz, Paul, Pert, & Pert, 1985) binding studies (Figure 11.5) identified two classes of putative nicotinic receptors in rat brain. One type bound radiolabeled ACh and nicotine with high affinity, and the other did not. The low-affinity receptors did, however, bind α-BTX and were hypothesized to represent a unique nAChR subtype. Protein purification methods provided some success in isolating the brain nicotine binding sites (Whiting & Lindstrom, 1986) and identified two subunits named α and β, based on size, as was done with *Torpedo* receptor subunits. At about the same time, the molecular cloning methods that had successfully delivered the muscle receptor subunits so that they could be studied outside the NMJ were used to find the nicotine receptors of the brain and autonomic ganglia. These efforts, largely spearheaded by Jim Boulter and Marc Ballivet, along with other members of the Heinemann and Patrick laboratories at the Salk Institute, identified an initial cadre of nAChR subunit genes that included three new putative alpha subunits designated α2, α3, and α4, as well as a novel non-alpha subunit designated β2 because it would substitute for the muscle β subunit when co-expressed with the other muscle subunits in *Xenopus* oocytes (Boulter et al., 1987; Deneris et al., 1987). The new clones that were identified as putative α subunits contained a pair of vicinal (adjacent) cysteines at a position homologous to similar residues in the muscle alpha subunit. These unusually adjacent cysteines are disulfide linked, and reduction of that bond was known to drastically reduce receptor function (Kao & Karlin, 1986). The expression of these genes in rat brain was mapped with *in situ* hybridization (Wada et al., 1989), and the overlapping expression of α4 and β2 expression corresponded well with the pattern of high-affinity nicotine binding sites previously described (Clarke et al., 1985). The predicted protein size of the α4 subunit was significantly larger than that of β2, and it was later confirmed that the protein that was previously identified as the brain receptor β subunit was actually coded for by the α4 gene (Whiting, Esch, Shimasaki, & Lindstrom, 1987). Cloning efforts ultimately identified nine neuronal α subunits expressed in terrestrial vertebrates

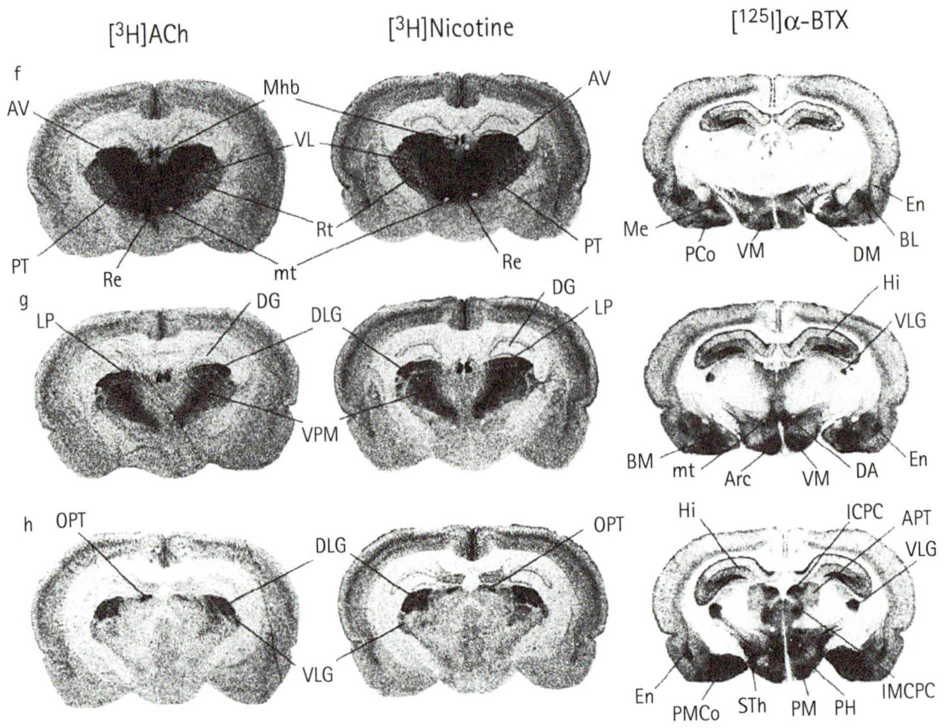

FIGURE 11.5 Autoradiographic comparison of [^{3}H]ACh, [^{3}H]nicotine, and [^{125}I]α-BTX binding in rat brain, from Clarke et al., *Journal of Neuroscience*. 1985. 5(5):1307–1315. Used with permission. APT, anterior pretectal area; Arc, arcuate hypothalamic nucleus; AV, anteroventral thalamic nucleus; BL, basolateral amygdaloid nucleus; BM, basomedial amygdaloid nucleus; DA, dorsal hypothalamic area; DCo, dorsal cochlear nucleus; DG, dentate gyrus; DLG, dorsal lateral geniculate nucleus; DM, dorsomedial hypothalamic nucleus; En, endopiriform nucleus; Hi, hippocampus; ICPC, lntracommissural nucleus of the posterior commissure; IMCPC, interstitial magnocellular nucleus of the posterior commissure; LP, lateral posterior thalamic nucleus (pulvinar); Me, medial amygdaloid nucleus; MHb, medial habenular nucleus; mt, mammillothalamic tract; OPT, olivary pretectal nucleus; PCo, posterior cortical amygdaloid nucleus; PH, posterior hypothalamic nucleus; PM, paramedian lobule; PMCo, Posteromedial cortical amygdaloid nucleus; PT, paratenial thalamic nucleus; Re, reuniens thalamic nucleus; Rt, reticular thalamic nucleus; STh, subthalamic nucleus; VL, ventrolateral thalamic nucleus; VLG, ventral lateral geniculate nucleus; VM, ventromedial hypothalamic nucleus; VPM, ventroposterior thalamic nucleus, medial part.

(α2–α10), although α8, which was originally cloned from chicken, has not been found in mammals. Three non-alpha subunits, designated β2 – β4, were also cloned (Figure 11.6). It is believed that all functional nAChR assemble as pentamers, and, as noted before, the majority of high-affinity nicotine receptors in rat brain contain α4 and β2 subunits. The β2 subunit can also be co-expressed with α2 or α3 subunits to make functionally and pharmacologically distinct subtypes (Patrick et al., 1989; Luetje & Patrick, 1991; Papke et al., 1989); likewise, the β4 subunit can co-assemble with α2, α3, or α4 subunits

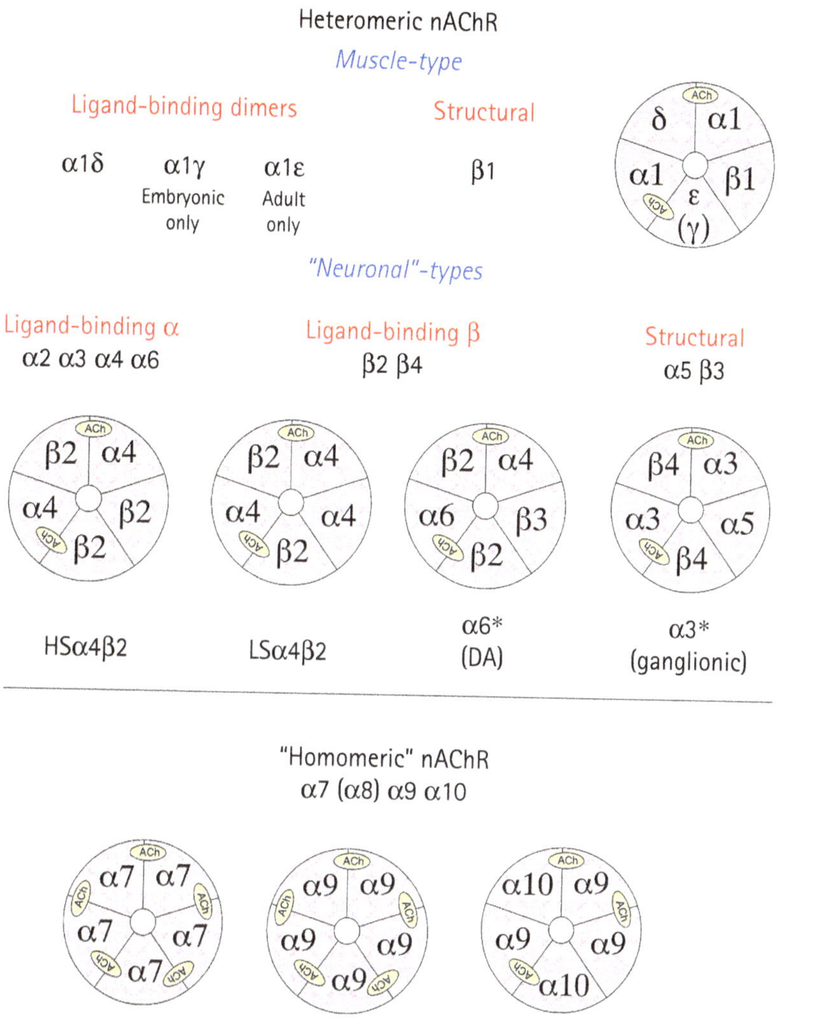

FIGURE 11.6 Subunits of nAChR and pentameric combinations. Shown are the known subunits of vertebrate nAChR, organized into three groups. On the top are subunits that are associated with receptors expressed in muscle cells. These form as pentamers with two ligand-binding sites at the interface between α1 and either α δ, γ, or ε subunits. The γ subunit is expressed in embryonic or denervated muscle as well as in the *Torpedo* electric organ. The β1 subunit is considered structural and does not contribute to agonist binding sites. In the middle are subunits that co-assemble to form neuronal heteromeric receptors that can bind ACh and nicotine with high affinity in the desensitized state. Agonist binding sites form primarily between the α and β subunits indicated, while the α5 and β3 subunits, like β1, are considered structural. Some pentameric combinations known to exist are shown below, including the high sensitivity (HS) and low sensitivity (LS) forms of α4β2, one of several possible combinations of α6-containing (α6*) receptors associated with dopaminergic neurons and one of several possible configurations of α3-containing (α3*) receptors found in autonomic ganglia and the medial habenula in the brain. At the bottom are subunits associated with α-BTX-sensitive receptors, most notably α7, which can form homomeric pentamers. Note that, to date, α8 subunits have only been reported in chicken.

(Duvoisin, Deneris, Patrick, & Heinemann, 1989; Papke & Heinemann, 1991). The α5, α6, and β3 subunits also can co-assemble with these functional subtypes, but details on their roles were only understood some years later. Some of this delay was because these subunits were not required for function or did not express well in *Xenopus* oocytes.

In the pairwise combinations of α2, α3, or α4 with either β2 or β4, two agonist binding sites form at the interface between α–β pairs, configurations that engage only four of the five subunits. This is similar to the situation in muscle receptors, where the ligand-binding domains are at interfaces between α1 and either γ, δ, or ε subunits. The β1 subunit does not contribute to binding sites and can be considered a "structural subunit," which is needed to complete the pentamer but not required for the ACh binding site. The neuronal β2 and β4 subunits, which do contribute to binding sites, can take the position of the structural subunit in muscle receptors. In a heteromeric neuronal receptor formed with, for example, α2 and β2, which one takes the position of the extra, structural subunit? Two different channel types, distinguishable by differences in conductance (33.6 +/− 1.8 pS and 15.5 +/− 1.3 pS) were observed in the very first single-channel recording from receptors known to contain just α2 and β2 subunits (Papke et al., 1989). It was hypothesized that these might arise from receptors formed with different subunit stoichiometry, either three α2s and two β2s, or the reverse. This hypothesis was supported by the observation that when β2 RNA was injected into oocytes at a nine-fold higher level than the α2 RNA, there was significantly increased occurrence of the smaller conductance channels that presumably had the α2(2)β2(3) composition. It was later shown that α5 subunits, which fail to form functional receptors in pairwise combinations, do in fact co-assemble with other α/β pairs and take the position of an obligatory structural subunit, with significant effects on receptor function (Gerzanich, Wang, Kuryatov, & Lindstrom, 1998). Likewise, it has since been shown that, in other receptors formed with single α and β subunit combinations, the structural subunit position can be taken by either an α or a β, with significant effect on receptor properties. This has been most extensively studied for α4β2 receptors, where the α4(2)β2(3) and α4(3)β2(2) configurations have been identified as high-sensitivity and low-sensitivity subtypes, respectively, based on the potency with which typical agonists like ACh and nicotine activate these receptors (Kuryatov, Onksen, & Lindstrom, 2008; Tapia, Kuryatov, & Lindstrom, 2007). Recently this has been shown to involve the presence of a novel low-affinity binding site at the α4/α4 interface in α4(3)β2(2) receptors (Wang et al., 2015).

Other puzzles and challenges were associated with α6- and β3-containing receptors, which were long considered "orphan" subunits, until it was shown that α6 could indeed form functional receptors and contribute (albeit inefficiently) to ACh binding sites (Gerzanich, Kuryatov, Anand, & Lindstrom, 1996; Kuryatov, Olale, Cooper, Choi, & Lindstrom, 2000). It was also shown that under certain expression conditions, β3 would assemble with other subunits as a structural subunit (Forsayeth & Kobrin, 1996). Interest in α6- and β3-containing receptors remains high because both of these subunits are expressed at high levels in dopamergic centers in the brain and are implicated in the reinforcing aspects of nicotine addiction (Grady et al., 2007). For many years, the pharmacology of putative α6 receptors was explored using artificial constructs that were chimeras of α6 and either α3 or α4 (Papke et al., 2008) since these expressed more efficiently in oocytes. Recently, improved expression of receptors containing complete α6 subunits has been achieved with subunit concatamers (Kuryatov & Lindstrom, 2011). The expression of β4 has also been reported in dopaminergic neurons in the brain (Azam, Winzer-Serhan, Chen, & Leslie, 2002), where it may combine with α6. However, the highest expression of β4 in the brain is in the medial habenula, a small nucleus remarkably high in the expression of multiple nAChR subtypes (Wada et al., 1989).

While α2-containing receptors show relatively low levels of expression in rodent brain (Wada et al., 1989), there is evidence that expression of α2-containing receptors is higher in primate brain (Han et al., 2000; Han et al., 2003). The nAChRs of autonomic ganglia contain predominately α3, β2, β4, and α5 subunits (David et al., 2010), so the co-expression of α3 and β4 is often used as model for these receptors, while α4 and β2 are used to model brain receptors. However, it should be noted that, while receptors specifically containing α3 and β4 are the primary receptor subtype on adrenal chromaffin cells (Sala, Nistri, & Criado, 2008), they account for only about 50% of the receptors on sympathetic neurons (David et al., 2010).

With α4β2-containing receptors largely understood to account for the high-affinity nicotine binding sites in brain by 1985, the α-BTX binding sites remained a mystery for a few years longer. It was once proposed that α5 expression might account for these receptors (Chini, Clementi, Hukovic, & Sher, 1992); however, the mystery was solved with the cloning of α7 (Seguela, Wadiche, Dinely-Miller, Dani, & Patrick, 1993). The α7 gene formed functional α-BTX-sensitive homopentameric receptors when expressed alone in *Xenopus* oocytes. A related gene, α8, was cloned from chick retina, which could form receptors co-assembled with chick α7 (Gotti et al., 1997); however, there is no α8 in mammals. Although receptors containing α7 and β2 have been reported (Liu, Huang, Shen, Steffensen, & Wu, 2012), the co-expression of these two genes generally just reduces α7 function, and β2 subunits do not contribute to the agonist activation sites in these unusual receptors.

The last two mammalian nAChR subunits cloned were α9 and α10, which are important for the function of the inner ear (Elgoyhen, Johnson, Boulter, Vetter, & Heinemann, 1994; Elgoyhen et al., 2001). These subunits function in *Xenopus* oocytes, but rather poorly compared to α7 and most of the heteromeric receptors. The α9 subunits can assemble into homopentamers, but function better when co-expressed with α10. Like the α7 receptors, the α9 receptors can be activated by choline as well as by ACh, and they are blocked by α-BTX. The complete family of mammalian nAChR subunit coding genes is shown in Figure 11.6.

Simple models for nAChR function derived from the NMJ work passably well to describe synaptic function in autonomic ganglia (Margiotta, Berg, & Dionne, 1987; Taylor, 1990), but they fail when it comes to describing nAChR function in the brain. In general, nAChR in the vertebrate brain are not postsynaptic, but rather are presynaptic, perisynaptic, or located on neuronal cell bodies, so they function to modulate the release of other neurotransmitters or of general neuronal excitability (Chesselet, 1984; McGehee, Heath, Gelber, Devay, & Role, 1995; Wonnacott, 1997; Wonnacott,

Irons, Rapier, Thorne, & Lunt, 1989). Likewise, the release of ACh in the brain is not rapid and highly localized as it is at the NMJ, but rather it is diffuse and associated with volume transmission, with concentrations of ACh rising and falling over the course of several milliseconds (Descarries, Gisiger, & Steriade, 1997). Although cholinergic fibers in the brain, most of which arise from forebrain nuclei, do not mediate point-to-point synaptic transmission as do glutamate receptors, the salience of signals mediated by other neurotransmitters can be increased or decreased based on whether that activity is in phase or out of phase with the oscillation of ACh (Jiang et al., 2016). Numerous neurological disorders, including dementia (Perry, Martin-Ruiz, & Court, 2001), depression (Picciotto, Brunzell, & Caldarone, 2002), and schizophrenia (Terry, 2008), are associated with genetic or age-dependent loss or alteration in brain nAChR.

Cys-Loop Superfamily of Ligand-Gated Ion Channels

The exploration of nAChR genes, and their ability to function in multiple combinations, foreshadowed the discovery that they were one branch in a large superfamily of pentameric ligand-gated ion channels (Betz, 1990; Stroud, McCarthy, & Shuster, 1990). The family clearly included the subunits of glycine and GABA receptors, and it was originally hypothesized that it might include, as more distantly related, glutamate receptors, which had just recently been cloned (Hollmann, O'Shea-Greenfield, Rogers, & Heinemann, 1989). This proved not to be the case, as the evolution of glutamate receptors could be traced back to bacterial amino acid-binding proteins and simple potassium channels (Wo & Oswald, 1995). The family of genes expressed in vertebrates was later shown to include one class of serotonin receptors (*5HT3*), and a zinc-activated channel (Davies, Wang, Hales, & Kirkness, 2003). All of the family members had the general structural features discussed later, with varying levels of conservation. One of the most conserved features among the vertebrate receptors is a sequence of 15 amino acids bracketed by disulfide-linked cysteines. This sequence has been shown to be of great functional importance since it is essential for the transduction of ligand binding to channel activation (Bouzat, Bartos, Corradi, & Sine, 2008). Based on the presence of this sequence, the collection of genes first identified in vertebrates are referred to as "the Cys-loop superfamily" of ligand-gated ion channels. In the current era of genome sequencing, it has become clear that this is a very large family of related genes coding ion channel receptors activated by a variety of chemical signals (Liebeskind, Hillis, & Zakon, 2015), all of which have residues homologous to the vertebrate Cys-loop, but not always with the eponymous disulfide, so alternative names have been proposed for the extended superfamily as simply "pentameric ligand-gated ion channels" (pLGIC), or more recently, the "Pro-loop ligand-gated ion channels" superfamily, recognizing that there is a conserved proline in the transduction loop, regardless of whether the cysteine residues are present (Jaiteh, Taly, & Henin, 2016).

Nicotinic receptor genes are particularly abundant in insects (Dent, 2010; Jones, Brown, & Sattelle, 2007) and other invertebrates such as the well-studied nematode *Caenorhabditis elegans* (Holden-Dye, Joyner, O'Connor, & Walker, 2013). As discussed before, in vertebrates, nAChR are essential mediators of synaptic function in the periphery, but they are less important in the CNS, where glutamate is the primary excitatory neurotransmitter. Knockout of either the α4 or β2 gene in mice essentially eliminates the high-affinity nAChR in brain, and although the behavioral effects of nicotine are reduced, the animals remain perfectly viable (Champtiaux & Changeux, 2004). In contrast, in insects and many other invertebrates, nAChR are essential for the function of the CNS, making them the primary targets for many types of insecticides, including organophosphates (irreversible AChE inhibitors) and neonicotinoids (Jones et al., 2007). Indeed, it is likely that nicotine itself evolved as a chemical defense against insects (Millar & Denholm, 2007).

The extended family of pentameric ligand-gated ion channels includes two analogs isolated from prokaryotes that are gated by protons: GLIC, from the cyanobacterium *Gloebacter violaceous* (Bocquet et al., 2007), and ELIC, from the Enterobacteria *Dickeya dadantii* (previously *Erwinia chrysanthemi*) (Hilf & Dutzler, 2008). Crystal structures have been published for both of these proteins, including one of ELIC in the putative open-channel conformation (Hilf & Dutzler, 2009). The signal-transducing loop in these proteins lacks the cysteine residues, and they have no intracellular domain, which undoubtedly contributed to the feasibility of obtaining crystal structures (see following discussion of crystal structures). It is tempting to speculate that these bacterial proteins may somehow be ancestral to the pLGIC superfamily of neurotransmitter receptors, in the manner that certain bacterial proteins are believed to have been at the root of the evolution of glutamate receptors (Wo & Oswald, 1995); however, there is no clear indication of a lineage that would support this speculation (Jaiteh et al., 2016). To date, GLIC and ELIC are the only known pLGIC from prokaryotes. There are no known pLGIC from protozoa, or porifera (sponges), and likewise, there are no known pLGIC from plants, although plants do have ACh and the enzymes for ACh synthesis and breakdown. Given these gaps in phylogeny, it seems likely that the prokaryotic pLGIC are ancestral to those in metazoa, but rather plausible that they were acquired by the bacterial species more recently through horizontal gene transfer.

nAChR Structure

There has been a punctate progression in our understanding of nAChR since the amino acid sequences became available for the *Torpedo* and muscle subunits. Features were identified in these sequences that have been noted to be either relatively well conserved or consistently variable among all of the Cys-loop proteins. The Kyte-Doolittle hydropathy plots (Kyte & Doolittle, 1982) in Figure 11.7 identify some of these features for the sequences of adult human muscle subunits, highlighted in α1. Note that the plots in Figure 11.7 include the initial hydrophobic signal sequences (SS), which are cleaved in the processing of the mature proteins. The amino terminal segment of the mature proteins is the primary extracellular domain of approximately 220 amino acids in length for all subunits. Within this domain is the highly conserved Cys-loop and the vicinal cysteines that are found only in the α subunits. The extracellular domain is followed by three closely spaced transmembrane (TM) domains that thread in and out of the membrane, so the short loop between TM1 and TM2 is intracellular, and the sequence between TM2 and TM3 is extracellular. In each subunit, there is a large and highly variable intracellular domain (ICD) between TM3 and TM4 (Stokes, Treinin, & Papke, 2015). The size differences noted in the first protein purification of the *Torpedo* receptor subunits are due to differences in this ICD. There is a short piece of extracellular sequence following TM4. Also shown in Figure 11.7 is the hydrophobicity profile of the human α4 subunit. While it contains the conserved features identified in the muscle subunit, it has an unusually large ICD.

While we have come to see X-ray resolved crystals as a sort of pinnacle for understanding protein structures, it should be noted that before any of the structures now available were obtained, a great deal of information anticipating those structures came from biochemical approaches and site-directed mutagenesis combined with functional studies. Those older data remain key for our interpretation of new data from X-ray crystal structures, since crystals are fundamentally static images, and receptors are dynamic and usually of little interest when frozen into their high-affinity desensitized states.

The first crystal structure related to pLGIC that became available was of a protein only distantly related to nAChR, an acetylcholine binding protein (AChBP) released from glial cells of a snail (*Lymnaea stagnalis*) (Brejc et al., 2001). This protein evolved to modulate neuronal nicotinic transmission (Smit et al., 2001). It forms as a homopentamer, like α7 nAChR. The first form isolated was shown to have only 24% sequence identity with the human α7 extracellular domain but conserved structural homology. The structures have subsequently been used for multiple homology models of various nAChR subtypes (Rucktooa, Smit, & Sixma, 2009) and combined with data from the *Torpedo* receptor to generate a model of nAChR, although absent a complete intracellular domain (Unwin, 2005). The AChBP was also used as a starting point for extensive site-directed mutagenesis to achieve a form much closer (64% identity) in sequence to actual α7 subunits (Li et al., 2011). An AChBP sequence was spliced onto the transmembrane domains of a 5HT3 receptor to form a novel ACh-gated channel. Only when amino-acid sequences of three loops in ACh-binding protein were changed to their 5-HT3A counterparts was ACh able to trigger the opening of the ion pore (Bouzat et al., 2004). These experiments confirmed the importance of both the Cys-loop and the short sequence between TM2 and TM3 for channel activation.

While a large body of useful information continues to come from the AChBP and related artificial proteins, arguably one of greatest advancements for structural studies

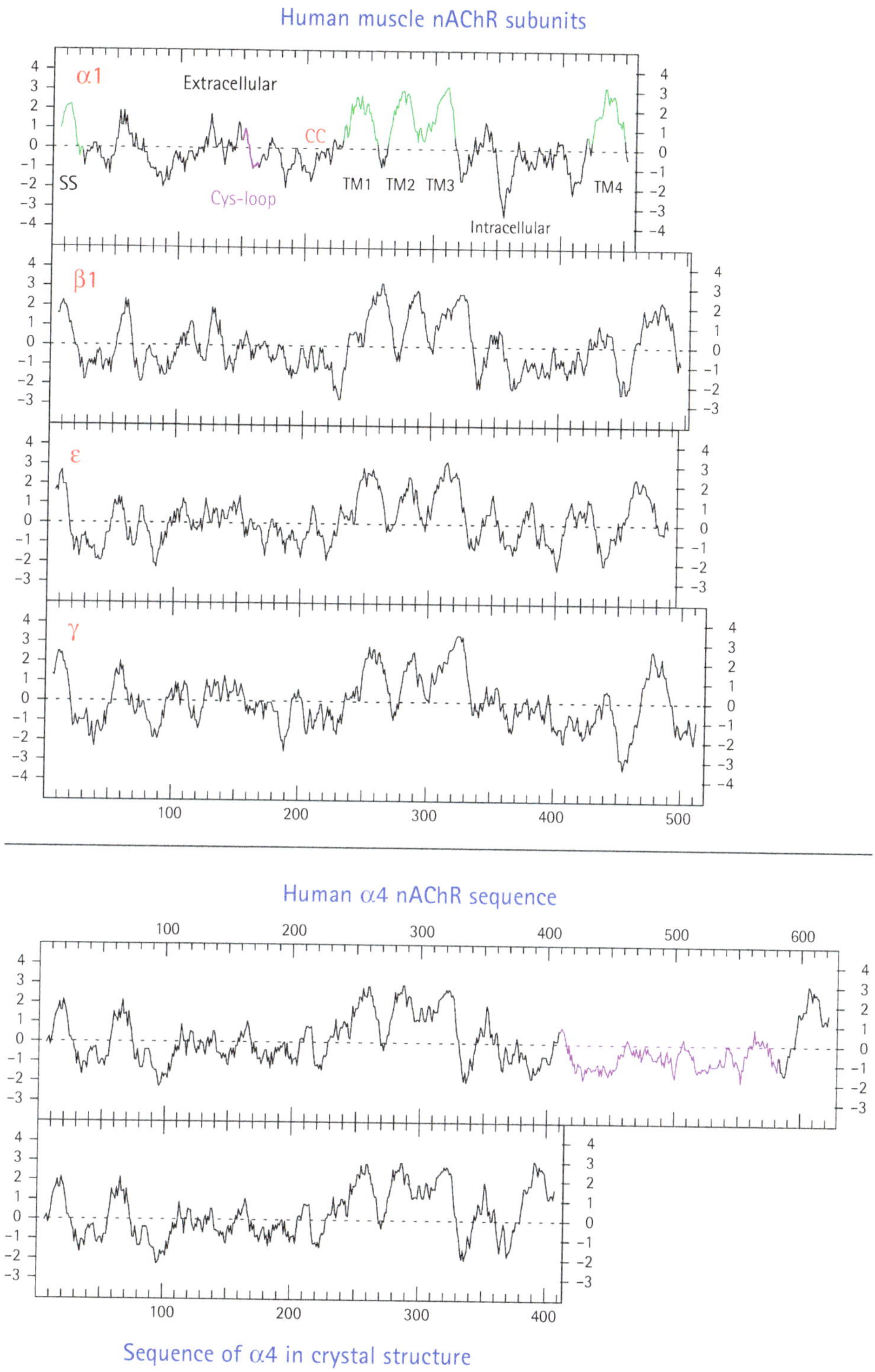

FIGURE 11.7 Hydrophobicity profiles of human nAChR subunits. Shown are Kyte-Doolittle plots for adult human muscle nAChR subunits deduced from the cloned cDNA sequences. The plots include the signal sequences (SS), which are cleaved from the mature protein. Other structural features are highlighted in the α1 plot, including the Cys-loop, the transmembrane (TM), and intracellular domains. Also shown is a plot for the human α4 sequence, highlighting (in purple) the very large intracellular domain that was deleted from the protein in order to get crystal structures (Morales-Perez et al., 2016) shown in Figure 11.8.

of nAChR was the crystallization of α4β2-related pentamers (Morales-Perez, Noviello, & Hibbs, 2016), containing two α4 and three β2 subunits and nicotine in the two agonist binding sites (as in the HSα4β2 shown in Figure 11.6). Crystallization of these subunits was only possible if the ICDs of α4 and β2 were deleted and short linkers substituted. The deleted part of α4 is indicated in Figure 11.7. Images based on the α4β2 structure are shown in Figure 11.8. While of higher resolution than earlier structures, for the most part, they support earlier models (Unwin, 2005) and are consistent with early biochemical studies and functional studies of site-directed mutants. The top view shows the pore-forming TM2 domain. The significance of this domain for the conductance pathway was first shown with labeling studies of *Torpedo* receptors using noncompetitive antagonists (Heidmann & Changeux, 1986) and later studied in greater detail with scanning cysteine mutation studies (Akabas, Stauffer, Xu, & Karlin, 1992). Residues in TM2 determine the selectivity for positive or negative ions, as in the cases of nAChR and 5HT3 versus GABA and glycine receptors, respectively. A relatively small number of mutations are required to reverse the charge selectivity of α7 nAChR (Galzi et al., 1992) or 5HT3 receptors from (Gunthorpe & Lummis, 2001) cationic to anionic, while reciprocal mutations reverse glycine receptors from anionic to cationic channels (Keramidas, Moorhouse, French, Schofield, & Barry, 2000).

Consistent with many previous studies (Papke, 2014), the agonist binding domains were at the interface between the α and β subunits. As predicted, each subunit contributes certain subdomains (Corringer, Le Novere, & Changeux, 2000). The α subunit side is considered the primary surface with A-, B-, and C-loop subdomains, with, as previously noted, the important vicinal cysteines located at the tip of the C-loop. On the complementary surface of the binding site, the β subunit contributes the D-, E-, and F-loop subdomains (Figure 11.8C). At the heart of the agonist binding site is a hydrophobic pocket created by several well-conserved aromatic residues (Bisson, Westera, Schubiger, & Scapozza, 2008). Activation is associated with the cation-pi bonds between the positively charged nitrogen of the agonist and these aromatic residues (Xiu, Puskar, Shanata, Lester, & Dougherty, 2009).

Before agonist binding, the C-loop is positioned like a relatively loose "lid" over the aromatic pocket, the "apo" position (Hansen et al., 2005), and the receptor has low affinity for the agonist. Single-channel studies indicate that ACh dissociation rates from the resting state and intraburst closed states of the receptor are much faster than the closing rate of the agonist-bound channel, at least for muscle-type receptors (Colquhoun & Ogden, 1988; Papke et al., 1988). The C-loop collapses down once the agonist is in the site, effectively closing over it, thereby increasing the affinity of the agonist binding. These conformational changes are consistent with the observation that activation of heteromeric nAChR requires the rapid application of relatively high concentrations of agonist and with low-affinity binding (Papke, Dwoskin, & Crooks, 2007), while it is well known that receptors in the desensitized state show a thousand-fold higher affinity for ACh and nicotine (Clarke et al., 1985; Reavill, Jenner, Kumar, & Stolerman, 1988). High affinity for agonist is a feature of the open state (agonist bound and C-loop closed) of the receptor.

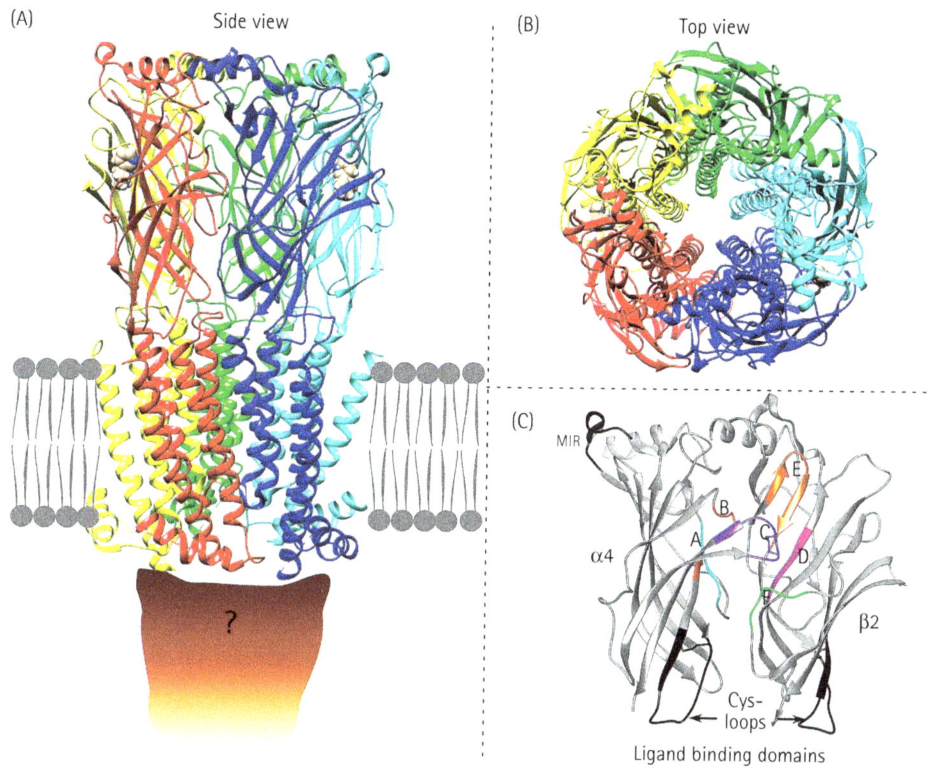

FIGURE 11.8 Images generated from the α4β2 crystal structure with nicotine in the ligand-binding domain (Morales-Perez et al., 2016). **(A)** A side view of the receptor with the likely location of the lipid bilayer represented in gray. Note that the coiled tails at the base of the transmembrane domains are a feature of the crystal where the natural intracellular domain was deleted and probably are not the same in the native receptor. The amorphous shape at the bottom represents the deleted intracellular domain of an α4 subunit, roughly to the scale of the rest of the protein. **(B)** Top view of the crystal showing the pentameric composition and the pore-forming TM2 domains in the center, presumably in a closed, desensitized conformation. The yellow and the dark blue subunits are α4 and the other three are β2 subunits. **(C)** Focus on the ligand-binding domain with highlights of the subdomains and other features.

(The author thanks Alican Gulsevin and Professor Nicole Horenstein for preparing the images used in this figure.)

It is, of course, reasonable that the AChBP and nAChR crystal structures with bound ligand should represent desensitized states/high-affinity states, since even under natural conditions, these are the most stable conformational states. The frozen images cannot readily capture the dynamic and transient condition of receptor activation. The best efforts to date to do so produced images of intermediate resolution of *Torpedo* receptors exposed to agonist and then rapidly frozen in liquid nitrogen (Miyazawa, Fujiyoshi, & Unwin, 2003; Unwin & Fujiyoshi, 2012). Receptor-containing tubes of membrane that had agonist successfully delivered were co-labeled with electron-dense ferritin particles and so could be distinguished from unstimulated tubes. This allowed image averaging of receptors in two different conformations. In the first publication of this method (Miyazawa et al., 2003), it was reported that channel opening was associated with a rotation of the TM2 helices in all five subunits at their point of closest intersubunit contact. In a later study with the same method, but theoretically higher resolution, it was proposed that upon ACh binding, the structural β subunits "tilt outward, destabilizing the arrangement of pore-lining helices, which in the closed channel bends inward symmetrically to form a central hydrophobic gate (Unwin & Fujiyoshi, 2012)." Studies of the prokaryotic channels (Bocquet et al., 2009) suggest a twisting mechanism for channel gating, more consistent with the early data from *Torpedo* channels than with the bulging beta subunit model later proposed.

Within the C-loop, there is a conserved aspartate residue positioned so that it could form a salt bridge with a lysine residue underneath the C-loop. These residues are present in both the AChBP and mammalian receptors. It was proposed that, for muscle-type receptors, such a salt bridge is present in the closed conformation of the channel, and that upon ACh binding, the salt bridge is broken during the conformational change associated with channel opening. These same residues are present in both neuronal α4 and α7 subunits and were confirmed to be essential for effective activation by ACh (Horenstein, McCormack, Stokes, Ren, & Papke, 2007).

Also shown in Figure 11.8C are the Cys-loops at the bottom of the extracellular domains and the main immunogenic region (MIR) at the top, a short, eight-amino-acid section that was identified as important for the etiology of the autoimmune disease myasthenia gravis (Tzartos, Kokla, Walgrave, & Conti-Tronconi, 1988), which targets nAChR of the NJM. The large majority of the autoantibodies in this disease recognize this sequence in the muscle α subunit, and monoclonal antibodies to this sequence are sufficient to induce an experimental mode for the disease (Tzartos et al., 1988). A curious observation is that there is great diversity in the amino acid composition of this region, across all nAChR subunits, greater than for any other region in the extracellular domain (Patrick, Neff, Dineley, & Char, 1996). Aside from its documented antigenicity in the case of the muscle α subunit, there is no known functional significance to this subdomain in any nAChR. It was proposed that synthetic peptides to these sequences might be sufficient to make nAChR subunit-specific antibodies, but the approach was not highly successful. It is nonetheless interesting to speculate that, as presented on the three-dimensional surface of the receptors, these epitopes may nonetheless act as molecular fingerprints.

The amino acid sequence of the TM2 domains is similar in all the mammalian nAChR subunits (Figure 11.9), consistent with its essential role in ion conduction. In general, neuronal nAChR have significantly higher calcium permeability than do muscle-type receptors (Francis & Papke, 1996), with α7-type receptors having particularly high calcium permeability, a feature determined by sequence in the TM2 domain (Bertrand, Galzi, Devillers-Thiery, Bertrand, & Changeux, 1993). Muscle nAChR have linear current–voltage relationships (Francis & Papke, 1996), while neuronal nAChR tend to be inward-rectifying due to a block of outward currents by intracellular polyamines (Haghighi & Cooper, 2000). There are rings of hydrophobic residues in

	9'
α1	MTLSISVLLSLTVFLLVIVE
β1	MGLSIFALLTLTVFLLLLAD
δ	TSVAISVLLAQSVFLLLISK
γ	CTVATNVLLAQTVFLFLVAK
ε	CTVSINVLLAQTVFLFLIAQ
α2	ITLCISVLLSLTVFLLLITE
α3	VTLCISVLLSLTVFLLVITE
α4	ITLCISVLLSLTVFLLLITE
α7	ISLGITVLLSLTVFMLLVAE
β2	MTLCISVLLALTVFLLLISK
β4	MTLCISVLLALTFFLLLISK
	6' 10'

FIGURE 11.9 Sequences in the pore-forming TM2 domains of nAChR. TM2 sequences are shown for the subunits indicated, aligned from the amino terminal intracellular side at left (1') to extracellular end at right (20'). The conduction-limiting leucine in the 9' position is indicated in red, and the residues that regulate sensitivity to noncompetitive antagonists at the 6' and 10' positions in the β subunits are in green.

TM2 that apparently occlude the conduction pathway in closed states of the receptor. Several studies have indicated a special role for the leucine in the 9′ position (referring to the ninth residue from the intracellular side of the predicted helix (Miller, 1989) in stabilizing a ligand-dependent closed state (Bertrand et al., 1992; Revah et al., 1991), so when this residue is mutated to the hydrophilic residue threonine, there is a large increase in channel function, and antagonists can behave as agonists. This mutation was first studied in α7 receptors, which have an intrinsically low open probability, but homologous mutations in other subunits, including subunits of GABAa receptors (Bianchi & Macdonald, 2001), have similar effects. Likewise, mutation of the 9′ residue to serine (Wang et al., 2014) or alanine (Fonck et al., 2009) also increases channel opening, but without as severe a change in pharmacology as with the L9′T mutation. Mice genetically engineered to have the L9′S gain-of-function mutation in their α6 subunits have been used to confirm the importance of α6-containing receptors in mid-brain dopamine release (Wang et al., 2014). The 6′ and 10′ TM2 (Figure 11.9 in green) residues have been shown to be important for determining the selectivity of local anesthetics (Papke, Horenstein, & Placzek, 2001) and other noncompetitive antagonists that are believed to be open-channel blockers. For example, reversing the sequence at these two sites between neuronal beta subunits and β1 prevents block of the neuronal receptor antagonists mecamylamine (Webster et al., 1999) and bis (2,2,6,6-tetramethyl- 4-piperidinyl) sebacate (BTMPS) (Francis, Choi, Horenstein, & Papke, 1998).

As noted previously, the functional roles played by the nAChR ICDs are largely unknown, and these domains elude the crystallographers. It has been proposed, with at least some evidence in the case of *Torpedo* receptors (Miyazawa, Fujiyoshi, Stowell, & Unwin, 1999), that portals for ion conduction are present in the ICD so that ion flow can be effectively directed under the membrane. Rings of negatively charged residues are present in the extracellular vestibule of the receptor and have been shown to be important for good ionic conductance since it is believed they can draw charge into the vestibule (Imoto et al., 1988; Konno, Busch, Von Kitzing, Imoto, & Wang, 1991). There is no evidence for intracellular vestibules, and, indeed, efficient conduction would best be promoted by transverse tunnels that could direct ion flow to the sub-membrane surface and discharge the electric field. Additional structure in the ICD would probably serve for sites of protein–protein interactions with intracellular scaffolding proteins or as intracellular signal transduction machinery, all of which could be modulated by protein phosphorylation (reviewed in Stokes et al., 2015).

nAChR-Associated Proteins

The highly ordered and structured extracellular and transmembrane domains of nAChR are in some way the exception rather than the rule for proteins, since in general, proteins need disorder and flexibility in order to work with other protein partners (Janin & Sternberg, 2013). The intrinsic disorder of the diverse nAChR ICDs suggest a large range of interdependent functions that are likely to occur between the receptors and intracellular proteins. The first putative intracellular protein partner to be identified for nAChR was a 43 kilodalton protein that co-purified with the subunits of the *Torpedo* nAChR (Porter & Froehner, 1985). This protein was not covalently attached to the AChR subunits and was stripped from the membranes with an alkaline wash. The protein was subsequently identified as *rapsyn*, part of an intracellular protein scaffold stabilizing receptors at the NMJ (Ramarao & Cohen, 1998). Other proteins, such as 14-3-3eta, important for α4-containing receptors (Jeanclos et al., 2001), have been identified as "chaperones," helping in the assembly and membrane insertion of nAChR. One important nAChR chaperone, revealed by an analysis of *C. elegans* mutants, is RIC-3. Nematodes with this mutation were resistant to pesticides that were AChE inhibitors because they were deficient in nAChR expression (Halevi et al., 2002). Prior to the discovery of RIC-3, the study of heterologously expressed α7 nAChR was almost exclusively conducted in *Xenopus* oocytes, because mammalian cell lines, such as HEK cells, which could be effectively transfected with heteromeric receptors (Stauderman et al., 1998), appeared incapable of expressing functional α7 receptors. This limitation was overcome when cells were co-transfected with RIC-3 (Williams et al., 2005). It has since been shown that RIC-3 can modulate the expression and subcellular localization of multiple nAChR subtypes (Treinin, 2008). Another recently reported α7 protein chaperone, NACHO, appears to be important for α7 function in brain (Gu et al., 2016). Although not yet characterized in detail, NACHO appears to be a more specific chaperone for α7 than is RIC-3, and to work through a different mechanism, since α7 mutants that are not upregulated by RIC-3 (Castillo et al., 2005) can still be functionally upregulated by NACHO (Papke, unpublished, 2017).

As discussed previously, snake neurotoxins have proven very important for the isolation and functional studies of nAChR. A search for related proteins in the mammalian genome brought to light a family of lymphocyte antigen 6 (Ly6) proteins with sequence homology to the three-finger snake neurotoxins. Following the initial identification of Lynx1 as an endogenous modulator of nAChR function (Miwa et al., 1999), numerous other related proteins have been shown to be positive or negative modulators of various types of nAChR (Tsetlin, 2015).

Just as nAChR function can be regulated by membrane-bound or intracellular protein partners, nAChR can also regulate intracellular signal transduction through a variety of proteins defining each receptor's interactome (McClure-Begley et al., 2013; Paulo, Brucker, & Hawrot, 2009), allowing nAChR to function as mediators of both electrical and metabolic signaling (Valbuena & Lerma, 2016). The ghost of Otto Loewi must have been smiling when Kevin Tracey and co-workers reported that electric stimulation of the vagus nerve could reduce the inflammatory response to the acute administration of bacterial endotoxin, due in large part to the release of ACh into the bloodstream. Surprisingly, although the cellular targets for the ACh released by the vagus were cells of the immune system, the molecular targets were shown to be α7 nAChR (Wang et al., 2003). The suppression of immune responses mediated by α7 nAChR through the cholinergic anti-inflammatory pathway appears to be independent of α7 ion channel activation, but rather associated with intracellular signal transduction (Rosas-Ballina & Tracey, 2009) and in particular the STAT3 pathway (de Jonge et al., 2005; Egea et al., 2015).

Pharmacology of nAChR and Therapeutics Indications

Where there are receptors, there are drug targets. The various nAChR subtypes each present multiple sites that can be differentially targeted by drugs. The ligand-binding domains can be targets for both agonists and antagonists, especially when such drugs can selectively target specific receptors. Modern derivatives of curare, a competitive antagonist of muscle nAChR, are used to produce muscle paralysis when required for surgical procedures. Succinylcholine, a neuromuscular agonist, is also a paralytic, although its mode of action is to produce chronic depolarization and receptor desensitization, similar to how high concentrations of nicotine can both activate and desensitize ganglionic receptors. Although ganglionic nAChR were once considered valid targets for the management of hypertension through the use of ganglionic blockers such as mecamylamine (McQueen & Smirk, 1957) and hexamethonium (Campbell & Robertson, 1950), interest in neuronal nAChR as therapeutic targets has shifted to the brain. As with any drug development program, it has been important to build from the basic nicotinic pharmacophore (Figure 11.1) and define the features of subtype-selective agonists, partial agonists, or antagonists. However, the assignment of specific nAChR as targets for specific indications has been difficult, except in regard to a basic split between the high-affinity heteromeric receptors and α7.

The most important indication related to the high-affinity α4-containing (α4*) nAChR undoubtedly is the management of nicotine addiction and dependence, due to the high incidence of smoking-related illnesses. The Centers for Disease Control and Prevention (CDC) reports that in the United States alone, there are on average more than 1,300 deaths per day due to smoking-related illnesses. High-affinity nAChR may also play roles in Alzheimer's disease and other dementias (Haydar & Dunlop, 2010), as well as depression (Mineur & Picciotto, 2010) and Parkinson's disease (Picciotto & Zoli, 2008). Additionally, a rare form of familial seizure disorder, autosomal-dominant nocturnal frontal lobe epilepsy (ADNFLE), is associated with various mutations in nAChR subunits of high-affinity heteromeric receptors (Becchetti, Aracri, Meneghini, Brusco, & Amadeo, 2015).

The behavioral compulsions associated with the habitual use of tobacco products arguably represent the subtlest and most pernicious of all drug addictions. In general, there are no overt signs of intoxication or incapacitation, and the health costs may take years or even decades to manifest. It is believed that the short-term rewarding (reinforcing) aspect of nicotine addiction comes from the facilitation of dopamine release in the midbrain, specifically in the circuit from the ventral tegmentum to the nucleus accumbens (Picciotto & Kenny, 2013). With habitual exposure to nicotine, the high-affinity nAChR increase in number in the brain, and additionally there may be a switch in the relative abundance of high-sensitivity (HS) compared to low-sensitivity (LS) forms of α4β2 receptors, although this effect, considered to be a preferential chemical chaperoning of HS α4β2 receptors by nicotine, has only been shown conclusively *in vitro* (Lester et al., 2009; Srinivasan et al., 2011). In any case, the changes in nAChR expression and function that occur with chronic nicotine use are believed to produce a dependence, so a smoker attempting to quit manifests a withdrawal syndrome that encourages relapse.

As noted previously, both α4 and α6-containing receptors, usually in combinations with β2, are strongly implicated as underlying nicotine addiction. Knockout of any of these three genes in mice is associated with a reduction in motivation to self-administer nicotine (Champtiaux & Changeux, 2004). Interestingly, the α5 gene has also been implicated in nicotine self-administration by animals, although in a qualitatively different manner. The selective knockout of α5 in the medial habenula, a part of the brain where it could assemble with a variety of different subunits, including α3 and β4, reduces the aversive effects of high dosages of nicotine, leading to increased self-administration. In humans, α5 polymorphisms have been identified that are associated with increased risk of nicotine dependence and subsequent cancer (Chen et al., 2009).

The oldest and most basic approach for aiding smoking cessation has been nicotine replacement therapy, substituting nicotine in alternative forms, such as a chewing gum or a patch, for the nicotine delivered by cigarettes. While these approaches do separate nicotine from other potential carcinogens in smoke, it is questionable whether they address nicotine addiction or merely assuage it with another manner of applying the drug. It can be argued that nicotine patches will not provide the short-term reinforcing effects that would come from the pulsatile delivery of nicotine to the dopamine-releasing centers in the brain. It also likely that the steady delivery of low concentrations of nicotine will sufficiently desensitize brain receptors so that they would respond less effectively to nicotine as delivered by a cigarette. Hypothetically, low concentrations of nicotine delivered from a transdermal patch may also help suppress withdrawal; however, studies show this is only partially true (Rose, Levin, Behm, Adivi, & Schur, 1990).

An alternative approach for smoking cessation therapies has been the development of drugs that are partial agonists of β2-containing nAChR, inspired by the properties of the plant alkaloid cytisine (Papke & Heinemann, 1994). Cytisine itself is being used as a smoking cessation aid in Europe (Etter, Lukas, Benowitz, West, & Dresler, 2008) and was the inspiration for the drug varenicline (Coe et al., 2005). At low concentrations, they blunt phasic activation of the high-affinity brain nAChR while providing low levels of steady-state activation (Papke, Trocme-Thibierge, Guendisch, Abbas Al Rubaiy, & Bloom, 2011). While these drugs are only weak activators of the β2-containing receptors, they are more efficacious agonists for other subtypes such as α3β4 and α7 nAChR (Mihalak, Carroll, & Luetje, 2006; Papke, Wecker, & Stitzel, 2010). This less-than-ideal profile encourages the further development of new partial agonists based on either cytisine (Mineur et al., 2009) or alternative scaffolds (Horenstein, Quadri, Stokes, Shoaib, & Papke, 2017; Tomassoli et al., 2011).

It has also been proposed that nAChR antagonists could be used for smoking cessation therapies. For example, the neuronal nicotinic antagonist mecamylamine used in conjunction with transdermal nicotine patches was more effective than patch therapy alone (Rose et al., 1994). The atypical antidepressant bupropion is also used for smoking cessation therapy, with an efficacy roughly equivalent to that of varenicline. Although the efficacy of bupropion is likely to come from the management of depression common during smoking withdrawal, it has also been reported to be a nAChR antagonist (Slemmer, Martin, & Damaj, 2000).

While most smoking therapeutics target the nAChR associated with dopamine release, it should be noted that individuals suffering from mental illness, and schizophrenics in particular, may represent a special population of smokers for whom nicotine is a sort of medication targeting mostly α7 receptors (Leonard, Mexal, & Freedman, 2007).

There is a general loss of cholinergic function and of high-affinity nAChR in Alzheimer's disease and other dementias, leading to the use of AChE inhibitors such as galantamine (Corey-Bloom, 2003) as therapeutics, although their efficacy is low. In contrast, depression may be associated with a state of cholinergic hyperfunction, so both partial agonists and nicotinic antagonists have been proposed as therapies to augment the efficacy of standard antidepressants (Mineur & Picciotto, 2010). An isomer of mecamylamine was advanced to Phase III clinical trials for depression (Lindsley, 2010) but was found to lack efficacy. Interestingly, mecamylamine, which was reported to reduce motor dysfunction in Tourette's syndrome (Sanberg, Shytle, & Silver, 1998), also failed when it was put into trials as a monotherapy (Silver et al., 2001). The balance between cholinergic and dopaminergic activity in the basal ganglia is important for control of motor function. It seems that the earlier positive results were probably due to the management of hypercholinergic activity that occurred as a side effect of the antipsychotic drugs (dopamine-receptor blockers) that the patients were taking as their standard therapy. Although the hypercholinergic states that sometimes develop in Parkinson's disease are treated with anti-muscarinics, there may be a connection between nAChR and the genesis of the disease, since there is a strong inverse relationship between smoking and the occurrence of the disease in later life (Li, Li, Liu, Shen, & Tang, 2015).

Most drugs that target Cys-loop receptors do so at either the agonist binding sites or the ion channel, but there is growing appreciation nAChR may also be targeted by drugs working at allosteric binding sites. Benzodiazepines and barbiturates, important drugs used as sedatives/hypnotics and as seizure medicines, are positive allosteric modulators (PAMs) of GABAa Cys-loop receptors. Like heteromeric nAChR, GABAa receptors typically have two binding sites for GABA at specific subunit interfaces. Benzodiazepines bind to alternative interfaces and can increase the affinity of efficacy of GABA. Similar allosteric sites have been identified on heteromeric nAChR (Wang & Lindstrom, 2017), suggesting a new direction for future development of therapeutics that target these receptor subtypes.

The therapeutic targeting of homomeric nAChR, α7 in particular, involves special challenges and presents potentially great opportunities. The kinetic models and energy landscapes described previously can only be effectively applied to heteromeric nAChR that manifest high-affinity desensitized states. Homomeric α7 receptors have five potential agonist binding sites, and maximal channel opening occurs when only one or two of those sites are occupied (Andersen, Corradi, Sine, & Bouzat, 2013; Williams et al., 2011). The binding of higher levels of agonist induces conformational changes to novel nonconducting states representing a sort of desensitization that is unique to α7 (Williams, Wang, & Papke, 2011a). The L9′T mutation discussed before largely eliminates this unique form of desensitization and produces very large increases in current for this receptor. The unique properties of α7 nAChR have also made them sensitive to types of PAMs that affect only α7 nAChR (Williams, Peng, Kimbrell, & Papke, 2012; Williams et al., 2011a; Williams, Wang, & Papke, 2011b). These α7-selective PAMs fall into two basic categories: Type I PAMs, which have a transient effect on increasing channel opening, and Type II PAMs, which destabilize or reverse α7 desensitization (Gronlien et al., 2007). These PAMs bind at a site within the α7 TM domains (Gill et al., 2011) and not in the extracellular domain as do PAMs for heteromeric receptors. Additionally, there are compounds that bind to both the TM PAM site and a novel site in the extracellular domain to produce both transient activation when applied alone and persistent potentiation of subsequent agonist applications (Horenstein et al., 2016). The first such ago-PAMs to be described were 4BP-TQS (Gill-Thind, Dhankher, D'Oyley, Sheppard,

& Millar, 2015) and its stereoisomer GAT107 (Papke et al., 2014). Subsequent work has identified an interesting range of properties for related compounds, including allosteric antagonism (Gill-Thind et al., 2015). The identification of these α7-selective PAMs and ago-PAMs increases the potential range of drug types that could be used for targeting this receptor, although it should be noted that most of the preclinical *in vitro* work with these compounds has been done at room temperature, and these compounds show a remarkable decrease in activity at body temperature (Williams et al., 2012).

In addition to having unique activation and desensitization properties, α7 receptors also have a unique pharmacophore. While the classic nicotinic pharmacophore (Figure 11.1) included both the charged nitrogen and a hydrogen bond acceptor, it was later shown that the charged nitrogen alone, in the form of tetramethylammonium (TMA) was sufficient to fully activate heteromeric neuronal and α7 nAChR (Papke, Bencherif, & Lippiello, 1996). In the same study it was also shown that choline, the ubiquitous ACh precursor, is an α7-selective agonist. TMA is just one example of a nonselective agonist that could be used as a scaffold and modified through either the addition of a hydroxyl, steric bulk, or a large hydrophobic group, to achieve α7-selective activation (Horenstein, Leonik, & Papke, 2008).

One of the first α7-selective partial agonists identified, GTS-21 (DMBXA), was proposed as a potential therapeutic for Alzheimer's disease based on its cytoprotective and cognition-enhancing effects in models of cholinergic hypo-function (Martin et al., 1994; Meyer et al., 1994). Since those studies, numerous other α7-selective agonists and partial agonists have been shown to be effective in animal models of learning and memory (Boess et al., 2007; Levin, McClernon, & Rezvani, 2006; Van Kampen et al., 2004). Ion channel activity was indicated to be particularly important for the pro-cognitive effects of α7-drugs, since in two separate studies (Briggs et al., 2009; Pieschl et al., 2017), NS6740, an apparent antagonist of α7-channel activation, was not only ineffective as a cognitive enhancer, it blocked the pro-cognitive effects of more efficacious agonists. Interestingly, NS6740 was shown not to be strictly an antagonist, but rather a very weak partial agonist that produced prolonged desensitization that could be partly reversed by co-application with a Type II PAM (Briggs et al., 2009), and when used in conjunction with the ago-PAM GAT107, NS6740 produced large persistent currents (Papke et al., 2017).

Following the reports of positive results with α7 drugs in animal models of learning and memory, several large companies initiated programs for α7 drug development. Some compounds made it into trials for Alzheimer's disease and attention-deficit hyperactivity disorder, but none were shown to be effective at safe doses. There is a partial gene duplication of α7 in humans that produces a short nonfunctional form of the receptor. The occurrence of this gene and other α7 polymorphisms have been linked to schizophrenia (Gault et al., 2003; Leonard et al., 2002). Limited trials are still ongoing with GTS-21 as a treatment for negative and cognitive symptoms of schizophrenia (Kem et al., 2017), although a similar trial with the potent α7-agonist TC-5619 (Hauser et al., 2009) was not successful (Walling et al., 2016).

Although the mindset of nAChR being ion channels and modulators of brain function encouraged an initial clinical focus on disorders of the CNS, discovery of the cholinergic anti-inflammatory pathway is promoting entirely new approaches for clinical development into areas like asthma (Assayag, Beaulieu, & Cormier, 2014), arthritis (van Maanen et al., 2009), and inflammatory pain (Alsharari, Freitas, & Damaj, 2013). The best drugs in these areas may not be channel activators, but rather silent agonists like NS6740 (Papke et al., 2015).

Concluding Remarks

Nicotinic acetylcholine receptors stand out from a large family of related proteins as being the essential initiators of every human movement, including the act of speaking. In the brain, they may help in the shaping of the words to be spoken, but in animals, they may be removed genetically from the brain with little apparent effect, except for loss of interest in nicotine.

For further reading, I recommend this excellent and comprehensive textbook dedicated just to nicotinic receptors:

> *Nicotinic Receptors*, edited by Robin Lester (2014), published by Humana Press, and available online at: http://www.springer.com/us/book/9781493911660.

Acknowledgements

Work in the Papke lab is supported by National Institutes of Health grant RO1-GM57481. I thank Alican Gulsevin and Professor Nicole Horenstein for preparing the images used in Figure 11.8. I also thank Clare Stokes for her careful proofreading and thoughtful comments.

References

Akabas, M. H., Stauffer, D. A., Xu, M., & Karlin, A. (1992). Acetylcholine receptor channel structure probed in cysteine-substitution mutants. *Science, 258*, 307–310.

Aldrich, R. W., Corey, D. P., & Stevens, C. F. (1983). A reinterpretation of mammalian sodium channel gating based on single channel recording. *Nature, 306*(5942), 436–441.

Alsharari, S. D., Freitas, K., & Damaj, M. I. (2013). Functional role of alpha7 nicotinic receptor in chronic neuropathic and inflammatory pain: Studies in transgenic mice. *Biochemical Pharmacology, 86*(8), 1201–1207.

Andersen, N., Corradi, J., Sine, S. M., & Bouzat, C. (2013). Stoichiometry for activation of neuronal alpha7 nicotinic receptors. *Proceedings of the National Academy of Sciences of the United States of America, 110*(51), 20819–20824.

Assayag, E. I., Beaulieu, M. J., & Cormier, Y. (2014). Bronchodilatory and anti-inflammatory effects of ASM-024, a nicotinic receptor ligand, developed for the treatment of asthma. *PLoS One, 9*(1), e86091. doi:10.1371/journal.pone.0086091

Azam, L., Winzer-Serhan, U. H., Chen, Y., & Leslie, F. M. (2002). Expression of neuronal nicotinic acetylcholine receptor subunit mRNAs within midbrain dopamine neurons. *Journal of Comparative Neurology, 444*(3), 260–274.

Ballivet, M., Patrick, J., Lee, J., & Heinemann, S. (1982). Molecular cloning of cDNA coding for the gamma subunit of *Torpedo* acetylcholine receptor. *Proceedings of the National Academy of Sciences of the United States of America, 79*, 4466–4470.

Bartol, T. M., Jr., Land, B. R., Salpeter, E. E., & Salpeter, M. M. (1991). Monte Carlo simulation of miniature endplate current generation in the vertebrate neuromuscular junction. *Biophysical Journal, 59*(6), 1290–1307.

Becchetti, A., Aracri, P., Meneghini, S., Brusco, S., & Amadeo, A. (2015). The role of nicotinic acetylcholine receptors in autosomal dominant nocturnal frontal lobe epilepsy. *Frontiers in Physiology, 6*, 22. doi:10.3389/fphys.2015.00022

Beers, W. H., & Reich, E. (1970). Structure and activity of acetylcholine. *Nature, 228*, 917–922.

Bertrand, D., Devillers-Thiéry, A., Revah, F., Galzi, J.-L., Hussy, N., Mulle, C., . . . Changeux, J.-P. (1992). Unconventional pharmacology of a neuronal nicotinic receptor mutated in the channel domain. *Proceedings of the National Academy of Sciences of the United States of America, 89*, 1261–1265.

Bertrand, D., Galzi, J. L., Devillers-Thiery, A., Bertrand, S., & Changeux, J.-P. (1993). Mutations at two distinct sites within the channel domain M2 alter calcium permeability of neuronal alpha 7 nicotinic receptor. *Proceedings of the National Academy of Sciences of the United States of America, 90*(15), 6971–6975.

Betz, H. (1990). Ligand gated channels in the brain: The amino acid receptor superfamily. *Neuron, 5*, 383–392.

Bianchi, M. T., & Macdonald, R. L. (2001). Mutation of the 9′ leucine in the GABA(A) receptor gamma2L subunit produces an apparent decrease in desensitization by stabilizing open states without altering desensitized states. *Neuropharmacology, 41*(6), 737–744.

Bisson, W. H., Westera, G., Schubiger, P. A., & Scapozza, L. (2008). Homology modeling and dynamics of the extracellular domain of rat and human neuronal nicotinic acetylcholine receptor subtypes alpha4beta2 and alpha7. *Journal of Molecular Modeling, 14*(10), 891–899.

Bocquet, N., Nury, H., Baaden, M., Le Poupon, C., Changeux, J. P., Delarue, M., & Corringer, P. J. (2009). X-ray structure of a pentameric ligand-gated ion channel in an apparently open conformation. *Nature, 457*(7225), 111–114.

Bocquet, N., Prado de Carvalho, L., Cartaud, J., Neyton, J., Le Poupon, C., Taly, A., . . . Corringer, P. J. (2007). A prokaryotic proton-gated ion channel from the nicotinic acetylcholine receptor family. *Nature, 445*(7123), 116–119.

Boess, F. G., De Vry, J., Erb, C., Flessner, T., Hendrix, M., Luithle, J., . . . Koenig, G. (2007). The novel alpha7 nicotinic acetylcholine receptor agonist N-[(3R)-1-azabicyclo[2.2.2]oct-3-yl]-7-[2-(methoxy)phenyl]-1-benzofuran-2-carboxamide improves working and recognition memory in rodents. *Journal of Pharmacology and Experimental Therapeutics, 321*(2), 716–725.

Boulter, J., Connolly, J., Deneris, E., Goldman, D., Heinemann, S., & Patrick, J. (1987). Functional expression of two neural nicotinic acetylcholine receptors from cDNA clones identifies a gene family. *Proceedings of the National Academy of Sciences of the United States of America, 84*, 7763–7767.

Bouzat, C., Bartos, M., Corradi, J., & Sine, S. M. (2008). The interface between extracellular and transmembrane domains of homomeric Cys-loop receptors governs open-channel lifetime and rate of desensitization. *Journal of Neuroscience, 28*(31), 7808–7819.

Bouzat, C., Gumilar, F., Spitzmaul, G., Wang, H. L., Rayes, D., Hansen, S. B., ... Sine, S. M. (2004). Coupling of agonist binding to channel gating in an ACh-binding protein linked to an ion channel. *Nature, 430*(7002), 896–900.

Brejc, K., van Dijk, W. J., Klaassen, R. V., Schuurmans, M., van Der Oost, J., Smit, A. B., & Sixma, T. K. (2001). Crystal structure of an ACh-binding protein reveals the ligand-binding domain of nicotinic receptors. *Nature, 411*(6835), 269–276.

Briggs, C. A., Gronlien, J. H., Curzon, P., Timmermann, D. B., Ween, H., Thorin-Hagene, K., ... Gopalakrishnan, M. (2009). Role of channel activation in cognitive enhancement mediated by alpha7 nicotinic acetylcholine receptors. *British Journal of Pharmacology, 158*(6), 1486–1494.

Campbell, A., & Robertson, E. (1950). Treatment of severe hypertension with hexamethonium bromide. *British Medical Journal, 2*(4683), 804–806.

Cash, D. E., Aoshima, H., Pasquale, E. B., & Hess, G. (1985). Acetylcholine-receptor-mediated ion fluxes in *Electrophorus electricus* and *Torpedo californica* membrane vesicles. *Reviews of Physiology, Biochemistry, and Pharmacology, 102*, 73–117.

Castillo, M., Mulet, J., Gutierrez, L. M., Ortiz, J. A., Castelan, F., Gerber, S., ... Criado, M. (2005). Dual role of the RIC-3 protein in trafficking of serotonin and nicotinic acetylcholine receptors. *Journal of Biological Chemistry, 280*(29), 27062–27068.

Champtiaux, N., & Changeux, J. P. (2004). Knockout and knockin mice to investigate the role of nicotinic receptors in the central nervous system. *Progress in Brain Research, 145*, 235–251.

Chen, X., Chen, J., Williamson, V. S., An, S. S., Hettema, J. M., Aggen, S. H., ... Kendler, K. S. (2009). Variants in nicotinic acetylcholine receptors alpha5 and alpha3 increase risks to nicotine dependence. *American Journal of Medical Genetics, Part B: Neuropsychiatric Genetics, 150 B*(7), 926–933.

Chesselet, M.-F. (1984). Presynaptic regulation of neurotransmitter release in the brain. *Neuroscience, 12*, 347–375.

Chini, B., Clementi, F., Hukovic, N., & Sher, E. (1992). Neuronal-type alpha-bungarotoxin receptors and the alpha 5-nicotinic receptor subunit gene are expressed in neuronal and nonneuronal human cell lines. *Proceedings of the National Academy of Sciences of the United States of America, 89*(5), 1572–1576.

Clarke, P. B. S., Schwartz, R. D., Paul, S. M., Pert, C. B., & Pert, A. (1985). Nicotinic binding in rat brain: Autoradiographic comparison of [^{3}H] acetylcholine [$_3$H] nicotine and [^{125}I]-alpha-bungarotoxin. *Journal of Neuroscience, 5*, 1307–1315.

Coe, J. W., Brooks, P. R., Vetelino, M. G., Wirtz, M. C., Arnold, E. P., Huang, J., ... O'Neill, B. T. (2005). Varenicline: An alpha4beta2 nicotinic receptor partial agonist for smoking cessation. *Journal of Medical Chemistry, 48*(10), 3474–3477.

Colquhoun, D., & Hawkes, A. G. (1983). The principles of the stochastic interpretation of ion channel mechanisms. In B. Sakmann & E. Neher (Eds.), *Single-channel recording* (pp. 135–175). New York: Plenum Press.

Colquhoun, D., & Ogden, D. C. (1988). Activation of ion channels in the frog end-plate by high concentrations of acetylcholine. *Journal of Physiology, 395*, 131–159.

Colquhoun, D., & Sakmann, B. (1985). Fast events in single-channel currents activated by acetylcholine and its analogues at the frog muscle end-plate. *Journal of Physiology, 369*, 501–557.

Corey-Bloom, J. (2003). Galantamine: A review of its use in Alzheimer's disease and vascular dementia. *International Journal of Clinical Practice, 57*(3), 219–223.

Corringer, P. J., Le Novere, N., & Changeux, J. P. (2000). Nicotinic receptors at the amino acid level. *Annual Review of Pharmacology and Toxicology, 40*, 431–458.

Dani, J. (1986). Ion-channel entrances influence permeation: Net charge size, shape, and binding considerations. *Biophysical Journal, 49*, 607–617.

David, R., Ciuraszkiewicz, A., Simeone, X., Orr-Urtreger, A., Papke, R. L., McIntosh, J. M., . . . Scholze, P. (2010). Biochemical and functional properties of distinct nicotinic acetylcholine receptors in the superior cervical ganglion of mice with targeted deletions of nAChR subunit genes. *European Journal of Neuroscience, 31*(6), 978–993.

Davies, P. A., Wang, W., Hales, T. G., & Kirkness, E. F. (2003). A novel class of ligand-gated ion channel is activated by Zn2+. *Journal of Biological Chemistry, 278*(2), 712–717.

de Jonge, W. J., van der Zanden, E. P., The, F. O., Bijlsma, M. F., van Westerloo, D. J., Bennink, R. J., . . . Boeckxstaens, G. E. (2005). Stimulation of the vagus nerve attenuates macrophage activation by activating the Jak2-STAT3 signaling pathway. *Nature Immunology, 6*(8), 844–851.

Deneris, E. S., Boulter, J., Connolly, J., Wada, K., Patrick, J., & Heinemann, S. (1987). Identification of a gene proposed to encode a non-alpha subunit of neuronal nicotinic acetylcholine receptors. *Clinical Chemistry, 35*(5), 731–737.

Dent, J. A. (2010). The evolution of pentameric ligand-gated ion channels. *Advances in Experimental Medicine and Biology, 683*, 11–23.

Descarries, L., Gisiger, V., & Steriade, M. (1997). Diffuse transmission by acetylcholine in the CNS. *Progress in Neurobiology, 53*(5), 603–625.

Dunn, S., & Raftery, M. (1982). Activation and desensitization of *Torpedo* acetylcholine receptor: Evidence for separate binding sites. *Proceedings of the National Academy of Sciences of the United States of America, 79*, 6757–6761.

Duvoisin, R. M., Deneris, E. S., Patrick, J., & Heinemann, S. (1989). The functional diversity of the neuronal nicotinic acetylcholine receptors is increased by a novel subunit: beta 4. *Neuron, 3*(4), 487–496.

Egea, J., Buendia, I., Parada, E., Navarro, E., Leon, R., & Lopez, M. G. (2015). Anti-inflammatory role of microglial alpha7 nAChRs and its role in neuroprotection. *Biochemical Pharmacology, 97*(4), 463–472.

Eldefrawi, M. E., Aronstam, R. S., Bakry, N. M., Eldefrawi, A. T., & Albuquerque, E. X. (1980). Activation, inactivation, and desensitization of acetylcholine receptor channel complex detected by binding of perhydrohistrionicotoxin. *Proceedings of the National Academy of Sciences of the United States of America, 77*(4), 2309–2313.

Elgoyhen, A. B., Johnson, D. S., Boulter, J., Vetter, D. E., & Heinemann, S. (1994). a9: An acetylcholine receptor with novel pharmacological properties expressed in rat cochlear hair cells. *Cell, 79*, 705–715.

Elgoyhen, A. B., Vetter, D. E., Katz, E., Rothlin, C. V., Heinemann, S. F., & Boulter, J. (2001). Alpha10: A determinant of nicotinic cholinergic receptor function in mammalian vestibular and cochlear mechanosensory hair cells. *Proceedings of the National Academy of Sciences of the United States of America, 98*(6), 3501–3506.

Etter, J. F., Lukas, R. J., Benowitz, N. L., West, R., & Dresler, C. M. (2008). Cytisine for smoking cessation: A research agenda. *Drug and Alcohol Dependency, 92*(1–3), 3–8.

Fatt, P., & Katz, B. (1951). An analysis of the endplate potential recorded with an intra-cellular electrode. *Journal of Physiology, 115*, 320–370.

Feldberg, W., & Fessard, A. (1942). The cholinergic nature of the nerves to the electric organ of the *Torpedo* (*Torpedo marmorata*). *Journal of Physiology, 101*(2), 200–216.

Fonck, C., Nashmi, R., Salas, R., Zhou, C., Huang, Q., De Biasi, M., . . . Lester, H. A. (2009). Demonstration of functional alpha4-containing nicotinic receptors in the medial habenula. *Neuropharmacology, 56*(1), 247–253.

Forsayeth, J. R., & Kobrin, E. (1996). Formation of Oligomers containing the beta3 and beta4 subunits of the rat nicotinic receptor. *Journal of Neuroscience, 17*(5), 1531–1538.

Francis, M. M., Choi, K. I., Horenstein, B. A., & Papke, R. L. (1998). Sensitivity to voltage-independent inhibition determined by pore-lining region of ACh receptor. *Biophysical Journal, 74*(5), 2306–2317.

Francis, M. M., & Papke, R. L. (1996). Muscle-type nicotinic acetylcholine receptor delta subunit determines sensitivity to noncompetitive inhibitors while gamma subunit regulates divalent permeability. *Neuropharmacology, 35*(11), 1547–1556.

Froehner, S. C. (1991). The submembrane machinery for nicotinic acetylcholine receptor clustering. *Journal of Cell Biology, 114*(1), 1–7.

Galzi, J.-L., Devillers-Thiery, A., Hussy, N., Bertrand, S., Changeux, J.-P., & Bertrand, D. (1992). Mutations in the channel domain of a neuronal nicotinic receptor convert ion selectivity from cationic to anionic. *Nature, 359*, 500–505.

Gault, J., Hopkins, J., Berger, R., Drebing, C., Logel, J., Walton, C., . . . Leonard, S. (2003). Comparison of polymorphisms in the alpha7 nicotinic receptor gene and its partial duplication in schizophrenic and control subjects. *American Journal of Medical Genetics, Part B: Neuropsychiatric Genetics, 123*(1), 39–49.

Gerzanich, V., Kuryatov, A., Anand, R., & Lindstrom, J. (1996). "Orphan" alpha6 nicotinic AChR subunit can form a functional heteromeric acetylcholine receptor. *Molecular Pharmacology, 51*, 320–327.

Gerzanich, V., Wang, F., Kuryatov, A., & Lindstrom, J. (1998). Alpha5 Subunit alters desensitization, pharmacology, Ca++ permeability and Ca++ modulation of human neuronal alpha 3 nicotinic receptors. *Journal of Pharmacology and Experimental Therapeutics, 286*(1), 311–320.

Gill, J. K., Savolainen, M., Young, G. T., Zwart, R., Sher, E., & Millar, N. S. (2011). Agonist activation of {alpha}7 nicotinic acetylcholine receptors via an allosteric transmembrane site. *Proceedings of the National Academy of Sciences of the United States of America, 108*(14), 5867–5872.

Gill-Thind, J. K., Dhankher, P., D'Oyley, J. M., Sheppard, T. D., & Millar, N. S. (2015). Structurally similar allosteric modulators of alpha7 nicotinic acetylcholine receptors exhibit five distinct pharmacological effects. *Journal of Biological Chemistry, 290*(6), 3552–3562.

Gotti, C., Moretti, M., Maggi, R., Longhi, R., Hanke, W., Klinke, N., & Clementi, F. (1997). Alpha7 and alpha8 nicotinic receptor subtypes immunopurified from chick retina have different immunological, pharmacological and functional properties. *European Journal of Neuroscience, 9*(6), 1201–1211.

Grady, S. R., Salminen, O., Laverty, D. C., Whiteaker, P., McIntosh, J. M., Collins, A. C., & Marks, M. J. (2007). The subtypes of nicotinic acetylcholine receptors on dopaminergic terminals of mouse striatum. *Biochemical Pharmacology, 74*(8), 1235–1246.

Gronlien, J. H., Haakerud, M., Ween, H., Thorin-Hagene, K., Briggs, C. A., Gopalakrishnan, M., & Malysz, J. (2007). Distinct profiles of alpha7 nAChR positive allosteric modulation revealed by structurally diverse chemotypes. *Molecular Pharmacology, 72*(3), 715–724.

Gu, S., Matta, J. A., Lord, B., Harrington, A. W., Sutton, S. W., Davini, W. B., & Bredt, D. S. (2016). Brain alpha7 nicotinic acetylcholine receptor assembly requires NACHO. *Neuron, 89*(5), 948–955.

Gunthorpe, M. J., & Lummis, S. C. (2001). Conversion of the ion selectivity of the 5-HT(3a) receptor from cationic to anionic reveals a conserved feature of the ligand-gated ion channel superfamily. *Journal of Biological Chemistry, 276*(24), 10977–10983.

Haghighi, A. P., & Cooper, E. (2000). A molecular link between inward rectification and calcium permeability of neuronal nicotinic acetylcholine alpha3beta4 and alpha4beta2 receptors. *Journal of Neuroscience, 20*(2), 529–541.

Halevi, S., McKay, J., Palfreyman, M., Yassin, L., Eshel, M., Jorgensen, E., & Treinin, M. (2002). The *C. elegans* ric-3 gene is required for maturation of nicotinic acetylcholine receptors. *EMBO Journal, 21*(5), 1012–1020.

Han, Z. Y., Le Novere, N., Zoli, M., Hill, J. A., Jr., Champtiaux, N., & Changeux, J. P. (2000). Localization of nAChR subunit mRNAs in the brain of *Macaca mulatta*. *European Journal of Neuroscience, 12*(10), 3664–3674.

Han, Z. Y., Zoli, M., Cardona, A., Bourgeois, J. P., Changeux, J. P., & Le Novere, N. (2003). Localization of [3H]nicotine, [3H]cytisine, [3H]epibatidine, and [125I]alpha-bungarotoxin binding sites in the brain of *Macaca mulatta*. *Journal of Comparative Neurology, 461*(1), 49–60.

Hansen, S. B., Sulzenbacher, G., Huxford, T., Marchot, P., Taylor, P., & Bourne, Y. (2005). Structures of Aplysia AChBP complexes with nicotinic agonists and antagonists reveal distinctive binding interfaces and conformations. *EMBO Journal, 24*(20), 3635–3646.

Hauser, T. A., Kucinski, A., Jordan, K. G., Gatto, G. J., Wersinger, S. R., Hesse, R. A., . . . Bencherif, M. (2009). TC-5619: An alpha7 neuronal nicotinic receptor-selective agonist that demonstrates efficacy in animal models of the positive and negative symptoms and cognitive dysfunction of schizophrenia. *Biochemical Pharmacology, 78*(7), 803–812.

Haydar, S. N., & Dunlop, J. (2010). Neuronal nicotinic acetylcholine receptors—targets for the development of drugs to treat cognitive impairment associated with schizophrenia and Alzheimer's disease. *Current Topics in Medicinal Chemistry, 10*(2), 144–152.

Heidmann, T., & Changeux, J.-P. (1986). Characterisation of the transient agonist-triggered state of the acetylcholine receptor rapidly labeled by the non-competitive blocker [^{3}H] chlorpromazine: Additional evidence for the open channel conformation. *Biochemistry, 25*, 6109–6113.

Hilf, R. J., & Dutzler, R. (2008). X-ray structure of a prokaryotic pentameric ligand-gated ion channel. *Nature, 452*(7185), 375–379.

Hilf, R. J., & Dutzler, R. (2009). Structure of a potentially open state of a proton-activated pentameric ligand-gated ion channel. *Nature, 457*(7225), 115–118.

Hodgkin, A. L., & Huxley, A. F. (1952). A quantitative description of membrane current and its application to conduction and excitation in nerve. *Journal of Physiology, 117*, 500–504.

Holden-Dye, L., Joyner, M., O'Connor, V., & Walker, R. J. (2013). Nicotinic acetylcholine receptors: A comparison of the nAChRs of *Caenorhabditis elegans* and parasitic nematodes. *Parasitology International, 62*(6), 606–615.

Hollmann, M., O'Shea-Greenfield, A., Rogers, S. W., & Heinemann, S. (1989). Cloning by functional expression of a member of the glutamate receptor family. *Nature, 342*, 643–648.

Horenstein, N. A., Leonik, F. M., & Papke, R. L. (2008). Multiple pharmacophores for the selective activation of nicotinic alpha7-type acetylcholine receptors. *Molecular Pharmacology, 74*(6), 1496–1511.

Horenstein, N. A., McCormack, T. J., Stokes, C., Ren, K., & Papke, R. L. (2007). Reversal of agonist selectivity by mutations of conserved amino acids in the binding site of nicotinic acetylcholine receptors. *Journal of Biological Chemistry, 282*(8), 5899–5909.

Horenstein, N. A., Papke, R. L., Kulkarni, A. R., Chaturbhuj, G. U., Stokes, C., Manther, K., & Thakur, G. A. (2016). Critical molecular determinants of alpha7 nicotinic acetylcholine

receptor allosteric activation: Separation of direct allosteric activation and positive allosteric modulation. *Journal of Biological Chemistry, 291*(10), 5049–5067.
Horenstein, N. A., Quadri, M., Stokes, C., Shoaib, M., & Papke, R. L. (2017). Cracking the betel nut: Cholinergic activity of Areca alkaloids and related compounds. *Nicotine and Tobacco Research*, in press.
Horn, R., & Vandenberg, C. A. (1984). Statistical properties of single sodium channels. *Journal of General Physiology, 84*(4), 505–534.
Hucho, F. (1979). Photoaffinity derivatives of alpha-bungarotoxin and alpha-Naja naja siamensis toxin. *FEBS Letters, 103*, 27–32.
Hucho, F. (1986). The nicotinic acetylcholine receptor and its ion channel. *European Journal of Biochemistry, 158*, 211–226.
Imoto, K., Busch, C., Sakmann, B., Mishina, M., Konno, T., Nakai, J., . . . Numa, S. (1988). Rings of negatively charged amino acids determine the acetylcholine receptors channel conductance. *Nature, 335*, 645–648.
Jaiteh, M., Taly, A., & Henin, J. (2016). Evolution of pentameric ligand-gated ion channels: Pro-loop receptors. *PLoS One, 11*(3), e0151934. doi:10.1371/journal.pone.0151934
Janin, J., & Sternberg, M. J. (2013). Protein flexibility, not disorder, is intrinsic to molecular recognition. *F1000 Biology Reports, 5*, 2.
Jeanclos, E. M., Lin, L., Treuil, M. W., Rao, J., DeCoster, M. A., & Anand, R. (2001). The chaperone protein 14-3-3eta interacts with the nicotinic acetylcholine receptor alpha 4 subunit. Evidence for a dynamic role in subunit stabilization. *Journal of Biological Chemistry, 276*(30), 28281–28290.
Jiang, L., Kundu, S., Lederman, J. D., Lopez-Hernandez, G. Y., Ballinger, E. C., Wang, S., . . . Role, L. W. (2016). Cholinergic signaling controls conditioned fear behaviors and enhances plasticity of cortical-amygdala circuits. *Neuron, 90*(5), 1057–1070.
Jones, A. K., Brown, L. A., & Sattelle, D. B. (2007). Insect nicotinic acetylcholine receptor gene families: From genetic model organism to vector, pest and beneficial species. *Invertebrate Neuroscience, 7*(1), 67–73.
Kao, P. N., & Karlin, A. (1986). Acetylcholine receptor binding site contains a disulfide cross-link between adjacent half-cystinyl residues. *Journal of Biological Chemistry, 261*(18), 8085–8088.
Karlin, A., Weill, C. L., McNamee, M. G., & Valderrama, R. (1976). Facets of the structures of acetylcholine receptors from Electrophorus and Torpedo. *Cold Spring Harbor Symposia on Quantitative Biology, 40*, 203–210.
Katz, B., & Miledi, R. (1972). The statistical nature of the acetylcholine potential and its molecular components. *Journal of Physiology, 224*, 665–699.
Katz, B., & Thesleff, S. (1957). A study of the "desensitization" produced by acetylcholine at the motor end-plate. *Journal of Physiology, 138*, 63–80.
Kem, W. R., Olincy, A., Johnson, L., Harris, J., Wagner, B. D., Buchanan, R. W., . . . Freedman, R. (2017). Pharmacokinetic limitations on effects of an alpha7 nicotinic receptor agonist in schizophrenia: Randomized trial with an extended release formulation. *Neuropsychopharmacology*, in press. doi: 10.1038/npp.2017.182
Keramidas, A., Moorhouse, A. J., French, C. R., Schofield, P. R., & Barry, P. H. (2000). M2 pore mutations convert the glycine receptor channel from being anion- to cation-selective. *Biophysical Journal, 79*(1), 247–259.
Konno, T., Busch, C., Von Kitzing, E., Imoto, K., & Wang, F. (1991). Rings of anionic amino acids as structural determinants of ion selectivity in the acetylcholine receptor channel. *Proceedings of the Royal Society of London, Series B, 244*, 69–79.

Kubo, T., Noda, M., Takai, T., Tanabe, T., Kayano, T., Shimizu, S., ... Numa, S. (1985). Primary structure of delta-subunit precursor of calf muscle acetylcholine receptor deduced from cDNA sequence. *European Journal of Biochemistry, 149*, 5–13.

Kuryatov, A., & Lindstrom, J. (2011). Expression of functional human alpha6beta2beta3* acetylcholine receptors in *Xenopus laevis* oocytes achieved through subunit chimeras and concatamers. *Molecular Pharmacology, 79*(1), 126–140.

Kuryatov, A., Olale, F., Cooper, J., Choi, C., & Lindstrom, J. (2000). Human alpha6 AChR subtypes: Subunit composition, assembly, and pharmacological responses. *Neuropharmacology, 39*(13), 2570–2590.

Kuryatov, A., Onksen, J., & Lindstrom, J. (2008). Roles of accessory subunits in alpha4beta2(*) nicotinic receptors. *Molecular Pharmacology, 74*(1), 132–143.

Kusano, K., Miledi, R., & Stinnakre, J. (1982). Cholinergic and catecholaminergic receptors in the *Xenopus* oocyte membrane. *Journal of Physiology, 328*, 143–170.

Kyte, J., & Doolittle, R. F. (1982). A simple method for displaying the hydropathic character of a protein. *Journal of Molecular Biology, 157*, 105–132.

Langley, J. N. (1905). On the reaction of cells and nerve-endings to certain poisons, chiefly as regards the reaction of striated muscle to nicotine and to curari. *Journal of Physiology, 33*, 374–413.

Langley, J. N., & Anderson, H. K. (1892). The action of Nicotin on the Ciliary Ganglion and on the endings of the Third Cranial Nerve. *Journal of Physiology, 13*(5), 460–468.

Leonard, S., Gault, J., Hopkins, J., Logel, J., Vianzon, R., Short, M., ... Freedman, R. (2002). Association of promoter variants in the alpha7 nicotinic acetylcholine receptor subunit gene with an inhibitory deficit found in schizophrenia. *Archives of General Psychiatry, 59*(12), 1085–1096.

Leonard, S., Mexal, S., & Freedman, R. (2007). Smoking, genetics and schizophrenia: Evidence for self medication. *Journal of Dual Diagnosis, 3*(3–4), 43–59.

Lester, H. A., Xiao, C., Srinivasan, R., Son, C. D., Miwa, J., Pantoja, R., ... Wang, J. C. (2009). Nicotine is a selective pharmacological chaperone of acetylcholine receptor number and stoichiometry. Implications for drug discovery. *The AAPS Journal, 11*(1), 167–177.

Levin, E. D., McClernon, F. J., & Rezvani, A. H. (2006). Nicotinic effects on cognitive function: Behavioral characterization, pharmacological specification, and anatomic localization. *Psychopharmacology (Berlin), 184*(3–4), 523–539.

Li, S. X., Huang, S., Bren, N., Noridomi, K., Dellisanti, C. D., Sine, S. M., & Chen, L. (2011). Ligand-binding domain of an alpha7-nicotinic receptor chimera and its complex with agonist. *Nature Neuroscience, 14*(10), 1253–1259.

Li, X., Li, W., Liu, G., Shen, X., & Tang, Y. (2015). Association between cigarette smoking and Parkinson's disease: A meta-analysis. *Archives of Gerontology and Geriatrics, 61*(3), 510–516.

Liebeskind, B. J., Hillis, D. M., & Zakon, H. H. (2015). Convergence of ion channel genome content in early animal evolution. *Proceedings of the National Academy of Sciences of the United States of America, 112*(8), E846–851.

Lindsley, C. W. (2010). (S)-(+)-mecamylamine (TC-5214): A neuronal nicotinic receptor modulator enters Phase III trials as an adjunct treatment for major depressive disorder (MDD). *ACS Chemical Neuroscience, 1*(8), 530–531.

Liu, Q., Huang, Y., Shen, J., Steffensen, S., & Wu, J. (2012). Functional alpha7beta2 nicotinic acetylcholine receptors expressed in hippocampal interneurons exhibit high sensitivity to pathological level of amyloid beta peptides. *BMC Neuroscience, 13*, 155.

Loewi, O. (1921). Über humorale Übertragbarkeit der Herznervenwirkung. I. *Pflügers Archive, 189*, 239–242.

Loewi, O. (1957). On the background of the discovery of neurochemical transmission. *Journal of Mt. Sinai Hospital, New York, 24*(6), 1014–1016.

Luetje, C. W., & Patrick, J. (1991). Both a- and b-subunits contribute to the agonist sensitivity of neuronal nicotinic acetylcholine receptors. *Journal of Neuroscience, 11*(3), 837–845.

Margiotta, J. F., Berg, D. K., & Dionne, V. E. (1987). The properties and regulation of functional acetylcholine receptors on chick ciliary ganglion neurons. *Journal of Neuroscience, 7*, 3612–3622.

Martin, E. J., Panikar, K. S., King, M. A., Deyrup, M., Hunter, B., Wang, G., & Meyer, E. (1994). Cytoprotective actions of 2,4-dimethoxybenzylidene anabaseine in differentiated PC12 cells and septal cholinergic cells. *Drug Development Research, 31*, 134–141.

McClure-Begley, T. D., Stone, K. L., Marks, M. J., Grady, S. R., Colangelo, C. M., Lindstrom, J. M., & Picciotto, M. R. (2013). Exploring the nicotinic acetylcholine receptor-associated proteome with iTRAQ and transgenic mice. *Genomics Proteomics Bioinformatics, 11*(4), 207–218.

McGehee, D. S., Heath, M. J. S., Gelber, S., Devay, P., & Role, L. W. (1995). Nicotine enhancement of fast excitatory synaptic transmission in CNS by presynaptic receptors. *Science, 269*, 1692–1696.

McQueen, E. G., & Smirk, F. H. (1957). Use of mecamylamine in the management of hypertension. *British Medical Journal, 1*(5016), 422–425.

Merlie, J. P., Isenberg, K. E., Russell, S. D., & Sanes, J. R. (1984). Denervation supersensitivity in skeletal muscle: Analysis with a cloned cDNA probe. *Journal of Cell Biology, 99*, 332–335.

Methfessel, C., Witzemann, V., Takahashi, T., Mishina, M., Numa, S., & Sakmann, B. (1986). Patch clamp measurements on *Xenopus laevis* oocytes: Currents through endogenous channels and implanted acetylcholine receptor and sodium channels. *Pflüegers Archiv: European Journal of Physiology, 407*, 577–588.

Meyer, E., deFiebre, C. M., Hunter, B., Simpkins, C. E., Frauworth, N., & deFiebre, N. C. (1994). Effects of anabaseine-related analogs on rat brain nicotinic receptor binding and on avoidance behaviors. *Drug Development Research, 31*, 127–134.

Mihalak, K. B., Carroll, F. I., & Luetje, C. W. (2006). Varenicline is a partial agonist at alpha4beta2 and a full agonist at alpha7 neuronal nicotinic receptors. *Molecular Pharmacology, 70*(3), 801–805.

Miledi, R., Parker, I., & Sumikawa, K. (1982a). Properties of acetylcholine receptors translated by cat muscle mRNA in *Xenopus* oocytes. *EMBO Journal, 1*(11), 1307–1312.

Miledi, R., Parker, I., & Sumikawa, K. (1982b). Synthesis of chick brain GABA receptors by frog oocytes. *Proceedings of the Royal Society of London: Series B, 216*, 509–515.

Millar, N. S., & Denholm, I. (2007). Nicotinic acetylcholine receptors: Targets for commercially important insecticides. *Invertebrate Neuroscience, 7*(1), 53–66.

Miller, C. (1989). Genetic manipulation of ion channels: A new approach to structure and mechanism. *Neuron, 2*(3), 1195–1205.

Mineur, Y. S., Eibl, C., Young, G., Kochevar, C., Papke, R. L., Gundisch, D., & Picciotto, M. R. (2009). Cytisine-based nicotinic partial agonists as novel antidepressant compounds. *Journal of Pharmacology and Experimental Therapeutics, 329*(1), 377–386.

Mineur, Y. S., & Picciotto, M. R. (2010). Nicotine receptors and depression: Revisiting and revising the cholinergic hypothesis. *Trends in Pharmacological Sciences, 31*(12), 580–586.

Mishina, M., Takai, T., Imoto, K., Noda, M., Takahashi, T., Numa, S., . . . Sakmann, B. (1986). Molecular distinction between fetal and adult forms of muscle acetylcholine receptor. *Nature, 321*, 406–411.

Miwa, J. M., Ibanez-Tallon, I., Crabtree, G. W., Sanchez, R., Sali, A., Role, L. W., & Heintz, N. (1999). Lynx1, an endogenous toxin-like modulator of nicotinic acetylcholine receptors in the mammalian CNS. *Neuron, 23*(1), 105–114.

Miyazawa, A., Fujiyoshi, Y., Stowell, M., & Unwin, N. (1999). Nicotinic acetylcholine receptor at 4.6 A resolution: Transverse tunnels in the channel wall. *Journal of Molecular Biology, 288*(4), 765–786.

Miyazawa, A., Fujiyoshi, Y., & Unwin, N. (2003). Structure and gating mechanism of the acetylcholine receptor pore. *Nature, 423*(6943), 949–955.

Morales-Perez, C. L., Noviello, C. M., & Hibbs, R. E. (2016). X-ray structure of the human alpha4beta2 nicotinic receptor. *Nature, 538*(7625), 411–415.

Neher, E., & Sakmann, B. (1976a). Noise analysis of drug induced voltage clamp currents in denervated frog muscle fibers. *Journal of Physiology, 258*, 705–729.

Neher, E., & Sakmann, B. (1976b). Single-channel currents recorded from membrane of denervated frog muscle fibres. *Nature, 260*(5554), 799–802.

Neher, E., & Steinbach, J. H. (1978). Local anaesthetics transiently block current through single acetylcholine receptor channels. *Journal of Physiology, 277*, 135–176.

Noda, M., Furutani, Y., Takahashi, H., Toyosato, M., Tanabe, T., Shimizu, S., . . . Numa, S. (1983). Cloning and sequence analysis of calf cDNA and human genomic DNA encoding alpha-subunit precursor of muscle acetylcholine receptor subunits. *Nature, 302*, 818–823.

Noda, M., Takahashi, H., Tanabe, T., Toyosato, M., Furutani, Y., Hirose, T., . . . Numa, S. (1982). Primary structure of alpha-subunit precursor of *Torpedo californica* acetylcholine receptor deduced from cDNA sequence. *Nature, 299*, 793–797.

Ortiz-Miranda, S., Lasalde, J., Pappone, P., & McNamee, M. (1997). Mutations in the M4 domain of the *Torpedo californica* nicotinic acetylcholine receptor alter channel opening and closing. *Journal of Membrane Biology, 158*(1), 17–30.

Oswald, R. E., & Changeux, J.-P. (1982). Crosslinking of alpha-bungarotoxin to the acetylcholine receptor from *Torpedo marmorata* by ultraviolet irradiation. *FEBS Letters, 139*, 225–229.

Papke, R. L. (1993). The kinetic properties of neuronal nicotinic receptors: Genetic basis of functional diversity. *Progress in Neurobiology, 41*, 509–531.

Papke, R. L. (2010). Tricks of perspective: Insights and limitations to the study of macroscopic currents for the analysis of nAChR activation and desensitization. *Journal of Molecular Neuroscience, 40*(1–2), 77–86.

Papke, R. L. (2014). Merging old and new perspectives on nicotinic acetylcholine receptors. *Biochemical Pharmacology, 89*(1), 1–11.

Papke, R. L., Bagdas, D., Kulkarni, A. R., Gould, T., AlSharari, S., Thakur, G. A., & Damaj, I. M. (2015). The analgesic-like properties of the alpha7 nAChR silent agonist NS6740 is associated with nonconducting conformations of the receptor. *Neuropharmacology, 91*, 34–42.

Papke, R. L., Bencherif, M., & Lippiello, P. (1996). An evaluation of neuronal nicotinic acetylcholine receptor activation by quaternary nitrogen compounds indicates that choline is selective for the a7 subtype. *Neuroscience Letters, 213*, 201–204.

Papke, R. L., Boulter, J., Patrick, J., & Heinemann, S. (1989). Single-channel currents of rat neuronal nicotinic acetylcholine receptors expressed in *Xenopus laevis* oocytes. *Neuron, 3*, 589–596.

Papke, R. L., Dwoskin, L. P., & Crooks, P. A. (2007). The pharmacological activity of nicotine and nornicotine on nAChRs subtypes: Relevance to nicotine dependence and drug discovery. *Journal of Neurochemistry, 101*(1), 160–167.

Papke, R. L., Dwoskin, L. P., Crooks, P. A., Zheng, G., Zhang, Z., McIntosh, J. M., & Stokes, C. (2008). Extending the analysis of nicotinic receptor antagonists with the study of alpha6 nicotinic receptor subunit chimeras. *Neuropharmacology, 54*(8), 1189–1200.

Papke, R. L., & Heinemann, S. F. (1991). The role of the b4 subunit in determining the kinetic properties of rat neuronal nicotinic acetylcholine a3-receptor. *Journal of Physiology (London), 440*, 95–112.

Papke, R. L., & Heinemann, S. F. (1994). The partial agonist properties of cytisine on neuronal nicotinic receptors containing the beta2 subunit. *Molecular Pharmacology, 45*, 142–149.

Papke, R. L., Horenstein, B. A., & Placzek, A. N. (2001). Inhibition of wild-type and mutant neuronal nicotinic acetylcholine receptors by local anesthetics. *Molecular Pharmacology, 60*(6), 1–10.

Papke, R. L., Horenstein, N. A., Kulkarni, A. R., Stokes, C., Corrie, L. W., Maeng, C. Y., & Thakur, G. A. (2014). The activity of GAT107, an allosteric activator and positive modulator of alpha7 nicotinic acetylcholine receptors (nAChR), is regulated by aromatic amino acids that span the subunit interface. *Journal of Biological Chemistry, 289*(7), 4515–4531.

Papke, R. L., Millhauser, G., Lieberman, Z., & Oswald, R. E. (1988). Relationships of agonist properties to the single channel kinetics of nicotinic acetylcholine receptors. *Biophysical Journal, 53*, 1–10.

Papke, R. L., & Smith-Maxwell, C. (2009). High throughput electrophysiology with *Xenopus* oocytes. *Combinatorial Chemistry & High Throughput Screening, 12*(1), 38–50.

Papke, R. L., Stokes, C., Damaj, M. I., Thakur, G. A., Manther, K., Treinin, M., . . . Horenstein, N. A. (2017). Persistent activation of alpha7 nicotinic ACh receptors associated with stable induction of different desensitized states. *British Journal of Pharmacology*, in press. doi: 10.1111/bph.13851. [Epub ahead of print]

Papke, R. L., Trocme-Thibierge, C., Guendisch, D., Abbas Al Rubaiy, S. A., & Bloom, S. A. (2011). Electrophysiological perspectives on the therapeutic use of nicotinic acetylcholine receptor partial agonists. *Journal of Pharmacology and Experimental Therapeutics, 337*(2), 367–379.

Papke, R. L., Wecker, L., & Stitzel, J. A. (2010). Activation and inhibition of mouse muscle and neuronal nicotinic acetylcholine receptors expressed in *Xenopus* oocytes. *Journal of Pharmacology and Experimental Therapeutics, 333*(2), 501–518.

Patrick, J., Boulter, J., Deneris, E., Wada, K., Wada, E., Connolly, J., . . . Heinemann, S. (1989). Structure and function of neuronal nicotinic acetylcholine receptors deduced from cDNA clones. *Progress in Brain Research. 79*, 27–33.

Patrick, J. W., Neff, S., Dineley, K., & Char, D. (1996). Immunological approaches to nicotine receptors. *NIDA—Research Monographs, 161*, 3–19.

Paulo, J., Brucker, W., & Hawrot, E. (2009). Proteomic analysis of an alpha7 nicotinic acetylcholine receptor interactome. *Journal of Proteome Research, 8*(4), 1849–1858.

Perry, E. K., Martin-Ruiz, C. M., & Court, J. A. (2001). Nicotinic receptor subtypes in human brain related to aging and dementia. *Alcohol, 24*(2), 63–68.

Picciotto, M. R., Brunzell, D. H., & Caldarone, B. J. (2002). Effect of nicotine and nicotinic receptors on anxiety and depression. *Neuroreport, 13*(9), 1097–1106.

Picciotto, M. R., & Kenny, P. J. (2013). Molecular mechanisms underlying behaviors related to nicotine addiction. *Cold Spring Harbor Perspectives in Medicine, 3*(1), a012112.

Picciotto, M. R., & Zoli, M. (2008). Neuroprotection via nAChRs: The role of nAChRs in neurodegenerative disorders such as Alzheimer's and Parkinson's disease. *Frontiers in Bioscience, 13*, 492–504.

Pieschl, R. L., Miller, R., Jones, K. M., Post-Munson, D. J., Chen, P., Newberry, K., ... Li, Y. W. (2017). Effects of BMS-902483, an alpha7 nicotinic acetylcholine receptor partial agonist, on cognition and sensory gating in relation to receptor occupancy in rodents. *European Journal of Pharmacology, 807*, 1–11.

Porter, S., & Froehner, S. (1985). Interaction of the 43K protein with components of *Torpedo* postsynaptic membranes. *Biochemistry, 24*, 425–432.

Ramarao, M. K., & Cohen, J. B. (1998). Mechanism of nicotinic acetylcholine receptor cluster formation by rapsyn. *Proceedings of the National Academy of Sciences of the United States of America, 95*(7), 4007–4012.

Reavill, C., Jenner, P., Kumar, R., & Stolerman, I. P. (1988). High affinity binding of [3H] (-)-nicotine to rat brain membranes and its inhibition by analogues of nicotine. *Neuropharmacology, 27*(3), 235–241.

Revah, F., Bertrand, D., Galzi, J.-L., Devillers-Thiery, A., Mulle, C., Hussy, N., ... Changeux, J.-P. (1991). Mutations in the channel domain alter desensitization of a neuronal nicotinic receptor. *Nature, 353*, 846–849.

Reynolds, J. A., & Karlin, A. (1978). Molecular weight in detergent solution of acetylcholine receptor from *Torpedo californica*. *Biochemistry, 17*(11), 2035–2038.

Rosas-Ballina, M., & Tracey, K. J. (2009). Cholinergic control of inflammation. *Journal of Internal Medicine, 265*(6), 663–679.

Rose, J. E., Behm, F. M., Westman, E. C., Levin, E. D., Stein, R. M., & Ripka, G. V. (1994). Mecamylamine combined with nicotine skin patch facilitates smoking cessation beyond nicotine patch treatment alone. *Clinical Pharmacology and Therapeutics, 56*(1), 86–99.

Rose, J. E., Levin, E. D., Behm, F. M., Adivi, C., & Schur, C. (1990). Transdermal nicotine facilitates smoking cessation. *Clinical Pharmacology and Therapeutics, 47*(3), 323–330.

Rucktooa, P., Smit, A. B., & Sixma, T. K. (2009). Insight in nAChR subtype selectivity from AChBP crystal structures. *Biochemical Pharmacology, 78*(7), 777–787.

Sala, F., Nistri, A., & Criado, M. (2008). Nicotinic acetylcholine receptors of adrenal chromaffin cells. *Acta Physiologica (Oxford), 192*(2), 203–212.

Salpeter, M. M. (1987). *The vertebrate neuromuscular junction*. New York: Alan R. Liss.

Sanberg, P. R., Shytle, R. D., & Silver, A. A. (1998). Treatment of Tourette's syndrome with mecamylamine. *Lancet, 352*, 705–706.

Seguela, P., Wadiche, J., Dinely-Miller, K., Dani, J. A., & Patrick, J. W. (1993). Molecular cloning, functional properties and distribution of rat brain alpha 7: A nicotinic cation channel highly permeable to calcium. *Journal of Neuroscience, 13*(2), 596–604.

Sigworth, F. J., & Sine, S. M. (1987). Data transformations for improved display and fitting of single-channel dwell time histograms. *Biophysical Journal, 52*(6), 1047–1054.

Silver, A. A., Shytle, R. D., Sheehan, K. H., Sheehan, D. V., Ramos, A., & Sanberg, P. R. (2001). Multicenter, double-blind, placebo-controlled study of mecamylamine monotherapy for Tourette's disorder. *Journal of the American Academy of Child and Adolescent Psychiatry, 40*(9), 1103–1110.

Sine, S., & Taylor, P. (1980). The relationship between agonist occupation and the permeability response of the cholinergic receptor revealed by bound cobra alpha-toxin. *Journal of Biological Chemistry, 255*, 10144–10156.

Sine, S. M., & Steinbach, J. H. (1984a). Activation of a nicotinic acetylcholine receptor. *Biophysical Journal, 45*, 175–185.

Sine, S. M., & Steinbach, J. H. (1984b). Agonists block currents through acetylcholine receptor channels. *Biophysical Journal, 46*(2), 277–283.

Sine, S. M., & Steinbach, J. H. (1987). Activation of acetylcholine receptors on clonal mammalian BC3H-1 cells by high concentrations of agonist. *Journal of Physiology, 385*, 325–359.

Slemmer, J. E., Martin, B. R., & Damaj, M. I. (2000). Bupropion is a nicotinic antagonist. *Journal of Pharmacology and Experimental Therapeutics, 295*(1), 321–327.

Smit, A. B., Syed, N. I., Schaap, D., van Minnen, J., Klumperman, J., Kits, K. S., . . . Geraerts, W. P. (2001). A glia-derived acetylcholine-binding protein that modulates synaptic transmission. *Nature, 411*(6835), 261–268.

Srinivasan, R., Pantoja, R., Moss, F. J., Mackey, E. D., Son, C. D., Miwa, J., & Lester, H. A. (2011). Nicotine up-regulates alpha4beta2 nicotinic receptors and ER exit sites via stoichiometry-dependent chaperoning. *Journal of General Physiology, 137*(1), 59–79.

Stauderman, K. A., Mahaffy, L. S., Akong, M., Veliçelebi, G., Chavez-Noriega, L. E., Crona, J. H., . . . Corey-Naeve, J. (1998). Characterization of human recombinant neuronal nicotinic acetylcholine receptor subunit combinations alpha2beta4, alpha3beta4 and alpha4beta4 stably expressed in HEK293 cells. *Journal of Pharmacology and Experimental Therapeutics, 284*(2), 777–789.

Stokes, C., Treinin, M., & Papke, R. L. (2015). Looking below the surface of nicotinic acetylcholine receptors. *Trends in Pharmacological Sciences, 36*(8), 514–523.

Stroud, R. M., McCarthy, M. P., & Shuster, M. (1990). Nicotinic acetylcholine receptor superfamily of ligand-gated ion channels. *Biochemistry, 29*, 11009–11023.

Takai, T., Noda, M., Mishina, M., Shimizu, S., Furutani, Y., Kayano, T., . . . Numa, S. (1985). Cloning sequencing and expression of cDNA for a novel subunit of acetylcholine receptor from calf muscle. *Nature, 315*, 761–764.

Tapia, L., Kuryatov, A., & Lindstrom, J. (2007). Ca2+ permeability of the (alpha4)3(beta2)2 stoichiometry greatly exceeds that of (alpha4)2(beta2)3 human acetylcholine receptors. *Molecular Pharmacology, 71*(3), 769–776.

Taylor, P. (1990). Neuromuscular junction and autonomic ganglia. In A. G. Gilman, T. W. Rall, A. S. Nies, & P. Taylor (Eds.), *The pharmacological basis for therapeutics* (8th ed., pp. 166–186). Elmsford, NY: Pergamon Press.

Terry, A. V., Jr. (2008). Role of the central cholinergic system in the therapeutics of schizophrenia. *Current Neuropharmacology, 6*(3), 286–292.

Tomassoli, I., Eibl, C., Wulf, M., Papke, R. L., Picciotto, M. R., & Gündisch, D. (2011). The twin drug approach for novel nicotinic acetylcholine receptor (nAChR) ligands: Synthesis and structure-affinity relationships. *Biochemical Pharmacology, 82*, 1023.

Treinin, M. (2008). RIC-3 and nicotinic acetylcholine receptors: Biogenesis, properties, and diversity. *Biotechnology Journal, 3*(12), 1539–1547.

Tsetlin, V. I. (2015). Three-finger snake neurotoxins and Ly6 proteins targeting nicotinic acetylcholine receptors: Pharmacological tools and endogenous modulators. *Trends in Pharmacological Sciences, 36*(2), 109–123.

Tzartos, S. J., Kokla, A., Walgrave, S. L., & Conti-Tronconi, B. M. (1988). Localization of the main immunogenic region of human muscle acetylcholine receptor to residues 67–76 of the alpha subunit. *Proceedings of the National Academy of Sciences of the United States of America, 85*, 2899–2903.

Unwin, M., Toyoshima, C., & Kubalek, E. (1988). Arrangement of the acetylcholine receptor subunits in the resting and desensitized states determined by cryoelectron microscopy of crystallized *Torpedo* postsynaptic membranes. *Journal of Cell Biology, 107*, 1123–1138.

Unwin, N. (1993). The nicotinic acetylcholine receptor at 9A resolution. *Journal of Molecular Biology, 229*(4), 1101–1124.

Unwin, N. (2005). Refined structure of the nicotinic acetylcholine receptor at 4A resolution. *Journal of Molecular Biology, 346*(4), 967–989.

Unwin, N., & Fujiyoshi, Y. (2012). Gating movement of acetylcholine receptor caught by plunge-freezing. *Journal of Molecular Biology, 422*(5), 617–634.

Valbuena, S., & Lerma, J. (2016). Non-canonical signaling, the hidden life of ligand-gated ion channels. *Neuron, 92*(2), 316–329.

Van Kampen, M., Selbach, K., Schneider, R., Schiegel, E., Boess, F., & Schreiber, R. (2004). AR-R 17779 improves social recognition in rats by activation of nicotinic alpha7 receptors. *Psychopharmacology (Berlin), 172*(4), 375–383.

Van Maanen, M. A., Lebre, M. C., van der Poll, T., LaRosa, G. J., Elbaum, D., Vervoordeldonk, M. J., & Tak, P. P. (2009). Stimulation of nicotinic acetylcholine receptors attenuates collagen-induced arthritis in mice. *Arthritis and Rheumatism, 60*(1), 114–122.

Wada, E., Wada, K., Boulter, J., Deneris, E., Heinemann, S., Patrick, J., & Swanson, L. W. (1989). Distribution of alpha2, alpha3, alpha4, and beta2 neuronal nicotinic receptor subunit mRNAs in the central nervous system: A hybridization histochemical study in the rat. *Journal of Comparative Neurology, 284*, 314–335.

Walling, D., Marder, S. R., Kane, J., Fleischhacker, W. W., Keefe, R. S., Hosford, D. A., . . . Dunbar, G. C. (2016). Phase 2 trial of an alpha-7 nicotinic receptor agonist (TC-5619) in negative and cognitive symptoms of schizophrenia. *Schizophrenia Bulletin, 42*(2), 335–343.

Wang, H., Yu, M., Ochani, M., Amella, C. A., Tanovic, M., Susarla, S., . . . Tracey, K. J. (2003). Nicotinic acetylcholine receptor alpha7 subunit is an essential regulator of inflammation. *Nature, 421*(6921), 384–388.

Wang, J., Kuryatov, A., Sriram, A., Jin, Z., Kamenecka, T. M., Kenny, P. J., & Lindstrom, J. (2015). An accessory agonist binding site promotes activation of alpha4beta2* nicotinic acetylcholine receptors. *Journal of Biological Chemistry, 290*(22), 13907–13918.

Wang, J., & Lindstrom, J. (2017). Orthosteric and allosteric potentiation of heteromeric neuronal nicotinic acetylcholine receptors. *British Journal of Pharmacology*, in press. doi: 10.1111/bph.13745. [Epub ahead of print]

Wang, Y., Lee, J. W., Oh, G., Grady, S. R., McIntosh, J. M., Brunzell, D. H., . . . Drenan, R. M. (2014). Enhanced synthesis and release of dopamine in transgenic mice with gain-of-function alpha6* nAChRs. *Journal of Neurochemistry, 129*(2), 315–327.

Weber, M., & Changeux, J.-P. (1974). Binding of *Naja nigricollis* [3H]alpha-toxin to membrane fragments from *Electrophorus* and *Torpedo* electric organs. I. Binding of the tritiated alpha-neurotoxin in the absence of effector. *Molecular Pharmacology, 10*, 1–14.

Webster, J. C., Francis, M. M., Porter, J. K., Robinson, G., Stokes, C., Horenstein, B., & Papke, R. L. (1999). Antagonist activities of mecamylamine and nicotine show reciprocal dependence on beta subunit sequence in the second transmembrane domain. *British Journal of Pharmacology, 127*, 1337–1348.

Whiting, P., Esch, F., Shimasaki, S., & Lindstrom, J. (1987). Neuronal nicotinic acetylcholine receptor b-subunit is coded for by the cDNA clone a_4. *FEBS Letters, 219*, 459–463.

Whiting, P. J., & Lindstrom, J. M. (1986). Purification and characterization of a nicotinic acetylcholine receptor from chick brain. *Biochemistry, 25*, 2082–2093.

Williams, D. K., Peng, C., Kimbrell, M. R., & Papke, R. L. (2012). The intrinsically low open probability of alpha7 nAChR can be overcome by positive allosteric modulation and serum factors leading to the generation of excitotoxic currents at physiological temperatures. *Molecular Pharmacology, 82*(4), 746–759.

Williams, D. K., Stokes, C., Horenstein, N. A., & Papke, R. L. (2011). The effective opening of nicotinic acetylcholine receptors with single agonist binding sites. *Journal of General Physiology, 137*(4), 369–384.

Williams, D. K., Wang, J., & Papke, R. L. (2011a). Investigation of the molecular mechanism of the alpha7 nAChR positive allosteric modulator PNU-120596 provides evidence for two distinct desensitized states. *Molecular Pharmacology, 80*(6), 1013–1032.

Williams, D. K., Wang, J., & Papke, R. L. (2011b). Positive allosteric modulators as an approach to nicotinic acetylcholine receptor-targeted therapeutics: Advantages and limitations. *Biochemical Pharmacology, 82*(8), 915–930.

Williams, M. E., Burton, B., Urrutia, A., Shcherbatko, A., Chavez-Noriega, L. E., Cohen, C. J., & Aiyar, J. (2005). RIC-3 promotes functional expression of the nicotinic acetylcholine receptor alpha7 subunit in mammalian cells. *Journal of Biological Chemistry, 280*(2), 1257–1263.

Wo, Z. G., & Oswald, R. E. (1995). Unraveling the modular design of glutamate-gated ion channels. *Trends in Neuroscience, 18*(4), 161–168.

Wonnacott, S. (1986). Alpha-bungarotoxin binds to low-affinity nicotine binding sites in rat brain. *Journal of Neurochemistry, 47*(6), 1706–1712.

Wonnacott, S. (1997). Presynaptic nicotinic ACh receptors. *Trends in Neuroscience, 20*(2), 92–98.

Wonnacott, S., Irons, J., Rapier, C., Thorne, B., & Lunt, G. G. (1989). Presynaptic modulation of transmitter release by nicotinic receptors. *Progress in Brain Research, 79*, 157–163.

Xiu, X., Puskar, N. L., Shanata, J. A., Lester, H. A., & Dougherty, D. A. (2009). Nicotine binding to brain receptors requires a strong cation-pi interaction. *Nature, 458*(7237), 534–537.

CHAPTER 12

$GABA_A$ RECEPTOR PHYSIOLOGY AND PHARMACOLOGY

MARTIN WALLNER, A. KERSTIN LINDEMEYER, AND RICHARD W. OLSEN

NONSTANDARD ABBREVIATIONS

GABA:	γ-aminobutyric acid;
$GABA_AR$:	$GABA_A$ receptor(s);
δ-$GABA_AR$:	δ subunit-containing $GABA_AR$;
[GABA]:	GABA concentration;
EtOH:	ethanol or alcohol;
PAM/NAM:	positive/negative allosteric modulator;
TM1/2/3/4:	transmembrane helix #1/2/3/4,
TMD:	transmembrane domain;
ECD:	extracellular domain;
CNS:	central nervous system;
NGFC:	neurogliaform cell;
BZ:	benzodiazepine;
DZ:	diazepam;
THIP:	4,5,6,7-tetrahydroisoxazolo(5,4-c)pyridin-3-ol, also known as gaboxadol;
IPSC:	inhibitory postsynaptic current

Introduction

γ-Aminobutyric acid (GABA) is the major inhibitory neurotransmitter in the vertebrate central nervous system, and it has a similar role in invertebrates or even simpler organisms. Neurotransmission mediated by GABA involves action on postsynaptic and extrasynaptic receptors of two main categories: $GABA_A$ and $GABA_B$ (Olsen & Li, 2012). $GABA_B$ receptors are G protein-coupled receptors in which the ligand response is mediated by a repertoire of slow cascades of events affecting the physiology of the postsynaptic cell, with details depending on the G protein partner. $GABA_A$ receptors ($GABA_A$Rs) are the subject of this chapter. They are heteromeric chloride channels mediating rapid inhibitory neurotransmission.

$GABA_A$Rs mediate rapid physiological inhibition by increasing membrane conductance with an equilibrium potential (E_{GABA}) near the resting level of ~ −70 mV. This conductance increase is due to Cl^- ion flux through the channel formed by the $GABA_A$R protein, opened by a conformational change favored by binding the neurotransmitter, GABA (Martin & Olsen, 2000; Roberts, Chase, & Tower, 1976), and usually is accompanied by hyperpolarization, increasing the firing threshold and reducing action potential initiation; i.e., neuronal inhibition. Reduction in excitatory potentials also can result from shunting Cl^- currents carried by $GABA_A$R channels, with or without hyperpolarization. Some $GABA_A$Rs are located on nerve terminals, excitatory or inhibitory, even GABAergic ones, and their activation (generally via GABAergic innervation) inhibits neurotransmitter release. On the other hand, increased Cl^- permeability could depolarize the target cell if the Cl^- gradient has higher intracellular Cl^-. Whether this can excite the cell to fire and/or activate Ca^{2+} entry is not clear (Zilberter, 2016), although proposed as physiologically relevant, especially in embryonic neurons (Ben-Ari, Gaiarsa, Tyzio, & Khazipov, 2007), suggesting that E_{GABA} is an important physiological entity dependent on the expression of various Cl^- pumps in the target cell, and their expression in various normal and disease states, such as epilepsy (Olsen & Li, 2012; Olsen & Spigelman, 2012). The $GABA_A$R protein has two binding sites for GABA, and the channel opening rate, closing rate, and desensitization in the continued presence of GABA have been described for many subtypes of $GABA_A$Rs; e.g., (Macdonald & Olsen, 1994).

$GABA_A$Rs are members of the Cys-loop pentameric ligand-gated ion channel superfamily, which includes the nicotinic acetylcholine receptors, glycine receptors, and 5-HT3 receptors (Galzi & Changeux, 1994). They comprise a family of over 20 $GABA_A$R subunit combinations called *subtypes*, constructed from a family of 19 subunit genes divided into eight classes according to sequence homology: α1-6, β1-3, γ1-3, δ, ε, θ, π, and ρ1-3 (Olsen & Sieghart, 2008). Most of the $GABA_A$R subunit genes are found organized in clusters on human chromosomes 4 (α4, β1, α2, γ1), 5 (α6, β2, α1, γ2), 15 (β3, α5, γ3), and X (θ, α3, ε) (McLean, Farb, & Russek, 1995; Russek, 1999; Simon, Wakimoto, Fujita, Lalande, & Barnard, 2004). In the forebrain, typically, 2 α and 2 β are present with either one γ or δ subunit in one pentamer. According to the subunit members present, distinct GABA$_A$R subtypes have different pharmacology, channel kinetic properties, and topography, with resulting functional and behavioral consequences (Olsen & Sieghart, 2008; Puia, Vicini, Seeburg, & Costa, 1991; Rudolph & Knoflach, 2011; Whiting et al., 1999). Because of the widespread presence of the GABA$_A$R subtypes, they are involved in virtually all CNS functions, with subtype dependence based on the neurocircuitry involved. This leads to an involvement in myriad neurological and psychiatric disorders and corresponding rich pharmacology. GABA$_A$Rs have been favorite research targets for years, first when discovered to play a role in the mechanisms of action of many useful neurotherapeutic agents, and then as the targets for new and better such agents.

This chapter gives a brief summary of the more important clinical agents acting as positive or negative allosteric modulators (PAMs and NAMs) of GABA$_A$Rs, examples of some more esoteric agents, giving some novel or curious insight into neuropharmacology, and an emphasis on agents studied in our lab. We also include an attempt to note interesting recently discovered drugs of interest. We apologize to those workers not mentioned: the subject is just too extensive for any review to be all-inclusive, and we must necessarily pick and choose selected drugs.

The emphasis of early work in the Olsen lab (Olsen, 1982; Olsen, Fischer, & Dunwiddie, 1986) depended on the development, over 50 years ago, of the technique of radioligand receptor binding (Yamamura, Enna, & Kuhar, 1978). Briefly, physicist Wally Gilbert reasoned that if the hypothetical *lac* repressor protein mediated regulation of expression of the multiple genes coded for by the *lac* operon cluster, as postulated by Jacob and Monod, who had already (with Lwoff) received the Nobel Prize for this insight, the only way to identify the repressor protein in a cell-free homogenate was to use its known property of binding the sugar lactose. When present in the medium, lactose stimulates gene expression of the *lac* operon genes, by releasing repression, present when lactose is not available (Jacob & Monod, 1961). Using equilibrium dialysis to quantitate binding of the radioactive analogue of lactose, [^{3}H]isopropyl-thio-β-galactoside, they demonstrated a lactose binding protein in *Salmonella* cells (as used by Jacob & Monod), present only when the cells had been grown in lactose medium, and eliminated by mutation of the *lac* repressor gene *i* (Gilbert & Muller-Hill, 1966); he then purified the lactose-binding protein coded for by the *lac* repressor gene, which also bound to the repressor site on the DNA at the initiation region of the *lac* operon.

This technique was rapidly applied to in vitro assays for receptors, transporters, and other binding proteins for hormones and neurotransmitters, as well as other ligands (e.g., Changeux, 1981; Cuatrecasas, 1974; Lefkowitz, 1975; Pert & Snyder, 1973). This allowed the identification of dozens of important receptors in cell-free assays of any chosen tissue homogenate, followed by characterization, counting, and structure-activity studies in addition to purification leading to biochemical/molecular analyses and cloning of the genes for the receptors. In the case of GABA$_A$R, the receptor was identified in mammalian brain homogenates using binding of radiolabeled [^{3}H] GABA itself (Zukin, Young, & Snyder, 1974). GABA$_A$Rs were (over time) also shown to possess binding sites for PAMs and NAMs, starting with the important neuropharmacological agents, the benzodiazepines (BZ), like diazepam (Valium®), as suggested by pharmacologists (Costa & Guidotti, 1979; Haefely, 1982). The binding sites were shown to mediate the drug actions; i.e., function as the drug's receptor (Braestrup & Squires, 1977; Möhler & Okada, 1977).

Our novel approach was to synthesize a radioligand analogue, [^{3}H]dihydropicrotoxin (DHP), of the universal GABA$_A$R channel blocker, the plant convulsant picrotoxinin (Olsen, 1982; Olsen, Ticku, & Miller, 1978; Ticku, Ban, & Olsen, 1978). This ligand bound to the "picrotoxin receptor" sites on the GABA$_A$R, providing another tool for in vitro studies, and leading to discovery of a wealth of ligands that competitively or allosterically modulated picrotoxin binding *if* their pharmacological effect on organisms, cells, and channels involved GABA$_A$Rs. Clearly, the agents binding to the picrotoxin site on GABA$_A$R had activity as PAMs or NAMs on GABA$_A$R function and pharmacology in vivo. We quickly showed that not only did structural analogues of picrotoxin inhibit [^{3}H] DHP binding, but also the cage convulsants synthesized by Casida (Casida et al., 1976) were found to bind to the picrotoxin site (Ticku & Olsen, 1979), and blocked GABA$_A$R currents (Bowery, Collins, & Hill, 1976); likewise, the barbiturates (allosterically) inhibited picrotoxin binding (Ticku & Olsen, 1978), as well as enhancing GABA$_A$R currents in neurons, and producing sedative and anticonvulsant efficacy in animals (Bowery & Dray, 1976; Nicoll, Eccles, Oshima, & Rubia, 1975; Ransom & Barker, 1975; Study & Barker, 1981).

This led to discovery that the barbiturates (Leeb-Lundberg, Snowman, & Olsen, 1980; Willow & Johnston, 1981) and related CNS depressants like etazolate (Leeb-Lundberg, Snowman, & Olsen, 1981; Supavilai, Mannonen, & Karobath, 1982) and neurosteroids enhance GABA$_A$R currents and modulate BZ binding to the BZ sites on GABA$_A$Rs, as well as the GABA binding site (Olsen & Snowman, 1982; Turner, Ransom, Yang, & Olsen, 1989; Willow & Johnston, 1981) and enhance the GABA enhancement of BZ binding to GABA$_A$Rs (Skolnick, Moncada, Barker, & Paul, 1981). We demonstrated for a large series of barbiturates and related compounds that the inhibition of picrotoxin binding in a chemically specific and stereospecific manner correlated with their potency as PAMs on GABA$_A$R currents and the potency as sedatives/hypnotics/anesthetics in organisms (Olsen et al., 1986). It eventually became clear that many general anesthetics mediate anesthetic efficacy, at least in part, by enhancing GABA$_A$R-mediated inhibition in the brain by binding directly to sites on the GABA$_A$R proteins that produce PAM efficacy and modulation of the various ligand binding sites on GABA$_A$Rs (Forman & Chin, 2008; Johnston, 1983; Mihic et al., 1997; Olsen, Sapp, Bureau, Turner, & Kokka, 1991; Peters, Kirkness, Callachan, Lambert, & Turner, 1988). In particular, the intravenous anesthetics (etomidate, propofol, barbiturates, and neurosteroid anesthetics (Harrison & Simmonds, 1984; Smith, 2003) act in a (stereo)chemically specific manner on GABA$_A$Rs, and also more potently than on other potential brain targets, including various ion channels (Olsen & Li, 2011; Olsen et al., 1991). There is also evidence that GABA$_A$Rs are important targets of volatile anesthetics and long-chain alcohols (Mihic et al., 1997) and mediate at least part of their anesthetic actions (Werner et al., 2011).

Specific binding sites for these anesthetics have been identified at subunit interfaces in the transmembrane domain (see Figure 12.1) of $GABA_AR$ proteins (Forman & Miller, 2016; Li et al., 2006; Olsen, 2015; Olsen, Chang, Li, Hanchar, & Wallner, 2004). Genetically engineered mice demonstrated that general anesthetic action not only requires $GABA_ARs$, but also requires specific subtypes of $GABA_ARs$ in specific brain locations (Jurd et al., 2003; Reynolds et al., 2003; Rudolph & Antkowiak, 2004); this appears to result from slight differences in the ligand binding sites in $GABA_AR$ subtypes (Chiara et al., 2013; reviewed in Olsen, 2015; Rudolph & Antkowiak, 2004). In addition, our lab has demonstrated that certain subtypes of $GABA_AR$, involved in tonic inhibitory currents via extrasynaptically localized membrane sites, are involved in low millimolar PAM action on $GABA_ARs$ by ethanol (EtOH) (Hanchar et al., 2006; Hanchar, Dodson, Olsen, Otis, & Wallner, 2005; Wallner, Hanchar, & Olsen, 2003; Wallner, Hanchar, & Olsen, 2006b). EtOH appears to bind to a modified GABA/BZ binding site in the extracellular domain of δ subunit-containing $GABA_ARs$ (Olsen et al., 2014; Wallner, Hanchar, & Olsen, 2014). A tentative structural model of the $GABA_ARs$ protein showing the binding sites for the major classes of ligands is shown in Figure 12.1, both the extracellular domain (ECD) sites for GABA, BZ, and EtOH, and the transmembrane domain (TMD) showing the sites for anesthetics and channel blockers (Olsen, 2015).

$GABA_A$ Receptor Pharmacology

$GABA_AR$ Binding Sites

$GABA_ARs$ exhibit multiple binding sites for a large variety of exogenous and endogenous ligands (Olsen, 2015; Sieghart, 2015). Ligand binding sites have been identified by radioligand binding, photolabeling studies, and mutagenesis, and the binding sites for BZs, GABA, and orthosteric ligands (muscimol and 4,5,6,7-tetrahydroisoxazolo(5,4-c)pyridin-3-ol [THIP]) have been confirmed using homology models of the extracellular domain of the acetylcholine binding protein (AChBP) (Brejc et al., 2001; Ernst, Brauchart, Boresch, & Sieghart, 2003; Ernst, Bruckner, Boresch, & Sieghart, 2005). Since then a number of structures of pentameric ligand-gated ion channels (pLGICs) that include membrane-spanning domains (TMD) have been solved. The first one was a low-resolution structure of the *Torpedo marmorata* nicotinic acetylcholine receptor (Unwin, 2005). This was soon followed by high-resolution structures of the prokaryotic *Erwinia chrysanthemi* pentameric ligand-gated ion channel (ELIC) (Hilf & Dutzler, 2008), the *Gloeobacter violaceus* pentameric ligand-gated ion channel homologue (GLIC) (Bocquet et al., 2009; Corringer et al., 2010), and the glutamate-gated anion-selective Cys-loop receptor GluCl found in *C. elegans* (Hibbs & Gouaux, 2011). Recently, a homopentameric human β3-$GABA_AR$ has been crystallized (Miller & Aricescu, 2014). After all, we are still lacking structural information about heteropentameric

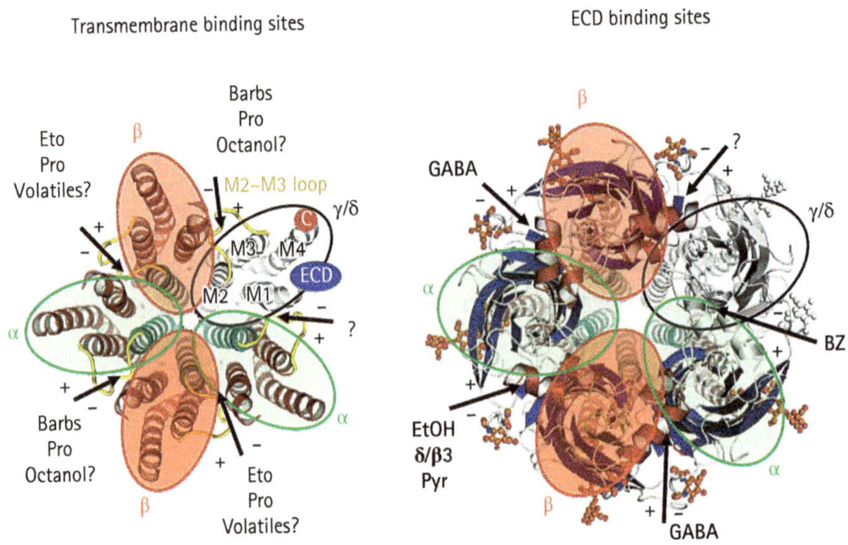

FIGURE 12.1 $GABA_AR$ ligand sites at subunit interfaces identified by mutagenesis and/or affinity labeling.

The left panel shows the transmembrane ligand binding sites, while the right panel shows the ECD binding sites. The protein structures are taken from the X-ray crystallography-derived structure of the recombinant mammalian homomeric β3-$GABA_AR$ (Miller & Aricescu, 2014), on which is displayed a homologous native $GABA_AR$ composed of an α_β_α_β_γ/δ heteropentamer (actual subunits arbitrary, no specific sequence implied, although they are all homologous to β3). The protein is viewed looking from the extracellular face, perpendicular to the cell membrane/synapse. Thus, the ECD pentamer on the right would actually be positioned directly on top of the TMD pentamer at left. Both portions indicate locations of ligand binding sites at subunit interfaces. The two α subunits are indicated by the green-shaded oval, the two β subunits by the pink-shaded ovals, and the one γ/δ subunit is indicated by the clear oval. An example C-terminus is indicated by a small red circled "C" at the bottom of the TMD of the γ/δ subunit; the TM1,2,3,4 domains are also labeled in this example subunit, and the N-terminus of the TMD of each subunit would attach to its ECD at the position indicated by the small blue oval "ECD." Ligand binding sites for the compounds listed (in shorthand) are indicated by arrows. Note that the heteropentameric proteins show several different but homologous subunit interfaces, so the pharmacological specificity varies with $GABA_AR$ subtype. The ligands named are BZ (benzodiazepines) and GABA, EtOH, and Pyr (pyrazoloquinolines) in the ECD. In the TMD, Eto (etomidate), Pro (propofol), octanol, volatiles, and barbs (barbiturates) binding sites are located.

$GABA_AR$s with ligands (BZ, anesthetics) docked into binding sites at extracellular and transmembrane subunit interfaces (see Figure 12.1). A study on homology models identified further putative binding sites for $GABA_AR$ allosteric ligands using multiple structure templates in addition to the structure of the β3 homopentamer (Puthenkalam et al., 2016). Recently, a drug-screening method on $GABA_AR$s has been developed that attaches a fluorophore in distinct $GABA_AR$ binding sites, using ligand-directed acyl imidazole chemistry. This method seems to be suitable for high-throughput screening, to be more sensitive than traditional radioligand binding approaches, and is able to detect subunit-specific drug candidates (Yamaura, Kiyonaka, Numata, Inoue, & Hamachi, 2016).

GABA Binding to GABA$_A$Rs

Binding of the inhibitory neurotransmitter GABA to the two extracellular domain β+/α− GABA$_A$R interfaces in the pentamer causes a conformational change and increases the permeability of chloride ions, thereby causing inhibition (Smith & Olsen, 1995). THIP, also known as gaboxadol, and muscimol are both conformationally restricted GABA analogs (see Figure 12.2) that bind to the GABA binding site and extrasynaptic δ-containing GABA$_A$Rs show particularly high affinity for all three ligands (Meera, Wallner, & Otis, 2011). A study using molecular dynamics simulation on α4βδ receptors suggests GABA is binding to the critical amino acid residues in the binding site at α+/β−, most of which are occluded by the C-loop, through an electrostatic funnel (Carpenter & Lightstone, 2016).

The Benzodiazepines (BZ) Binding Site and Subtype-Selective Modulators

Benzodiazepines (BZs), the most widely used class of anxiolytic and sedative/hypnotic prescription drugs, are GABA$_A$R modulators that bind to the α+/γ- interface and allosterically modulate GABA$_A$R function (Möhler, 1982; Rudolph & Möhler, 2004; Sigel & Buhr, 1997; Sigel & Luscher, 2011). Interestingly, the amino acids involved in the binding site for BZs involve homologous residues and microdomains that produce the GABA agonist/antagonist sites at other subunit interfaces; namely, the GABA binding site at the β+/α− interface (Ernst et al., 2003; Sigel & Buhr, 1997; Smith & Olsen, 1995). Classical BZs, like diazepam (DZ), target non-selectively α1, α2, α3, and α5 subunit-containing GABA$_A$Rs, thereby displaying anxiolytic, muscle relaxant, anticonvulsant, and anterograde amnestic properties. These features are

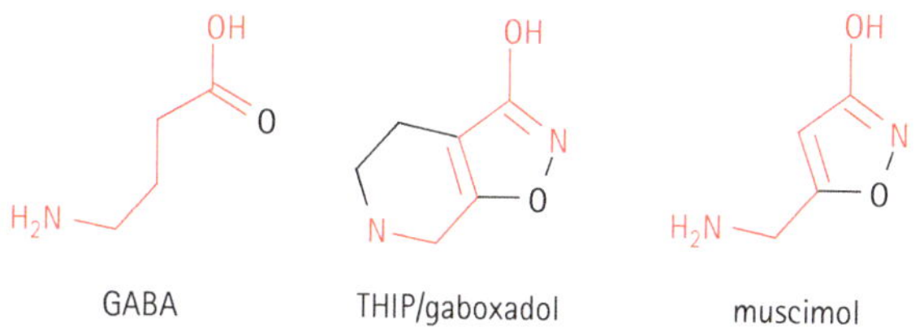

FIGURE 12.2 GABA and the GABA analogs THIP and muscimol.
Backbone structures in the conformationally restricted GABA analogs THIP and muscimol are drawn in red.

related to certain receptor subtypes, which are found in different brain regions and networks. Studies on mice with α1 subunits rendered DZ-insensitive by the H101R point mutation did not show BZ-induced sedation or anterograde amnesia. They exhibit reduced anticonvulsant action, but no changes in anxiolytic, myorelaxant, motor-impairing, or EtOH-enhanced behavior in response to BZs (McKernan et al., 2000; Rudolph et al., 1999). Mice expressing the DZ-insensitive α2(H101R) mutant are resistant to anxiolytic-like effects of BZs, whereas sedation was present (Löw et al., 2000). These animals also showed strongly reduced BZ-induced myorelaxation, while only higher doses were needed for this effect in α3(H101R) (Crestani et al., 2001). Using cell-type and region-specific conditional α2 knock-out mice, it has been shown that α2 modulates anxiety and fear through different brain circuits in the hippocampus (Engin et al., 2016). The α2 subunit is implicated in reinforcing responses to cocaine in the nucleus accumbens (Dixon et al., 2010) and alcohol in the central amygdala (Liu et al., 2011). An important role of α2 in mediating antihyperalgesia in spinal cord has recently been shown in point-mutated mice (Ralvenius, Benke, Acuna, Rudolph, & Zeilhofer, 2015), indicating that subtype-selective PAMs could constitute a rational approach to the treatment of chronic pain syndromes (Zeilhofer, Ralvenius, & Acuna, 2015). The α5 subunit is involved in cognition (Collinson et al., 2002) and the production of amnesia-inducing actions of general anesthetics, localized to the hippocampal CA1 region (Cheng et al., 2006); also, stimulation of α5-containing receptors mediates tolerance to sedation induced by DZ (van Rijnsoever et al., 2004). Of major interest is the discovery of subtype-specific BZs for the development of therapeutics with minimal side effects (for example non-sedative, non-addictive anxiolytic drugs that would have higher efficacy at α2 and α3 compared to α1 and α5 subunits—Rudolph & Knoflach, 2011; Tan, Rudolph, & Lüscher, 2011—which might be only partially possible; Chagraoui, Skiba, Thuillez, & Thibaut, 2016). Selective BZs could have an indication as novel therapeutics for the treatment of depression (Lüscher, Shen, & Sahir, 2011), schizophrenia, chronic pain (Engin, Liu, & Rudolph, 2012), cognitive impairment, stroke (Rudolph & Knoflach, 2011), and drug abuse (Dixon et al., 2010; Lindemeyer et al., 2017; Liu et al., 2011).

Zolpidem (an imidazopyridine) and other clinically used hypnotics like zopiclone (a cyclopyrrolone), and zaleplon (a pyrazolopyrimidine), as well as beta-carbolines, triazolopyridazines, and quinolones, show a higher affinity for α1-containing receptors than for α2- or α3-containing subtypes, while they have no effect on α5-containing $GABA_A$Rs (Crestani, Martin, Mohler, & Rudolph, 2000; Olsen & Sieghart, 2008, 2009).

Selective positive allosteric modulators (PAMs) for α2 (and possibly α3), like the BZ-derivative TPA023 (MK0777) (Atack, Wafford, et al., 2006) are anxiolytic, but not sedative, and also showed enhancement of cognition in schizophrenia (Atack, 2010). TPA023 did not make it past phase II clinical trials due to ocular cataract formation as a side effect (Möhler, 2011). Other drugs with a similar pre-clinical profile, like MRK-409 or ocinaplon, also have been stopped in clinical trials due to various other side effects (Rudolph & Möhler, 2014). Recently imidazobenzodiazepine oxazole derivatives have been synthesized that show some α2/α3 selectivity (Poe et al., 2016).

Selective *inverse agonists* at the α5 subtype, a subunit that is highly expressed in the hippocampus CA1 field, and implicated in mediating an unconventional form of tonic GABAergic inhibition mediated by α5βγ receptors, have been suggested to enhance cognition (spatial memory), without anxiogenic or proconvulsive components (Sternfeld et al., 2004). Such α5 inverse agonists also reverse the amnestic effects of BZs, general anesthetics (Antkowiak & Rudolph, 2016; Cheng et al., 2006), and alcohol (Nutt, Besson, Wilson, Dawson, & Lingford-Hughes, 2007). They may be beneficial for treating autism, Down syndrome, affective disorders, and schizophrenia (Rudolph & Möhler, 2014). Clinical trials for the α5 inverse agonist cognitive enhancer L-655,708 had to be stopped due to toxic side effects (Atack, Bayley, et al., 2006; Dawson et al., 2006).

Flumazenil (Ro15-1788), first introduced by Hoffman-LaRoche under the brand name Anexate®, a competitive antagonist with a high binding affinity to the BZ site of GABA$_A$Rs (Hunkeler et al., 1981), blocks binding of BZs and other substances that bind the BZ sites, such as zolpidem, and serves as an antidote for BZ overdoses.

Of the five different subunit interfaces, the two α-/β+ sites bind GABA and other orthosteric ligands (THIP, muscimol, bicuculline), and the α+/γ- site binds BZs (see Figure 12.1); it may in hindsight not be surprising that in the last 10 years, there have been reports that certain BZs and structurally unrelated BZ-site ligands can also bind with high affinity to other subunit interfaces and act as allosteric modulators at (in particular) the α+/β- site, which shares the α+ interface with the classical BZ binding site (see Figure 12.1). Such alternative subunit interfaces, like α+/β– harbor modified GABA/BZ binding sites that may bind some drugs, including pyrazoloquinolones (Baur et al., 2008; Maldifassi, Baur, & Sigel, 2016b; Ramerstorfer et al., 2011; Sieghart, 2015; Sieghart, Ramerstorfer, Sarto-Jackson, Varagic, & Ernst, 2012), and the α4/6+/β3- site, have been implicated in low-dose EtOH actions (see section "Tonic GABA$_A$Rs as Low-Dose Alcohol Targets).

Channel Pore Blocker

Unlike bicuculline, which is a GABA competitive antagonist at the GABA site in the extracellular domain of the GABA$_A$R protein (Curtis, Duggan, Felix, & Johnston, 1970), picrotoxinin and other structurally unrelated agents, like insecticides such as lindane and fipronil, block GABA$_A$Rs non-competitively at a binding side inside the chloride channel and display convulsive properties (Chen, Durkin, & Casida, 2006; Inoue & Akaike, 1988; Olsen, 2006). In fact, picrotoxin has been shown to bind at the narrowest part of the ion channel pore in a picrotoxin-co-crystal structure with the GluCl glutamate-gated chloride channel (Hibbs & Gouaux, 2011) at a location determined by amino acid residues that, when mutated, lead to insecticide-resistant insect homomeric GABA$_A$Rs (ffrench-Constant, Rocheleau, Steichen, & Chalmers, 1993).

Jayakar et al. (2014) demonstrated that some chemical analogues of general anesthetics with convulsant efficacy may mediate this action via the general anesthetic binding sites in the transmembrane domain, identified by photoaffinity-labeling with chemical analogues of etomidate, propofol, and barbiturates (Chiara et al., 2016; Li et al., 2006) (see below sections on anesthetics). There is evidence that α-thujone, a monoterpene compound found in the original recipes of absinthe, is acting on $GABA_AR$ by blocking the channel pore, thus having excitatory, possibly proconvulsant properties Hold, Sirisoma, Ikeda, Narahashi, & Casida, 2000, and for *fin de siècle* late 19th century Parisians, the green α-thujone-containing juice blocked sedation by alcohol while sparing activation of creative brain activity (Olsen, 2000).

Neuroactive Steroids

Certain endogenous neurosteroids have been identified as $GABA_AR$ modulators (Belelli & Lambert, 2005; Olsen & Sapp, 1995). The group of steroids active on the nervous system are classified as *neuroactive steroids*, composed of endogenous steroids, many of them hormones and hormone metabolites, and synthetic analogues. The neurosteroids, allopregnanolone (progesterone metabolite) and tetrahydro-deoxycorticosterone (cortisol metabolite), are established $GABA_AR$ PAMs (Majewska, Harrison, Schwartz, Barker, & Paul, 1986), based on the long-known anesthetic efficacy of progesterone (Selye, 1942) and the clinical utilization of its analogue alphaxalone as an intravenous short-acting general anesthetic (Antkowiak & Rudolph, 2016; Harrison & Simmonds, 1984). The neurosteroid ganaxolone (synthetic non-metabolizeable analog of allopregnanolone) and other neurosteroids are potential clinical therapies for seizure disorders/epilepsy (Mula, 2016; Zaccara & Schmidt, 2016) and for the treatment of anxiety and attention deficits in children with Fragile X syndrome, one type of autism spectrum disorder (ASD) (Braat & Kooy, 2015). Steroid binding sites on $GABA_ARs$ have been proposed, based on site-directed mutagenesis alone, and a homology model derived from the *Torpedo* nAChR structural model (Unwin, 2005), to bind the $GABA_AR$ at two sites in the transmembrane domain: α1M236 and βY284 initiate direct activation, whereas αQ241 and αN407 mediate the potentiation of responses to GABA by neurosteroids. These ligands have the highest activity on $GABA_ARs$ containing δ or α5 subunits (Hosie, Wilkins, da Silva, & Smart, 2006; Hosie, Wilkins, & Smart, 2007). The neuroactive steroid-binding sites are in the TMD but apparently are not the same as etomidate, because steroids enhance etomidate binding (Li, Chiara, Cohen, & Olsen, 2009) indicating that the steroid interaction of Hosie et al. 2006 with the etomidate binding residue α1M236 identified by affinity labeling must be allosteric. Furthermore, generally accepted helical arrangements for the transmembrane domains of $GABA_ARs$ based on templates of $GABA_AR$ and related LGIC crystal structures do not allow contact of any of the etomidate/propofol binding sites identified by affinity labeling with the residues shown by mutagenesis by Hosie et al. (2006), nor the βN235 discovered using mouse point mutations knock-ins in β3 (Jurd et al., 2003) or β2 (Reynolds et al., 2003; Rudolph & Antkowiak, 2004) of a "Mihic residue" identified as critical for action of anesthetics to modulate $GABA_ARs$ (Mihic et al., 1997). Of course, the native protein may be more flexible than the crystal structure shows, and may explain the binding of steroids. A successful steroid anesthetic affinity labeling was achieved, showing stereo-selective covalent attachment of an azi-steroid to F301, identified by mass spectrometry, in the TM3 domain of recombinant homomeric β3-GABA$_A$R (Chen et al., 2012). Alternatively, Akk et al. (2005) suggest that steroids bind numerous low-affinity sites at the membrane lipid–GABA$_A$R protein interface and modulate its kinetics of opening and closing. In other words, because of their hydrophobicity, they accumulate at the membrane/ion channel interface in concentrations higher than found in the medium/recording solution and appear more potent than they actually are. This theory requires further study for validation or refutation.

Anesthetics

GABA$_A$Rs are targets of volatile anesthetics, the intravenous anesthetics propofol and etomidate, as well as barbiturates, steroid anesthetics and EtOH (Franks & Lieb, 1994; Olsen & Li, 2011). Anesthetic binding sites on the GABA$_A$R can be identified using site-directed mutagenesis (Mihic et al., 1997), substituted cysteine modification protection (SCAMP) (Forman & Miller, 2016), or photoaffinity labeling (Forman & Miller, 2011; Olsen & Li, 2011).

Mihic et al. (1997) used site-directed mutagenesis to show that certain residues in the TMD of both GABA$_A$Rs, and the related inhibitory glycine receptor ligand-gated chloride channels (GlyRs), are critical for modulation by volatile anesthetics and long-chain alcohols. By replacing the residues in GABA$_A$Rs and GlyRs with the corresponding amino acids in the anesthetic-insensitive GABA$_A$R rho1 subunit, they identified two residues, one in TM2 (α1S270 or α2S270, or corresponding β1S265 or β3N265) and one in TM3 (α1A291 or corresponding β1M286), that were essential for modulation by the two types of anesthetic. The authors argued that these residues form an aqueous pocket on the exterior of the TMD ion channel domain that can perturb channel gating by GABA. The size of the active anesthetics shows a "cut-off" in maximum volume that is dependent on the volume of the amino acid side chains in the pocket (Jenkins et al., 2001; Koltchine et al., 1999; Wick et al., 1998), which is also modulated by intravenous general anesthetics propofol and chemical analogues (Krasowski, Nishikawa, Nikolaeva, Lin, & Harrison, 2001). Mutation of these residues to cysteine and alkylation inhibited anesthetic modulation, and the treatment with a sulfhydryl residue alkylating reagent could be blocked by including excess free anesthetics, such as propofol and etomidate, in the reaction; in addition, it was shown that the arrangement of TM1-TM2-TM3 in the membrane could be modeled in a manner consistent with agonist and modulator-sensitive conformational changes (Bali & Akabas, 2004). Mutations of the two critical residues was shown to prevent anesthetic modulation in vitro (Krasowski & Harrison, 2000; Ueno, Wick, Ye, Harrison, & Harris, 1999) and in vivo (Werner et al., 2011). Whereas these residues mediated modulation by long-chain alcohol anesthetics, including EtOH, but only at ≥100 mM, it was argued that EtOH might act on some other GABA$_A$R subtypes at <100 mM (Lobo & Harris, 2008). Furthermore, mutation of β265 to cysteine allowed covalent attachment by an alkylating analogue of n-propanol (Mascia, Trudell, & Harris, 2000), suggesting that propanol and EtOH could bind to this site at pharmacologically relevant intoxicating, less than anesthetic, concentrations (10–40 mM). Note that both of the "Mihic residues" were independently verified as anesthetic targets: TM3 residue βM286 was later identified by affinity labeling with an azi-etomidate reagent as a binding site for that anesthetic (Li et al., 2006), and the TM2 residue β3N265 was shown by mouse knock-in engineering to be critical for in vitro (Siegwart, Jurd, & Rudolph, 2002) and in vivo action of etomidate, propofol, and to a much lesser extent, also the volatile anesthetics enflurane and halothane (Jurd et al., 2003).

Photolabeling with the etomidate analog azietomidate revealed a binding of $GABA_AR$ at the β+/α– interface (α1M236 and β3M268) (Li et al., 2006). This binding site for the intravenous general anesthetics etomidate and propofol that appears to be in TM2 of β subunits seems to be close by, or overlapping with, the binding site(s) of volatile anesthetics (Li, Chiara, Cohen, & Olsen, 2010; Rudolph & Antkowiak, 2004). For propofol, multiple binding sites exist at the $GABA_AR$ (Jayakar et al., 2014). A recent study using subunit concatenations with or without β2N265I mutation suggests propofol acting predominantly on the γβ+/α–β and γ+/β– interfaces, while propofol predominantly acts on γβ+/α–β, αβ+/α–γ, and γ+/β– interfaces at the α1β2γ2 $GABA_AR$ (Maldifassi, Baur, & Sigel, 2016a).

Barbiturate affinity and efficacy depend on the subunit composition, but the α subunit seems to be more important than β (Thompson, Whiting, & Wafford, 1996). More recently, photoaffinity labeling followed by protein sequencing has shown barbiturate binding at the α+/β– and γ+/β– interfaces in the TMD, revealing two structurally related but pharmacologically distinct classes of general anesthetic binding sites in the TMD (Chiara et al., 2013). Such subunit-interface binding sites have already been described for numerous categories of ligands, especially GABA and BZ, in the ECD (Olsen, 2015). Maldifassi et al. (2016a) suggesting predominant binding of propofol, etomidate and barbiturate at the αβ+/α–γ interface and in addition to the α+/β– and/or α+/γ– TMD interfaces in α1β2γ2. Photolabeling studies also suggest binding sites for etomidate and barbiturates at α4β3δ $GABA_AR$ subtypes at the β+/α–, and ("most likely") at the β+/β– TMD interfaces, respectively, which were not binding alphaxalone or DS2 (Chiara et al., 2016).

Flavonoids

Flavonoids are found in many plants and a few microorganisms. They have first been discovered as modulators of the BZ-site of $GABA_ARs$, but the action of different compounds within this group seems to exceed binding to only this site, potentially acting at more than one additional binding site on $GABA_ARs$. Flavonoids can act as either positive, negative, or neutralizing on $GABA_ARs$, or directly as allosteric agonists (Hanrahan, Chebib, & Johnston, 2015; Marder & Paladini, 2002; Wasowski & Marder, 2012). Flavonoids share the basic structure of a phenylbenzopyran; most commonly of a flavan (2-phenylchromane). Subgroups include flavanoles, flavanones, flavanonole, flavones, flavonoles, and isoflavones. Among these groups, flavones and isoflavones especially have been found to interact with the BZ binding site of $GABA_ARs$. This is probably due to their rigid, conjugated structure (Hanrahan, Chebib, & Johnston, 2011). Structure-activity experiments have shown that flavones are more potent on BZ radioligand binding than their flavanone or flavonol counterparts; glycosylation had a negative influence on binding (Wang et al., 2002), but might be more potent in vivo; the modification is probably irrelevant to the efficacy. Flavones with a high affinity at the BZ binding pocket of the $GABA_AR$ have been synthesized according to the pharmacophore model by Cook (Dekermendjian et al., 1999). A method for the synthesis of a series of flavone scaffolds has been described (Yao, Song, Wang, Dixon, & Lam, 2007). Flavonoid structures may be useful as lead scaffolds for the development of new potent and selective $GABA_AR$ drugs.

Relevant is the position 6 on flavones for their effects on $GABA_ARs$. Generally, flavonoids can interact with $GABA_ARs$ in a flumazenil-sensitive or flumazenil-insensitive manner (Hanrahan et al., 2015). Some flavonoids show subtype-selectivity like 6,2'-dihydroxyflavone (Wang et al., 2007) or the flavan-3-ol ester Fa131 (Fernandez, Mewett, Hanrahan, Chebib, & Johnston, 2008). The flavone hispidulin, in addition to the other $GABA_AR$ subtypes, also enhanced $\alpha6\beta2\gamma2$ $GABA_ARs$ (Kavvadias et al., 2004) and showed lowered seizure susceptibility. This study by Kavvadias also demonstrates that hispidulin crosses the blood–brain barrier unmodified through passive diffusion transcellularly through endothelial cells. 6-Hydroxyflavones have been shown to be anxiolytic but not sedative at anxiolytic dose in animal models. They showed a stronger potentiation on $\alpha2$ and $\alpha3$ subtypes than $\alpha1$ or $\alpha5$, measured by patch-clamp recordings on recombinant receptors expressed in HEK293 cells (Ren et al., 2010). (+)-Catechin, a natural flavan-3-ol, is an agonist at $\alpha4\beta3\delta$ $GABA_ARs$ expressed in *Xenopus laevis* oocytes, while (+)-taxifolin was a negative modulator (Eghorn et al., 2014). The flavonoid dihydromyricetin (DHM), isolated from the plant *Hovenia dulcis,* potentiates synaptic and extrasynaptic $GABA_ARs$ in dentate gyrus granule cells in a flumazenil-sensitive manner. DHM counteracts alcohol intoxication and ameliorates withdrawal symptoms and alcohol craving in rats (Shen et al., 2012). It is on the market as an anti-hangover nutraceutical. DHM also improved behavioral deficits and neuropathology in a mouse model of Alzheimer's disease (Liang, Lopez-Valdes, et al., 2014).

Cannabinoids

Endocannabinoids like 2-arachidonylglycerol (2-AG) (Sigel et al., 2011) and N-arachidonylglycine (Baur, Kielar, et al., 2013), a lipid with analgesic features, have been identified as positive $GABA_AR$ modulators, at $GABA_AR$ subtypes (Baur, Gertsch, & Sigel, 2013; Sigel et al., 2011). Cannabidiol (CBD), the major non-intoxicating component of cannabis, and a structural isomer of the main psychoactive component Δ^9THC (Δ^9-tetrahydrocannabinol), has anxiolytic, sedative, and anticonvulsant effects and has been suggested for treating pediatric epilepsies such as Dravet syndrome. Since CBD has low affinity for the main cannabinoid (CB) receptor, and exhibits an activity profile similar to that of GABA PAMs enhancing the activity of $GABA_AR$-subtypes, effects on $GABA_AR$ might be in part responsible for beneficial anticonvulsant and anxiolytic CBD effects.

In contrast to studies on recombinant receptors where 2-AG increases $GABA_AR$ activity at low non-saturating 1 μM GABA concentrations, 2-AG leads to a decrease in GABA peak currents in neurons evoked by high saturating (1 mM) GABA concentrations that mimic synaptic GABA transmission. Therefore the functional impact of endo- and perhaps also phytocannabinoids like CBD and Δ^9THC on $GABA_AR$ activity could depend on the physiological context of $GABA_AR$ activation (Golovko et al., 2015).

Neurotransmitter, Amino Acids

The neurotransmitter dopamine (0.1–10 mM) has been reported to act on $GABA_AR$s. The effects of dopamine were opposing, such as it inhibited tonic striatal GABA-evoked currents, but also directly activated recombinant $GABA_AR$s composed of α1β3γ2, α1β2γ2, and α5β3γ2 expressed in HEK293 cells (Hoerbelt, Lindsley, & Fleck, 2015). Dopamine effects did not require the presence of α subunits. Histamine modulates heteromultimeric $GABA_AR$s in a γ2-controlled manner, shown on recombinant α1β2γ2 $GABA_AR$s expressed in Xenopus oocytes, where 1 mM histamine doubled the response of 3 μM GABA, while it directly activated β2 or β3 homomultimeric channels with an EC_{50} of 212 μM and 174 μM, respectively (Saras et al., 2008). At physiological concentrations (10–100 μM), taurine, which is highly abundant in the brain, acts as an agonist on extrasynaptic α4β2δ $GABA_AR$s in the thalamus and is much more potent on that subtype in a recombinant system than the α1β2γ2 combination (Jia, Yue, et al., 2008). Another study shows highest efficacy when the β1 subunit is present (Kletke, Gisselmann, May, Hatt, & Sergeeva, 2013).

β-Selective Compounds

Loreclazole (Wingrove, Wafford, Bain, & Whiting, 1994), etomidate (Hill-Venning, Belelli, Peters, & Lambert, 1997), and propofol effects are essentially highly selective for β2 and β3 subunit–containing receptors, an effect that is due to a single amino-acid β2/3N265 over β1S265 (Belelli, Lambert, Peters, Wafford, & Whiting, 1997). In knock-in animals carrying a mutated etomidate- and propofol-insensitive β3N265M subunit, mice become highly resistant to etomidate and propofol anesthesia (Jurd et al., 2003), demonstrating that β3 subunits are critical for mediating anesthetic effects. Surprisingly, a mouse with the same mutation in the β2 subunit (β2-N265M) only showed a lack of sedative effects, leading to the conclusion that anesthesia and sedation are mediated by distinct $GABA_AR$ isoforms (Reynolds et al., 2003). Fragrant dioxane derivatives (FDDs) prefer β1 to β2 and β3 (Sergeeva et al., 2010). γ-Hydroxybutyric acid (GHB) also shows β-selectivity. GHB activates α4β1δ $GABA_ARs$ with high affinity, whereas α4β2/3δ $GABA_ARs$ are low-affinity targets (Absalom et al., 2012). Salicylidene salicylhydrazide is a β1 selective negative allosteric modulator (Thompson et al., 2004). Beta subunit selectivity has also been reported for low-dose alcohol effects on δ subunit-containing $GABA_ARs$, with the β3 subunit required for effects at alcohol concentrations below the legal U.S. driving limit of ~16 mM EtOH blood alcohol (Wallner et al., 2003), and it has been shown that high alcohol sensitivity requires β3Y66, which is β2S66 and β1A66 (see section "$GABA_ARs$ as EtOH Targets" below).

Three Forms of GABAergic Inhibition—Synaptic, Extrasynaptic and Intermediate ($GABA_{A,slow}$)—Lead to Functional Selectivity to GABAergic Modulators

Classic synaptic GABAergic inhibition is the result of an extremely rapid ($\tau < 1$ ms) transient activation of postsynaptic $GABA_ARs$ in response to vesicular presynaptic GABA release with typically very high saturating [GABA] (~1 mM), which is in part due to the limited volume of the synaptic cleft. Fast receptor desensitization and deactivation contributes to the very short duration of GABAergic inhibitory postsynaptic currents (IPSCs) with deactivation due to quick GABA removal by diffusion and GABA reuptake mechanisms.

In contrast, *tonic inhibition* is mediated by highly GABA-sensitive extrasynaptic δ subunit-containing $GABA_AR$ which shows persistent activation with little desensitization at low extracellular ambient [GABA], generally thought to be in the range of 50 nM—1 μM (Belelli et al., 2009; Farrant & Nusser, 2005; Glaeser & Hare, 1975; Santhakumar, Hanchar, Wallner, Olsen, & Otis, 2006). While extrasynaptic receptors, despite the large nonsynaptic area in the neuronal plasma membrane, have a very low density and thus make up only a relatively small fraction of total $GABA_ARs$ in the brain, because of their persistent activity, it has been suggested that in fact tonic inhibition has a stronger influence on dampening neuronal excitability when compared to synaptic inhibition (Brickley & Mody, 2012).

In addition to fast synaptic and persistent tonic inhibition, a third *intermediate form of inhibitory control* is mediated by prolonged ($\tau \sim 50$ ms) GABA transients produced by nonsynaptic vesicular GABA release from neurogliaform cells (NGFC) that produce "clouds" of GABA, with GABA concentrations sufficiently high to activate not only extrasynaptic δ-containing $GABA_ARs$, but also less GABA-sensitive γ2-containing $GABA_ARs$ and metabotropic $GABA_BRs$. $GABA_BRs$ that mediate homo- and heterosynaptic depression (Capogna & Pearce, 2011; Olah et al., 2009; Overstreet-Wadiche & McBain, 2015; Szabadics, Tamas, & Soltesz, 2007). The vesicular GABA release of NGFCs leads to prolonged GABA transients that evoke slow inhibitory postsynaptic currents (IPSC), also known as $GABA_{A,slow}$ IPSCs. In hippocampal CA1 neurons, $GABA_{A,slow}$ IPSCs are mediated by α5βγ2 receptors, while in other brain regions they are mediated by α1,2,3βγ2 receptors (Overstreet-Wadiche & McBain, 2015). Fast synaptic GABA currents ($GABA_{A,fast}$), where $GABA_A$Rs are generally saturated by GABA, cannot exhibit PAM-induced further increase in peak GABA currents due to the ceiling effect at saturating [GABA]. In contrast, application of GABA PAMs, as well as GABA reuptake inhibitors (like NO711), lead to an increased amplitude of $GABA_{A,slow}$ currents, indicating that vesicular NGFC GABA release produces sub-saturating [GABA] for these particular GABA receptor subtypes (Prenosil et al., 2006). Because of the functional difference in the potential of PAMs to increase $GABA_{A,slow}$, but not $GABA_{A,fast}$ peak currents, non-saturated $GABA_{A,slow}$ currents could make important contributions to the behavioral effects of GABA PAMs (anesthetics and BZs).

Physiology and Unique Pharmacology of Extrasynaptic δ-$GABA_A$Rs

In terms of subunit composition, most $GABA_A$Rs that mediate tonic currents are composed of α4, β, and δ subunits in the forebrain, with the cerebellar granule cells' specific α6 subunit substituting for the α4 subunit in the cerebellar granular cells (Brickley, Revilla, Cull-Candy, Wisden, & Farrant, 2001; Nusser et al., 1999). $GABA_A$R δ subunits have an almost exclusive association with α4 and α6 subunits (Jones et al., 1997; Sur et al., 1999), with the exception of hippocampal interneurons, where δ subunits can also associate with α1 subunits to generate highly alcohol-sensitive tonic currents (Glykys et al., 2007). It is generally thought that in δ-$GABA_A$Rs, the δ subunit substitutes for the single γ subunit in a αβαβγ/δ heteropentameric complex (Olsen & Sieghart, 2008). Since γ subunits are required to form classical BZ α+/γ- binding sites, extrasynaptic δ-receptors are generally insensitive to most classical BZ ligands like DZ, although it has been shown that a single amino acid change in the δ subunit (δH68A), from histidine in the δ subunit to the homologous alanine found in γ, confers diazepam sensitivity to recombinantly expressed α4β3δH68A receptors (Meera, Olsen, Otis, & Wallner, 2010), supporting the notion that homologous BZ-ligand binding pockets exist at subunit interfaces other than the classical BZ site at the α+/γ- subunit interface in the ECD.

Tonic Current-Carrying $GABA_A$Rs as Low-Dose Alcohol (EtOH) Targets

There have been over four decades of detailed positive evidence, predominantly from rat models, for a role of $GABA_A$Rs in mediating the anxiolytic, mood-enhancing alcohol effect that we experience during social alcohol consumption with blood alcohol levels usually below the U.S. 17 mM legal driving limit (Aguayo, Peoples, Yeh, & Yevenes, 2002; Olsen, Hanchar, Meera, & Wallner, 2007; Suzdak et al., 1986; Wallner, Hanchar, & Olsen, 2006a; Wallner & Olsen, 2008). While many proteins, including GABA$_A$Rs (Mihic et al., 1997), are changed in their function at high >50 mM ethanol (EtOH) concentrations, tonic GABA currents are uniquely sensitive to alcohol, with essential consensus that tonic currents mediated by δ-GABA$_A$Rs in neurons are increased in their activity by relevant low (<30 mM) EtOH concentrations (Carta, Mameli, & Valenzuela, 2004; Centanni, Burnett, Trantham-Davidson, & Chandler, 2017; Fleming, Wilson, & Swartzwelder, 2007; Hanchar et al., 2005; Herman, Contet, Justice, Vale, & Roberto, 2013; Herman & Roberto, 2016; Jia, Chandra, Homanics, & Harrison, 2008; Jia, Pignataro, & Harrison, 2007; Liang et al., 2006; Mody, Glykys, & Wei, 2007; Santhakumar, Meera, Karakossian, & Otis, 2013; Santhakumar, Wallner, & Otis, 2007; Wei, Faria, & Mody, 2004). High EtOH sensitivity (≤10 mM) has also been reported in recombinantly expressed α4/6βδ receptors (Sundstrom-Poromaa et al., 2002), with significant β3 selectivity (Wallner et al., 2003).

Three lines of evidence point towards a role of a homologous BZ-like site that mediates low-dose EtOH actions.

(1) The α6R100Q mutation changes one of the most influential BZ-binding site residues at the α+ interface (histidine (H) in α1,2,3,5, arginine (R) in α4,6) and is a naturally occurring polymorphism initially found in alcohol non-tolerant (ANT) rats that show increased alcohol-induced motor impairment (Korpi, Kleingoor, Kettenmann, & Seeburg, 1993). The α6R100Q mutation leads to increased EtOH sensitivity when expressed with β3 (but not with β1 or β2) and δ subunits, increases EtOH effects in slices at concentrations as low as 10 mM, and leads to increased EtOH-induced motor coordination in rats homozygous for the α6R100Q mutation (Hanchar et al., 2005). The α6R100Q polymorphism was also found to be enriched in two independently generated alcohol non-EtOH-preferring rat lines (Carr, Spence, Peter Eriksson, Lumeng, & Li, 2003; Saba et al., 2001).

(2) The (imidazo)benzodiazepine Ro15-4513 (a close structural analog of flumazenil), reported as a behavioral alcohol antagonist in rats (Suzdak et al., 1986), reverses low-concentration EtOH enhancement of α4/6β3δ receptors (Wallner et al., 2006b) in a competitive manner; [^{3}H]Ro15-4513 binds to α4/6β3δ receptors but not to α4/6β3γ2 receptors in a manner inhibited by EtOH (10–20 mM) (Hanchar et al., 2006).

(3) The reported higher alcohol sensitivity of α4/6β3δ GABA$_A$Rs is due to an amino acid difference in the three mammalian β subunits β1S66, β2A66, β3Y66 (Wallner et al., 2014). Position 66 in GABA$_A$R beta subunits is homologous to position γ2A79, which lines the classical BZ α+/γ- binding site (Kucken, Teissere, Seffinga-Clark, Wagner, & Czajkowski, 2003). These findings led us to a model where the EtOH site and the overlapping Ro15-4513 site is located at the α4,6+/β3- interface in extrasynaptic δ-GABA$_A$Rs (Wallner et al., 2014) (see Figure 12.1, ECD).

One of the rationales that prompted us to study low-dose EtOH effects on δ-GABA$_A$Rs and tonic currents was the remarkable observation that rats that have been treated with alcohol have a remarkable down-regulation of GABA$_A$R δ subunits (Gonzalez et al., 2012), which is consistent with δ-GABA$_A$Rs being a direct alcohol target, and consistent with behavioral tolerance to EtOH in alcohol-treated rats (for further details, see section on GABA$_A$R plasticity) (Cagetti, Liang, Spigelman, & Olsen, 2003).

The GABA Analogs THIP and Muscimol Are δ-GABA$_A$R Selective Ligands

Another marked difference is sensitivity to GABA, which is much higher in the presence of the δ subunit, to allow for activation of extrasynaptic α4/6βδ receptors by low tonic (up to ~1 μM) GABA concentrations. This unusually high GABA sensitivity of δ-GABA$_A$Rs also extends to higher affinity of the conformationally restricted GABA analogs THIP/gaboxadol and muscimol (see Figure 12.2). Co-expression of δ subunits with α4/6 and β3 subunits leads to dramatically increased THIP sensitivity, with an EC_{50} of around 30 nM (Meera et al., 2011), which is in excellent agreement with the observed binding affinity of [^{3}H]THIP (equilibrium K_D of 38 nM) in cerebellar membranes (Friemel et al., 2007). Similarly, high-affinity muscimol binding is absent in the forebrain of δ-KO mice (Mihalek et al., 1999), and, consistent with the much higher THIP and muscimol affinity, δ-KO mice show dramatically reduced behavioral muscimol and THIP sensitivity (Boehm, Homanics, Blednov, & Harris, 2006; Chandra et al., 2010). This shows that, despite the fairly low abundance (5–10% of total GABA$_A$Rs) of δ-GABA$_A$Rs overall, muscimol and THIP are highly selective for extrasynaptic δ-GABA$_A$Rs. While THIP/gaboxadol was tried unsuccessfully (due to its hallucinogenic side effects) in clinical trials as a sleeping pill, muscimol is, along with ibotenic acid, the major active ingredient in the hallucinogenic mushroom *Amanita muscaria* (DeFeudis, 1980).

Low GABA Efficacy of Recombinantly Expressed δ-GABA$_A$Rs Might Not Be a Feature of Tonic GABA Currents in Neurons

There are numerous reports that recombinant δ-GABA$_A$R expression (in oocytes and human embryonic kidney [HEK] cells) produces GABA currents with very low current amplitudes, even with application of saturating [GABA]. However, co-application of relevant concentrations of anesthetics (neurosteroids like tetrahydroxycorticosterone [THDOC] and ganaxolone, propofol, and etomidate, but also barbiturates and etazolate) leads to rather dramatic increases (often more than 1000%, depending on exact subunit composition and anesthetic used) in current amplitudes (Meera, Olsen, Otis, & Wallner, 2009; Wallner et al., 2003; Wohlfarth, Bianchi, & Macdonald, 2002;

Zheleznova, Sedelnikova, & Weiss, 2008). In many cases, this is a feature not only of αβδ, but also of functionally expressed "binary" αβ GABA$_A$Rs, and therefore predominantly an indication that co-expression of the γ subunit (most frequently γ2) increases GABA efficacy rather than the δ subunit decreasing efficacy (Meera et al., 2009).

More recently, two related compounds called delta-specific 1 and 2 (DS1 and DS2), which showed δ-GABA$_A$R specificity, have been developed. Unlike other GABAergic anesthetics such as etomidate propofol and barbiturates, DS1 and DS2 current increases are dependent on δ-subunit expression and do not enhance GABA currents in "binary" GABA$_A$Rs formed by α and β subunits alone. (Jensen et al., 2013; Wafford et al., 2009). GABA current increases seen with DS2 are therefore an indication that at least some of the functional GABA$_A$Rs formed in recombinant systems are indeed ones that have δ subunits incorporated and are not due to receptors formed by α and β subunits alone. Problems with δ subunit incorporation into recombinant GABA$_A$Rs in many studies are a likely reason why there has been so much controversy about the EtOH sensitivity of recombinantly expressed δ-GABA$_A$Rs (Borghese & Harris, 2007; Lovinger & Homanics, 2007; Meera et al., 2010) and why the high THIP (and muscimol) sensitivity of recombinant δ-GABA$_A$Rs was missed in many studies (Meera et al., 2011).

While GABAergic anesthetics (propofol, etomidate, barbiturates, THDOC/ganaxalone) produce incredible increases in GABA efficacy in recombinantly expressed δ-GABA$_A$Rs, with often more than tenfold increases in current amplitudes at saturating [GABA], this dramatic increase is not seen in native δ-GABA$_A$Rs that mediate tonic GABA currents. This also is the case for DS2, where application of only very high (10 μM) [DS2] leads to measureable increases in native tonic GABA currents at usually sub-saturating GABA concentrations (Jensen et al., 2013), but nowhere close to the extent (10–30-fold increases) seen with low-dose ≤1 μM DS2 application on recombinant αβδ currents. This suggest that the incredible increase in efficacy seen with PAMs like etomidate, propofol, barbiturate, and neurosteroid anesthetics is likely to be an artifact of recombinant subunit expression. This is consistent with the observation that δ-KO mice only show a significant decrease in behavioral (loss of righting reflex) sensitivity to steroid anesthetics (alphaxalone, pregnanolone) but not to barbiturates, etomidate, or propofol (Mihalek et al., 1999). This contrasts with a drastically reduced behavioral sensitivity of δ-KO mice to gaboxadol/THIP (Boehm et al., 2006) and muscimol (Chandra et al., 2010).

Measurement of Tonic Currents Mediated by α5 Subunit-Containing Receptors Requires Increasing the Ambient GABA Concentration

There is also evidence that extrasynaptic α5-containing GABA$_A$Rs can contribute to tonic currents, although most (but not all) of these experiments were performed either with elevated GABA (e.g., 5 μM) in the recording solution, or after artificially increasing ambient [GABA] using the GABA transaminase inhibitor (anticonvulsant drug) vigabatrin (Caraiscos et al., 2004; Wu, Wang, & Richerson, 2003). Under such increased [GABA] conditions, α5 subunit-containing $GABA_A$Rs have been shown in brain slices and cultured neurons to mediate tonic inhibition in hippocampal CA1/CA3 pyramidal cells, and, to a lesser extent also in dentate gyrus granule cells and molecular layer interneurons. Under these conditions with artificially elevated ambient [GABA], mice lacking α5 subunits show reduced tonic inhibition. In addition, tonic currents present under these conditions in neurons can be partially blocked by the α5 selective benzodiazepine inverse agonist L655,708, with essentially total lack of tonic currents in mice lacking both α5 and δ subunits (Glykys, Mann, & Mody, 2008). However, since recombinant α5βγ2 receptors are activated only at fairly high [GABA] >>1 μM, and only have a slightly higher GABA sensitivity when compared to, for instance, abundant synaptic α1β2γ2 receptors, it remains an open question if, under physiological ambient GABA concentrations, extrasynaptic α5-containing $GABA_A$Rs, thought to be composed of α5β3γ2 subunits, can respond to the low ≤1 μM [GABA] concentrations that usually occur in the extrasynaptic space. A possible solution to this problem is that vesicular GABA release from NGFCs can lead to local GABA "clouds" where the [GABA] is sufficiently high to not only open classical δ-$GABA_A$Rs but also extrasynaptic BZ-sensitive α5βγ2 $GABA_A$Rs (Overstreet-Wadiche & McBain, 2015).

The Functional Distinction Between Tonic and Phasic Currents Might Be Shifted Towards Tonic Inhibition by $GABA_A$R PAMs

Synaptic $GABA_A$Rs are thought to be initially inserted into the plasma membrane outside of synapses and move by lateral membrane-bound diffusion into the postsynaptic membrane, where they are clustered by anchoring and scaffolding molecules; e.g., gephryin (for review, see Jacob, Moss, & Jurd, 2008). Therefore, besides α5-containing receptors, a substantial portion of so-called synaptic receptors might therefore be located extrasynaptically, and it is possible that even clustered synaptic $GABA_A$Rs could be exposed to low ambient ≤1 μM [GABA]. It is tempting to speculate that evolution has "designed" most synaptic γ subunit-containing receptors so that they show little or no activity at low ambient [GABA], to allow extrasynaptic and synaptic receptors to act largely as two independent inhibitory systems: synaptic receptors aid in fast neuronal computation in point-to-point communication tasks, whereas highly GABA-sensitive extrasynaptic $GABA_A$Rs are slowly acting systems designed to set neuronal excitability in a more general way.

A common feature of $GABA_A$R PAMs like BZs, barbiturates, and other general anesthetics (propofol, etomidate, and steroid anesthetics like THDOC, [allo] pregnanolone, alphaxalone), is that they lead to activation of essentially all $GABA_A$R subtypes, and this is generally due to an increase in GABA-sensitivity (i.e., left-shift of the GABA concentration–response curve) (Feng, Jounaidi, Haburcak, Yang, & Forman, 2014; Rüsch, Zhong, & Forman, 2004). Such increased GABA sensitivity of classical "synaptic" receptors in the presence of PAMs might lower the threshold for GABA activation so that, after pharmacological activation, these rather abundant "synaptic" GABA$_A$Rs may now respond to ambient [GABA], and even more so to non-vesicular GABA release from, for example, NGF cells, resulting in sustained GABA currents. Of course, GABA$_A$R subtypes with higher GABA sensitivity that are already closer to the activation threshold where ambient GABA would lead to receptor activation would be particularly sensitive to being shifted into tonic mode by PAMs. It is worth noting in this context that α1/5β3γ2 receptors have been shown to have a higher GABA sensitivity when compared to α1β2γ2 receptors (Caraiscos et al., 2004), and this increased GABA sensitivity of β3-containing GABA$_A$Rs might help explain why β3 subunits mediate actions of etomidate and propofol that are more critical to general anesthesia, whereas β2 subunits mediate only sedative effects (Jurd et al., 2003; Reynolds et al., 2003).

An important aspect of the physiology of tonic GABA currents is the source and regulation of ambient [GABA]. Diffusion of GABA after synaptic release ("spill-over") almost certainly is an important source (Glykys & Mody, 2007), whereas others have provided evidence that such small amounts of GABA could also be the result of GABA transporters running in reverse (Ransom, Tao, Wu, Spain, & Richerson, 2013; Richerson & Wu, 2003), depending on GABA gradients and the other ion gradients coupled to GABA transport. Vesicular GABA release from NGF cells might be another important source for ambient GABA (Overstreet-Wadiche & McBain, 2015)

GABA$_A$R Plasticity in Epilepsy and Alcohol-Induced Hyperexcitability

GABA$_A$R plasticity, defined as changes in GABA$_A$R subunit abundance and distribution, has been observed in numerous hyperexcitability animal models involving drug withdrawal and/or in induced epilepsy (Lagrange, Botzolakis, & Macdonald, 2007; Olsen & Spigelman, 2012; Schwarzer et al., 1997). The GABA$_A$R α4 subunit is up-regulated in neurosteroid withdrawal, accompanied by increased anxiety and tolerance to BZs (Smith & Gong, 2005), as well as in puberty and the menstrual cycle (Afroz, Parato, Shen, & Smith, 2016; Maguire, Stell, Rafizadeh, & Mody, 2005), and in pregnancy (Licheri et al., 2015; Maguire & Mody, 2008). The α4 subunit is also elevated in several models of epilepsy (Brooks-Kayal & Russek, 2012; Joshi & Kapur, 2012; Mody, 2012). There are two possible major principles involved: (1) GABA$_A$Rs are abnormal in epileptic individuals, as either a cause or a result of seizures, and (2) seizures probably induce plastic changes in GABA$_A$Rs. Numerous other changes in GABA$_A$R subunit composition, including decreased α1 and δ subunits, have been documented by immunohistochemical studies in epileptic hippocampal formation and neocortex (Houser, Zhang, & Peng, 2012).

Plastic changes in $GABA_AR$ expression such as altered subunit composition and localization, both regional and subcellular, have been reported in brains of animals, or primary cultured neurons, treated with acute and chronic EtOH (Follesa, Biggio, Caria, Gorini, & Biggio, 2004; Kumar, Fleming, & Morrow, 2004; Mhatre & Ticku, 1992; Olsen & Spigelman, 2012; Sanna et al., 2011).

$GABA_AR$ δ subtypes, once activated by low doses of EtOH, have been shown to undergo rapid internalization within 10 minutes after EtOH exposure, mediated by clathrin adaptor binding to sites in the intracellular domain of δ subunits (Gonzalez, Moss, & Olsen, 2012; Liang et al., 2007; Shen et al., 2011). This is an example of the well-known principle of use-dependent down-regulation of synaptic transmission at the receptor level; that is, receptor function is reduced (Gallager, Lakoski, Gonsalves, & Rauch, 1984). For example, the $GABA_AR$ plasticity seen in status epilepticus has been shown to involve overstimulation of synaptic γ2-containing $GABA_ARs$ and their phosphorylation-dependent internalization (Terunuma et al., 2008), leading to compensatory increase in other $GABA_AR$ subtypes, including extrasynaptic ones (Joshi & Kapur, 2012). Excessive GABA levels or efficacy induced by PAMs such as EtOH will down-regulate target $GABA_ARs$ and are associated with acute EtOH tolerance. Repeated exposure triggers long-lasting plastic changes in synaptic and extrasynaptic $GABA_AR$ subunit composition in the CNS along with alterations in $GABA_AR$-mediated synaptic and extrasynaptic currents, pharmacology, and behavior, which become irreversible after chronic intermittent ethanol (CIE) administration (60-dose cycles of intoxication and withdrawal) (Cagetti et al., 2003; Liang, Cagetti, Olsen, & Spigelman, 2004; Liang, Spigelman, & Olsen, 2009; Liang et al., 2007; Liang et al., 2006; Lindemeyer et al., 2017). Although most measurements were made in the hippocampal formation, changes in EtOH-sensitive GABAARs throughout the CNS are likely, thus affecting many behaviors (Diaz, Christian, Anderson, & McCool, 2011; Liang, Lindemeyer, et al., 2014; Lindemeyer et al., 2014). These alterations are consistent with alcohol dependence, including increased anxiety and seizure susceptibility, tolerance to the sedative effects of EtOH and other GABAergic modulators, and increased consumption (O'Dell, Roberts, Smith, & Koob, 2004; Olsen & Spigelman, 2012; Rimondini, Sommer, & Heilig, 2003; Simms et al., 2008).

References

Absalom, N., Eghorn, L. F., Villumsen, I. S., Karim, N., Bay, T., Olsen, J. V., . . . Wellendorph, P. (2012). a4bd $GABA_A$ receptors are high-affinity targets for g-hydroxybutyric acid (GHB). *Proceedings of the National Academy of Sciences of the United States of America, 109*(33), 13404–13409.

Afroz, S., Parato, J., Shen, H., & Smith, S. S. (2016). Synaptic pruning in the female hippocampus is triggered at puberty by extrasynaptic $GABA_A$ receptors on dendritic spines. *Elife*, 5. doi: 10.7554/eLife.15106

Aguayo, L. G., Peoples, R. W., Yeh, H. H., & Yevenes, G. E. (2002). $GABA_A$ receptors as molecular sites of ethanol action. Direct or indirect actions? *Current Topics in Medicinal Chemistry*, 2(8), 869–885.

Akk, G., Shu, H. J., Wang, C., Steinbach, J. H., Zorumski, C. F., Covey, D. F., & Mennerick, S. (2005). Neurosteroid access to the GABA$_A$ receptor. *Journal of Neuroscience, 25*(50), 11605–11613.

Antkowiak, B., & Rudolph, U. (2016). New insights in the systemic and molecular underpinnings of general anesthetic actions mediated by GABA$_A$ receptors. *Current Opinion in Anaesthesiology, 29*(4), 447–453.

Atack, J. R. (2010). GABA$_A$ receptor alpha2/alpha3 subtype-selective modulators as potential nonsedating anxiolytics. *Current Topics in Behavioral Neurosciences, 2*, 331–360.

Atack, J. R., Bayley, P. J., Seabrook, G. R., Wafford, K. A., McKernan, R. M., & Dawson, G. R. (2006). L-655,708 enhances cognition in rats but is not proconvulsant at a dose selective for a5-containing GABA$_A$ receptors. *Neuropharmacology, 51*(6), 1023–1029.

Atack, J. R., Wafford, K. A., Tye, S. J., Cook, S. M., Sohal, B., Pike, A., . . . McKernan, R. M. (2006). TPA023 [7-(1,1-dimethylethyl)-6-(2-ethyl-2H-1,2,4-triazol-3-ylmethoxy)-3-(2-fluorophenyl)-1,2,4-triazolo[4,3-b]pyridazine], an agonist selective for alpha2- and alpha3-containing GABA$_A$ receptors, is a nonsedating anxiolytic in rodents and primates. *Journal of Pharmacology and Experimental Therapeutics, 316*(1), 410–422.

Bali, M., & Akabas, M. H. (2004). Defining the propofol binding site location on the GABA$_A$ receptor. *Molecular Pharmacology, 65*(1), 68–76.

Baur, R., Gertsch, J., & Sigel, E. (2013). Do N-arachidonyl-glycine (NA-glycine) and 2-arachidonoyl glycerol (2-AG) share mode of action and the binding site on the b2 subunit of GABA$_A$ receptors? *Peer Journal 1*, e149.

Baur, R., Kielar, M., Richter, L., Ernst, M., Ecker, G. F., & Sigel, E. (2013). Molecular analysis of the site for 2-arachidonylglycerol (2-AG) on the beta(2) subunit of GABA$_A$ receptors. *Journal of Neurochemistry, 126*(1), 29–36.

Baur, R., Tan, K. R., Luscher, B. P., Gonthier, A., Goeldner, M., & Sigel, E. (2008). Covalent modification of GABA$_A$ receptor isoforms by a diazepam analogue provides evidence for a novel benzodiazepine binding site that prevents modulation by these drugs. *Journal of Neurochemistry, 106*(6), 2353–2363.

Belelli, D., Harrison, N. L., Maguire, J., Macdonald, R. L., Walker, M. C., & Cope, D. W. (2009). Extrasynaptic GABA$_A$ receptors: Form, pharmacology, and function. *Journal of Neuroscience, 29*(41), 12757–12763.

Belelli, D., & Lambert, J. J. (2005). Neurosteroids: Endogenous regulators of the GABA$_A$ receptor. *Nature Reviews Neuroscience, 6*(7), 565–575.

Belelli, D., Lambert, J. J., Peters, J. A., Wafford, K., & Whiting, P. J. (1997). The interaction of the general anesthetic etomidate with the GABA$_A$ receptor is influenced by a single amino acid. *Proceedings of the National Academy of Sciences of the United States of America, 94*(20), 11031–11036.

Ben-Ari, Y., Gaiarsa, J. L., Tyzio, R., & Khazipov, R. (2007). GABA: A pioneer transmitter that excites immature neurons and generates primitive oscillations. *Physiological Reviews, 87*(4), 1215–1284.

Bocquet, N., Nury, H., Baaden, M., Le Poupon, C., Changeux, J.-P., Delarue, M., & Corringer, P. J. (2009). X-ray structure of a pentameric ligand-gated ion channel in an apparently open conformation. *Nature, 457*(7225), 111–114.

Boehm, S. L., 2nd, Homanics, G. E., Blednov, Y. A., & Harris, R. A. (2006). d-Subunit containing GABA$_A$ receptor knockout mice are less sensitive to the actions of 4,5,6,7-tetrahydroisoxazolo-[5,4-c]pyridin-3-ol (THIP). *European Journal of Pharmacology, 541*(3), 158–162.

Borghese, C. M., & Harris, R. A. (2007). Studies of ethanol actions on recombinant d-containing $GABA_A$ receptors yield contradictory results. *Alcohol, 41*(3), 155–162.

Bowery, N. G., Collins, J. F., & Hill, R. G. (1976). Bicyclic phosphorus esters that are potent convulsants and GABA antagonists. *Nature, 261*(5561), 601–603.

Bowery, N. G., & Dray, A. (1976). Barbiturate reversal of amino acid antagonism produced by convulsant agents. *Nature, 264*(5583), 276–278.

Braat, S., & Kooy, R. F. (2015). Insights into $GABA_A$ergic system deficits in fragile X syndrome lead to clinical trials. *Neuropharmacology, 88*, 48–54.

Braestrup, C., & Squires, R. F. (1977). Specific benzodiazepine receptors in rat brain characterized by high-affinity [^{3}H]diazepam binding. *Proceedings of the National Academy of Sciences of the United States of America, 74*(9), 3805–3809.

Brejc, K., van Dijk, W. J., Klaassen, R. V., Schuurmans, M., van Der Oost, J., Smit, A. B., & Sixma, T. K. (2001). Crystal structure of an ACh-binding protein reveals the ligand-binding domain of nicotinic receptors. *Nature, 411*(6835), 269–276.

Brickley, S. G., & Mody, I. (2012). Extrasynaptic $GABA_A$ receptors: Their function in the CNS and implications for disease. *Neuron, 73*(1), 23–34.

Brickley, S. G., Revilla, V., Cull-Candy, S. G., Wisden, W., & Farrant, M. (2001). Adaptive regulation of neuronal excitability by a voltage-independent potassium conductance. *Nature, 409*(6816), 88–92.

Brooks-Kayal, A. R., & Russek, S. J. (2012). Regulation of $GABA_A$ receptor gene expression and epilepsy. In J. L. Noebels, M. Avoli, M. A. Rogawski, R. W. Olsen, & A. V. Delgado-Escueta (Eds.), *Jasper's Basic Mechanisms of the Epilepsies* (pp. 574–580). New York, NY: Oxford University Press.

Cagetti, E., Liang, J., Spigelman, I., & Olsen, R. W. (2003). Withdrawal from chronic intermittent ethanol treatment changes subunit composition, reduces synaptic function, and decreases behavioral responses to positive allosteric modulators of $GABA_A$ receptors. *Molecular Pharmacology, 63*(1), 53–64.

Capogna, M., & Pearce, R. A. (2011). $GABA_{A,slow}$: Causes and consequences. *Trends in Neurosciences, 34*(2), 101–112.

Caraiscos, V. B., Elliott, E. M., You-Ten, K. E., Cheng, V. Y., Belelli, D., Newell, J. G., . . . Orser, B. A. (2004). Tonic inhibition in mouse hippocampal CA1 pyramidal neurons is mediated by a5 subunit-containing $GABA_A$ receptors. *Proceedings of the National Academy of Sciences of the United States of America, 101*(10), 3662–3667.

Carpenter, T. S., & Lightstone, F. C. (2016). An electrostatic funnel in the GABA-binding pathway. *PLoS Computational Biology, 12*(4), e1004831.

Carr, L. G., Spence, J. P., Peter Eriksson, C. J., Lumeng, L., & Li, T. K. (2003). AA and ANA rats exhibit the R100Q mutation in the $GABA_A$ receptor a6 subunit. *Alcohol, 31*(1–2), 93–97.

Carta, M., Mameli, M., & Valenzuela, C. F. (2004). Alcohol enhances GABAergic transmission to cerebellar granule cells via an increase in Golgi cell excitability. *Journal of Neuroscience, 24*(15), 3746–3751.

Casida, J. E., Eto, M., Moscioni, A. D., Engel, J. L., Milbrath, D. S., & Verkade, J. G. (1976). Structure-toxicity relationships of 2,6,7-trioxabicyclo(2.2.2)octanes and related compounds. *Toxicology and Applied Pharmacology, 36*(2), 261–279.

Centanni, S. W., Burnett, E. J., Trantham-Davidson, H., & Chandler, L. J. (2017). Loss of d-GABA receptor-mediated tonic currents in the adult prelimbic cortex following adolescent alcohol exposure. *Addiction Biology, 22*(3), 616–628.

Chagraoui, A., Skiba, M., Thuillez, C., & Thibaut, F. (2016). To what extent is it possible to dissociate the anxiolytic and sedative/hypnotic properties of $GABA_A$ receptor modulators? *Progress in Neuro-Psychopharmacology and Biological Psychiatry, 71*, 189–202.
Chandra, D., Halonen, L. M., Linden, A. M., Procaccini, C., Hellsten, K., Homanics, G. E., & Korpi, E. R. (2010). Prototypic $GABA_A$ receptor agonist muscimol acts preferentially through forebrain high-affinity binding sites. *Neuropsychopharmacology, 35*(4), 999–1007.
Changeux, J. P. (1981). *The Harvey Lectures.* Amsterdam, Netherlands: Elsevier Science & Technology Books.
Chen, L., Durkin, K. A., & Casida, J. E. (2006). Structural model for $GABA_A$ receptor noncompetitive antagonist binding: Widely diverse structures fit the same site. *Proceedings of the National Academy of Sciences of the United States of America, 103*(13), 5185–5190.
Chen, Z.-W., Manion, B., Townsend, R. R., Reichert, D. E., Covey, D. F., Steinbach, J. H., . . . Evers, A. S. (2012). Neurosteroid analog photolabeling of a site in the third transmembrane domain of the b3 subunit of the $GABA_A$ receptor. *Molecular Pharmacology, 82*(3), 408–419.
Cheng, V. Y., Martin, L. J., Elliott, E. M., Kim, J. H., Mount, H. T., Taverna, F. A., . . . Orser, B. A. (2006). Alpha5 $GABA_A$ receptors mediate the amnestic but not sedative-hypnotic effects of the general anesthetic etomidate. *Journal of Neuroscience, 26*(14), 3713–3720.
Chiara, D. C., Jayakar, S. S., Zhou, X., Zhang, X., Savechenkov, P. Y., Bruzik, K. S., . . . Cohen, J. B. (2013). Specificity of intersubunit general anesthetic-binding sites in the transmembrane domain of the human a1b3g2 $GABA_A$ receptor. *Journal of Biological Chemistry, 288*(27), 19343–19357.
Chiara, D. C., Jounaidi, Y., Zhou, X., Savechenkov, P. Y., Bruzik, K. S., Miller, K. W., & Cohen, J. B. (2016). General anesthetic binding sites in human alpha4beta3delta $GABA_A$ receptors ($GABA_A$Rs). *Journal of Biological Chemistry, 291*(51), 26529–26539.
Collinson, N., Kuenzi, F. M., Jarolimek, W., Maubach, K. A., Cothliff, R., Sur, C., . . . Rosahl, T. W. (2002). Enhanced learning and memory and altered GABAergic synaptic transmission in mice lacking the a5 subunit of the $GABA_A$ receptor. *Journal of Neuroscience, 22*(13), 5572–5580.
Corringer, P. J., Baaden, M., Bocquet, N., Delarue, M., Dufresne, V., Nury, H., . . . Van Renterghem, C. (2010). Atomic structure and dynamics of pentameric ligand-gated ion channels: New insight from bacterial homologues. *Journal of Physiology, 588*(Pt 4), 565–572.
Costa, E., & Guidotti, A. (1979). Molecular mechanisms in the receptor action of benzodiazepines. *Annual Review of Pharmacology and Toxicology, 19*, 531–545.
Crestani, F., Low, K., Keist, R., Mandelli, M., Möhler, H., & Rudolph, U. (2001). Molecular targets for the myorelaxant action of diazepam. *Molecular Pharmacology, 59*(3), 442–445.
Crestani, F., Martin, J. R., Mohler, H., & Rudolph, U. (2000). Mechanism of action of the hypnotic zolpidem in vivo. *British Journal of Pharmacology, 131*(7), 1251–1254.
Cuatrecasas, P. (1974). Membrane receptors. *Annual Review of Biochemistry, 43*(0), 169–214.
Curtis, D. R., Duggan, A. W., Felix, D., & Johnston, G. A. (1970). Bicuculline and central GABA receptors. *Nature, 228*(5272), 676–677.
Dawson, G. R., Maubach, K. A., Collinson, N., Cobain, M., Everitt, B. J., MacLeod, A. M., . . . Atack, J. R. (2006). An inverse agonist selective for a5 subunit-containing $GABA_A$ receptors enhances cognition. *Journal of Pharmacology and Experimental Therapeutics, 316*(3), 1335–1345.
DeFeudis, F. V. (1980). Physiological and behavioral studies with muscimol. *Neurochemical Research, 5*(10), 1047–1068.
Dekermendjian, K., Kahnberg, P., Witt, M. R., Sterner, O., Nielsen, M., & Liljefors, T. (1999). Structure-activity relationships and molecular modeling analysis of flavonoids binding

to the benzodiazepine site of the rat brain $GABA_A$ receptor complex. *Journal of Medicinal Chemistry*, *42*(21), 4343–4350.

Diaz, M. R., Christian, D. T., Anderson, N. J., & McCool, B. A. (2011). Chronic ethanol and withdrawal differentially modulate lateral/basolateral amygdala paracapsular and local GABAergic synapses. *Journal of Pharmacology and Experimental Therapeutics*, *337*(1), 162–170.

Dixon, C. I., Morris, H. V., Breen, G., Desrivieres, S., Jugurnauth, S., Steiner, R. C., . . . Stephens, D. N. (2010). Cocaine effects on mouse incentive-learning and human addiction are linked to alpha2 subunit-containing $GABA_A$ receptors. *Proceedings of the National Academy of Sciences of the United States of America*, *107*(5), 2289–2294.

Eghorn, L. F., Hoestgaard-Jensen, K., Kongstad, K. T., Bay, T., Higgins, D., Frolund, B., & Wellendorph, P. (2014). Positive allosteric modulation of the GHB high-affinity binding site by the $GABA_A$ receptor modulator monastrol and the flavonoid catechin. *European Journal of Pharmacology*, *740*, 570–577.

Engin, E., Liu, J., & Rudolph, U. (2012). Alpha2-containing $GABA_A$ receptors: A target for the development of novel treatment strategies for CNS disorders. *Pharmacology and Therapeutics*, *136*(2), 142–152.

Engin, E., Smith, K. S., Gao, Y., Nagy, D., Foster, R. A., Tsvetkov, E., . . . Rudolph, U. (2016). Modulation of anxiety and fear via distinct intrahippocampal circuits. *Elife*, *5*, e14120.

Ernst, M., Brauchart, D., Boresch, S., & Sieghart, W. (2003). Comparative modeling of $GABA_A$ receptors: Limits, insights, future developments. *Neuroscience*, *119*(4), 933–943.

Ernst, M., Bruckner, S., Boresch, S., & Sieghart, W. (2005). Comparative models of $GABA_A$ receptor extracellular and transmembrane domains: Important insights in pharmacology and function. *Molecular Pharmacology*, *68*(5), 1291–1300.

Farrant, M., & Nusser, Z. (2005). Variations on an inhibitory theme: Phasic and tonic activation of $GABA_A$ receptors. *Nature Reviews Neuroscience*, *6*(3), 215–229.

Feng, H. J., Jounaidi, Y., Haburcak, M., Yang, X., & Forman, S. A. (2014). Etomidate produces similar allosteric modulation in a1b3d and a1b3g2L $GABA_A$ receptors. *British Journal of Pharmacology*, *171*(3), 789–798.

Fernandez, S. P., Mewett, K. N., Hanrahan, J. R., Chebib, M., & Johnston, G. A. (2008). Flavan-3-ol derivatives are positive modulators of $GABA_A$ receptors with higher efficacy for the alpha(2) subtype and anxiolytic action in mice. *Neuropharmacology*, *55*(5), 900–907.

ffrench-Constant, R. H., Rocheleau, T. A., Steichen, J. C., & Chalmers, A. E. (1993). A point mutation in a *Drosophila* GABA receptor confers insecticide resistance. *Nature*, *363*(6428), 449–451.

Fleming, R. L., Wilson, W. A., & Swartzwelder, H. S. (2007). Magnitude and ethanol sensitivity of tonic $GABA_A$ receptor-mediated inhibition in dentate gyrus changes from adolescence to adulthood. *Journal of Neurophysiology*, *97*(5), 3806–3811.

Follesa, P., Biggio, F., Caria, S., Gorini, G., & Biggio, G. (2004). Modulation of $GABA_A$ receptor gene expression by allopregnanolone and ethanol. *European Journal of Pharmacology*, *500*(1–3), 413–425.

Forman, S. A., & Chin, V. A. (2008). General anesthetics and molecular mechanisms of unconsciousness. *International Anesthesiology Clinics*, *46*(3), 43–53.

Forman, S. A., & Miller, K. W. (2011). Anesthetic sites and allosteric mechanisms of action on Cys-loop ligand-gated ion channels. *Canadian Journal of Anaesthia*, *58*(2), 191–205.

Forman, S. A., & Miller, K. W. (2016). Mapping general anesthetic sites in heteromeric GABA$_A$ receptors reveals a potential for targeting receptor subtypes. *Anesthesia and Analgesia, 123*(5), 1263–1273.

Franks, N. P., & Lieb, W. R. (1994). Molecular and cellular mechanisms of general anaesthesia. *Nature, 367*(6464), 607–614.

Friemel, A., Ebert, B., Hutson, P. H., Brust, P., Nieber, K., & Deuther-Conrad, W. (2007). Postnatal development and kinetics of [^{3}H]gaboxadol binding in rat brain: In vitro homogenate binding and quantitative autoradiography. *Brain Research, 1170*, 39–47.

Gallager, D. W., Lakoski, J. M., Gonsalves, S. F., & Rauch, S. L. (1984). Chronic benzodiazepine treatment decreases postsynaptic GABA sensitivity. *Nature, 308*(5954), 74–77.

Galzi, J.-L., & Changeux, J.-P. (1994). Ligand-gated ion channels as unconventional allosteric proteins. *Current Opinion in Structural Biology, 4*, 554–565.

Gilbert, W., & Muller-Hill, B. (1966). Isolation of the lac repressor. *Proceedings of the National Academy of Sciences of the United States of America, 56*(6), 1891–1898.

Glaeser, B. S., & Hare, T. A. (1975). Measurement of GABA in human cerebrospinal fluid. *Biochemical Medicine, 12*(3), 274–282.

Glykys, J., Mann, E. O., & Mody, I. (2008). Which GABA$_A$ receptor subunits are necessary for tonic inhibition in the hippocampus? *Journal of Neuroscience, 28*(6), 1421–1426.

Glykys, J., & Mody, I. (2007). The main source of ambient GABA responsible for tonic inhibition in the mouse hippocampus. *Journal of Physiology, 582*(Pt 3), 1163–1178.

Glykys, J., Peng, Z., Chandra, D., Homanics, G. E., Houser, C. R., & Mody, I. (2007). A new naturally occurring GABA$_A$ receptor subunit partnership with high sensitivity to ethanol. *Nature Neuroscience, 10*(1), 40–48.

Golovko, T., Min, R., Lozovaya, N., Falconer, C., Yatsenko, N., Tsintsadze, T., . . . Burnashev, N. (2015). Control of inhibition by the direct action of cannabinoids on GABA$_A$ receptors. *Cerebral Cortex, 25*(9), 2440–2455.

Gonzalez, C., Moss, S. J., & Olsen, R. W. (2012). Ethanol promotes clathrin adaptor-mediated endocytosis via the intracellular domain of delta-containing GABA$_A$ receptors. *Journal of Neuroscience, 32*(49), 17874–17881.

Haefely, W. (1982). Neurophysiology of benzodiazepines: summary. In E. Usdin, P. Skolnick, J. F. Tallman, D. Greenblatt, S. M. Paul (Eds.), *Pharmacology of Benzodiazepines* (pp. 509–516). Macmillan Press, London UK

Hanchar, H. J., Chutsrinopkun, P., Meera, P., Supavilai, P., Sieghart, W., Wallner, M., & Olsen, R. W. (2006). Ethanol potently and competitively inhibits binding of the alcohol antagonist Ro15-4513 to a4/6b3d GABA$_A$ receptors. *Proceedings of the National Academy of Sciences of the United States of America, 103*(22), 8546–8550.

Hanchar, H. J., Dodson, P. D., Olsen, R. W., Otis, T. S., & Wallner, M. (2005). Alcohol-induced motor impairment caused by increased extrasynaptic GABA$_A$ receptor activity. *Nature Neuroscience, 8*(3), 339–345.

Hanrahan, J. R., Chebib, M., & Johnston, G. A. (2011). Flavonoid modulation of GABA$_A$ receptors. *British Journal of Pharmacology, 163*(2), 234–245.

Hanrahan, J. R., Chebib, M., & Johnston, G. A. (2015). Interactions of flavonoids with ionotropic GABA receptors. *Advances in Pharmacology, 72*, 189–200.

Harrison, N. L., & Simmonds, M. A. (1984). Modulation of the GABA receptor complex by a steroid anaesthetic. *Brain Research, 323*(2), 287–292.

Herman, M. A., Contet, C., Justice, N. J., Vale, W., & Roberto, M. (2013). Novel subunit-specific tonic GABA currents and differential effects of ethanol in the central amygdala of CRF receptor-1 reporter mice. *Journal of Neuroscience, 33*(8), 3284–3298.

Herman, M. A., & Roberto, M. (2016). Cell-type-specific tonic GABA signaling in the rat central amygdala is selectively altered by acute and chronic ethanol. *Addiction Biology, 21*(1), 72–86.

Hibbs, R. E., & Gouaux, E. (2011). Principles of activation and permeation in an anion-selective Cys-loop receptor. *Nature, 474*(7249), 54–60.

Hilf, R. J., & Dutzler, R. (2008). X-ray structure of a prokaryotic pentameric ligand-gated ion channel. *Nature, 452*(7185), 375–379.

Hill-Venning, C., Belelli, D., Peters, J. A., & Lambert, J. J. (1997). Subunit-dependent interaction of the general anaesthetic etomidate with the gamma-aminobutyric acid type A receptor. *British Journal of Pharmacology, 120*(5), 749–756.

Hoerbelt, P., Lindsley, T. A., & Fleck, M. W. (2015). Dopamine directly modulates $GABA_A$ receptors. *Journal of Neuroscience, 35*(8), 3525–3536.

Hold, K. M., Sirisoma, N. S., Ikeda, T., Narahashi, T., & Casida, J. E. (2000). A-thujone (the active component of absinthe): $GABA_A$ receptor modulation and metabolic detoxification. *Proceedings of the National Academy of Sciences of the United States of America, 97*(8), 3826–3831.

Hosie, A. M., Wilkins, M. E., da Silva, H. M., & Smart, T. G. (2006). Endogenous neurosteroids regulate $GABA_A$ receptors through two discrete transmembrane sites. *Nature, 444*(7118), 486–489.

Hosie, A. M., Wilkins, M. E., & Smart, T. G. (2007). Neurosteroid binding sites on $GABA_A$ receptors. *Pharmacology and Therapeutics, 116*(1), 7–19.

Houser, C. R., Zhang, N., & Peng, Z. (2012). Alterations in the distribution of $GABA_A$ receptors in epilepsy. In J. L. Noebels, M. Avoli, M. A. Rogawski, R. W. Olsen, & A. V. Delgado-Escueta (Eds.), *Jasper's Basic Mechanisms of the Epilepsies* (pp. 532–544). Oxford University Press, NY.

Hunkeler, W., Möhler, H., Pieri, L., Polc, P., Bonetti, E. P., Cumin, R., . . . Haefely, W. (1981). Selective antagonists of benzodiazepines. *Nature, 290*(5806), 514–516.

Inoue, M., & Akaike, N. (1988). Blockade of gamma-aminobutyric acid-gated chloride current in frog sensory neurons by picrotoxin. *Neuroscience Research, 5*(5), 380–394.

Jacob, F., & Monod, J. (1961). Genetic regulatory mechanisms in the synthesis of proteins. *Journal of Molecular Biology, 3*, 318–356.

Jacob, T. C., Moss, S. J., & Jurd, R. (2008). $GABA_A$ receptor trafficking and its role in the dynamic modulation of neuronal inhibition. *Nature Reviews Neuroscience, 9*(5), 331–343.

Jayakar, S. S., Zhou, X., Chiara, D. C., Dostalova, Z., Savechenkov, P. Y., Bruzik, K. S., . . . Cohen, J. B. (2014). Multiple propofol-binding sites in a $GABA_AR$ identified using a photoreactive propofol analog. *Journal of Biological Chemistry, 289*(40), 27456–27468.

Jenkins, A., Greenblatt, E. P., Faulkner, H. J., Bertaccini, E., Light, A., Lin, A., . . . Harrison, N. L. (2001). Evidence for a common binding cavity for three general anesthetics within the $GABA_A$ receptor. *Journal of Neuroscience, 21*(6), RC136.

Jensen, M. L., Wafford, K. A., Brown, A. R., Belelli, D., Lambert, J. J., & Mirza, N. R. (2013). A study of subunit selectivity, mechanism and site of action of the d selective compound 2 (DS2) at human recombinant and rodent native $GABA_A$ receptors. *British Journal of Pharmacology, 168*(5), 1118–1132.

Jia, F., Chandra, D., Homanics, G. E., & Harrison, N. L. (2008). Ethanol modulates synaptic and extrasynaptic $GABA_A$ receptors in the thalamus. *Journal of Pharmacology and Experimental Therapeutics, 326*(2), 475–482.

Jia, F., Pignataro, L., & Harrison, N. L. (2007). $GABA_A$ receptors in the thalamus: A4 subunit expression and alcohol sensitivity. *Alcohol, 41*(3), 177–185.

Jia, F., Yue, M., Chandra, D., Keramidas, A., Goldstein, P. A., Homanics, G. E., & Harrison, N. L. (2008). Taurine is a potent activator of extrasynaptic $GABA_A$ receptors in the thalamus. *Journal of Neuroscience, 28*(1), 106–115.

Johnston, G. A. (1983). Regulation of GABA receptors by barbiturates and by related sedative-hypnotic and anticonvulsant drugs. In S. J. Enna (Ed.), *The GABA Receptors* (pp. 108–128). Totowa, NJ: Humana Press.

Jones, A., Korpi, E. R., McKernan, R. M., Pelz, R., Nusser, Z., Makela, R., . . . Wisden, W. (1997). Ligand-gated ion channel subunit partnerships: $GABA_A$ receptor a6 subunit gene inactivation inhibits d subunit expression. *Journal of Neuroscience, 17*(4), 1350–1362.

Joshi, S., & Kapur, J. (2012). $GABA_A$ receptor plasticity during status epilepticus. In J. L. Noebels, M. Avoli, M. A. Rogawski, R. W. Olsen, & A. V. Delgado-Escueta (Eds.), *Jasper's Basic Mechanisms of the Epilepsies* (pp. 545–554). Oxford University Press, NY.

Jurd, R., Arras, M., Lambert, S., Drexler, B., Siegwart, R., Crestani, F., . . . Rudolph, U. (2003). General anesthetic actions in vivo strongly attenuated by a point mutation in the $GABA_A$ receptor b3 subunit. *The FASEB Journal, 17*(2), 250–252.

Kavvadias, D., Sand, P., Youdim, K. A., Qaiser, M. Z., Rice-Evans, C., Baur, R., . . . Schreier, P. (2004). The flavone hispidulin, a benzodiazepine receptor ligand with positive allosteric properties, traverses the blood–brain barrier and exhibits anticonvulsive effects. *British Journal of Pharmacology, 142*(5), 811–820.

Kletke, O., Gisselmann, G., May, A., Hatt, H., & Sergeeva, O. (2013). Partial agonism of taurine at gamma-containing native and recombinant $GABA_A$ receptors. *PloS One, 8*(4), e61733.

Koltchine, V. V., Finn, S. E., Jenkins, A., Nikolaeva, N., Lin, A., & Harrison, N. L. (1999). Agonist gating and isoflurane potentiation in the human $GABA_A$ receptor determined by the volume of a second transmembrane domain residue. *Molecular Pharmacology, 56*(5), 1087–1093.

Korpi, E. R., Kleingoor, C., Kettenmann, H., & Seeburg, P. H. (1993). Benzodiazepine-induced motor impairment linked to point mutation in cerebellar $GABA_A$ receptor. *Nature, 361*(6410), 356–359.

Krasowski, M. D., & Harrison, N. L. (2000). The actions of ether, alcohol and alkane general anaesthetics on $GABA_A$ and glycine receptors and the effects of TM2 and TM3 mutations. *British Journal of Pharmacology, 129*(4), 731–743.

Krasowski, M. D., Nishikawa, K., Nikolaeva, N., Lin, A., & Harrison, N. L. (2001). Methionine 286 in transmembrane domain 3 of the $GABA_A$ receptor b subunit controls a binding cavity for propofol and other alkylphenol general anesthetics. *Neuropharmacology, 41*(8), 952–964.

Kucken, A. M., Teissere, J. A., Seffinga-Clark, J., Wagner, D. A., & Czajkowski, C. (2003). Structural requirements for imidazobenzodiazepine binding to $GABA_A$ receptors. *Molecular Pharmacology, 63*(2), 289–296.

Kumar, S., Fleming, R. L., & Morrow, A. L. (2004). Ethanol regulation of $GABA_A$ receptors: Genomic and nongenomic mechanisms. *Pharmacology and Therapeutics, 101*(3), 211–226.

Lagrange, A. H., Botzolakis, E. J., & Macdonald, R. L. (2007). Enhanced macroscopic desensitization shapes the response of a4 subtype-containing $GABA_A$ receptors to synaptic and extrasynaptic GABA. *Journal of Physiology, 578*(Pt 3), 655–676.

Leeb-Lundberg, F., Snowman, A., & Olsen, R. W. (1980). Barbiturate receptor sites are coupled to benzodiazepine receptors. *Proceedings of the National Academy of Sciences of the United States of America*, *77*(12), 7468–7472.

Leeb-Lundberg, F., Snowman, A., & Olsen, R. W. (1981). Perturbation of benzodiazepine receptor binding by pyrazolopyridines involves picrotoxinin/barbiturate receptor sites. *Journal of Neuroscience*, *1*(5), 471–477.

Lefkowitz, R. J. (1975). Identification of adenylate cyclase-coupled b-adrenergic receptors with radiolabeled b-adrenergic antagonists. *Biochemical Pharmacology*, *24*(18), 1651–1658.

Li, G.-D., Chiara, D. C., Cohen, J. B., & Olsen, R. W. (2010). Numerous classes of general anesthetics inhibit etomidate binding to $GABA_A$ receptors. *Journal of Biological Chemistry*, *285*(12), 8615–8620.

Li, G.-D., Chiara, D. C., Sawyer, G. W., Husain, S. S., Olsen, R. W., & Cohen, J. B. (2006). Identification of a $GABA_A$ receptor anesthetic binding site at subunit interfaces by photolabeling with an etomidate analog. *Journal of Neuroscience*, *26*(45), 11599–11605.

Li, G. D., Chiara, D. C., Cohen, J. B., & Olsen, R. W. (2009). Neurosteroids allosterically modulate binding of the anesthetic etomidate to gamma-aminobutyric acid type A receptors. *Journal of Biological Chemistry*, *284*(18), 11771–11775.

Liang, J., Cagetti, E., Olsen, R. W., & Spigelman, I. (2004). Altered pharmacology of synaptic and extrasynaptic $GABA_A$ receptors on CA1 hippocampal neurons is consistent with subunit changes in a model of alcohol withdrawal and dependence. *Journal of Pharmacology and Experimental Therapeutics*, *310*(3), 1234–1245.

Liang, J., Lindemeyer, A. K., Suryanarayanan, A., Meyer, E. M., Marty, V. N., Ahmad, S. O., . . . Spigelman, I. (2014). Plasticity of $GABA_A$ receptor-mediated neurotransmission in the nucleus accumbens of alcohol-dependent rats. *Journal of Neurophysiology*, *112*(1), 39–50.

Liang, J., Lopez-Valdes, H. E., Martinez-Coria, H., Lindemeyer, A. K., Shen, Y., Shao, X. M., & Olsen, R. W. (2014). Dihydromyricetin ameliorates behavioral deficits and reverses neuropathology of transgenic mouse models of Alzheimer's disease. *Neurochemical Research*, *39*(6), 1171–1181.

Liang, J., Spigelman, I., & Olsen, R. W. (2009). Tolerance to sedative/hypnotic actions of GABAergic drugs correlates with tolerance to potentiation of extrasynaptic tonic currents of alcohol-dependent rats. *Journal of Neurophysiology*, *102*(1), 224–233.

Liang, J., Suryanarayanan, A., Abriam, A., Snyder, B., Olsen, R. W., & Spigelman, I. (2007). Mechanisms of reversible $GABA_A$ receptor plasticity after ethanol intoxication. *Journal of Neuroscience*, 27(45), 12367–12377.

Liang, J., Zhang, N., Cagetti, E., Houser, C. R., Olsen, R. W., & Spigelman, I. (2006). Chronic intermittent ethanol-induced switch of ethanol actions from extrasynaptic to synaptic hippocampal $GABA_A$ receptors. *Journal of Neuroscience*, *26*(6), 1749–1758.

Licheri, V., Talani, G., Gorule, A. A., Mostallino, M. C., Biggio, G., & Sanna, E. (2015). Plasticity of $GABA_A$ receptors during pregnancy and postpartum period: From gene to function. *Neural Plasticity*, *2015*, 170435.

Lindemeyer, A. K., Liang, J., Marty, V. N., Meyer, E. M., Suryanarayanan, A., Olsen, R. W., & Spigelman, I. (2014). Ethanol-induced plasticity of $GABA_A$ receptors in the basolateral amygdala. *Neurochemical Research*, 39(6), 1162–1170.

Lindemeyer, A. K., Shen, Y., Yazdani, F., Shao, X. M., Davies, D. L., Spigelman, I., . . . Liang, J. (2017). Alpha 2 subunit-containing $GABA_A$ receptor subtypes are up-regulated and contribute to alcohol-induced functional plasticity in rat hippocampus. *Molecular Pharmacology*, 92(2), 101–102.

Liu, J., Yang, A. R., Kelly, T., Puche, A., Esoga, C., June, H. L., Jr., . . . Aurelian, L. (2011). Binge alcohol drinking is associated with GABA$_A$ a2-regulated Toll-like receptor 4 (TLR4) expression in the central amygdala. *Proceedings of the National Academy of Sciences of the United States of America*, *108*(11), 4465–4470.

Lobo, I. A., & Harris, R. A. (2008). GABA$_A$ receptors and alcohol. *Pharmacology, Biochemistry and Behavior*, *90*(1), 90–94.

Lovinger, D. M., & Homanics, G. E. (2007). Tonic for what ails us? High-affinity GABA$_A$ receptors and alcohol. *Alcohol*, *41*(3), 139–143.

Löw, K., Crestani, F., Keist, R., Benke, D., Brunig, I., Benson, J. A., . . . Rudolph, U. (2000). Molecular and neuronal substrate for the selective attenuation of anxiety. *Science*, *290*(5489), 131–134.

Lüscher, B., Shen, Q., & Sahir, N. (2011). The GABAergic deficit hypothesis of major depressive disorder. *Molecular Psychiatry*, *16*(4), 383–406.

Macdonald, R. L., & Olsen, R. W. (1994). GABA$_A$ receptor channels. *Annual Review of Neuroscience*, *17*, 569–602.

Maguire, J., & Mody, I. (2008). GABA$_A$R plasticity during pregnancy: Relevance to postpartum depression. *Neuron*, *59*(2), 207–213.

Maguire, J. L., Stell, B. M., Rafizadeh, M., & Mody, I. (2005). Ovarian cycle-linked changes in GABA$_A$ receptors mediating tonic inhibition alter seizure susceptibility and anxiety. *Nature Neuroscience*, *8*(6), 797–804.

Majewska, M. D., Harrison, N. L., Schwartz, R. D., Barker, J. L., & Paul, S. M. (1986). Steroid hormone metabolites are barbiturate-like modulators of the GABA receptor. *Science*, 232(4753), 1004–1007.

Maldifassi, M. C., Baur, R., & Sigel, E. (2016a). Functional sites involved in modulation of the GABA$_A$ receptor channel by the intravenous anesthetics propofol, etomidate and pentobarbital. *Neuropharmacology*, *105*, 207–214.

Maldifassi, M. C., Baur, R., & Sigel, E. (2016b). Molecular mode of action of CGS 9895 at a1b2g2 GABA$_A$ receptors. *Journal of Neurochemistry*, *138*(5), 722–730.

Marder, M., & Paladini, A. C. (2002). GABA$_A$ receptor ligands of flavonoid structure. *Current Topics in Medicinal Chemistry*, 2(8), 853–867.

Martin, D. L., & Olsen, R. W. (2000). *GABA in the Nervous System: The View at Fifty Years* (1st ed.). Philadelphia, PA: Lippincott Williams & Wilkins Publishers.

Mascia, M. P., Trudell, J. R., & Harris, R. A. (2000). Specific binding sites for alcohols and anesthetics on ligand-gated ion channels. *Proceedings of the National Academy of Sciences of the United States of America*, *97*(16), 9305–9310.

McKernan, R. M., Rosahl, T. W., Reynolds, D. S., Sur, C., Wafford, K. A., Atack, J. R., . . . Whiting, P. J. (2000). Sedative but not anxiolytic properties of benzodiazepines are mediated by the GABA$_A$ receptor a1 subtype. *Nature Neuroscience*, 3(6), 587–592.

McLean, P. J., Farb, D. H., & Russek, S. J. (1995). Mapping of the a4 subunit gene (GABRA4) to human chromosome 4 defines an a2-a4-b1-g1 gene cluster: Further evidence that modern GABA$_A$ receptor gene clusters are derived from an ancestral cluster. *Genomics*, *26*(3), 580–586.

Meera, P., Olsen, R. W., Otis, T. S., & Wallner, M. (2009). Etomidate, propofol and the neurosteroid THDOC increase the GABA efficacy of recombinant a4b3d and a4b3 GABA$_A$ receptors expressed in HEK cells. *Neuropharmacology*, *56*(1), 155–160.

Meera, P., Olsen, R. W., Otis, T. S., & Wallner, M. (2010). Alcohol- and alcohol antagonist-sensitive human GABA$_A$ receptors: Tracking δ subunit incorporation into functional receptors. *Molecular Pharmacology*, *78*, 918–924.

Meera, P., Wallner, M., & Otis, T. S. (2011). Molecular basis for the high THIP/gaboxadol sensitivity of extrasynaptic $GABA_A$ receptors. *Journal of Neurophysiology, 106*(4), 2057–2064.

Mhatre, M. C., & Ticku, M. K. (1992). Chronic ethanol administration alters $GABA_A$ receptor gene expression. *Molecular Pharmacology, 42*(3), 415–422.

Mihalek, R. M., Banerjee, P. K., Korpi, E. R., Quinlan, J. J., Firestone, L. L., Mi, Z. P., ... Homanics, G. E. (1999). Attenuated sensitivity to neuroactive steroids in $GABA_A$ receptor d subunit knockout mice. *Proceedings of the National Academy of Sciences of the United States of America, 96*(22), 12905–12910.

Mihic, S. J., Ye, Q., Wick, M. J., Koltchine, V. V., Krasowski, M. D., Finn, S. E., ... Harrison, N. L. (1997). Sites of alcohol and volatile anaesthetic action on $GABA_A$ and glycine receptors. *Nature, 389*(6649), 385–389.

Miller, P. S., & Aricescu, A. R. (2014). Crystal structure of a human GABA receptor. *Nature, 512*, 270–275.

Mody, I. (2012). Plasticity of $GABA_A$ receptors relevant to neurosteroid actions. In J. L. Noebels, M. Avoli, M. A. Rogawski, R. W. Olsen, & A. V. Delgado-Escueta (Eds.), *Jasper's Basic Mechanisms of the Epilepsies* (pp. 555–561). New York, NY: Oxford University Press.

Mody, I., Glykys, J., & Wei, W. (2007). A new meaning for "gin & tonic": Tonic inhibition as the target for ethanol action in the brain. *Alcohol, 41*(3), 145–153.

Möhler, H. (1982). Benzodiazepine receptors: Differential interaction of benzodiazepine agonists and antagonists after photoaffinity labeling with flunitrazepam. *European Journal of Pharmacology, 80*(4), 435–436.

Möhler, H. (2011). The rise of a new GABA pharmacology. *Neuropharmacology, 60*(7–8), 1042–1049.

Möhler, H., & Okada, T. (1977). Benzodiazepine receptor: Demonstration in the central nervous system. *Science, 198*(4319), 849–851.

Mula, M. (2016). Investigational new drugs for focal epilepsy. *Expert Opinion on Investigational Drugs, 25*(1), 1–5.

Nicoll, R. A., Eccles, J. C., Oshima, T., & Rubia, F. (1975). Prolongation of hippocampal inhibitory postsynaptic potentials by barbiturates. *Nature, 258*(5536), 625–627.

Nusser, Z., Ahmad, Z., Tretter, V., Fuchs, K., Wisden, W., Sieghart, W., & Somogyi, P. (1999). Alterations in the expression of $GABA_A$ receptor subunits in cerebellar granule cells after the disruption of the a6 subunit gene. *European Journal of Neuroscience, 11*(5), 1685–1697.

Nutt, D. J., Besson, M., Wilson, S. J., Dawson, G. R., & Lingford-Hughes, A. R. (2007). Blockade of alcohol's amnestic activity in humans by an a5 subtype benzodiazepine receptor inverse agonist. *Neuropharmacology, 53*(7), 810–820.

O'Dell, L. E., Roberts, A. J., Smith, R. T., & Koob, G. F. (2004). Enhanced alcohol self-administration after intermittent versus continuous alcohol vapor exposure. *Alcoholism, Clinical and Experimental Research, 28*(11), 1676–1682.

Olah, S., Fule, M., Komlosi, G., Varga, C., Baldi, R., Barzo, P., & Tamas, G. (2009). Regulation of cortical microcircuits by unitary GABA-mediated volume transmission. *Nature, 461*(7268), 1278–1281.

Olsen, R. W. (1982). Drug interactions at the GABA receptor-ionophore complex. *Annual Review of Pharmacology and Toxicology, 22*, 245–277.

Olsen, R. W. (2000). Absinthe and $GABA_A$ receptors. *Proceedings of the National Academy of Sciences of the United States of America, 97*(9), 4417–4418.

Olsen, R. W. (2006). Picrotoxin-like channel blockers of $GABA_A$ receptors. *Proceedings of the National Academy of Sciences of the United States of America, 103*(16), 6081–6082.

Olsen, R. W. (2015). Allosteric ligands and their binding sites define GABA$_A$ receptor subtypes. *Advances in Pharmacology, 73*, 167–202.
Olsen, R. W., Chang, C. S., Li, G., Hanchar, H. J., & Wallner, M. (2004). Fishing for allosteric sites on GABA$_A$ receptors. *Biochemical Pharmacology, 68*(8), 1675–1684.
Olsen, R. W., Fischer, J. B., & Dunwiddie, T. V. (1986). Barbiturate enhancement of GABA receptor binding and function as a mechanism of anesthesia. In S. Roth & K. W. Miller (Eds.), *Molecular and Cellular Mechanisms of Anaesthetics* (pp. 165–177). New York: Plenum Publishing Corporation.
Olsen, R. W., Hanchar, H. J., Meera, P., & Wallner, M. (2007). GABA$_A$ receptor subtypes: The "one glass of wine" receptors. *Alcohol, 41*(3), 201–209.
Olsen, R. W., & Li, G.-D. (2011). GABA$_A$ receptors as molecular targets of general anesthetics: Identification of binding sites provides clues to allosteric modulation. *Canadian Journal of Anaesthia, 58*(2), 206–215.
Olsen, R. W., & Li, G.-D. (2012). GABA. In S. Brady, G. Siegel, R. W. Albers, & D. Price (Eds.), *Basic Neurochemistry: Molecular, Cellular and Medical Aspects* (Vol. 8; pp. 367–376): Burlington, VT: Academic Press.
Olsen, R. W., Li, G.-D., Wallner, M., Trudell, J. R., Bertaccini, E. J., Lindahl, E., . . . Davies, D. L. (2014). Structural models of ligand-gated ion channels: Sites of action for anesthetics and ethanol. *Alcoholism: Clinical and Experimental Research, 38*(3), 595–603.
Olsen, R. W., Sapp, D. M., Bureau, M. H., Turner, D. M., & Kokka, N. (1991). Allosteric actions of central nervous system depressants including anesthetics on subtypes of the inhibitory GABA$_A$ receptor-chloride channel complex. *Annals of the New York Academy of Sciences, 625*, 145–154.
Olsen, R. W., & Sapp, D. W. (1995). Neuroactive steroid modulation of GABA$_A$ receptors. In G. Biggio (Ed.), *Advances in Biochemical Psychopharmacology* (Vol. 48, pp. 57–74). New York: Raven Press.
Olsen, R. W., & Sieghart, W. (2008). International Union of Pharmacology. LXX. Subtypes of GABA$_A$ receptors: Classification on the basis of subunit composition, pharmacology, and function. Update. *Pharmacological Reviews, 60*(3), 243–260.
Olsen, R. W., & Sieghart, W. (2009). GABA$_A$ receptors: Subtypes provide diversity of function and pharmacology. *Neuropharmacology, 56*(1), 141–148.
Olsen, R. W., & Snowman, A. M. (1982). Chloride-dependent enhancement by barbiturates of GABA receptor binding. *Journal of Neuroscience, 2*(12), 1812–1823.
Olsen, R. W., & Spigelman, I. (2012). GABA$_A$ receptor plasticity in alcohol withdrawal. In J. L. Noebels, M. Avoli, M. A. Rogawski, R. W. Olsen, & A. V. Delgado-Escueta (Eds.), *Jasper's Basic Mechanisms of the Epilepsies* (4th ed.; pp. 562–573). New York, NY: Oxford University Press.
Olsen, R. W., Ticku, M. K., & Miller, T. (1978). Dihydropicrotoxinin binding to crayfish muscle sites possibly related to GABA receptor-ionophores. *Molecular Pharmacology, 14*(3), 381–390.
Overstreet-Wadiche, L., & McBain, C. J. (2015). Neurogliaform cells in cortical circuits. *Nature Reviews Neuroscience, 16*(8), 458–468.
Pert, C. B., & Snyder, S. H. (1973). Properties of opiate-receptor binding in rat brain. *Proceedings of the National Academy of Sciences of the United States of America, 70*(8), 2243–2247.
Peters, J. A., Kirkness, E. F., Callachan, H., Lambert, J. J., & Turner, A. J. (1988). Modulation of the GABA$_A$ receptor by depressant barbiturates and pregnane steroids. *British Journal of Pharmacology, 94*(4), 1257–1269.

Poe, M. M., Methuku, K. R., Li, G., Verma, A. R., Teske, K. A., Stafford, D. C., . . . Schkeryantz, J. M. (2016). Synthesis and characterization of a novel gamma-aminobutyric acid type A ($GABA_A$) receptor ligand that combines outstanding metabolic stability, pharmacokinetics, and anxiolytic efficacy. *Journal of Medicinal Chemistry*, *59*(23), 10800–10806.

Prenosil, G. A., Schneider Gasser, E. M., Rudolph, U., Keist, R., Fritschy, J. M., & Vogt, K. E. (2006). Specific subtypes of $GABA_A$ receptors mediate phasic and tonic forms of inhibition in hippocampal pyramidal neurons. *Journal of Neurophysiology*, *96*(2), 846–857.

Puia, G., Vicini, S., Seeburg, P. H., & Costa, E. (1991). Influence of recombinant $GABA_A$ receptor subunit composition on the action of allosteric modulators of GABA-gated Cl- currents. *Molecular Pharmacology*, *39*(6), 691–696.

Puthenkalam, R., Hieckel, M., Simeone, X., Suwattanasophon, C., Feldbauer, R. V., Ecker, G. F., & Ernst, M. (2016). Structural studies of $GABA_A$ receptor binding sites: Which experimental structure tells us what? *Frontiers in Molecular Neuroscience*, *9*, 44.

Ralvenius, W. T., Benke, D., Acuna, M. A., Rudolph, U., & Zeilhofer, H. U. (2015). Analgesia and unwanted benzodiazepine effects in point-mutated mice expressing only one benzodiazepine-sensitive $GABA_A$ receptor subtype. *Nature Communications*, *6*, 6803.

Ramerstorfer, J., Furtmuller, R., Sarto-Jackson, I., Varagic, Z., Sieghart, W., & Ernst, M. (2011). The $GABA_A$ receptor a+b− interface: A novel target for subtype selective drugs. *Journal of Neuroscience*, *31*(3), 870–877.

Ransom, B. R., & Barker, J. L. (1975). Pentobarbital modulates transmitter effects on mouse spinal neurones grown in tissue culture. *Nature*, *254*(5502), 703–705.

Ransom, C. B., Tao, W., Wu, Y., Spain, W. J., & Richerson, G. B. (2013). Rapid regulation of tonic GABA currents in cultured rat hippocampal neurons. *Journal of Neurophysiology*, *109*(3), 803–812.

Ren, L., Wang, F., Xu, Z., Chan, W. M., Zhao, C., & Xue, H. (2010). $GABA_A$ receptor subtype selectivity underlying anxiolytic effect of 6-hydroxyflavone. *Biochemical Pharmacology*, *79*(9), 1337–1344.

Reynolds, D. S., Rosahl, T. W., Cirone, J., O'Meara, G. F., Haythornthwaite, A., Newman, R. J., . . . Wafford, K. A. (2003). Sedation and anesthesia mediated by distinct $GABA_A$ receptor isoforms. *Journal of Neuroscience*, *23*(24), 8608–8617.

Richerson, G. B., & Wu, Y. (2003). Dynamic equilibrium of neurotransmitter transporters: Not just for reuptake anymore. *Journal of Neurophysiology*, *90*(3), 1363–1374.

Rimondini, R., Sommer, W., & Heilig, M. (2003). A temporal threshold for induction of persistent alcohol preference: Behavioral evidence in a rat model of intermittent intoxication. *Journal of Studies on Alcohol*, *64*(4), 445–449.

Roberts, E., Chase, T. N., & Tower, D. (Eds.). (1976). *GABA in Nervous System Function.* New York: Raven Press.

Rudolph, U., & Antkowiak, B. (2004). Molecular and neuronal substrates for general anaesthetics. *Nature Reviews Neuroscience*, *5*(9), 709–720.

Rudolph, U., Crestani, F., Benke, D., Brunig, I., Benson, J. A., Fritschy, J. M., . . . Möhler, H. (1999). Benzodiazepine actions mediated by specific $GABA_A$ receptor subtypes. *Nature*, *401*(6755), 796–800.

Rudolph, U., & Knoflach, F. (2011). Beyond classical benzodiazepines: Novel therapeutic potential of $GABA_A$ receptor subtypes. *Nature Reviews Drug Discovery*, *10*(9), 685–697.

Rudolph, U., & Möhler, H. (2004). Analysis of $GABA_A$ receptor function and dissection of the pharmacology of benzodiazepines and general anesthetics through mouse genetics. *Annual Review of Pharmacology and Toxicology*, *44*, 475–498.

Rudolph, U., & Möhler, H. (2014). GABA$_A$ receptor subtypes: Therapeutic potential in Down syndrome, affective disorders, schizophrenia, and autism. *Annual Review of Pharmacology and Toxicology*, *54*, 483–507.

Rüsch, D., Zhong, H., & Forman, S. A. (2004). Gating allosterism at a single class of etomidate sites on a1b2g2L GABA$_A$ receptors accounts for both direct activation and agonist modulation. *Journal of Biological Chemistry*, *279*(20), 20982–20992.

Russek, S. J. (1999). Evolution of GABA$_A$ receptor diversity in the human genome. *Gene*, *227*(2), 213–222.

Saba, L., Porcella, A., Congeddu, E., Colombo, G., Peis, M., Pistis, M., . . . Pani, L. (2001). The R100Q mutation of the GABA$_A$ a6 receptor subunit may contribute to voluntary aversion to ethanol in the sNP rat line. *Brain Research: Molecular Brain Research*, *87*(2), 263–270.

Sanna, E., Talani, G., Obili, N., Mascia, M. P., Mostallino, M. C., Secci, P. P., . . . Follesa, P. (2011). Voluntary ethanol consumption induced by social isolation reverses the increase of alpha4/delta GABA$_A$ receptor gene expression and function in the hippocampus of C57BL/6J mice. *Frontiers in Neuroscience*, *5*, 15.

Santhakumar, V., Hanchar, H. J., Wallner, M., Olsen, R. W., & Otis, T. S. (2006). Contributions of the GABA$_A$ receptor a6 subunit to phasic and tonic inhibition revealed by a naturally occurring polymorphism in the a6 gene. *Journal of Neuroscience*, *26*(12), 3357–3364.

Santhakumar, V., Meera, P., Karakossian, M. H., & Otis, T. S. (2013). A reinforcing circuit action of extrasynaptic GABA$_A$ receptor modulators on cerebellar granule cell inhibition. *PLoS One*, *8*(8), e72976.

Santhakumar, V., Wallner, M., & Otis, T. (2007). Ethanol acts directly on extrasynaptic subtypes of GABA$_A$ receptors to increase tonic inhibition. *Alcohol*, *41*(3), 211–221.

Saras, A., Gisselmann, G., Vogt-Eisele, A. K., Erlkamp, K. S., Kletke, O., Pusch, H., & Hatt, H. (2008). Histamine action on vertebrate GABA$_A$ receptors: Direct channel gating and potentiation of GABA responses. *Journal of Biological Chemistry*, *283*(16), 10470–10475.

Schwarzer, C., Tsunashima, K., Wanzenbock, C., Fuchs, K., Sieghart, W., & Sperk, G. (1997). GABA$_A$ receptor subunits in the rat hippocampus II: Altered distribution in kainic acid-induced temporal lobe epilepsy. *Neuroscience*, *80*(4), 1001–1017.

Selye, H. (1942). Correlations between the chemical structure and pharmacological actions of the steroids. *Endocrinology*, *30*, 437–442.

Sergeeva, O. A., Kletke, O., Kragler, A., Poppek, A., Fleischer, W., Schubring, S. R., . . . Hatt, H. (2010). Fragrant dioxane derivatives identify beta1-subunit-containing GABA$_A$ receptors. *Journal of Biological Chemistry*, *285*(31), 23985–23993.

Shen, Y., Lindemeyer, A. K., Gonzalez, C., Shao, X. M., Spigelman, I., Olsen, R. W., & Liang, J. (2012). Dihydromyricetin as a novel anti-alcohol intoxication medication. *Journal of Neuroscience*, *32*(1), 390–401.

Shen, Y., Lindemeyer, A. K., Spigelman, I., Sieghart, W., Olsen, R. W., & Liang, J. (2011). Plasticity of GABA$_A$ receptors after ethanol pre-exposure in cultured hippocampal neurons. *Molecular Pharmacology*, *79*(3), 432–442.

Sieghart, W. (2015). Allosteric modulation of GABA$_A$ receptors via multiple drug-binding sites. *Advances in Pharmacology*, *72*, 53–96.

Sieghart, W., Ramerstorfer, J., Sarto-Jackson, I., Varagic, Z., & Ernst, M. (2012). A novel GABA$_A$ receptor pharmacology: Drugs interacting with the a(+) b(−) interface. *British Journal of Pharmacology*, *166*(2), 476–485.

Siegwart, R., Jurd, R., & Rudolph, U. (2002). Molecular determinants for the action of general anesthetics at recombinant a2b3g2 $GABA_A$ receptors. *Journal of Neurochemistry, 80*(1), 140–148.

Sigel, E., Baur, R., Racz, I., Marazzi, J., Smart, T. G., Zimmer, A., & Gertsch, J. (2011). The major central endocannabinoid directly acts at $GABA_A$ receptors. *Proceedings of the National Academy of Sciences of the United States of America, 108*(44), 18150–18155.

Sigel, E., & Buhr, A. (1997). The benzodiazepine binding site of $GABA_A$ receptors. *Trends in Pharmacological Science, 18*(11), 425–429.

Sigel, E., & Luscher, B. P. (2011). A closer look at the high affinity benzodiazepine binding site on $GABA_A$ receptors. *Current Topics in Medicinal Chemistry, 11*(2), 241–246.

Simms, J. A., Steensland, P., Medina, B., Abernathy, K. E., Chandler, L. J., Wise, R., & Bartlett, S. E. (2008). Intermittent access to 20% ethanol induces high ethanol consumption in Long-Evans and Wistar rats. *Alcoholism, Clinical and Experimental Research, 32*(10), 1816–1823.

Simon, J., Wakimoto, H., Fujita, N., Lalande, M., & Barnard, E. A. (2004). Analysis of the set of $GABA_A$ receptor genes in the human genome. *Journal of Biological Chemistry, 279*(40), 41422–41435.

Skolnick, P., Moncada, V., Barker, J. L., & Paul, S. M. (1981). Pentobarbital: Dual actions to increase brain benzodiazepine receptor affinity. *Science, 211*(4489), 1448–1450.

Smith, G. B., & Olsen, R. W. (1995). Functional domains of $GABA_A$ receptors. *Trends in Pharmacological Science, 16*(5), 162–168.

Smith, S. S. (Ed.). (2003). *Neurosteroid Effects in the Central Nervous System: The Role of the $GABA_A$ Receptor*. Boca Raton, FL: CRC Press.

Smith, S. S., & Gong, Q. H. (2005). Neurosteroid administration and withdrawal alter $GABA_A$ receptor kinetics in CA1 hippocampus of female rats. *Journal of Physiology, 564*(Pt 2), 421–436.

Sternfeld, F., Carling, R. W., Jelley, R. A., Ladduwahetty, T., Merchant, K. J., Moore, K. W., ... MacLeod, A. M. (2004). Selective, orally active $GABA_A$ alpha5 receptor inverse agonists as cognition enhancers. *Journal of Medicinal Chemistry, 47*(9), 2176–2179.

Study, R. E., & Barker, J. L. (1981). Diazepam and (–)-pentobarbital: Fluctuation analysis reveals different mechanisms for potentiation of g-aminobutyric acid responses in cultured central neurons. *Proceedings of the National Academy of Sciences of the United States of America, 78*(11), 7180–7184.

Sundstrom-Poromaa, I., Smith, D. H., Gong, Q. H., Sabado, T. N., Li, X., Light, A., ... Smith, S. S. (2002). Hormonally regulated a4b2d $GABA_A$ receptors are a target for alcohol. *Nature Neuroscience, 5*(8), 721–722.

Supavilai, P., Mannonen, A., & Karobath, M. (1982). Modulation of GABA binding sites by CNS depressants and CNS convulsants. *Neurochemistry International, 4*(4), 259–268.

Sur, C., Farrar, S. J., Kerby, J., Whiting, P. J., Atack, J. R., & McKernan, R. M. (1999). Preferential coassembly of a4 and d subunits of the $GABA_A$ receptor in rat thalamus. *Molecular Pharmacology, 56*(1), 110–115.

Suzdak, P. D., Glowa, J. R., Crawley, J. N., Schwartz, R. D., Skolnick, P., & Paul, S. M. (1986). A selective imidazobenzodiazepine antagonist of ethanol in the rat. *Science, 234*(4781), 1243–1247.

Szabadics, J., Tamas, G., & Soltesz, I. (2007). Different transmitter transients underlie presynaptic cell type specificity of $GABA_{A,slow}$ and $GABA_{A,fast}$. *Proceedings of the National Academy of Sciences of the United States of America, 104*(37), 14831–14836.

Tan, K. R., Rudolph, U., & Lüscher, C. (2011). Hooked on benzodiazepines: GABA$_A$ receptor subtypes and addiction. *Trends in Neurosciences*, *34*(4), 188–197.

Terunuma, M., Xu, J., Vithlani, M., Sieghart, W., Kittler, J., Pangalos, M., ... Moss, S. J. (2008). Deficits in phosphorylation of GABA$_A$ receptors by intimately associated protein kinase C activity underlie compromised synaptic inhibition during status epilepticus. *Journal of Neuroscience*, *28*(2), 376–384.

Thompson, S. A., Wheat, L., Brown, N. A., Wingrove, P. B., Pillai, G. V., Whiting, P. J., ... Wafford, K. A. (2004). Salicylidene salicylhydrazide, a selective inhibitor of beta 1-containing GABAA receptors. *British Journal of Pharmacology*, *142*(1), 97–106.

Thompson, S. A., Whiting, P. J., & Wafford, K. A. (1996). Barbiturate interactions at the human GABAA receptor: Dependence on receptor subunit combination. *British Journal of Pharmacology*, *117*(3), 521–527.

Ticku, M. K., Ban, M., & Olsen, R. W. (1978). Binding of [^{3}H]a-dihydropicrotoxinin, a GABA synaptic antagonist, to rat brain membranes. *Molecular Pharmacology*, *14*(3), 391–402.

Ticku, M. K., & Olsen, R. W. (1978). Interaction of barbiturates with dihydropicrotoxinin binding sites related to the GABA receptor-ionophore system. *Life Sciences*, *22*(18), 1643–1651.

Ticku, M. K., & Olsen, R. W. (1979). Cage convulsants inhibit picrotoxinin binding. *Neuropharmacology*, *18*(3), 315–318.

Turner, D. M., Ransom, R. W., Yang, J. S., & Olsen, R. W. (1989). Steroid anesthetics and naturally occurring analogs modulate the GABA receptor complex at a site distinct from barbiturates. *Journal of Pharmacology and Experimental Therapeutics*, *248*(3), 960–966.

Ueno, S., Wick, M. J., Ye, Q., Harrison, N. L., & Harris, R. A. (1999). Subunit mutations affect ethanol actions on GABA$_A$ receptors expressed in Xenopus oocytes. *British Journal of Pharmacology*, *127*(2), 377–382.

Unwin, N. (2005). Refined structure of the nicotinic acetylcholine receptor at 4Å resolution. *Journal of Molecular Biology*, *346*(4), 967–989.

van Rijnsoever, C., Tauber, M., Choulli, M. K., Keist, R., Rudolph, U., Möhler, H., ... Crestani, F. (2004). Requirement of a5-GABA$_A$ receptors for the development of tolerance to the sedative action of diazepam in mice. *Journal of Neuroscience*, *24*(30), 6785–6790.

Wafford, K. A., van Niel, M. B., Ma, Q. P., Horridge, E., Herd, M. B., Peden, D. R., ... Lambert, J. J. (2009). Novel compounds selectively enhance d subunit containing GABA$_A$ receptors and increase tonic currents in thalamus. *Neuropharmacology*, *56*(1), 182–189.

Wallner, M., Hanchar, H. J., & Olsen, R. W. (2003). Ethanol enhances a4b3d and a6b3d GABA$_A$ receptors at low concentrations known to affect humans. *Proceedings of the National Academy of Sciences of the United States of America*, *100*(25), 15218–15223.

Wallner, M., Hanchar, H. J., & Olsen, R. W. (2006a). Low dose acute alcohol effects on GABA$_A$ receptor subtypes. *Pharmacology and Therapeutics*, *112*, 513–528.

Wallner, M., Hanchar, H. J., & Olsen, R. W. (2006b). Low dose alcohol actions on a4b3d GABA$_A$ receptors are reversed by the behavioral alcohol antagonist Ro15-4513. *Proceedings of the National Academy of Sciences of the United States of America*, *103*(22), 8540–8545.

Wallner, M., Hanchar, H. J., & Olsen, R. W. (2014). Alcohol selectivity of b3-containing GABA$_A$ receptors: Evidence for a unique extracellular alcohol/imidazobenzodiazepine Ro15-4513 binding site at the a+b– subunit interface in ab3d GABA$_A$ receptors. *Neurochemical Research*, *39*, 1118–1126.

Wallner, M., & Olsen, R. W. (2008). Physiology and pharmacology of alcohol: The imidazobenzodiazepine alcohol antagonist site on subtypes of $GABA_A$ receptors as an opportunity for drug development? *British Journal of Pharmacology, 154*(2), 288–298.

Wang, F., Xu, Z., Yuen, C. T., Chow, C. Y., Lui, Y. L., Tsang, S. Y., & Xue, H. (2007). 6,2′-Dihydroxyflavone, a subtype-selective partial inverse agonist of $GABA_A$ receptor benzodiazepine site. *Neuropharmacology, 53*(4), 574–582.

Wang, H., Hui, K. M., Chen, Y., Xu, S., Wong, J. T., & Xue, H. (2002). Structure-activity relationships of flavonoids, isolated from *Scutellaria baicalensis*, binding to benzodiazepine site of $GABA_A$ receptor complex. *Planta Medica, 68*(12), 1059–1062.

Wasowski, C., & Marder, M. (2012). Flavonoids as $GABA_A$ receptor ligands: The whole story? *Journal of Experimental Pharmacology, 4*, 9–24.

Wei, W., Faria, L. C., & Mody, I. (2004). Low ethanol concentrations selectively augment the tonic inhibition mediated by d subunit-containing $GABA_A$ receptors in hippocampal neurons. *Journal of Neuroscience, 24*(38), 8379–8382.

Werner, D. F., Swihart, A., Rau, V., Jia, F., Borghese, C. M., McCracken, M. L., . . . Homanics, G. E. (2011). Inhaled anesthetic responses of recombinant receptors and knockin mice harboring a2(S270H/L277A) $GABA_A$ receptor subunits that are resistant to isoflurane. *Journal of Pharmacology and Experimental Therapeutics, 336*(1), 134–144.

Whiting, P. J., Bonnert, T. P., McKernan, R. M., Farrar, S., Le Bourdelles, B., Heavens, R. P., . . . Wafford, K. A. (1999). Molecular and functional diversity of the expanding $GABA_A$ receptor gene family. *Annals of the New York Academy of Sciences, 868*, 645–653.

Wick, M. J., Mihic, S. J., Ueno, S., Mascia, M. P., Trudell, J. R., Brozowski, S. J., . . . Harris, R. A. (1998). Mutations of gamma-aminobutyric acid and glycine receptors change alcohol cutoff: Evidence for an alcohol receptor? *Proceedings of the National Academy of Sciences of the United States of America, 95*(11), 6504–6509.

Willow, M., & Johnston, G. A. (1981). Enhancement by anesthetic and convulsant barbiturates of GABA binding to rat brain synaptosomal membranes. *Journal of Neuroscience, 1*(4), 364–367.

Wingrove, P. B., Wafford, K. A., Bain, C., & Whiting, P. J. (1994). The modulatory action of loreclezole at the $GABA_A$ receptor is determined by a single amino acid in the b2 and b3 subunit. *Proceedings of the National Academy of Sciences of the United States of America, 91*(10), 4569–4573.

Wohlfarth, K. M., Bianchi, M. T., & Macdonald, R. L. (2002). Enhanced neurosteroid potentiation of ternary $GABA_A$ receptors containing the d subunit. *Journal of Neuroscience, 22*(5), 1541–1549.

Wu, Y., Wang, W., & Richerson, G. B. (2003). Vigabatrin induces tonic inhibition via GABA transporter reversal without increasing vesicular GABA release. *Journal of Neurophysiology, 89*(4), 2021–2034.

Yamamura, H. I., Enna, S. J., & Kuhar, M. J. (Eds.). (1978). *Neurotransmitter Receptor Binding*. New York: Raven.

Yamaura, K., Kiyonaka, S., Numata, T., Inoue, R., & Hamachi, I. (2016). Discovery of allosteric modulators for $GABA_A$ receptors by ligand-directed chemistry. *Nature Chemical Biology, 12*(10), 822–830.

Yao, N., Song, A., Wang, X., Dixon, S., & Lam, K. S. (2007). Synthesis of flavonoid analogues as scaffolds for natural product-based combinatorial libraries. *Journal of Combinatorial Chemistry, 9*(4), 668–676.

Zaccara, G., & Schmidt, D. (2016). Do traditional anti-seizure drugs have a future? A review of potential anti-seizure drugs in clinical development. *Pharmacological Research, 104*, 38–48.
Zeilhofer, H. U., Ralvenius, W. T., & Acuna, M. A. (2015). Restoring the spinal pain gate: GABA$_A$ receptors as targets for novel analgesics. *Advances in Pharmacology, 73*, 71–96.
Zheleznova, N., Sedelnikova, A., & Weiss, D. S. (2008). a1b2d, a silent GABA$_A$ receptor: Recruitment by tracazolate and neurosteroids. *British Journal of Pharmacology, 153*(5), 1062–1071.
Zilberter, M. (2016). Reality of inhibitory GABA in neonatal brain: Time to rewrite the textbooks? *Journal of Neuroscience, 36*(40), 10242–10244.
Zukin, S. R., Young, A. B., & Snyder, S. H. (1974). GABA binding to receptor sites in the rat central nervous system. *Proceedings of the National Academy of Sciences of the United States of America, 71*(12), 4802–4807.

CHAPTER 13

P2X RECEPTORS

ANNETTE NICKE, THOMAS GRUTTER,
AND TERRANCE M. EGAN

INTRODUCTION

THERE may be no molecule more versatile than adenosine triphosphate (ATP). In large part, this is because ATP provides a high-energy phosphate bond that is the primary currency of energy transfer in most living cells, while also serving as an essential substrate in intracellular signaling pathways and a synthetic building block of nucleic acids. More recently appreciated is its role as an intermediary of cellular communication (Burnstock, Campbell, Satchell, & Smythe, 1970; Holton & Holton, 1954). ATP is released from a variety of cells where it acts on two classes of membrane-bound purinergic receptor families (Burnstock & Kennedy, 1985): G-protein-coupled metabotropic P2Y receptors (von Kugelgen & Harden, 2011) and ligand-gated ion channels called *P2X receptors*. The P2X family is composed of seven eukaryotic gene products (P2X1R–P2X7Rs) that target cell surface membranes (Khakh & North, 2012) and intracellular vesicles (Fountain et al., 2007; Parkinson et al., 2014; Qureshi, Paramasivam, Yu, & Murrell-Lagnado, 2007; Toulme et al., 2010). Six of the seven P2XR subunits form functional homotrimeric receptors (P2X6R is the exception), while all but the P2X7 subunit form heteromeric complexes with other family members (Torres, Egan, & Voigt, 1999).

In this review, we discuss how the tight binding of pericellular ATP to an extracellular site on P2XRs causes cations to flow across cell surface membranes, and in so doing, initiates or modulates a number of critical physiological and pathophysiological processes.

AN OVERVIEW OF STRUCTURE

P2X2Rs are made of three subunits that form an ion-conducting channel (Kaczmarek-Hajek, Lorinczi, Hausmann, & Nicke, 2012). Each subunit has cytoplasmic N- and

C-termini, two transmembrane helices (TM1 and TM2), and a large extracellular loop that forms interfacial ATP binding pockets located about 40 Å from the extracellular surface of the membrane (Figure 13.1A–C). The binding sites are lined by basic and polar residues of adjoining subunits that recognize ATP through extensive intersubunit interactions (Evans, 2009; Hattori & Gouaux, 2012). Agonist occupation sets in motion a conformational change that opens the channel gate in the transmembrane pore, relieving the constriction that limits ion transport across the membrane (Mansoor et al., 2016). Remarkably in contrast to other classes of neurotransmitter-gated ion channels, extracellular cations enter the channel through fenestrations that open near the outer leaflet of the lipid bilayer, and cross the membrane through a pore lined by TM2 (Habermacher, Dunning, Chataigneau, & Grutter, 2016; Samways, Li, & Egan, 2014).

Channel Gating

P2XRs cycle between different functional states in response to ATP. Agonist binding causes first the receptor to rapidly switch from a closed, resting state to an open, conducting state, which allows cations to flow across the membrane. This activity, also known as *gating*, is a very fast process that usually takes place on the millisecond time scale. In all P2XRs except the P2X7R, sustained application of ATP then leads to the formation of a third state, named the *desensitized state*, in which the pore is inactivated. This temporary inactivation terminates ion flux despite the fact that ATP remains bound to the receptor. The kinetics of desensitization vary considerably between P2XR subtypes, ranging from very fast desensitization (hundreds of milliseconds) for P2X1Rs and P2X3Rs to slow desensitization (several seconds) for P2X2Rs, P2X4Rs, and P2X5Rs. Dissociation of ATP from desensitized and/or open states, in turn, reverts the channel to the initial closed, resting state, and allows the receptor to be activated again. These different functional states are therefore closely linked and form a gating cycle that shapes P2XR signaling.

The molecular mechanism underlying the gating cycle remained unclear for many years. However, recent X-ray structures of P2XRs from different species, including humans, solved in closed, open, and desensitized states provided the first "molecular movie" of how ATP-binding opens and desensitizes the channel (Hattori & Gouaux, 2012; Karasawa & Kawate, 2016; Kasuya et al., 2016; Kawate, Michel, Birdsong, & Gouaux, 2009; Mansoor et al., 2016). Supported by various functional studies using the substituted-cysteine accessibility method (Egan, Haines, & Voigt, 1998; Haines, Voigt, Migita, Torres, & Egan, 2001; L. H. Jiang, Rassendren, Spelta, Surprenant, & North, 2001; Kracun, Chaptal, Abramson, & Khakh, 2010; Li, Chang, Silberberg, & Swartz, 2008; Rassendren, Buell, Newbolt, North, & Surprenant, 1997; Samways et al., 2014), engineered disulphide bonds or metal bridging (L. H. Jiang et al., 2001; Kawate, Robertson, Li, Silberberg, & Swartz, 2011; Kowalski et al., 2014; Li, Kawate, Silberberg, & Swartz, 2010; Marquez-Klaka, Rettinger, & Nicke, 2009; Roberts et al., 2012; Stelmashenko, Compan, Browne, & North,

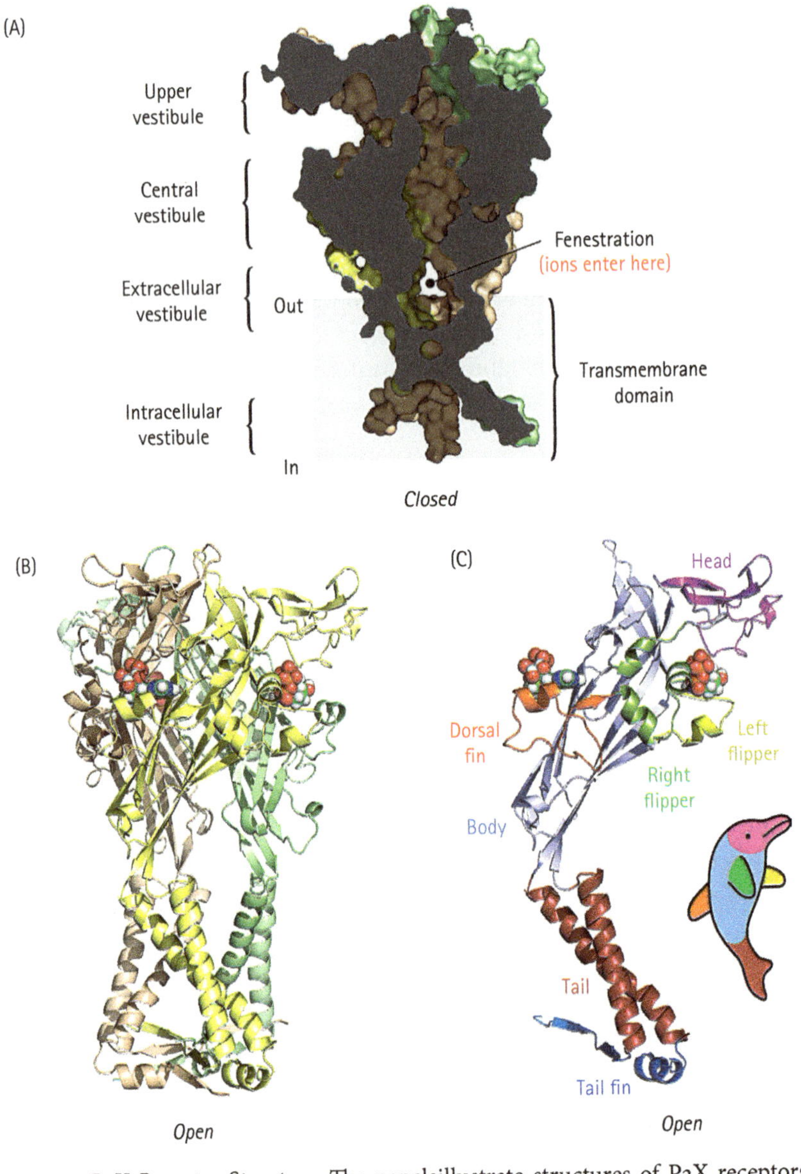

FIGURE 13.1 *P2X Receptor Structure*. The panelsillustrate structures of P2X receptors determined by crystallography. (A) A sagittal slice of the closed state of the zebrafish P2X4R. The oligomeric receptor forms four vestibules and a transmembrane pore. In this and all panels, the terminal cytoplasmic domains are truncated to varying degrees and not shown. (B) The open state of the human P2X3R. Like all P2XRs, the human P2X3R is made of three identical subunits that form a transmembrane pore. When ATP bindsat the extracellular subunit interfaces, the helical transmembrane domains undergo a conformation change thatrelieves the narrow constriction that impedes the flow of ions through the pore. (C) A single human P2X3R subunit forms a dolphin-shaped structure.

2014; Stephan et al., 2016; Zhao et al., 2014), and voltage-clamp fluorometry (Fryatt & Evans, 2014; Lorinczi et al., 2012), these structures reveal the key steps of the gating cycle. Further, recent approaches based on chemical optogenetics offer innovative ways of probing the molecular motions of P2XR gating (Browne et al., 2014; Habermacher, Dunning, et al., 2016; Lemoine et al., 2013). Here, we will focus on molecular changes that occur during gating in light of these experimental techniques.

Tightening of the "Binding Jaw" Is Vital for Agonist Action

The first evidence supporting an interfacial ATP-binding site came from early biochemical and mutational studies (Marquez-Klaka, Rettinger, Bhargava, Eisele, & Nicke, 2007; Marquez-Klaka et al., 2009; Wilkinson, Jiang, Surprenant, & North, 2006). These studies were later confirmed by other experiments (R. Jiang et al., 2011), including crystal structures of ATP-bound states (Hattori & Gouaux, 2012; Karasawa & Kawate, 2016; Kasuya et al., 2016; Mansoor et al., 2016). There is now convincing evidence that ATP binding induces the closure of the cavity that is formed by the head domain of one subunit and the dorsal fin of another (Figure 13.1B,C). The closure of this binding cavity, sometimes referred to as the "binding jaw," seems to be an essential motion that triggers the initial steps of the gating process.

Several experimental data, including protein engineering (R. Jiang, Taly, & Grutter, 2013), optical readout (Lörinczi et al., 2012), molecular modelling (Du, Dong, & Zhou, 2012; R. Jiang et al., 2013), and structural data (Hattori & Gouaux, 2012), have revealed the dynamic nature of the loops that frame the ATP-binding sites. For a detailed description of these motions, the reader is invited to read other reviews (Habermacher, Dunning, et al., 2016; R. Jiang et al., 2013). Briefly, it is suggested that two existent histidine residues, which are involved in the potentiation of the ATP response and located at the mouth of the ATP-binding site (His^{120} at the head domain and His^{213} in the dorsal fin), come closer together during ATP activation (R. Jiang et al., 2013; Nagaya, Tittle, Saar, Dellal, & Hume, 2005). By engineering zinc-binding sites to other binding loops (R. Jiang et al., 2012; Zhao et al., 2014), this tightening motion was extended to the entire binding jaw, suggesting that this movement is critical in the initial steps of ATP activation.

The crystal structures in apo and ATP-bound states from zebrafish P2X4R (zfP2X4R) and human P2X3R (hP2X3R) confirm the overall closing of the binding cleft (Hattori & Gouaux, 2012; Mansoor et al., 2016). However, close inspection of the movement of binding loops upon ATP binding in zfP2X4Rs and hP2X3Rs reveals striking differences. For instance, residues Glu^{112} and Thr^{202} of hP2X3R, which are equivalent to the zinc-potentiating residues His^{120} and His^{213} from rP2X2R, move apart upon ATP binding, while the equivalent residues in the zfP2X4R (Pro^{125} and His^{219}) move closer together. Conversely, the distance separating the residues Glu^{125} and Thr^{202}, located in the mouth of the binding jaw of the hP2X3R, is consistently reduced in the ATP-bound state relative to the apo state, a change not observed for the distance separating the equivalent residues in zfP2X4R (Asn^{140} and His^{219}). These data suggest that, although there is an overall closure of the binding jaw, this movement is not necessarily associated with a concerted tightening motion of all the binding loops, and subtle, yet significant, differences may exist between different P2XRs.

While the molecular details underlying agonist action seem to be disentangled, the reason why competitive antagonists fail to open the channel has remained elusive. However, the recent X-ray structures of two representative classes of P2XR competitive antagonists, TNP-ATP and A-317491, bound to hP2X3R, now provide new clues (Mansoor et al., 2016). Direct comparison of the structures bound to these competitive antagonists to those bound to ATP shows a completely different binding mode of the former. Although these molecules all compete for the same interfacial orthosteric site, the orientation and penetration into the binding cleft are strikingly different. It seems that TNP-ATP and A-317491 adopt a Y-shaped structure upon binding, while ATP adopts an unusual U-shaped form. These Y-shaped conformations allow the molecules to bind deeper within the binding pocket formed by the "left flipper" of one subunit and the dorsal fin of another. By more deeply occupying this pocket, TNP-ATP and A-317491 prevent ATP-induced closure of the binding jaw that is necessary to initiate pore opening. As a result, the distance separating Glu^{125} and Thr^{202} in the hP2X3R, which monitors jaw closure in the ATP-bound state, remains similar to that of the apo state. Thus, in agreement with previous work (Huang et al., 2014; R. Jiang et al., 2013; R. Jiang et al., 2012), antagonists seem to bind to and stabilize the apo state of the receptor, preventing the opening of the pore.

Illuminating Pore Expansion During ATP Activation

What happens following tightening of the binding jaw? Studies using cysteine-engineered receptors to lock gating by disulphide bridges revealed a global reorganization of the subunit interface (L. H. Jiang et al., 2003; R. Jiang et al., 2010; Roberts et al., 2012; Stelmashenko et al., 2014). This reorganization was further confirmed by the crystal structures of the ATP-bound state (Hattori & Gouaux, 2012; Mansoor et al., 2016), which show that agonist binding induces an expansion of the lateral fenestrations through an outward flexing of the lower body domain. Because the β-sheets of the lower body domain are directly connected to the transmembrane helices, their outward movement induces an outward displacement of the outer ends of the transmembrane domain. This expansion, in turn, causes the physical opening of the pore.

Although appealing, this mechanism, while observed in crystals, has to be demonstrated by additional experiments that monitor structural changes in a native environment. This is vital because the structure of the ATP-bound receptor, resolved in a non-native environment, revealed striking gaps between transmembrane helices that seem incompatible with the conduction of ions (Heymann et al., 2013).

To demonstrate such outward expansion of TM helices, two independent groups designed photoswitchable cross-linkers, named BMA (Bis-Maleimido Azobenzene) (Browne et al., 2014) and MAM (Maleimide Azobenzene Maleimide) (Habermacher, Martz, et al., 2016), which were covalently tethered into the P2XR pore and used as mechanical "nano-tweezers" able to expand or shrink the channel in response to light. The photoswitchable nature of these molecules originates from a central azobenzene core, which isomerizes between defined *trans* and *cis* isomers upon light irradiation at a particular wavelength, thus changing the length of the molecule. At both extremities of the azobenzene two maleimides are covalently attached that allow tethering to two engineered cysteines via a Michael addition reaction. When properly attached at the outer ends of TM2 helices of the rP2X2R at residue I328C or P329C, light-controlled toggling between *cis* and *trans* isomers acts to bring the subunits closer or further apart, thus closing or opening the channel. Importantly, the light-gated motions do not introduce disorder in the structure of the channel (Browne et al., 2014; Habermacher, Martz, et al., 2016), in contrast to other studies that used bridged azobenzene to control the folding/unfolding of peptides (Beharry & Woolley, 2011).

Interpreted in the light of the zfP2X4R X-ray structures with molecular dynamics simulations, one of these molecular rulers, MAM, revealed that the extent to which TM2 expands is less than initially suggested by the crystal structure (Habermacher, Martz, et al., 2016). As a result, the large gaps seen at the subunit interface in crystals diminished in the new modelled structure, so that the open pore became more compatible with the conduction of ions. Of great interest, the recently solved ATP-bound hP2X3R structure, in which the outer ends seem less extended than those of the more truncated zfP2X4R, seems more compatible with ion conduction (Mansoor et al., 2016). Therefore, these photoswitchable cross-linkers are useful not only for monitoring movements in native membranes, but also for refining current structures.

These engineering studies reveal that the external region of the pore is particularly effective for manipulating the molecular motions involved in channel opening. In line with this idea, other work showed that tethering of simple hydrophobic chemicals at the same region also allows the channel to be strikingly gated in the absence of ATP. For instance, tethering hydrophobic methanethiosulfonates (MTS) compounds at I328C gates the rP2X2R channel without ATP, suggesting that attachment of lipophilic MTS reagents, most likely by intercalation between separate transmembrane domains, is sufficient for channel opening (Rothwell, Stansfeld, Bragg, Verkhratsky, & North, 2014). Other work showed that tethering a photoswitchable reagent at the same I328C mutant reprogrammed the gating machinery of a rP2X2R receptor into a light-gated ion channel (Lemoine et al., 2013). This method, named *optogating*, employs the photoswitch MEA-TMA. This molecule is closely related to MAM and BMA, in that it comprises an azobenzene core, but differs at the extremities, comprising only one maleimide moiety for cysteine attachment, and at the other end a positively charged trimethyl ammonium group that is important for efficient labelling. Irradiation in the visible or near-ultraviolet (UV) light causes the tethered MEA-TMA to undergo isomerization to the *trans* or *cis* isomer, respectively, which in turn induces an opening or closing of the channel. The mechanism of MEA-TMA gating is currently unknown, but it probably shares similar features with that underlying the action of hydrophobic MTS compounds. However, unlike the lipophilic MTS reagents that constitutively gated the channel upon tethering without external control, gating by MEA-TMA can be dynamically controlled by light, thus making optogating a powerful strategy to remotely control P2X signaling with exquisite temporal precision.

Other Gating Motions of TM2 Helices

The use of these photoswitchable tweezers has improved the X-ray structure of the ATP-bound state. These molecules thus are valuable tools to monitor other gating motions of the receptor. When attached between I328C from one subunit and S345C from another in P2X2Rs, the *cis* isomer of MAM induces pore opening, suggesting that the length of tethered MAM molecule shortens during activation (Habermacher, Martz, et al., 2016). To explain this result, it was proposed that all three TM2 helices, which are straight in the closed state, bend around a highly conserved and flexible glycine residue in the open state (Gly^{342} in the rP2X2R). In line with this hypothesis, the recent X-ray structure of the hP2X3R reveals a striking structural change at the level of the conserved glycine residue in TM2, in which the structure of the helix switches from a regular alpha-helix to a stretched 3_{10}-helix (Mansoor et al., 2016). This local rearrangement of the structure bends all three TM2 helices, allowing the flow of ions, in agreement with the motion anticipated earlier by experiments using with the photoswitchable MAM.

A Cytoplasmic Cap to Stabilize the Open State Structure?

Contrary to the extracellular and transmembrane domains, for which much has been elucidated about the structure and gating, less information is available for the intracellular domain. As these domains are recognized to play critical roles in ATP gating (Allsopp & Evans, 2011; Allsopp, Farmer, Fryatt, & Evans, 2013; Boue-Grabot, Archambault, & Seguela, 2000; Robinson & Murrell-Lagnado, 2013), it is thus important to investigate how they modulate the function of P2XRs. A significant contribution to the field was made recently with the structure determination of the hP2X3R, which revealed, for the first time, the nearly complete structure of the intracellular domain in the ATP-bound state (Mansoor et al., 2016). The structure shows that this domain includes elements of secondary structure from both termini, including two sequential beta-strands from the N-terminus and a beta-strand from the C-terminus. All these strands are knitted together by extensive contact to form a network of three beta-sheets that sit beneath the transmembrane domain, capping the cytoplasmic surface of the pore (Figure 13.1B). The role of this cytoplasmic cap is unclear, but it is suggested that the formation of this cap has a central role in the stability of the open state and provides cytoplasmic fenestrations through which ions exit the pore. Strikingly, this cap was not resolved in the structure of either apo or desensitized state, suggesting that it is structured only in the open state. The reasons are unclear, but one possibility is that the structure of the cap is likely to be disassembled in the apo and desensitized states. In such a scenario, this three-beta-sheet network would be transient, a hypothesis that deserves to be tested by additional experiments.

Ion Conduction

All seven P2XRs are permeable to Na^+, K^+, and Ca^{2+} (Egan, Samways, & Li, 2006). The inward flow of Na^+ and Ca^{2+} depolarizes the membrane and increases the intracellular free Ca^{2+} concentration ($[Ca^{2+}]_i$). In excitable cells, the rise in $[Ca^{2+}]_i$ is the product of two distinct currents. The first results from direct Ca^{2+} permeation through the P2X channel (Benham & Tsien, 1987), and the second reflects activation of voltage-gated Ca^{2+} channels by the depolarizing effect of Na^+ influx (Gu & MacDermott, 1997). In non-excitable cells that lack voltage-gated Ca^{2+} channels, the Ca^{2+} that transits the pore is the main signal by which P2XRs modulate cellular function (Stojilkovic et al., 2005). The resulting rise in $[Ca^{2+}]_i$ is large enough to trigger transmitter release from neurons (Pankratov, Lalo, Verkhratsky, & North, 2006); hormone release from endocrine glands (Troadec, Thirion, Nicaise, Lemos, & Dayanithi, 1998); cytokine release from immune cells (Di Virgilio, 2007; Guerra, Gavala, Chung, & Bertics, 2007); and contraction of smooth (Brain, Jackson, Trout, & Cunnane, 2002; Gitterman & Evans, 2001; Ramme, Regenold, Starke, Busse, & Illes, 1987), skeletal (Buvinic et al., 2009), and cardiac (Shen, Pappano, & Liang, 2006) muscle. It also initiates feedback modulation of key properties of the P2XR itself, including receptor desensitization (Cook, Rodland, & McCleskey, 1998; Gu & MacDermott, 1997), current facilitation (Roger, Pelegrin, & Surprenant, 2008), and the lateral mobility of the receptor in the membrane (Toulme & Khakh, 2012). Even K^+ efflux plays a physiological role, as ATP-dependent depletion of cytosolic K^+ through activated P2X7Rs may initiate NLRP3 activation of downstream inflammasome signaling and pyroptosis in immune cells (Katsnelson, Rucker, Russo, & Dubyak, 2015; Perregaux & Gabel, 1994).

The Conduction Pathway

The original zfP2X4R crystal structure, when imaged in the closed state, showed two possible permeation pathways (Kawate et al., 2009). The first, called the "central pathway," runs along the threefold axis of symmetry, encompassing three wide vestibules with significant acidic character, leading to a transmembrane pore predominately lined by hydrophobic amino acids (Figure 13.1A). By analogy to other classes of channels, it was originally thought that ions would enter the central pathway through an opening at the top, and traverse the length of the protein before reaching the transmembrane pore. In the closed state, the central pathway contains multiple constrictions in the extracellular domain that must widen during gating to allow unrestricted flow of permeating cations to the transmembrane pore; in theory, these constrictions could be the gate(s) that open when the channel activates. The second route, called the "lateral pathway," is significantly shorter. Here, cations bypass the upper and central vestibules by entering the extracellular vestibule through three short (~15 Å deep) but wide (~8 Å) fenestrations located close to the outer leaflet of the membrane. In comparison to the central pathway, the lateral pathway is unobstructed. Whether cations use the central pathway, the lateral pathways, or both was at first uncertain, although persuasive evidence in the form of computational studies, disulphide cross-linking, cysteine scanning mutagenesis, and single-channel current recordings strongly favored the lateral pathway (Kawate et al., 2011; Samways, Khakh, Dutertre, & Egan, 2011). The definitive answer came in the form of the open state structures of hP2X3R and zfP2X4R, which show that the central pathway remains constricted during gating and therefore cannot serve as a conduit for ion transport. In contrast, the fenestrations of the lateral pathway flex and widen during receptor activation, providing an unbarred pathway for ions to enter the extracellular vestibule near the entrance to the transmembrane pore (Hattori & Gouaux, 2012; Mansoor et al., 2016).

The most complete picture of the transmembrane domain comes from the hP2X3R structure (Mansoor et al., 2016). Within the plane of the membrane, the pore is made of the α-helical TM2s of the three constituent subunits. In the closed state, the channel gate is formed by Ile^{323}, Val^{326}, and Thr^{330}, which form a constriction that is too narrow (<1 Å) to pass dehydrated Na^+. When ATP occupies the binding pocket, Ile^{323} and Val^{326} rotate outward, and the lumen forms a continuous open channel wide enough (~3 Å) to allow passage of partially dehydrated cations. Thr^{330} (and Ser^{331}) now sit at the narrowest point in the open pore, making them prime candidates to form the ion selectivity filter. Interestingly, the ion channel gate and selectivity filter are formed from the same subset of residues within the transmembrane pore, a situation anticipated from the results of functional studies (Migita, Haines, Voigt, & Egan, 2001; Pippel et al., 2017). After cations pass the filter/gate, they exit the pore in an unexpected manner. The orifice at the bottom of the threefold axis of symmetry is too small for ion transport. However, an alternative pathway is formed from parts of the cytoplasmic cap and TM2 helices within the boundary of the membrane, where channel protein and polar lipid head groups form rivulets for ions and water. Although the hypothesis that cations use this pathway to exit the pore is still unproven, it is strongly supported by molecular dynamics simulations (Mansoor et al., 2016).

Selectivity Filters

Most, but not all, P2X receptors prefer cations to anions by a factor of >10. The exception is the P2X5R, which in some species shows significant Cl^- permeability ($P_{Cl}/P_{Na} \approx 0.5$) (Bo et al., 2003; Thomas & Hume, 1990). The molecular basis of the preference for cations is not fully characterized, although structural and functional experiments suggest two sites where selection might occur. The first site involves acidic amino acids in the central cavity and extracellular fenestrations (Kawate et al., 2009; Samways & Egan, 2007). Movement through the fenestrations is enhanced by the electrostatic pull of the negative charge, which serves to concentrate cations near the top of the transmembrane pore while repelling anions. The enhanced electrochemical gradient then drives the cations through the open pore. In the case of the hP2X5R, the fenestrations exhibit positive electrostatic potential, which might explain their unusually high Cl^- permeability (Kawate et al., 2011). The second site is the transmembrane selectivity filter formed by Thr^{330}, which occupies the narrowest point in the conduction pathway of the open hP2X3R (Mansoor et al., 2016). Mutating threonine to lysine at the homologous site in the rP2X2R produces a channel permeable to anions, showing that Cl^- is able to move at least this far into the channel where it senses the local electrostatic environment (Browne et al., 2011), a hypothesis supported by recent experiments using anionic and cationic thiol-reactive compounds and cysteine substituted mutants of the hP2X7R (Pippel et al., 2017).

Do P2XRs show a preference for one physiological cation over another? The answer is yes, in that most P2XRs favor Ca^{2+}, the ion most responsible for triggering the physiological outcomes of receptor activation. Preference for Ca^{2+} is typically calculated from the reversal potential of the ATP-gated current (Hille, 2001), and quantified as Ca^{2+} permeability relative to a reference monovalent cation ion like Na^+ (P_{Ca}/P_{Na}). In the case of P2XRs, P_{Ca}/P_{Na} ranges from ~1 to ~13, depending on species, splice variant, and subunit composition of the trimeric receptor (Egan et al., 2006; Jarvis & Khakh, 2009; Liang et al., 2015). A more accurate measure of the significance of Ca^{2+} is the fractional Ca^{2+} current (also called the "Pf%"), defined as the fraction of the total ATP-gated current carried by Ca^{2+} (Neher, 1995). Unlike reversal potential-based methods of determining P_{Ca}/P_{Na}, Pf% is measured at resting membrane potentials, uses physiological solutions, and does not depend on the Goldman-Hodgkin-Katz assumptions (Frings et al., 2000; Neher, 1995). The Pf% of P2XRs varies from ~3% for receptors that show almost no preference for Ca^{2+} (homomeric rP2X3R, the murine P2X7kR splice variant, and heteromeric rP2X2/3R and rP2X1/5R are examples), to values as high as ~16% for hP2X4Rs and ~20% for zfP2X7Rs (Egan & Khakh, 2004; Liang et al., 2015; Samways & Egan, 2007).

Where does Ca^{2+} selection occur? It probably happens at the same two sites linked to cation versus anion selection we described before. First, the electrostatic potential of the acidic amino acids of the extracellular fenestrations exerts a greater pull on divalent than on monovalent cations, and thus preferentially concentrate Ca^{2+}. Notably, not all P2XRs contain the appropriate acidic fenestral residues, which might explain the range of Pf%s measured from different family members. For example, removing the fixed charge of key sites in the extracellular fenestrations of the highly Ca^{2+} permeable P2X4R significantly reduces Pf% from ~16% to 9%, while adding charge to the less permeable P2X2R increases Pf% from ~6% to ~13% (Samways & Egan, 2007). Second, mutating Thr^{339} and Ser^{340} within the transmembrane pore of the rP2X2R (equivalent to Thr^{330} and Ser^{331} of the hP2X3R) significantly alters Ca^{2+} flux (Egan & Khakh, 2004; Migita et al., 2001). Based on the crystal structure of the hP2X3R, these residues face the lumen of the open channel at its narrowest point, putting them in position to interact with Ca^{2+} as it moves through the pore (Mansoor et al., 2016). That they do play a role is supported by abundant but indirect evidence; however, the mechanism by which they influence current is still poorly understood (Samways et al., 2014). A simple hypothesis is that the side chains of the polar residues face into the pore allowing hydroxyl oxygens to solvate Ca^{2+}. In support of this hypothesis is the fact that the effect of the mutation is predicted by the nature of the change in the character of the side-chain (Browne et al., 2011; Egan & Khakh, 2004; Migita et al., 2001). However, highly Ca^{2+} permeable P2X1Rs and P2X4Rs contain hydrophobic residues at these sites, casting doubt on the significance of the side chains and implying that backbone carbonyl oxygens interact with Ca^{2+}. It is possible that both the side chains and the carbonyl oxygens influence permeability in a manner that depends on the subunit composition and the state of the channel. Future experiments using unnatural amino acid substitutions to probe the contribution of backbone carbonyls to selection may help to bring clarity to what remains a clouded picture.

Time-Dependent Changes in Reversal Potentials

A core property of all ligand-gated ion channels is their ability to alter membrane permeability when bound to an agonist. P2XRs are no different, as they allow small physiological cations (Na^+, K^+, and Ca^{2+}) to pass into and out of cells. In addition, some P2XRs (P2X2R, P2X4R, and P2X7R) have the unusual property of showing a time-dependent shift in the E_{rev} of their ligand-gated current when the plasmalemmal membrane separates cations with different permeabilities (Khakh, Bao, Labarca, & Lester, 1999; Surprenant, Rassendren, Kawashima, North, & Buell, 1996; Virginio, MacKenzie, Rassendren, North, & Surprenant, 1999). When Na^+ is the sole permeable intracellular cation and $NMDG^+$ (*N*-methyl-D-glucamine, a synthetic organic cation) is the sole permeable extracellular ion, the $E_{rev,}$ called the “bi-ionic potential,” is defined by the following equation:

$$E_{rev} = (RT/F * ln(P_{NMDG} * [NMDG]_{out})/(P_{Na} * [Na]_{in})$$

where R, T, and F are constants, and P_{NMDG} and P_{Na} are the permeabilities of the two cations. The equation assumes that the concentrations of ions on either side of the membrane are static, and that a shift in E_{rev} signals a change in membrane permeability to one or both ions. If so, then the positive shift in E_{rev} measured during long applications of ATP must result from a time-dependent change in the relative permeability (P_{NMDG}/P_{Na}) of $NMDG^+$ and Na^+. The underlying cause of the shift in E_{rev} is an increase in $NMDG^+$ permeability (i.e., P_{NMDG}/P_{Na} changes from ~0.05 to ~0.6), thought to result from a gradual dilation of a narrow part of the pore. This phenomenon, called “pore dilation,” is supported by experiments using electrophysiology (Bean, 2015), measurements of dye uptake by fluorescence microscopy (Ferreira, Pereira, & Faria, 2015), and atomic force microscopy (Shinozaki et al., 2009). However, inconsistences in the theory of pore dilation exist. Chief amongst these is the inability to measure an increase in ATP-gated single-channel conductance as expected for a widening pore (Riedel, Schmalzing, &

Markwardt, 2007). The alternative hypothesis is that the positive shift in E_{rev} reflects changes in ion concentrations. If the current is large enough and the agonist is applied for more than a few seconds, then a low but unchanging constitutive permeability to $NMDG^+$ ($P_{NMDG}/P_{Na} \approx 0.06$) could disrupt the ionic gradients across the membrane in a manner that would shift E_{rev} towards 0 mV. Indeed, using a mix of electrophysiology and mathematical modeling, Li et al. were able to convincingly demonstrate that cytoplasmic $NMDG^+$ accumulation and Na^+ depletion do in fact occur (Li, Toombes, Silberberg, & Swartz, 2015). The positive correlation of empirical and modeled data led the authors to conclude that the time-dependent shift in the E_{rev} of the ATP-gated current is an artifact of the method used to measure permeability, exclusively caused by changes in the concentrations of cytoplasmic cations. This epochal shift in understanding impacts more than just ATP-gated P2XRs, because other types of ligand-gated ion channels display similar time-dependent changes in E_{rev} (Baconguis & Gouaux, 2012; Banke, Chaplan, & Wickenden, 2010; Chung, Guler, & Caterina, 2008). If true, then it also necessitates reinvestigation of the reported role of genuine pore dilation in a number of pathophysiological processes (Bautista & Julius, 2008; Sorge et al., 2012; Wei, Caseley, Li, & Jiang, 2016). The jury is still out, and additional experiments, including the reexamination of published results, are needed to determine whether or not genuine pore dilation and/or the redistribution of ions is the ultimate cause of the change in E_{rev}.

In line with the work of Li et al. (2015), a very recent study combining single-channel recordings, the use of mechanical "nano-tweezers," and molecular modeling have shown that the motions leading to Na^+ flow are very similar to those that drive $NMDG^+$ permeability, suggesting that once the pore is rapidly opened by ATP, both $NMDG^+$ and Na^+ ions can flow through a similar open state, yet at different rates (Harkat et al., 2017). Indeed, $NMDG^+$ stably "percolates" through the open pore, but much slower than Na^+. The reason for this seems to be related to the fact that $NMDG^+$ needs to "snake" through the pore in a fully linear conformation with the positively charged head group pointing downwards along the electrochemical gradient, a process that takes a considerable amount of time. The authors also identify a polyamine (spermidine) as a natural cation able to permeate the P2X pore. As polyamines are known to modulate a number of ion channels, including synaptic *N*-methyl D-aspartate (NMDA) receptors (Mony, Zhu, Carvalho, & Paoletti, 2011), the permeability of the P2X pore to spermidine provides new insights into the physiological function of P2X receptors.

Physiological Function

P2XRs are found on all types of mammalian cells, where they serve as a receptor for extracellular ATP, an autocrine/paracrine signaling molecule, and fast synaptic transmitter. For example, ATP is released from specialized sensory cells like taste buds and stressed or damaged cells through canonical and non-canonical pathways, and then acts on the P2XRs of primary afferent nerve terminals to transmit sensations of taste, bladder filling, and pain. In neurons of the peripheral nervous system, vesicular ATP is synaptically released from sympathetic and parasympathetic nerve endings and functions as an excitatory co-transmitter on smooth muscle cells, for example in arteries, bladder, and *vas deferens* (Kennedy, 2015). Evidence that ATP functions as an excitatory transmitter at postsynaptic P2XRs is controversial in the central nervous system (CNS), where the major source of ATP seems to be astrocytes and not neurons, and where ATP predominately acts as a neuromodulator and gliotransmitter (Khakh & North, 2012; Pougnet et al., 2014).

Heteromerization between subunits is well characterized in vitro (Torres et al., 1999). However, evidence of heteromerization in native preparations is limited, with only P2X2/3Rs and P2X1/5Rs conclusively characterized *in vivo* so far (Saul, Hausmann, Kless, & Nicke, 2013). In the remainder of this review, we focus on *in vivo* data obtained from genetically modified mice to describe significant physiological functions of P2XR subtypes. In many cases, these are supported by excellent pharmacological studies from a variety of species and model systems, although these experiments, as well as many studies describing additional physiological functions, are not described here.

P2X1R

The P2X1R shows fast desensitization upon agonist activation, which often masks the significance of its actions on tissue or cell preparations that release ATP. On the other hand, the rapid desensitization is also a distinguishing characteristic that suggests a P2X1R-specific response. P2X3Rs also show rapid desensitization, so additional experiments using P2X1R- or P2X3R-selective antagonists (Khakh & North, 2012) must be performed to conclusively identify the responsible receptor. P2X1Rs are found on smooth muscle cells of the *vas deferens*, bladder, and arteries, as well as on platelets, immune cells, and certain neurons; for example, of superior cervical ganglia (Kaczmarek-Hajek et al., 2012). P2X1R activation causes membrane depolarization and contraction in smooth muscle cells and accounts for up to 60% of the contractile response to sympathetic nerve stimulation in the *vas deferens*. As a physiological consequence of its deletion, sperm numbers in the ejaculate of $P2X1^{-/-}$ mice are reduced, and fertility is decreased by 90% (Mulryan et al., 2000). The residual contractile response is due to α1-receptor activation by co-released noradrenaline, and the respective α1/P2X1 double knockout mouse is 100% infertile (White et al., 2013). Smooth muscle P2X1Rs are also involved in renal autoregulation and prostatic contraction (Guan, Osmond, & Inscho, 2007; Ventura et al., 2011).

In platelets, P2X1R-induced increase of intracellular Ca^{2+} amplifies signals for thrombus formation via platelet G-protein coupled and tyrosine kinase receptors (Mahaut-Smith, Jones, & Evans, 2011; Oury, Lecut, Hego, Wera, & Delierneux, 2015). Transgenic mice that overexpress P2X1Rs in platelets display a mildly prothrombotic phenotype (Oury et al., 2003), and P2X1 knockout mice show reduced aggregation and decreased mortality in a model of systemic thromboembolism (Hechler et al., 2003).

Neutrophil P2X1Rs play a role in recruitment and homeostatic regulation of neutrophils, and P2X1R deletion leads to overactivation of neutrophils upon lipopolysacharide LPS injection. Furthermore, placing wild-type neutrophils into $P2X1^{-/-}$ mice has shown a contribution to thrombus formation by the induction of fibrin formation (Darbousset et al., 2014; Lecut et al., 2012). A physiological role of P2X1Rs in the nervous system is not well defined, and no behavioral phenotype is known.

P2X2R

The slowly desensitizing P2X2R is abundantly expressed in the central and peripheral nervous system as well as in multiple non-neuronal tissues. In spite of its wide distribution, $P2X2^{-/-}$ mice show a moderate phenotype with impaired neurotransmission in the myenteric plexus (reduced peristalsis *in vitro*), pelvic afferent nerves (bladder hyporeflexia), and carotid sinus nerve (reduced ventilatory responses to hypoxia), and showed almost no ATP responses in coeliac and superior cervical ganglia (Cockayne et al., 2005; Ren et al., 2003; Rong et al., 2003). Cochlear P2X2Rs mediate adaptation to elevated sound levels, and $P2X2^{-/-}$ mice and humans carrying a dominant negative P2X2 point mutation suffer from progressive hearing loss (Housley et al., 2013; Yan et al., 2013). P2X2R subunits form slowly desensitizing P2X2/3R heterotrimers with subunits of the fast desensitizing P2X3R that contribute to P2X2R-mediated sensory transmission in nodose ganglia and, to a lesser extent, to P2X3R-mediated transmission in the dorsal root ganglia, where they account for the persistent but not the acute response in a formalin model of acute inflammatory pain (Cockayne et al., 2005).

P2X3Rs

Like the P2X2R, the P2X3R plays a crucial role in somatic and visceral sensory function, but it has a more restricted expression pattern in primary sensory neurons (Abdulqawi et al., 2015; Ford et al., 2015).

The contribution of P2X3Rs to pain was demonstrated using two different strains of $P2X3^{-/-}$ mice. Using these models, a role for P2X3Rs in agonist-evoked nocifensive responses and both the acute (as the homomer) and the persistent (as a component of the P2X2/3 heteromer) phase of formalin-induced pain was demonstrated. In contrast, the knockout mice had normal responses to noxious thermal and mechanical stimuli (Cockayne et al., 2005; Cockayne et al., 2000; Souslova et al., 2000). However, thermal hyporesponsiveness to non-noxious temperatures was observed in *in vivo* recordings (Souslova et al 2000). In addition, the mice showed urinary bladder hyporeflexia with decreased voiding frequency (Cockayne et al., 2000), as well as impaired peristalsis in the small intestine (Bian et al., 2003), blunted responses of gastric vagal afferents (McIlwrath, Davis, & Bielefeldt, 2009), and altered synaptic plasticity in the hippocampus (Wang et al., 2006), in *in vitro* studies.

Urothelium-released ATP is generally considered to be important for the P2X-induced micturition reflex, but a recent study in which younger P2X2$^{-/-}$ and P2X3$^{-/-}$ mice and different *in vivo* methods were used suggests that ATP signaling is not essential under normal physiological conditions but increased and particularly relevant under pathological conditions (Takezawa et al., 2016), supporting the importance of P2X2Rs and P2X3Rs as therapeutic targets for bladder disorders.

Interestingly, P2X2/P2X3 double knockout mice show about 90% lethality during the early postnatal phase. Death usually occurs as the result of bacterial bronchial pneumonia and multiple developmental abnormalities (Cockayne et al., 2005). Surviving mice appear normal. In addition to the impairments described for the respective single knockout animals, the double knockouts show a loss of gustatory nerve responses, showing that ATP, acting on P2X2/3Rs, is the primary neurotransmitter in taste buds (Finger et al., 2005). P2X2/3Rs are also found on primary afferent neurons in airways, and recent data from a phase 2 study suggest that P2X3R antagonists like AF-219 are promising antitussives (Abdulqawi et al., 2015).

P2X4Rs

The P2X4R shows intermediate desensitization and a wider expression profile than other family members (Kaczmarek-Hajek et al., 2012). Unfortunately, its characterization in native tissues has so far been hampered by a lack of selective ligands. Instead, the ability of the positive allosteric modulator ivermectin to potentiate ATP-gated currents is regarded as a defining characteristic (Khakh & North, 2012), although even these results should be considered with caution (Norenberg et al., 2012). Interestingly, this subtype undergoes dynamic trafficking, and a significant fraction of the receptor is localized to intracellular compartments/lysosomes (Huang et al., 2014; Murrell-Lagnado & Qureshi, 2008; Xu et al., 2014). Its intracellular location raises the following question: Do cytoplasmic ATP and intracellular P2X4Rs play a meaningful role in cellular physiology? The answer is: Maybe so! For example, the distantly related P2XR isoforms of *D. discoideum* are localized to and functional in intracellular contractile vacuoles (Fountain et al., 2007; Parkinson et al., 2014), and P2X4Rs on storage organelles for lung surfactant produce a localized Ca^{2+} influx at the site of vesicle fusion with the plasma membrane that promotes surfactant secretion (Miklavc, Thompson, & Frick, 2013).

Three P2X4$^{-/-}$ knockout animals have been generated by different laboratories (Brone, Moechars, Marrannes, Mercken, & Meert, 2007; Sim et al., 2006; Yamamoto et al., 2006), and they confirm the previously suspected role of spinal microglial P2X4Rs in chronic inflammatory and neuropathic pain (Tsuda et al., 2009; Ulmann, Hirbec, & Rassendren, 2010). Using these animals, P2X4Rs were shown to modulate release of brain-derived neurotrophic factor (BDNF) from microglia and increase synthesis of prostaglandin E2 (PGE2) by macrophages. Moreover, endothelial P2X4Rs play a role in flow-induced Ca^{2+} influx, production of vasodilatory nitric oxide (NO) and vascular remodeling, traits that could explain the elevated blood pressure recorded from $P2X4^{-/-}$ mice (Yamamoto et al., 2006).

Transgenic mice overexpressing the human P2X4R in cardiomyocytes exhibit increased myocyte contractility (Hu et al., 2001), suggesting that it is a possible target to improve heart performance following cardiomyopathy and myocardial infarction (Shen, Shutt, Agosto, Pappano, & Liang, 2009; Sonin et al., 2008), whereas cardiac-specific P2X4 deletion resulted in a more severe heart failure phenotype after infarction (Yang et al., 2014).

In the CNS, P2X4Rs are localized to perisynaptic positions (Rubio & Soto, 2001). $P2X4^{-/-}$ mice revealed altered regulation of hippocampal synaptic plasticity (Sim et al., 2006) and perceptual and sociocommunicative deficits (Wyatt et al., 2013). A recently generated P2X4R reporter mouse model showed high P2X4R expression in the hypothalamic arcuate nucleus where presynaptic P2X4Rs regulate GABA release that might be involved in feeding control (Xu et al., 2014).

P2X5R

P2X5R subunits appear to be widely distributed in murine tissues and are implicated in cell differentiation and cancer. In contrast to frog, chick, or zebrafish isoforms, which are efficiently expressed, rodent isoforms give small current responses when heterologously expressed (Kaczmarek-Hajek et al., 2012). In complex with acid-sensing ion channel subtype 3 ASIC3 channels, they are suggested to form a coincidence detector for ATP and H^+ in sensory neurons in muscle tissue (Birdsong et al., 2010). In humans, but not in other species, a predominant single nucleotide polymorphism (SNP) leads to a non-functional P2X5R splice variant in which the second TM domain is missing. Whether or not this isoform traffics to the plasma membrane is still questioned (Kotnis et al., 2010). The prevalence of the non-functional isoform suggests that it does not fulfil an important physiological function in humans. However, a recent study identified the P2X5R as a surface marker for human and mouse brown adipose cells (Ussar et al., 2014).

P2X6Rs

P2X6 subunits are unable to homo-oligomerize efficiently. However, their expression overlaps to a large extent with P2X4 subunits, suggesting the formation of P2X4/6R heteromultimers. Although P2X4/6R formation has been shown biochemically with heterologously expressed subunits, evidence for their existence in native tissues is sparse, and no physiological function is known (Saul et al., 2013). A $P2X6^{-/-}$ mouse was recently generated, but no obvious phenotype was identified, in a study focused on renal electrolyte handling (de Baaij et al., 2016).

P2X7Rs

P2X7Rs are expressed in cells of hematopoietic origin, different types of glia cells, endothelia, and epithelia (Kaczmarek-Hajek et al., 2012). Although P2X7 transcripts are found in neurons, the specific localization and function of neuronal P2X7Rs remains controversial (Anderson & Nedergaard, 2006; Illes, Khan & Rubini, 2017; Miras-Portugal et al., 2017; Sperlagh & Illes, 2014). The P2X7R differs structurally and functionally from all other P2X subtypes. It contains a particularly long C-terminal sequence that constitutes about 40% of the whole protein and is supposed to be localized intracellularly. Compared to other P2XRs, it has an at least 10- to 100-fold lower sensitivity to ATP, which acts as a partial agonist at this receptor, suggesting that it functions as a "danger signal" detector in situations like inflammation and cellular injury where unusually high concentrations of extracellular ATP are present. In rodents, but not humans, a more sensitive splice variant exists that is activated by covalent modification by ADP ribosylation (Schwarz et al., 2012; Xu et al., 2012). Activation of the P2X7R mediates downstream signaling cascades that trigger cytokine release, plasma membrane reorganization, increase of membrane permeability, ectodomain shedding, and apoptosis (Bartlett, Stokes, & Sluyter, 2014). Several of these long-term effects probably result from a P2X7R-dependent influx of Ca^{2+}, although complete descriptions of the signaling pathways are lacking.

Several laboratories have investigated the role of P2X7Rs in inflammasome assembly and the subsequent maturation and release of interleukin-1-beta (IL-1β) in macrophages and other immune cells. Here, ATP-evoked K^{+}-efflux probably functions as a co-stimulatory signal to Toll-like receptor activation (Dubyak, 2012; Di Virgilio et al, 2017). Data from three $P2X7^{-/-}$ mouse lines have been published so far (Basso et al., 2009; Chessell et al., 2005; Sim, Young, Sung, North, & Surprenant, 2004; Solle et al., 2001). Although two of these lines were shown to express alternative splice variants and/or incomplete receptor transcripts (Kaczmarek-Hajek et al., 2012), all three demonstrated a role for P2X7Rs in inflammatory and neuropathic pain. Also, roles in bone formation and T-cell differentiation have been shown (Ke et al, 2003; Rissiek et al, 2015). In agreement with a pathophysiological role, P2X7R blockade/deletion ameliorates tissue damage and has beneficial effects in numerous pathophysiological and inflammatory conditions, including autoimmune diseases such as rheumatoid arthritis and inflammatory bowel disease, epilepsy and other neurological diseases, and cancer (Bartlett, Stokes, & Sluyter, 2014; Bhattacharya & Biber, 2016; Di Virgilio & Adinolfi, 2017; Eser et al., 2015; Kaczmarek-Hajek et al., 2012; Roger et al., 2015; Sperlagh and Illes, 2014). In addition, several SNPs are identified in humans and rodents that are associated with susceptibility to infections, pain sensitivity, osteoporosis, and mood disorders, to only name a few (Bartlett et al., 2014). Although importance of the P2X7R as a potential drug target drives the development of a number of selective antagonists, so far none of them have demonstrated significant therapeutic efficacy in clinical studies (Park & Kim, 2017; Pevarello et al., 2017).

Summary

The use of P2XR knockout mice helped to overcome the complexities and deficiencies of the current state of P2XR pharmacology, and they continue to provide crucial tools to decipher P2XR physiological and pathophysiological functions. The mostly mild knockout mouse phenotypes (at least in single-knockout animals) suggest modulatory physiological roles rather than important physiological functions. Since upregulation of P2XRs and/or increased ATP levels are generally observed in pathophysiological conditions, P2XRs are assumed to represent pathology-specific drug targets. Besides possible compensatory effects in the knockout mice, the limited transferability of mouse data to humans remains a major challenge, in particular where the immune system is involved. Another problem is, as in other fields of research, the lack of selective antibodies and the uncritical use of the available antibodies. This is particularly evident in the case of the P2X7 subtype where the exact localization as well as the contribution of neuronal and/or non-neuronal P2X7Rs to pathologies remain debated (Anderson & Nedergaard, 2006; Sperlagh & Illes, 2014). To resolve these issues, better and species-specific antibodies and nanobodies and novel conditional and reporter mouse models are being developed. In addition, human models such as organoids from differentiated stem cells are emerging as novel test systems.

Acknowledgements

We thank Eric Gouaux and the Gouaux laboratory for the PDBs of the assembled hP2X3R structures and the dolphin cartoon of Figure 13.1C. Work in the authors' laboratories is supported by grants from the Deutsche Forschungsgemeinschaft (NI 592/7-1 to Annette Nicke), Agence Nationale de la Recherche (ANR-14-CE11-0004-01 to Thomas Grutter), and the National Institutes of Health (R01GM112188 to Terrance M. Egan).

References

Abdulqawi, R., Dockry, R., Holt, K., Layton, G., McCarthy, B. G., Ford, A. P., & Smith, J. A. (2015). P2X3 receptor antagonist (AF-219) in refractory chronic cough: a randomised, double-blind, placebo-controlled phase 2 study. *Lancet, 385*(9974), 1198–1205.

Allsopp, R. C., & Evans, R. J. (2011). The intracellular amino terminus plays a dominant role in desensitization of ATP-gated P2X receptor ion channels. *Journal of Biological Chemistry, 286*(52), 44691–44701.

Allsopp, R. C., Farmer, L. K., Fryatt, A. G., & Evans, R. J. (2013). P2X receptor chimeras highlight roles of the amino terminus to partial agonist efficacy, the carboxyl terminus to recovery from desensitization, and independent regulation of channel transitions. *Journal of Biological Chemistry, 288*(29), 21412–21421.

Anderson, C. M., & Nedergaard, M. (2006). Emerging challenges of assigning P2X7 receptor function and immunoreactivity in neurons. *Trends in Neurosciences, 29*(5), 257–262.

Baconguis, I., & Gouaux, E. (2012). Structural plasticity and dynamic selectivity of acid-sensing ion channel-spider toxin complexes. *Nature*, *489*(7416), 400–405.

Banke, T. G., Chaplan, S. R., & Wickenden, A. D. (2010). Dynamic changes in the TRPA1 selectivity filter lead to progressive but reversible pore dilation. *American Journal of Physiology–Cell Physiology*, *298*(6), C1457–C1468.

Bartlett, R., Stokes, L., & Sluyter, R. (2014). The P2X7 receptor channel: recent developments and the use of P2X7 antagonists in models of disease. *Journal of Molecular Biology*, *66*(3), 638–675.

Basso, A. M., Bratcher, N. A., Harris, R. R., Jarvis, M. F., Decker, M. W., & Rueter, L. E. (2009). Behavioral profile of P2X7 receptor knockout mice in animal models of depression and anxiety: relevance for neuropsychiatric disorders. *Behavioural Brain Research*, *198*(1), 83–90.

Bautista, D., & Julius, D. (2008). Fire in the hole: pore dilation of the capsaicin receptor TRPV1. *Nature Neuroscience*, *11*(5), 528–529.

Bean, B. P. (2015). Pore dilation reconsidered. *Nature Neuroscience*, *18*(11), 1534–1535.

Beharry, A. A., & Woolley, G. A. (2011). Azobenzene photoswitches for biomolecules. *Chemical Society Reviews*, *40*(8), 4422–4437.

Benham, C. D., & Tsien, R. W. (1987). A novel receptor-operated Ca2+-permeable channel activated by ATP in smooth muscle. *Nature*, *328*(6127), 275–278.

Bhattacharya, A., & Biber, K. (2016). The microglial ATP-gated ion channel P2X7 as a CNS drug target. *Glia*, *64*(10), 1772–1787.

Bian, X., Ren, J., DeVries, M., Schnegelsberg, B., Cockayne, D. A., Ford, A. P., & Galligan, J. J. (2003). Peristalsis is impaired in the small intestine of mice lacking the P2X3 subunit. *Journal of Physiology*, *551*(Pt 1), 309–322.

Birdsong, W. T., Fierro, L., Williams, F. G., Spelta, V., Naves, L. A., Knowles, M., ... McCleskey, E. W. (2010). Sensing muscle ischemia: coincident detection of acid and ATP via interplay of two ion channels. *Neuron*, *68*(4), 739–749.

Bo, X., Jiang, L. H., Wilson, H. L., Kim, M., Burnstock, G., Surprenant, A., & North, R. A. (2003). Pharmacological and biophysical properties of the human P2X5 receptor. *Molecular Pharmacology*, *63*(6), 1407–1416.

Boue-Grabot, E., Archambault, V., & Seguela, P. (2000). A protein kinase C site highly conserved in P2X subunits controls the desensitization kinetics of P2X(2) ATP-gated channels. *Journal of Biological Chemistry*, *275*(14), 10190–10195.

Brain, K. L., Jackson, V. M., Trout, S. J., & Cunnane, T. C. (2002). Intermittent ATP release from nerve terminals elicits focal smooth muscle Ca2+ transients in mouse vas deferens. *Journal of Physiology*, *541*(Pt 3), 849–862.

Brone, B., Moechars, D., Marrannes, R., Mercken, M., & Meert, T. (2007). P2X currents in peritoneal macrophages of wild type and P2X4 -/- mice. *Immunology Letters*, *113*(2), 83–89.

Browne, L. E., Cao, L., Broomhead, H. E., Bragg, L., Wilkinson, W. J., & North, R. A. (2011). P2X receptor channels show threefold symmetry in ionic charge selectivity and unitary conductance. *Nature Neuroscience*, *14*(1), 17–18.

Browne, L. E., Nunes, J. P., Sim, J. A., Chudasama, V., Bragg, L., Caddick, S., & North, R. A. (2014). Optical control of trimeric P2X receptors and acid-sensing ion channels. *Proceedings of the National Academy of Sciences of the United States of America*, *111*(1), 521–526.

Burnstock, G., Campbell, G., Satchell, D., & Smythe, A. (1970). Evidence that adenosine triphosphate or a related nucleotide is the transmitter substance released by non-adrenergic inhibitory nerves in the gut. *British Journal of Pharmacology*, *40*(4), 668–688.

Burnstock, G., & Kennedy, C. (1985). Is there a basis for distinguishing two types of P2-purinoceptor? *General Pharmacology*, *16*(5), 433–440.

Buvinic, S., Almarza, G., Bustamante, M., Casas, M., Lopez, J., Riquelme, M., . . . Jaimovich, E. (2009). ATP released by electrical stimuli elicits calcium transients in skeletal muscle. *Journal of Biological Chemistry*, *284*(50), 34490–34505.

Chessell, I. P., Hatcher, J. P., Bountra, C., Michel, A. D., Hughes, J. P., Green, P., . . . Buell, G. N. (2005). Disruption of the P2X7 purinoceptor gene abolishes chronic inflammatory and neuropathic pain. *Pain*, *114*(3), 386–396.

Chung, M. K., Guler, A. D., & Caterina, M. J. (2008). TRPV1 shows dynamic ionic selectivity during agonist stimulation. *Nature Neuroscience*, *11*(5), 555–564.

Cockayne, D. A., Dunn, P. M., Zhong, Y., Rong, W., Hamilton, S. G., Knight, G. E., . . . Ford, A. P. (2005). P2X2 knockout mice and P2X2/P2X3 double knockout mice reveal a role for the P2X2 receptor subunit in mediating multiple sensory effects of ATP. *Journal of Physiology*, *567*(Pt 2), 621–639.

Cockayne, D. A., Hamilton, S. G., Zhu, Q. M., Dunn, P. M., Zhong, Y., Novakovic, S., . . . Ford, A. P. (2000). Urinary bladder hyporeflexia and reduced pain-related behaviour in P2X3-deficient mice. *Nature*, *407*(6807), 1011–1015.

Cook, S. P., Rodland, K. D., & McCleskey, E. W. (1998). A memory for extracellular Ca2+ by speeding recovery of P2X receptors from desensitization. *Journal of Neuroscience*, *18*(22), 9238–9244.

Darbousset, R., Delierneux, C., Mezouar, S., Hego, A., Lecut, C., Guillaumat, I., . . . Dubois, C. (2014). P2X1 expressed on polymorphonuclear neutrophils and platelets is required for thrombosis in mice. *Blood*, *124*(16), 2575–2585.

de Baaij, J. H., Kompatscher, A., Viering, D. H., Bos, C., Bindels, R. J., & Hoenderop, J. G. (2016). P2X6 knockout mice exhibit normal electrolyte homeostasis. *PLoS One*, *11*(6), e0156803.

Di Virgilio, F. (2007). Purinergic signalling in the immune system. A brief update. *Purinergic Signalling*, *3*(1–2), 1–3.

Di Virgilio, F., & Adinolfi, E. (2017). Extracellular purines, purinergic receptors and tumor growth. *Oncogene*, *36*(3), 293–303.

Di Virgilio, F., Dal Ben, D., Sarti, A. C., Giuliani, A. L., & Falzoni, S. (2017). The P2X7 Receptor in Infection and Inflammation. *Immunity*, *47*(1), 15–31.

Du, J., Dong, H., & Zhou, H. X. (2012). Gating mechanism of a P2X4 receptor developed from normal mode analysis and molecular dynamics simulations. *Proceedings of the National Academy of Sciences of the United States of America*, *109*(11), 4140–4145.

Dubyak, G. R. (2012). P2X7 receptor regulation of non-classical secretion from immune effector cells. *Cellular Microbiology*, *14*(11), 1697–1706.

Egan, T. M., Haines, W. R., & Voigt, M. M. (1998). A domain contributing to the ion channel of ATP-gated P2X2 receptors identified by the substituted cysteine accessibility method. *Journal of Neuroscience*, *18*(7), 2350–2359.

Egan, T. M., & Khakh, B. S. (2004). Contribution of calcium ions to P2X channel responses. *Journal of Neuroscience*, *24*(13), 3413–3420.

Egan, T. M., Samways, D. S., & Li, Z. (2006). Biophysics of P2X receptors. *Pflugers Archiv*, *452*(5), 501–512.

Eser, A., Colombel, J. F., Rutgeerts, P., Vermeire, S., Vogelsang, H., Braddock, M., . . . Reinisch, W. (2015). Safety and efficacy of an oral inhibitor of the purinergic receptor P2X7 in adult patients with moderately to severely active Crohn's disease: a randomized placebo-controlled, double-blind, phase IIa study. *Inflammatory Bowel Disease*, *21*(10), 2247–2253.

Evans, R. J. (2009). Orthosteric and allosteric binding sites of P2X receptors. *European Biophysics Journal*, *38*(3), 319–327.

Ferreira, L., Pereira, L., & Faria, R. (2015). Fluorescent dyes as a reliable tool in P2X7 receptor-associated pore studies. *Journal of Bioenergetics and Biomembranes*, *47*(4), 283–307.

Finger, T. E., Danilova, V., Barrows, J., Bartel, D. L., Vigers, A. J., Stone, L., . . . Kinnamon, S. C. (2005). ATP signaling is crucial for communication from taste buds to gustatory nerves. *Science*, *310*(5753), 1495–1499.

Ford, A. P., Undem, B. J., Birder, L. A., Grundy, D., Pijacka, W., & Paton, J. F. (2015). P2X3 receptors and sensitization of autonomic reflexes. *Autonomic Neuroscience*, *191*, 16–24.

Fountain, S. J., Parkinson, K., Young, M. T., Cao, L., Thompson, C. R., & North, R. A. (2007). An intracellular P2X receptor required for osmoregulation in *Dictyostelium discoideum*. *Nature*, *448*(7150), 200–203.

Frings, S., Hackos, D. H., Dzeja, C., Ohyama, T., Hagen, V., Kaupp, U. B., & Korenbrot, J. I. (2000). Determination of fractional calcium ion current in cyclic nucleotide-gated channels. *Methods in Enzymology*, *315*, 797–817.

Fryatt, A. G., & Evans, R. J. (2014). Kinetics of conformational changes revealed by voltage-clamp fluorometry give insight to desensitization at ATP-gated human P2X1 receptors. *Molecular Pharmacology*, *86*(6), 707–715.

Gitterman, D. P., & Evans, R. J. (2001). Nerve evoked P2X receptor contractions of rat mesenteric arteries; dependence on vessel size and lack of role of L-type calcium channels and calcium induced calcium release. *British Journal of Pharmacology*, *132*(6), 1201–1208.

Gu, J. G., & MacDermott, A. B. (1997). Activation of ATP P2X receptors elicits glutamate release from sensory neuron synapses. *Nature*, *389*(6652), 749–753.

Guan, Z., Osmond, D. A., & Inscho, E. W. (2007). P2X receptors as regulators of the renal microvasculature. *Trends in Pharmacological Sciences*, *28*(12), 646–652.

Guerra, A. N., Gavala, M. L., Chung, H. S., & Bertics, P. J. (2007). Nucleotide receptor signalling and the generation of reactive oxygen species. *Purinergic Signalling*, *3*(1–2), 39–51.

Habermacher, C., Dunning, K., Chataigneau, T., & Grutter, T. (2016). Molecular structure and function of P2X receptors. *Neuropharmacology*, *104*, 18–30.

Habermacher, C., Martz, A., Calimet, N., Lemoine, D., Peverini, L., Specht, A., . . . Grutter, T. (2016). Photo-switchable tweezers illuminate pore-opening motions of an ATP-gated P2X ion channel. *Elife*, *5*, e11050.

Haines, W. R., Voigt, M. M., Migita, K., Torres, G. E., & Egan, T. M. (2001). On the contribution of the first transmembrane domain to whole-cell current through an ATP-gated ionotropic P2X receptor. *Journal of Neuroscience*, *21*(16), 5885–5892.

Harkat, M., Peverini, L., Cerdan, A. H., Dunning, K., Beudez, J., Martz, A., . . . Grutter, T. (2017). On the permeation of large organic cations through the pore of ATP-gated P2X receptors. *Proceedings of the National Academy of Sciences of the United States of America*, *114*(19), E3786–E3795.

Hattori, M., & Gouaux, E. (2012). Molecular mechanism of ATP binding and ion channel activation in P2X receptors. *Nature*, *485*(7397), 207–212.

Hechler, B., Lenain, N., Marchese, P., Vial, C., Heim, V., Freund, M., . . . Gachet, C. (2003). A role of the fast ATP-gated P2X1 cation channel in thrombosis of small arteries in vivo. *Journal of Experimental Medicine*, *198*(4), 661–667.

Heymann, G., Dai, J., Li, M., Silberberg, S. D., Zhou, H. X., & Swartz, K. J. (2013). Inter- and intrasubunit interactions between transmembrane helices in the open state of P2X receptor channels. *Proceedings of the National Academy of Sciences of the United States of America, 110*(42), E4045–4054.

Hille, B. (2001). *Ion channels of excitable membranes.* Third edition. Sunderland, MA: Sinauer Associates.

Holton, F. A., & Holton, P. (1954). The capillary dilator substances in dry powders of spinal roots: a possible role of adenosine triphosphate in chemical transmission from nerve endings. *Journal of Physiology (London), 126*, 124–140.

Housley, G. D., Morton-Jones, R., Vlajkovic, S. M., Telang, R. S., Paramananthasivam, V., Tadros, S. F., . . . Ryan, A. F. (2013). ATP-gated ion channels mediate adaptation to elevated sound levels. *Proceedings of the National Academy of Sciences of the United States of America, 110*(18), 7494–7499.

Hu, B., Mei, Q. B., Yao, X. J., Smith, E., Barry, W. H., & Liang, B. T. (2001). A novel contractile phenotype with cardiac transgenic expression of the human P2X4 receptor. *Federation of American Societies for Experimental Biology Journal, 15*(14), 2739–2741.

Huang, L. D., Fan, Y. Z., Tian, Y., Yang, Y., Liu, Y., Wang, J., . . . Yu, Y. (2014). Inherent dynamics of head domain correlates with ATP-recognition of P2X4 receptors: insights gained from molecular simulations. *PLoS One, 9*(5), e97528.

Illes P, Khan TM & Rubini P (2017). Neuronal P2X7 receptors revisited: Do they really exist? *Journal of Neuroscience, 30*, 7049–7062.

Jarvis, M. F., & Khakh, B. S. (2009). ATP-gated P2X cation-channels. *Neuropharmacology, 56*(1), 208–215.

Jiang, L. H., Kim, M., Spelta, V., Bo, X., Surprenant, A., & North, R. A. (2003). Subunit arrangement in P2X receptors. *Journal of Neuroscience, 23*(26), 8903–8910.

Jiang, L. H., Rassendren, F., Spelta, V., Surprenant, A., & North, R. A. (2001). Amino acid residues involved in gating identified in the first membrane-spanning domain of the rat P2X(2) receptor. *Journal of Biological Chemistry, 276*(18), 14902–14908.

Jiang, R., Lemoine, D., Martz, A., Taly, A., Gonin, S., Prado de Carvalho, L., . . . Grutter, T. (2011). Agonist trapped in ATP-binding sites of the P2X2 receptor. *Proceedings of the National Academy of Sciences of the United States of America, 108*(22), 9066–9071.

Jiang, R., Martz, A., Gonin, S., Taly, A., de Carvalho, L. P., & Grutter, T. (2010). A putative extracellular salt bridge at the subunit interface contributes to the ion channel function of the ATP-gated P2X2 receptor. *Journal of Biological Chemistry, 285*(21), 15805–15815.

Jiang, R., Taly, A., & Grutter, T. (2013). Moving through the gate in ATP-activated P2X receptors. *Trends in Biochemical Science, 38*(1), 20–29.

Jiang, R., Taly, A., Lemoine, D., Martz, A., Cunrath, O., & Grutter, T. (2012). Tightening of the ATP-binding sites induces the opening of P2X receptor channels. *The EMBO Journal, 31*, 2134–2143.

Kaczmarek-Hajek, K., Lorinczi, E., Hausmann, R., & Nicke, A. (2012). Molecular and functional properties of P2X receptors—recent progress and persisting challenges. *Purinergic Signalling, 8*(3), 375–417.

Karasawa, A., & Kawate, T. (2016). Structural basis for subtype-specific inhibition of the P2X7 receptor. *Elife, 5*, e22153.

Kasuya, G., Fujiwara, Y., Takemoto, M., Dohmae, N., Nakada-Nakura, Y., Ishitani, R., . . . Nureki, O. (2016). Structural insights into divalent cation modulations of ATP-gated P2X receptor channels. *Cell Reports, 14*(4), 932–944.

Katsnelson, M. A., Rucker, L. G., Russo, H. M., & Dubyak, G. R. (2015). K+ efflux agonists induce NLRP3 inflammasome activation independently of Ca2+ signaling. *Journal of Immunology, 194*(8), 3937–3952.

Kawate, T., Michel, J. C., Birdsong, W. T., & Gouaux, E. (2009). Crystal structure of the ATP-gated P2X(4) ion channel in the closed state. *Nature, 460*(7255), 592–598.

Kawate, T., Robertson, J. L., Li, M., Silberberg, S. D., & Swartz, K. J. (2011). Ion access pathway to the transmembrane pore in P2X receptor channels. *Journal of General Physiology, 137*(6), 579–590.

Ke, H. Z., Qi, H., Weidema, A. F., Zhang, Q., Panupinthu, N., Crawford, D. T., . . . Thompson, D. D. (2003). Deletion of the P2X7 nucleotide receptor reveals its regulatory roles in bone formation and resorption. *Molecular Endocrinology, 17*(7), 1356–1367.

Kennedy, C. (2015). ATP as a cotransmitter in the autonomic nervous system. *Autonomic Neuroscience, 191*, 2–15.

Khakh, B. S., Bao, X. R., Labarca, C., & Lester, H. A. (1999). Neuronal P2X transmitter-gated cation channels change their ion selectivity in seconds. *Nature Neuroscience, 2*(4), 322–330.

Khakh, B. S., & North, R. A. (2012). Neuromodulation by extracellular ATP and P2X receptors in the CNS. *Neuron, 76*(1), 51–69.

Kotnis, S., Bingham, B., Vasilyev, D. V., Miller, S. W., Bai, Y., Yeola, S., . . . Whiteside, G. T. (2010). Genetic and functional analysis of human P2X5 reveals a distinct pattern of exon 10 polymorphism with predominant expression of the nonfunctional receptor isoform. *Molecular Pharmacology, 77*(6), 953–960.

Kowalski, M., Hausmann, R., Dopychai, A., Grohmann, M., Franke, H., Nieber, K., . . . Riedel, T. (2014). Conformational flexibility of the agonist binding jaw of the human P2X3 receptor is a prerequisite for channel opening. *British Journal of Pharmacology, 171*(22), 5093–5112.

Kracun, S., Chaptal, V., Abramson, J., & Khakh, B. S. (2010). Gated access to the pore of a P2X receptor: structural implications for closed–open transitions. *Journal of Biological Chemistry, 285*(13), 10110–10121.

Lecut, C., Faccinetto, C., Delierneux, C., van Oerle, R., Spronk, H. M., Evans, R. J., . . . Oury, C. (2012). ATP-gated P2X1 ion channels protect against endotoxemia by dampening neutrophil activation. *Journal of Thrombosis and Haemostasis, 10*(3), 453–465.

Lemoine, D., Habermacher, C., Martz, A., Mery, P. F., Bouquier, N., Diverchy, F., . . . Grutter, T. (2013). Optical control of an ion channel gate. *Proceedings of the National Academy of Sciences of the United States of America, 110*(51), 20813–20818.

Li, M., Chang, T. H., Silberberg, S. D., & Swartz, K. J. (2008). Gating the pore of P2X receptor channels. *Nature Neuroscience, 11*(8), 883–887.

Li, M., Kawate, T., Silberberg, S. D., & Swartz, K. J. (2010). Pore-opening mechanism in trimeric P2X receptor channels. *Nature Communications, 1*, 44.

Li, M., Toombes, G. E., Silberberg, S. D., & Swartz, K. J. (2015). Physical basis of apparent pore dilation of ATP-activated P2X receptor channels. *Nature Neuroscience, 18*(11), 1577–1583.

Liang, X., Samways, D. S., Wolf, K., Bowles, E. A., Richards, J. P., Bruno, J., . . . Egan, T. M. (2015). Quantifying Ca2+ current and permeability in ATP-gated P2X7 receptors. *Journal of Biological Chemistry, 290*(12), 7930–7942.

Lorinczi, E., Bhargava, Y., Marino, S. F., Taly, A., Kaczmarek-Hajek, K., Barrantes-Freer, A., . . . Nicke, A. (2012). Involvement of the cysteine-rich head domain in activation and desensitization of the P2X1 receptor. *Proceedings of the National Academy of Sciences of the United States of America, 109*(28), 11396–11401.

Mahaut-Smith, M. P., Jones, S., & Evans, R. J. (2011). The P2X1 receptor and platelet function. *Purinergic Signalling, 7*(3), 341–356.

Mansoor, S. E., Lu, W., Oosterheert, W., Shekhar, M., Tajkhorshid, E., & Gouaux, E. (2016). X-ray structures define human P2X(3) receptor gating cycle and antagonist action. *Nature, 538*(7623), 66–71.

Marquez-Klaka, B., Rettinger, J., Bhargava, Y., Eisele, T., & Nicke, A. (2007). Identification of an intersubunit cross-link between substituted cysteine residues located in the putative ATP binding site of the P2X1 receptor. *Journal of Neuroscience, 27*(6), 1456–1466.

Marquez-Klaka, B., Rettinger, J., & Nicke, A. (2009). Inter-subunit disulfide cross-linking in homomeric and heteromeric P2X receptors. *European Biophysics Journal, 38*(3), 329–338.

McIlwrath, S. L., Davis, B. M., & Bielefeldt, K. (2009). Deletion of P2X3 receptors blunts gastro-oesophageal sensation in mice. *Neurogastroenterology and Motility, 21*(8), 890–e866.

Migita, K., Haines, W. R., Voigt, M. M., & Egan, T. M. (2001). Polar residues of the second transmembrane domain influence cation permeability of the ATP-gated P2X(2) receptor. *Journal of Biological Chemistry, 276*(33), 30934–30941.

Miklavc, P., Thompson, K. E., & Frick, M. (2013). A new role for P2X4 receptors as modulators of lung surfactant secretion. *Frontiers in Cellular Neuroscience, 7*, 171.

Miras-Portugal MT, Sebastian-Serrano A, de Diego Garcia L, & Diaz-Hernandez M (2017) Neuronal P2X7 Receptor: Involvement in Neuronal Physiology and Pathology. *Journal of Neuroscience, 37*(30), 7063–7072.

Mony, L., Zhu, S., Carvalho, S., & Paoletti, P. (2011). Molecular basis of positive allosteric modulation of GluN2B NMDA receptors by polyamines. *The EMBO Journal, 30*(15), 3134–3146.

Mulryan, K., Gitterman, D. P., Lewis, C. J., Vial, C., Leckie, B. J., Cobb, A. L., . . . Evans, R. J. (2000). Reduced vas deferens contraction and male infertility in mice lacking P2X1 receptors. *Nature, 403*(6765), 86–89.

Murrell-Lagnado, R. D., & Qureshi, O. S. (2008). Assembly and trafficking of P2X purinergic receptors (review). *Molecular Membrane Biology, 25*(4), 321–331.

Nagaya, N., Tittle, R. K., Saar, N., Dellal, S. S., & Hume, R. I. (2005). An intersubunit zinc binding site in rat P2X2 receptors. *Journal of Biological Chemistry, 280*(28), 25982–25993.

Neher, E. (1995). The use of fura-2 for estimating Ca buffers and Ca fluxes. *Neuropharmacology, 34*(11), 1423–1442.

Norenberg, W., Sobottka, H., Hempel, C., Plotz, T., Fischer, W., Schmalzing, G., & Schaefer, M. (2012). Positive allosteric modulation by ivermectin of human but not murine P2X7 receptors. *British Journal of Pharmacology, 167*(1), 48–66.

Oury, C., Kuijpers, M. J., Toth-Zsamboki, E., Bonnefoy, A., Danloy, S., Vreys, I., . . . Hoylaerts, M. F. (2003). Overexpression of the platelet P2X1 ion channel in transgenic mice generates a novel prothrombotic phenotype. *Blood, 101*(10), 3969–3976.

Oury, C., Lecut, C., Hego, A., Wera, O., & Delierneux, C. (2015). Purinergic control of inflammation and thrombosis: role of P2X1 receptors. *Computational and Structural Biotechnology Journal, 13*, 106–110.

Pankratov, Y., Lalo, U., Verkhratsky, A., & North, R. A. (2006). Vesicular release of ATP at central synapses. *Pflugers Archiv, 452*(5), 589–597.

Park, J. H., & Kim, Y. C. (2017). P2X7 receptor antagonists: a patent review (2010-2015). *Expert Opinion on Therapeutic Patients, 27*(3), 257–267.

Parkinson, K., Baines, A. E., Keller, T., Gruenheit, N., Bragg, L., North, R. A., & Thompson, C. R. (2014). Calcium-dependent regulation of Rab activation and vesicle fusion by an intracellular P2X ion channel. *Nature Cell Biology, 16*(1), 87–98.

Perregaux, D., & Gabel, C. A. (1994). Interleukin-1 beta maturation and release in response to ATP and nigericin. Evidence that potassium depletion mediated by these agents is a necessary and common feature of their activity. *Journal of Biological Chemistry*, *269*(21), 15195–15203.

Pevarello, P., Bovolenta, S., Tarroni, P., Za, L., Severi, E., Torino, D., & Vitalone, R. (2017). P2X7 antagonists for CNS indications: recent patent disclosures. *Pharmaceutical Patent Analyst*, *6*(2), 61–76.

Pippel, A., Stolz, M., Woltersdorf, R., Kless, A., Schmalzing, G., & Markwardt, F. (2017). Localization of the gate and selectivity filter of the full-length P2X7 receptor. *Proceedings of the National Academy of Sciences of the United States of America*, *114*(11), E2156–E2165.

Pougnet, J. T., Toulme, E., Martinez, A., Choquet, D., Hosy, E., & Boue-Grabot, E. (2014). ATP P2X receptors downregulate AMPA receptor trafficking and postsynaptic efficacy in hippocampal neurons. *Neuron*, *83*(2), 417–430.

Qureshi, O. S., Paramasivam, A., Yu, J. C., & Murrell-Lagnado, R. D. (2007). Regulation of P2X4 receptors by lysosomal targeting, glycan protection and exocytosis. *Journal of Cell Science*, *120*(Pt 21), 3838–3849.

Ramme, D., Regenold, J. T., Starke, K., Busse, R., & Illes, P. (1987). Identification of the neuroeffector transmitter in jejunal branches of the rabbit mesenteric artery. *Naunyn Schmiedeberg's Archives of Pharmacology*, *336*(3), 267–273.

Rassendren, F., Buell, G., Newbolt, A., North, R. A., & Surprenant, A. (1997). Identification of amino acid residues contributing to the pore of a P2X receptor. *The EMBO Journal*, *16*(12), 3446–3454.

Ren, J., Bian, X., DeVries, M., Schnegelsberg, B., Cockayne, D. A., Ford, A. P., & Galligan, J. J. (2003). P2X2 subunits contribute to fast synaptic excitation in myenteric neurons of the mouse small intestine. *Journal of Physiology*, *552*(Pt 3), 809–821.

Riedel, T., Schmalzing, G., & Markwardt, F. (2007). Influence of extracellular monovalent cations on pore and gating properties of P2X7 receptor-operated single-channel currents. *Biophysical Journal*, *93*(3), 846–858.

Rissiek, B., Haag, F., Rissiek, B., Haag, F., Boyer, O., Koch-Nolte, F., & Adriouch, S. (2015). P2X7 on Mouse T Cells: One Channel, Many Functions. *Frontiers in* Immunology, 6, 204. doi: 10.3389/fimmu.2015.00204. eCollection.

Roberts, J. A., Allsopp, R. C., El Ajouz, S., Vial, C., Schmid, R., Young, M. T., & Evans, R. J. (2012). Agonist binding evokes extensive conformational changes in the extracellular domain of the ATP-gated human P2X1 receptor ion channel. *Proceedings of the National Academy of Sciences of the United States of America*, *109*(12), 4663–4667.

Robinson, L. E., & Murrell-Lagnado, R. D. (2013). The trafficking and targeting of P2X receptors. *Frontiers in Cellular Neuroscience*, *7*, 233.

Roger, S., Jelassi, B., Couillin, I., Pelegrin, P., Besson, P., & Jiang, L. H. (2015). Understanding the roles of the P2X7 receptor in solid tumour progression and therapeutic perspectives. *Biochimica et Biophysica Acta*, *1848*(10 Pt B), 2584–2602.

Roger, S., Pelegrin, P., & Surprenant, A. (2008). Facilitation of P2X7 receptor currents and membrane blebbing via constitutive and dynamic calmodulin binding. *Journal of Neuroscience*, *28*(25), 6393–6401.

Rong, W., Gourine, A. V., Cockayne, D. A., Xiang, Z., Ford, A. P., Spyer, K. M., & Burnstock, G. (2003). Pivotal role of nucleotide P2X2 receptor subunit of the ATP-gated ion channel mediating ventilatory responses to hypoxia. *Journal of Neuroscience*, *23*(36), 11315–11321.

Rothwell, S. W., Stansfeld, P. J., Bragg, L., Verkhratsky, A., & North, R. A. (2014). Direct gating of ATP-activated ion channels (P2X2 receptors) by lipophilic attachment at the outer end of the second transmembrane domain. *Journal of Biological Chemistry, 289*(2), 618–626.

Rubio, M. E., & Soto, F. (2001). Distinct localization of P2X receptors at excitatory postsynaptic specializations. *Journal of Neuroscience, 21*(2), 641–653.

Samways, D. S., & Egan, T. M. (2007). Acidic amino acids impart enhanced Ca^{2+} permeability and flux in two members of the ATP-gated P2X receptor family. *Journal of General Physiology, 129*(3), 245–256.

Samways, D. S., Khakh, B. S., Dutertre, S., & Egan, T. M. (2011). Preferential use of unobstructed lateral portals as the access route to the pore of human ATP-gated ion channels (P2X receptors). *Proceedings of the National Academy of Sciences of the United States of America, 108*(33), 13800–13805.

Samways, D. S., Li, Z., & Egan, T. M. (2014). Principles and properties of ion flow in P2X receptors. *Frontiers in Cellular Neuroscience, 8*, 6.

Saul, A., Hausmann, R., Kless, A., & Nicke, A. (2013). Heteromeric assembly of P2X subunits. *Frontiers in Cellular Neuroscience, 7*, 250.

Schwarz et al. 2012, Alternative splicing of the N-terminal cytosolic and transmembrane domains of P2X7 controls gating of the ion channel by ADP-ribosylation. *PLOS One, 7*, e41269.

Shen, J. B., Pappano, A. J., & Liang, B. T. (2006). Extracellular ATP-stimulated current in wild-type and P2X4 receptor transgenic mouse ventricular myocytes: implications for a cardiac physiologic role of P2X4 receptors. *Federation of American Societies for Experimental Biology Journal, 20*(2), 277–284.

Shen, J. B., Shutt, R., Agosto, M., Pappano, A., & Liang, B. T. (2009). Reversal of cardiac myocyte dysfunction as a unique mechanism of rescue by P2X4 receptors in cardiomyopathy. *American Journal of Physiology–Heart and Circulatory Physiology, 296*(4), H1089–H1095.

Shinozaki, Y., Sumitomo, K., Tsuda, M., Koizumi, S., Inoue, K., & Torimitsu, K. (2009). Direct observation of ATP-induced conformational changes in single P2X4 receptors. *PLoS Biology, 7*(5), e103.

Sim, J. A., Chaumont, S., Jo, J., Ulmann, L., Young, M. T., Cho, K., . . . Rassendren, F. (2006). Altered hippocampal synaptic potentiation in P2X4 knock-out mice. *Journal of Neuroscience, 26*(35), 9006–9009.

Sim, J. A., Young, M. T., Sung, H. Y., North, R. A., & Surprenant, A. (2004). Reanalysis of P2X7 receptor expression in rodent brain. *Journal of Neuroscience, 24*(28), 6307–6314.

Solle, M., Labasi, J., Perregaux, D. G., Stam, E., Petrushova, N., Koller, B. H., . . . Gabel, C. A. (2001). Altered cytokine production in mice lacking P2X(7) receptors. *Journal of Biological Chemistry, 276*(1), 125–132.

Sonin, D., Zhou, S. Y., Cronin, C., Sonina, T., Wu, J., Jacobson, K. A., . . . Liang, B. T. (2008). Role of P2X purinergic receptors in the rescue of ischemic heart failure. *American Journal of Physiology–Heart and Circulatory Physiology, 295*(3), H1191–H1197.

Sorge, R. E., Trang, T., Dorfman, R., Smith, S. B., Beggs, S., Ritchie, J., . . . Mogil, J. S. (2012). Genetically determined P2X7 receptor pore formation regulates variability in chronic pain sensitivity. *Nature Medicine, 18*(4), 595–599.

Souslova, V., Cesare, P., Ding, Y., Akopian, A. N., Stanfa, L., Suzuki, R., . . . Wood, J. N. (2000). Warm-coding deficits and aberrant inflammatory pain in mice lacking P2X3 receptors. *Nature, 407*(6807), 1015–1017.

Sperlagh, B., & Illes, P. (2014). P2X7 receptor: an emerging target in central nervous system diseases. *Trends in Pharmacological Sciences, 35*(10), 537–547.

Stelmashenko, O., Compan, V., Browne, L. E., & North, R. A. (2014). Ectodomain movements of an ATP-gated ion channel (P2X2 receptor) probed by disulfide locking. *Journal of Biological Chemistry, 289*(14), 9909–9917.

Stephan, G., Kowalski-Jahn, M., Zens, C., Schmalzing, G., Illes, P., & Hausmann, R. (2016). Inter-subunit disulfide locking of the human P2X3 receptor elucidates ectodomain movements associated with channel gating. *Purinergic Signalling, 12*(2), 221–233.

Stojilkovic, S. S., Tomic, M., He, M. L., Yan, Z., Koshimizu, T. A., & Zemkova, H. (2005). Molecular dissection of purinergic P2X receptor channels. *Annals of the New York Academy of Sciences, 1048*, 116–130.

Surprenant, A., Rassendren, F., Kawashima, E., North, R. A., & Buell, G. (1996). The cytolytic P2Z receptor for extracellular ATP identified as a P2X receptor (P2X7). *Science, 272*(5262), 735–738.

Takezawa, K., Kondo, M., Kiuchi, H., Ueda, N., Soda, T., Fukuhara, S., . . . Shimada, S. (2016). Authentic role of ATP signaling in micturition reflex. *Scientific Reports, 6*, 19585.

Thomas, S. A., & Hume, R. I. (1990). Permeation of both cations and anions through a single class of ATP-activated ion channels in developing chick skeletal muscle. *Journal of General Physiology, 95*(4), 569–590.

Torres, G. E., Egan, T. M., & Voigt, M. M. (1999). Hetero-oligomeric assembly of P2X receptor subunits. Specificities exist with regard to possible partners. *Journal of Biological Chemistry, 274*(10), 6653–6659.

Toulme, E., Garcia, A., Samways, D., Egan, T. M., Carson, M. J., & Khakh, B. S. (2010). P2X4 receptors in activated C8-B4 cells of cerebellar microglial origin. *Journal of General Physiology, 135*(4), 333–353.

Toulme, E., & Khakh, B. S. (2012). Imaging P2X4 receptor lateral mobility in microglia: regulation by calcium and p38 MAPK. *Journal of Biological Chemistry.*

Troadec, J. D., Thirion, S., Nicaise, G., Lemos, J. R., & Dayanithi, G. (1998). ATP-evoked increases in [Ca2+]i and peptide release from rat isolated neurohypophysial terminals via a P2X2 purinoceptor. *Journal of Physiology, 511*(Part 1), 89–103.

Tsuda, M., Kuboyama, K., Inoue, T., Nagata, K., Tozaki-Saitoh, H., & Inoue, K. (2009). Behavioral phenotypes of mice lacking purinergic P2X4 receptors in acute and chronic pain assays. *Molecular Pain, 5*, 28.

Ulmann, L., Hirbec, H., & Rassendren, F. (2010). P2X4 receptors mediate PGE2 release by tissue-resident macrophages and initiate inflammatory pain. *The EMBO Journal, 29*(14), 2290–2300.

Ussar, S., Lee, K. Y., Dankel, S. N., Boucher, J., Haering, M. F., Kleinridders, A., . . . Kahn, C. R. (2014). ASC-1, PAT2, and P2RX5 are cell surface markers for white, beige, and brown adipocytes. *Science Translational Medicine, 6*(247), 247ra103.

Ventura, S., Oliver, V., White, C. W., Xie, J. H., Haynes, J. M., & Exintaris, B. (2011). Novel drug targets for the pharmacotherapy of benign prostatic hyperplasia (BPH). *British Journal of Pharmacology, 163*(5), 891–907.

Virginio, C., MacKenzie, A., Rassendren, F. A., North, R. A., & Surprenant, A. (1999). Pore dilation of neuronal P2X receptor channels. *Nature Neuroscience, 2*(4), 315–321.

von Kugelgen, I., & Harden, T. K. (2011). Molecular pharmacology, physiology, and structure of the P2Y receptors. *Advances in Pharmacology, 61*, 373–415.

Wang, Y., Mackes, J., Chan, S., Haughey, N. J., Guo, Z., Ouyang, X., . . . Mattson, M. P. (2006). Impaired long-term depression in P2X3 deficient mice is not associated with a spatial learning deficit. *Journal of Neurochemistry*, *99*(5), 1425–1434.

Wei, L., Caseley, E., Li, D., & Jiang, L. H. (2016). ATP-induced P2X receptor-dependent large pore formation: How much do we know? *Frontiers in Pharmacology*, *7*, 5.

White, C. W., Choong, Y. T., Short, J. L., Exintaris, B., Malone, D. T., Allen, A. M., . . . Ventura, S. (2013). Male contraception via simultaneous knockout of alpha1A-adrenoceptors and P2X1-purinoceptors in mice. *Proceedings of the National Academy of Sciences of the United States of America*, *110*(51), 20825–20830.

Wilkinson, W. J., Jiang, L. H., Surprenant, A., & North, R. A. (2006). Role of ectodomain lysines in the subunits of the heteromeric P2X2/3 receptor. *Molecular Pharmacology*, *70*(4), 1159–1163.

Wyatt, L. R., Godar, S. C., Khoja, S., Jakowec, M. W., Alkana, R. L., Bortolato, M., & Davies, D. L. (2013). Sociocommunicative and sensorimotor impairments in male P2X4-deficient mice. *Neuropsychopharmacology*, *38*(10), 1993–2002.

Xu, J., Chai, H., Ehinger, K., Egan, T. M., Srinivasan, R., Frick, M., & Khakh, B. S. (2014). Imaging P2X4 receptor subcellular distribution, trafficking, and regulation using P2X4-pHluorin. *Journal of General Physiology*, *144*(1), 81–104.

Xu, X. J., Boumechache, M., Robinson, L. E., Marschall, V., Gorecki, D. C., Masin, M., & Murrell-Lagnado, R. D. (2012). Splice variants of the P2X7 receptor reveal differential agonist dependence and functional coupling with pannexin-1. *Journal of Cell Science*, *125*(Part 16), 3776–3789.

Yamamoto, K., Sokabe, T., Matsumoto, T., Yoshimura, K., Shibata, M., Ohura, N., . . . Ando, J. (2006). Impaired flow-dependent control of vascular tone and remodeling in P2X4-deficient mice. *Nature Medicine*, *12*(1), 133–137.

Yan, D., Zhu, Y., Walsh, T., Xie, D., Yuan, H., Sirmaci, A., . . . Liu, X. Z. (2013). Mutation of the ATP-gated P2X(2) receptor leads to progressive hearing loss and increased susceptibility to noise. *Proceedings of the National Academy of Sciences of the United States of America*, *110*(6), 2228–2233.

Yang, T., Shen, J. B., Yang, R., Redden, J., Dodge-Kafka, K., Grady, J., . . . Liang, B. T. (2014). Novel protective role of endogenous cardiac myocyte P2X4 receptors in heart failure. *Circulation: Heart Failure*, *7*(3), 510–518.

Zhao, W. S., Wang, J., Ma, X. J., Yang, Y., Liu, Y., Huang, L. D., . . . Yu, Y. (2014). Relative motions between left flipper and dorsal fin domains favour P2X4 receptor activation. *Nature Communications*, *5*, 4189.

CHAPTER 14

LARGE CONDUCTANCE POTASSIUM CHANNELS IN THE NERVOUS SYSTEM

WILLY CARRASQUEL-URSULAEZ, YENISLEIDY LORENZO, FELIPE ECHEVERRIA, AND RAMON LATORRE

INTRODUCTION

THE Slo family of K^+ channels is distinguished by its unusually large single-channel conductance. It is named after *Slowpoke*, the gene that encodes this channel in *Drosophila* (Atkinson, Robertson, & Ganetzky, 1991). In the mammalian genome, there are four genes encoding Slo channels: *Slo1* (*K_{Ca}1.1; KCNMA1* in mammals); two highly similar *Slo2* paralogues, *Slo2.1* (*K_{Na}1.2, KCNT2, Slick*, formerly K_{Ca}4.2 [Kaczmarek et al., 2017]) and *Slo2.2* (*K_{Na}1.1, KCNT1, Slack*, formerly K_{Ca}4.2 [Kaczmarek et al., 2017]); and Slo3 (*K_{Ca}5.1, KCNU1*). Although all four channels show structural similarities, their activity is modulated by different internal cations: Slo1 is activated by Ca^{2+}, Slo2.1 and Slo2.2 are both activated by Na^+, and Slo3 opening is mediated by alkalization. Certain human Slo3 variants can also be activated by Ca^{2+} (Brenker et al., 2014; Geng et al., 2017). Slo3 is found exclusively in sperm cells (Schreiber et al., 1998); we will therefore not discuss this channel further. Here, we highlight the distribution and physiological roles of the Slo family in the nervous system. Importantly, our discussion takes into account recent advances in structural data that have provided crucial insights into the gating mechanisms of these channels.

Slo1 (BK) Channel

The Slo1 channel, also named Big K (BK) (Blatz AL, 1987) or Maxi K (Latorre & Miller, 1983), is the most studied member of the Slo family. In mammals, BK is encoded by a single gene (*Slo1* or *KCNMA1*) (Butler, Tsunoda, McCobb, Wei, & Salkoff, 1993). At ~250 pS in symmetrical 100 mM K^+, its single-channel conductance is the largest within the family (Table 14.1) (Latorre, Vergara, & Hidalgo, 1982; Marty, 1981; Pallotta, Magleby, & Barrett, 1981). Although BK is the product of a single gene, a large functional diversity is achieved by alternative splicing (Fodor & Aldrich, 2009), metabolic regulation (Sitdikova, Fuchs, Kainz, Weiger, & Hermann, 2014), and association with accessory subunits (Behrens et al., 2000; Brenner, Jegla, Wickenden, Liu, & Aldrich, 2000; Knaus et al., 1994; Meera, Wallner, & Toro, 2000; Wallner, Meera, & Toro, 1999; Xia, Ding, Zeng, Duan, & Lingle, 2000; Yan & Aldrich, 2010, 2012).

The activation of BK leads to a large efflux of K^+ ions, thus hyperpolarizing the cellular membrane. Combined with its ability to act as a sensor of changes in the internal Ca^{2+} concentration ($[Ca^{2+}]_i$), this allows BK to control the activity of excitable cells. For example, BK channel activation—mediated by the Ca^{2+} influx through voltage-dependent Ca^{2+} (Cav) channels—tends to drive excitable cells toward their resting state. In the central nervous system (CNS), BK influences the shape, frequency, and propagation of action potentials (APs), as well as neurotransmitter release. BK is also expressed in the nuclear envelope of neurons, where it modifies gene transcription and neuronal morphology (Li & Gao, 2016). Moreover, BK expression in non-neuronal cells, like astrocytes or vascular smooth muscle cells, influences brain activity by regulating cerebral blood flow (Filosa et al., 2006).

Structural Features

Functional BK channels are homotetramers of the pore-forming α subunit (Hite, Tao, & MacKinnon, 2017; Quirk & Reinhart, 2001; Shen et al., 1994; Tao, Hite, & MacKinnon, 2017; Wang & Sigworth, 2009). Each of the four α subunits contains seven transmembrane (TM) segments, S0 to S6, and a large C-terminal domain (CTD) that comprises about two-thirds of the protein (Figure 14.1A). The TM region is composed of two distinct functional domains, the voltage sensor domain (VSD) and the pore domain (PD) (Figure 14.1A).

Several structural features distinguish BK from the family of voltage-dependent K^+ (Kv) channels. Due to the presence of an additional TM segment in BK (S0), the channel's N-terminus points towards the external side of the cell (Figure 14.1A)

Table 14.1 Slo Channels Family

	Alternative names	Gene (human)	Chromosomal location	Conductance (symmetrical K^+)	Accessory subunits
Slo1	BK, K_{Ca}, Maxi-K, $K_{Ca}1.1$	*KCNMA1*	10q22	100–270 pS	β1–4, γ1–4
Slo2.1	Slick, $K_{Na}1.2$	*KCNT2*	1q31.3	60–140 pS	Not reported
Slo2.2	Slack, $K_{Na}1.1$	*KCNT1*	9q34.3	100–180 pS	Not reported
Slo3	K^+ large conductance pH-sensitive channel, $K_{Ca}5.1$	*KCNU1*	8p11.2	70–100 pS	γ2

(Hite et al., 2017; Tao et al., 2017; Wallner, Meera, & Toro, 1996). Furthermore, in BK, the VSD interacts directly with the PD of the same subunit, whereas the VSD of Kv channels makes contact with the PD of the adjacent subunit (Figure 14.1B–C) (Hite et al., 2017; Tao et al., 2017). Finally, a large S6-RCK1 linker allows the VSD to contact the CTD from a neighboring subunit (Figure 14.1B) (Hite et al., 2017; Tao et al., 2017).

The Voltage Sensor Domain (VSD)

The VSD is formed by TM segments S0 to S4 of each subunit (Figure 14.1A and D). Like Kv channels, BK has a positively charged S4 (Figure 14.2) (Adelman et al., 1992; Atkinson et al., 1991; Stefani et al., 1997). However, there are some functional differences between the S4 of BK and those of Kv channels. In BK, the number of gating charges per channel (i.e., the number of elemental charges that move upon activation of each voltage sensor) is about four (Stefani et al., 1997); i.e., a third of the number of gating charges of Kv channels (Aggarwal & MacKinnon, 1996; Schoppa, McCormack, Tanouye, & Sigworth, 1992; Seoh, Sigg, Papazian, & Bezanilla, 1996). Unlike Kv channels, where voltage-sensing residues reside in S4, BK channel voltage-sensing residues are located in TM segments S2, S3, and S4 (D153 and R168 in S2; D186 in S3; and R213 in S4)[1] (Figure 14.1D) (Diaz et al., 1998; Ma, Lou, & Horrigan, 2006).

The C-Terminal (Ca^{2+} Sensor) Domain (CTD)

The CTD consists of a tetrameric gating ring comprising two tandem regulators of K^+ conductance (RCK) domains (Jiang, Pico, Cadene, Chait, & MacKinnon, 2001), RCK1 and RCK2. Each of these contains a high-affinity Ca^{2+} binding site (Figure 14.1A and E) (Bian, Favre, & Moczydlowski, 2001; Cox, 2005; Latorre & Brauchi, 2006; Schreiber & Salkoff, 1997; Yuan, Leonetti, Hsiung, & MacKinnon, 2011; Yuan, Leonetti, Pico, Hsiung, &

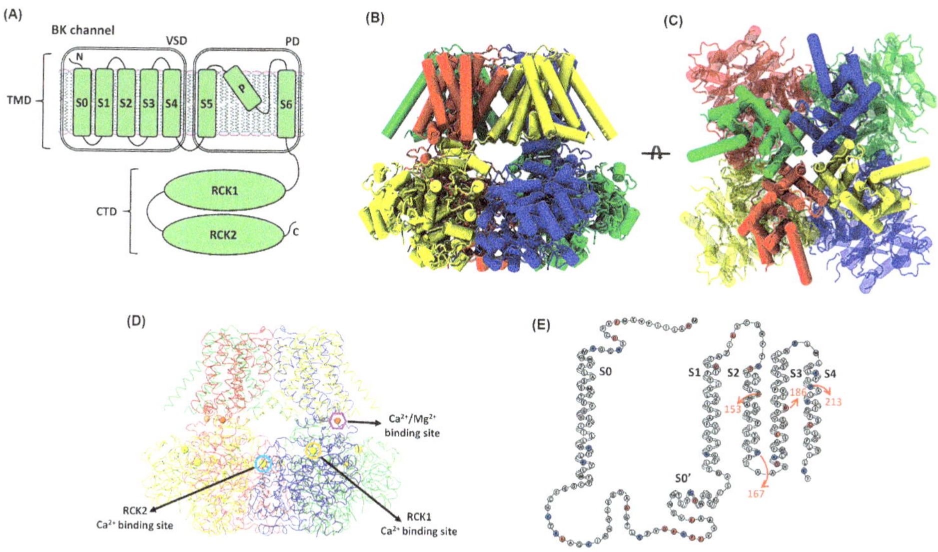

FIGURE 14.1 Structure of BK channel.
(*A*) Topology of the BK channels. The seven α-helical transmembrane segments are represented as vertical rectangles. S0 to S4 form the voltage sensor domain (VSD). The pore domain (PD) consist of S5, the pore helix (P), and S6. VSD and PD form the transmembrane domain (TMD). The C-terminal domain (CTD) consist of two RCK domains (horizontal ovals). N terminal is extracellular. (*B*) Side and (*C*) top views of the full tetrameric channel, respectively. Each subunit is represented by a different color. In (*C*), CTD is transparent. Note that VSD and PD are not domain swapped, while the VSD contacts the CTD from a neighboring subunit. (*D*) BK channel ribbon representation depicting the three divalent cation binding sites. All sites are repeated four times due to the tetrameric symmetry of the channel. The location of one high-affinity Ca^{2+} binding site in RCK1, one high-affinity Ca^{2+} binding site in the RCK2, and one low-affinity Ca^{2+} and Mg^{2+} binding site are indicated by the heptagon, the octagon, and the hexagon, respectively. The spheres represent Ca^{2+} and Mg^{2+} ions, as noted. (*E*) The primary and secondary structure and the topology of the BK VSD. Arrows and position numbers indicate voltage-sensing residues located in the S2, S3, and S4 transmembrane segments. The coordinates of the BK channel and the topology of the VSD were obtained from the *Aplysia californica* Slo1 structure (Protein Data Bank [PDB] ID: 5TJI) (Hite et al., 2017). Rendering was made using the Visual Molecular Dynamics, Version 1.9.2 (Humphrey, Dalke, & Schulten, 1996).

MacKinnon, 2010; Zeng, Xia, & Lingle, 2005). A less selective, low-affinity divalent cation (Mg^{2+}/Ca^{2+}) binding site is located at the interface between the VSD and the CTD (Figure 14.1E) (Tao et al., 2017; H. Yang et al., 2007; X. Zhang, Solaro, & Lingle, 2001). All three binding sites have been described in atomic detail for *Aplysia* BK. In addition, the RCK2 site has been also described in the BK zebrafish (Tao et al., 2017; Yuan et al., 2011). The broad concentration range characterizing the Ca^{2+}-dependence of BK currents is explained by the existence of multiple regulatory Ca^{2+} binding sites, each with different Ca^{2+}-affinities.

The Pore Domain (PD) and the Gate

Like in other K^+ channels, the PD of the BK channel is formed by two TM segments from each subunit, S5 and S6, and the pore helix (Figure 14.1A). A signature sequence, TVGYGD, which is characteristic of the selectivity filter of many K^+ channels, is located between the pore helix and S6 (Figure 14.1A and 14.3A) (Heginbotham, Lu, Abramson, & MacKinnon, 1994). Compared to other K^+ channels, including Kv channels, where the gate is a "bundle crossing" formed by four S6 pore-lining helices, available evidence indicates that in BK, the gate is in, or in the neighborhood of, the selectivity filter (del Camino & Yellen, 2001; Doyle et al., 1998; Holmgren, Smith, & Yellen, 1997; Li & Aldrich, 2006; Liu, Holmgren, Jurman, & Yellen, 1997; Rothberg, Shin, Phale, & Yellen, 2002; Shin, Rothberg, & Yellen, 2001; Thompson & Begenisich, 2012; Wilkens & Aldrich, 2006).

Molecular Determinants of the Large Conductance

Although the amino acid composition of the selectivity filter is identical in BK and Kv channels, the unitary conductance of BK exceeds that of Kv channels by an order of magnitude. Paradoxically, in spite of having a large conductance, BK is the most selective K^+ channel currently known (Latorre et al., 2017). The mechanism underlying this unusually large conductance has not yet been fully resolved, but several structural features of the BK conduction system may contribute to a possible explanation at the molecular level. In BK, a ring of four negatively charged residues (D292) in the external entrance of the selectivity filter and, unlike Kv channels, a ring of eight negative charges (E321 and E324) surround the entrance of the internal vestibule (the aqueous region underneath the selectivity filter) (Figure 14.3A–B) (Carrasquel-Ursulaez et al.,

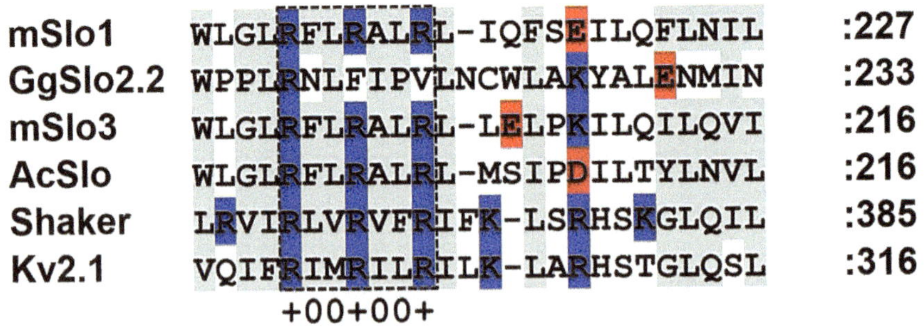
mSlo1 WLGLRFLRALRL-IQFSEILQFLNIL :227
GgSlo2.2 WPPLRNLFIPVLNCWLAKYALENMIN :233
mSlo3 WLGLRFLRALRL-LELPKILQILQVI :216
AcSlo WLGLRFLRALRL-MSIPDILTYLNVL :216
Shaker LRVIRLVRVFRIFK-LSRHSKGLQIL :385
Kv2.1 VQIFRIMRILRILK-LARHSTGLQSL :316
+00+00+

FIGURE 14.2 Alignment of the S4 transmembrane segment of several K^+ channels.
Mouse BK channel (mSlo1), chicken Slack channel (GgSlo2.2), mouse Slo3 channel (mSlo3), *Aplysia californica* BK channel (AcSlo), Shaker and Kv2.1 S4 TM segments were aligned. The numbers indicates the position in the primary structure of the last residue in each sequence. Positively and negatively charged residues are highlighted. Identical or conserved residues in the different channels are shown in gray. The typical sequence + 00 + 00 + (only absent in Slo2.2) is enclosed between a discontinuous line rectangle.

2015). The neutralization of the external ring (Figure 14.3A–C; D292N) has been found to reduce single-channel conductance by around 40%, due to a decrease in the negative surface-charge density in that region (Haug, Olcese, Toro, & Stefani, 2004; Haug, Sigg, et al., 2004). Notably, the corresponding aspartate residue is conserved in Shaker (D447), and its neutralization (D447N) abolishes the ionic conductance of the channel while retaining normal gating currents (Hurst, Toro, & Stefani, 1996). Neutralization of the internal ring (Figure 14.3A–B and D; E321N/E324N double mutant) reduces the unitary conductance by half (Brelidze, Niu, & Magleby, 2003; Nimigean, Chappie, & Miller, 2003). At very high symmetrical K^+ concentrations, the maximal conductance is identical in wild-type (WT) and double mutant channels, indicating that the negative charges increase the K^+ concentration in the internal vestibule through long-range electrostatic interactions. The electrostatic potential generated by the negatively charged ring is enough to raise the K^+ concentration at the channel entrance about two-fold (Carvacho et al., 2008).

Substitution of the hydrophobic F315 residue located in the S6 domain by alanine, isoleucine, or tyrosine has been shown to reduce single-channel conductance (Carrasquel-Ursulaez et al., 2015; Lippiat, Standen, & Davies, 2000). F315 interacts with L312 of the contiguous subunit (Figure 14.3A and E) (Carrasquel-Ursulaez et al., 2015; Hite et al., 2017; Wu et al., 2009); this interaction constitutes a stable "hydrophobic ring" through which potassium ions have to pass. As a consequence, the distortion of the geometry of this ring has important effects on the unitary conductance (Carrasquel-Ursulaez et al., 2015).

Another important factor in determining channel conductance is the large size of the BK internal cavity (Figure 14.3B). The radius of the internal entrance to the BK channel is ~20 Å and thus wider than in Kv channels (~10 Å) (Hite et al., 2017; Jiang et al., 2002; Long, Tao, Campbell, & MacKinnon, 2007; Naranjo, Moldenhauer, Pincuntureo, & Diaz-Franulic, 2016; Tao et al., 2017; Zhou, Xia, & Lingle, 2011). MthK, a bacterial K^+ channel of large unitary conductance, also has a wide (~15 Å) internal vestibule (Jiang et al., 2002). Thus, in BK and in MthK, most of the electric field drops in the selectivity filter (Carvacho et al., 2008; Diaz-Franulic, Sepulveda, Navarro-Quezada, Gonzalez-Nilo, & Naranjo, 2015; Jiang et al., 2002). In Shaker K^+ channels, most (92%) of the electrical resistance resides outside the selectivity filter, while in BK, the selectivity filter represents 50% of the total resistance of the conductive pathway (Brelidze & Magleby, 2005; Diaz-Franulic et al., 2015; Geng, Niu, & Magleby, 2011).

Allosteric Nature of Ca^{2+}- and Voltage-Dependent BK Channel Activation

Several kinetic models have been developed to understand the allosteric nature of the BK channel. Among these, the Horrigan and Aldrich (H-A) allosteric model stands out due to its optimum balance between simplicity and predictive capabilities (Horrigan & Aldrich, 2002). Latorre et al. (Latorre et al., 2017) show a rational approach to the experimental study of BK channels using the H-A and other kinetic models.

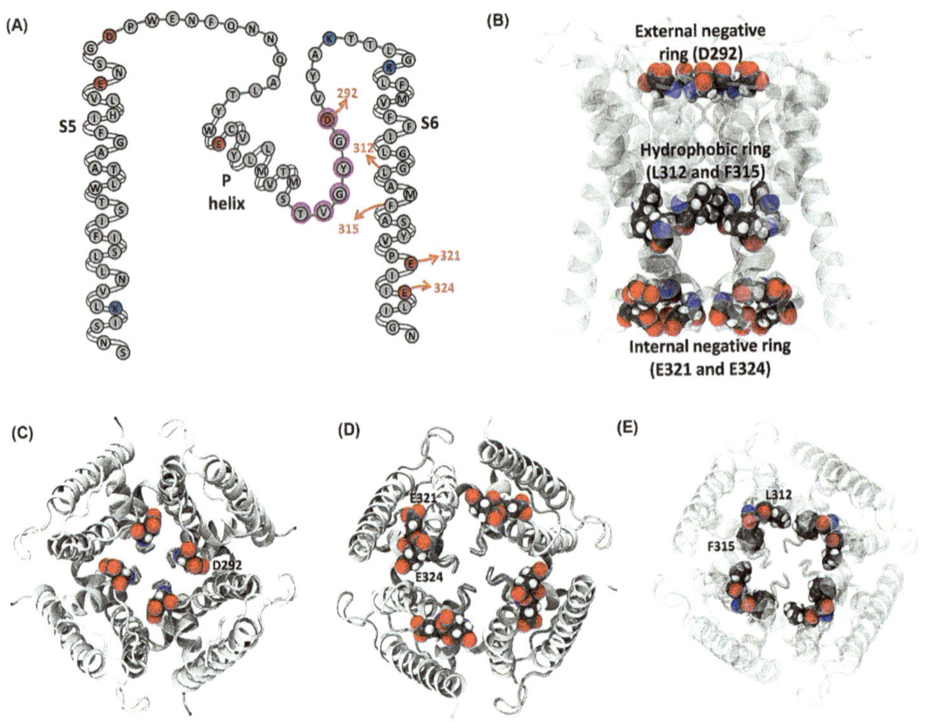

FIGURE 14.3 The BK channel pore domain.
(*A*) Primary and secondary structure and the topology of the pore domain. Residues within the characteristic sequence of K^+ channels selectivity filter, TVGYGD, are highlighted. Residues involved in determining the large conductance of the channel are indicated with arrows and position numbers. (*B*) Side view of the pore domain. The structures defined by the crucial residues shown in (*A*) are highlighted. (*C*) Top view of the BK channel PD. Four aspartate residues (D292) are located at the outer entrance of the BK channel forming the negatively-charged ring. (*D*) Bottom view of the BK channel PD. The internal negative ring contains eight glutamate residues (E321 and E324), which are located at the inner entrance of the BK channel. (*E*) Bottom view of the BK channel PD. Eight hydrophobic residues (L312 and F315) form a hydrophobic ring. Atomic coordinates were taken from the BK channel homology model made from *Aplysia* Slo1 structure (PDB ID: 5TJI) (Hite et al., 2017). Amino acid atoms are shown as spheres using this color code: C = gray, N = blue, O = red H = white. Illustrations were prepared using the Visual Molecular Dynamics, Version 1.9.2 (Humphrey et al., 1996).

Allosteric Coupling Between VSD and PD

While it is as yet unclear how the voltage-driven conformational changes of the VSD are coupled with the opening of the pore, the S6 segment appears to play an important role. It is known that mutations in this segment greatly modify the coupling between VSD and PD (Carrasquel-Ursulaez et al., 2015; X. Chen, Yan, & Aldrich, 2014; Lippiat et al., 2000; Wang & Brenner, 2006; Wu et al., 2009; Zhou et al., 2011), and molecular modeling suggests that a "hydrophobic ring" may act as an integration node between the allosteric signals from both domains (Carrasquel-Ursulaez et al., 2015).

Allosteric Coupling Between CTD and PD

Ca^{2+} binding alters the conformation of the gating ring (Ghatta, Nimmagadda, Xu, & O'Rourke, 2006; Hite et al., 2017; Javaherian et al., 2011; Miranda et al., 2013; Savalli, Pantazis, Yusifov, Sigg, & Olcese, 2012; Tao et al., 2017; Wu, Yang, Ye, & Jiang, 2010; Yuan et al., 2011; Yuan et al., 2010; Yusifov, Javaherian, Pantazis, Gandhi, & Olcese, 2010; Yusifov, Savalli, Gandhi, Ottolia, & Olcese, 2008). It is thought that the allosteric coupling between the CTD and the pore is mediated by the S6-RCK1 linker that connects these two domains, because its shortening increases channel activity, while its elongation decreases Ca^{2+} sensitivity (Niu, Qian, & Magleby, 2004).

Upon Ca^{2+} binding, the structure formed by the four RCK1 domains undergoes an expansion of 10–14 Å (Hite et al., 2017; Yuan et al., 2011). Even larger rearrangements of the gating ring have been proposed by other authors (Javaherian et al., 2011; Miranda et al., 2013). Regardless of the magnitude of Ca^{2+} binding-induced conformational changes, the expansion of the gating ring is transmitted to the pore domain in two ways (Hite et al., 2017). The RCK1 domain exerts a pulling force on the covalently attached S6-RCK1 linker and also on the non-covalent interaction between the interfaces of the RCK1 and the VSD, thus separating the S6 TM segments.

Allosteric Coupling Between VSD and CTD

There is a strong functional coupling between the CTD and the VSD (Carrasquel-Ursulaez et al., 2015; Savalli et al., 2012). The coupling between the site located on RCK1 and the VSD is stronger than that between the site located on RCK2 and the VSD (Savalli et al., 2012). Ca^{2+} binding to the RCK1 site has been found to strongly stabilize the active configuration of the VSD (Savalli et al., 2012). These differences may be due to the close physical vicinity of RCK1 and VSD (Hite et al., 2017; Tao et al., 2017; H. Yang et al., 2008).

BK Channel Expression Pattern in the Central Nervous System

BK is widely distributed in different types of excitable and non-excitable cells, where it is involved in a large variety of physiological processes (Latorre et al., 2017). In the CNS, BK is expressed in the olfactory bulb, cortex, basal ganglia, hippocampus, thalamus, habenulointerpenduncular tract, cerebellum, substantia nigra, pontine nuclei, caudate putamen, pallidal areas, vestibular nuclei, and spinal cord (Chang, Dworetzky, Wang, & Goldstein, 1997; Grunnet & Kaufmann, 2004; Hu et al., 2001; Knaus et al., 1996;

Misonou et al., 2006; Sailer et al., 2006; U. Sausbier et al., 2006; Tseng-Crank et al., 1994; Wanner et al., 1999). At the ultrastructural level, BK distribution varies greatly between neuronal types. It is commonly expressed in axons and terminals, but in some neurons, it is also present in soma and dendrites (Hu et al., 2001; Kaufmann et al., 2009; Knaus et al., 1996; Misonou et al., 2006; Sausbier et al., 2006). We will discuss the details of the distribution and function of BK channels later, in the two sections "The Role of (α+β)BK Channels in the Nervous System" and "BK Channels and Channelopathies."

Ca^{2+} Sources

Except when co-expressed with γ1 or γ2 subunits (J. Yan & Aldrich, 2010, 2012), BK requires micromolar internal concentrations of Ca^{2+} in order to be able to modify the membrane potential (Berkefeld, Fakler, & Schulte, 2010). Compared with other Ca^{2+}-activated K^+ channels, such as SK channels, the affinity of BK to Ca^{2+} is low, implying that only BK channels located very close to the active sources of Ca^{2+} influx will become activated. The levels of $[Ca^{2+}]_i$ required for BK activation are reached almost exclusively in local Ca^{2+}-signaling nanodomains centered around Cav channels (Augustine, Santamaria, & Tanaka, 2003; Fakler & Adelman, 2008). BK channels not tightly associated with Cav channels may play a role in Ca^{2+} sensing under conditions where $[Ca^{2+}]_i$ rises above the physiological range (Hu et al., 2001; Runden-Pran, Haug, Storm, & Ottersen, 2002).

In neurons, the increase in cytosolic Ca^{2+} results from Ca^{2+} entry through Cav channels. Functional studies suggest that there is a tight spatial coupling between BK and Cav channels in these cells. For instance, BAPTA (a fast-binding Ca^{2+} chelator), but not EGTA (a slow-binding Ca^{2+} chelator), can disrupt the BK–Cav interaction and suppress BK currents (Edgerton & Reinhart, 2003; Lancaster & Nicoll, 1987; Muller, Kukley, Uebachs, Beck, & Dietrich, 2007; Robitaille, Garcia, Kaczorowski, & Charlton, 1993; Storm, 1987; Sun, Gu, & Haddad, 2003). In the rat brain, the coupling between N-type Ca^{2+} channels and BK channels is completely resistant even to BAPTA buffering, suggesting a distance of only few tens of nanometers between the pores of both channels (Loane, Lima, & Marrion, 2007). Further direct evidence of the physical association between Cav and BK channels has been gathered using co-immunoprecipitation, mass spectrometry, and high-resolution imaging techniques (Bao, de Jong, Alevra, & Schild, 2015; Berkefeld et al., 2006; Grunnet & Kaufmann, 2004; Indriati et al., 2013; Kaufmann et al., 2009; Loane et al., 2007; Samaranayake, Saunders, Greene, & Navaratnam, 2004).

In the nervous system, BK co-localizes with L, P/Q, T, and N-type Cav channels (the only exception being R-type channels) (Berkefeld et al., 2006; Grunnet & Kaufmann, 2004). Because these different Cav channels vary in their activation characteristics, they modulate BK activation by voltage. Therefore, Cav channels increase the functional diversity of BK channels, thus determining their specific effects on neuronal function (Berkefeld & Fakler, 2008, 2013; Berkefeld et al., 2010).

The Cav channels' effect of on BK opening depends on the subcellular location of each channel. For example, P/Q and N-type Cav channels are fast-activating. Ca^{2+} entry through these types of Cav channels can thus reduce neurotransmitter release through the BK channel-dependent shortening of AP duration (Castillo, Weisskopf, & Nicoll, 1994; Kulik et al., 2004; Pelkey, Topolnik, Lacaille, & McBain, 2006; Wu, Westenbroek, Borst, Catterall, & Sakmann, 1999). The slowly-activating L- and T-type Cav channels are located mainly in dendrites and soma, where they regulate the role of BK in synaptic summation and AP propagation (Dobremez et al., 2005; Engbers, Anderson, Zamponi, & Turner, 2013; Obermair, Szabo, Bourinet, & Flucher, 2004; Rehak et al., 2013).

The association and coupling of BK with specific Cav channel types vary across different brain regions (Berkefeld & Fakler, 2008). In hippocampal CA1 neurons, N-type Cav channels and BK channels are very close (Marrion & Tavalin, 1998). In cortical pyramidal neurons, both N- and L-type Cav channels provide the Ca^{2+} that activates BK channels (Sun et al., 2003). In cerebellar neurons, low-voltage-activated T-type Cav channels are associated with BK channels (Engbers et al., 2013).

It has been reported that neuronal BK can also be activated by Ca^{2+} from internal sources. In Purkinje cells and dentate gyrus granule cells, some apical extensions of the smooth endoplasmic reticulum are located underneath clusters of BK channels (Kaufmann et al., 2009; Kaufmann, Kasugai, Ferraguti, & Storm, 2010). These cisterns may serve as an intracellular source of Ca^{2+} sparks, opening coordinately clustered BK channels (Petersen, Tepikin, & Park, 2001).

Modulation by Accessory Subunits Generates BK Channel Diversity

BK channels are characterized by a large phenotypical diversity. One important source of this diversity is alternative splicing, which generates BK channels of different kinetics and Ca^{2+} sensitivities (Fodor & Aldrich, 2009; Latorre et al., 2017). Nevertheless, the differences between splice variants are modest compared with the effect of co-expression with accessory β and γ subunits. In this chapter, we will focus on the interaction between BK and its neuron-expressed accessory subunits.

β Subunit Family

Four members of the β subunit family (β1–β4), encoded by four different genes (*KCNMB1–4*), have been cloned and characterized. Their secondary structure is conserved and consists of two TM segments (TM1 and TM2) connected by a large external loop, with both N- and C-termini pointing towards the cytoplasmic side (Figure 14.4) (Hoshi, Pantazis, & Olcese, 2013; Orio, Rojas, Ferreira, & Latorre, 2002;

Torres, Granados, & Latorre, 2014; Torres, Morera, Carvacho, & Latorre, 2007). Two members of this family, β2 and β4, are highly expressed in neurons (Brenner et al., 2000), where they determine the specific role of BK channels (Wang, Jaffe, & Brenner, 2014). The expression of the other two members of the β subunit family, β1 and β3, in the nervous system is weak (Behrens et al., 2000; Brenner et al., 2000; Poulsen et al., 2009; Uebele et al., 2000). In particular, β1 expression is highly abundant in cerebellar artery myocytes but undetectable in the brain, with the exception of certain hypothalamic regions, cerebellar Purkinje cells, and several brainstem nuclei (Behrens et al., 2000; Brenner et al., 2000; Chang et al., 1997; Z. Jiang, Wallner, Meera, & Toro, 1999; Salzmann et al., 2010; Tseng-Crank et al., 1994; Weiger et al., 2000).

γ Subunit Family

The other family of proteins modulating BK channel voltage and Ca^{2+} sensitivity is that of γ accessory subunits. Four γ subunits have been described (γ1–γ4), each of which is encoded by a different gene (γ1: *LRRC26*, γ2: *LRRC52*, γ3: *LRRC55*, and γ4: *LRRC38*) (J. Yan & Aldrich, 2012). They share most of their structural features: a single TM segment, a large leucine rich repeat (LRR) domain, and a short C-terminus (Figure 14.4). The most marked functional effect of the γ subunits on BK consists of a leftward shift of the channel's conductance-voltage curve in the absence of internal Ca^{2+} (J. Yan & Aldrich, 2010, 2012; C. Yang, Zeng, Zhou, Xia, & Lingle, 2011). Yan and Aldrich (2012) reported high levels of γ3 mRNA of in the brain, while γ1 and γ4 are expressed at lower levels, and γ2 is absent from this tissue. On the other hand, Yang et al. (2017) found an absence of *LRRC26* promoter activity in excitable cells, including neurons and smooth muscle (C. Yang et al., 2017). The γ1 subunit is expressed in cerebral arterial myocytes, where it elevates voltage- and Ca^{2+}-sensitivity of the BK channel, contributing to a decrease in myogenic tone and promoting vasodilation (Evanson, Bannister, Leo, & Jaggar, 2014). The role of γ subunits in neurons remains elusive.

The Role of (α+β)BK Channels in the Nervous System

Although neuronal BK formed by the α subunit alone can be functional, the contribution of BK to AP repolarization and to the firing rate is determined by the presence of two members of the β family of accessory subunits (Wang et al., 2014), β2 and β4, which are highly expressed in central neurons (Brenner et al., 2000). The interaction between α subunits and these accessory subunits defines the three BK subtypes found in central neurons: inactivating (α+β2)BK channels, non-inactivating fast-gated iberiotoxin-sensitive type I αBK (α alone) channels and non-inactivating slow-gated iberiotoxin-insensitive type II (α+β4)BK channels (Table 14.2) (Wang et al., 2014). In addition, some splice variants of the β3 subunit can inactivate BK; however, β3 expression in the brain is weak (Hu et al., 2003; Uebele et al., 2000; Wallner et al., 1999; Xia,

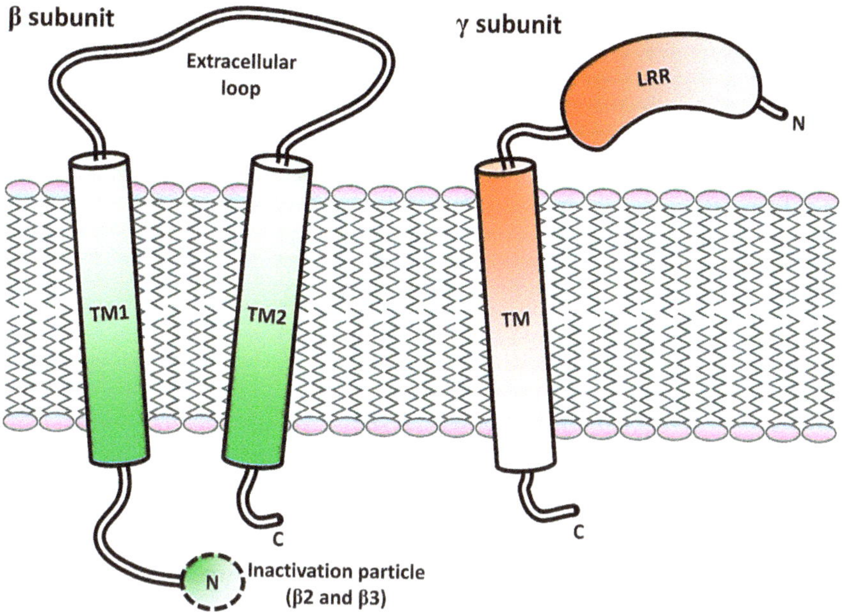

FIGURE 14.4 Topology and general structural features of β and γ families of BK accessory subunits.
All β-subunits have N- and C-terminus facing the intracellular side and two transmembrane segments linked by a large extracellular loop. β2 and β3 containing additional residues in their N-termini, which constitute the particle of inactivation (broken circle). γ- subunits contains only one transmembrane segment and an extracellular LRR domain. The N-terminus and the C-terminus are facing the extracellular and intracellular sides, respectively.

Ding, & Lingle, 1999). As we will see, some findings indicate an expression of (α+β1)BK channels in the nervous system.

The kinetics of the BK currents is a determinant of AP shape. As a general rule, faster activation kinetics produces faster repolarization, and slower deactivation kinetics increases the magnitude of after-hyperpolarization (AHP) (Figure 14.5). These effects have implications beyond AP shape: for example, an increase in AHP can help Na^+

Table 14.2 Neuronal BK Channel Types

Neuronal BK channel type	Subunit composition	Activation kinetic	Deactivation kinetic	Paxilline sensitivity	Iberiotoxin sensitivity	Martentoxin sensitivity
Non-inactivating (Type I)	α alone	Fast	Fast	Yes	Yes	Not
Inactivating	α+β2	Slow	Fast	Yes	Yes	Not
Non-inactivating (Type II)	α+β4	Slower	Slow	Yes	Not	Yes

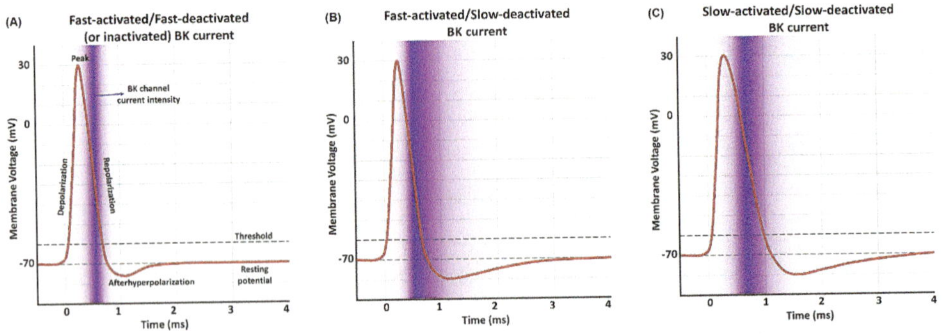

FIGURE 14.5 Modulation of the action potential shape by BK channels with three different kinetic characteristics.
(*A*) Fast-activated/Fast-deactivated (or inactivated), (*B*) Fast-activated/Slow-deactivated, (*C*) Slow-activated/Slow-deactivated. The gradient of intensity represents the time course of Slo1 current intensity.

channels recover from inactivation, thus promoting the occurrence of another AP and increasing AP frequency.

(α+β1)BK Channels

While ethanol can increase αBK channel activity (Crowley, Treistman, & Dopico, 2003; Chu, Dopico, Lemos, & Treistman, 1998; Dopico, Lemos, & Treistman, 1996; Knott, Dopico, Dayanithi, Lemos, & Treistman, 2002), the association with β1 renders BK insensitive to alcohol-induced potentiation (Feinberg-Zadek & Treistman, 2007). This pharmacological signature has been used to identify putative (α+β1)BK channels in neurons. β1 protein, β1 mRNA, and BK channels lacking ethanol-induced activation have been found in the soma of supraoptic magnocellular neurons of the rat and the soma of medium spiny neurons of the rat nucleus accumbens (Dopico, Widmer, Wang, Lemos, & Treistman, 1999; Martin et al., 2004; Wynne, Puig, Martin, & Treistman, 2009).

(α+β2)BK Channels

The co-expression of β2 and α subunits results in inactivating BK channels (Table 14.2). In this case, the inactivating "particle" is the N-terminus of the β2 subunit, thus resembling the ball-and-chain-type inactivation proposed by Armstrong and Bezanilla for Na^+ channels (Bezanilla & Armstrong, 1977). The β2 N-terminus (Wallner et al., 1999; Xia et al., 1999) consists of a pore-blocking "ball domain" (first 17 residues), and a "chain domain" that connects the ball to TM1 (Figure 14.4) (Bentrop, Beyermann, Wissmann, & Fakler, 2001). Moreover, β2 increases the apparent Ca^{2+} sensitivity and slows down the gating kinetics of BK (Brenner et al., 2000; Wallner et al., 1999; Xia et al.,

1999). In neurons, (α+β2)BK channels rapidly activate and inactivate. Due to a fast repolarization, this type of current decreases AP duration (Figure 14.5A), but the fast inactivation leads to a moderation of the AHP (Figure 14.5A).

β2 Expression Patterns and the Role of (α+β2)BK Channels in Neurons

β2 is expressed throughout the brain, although to a lesser extent than β4 (Behrens et al., 2000; Brenner et al., 2000; Weiger et al., 2000). Inactivating BK channels have been detected in the hippocampus, neocortex, lateral amygdala pyramidal neurons, cerebellar Purkinje cells, dorsal root ganglion (DRG) neurons, and suprachiasmatic nucleus (Faber & Sah, 2003; Haghdoost-Yazdi, Janahmadi, & Behzadi, 2008; Hicks & Marrion, 1998; Ikemoto, Ono, Yoshida, & Akaike, 1989; Li et al., 2007; McLarnon, 1995; Sun et al., 2003). In CA1 neurons, BK channels contribute to the repolarization of the membrane only after the first few of a train of APs, resulting in a frequency-dependent spike broadening (Faber & Sah, 2003; Shao, Halvorsrud, Borg-Graham, & Storm, 1999). In rat lateral amygdala projection neurons, the frequency dependence of AP broadening is partly due to BK inactivation (Faber & Sah, 2003). Mouse neocortical pyramidal neurons (which express both inactivating and non-inactivating BK channels) displayed AP broadening after blockade of inactivating BK by iberiotoxin (Sun et al., 2003).

Cerebellar Purkinje cells, which express inactivating and non-inactivating BK channels in their soma and dendrites (Widmer, Rowe, & Shipston, 2003), exhibit both tonic and bursting modes in their firing of APs (Cingolani, Gymnopoulos, Boccaccio, Stocker, & Pedarzani, 2002; Hockberger, Tseng, & Connor, 1989). Bursts containing a variable number of Na^+ spikes riding on a Ca^{2+} spike are called Na^+-Ca^{2+} bursts (Cavelier, Pouille, Desplantez, Beekenkamp, & Bossu, 2002; McKay & Turner, 2004). The Na^+-Ca^{2+} burst mode can be artificially induced by application of 4-aminopyridine, a blocker of Kv channels (Yazdi, Janahmadi, & Behzadi, 2007). A blocking of iberiotoxin-sensitive (α+β2)BK channels does not affect the Na^+ spike in the tonic firing, but changes the shape of Na^+ and Ca^{2+} spikes in the bursts induced by 4-aminopyridine (Haghdoost-Yazdi et al., 2008). During burst firing, activation of (α+β2)BK channels enhances a fast AHP while decreasing the duration of Na^+ and Ca^{2+} spikes (Haghdoost-Yazdi et al., 2008).

In the suprachiasmatic nuclei (SCN), the master pacemaker that drives the circadian control of physiology and behavior in mammals, inactivating BK currents predominate during the day (Whitt, Montgomery, & Meredith, 2016). Conversely, inactivating BK current decreases during the night (Whitt et al., 2016). This diurnal variation of BK currents features is abolished in β2 knockout mice, suggesting that inactivation is used as a switch to regulate circadian variations in SCN excitability (Whitt et al., 2016).

(α+β2)BK channels are involved in modulating nociceptive information from the peripheral to the central nervous system (Li et al., 2007). In rat lumbar L4–L6 DRGs, β2 was identified as the molecular agent responsible for the inactivation of BK channels currents in small sensory neurons (Li et al., 2007). In these neurons, the blockade of inactivating BK currents caused AP broadening, an increase in the firing frequency, and a reduced AHP.

(α+β4)BK Channels

β4 slows down both activation and deactivation kinetics of BK, in addition to producing complex effects on its Ca^{2+} sensitivity (Behrens et al., 2000; Brenner et al., 2000; Weiger et al., 2000). At a $[Ca^{2+}]_i$ below 7 μM, (α+β4)BK channels are less Ca^{2+} sensitive than BK channels formed by α subunits only. At higher $[Ca^{2+}]_i$, channel activity is enhanced in the presence of β4, compared to αBK channels (Brenner et al., 2000; Ha, Heo, & Park, 2004; Wang, Rothberg, & Brenner, 2006). (α+β4)BK channels are characterized by several unique pharmacological properties, including a resistance to blockade by charybdotoxin or iberiotoxin (Meera et al., 2000). The slow kinetics of the (α+β4)BK gating is relevant in neurons because it reduces BK contribution to AP repolarization, resulting in AP broadening and an increase in AHP (Figure 15.5C) (Brenner et al., 2005).

β4 Expression Patterns and the Role of (α+β4)BK Channels in the Neuronal Soma

β4 is the most highly expressed β subunit in brain and spinal cord neurons (Behrens et al., 2000; Brenner et al., 2000; Weiger et al., 2000). Its distribution overlaps with that of the BK α subunit, and it has been found in neurohypophysial terminals, trigeminal ganglia, the posterior pituitary, pyramidal neurons of the cortex, CA3 pyramidal neurons, hippocampus dentate gyrus granule cells, olfactory bulb, paraventricular nucleus of the hypothalamus, supraoptic nucleus, facial and trigeminal motor nuclei, and cerebellar Purkinje cells (Benton, Lewis, Bant, & Raman, 2013; Bielefeldt, Rotter, & Jackson, 1992; Brenner et al., 2005; Deng et al., 2013; Dopico et al., 1999; Lein et al., 2007; Petrik & Brenner, 2007; Samengo, Curro, Barrese, Taglialatela, & Martire, 2014; Shruti, Clem, & Barth, 2008; Shruti et al., 2012; Weiger et al., 2000). It should be noted that β4 is also present in neuronal mitochondria, with higher levels in neurons of the thalamus and brainstem (Piwonska, Wilczek, Szewczyk, & Wilczynski, 2008).

Mouse hippocampal CA3 pyramidal cells strongly express β4. However, the addition of paxilline, which can block (α+β4)BK channels, and iberiotoxin causes the same effect, suggesting that (α+β4)BK channels do not participate in generating the macroscopic ionic currents (Shruti et al., 2012). Interestingly, in the CA3 neurons of β4 knockout mice, BK currents are increased (Shruti et al., 2012); moreover, it is known that in this type of neuron, β4 promotes the retention of the α subunit in the endoplasmic reticulum via a C-terminal retention sequence (L. Chen et al., 2013; Shruti et al., 2012). Taken together, these results suggest that in CA3 neurons, β4 expression represses forward trafficking of the BK channel α subunit towards the cell membrane (Shruti et al., 2012). (α+β4)BK channels are also expressed in the trigeminal ganglion, which projects primary terminal afferents into the trigeminal caudal nucleus (TCN) (Samengo et al., 2014; Wulf-Johansson et al., 2010). In the TCN, (α+β4)BK channels modulate glutamate release from synaptosomes, a process which is inhibited by BK channel activation and stimulated by BK blocking (Samengo et al., 2014; Wulf-Johansson et al., 2010). In cerebellar Purkinje neurons, there is a mixture of iberiotoxin-sensitive and

-insensitive BK channels (Benton et al., 2013). The presence of iberiotoxin-insensitive BK channels is consistent with a co-expression with β4 in these cells (Petrik & Brenner, 2007). The slow-gating of the iberiotoxin-insensitive BK channels sustains the activity of these channels during the AHP and contributes to a persistent interspike conductance (Benton et al., 2013).

In hippocampus dentate gyrus granule neurons, knockout of the β4 subunit transforms slow-gated type-II into fast-gated type I BK channels (Wang et al., 2016). The lack of β4 leads to decreased AP duration, thus diminishing AHP amplitude and augmenting spike frequency, resulting in spontaneous temporal lobe seizures (Brenner et al., 2005). These observations strongly suggest that β4 protects dentate gyrus cells against hyperexcitability (Wang et al., 2016).

The Role of (α+β4)BK Channels in Nerve Terminals

(α+β4)BK channels are expressed in posterior pituitary nerve terminals (Bielefeldt & Jackson, 1993; Bielefeldt et al., 1992; Brenner et al., 2005). In these terminals, BK activation inhibits high-frequency AP firing in response to an increased Ca^{2+} influx through L-type Ca_V channels (Bielefeldt & Jackson, 1993). The fiber terminals of hippocampal dentate gyrus granule cells (mossy fiber boutons) express a mixture of iberiotoxin-sensitive (fast-gated) and iberiotoxin-insensitive (slow-gated) BK channels, suggesting that the β4 subunit is trafficked to presynaptic locations (Alle, Kubota, & Geiger, 2011). However, the fast-gated BK channel subtype does not contribute to presynaptic terminal AP repolarization, which is dominated by K_V3 channels in these cells (Alle et al., 2011). In contrast to somatic BK channels in the same neurons, in mossy fibers, both the fast-gated and the slow-gated BK components were sensitive to a slow calcium buffer (EGTA) (Alle et al., 2011; Muller et al., 2007). Therefore slow-activation of presynaptic BK channels might also be due to the lack of a neighboring calcium source, which is required for fast BK channel activation. All these findings indicate that presynaptic BK channels in mossy fibers are restricted to limiting additional increases of the AHP in the case of Kv3 hypofunction.

BK Channels in Sensory Receptors

Hearing

It has been suggested that in the cochlea of birds, reptiles, and amphibians, the tonotopic map is defined by the interplay between BK and Cav channels of sensory hair cells (Fettiplace & Fuchs, 1999). Since β subunits are mainly expressed in low-frequency hair cells, the differential distribution and gating features of αBK, (α + β1)BK, and (α + β4)BK channels could determine the electrical tuning of hair cells located along the cochlea (Ramanathan, Michael, Jiang, Hiel, & Fuchs, 1999). In particular, the β1 subunit is abundantly expressed in the low-frequency apex of the cochlea (Miranda-Rottmann, Kozlov, & Hudspeth, 2010). BK channels are also expressed in the inner and outer hair cells of mammals, where mRNA has been identified for the α, β1, and β4 subunits (Brandle et al., 2001; Langer, Grunder, & Rusch, 2003; Pyott & Duncan, 2016; Skinner et al., 2003). In the cochlea of mice affected by age-related hearing loss, BK expression gradually decreases with age (Pan et al., 2016). Surprisingly, α, β1, or β4 knockout mice have normal cochlear function, suggesting that BK channels do not play an essential role in this physiological process (Pyott et al., 2007). Nevertheless, α knockout mice are more resistant to noise-induced hearing loss (Pyott et al., 2007), and, as suggested by specifically deleting the α subunit in the inner and outer hair cells of mice, the role of BK channel could lie in defining the temporal fine structure and dynamic range of auditory information (Kurt et al., 2012). On the other hand, BK blockade on the dural surface of the mouse auditory midbrain has been found to suppress the neural activity evoked by sound in the inferior colliculus (Scott et al., 2017), where BK modulates both excitatory and inhibitory inputs (Sivaramakrishnan & Oliver, 2001).

In addition to their essential role in the structural maturation of efferent outer hair cell synapses, BK channels are functionally coupled to ionotropic acetylcholine receptors (AChR) (Rohmann, Wersinger, Braude, Pyott, & Fuchs, 2015). The outer hair cells of the mammalian cochlea are inhibited by cholinergic neurons of the olivary complex of the brainstem, thus improving acoustic sensitivity. This inhibition results in the activation of AchRs, through which calcium enters to activate SK and BK channels, leading to a hyperpolarization of the hair cells (Engel et al., 2006; Evans, 1996; Fuchs & Murrow, 1992; Housley & Ashmore, 1991; Maison, Pyott, Meredith, & Liberman, 2013; Shigemoto & Ohmori, 1991; Wersinger, McLean, Fuchs, & Pyott, 2010).

Vision

BK currents have been described in several retinal cell types, including rod and cone photoreceptors, and pigment epithelial, bipolar, amacrine, and ganglion cells (Barnes & Hille, 1989; Burrone & Lagnado, 1997; Grimes, Li, Chavez, & Diamond, 2009; Henne & Jeserich, 2004; Mitra & Slaughter, 2002b; Pelucchi, Grimaldi, & Moriondo, 2008; Sakaba, Ishikane, & Tachibana, 1997; Tanimoto et al., 2012; Tao & Kelly, 1996; Wang, Robinson, & Chalupa, 1998; Wimmers, Halsband, Seyler, Milenkovic, & Strauss, 2008; Xu & Slaughter, 2005). In some amacrine cells, spontaneous miniature outwards currents are mediated by BK channels. These currents act as synaptic noise suppressors that help keep certain glutamatergic synapses quiescent (Mitra & Slaughter, 2002a, 2002b). BK channels also prevent the activation of Cav channels in A17 amacrine cells, limiting gamma-aminobutyric acid (GABA) release onto rod bipolar cells (Grimes et al., 2009). This is in line with the results from experiments on BK knockout mice, whose rod bipolar cell activity *in vivo* has been found to be reduced at intermediate lighting stimulus condition (Tanimoto et al., 2012).

In the ribbon synapse of salamander rod photoreceptors, BK channel activity leads to an unexpected increase in Ca^{2+}-mediated neurotransmitter release (Xu & Slaughter, 2005). In order to explain this counterintuitive observation, Xu and Slaughter (2005) hypothesize that K^+ efflux through the BK channel produces an accumulation of K^+ in the synaptic cleft, which then leads to an increase in Ca^{2+} currents. In fact, Ca^{2+} currents increase with extracellular K^+ in a concentration-dependent manner, most likely due to the existence of an external K^+ binding site in the Cav channel that enhances channel activity.

Photoreceptor shedding and the phagocytosis of shed photoreceptor outer segments are synchronized in a process whose disruption results in blindness (Strauss, 2005). In porcine retinal pigment epithelium (RPE) cells, these two processes are regulated by BK and Cav1.3 channels (Muller, Mas Gomez, Ruth, & Strauss, 2014). In particular, BK channel blockade was shown to diminish phagocytosis in RPE cell cultures (Muller et al., 2014). Although the RPE cells of BK knockout mice retained their phagocytic capacity *in vivo*, the circadian synchronization of phagocytosis and photoreceptor shedding was lost (Muller et al., 2014).

BK Channel and Channelopathies

As it is summarized in Table 14.3, BK channels are involved with several nervous system diseases.

Epilepsy

A single amino acid substitution in the BK channel's RCK1 domain, D369G (D434G in humans), increases neuronal activity by causing BK channel hyperactivity (Du et al., 2005; Wang, Rothberg, & Brenner, 2009). The physiological consequence of this mutation is to promote high-frequency firing by accelerating AP repolarization and increasing the AHP, which can speed up the recovery of inactivated voltage-dependent Na^+ (Nav) channels. The effects of the mutation on AP shape are similar to what is shown in Figure 14.5B.

In pyramidal neurons of the mouse somatosensory cortex, BK currents increased 24 hours after generalized seizures induced by picrotoxin, a convulsive drug (Shruti et al., 2008). This rise in BK channel activity is related to increases in evoked firing rates, which can be returned to normal levels by blocking BK channels (Shruti et al., 2008). Administration of paxilline 24 hours after seizure induction prevented seizures after a second exposure to picrotoxin, but this anticonvulsant effect did not extend to mice without previous seizure experience (Sheehan, Benedetti, & Barth, 2009). An increase in BK transcription and/or translation, and the post-translational modifications of previously existing BK channels have been postulated as possible mechanisms mediating seizure-induced potentiation of the BK channel activity (Contet, Goulding, Kuljis, & Barth, 2016).

β3 and β4 accessory subunits are associated with certain nervous system disorders. Certain patients affected by the dup(3q) syndrome, which is characterized by mental and growth retardation, carry a duplication of the β3 gene *KCNMB3* (Riazi et al., 1999). In addition, a truncation in *KCNMB3* confers susceptibility to idiopathic generalized

Table 14.3 BK Channel Implication in Neuronal Diseases

Diseases	Evidence of BK channel implication	References
Epileptic disorders	Point mutations in the α subunit: D369G (D434G in human)	Wang et al. (2009), Du et al. (2005)
	BK blockade prevents seizure after second picrotoxin exposure	Shruti et al. (2008), Sheehan et al. (2009)
	KCNMB3 duplication	Riazi et al. (1999)
	KCNMB3 truncation	Lorenz et al. (2007)
	KCNMB4 knockout	Brenner et al. (2005)
	KCNMB4 single-nucleotide polymorphism	Cavalleri et al. (2007)
Ataxia	*KCNMA1* knockout	Imlach et al. (2008), Meredith et al. (2004), Sausbier et al. (2004), Typlt et al. (2013)
	Point mutations in human α subunit: D434G, E884K, and N1053S	Du et al. (2005), Zhang et al. (2015)
Adult-onset neuronal ceroid lipofuscinosis	Elevation of BK functional expression as a result of CSPα mutations or knockout	Fernandez-Chacon et al. (2004), Noskova et al. (2011), Velinov et al. (2012), Kyle et al. (2013), Ahrendt et al. (2014), Donnelier et al. (2015)
Alzheimer's disease	*KCNMA1* single-nucleotide polymorphism	Burns et al. (2011), Beecham et al. (2009)
	Not clear yet	Chishti et al. (2001), Jafari et al. (2015), Wang et al. (2015), Wang et al. (2015), Yamamoto et al. (2011), Ye et al. (2010)
Mental retardation	*KCNMA1* de novo mutations	Zhang et al. (2015), Laumonnier et al. (2006)
	Nonsense mutation in the gene encoding cereblon, which suppresses BK channels expression	Liu et al. (2014), Jo et al. (2005), Higgins et al. (2008)
	Transcriptional silencing of gene encoding FMRP, which interacts with β4	Deng et al. (2013)
Ethanol addiction	Not clear yet	Edenberg et al. (2010), Han et al. (2013), Kendler et al. (2011), Schuckit et al. (2005)
	Ethanol-induced degradation of ethanol-sensitive BK splicing variants and expression of ethanol-insensitive STREX variant	Pietrzykowski et al. (2008), Velazquez-Marrero et al. (2011), Velazquez-Marrero et al. (2014)
	Ethanol-induced reduction in BK trafficking to the membrane	Palacio et al. (2015), Pietrzykowski et al. (2004), Velazquez-Marrero et al. (2011)
	β1 and β4 control ethanol tolerance to the molecular, cellular, and behavioral levels	Kreifeldt et al. (2015), Kreifeldt et al. (2015), Martin et al. (2008)

Table 14.3 Continued

Diseases	Evidence of BK channel implication	References
Pain disorders	BK-mediated effects of analgesic compounds	Baillie et al. (2015), Gruss et al. (2001), Gruss et al. (2002)
	Reduction in BK expression in DRG due to peripheral nerve injury, which increases excitatory transmission in the dorsal horn	Furukawa et al. (2008), Chen et al. (2009), Cao et al. (2012), Sarantopoulos et al. (2007), Song & Marvizon (2005), Shoudai et al. (2007)
	Increase in BK expression in dorsal horn interneurons due to peripheral nerve injury, which suppress inhibitory signaling	Chen et al. (2009)
	BK knockout in DRG neurons induced nociceptive behavior	Lu et al. (2014)
	Blockade of spinal BK channels induces mechanical hyperalgesia	Chen et al. (2009)
	Exposition of juxtaparanodal BK channels during spinal cord compression injury, which impair axonal conduction	Ye et al. (2012)
Ischemia	BK channels blockade or genetic ablation increase neuronal death during ischemia	Liao et al. (2010), Katsuki et al. (2005), Runden-Pran et al. (2002)
	BK openers reduced cortical infarcts in rats after middle cerebral artery occlusion	Gribkoff et al. (2001)

epilepsy (Lorenz, Heils, Kasper, & Sander, 2007). How β3 mutations alter neuronal function is difficult to explain because of the low levels of this subunit in the brain (Behrens et al., 2000; Brenner et al., 2000; Poulsen et al., 2009; Uebele et al., 2000). β4 knockout reduces AP and AHP duration and increases firing rates in mice dentate gyrus granule cells. Mice lacking β4 exhibit spontaneous nonconvulsive seizures (Brenner et al., 2005). In addition, a single nucleotide polymorphism (SNP) in *KCNMB4* confers predisposition to mesial temporal lobe epilepsy (Cavalleri et al., 2007).

Ataxia

In line with the high levels of BK expression in the cerebellum, BK channel dysfunction is implicated in movement disorders. BK knockout mice exhibit a moderate ataxic phenotype (Imlach et al., 2008; Meredith, Thorneloe, Werner, Nelson, & Aldrich, 2004; Sausbier et al., 2004; Typlt et al., 2013). BK inhibitors, such as paxilline or lolitrem B, induce tremors in WT but not BK knockout mice (Imlach et al., 2008). Interestingly, β4 appears to be necessary for toxin-induced ataxia, although the molecular basis for this need is unclear (Imlach et al., 2008). Conversely, the gain-of-function hBK mutation *D434G* (*D369G* in mouse) (Du et al., 2005) and two hBK gating ring mutations (*E884K* and *N1053S*) are all associated with paroxysmal nonkinesigenic dyskinesia (Z. B. Zhang, Tian, Gao, Jiang, & Wu, 2015). The latter two residues are conserved in several mammalian species, but their mutational effects remain unknown (Z. B. Zhang et al., 2015).

Adult-Onset Neuronal Ceroid Lipofuscinosis

Mutations in the gene encoding Cysteine String Protein α (CSPα), a chaperone for proteins involved in presynaptic neurotransmitter vesicle trafficking (Chamberlain & Burgoyne, 2000), cause adult-onset ceroid lipofuscinosis (ANCL). ANCL is a neurodegenerative disorder characterized by the accumulation of lipofuscin, an autofluorescent lipopigment, in neuronal lysosomes. It causes seizures, ataxia, and dementia (Fernandez-Chacon et al., 2004; Noskova et al., 2011; Velinov et al., 2012). In CSPα knockout mice, functional expression levels of brain synaptosomal BK are highly elevated, without a concurrent alteration of BK α mRNA levels (Ahrendt, Kyle, Braun, & Braun, 2014; Kyle, Ahrendt, Braun, & Braun, 2013). BK was furthermore found to be abundantly expressed in the postmortem cortex of an ANCL patient (Donnelier et al., 2015). Together, these findings suggest that, under normal conditions, CSPα reduces BK α trafficking from the endoplasmic reticulum to the presynaptic terminals membrane. Alternatively, it may increase the rate of BK channel removal from the plasma membrane. An increase in BK channel density following CSPα malfunction may contribute to the onset of ANCL.

Alzheimer's Disease (AD)

A SNP in the *KCNMA1* gene has been found to be associated with late-onset risk of AD in humans (Beecham et al., 2009; Burns et al., 2011); however, the functional effects of this mutation are unknown. Some experiments on rodents point towards an intricate, though as yet unclear, role for BK in AD pathogenesis (Chishti et al., 2001; Jafari et al., 2015; Wang et al., 2015; Wang et al., 2015; Yamamoto et al., 2011; Ye, Jalini, Mylvaganam, & Carlen, 2010).

Mental Retardation

Both decreased and increased BK channel activity may cause mental retardation. Several cases of mental retardation have been linked to de novo *KCNMA1* mutations that decrease BK currents (Laumonnier et al., 2006; Z. B. Zhang et al., 2015), as well as to a nonsense mutation in the gene encoding cereblon, which under normal conditions suppresses the surface expression of neuronal BK channels by promoting its ubiquitination (Higgins, Hao, Kosofsky, & Rajadhyaksha, 2008; Jo, Lee, Song, Jung, & Park, 2005; Liu et al., 2014). A link between BK and mental retardations has been established through studies on the Fragile X Mental Retardation Protein (FMRP), which is essential for cognitive development due to its role as a regulator of local protein synthesis in dendrites (Bassell & Warren, 2008). FMRP loss, by transcriptional silencing of the *Fmr1* gene, causes Fragile X syndrome (FXS), the most common inheritable cause of human mental disability, which is also linked to autism (Wijetunge, Chattarji, Wyllie, & Kind, 2013). In CA3 hippocampal neurons, FMRP interacts with the β4 subunit to reduce BK Ca^{2+} sensitivity, thus causing AP broadening (Deng et al., 2013). Likewise, AP broadening caused by the loss of FMRP also occurs in cortical pyramidal neurons (Deng et al., 2013). These effects on AP duration lead to elevated presynaptic Ca^{2+} influx and neurotransmitter release.

Ethanol Addiction

It is a topic of debate whether the interaction of ethanol with BK channels directly contributes to the intoxicating effects of ethanol, or whether it acts on the adaptive mechanisms that generate chronic tolerance or drinking escalation (Contet et al., 2016). In several studies, *KCNMA1* and *KCNMB1* have been associated with a risk of developing ethanol dependence (Edenberg et al., 2010; Han et al., 2013; Kendler et al., 2011; Schuckit et al., 2005). The BK channel is a known molecular target of ethanol (Dopico et al., 1996). Whether ethanol acts as BK activator or inhibitor depends on several factors, such as $[Ca^{2+}]_i$ or $[Mg^{2+}]_i$. For example, co-expression with the β1 subunit simultaneously yields ethanol-resistant (α+β1)BK channels in neurons and ethanol-inhibited (α+β1)BK channels in cerebral artery myocytes (Bukiya, Liu, & Dopico, 2009; Martin et al., 2004; Wynne et al., 2009). At low $[Mg^{2+}]_i$, ethanol is a BK channel inhibitor, which is relevant because chronic ethanol consumption leads to Mg^{2+} depletion (Marrero, Treistman, & Lemos, 2015; Romani, 2008). The effects of ethanol are strongly dependent on the BK splice variant, and some variants are more ethanol-sensitive than others. In the adult mammalian brain, ethanol causes an increase in microRNA (miR-9) expression, whereby only BK mRNA containing miR-9 binding sites are degraded, and the expression of other ethanol-insensitive stress-axis hormones regulated (STREX) exon containing isoforms is favored (Pietrzykowski et al., 2008; Velazquez-Marrero et al., 2011). This microRNA-mediated BK α isoform switch contributes to the rapid ethanol-desensitization of BK channels in the supraoptic nucleus and striatal neurons (Pietrzykowski et al., 2008; Velazquez-Marrero et al., 2011). The rapid desensitization is also partially dependent on the phosphorylation state of the BK channel (Velazquez-Marrero, Seale, Treistman, & Martin, 2014). In addition, neuronal exposure to ethanol over a period of six hours has been shown to reduce the cell surface expression of BK, thus indicating that ethanol can modulate BK channel trafficking to the cell membrane (Palacio et al., 2015; Pietrzykowski et al., 2004; Velazquez-Marrero et al., 2011). The precise subunit composition of the BK channels also plays an important role in the adaptive response to acute and chronic alcohol exposure in mammals. Drinking escalation was accelerated and the chronic tolerance to ethanol-induced sedation and hypothermia was reduced in β1 knockout ethanol-dependent mice (Kreifeldt, Cates-Gatto, Roberts, & Contet, 2015; Kreifeldt, Le, Treistman, Koob, & Contet, 2013). In contrast, β4 knockout mice exhibit tolerance to the hypolocomotor effect of ethanol, and the drinking escalation in ethanol-dependent individuals is attenuated (Kreifeldt et al., 2013; Martin et al., 2008). More precisely, while ethanol-induced BK current potentiation has a desensitizing effect on the striatal spiny neurons of WT mice, such a desensitization does not occur in the same neurons of β4 knockout animals (Martin et al., 2008). These findings indicate that β1 and β4 BK accessory subunits control ethanol tolerance at the molecular, cellular, and behavioral levels.

Pain Disorders

BK channels are implicated in the flow of sensory information along the spinal cord, where this channel is expressed in subsets of small- and medium-diameter DRG neurons (Li et al., 2007; Lu et al., 2014; Sarantopoulos et al., 2007; Scholz, Gruss, & Vogel, 1998; Shieh et al., 2007; X. F. Zhang, Gopalakrishnan, & Shieh, 2003; X. L. Zhang, Mok, Katz, & Gold, 2010). Specifically, BK channels are highly expressed in isolectin-B4 positive small-diameter DRG neurons, which are nociceptive primary afferents (Gruss et al., 2001; X. L. Zhang et al., 2010). In addition, it has been reported that the effects of some analgesic compounds are mediated by BK channels in these and other cells (Baillie, Schmidhammer, & Mulligan, 2015; Gruss, Hempelmann, & Scholz, 2002; Gruss et al., 2001). Peripheral nerve injury (a preclinical model of neuropathic pain) reduces BK channel expression and activity in small and medium DRG neurons, leading to an increase in excitatory transmission in the superficial dorsal horn (Cao, Chen, Li, & Pan, 2012; S. R. Chen, Cai, & Pan, 2009; Furukawa, Takasusuki, Fukushima, & Hori, 2008; Sarantopoulos et al., 2007). In the dorsal horn presynaptic terminals of rodent models, BK inhibits transmitter release from primary afferents and inhibitory interneurons (Furukawa et al., 2008; Song & Marvizon, 2005). BK reduces spontaneous glycinergic transmission in the sacral dorsal commissural nucleus and mediates the inhibitory effect of N-methyl-D-aspartate (NMDA) on opioid release (Shoudai et al., 2007; Song & Marvizon, 2005). In contrast with DRG neurons, where ligation reduces BK expression, BK expression is increased in dorsal horn interneurons (S. R. Chen et al., 2009). BK can suppress inhibitory signaling within the dorsal horn, which can potentiate nociception in neuropathic pain. Knockout of BK in DRG neurons induces a nociceptive behavior in models of chronic inflammatory pain (Lu et al., 2014). Similarly, the blockade of spinal BK channels with iberiotoxin has been found to induce mechanical hyperalgesia (S. R. Chen et al., 2009).

BK is expressed underneath the myelin sheath of the juxtaparanodal sites of spinal cord white matter axons, where, under normal conditions, it does not modulate axonal conduction (Ye et al., 2012). However, in a model of spinal cord compression injury, demyelination exposes juxtaparanodal BK channels, and the activity of these exposed channels impairs axonal conduction (Ye et al., 2012).

Ischemia

During a cerebral ischemic stroke, BK channels take on a protective role. BK knockout mice are characterized by larger infarct volumes, more important neurological deficits, and lower levels of survival than their WT counterparts (Liao et al., 2010). Furthermore, blockade or genetic ablation of the BK channel leads to an increase in NMDA-induced neurotoxicity and neuronal death in ischemia-like conditions (Katsuki, Shinohara, Fujimoto, Kume, & Akaike, 2005; Liao et al., 2010; Runden-Pran et al., 2002). Conversely, administration of BK-openers following middle cerebral artery occlusion was found to reduce cortical infarct in rats (Gribkoff et al., 2001).

Slo2 Channels

Na^+ activated K^+ (K_{Na}) currents mediated by members of the Slo2 channel family have been described for the neurons of several species (Bader, Bernheim, & Bertrand, 1985; Bischoff, Vogel, & Safronov, 1998; Dale, 1993; Dryer, 1991; Dryer, Fujii, & Martin, 1989; Egan, Dagan, Kupper, & Levitan, 1992; Gao et al., 2008; Haimann & Bader, 1989; Haimann, Magistretti, & Pozzi, 1992; Hartung, 1985; Nuwer, Picchione, & Bhattacharjee, 2010; Safronov & Vogel, 1996; Saito & Wu, 1991; Schwindt, Spain, & Crill, 1989; Tamsett, Picchione, & Bhattacharjee, 2009). These K_{Na} channels are encoded by the genes *Slo2.1* (*KCNT2, KNa1.2, Slick*) and *Slo2.2* (*KCNT1, KNa1.1, Slack*) (Bhattacharjee et al., 2003; Joiner et al., 1998; Yuan et al., 2003). The first member of the family to be described was the Sequence Like A Calcium-activated K^+ Channel, or Slack. Its name originates from the fact that its S6 and part of its PD are similar to the corresponding domains of BK (Joiner et al., 1998). The other Slo2 channel, Slick, or Sequence Like an Intermediate Conductance K^+ Channel, takes its name from its close relationship to the Slack channel and makes reference to the fact that the single-channel conductance of the Slack channel falls between that of BK and the typical conductance of other K^+ channels (Table 14.1) (Bhattacharjee et al., 2003). Both Slick and Slack channels are also activated by cytosolic Cl^- (Bhattacharjee et al., 2003; Yuan et al., 2000); under physiological conditions, Slick is the more Cl^--sensitive of the two. There is a ~74% identity between Slick and

Slack, with the greatest divergence between their distal C-terminal regions, which explains why the modulating effects of various intracellular factors on the two channels differ (Bhattacharjee & Kaczmarek, 2005; Santi et al., 2006; Tejada, Hashem, Calloe, & Klaerke, 2017; Tejada et al., 2014).

The unitary conductance of the Slick channel (KCNT2 in humans) is ~140 pS in symmetrical 130 mM K^+. Its half-activation concentration is 89 mM Na^+, but a basal level of channel activity can be observed even in the absence of internal Na^+ (Bhattacharjee et al., 2003).

The unitary conductance of the Slack channel (KCNT1 in humans) ranges from 165 to 183 pS in symmetrical concentrations of K^+ varying from 130 to 160 mM K^+ (Bhattacharjee et al., 2003; Yuan et al., 2003). This value is strongly dependent on the ionic composition of the solution and is reduced to 25–90 pS under different conditions (Joiner et al., 1998; Yuan et al., 2003). In excised patches, the internal half-activation Na^+ concentration of Slack channels is ~40 mM (Bhattacharjee et al., 2003; Yuan et al., 2003). Unlike Slick, Slack channel activation has an absolute requirement for internal Na^+ (Bhattacharjee et al., 2003). In the rat, five different *Slack* transcripts have been described, resulting in Slack channels with different cytoplasmic amino termini (Brown et al., 2008). Two of these transcripts, *Slack*-A and *Slack*-B, are functionally expressed, and while both isoforms have the same unitary conductance, their activation kinetics differ greatly. Slack-B has a long N-terminal domain, making it the largest known K^+ channel subunit. Slack-B channels activate slowly over hundreds of milliseconds, while Slack-A rapidly activates upon depolarization (Bhattacharjee et al., 2003; Joiner et al., 1998; Yuan et al., 2003). Slick and Slack-B, but not Slack-A, subunits can form heteromeric channels that differ in their properties from both homomers, thus increasing channel diversity (H. Chen et al., 2009). Current activation is markedly slower for Slick/Slack-B than for homomeric Slick or Slack-B, and the unitary conductance of the Slack-B/Slick heteromers is intermediate between that of Slack and Slick (H. Chen et al., 2009).

Common Structural Features

The structural features of Slo2 channels are similar to those of the Slo1 channel, except that Slo2 channels lack the S0 TM segment (Figure 14.1A–C). The structure of Slack has been determined by cryo-electron microscopy technique, both open and in multiple closed conformations (Hite & MacKinnon, 2017; Hite et al., 2015).

The Pore Domain and the Gate

The selectivity filter is similar to that of BK and other K^+ channels (Heginbotham et al., 1994; Hite & MacKinnon, 2017; Hite et al., 2015). Whether it is the bundle crossing, as in Kv channels, or the selectivity filter itself, as in BK, that serves as the gate to K^+ passage in Slo2 channels is still a matter of debate.

The substitution of two conserved leucine residues in Slick's S6 segment, near the bottom of the selectivity filter, by asparagine (L267N and L270N) stabilizes the channel in its open state, suggesting the presence of a putative bundle crossing (Giese, Gardner, Hansen, & Sanguinetti, 2017). However, in the BK channel, where the gate resides in the selectivity filter, some S6 mutations also produce constitutively open channels (X. Chen et al., 2014). Moreover, internal verapamil was found to block the Slick channel in an activation-independent manner, indicating that the compound has access to its binding site in the central channel cavity even when the channel is closed. This represents strong evidence against the presence in Slick channels of a bundle crossing gate similar to those found in Kv channels (Garg, Gardner, Garg, & Sanguinetti, 2013). In addition, it has been determined that the two S6 residues P271 and E275 maintain the inner pore in an open configuration, thus preventing the formation of a tight bundle crossing (Garg et al., 2013). Conversely, hydrophobic interaction between the pore helix residue F240 and L209 in the S5 TM segment stabilizes the closed state, suggesting that the gate is located in the PD (Garg et al., 2013; Suzuki, Hansen, & Sanguinetti, 2016).

In the absence of Na^+, the structure of chicken Slack shows an apparent bundle crossing resulting from the interaction of its S6 segments. The narrowest point, which has a diameter of 4-6 Å, is formed by a ring of methionines (M333) near the carboxyl end of S6 (Hite & MacKinnon, 2017; Hite et al., 2015). Hite et al. (2015) argue that this structure is compatible with a closed conformation, since the diameter of this pore is smaller than that of a hydrated K^+ (~8 Å), and the energy barrier for incoming K^+ ions is likely to be further raised by the presence of positively charged residues in the neighborhood of M333. However, Giese et al. (2017) have recently produced functional evidence against the presence of a bundle crossing gate in the Slack channel. They found that the substitution of the human homologue to M333, M354, for glutamate did not enhance activation, as would be expected if M354 were part of a hydrophobic gate (Giese et al., 2017). Interestingly, Hite and MacKinnon (2017) found that the Slack channel exists in multiple Na^+-independent closed conformations, but that there is a single open conformation, formed from the ensemble of closed conformations in a Na^+-dependent manner (Hite & MacKinnon, 2017).

Absence of a Canonical Voltage Sensor Domain

While the intrinsic voltage dependence of both Slick and Slack is weak, the S4 segments of both channels lack the canonical gating charges that are associated with voltage sensing in K_V and BK channels (Figure 14.2) (Y. Yan, Yang, Bian, & Sigworth, 2012; Yuan et al., 2003). Neutralization of charged residues in the S1–S4 TM domain does not modify the voltage dependence of Slick channel activation, indicating that the voltage sensor resides in a different region of the channel (Dai, Garg, & Sanguinetti, 2010). In addition, the absence of a conformational change within the S1–S4 TM domain upon Slack channel opening is consistent with this channel's being a purely ligand-gated K^+

channel (Hite & MacKinnon, 2017). Where Slo2 intrinsic voltage dependence resides is still to be determined.

The C-Terminal (Na^+ Sensor) Domain

Screening Slick and Slack for potential Na^+ coordination motifs (DX(R/K)XXH) identified a conserved site in the RCK2 domain of both channels. Charge reversion of the initial aspartate (D757 in Slick and D818 in Slack) greatly attenuated channel activation by cytosolic Na^+ (Thomson, Hansen, & Sanguinetti, 2015; Z. Zhang, Rosenhouse-Dantsker, Tang, Noskov, & Logothetis, 2010).

Structural studies of Slack revealed that, similar to BK channel activation by Ca^{2+}, cytoplasmic Na^+ binding causes an expansion of the Slack gating ring, which displaces the S6 and thus opens the channel by exerting force via the S6–RCK1 linker (Hite & MacKinnon, 2017). As a consequence of the displacement of the S6 helices, the internal channel vestibule expands from 6 Å at the level of M333 (in chicken Slack) in the closed state, to about 20 Å in the open state (Hite & MacKinnon, 2017). The Slo2 RCK2 domain has a region resembling the "calcium bowl" of BK, but containing positively rather than negatively charged residues. Although this region had been suggested to comprise a putative "chloride bowl" (Yuan et al., 2000; Yuan et al., 2003), neutralization of charges in this region did not affect the Na^+ or Cl^- sensitivity of Slick and Slack (Bhattacharjee et al., 2003).

Na^+ Entry Pathways

As previously stated, Slo2 channels are activated by cytosolic Na^+. The estimated cytosolic Na^+ concentration in resting unstimulated neurons is between 4 mM and 15 mM (Rose, 2002), potentially high enough to induce a small activation of Slo2 channels in neurons under resting conditions (Tamsett et al., 2009). Several physiological processes can increase internal Na^+ concentrations beyond this range, thus increasing Slo2 opening probabilities. A brief and localized rise in internal Na^+ levels can occur with the influx of Na^+ entry through Nav channels during an AP and through postsynaptic ionotropic ligand-gated receptors, such as α-amino-3-hydroxy-5-methyl-4-isoxazolepropionic acid (AMPA) and NMDA glutamate receptors (Nanou & El Manira, 2007; Nanou et al., 2008; Rose & Ransom, 1997; Uchino et al., 2003; Zamalloa, Bailey, & Pineda, 2009). Although in most neurons, the AP is driven by rapidly inactivating Nav currents, a small component of the Na^+ current (I_{NaP}) through persistent Na^+ channels does not inactivate during maintained depolarization. This Na^+ entry through I_{NaP} channels is responsible for a sustained elevation of cytosolic Na^+ levels (Rose & Ransom, 1997), which activates Slo2 channels (Budelli et al., 2009; Hage & Salkoff, 2012). Although I_{NaP} are small, the proximity between Slo2 and persistent Na^+ channels allows their functional coupling. Since inhibition of I_{NaP} also eliminates Slo2 currents, both channels appear to co-localize (Budelli et al., 2009; Hage & Salkoff, 2012). In addition, repetitive tetanic stimulation can raise local Na^+ concentrations to 45-100 mM in restricted compartments such as dendritic spines, which may activate Slo2 channels (Rose & Konnerth, 2001; Zhong, Beaumont, & Zucker, 2001).

Differential Regulation of Slo2 by Intracellular Factors

Nicotinamide Adenine Dinucleotide (NAD^+)

Neuronal Slo2 channel activity is reduced upon excision of membrane patches (Haimann et al., 1992; Tamsett et al., 2009). This loss of activity appears result from the loss of nicotinamide adenine dinucleotide (NAD^+) because both Slo2 subunits have a putative NAD^+-binding site within their RCK2 domain (Tamsett et al., 2009). A single amino acid substitution within the putative NAD^+-binding site of Slack, G792A, abolished the ability of NAD^+ to potentiate Slack activity (Tamsett et al., 2009). Moreover, application of NAD^+ to the cytoplasmic face of K_{Na} channels excised from DRG neurons, which express Slack channels, increases the K_{Na} channel open probability and reduces the Na^+ concentration required for half activation from ~50 mM to ~17 mM (Tamsett et al., 2009).

Adenosine Trisphosphate (ATP)

A potentially distinctive feature of Slick channels is their inhibition by cytosolic ATP. Slick current is reduced by ~80% in the presence of 5 mM ATP at the cytoplasmic face of excised patches from chinese hamster ovary CHO cells heterologously expressing human Slick (hSlick). These results suggest that Slick channel activity increases during periods of high metabolic demand when cellular ATP levels are low (Bhattacharjee et al., 2003; Bhattacharjee & Kaczmarek, 2005; Kaczmarek, 2013; B. Yang, Desai, & Kaczmarek, 2007). A consensus ATP-binding site (GXXXXGKT) near the C-terminal of the RCK2 domain is thought to be responsible for ATP sensitivity (Bhattacharjee et al., 2003). The mutation of the first glycine residue in this consensus sequence (G1032S) renders the Slick channel insensitive to internal ATP (Bhattacharjee et al., 2003). It is worth noting that there are conflicting reports in the literature regarding the inhibitory effect of internal ATP on the Slick channel. Slick channels from rat striatal neurons were not inhibited by ATP when they were heterologously expressed in HEK293T cells (Berg, Sen, & Bayliss, 2007) or in excised patches from *Xenopus* oocytes (Berg et al., 2007; Garg & Sanguinetti, 2014). The precise conditions under which ATP regulates Slick channels have thus not yet been completely established. In particular, Garg and Sanguinetti (2014) argued that during ischemia, it is the increase in internal Na^+ and Cl^- concentrations, and not the drop in ATP, that is responsible for Slick activation.

Protein Kinase C (PKC) and G Proteins

Slick channel activity is modulated by protein kinase C (PKC), whose presence produces a decrease in the current amplitude of Slick (Santi et al., 2006). The specific site of Slick phosphorylation by PKC remains to be identified. Chimeric channels containing the

Slick CTD and the Slack TM region are inhibited by the application of a PKC activator, indicating that the PKC sites reside in the CTD (Santi et al., 2006). In contrast to Slick channels, treatment of Slack channels with PKC activators leads to an increase in current amplitude and a slowing of the activation rate (Barcia et al., 2012; Santi et al., 2006). This effect on Slack is due to the phosphorylation of a single residue, S407, which is located in the "hinge" region of the C-terminus, between S6 and the first RCK domain (Barcia et al., 2012). Slack-B/Slick heteromeric channels respond to PKC activation in quite a different manner from the response of either subunit expressed alone (H. Chen et al., 2009). Here, PKC activators suppress current by ~90%, which is greater than the degree of inhibition measured for Slick homotetrameric channels (~50%). This effect cannot be explained by any linear combination of Slick or Slack-B subunits, because PKC potentiates Slack-B channel activity (Barcia et al., 2012).

Throughout the nervous system, there is widespread co-localization of both Slo2 channels with the $G\alpha_q$-protein coupled M1 muscarinic receptor and the mGluR1 metabotropic glutamate receptor (Santi et al., 2006). Activation of these receptors suppresses Slick currents while increasing Slack currents, an effect that is consistent with the modulation of Slick and Slack by PKC (Santi et al., 2006). Differential receptor modulation can thus distinguish between Slick and Slack. However, when *Slack* cDNA is transfected into a mammalian expression system (HEK293 cells), the resulting Slack channels are also inhibited by $G\alpha_q$-coupled receptors (Berg et al., 2007). The reasons behind this inconsistent behavior between different expression systems are at present unclear.

Protein-Kinase A (PKA)

K_{Na} currents have been recorded and Slack subunits are expressed in primary afferent nociceptor neurons located in the DRG (Gao et al., 2008; Nuwer et al., 2010; Tamsett et al., 2009). Some inflammatory substances that activate the PKA pathway markedly increase the excitability of these neurons, sensitizing them to thermal and mechanical stimulation. These PKA effects can be attributed to a decrease in Slack current (Nuwer et al., 2010). Unlike PKC, the effects of PKA on Slack channels are indirect and do not involve phosphorylation of the channel subunit. Instead, the modulatory effects of PKA are due to a fast internalization of the channel from the plasma membrane into internal organelles (Nuwer, Picchione, & Bhattacharjee, 2009; Nuwer et al., 2010).

Phosphatidylinositol 4,5-bisphosphate (PIP2)

PIP2 activates both Slick and Slack channels (Tejada, Jensen, & Klaerke, 2012). Activating PIP2 effects appears to be mediated by direct interaction with lysine residues K306 in Slick and K339 in Slack, located at the proximal C-terminal of their respective channels (Tejada et al., 2012).

Estradiol

17β-estradiol is able to activate *Slack* channels reconstituted into lipid bilayers from extracts of rat cortex (L. Zhang et al., 2005). A similar effect was reported in

Slack-expressing HEK-293 cells, presumably by direct binding of 17β-estradiol to the channel (L. Zhang et al., 2005).

Cell Volume

Slick, but not Slack-B, channels are cell volume–sensitive (Tejada et al., 2014). The sensitivity of the heteromeric channels is proportional to the number of Slick subunits in the tetramer (Tejada et al., 2017).

Expression Patterns of Slo2 Channels

Slick mRNA is widely distributed throughout the body; its highest levels are found in the brain (Bhattacharjee et al., 2003). Slick protein expression is found at high levels in the olfactory bulb, midbrain, brainstem, hippocampus, and throughout the cerebral cortex, with high expression levels in primary somatosensory and visual regions (Bhattacharjee, von Hehn, Mei, & Kaczmarek, 2005). Likewise, high levels of *Slick* expression have been observed in the neurons of sensory pathways, including those in the auditory brainstem (Bhattacharjee et al., 2005). In the peripheral nervous system, Slick channel is expressed in small- and medium-sized DRG neurons containing calcitonin gene-related peptide (CGRP). Slick channels are synthetized in the soma and transported to the periphery via CGRP-containing large dense core vesicles (Tomasello, Hurley, Wrabetz, & Bhattacharjee, 2017).

Two different *Slack* transcripts (of about 4.5 kb and 7.5 kb) are abundantly expressed in the rat brain and kidney (Joiner et al., 1998). In the adult rat brain, *Slack* mRNA is highly expressed in neurons, without evident expression in glial cells. The mRNA of both isoforms (*Slack*-A and *Slack*-B) is found throughout the brain, but the highest levels of both are detected in the brainstem and olfactory bulb (Brown et al., 2008; Joiner et al., 1998).

The Slack-B protein is detected in brainstem and olfactory bulb (Bhattacharjee et al., 2005). Abundant transcripts are found in the red nucleus, deep cerebellar nuclei, vestibular and oculomotor nuclei, and in the trigeminal system and reticular formation, where they are expressed in both cell bodies and axonal fibers. In addition, the Slack-B protein was found in the auditory brainstem nuclei and in neurons of the thalamus, substantia nigra, and amygdala. The only region of the cerebral cortex where the *Slack*-B isoform has been found is the frontal cortex (Bhattacharjee et al., 2005). Of note, there is a widespread co-localization of Slick and Slack-B, meaning that the K_{Na} channels in the olfactory bulb and brainstem are likely to be Slick/Slack-B heteromers (Bhattacharjee et al., 2005; H. Chen et al., 2009).

Brain regions such as the dendrites of hippocampal neurons and olfactory bulb glomeruli express further Slack isoforms, different from the Slack-B protein (Brown et al., 2008). *Slack* is also expressed in neurons of the peripheral nervous system. Slack protein is found in both the soma and the axonal tracts of neurons of the DRG (Nuwer et al., 2010; Tamsett et al., 2009). *Slack* expression in axonal tracts has also been confirmed by electrophysiological experiments on the peripheral axons of *Xenopus* neurons, where it is co-expressed with Nav channels (Koh, Jonas, & Vogel, 1994).

The Role of Slo2 Channels in the Nervous System

Slo2 channels are widely expressed in the nervous system, where they contribute to the slow repolarization following an AP (Bhattacharjee et al., 2003). As previously stated, activation of K_{Na} currents requires a rise in cytosolic Na^+ concentration, which results from different mechanisms of Na^+ influx, under physiological or pathophysiological conditions (Dryer, 1994; Dryer, 2003; Dryer et al., 1989; Uchino et al., 2003). Therefore, Slo2 channels might either play a direct role in normal AP discharges, or they may facilitate a kind of "reserve conductance," thus playing an important protective role under pathological conditions such as ischemia or hypoxia, where $[Na^+]_i$ is increased (Dryer, 1994; Dryer, 2003). The K_{Na} conductance through Slo2 channels has several functions in neurons, including the stabilization of the resting membrane potential and the shaping and adaptation of firing patterns through the generation of slow AHPs and depolarizing afterpotentials (Gao et al., 2008; Markham, Kaczmarek, & Zakon, 2013; B. Yang et al., 2007; Yuan et al., 2003; Y. Zhang et al., 2012).

The suppression of Slack channel translation in mitral cells of the olfactory bulb and medium spiny neurons of the striatum revealed that Slack channels are responsible for the total delayed outward currents under physiological conditions in both cells (Budelli et al., 2009; Lu, Das, Fadool, & Kaczmarek, 2010). K_{Na} currents can be activated by a single AP in some neuronal types, such as hippocampal CA1 neurons and neocortical pyramidal neurons. In these cells, K_{Na} currents contribute to AP repolarization or to the depolarizing afterpotential, which is generated by I_{NaP} (Franceschetti et al., 2003; Liu & Stan Leung, 2004). In a variety of neuronal types, the activation of K_{Na} currents following an AP burst produces a slow afterhyperpolarization (sAHP) (Descalzo, Nowak, Brumberg, McCormick, & Sanchez-Vives, 2005; Foehring, Schwindt, & Crill, 1989; Franceschetti et al., 2003; Kim & McCormick, 1998; Kubota & Saito, 1991; Sanchez-Vives, Nowak, & McCormick, 2000; Sandler, Puil, & Schwarz, 1998; Schwindt et al., 1989). This sAHP influences the timing of regular AP bursts (Kim & McCormick, 1998). In addition, the slow activation of K_{Na} channels during AP bursts leads to an adaptation of the firing rate (Nuwer et al., 2010). This adaptation can be predicted by a numerical simulation of participating K_{Na} currents (Benda, Maler, & Longtin, 2010; Brown et al., 2008). Neurons in which K^+ currents are dominated by a Slack-A–like current can adapt very rapidly to repeated or maintained stimulation (Brown et al., 2008). In contrast, Slack-B currents can allow neurons to fire rhythmically during maintained stimulation and enable the variation of the adaptation rate with the intensity of stimulation (Brown et al., 2008).

In locomotion-controlling neurons of the lamprey spinal cord (Nanou & El Manira, 2007; Nanou et al., 2008), two distinct components of K_{Na} current are triggered by a single AP firing: a fast transient component and a slower sustained component

(Cangiano, Wallen, & Grillner, 2002; Hess, Nanou, & El Manira, 2007; Wallen et al., 2007). Both these components flow through different Na^+ entry pathways with different tetrodotoxin (TTX) sensitivities. The fast component is activated by Na^+ influx through TTX-sensitive Na^+ channels during the AP and determines AP duration and amplitude (Hess et al., 2007). The sustained component is activated by an independent TTX-insensitive Na^+ entry pathway and regulates the sAHP during repetitive bursting (Hess et al., 2007).

The neurons of the medial nucleus of the trapezoid body (MNTB) in the auditory brainstem fire at high frequency (Paolini, FitzGerald, Burkitt, & Clark, 2001). To facilitate an adequate transmission of information about the position of the source of a sound stimulus, accurate AP timing is paramount in these neurons. *Slack* and *Slick* are both highly expressed in MNTB neurons (Bhattacharjee, Gan, & Kaczmarek, 2002; Bhattacharjee et al., 2005; Brown et al., 2010), where they may play the role of K_{Na} channels in neurons that fire at high rates and with high temporal accuracy (B. Yang et al., 2007). K_{Na} channels co-localize with AMPA receptors and voltage-dependent Na^+ channels in the MNTB neuron soma (B. Yang et al., 2007). K_{Na} channel activation by an increase in $[Na^+]_i$ or through pharmacological Slack activators improves the temporal accuracy of firing. This is due to the voltage dependence of K_{Na} channels, which can open at voltages near the resting potential of MNTB neurons, thus increasing the resting conductance and decreasing the membrane time constant (B. Yang et al., 2007).

Slo2 Channels and Neuronal Diseases

Small changes in the function of Slo2 channels can produce large changes in neuronal development and learning capabilities, as it is summarized in Table 14.4 (Barcia et al., 2012; Bausch et al., 2015; Heron et al., 2012).

Fragile X Syndrome

The cytoplasmic C-terminus of the Slack subunit interacts with the RNA-binding protein FMRP (Brown et al., 2010), a potent activator of Slack channel activity (Brown et al., 2010). It is known that mutations in human *Slack* have serious adverse effects on intellectual development (Barcia et al., 2012; Heron et al., 2012).

Autosomal Dominant Nocturnal Frontal Lobe Epilepsy

The Slack channel is implicated in autosomal dominant nocturnal frontal lobe epilepsy (ADNFLE) and inherited intellectual and psychiatric disease (Heron et al., 2012). Human ADNFLE patients from three different families carried missense mutations in the *Slack* gene (Heron et al., 2012).

Epilepsy

Certain de novo mutations in the human Slick and Slack channels cause epileptic disorders. A mutation of a residue located in the P-helix of the Slick channel (F240L)

Table 14.4 Slo2 Channels Implication in Neuronal Diseases

Diseases	Evidence of Slo2 channel implication	References
Mental retardation	Slack interacts with FMRP, which is a potent Slack activator	Brown et al. (2010)
	Slack mutations produce defects in intellectual development in humans	Barcia et al. (2012), Heron et al. (2012)
Autosomal-dominant nocturnal frontal lobe epilepsy (ADNFLE)	Three non-related ADNFLE patients displayed *Slack* gene missense mutations	Heron et al. (2012)
Epilepsy	De novo mutation in Slick P-helix (F240L) in an early infantile encephalopathy patient	Gururaj et al. (2017)
	Several patients suffering epileptic encephalopathies of the infancy displayed mutations in Slack channels	Barcia et al. (2012)
Learning defects	*Slack* knockout mice lose cognitive flexibility, show inability to adapt fast to new situations, and defects during reversal learning tasks	Bausch et al. (2015)
Ischemia	Nematodes lacking *Slack* orthologue are more sensitive to hypoxic death	Yuan et al. (2003)
	Slack activation in *Xenopus* oocytes exposed to ischemia/hypoxia-like conditions	Ruffin et al. (2008)
Pain disorders	Thermal hyperalgesia and increased AP peak in Slick KO mice	Tomasello et al. (2017)
	Slack KO increase sensitivity to neuropathic pain in mice	Lu et al. (2015)
	Slack regulates synaptic transmission at the superficial, pain-processing lamina of the dorsal horn. Slack KO mice have an enhanced response to thermal nociceptive stimuli	Evely et al. (2017)

was found in a patient with early infantile epileptic encephalopathy (Gururaj, Palmer, et al., 2017). This change-of-function mutation reversed the Cl^--sensitivity and made the channel more selective for Na^+ than K^+ (Gururaj, Palmer, et al., 2017). Several gain-of-function mutations in the CTD-coding region of the *Slack* gene (which produce channels constitutively active) were found in six of twelve unrelated patients suffering from malignant migrating partial seizures of infancy, a rare epileptic encephalopathy of infancy (Barcia et al., 2012).

Learning Defects

Bausch et al. (2015) (Bausch et al., 2015) studied the Slack-dependent aspects of the memory and learning in *Slack* knockout mice (Bausch et al., 2015). Although these mice showed normal working memory, reference memory, and cerebellar control of movement, and almost normal locomotor activity, they lost cognitive flexibility, and showed an atypical initial response to novelty (an inability for the fast adaptation to new situations), in addition to defects during reversal learning tasks (Bausch et al., 2015).

Ischemia

It has been suggested that the activation of K_{Na} channels and the ensuing hyperpolarization of cells could act as a protective mechanism against hypoxia, when levels of internal Na^+ and Cl^-, H^+ and CO_2 rise (Kameyama et al., 1984). In support of this hypothesis, nematodes lacking the *Slack* orthologue are significantly more sensitive to hypoxic death than their WT counterparts (Yuan et al., 2003). Additionally, Slack activity increases in *Xenopus* oocytes exposed to increasing concentrations of H^+ and CO_2, demonstrating that an increase in internal Na^+ concentration activates Slack under cellular conditions similar to those in ischemia/hypoxia (Ruffin et al., 2008).

Pain Disorders

Several studies indicate that Slo2 channels are implicated in pain processing, but the precise mechanism and the levels within the sensory pathway where Slo2 channels participate still need to be clarified. Slick KO mice exhibit thermal hyperalgesic behavior and increase in Ca^{2+} AP peak in DRG neurons (Tomasello et al., 2017). Due to the presynaptic localization of Slick channels, it is hypothesized that this increase in AP peak, without a change in repolarization time constant, would lead to Ca^{2+} accumulation during AP repolarization, which can result in elevated neurotransmitter release (Tomasello et al., 2017).

Slack manipulation, by either global or sensory DRG neuron-specific KO, increases sensitivity to neuropathic, but not acute nociceptive or inflammatory, pain in mice (Lu et al., 2015). In DRG neurons, the Slack function relies strongly on a PKA-modulated channel density. (Nuwer et al., 2010). The PKA activation diminishes Slack density by promoting its internalization, a process carried out by Adaptor Protein-2 via clathrin-mediated endocytosis (Gururaj, Evely, et al., 2017).

At the level of the superficial, pain-processing lamina of the dorsal horn, Slack channels regulate synaptic transmission (Evely et al., 2017). In contrast to previous results (Lu et al., 2015; Martinez-Espinosa et al., 2015), Evely et al. (2017) found that Slack KO mice have an enhanced nociceptive response to thermal stimuli, suggesting that Slack regulates the nociceptive thermal threshold (Evely et al., 2017).

Conclusions

The Slo channel family is a key player in the physiology of the nervous system. Despite being encoded by only a handful of genes, this channel family is characterized by a remarkable functional diversity, which is attained through alternative splicing, the association with accessory subunits, heteromer formation, and metabolic modification. This diversity enables Slo channels to perform specific functions in each cell type. The recent elucidation of their various structures has significantly deepened our knowledge about the specific structural determinants of the Slo channel function. However, the answers to several important problems, such as that of the molecular determinants of these channel's unusually large conductance, remain elusive.

Future Directions

Several key questions currently remain unanswered about the Slo channel family and its role in the nervous system. Among the major recent advances of this field was the determination of the structures of Slo1 and Slo2.2. Without doubt, these structures will greatly help the design of new and more potent agonists and inhibitors of these channels in a major step towards a structurally oriented Slo channel pharmacology. In addition, the newly available information on their structures will help reveal the molecular determinants of their large conductance and exquisitely high selectivity. It seems likely that these open questions can be resolved by focusing future studies on the molecular characteristics of the internal vestibule. Finally, the precise role of Slo channels in neuronal diseases should be further pinpointed, in order to aid the rational approach to the treatment of these ailments.

Acknowledgements

This work was supported by Fondo Nacional de Desarrollo Científico yTenológico (FONDECYT) Grant No. 1150273 to RL, Comisión Nacional de Investigación Científica y Tecnológica (CONICYT) doctoral fellowship to YL, and the Air Force Office of Scientific Research under award number FA9550-16-1-0384 to RL. The Centro Interdisciplinario de Neurociencia de Valparaíso is a Millennium Institute (P09-022-F) supported by the Millennium Scientific Initiative of Ministerio de Economía, Fomento y Turismo of Chile.

Note

1. Throughout the chapter, the numbering of the residues corresponds to the murine Slo channel, unless otherwise stated.

References

Adelman, J. P., Shen, K. Z., Kavanaugh, M. P., Warren, R. A., Wu, Y. N., Lagrutta, A., ... North, R. A. (1992). Calcium-activated potassium channels expressed from cloned complementary DNAs. *Neuron*, *9*(2), 209–216.

Aggarwal, S. K., & MacKinnon, R. (1996). Contribution of the S4 segment to gating charge in the Shaker K+ channel. *Neuron*, *16*(6), 1169–1177.

Ahrendt, E., Kyle, B., Braun, A. P., & Braun, J. E. (2014). Cysteine string protein limits expression of the large conductance, calcium-activated K(+) (BK) channel. *PLoS One*, *9*(1), e86586. doi:10.1371/journal.pone.0086586

Alle, H., Kubota, H., & Geiger, J. R. (2011). Sparse but highly efficient Kv3 outpace BKCa channels in action potential repolarization at hippocampal mossy fiber boutons. *Journal of Neuroscience*, *31*(22), 8001–8012. doi:10.1523/JNEUROSCI.0972-11.2011

Atkinson, N. S., Robertson, G. A., & Ganetzky, B. (1991). A component of calcium-activated potassium channels encoded by the *Drosophila* Slo locus. *Science*, *253*(5019), 551–555.

Augustine, G. J., Santamaria, F., & Tanaka, K. (2003). Local calcium signaling in neurons. *Neuron*, *40*(2), 331–346.

Bader, C. R., Bernheim, L., & Bertrand, D. (1985). Sodium-activated potassium current in cultured avian neurones. *Nature*, *317*(6037), 540–542.

Baillie, L. D., Schmidhammer, H., & Mulligan, S. J. (2015). Peripheral mu-opioid receptor mediated inhibition of calcium signaling and action potential-evoked calcium fluorescent transients in primary afferent CGRP nociceptive terminals. *Neuropharmacology*, *93*, 267–273. doi:10.1016/j.neuropharm.2015.02.011

Bao, G., de Jong, D., Alevra, M., & Schild, D. (2015). Ca(2+)-BK channel clusters in olfactory receptor neurons and their role in odour coding. *European Journal of Neuroscience*, *42*(11), 2985–2995. doi:10.1111/ejn.13095

Barcia, G., Fleming, M. R., Deligniere, A., Gazula, V. R., Brown, M. R., Langouet, M., . . . Nabbout, R. (2012). De novo gain-of-function KCNT1 channel mutations cause malignant migrating partial seizures of infancy. *Nature Genetics*, *44*(11), 1255–1259. doi:10.1038/ng.2441

Barnes, S., & Hille, B. (1989). Ionic channels of the inner segment of tiger salamander cone photoreceptors. *Journal of General Physiology*, *94*(4), 719–743.

Bassell, G. J., & Warren, S. T. (2008). Fragile X syndrome: Loss of local mRNA regulation alters synaptic development and function. *Neuron*, *60*(2), 201–214. doi:10.1016/j.neuron.2008.10.004

Bausch, A. E., Dieter, R., Nann, Y., Hausmann, M., Meyerdierks, N., Kaczmarek, L. K., . . . Lukowski, R. (2015). The sodium-activated potassium channel Slack is required for optimal cognitive flexibility in mice. *Learning & Memory*, *22* (7), 323–335. doi:10.1101/lm.037820.114

Beecham, G. W., Martin, E. R., Li, Y. J., Slifer, M. A., Gilbert, J. R., Haines, J. L., & Pericak-Vance, M. A. (2009). Genome-wide association study implicates a chromosome 12 risk locus for late-onset Alzheimer disease. *American Journal of Human Genetics*, *84*(1), 35–43. doi:10.1016/j.ajhg.2008.12.008

Behrens, R., Nolting, A., Reimann, F., Schwarz, M., Waldschutz, R., & Pongs, O. (2000). hKCNMB3 and hKCNMB4, cloning and characterization of two members of the large-conductance calcium-activated potassium channel beta subunit family. *FEBS Letters*, *474* (1), 99–106.

Benda, J., Maler, L., & Longtin, A. (2010). Linear versus nonlinear signal transmission in neuron models with adaptation currents or dynamic thresholds. *Journal of Neurophysiology*, *104*(5), 2806–2820. doi:10.1152/jn.00240.2010

Benton, M. D., Lewis, A. H., Bant, J. S., & Raman, I. M. (2013). Iberiotoxin-sensitive and -insensitive BK currents in Purkinje neuron somata. *Journal of Neurophysiology*, *109*(10), 2528–2541. doi:10.1152/jn.00127.2012

Bentrop, D., Beyermann, M., Wissmann, R., & Fakler, B. (2001). NMR structure of the "ball-and-chain" domain of KCNMB2, the beta 2-subunit of large conductance Ca2+- and voltage-activated potassium channels. *Journal of Biological Chemistry, 276*(45), 42116–42121. doi:10.1074/jbc.M107118200

Berg, A. P., Sen, N., & Bayliss, D. A. (2007). TrpC3/C7 and Slo2.1 are molecular targets for metabotropic glutamate receptor signaling in rat striatal cholinergic interneurons. *Journal of Neuroscience, 27*(33), 8845–8856. doi:10.1523/JNEUROSCI.0551-07.2007

Berkefeld, H., & Fakler, B. (2008). Repolarizing responses of BKCa-Cav complexes are distinctly shaped by their Cav subunits. *Journal of Neuroscience, 28*(33), 8238–8245. doi:28/33/8238 [pii] 10.1523/JNEUROSCI.2274-08.2008

Berkefeld, H., & Fakler, B. (2013). Ligand-gating by Ca2+ is rate limiting for physiological operation of BK(Ca) channels. *Journal of Neuroscience, 33*(17), 7358–7367. doi:10.1523/JNEUROSCI.5443-12.2013

Berkefeld, H., Fakler, B., & Schulte, U. (2010). Ca2+-activated K+ channels: From protein complexes to function. *Physiological Reviews, 90*(4), 1437–1459. doi:10.1152/physrev.00049.2009

Berkefeld, H., Sailer, C. A., Bildl, W., Rohde, V., Thumfart, J. O., Eble, S., . . . Fakler, B. (2006). BKCa-Cav channel complexes mediate rapid and localized Ca2+-activated K+ signaling. *Science, 314*(5799), 615–620.

Bezanilla, F., & Armstrong, C. M. (1977). Inactivation of the sodium channel. I. Sodium current experiments. *Journal of General Physiology, 70*(5), 549–566.

Bhattacharjee, A., Gan, L., & Kaczmarek, L. K. (2002). Localization of the Slack potassium channel in the rat central nervous system. *Journal of Comparative Neurology, 454*(3), 241–254. doi:10.1002/cne.10439

Bhattacharjee, A., Joiner, W. J., Wu, M., Yang, Y., Sigworth, F. J., & Kaczmarek, L. K. (2003). Slick (Slo2.1), a rapidly-gating sodium-activated potassium channel inhibited by ATP. *Journal of Neuroscience, 23*(37), 11681–11691.

Bhattacharjee, A., & Kaczmarek, L. K. (2005). For K+ channels, Na+ is the new Ca2+. *Trends in Neurosciences, 28*(8), 422–428. doi:10.1016/j.tins.2005.06.003

Bhattacharjee, A., von Hehn, C. A., Mei, X., & Kaczmarek, L. K. (2005). Localization of the Na+-activated K+ channel Slick in the rat central nervous system. *Journal of Comparative Neurology, 484*(1), 80–92. doi:10.1002/cne.20462

Bian, S., Favre, I., & Moczydlowski, E. (2001). Ca2+-binding activity of a COOH-terminal fragment of the Drosophila BK channel involved in Ca2+-dependent activation. *Proceedings of the National Academy of Sciences of the United States of America, 98*(8), 4776–4781. doi:10.1073/pnas.081072398

Bielefeldt, K., & Jackson, M. B. (1993). A calcium-activated potassium channel causes frequency-dependent action-potential failures in a mammalian nerve terminal. *Journal of Neurophysiology, 70*(1), 284–298.

Bielefeldt, K., Rotter, J. L., & Jackson, M. B. (1992). Three potassium channels in rat posterior pituitary nerve terminals. *Journal of Physiology, 458*, 41–67.

Bischoff, U., Vogel, W., & Safronov, B. V. (1998). Na+-activated K+ channels in small dorsal root ganglion neurones of rat. *Journal of Physiology, 510 (Pt 3)*, 743–754.

Blatz AL, M. K. (1987). Calcium-activated potassium channels. *Trends in Neurosciences, 10*, 463–467.

Brandle, U., Frohnmayer, S., Krieger, T., Zenner, H. P., Ruppersberg, J. P., & Maassen, M. M. (2001). Expression of Ca(2+)-activated K(+) channel subunits and splice variants in the rat cochlea. *Hearing Research, 161*(1–2), 23–28.

Brelidze, T. I., & Magleby, K. L. (2005). Probing the geometry of the inner vestibule of BK channels with sugars. *Journal of General Physiology, 126*(2), 105–121. doi:10.1085/jgp.200509286

Brelidze, T. I., Niu, X., & Magleby, K. L. (2003). A ring of eight conserved negatively charged amino acids doubles the conductance of BK channels and prevents inward rectification. *Proceedings of the National Academy of Sciences of the United States of America, 100*(15), 9017–9022. doi:10.1073/pnas.1532257100

Brenker, C., Zhou, Y., Muller, A., Echeverry, F. A., Trotschel, C., Poetsch, A., . . . Strunker, T. (2014). The Ca2+-activated K+ current of human sperm is mediated by Slo3. *Elife, 3*, e01438. doi:10.7554/eLife.01438

Brenner, R., Chen, Q. H., Vilaythong, A., Toney, G. M., Noebels, J. L., & Aldrich, R. W. (2005). BK channel beta4 subunit reduces dentate gyrus excitability and protects against temporal lobe seizures. *Nature Neuroscience, 8*(12), 1752–1759. doi:10.1038/nn1573

Brenner, R., Jegla, T. J., Wickenden, A., Liu, Y., & Aldrich, R. W. (2000). Cloning and functional characterization of novel large conductance calcium-activated potassium channel beta subunits, hKCNMB3 and hKCNMB4. *Journal of Biological Chemistry, 275*(9), 6453–6461.

Brown, M. R., Kronengold, J., Gazula, V. R., Chen, Y., Strumbos, J. G., Sigworth, F. J., . . . Kaczmarek, L. K. (2010). Fragile X mental retardation protein controls gating of the sodium-activated potassium channel Slack. *Nature Neuroscience, 13*(7), 819–821. doi:10.1038/nn.2563

Brown, M. R., Kronengold, J., Gazula, V. R., Spilianakis, C. G., Flavell, R. A., von Hehn, C. A., . . . Kaczmarek, L. K. (2008). Amino-termini isoforms of the Slack K+ channel, regulated by alternative promoters, differentially modulate rhythmic firing and adaptation. *Journal of Physiology, 586*(21), 5161–5179. doi:10.1113/jphysiol.2008.160861

Budelli, G., Hage, T. A., Wei, A., Rojas, P., Jong, Y. J., O'Malley, K., & Salkoff, L. (2009). Na+-activated K+ channels express a large delayed outward current in neurons during normal physiology. *Nature Neuroscience, 12*(6), 745–750. doi:10.1038/nn.2313

Bukiya, A. N., Liu, J. X., & Dopico, A. M. (2009). The BK channel accessory beta(1) subunit determines alcohol-induced cerebrovascular constriction. *FEBS Letters, 583* (17), 2779–2784. doi:10.1016/j.febslet.2009.07.019

Burns, L. C., Minster, R. L., Demirci, F. Y., Barmada, M. M., Ganguli, M., Lopez, O. L., . . . Kamboh, M. I. (2011). Replication study of genome-wide associated SNPs with late-onset Alzheimer's disease. *American Journal of Medical Genetics Part B, Neuropsychiatric Genetics, 156B*(4), 507–512. doi:10.1002/ajmg.b.31194

Burrone, J., & Lagnado, L. (1997). Electrical resonance and Ca2+ influx in the synaptic terminal of depolarizing bipolar cells from the goldfish retina. *Journal of Physiology, 505* (*Pt 3*), 571–584.

Butler, A., Tsunoda, S., McCobb, D. P., Wei, A., & Salkoff, L. (1993). mSlo, a complex mouse gene encoding "maxi" calcium-activated potassium channels. *Science, 261*(5118), 221–224.

Cangiano, L., Wallen, P., & Grillner, S. (2002). Role of apamin-sensitive k(ca) channels for reticulospinal synaptic transmission to motoneuron and for the afterhyperpolarization. *Journal of Neurophysiology, 88*(1), 289–299.

Cao, X. H., Chen, S. R., Li, L., & Pan, H. L. (2012). Nerve injury increases brain-derived neurotrophic factor levels to suppress BK channel activity in primary sensory neurons. *Journal of Neurochemistry, 121*(6), 944–953. doi:10.1111/j.1471-4159.2012.07736.x

Carrasquel-Ursulaez, W., Contreras, G. F., Sepulveda, R. V., Aguayo, D., Gonzalez-Nilo, F., Gonzalez, C., & Latorre, R. (2015). Hydrophobic interaction between contiguous residues in the S6 transmembrane segment acts as a stimuli integration node in the BK channel. *Journal of General Physiology, 145*(1), 61–74. doi:10.1085/jgp.201411194

Carvacho, I., Gonzalez, W., Torres, Y. P., Brauchi, S., Alvarez, O., Gonzalez-Nilo, F. D., & Latorre, R. (2008). Intrinsic electrostatic potential in the BK channel pore: Role in determining single channel conductance and block. *Journal of General Physiology, 131*(2), 147–161. doi:10.1085/jgp.200709862

Castillo, P. E., Weisskopf, M. G., & Nicoll, R. A. (1994). The role of Ca2+ channels in hippocampal mossy fiber synaptic transmission and long-term potentiation. *Neuron, 12*(2), 261–269.

Cavalleri, G. L., Weale, M. E., Shianna, K. V., Singh, R., Lynch, J. M., Grinton, B., ... Goldstein, D. B. (2007). Multicentre search for genetic susceptibility loci in sporadic epilepsy syndrome and seizure types: A case-control study. *Lancet Neurology, 6* (11), 970–980. doi:10.1016/S1474-4422(07)70247-8

Cavelier, P., Pouille, F., Desplantez, T., Beekenkamp, H., & Bossu, J. L. (2002). Control of the propagation of dendritic low-threshold Ca(2+) spikes in Purkinje cells from rat cerebellar slice cultures. *Journal of Physiology, 540*(Pt 1), 57–72.

Cingolani, L. A., Gymnopoulos, M., Boccaccio, A., Stocker, M., & Pedarzani, P. (2002). Developmental regulation of small-conductance Ca2+-activated K+ channel expression and function in rat Purkinje neurons. *Journal of Neuroscience, 22*(11), 4456–4467. doi:20026415

Contet, C., Goulding, S. P., Kuljis, D. A., & Barth, A. L. (2016). BK channels in the central nervous system. *International Review of Neurobiology, 128*, 281–342. doi:10.1016/bs.irn.2016.04.001

Cox, D. H. (2005). The BKCa channel's Ca2+-binding sites, multiple sites, multiple ions. *Journal of General Physiology, 125*(3), 253–255. doi:10.1085/jgp.200509270

Crowley, J. J., Treistman, S. N., & Dopico, A. M. (2003). Cholesterol antagonizes ethanol potentiation of human brain BKCa channels reconstituted into phospholipid bilayers. *Molecular Pharmacology, 64*(2), 365–372. doi:10.1124/mol.64.2.365

Chamberlain, L. H., & Burgoyne, R. D. (2000). Cysteine-string protein: The chaperone at the synapse. *Journal of Neurochemistry, 74*(5), 1781–1789.

Chang, C. P., Dworetzky, S. I., Wang, J., & Goldstein, M. E. (1997). Differential expression of the alpha and beta subunits of the large-conductance calcium-activated potassium channel: Implication for channel diversity. *Brain Research, Molecular Brain Research, 45*(1), 33–40.

Chen, H., Kronengold, J., Yan, Y., Gazula, V. R., Brown, M. R., Ma, L., ... Kaczmarek, L. K. (2009). The N-terminal domain of Slack determines the formation and trafficking of Slick/Slack heteromeric sodium-activated potassium channels. *Journal of Neuroscience, 29*(17), 5654–5665. doi:10.1523/JNEUROSCI.5978-08.2009

Chen, L., Bi, D., Tian, L., McClafferty, H., Steeb, F., Ruth, P., ... Shipston, M. J. (2013). Palmitoylation of the beta4-subunit regulates surface expression of large conductance calcium-activated potassium channel splice variants. *Journal of Biological Chemistry, 288*(18), 13136–13144. doi:10.1074/jbc.M113.461830

Chen, S. R., Cai, Y. Q., & Pan, H. L. (2009). Plasticity and emerging role of BKCa channels in nociceptive control in neuropathic pain. *Journal of Neurochemistry, 110*(1), 352–362. doi:10.1111/j.1471-4159.2009.06138.x

Chen, X., Yan, J., & Aldrich, R. W. (2014). BK channel opening involves side-chain reorientation of multiple deep-pore residues. *Proceedings of the National Academy of Sciences of the United States of America, 111*(1), E79–88. doi:10.1073/pnas.1321697111

Chishti, M. A., Yang, D. S., Janus, C., Phinney, A. L., Horne, P., Pearson, J., . . . Westaway, D. (2001). Early-onset amyloid deposition and cognitive deficits in transgenic mice expressing a double mutant form of amyloid precursor protein 695. *Journal of Biological Chemistry, 276*(24), 21562–21570. doi:10.1074/jbc.M100710200

Chu, B., Dopico, A. M., Lemos, J. R., & Treistman, S. N. (1998). Ethanol potentiation of calcium-activated potassium channels reconstituted into planar lipid bilayers. *Molecular Pharmacology, 54*(2), 397–406.

Dai, L., Garg, V., & Sanguinetti, M. C. (2010). Activation of Slo2.1 channels by niflumic acid. *Journal of General Physiology, 135*(3), 275–295. doi:10.1085/jgp.200910316

Dale, N. (1993). A large, sustained Na(+)- and voltage-dependent K+ current in spinal neurons of the frog embryo. *Journal of Physiology, 462*, 349–372.

del Camino, D., & Yellen, G. (2001). Tight steric closure at the intracellular activation gate of a voltage-gated K(+) channel. *Neuron, 32*(4), 649–656.

Deng, P. Y., Rotman, Z., Blundon, J. A., Cho, Y., Cui, J., Cavalli, V., . . . Klyachko, V. A. (2013). FMRP regulates neurotransmitter release and synaptic information transmission by modulating action potential duration via BK channels. *Neuron, 77*(4), 696–711. doi:10.1016/j.neuron.2012.12.018

Descalzo, V. F., Nowak, L. G., Brumberg, J. C., McCormick, D. A., & Sanchez-Vives, M. V. (2005). Slow adaptation in fast-spiking neurons of visual cortex. *Journal of Neurophysiology, 93*(2), 1111–1118. doi:10.1152/jn.00658.2004

Diaz-Franulic, I., Sepulveda, R. V., Navarro-Quezada, N., Gonzalez-Nilo, F., & Naranjo, D. (2015). Pore dimensions and the role of occupancy in unitary conductance of Shaker K channels. *Journal of General Physiology, 146*(2), 133–146. doi:10.1085/jgp.201411353

Diaz, L., Meera, P., Amigo, J., Stefani, E., Alvarez, O., Toro, L., & Latorre, R. (1998). Role of the S4 segment in a voltage-dependent calcium-sensitive potassium (hSlo) channel. *Journal of Biological Chemistry, 273*(49), 32430–32436.

Dobremez, E., Bouali-Benazzouz, R., Fossat, P., Monteils, L., Dulluc, J., Nagy, F., & Landry, M. (2005). Distribution and regulation of L-type calcium channels in deep dorsal horn neurons after sciatic nerve injury in rats. *European Journal of Neuroscience, 21*(12), 3321–3333. doi:10.1111/j.1460-9568.2005.04177.x

Donnelier, J., Braun, S. T., Dolzhanskaya, N., Ahrendt, E., Braun, A. P., Velinov, M., & Braun, J. E. A. (2015). Increased expression of the large conductance, calcium-activated K+ (BK) channel in adult-onset neuronal ceroid lipofuscinosis. *PLoS One, 10*(4). doi:ARTN e0125205 10.1371/journal.pone.0125205

Dopico, A. M., Lemos, J. R., & Treistman, S. N. (1996). Ethanol increases the activity of large conductance, Ca(2+)-activated K+ channels in isolated neurohypophysial terminals. *Molecular Pharmacology, 49*(1), 40–48.

Dopico, A. M., Widmer, H., Wang, G., Lemos, J. R., & Treistman, S. N. (1999). Rat supraoptic magnocellular neurones show distinct large conductance, Ca2+-activated K+ channel subtypes in cell bodies versus nerve endings. *Journal of Physiology, 519 Pt 1*, 101–114.

Doyle, D. A., Morais Cabral, J., Pfuetzner, R. A., Kuo, A., Gulbis, J. M., Cohen, S. L., . . . MacKinnon, R. (1998). The structure of the potassium channel: Molecular basis of K+ conduction and selectivity. *Science, 280*(5360), 69–77.

Dryer, S. E. (1991). Na(+)-activated K+ channels and voltage-evoked ionic currents in brain stem and parasympathetic neurones of the chick. *Journal of Physiology, 435*, 513–532.

Dryer, S. E. (1994). Na +activated K + channels: A new family of large-conductance ion channels. *Trends in Neurosciences, 17*(4).

Dryer, S. E. (2003). Molecular identification of the Na+-activated K+ channel. *Neuron, 37*(6), 727–730.

Dryer, S. E., Fujii, J. T., & Martin, A. R. (1989). A Na+-activated K+ current in cultured brain stem neurones from chicks. *Journal of Physiology, 410*, 283–296.

Du, W., Bautista, J. F., Yang, H., Diez-Sampedro, A., You, S. A., Wang, L., . . . Wang, Q. K. (2005). Calcium-sensitive potassium channelopathy in human epilepsy and paroxysmal movement disorder. *Nature Genetics, 37*(7), 733–738. doi:10.1038/ng1585

Edenberg, H. J., Koller, D. L., Xuei, X., Wetherill, L., McClintick, J. N., Almasy, L., . . . Foroud, T. (2010). Genome-wide association study of alcohol dependence implicates a region on chromosome 11. *Alcoholism—Clinical and Experimental Research, 34*(5), 840–852. doi:10.1111/j.1530-0277.2010.01156.x

Edgerton, J. R., & Reinhart, P. H. (2003). Distinct contributions of small and large conductance Ca2+-activated K+ channels to rat Purkinje neuron function. *Journal of Physiology, 548*(Pt 1), 53–69. doi:10.1113/jphysiol.2002.027854

Egan, T. M., Dagan, D., Kupper, J., & Levitan, I. B. (1992). Properties and rundown of sodium-activated potassium channels in rat olfactory bulb neurons. *Journal of Neuroscience, 12*(5), 1964–1976.

Engbers, J. D., Anderson, D., Zamponi, G. W., & Turner, R. W. (2013). Signal processing by T-type calcium channel interactions in the cerebellum. *Frontiers in Cellular Neuroscience, 7*, 230. doi:10.3389/fncel.2013.00230

Engel, J., Braig, C., Ruttiger, L., Kuhn, S., Zimmermann, U., Blin, N., . . . Knipper, M. (2006). Two classes of outer hair cells along the tonotopic axis of the cochlea. *Neuroscience, 143*(3), 837–849. doi:10.1016/j.neuroscience.2006.08.060

Evans, M. G. (1996). Acetylcholine activates two currents in guinea-pig outer hair cells. *Journal of Physiology, 491 (Pt 2)*, 563–578.

Evanson, K. W., Bannister, J. P., Leo, M. D., & Jaggar, J. H. (2014). LRRC26 is a functional BK channel auxiliary gamma subunit in arterial smooth muscle cells. *Circulation Research, 115*(4), 423–431. doi:10.1161/CIRCRESAHA.115.303407

Evely, K. M., Pryce, K. D., Bausch, A. E., Lukowski, R., Ruth, P., Haj-Dahmane, S., & Bhattacharjee, A. (2017). [EXPRESS] Slack KNa channels influence dorsal horn synapses and nociceptive behavior. *Molecular Pain, 13*, 1744806917714342. doi:10.1177/1744806917714342

Faber, E. S., & Sah, P. (2003). Ca2+-activated K+ (BK) channel inactivation contributes to spike broadening during repetitive firing in the rat lateral amygdala. *Journal of Physiology, 552*(Pt 2), 483–497. doi:10.1113/jphysiol.2003.050120

Fakler, B., & Adelman, J. P. (2008). Control of K(Ca) channels by calcium nano/microdomains. *Neuron, 59*(6), 873–881. doi:10.1016/j.neuron.2008.09.001

Feinberg-Zadek, P. L., & Treistman, S. N. (2007). Beta-subunits are important modulators of the acute response to alcohol in human BK channels. *Alcoholism: Clinical & Experimental Research, 31*(5), 737–744. doi:10.1111/j.1530-0277.2007.00371.x

Fernandez-Chacon, R., Wolfel, M., Nishimune, H., Tabares, L., Schmitz, F., Castellano-Munoz, M., . . . Sudhof, T. C. (2004). The synaptic vesicle protein CSP alpha prevents presynaptic degeneration. *Neuron, 42*(2), 237–251.

Fettiplace, R., & Fuchs, P. A. (1999). Mechanisms of hair cell tuning. *Annual Review of Physiology, 61*, 809–834. doi:10.1146/annurev.physiol.61.1.809

Filosa, J. A., Bonev, A. D., Straub, S. V., Meredith, A. L., Wilkerson, M. K., Aldrich, R. W., & Nelson, M. T. (2006). Local potassium signaling couples neuronal activity to vasodilation in the brain. *Nature Neuroscience, 9*(11), 1397–1403. doi:10.1038/nn1779

Fodor, A. A., & Aldrich, R. W. (2009). Convergent evolution of alternative splices at domain boundaries of the BK channel. *Annual Review of Physiology, 71*, 19–36. doi:10.1146/annurev.physiol.010908.163124

Foehring, R. C., Schwindt, P. C., & Crill, W. E. (1989). Norepinephrine selectively reduces slow Ca2+- and Na+-mediated K+ currents in cat neocortical neurons. *Journal of Neurophysiology, 61*(2), 245–256.

Franceschetti, S., Lavazza, T., Curia, G., Aracri, P., Panzica, F., Sancini, G., . . . Magistretti, J. (2003). Na+-activated K+ current contributes to postexcitatory hyperpolarization in neocortical intrinsically bursting neurons. *Journal of Neurophysiology, 89*(4), 2101–2111. doi:10.1152/jn.00695.2002

Fuchs, P. A., & Murrow, B. W. (1992). Cholinergic inhibition of short (outer) hair cells of the chick's cochlea. *Journal of Neuroscience, 12*(3), 800–809.

Furukawa, N., Takasusuki, T., Fukushima, T., & Hori, Y. (2008). Presynaptic large-conductance calcium-activated potassium channels control synaptic transmission in the superficial dorsal horn of the mouse. *Neuroscience Letters, 444* (1), 79–82. doi:10.1016/j.neulet.2008.08.022

Gao, S. B., Wu, Y., Lu, C. X., Guo, Z. H., Li, C. H., & Ding, J. P. (2008). Slack and Slick KNa channels are required for the depolarizing afterpotential of acutely isolated, medium diameter rat dorsal root ganglion neurons. *Acta Pharmacologica Sinica, 29*(8), 899–905. doi:10.1111/j.1745-7254.2008.00842.x

Garg, P., Gardner, A., Garg, V., & Sanguinetti, M. C. (2013). Structural basis of ion permeation gating in Slo2.1 K+ channels. *Journal of General Physiology, 142*(5), 523–542. doi:10.1085/jgp.201311064

Garg, P., & Sanguinetti, M. C. (2014). Intracellular ATP does not inhibit Slo2.1 K+ channels. *Physiological Reports, 2*(9). doi:10.14814/phy2.12118

Geng, Y., Ferreira, J. J., Dzikunu, V., Butler, A., Lybaert, P., Yuan, P., . . . Santi, C. M. (2017). A genetic variant of the sperm-specific SLO3 K+ channel has altered pH and Ca2+ sensitivities. *Journal of Biological Chemistry, 292*(21), 8978–8987. doi:10.1074/jbc.M117.776013

Geng, Y., Niu, X., & Magleby, K. L. (2011). Low resistance, large dimension entrance to the inner cavity of BK channels determined by changing side-chain volume. *Journal of General Physiology, 137*(6), 533–548. doi:10.1085/jgp.201110616

Ghatta, S., Nimmagadda, D., Xu, X., & O'Rourke, S. T. (2006). Large-conductance, calcium-activated potassium channels: Structural and functional implications. *Pharmacology & Therapeutics, 110*(1), 103–116. doi:S0163-7258(05)00226-3 [pii] 10.1016/j.pharmthera.2005.10.007

Giese, M. H., Gardner, A., Hansen, A., & Sanguinetti, M. C. (2017). Molecular mechanisms of Slo2 K+ channel closure. *Journal of Physiology, 595*(7), 2321–2336. doi:10.1113/JP273225

Gribkoff, V. K., Starrett, J. E., Jr., Dworetzky, S. I., Hewawasam, P., Boissard, C. G., Cook, D. A., . . . Yeola, S. W. (2001). Targeting acute ischemic stroke with a calcium-sensitive opener of maxi-K potassium channels. *Nature Medicine, 7*(4), 471–477. doi:10.1038/86546

Grimes, W. N., Li, W., Chavez, A. E., & Diamond, J. S. (2009). BK channels modulate pre- and postsynaptic signaling at reciprocal synapses in retina. *Nature Neuroscience, 12*(5), 585–592. doi:10.1038/nn.2302

Grunnet, M., & Kaufmann, W. A. (2004). Coassembly of big conductance Ca2+-activated K+ channels and L-type voltage-gated Ca2+ channels in rat brain. *Journal of Biological Chemistry, 279*(35), 36445–36453. doi:10.1074/jbc.M402254200

Gruss, M., Hempelmann, G., & Scholz, A. (2002). Trichloroethanol alters action potentials in a subgroup of primary sensory neurones. *Neuroreport, 13*(6), 853–856.

Gruss, M., Henrich, M., Konig, P., Hempelmann, G., Vogel, W., & Scholz, A. (2001). Ethanol reduces excitability in a subgroup of primary sensory neurons by activation of BK(Ca) channels. *European Journal of Neuroscience, 14*(8), 1246–1256.

Gururaj, S., Evely, K. M., Pryce, K. D., Li, J., Qu, J., & Bhattacharjee, A. (2017). Protein Kinase A induced internalization of Slack channels from the neuronal membrane occurs by Adaptor Protein-2/Clathrin mediated endocytosis. *Journal of Biological Chemistry*. doi:10.1074/jbc.M117.804716

Gururaj, S., Palmer, E. E., Sheehan, G. D., Kandula, T., Macintosh, R., Ying, K., . . . Bhattacharjee, A. (2017). A de novo mutation in the sodium-activated potassium channel KCNT2 alters ion selectivity and causes epileptic encephalopathy. *Cell Reports, 21* (4), 926–933. doi:10.1016/j.celrep.2017.09.088

Ha, T. S., Heo, M. S., & Park, C. S. (2004). Functional effects of auxiliary beta4-subunit on rat large-conductance Ca(2+)-activated K(+) channel. *Biophysical Journal, 86* (5), 2871–2882. doi:10.1016/S0006-3495(04)74339-8

Hage, T. A., & Salkoff, L. (2012). Sodium-activated potassium channels are functionally coupled to persistent sodium currents. *Journal of Neuroscience, 32*(8), 2714–2721. doi:10.1523/JNEUROSCI.5088-11.2012

Haghdoost-Yazdi, H., Janahmadi, M., & Behzadi, G. (2008). Iberiotoxin-sensitive large conductance Ca2+ -dependent K+ (BK) channels regulate the spike configuration in the burst firing of cerebellar Purkinje neurons. *Brain Research, 1212* , 1–8. doi:10.1016/j.brainres.2008.03.030

Haimann, C., & Bader, C. R. (1989). Sodium-activated potassium channel in avian sensory neurons. *Cell Biology International Reports, 13*(12), 1133–1139.

Haimann, C., Magistretti, J., & Pozzi, B. (1992). Sodium-activated potassium current in sensory neurons: A comparison of cell-attached and cell-free single-channel activities. *Pflugers Archives, 422* (3), 287–294.

Han, S. Z., Yang, B. Z., Kranzler, H. R., Liu, X. M., Zhao, H. Y., Farrer, L. A., . . . Gelemter, J. (2013). Integrating GWASs and human protein interaction networks identifies a gene subnetwork underlying alcohol dependence. *American Journal of Human Genetics, 93*(6), 1027–1034. doi:10.1016/j.ajhg.2013.10.021

Hartung, K. (1985). Potentiation of a transient outward current by Na+ influx in crayfish neurones. *Pflugers Archives, 404* (1), 41–44.

Haug, T., Olcese, R., Toro, L., & Stefani, E. (2004). Regulation of K+ flow by a ring of negative charges in the outer pore of BKCa channels. Part II: Neutralization of aspartate 292 reduces long channel openings and gating current slow component. *Journal of General Physiology, 124*(2), 185–197. doi:10.1085/jgp.200308950

Haug, T., Sigg, D., Ciani, S., Toro, L., Stefani, E., & Olcese, R. (2004). Regulation of K+ flow by a ring of negative charges in the outer pore of BKCa channels. Part I: Aspartate 292 modulates K+ conduction by external surface charge effect. *Journal of General Physiology, 124*(2), 173–184. doi:10.1085/jgp.200308949

Heginbotham, L., Lu, Z., Abramson, T., & MacKinnon, R. (1994). Mutations in the K+ channel signature sequence. *Biophysical Journal, 66* (4), 1061–1067. doi:10.1016/S0006-3495(94)80887-2

Henne, J., & Jeserich, G. (2004). Maturation of spiking activity in trout retinal ganglion cells coincides with upregulation of Kv3.1- and BK-related potassium channels. *Journal of Neuroscience Research, 75* (1), 44–54. doi:10.1002/jnr.10830

Heron, S. E., Smith, K. R., Bahlo, M., Nobili, L., Kahana, E., Licchetta, L., . . . Dibbens, L. M. (2012). Missense mutations in the sodium-gated potassium channel gene KCNT1 cause severe autosomal dominant nocturnal frontal lobe epilepsy. *Nature Genetics, 44*(11), 1188–1190. doi:10.1038/ng.2440

Hess, D., Nanou, E., & El Manira, A. (2007). Characterization of Na+-activated K+ currents in larval lamprey spinal cord neurons. *Journal of Neurophysiology, 97*(5), 3484–3493. doi:10.1152/jn.00742.2006

Hicks, G. A., & Marrion, N. V. (1998). Ca2+-dependent inactivation of large conductance Ca2+ -activated K+ (BK) channels in rat hippocampal neurones produced by pore block from an associated particle. *Journal of Physiology, 508 (Pt 3)*, 721–734.

Higgins, J. J., Hao, J., Kosofsky, B. E., & Rajadhyaksha, A. M. (2008). Dysregulation of large-conductance Ca(2+)-activated K(+) channel expression in nonsyndromal mental retardation due to a cereblon p.R419X mutation. *Neurogenetics, 9*(3), 219–223. doi:10.1007/s10048-008-0128-2

Hite, R. K., & MacKinnon, R. (2017). Structural titration of Slo2.2, a Na+-dependent K+ channel. *Cell, 168*(3), 390–399 e311. doi:10.1016/j.cell.2016.12.030

Hite, R. K., Tao, X., & MacKinnon, R. (2017). Structural basis for gating the high-conductance Ca2+-activated K+ channel. *Nature, 541*(7635), 52–57. doi:10.1038/nature20775

Hite, R. K., Yuan, P., Li, Z., Hsuing, Y., Walz, T., & MacKinnon, R. (2015). Cryo-electron microscopy structure of the Slo2.2 Na(+)-activated K(+) channel. *Nature, 527*(7577), 198–203. doi:10.1038/nature14958

Hockberger, P. E., Tseng, H. Y., & Connor, J. A. (1989). Development of rat cerebellar Purkinje cells: Electrophysiological properties following acute isolation and in long-term culture. *Journal of Neuroscience, 9*(7), 2258–2271.

Holmgren, M., Smith, P. L., & Yellen, G. (1997). Trapping of organic blockers by closing of voltage-dependent K+ channels: Evidence for a trap door mechanism of activation gating. *Journal of General Physiology, 109*(5), 527–535.

Horrigan, F. T., & Aldrich, R. W. (2002). Coupling between voltage sensor activation, Ca2+ binding and channel opening in large conductance (BK) potassium channels. *Journal of General Physiology, 120*(3), 267–305.

Hoshi, T., Pantazis, A., & Olcese, R. (2013). Transduction of voltage and Ca2+ signals by Slo1 BK channels. *Physiology, 28*(3), 172–189. doi:10.1152/physiol.00055.2012

Housley, G. D., & Ashmore, J. F. (1991). Direct measurement of the action of acetylcholine on isolated outer hair cells of the guinea pig cochlea. *Proceedings. Biological Sciences, 244*(1310), 161–167. doi:10.1098/rspb.1991.0065

Hu, H., Shao, L. R., Chavoshy, S., Gu, N., Trieb, M., Behrens, R., . . . Storm, J. F. (2001). Presynaptic Ca2+-activated K+ channels in glutamatergic hippocampal terminals and their role in spike repolarization and regulation of transmitter release. *Journal of Neuroscience, 21*(24), 9585–9597. doi:21/24/9585 [pii]

Hu, S., Labuda, M. Z., Pandolfo, M., Goss, G. G., McDermid, H. E., & Ali, D. W. (2003). Variants of the KCNMB3 regulatory subunit of maxi BK channels affect channel inactivation. *Physiological Genomics, 15*(3), 191–198. doi:10.1152/physiolgenomics.00110.2003

Humphrey, W., Dalke, A., & Schulten, K. (1996). VMD: Visual molecular dynamics. *Journal of Molecular Graphics, 14*(1), 33–38, 27–38.

Hurst, R. S., Toro, L., & Stefani, E. (1996). Molecular determinants of external barium block in Shaker potassium channels. *FEBS Letters, 388* (1), 59–65.

Ikemoto, Y., Ono, K., Yoshida, A., & Akaike, N. (1989). Delayed activation of large-conductance Ca2+-activated K channels in hippocampal neurons of the rat. *Biophysical Journal, 56* (1), 207–212. doi:10.1016/S0006-3495(89)82665-7

Imlach, W. L., Finch, S. C., Dunlop, J., Meredith, A. L., Aldrich, R. W., & Dalziel, J. E. (2008). The molecular mechanism of "ryegrass staggers," a neurological disorder of K+ channels. *Journal of Pharmacology & Experimental Therapeutics, 327*(3), 657–664. doi:10.1124/jpet.108.143933

Indriati, D. W., Kamasawa, N., Matsui, K., Meredith, A. L., Watanabe, M., & Shigemoto, R. (2013). Quantitative localization of Cav2.1 (P/Q-type) voltage-dependent calcium channels in Purkinje cells: Somatodendritic gradient and distinct somatic coclustering with calcium-activated potassium channels. *Journal of Neuroscience, 33*(8), 3668–3678. doi:10.1523/JNEUROSCI.2921-12.2013

Jafari, A., Noursadeghi, E., Khodagholi, F., Saghiri, R., Sauve, R., Aliaghaei, A., & Eliassi, A. (2015). Brain mitochondrial ATP-insensitive large conductance Ca(+)(2)-activated K(+) channel properties are altered in a rat model of amyloid-beta neurotoxicity. *Experimental Neurology, 269*, 8–16. doi:10.1016/j.expneurol.2014.12.024

Javaherian, A. D., Yusifov, T., Pantazis, A., Franklin, S., Gandhi, C. S., & Olcese, R. (2011). Metal-driven operation of the human large-conductance voltage- and Ca2+-dependent potassium channel (BK) gating ring apparatus. *Journal of Biological Chemistry, 286*(23), 20701–20709. doi:10.1074/jbc.M111.235234

Jiang, Y., Lee, A., Chen, J., Cadene, M., Chait, B. T., & MacKinnon, R. (2002). The open pore conformation of potassium channels. *Nature, 417*(6888), 523–526. doi:10.1038/417523a

Jiang, Y., Pico, A., Cadene, M., Chait, B. T., & MacKinnon, R. (2001). Structure of the RCK domain from the E. coli K+ channel and demonstration of its presence in the human BK channel. *Neuron, 29*(3), 593–601.

Jiang, Z., Wallner, M., Meera, P., & Toro, L. (1999). Human and rodent MaxiK channel beta-subunit genes: Cloning and characterization. *Genomics, 55*(1), 57–67. doi:10.1006/geno.1998.5627

Jo, S., Lee, K. H., Song, S., Jung, Y. K., & Park, C. S. (2005). Identification and functional characterization of cereblon as a binding protein for large-conductance calcium-activated potassium channel in rat brain. *Journal of Neurochemistry, 94*(5), 1212–1224. doi:10.1111/j.1471-4159.2005.03344.x

Joiner, W. J., Tang, M. D., Wang, L. Y., Dworetzky, S. I., Boissard, C. G., Gan, L., . . . Kaczmarek, L. K. (1998). Formation of intermediate-conductance calcium-activated potassium channels by interaction of Slack and Slo subunits. *Nature Neuroscience, 1*(6), 462–469. doi:10.1038/2176

Kaczmarek, L. K. (2013). Slack, Slick and sodium-activated potassium channels. *ISRN Neuroscience, 2013* (2013). doi:10.1155/2013/354262

Kaczmarek, L. K., Aldrich, R. W., Chandy, K. G., Grissmer, S., Wei, A. D., & Wulff, H. (2017). International Union of Basic and Clinical Pharmacology. C. Nomenclature and properties of calcium-activated and sodium-activated potassium channels. *Pharmacological Reviews, 69*(1), 1–11. doi:10.1124/pr.116.012864

Kameyama, M., Kakei, M., Sato, R., Shibasaki, T., Matsuda, H., & Irisawa, H. (1984). Intracellular Na+ activates a K+ channel in mammalian cardiac cells. *Nature, 309*(5966), 354–356.

Katsuki, H., Shinohara, A., Fujimoto, S., Kume, T., & Akaike, A. (2005). Tetraethylammonium exacerbates ischemic neuronal injury in rat cerebrocortical slice cultures. *European Journal of Pharmacology, 508*(1–3), 85–91. doi:10.1016/j.ejphar.2004.11.058

Kaufmann, W. A., Ferraguti, F., Fukazawa, Y., Kasugai, Y., Shigemoto, R., Laake, P., ... Ottersen, O. P. (2009). Large-conductance calcium-activated potassium channels in Purkinje cell plasma membranes are clustered at sites of hypolemmal microdomains. *Journal of Comparative Neurology*, *515*(2), 215–230. doi:10.1002/cne.22066

Kaufmann, W. A., Kasugai, Y., Ferraguti, F., & Storm, J. F. (2010). Two distinct pools of large-conductance calcium-activated potassium channels in the somatic plasma membrane of central principal neurons. *Neuroscience*, *169*(3), 974–986. doi:10.1016/j.neuroscience.2010.05.070

Kendler, K. S., Kalsi, G., Holmans, P. A., Sanders, A. R., Aggen, S. H., Dick, D. M., ... Gejman, P. V. (2011). Genomewide association analysis of symptoms of alcohol dependence in the Molecular Genetics of Schizophrenia (MGS2) control sample. *Alcoholism—Clinical and Experimental Research*, *35*(5), 963–975. doi:10.1111/j.1530-0277.2010.01427.x

Kim, U., & McCormick, D. A. (1998). Functional and ionic properties of a slow afterhyperpolarization in ferret perigeniculate neurons in vitro. *Journal of Neurophysiology*, *80*(3), 1222–1235.

Knaus, H. G., Folander, K., Garcia-Calvo, M., Garcia, M. L., Kaczorowski, G. J., Smith, M., & Swanson, R. (1994). Primary sequence and immunological characterization of beta-subunit of high conductance Ca(2+)-activated K+ channel from smooth muscle. *Journal of Biological Chemistry*, *269*(25), 17274–17278.

Knaus, H. G., Schwarzer, C., Koch, R. O., Eberhart, A., Kaczorowski, G. J., Glossmann, H., ... Sperk, G. (1996). Distribution of high-conductance Ca(2+)-activated K+ channels in rat brain: Targeting to axons and nerve terminals. *Journal of Neuroscience*, *16*(3), 955–963.

Knott, T. K., Dopico, A. M., Dayanithi, G., Lemos, J., & Treistman, S. N. (2002). Integrated channel plasticity contributes to alcohol tolerance in neurohypophysial terminals. *Molecular Pharmacology*, *62*(1), 135–142.

Koh, D. S., Jonas, P., & Vogel, W. (1994). Na(+)-activated K+ channels localized in the nodal region of myelinated axons of Xenopus. *Journal of Physiology*, *479* (*Pt 2*), 183–197.

Kreifeldt, M., Cates-Gatto, C., Roberts, A. J., & Contet, C. (2015). BK Channel beta 1 subunit contributes to behavioral adaptations elicited by chronic intermittent ethanol exposure. *Alcoholism—Clinical and Experimental Research*, *39*(12), 2394–2402. doi:10.1111/acer.12911

Kreifeldt, M., Le, D., Treistman, S. N., Koob, G. F., & Contet, C. (2013). BK channel beta1 and beta4 auxiliary subunits exert opposite influences on escalated ethanol drinking in dependent mice. *Frontiers in Integrative Neuroscience*, *7*, 105. doi:10.3389/fnint.2013.00105

Kubota, M., & Saito, N. (1991). Sodium- and calcium-dependent conductances of neurones in the zebra finch hyperstriatum ventrale pars caudale in vitro. *Journal of Physiology*, *440*, 131–142.

Kulik, A., Nakadate, K., Hagiwara, A., Fukazawa, Y., Lujan, R., Saito, H., ... Shigemoto, R. (2004). Immunocytochemical localization of the alpha 1A subunit of the P/Q-type calcium channel in the rat cerebellum. *European Journal of Neuroscience*, *19*(8), 2169–2178. doi:10.1111/j.0953-816X.2004.03319.x

Kurt, S., Sausbier, M., Ruttiger, L., Brandt, N., Moeller, C. K., Kindler, J., ... Schulze, H. (2012). Critical role for cochlear hair cell BK channels for coding the temporal structure and dynamic range of auditory information for central auditory processing. *The FASEB Journal*, *26*(9), 3834–3843. doi:10.1096/fj.11-200535

Kyle, B. D., Ahrendt, E., Braun, A. P., & Braun, J. E. (2013). The large conductance, calcium-activated K+ (BK) channel is regulated by cysteine string protein. *Scientific Reports*, *3* , 2447. doi:10.1038/srep02447

Lancaster, B., & Nicoll, R. A. (1987). Properties of two calcium-activated hyperpolarizations in rat hippocampal neurones. *Journal of Physiology*, *389*, 187–203.

Langer, P., Grunder, S., & Rusch, A. (2003). Expression of Ca2+-activated BK channel mRNA and its splice variants in the rat cochlea. *Journal of Comparative Neurology*, *455*(2), 198–209. doi:10.1002/cne.10471

Latorre, R., & Brauchi, S. (2006). Large conductance Ca2+-activated K+ (BK) channel: Activation by Ca2+ and voltage. *Biological Research*, *39*(3), 385–401. doi:/S0716-97602006000300003

Latorre, R., Castillo, K., Carrasquel-Ursulaez, W., Sepulveda, R. V., Gonzalez-Nilo, F., Gonzalez, C., & Alvarez, O. (2017). Molecular determinants of BK channel functional diversity and functioning. *Physiological Reviews*, *97*(1), 39–87. doi:10.1152/physrev.00001.2016

Latorre, R., & Miller, C. (1983). Conduction and selectivity in potassium channels. *Journal of Membrane Biology*, *71*(1–2), 11–30.

Latorre, R., Vergara, C., & Hidalgo, C. (1982). Reconstitution in planar lipid bilayers of a Ca2+ -dependent K+ channel from transverse tubule membranes isolated from rabbit skeletal muscle. *Proceedings of the National Academy of Sciences of the United States of America*, *79*(3), 805–809.

Laumonnier, F., Roger, S., Guerin, P., Molinari, F., M'Rad, R., Cahard, D., . . . Briault, S. (2006). Association of a functional deficit of the BKCa channel, a synaptic regulator of neuronal excitability, with autism and mental retardation. *American Journal of Psychiatry*, *163*(9), 1622–1629. doi: 10.1176/appi.ajp.163.9.1622

Lein, E. S., Hawrylycz, M. J., Ao, N., Ayres, M., Bensinger, A., Bernard, A., . . . Jones, A. R. (2007). Genome-wide atlas of gene expression in the adult mouse brain. *Nature*, *445*(7124), 168–176. doi:10.1038/nature05453

Li, B., & Gao, T. M. (2016). Functional role of mitochondrial and nuclear BK channels. *International Review of Neurobiology*, *128*, 163–191. doi:10.1016/bs.irn.2016.03.018

Li, W., & Aldrich, R. W. (2006). State-dependent block of BK channels by synthesized shaker ball peptides. *Journal of General Physiology*, *128*(4), 423–441. doi:10.1085/jgp.200609521

Li, W., Gao, S. B., Lv, C. X., Wu, Y., Guo, Z. H., Ding, J. P., & Xu, T. (2007). Characterization of voltage-and Ca2+-activated K+ channels in rat dorsal root ganglion neurons. *Journal of Cellular Physiology*, *212*(2), 348–357. doi:10.1002/jcp.21007

Liao, Y., Kristiansen, A. M., Oksvold, C. P., Tuvnes, F. A., Gu, N., Runden-Pran, E., . . . Storm, J. F. (2010). Neuronal Ca2+-activated K+ channels limit brain infarction and promote survival. *PLoS One*, *5*(12), e15601. doi:10.1371/journal.pone.0015601

Lippiat, J. D., Standen, N. B., & Davies, N. W. (2000). A residue in the intracellular vestibule of the pore is critical for gating and permeation in Ca2+-activated K+ (BKCa) channels. *Journal of Physiology*, *529 Pt 1*, 131–138.

Liu, J. Y., Ye, J., Zou, X. L., Xu, Z. H., Feng, Y., Zou, X. X., . . . Cang, Y. (2014). CRL4A(CRBN) E3 ubiquitin ligase restricts BK channel activity and prevents epileptogenesis. *Nature Communications*, *5*. doi:10.1038/Ncomms4924

Liu, X., & Stan Leung, L. (2004). Sodium-activated potassium conductance participates in the depolarizing afterpotential following a single action potential in rat hippocampal CA1 pyramidal cells. *Brain Research*, *1023* (2), 185–192. doi:10.1016/j.brainres.2004.07.017

Liu, Y., Holmgren, M., Jurman, M. E., & Yellen, G. (1997). Gated access to the pore of a voltage-dependent K+ channel. *Neuron*, *19*(1), 175–184.

Loane, D. J., Lima, P. A., & Marrion, N. V. (2007). Co-assembly of N-type Ca2+ and BK channels underlies functional coupling in rat brain. *Journal of Cell Science*, *120*(Pt 6), 985–995. doi:10.1242/jcs.03399

Long, S. B., Tao, X., Campbell, E. B., & MacKinnon, R. (2007). Atomic structure of a voltage-dependent K+ channel in a lipid membrane-like environment. *Nature, 450*(7168), 376–382. doi:10.1038/nature06265

Lorenz, S., Heils, A., Kasper, J. M., & Sander, T. (2007). Allelic association of a truncation mutation of the KCNMB3 gene with idiopathic generalized epilepsy. *American Journal of Medical Genetics Part B, Neuropsychiatric Genetics, 144B*(1), 10–13. doi:10.1002/ajmg.b.30369

Lu, R., Bausch, A. E., Kallenborn-Gerhardt, W., Stoetzer, C., Debruin, N., Ruth, P., . . . Schmidtko, A. (2015). Slack channels expressed in sensory neurons control neuropathic pain in mice. *Journal of Neuroscience, 35*(3), 1125–1135. doi:10.1523/JNEUROSCI.2423-14.2015

Lu, R., Lukowski, R., Sausbier, M., Zhang, D. D., Sisignano, M., Schuh, C. D., ... Schmidtko, A. (2014). BKCa channels expressed in sensory neurons modulate inflammatory pain in mice. *Pain, 155*(3), 556–565. doi:10.1016/j.pain.2013.12.005

Lu, S., Das, P., Fadool, D. A., & Kaczmarek, L. K. (2010). The slack sodium-activated potassium channel provides a major outward current in olfactory neurons of Kv1.3-/- super-smeller mice. *Journal of Neurophysiology, 103*(6), 3311–3319. doi:10.1152/jn.00607.2009

Ma, Z., Lou, X. J., & Horrigan, F. T. (2006). Role of charged residues in the S1-S4 voltage sensor of BK channels. *Journal of General Physiology, 127*(3), 309–328. doi:10.1085/jgp.200509421

Maison, S. F., Pyott, S. J., Meredith, A. L., & Liberman, M. C. (2013). Olivocochlear suppression of outer hair cells in vivo: evidence for combined action of BK and SK2 channels throughout the cochlea. *Journal of Neurophysiology, 109*(6), 1525–1534. doi:10.1152/jn.00924.2012

Markham, M. R., Kaczmarek, L. K., & Zakon, H. H. (2013). A sodium-activated potassium channel supports high-frequency firing and reduces energetic costs during rapid modulations of action potential amplitude. *Journal of Neurophysiology, 109*(7), 1713–1723. doi:10.1152/jn.00875.2012

Marrero, H. G., Treistman, S. N., & Lemos, J. R. (2015). Ethanol effect on BK channels is modulated by magnesium. *Alcoholism—Clinical and Experimental Research, 39*(9), 1671–1679. doi:10.1111/acer.12821

Marrion, N. V., & Tavalin, S. J. (1998). Selective activation of Ca2+-activated K+ channels by co-localized Ca2+ channels in hippocampal neurons. *Nature, 395*(6705), 900–905. doi:10.1038/27674

Martin, G., Puig, S., Pietrzykowski, A., Zadek, P., Emery, P., & Treistman, S. (2004). Somatic localization of a specific large-conductance calcium-activated potassium channel subtype controls compartmentalized ethanol sensitivity in the nucleus accumbens. *Journal of Neuroscience, 24*(29), 6563–6572. doi:10.1523/JNEUROSCI.0684-04.2004

Martin, G. E., Hendrickson, L. M., Penta, K. L., Friesen, R. M., Pietrzykowski, A. Z., Tapper, A. R., & Treistman, S. N. (2008). Identification of a BK channel auxiliary protein controlling molecular and behavioral tolerance to alcohol. *Proceedings of the National Academy of Sciences of the United States of America, 105*(45), 17543–17548. doi:10.1073/pnas.0801068105

Martinez-Espinosa, P. L., Wu, J., Yang, C., Gonzalez-Perez, V., Zhou, H., Liang, H., ... Lingle, C. J. (2015). Knockout of Slo2.2 enhances itch, abolishes KNa current, and increases action potential firing frequency in DRG neurons. *Elife, 4*. doi:10.7554/eLife.10013

Marty, A. (1981). Ca-dependent K channels with large unitary conductance in chromaffin cell membranes. *Nature, 291*(5815), 497–500.

McKay, B. E., & Turner, R. W. (2004). Kv3 K+ channels enable burst output in rat cerebellar Purkinje cells. *European Journal of Neuroscience, 20*(3), 729–739. doi:10.1111/j.1460-9568.2004.03539.x

McLarnon, J. G. (1995). Inactivation of a high conductance calcium dependent potassium current in rat hippocampal neurons. *Neuroscience Letters, 193* (1), 5–8.

Meera, P., Wallner, M., & Toro, L. (2000). A neuronal beta subunit (KCNMB4) makes the large conductance, voltage- and Ca2+-activated K+ channel resistant to charybdotoxin and iberiotoxin. *Proceedings of the National Academy of Sciences of the United States of America, 97*(10), 5562–5567. doi:10.1073/pnas.100118597

Meredith, A. L., Thorneloe, K. S., Werner, M. E., Nelson, M. T., & Aldrich, R. W. (2004). Overactive bladder and incontinence in the absence of the BK large conductance Ca2+-activated K+ channel. *Journal of Biological Chemistry, 279*(35), 36746–36752. doi:10.1074/jbc.M405621200

Miranda-Rottmann, S., Kozlov, A. S., & Hudspeth, A. J. (2010). Highly specific alternative splicing of transcripts encoding BK channels in the chicken's cochlea is a minor determinant of the tonotopic gradient. *Molecular & Cellular Biology, 30*(14), 3646–3660. doi:MCB.00073-10 [pii] 10.1128/MCB.00073-10

Miranda, P., Contreras, J. E., Plested, A. J., Sigworth, F. J., Holmgren, M., & Giraldez, T. (2013). State-dependent FRET reports calcium- and voltage-dependent gating-ring motions in BK channels. *Proceedings of the National Academy of Sciences of the United States of America, 110*(13), 5217–5222. doi:10.1073/pnas.1219611110

Misonou, H., Menegola, M., Buchwalder, L., Park, E. W., Meredith, A., Rhodes, K. J., . . . Trimmer, J. S. (2006). Immunolocalization of the Ca2+-activated K+ channel Slo1 in axons and nerve terminals of mammalian brain and cultured neurons. *Journal of Comparative Neurology, 496*(3), 289–302. doi:10.1002/cne.20931

Mitra, P., & Slaughter, M. M. (2002a). Calcium-induced transitions between the spontaneous miniature outward and the transient outward currents in retinal amacrine cells. *Journal of General Physiology, 119*(4), 373–388.

Mitra, P., & Slaughter, M. M. (2002b). Mechanism of generation of spontaneous miniature outward currents (SMOCs) in retinal amacrine cells. *Journal of General Physiology, 119*(4), 355–372.

Muller, A., Kukley, M., Uebachs, M., Beck, H., & Dietrich, D. (2007). Nanodomains of single Ca2+ channels contribute to action potential repolarization in cortical neurons. *Journal of Neuroscience, 27*(3), 483–495. doi:10.1523/JNEUROSCI.3816-06.2007

Muller, C., Mas Gomez, N., Ruth, P., & Strauss, O. (2014). CaV1.3 L-type channels, maxiK Ca(2+)-dependent K(+) channels and bestrophin-1 regulate rhythmic photoreceptor outer segment phagocytosis by retinal pigment epithelial cells. *Cell Signaling, 26*(5), 968–978. doi:10.1016/j.cellsig.2013.12.021

Nanou, E., & El Manira, A. (2007). A postsynaptic negative feedback mediated by coupling between AMPA receptors and Na+-activated K+ channels in spinal cord neurones. *European Journal of Neuroscience, 25*(2), 445–450. doi:10.1111/j.1460-9568.2006.05287.x

Nanou, E., Kyriakatos, A., Bhattacharjee, A., Kaczmarek, L. K., Paratcha, G., & El Manira, A. (2008). Na+-mediated coupling between AMPA receptors and KNa channels shapes synaptic transmission. *Proceedings of the National Academy of Sciences of the United States of America, 105*(52), 20941–20946. doi:10.1073/pnas.0806403106

Naranjo, D., Moldenhauer, H., Pincuntureo, M., & Diaz-Franulic, I. (2016). Pore size matters for potassium channel conductance. *Journal of General Physiology, 148*(4), 277–291. doi:10.1085/jgp.201611625

Nimigean, C. M., Chappie, J. S., & Miller, C. (2003). Electrostatic tuning of ion conductance in potassium channels. *Biochemistry, 42*(31), 9263–9268. doi:10.1021/bi0348720

Niu, X., Qian, X., & Magleby, K. L. (2004). Linker-gating ring complex as passive spring and Ca(2+)-dependent machine for a voltage- and Ca(2+)-activated potassium channel. *Neuron, 42*(5), 745–756. doi:10.1016/j.neuron.2004.05.001

Noskova, L., Stranecky, V., Hartmannova, H., Pristoupilova, A., Baresova, V., Ivanek, R., . . . Kmoch, S. (2011). Mutations in DNAJC5, encoding cysteine-string protein alpha, cause autosomal-dominant adult-onset neuronal ceroid lipofuscinosis. *American Journal of Human Genetics, 89*(2), 241–252. doi:10.1016/j.ajhg.2011.07.003

Nuwer, M. O., Picchione, K. E., & Bhattacharjee, A. (2009). cAMP-dependent kinase does not modulate the Slack sodium-activated potassium channel. *Neuropharmacology, 57*(3), 219–226. doi:10.1016/j.neuropharm.2009.06.006

Nuwer, M. O., Picchione, K. E., & Bhattacharjee, A. (2010). PKA-induced internalization of slack KNa channels produces dorsal root ganglion neuron hyperexcitability. *Journal of Neuroscience, 30*(42), 14165–14172. doi:10.1523/JNEUROSCI.3150-10.2010

Obermair, G. J., Szabo, Z., Bourinet, E., & Flucher, B. E. (2004). Differential targeting of the L-type Ca2+ channel alpha 1C (CaV1.2) to synaptic and extrasynaptic compartments in hippocampal neurons. *European Journal of Neuroscience, 19*(8), 2109–2122. doi:10.1111/j.0953-816X.2004.03272.x

Orio, P., Rojas, P., Ferreira, G., & Latorre, R. (2002). New disguises for an old channel: MaxiK channel beta-subunits. *News in Physiological Sciences, 17*, 156–161.

Palacio, S., Velazquez-Marrero, C., Marrero, H. G., Seale, G. E., Yudowski, G. A., & Treistman, S. N. (2015). Time-dependent effects of ethanol on BK channel expression and trafficking in hippocampal neurons. *Alcoholism—Clinical and Experimental Research, 39*(9), 1619–1631. doi:10.1111/acer.12808

Pallotta, B. S., Magleby, K. L., & Barrett, J. N. (1981). Single channel recordings of Ca2+-activated K+ currents in rat muscle cell culture. *Nature, 293*(5832), 471–474.

Pan, C., Chu, H., Lai, Y., Liu, Y., Sun, Y., Du, Z., . . . Tao, Y. (2016). Down-regulation of the large conductance Ca(2+)-activated K(+) channel expression in C57BL/6J cochlea. *Acta Oto-Laryngologica, 136*(9), 875–878. doi:10.3109/00016489.2016.1168941

Paolini, A. G., FitzGerald, J. V., Burkitt, A. N., & Clark, G. M. (2001). Temporal processing from the auditory nerve to the medial nucleus of the trapezoid body in the rat. *Hearing Research, 159* (1–2), 101–116.

Pelkey, K. A., Topolnik, L., Lacaille, J. C., & McBain, C. J. (2006). Compartmentalized Ca(2+) channel regulation at divergent mossy-fiber release sites underlies target cell-dependent plasticity. *Neuron, 52*(3), 497–510. doi:10.1016/j.neuron.2006.08.032

Pelucchi, B., Grimaldi, A., & Moriondo, A. (2008). Vertebrate rod photoreceptors express both BK and IK calcium-activated potassium channels, but only BK channels are involved in receptor potential regulation. *Journal of Neuroscience Research, 86* (1), 194–201. doi:10.1002/jnr.21467

Petersen, O. H., Tepikin, A., & Park, M. K. (2001). The endoplasmic reticulum: One continuous or several separate Ca(2+) stores? *Trends in Neurosciences, 24*(5), 271–276.

Petrik, D., & Brenner, R. (2007). Regulation of STREX exon large conductance, calcium-activated potassium channels by the beta4 accessory subunit. *Neuroscience, 149*(4), 789–803. doi:10.1016/j.neuroscience.2007.07.066

Pietrzykowski, A. Z., Friesen, R. M., Martin, G. E., Puig, S. I., Nowak, C. L., Wynne, P. M., . . . Treistman, S. N. (2008). Posttranscriptional regulation of BK channel splice variant stability by miR-9 underlies neuroadaptation to alcohol. *Neuron, 59*(2), 274–287. doi:10.1016/j.neuron.2008.05.032

Pietrzykowski, A. Z., Martin, G. E., Puig, S. I., Knott, T. K., Lemos, J. R., & Treistman, S. N. (2004). Alcohol tolerance in large-conductance, calcium-activated potassium channels of CNS terminals is intrinsic and includes two components: Decreased ethanol potentiation and decreased channel density. *Journal of Neuroscience*, *24*(38), 8322–8332. doi:10.1523/Jneurosci.1536-04.2004

Piwonska, M., Wilczek, E., Szewczyk, A., & Wilczynski, G. M. (2008). Differential distribution of Ca2+-activated potassium channel beta4 subunit in rat brain: Immunolocalization in neuronal mitochondria. *Neuroscience*, *153*(2), 446–460. doi:10.1016/j.neuroscience.2008.01.050

Poulsen, A. N., Wulf, H., Hay-Schmidt, A., Jansen-Olesen, I., Olesen, J., & Klaerke, D. A. (2009). Differential expression of BK channel isoforms and beta-subunits in rat neuro-vascular tissues. *Biochimica Biophysica Acta*, *1788*(2), 380–389. doi:10.1016/j.bbamem.2008.10.001

Pyott, S. J., & Duncan, R. K. (2016). BK channels in the vertebrate inner ear. *International Review of Neurobiology*, *128*, 369–399. doi:10.1016/bs.irn.2016.03.016

Pyott, S. J., Meredith, A. L., Fodor, A. A., Vazquez, A. E., Yamoah, E. N., & Aldrich, R. W. (2007). Cochlear function in mice lacking the BK channel alpha, beta1, or beta4 subunits. *Journal of Biological Chemistry*, *282*(5), 3312–3324. doi:10.1074/jbc.M608726200

Quirk, J. C., & Reinhart, P. H. (2001). Identification of a novel tetramerization domain in large conductance K(ca) channels. *Neuron*, *32*(1), 13–23.

Ramanathan, K., Michael, T. H., Jiang, G. J., Hiel, H., & Fuchs, P. A. (1999). A molecular mechanism for electrical tuning of cochlear hair cells. *Science*, *283*(5399), 215–217.

Rehak, R., Bartoletti, T. M., Engbers, J. D., Berecki, G., Turner, R. W., & Zamponi, G. W. (2013). Low voltage activation of KCa1.1 current by Cav3-KCa1.1 complexes. *PLoS One*, *8*(4), e61844. doi:10.1371/journal.pone.0061844

Riazi, M. A., Brinkman-Mills, P., Johnson, A., Naylor, S. L., Minoshima, S., Shimizu, N., . . . McDermid, H. E. (1999). Identification of a putative regulatory subunit of a calcium-activated potassium channel in the dup(3q) syndrome region and a related sequence on 22q11.2. *Genomics*, *62*(1), 90–94. doi:10.1006/geno.1999.5975

Robitaille, R., Garcia, M. L., Kaczorowski, G. J., & Charlton, M. P. (1993). Functional colocalization of calcium and calcium-gated potassium channels in control of transmitter release. *Neuron*, *11*(4), 645–655.

Rohmann, K. N., Wersinger, E., Braude, J. P., Pyott, S. J., & Fuchs, P. A. (2015). Activation of BK and SK channels by efferent synapses on outer hair cells in high-frequency regions of the rodent cochlea. *Journal of Neuroscience*, *35*(5), 1821–1830. doi:10.1523/JNEUROSCI.2790-14.2015

Romani, A. M. P. (2008). Magnesium homeostasis and alcohol consumption. *Magnesium Research*, *21*(4), 197–204.

Rose, C. R. (2002). Na+ signals at central synapses. *Neuroscientist*, *8*(6), 532–539. doi:10.1177/1073858402238512

Rose, C. R., & Konnerth, A. (2001). NMDA receptor-mediated Na+ signals in spines and dendrites. *Journal of Neuroscience*, *21*(12), 4207–4214.

Rose, C. R., & Ransom, B. R. (1997). Regulation of intracellular sodium in cultured rat hippocampal neurones. *Journal of Physiology*, *499* (*Pt* 3), 573–587.

Rothberg, B. S., Shin, K. S., Phale, P. S., & Yellen, G. (2002). Voltage-controlled gating at the intracellular entrance to a hyperpolarization-activated cation channel. *Journal of General Physiology*, *119*(1), 83–91.

Ruffin, V. A., Gu, X. Q., Zhou, D., Douglas, R. M., Sun, X., Trouth, C. O., & Haddad, G. G. (2008). The sodium-activated potassium channel Slack is modulated by hypercapnia and acidosis. *Neuroscience, 151*(2), 410–418. doi:10.1016/j.neuroscience.2007.10.031

Runden-Pran, E., Haug, F. M., Storm, J. F., & Ottersen, O. P. (2002). BK channel activity determines the extent of cell degeneration after oxygen and glucose deprivation: A study in organotypical hippocampal slice cultures. *Neuroscience, 112*(2), 277–288.

Safronov, B. V., & Vogel, W. (1996). Properties and functions of Na(+)-activated K+ channels in the soma of rat motoneurones. *Journal of Physiology, 497 (Pt 3)*, 727–734.

Sailer, C. A., Kaufmann, W. A., Kogler, M., Chen, L., Sausbier, U., Ottersen, O. P., . . . Knaus, H. G. (2006). Immunolocalization of BK channels in hippocampal pyramidal neurons. *European Journal of Neuroscience, 24*(2), 442–454. doi:10.1111/j.1460-9568.2006.04936.x

Saito, M., & Wu, C. F. (1991). Expression of ion channels and mutational effects in giant Drosophila neurons differentiated from cell division-arrested embryonic neuroblasts. *Journal of Neuroscience, 11*(7), 2135–2150.

Sakaba, T., Ishikane, H., & Tachibana, M. (1997). Ca2+ -activated K+ current at presynaptic terminals of goldfish retinal bipolar cells. *Neuroscience Research, 27*(3), 219–228.

Salzmann, M., Seidel, K. N., Bernard, R., Pruss, H., Veh, R. W., & Derst, C. (2010). BKbeta1 subunits contribute to BK channel diversity in rat hypothalamic neurons. *Cellular & Molecular Neurobiology, 30*(6), 967–976. doi:10.1007/s10571-010-9527-7

Samaranayake, H., Saunders, J. C., Greene, M. I., & Navaratnam, D. S. (2004). Ca(2+) and K(+) (BK) channels in chick hair cells are clustered and colocalized with apical-basal and tonotopic gradients. *Journal of Physiology, 560*(Pt 1), 13–20. doi:10.1113/jphysiol.2004.069856

Samengo, I., Curro, D., Barrese, V., Taglialatela, M., & Martire, M. (2014). Large conductance calcium-activated potassium channels: Their expression and modulation of glutamate release from nerve terminals isolated from rat trigeminal caudal nucleus and cerebral cortex. *Neurochemical Research, 39*(5), 901–910. doi:10.1007/s11064-014-1287-1

Sanchez-Vives, M. V., Nowak, L. G., & McCormick, D. A. (2000). Cellular mechanisms of long-lasting adaptation in visual cortical neurons in vitro. *Journal of Neuroscience, 20*(11), 4286–4299.

Sandler, V. M., Puil, E., & Schwarz, D. W. (1998). Intrinsic response properties of bursting neurons in the nucleus principalis trigemini of the gerbil. *Neuroscience, 83*(3), 891–904.

Santi, C. M., Ferreira, G., Yang, B., Gazula, V. R., Butler, A., Wei, A., . . . Salkoff, L. (2006). Opposite regulation of Slick and Slack K+ channels by neuromodulators. *Journal of Neuroscience, 26*(19), 5059–5068. doi:10.1523/JNEUROSCI.3372-05.2006

Sarantopoulos, C. D., McCallum, J. B., Rigaud, M., Fuchs, A., Kwok, W. M., & Hogan, Q. H. (2007). Opposing effects of spinal nerve ligation on calcium-activated potassium currents in axotomized and adjacent mammalian primary afferent neurons. *Brain Research, 1132* (1), 84–99. doi:10.1016/j.brainres.2006.11.055

Sausbier, M., Hu, H., Arntz, C., Feil, S., Kamm, S., Adelsberger, H., . . . Ruth, P. (2004). Cerebellar ataxia and Purkinje cell dysfunction caused by Ca2+-activated K+ channel deficiency. *Proceedings of the National Academy of Sciences of the United States of America, 101*(25), 9474–9478. doi:10.1073/pnas.0401702101

Sausbier, U., Sausbier, M., Sailer, C. A., Arntz, C., Knaus, H. G., Neuhuber, W., & Ruth, P. (2006). Ca2+ -activated K+ channels of the BK-type in the mouse brain. *Histochemistry & Cell Biology, 125*(6), 725–741. doi:10.1007/s00418-005-0124-7

Savalli, N., Pantazis, A., Yusifov, T., Sigg, D., & Olcese, R. (2012). The contribution of RCK domains to human BK channel allosteric activation. *Journal of Biological Chemistry, 287*(26), 21741–21750. doi:10.1074/jbc.M112.346171

Scott, L. L., Brecht, E. J., Philpo, A., Iyer, S., Wu, N. S., Mihic, S. J., ... Walton, J. P. (2017). A novel BK channel-targeted peptide suppresses sound evoked activity in the mouse inferior colliculus. *Scientific Reports, 7*, 42433. doi:10.1038/srep42433

Scholz, A., Gruss, M., & Vogel, W. (1998). Properties and functions of calcium-activated K+ channels in small neurones of rat dorsal root ganglion studied in a thin slice preparation. *Journal of Physiology, 513 (Pt 1)*, 55–69.

Schoppa, N. E., McCormack, K., Tanouye, M. A., & Sigworth, F. J. (1992). The size of gating charge in wild-type and mutant Shaker potassium channels. *Science, 255*(5052), 1712–1715.

Schreiber, M., & Salkoff, L. (1997). A novel calcium-sensing domain in the BK channel. *Biophysical Journal, 73* (3), 1355–1363. doi:10.1016/S0006-3495(97)78168-2

Schreiber, M., Wei, A., Yuan, A., Gaut, J., Saito, M., & Salkoff, L. (1998). Slo3, a novel pH-sensitive K+ channel from mammalian spermatocytes. *Journal of Biological Chemistry, 273*(6), 3509–3516.

Schuckit, M. A., Wilhelmsen, K., Smith, T. L., Feiler, H. S., Lind, P., Lange, L. A., & Kalmijn, J. (2005). Autosomal linkage analysis for the level of response to alcohol. *Alcoholism—Clinical and Experimental Research, 29*(11), 1976–1982. doi:10.1097/01.alc.0000187598.82921.27

Schwindt, P. C., Spain, W. J., & Crill, W. E. (1989). Long-lasting reduction of excitability by a sodium-dependent potassium current in cat neocortical neurons. *Journal of Neurophysiology, 61*(2), 233–244.

Seoh, S. A., Sigg, D., Papazian, D. M., & Bezanilla, F. (1996). Voltage-sensing residues in the S2 and S4 segments of the Shaker K+ channel. *Neuron, 16*(6), 1159–1167.

Shao, L. R., Halvorsrud, R., Borg-Graham, L., & Storm, J. F. (1999). The role of BK-type Ca2+-dependent K+ channels in spike broadening during repetitive firing in rat hippocampal pyramidal cells. *The Journal of Physiology, 521 Pt 1*, 135–146.

Sheehan, J. J., Benedetti, B. L., & Barth, A. L. (2009). Anticonvulsant effects of the BK-channel antagonist paxilline. *Epilepsia, 50*(4), 711–720. doi:10.1111/j.1528-1167.2008.01888.x

Shen, K. Z., Lagrutta, A., Davies, N. W., Standen, N. B., Adelman, J. P., & North, R. A. (1994). Tetraethylammonium block of Slowpoke calcium-activated potassium channels expressed in Xenopus oocytes: Evidence for tetrameric channel formation. *Pflugers Archives, 426* (5), 440–445.

Shieh, C. C., Turner, S. C., Zhang, X. F., Milicic, I., Parihar, A., Jinkerson, T., ... Gopalakrishnan, M. (2007). A-272651, a nonpeptidic blocker of large-conductance Ca2+-activated K+ channels, modulates bladder smooth muscle contractility and neuronal action potentials. *British Journal of Pharmacology, 151*(6), 798–806. doi:10.1038/sj.bjp.0707278

Shigemoto, T., & Ohmori, H. (1991). Muscarinic receptor hyperpolarizes cochlear hair cells of chick by activating Ca(2+)-activated K+ channels. *Journal of Physiology, 442*, 669–690.

Shin, K. S., Rothberg, B. S., & Yellen, G. (2001). Blocker state dependence and trapping in hyperpolarization-activated cation channels: Evidence for an intracellular activation gate. *Journal of General Physiology, 117*(2), 91–101.

Shoudai, K., Nonaka, K., Maeda, M., Wang, Z. M., Jeong, H. J., Higashi, H., ... Akaike, N. (2007). Effects of various K+ channel blockers on spontaneous glycine release at rat spinal neurons. *Brain Research, 1157*, 11–22. doi:10.1016/j.brainres.2006.09.097

Shruti, S., Clem, R. L., & Barth, A. L. (2008). A seizure-induced gain-of-function in BK channels is associated with elevated firing activity in neocortical pyramidal neurons. *Neurobiology of Disease, 30*(3), 323–330. doi:10.1016/j.nbd.2008.02.002

Shruti, S., Urban-Ciecko, J., Fitzpatrick, J. A., Brenner, R., Bruchez, M. P., & Barth, A. L. (2012). The brain-specific Beta4 subunit downregulates BK channel cell surface expression. *PLoS One, 7*(3), e33429. doi:10.1371/journal.pone.0033429

Sitdikova, G. F., Fuchs, R., Kainz, V., Weiger, T. M., & Hermann, A. (2014). Phosphorylation of BK channels modulates the sensitivity to hydrogen sulfide (H2S). *Frontiers in Physiology, 5*, 431. doi:10.3389/fphys.2014.00431

Sivaramakrishnan, S., & Oliver, D. L. (2001). Distinct K currents result in physiologically distinct cell types in the inferior colliculus of the rat. *Journal of Neuroscience, 21*(8), 2861–2877.

Skinner, L. J., Enee, V., Beurg, M., Jung, H. H., Ryan, A. F., Hafidi, A., . . . Dulon, D. (2003). Contribution of BK Ca2+-activated K+ channels to auditory neurotransmission in the Guinea pig cochlea. *Journal of Neurophysiology, 90*(1), 320–332. doi:10.1152/jn.01155.2002

Song, B., & Marvizon, J. C. (2005). N-methyl-D-aspartate receptors and large conductance calcium-sensitive potassium channels inhibit the release of opioid peptides that induce mu-opioid receptor internalization in the rat spinal cord. *Neuroscience, 136*(2), 549–562. doi:10.1016/j.neuroscience.2005.08.032

Stefani, E., Ottolia, M., Noceti, F., Olcese, R., Wallner, M., Latorre, R., & Toro, L. (1997). Voltage-controlled gating in a large conductance Ca2+-sensitive K+channel (hslo). *Proceedings of the National Academy of Sciences of the United States of America, 94*(10), 5427–5431.

Storm, J. F. (1987). Intracellular injection of a Ca2+ chelator inhibits spike repolarization in hippocampal neurons. *Brain Research, 435* (1–2), 387–392.

Strauss, O. (2005). The retinal pigment epithelium in visual function. *Physiological Reviews, 85*(3), 845–881. doi:10.1152/physrev.00021.2004

Sun, X., Gu, X. Q., & Haddad, G. G. (2003). Calcium influx via L- and N-type calcium channels activates a transient large-conductance Ca2+-activated K+ current in mouse neocortical pyramidal neurons. *Journal of Neuroscience, 23*(9), 3639–3648.

Suzuki, T., Hansen, A., & Sanguinetti, M. C. (2016). Hydrophobic interactions between the S5 segment and the pore helix stabilizes the closed state of Slo2.1 potassium channels. *Biochimica Biophysica Acta, 1858*(4), 783–792. doi:10.1016/j.bbamem.2015.12.024

Tamsett, T. J., Picchione, K. E., & Bhattacharjee, A. (2009). NAD+ activates KNa channels in dorsal root ganglion neurons. *Journal of Neuroscience, 29*(16), 5127–5134. doi:10.1523/JNEUROSCI.0859-09.2009

Tanimoto, N., Sothilingam, V., Euler, T., Ruth, P., Seeliger, M. W., & Schubert, T. (2012). BK channels mediate pathway-specific modulation of visual signals in the in vivo mouse retina. *Journal of Neuroscience, 32*(14), 4861–4866. doi:10.1523/JNEUROSCI.4654-11.2012

Tao, Q., & Kelly, M. E. (1996). Calcium-activated potassium current in cultured rabbit retinal pigment epithelial cells. *Current Eye Research, 15*(3), 237–246.

Tao, X., Hite, R. K., & MacKinnon, R. (2017). Cryo-EM structure of the open high-conductance Ca2+-activated K+ channel. *Nature, 541*(7635), 46–51. doi:10.1038/nature20608

Tejada, M. A., Hashem, N., Calloe, K., & Klaerke, D. A. (2017). Heteromeric Slick/Slack K+ channels show graded sensitivity to cell volume changes. *PLoS One, 12*(2), e0169914. doi:10.1371/journal.pone.0169914

Tejada, M. A., Stople, K., Hammami Bomholtz, S., Meinild, A. K., Poulsen, A. N., & Klaerke, D. A. (2014). Cell volume changes regulate slick (Slo2.1), but not slack (Slo2.2) K+ channels. *PLoS One, 9*(10), e110833. doi:10.1371/journal.pone.0110833

Tejada, M. d. l. A., Jensen, L. J., & Klaerke, D. A. (2012). PIP(2) modulation of Slick and Slack K(+) channels. *Biochemical & Biophysical Research Communications*, *424*(2), 208–213. doi:10.1016/j.bbrc.2012.06.038

Thompson, J., & Begenisich, T. (2012). Selectivity filter gating in large-conductance Ca(2+)-activated K+ channels. *Journal of General Physiology*, *139*(3), 235–244. doi:10.1085/jgp.201110748

Thomson, S. J., Hansen, A., & Sanguinetti, M. C. (2015). Identification of the intracellular Na+ sensor in Slo2.1 potassium channels. *Journal of Biological Chemistry*, *290*(23), 14528–14535. doi:10.1074/jbc.M115.653089

Tomasello, D. L., Hurley, E., Wrabetz, L., & Bhattacharjee, A. (2017). Slick (Kcnt2) sodium-activated potassium channels limit peptidergic nociceptor excitability and hyperalgesia. *Journal of Experimental Neuroscience*, *11*, 1179069517726996. doi:10.1177/1179069517726996

Torres, Y. P., Granados, S. T., & Latorre, R. (2014). Pharmacological consequences of the coexpression of BK channel alpha and auxiliary beta subunits. *Frontiers in Physiology*, *5*, 383. doi:10.3389/fphys.2014.00383

Torres, Y. P., Morera, F. J., Carvacho, I., & Latorre, R. (2007). A marriage of convenience: beta-subunits and voltage-dependent K+ channels. *Journal of Biological Chemistry*, *282*(34), 24485–24489. doi:10.1074/jbc.R700022200

Tseng-Crank, J., Foster, C. D., Krause, J. D., Mertz, R., Godinot, N., DiChiara, T. J., & Reinhart, P. H. (1994). Cloning, expression, and distribution of functionally distinct Ca(2+)-activated K+ channel isoforms from human brain. *Neuron*, *13*(6), 1315–1330.

Typlt, M., Mirkowski, M., Azzopardi, E., Ruettiger, L., Ruth, P., & Schmid, S. (2013). Mice with deficient BK channel function show impaired prepulse inhibition and spatial learning, but normal working and spatial reference memory. *PLoS One*, *8*(11), e81270. doi:10.1371/journal.pone.0081270

Uchino, S., Wada, H., Honda, S., Hirasawa, T., Yanai, S., Nakamura, Y., . . . Kohsaka, S. (2003). Slo2 sodium-activated K+ channels bind to the PDZ domain of PSD-95. *Biochemical & Biophysical Research Communications*, *310*(4), 1140–1147.

Uebele, V. N., Lagrutta, A., Wade, T., Figueroa, D. J., Liu, Y., McKenna, E., . . . Swanson, R. (2000). Cloning and functional expression of two families of beta-subunits of the large conductance calcium-activated K+ channel. *Journal of Biological Chemistry*, *275*(30), 23211–23218. doi:10.1074/jbc.M910187199

Velazquez-Marrero, C., Seale, G. E., Treistman, S. N., & Martin, G. E. (2014). Large conductance voltage- and Ca2+-gated potassium (BK) channel beta 4 subunit influences sensitivity and tolerance to alcohol by altering its response to kinases. *Journal of Biological Chemistry*, *289*(42), 29261–29272. doi:10.1074/jbc.M114.604306

Velazquez-Marrero, C., Wynne, P., Bernardo, A., Palacio, S., Martin, G., & Treistman, S. N. (2011). The relationship between duration of initial alcohol exposure and persistence of molecular tolerance is markedly nonlinear. *Journal of Neuroscience*, *31*(7), 2436–2446. doi:10.1523/Jneurosci.5429-10.2011

Velinov, M., Dolzhanskaya, N., Gonzalez, M., Powell, E., Konidari, I., Hulme, W., . . . Zuchner, S. (2012). Mutations in the gene DNAJC5 cause autosomal dominant Kufs disease in a proportion of cases: Study of the Parry family and 8 other families. *PLoS One*, *7*(1), e29729. doi:10.1371/journal.pone.0029729

Wallen, P., Robertson, B., Cangiano, L., Low, P., Bhattacharjee, A., Kaczmarek, L. K., & Grillner, S. (2007). Sodium-dependent potassium channels of a Slack-like subtype contribute to the

slow afterhyperpolarization in lamprey spinal neurons. *Journal of Physiology, 585*(Pt 1), 75–90. doi:10.1113/jphysiol.2007.138156
Wallner, M., Meera, P., & Toro, L. (1996). Determinant for beta-subunit regulation in high-conductance voltage-activated and Ca(2+)-sensitive K+ channels: an additional transmembrane region at the N terminus. *Proceedings of the National Academy of Sciences of the United States of America, 93*(25), 14922–14927.
Wallner, M., Meera, P., & Toro, L. (1999). Molecular basis of fast inactivation in voltage and Ca2+-activated K+ channels: A transmembrane beta-subunit homolog. *Proceedings of the National Academy of Sciences of the United States of America, 96*(7), 4137–4142.
Wang, B., & Brenner, R. (2006). An S6 mutation in BK channels reveals beta1 subunit effects on intrinsic and voltage-dependent gating. *Journal of General Physiology, 128*(6), 731–744. doi:10.1085/jgp.200609596
Wang, B., Bugay, V., Ling, L., Chuang, H. H., Jaffe, D. B., & Brenner, R. (2016). Knockout of the BK beta4-subunit promotes a functional coupling of BK channels and ryanodine receptors that mediate a fAHP-induced increase in excitability. *Journal of Neurophysiology, 116*(2), 456–465. doi:10.1152/jn.00857.2015
Wang, B., Jaffe, D. B., & Brenner, R. (2014). Current understanding of iberiotoxin-resistant BK channels in the nervous system. *Frontiers in physiology*, 5, 382. doi:10.3389/fphys.2014.00382
Wang, B., Rothberg, B. S., & Brenner, R. (2006). Mechanism of beta4 subunit modulation of BK channels. *Journal of General Physiology, 127*(4), 449–465. doi:10.1085/jgp.200509436
Wang, B., Rothberg, B. S., & Brenner, R. (2009). Mechanism of increased BK channel activation from a channel mutation that causes epilepsy. *Journal of General Physiology, 133*(3), 283–294. doi:10.1085/jgp.200810141
Wang, F., Zhang, Y., Wang, L., Sun, P., Luo, X., Ishigaki, Y., . . . Kato, N. (2015). Improvement of spatial learning by facilitating large-conductance calcium-activated potassium channel with transcranial magnetic stimulation in Alzheimer's disease model mice. *Neuropharmacology, 97*, 210–219. doi:10.1016/j.neuropharm.2015.05.027
Wang, G. Y., Robinson, D. W., & Chalupa, L. M. (1998). Calcium-activated potassium conductances in retinal ganglion cells of the ferret. *Journal of Neurophysiology, 79*(1), 151–158.
Wang, L., Kang, H., Li, Y., Shui, Y., Yamamoto, R., Sugai, T., & Kato, N. (2015). Cognitive recovery by chronic activation of the large-conductance calcium-activated potassium channel in a mouse model of Alzheimer's disease. *Neuropharmacology, 92*, 8–15. doi:10.1016/j.neuropharm.2014.12.033
Wang, L., & Sigworth, F. J. (2009). Structure of the BK potassium channel in a lipid membrane from electron cryomicroscopy. *Nature, 461*(7261), 292–295. doi:10.1038/nature08291
Wanner, S. G., Koch, R. O., Koschak, A., Trieb, M., Garcia, M. L., Kaczorowski, G. J., & Knaus, H. G. (1999). High-conductance calcium-activated potassium channels in rat brain: Pharmacology, distribution, and subunit composition. *Biochemistry, 38*(17), 5392–5400. doi:10.1021/bi983040c
Weiger, T. M., Holmqvist, M. H., Levitan, I. B., Clark, F. T., Sprague, S., Huang, W. J., . . . Curtis, R. (2000). A novel nervous system beta subunit that downregulates human large conductance calcium-dependent potassium channels. *The Journal of Neuroscience: The Official Journal of the Society for Neuroscience, 20*(10), 3563–3570.
Wersinger, E., McLean, W. J., Fuchs, P. A., & Pyott, S. J. (2010). BK channels mediate cholinergic inhibition of high frequency cochlear hair cells. *PLoS One, 5*(11), e13836. doi:10.1371/journal.pone.0013836

Whitt, J. P., Montgomery, J. R., & Meredith, A. L. (2016). BK channel inactivation gates day-time excitability in the circadian clock. *Nature Communications*, *7*, 10837. doi:10.1038/ncomms10837

Widmer, H. A., Rowe, I. C., & Shipston, M. J. (2003). Conditional protein phosphorylation regulates BK channel activity in rat cerebellar Purkinje neurons. *Journal of Physiology*, *552*(Pt 2), 379–391. doi:10.1113/jphysiol.2003.046441

Wijetunge, L. S., Chattarji, S., Wyllie, D. J., & Kind, P. C. (2013). Fragile X syndrome: From targets to treatments. *Neuropharmacology*, *68*, 83–96. doi:10.1016/j.neuropharm.2012.11.028

Wilkens, C. M., & Aldrich, R. W. (2006). State-independent block of BK channels by an intracellular quaternary ammonium. *Journal of General Physiology*, *128*(3), 347–364. doi:10.1085/jgp.200609579

Wimmers, S., Halsband, C., Seyler, S., Milenkovic, V., & Strauss, O. (2008). Voltage-dependent Ca2+ channels, not ryanodine receptors, activate Ca2+-dependent BK potassium channels in human retinal pigment epithelial cells. *Molecular Vision*, *14*, 2340–2348.

Wu, L. G., Westenbroek, R. E., Borst, J. G., Catterall, W. A., & Sakmann, B. (1999). Calcium channel types with distinct presynaptic localization couple differentially to transmitter release in single calyx-type synapses. *Journal of Neuroscience*, *19*(2), 726–736.

Wu, Y., Xiong, Y., Wang, S., Yi, H., Li, H., Pan, N., ... Ding, J. (2009). Intersubunit coupling in the pore of BK channels. *Journal of Biological Chemistry*, *284*(35), 23353–23363. doi:10.1074/jbc.M109.027789

Wu, Y., Yang, Y., Ye, S., & Jiang, Y. (2010). Structure of the gating ring from the human large-conductance Ca(2+)-gated K(+) channel. *Nature*, *466*(7304), 393–397. doi:10.1038/nature09252

Wulf-Johansson, H., Amrutkar, D. V., Hay-Schmidt, A., Poulsen, A. N., Klaerke, D. A., Olesen, J., & Jansen-Olesen, I. (2010). Localization of large conductance calcium-activated potassium channels and their effect on calcitonin gene-related peptide release in the rat trigemino-neuronal pathway. *Neuroscience*, *167*(4), 1091–1102. doi:10.1016/j.neuroscience.2010.02.063

Wynne, P. M., Puig, S. I., Martin, G. E., & Treistman, S. N. (2009). Compartmentalized beta subunit distribution determines characteristics and ethanol sensitivity of somatic, dendritic, and terminal large-conductance calcium-activated potassium channels in the rat central nervous system. *Journal of Pharmacology & Experimental Therapeutics*, *329*(3), 978–986. doi:10.1124/jpet.108.146175

Xia, X. M., Ding, J. P., & Lingle, C. J. (1999). Molecular basis for the inactivation of Ca2+- and voltage-dependent BK channels in adrenal chromaffin cells and rat insulinoma tumor cells. *Journal of Neuroscience*, *19*(13), 5255–5264.

Xia, X. M., Ding, J. P., Zeng, X. H., Duan, K. L., & Lingle, C. J. (2000). Rectification and rapid activation at low Ca2+ of Ca2+-activated, voltage-dependent BK currents: Consequences of rapid inactivation by a novel beta subunit. *Journal of Neuroscience*, *20*(13), 4890–4903.

Xu, J. W., & Slaughter, M. M. (2005). Large-conductance calcium-activated potassium channels facilitate transmitter release in salamander rod synapse. *Journal of Neuroscience*, *25*(33), 7660–7668. doi:10.1523/JNEUROSCI.1572-05.2005

Yamamoto, K., Ueta, Y., Wang, L., Yamamoto, R., Inoue, N., Inokuchi, K., ... Kato, N. (2011). Suppression of a neocortical potassium channel activity by intracellular amyloid-beta and its rescue with Homer1a. *Journal of Neuroscience*, *31*(31), 11100–11109. doi:10.1523/JNEUROSCI.6752-10.2011

Yan, J., & Aldrich, R. W. (2010). LRRC26 auxiliary protein allows BK channel activation at resting voltage without calcium. *Nature*, *466*(7305), 513–516. doi:10.1038/nature09162

Yan, J., & Aldrich, R. W. (2012). BK potassium channel modulation by leucine-rich repeat-containing proteins. *Proceedings of the National Academy of Sciences of the United States of America, 109*(20), 7917–7922. doi:10.1073/pnas.1205435109

Yan, Y., Yang, Y., Bian, S., & Sigworth, F. J. (2012). Expression, purification and functional reconstitution of slack sodium-activated potassium channels. *Journal of Membrane Biology, 245*(11), 667–674. doi:10.1007/s00232-012-9425-7

Yang, B., Desai, R., & Kaczmarek, L. K. (2007). Slack and Slick K(Na) channels regulate the accuracy of timing of auditory neurons. *Journal of Neuroscience, 27*(10), 2617–2627. doi:10.1523/JNEUROSCI.5308-06.2007

Yang, C., Gonzalez-Perez, V., Mukaibo, T., Melvin, J. E., Xia, X. M., & Lingle, C. J. (2017). Knockout of the LRRC26 subunit reveals a primary role of LRRC26-containing BK channels in secretory epithelial cells. *Proceedings of the National Academy of Sciences of the United States of America, 114*(18), E3739–E3747. doi:10.1073/pnas.1703081114

Yang, C., Zeng, X. H., Zhou, Y., Xia, X. M., & Lingle, C. J. (2011). LRRC52 (leucine-rich-repeat-containing protein 52), a testis-specific auxiliary subunit of the alkalization-activated Slo3 channel. *Proceedings of the National Academy of Sciences of the United States of America, 108*(48), 19419–19424. doi:10.1073/pnas.1111104108

Yang, H., Hu, L., Shi, J., Delaloye, K., Horrigan, F. T., & Cui, J. (2007). Mg2+ mediates interaction between the voltage sensor and cytosolic domain to activate BK channels. *Proceedings of the National Academy of Sciences of the United States of America, 104*(46), 18270–18275. doi:10.1073/pnas.0705873104

Yang, H., Shi, J., Zhang, G., Yang, J., Delaloye, K., & Cui, J. (2008). Activation of Slo1 BK channels by Mg2+ coordinated between the voltage sensor and RCK1 domains. *Nature Structural & Molecular Biology, 15*(11), 1152–1159. doi:10.1038/nsmb.1507

Yazdi, H. H., Janahmadi, M., & Behzadi, G. (2007). The role of small-conductance Ca2+-activated K+ channels in the modulation of 4-aminopyridine-induced burst firing in rat cerebellar Purkinje cells. *Brain Research, 1156* , 59–66. doi:10.1016/j.brainres.2007.04.031

Ye, H., Buttigieg, J., Wan, Y., Wang, J., Figley, S., & Fehlings, M. G. (2012). Expression and functional role of BK channels in chronically injured spinal cord white matter. *Neurobiology of Disease, 47*(2), 225–236. doi:10.1016/j.nbd.2012.04.006

Ye, H., Jalini, S., Mylvaganam, S., & Carlen, P. (2010). Activation of large-conductance Ca(2+)-activated K(+) channels depresses basal synaptic transmission in the hippocampal CA1 area in APP (swe/ind) TgCRND8 mice. *Neurobiology of Aging, 31*(4), 591–604. doi:10.1016/j.neurobiolaging.2008.05.012

Yuan, A., Dourado, M., Butler, A., Walton, N., Wei, A., & Salkoff, L. (2000). SLO-2, a K+ channel with an unusual Cl- dependence. *Nature Neuroscience, 3*(8), 771–779. doi:10.1038/77670

Yuan, A., Santi, C. M., Wei, A., Wang, Z. W., Pollak, K., Nonet, M., . . . Salkoff, L. (2003). The sodium-activated potassium channel is encoded by a member of the Slo gene family. *Neuron, 37*(5), 765–773.

Yuan, P., Leonetti, M. D., Hsiung, Y., & MacKinnon, R. (2011). Open structure of the Ca2+ gating ring in the high-conductance Ca2+-activated K+ channel. *Nature, 481*(7379), 94–97. doi:10.1038/nature10670

Yuan, P., Leonetti, M. D., Pico, A. R., Hsiung, Y., & MacKinnon, R. (2010). Structure of the human BK channel Ca2+-activation apparatus at 3.0 A resolution. *Science, 329*(5988), 182–186. doi:10.1126/science.1190414

Yusifov, T., Javaherian, A. D., Pantazis, A., Gandhi, C. S., & Olcese, R. (2010). The RCK1 domain of the human BKCa channel transduces Ca2+ binding into structural rearrangements. *Journal of General Physiology, 136*(2), 189–202. doi:10.1085/jgp.200910374

Yusifov, T., Savalli, N., Gandhi, C. S., Ottolia, M., & Olcese, R. (2008). The RCK2 domain of the human BKCa channel is a calcium sensor. *Proceedings of the National Academy of Sciences of the United States of America, 105*(1), 376–381. doi:0705261105 [pii] 10.1073/pnas.0705261105

Zamalloa, T., Bailey, C. P., & Pineda, J. (2009). Glutamate-induced post-activation inhibition of locus coeruleus neurons is mediated by AMPA/kainate receptors and sodium-dependent potassium currents. *British Journal of Pharmacology, 156*(4), 649–661. doi:10.1111/j.1476-5381.2008.00004.x

Zeng, X. H., Xia, X. M., & Lingle, C. J. (2005). Divalent cation sensitivity of BK channel activation supports the existence of three distinct binding sites. *Journal of General Physiology, 125*(3), 273–286.

Zhang, L., Sukhareva, M., Barker, J. L., Maric, D., Hao, Y., Chang, Y. H., . . . Rubinow, D. R. (2005). Direct binding of estradiol enhances Slack (sequence like a calcium-activated potassium channel) channels' activity. *Neuroscience, 131*(2), 275–282. doi:10.1016/j.neuroscience.2004.10.042

Zhang, X., Solaro, C. R., & Lingle, C. J. (2001). Allosteric regulation of BK channel gating by Ca(2+) and Mg(2+) through a nonselective, low affinity divalent cation site. *Journal of General Physiology, 118*(5), 607–636.

Zhang, X. F., Gopalakrishnan, M., & Shieh, C. C. (2003). Modulation of action potential firing by iberiotoxin and NS1619 in rat dorsal root ganglion neurons. *Neuroscience, 122*(4), 1003–1011.

Zhang, X. L., Mok, L. P., Katz, E. J., & Gold, M. S. (2010). BKCa currents are enriched in a subpopulation of adult rat cutaneous nociceptive dorsal root ganglion neurons. *European Journal of Neuroscience, 31*(3), 450–462. doi:10.1111/j.1460-9568.2009.07060.x

Zhang, Y., Brown, M. R., Hyland, C., Chen, Y., Kronengold, J., Fleming, M. R., . . . Kaczmarek, L. K. (2012). Regulation of neuronal excitability by interaction of fragile X mental retardation protein with slack potassium channels. *The Journal of Neuroscience: The Official Journal of the Society for Neuroscience, 32*(44), 15318–15327. doi: 10.1523/JNEUROSCI.2162-12.2012.

Zhang, Z., Rosenhouse-Dantsker, A., Tang, Q. Y., Noskov, S., & Logothetis, D. E. (2010). The RCK2 domain uses a coordination site present in Kir channels to confer sodium sensitivity to Slo2.2 channels. *Journal of Neuroscience, 30*(22), 7554–7562. doi:10.1523/JNEUROSCI.0525-10.2010

Zhang, Z. B., Tian, M. Q., Gao, K., Jiang, Y. W., & Wu, Y. (2015). De novo KCNMA1 mutations in children with early-onset paroxysmal dyskinesia and developmental delay. *Movement Disorders, 30*(9), 1290–1292. doi:10.1002/mds.26216

Zhong, N., Beaumont, V., & Zucker, R. S. (2001). Roles for mitochondrial and reverse mode Na+/Ca2+ exchange and the plasmalemma Ca2+ ATPase in post-tetanic potentiation at crayfish neuromuscular junctions. *Journal of Neuroscience, 21*(24), 9598–9607.

Zhou, Y., Xia, X. M., & Lingle, C. J. (2011). Cysteine scanning and modification reveal major differences between BK channels and Kv channels in the inner pore region. *Proceedings of the National Academy of Sciences of the United States of America, 108*(29), 12161–12166. doi:10.1073/pnas.1104150108

CHAPTER 15

HYPERPOLARIZATION-ACTIVATED CYCLIC NUCLEOTIDE-GATED CHANNELS

ALESSIO MASI, MARIA NOVELLA ROMANELLI, GUIDO MANNAIONI, AND ELISABETTA CERBAI

Hyperpolarization-activated, cyclic nucleotide-gated (HCN) channels belong to the voltage-gated pore-loop channels family. The HCN subfamily comprises four subunits, encoded by four distinct genes, named HCN1, 2, 3, and 4. All subunits are made up of six transmembrane domains (S_1–S_6), a voltage-sensing segment in S_4, and the pore-loop element between S_4 and S_5. Functional channels are typically composed of homo- or hetero-tetramers. Altogether, these traits match the identikit of six-transmembrane domain, voltage-gated K^+ channels. What sets HCN channels apart is their peculiar ionic selectivity and their unusual gating properties. In fact, unlike their close relatives, HCN channels are activated by hyperpolarization and permeable to both Na^+ and K^+. These properties, first described in the late 1970s by Brown, DiFrancesco, and Noble in heart muscle cells, were seen as pretty unusual features, even funny, thus the current was nicknamed I_f, where "f" stands for "funny." Since then, the physiological relevance of HCN current has been well established in the heart, where it is now the target of a drug with therapeutic use. In the meantime, HCN channels have been discovered in several neuronal types in both central and peripheral districts of the nervous system and studied extensively with multiple approaches. The interest of neurophysiologists in HCN current was prompted by the unique biophysical features, which appeared suited to explain some remarkable electrical properties of single neurons and complex neural systems, such as intrinsically generated rhythmic firing and network oscillations. Two decades of extraordinary experimental effort have now established that HCN channels are key elements of neuronal physiology. In the attempt to review the role of HCN channels in the normal operation of neurons, as well as in pathological states, this chapter will start by discussing general molecular and biophysical concepts, will then focus on the most prominent and well-documented functions at cellular and network levels in physiological conditions, and will finally overview some remarkable examples of HCN channel–related pathologies affecting nerve cells.

Basic Facts About HCN Channels in the Nervous System

To understand the multiple roles played by HCN channels in physiological and pathological states of the nervous system, it is necessary to start from the expression pattern of the four pore-forming subunits and to overview the defining molecular and biophysical properties of HCN channels in nerve cells.

Expression Pattern at Regional, Cellular, and Subcellular Level

Available data on HCN channel expression at the supracellular level comes from in situ hybridization and immunohistochemistry studies performed in brain sections from both adult and developing rodent brains (Bender et al., 2001; Monteggia, Eisch, Tang, Kaczmarek, & Nestler, 2000; Moosmang, Biel, Hofmann, & Ludwig, 1999; Notomi & Shigemoto, 2004; Santoro et al., 2000). Single-cell reverse transcription-polymerase chain reaction (RT-PCR) and electron microscopy have provided further details on the expression pattern of HCN transcripts and proteins with cellular and subcellular resolution (Bender et al., 2007; Brewster et al., 2007; Dufour, Woodhouse, & Goaillard, 2014; Franz, Liss, Neu, & Roeper, 2000; Huang et al., 2011; Luján, Albasanz, Shigemoto, & Juiz, 2005; Paspalas, Wang, & Arnsten, 2013; Ramakrishnan, Drescher, Khan, Hatfield, & Drescher, 2012). All four HCN isoforms are expressed in the brain, with the relative abundance of each subunit relating to the area, neuronal type, and cellular compartment. Despite the different methodologies and animal models employed, all reports addressing HCN distribution show fairly consistent results. In the central nervous system (CNS), HCN1 is highly expressed in the neocortex, hippocampus, cerebellar cortex, brainstem, and spinal cord. HCN2 has a diffuse expression pattern across the CNS, with highest density in thalamic and brainstem nuclei. Conversely, HCN4 expression is highly localized, with few areas exhibiting strong expression levels, such as the olfactory bulb and the thalamus, with a distribution topography that appears complementary to that of HCN1. The expression of HCN3 is sparse throughout the brain. All isoforms are present in the retina (Fyk-Kolodziej & Pourcho, 2007; Muller et al., 2003). In the peripheral nervous system (PNS), all HCN subunits are expressed. HCN1 is the isoform with highest expression in dorsal root ganglia (Chaplan et al., 2003), although a prominent function for HCN2 in the transmission of painful stimuli has also been reported (Emery, Young, Berrocoso, Chen, & McNaughton, 2011). Last but not least, HCN current has been recorded in many ganglion neurons of the autonomic nervous system (Doan et al., 2004; Galligan, Tatsumi, Shen, Surprenant, & North, 1990; Kullmann et al., 2016; Lamas, 1998; Zhang & Cuevas, 2002).

Biophysical Features of Neuronal HCN Channels

As stated before, this chapter will attempt to explain how HCN channels impart specific electrical properties to the single neuron, and how these impinge on neuronal input–output properties, synaptic transmission, and the operation of neuronal ensembles. Therefore, only some essential concepts on basic HCN channels' biophysics will be provided here as tools for the reader to fully grasp the neurophysiological significance of HCN current. The brave reader who is willing to penetrate the multitude of studies that have addressed this matter over the past 40 years will find full satisfaction browsing the reference section of some excellent reviews (Biel, Wahl-Schott, Michalakis, & Zong, 2009; He, Chen, Li, & Hu, 2014; Sartiani, Mannaioni, Masi, Novella Romanelli, & Cerbai, 2017).

As mentioned, HCN subunits display all the standard structural features of voltage-gated K^+ channels (K_V), with which they share remarkable aminoacidic sequence homology, especially in functional domains. Nonetheless, HCN channels have a few remarkable peculiarities. These lie essentially in (i) the function of the voltage sensor, (ii) the selectivity filter, (iii) the absence of an inactivation gate, and (iv) the presence of a highly specialized cyclic nucleotide binding domain (CNBD) at the C-terminus. Universally present in voltage-gate ion channels, the voltage sensor is a typical α helix motif enriched with positively charged aminoacidic residues (Lys, Arg). These residues enable the segment to respond with outward or inward displacement following changes in magnitude of the membrane potential (V_M). In "classic" depolarization-activated channels, depolarization causes outward motion of the voltage sensor and opening of the pore (Bahring, Barghaan, Westermeier, & Wollberg, 2012; Yellen, 1998). In contrast, HCN channels are stabilized in the closed state by depolarization ($V_M \geq -50$ mV), whereas hyperpolarization ($V_M \leq -60$ mV) causes opening of the pore (Prole & Yellen, 2006). Such a striking exception from the rule seems to depend on the uniquely large size of the S_4 helix. In depolarization, the S_4 helix compresses the S_5–S_6 helices, thus causing pore obstruction, while hyperpolarization causes an inward shift of the S_4 sensor and unblocking of the pore. Once the channel is open, both Na^+ and K^+ ions are allowed through the pore. Such loose selectivity is due to the presence of only two binding sites for K^+, as opposed to the four present in K_V channels (Lee & MacKinnon, 2017). Because of this structural feature, K^+:Na^+ selectivity drops from 1000:1 to 4:1. Although a certain preference for K^+ over Na^+ is maintained, since HCN channels open at potentials that are closer to the equilibrium potential of K^+ (~ −90 mV) than to that of Na^+ (~ +60 mV), the electrochemical gradient driving Na^+ inside is much greater than that driving K^+ outside when channels are open, with the consequence that HCN current is, under physiological conditions, an inward depolarizing current.

The half-activation potential ($V_{1/2}$, an index of voltage dependence) of HCN channels lies between −75 and −90 mV and varies based on the intrinsic properties of HCN isoforms, host cell type, and experimental conditions. Gating kinetics—i.e., the speed at which the channel opens when the membrane is hyperpolarized—is also voltage dependent, with greater hyperpolarization leading to faster activation. In this respect, compared to the majority of voltage-gated ion channels, HCN channels are exceptionally slow. In fact, with the exception of a minor initial component termed "instantaneous current," which reaches steady-state amplitude within few milliseconds and whose molecular aspects are still unclear, the main component takes from hundreds of milliseconds to several seconds to reach the steady state. Gating kinetics are isoform and voltage dependent, with stronger hyperpolarization pulses leading to faster gating. In this regard, HCN1 is the isoform showing the fastest kinetics, reaching full activation in fewer than 100 milliseconds. Finally, HCN current does not inactivate—i.e., the current maintains steady-state amplitude until the membrane potential returns to depolarized values. This feature is also very unusual among voltage-gated channels, and it entails a number of important consequences, which will be discussed later in the chapter.

Functional Modulation by Multiple Heterogeneous Mechanisms

HCN channels are true voltage-gated channels. In other words, voltage is always sufficient to gate the channel. Nevertheless, gating properties may be modulated by several physiological mechanisms. These include covalent modifications, physical interaction with scaffold proteins and lipidic membrane constituents, and sensitivity to protons and other ions (He et al., 2014; Sartiani et al., 2017). This chapter will mainly focus on the examples for which strong evidence of physiological relevance in a native neuronal environment has been provided.

The modulation by intracellular cyclic nucleotides is a defining feature of HCN channels. The CNBD is highly conserved across isoforms and exerts a steric blocking action on the channel pore in the absence of the ligand. Cyclic adenosine monophosphate (cAMP) binding relieves self-inhibition, resulting in a positive shift of the activation curve and acceleration of gating kinetics. The overall effect exerted by cAMP-dependent modulation is an increase in current availability. Of note, cAMP promotes channel opening by increasing its voltage sensitivity; thus the same effect can be produced experimentally by increasing the amplitude of the hyperpolarizing test stimulus (Biel et al., 2009). Sensitivity to cAMP concentration, as well as the magnitude of the $V_{1/2}$ shift ("intrinsic activity"), varies to a large extent across isoforms: HCN2 and 4 are the most responsive (10–20 mV shift), while HCN1 is much less responsive (2–6 mV shift). As a striking exception, HCN3, although equipped with a normal CNBD, is not activated by cAMP. Rather, there is evidence for a negative shift of $V_{1/2}$ in this isoform. Cyclic guanosine monophosphate (cGMP) and cyclic cytosine monophosphate (cCMP) have also been reported to activate HCN channels, yet with lower potency and intrinsic activity. It is important to point out that a large majority of the studies addressing basic HCN channels' properties were carried out in heterologous cellular systems. Hence, the actual significance of HCN current modulation by cyclic nucleotides in native conditions is still incompletely understood.

HCN channels are sensitive to lipidic acids, normal constituents of the plasma membrane. Phosphatidylinositol 4,5 bisphosphate (PIP_2) depolarizes the $V_{1/2}$ by ~20 mV, thus greatly contributing to the physiological relevance of HCN current. The molecular mechanism underlying PIP_2 action on HCN channels is unknown, although it is certainly independent from the presence of cyclic nucleotides or the CNBD itself, as shown by experiments performed in CNBD-lacking mutants. In dopaminergic neurons of the *substantia nigra pars compacta*, expressing mainly HCN2 and 4, pharmacological inhibition of PIP_2 synthesis shifts the $V_{1/2}$ from −65 to −77 mV and reduces the frequency of autonomous firing (Zolles et al., 2006).

The presence of a protonable histidine residue near the S_4 helix makes murine HCN2 sensitive to intracellular pH. Acidic (pH~6) and alkaline (pH~9) environments affect gating by shifting $V_{1/2}$ to more negative or less negative potentials, respectively. This mechanism is thought to apply to other HCN isoforms and to contribute to the physiological role of these in the regulation of respiratory rhythm (Hawkins et al., 2015).

Several transmembrane and cytosolic proteins have been shown to interact with HCN channels and to modulate their expression and function. In nerve cells, the most important among these ancillary proteins is the tetratricopeptide repeat-containing Rab8b-interacting protein (TRIP8b). TRIP8b is a brain-specific cytosolic scaffold protein capable of interacting with the C-terminus domain of HCN channels. TRIP8b has several N-terminal splice variants that determine the heterogeneous subcellular localization of HCN channels. TRIP8b global knock-out (KO) mice display alterations in thalamic excitability, partly resembling the epileptic phenotype of HCN2 KO, but without cardiac phenotype. In the hippocampus, TRIP8b deletion causes overall reduction of HCN1 expression levels with disruption of the distinctive somatodendritic expression gradient (Lewis, Estep, & Chetkovich, 2010). HCN channel topology and basic current properties are illustrated in Figure 15.1.

Properties and Function of HCN Channels in the Nervous System

This section will overview the contribution of HCN channels to critical neuronal functions such as the regulation of resting membrane potential and intrinsic excitability, generation of rhythm, synaptic excitability. The influence of HCN current on the normal operation of neurons is determined by subunit stoichiometry, interplay with other membrane conductances and action of modulatory mechanisms.

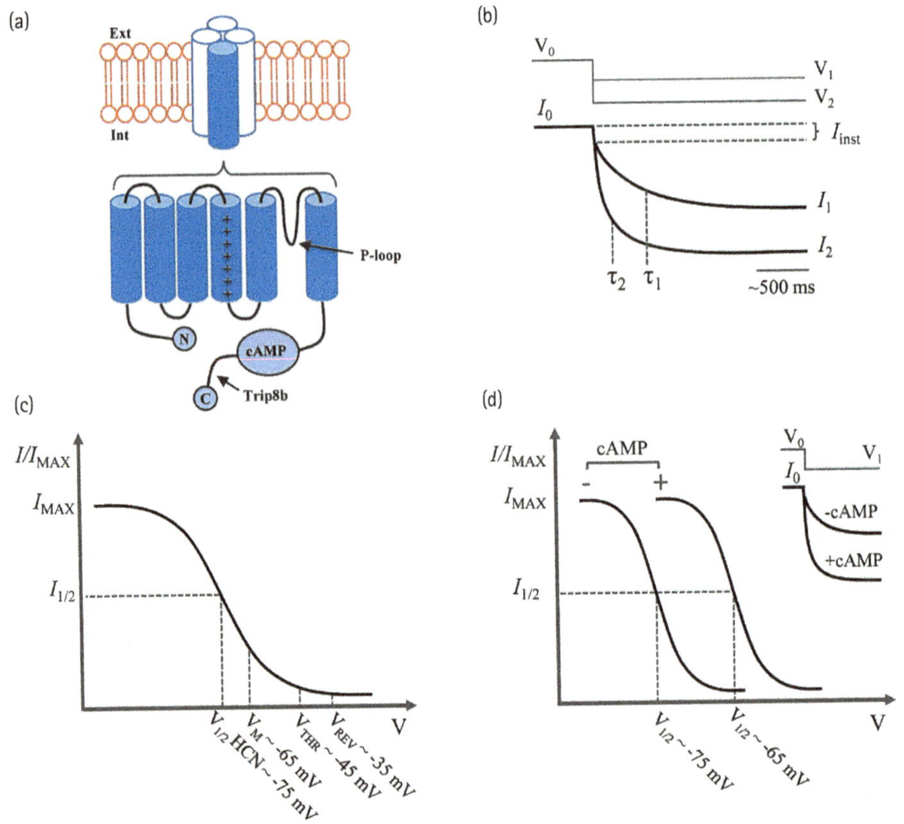

FIGURE 15.1 A. HCN channels are present at cell surface as homo- or hetero-tetrameric structures. Individual subunits are formed by 6 transmembrane domains (1–6, *from left to right*). Voltage-sensing domain is located in S_4, the pore-loop between S_5 and S_6, while most regulatory domains are found in the intracellular tail at C-terminus. **B.** Basic electrophysiological properties of HCN current. Negative voltage steps elicit an instantaneous current (I_{inst}), followed by a main, voltage-dependent, slow component. Gating kinetics (τ) are related to magnitude of voltage pulse. **C.** Normalized current-voltage curve highlighting $V_{1/2}$ and V_{REV} of HCN current in relation to critical membrane potential values such as resting (V_M) and AP threshold (V_{THR}). **D.** Intracellular cAMP concentration modulates the availability of HCN current by shifting the activation curve towards a more negative (-cAMP) or more positive (+cAMP) range of potentials. In addition, cAMP accelerates gating kinetics (*right hand*).

Neurophysiological Implications of the Unique Biophysical Properties of HCN Channels

A large body of literature is available on the role of HCN channels in nerve cells. The basic aspects of HCN channels' physiology in the CNS have been the topic of some recent, exhaustive reviews and will not be treated here in detail (He et al., 2014; Shah, 2016). However, in order to fully comprehend the relevance of HCN current in the physiology and pathology of the nervous system, it is necessary to review its basic biophysical properties proper, and in the context of neuronal and network excitability. In essence, these can be narrowed down to the following hallmarks: permeability to both Na^+ and K^+ (a); reversal potential (V_{REV}) lying close to action potential (AP) threshold (b) and at the bottom of the activation curve (*c*); slow gating kinetics (d). Because of properties *a* and *b*, HCN channels mediate a tonic inward current, which, at potentials below ~50 mV, depolarizes the membrane, promoting intrinsic AP firing. In the same conditions, however, HCN current diminishes membrane resistance, thus opposing synaptically generated depolarizations. This dual action accounts for the complex consequences of HCN channels' mutations for neuronal excitability reported in cellular and animal models. Property *c* implies that HCN current drives the membrane potential towards values where HCN channels are closed. In other words, HCN current is self-limiting, thus exerting a stabilizing force at subthreshold potentials that actively opposes both downward and upward shifts. Finally, property *d* confers a preference for low-frequency inputs to the membrane potential–stabilizing effect of HCN current. In other words, HCN current will effectively oppose, and thus filter out, slow (or low-frequency) voltage oscillations, while fast (or high-frequency) oscillations will escape filtering. In electronics, this process is called "high-pass filtering," and in neurons, HCN current serves as a high-pass filter.

The concepts outlined here concern general and constitutive aspects of neuronal HCN channels. The resulting impact on the electrical activity of the neuron varies greatly, based on membrane expression levels and relative HCN subunit prevalence. Furthermore, the influence of many heterogeneous factors, including ionic species, second messengers, membrane lipids, and auxiliary/scaffolding proteins, shapes the activity of HCN channels and their influence on intrinsic excitability and synaptic transmission. Because of these regulatory mechanisms, HCN channels may serve as final effectors of important signaling pathways initiated by circulating or synaptically released chemical messengers. The main contributions of HCN current to resting membrane potential, AP firing frequency, and synaptic excitability are summarized in Figure 15.2.

Regulation of Resting Membrane Potential and Intrinsic Excitability

The influence exerted by HCN current on the intrinsic excitability of neurons is complex. At –65 mV, a fraction of HCN channels are constitutively open (Kase & Imoto, 2012), sustaining a persistent Na^+ inflow that keeps the membrane relatively depolarized and opposes hyperpolarizing inputs. Vice-versa, depolarization causes channel deactivation, and the resulting cessation of the inward current has a hyperpolarizing effect. As a result, HCN current acts as a membrane potential stabilizer in the subthreshold range. Although blockage of HCN current hyperpolarizes the neuron and thereby reduces its

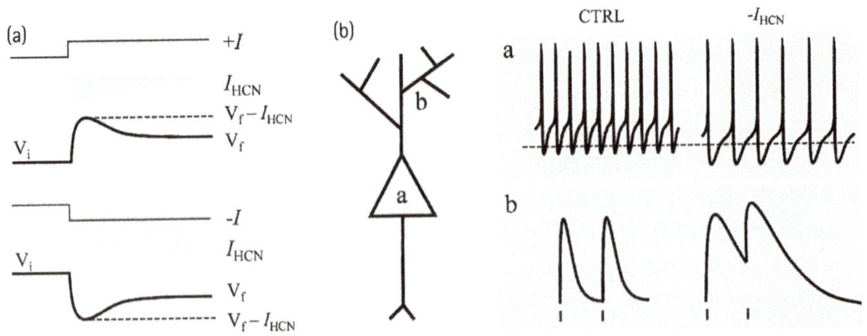

FIGURE 15.2 A. Stabilizing action exerted by HCN current on membrane potential following voltage perturbations. *Top*, injection of positive current +*I* causes fast depolarization of the membrane potential. However, deactivation of HCN current causes a rebound effect towards more negative potential V_1. Symmetrically, activation of HCN current promotes a positive rebound of membrane potential V_2 from initial hyperpolarization following injection of negative current −*I* (*bottom*). B. HCN current at somatic level (a) promotes firing in autonomous spiking neurons. Removal of HCN current hyperpolarizes the neuron and reduces firing rate. At somatodendritic level (b), HCN current accelerates the decay kinetics of EPSPs, thus promoting the ability of the neuron to resolve individual synaptic events (*left*). After removal of HCN current (*right*), temporal precision is impaired and summation of EPSPs occurs.

intrinsic excitability, for the same reason, the response to depolarizing synaptic inputs becomes potentiated. This dual action exerted by HCN current on overall neuronal excitability has been clearly established in many neuronal types and brain areas, including the thalamus, the hippocampus, and the cerebellum, and it has important repercussions on higher-order brain functions such as learning and memory, control of circadian rhythm, and sleep–wakefulness state (Gasparini & DiFrancesco, 1997; Maccaferri, Mangoni, Lazzari, & DiFrancesco, 1993; McCormick & Pape, 1990; Nolan, Dudman, Dodson, & Santoro, 2007).

In the axon initial segment of medial superior olive principal neurons, HCN1 current reduces spike probability by elevating the firing threshold. Serotonin 5-HT_{1A} receptor engagement shifts HCN channel $V_{1/2}$ in the hyperpolarizing direction, thereby relieving the brake on spike probability (Ko, Rasband, Meseguer, Kramer, & Golding, 2016). Serotonin-dependent modulation of HCN current is also involved in the control of respiratory rhythm in the retrotrapezoid nucleus (RTN). Here, activation of the 5-HT_7 serotonin receptor causes a cAMP-dependent depolarizing shift in $V_{1/2}$, increasing the firing rate of RTN neurons in vitro and the frequency of respiratory rhythm in vivo (Hawkins et al., 2015).

The ability of HCN channels to set the membrane potential plays an important role in the physiology of the retina. Early in situ hybridization studies have shown high levels of HCN1 in retinal photoreceptors (Moosmang et al., 2001). Later, functional studies have uncovered an important role exerted by HCN current in the regulation of the electrical events following phototransduction. In the outer segment of rod photoreceptors, a light flash shuts the inward, depolarizing ("dark") current mediated by cyclic nucleotide-gated (CNG) channels, leading to hyperpolarization of the membrane. This causes a strong reduction in synaptic release at the rod–bipolar cell synapse. Hyperpolarization also activates HCN1 current located in the membrane of the inner segment, in turn promoting the rebound of membrane potential back to the resting, depolarized state. Thus, synaptic activity recovers, reducing rod output and preventing saturation of the downstream retinal network (Seeliger et al., 2011). This regulatory action on photoreceptor function is at the basis of the side effects produced by ivabradine, an HCN channel blocker with medical applications in cardiovascular pathologies (Cangiano, Gargini, Della Santina, Demontis, & Cervetto, 2007).

Rhythmogenesis

In a functional interplay with other membrane ionic mechanisms, the activation-deactivation cycle of HCN channels helps setting the pace of subthreshold membrane oscillations, which determines AP discharge rate and, ultimately, neuronal output. HCN current cooperates with a persistent, subthreshold Na^+ current in setting a 4–10 Hz ("theta-like") rhythm in pyramidal and stellate cells of the entorhinal cortex (EC) layer II (Alonso & Llinas, 1989). Both cell types in this area are critically involved during spatial navigation tasks. These cells are named "grid cells," because they show periodic, hexagon-shaped firing locations that scale up progressively along the dorsal–ventral axis of the medial entorhinal cortex (Moser, Rowland, & Moser, 2015). In this structure, relative HCN1/HCN2 expression ratio and oscillation frequency decrease following a dorsal–ventral gradient (Notomi & Shigemoto, 2004). Such a dorsal–ventral frequency gradient correlates with the different time constants of HCN1 and HCN2 isoforms. In HCN1 KO mice, the dorsal–ventral gradient of the grid pattern is preserved, but the size and spacing of the grid fields, as well as the period of the accompanying theta modulation, are expanded (Giocomo & Hasselmo, 2009; Giocomo et al., 2011).

The firing pattern of thalamocortical neurons is governed by HCN current, here mediated by HCN2 channels. Adrenergic and serotonergic stimulation causes cAMP-mediated modulation of HCN current. During wakefulness and REM sleep, thalamocortical neurons fire in a single-spike, or "transmission," mode. In this state, information is effectively transmitted to the cortex. During certain physiological and pathological states, such as non-REM sleep and absence seizures, thalamocortical neurons fire in "burst" mode, and the transfer of signals to the cortex is thought to be less effective (McCormick & Pape, 1990). In agreement, global HCN2 deletion causes absence epilepsy with typical "spike-and-wave" discharges in EEG recordings (Ludwig et al., 2003).

A similar role has been described for HCN3 in leaflet neurons of the intergeniculate nucleus, a retino-recipient thalamic structure implicated in orchestrating the circadian rhythm. Here, HCN3-mediated current drives low-threshold burst firing and spontaneous oscillations and is bi-directionally modulated by PIP_2. Depletion of PIP_2 or pharmacological block of HCN current results in profound inhibition of excitability (Ying et al., 2011).

Synaptic Excitability and Plasticity

Due to the properties described previously, HCN current shapes the amplitude and temporal dynamics of synaptic potentials. This deeply affects the integrative properties of the somatodendritic compartment and the ability of the neuron to express different forms of synaptic plasticity.

Functional HCN channels are strongly expressed along the dendritic arborisation of neocortical and hippocampal neurons with an increasing soma-to-dendrites expression gradient (Bender et al., 2001; Harnett, Magee, & Williams, 2015; Lorincz, Notomi, Tamas, Shigemoto, & Nusser, 2002). Here, HCN current constitutes a shunt conductance accelerating the decay of excitatory post-synaptic potentials (EPSPs). As a result, temporal summation during EPSP sequences is reduced, and the ability of the neuron to resolve individual inputs at the somatic level is enhanced. This function, first discovered and characterized in pyramidal neurons of the hippocampal CA1 region and somatosensory cortex (Berger & Luscher, 2003; Magee, 1998, 1999; Williams & Stuart, 2000), has since been described in subcortical structures (Carbone, Costa, Provensi, Mannaioni, & Masi, 2017; Engel & Seutin, 2015; Masi et al., 2015; Ying et al., 2007). In the dendritic compartment of CA1 pyramidal neurons, the shunting effect exerted by HCN current limits voltage-dependent Ca^{2+} entry, with important consequences for synaptic excitability (Tsay, Dudman, & Siegelbaum, 2007). The non-uniform distribution of HCN current is also responsible for a phenomenon termed "site independence" of synaptic potentials, whereby the decay time of distally generated EPSPs is similar to that of proximally generated EPSPs (Williams & Stuart, 2000).

HCN current accelerates the decay time of both forward- and back-propagating depolarization waves, thus increasing the precision with which the dendrite detects the coincidence of salient electrical events. These include the coincidence of EPSPs and APs, an event leading to synaptic potentiation and, presumably, memory formation (Pavlov, Scimemi, Savtchenko, Kullmann, & Walker, 2011). In several brain areas, HCN function constrains long-term potentiation, thus affecting the associated cognitive functions. In keeping with brain-slice physiology experiments, transgenic animals have provided evidence of a role for HCN current in cognition. Indeed, HCN1 KO mice show improved hippocampal-dependent learning and memory performance. In these mice, proximal CA3–CA1 synapses function normally, whereas distal EC Layer III-CA1 contacts are potentiated, as predicted by stronger expression of HCN channels in this compartment in normal conditions (Nolan et al., 2004).

It has been suggested that HCN current affects higher cortical functions such as cognition, movement planning, and execution by setting the strength of connectivity within local microcircuits. HCN current is abundantly expressed in dorsal–lateral prefrontal cortex (dlPFC) layer III neurons, a population crucially involved in spatial working memory. Working memory is a short-term memory form engaged during the execution of complex cognitive tasks that requires the transient activation of intracortical microcircuits. In dlPFC layer III neurons, HCN channels are modulated by the opposite action of α_2-adrenergic (α_2-AR) and type-1 dopaminergic receptor (D_1R) stimulation of cAMP signaling. When α_2-AR stimulation inhibits the cAMP-HCN channel pathway, neurons are more excitable and more tightly connected to recurring microcircuit activity, thus performance is optimal. In contrast, negative affective states induced by stress cause a D_1R-mediated enhancement of the cAMP-HCN pathway, leading to a functional disconnection from the local network and impairment in working memory performance (Arnsten & Jin, 2014; M. Wang et al., 2007).

In the primary motor cortex of the mouse, HCN expression is specifically elevated in corticospinal neurons of layer V. HCN current confers to these neurons a 4 Hz-resonance preference and gates synaptic inputs from layer II/III pyramidal neurons, determining the efficacy of signal transmission between the two layers and thus the transmission efficacy of motor inputs. Also in this context, α_2-AR stimulation modulates HCN function (Sheets et al., 2011).

Resonance Properties

Dendritic HCN current confers to the neuron specific "resonance" properties; i.e., the ability to respond preferentially to inputs at a certain frequency, and determines the extent to which the neuron takes part to synchronous network activity. Due to its slow gating kinetics, HCN current has high-pass filtering properties, which, combined to the low-pass filtering action exerted by the membrane time constant, contribute to impart a resonance frequency in the 1–10 Hz range to sensorimotor neurons (Hutcheon, Miura, & Puil, 1996). In the hippocampus, the power of theta oscillation and the strength of the perforant path-CA1 pyramidal neuron synapses, both influenced by HCN1 function, are required for hippocampal-dependent learning and memory storage (Nolan et al., 2004). The ability to resonate in an HCN-dependent manner has been reported for other areas (Borel, Guadagna, Jang, Kwag, & Paulsen, 2013; Ulrich, 2002; Wang et al., 2006; Xue et al., 2012). Although the role of HCN current in oscillating activity and resonance properties has been clearly established at single cell or local network level, its actual significance for higher-order functions remains undetermined.

Neurotransmitter Release

HCN current at synaptic terminals has been reported to control the efficacy of vesicle release. HCN1 and HCN2 isoforms are present at GABAergic terminals of pallidal axon collaterals (Boyes, Bolam, Shigemoto, & Stanford, 2007) and at glutamatergic terminals making contacts onto EC layer III pyramidal neurons (Huang et al., 2011). In both cases, pharmacological or genetic KO of HCN function leads to elevation of spontaneous synaptic release. Huang et al. (2011) have suggested that HCN current exerts this inhibitory action by setting the potential to a value where N-type $Ca_V3.2$ Ca^{2+} channels are in a partially inactivated state. HCN deletion hyperpolarizes the terminal and removes Ca^{2+} channel inactivation. Thus, when the AP invades the terminal, a larger amount of Ca^{2+} flows in through $Ca_V3.2$ Ca^{2+} channels. A follow-up study by the same authors shows that HCN1 channels restrict the rate of exocytosis from a subset of cortical synaptic terminals within the EC and constrain spontaneous as well as evoked release (Huang, Li, Aguado, Lujan, & Shah, 2017).

HCN Channels in Pathological States of the Central Nervous System

To date, mutations in HCN genes have been clearly reported as primary causes of rare epilepsy forms. Furthermore, multiple lines of evidence suggest that acquired HCN current defects are associated to other pathological states as diverse as Parkinson's disease, neuropathic pain, substance use disorder. This last section will present some remarkable examples of HCN-associated pathologies and the mechanisms linking HCN abnormalities to disease.

Epilepsy

Epilepsy is the neurological disorder with strongest correlation with genetic or epigenetic alterations in HCN channels. In most HCN-linked epilepsies, increased neuronal excitability is associated with loss-of-function mutations in HCN1 and HCN2 isoforms (DiFrancesco & DiFrancesco, 2015; Nava et al., 2014). HCN1 null mice show increased susceptibility to kaynic acid-induced seizures. In addition, these mice show increased synaptic excitability of cortical neurons, in spite of the more negative membrane potential (Huang, Walker, & Shah, 2009). In contrast, HCN2 KO mice show absence seizures (Ludwig et al., 2003). Evidence of the role of HCN mutations in the pathophysiology of epilepsy are corroborated by genetic studies on human subjects. Next generation sequencing analysis on epileptic individuals point to a strong association between single-nucleotide mutations in HCN2 gene, leading to a loss-of-function phenotype, and idiopathic generalized epilepsy (DiFrancesco et al., 2011; Tang, Sander, Craven, Hempelmann, & Escayg, 2008). Another study, however, reported a significant association between febrile seizures and a single-point, gain-of-function mutation in HCN2 (Dibbens et al., 2010).

HCN loss of function phenotype may also result as a consequence of changes in the expression of auxiliary proteins, such as TRIP8b, that normally control channel surface expression or correct subcellular targeting. TRIP8b KO mice show spontaneous spike-wave discharges on EEG, the electrographic hallmark of absence seizures, which resemble the phenotype of global HCN2 deletion. At cellular and molecular levels, HCN2 channels are strongly reduced in thalamic-projecting cortical layer 5b neurons and thalamic relay neurons, but unaltered in inhibitory neurons of the reticular thalamic nucleus (Heuermann et al., 2016).

Interestingly, there is also evidence of the opposite relationship between HCN channels' expression and epilepsy, as a number of studies have reported remodeling of HCN1 and 2 expression following experimental seizure induction. In pyramidal neurons of the hippocampus, HCN1 expression goes down, while HCN2 expression goes up, after febrile seizures (Brewster et al., 2002). Abnormal HCN channels' expression has been detected in autopsy specimens from individuals suffering from temporal lobe epilepsy (Bender et al., 2003). The significance of such neuroadaptative change triggered by epileptic states and involving HCN expression remains to be clarified.

In summary, a relatively large number of preclinical and clinical studies support the evidence linking HCN channelopathy to epilepsy. However, they do not unambiguously point to a straightforward mechanistic link between HCN alteration and clinical manifestations. For these reasons, the exploitation of HCN channels as potential targets for antiepileptic medications requires further preclinical advancements.

Parkinson's Disease

Parkinson's disease (PD) is caused by massive degeneration of nigrostriatal dopaminergic (DA) neurons. *Substantia nigra pars compacta* (SNc) DA neurons express high levels of HCN2 and 4 (Dufour et al., 2014; Neu et al., 2002). Preclinical evidence has highlighted a correlative link between HCN loss of function and selective vulnerability of SNc DA neurons in PD models linked to mitochondrial failure. The Mitopark mouse, a transgenic PD model based on a DA neuron-targeted mitochondrial defect leading to selective nigrostriatal degeneration and PD-like phenotype, shows reduced HCN current density in SNc DA neurons (Branch et al., 2016; Good et al., 2011). Of note, HCN downregulation is an early event preceding nigrostriatal degeneration (Good et al., 2011). With brain slice patch clamp recordings, it has been reported that 1-Methyl-4-phenylpyridinium (MPP^+), a mitochondrial toxin able to produce selective nigrostriatal degeneration, causes HCN current inhibition in SNc DA neurons. MPP^+-induced HCN block leads to increased synaptic excitability (Masi, Narducci, Landucci, Moroni, & Mannaioni, 2013). In addition, in agreement with respective HCN expression levels, the effect of HCN block on synaptic excitability and Ca^{2+} entry is significantly greater in SNc compared to other DA subsets (Masi et al., 2015) and pharmacological blockade of HCN channels in vivo causes selective nigrostriatal DA degeneration (Carbone et al., 2017). These findings point to a significant correlation between altered HCN current and nigrostriatal vulnerability, with potential implications for the pathogenesis of PD and the development of disease-modifying medications.

Neuropathy

The precise role of HCN current in physiological somatosensory transmission is imperfectly understood. Indeed, healthy sensory neurons seem largely unaffected by pharmacological HCN blockade. Consistently, safety studies with HCN blocker ivabradine did not report any adverse reactions such as dysesthesias and paresthesias (Herrmann, Schnorr, & Ludwig, 2015). Therefore, it appears that HCN current acquires major functional relevance in pathological states. Multiple studies have suggested that HCN channels undergo changes in expression or function following neuronal damage or in experimental models of inflammation (Acosta et al., 2012; Chaplan et al., 2003; Jiang, Sun, Tu, & Wan, 2008; Papp, Holló, & Antal, 2010; Schnorr et al., 2014; Weng, Smith, Sathish, & Djouhri, 2012). These reports point to an association between HCN current upregulation and pathological pain conditions such as allodynia and hyperalgesia (Herrmann et al., 2015). HCN current upregulation seems to depend on multiple factors such as increased expression or functional modification (Descoeur et al., 2011; Papp et al., 2010; Resta et al., 2018; Schnorr et al., 2014), PIP_2-mediated modulation (Pian, Bucchi, Decostanzo, Robinson, & Siegelbaum, 2007), protein kinase A (PKA) over-activation (Cheng & Zhou, 2013), and elevated intracellular levels of cAMP triggered by inflammatory mediators (Jafri & Weinreich, 1998; Momin, Cadiou, Mason, & McNaughton, 2008; Resta et al., 2016). Some of the works cited here suggest that targeting pathological HCN current alterations in sensory neurons with selective blockers devoid of activity on cardiac isoforms can be seen as a therapeutic avenue in the treatment of pathological pain.

Retinal Pathology

The first evidence of an HCN-related pathology in the retina was the discovery that prolonged treatment with ivabradine, an HCN inhibitor used in the treatment of angina pectoris and chronic heart failure, causes visual disturbances described as sensations of enhanced brightness, called "phosphenes" (Cervetto, Demontis, & Gargini, 2007). These symptoms revert following drug discontinuation and do not lead to modification of overall retinal morphology or HCN expression in animals (Della Santina et al., 2010).

Other reports have addressed the role of HCN in the pathophysiology of degenerative diseases of the retina. In one study, the absence of HCN1 in cyclic nucleotide-gated channel β1 (CNGB1)-KO mice exacerbated photoreceptor degeneration, thus suggesting that intact HCN1 functions may have a pro-survival role in this cell type (Schon et al., 2016).

Addiction

Natural and artificial rewarding stimuli cause activation of the ventral tegmental area (VTA) DA system. These neurons show slow, autonomous, regular firing, which is transiently interrupted by bursts or pauses during reward or punishment (Schultz, 2016). The antagonism of the excitatory action exerted by psychostimulants on the reward DA system is considered the most promising strategy to develop pharmacological therapies to treat addiction. In this respect, HCN current contributes to shape the firing pattern of VTA DA both in physiological conditions and in the presence of substances of abuse such as ethanol (Appel, Liu, McElvain, & Brodie, 2003; Okamoto, Harnett, & Morikawa, 2006), cocaine (Arencibia-Albite, Vazquez-Torres, & Jimenez-Rivera, 2017; Goertz et al., 2015), and methamphetamine (Gonzalez et al., 2016). The development of subunit- or cell type–specific HCN blockers or modulators may control the rewarding actions of these substances (Novella Romanelli et al., 2016). However, there is also convincing evidence that one of the long-term effects of drug abuse is the induction of a negative hypodopaminergic state that substance intake can transiently compensate for in drug addicts (Diana, 2011; Koob & Volkow, 2010). In this respect, HCN current density and spontaneous firing frequency of VTA DA neurons from mice previously exposed to a voluntary drinking paradigm is inversely correlated to ethanol intake (Juarez et al., 2017). Prospectively, restoring the normal tone of the DA system in drug abusers through functional upregulation of HCN current may be effective in relieving craving or preventing relapse. To achieve this aim, however, it will be necessary to develop compounds able to enhance HCN current with subunit and neuronal type specificity (Novella Romanelli et al., 2016).

References

Acosta, C., McMullan, S., Djouhri, L., Gao, L., Watkins, R., Berry, C., . . . Lawson, S. N. (2012). HCN1 and HCN2 in rat DRG neurons: Levels in nociceptors and non-nociceptors, NT3-dependence and influence of CFA-induced skin inflammation on HCN2 and NT3 expression. *PLoS One, 7*(12), e50442. Retrieved from http://www.ncbi.nlm.nih.gov/pubmed/23236374. doi:10.1371/journal.pone.0050442

Alonso, A., & Llinas, R. R. (1989). Subthreshold Na^+-dependent theta-like rhythmicity in stellate cells of entorhinal cortex layer II. *Nature, 342*(6246), 175–177. doi:10.1038/342175a0

Appel, S. B., Liu, Z., McElvain, M. A., & Brodie, M. S. (2003). Ethanol excitation of dopaminergic ventral tegmental area neurons is blocked by quinidine. *Journal of Pharmacology and Experimental Therapeutics, 306*(2), 437–446. Retrieved from http://www.ncbi.nlm.nih.gov/pubmed/12721326. doi:10.1124/jpet.103.050963

Arencibia-Albite, F., Vazquez-Torres, R., & Jimenez-Rivera, C. A. (2017). Cocaine sensitization increases subthreshold activity in dopamine neurons from the ventral tegmental area. *Journal of Neurophysiology, 117*(2), 612–623. doi:10.1152/jn.00465.2016

Arnsten, A. F., & Jin, L. E. (2014). Molecular influences on working memory circuits in dorsolateral prefrontal cortex. *Progress in Molecular Biology and Translational Science, 122*, 211–231. Retrieved from http://www.ncbi.nlm.nih.gov/pubmed/24484703. doi:10.1016/B978-0-12-420170-5.00008-8

Bahring, R., Barghaan, J., Westermeier, R., & Wollberg, J. (2012). Voltage sensor inactivation in potassium channels. *Frontiers in Pharmacology, 3*, 100. doi:10.3389/fphar.2012.00100

Bender, R. A., Brewster, A., Santoro, B., Ludwig, A., Hofmann, F., Biel, M., & Baram, T. Z. (2001). Differential and age-dependent expression of hyperpolarization-activated, cyclic nucleotide-gated cation channel isoforms 1–4 suggests evolving roles in the developing rat hippocampus. *Neuroscience, 106*(4), 689–698. Retrieved from http://www.ncbi.nlm.nih.gov/pubmed/11682156.

Bender, R. A., Kirschstein, T., Kretz, O., Brewster, A. L., Richichi, C., Rüschenschmidt, C., ... Baram, T. Z. (2007). Localization of HCN1 channels to presynaptic compartments: Novel plasticity that may contribute to hippocampal maturation. *Journal of Neuroscience, 27*(17), 4697–4706. Retrieved from http://www.ncbi.nlm.nih.gov/pubmed/17460082. doi:10.1523/JNEUROSCI.4699-06.2007

Bender, R. A., Soleymani, S. V., Brewster, A. L., Nguyen, S. T., Beck, H., Mathern, G. W., & Baram, T. Z. (2003). Enhanced expression of a specific hyperpolarization-activated cyclic nucleotide-gated cation channel (HCN) in surviving dentate gyrus granule cells of human and experimental epileptic hippocampus. *Journal of Neuroscience, 23*(17), 6826–6836. Retrieved from http://www.ncbi.nlm.nih.gov/pubmed/12890777.

Berger, T., & Luscher, H. R. (2003). Timing and precision of spike initiation in layer V pyramidal cells of the rat somatosensory cortex. *Cerebral Cortex, 13*(3), 274–281. doi:10.1093/cercor/13.3.274

Biel, M., Wahl-Schott, C., Michalakis, S., & Zong, X. (2009). Hyperpolarization-activated cation channels: From genes to function. *Physiological Reviews, 89*(3), 847–885. Retrieved from http://www.ncbi.nlm.nih.gov/pubmed/19584315. doi:10.1152/physrev.00029.2008

Borel, M., Guadagna, S., Jang, H. J., Kwag, J., & Paulsen, O. (2013). Frequency dependence of CA3 spike phase response arising from h-current properties. *Frontiers in Cellular Neuroscience, 7*, 263. doi:10.3389/fncel.2013.00263

Boyes, J., Bolam, J. P., Shigemoto, R., & Stanford, I. M. (2007). Functional presynaptic HCN channels in the rat globus pallidus. *European Journal of Neuroscience, 25*(7), 2081–2092. Retrieved from http://www.ncbi.nlm.nih.gov/pubmed/17439493. doi:10.1111/j.1460-9568.2007.05463.x

Branch, S. Y., Chen, C., Sharma, R., Lechleiter, J. D., Li, S., & Beckstead, M. J. (2016). Dopaminergic neurons exhibit an age-dependent decline in electrophysiological parameters in the MitoPark mouse model of Parkinson's disease. *Journal of Neuroscience, 36*(14), 4026–4037. doi:10.1523/jneurosci.1395-15.2016

Brewster, A., Bender, R. A., Chen, Y., Dube, C., Eghbal-Ahmadi, M., & Baram, T. Z. (2002). Developmental febrile seizures modulate hippocampal gene expression of hyperpolarization-activated channels in an isoform- and cell-specific manner. *Journal of Neuroscience, 22*(11), 4591–4599. Retrieved from http://www.ncbi.nlm.nih.gov/pubmed/12040066.

Brewster, A. L., Chen, Y., Bender, R. A., Yeh, A., Shigemoto, R., & Baram, T. Z. (2007). Quantitative analysis and subcellular distribution of mRNA and protein expression of the hyperpolarization-activated cyclic nucleotide-gated channels throughout development in rat hippocampus. *Cerebral Cortex, 17*(3), 702–712. Retrieved from http://www.ncbi.nlm.nih.gov/pubmed/16648453. doi:10.1093/cercor/bhk021

Cangiano, L., Gargini, C., Della Santina, L., Demontis, G. C., & Cervetto, L. (2007). High-pass filtering of input signals by the Ih current in a non-spiking neuron, the retinal rod bipolar cell. *PLoS One, 2*(12), e1327. Retrieved from http://www.ncbi.nlm.nih.gov/pubmed/18091997. doi:10.1371/journal.pone.0001327

Carbone, C., Costa, A., Provensi, G., Mannaioni, G., & Masi, A. (2017). The hyperpolarization-activated current determines synaptic excitability, calcium activity and specific viability of substantia nigra dopaminergic neurons. *Frontiers in Cellular Neuroscience, 11*, 187. doi:10.3389/fncel.2017.00187

Cervetto, L., Demontis, G. C., & Gargini, C. (2007). Cellular mechanisms underlying the pharmacological induction of phosphenes. *British Journal of Pharmacology, 150*(4), 383–390. doi:10.1038/sj.bjp.0706998

Chaplan, S. R., Guo, H. Q., Lee, D. H., Luo, L., Liu, C., Kuei, C., . . . Dubin, A. E. (2003). Neuronal hyperpolarization-activated pacemaker channels drive neuropathic pain. *Journal of Neuroscience, 23*(4), 1169–1178. Retrieved from http://www.ncbi.nlm.nih.gov/pubmed/12598605.

Cheng, Q., & Zhou, Y. (2013). Novel role of KT5720 on regulating hyperpolarization-activated cyclic nucleotide-gated channel activity and dorsal root ganglion neuron excitability. *DNA and Cell Biology, 32*(6), 320–328. Retrieved from http://www.ncbi.nlm.nih.gov/pubmed/23713946. doi:10.1089/dna.2013.2021

Della Santina, L., Bouly, M., Asta, A., Demontis, G. C., Cervetto, L., & Gargini, C. (2010). Effect of HCN channel inhibition on retinal morphology and function in normal and dystrophic rodents. *Investigative Ophthalmology and Visual Science, 51*(2), 1016–1023. Retrieved from http://www.ncbi.nlm.nih.gov/pubmed/19741244. doi:10.1167/iovs.09-3680

Descoeur, J., Pereira, V., Pizzoccaro, A., Francois, A., Ling, B., Maffre, V., . . . Bourinet, E. (2011). Oxaliplatin-induced cold hypersensitivity is due to remodeling of ion channel expression in nociceptors. *EMBO Molecular Medicine, 3*(5), 266–278. Retrieved from http://www.ncbi.nlm.nih.gov/pubmed/21438154. doi:10.1002/emmm.201100134

Diana, M. (2011). The dopamine hypothesis of drug addiction and its potential therapeutic value. *Frontiers in Psychiatry, 2*. doi:10.3389/fpsyt.2011.00064

Dibbens, L. M., Reid, C. A., Hodgson, B., Thomas, E. A., Phillips, A. M., Gazina, E., . . . Petrou, S. (2010). Augmented currents of an HCN2 variant in patients with febrile seizure syndromes. *Annals of Neurology, 67*(4), 542–546. doi:10.1002/ana.21909

DiFrancesco, J. C., Barbuti, A., Milanesi, R., Coco, S., Bucchi, A., Bottelli, G., . . . DiFrancesco, D. (2011). Recessive loss-of-function mutation in the pacemaker HCN2 channel causing increased neuronal excitability in a patient with idiopathic generalized epilepsy. *Journal of Neuroscience, 31*(48), 17327–17337. Retrieved from http://www.ncbi.nlm.nih.gov/pubmed/22131395. doi:10.1523/JNEUROSCI.3727-11.2011

DiFrancesco, J. C., & DiFrancesco, D. (2015). Dysfunctional HCN ion channels in neurological diseases. *Frontiers in Cellular Neuroscience, 6*, 174. Retrieved from http://www.ncbi.nlm.nih.gov/pubmed/25805968. doi:10.3389/fncel.2015.00071

Doan, T. N., Stephans, K., Ramirez, A. N., Glazebrook, P. A., Andresen, M. C., & Kunze, D. L. (2004). Differential distribution and function of hyperpolarization-activated channels in sensory neurons and mechanosensitive fibers. *Journal of Neuroscience, 24*(13), 3335–3343. Retrieved from http://www.ncbi.nlm.nih.gov/pubmed/15056713. doi:10.1523/JNEUROSCI.5156-03.2004

Dufour, M. A., Woodhouse, A., & Goaillard, J. M. (2014). Somatodendritic ion channel expression in substantia nigra pars compacta dopaminergic neurons across postnatal development. *Journal of Neuroscience Research, 92*(8), 981–999. Retrieved from http://www.ncbi.nlm.nih.gov/pubmed/24723263. doi:10.1002/jnr.23382

Emery, E. C., Young, G. T., Berrocoso, E. M., Chen, L., & McNaughton, P. A. (2011). HCN2 ion channels play a central role in inflammatory and neuropathic pain. *Science, 333*(6048), 1462–1466. Retrieved from http://www.ncbi.nlm.nih.gov/pubmed/21903816. doi:10.1126/science.1206243

Engel, D., & Seutin, V. (2015). High dendritic expression of Ih in the proximity of the axon origin controls the integrative properties of nigral dopamine neurons. *Journal of Physiology, 593*(22), 4905–4922. doi:10.1113/jp271052

Franz, O., Liss, B., Neu, A., & Roeper, J. (2000). Single-cell mRNA expression of HCN1 correlates with a fast gating phenotype of hyperpolarization-activated cyclic nucleotide-gated ion channels (Ih) in central neurons. *European Journal of Neuroscience, 12*(8), 2685–2693. Retrieved from http://www.ncbi.nlm.nih.gov/pubmed/10971612.

Fyk-Kolodziej, B., & Pourcho, R. G. (2007). Differential distribution of hyperpolarization-activated and cyclic nucleotide-gated channels in cone bipolar cells of the rat retina. *Journal of Comparative Neurology, 501*(6), 891–903. Retrieved from http://www.ncbi.nlm.nih.gov/pubmed/17311321. doi:10.1002/cne.21287

Galligan, J. J., Tatsumi, H., Shen, K. Z., Surprenant, A., & North, R. A. (1990). Cation current activated by hyperpolarization (IH) in guinea pig enteric neurons. *American Journal of Physiology, 259*(6 Pt 1), G966–G972. doi:10.1152/ajpgi.1990.259.6.g966

Gasparini, S., & DiFrancesco, D. (1997). Action of the hyperpolarization-activated current (Ih) blocker ZD 7288 in hippocampal CA1 neurons. *Pflugers Archiv, 435*(1), 99–106. doi:10.1007/s004240050488

Giocomo, L. M., & Hasselmo, M. E. (2009). Knock-out of HCN1 subunit flattens dorsal-ventral frequency gradient of medial entorhinal neurons in adult mice. *Journal of Neuroscience, 29*(23), 7625–7630. doi:10.1523/jneurosci.0609-09.2009

Giocomo, L. M., Hussaini, S. A., Zheng, F., Kandel, E. R., Moser, M. B., & Moser, E. I. (2011). Grid cells use HCN1 channels for spatial scaling. *Cell, 147*(5), 1159–1170. Retrieved from http://www.ncbi.nlm.nih.gov/pubmed/22100643. doi:10.1016/j.cell.2011.08.051

Goertz, R. B., Wanat, M. J., Gomez, J. A., Brown, Z. J., Phillips, P. E., & Paladini, C. A. (2015). Cocaine increases dopaminergic neuron and motor activity via midbrain alpha1 adrenergic signaling. *Neuropsychopharmacology, 40*(5), 1151–1162. doi:10.1038/npp.2014.296

Gonzalez, B., Rivero-Echeto, C., Muniz, J. A., Cadet, J. L., Garcia-Rill, E., Urbano, F. J., & Bisagno, V. (2016). Methamphetamine blunts Ca(2^+) currents and excitatory synaptic transmission through D1/5 receptor-mediated mechanisms in the mouse medial prefrontal cortex. *Addiction Biology, 21*(3), 589–602. doi:10.1111/adb.12249

Good, C. H., Hoffman, A. F., Hoffer, B. J., Chefer, V. I., Shippenberg, T. S., Bäckman, C. M., . . . Lupica, C. R. (2011). Impaired nigrostriatal function precedes behavioral deficits in a genetic mitochondrial model of Parkinson's disease. *FASEB Journal, 25*(4), 1333–1344. Retrieved from http://www.ncbi.nlm.nih.gov/pubmed/21233488. doi:10.1096/fj.10-173625

Harnett, M. T., Magee, J. C., & Williams, S. R. (2015). Distribution and function of HCN channels in the apical dendritic tuft of neocortical pyramidal neurons. *Journal of Neuroscience, 35*(3), 1024–1037. Retrieved from http://www.ncbi.nlm.nih.gov/pubmed/25609619. doi:10.1523/JNEUROSCI.2813-14.2015

Hawkins, V. E., Hawryluk, J. M., Takakura, A. C., Tzingounis, A. V., Moreira, T. S., & Mulkey, D. K. (2015). HCN channels contribute to serotonergic modulation of ventral surface chemosensitive neurons and respiratory activity. *Journal of Neurophysiology, 113*(4), 1195–1205. Retrieved from http://www.ncbi.nlm.nih.gov/pubmed/25429115. doi:10.1152/jn.00487.2014

He, C., Chen, F., Li, B., & Hu, Z. (2014). Neurophysiology of HCN channels: From cellular functions to multiple regulations. *Progress in Neurobiology, 112*, 1–23. doi:10.1016/j.pneurobio.2013.10.001

Herrmann, S., Schnorr, S., & Ludwig, A. (2015). HCN channels—modulators of cardiac and neuronal excitability. *International Journal of Molecular Science, 16*(1), 1429–1447. Retrieved from http://www.ncbi.nlm.nih.gov/pubmed/25580535. doi:10.3390/ijms16011429

Heuermann, R. J., Jaramillo, T. C., Ying, S. W., Suter, B. A., Lyman, K. A., Han, Y., ... Chetkovich, D. M. (2016). Reduction of thalamic and cortical Ih by deletion of TRIP8b produces a mouse model of human absence epilepsy. *Neurobiology of Disease, 85*, 81–92. doi:10.1016/j.nbd.2015.10.005

Huang, Z., Li, G., Aguado, C., Lujan, R., & Shah, M. M. (2017). HCN1 channels reduce the rate of exocytosis from a subset of cortical synaptic terminals. *Scientific Reports, 7*, 40257. doi:10.1038/srep40257

Huang, Z., Lujan, R., Kadurin, I., Uebele, V. N., Renger, J. J., Dolphin, A. C., & Shah, M. M. (2011). Presynaptic HCN1 channels regulate Cav3.2 activity and neurotransmission at select cortical synapses. *Nature Neuroscience, 14*(4), 478–486. Retrieved from http://www.ncbi.nlm.nih.gov/pubmed/21358644. doi:10.1038/nn.2757

Huang, Z., Walker, M. C., & Shah, M. M. (2009). Loss of dendritic HCN1 subunits enhances cortical excitability and epileptogenesis. *Journal of Neuroscience, 29*(35), 10979–10988. Retrieved from http://www.ncbi.nlm.nih.gov/pubmed/19726656. doi:10.1523/JNEUROSCI.1531-09.2009

Hutcheon, B., Miura, R. M., & Puil, E. (1996). Models of subthreshold membrane resonance in neocortical neurons. *Journal of Neurophysiology, 76*(2), 698–714. Retrieved from doi: 10.1152/jn.1996.76.2.698

Jafri, M. S., & Weinreich, D. (1998). Substance P regulates Ih via a NK-1 receptor in vagal sensory neurons of the ferret. *Journal of Neurophysiology, 79*(2), 769–777. doi:10.1152/jn.1998.79.2.769

Jiang, Y. Q., Sun, Q., Tu, H. Y., & Wan, Y. (2008). Characteristics of HCN channels and their participation in neuropathic pain. *Neurochemical Research, 33*(10), 1979–1989. Retrieved from http://www.ncbi.nlm.nih.gov/pubmed/18461446. doi:10.1007/s11064-008-9717-6

Juarez, B., Morel, C., Ku, S. M., Liu, Y., Zhang, H., Montgomery, S., ... Han, M. H. (2017). Midbrain circuit regulation of individual alcohol drinking behaviors in mice. *Nature Communications, 8*(1), 2220. doi:10.1038/s41467-017-02365-8

Kase, D., & Imoto, K. (2012). The role of HCN channels on membrane excitability in the nervous system. *Journal of Signal Transduction, 2012*, 619747. Retrieved from http://www.ncbi.nlm.nih.gov/pubmed/22934165. doi:10.1155/2012/619747

Ko, K. W., Rasband, M. N., Meseguer, V., Kramer, R. H., & Golding, N. L. (2016). Serotonin modulates spike probability in the axon initial segment through HCN channels. *Nature Neuroscience, 19*(6), 826–834. doi:10.1038/nn.4293

Koob, G. F., & Volkow, N. D. (2010). Neurocircuitry of addiction. *Neuropsychopharmacology, 35*(1): 217–238. doi 10.1038/npp.2009.110.

Kullmann, P. H., Sikora, K. M., Clark, K. L., Arduini, I., Springer, M. G., & Horn, J. P. (2016). HCN hyperpolarization-activated cation channels strengthen virtual nicotinic EPSPs and thereby elevate synaptic amplification in rat sympathetic neurons. *Journal of Neurophysiology, 116*(2), 438–447. doi:10.1152/jn.00223.2016

Lamas, J. A. (1998). A hyperpolarization-activated cation current (Ih) contributes to resting membrane potential in rat superior cervical sympathetic neurones. *Pflugers Archiv, 436*(3), 429–435. doi:10.1007/s004240050653.

Lee, C. H., & MacKinnon, R. (2017). Structures of the human HCN1 hyperpolarization-activated channel. *Cell, 168*(1–2), 111–120. e111. doi:10.1016/j.cell.2016.12.023

Lewis, A. S., Estep, C. M., & Chetkovich, D. M. (2010). The fast and slow ups and downs of HCN channel regulation. *Channels (Austin)*, *4*(3), 215–231. Retrieved from http://www.ncbi.nlm.nih.gov/pubmed/20305382.

Lorincz, A., Notomi, T., Tamas, G., Shigemoto, R., & Nusser, Z. (2002). Polarized and compartment-dependent distribution of HCN1 in pyramidal cell dendrites. *Nature Neuroscience*, *5*(11), 1185–1193. doi:10.1038/nn962

Ludwig, A., Budde, T., Stieber, J., Moosmang, S., Wahl, C., Holthoff, K., ... Hofmann, F. (2003). Absence epilepsy and sinus dysrhythmia in mice lacking the pacemaker channel HCN2. *EMBO Journal*, *22*(2), 216–224. Retrieved from http://www.ncbi.nlm.nih.gov/pubmed/12514127. doi:10.1093/emboj/cdg032

Luján, R., Albasanz, J. L., Shigemoto, R., & Juiz, J. M. (2005). Preferential localization of the hyperpolarization-activated cyclic nucleotide-gated cation channel subunit HCN1 in basket cell terminals of the rat cerebellum. *European Journal of Neuroscience*, *21*(8), 2073–2082. Retrieved from http://www.ncbi.nlm.nih.gov/pubmed/15869503. doi:10.1111/j.1460-9568.2005.04043.x

Maccaferri, G., Mangoni, M., Lazzari, A., & DiFrancesco, D. (1993). Properties of the hyperpolarization-activated current in rat hippocampal CA1 pyramidal cells. *Journal of Neurophysiology*, *69*(6), 2129–2136. doi:10.1152/jn.1993.69.6.2129.

Magee, J. C. (1998). Dendritic hyperpolarization-activated currents modify the integrative properties of hippocampal CA1 pyramidal neurons. *Journal of Neuroscience*, *18*(19), 7613–7624.

Magee, J. C. (1999). Dendritic Ih normalizes temporal summation in hippocampal CA1 neurons. *Nature Neuroscience*, *2*(9), 848. doi:10.1038/12229

Masi, A., Narducci, R., Landucci, E., Moroni, F., & Mannaioni, G. (2013). MPP($^+$)-dependent inhibition of Ih reduces spontaneous activity and enhances EPSP summation in nigral dopamine neurons. *British Journal of Pharmacology*, *169*(1), 130–142. Retrieved from http://www.ncbi.nlm.nih.gov/pubmed/23323755. doi:10.1111/bph.12104

Masi, A., Narducci, R., Resta, F., Carbone, C., Kobayashi, K., & Mannaioni, G. (2015). Differential contribution of Ih to the integration of excitatory synaptic inputs in substantia nigra pars compacta and ventral tegmental area dopaminergic neurons. *European Journal of Neuroscience*, *42*(9), 2699–2706. Retrieved from http://www.ncbi.nlm.nih.gov/pubmed/26354486. doi:10.1111/ejn.13066

McCormick, D. A., & Pape, H. C. (1990). Properties of a hyperpolarization-activated cation current and its role in rhythmic oscillation in thalamic relay neurones. *Journal of Physiology*, *431*, 291–318. doi:10.1113/jphysiol.1990.sp018331.

Momin, A., Cadiou, H., Mason, A., & McNaughton, P. A. (2008). Role of the hyperpolarization-activated current Ih in somatosensory neurons. *Journal of Physiology*, *586*(24), 5911–5929. doi:10.1113/jphysiol.2008.163154

Monteggia, L. M., Eisch, A. J., Tang, M. D., Kaczmarek, L. K., & Nestler, E. J. (2000). Cloning and localization of the hyperpolarization-activated cyclic nucleotide-gated channel family in rat brain. *Brain Research: Molecular Brain Research*, *81*(1–2), 129–139. Retrieved from http://www.ncbi.nlm.nih.gov/pubmed/11000485.

Moosmang, S., Biel, M., Hofmann, F., & Ludwig, A. (1999). Differential distribution of four hyperpolarization-activated cation channels in mouse brain. *Biological Chemistry*, *380*(7–8), 975–980. Retrieved from http://www.ncbi.nlm.nih.gov/pubmed/10494850. doi:10.1515/BC.1999.121

Moosmang, S., Stieber, J., Zong, X., Biel, M., Hofmann, F., & Ludwig, A. (2001). Cellular expression and functional characterization of four hyperpolarization-activated pacemaker channels in cardiac and neuronal tissues. *European Journal of Biochemistry, 268*(6), 1646–1652. Retrieved from http://www.ncbi.nlm.nih.gov/pubmed/11248683.

Moser, M. B., Rowland, D. C., & Moser, E. I. (2015). Place cells, grid cells, and memory. *Cold Spring Harbor Perspectives in Biology, 7*(2), a021808. doi:10.1101/cshperspect.a021808

Muller, F., Scholten, A., Ivanova, E., Haverkamp, S., Kremmer, E., & Kaupp, U. B. (2003). HCN channels are expressed differentially in retinal bipolar cells and concentrated at synaptic terminals. *European Journal of Neuroscience, 17*(10), 2084–2096. doi:10.1046/j.1460-9568.2003.02634

Nava, C., Dalle, C., Rastetter, A., Striano, P., de Kovel, C. G., Nabbout, R., . . . Consortium, E. R. (2014). De novo mutations in HCN1 cause early infantile epileptic encephalopathy. *Nature Genetics, 46*(6), 640–645. Retrieved from http://www.ncbi.nlm.nih.gov/pubmed/24747641. doi:10.1038/ng.2952

Neu, A., Neuhoff, H., Trube, G., Fehr, S., Ullrich, K., Roeper, J., & Isbrandt, D. (2002). Activation of GABA(A) receptors by guanidinoacetate: A novel pathophysiological mechanism. *Neurobiology of Disease, 11*(2), 298–307. Retrieved from http://www.ncbi.nlm.nih.gov/pubmed/12505422.

Nolan, M. F., Dudman, J. T., Dodson, P. D., & Santoro, B. (2007). HCN1 channels control resting and active integrative properties of stellate cells from layer II of the entorhinal cortex. *Journal of Neuroscience, 27*(46), 12440–12451. doi:10.1523/jneurosci.2358-07.2007

Nolan, M. F., Malleret, G., Dudman, J. T., Buhl, D. L., Santoro, B., Gibbs, E., . . . Morozov, A. (2004). A behavioral role for dendritic integration: HCN1 channels constrain spatial memory and plasticity at inputs to distal dendrites of CA1 pyramidal neurons. *Cell, 119*(5), 719–732. doi:10.1016/j.cell.2004.11.020

Notomi, T., & Shigemoto, R. (2004). Immunohistochemical localization of Ih channel subunits, HCN1–4, in the rat brain. *Journal of Comparative Neurology, 471*(3), 241–276. doi:10.1002/cne.11039

Novella Romanelli, M., Sartiani, L., Masi, A., Mannaioni, G., Manetti, D., Mugelli, A., & Cerbai, E. (2016). HCN channels modulators: The need for selectivity. *Current Topics in Medicinal Chemistry, 16*(16), 1764–1791. doi:10.2174/1568026616999160315130832.

Okamoto, T., Harnett, M. T., & Morikawa, H. (2006). Hyperpolarization-activated cation current (Ih) is an ethanol target in midbrain dopamine neurons of mice. *Journal of Neurophysiology, 95*(2), 619–626. doi:10.1152/jn.00682.2005

Papp, I., Holló, K., & Antal, M. (2010). Plasticity of hyperpolarization-activated and cyclic nucleotide-gated cation channel subunit 2 expression in the spinal dorsal horn in inflammatory pain. *European Journal of Neuroscience, 32*(7), 1193–1201. Retrieved from http://www.ncbi.nlm.nih.gov/pubmed/20726890. doi:10.1111/j.1460-9568.2010.07370.x

Paspalas, C. D., Wang, M., & Arnsten, A. F. (2013). Constellation of HCN channels and cAMP regulating proteins in dendritic spines of the primate prefrontal cortex: Potential substrate for working memory deficits in schizophrenia. *Cerebral Cortex, 23*(7), 1643–1654. Retrieved from http://www.ncbi.nlm.nih.gov/pubmed/22693343. doi:10.1093/cercor/bhs152

Pavlov, I., Scimemi, A., Savtchenko, L., Kullmann, D. M., & Walker, M. C. (2011). I(h)-mediated depolarization enhances the temporal precision of neuronal integration. *Nature Communications, 2*, 199. Retrieved from https://www.ncbi.nlm.nih.gov/pubmed/21326231. doi:10.1038/ncomms1202

Pian, P., Bucchi, A., Decostanzo, A., Robinson, R. B., & Siegelbaum, S. A. (2007). Modulation of cyclic nucleotide-regulated HCN channels by PIP(2) and receptors coupled to phospholipase C. *Pflugers Archiv, 455*(1), 125–145. Retrieved from http://www.ncbi.nlm.nih.gov/pubmed/17605039. doi:10.1007/s00424-007-0295-2

Prole, D. L., & Yellen, G. (2006). Reversal of HCN channel voltage dependence via bridging of the S4–S5 linker and Post-S6. *Journal of General Physiology, 128*(3), 273–282. doi:10.1085/jgp.200609590

Ramakrishnan, N. A., Drescher, M. J., Khan, K. M., Hatfield, J. S., & Drescher, D. G. (2012). HCN1 and HCN2 proteins are expressed in cochlear hair cells: HCN1 can form a ternary complex with protocadherin 15 CD3 and F-actin-binding filamin A or can interact with HCN2. *Journal of Biological Chemistry, 287*(45), 37628–37646. Retrieved from http://www.ncbi.nlm.nih.gov/pubmed/22948144. doi:10.1074/jbc.M112.375832

Resta, F., Masi, A., Sili, M., Laurino, A., Moroni, F., & Mannaioni, G. (2016). Kynurenic acid and zaprinast induce analgesia by modulating HCN channels through GPR35 activation. *Neuropharmacology, 108*, 136–143. doi:10.1016/j.neuropharm.2016.04.038

Resta, F., Micheli, L., Laurino, A., Spinelli, V., Mello, T., Sartiani, L., … Masi, A. (2018). Selective HCN1 block as a strategy to control oxaliplatin-induced neuropathy. *Neuropharmacology, 131*, 403–413. doi:10.1016/j.neuropharm.2018.01.014

Santoro, B., Chen, S., Luthi, A., Pavlidis, P., Shumyatsky, G. P., Tibbs, G. R., & Siegelbaum, S. A. (2000). Molecular and functional heterogeneity of hyperpolarization-activated pacemaker channels in the mouse CNS. *Journal of Neuroscience, 20*(14), 5264–5275. Retrieved from http://www.ncbi.nlm.nih.gov/pubmed/10884310.

Sartiani, L., Mannaioni, G., Masi, A., Novella Romanelli, M., & Cerbai, E. (2017). The hyperpolarization-activated cyclic nucleotide-gated channels: From biophysics to pharmacology of a unique family of ion channels. *Pharmacological Reviews, 69*(4), 354–395. doi:10.1124/pr.117.014035

Schnorr, S., Eberhardt, M., Kistner, K., Rajab, H., Käßer, J., Hess, A., … Herrmann, S. (2014). HCN2 channels account for mechanical (but not heat) hyperalgesia during long-standing inflammation. *Pain, 155*(6), 1079–1090. Retrieved from http://www.ncbi.nlm.nih.gov/pubmed/24525276. doi:10.1016/j.pain.2014.02.006

Schon, C., Asteriti, S., Koch, S., Sothilingam, V., Garcia Garrido, M., Tanimoto, N., … Michalakis, S. (2016). Loss of HCN1 enhances disease progression in mouse models of CNG channel-linked retinitis pigmentosa and achromatopsia. *Human Molecular Genetics, 25*(6), 1165–1175. doi:10.1093/hmg/ddv639

Schultz, W. (2016). Dopamine reward prediction error coding. *Dialogues in Clinical Neuroscience, 18*(1), 23–32.

Seeliger, M. W., Brombas, A., Weiler, R., Humphries, P., Knop, G., Tanimoto, N., & Muller, F. (2011). Modulation of rod photoreceptor output by HCN1 channels is essential for regular mesopic cone vision. *Nature Communications, 2*, 532. doi:10.1038/ncomms1540

Shah, M. M. (2016). Hyperpolarization-activated cyclic nucleotide-gated channel currents in neurons. *Cold Spring Harbor Protocols, 2016*(7). doi:10.1101/pdb.top087346

Sheets, P. L., Suter, B. A., Kiritani, T., Chan, C. S., Surmeier, D. J., & Shepherd, G. M. (2011). Corticospinal-specific HCN expression in mouse motor cortex: I(h)-dependent synaptic integration as a candidate microcircuit mechanism involved in motor control. *Journal of Neurophysiology, 106*(5), 2216–2231. Retrieved from http://www.ncbi.nlm.nih.gov/pubmed/21795621. doi:10.1152/jn.00232.2011

Tang, B., Sander, T., Craven, K. B., Hempelmann, A., & Escayg, A. (2008). Mutation analysis of the hyperpolarization-activated cyclic nucleotide-gated channels HCN1 and HCN2 in idiopathic generalized epilepsy. *Neurobiology of Disease, 29*(1), 59–70. doi:10.1016/j.nbd.2007.08.006

Tsay, D., Dudman, J. T., & Siegelbaum, S. A. (2007). HCN1 channels constrain synaptically evoked $Ca2^+$ spikes in distal dendrites of CA1 pyramidal neurons. *Neuron, 56*(6), 1076–1089. Retrieved from http://www.ncbi.nlm.nih.gov/pubmed/18093528. doi:10.1016/j.neuron.2007.11.015

Ulrich, D. (2002). Dendritic resonance in rat neocortical pyramidal cells. *Journal of Neurophysiology, 87*(6), 2753–2759. doi:10.1152/jn.2002.87.6.2753.

Wang, M., Ramos, B. P., Paspalas, C. D., Shu, Y., Simen, A., Duque, A., . . . Arnsten, A. F. (2007). Alpha2A-adrenoceptors strengthen working memory networks by inhibiting cAMP-HCN channel signaling in prefrontal cortex. *Cell, 129*(2), 397–410. Retrieved from http://www.ncbi.nlm.nih.gov/pubmed/17448997. doi:10.1016/j.cell.2007.03.015

Wang, W. T., Wan, Y. H., Zhu, J. L., Lei, G. S., Wang, Y. Y., Zhang, P., & Hu, S. J. (2006). Theta-frequency membrane resonance and its ionic mechanisms in rat subicular pyramidal neurons. *Neuroscience, 140*(1), 45–55. doi:10.1016/j.neuroscience.2006.01.033

Weng, X., Smith, T., Sathish, J., & Djouhri, L. (2012). Chronic inflammatory pain is associated with increased excitability and hyperpolarization-activated current (Ih) in C- but not Aδ-nociceptors. *Pain, 153*(4), 900–914. Retrieved from http://www.ncbi.nlm.nih.gov/pubmed/22377439. doi:10.1016/j.pain.2012.01.019

Williams, S. R., & Stuart, G. J. (2000). Site independence of EPSP time course is mediated by dendritic I(h) in neocortical pyramidal neurons. *Journal of Neurophysiology, 83*(5), 3177–3182. doi:10.1152/jn.2000.83.5.3177.

Xue, W. N., Wang, Y., He, S. M., Wang, X. L., Zhu, J. L., & Gao, G. D. (2012). SK- and h-current contribute to the generation of theta-like resonance of rat substantia nigra pars compacta dopaminergic neurons at hyperpolarized membrane potentials. *Brain Structure and Function, 217*(2), 379–394. Retrieved from http://www.ncbi.nlm.nih.gov/pubmed/22108680. doi:10.1007/s00429-011-0361-6

Yellen, G. (1998). The moving parts of voltage-gated ion channels. *Quarterly Reviews of Biophysics, 31*(3), 239–295.

Ying, S. W., Jia, F., Abbas, S. Y., Hofmann, F., Ludwig, A., & Goldstein, P. A. (2007). Dendritic HCN2 channels constrain glutamate-driven excitability in reticular thalamic neurons. *Journal of Neuroscience, 27*(32), 8719–8732. Retrieved from http://www.ncbi.nlm.nih.gov/pubmed/17687049. doi:10.1523/JNEUROSCI.1630-07.2007

Ying, S. W., Tibbs, G. R., Picollo, A., Abbas, S. Y., Sanford, R. L., Accardi, A., . . . Goldstein, P. A. (2011). PIP2-mediated HCN3 channel gating is crucial for rhythmic burst firing in thalamic intergeniculate leaflet neurons. *Journal of Neuroscience, 31*(28), 10412–10423. Retrieved from http://www.ncbi.nlm.nih.gov/pubmed/21753018. doi:10.1523/JNEUROSCI.0021-11.2011

Zhang, H., & Cuevas, J. (2002). Sigma receptors inhibit high-voltage-activated calcium channels in rat sympathetic and parasympathetic neurons. *Journal of Neurophysiology, 87*(6), 2867–2879. Retrieved from http://www.ncbi.nlm.nih.gov/pubmed/12037190.

Zolles, G., Klöcker, N., Wenzel, D., Weisser-Thomas, J., Fleischmann, B. K., Roeper, J., & Fakler, B. (2006). Pacemaking by HCN channels requires interaction with phosphoinositides. *Neuron, 52*(6), 1027–1036. Retrieved from http://www.ncbi.nlm.nih.gov/pubmed/17178405. doi:10.1016/j.neuron.2006.12.005

SECTION 4

OTHER CHANNELS

CHAPTER 16

TANDEM PORE DOMAIN POTASSIUM CHANNELS

DOUGLAS A. BAYLISS

Introduction

In the nervous system, maintaining a negative resting membrane potential—i.e., where the voltage across the plasma membrane is more negative on the intracellular side—is vital for cell function. For neurons, the electrochemical driving force provided by this negative potential allows for ion flux that underlies the rapid electrical and chemical signaling that drives all manner of behaviors; for glial cells, it permits the transmembrane transport of ions and signaling molecules that supports their various homeostatic functions.

A stable negative resting membrane potential represents a steady state condition, at which there is no *net* current due to ionic flux across the membrane resistance. This is not an equilibrium. The major ionic species (K^+, Na^+, Ca^{2+}, Cl^-, HCO_3^-, etc.) are differentially distributed across the cell membrane, with Nernst equilibrium potentials (E_{ion}) that are not identical to the resting membrane potential (E_m). Thus, a basal current for each species arises at steady state from ionic flux determined by the resulting electrochemical gradient (E_m–E_{ion}) across a finite membrane conductance, as defined by Ohm's Law—essentially, there is some "leak" current for each ionic species. Additional currents derive from electrogenic ion pumps and transporters that maintain steady state ionic gradients. In sum, the steady state E_m results from combined contributions of all ion currents, with the E_m tending toward the E_{ion} of the particular ion with the greatest relative resting conductance. For most cells, E_K is a negative value (~–90 mV) due to the ~40-fold higher concentration of K^+ ions inside the cell, and, because the resting conductance is highest for K^+ ions, the cell obtains a negative resting E_m near E_K.

This basic concept—that cells possess a dominant "leak" K^+ conductance under basal conditions—has been long understood. However, specific K^+ channel candidates that might mediate this "leak" K^+ conductance became available only with the advent of molecular cloning, initially with the identification of the voltage-gated (K_V) and inwardly rectifying (Kir) family of K^+ channels. Indeed, at membrane potentials near the resting E_m, some K_V channels display so-called window currents (persistent currents arising from overlap of activation and inactivation voltage ranges), and some Kir channels conduct constitutive current. Although these K_V and Kir channels could conceivably underlie some "leak" K^+ conductance, their voltage-dependent characteristics do not match the relatively voltage-independent characteristics expected of a classical "leak" conductance. Moreover, this more classical type of "leak" conductance had been observed in various cell types, often as a voltage-independent K^+ current modulated by various physicochemical factors, neuromodulators, and clinically relevant drugs, but insensitive to classical blockers of K_V and Kir channels (e.g., TEA [tetraethylammonium]) (Franks & Lieb, 1988; Nicoll, 1988; Nicoll, Malenka, & Kauer, 1990). A molecular basis for this alternative, more classical "leak" K^+ conductance was ultimately realized with the cloning of the well-modulated and relatively voltage-independent K2P channels (Goldstein, Bockenhauer, O'Kelly, & Zilberberg, 2001; Goldstein, Price, Rosenthal, & Pausch, 1996; Goldstein, Wang, Ilan, & Pausch, 1998; Lesage, Guillemare, et al., 1996; Lesage & Lazdunski, 2000). It is now clear that this unique, structurally distinct family of "leak" K^+ channels plays an important role in cellular and integrative physiological functions mediated by the brain and peripheral nervous system.

General Features of K2P Channels

The *KCNK* gene family encodes the "two pore-domain" K^+ channels (K2P) that present unique structural and functional characteristics, distinctly different from those of K_V and Kir channels. The mammalian *KCNK* gene family includes 15 members, which are distributed among six clades based on both sequence homology and shared functional properties of the encoded K2P channels (Fig. 16.1A). A numerical International Union of Basic and Clinical Pharmacology (IUPHAR) nomenclature was established for K2P channel subunits, which was based on the Human Genome Organization (HUGO) naming of the corresponding genes according to the chronological order of their identification (Goldstein et al., 2005). In addition to these formal naming schemes, the channels are often referred to by a set of acronyms that were derived from some salient physiological or pharmacological property (Enyedi & Czirjak, 2010; Feliciangeli, Chatelain, Bichet, & Lesage, 2015). Due to vagaries inherent in the naming process, including later recognition that channel genes originally found in other species were actually orthologues of a differently named human gene, the family now includes 15 genes (and corresponding channels) that are numbered from 1–18, where 8, 11, and 14 are omitted. Expression of K2P subunits individually in heterologous systems yielded functional channel activity for all but three (*KCNK7*, *KCNK15*, *KCNK12*); these "silent" genes are nevertheless expressed in some tissues, including brain, and recent work suggests that their activity may depend in certain circumstances on co-expression of another K2P subunit (Blin et al., 2014) Renigunta, Zou, Kling, Schlichthorl, & Daut, 2014).

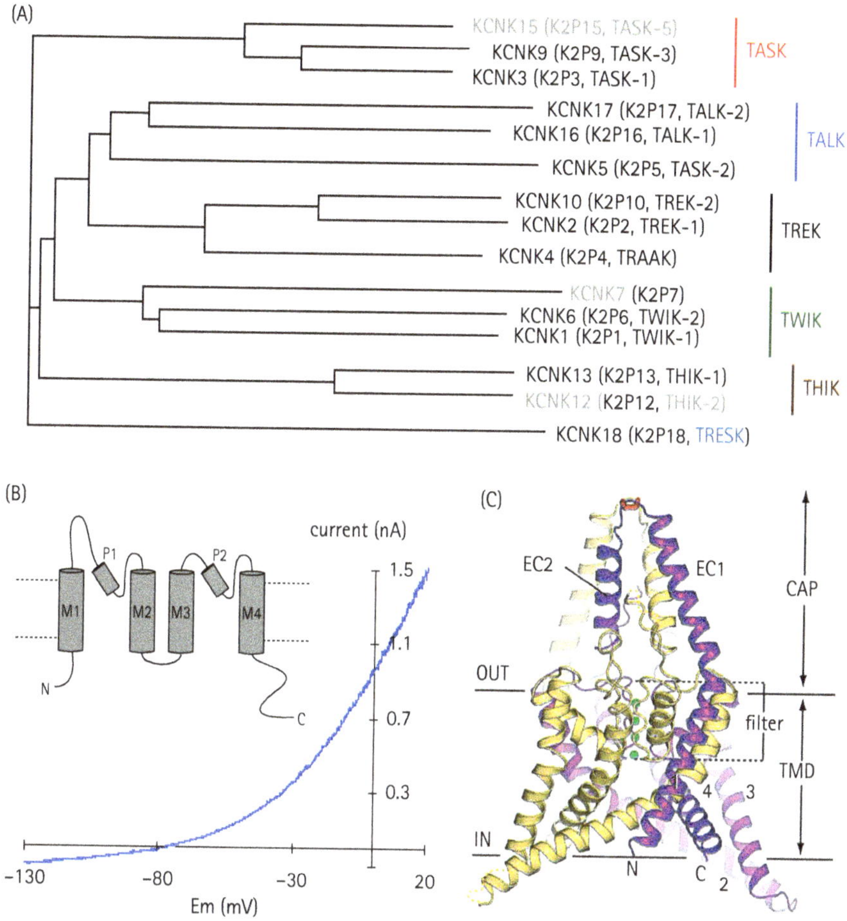

FIGURE 16.1 **The *KCNK* gene family. A.** Phylogeny analysis based on human K2P channel subunits. **B.** Weakly rectifying whole cell current-voltage (*I*–V) relationship obtained from mouse K2P2 expressed in HEK293 cells (adapted from Murbartian et al., 2005). *Inset*: Schematic of the general topology of K2P channel subunits. **C.** Structure of K2P10 determined by X-ray crystallography (in the "up-state" configuration). Note the prominent extracellular cap domain, and the fenestration through the channel vestibule.

From Dong et al., 2015. Reprinted with permission from AAAS.

Unique Structural Considerations

The primary structure of K2P channel subunits predicted a topology comprising cytoplasmic N- and C-termini, four transmembrane domains (M1 to M4), an extended M1P1 extracellular domain, a cytoplasmic M2M3 region, and two reentrant pore loops (P1, P2) (Fig. 1B, inset). The existence of two pore (P) loops in each subunit, rather than the single P-loop typical of other K^+ channel subunits, accounts for their designation as two pore-domain K^+ channels (K2P channels). For K_V and Kir channels, the central pore of the protein is formed from four P-loops, one from each subunit, in a tetrameric configuration (Jiang et al., 2003). For K2P channels, therefore, a dimeric conformation in which each of the two subunits contributes two P-loops, allows assembly of a comparable *pseudo*tetrameric channel pore (Brohawn, Campbell, & MacKinnon, 2013, 2014; Brohawn, del Marmol, & MacKinnon, 2012; Dong et al., 2015; Lolicato et al., 2017; McClenaghan et al., 2016; Miller & Long, 2012). It is now clear that K2P subunits can form channels as homodimers and heterodimers, combining with closely related subunits within their particular clade to yield channels with distinct properties (e.g., K2P3:K2P9; K2P2:K2P10) (Berg, Talley, Manger, & Bayliss, 2004; Blin et al., 2016; Kang, Han, Talley, Bayliss, & Kim, 2004; Lengyel, Czirjak, & Enyedi, 2016; Levitz et al., 2016)and/or allowing functional expression of otherwise silent subunits (K2P12 unsilenced by K2P13) (Blin et al., 2014) Renigunta et al., 2014); there have also been reports of heterodimeric channels formed between K2P subunits from different subgroups (K2P1:K2P3 or K2P1:K2P9; K2P1:K2P2) (Hwang et al., 2014; Wang et al., 2016; Zhou et al., 2009).

The characteristics expected for these channels from early biochemical and electrophysiological studies were subsequently verified by X-ray crystallographic analysis obtained for three different K2P channels (K2P1, K2P2 K2P4, K2P10) (Fig. 1C); these K2P channel structures revealed additional interesting features, some expected and others that had not been anticipated (Brohawn et al., 2013; Brohawn, Campbell, et al., 2014; Brohawn et al., 2012; Dong et al., 2015; Lolicato et al., 2017; McClenaghan et al., 2016; Miller & Long, 2012).

A large extracellular cap domain, not present in other K^+ channels, is formed by the extended M1P1 loop. In the channel structures, the M1P1 domains of the two subunits were covalently linked at the peak of the cap via a cysteine–disulfide bridge, a covalent link that may stabilize the dimer, as predicted based on earlier biochemical and functional studies of K2P1 (Hwang et al., 2014; Lesage, Reyes, et al., 1996). The cap appears in all structures to date, but the disulfide is unlikely to be found in all K2P channels, because the relevant cysteine residues are not uniformly present (e.g., there is no analogous M1P1 cysteine in K2P3 or K2P9 subunits) (Goldstein et al., 2016). At the peak of the cap, a "domain swapping" is observed between the two subunits, such that outer pore helices from one subunit envelop inner pore helices from the other subunit; whether there is any functional significance to this unique arrangement remains to be clarified, but it has been observed in structures from different K2P channels. The extracellular cap blocks access of ions to the pore from directly overhead, but it overlays two pathways lined with negatively charged residues that are large enough to funnel hydrated K^+ ions toward the pore.

The pore resembles that of other K^+ channels even though it derives in K2P channels from a pseudotetrameric arrangement in which non-identical residues from P1 and P2 pore domains comprise the selectivity filter, and is formed within the unique domain-swapping arrangement mentioned here. A large vestibule below the selectivity filter is also characteristic of other K^+ channels, although a unique feature is apparent in the K2P structure: a fenestration lined by hydrophobic residues provides access between the lipid bilayer and the vestibule, which may be occupied by lipids or ligands that modulate channel gating. The fenestration is apparent in K2P channel structures in a "down-state," but not in an alternative "up-state," reflecting conformational rearrangements around a hinge glycine in the subunit TM4 domains (Brohawn et al., 2013; Brohawn, Campbell, et al., 2014; Brohawn et al., 2012; Dong et al., 2015; Lolicato et al., 2017; McClenaghan et al., 2016; Miller & Long, 2012). The relationship between the down and up conformational states (i.e., either with or without the fenestration) and the conductive state of the channels remains to be firmly established.

At the cytoplasmic end of the pore, K2P channels do not possess the classical bundle-crossing gate that physically impedes ionic current in other K^+ channels; rather, functional studies indicate that K2P channels remain accessible to cytoplasmic ions up to the level of the selectivity filter, even when the channel is closed (Piechotta et al., 2011). Thus, gating in K2P channels takes place at the selectivity filter itself.

A cytoplasmic domain emerging at the end of TM4 appears to run antiparallel to the lipid bilayer, and provides a structural link between subunits. Interestingly, this C-helix region of the channel is also implicated as a functionally important site for effects of various factors that modulate K2P channel function. At this point, defining a structural basis for coupling actions of this domain to selectivity-filter gating remains an area of active research (Bagriantsev, Clark, & Minor, 2012; Bagriantsev, Peyronnet, Clark, Honore, & Minor, 2011; Chemin, Patel, Duprat, Lauritzen, et al., 2005; Honore, 2007; Honore, Maingret, Lazdunski, & Patel, 2002; Lolicato et al., 2017).

Unique Functional Characteristics

Heterologous expression of K2P channel subunits revealed properties that were unlike those of the previously identified K_V or Kir channels (Enyedi & Czirjak, 2010; Feliciangeli et al., 2015; Goldstein et al., 2001; Goldstein et al., 1998; Honore, 2007; Kim, 2005; Renigunta, Schlichthorl, & Daut, 2015; Talley, Sirois, Lei, & Bayliss, 2003). K2P channels invariably generated K-selective currents across a wide range of membrane potentials, including at negative potentials where neurons find themselves at rest or between action potentials (Fig. 1B). Thus, these K2P channels appeared well suited to provide a background K^+ conductance to establish this negative resting membrane potential. In addition, a strong instantaneous K2P current component accompanied step changes in voltage, and the associated current-voltage plots typically deviated only weakly from an Ohmic (i.e., linear) relationship, especially in symmetrical K^+ solutions. A more prominent, though still weak, rectification was obtained in physiological K^+ solutions that could often be approximated by the Goldman-Hodgkin-Katz (GHK) equation that accounts for effects of asymmetrical K^+ concentrations on ionic current. These latter considerations implied that K2P channels provided a background K^+ current that also fit a classical expectation for a "leak" or "open-rectifier" current.

This characterization remains largely accurate in a description of general K2P channel properties, but it obscures some important details (see Renigunta et al., 2015). First, unitary recordings revealed that single K2P channels, like other self-respecting ion channels, undergo stochastic transitions between closed and open states; they are not simply open,

K^+-selective membrane pores. Second, some K2P channel currents show inward rectification rather than outward rectification (Lesage, Guillemare, et al., 1996), indicating that that they do not dutifully follow the dictates of Ohm's Law for a voltage-independent open pore. Third, K2P channels show additional voltage dependence, both in the form of a time-dependent component of whole cell current that develops following changes in voltage step, and in voltage-dependent changes in single-channel open probability (P_O) (Kim, Bang, & Kim, 1999; Renigunta et al., 2015; Schewe et al., 2016). The basis for this voltage-dependence in P_O is unclear, since these channels do not possess a charged TM segment like the S4 domain that mediates voltage sensing in K_V and other voltage-dependent ion channels; recent work suggests a novel ion-flux gating mechanism that derives from ion movement within the membrane electric field (Schewe et al., 2016). Thus, the whole cell GHK rectification observed in some K2P channels is actually a fortuitous consequence of the combination of single-channel rectification and voltage-dependent gating (Kim et al., 1999; Renigunta et al., 2015). These more complex underlying channel properties notwithstanding, the observation that K2P channels provide weakly rectifying macroscopic K^+ currents across a wide range of membrane potentials remains valid, and they can thereby provide a "leak" K^+ current in the cell membrane.

The classical leak current designation, at least as incorporated into equivalent electrical circuit models, envisions a static background conductance. However, several different modulatory mechanisms that impinge on K2P channels were identified essentially immediately upon their functional expression. It was also subsequently recognized that these channels may be molecular substrates for the various non-K_V and non-Kir leak K^+ currents that had been recorded in native cells upon stimulation by various physicochemical factors, neuromodulators, and clinically relevant drugs.

Here, we discuss the principal defining modulatory characteristics of the major subgroups of K2P channels, especially as they relate to specific nervous system functions of those channels. In some cases, the channels are most prominently expressed outside of the nervous system, and we touch on those only briefly for completeness. A number of additional, exhaustive reviews on K2P channel structure and function are also available (Enyedi & Czirjak, 2010; Feliciangeli et al., 2015; Goldstein et al., 2001; Goldstein et al., 1998; Honore, 2007; Kim, 2005; Renigunta et al., 2015; Sepulveda, Pablo Cid, Teulon, & Niemeyer, 2015; Talley et al., 2003).

Properties of K2P Channel Subgroups

TWIK Family

The TWIK (_T_andem of P domains in a _w_eak _i_nwardly rectifying _K_$^{\pm}$ channel) family includes K2P1 (TWIK-1), K2P6 (TWIK-2) and K2P7 (*KCNK7*). Among this group, small currents were initially obtained following heterologous expression of cloned K2P1 and K2P6 subunits in Xenopus oocytes (Lesage, Guillemare, et al., 1996; Patel et al.,

2000; Salinas et al., 1999), both of which yielded weakly inwardly rectifying current-voltage (*I*-V) relationships that account for their common name; the currents from K2P6 were notable for their slow inactivation, an uncommon property in K2P channels (Patel et al., 2000). Functional currents have not yet been obtained from expression of recombinant *KCNK7* subunits. K2P1 and the (so far) non-functional *KCNK7* are both expressed throughout the nervous system (Lesage et al., 1997; Medhurst et al., 2001; Talley, Solorzano, Lei, Kim, & Bayliss, 2001); K2P6 is found mostly in the periphery (Medhurst et al., 2001), and will not be considered further.

K2P1 was the founding member of mammalian K2P channel family (Lesage, Guillemare, et al., 1996), and provided one of the initial high-resolution K2P structures (Miller & Long, 2012). However, there remains significant controversy and confusion as to the specific channel properties and physiological roles of K2P1. This uncertainty has arisen, in large part, from difficulties encountered by many groups in recording K2P1 channel currents in heterologous expression systems. In addition, knockout mice have not yet yielded any major insights into the K2P1 functions within the nervous system.

Various mechanisms have been advanced to explain the diminutive K2P1 channel currents. A dynamic covalent modification of membrane bound K2P1, involving the addition and removal of a small ubiquitin-like modulator (SUMO) adducts at Lys-274, was proposed to silence and activate the channel, respectively (Plant et al., 2010; Rajan, Plant, Rabin, Butler, & Goldstein, 2005); the de-sumoylated K2P1 channels showed GHK rectification rather than inward rectification (Rajan et al., 2005). The supporting evidence for this SUMO-based mechanism is compelling, although other laboratories have found the results difficult to replicate (Feliciangeli et al., 2007). An alternative (or additional) explanation for small K2P1 currents envisages constitutive, brisk, endocytic trafficking of the channel (Bichet et al., 2015; Feliciangeli et al., 2010). It was found that rapid retrieval of K2P1 channels from the plasma membrane occurred via a dynamin-mediated process requiring a C terminal di-isoleucine repeat of K2P1 (Feliciangeli et al., 2010); the process is regulated by Gαi/o-coupled receptors, which act to enhance membrane levels of K2P1 (Feliciangeli et al., 2010; Wang et al., 2016). Finally, a novel hydrophobic gating mechanism was suggested from molecular dynamic simulations based on the K2P1 structure (Aryal, Abd-Wahab, Bucci, Sansom, & Tucker, 2014). According to this model, a hydrophobic cuff within the pore leads to de-wetting of the inner cavity to impede ion permeation (Aryal et al., 2014); accordingly, in silico substitutions of polar residues at sites within that cuff were associated with water retention in the simulations (i.e., Leu-146, Leu-261), while, experimentally, the corresponding mutations yielded enhanced K2P1 currents (Aryal et al., 2014). Some combination of these mechanisms may account for the low activity of K2P1 channels in heterologous systems, and regulation of these dynamic processes could modulate K2P1 channel activity in native contexts.

In cerebellar granule neurons, for example, a fraction of background current was attributed to heterodimeric K2P1:K2P3 or K2P1:K2P9 channels that were silenced by sumoylation of the K2P1 subunit in the channels (Plant, Zuniga, Araki, Marks, & Goldstein, 2012). In some studies, a fraction of the large, linear, background K current found in astrocytes was attributed to K2P1 channels or to a K2P1:K2P2 heterodimer (Hwang et al., 2014; Wang et al., 2016; Zhou et al., 2009), with the latter reported to display a G protein (Gβγ) activation that permits astrocytic glutamate release via the channel (presumably via some ill-defined pore dilation mechanism that allows permeation by the large, anionic glutamate molecule through previously K^+-selective pores) (Hwang et al., 2014; Woo et al., 2012). This work will need to be reconciled with other reports that K2P1 and K2P2 do not co-assemble (Blin et al., 2016; Plant et al., 2012), and with recent data from knockout mice indicating that astrocytic leak K^+ currents do not require expression of either K2P1 or K2P2 subunits (Du et al., 2016); the latter may in part reflect a need for Gαi/o-coupled receptor activation to overcome intracellular retention of K2P1 in astrocytes (Wang et al., 2016).

A further unique feature described for K2P1 homodimeric channels involves dynamic ionic selectivity. Specifically, under conditions of low extracellular K^+ or low pH, the channel slowly obtains a significant Na^+ permeability that recovers with a similar slow time course ($\tau \sim$ 4–6 mins) (Chatelain et al., 2012; Chen, Chatelain, & Lesage, 2014; Ma, Zhang, & Chen, 2011). A conformational rearrangement within the K2P1 channel probably accounts for this dynamic ionic selectivity, perhaps due to actions on K2P1-specific residues within the selectivity filter or pore (Chatelain et al., 2012; Chen et al., 2014; Ma et al., 2011). The consequence is that K2P1 channels may have either a hyperpolarizing or a depolarizing influence on cell membranes, depending on the prevailing conditions. For example, removing extracellular K^+ leads to a "paradoxical depolarization" in cardiac myocytes that is dependent on K2P1 expression, and attributed to its enhanced Na permeation in lower [K] (Chen et al., 2014; Ma et al., 2011). In addition, the membrane potential of astrocytes (and pancreatic β cells) from K2P1-knockout mice are more hyperpolarized than their wild-type counterparts, implying that the channel may usually provide a depolarizing current in these cells (Chatelain et al., 2012; Wang et al., 2016). This could be the case if K2P1, with a prominent Na^+ selectivity in acidic endosomes, is recycled to the cell surface, and then rapidly retrieved again from the membrane before K^+ selectivity can be restored by the more alkaline extracellular pH (Chatelain et al., 2012).

In sum, K2P1 and *KCNK7* (but not K2P6) are widely expressed in the nervous system. The conditions that support functional channel activity of *KCNK7* have not yet been defined. The properties of K2P1 have also been difficult to ascertain, due primarily to low functional channel expression in heterologous systems. Nonetheless, a number of distinctive features of K2P1 have been observed (e.g., sumoylation, receptor-modulated rapid endocytosis, pore de-wetting, dynamic ionic selectivity); these unique channel mechanisms imply that contributions of K2P1 to native neuronal and astrocytic membrane current may depend on prevailing physiological conditions, and that K2P1 may not always generate a K^+-selective current.

TREK Family

The TREK (*T*WIK-*re*lated *K*± channel) family includes three subunits: K2P2 (TREK-1), K2P4 (TRAAK; *T*WIK-*r*elated *a*rachidonic *a*cid-activated *K*± channel), and K2P10

(TREK-2). The members of this K2P subgroup express well in heterologous systems and are arguably the best-studied group of K2P channels. These channels are considered as polymodal signal integrators since they are modulated by a whole host of different factors: temperature, stretch, pH, bioactive lipids, receptor-activated signaling pathways, and clinically useful drugs, like inhaled anesthetics and antidepressants (Honore, 2007).

The channels are active at negative membrane potentials and generally K^+-selective (with the exception of some structural variants of TREK-1); they activate with modest time-dependence upon depolarization to yield an outwardly rectifying *I*–V relationship, which reflects a voltage-dependent single-channel open probability superimposed on either outwardly rectifying (K2P2) or inwardly rectifying unitary conductance (K2P4, K2P10) (Blin et al., 2016). They are widely expressed throughout the brain and are found in neurons of sensory and autonomic ganglia (Aller & Wisden, 2008; Gu et al., 2002; Kang & Kim, 2006; Medhurst et al., 2001; Talley et al., 2001).

Structural diversity among this K2P channel family arises from some common mechanisms, but also from some more unusual processes. First, splice variants with different expression patterns but similar functional properties were identified for K2P2 and K2P10 (Gu et al., 2002; Mirkovic & Wickman, 2011; Rinne et al., 2014). For K2P2, some alternative splice variants yield truncated forms of the protein that act in dominant negative fashion on full-length K2P2, probably by interfering with surface expression (Rinne et al., 2014; Veale, Rees, Mathie, & Trapp, 2010); for K2P10, the splice variants are differentially compatible with an alternative translation initiation described later (Staudacher et al., 2011). In addition, all members of this K2P clade can mix and match to form heterodimers with unique properties (Blin et al., 2016; Lengyel et al., 2016; Levitz et al., 2016). As mentioned before, there is also some evidence for promiscuity in these pairings, with K2P2 identified as coexisting with K2P1 in astrocyte channels (Hwang et al., 2014; Wang et al., 2016; Zhou et al., 2009); however, this has not been independently confirmed (Blin et al., 2016; Du et al., 2016; Plant et al., 2012).

A more surprising mechanism for generating channel diversity was discovered for both K2P2 and K2P10—specifically, alternative translation initiation (ATI). For these subunits, a suboptimal Kozak sequence around the most 5' initiator methionine position allows ribosomal scanning to skip those start sites and initiate translation from a downstream methionine. This ATI process may be differentially regulated in the brain during development (Thomas, Plant, Wilkens, McCrossan, & Goldstein, 2008), and the N-terminally truncated ATI variants present with different single-channel properties (especially, distinct unitary conductance) (Simkin, Cavanaugh, & Kim, 2008; Thomas et al., 2008). For K2P2, the foreshortened ATI variant also exhibits an enhanced Na^+ selectivity, such that its expression would yield a cationic "leak" channel promoting membrane depolarization (increased excitability), as opposed to the hyperpolarizing influence of the full-length K2P2 variant (Thomas et al., 2008). This change in ionic selectivity was not observed in ATI variants of K2P10 (Simkin et al., 2008).

The documented mechanisms of modulation for this well-studied subgroup of K2P channels are the most numerous and diverse; interestingly, although diverse, many of these modulatory mechanisms converge on the proximal C terminal domain that runs antiparallel to, and probably interacts with, the inner leaflet of the membrane bilayer to provide a contact point for transduction of conformational changes to affect the selectivity filter "gate" (Bagriantsev et al., 2012; Bagriantsev et al., 2011; Honore, 2007; Lolicato et al., 2017; Patel & Honore, 2002).

In terms of physicochemical mechanisms, all three channels in this subgroup are activated by membrane stretch and by increases in temperature, involving the proximal C-terminal region (Honore, 2007; Patel & Honore, 2002). The activation by temperature is apparent from ~24°C to 42°C, with a Q_{10}~10 (Kang, Choe, & Kim, 2005; Maingret, Lauritzen, et al., 2000). The effects of negative pressure (or convex membrane curvature) are observed on channels expressed in cell-free patches or purified and incorporated into proteoliposomes (Brohawn, Campbell, et al., 2014; Brohawn, Su, & MacKinnon, 2014; Kang, Choe, Cavanaugh, & Kim, 2007; Maingret, Fosset, Lesage, Lazdunski, & Honore, 1999; Maingret, Patel, Lesage, Lazdunski, & Honore, 1999; Patel et al., 1998) (Berrier et al., 2013), suggesting direct effects of mechanical distortion on channel function; on the other hand, the modulation by temperature is lost in cell-free patches (Kang et al., 2005; Maingret, Lauritzen, et al., 2000).

Intracellular acidification strongly activates K2P2 and K2P10, via protonation of a glutamate residue located in the proximal C-terminal region of the channel (Kang et al., 2007; Maingret, Patel, et al., 1999); when activated by intracellular acidification, the effects of stretch and temperature are occluded (Honore et al., 2002). The channels are also sensitive to extracellular protons, via a titratable M1P1 loop histidine residue common to all these channels. However, acidification inhibits K2P2 and K2P4 while it activates K2P10; this differential modulation reflects electrostatic interactions of the common protonated histidine with nearby residues that are of opposite charge in the two channels (negative in K2P2 vs. neutral or positive in K2P10) (Sandoz, Douguet, Chatelain, Lazdunski, & Lesage, 2009) (Lesage & Barhanin, 2011).

The channels are also activated via proximal C terminal interactions with other proteins and various classes of lipids. For lipids, this includes polyunsaturated fatty acids (PUFAs), lysophospholipids (e.g., lysophosphatidylcholine), and phosphatidylinositol-4,5-bisphosphate (PIP2); the effects of lysophospholipids are apparently indirect, whereas activation by PUFAs and PIP2 is thought to be direct (Chemin, Patel, Duprat, Zanzouri, et al., 2005; Chemin, Patel, Duprat, Lauritzen, et al., 2005; Maingret, Patel, Lesage, Lazdunski, & Honore, 2000; Patel, Lazdunski, & Honore, 2001). The actions of arachidonate do not require breakdown via lipoxygenase or cyclooxygenase, but metabolism of PIP2 by phospholipase C (PLC) may contribute, in part, to reduced channel activity downstream of Gαq-coupled receptor signaling (Chemin et al., 2003; Chemin et al., 2007). Multiple proteins can interact within or near the proximal C terminal regulatory domain in K2P2 and K2P10 (but not K2P4) to affect channel function: A-kinase anchoring protein, AKAP150, binds to this region and fully activates the channels, occluding effects of other activators (arachidonate, stretch, pH) (Sandoz et al., 2006); microtubule-associated protein 2 (Mtap2) can also bind simultaneously to enhance surface expression (Sandoz et al., 2008). Together, the protein complexes aggregate signaling molecules to support dynamic modulation of channel function.

Physiologically, a number of G protein-coupled receptors (GPCRs) regulate activity of K2P2 and K2P10, but not K2P4; specifically, the channels are inhibited by Gαq- and Gαs-coupled receptors, and activated by Gαi-coupled receptors. For Gαq-coupled receptors, channel inhibition probably results from a combination of PLC-mediated breakdown of PIP2, a channel activator (Chemin et al., 2007), and a sequential mechanism whereby protein kinase C (PKC) phosphorylates K2P2 at Ser-333 to enable a subsequent inhibitory modification at Ser-300 within the proximal C terminal regulatory region (Murbartian, Lei, Sando, & Bayliss, 2005); in K2P10, the corresponding sites (Ser-326, Ser-359) contribute to GαqPCR modulation by protein kinase A (PKA) and PKC (Gu et al., 2002; Kang, Han, & Kim, 2006; Lesage, Terrenoire, Romey, & Lazdunski, 2000). Similarly, Gαs-coupled receptors mediate channel inhibition via PKA modification of the same phosphorylation sites (Murbartian et al., 2005), as does receptor-independent channel inhibition by the metabolic sensor, adenosine monophosphate–activated protein kinase (AMP kinase) (Kreneisz, Benoit, Bayliss, & Mulkey, 2009; Patel et al., 1998). Conversely, activation of K2P2 and K2P10 by Gαi-coupled receptors involves dephosphorylation at these sites, implying a constitutive level of phosphorylation-mediated channel inhibition under basal conditions (Cain, Meadows, Dunlop, & Bushell, 2008; Lesage et al., 2000).

Interestingly, various "off-target" actions of pharmacological agents on these channels might contribute to their clinically relevant therapeutic actions. For example, inhalational anesthetics activate K2P2 and K2P10 (but not K2P4) (Patel et al., 1999), providing ion channel substrates for the "leak" K^+ channels underlying anesthetic-evoked membrane hyperpolarization in various neurons (along with K2P3, K2P9) (Franks & Lieb, 1988; Nicoll & Madison, 1982). In addition, these same channels are inhibited in a state-dependent manner by antidepressants like fluoxetine and paroxetine (Heurteaux et al., 2006; Kennard et al., 2005); the state-dependent binding of norfluoxetine, localized to the intramembrane fenestration in structural studies of K2P10, has been exploited to suggest structurally distinct open states associated with various activation mechanisms (Dong et al., 2015; McClenaghan et al., 2016). Further in this respect, K2P2 and K2P10 are inhibited by spadin, a peptide product from the neurotensin 3 receptor propeptide that itself has antidepressant actions (Mazella et al., 2010). These observations suggest a role for K2P2 and K2P10 in mediating effects of anesthetic and antidepressant drugs.

The widespread expression and extensive modulatory potential of this subgroup of channels suggested possible contributions to various physiological and pharmacological mechanisms, possibilities that have been tested using currently available genetic mouse models. With respect to the pharmacology discussed before, genetic deletion of K2P2 yielded reduced sensitivity to inhalational anesthetics and a depression-resistant phenotype (Heurteaux et al., 2004; Heurteaux et al., 2006); in K2P2-deleted mice, the antidepressant effects of fluoxetine and paroxetine were occluded, and the firing behavior of serotonergic neurons under baseline conditions in vivo resembled that of mice treated with antidepressants (Heurteaux et al., 2006).

The polymodal nature of channel activation, together with expression in sensory neurons, was reflected in exaggerated responses to mechanical and thermal stimuli in mice deleted for these channels, consistent with removal of an activated "leak" K^+ channel that serves as an excitability "brake" in those cells (Alloui et al., 2006; Noel et al., 2009; Pereira et al., 2014). Likewise, acute siRNA-depletion of K2P10 from nociceptors caused membrane depolarization, enhanced spontaneous firing, and was associated with increased spontaneous pain in neuropathy models (Acosta et al., 2014). A contribution of K2P2 to generally limiting neuronal excitability was revealed by an increased seizure and ischemia susceptibility in K2P2-deleted mice (Heurteaux et al., 2004); moreover, the neuroprotection typically provided by PUFAs and lysophospholipids after seizure or ischemia was abrogated in K2P2-knockout mice, implicating lipid activation of the channel in this form of neuroprotection (Heurteaux et al., 2004).

In sum, K2P2, K2P4, and K2P10 represent a subfamily of channels with unique structural and functional features, some shared and some distinct, that underlie their characteristic polymodal regulation. These broad regulatory properties, along with expression throughout the central and peripheral nervous system, allow these specific K2P channels to serve in various physiological and pharmacological processes. Most notably, they act as a "tunable" excitability brake in peripheral and central neurons, where they contribute to neuroprotection and modulate the effects of sensory inputs. Surprisingly, K2P2 and K2P10 are "off-target" substrates for clinically relevant pharmacological agents that nevertheless may mediate at least some of their salutary actions (e.g., effects of anesthetics or antidepressant drugs). It remains to be seen if any of these varied roles can be exploited with specific compounds in novel ways for therapeutic advantage (e.g., see Vivier et al., 2017).

TASK Family

The TASK (_T_WIK-related _a_cid-_s_ensitive _K_$^{\pm}$ channel) family includes three subunits. The two functional subunits, K2P3 (TASK-1) and K2P9 (TASK-3), are widely expressed in the nervous system (Fig. 16.2A) and a number of other tissues (Aller & Wisden, 2008; Medhurst et al., 2001; Talley et al., 2001), where they produce K^+-selective background currents that are inhibited by extracellular acidification; these channels are also targets for inhibition by Gαq-coupled receptors, and for activation by inhalational anesthetics (Bayliss, Sirois, & Talley, 2003; Enyedi & Czirjak, 2010; Kim, 2005; Lesage & Barhanin, 2011). K2P15 (TASK-5) is a silent subunit that is differentially expressed in the brain, especially in auditory nuclei (Ashmole, Goodwin, & Stanfield, 2001; Karschin et al., 2001; Kim & Gnatenco, 2001); we will not consider it further here.

For K2P3 and K2P9, inward rectification of single-channel conductance coupled with modest voltage- and time-dependence of channel-open probability yields whole cell pH- and anesthetic-sensitive I–V relationships that are reasonably well fitted in physiological solutions by the GHK equation (Kim, 2005; Renigunta et al., 2015) (Fig. 16.2B, 16.2C). The channel kinetics of K2P3 and K2P9 are similar, but single-channel conductance is larger and divalent-sensitive for K2P9 (36 pS vs. 100 pS, in 1 mM and 0 mM divalents), and smaller and divalent-insensitive for K2P3 (~14 pS) (Kim, 2005). For both, proton inhibition requires a histidine residue near the first pore selectivity filter (His-98); despite this shared titratable residue, the two channels show notably different pH sensitivity that is conferred by additional divergent nearby residues (pKa for K2P3 ~7.4; for K2P9 ~6.7) (Kim, Bang, & Kim, 2000; Lopes, Zilberberg, & Goldstein, 2001; Rajan et al., 2000).

Surface expression of K2P3 and K2P9 is facilitated by protein–protein interactions, including by C-terminal binding to 14-3-3 proteins. This binding is dependent on PKA- or RSK2-mediated phosphorylation of the channel, and leads to release of β-coatamer binding to an N-terminal endoplasmic reticulum (ER) retention signal (Kilisch, Lytovchenko, Schwappach, Renigunta, & Daut, 2015; O'Kelly, Butler, Zilberberg, & Goldstein, 2002; Rajan et al., 2002; Renigunta et al., 2006; Zuzarte et al., 2009).

Structural diversity for K2P3 and K2P9 arises from heteromeric channel assembly; there are no splice variants produced from their cognate two-exon genes. The expression patterns for K2P3 and K2P9 are widespread, but distinct (Aller & Wisden, 2008; Medhurst et al., 2001; Talley et al., 2001); in some cell types, there is expression of only one subunit (e.g., K2P9 in striatal cholinergic cells) (Berg & Bayliss, 2007; Talley et al., 2001), whereas overlapping expression of both subunits is observed in other cell groups (e.g., cerebellar granule neurons, motor and sensory neurons, carotid body glomus cells, adrenocortical cells) (Aller & Wisden, 2008; Czirjak & Enyedi, 2002; Czirjak, Fischer, Spat, Lesage, & Enyedi, 2000; Davies et al., 2008; Heitzmann et al., 2008; Kim, Cavanaugh, Kim, & Carroll, 2009; Medhurst et al., 2001; Penton et al., 2012; Talley et al., 2001). As might be expected from this distribution, K2P3 and K2P9 subunits form homodimeric and heterodimeric channels, which have been identified in a number of native contexts based on conformation-specific constellations of single-channel and pharmacological properties (Berg et al., 2004; Kang, Han, et al., 2004; Kim et al., 2009) (Fig. 16.2B). In addition, as mentioned previously, heterodimers of K2P3 or K2P9 with K2P1 have also been observed in cerebellar granule neurons, where they were silenced by sumoylation of the K2P1 subunit (Plant et al., 2012).

A major regulatory feature, described in early studies of K2P3 and K2P9 channels, involves their inhibition downstream of Gαq-coupled receptors (Czirjak, Petheo, Spat, & Enyedi, 2001; Millar et al., 2000; Talley, Lei, Sirois, & Bayliss, 2000). Indeed, this receptor-mediated neuronal "leak" K^+ current inhibition was well known in neurons (Nicoll, 1988; Nicoll et al., 1990), and it now appears that these channels are a likely substrate for that current in various cell types (Enyedi & Czirjak, 2010). A number of signaling pathways have been described for this receptor-mediated mechanism, including direct inhibition by the activated Gαq subunit (Chen et al., 2006) and either PIP_2 depletion (Lopes et al., 2005) or diacylglycerol (DAG) production (Lindner, Leitner, Halaszovich, Hammond, & Oliver, 2011; Wilke et al., 2014) that would occur via Gαq-stimulated PLC activity (Czirjak et al., 2001; Wilke et al., 2014); the relative contributions

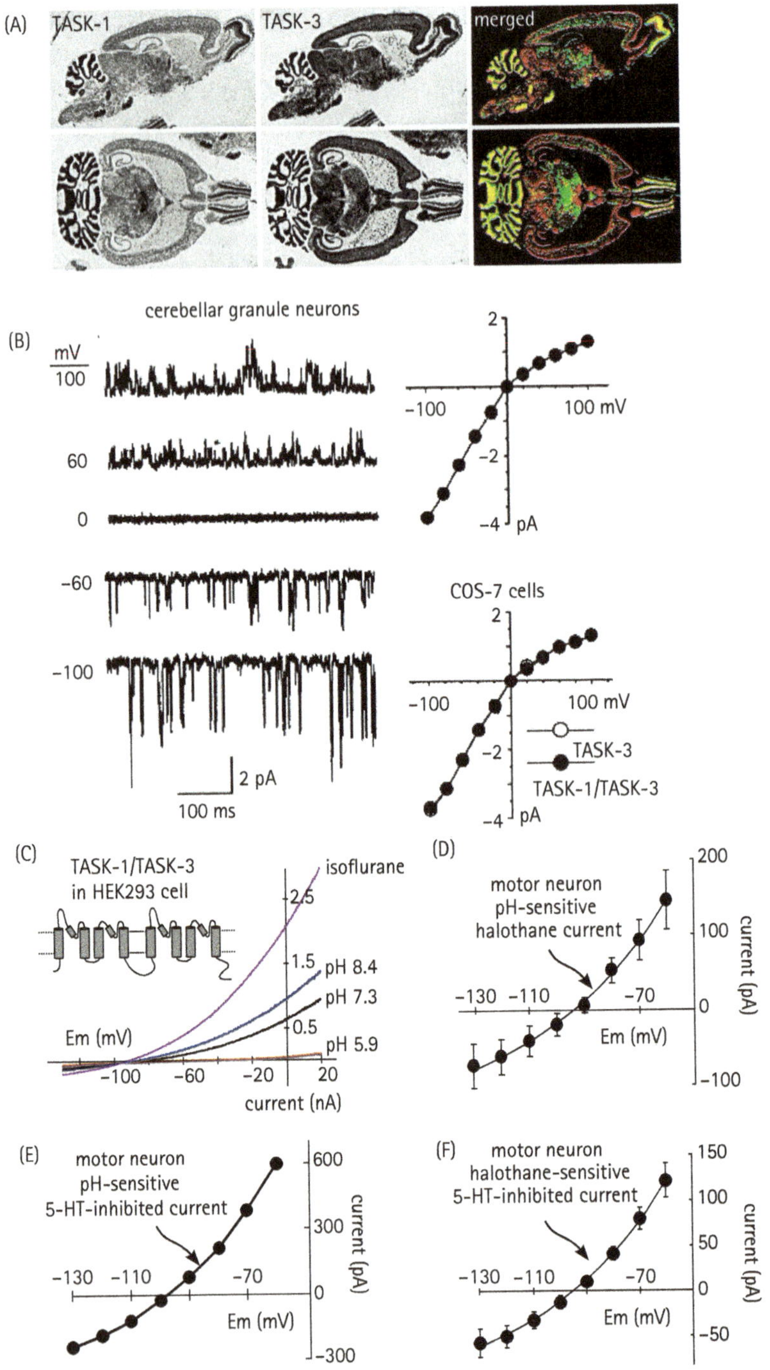

FIGURE 16.2 **TASK channels: Distribution and modulation. A.** In situ hybridization analysis of K2P3 and K2P9 expression in rat brain; note the differential but overlapping pattern of expression (adapted with permission from Talley et al., 2003). **B.** Single-channel TASK-like currents from cerebellar granule cells, which express both K2P3 and K2P9, display properties similar to

of these pathways may be context-dependent (Mathie, 2007). Gαq-mediated channel modulation may be opposed by PKC phosphorylation (Veale et al., 2007), but there is little evidence for other effects of channel phosphorylation, or for modulation of K2P3 or K2P9 by Gαs- or Gαi-coupled receptors. As with K2P2 channels, a proximal C-terminal region is also required for receptor-mediated inhibition of K2P3 and K2P9 (Talley & Bayliss, 2002).

Interestingly, this proximal C-terminal region is also required for activation of these channels by inhalational anesthetics (Patel et al., 1999; Talley & Bayliss, 2002), another well-known form of regulation for K2P3 and K2P9; residues critical for anesthetic activation have also been identified in regions outside from the proximal C terminus (Andres-Enguix et al., 2007). It remains to be determined whether these functionally important channel regions represent true binding sites or domains that transduce gating information from remote binding sites. Among these different forms of regulation, channel inhibition appears to prevail since anesthetic activation is completely abrogated in channels simultaneously inhibited by either extracellular acidification or Gαq receptor signaling (Meuth et al., 2003; Sirois, Lynch, & Bayliss, 2002; Talley & Bayliss, 2002); the joint pH-, anesthetic-, and transmitter-sensitive components of native motoneuronal currents present with TASK-like *I*–V characteristics (Fig. 2D-F), and they are eliminated in K2P3:K2P9 double knockout mice (Berg et al., 2004; Lazarenko, Willcox, et al., 2010; Sirois, Lei, Talley, Lynch, & Bayliss, 2000; Sirois et al., 2002).

K2P3 and K2P9 are also inhibited by hypoxia (low oxygen tension), an effect that was initially noted in presumed native correlates of these channels recorded in oxygen-sensitive glomus cells of the carotid body (Buckler, 1997, 2015; Buckler, Williams, & Honore, 2000); these channels were also implicated in O_2-mediated contraction of pulmonary arterial smooth muscle cells (Gurney et al., 2003; Olschewski et al., 2006), and mutations in K2P3 are associated with pulmonary hypertension in humans (Lambert et al., 2018). The native O_2-sensitive channels in peripheral arterial chemoreceptors were subsequently identified definitively as K2P3 and K2P9, first based on detailed

---

recombinant linked K2P3:K2P9 channels expressed in COS-7 cells. Note the voltage-dependent gating and the prominent inward rectification in the single-channel records of native and recombinant channels (adapted from Kang, Han, et al., 2003, with permission from John Wiley & Sons). **C.** Weak outward rectification of pH- and anesthetic-sensitive whole cell currents from recombinant K2P3:K2P9 channels expressed in HEK293 cells; the inset provides a cartoon of the concatenated TASK-1/TASK-3 construct (adapted from Berg et al., 2004). **D–F.** As observed with recombinant K2P3:K2P9 channels, native TASK-like currents in hypoglossal motoneurons are jointly regulated by extracellular pH, inhalational anesthetics, and neurotransmitters (5-HT, serotonin); the weakly rectifying whole cell *I*–V relationships represent the pH-sensitive components of anesthetic-activated (**D**; adapted from Sirois et al., 2000) and 5-HT inhibited current (**E**; adapted from Talley et al., 2000, with permission from Elsevier), and the halothane-sensitive component of 5-HT inhibited current (**F**; adapted from Sirois et al., 2002, with permission from John Wiley & Sons). These TASK-like current components were eliminated in motoneurons from K2P3:K2P9 (TASK) knockout mice (Lazarenko, Willcox, et al., 2010).

comparisons to single-channel properties of the cloned channels (Buckler et al., 2000; Kim et al., 2009), and then by using knockout mice (Turner & Buckler, 2013). These analyses concluded that the predominant native form of the O_2-sensitive channel in carotid body glomus cells is the K2P3:K2P9 heterodimer (Kim et al., 2009; Turner & Buckler, 2013). On the other hand, deletion of K2P3 and K2P9 had little to no effect on pulmonary arterial smooth muscle cells, and it did not disrupt hypoxic vasoconstriction (Manoury, Lamalle, Oliveira, Reid, & Gurney, 2011; Murtaza et al., 2017); this may reflect some species-dependent differences in K2P channel expression that are peculiar to mice (Manoury et al., 2011; Murtaza et al., 2017). The mechanism for channel inhibition by hypoxia remains unsettled, but it is likely to be indirect and mediated by metabolic changes associated with reduced O_2 levels in glomus cells (Buckler, 2015; Kim, 2013).

With respect to physiological roles for K2P3 and K2P9 channels, their demonstrated pH and O_2 sensitivity, together with expression in carotid body and brainstem regions associated with respiratory control, suggested that they could be involved in chemoreceptor-mediated stimulation of breathing (i.e., by hypoxia and CO_2-induced acidosis). However, this is not supported by data from knockout mice. In the case of carotid body chemoreceptors, hypoxia-stimulated breathing was normal despite reduced K2P3 and K2P9 channel activity (Ortega-Saenz et al., 2010; Trapp, Aller, Wisden, & Gourine, 2008); thus, although the O_2-sensitive K2P3:K2P9 channels are expressed in glomus cells (Kim et al., 2009; Turner & Buckler, 2013), compensatory changes were apparent in these global knockouts, and multiple or redundant mechanisms probably contribute to carotid body stimulation of breathing (Buckler, 2015). Notably, pharmacological blockers of K2P3 and K2P9 can acutely stimulate breathing in wild-type mice (e.g., PK-THPP, A1899), and the respiratory stimulant doxapram inhibits these channels (Cotten, 2013; Cotten, Keshavaprasad, Laster, Eger, & Yost, 2006). There was also no effect of K2P3 and K2P9 deletion on CO_2-induced stimulation of breathing, despite widespread expression of these pH-sensitive channels in brainstem respiratory neurons (Bayliss, Barhanin, Gestreau, & Guyenet, 2015; Bayliss, Talley, Sirois, & Lei, 2001; Mulkey et al., 2007). In the specific case of respiratory chemoreceptors located in the brainstem retrotrapezoid nucleus, the pH-sensitive mechanisms involve other channels (e.g., K2P5, see TALK Family, below) (Bayliss et al., 2015; Gestreau et al., 2010; Kumar et al., 2015; Wang et al., 2013). Thus, the neurobiological consequences of the pH and O_2 sensitivity of these channels, K2P3 and K2P5, remain to be clarified.

Physiological data from mice deleted for K2P3 and/or K2P9 have revealed prominent roles in regulation of activity, sleep-wake patterning, and responses to anesthetics. Specifically, K2P9-knockout mice have markedly elevated nocturnal activity and fragmented rapid-eye-movement (REM) sleep, effects that are largely reproduced by a K2P9 blocker compound (Coburn et al., 2012; Pang et al., 2009); mice deleted for K2P3 and K2P9 are less sensitive to sedating, hypnotic, and immobilizing effects of inhaled anesthetics (Lazarenko, Willcox, et al., 2010; Linden et al., 2006; Linden et al., 2007). The precise mechanisms for these effects remains unclear. For activity and sleep regulation, this may reflect particularly high expression levels and state-dependent modulation in various aminergic neurons of the reticular activating system (e.g., noradrenergic locus coeruleus, serotonergic raphe, histaminergic tuberomammillary, cholinergic basal forebrain nuclei) (Karschin et al., 2001; Marinc, Preisig-Muller, Pruss, Derst, & Veh, 2011; Sirois et al., 2000; Steinberg, Wafford, Brickley, Franks, & Wisden, 2015; Talley et al., 2001; Vu, Du, Bayliss, & Horner, 2015; Washburn, Sirois, Talley, Guyenet, & Bayliss, 2002). In addition, Gαq receptor inhibition of K2P3 and K2P9, and the accompanying membrane depolarization, is a major mechanism for a state-dependent switch from sleep-related burst firing to wake-related tonic firing in thalamocortical relay neurons (Bista et al., 2015; Bista et al., 2012). Conversely, activation of K2P3 and K2P9 channels in these same aminergic and thalamocortical cell groups may contribute to the sedating and hypnotic effects of inhalational anesthetics (Budde et al., 2008; Meuth et al., 2003; Sirois et al., 2000; Washburn et al., 2002). Immobilizing anesthetic actions mediated by K2P3 and K2P9 probably arise from anesthetic-induced hyperpolarization in motoneurons, since deletion of K2P3 and K2P9 from cholinergic neurons recapitulates the decreased anesthetic sensitivity of global knockout mice (Lazarenko, Willcox, et al., 2010). These channels also contribute to the neuroprotection in mouse models of stroke (Meuth et al., 2009), including the neuroprotective effects of inhalational anesthetics (C. Yao et al., 2017).

As with other K2P channels, K2P3 and K2P9 are expressed in peripheral sensory neurons where they can act as a modifiable brake on cell excitability. These channels, along with K2P18, are inhibited by a number of pungent agents (e.g., hydroxy-α-sanshool, from Szechuan peppercorns; piperines, from black peppercorns), providing a TRP (transient receptor potential) channel-independent mechanism for detecting chemicals that yield a distinct sensation (i.e., tingling paresthesia, numbing) (Bautista et al., 2008; Beltran et al., 2017). In addition, K2P9 is preferentially expressed at high levels in cold- and menthol-sensitive TRPM8-containing sensory neurons, and cold hypersensitivity is observed in mice deleted for K2P9 (Morenilla-Palao et al., 2014). Outside the nervous system, K2P3 and K2P9 are expressed in the heart, where they can contribute to action potential repolarization (Decher, Kiper, Rolfes, Schulze-Bahr, & Rinne, 2015; Rinne et al., 2015). K2P3 and K2P9 are also expressed in the mouse adrenal cortex, with K2P9 selectively localized to the zona glomerulosa, and expressed in mitochondria (Davies et al., 2008; Guagliardo et al., 2012; J. Yao et al., 2017). Genetic deletion of K2P3 disrupts zonation in female mice (Heitzmann et al., 2008); in male mice, deletion of K2P3 and/or K2P9 leads to a continuum of aldosteronism with elevated blood pressure (Davies et al., 2008; Guagliardo et al., 2012; Heitzmann et al., 2008; Penton et al., 2012). Finally, a role for these channels has been suggested for metabolic functions, such as dampening of lipolysis and thermogenesis in adipocytes (Chen et al., 2017; Pisani et al., 2016) and control of glucagon and insulin secretion from pancreatic α and β cells (Dadi, Luo, Vierra, & Jacobson, 2015; Dadi, Vierra, & Jacobson, 2014).

Although mice deleted for K2P3 and K2P9 have relatively minor impairments in the absence of physiological challenge (Lazarenko, Willcox, et al., 2010; Linden et al., 2006; Linden et al., 2007), the channels have been implicated in normal neuronal development. For example, in the cerebellum and hippocampus, K2P9 may contribute to apoptosis and cell proliferation, respectively. Moreover, invalidation of these channels affects leak conductance and firing properties in these and other neurons (Aller et al., 2005; Brickley et al., 2007; Gonzalez et al., 2009). Missense mutations of *KCNK9* that strongly reduce K2P9 currents lead to a mental retardation dysmorphism (Birk-Barel) syndrome in human patients (Barel et al., 2008; Veale, Hassan, Walsh, Al-Moubarak, & Mathie, 2014). Interestingly, gene deletion of a GABA receptor subunit that typically underlies a constitutive outward current in cerebellar cells leads to a compensatory upregulation of K2P3 and K2P9, seemingly to preserve normal neuronal excitability (Brickley, Revilla, Cull-Candy, Wisden, & Farrant, 2001). Thus, as principal determinants of membrane potential and excitability in many neurons, there must be substantial developmental pressure for homeostatic compensatory mechanisms to protect neurons from K2P channel dysfunction.

In sum, K2P3 and K2P9 are expressed widely throughout the nervous system, with particularly prominent expression in brainstem aminergic neurons and motoneurons; they are capable of forming homomeric and heteromeric channels that generate background K^+ currents, which are renowned for their inhibition by protons, Gαq-coupled receptors and hypoxia, and for their activation by inhalational anesthetics. As such, K2P3 and K2P9 account for multiple instances of receptor-mediated neuromodulation involving "leak" K^+ channel inhibition, effects that may be particularly important for regulating arousal state-dependent activity and sleep patterning. The physiological relevance of channel-intrinsic pH sensitivity and indirect hypoxia sensitivity remains unclear, but the latter is likely to be involved in breathing regulation by carotid body oxygen chemoreceptors. Activation of K2P3 and K2P9 by inhalational anesthetics contributes to multiple clinically relevant actions (sedation, hypnosis, immobilization, neuroprotection); in sensory neurons, K2P9 depresses cold sensitivity, and K2P3 and K2P9 are inhibited by pungent agents that evoke tingling sensations. During neuronal development, K2P3 and/or K2P9 can homeostatically regulate cell excitability or influence cell proliferation and apoptosis; loss-of-function mutations in K2P9 are associated with the Birk-Barel syndrome. There has been some success in developing compounds specific for K2P3 and K2P9; activators and inhibitors targeting these channels will provide useful complements to genetic studies for experimental analysis of their physiological function, with potential therapeutic utility.

TALK Family

The TALK (TWIK-related alkaline pH-activated $K^{\pm}$ channel) family comprises three subunits: K2P5 (TASK-2), K2P16 (TALK-1) and K2P17 (TALK-2). The members of this K2P subgroup are pH-dependent over a broad range and, as suggested by their name, display increasing activity as extracellular pH becomes progressively alkalized (Lesage & Barhanin, 2011). In contrast to other pH-sensitive K2P channels, for which histidine residues located in either the M1P1 loop or the pore are implicated (TREK and TASK families) (Lesage & Barhanin, 2011; Sepulveda et al., 2015), the basis for pH sensitivity in TALK family channels appears to involve positively charged residues in the P2M4 loop (e.g., Arg224 in K2P5, Lys242 in K2P17) (Niemeyer et al., 2007; Sepulveda et al., 2015). The channels are K^+-selective, active at negative membrane potentials, and they activate with some time-dependence upon depolarization to yield a modest, outwardly rectifying *I*–V relationship (Duprat, Girard, Jarretou, & Lazdunski, 2005; Girard et al., 2001; Han, Kang, & Kim, 2003; Kang & Kim, 2004; Reyes et al., 1998) that contrasts with the inward rectification observed for each of these K2P subunits at the single-channel level (Han et al., 2003; Kang & Kim, 2004). Notably, this subgroup of channels can be strongly activated by nitric oxide (NO) and specific reactive oxygen species (ROS) (Duprat et al., 2005; Gestreau et al., 2010); they also may be receptor modulated, since K2P5 can be inhibited by G protein βγ subunits (Anazco et al., 2013).

In general, this subgroup of K2P channels is thought to be most highly expressed in peripheral tissues (e.g., pancreas, kidney, cochlea) (Lesage & Barhanin, 2011), where the channels have been found to play critical roles in secretory processes. For example, deletion of K2P5 in mice disrupts HCO_3^- reabsorption from the kidney to precipitate a metabolic acidosis (Warth et al., 2004); loss of K2P5 also interferes with K^+ recycling by cochlear cells, resulting in progressive deafness (Cazals et al., 2015); and genetic ablation of K2P16 depolarizes pancreatic β cells to increase glucose-stimulated insulin secretion (Vierra et al., 2015). Since expression of this K2P subgroup has been considered to be predominantly peripheral (Lesage & Barhanin, 2011), possible roles for these channels in neuronal function remain largely unexplored—with the following notable exception.

A highly restricted brainstem expression of K2P5 was uncovered by using β-galactosidase staining in a mouse expressing LacZ from a gene trap embedded in the *Kcnk5* locus (Gestreau et al., 2010). Among the brainstem cell groups found to be positive for expression of this pH-sensitive K2P5 channel was the so-called retrotrapezoid nucleus (RTN), which contains a set of neurons that are responsible for sensing changes in CO_2/H^+ for the purpose of regulating breathing (Gestreau et al., 2010; Wang et al., 2013)—a homeostatic reflex that rapidly regulates CO_2 levels and acid–base balance (Bayliss et al., 2015; Guyenet & Bayliss, 2015). Accordingly, genetic deletion of K2P5 in mice eliminated pH sensitivity in a subpopulation of RTN neurons (Wang et al., 2013), and it strongly blunted CO_2-induced stimulation of breathing (Gestreau et al., 2010); a second receptor-mediated inhibition of an unidentified background K^+ channel accounted for much of the residual neuronal pH sensitivity and breathing stimulation by CO_2 (Guyenet & Bayliss, 2015; Kumar et al., 2015). Interestingly, then, via expression in these brainstem RTN neurons and in the kidney, K2P5 contributes to both acute and chronic mechanisms of acid-base regulation (Bayliss et al., 2015; Warth et al., 2004). A separate effect of K2P5 deletion on the respiratory response to hypoxia was also observed, prompting the interesting speculation that this might reflect a ROS-mediated activation of K2P5 (Gestreau et al., 2010); this latter hypothesis remains untested.

In sum, the TALK family of K2P channels generates outwardly rectifying "leak" K^+ currents that are strongly activated by alkalization over a broad pH range, and by NO or ROS. Although these subunits are typically considered to be important in peripheral organs, there may be discrete sites of expression for this subgroup within the nervous system that could mediate important physiological functions; this is the case for K2P5, which provides a molecular substrate for pH-dependent regulation of breathing by virtue of its expression in RTN neurons. Similar discoveries of other neuronal functions may yet be made for K2P16 and K2P17.

THIK Family

The THIK (*T*andem pore domain *h*alothane-*i*nhibited *K*± channel) family includes K2P12 (THIK-2) and K2P13 (THIK-1), both of which show differential, but overlapping, expression in the brain (Rajan et al., 2001). Initial recordings from these subunits revealed K^+ selective currents and GHK-type rectification for K2P13 (Rajan et al., 2001); unlike other K2P channels mentioned previously, the K2P13 channel currents are inhibited, rather than activated, by the inhalational anesthetic halothane (Rajan et al., 2001). On the other hand, K2P12 failed to generate any currents, either when expressed alone or together with K2P13 in various heterologous expression systems (Blin et al., 2014; Chatelain et al., 2013; Rajan et al., 2001; Renigunta et al., 2014). However, more recent work indicates that K2P12 can be trafficked to the cell surface to generate halothane-inhibited K^+ currents, similar to those mediated by K2P13, if strong N-terminal ER retention signals are disrupted by mutation or masked by co-expressed K2P13 (Blin et al., 2014; Chatelain et al., 2013; Renigunta et al., 2014). The shared properties of whole cell currents from homomeric K2P12 and K2P13 channels are reflected in similar current properties obtained from the corresponding heteromeric K2P12:K2P13 channels. Nonetheless, as observed for other K2P heterodimers, the single-channel conductance of the K2P12:K2P13 heterodimers (~3.5 pS) could be distinguished from that of either of the homomeric channels (K2P13: ~5 pS; K2P12: ~2.5 pS), despite the short, flickery openings for these very small conductance K2P channels (Blin et al., 2014; Kang, Hogan, & Kim, 2014).

There has been relatively little information regarding native correlates or physiological functions for K2P12- and/or K2P13-containing channels. In microglia, ATP and P2Y purinergic receptors activate a tonic background K^+ current with pharmacological characteristics similar to K2P13 that is absent in *Kcnk13* knockout mice (Madry et al., 2018). Interestingly, either pharmacological or genetic interference with K2P13 channel function reduced microglia surveillance and inflammatory cytyokine release, with an important role for tonic K2P13 activity in surveillance function and for ATP-stimulated channel activity in inflammasone activation (Madry et al., 2018).

The defining characteristic that inspired the common name for this group of K2P channels is their inhibition by inhalational anesthetics (Rajan et al., 2001). Accordingly, a native halothane-inhibited Purkinje cell current was attributed to K2P13, based on its expression in cerebellar Purkinje neurons (Bushell, Clarke, Mathie, & Robertson, 2002; Ishii, Nakajo, Yanagawa, & Kubo, 2010); interestingly, this Purkinje cell current and recombinant K2P13 were both found to be permeable to Cs^+ and activated by G protein $\beta\gamma$ subunits, suggesting an effector role downstream of $GABA_B$ receptor signaling in those neurons (Ishii et al., 2010). Respiratory chemoreceptor neurons also express K2P13 and a native halothane-inhibited "leak" K^+ current (Lazarenko, Fortuna, et al., 2010); the excitatory influence of this K2P13-mediated modulation on these RTN neurons, which provide a major drive for breathing, may contribute to maintained respiration during general anesthesia, even as other motor systems are strongly suppressed (Lazarenko, Fortuna, et al., 2010). Likewise, a halothane-inhibited "leak" K^+ current was also found to provide a direct excitation of sleep-active neurons of the ventrolateral preoptic nucleus (VLPO) (Moore et al., 2012); although not directly linked to K2P12 or K2P13, it was suggested that activation of those neurons via this mechanism might coopt this component of the physiological sleep network to facilitate the loss of consciousness associated with those anesthetic drugs (Moore et al., 2012).

A role for K2P13 in mediating a hypoxia-sensitive K^+ current in a subset of nitric oxide-synthase positive glossopharyngeal sensory neurons was suggested based on the demonstration that the neuronal hypoxia-sensitive current was occluded by halothane (Campanucci, Fearon, & Nurse, 2003), and that recombinant K2P13 was modestly inhibited by hypoxia after heterologous expression in HEK293 cells (Campanucci et al., 2003; Fearon et al., 2006; but see Kang et al., 2014). A small conductance halothane-inhibited channel with short, flickery openings reminiscent of K2P13 was observed in trigeminal ganglion neurons (Kang et al., 2014); like recombinant K2P13, this native channel was activated by arachidonic acid and strongly inhibited by cold (Kang et al., 2014), perhaps contributing to temperature-sensitive K^+ channels previously recorded in cold-activated sensory neurons (Madrid, de la Pena, Donovan-Rodriguez, Belmonte, & Viana, 2009; Viana, de la Pena, & Belmonte, 2002).

In sum, K2P12 and K2P13 can generate weakly rectifying background K^+ currents in physiological conditions; these K2P channels are notable for their very small unitary conductance, flickery activity, and inhibition by halothane and other inhaled anesthetics. By masking a powerful intracellular retention signal on K2P12, the related K2P13 may permit surface expression of heterodimeric K2P12:K2P13 channels at sites where the subunits are co-expressed. At present, only circumstantial evidence attributes native currents to these channels, with possible roles proposed in mediating specific effects of anesthetics or contributing to cold sensation.

TRESK Family

The TRESK (<u>T</u>WIK-<u>re</u>lated <u>s</u>pinal cord <u>K</u>$^{\pm}$ channel) family includes a single member K2P18 (TRESK), which obtained its moniker because it was originally cloned from spinal cord cDNA and because initial observations suggested a predominantly spinal expression pattern (Sano et al., 2003). However, this selective expression has not been borne out in subsequent quantitative analyses, which find a more widespread localization pattern and particularly high levels of K2P18 transcripts in sensory ganglia (Bautista et al., 2008; Cadaveira-Mosquera et al., 2012; Czirjak, Toth, & Enyedi, 2004; Dobler et al., 2007; Kang & Kim, 2006; Keshavaprasad et al., 2005). The major structural distinctions of K2P18, by comparison to other K2P channels, are an especially long

M2M3 intracellular loop domain and short C-terminal region; these characteristics are preserved among species orthologues, which otherwise show particularly low primary sequence conservation (Kang, Mariash, & Kim, 2004; Sano et al., 2003). In heterologous expression systems, K2P18 channels generate a nearly instantaneous, weakly rectifying K^+-selective whole cell current (Kang, Mariash, et al., 2004; Sano et al., 2003); the underlying channels exhibit an asymmetrical gating behavior, with low conductance openings (<20 pS) that are well resolved at depolarized potentials but that exhibited short bursts of highly flickery activity at negative membrane potentials (Czirjak et al., 2004; Kang, Mariash, et al., 2004).

The major regulatory features identified for K2P18 involve actions at the extended intracellular M2M3 loop unique to this channel. In particular, K2P18 is activated by Gαq-coupled receptor signaling via de-phosphorylation of a regulatory cluster of serine residues in this loop by calcineurin, a calcium-dependent phosphatase (Czirjak & Enyedi, 2010; Czirjak et al., 2004); basal inhibitory phosphorylation at those sites may be mediated by microtubule affinity-regulating kinases (MARKs) (Braun, Nemcsics, Enyedi, & Czirjak, 2011). In addition, the M2M3 loop coordinates various additional protein interactions (e.g., with 14-3-3 proteins) that can modulate phosphorylation-based regulation of the channel (Czirjak, Vuity, & Enyedi, 2008). Modest extracellular pH sensitivity was observed for mouse K2P18 and attributed to a pore-adjacent histidine (Kang, Mariash, et al., 2004; Keshavaprasad et al., 2005); the absence of that residue appears to account for the lack of pH sensitivity in the human K2P18 orthologue (Dobler et al., 2007). Although K2P18 is expressed in sensory ganglia and activated by inhalational anesthetics (Keshavaprasad et al., 2005; Liu, Au, Zou, Cotten, & Yost, 2004), gene deletion had negligible effects on anesthetic sensitivity in mice (Chae et al., 2010).

The high levels of K2P18 expression in sensory neurons suggested a role in regulating their excitability and associated physiological functions. Indeed, K2P18 accounts for a substantial fraction of background K^+ current in sensory neurons of dorsal root and trigeminal ganglia, and interfering with K2P18 channel expression or function increases sensory neuron excitability (Dobler et al., 2007). For example, mechanical hypersensitivity was evoked by reduced expression of K2P18 in DRG neurons, produced either by siRNA treatment or following nerve injury in a neuropathic pain model (Tulleuda et al., 2011; Zhou, Yang, Zhong, & Wang, 2013). The pungent ingredient in Szechuan peppers, α-hydroxy sanshool, causes a tingling sensation in the mouth and evokes aversive responses in behavioral experiments; this was attributed to inhibition by sanshool of K2P18 in sensory neurons (Bautista et al., 2008), although contributions from various additional K2P and other channels are also likely (e.g., K2P3, K2P9, TRPV1, NaV1.7) (Bautista et al., 2008; Koo et al., 2007; Riera et al., 2009; Tsunozaki et al., 2013).

A specific *KCNK18* frameshift mutation (F139WfsX24) was discovered in patients experiencing migraine with aura; the truncated protein product acts as a dominant negative in trigeminal ganglion neurons, and increased excitability due to loss of K2P18 function in pain-associated trigeminal neurons is a plausible mechanism for the debilitating migraine headaches (Lafreniere et al., 2010). Notably, all the patients carrying this mutation also reported aura symptoms, suggesting that disrupting K2P18 could contribute to that aberrant cortical behavior as well, and implying a normal role for K2P18 in limiting excitability of central neurons. It is important to note that although this F139WfsX24 variant was clearly linked to migraine with aura, other loss-of-function mutations in K2P18 did not show any genetic association with migraine (Andres-Enguix et al., 2012), suggesting a more complex and multifactorial etiology for K2P18 contributions to this disorder.

In sum, K2P18 stands apart from the other K2P channels. It displays "leak" type whole cell K^+ currents typical of the family that nonetheless belie unusual asymmetrical single-channel properties. It has an unusually short C-terminal region and long intracellular loop domain, the latter serving as a regulatory site for receptor- and calcineurin-mediated channel activation. Expression of K2P18 is particularly high in sensory ganglia, where it normally contributes to dampening neuronal excitability; decreased expression after nerve injury may underlie the sensory neuron hyperexcitability associated with neuropathic pain, and dominant negative mutations of the channel are implicated in a subset of patients who experience painful migraine, with aura.

Summary

The K2P channel family presents structural features and functional properties that are distinctly different from those of other K^+ channels. K2P channels assemble as homo- or heterodimers, bringing together the two pore-domains from each subunit to form a pseudotetrameric K^+-selective pore. Further unique features include a large extracellular cap structure that overlies a pair of electronegative pathways for K^+ ions to reach the selectivity filter, and a prominent hydrophobic fenestration that links the membrane bilayer to the inner channel vestibule. Characteristic weakly rectifying K2P currents are present over wide voltage ranges, including at resting membrane potentials (i.e., "leak" currents), but these macroscopic current profiles are typically the product of underlying single-channel rectification and modest voltage-dependent gating. The channel gate is localized to the selectivity filter, rather than at an intracellular bundle crossing, and gating is dynamically regulated by convergent effects of multiple channel-specific modulators, in many cases via actions on a proximal C-terminal helix region.

Modulatory mechanisms peculiar to each channel subtype allow adjustment of membrane potential and electroresponsive properties in K2P-expressing cells by a variety of physicochemical factors, neurochemicals, and clinically relevant drugs. The TWIK family of channels exemplifies multiple novel regulatory features, including reversible SUMOlyation, rapid endocytic recycling, pore de-wetting, and dynamic ion selectivity. TREK channel variants, with distinct unitary properties arising from differential splicing and alternative translation initiation, are polymodal signal integrators: channel activity is regulated by an incredibly diverse set of factors (e.g., stretch, pH, temperature, lipid and protein interactions, inhalational anesthetics, antidepressants, and by up- and down-modulation by GPCRs). TASK channels are activated by inhalational anesthetics and inhibited by GαqPCRs, low oxygen tension and protons; they may contribute to hypoxia-stimulated breathing, sleep patterning, arousal state, and anesthetic actions. TALK channels are activated by alkaline pH, nitric oxide, and reactive oxygen species; they are expressed most prominently outside the central nervous system, but TASK-2 contributes to CO_2-regulated breathing by the brainstem. In contradistinction to other K2P channels, THIK channels are inhibited rather than activated by inhalational anesthetics; heterodimerization with K2P13 allows surface trafficking to unsilence K2P12. The single TRESK channel subunit is the most diverse, presenting an extended cytoplasmic loop domain that mediates dynamic, phosphorylation-dependent channel modulation; mutations in K2P18 are associated with migraines, with aura.

In sum, K2P channels underlie background K^+ currents that provide a tunable brake on cellular excitability. These channels contribute particularly prominently to sensory neuron function and to regulation of brain state, and, surprisingly, they have been associated with clinical actions of various well-known drugs (e.g., anesthetics and antidepressants). There are currently few subunit-selective drugs targeting individual types of K2P channels, but such compounds may find use in treating dysfunctions of the myriad systems in which these channels are now appearing to play a key role.

References

Acosta, C., Djouhri, L., Watkins, R., Berry, C., Bromage, K., & Lawson, S. N. (2014). TREK2 expressed selectively in IB4-binding C-fiber nociceptors hyperpolarizes their membrane potentials and limits spontaneous pain. *Journal of Neuroscience*, *34*(4), 1494–1509. doi:10.1523/JNEUROSCI.4528-13.2014

Aller, M. I., Veale, E. L., Linden, A. M., Sandu, C., Schwaninger, M., Evans, L. J., . . . Brickley, S. G. (2005). Modifying the subunit composition of TASK channels alters the modulation of a leak conductance in cerebellar granule neurons. *Journal of Neuroscience*, *25*(49), 11455–11467. doi:10.1523/JNEUROSCI.3153-05.2005

Aller, M. I., & Wisden, W. (2008). Changes in expression of some two-pore domain potassium channel genes (KCNK) in selected brain regions of developing mice. *Neuroscience*, *151*(4), 1154–1172. doi:10.1016/j.neuroscience.2007.12.011

Alloui, A., Zimmermann, K., Mamet, J., Duprat, F., Noel, J., Chemin, J., . . . Lazdunski, M. (2006). TREK-1, a K+ channel involved in polymodal pain perception. *EMBO Journal*, *25* (11), 2368–2376. doi:10.1038/sj.emboj.7601116

Anazco, C., Pena-Munzenmayer, G., Araya, C., Cid, L. P., Sepulveda, F. V., & Niemeyer, M. I. (2013). G protein modulation of K2P potassium channel TASK-2: A role of basic residues in the C terminus domain. *Pflügers Archiv*, *465*(12), 1715–1726. doi:10.1007/s00424-013-1314-0

Andres-Enguix, I., Caley, A., Yustos, R., Schumacher, M. A., Spanu, P. D., Dickinson, R., . . . Franks, N. P. (2007). Determinants of the anesthetic sensitivity of two-pore domain acid-sensitive potassium channels: Molecular cloning of an anesthetic-activated potassium channel from *Lymnaea stagnalis*. *Journal of Biological Chemistry*, *282*(29), 20977–20990. doi:10.1074/jbc.M610692200

Andres-Enguix, I., Shang, L., Stansfeld, P. J., Morahan, J. M., Sansom, M. S., Lafreniere, R. G., ... Tucker, S. J. (2012). Functional analysis of missense variants in the TRESK (KCNK18) K channel. *Scientific Reports*, *2*, 237. doi:10.1038/srep00237

Aryal, P., Abd-Wahab, F., Bucci, G., Sansom, M. S., & Tucker, S. J. (2014). A hydrophobic barrier deep within the inner pore of the TWIK-1 K2P potassium channel. *Nature Communications*, *5*, 4377. doi:10.1038/ncomms5377

Ashmole, I., Goodwin, P. A., & Stanfield, P. R. (2001). TASK-5, a novel member of the tandem pore K+ channel family. *Pflügers Archiv*, *442*(6), 828–833.

Bagriantsev, S. N., Clark, K. A., & Minor, D. L., Jr. (2012). Metabolic and thermal stimuli control K(2P)2.1 (TREK-1) through modular sensory and gating domains. *EMBO Journal*, *31* (15), 3297–3308. doi:10.1038/emboj.2012.171

Bagriantsev, S. N., Peyronnet, R., Clark, K. A., Honore, E., & Minor, D. L., Jr. (2011). Multiple modalities converge on a common gate to control K2P channel function. *EMBO Journal*, *30* (17), 3594–3606. doi:10.1038/emboj.2011.230

Barel, O., Shalev, S. A., Ofir, R., Cohen, A., Zlotogora, J., Shorer, Z., ... Birk, O. S. (2008). Maternally inherited Birk Barel mental retardation dysmorphism syndrome caused by a mutation in the genomically imprinted potassium channel KCNK9. *American Journal of Human Genetics*, *83*(2), 193–199. doi:10.1016/j.ajhg.2008.07.010

Bautista, D. M., Sigal, Y. M., Milstein, A. D., Garrison, J. L., Zorn, J. A., Tsuruda, P. R., ... Julius, D. (2008). Pungent agents from Szechuan peppers excite sensory neurons by inhibiting two-pore potassium channels. *Nature Neuroscience*, *11*(7), 772–779. doi:10.1038/nn.2143

Bayliss, D. A., Barhanin, J., Gestreau, C., & Guyenet, P. G. (2015). The role of pH-sensitive TASK channels in central respiratory chemoreception. *Pflügers Archiv*, *467*(5), 917–929. doi:10.1007/s00424-014-1633-9

Bayliss, D. A., Sirois, J. E., & Talley, E. M. (2003). The TASK family: Two-pore domain background K+ channels. *Molecular Interventions*, *3*(4), 205–219. doi:10.1124/mi.3.4.205

Bayliss, D. A., Talley, E. M., Sirois, J. E., & Lei, Q. (2001). TASK-1 is a highly modulated pH-sensitive "leak" K(+) channel expressed in brainstem respiratory neurons. *Respiration Physiology*, *129*(1–2), 159–174.

Beltran, L. R., Dawid, C., Beltran, M., Levermann, J., Titt, S., Thomas, S., ... Hatt, H. (2017). The effect of pungent and tingling compounds from Piper nigrum L. on background K(+) currents. *Frontiers in Pharmacology*, *8*, 408. doi:10.3389/fphar.2017.00408

Berg, A. P., & Bayliss, D. A. (2007). Striatal cholinergic interneurons express a receptor-insensitive homomeric TASK-3-like background K+ current. *Journal of Neurophysiology*, *97*(2), 1546–1552. doi:10.1152/jn.01090.2006

Berg, A. P., Talley, E. M., Manger, J. P., & Bayliss, D. A. (2004). Motoneurons express heteromeric TWIK-related acid-sensitive K+ (TASK) channels containing TASK-1 (KCNK3) and TASK-3 (KCNK9) subunits. *Journal of Neuroscience*, *24*(30), 6693–6702. doi:10.1523/JNEUROSCI.1408-04.2004

Berrier, C., Pozza, A., de Lacroix de Lavalette, A., Chardonnet, S., Mesneau, A., Jaxel, C., ... Ghazi, A. (2013). The purified mechanosensitive channel TREK-1 is directly sensitive to membrane tension. *Journal of Biological Chemistry*, *288*(38), 27307–27314. doi:10.1074/jbc.M113.478321

Bichet, D., Blin, S., Feliciangeli, S., Chatelain, F. C., Bobak, N., & Lesage, F. (2015). Silent but not dumb: How cellular trafficking and pore gating modulate expression of TWIK1 and THIK2. *Pflügers Archiv*, *467*(5), 1121–1131. doi:10.1007/s00424-014-1631-y

Bista, P., Cerina, M., Ehling, P., Leist, M., Pape, H. C., Meuth, S. G., & Budde, T. (2015). The role of two-pore-domain background K(+) (K(2)p) channels in the thalamus. *Pflügers Archiv*, *467*(5), 895–905. doi:10.1007/s00424-014-1632-x

Bista, P., Meuth, S. G., Kanyshkova, T., Cerina, M., Pawlowski, M., Ehling, P., . . . Budde, T. (2012). Identification of the muscarinic pathway underlying cessation of sleep-related burst activity in rat thalamocortical relay neurons. *Pflügers Archiv*, *463*(1), 89–102. doi:10.1007/s00424-011-1056-9

Blin, S., Ben Soussia, I., Kim, E. J., Brau, F., Kang, D., Lesage, F., & Bichet, D. (2016). Mixing and matching TREK/TRAAK subunits generate heterodimeric K2P channels with unique properties. *Proceedings of the National Academy of Sciences of the United States of America*, *113*(15), 4200–4205. doi:10.1073/pnas.1522748113

Blin, S., Chatelain, F. C., Feliciangeli, S., Kang, D., Lesage, F., & Bichet, D. (2014). Tandem pore domain halothane-inhibited K+ channel subunits THIK1 and THIK2 assemble and form active channels. *Journal of Biological Chemistry*, *289*(41), 28202–28212. doi:10.1074/jbc.M114.600437

Braun, G., Nemcsics, B., Enyedi, P., & Czirjak, G. (2011). TRESK background K(+) channel is inhibited by PAR-1/MARK microtubule affinity-regulating kinases in Xenopus oocytes. *PLoS One*, *6*(12), e28119. doi:10.1371/journal.pone.0028119

Brickley, S. G., Aller, M. I., Sandu, C., Veale, E. L., Alder, F. G., Sambi, H., . . . Wisden, W. (2007). TASK-3 two-pore domain potassium channels enable sustained high-frequency firing in cerebellar granule neurons. *Journal of Neuroscience*, *27*(35), 9329–9340. doi:10.1523/JNEUROSCI.1427-07.2007

Brickley, S. G., Revilla, V., Cull-Candy, S. G., Wisden, W., & Farrant, M. (2001). Adaptive regulation of neuronal excitability by a voltage-independent potassium conductance. *Nature*, *409*(6816), 88–92. doi:10.1038/35051086

Brohawn, S. G., Campbell, E. B., & MacKinnon, R. (2013). Domain-swapped chain connectivity and gated membrane access in a Fab-mediated crystal of the human TRAAK K+ channel. *Proceedings of the National Academy of Sciences of the United States of America*, *110*(6), 2129–2134. doi:10.1073/pnas.1218950110

Brohawn, S. G., Campbell, E. B., & MacKinnon, R. (2014). Physical mechanism for gating and mechanosensitivity of the human TRAAK K+ channel. *Nature*, *516*(7529), 126–130. doi:10.1038/nature14013

Brohawn, S. G., del Marmol, J., & MacKinnon, R. (2012). Crystal structure of the human K2P TRAAK, a lipid- and mechano-sensitive K+ ion channel. *Science*, *335*(6067), 436–441. doi:10.1126/science.1213808

Brohawn, S. G., Su, Z., & MacKinnon, R. (2014). Mechanosensitivity is mediated directly by the lipid membrane in TRAAK and TREK1 K+ channels. *Proceedings of the National Academy of Sciences of the United States of America*, *111*(9), 3614–3619. doi:10.1073/pnas.1320768111

Buckler, K. J. (1997). A novel oxygen-sensitive potassium current in rat carotid body type I cells. *Journal of Physiology*, *498* (*Pt 3*), 649–662.

Buckler, K. J. (2015). TASK channels in arterial chemoreceptors and their role in oxygen and acid sensing. *Pflügers Archiv*, *467*(5), 1013–1025. doi:10.1007/s00424-015-1689-1

Buckler, K. J., Williams, B. A., & Honore, E. (2000). An oxygen-, acid- and anaesthetic-sensitive TASK-like background potassium channel in rat arterial chemoreceptor cells. *Journal of Physiology*, *525 Pt 1*, 135–142.

Budde, T., Coulon, P., Pawlowski, M., Meuth, P., Kanyshkova, T., Japes, A., . . . Pape, H. C. (2008). Reciprocal modulation of I (h) and I (TASK) in thalamocortical relay neurons by halothane. *Pflügers Archiv, 456*(6), 1061–1073. doi:10.1007/s00424-008-0482-9

Bushell, T., Clarke, C., Mathie, A., & Robertson, B. (2002). Pharmacological characterization of a non-inactivating outward current observed in mouse cerebellar Purkinje neurones. *British Journal of Pharmacology, 135*(3), 705–712. doi:10.1038/sj.bjp.0704518

Cadaveira-Mosquera, A., Perez, M., Reboreda, A., Rivas-Ramirez, P., Fernandez-Fernandez, D., & Lamas, J. A. (2012). Expression of K2P channels in sensory and motor neurons of the autonomic nervous system. *Journal of Molecular Neuroscience, 48*(1), 86–96. doi:10.1007/s12031-012-9780-y

Cain, S. M., Meadows, H. J., Dunlop, J., & Bushell, T. J. (2008). mGlu4 potentiation of K(2P)2.1 is dependent on C-terminal dephosphorylation. *Molecular and Cellular Neuroscience, 37*(1), 32–39. doi:10.1016/j.mcn.2007.08.009

Campanucci, V. A., Fearon, I. M., & Nurse, C. A. (2003). A novel O2-sensing mechanism in rat glossopharyngeal neurones mediated by a halothane-inhibitable background K+ conductance. *Journal of Physiology, 548*(Pt 3), 731–743. doi:10.1113/jphysiol.2002.035998

Cazals, Y., Bevengut, M., Zanella, S., Brocard, F., Barhanin, J., & Gestreau, C. (2015). KCNK5 channels mostly expressed in cochlear outer sulcus cells are indispensable for hearing. *Nature Communications, 6*, 8780. doi:10.1038/ncomms9780

Chae, Y. J., Zhang, J., Au, P., Sabbadini, M., Xie, G. X., & Yost, C. S. (2010). Discrete change in volatile anesthetic sensitivity in mice with inactivated tandem pore potassium ion channel TRESK. *Anesthesiology, 113*(6), 1326–1337. doi:10.1097/ALN.0b013e3181f90ca5

Chatelain, F. C., Bichet, D., Douguet, D., Feliciangeli, S., Bendahhou, S., Reichold, M., . . . Lesage, F. (2012). TWIK1, a unique background channel with variable ion selectivity. *Proceedings of the National Academy of Sciences of the United States of America, 109*(14), 5499–5504. doi:10.1073/pnas.1201132109

Chatelain, F. C., Bichet, D., Feliciangeli, S., Larroque, M. M., Braud, V. M., Douguet, D., & Lesage, F. (2013). Silencing of the tandem pore domain halothane-inhibited K+ channel 2 (THIK2) relies on combined intracellular retention and low intrinsic activity at the plasma membrane. *Journal of Biological Chemistry, 288*(49), 35081–35092. doi:10.1074/jbc.M113.503318

Chemin, J., Girard, C., Duprat, F., Lesage, F., Romey, G., & Lazdunski, M. (2003). Mechanisms underlying excitatory effects of group I metabotropic glutamate receptors via inhibition of 2P domain K+ channels. *EMBO Journal, 22* (20), 5403–5411. doi:10.1093/emboj/cdg528

Chemin, J., Patel, A., Duprat, F., Zanzouri, M., Lazdunski, M., & Honore, E. (2005). Lysophosphatidic acid-operated K+ channels. *Journal of Biological Chemistry, 280*(6), 4415–4421. doi:10.1074/jbc.M408246200

Chemin, J., Patel, A. J., Duprat, F., Lauritzen, I., Lazdunski, M., & Honore, E. (2005). A phospholipid sensor controls mechanogating of the K+ channel TREK-1. *EMBO Journal, 24* (1), 44–53. doi:10.1038/sj.emboj.7600494

Chemin, J., Patel, A. J., Duprat, F., Sachs, F., Lazdunski, M., & Honore, E. (2007). Up- and down-regulation of the mechano-gated K(2P) channel TREK-1 by PIP (2) and other membrane phospholipids. *Pflügers Archiv, 455*(1), 97–103. doi:10.1007/s00424-007-0250-2

Chen, H., Chatelain, F. C., & Lesage, F. (2014). Altered and dynamic ion selectivity of K+ channels in cell development and excitability. *Trends in Pharmacological Sciences, 35*(9), 461–469. doi:10.1016/j.tips.2014.06.002

Chen, X., Talley, E. M., Patel, N., Gomis, A., McIntire, W. E., Dong, B., . . . Bayliss, D. A. (2006). Inhibition of a background potassium channel by Gq protein alpha-subunits. *Proceedings of the National Academy of Sciences of the United States of America, 103*(9), 3422–3427. doi:10.1073/pnas.0507710103

Chen, Y., Zeng, X., Huang, X., Serag, S., Woolf, C. J., & Spiegelman, B. M. (2017). Crosstalk between KCNK3-mediated ion current and adrenergic signaling regulates adipose thermogenesis and obesity. *Cell, 171*(4), 836–848 e813. doi:10.1016/j.cell.2017.09.015

Coburn, C. A., Luo, Y., Cui, M., Wang, J., Soll, R., Dong, J., . . . Renger, J. J. (2012). Discovery of a pharmacologically active antagonist of the two-pore-domain potassium channel K2P9.1 (TASK-3). *ChemMedChem, 7*(1), 123–133. doi:10.1002/cmdc.201100351

Cotten, J. F. (2013). TASK-1 (KCNK3) and TASK-3 (KCNK9) tandem pore potassium channel antagonists stimulate breathing in isoflurane-anesthetized rats. *Anesthesia and Analgesia, 116*(4), 810–816. doi:10.1213/ANE.0b013e318284469d

Cotten, J. F., Keshavaprasad, B., Laster, M. J., Eger, E. I., 2nd, & Yost, C. S. (2006). The ventilatory stimulant doxapram inhibits TASK tandem pore (K2P) potassium channel function but does not affect minimum alveolar anesthetic concentration. *Anesthesia and Analgesia, 102*(3), 779–785. doi:10.1213/01.ane.0000194289.34345.63

Czirjak, G., & Enyedi, P. (2002). TASK-3 dominates the background potassium conductance in rat adrenal glomerulosa cells. *Molecular Endocrinology, 16*(3), 621–629. doi:10.1210/mend.16.3.0788

Czirjak, G., & Enyedi, P. (2010). TRESK background K(+) channel is inhibited by phosphorylation via two distinct pathways. *Journal of Biological Chemistry, 285*(19), 14549–14557. doi:10.1074/jbc.M110.102020

Czirjak, G., Fischer, T., Spat, A., Lesage, F., & Enyedi, P. (2000). TASK (TWIK-related acid-sensitive K+ channel) is expressed in glomerulosa cells of rat adrenal cortex and inhibited by angiotensin II. *Molecular Endocrinology, 14*(6), 863–874. doi:10.1210/mend.14.6.0466

Czirjak, G., Petheo, G. L., Spat, A., & Enyedi, P. (2001). Inhibition of TASK-1 potassium channel by phospholipase C. *American Journal of Physiology–Cell Physiology, 281*(2), C700–708. doi:10.1152/ajpcell.2001.281.2.C700

Czirjak, G., Toth, Z. E., & Enyedi, P. (2004). The two-pore domain K+ channel, TRESK, is activated by the cytoplasmic calcium signal through calcineurin. *Journal of Biological Chemistry, 279*(18), 18550–18558. doi:10.1074/jbc.M312229200

Czirjak, G., Vuity, D., & Enyedi, P. (2008). Phosphorylation-dependent binding of 14-3-3 proteins controls TRESK regulation. *Journal of Biological Chemistry, 283*(23), 15672–15680. doi:10.1074/jbc.M800712200

Dadi, P. K., Luo, B., Vierra, N. C., & Jacobson, D. A. (2015). TASK-1 potassium channels limit pancreatic alpha-cell calcium influx and glucagon secretion. *Molecular Endocrinology, 29*(5), 777–787. doi:10.1210/me.2014-1321

Dadi, P. K., Vierra, N. C., & Jacobson, D. A. (2014). Pancreatic beta-cell-specific ablation of TASK-1 channels augments glucose-stimulated calcium entry and insulin secretion, improving glucose tolerance. *Endocrinology, 155*(10), 3757–3768. doi:10.1210/en.2013-2051

Davies, L. A., Hu, C., Guagliardo, N. A., Sen, N., Chen, X., Talley, E. M., . . . Barrett, P. Q. (2008). TASK channel deletion in mice causes primary hyperaldosteronism. *Proceedings of the National Academy of Sciences of the United States of America, 105*(6), 2203–2208. doi:10.1073/pnas.0712000105

Decher, N., Kiper, A. K., Rolfes, C., Schulze-Bahr, E., & Rinne, S. (2015). The role of acid-sensitive two-pore domain potassium channels in cardiac electrophysiology: Focus on arrhythmias. *Pflügers Archiv, 467*(5), 1055–1067. doi:10.1007/s00424-014-1637-5

Dobler, T., Springauf, A., Tovornik, S., Weber, M., Schmitt, A., Sedlmeier, R., . . . Doring, F. (2007). TRESK two-pore-domain K+ channels constitute a significant component of background potassium currents in murine dorsal root ganglion neurones. *Journal of Physiology, 585*(Pt 3), 867–879. doi:10.1113/jphysiol.2007.145649

Dong, Y. Y., Pike, A. C., Mackenzie, A., McClenaghan, C., Aryal, P., Dong, L., . . . Carpenter, E. P. (2015). K2P channel gating mechanisms revealed by structures of TREK-2 and a complex with Prozac. *Science, 347*(6227), 1256–1259. doi:10.1126/science.1261512

Du, Y., Kiyoshi, C. M., Wang, Q., Wang, W., Ma, B., Alford, C. C., . . . Zhou, M. (2016). Genetic deletion of TREK-1 or TWIK-1/TREK-1 potassium channels does not alter the basic electrophysiological properties of mature hippocampal astrocytes in situ. *Frontiers in Cellular Neuroscience, 10*, 13. doi:10.3389/fncel.2016.00013

Duprat, F., Girard, C., Jarretou, G., & Lazdunski, M. (2005). Pancreatic two P domain K+ channels TALK-1 and TALK-2 are activated by nitric oxide and reactive oxygen species. *Journal of Physiology, 562*(Pt 1), 235–244. doi:10.1113/jphysiol.2004.071266

Enyedi, P., & Czirjak, G. (2010). Molecular background of leak K+ currents: Two-pore domain potassium channels. *Physiological Reviews, 90*(2), 559–605. doi:10.1152/physrev.00029.2009

Fearon, I. M., Campanucci, V. A., Brown, S. T., Hudasek, K., O'Kelly, I. M., & Nurse, C. A. (2006). Acute hypoxic regulation of recombinant THIK-1 stably expressed in HEK293 cells. *Advances in Experimental Medicine and Biology, 580*, 203–208; discussion 351–209. doi:10.1007/0-387-31311-7_31

Feliciangeli, S., Bendahhou, S., Sandoz, G., Gounon, P., Reichold, M., Warth, R., . . . Lesage, F. (2007). Does sumoylation control K2P1/TWIK1 background K+ channels? *Cell, 130*(3), 563–569. doi:10.1016/j.cell.2007.06.012

Feliciangeli, S., Chatelain, F. C., Bichet, D., & Lesage, F. (2015). The family of K2P channels: Salient structural and functional properties. *Journal of Physiology, 593*(12), 2587–2603. doi:10.1113/jphysiol.2014.287268

Feliciangeli, S., Tardy, M. P., Sandoz, G., Chatelain, F. C., Warth, R., Barhanin, J., . . . Lesage, F. (2010). Potassium channel silencing by constitutive endocytosis and intracellular sequestration. *Journal of Biological Chemistry, 285*(7), 4798–4805. doi:10.1074/jbc.M109.078535

Franks, N. P., & Lieb, W. R. (1988). Volatile general anaesthetics activate a novel neuronal K+ current. *Nature, 333*(6174), 662–664. doi:10.1038/333662a0

Gestreau, C., Heitzmann, D., Thomas, J., Dubreuil, V., Bandulik, S., Reichold, M., . . . Barhanin, J. (2010). Task2 potassium channels set central respiratory CO2 and O2 sensitivity. *Proceedings of the National Academy of Sciences of the United States of America, 107*(5), 2325–2330. doi:10.1073/pnas.0910059107

Girard, C., Duprat, F., Terrenoire, C., Tinel, N., Fosset, M., Romey, G., . . . Lesage, F. (2001). Genomic and functional characteristics of novel human pancreatic 2P domain K(+) channels. *Biochemical and Biophysical Research Communications, 282*(1), 249–256. doi:10.1006/bbrc.2001.4562

Goldstein, M., Rinne, S., Kiper, A. K., Ramirez, D., Netter, M. F., Bustos, D., . . . Decher, N. (2016). Functional mutagenesis screens reveal the "cap structure" formation in disulfide-bridge free TASK channels. *Scientific Reports, 6*, 19492. doi:10.1038/srep19492

Goldstein, S. A., Bayliss, D. A., Kim, D., Lesage, F., Plant, L. D., & Rajan, S. (2005). International Union of Pharmacology. LV. Nomenclature and molecular relationships of two-P potassium channels. *Pharmacological Reviews*, *57*(4), 527–540. doi:10.1124/pr.57.4.12

Goldstein, S. A., Bockenhauer, D., O'Kelly, I., & Zilberberg, N. (2001). Potassium leak channels and the KCNK family of two-P-domain subunits. *Nature Reviews Neuroscience*, *2*(3), 175–184. doi:10.1038/35058574

Goldstein, S. A., Price, L. A., Rosenthal, D. N., & Pausch, M. H. (1996). ORK1, a potassium-selective leak channel with two pore domains cloned from *Drosophila melanogaster* by expression in *Saccharomyces cerevisiae*. *Proceedings of the National Academy of Sciences of the United States of America*, *93*(23), 13256–13261.

Goldstein, S. A., Wang, K. W., Ilan, N., & Pausch, M. H. (1998). Sequence and function of the two P domain potassium channels: Implications of an emerging superfamily. *Journal of Molecular Medicine (Berlin)*, *76*(1), 13–20.

Gonzalez, J. A., Jensen, L. T., Doyle, S. E., Miranda-Anaya, M., Menaker, M., Fugger, L., . . . Burdakov, D. (2009). Deletion of TASK1 and TASK3 channels disrupts intrinsic excitability but does not abolish glucose or pH responses of orexin/hypocretin neurons. *European Journal of Neuroscience*, *30*(1), 57–64. doi:10.1111/j.1460-9568.2009.06789.x

Gu, W., Schlichthorl, G., Hirsch, J. R., Engels, H., Karschin, C., Karschin, A., . . . Daut, J. (2002). Expression pattern and functional characteristics of two novel splice variants of the two-pore-domain potassium channel TREK-2. *Journal of Physiology*, *539*(Pt 3), 657–668.

Guagliardo, N. A., Yao, J., Hu, C., Schertz, E. M., Tyson, D. A., Carey, R. M., . . . Barrett, P. Q. (2012). TASK-3 channel deletion in mice recapitulates low-renin essential hypertension. *Hypertension*, *59*(5), 999–1005. doi:10.1161/HYPERTENSIONAHA.111.189662

Gurney, A. M., Osipenko, O. N., MacMillan, D., McFarlane, K. M., Tate, R. J., & Kempsill, F. E. (2003). Two-pore domain K channel, TASK-1, in pulmonary artery smooth muscle cells. *Circulation Research*, *93*(10), 957–964. doi:10.1161/01.RES.0000099883.68414.61

Guyenet, P. G., & Bayliss, D. A. (2015). Neural control of breathing and CO2 homeostasis. *Neuron*, *87*(5), 946–961. doi:10.1016/j.neuron.2015.08.001

Han, J., Kang, D., & Kim, D. (2003). Functional properties of four splice variants of a human pancreatic tandem-pore K+ channel, TALK-1. *American Journal of Physiology–Cell Physiology*, *285*(3), C529–538. doi:10.1152/ajpcell.00601.2002

Heitzmann, D., Derand, R., Jungbauer, S., Bandulik, S., Sterner, C., Schweda, F., . . . Barhanin, J. (2008). Invalidation of TASK1 potassium channels disrupts adrenal gland zonation and mineralocorticoid homeostasis. *EMBO Journal*, *27* (1), 179–187. doi:10.1038/sj.emboj.7601934

Heurteaux, C., Guy, N., Laigle, C., Blondeau, N., Duprat, F., Mazzuca, M., . . . Lazdunski, M. (2004). TREK-1, a K+ channel involved in neuroprotection and general anesthesia. *EMBO Journal*, *23* (13), 2684–2695. doi:10.1038/sj.emboj.7600234

Heurteaux, C., Lucas, G., Guy, N., El Yacoubi, M., Thummler, S., Peng, X. D., . . . Lazdunski, M. (2006). Deletion of the background potassium channel TREK-1 results in a depression-resistant phenotype. *Nature Neuroscience*, *9*(9), 1134–1141. doi:10.1038/nn1749

Honore, E. (2007). The neuronal background K2P channels: Focus on TREK1. *Nature Reviews Neuroscience*, *8*(4), 251–261. doi:10.1038/nrn2117

Honore, E., Maingret, F., Lazdunski, M., & Patel, A. J. (2002). An intracellular proton sensor commands lipid- and mechano-gating of the K(+) channel TREK-1. *EMBO Journal*, *21* (12), 2968–2976. doi:10.1093/emboj/cdf288

Hwang, E. M., Kim, E., Yarishkin, O., Woo, D. H., Han, K. S., Park, N., . . . Park, J. Y. (2014). A disulphide-linked heterodimer of TWIK-1 and TREK-1 mediates passive conductance in astrocytes. *Nature Communications*, 5, 3227. doi:10.1038/ncomms4227

Ishii, H., Nakajo, K., Yanagawa, Y., & Kubo, Y. (2010). Identification and characterization of Cs(+)—permeable K(+) channel current in mouse cerebellar Purkinje cells in lobules 9 and 10 evoked by molecular layer stimulation. *European Journal of Neuroscience*, *32*(5), 736–748. doi:10.1111/j.1460-9568.2010.07336.x

Jiang, Y., Lee, A., Chen, J., Ruta, V., Cadene, M., Chait, B. T., & MacKinnon, R. (2003). X-ray structure of a voltage-dependent K+ channel. *Nature*, *423*(6935), 33–41. doi:10.1038/nature01580

Kang, D., Choe, C., Cavanaugh, E., & Kim, D. (2007). Properties of single two-pore domain TREK-2 channels expressed in mammalian cells. *Journal of Physiology*, *583*(Pt 1), 57–69. doi:10.1113/jphysiol.2007.136150

Kang, D., Choe, C., & Kim, D. (2005). Thermosensitivity of the two-pore domain K+ channels TREK-2 and TRAAK. *Journal of Physiology*, *564*(Pt 1), 103–116. doi:10.1113/jphysiol.2004.081059

Kang, D., Han, J., & Kim, D. (2006). Mechanism of inhibition of TREK-2 (K2P10.1) by the Gq-coupled M3 muscarinic receptor. *American Journal of Physiology–Cell Physiology*, *291*(4), C649–656. doi:10.1152/ajpcell.00047.2006

Kang, D., Han, J., Talley, E. M., Bayliss, D. A., & Kim, D. (2004). Functional expression of TASK-1/TASK-3 heteromers in cerebellar granule cells. *Journal of Physiology*, *554*(Pt 1), 64–77. doi:10.1113/jphysiol.2003.054387

Kang, D., Hogan, J. O., & Kim, D. (2014). THIK-1 (K2P13.1) is a small-conductance background K(+) channel in rat trigeminal ganglion neurons. *Pflügers Archiv*, *466*(7), 1289–1300. doi:10.1007/s00424-013-1358-1

Kang, D., & Kim, D. (2004). Single-channel properties and pH sensitivity of two-pore domain K+ channels of the TALK family. *Biochemical and Biophysical Research Communications*, *315*(4), 836–844. doi:10.1016/j.bbrc.2004.01.137

Kang, D., & Kim, D. (2006). TREK-2 (K2P10.1) and TRESK (K2P18.1) are major background K+ channels in dorsal root ganglion neurons. *American Journal of Physiology–Cell Physiology*, *291*(1), C138–146. doi:10.1152/ajpcell.00629.2005

Kang, D., Mariash, E., & Kim, D. (2004). Functional expression of TRESK-2, a new member of the tandem-pore K+ channel family. *Journal of Biological Chemistry*, *279*(27), 28063–28070. doi:10.1074/jbc.M402940200

Karschin, C., Wischmeyer, E., Preisig-Muller, R., Rajan, S., Derst, C., Grzeschik, K. H., . . . Karschin, A. (2001). Expression pattern in brain of TASK-1, TASK-3, and a tandem pore domain K(+) channel subunit, TASK-5, associated with the central auditory nervous system. *Molecular and Cellular Neuroscience*, *18*(6), 632–648. doi:10.1006/mcne.2001.1045

Kennard, L. E., Chumbley, J. R., Ranatunga, K. M., Armstrong, S. J., Veale, E. L., & Mathie, A. (2005). Inhibition of the human two-pore domain potassium channel, TREK-1, by fluoxetine and its metabolite norfluoxetine. *British Journal of Pharmacology*, *144*(6), 821–829. doi:10.1038/sj.bjp.0706068

Keshavaprasad, B., Liu, C., Au, J. D., Kindler, C. H., Cotten, J. F., & Yost, C. S. (2005). Species-specific differences in response to anesthetics and other modulators by the K2P channel TRESK. *Anesthesia and Analgesia*, *101*(4), 1042–1049, Table of Contents. doi:10.1213/01.ane.0000168447.87557.5a

Kilisch, M., Lytovchenko, O., Schwappach, B., Renigunta, V., & Daut, J. (2015). The role of protein–protein interactions in the intracellular traffic of the potassium channels TASK-1 and TASK-3. *Pflügers Archiv*, *467*(5), 1105–1120. doi:10.1007/s00424-014-1672-2

Kim, D. (2005). Physiology and pharmacology of two-pore domain potassium channels. *Current Pharmaceutical Design*, *11*(21), 2717–2736.

Kim, D. (2013). K(+) channels in O(2) sensing and postnatal development of carotid body glomus cell response to hypoxia. *Respiration Physiology and Neurobiology*, *185*(1), 44–56. doi:10.1016/j.resp.2012.07.005

Kim, D., Cavanaugh, E. J., Kim, I., & Carroll, J. L. (2009). Heteromeric TASK-1/TASK-3 is the major oxygen-sensitive background K+ channel in rat carotid body glomus cells. *Journal of Physiology*, *587*(Pt 12), 2963–2975. doi:10.1113/jphysiol.2009.171181

Kim, D., & Gnatenco, C. (2001). TASK-5, a new member of the tandem-pore K(+) channel family. *Biochemical and Biophysical Research Communications*, *284*(4), 923–930. doi:10.1006/bbrc.2001.5064

Kim, Y., Bang, H., & Kim, D. (1999). TBAK-1 and TASK-1, two-pore K(+) channel subunits: Kinetic properties and expression in rat heart. *American Journal of Physiology*, *277*(5 Pt 2), H1669–1678.

Kim, Y., Bang, H., & Kim, D. (2000). TASK-3, a new member of the tandem pore K(+) channel family. *Journal of Biological Chemistry*, *275*(13), 9340–9347.

Koo, J. Y., Jang, Y., Cho, H., Lee, C. H., Jang, K. H., Chang, Y. H., … Oh, U. (2007). Hydroxy-alpha-sanshool activates TRPV1 and TRPA1 in sensory neurons. *European Journal of Neuroscience*, *26*(5), 1139–1147. doi:10.1111/j.1460-9568.2007.05743.x

Kreneisz, O., Benoit, J. P., Bayliss, D. A., & Mulkey, D. K. (2009). AMP-activated protein kinase inhibits TREK channels. *Journal of Physiology*, *587*(Pt 24), 5819–5830. doi:10.1113/jphysiol.2009.180372

Kumar, N. N., Velic, A., Soliz, J., Shi, Y., Li, K., Wang, S., … Bayliss, D. A. (2015). PHYSIOLOGY. Regulation of breathing by CO(2) requires the proton-activated receptor GPR4 in retrotrapezoid nucleus neurons. *Science*, *348*(6240), 1255–1260. doi:10.1126/science.aaa0922

Lafreniere, R. G., Cader, M. Z., Poulin, J. F., Andres-Enguix, I., Simoneau, M., Gupta, N., … Rouleau, G. A. (2010). A dominant-negative mutation in the TRESK potassium channel is linked to familial migraine with aura. *Nature Medicine*, *16*(10), 1157–1160. doi:10.1038/nm.2216

Lambert, M., Boet, A., Rucker-Martin, C., Mendes-Ferreira, P., Capuano, V., Hatem, S., … Antigny, F. (2018). Loss of KCNK3 is a hallmark of RV hypertrophy/dysfunction associated with pulmonary hypertension. *Cardiovascular Research 114* (6): 880–893. doi:10.1093/cvr/cvy016

Lazarenko, R. M., Fortuna, M. G., Shi, Y., Mulkey, D. K., Takakura, A. C., Moreira, T. S., … Bayliss, D. A. (2010). Anesthetic activation of central respiratory chemoreceptor neurons involves inhibition of a THIK-1-like background K(+) current. *Journal of Neuroscience*, *30*(27), 9324–9334. doi:10.1523/JNEUROSCI.1956-10.2010

Lazarenko, R. M., Willcox, S. C., Shu, S., Berg, A. P., Jevtovic-Todorovic, V., Talley, E. M., … Bayliss, D. A. (2010). Motoneuronal TASK channels contribute to immobilizing effects of inhalational general anesthetics. *Journal of Neuroscience*, *30*(22), 7691–7704. doi:10.1523/JNEUROSCI.1655-10.2010

Lengyel, M., Czirjak, G., & Enyedi, P. (2016). Formation of functional heterodimers by TREK-1 and TREK-2 two-pore domain potassium channel subunits. *Journal of Biological Chemistry*, *291*(26), 13649–13661. doi:10.1074/jbc.M116.719039

Lesage, F., & Barhanin, J. (2011). Molecular physiology of pH-sensitive background K(2P) channels. *Physiology (Bethesda)*, *26*(6), 424–437. doi:10.1152/physiol.00029.2011

Lesage, F., Guillemare, E., Fink, M., Duprat, F., Lazdunski, M., Romey, G., & Barhanin, J. (1996). TWIK-1, a ubiquitous human weakly inward rectifying K+ channel with a novel structure. *EMBO Journal*, *15*(5), 1004–1011.

Lesage, F., Lauritzen, I., Duprat, F., Reyes, R., Fink, M., Heurteaux, C., & Lazdunski, M. (1997). The structure, function and distribution of the mouse TWIK-1 K+ channel. *FEBS Letters*, *402*(1), 28–32.

Lesage, F., & Lazdunski, M. (2000). Molecular and functional properties of two-pore-domain potassium channels. *American Journal of Physiology–Renal Physiology*, *279*(5), F793–801. doi:10.1152/ajprenal.2000.279.5.F793

Lesage, F., Reyes, R., Fink, M., Duprat, F., Guillemare, E., & Lazdunski, M. (1996). Dimerization of TWIK-1 K+ channel subunits via a disulfide bridge. *EMBO Journal*, *15* (23), 6400–6407.

Lesage, F., Terrenoire, C., Romey, G., & Lazdunski, M. (2000). Human TREK2, a 2P domain mechano-sensitive K+ channel with multiple regulations by polyunsaturated fatty acids, lysophospholipids, and Gs, Gi, and Gq protein-coupled receptors. *Journal of Biological Chemistry*, *275*(37), 28398–28405. doi:10.1074/jbc.M002822200

Levitz, J., Royal, P., Comoglio, Y., Wdziekonski, B., Schaub, S., Clemens, D. M., . . . Sandoz, G. (2016). Heterodimerization within the TREK channel subfamily produces a diverse family of highly regulated potassium channels. *Proceedings of the National Academy of Sciences of the United States of America*, *113*(15), 4194–4199. doi:10.1073/pnas.1522459113

Linden, A. M., Aller, M. I., Leppa, E., Vekovischeva, O., Aitta-Aho, T., Veale, E. L., . . . Korpi, E. R. (2006). The in vivo contributions of TASK-1-containing channels to the actions of inhalation anesthetics, the alpha(2) adrenergic sedative dexmedetomidine, and cannabinoid agonists. *Journal of Pharmacology and Experimental Therapeutics*, *317*(2), 615–626. doi:10.1124/jpet.105.098525

Linden, A. M., Sandu, C., Aller, M. I., Vekovischeva, O. Y., Rosenberg, P. H., Wisden, W., & Korpi, E. R. (2007). TASK-3 knockout mice exhibit exaggerated nocturnal activity, impairments in cognitive functions, and reduced sensitivity to inhalation anesthetics. *Journal of Pharmacology and Experimental Therapeutics*, *323*(3), 924–934. doi:10.1124/jpet.107.129544

Lindner, M., Leitner, M. G., Halaszovich, C. R., Hammond, G. R., & Oliver, D. (2011). Probing the regulation of TASK potassium channels by PI4,5P(2) with switchable phosphoinositide phosphatases. *Journal of Physiology*, *589*(Pt 13), 3149–3162. doi:10.1113/jphysiol.2011.208983

Liu, C., Au, J. D., Zou, H. L., Cotten, J. F., & Yost, C. S. (2004). Potent activation of the human tandem pore domain K channel TRESK with clinical concentrations of volatile anesthetics. *Anesthesia and Analgesia*, *99*(6), 1715–1722, Table of Contents. doi:10.1213/01.ANE.0000136849.07384.44

Lolicato, M., Arrigoni, C., Mori, T., Sekioka, Y., Bryant, C., Clark, K. A., & Minor, D. L., Jr. (2017). K2P2.1 (TREK-1)-activator complexes reveal a cryptic selectivity filter binding site. *Nature*, *547*(7663), 364–368. doi:10.1038/nature22988

Lopes, C. M., Rohacs, T., Czirjak, G., Balla, T., Enyedi, P., & Logothetis, D. E. (2005). PIP2 hydrolysis underlies agonist-induced inhibition and regulates voltage gating of two-pore domain K+ channels. *Journal of Physiology*, *564*(Pt 1), 117–129. doi:10.1113/jphysiol.2004.081935

Lopes, C. M., Zilberberg, N., & Goldstein, S. A. (2001). Block of Kcnk3 by protons. Evidence that 2-P-domain potassium channel subunits function as homodimers. *Journal of Biological Chemistry*, *276*(27), 24449–24452. doi:10.1074/jbc.C100184200

Ma, L., Zhang, X., & Chen, H. (2011). TWIK-1 two-pore domain potassium channels change ion selectivity and conduct inward leak sodium currents in hypokalemia. *Science Signaling*, *4*(176), ra37. doi:10.1126/scisignal.2001726

Madrid, R., de la Pena, E., Donovan-Rodriguez, T., Belmonte, C., & Viana, F. (2009). Variable threshold of trigeminal cold-thermosensitive neurons is determined by a balance between TRPM8 and Kv1 potassium channels. *Journal of Neuroscience*, *29*(10), 3120–3131. doi:10.1523/JNEUROSCI.4778-08.2009

Madry, C., Kyrargyri, V., Arancibia-Cárcamo, I.L., Jolivet, R., Kohsaka, S., Bryan, R.M., Attwell, D. (2018). Microglial ramification, surveillance, and interleukin-1β release are regulated by the two-pore domain K+ channel THIK-1. *Neuron*. *97*(2), 299–312. doi: 10.1016/j.neuron.2017.12.002.

Maingret, F., Fosset, M., Lesage, F., Lazdunski, M., & Honore, E. (1999). TRAAK is a mammalian neuronal mechano-gated K+ channel. *Journal of Biological Chemistry*, *274*(3), 1381–1387.

Maingret, F., Lauritzen, I., Patel, A. J., Heurteaux, C., Reyes, R., Lesage, F., . . . Honore, E. (2000). TREK-1 is a heat-activated background K(+) channel. *EMBO Journal*, *19* (11), 2483–2491. doi:10.1093/emboj/19.11.2483

Maingret, F., Patel, A. J., Lesage, F., Lazdunski, M., & Honore, E. (1999). Mechano- or acid stimulation, two interactive modes of activation of the TREK-1 potassium channel. *Journal of Biological Chemistry*, *274*(38), 26691–26696.

Maingret, F., Patel, A. J., Lesage, F., Lazdunski, M., & Honore, E. (2000). Lysophospholipids open the two-pore domain mechano-gated K(+) channels TREK-1 and TRAAK. *Journal of Biological Chemistry*, *275*(14), 10128–10133.

Manoury, B., Lamalle, C., Oliveira, R., Reid, J., & Gurney, A. M. (2011). Contractile and electrophysiological properties of pulmonary artery smooth muscle are not altered in TASK-1 knockout mice. *Journal of Physiology*, *589*(Pt 13), 3231–3246. doi:10.1113/jphysiol.2011.206748

Marinc, C., Preisig-Muller, R., Pruss, H., Derst, C., & Veh, R. W. (2011). Immunocytochemical localization of TASK-3 (K(2P)9.1) channels in monoaminergic and cholinergic neurons. *Cellular and Molecular Neurobiology*, *31*(2), 323–335. doi:10.1007/s10571-010-9625-6

Mathie, A. (2007). Neuronal two-pore-domain potassium channels and their regulation by G protein-coupled receptors. *Journal of Physiology*, *578*(Pt 2), 377–385. doi:10.1113/jphysiol.2006.121582

Mazella, J., Petrault, O., Lucas, G., Deval, E., Beraud-Dufour, S., Gandin, C., . . . Borsotto, M. (2010). Spadin, a sortilin-derived peptide, targeting rodent TREK-1 channels: A new concept in the antidepressant drug design. *PLoS Biology*, *8*(4), e1000355. doi:10.1371/journal.pbio.1000355

McClenaghan, C., Schewe, M., Aryal, P., Carpenter, E. P., Baukrowitz, T., & Tucker, S. J. (2016). Polymodal activation of the TREK-2 K2P channel produces structurally distinct open states. *Journal of General Physiology*, *147*(6), 497–505. doi:10.1085/jgp.201611601

Medhurst, A. D., Rennie, G., Chapman, C. G., Meadows, H., Duckworth, M. D., Kelsell, R. E., . . . Pangalos, M. N. (2001). Distribution analysis of human two pore domain potassium channels in tissues of the central nervous system and periphery. *Brain Research, Molecular Brain Research*, *86*(1–2), 101–114.

Meuth, S. G., Budde, T., Kanyshkova, T., Broicher, T., Munsch, T., & Pape, H. C. (2003). Contribution of TWIK-related acid-sensitive K+ channel 1 (TASK1) and TASK3 channels to the control of activity modes in thalamocortical neurons. *Journal of Neuroscience*, *23*(16), 6460–6469.

Meuth, S. G., Kleinschnitz, C., Broicher, T., Austinat, M., Braeuninger, S., Bittner, S., ... Wiendl, H. (2009). The neuroprotective impact of the leak potassium channel TASK1 on stroke development in mice. *Neurobiology of Disease*, *33*(1), 1–11. doi:10.1016/j.nbd.2008.09.006

Millar, J. A., Barratt, L., Southan, A. P., Page, K. M., Fyffe, R. E., Robertson, B., & Mathie, A. (2000). A functional role for the two-pore domain potassium channel TASK-1 in cerebellar granule neurons. *Proceedings of the National Academy of Sciences of the United States of America*, *97*(7), 3614–3618. doi:10.1073/pnas.050012597

Miller, A. N., & Long, S. B. (2012). Crystal structure of the human two-pore domain potassium channel K2P1. *Science*, *335*(6067), 432–436. doi:10.1126/science.1213274

Mirkovic, K., & Wickman, K. (2011). Identification and characterization of alternative splice variants of the mouse Trek2/Kcnk10 gene. *Neuroscience*, *194*, 11–18. doi:10.1016/j.neuroscience.2011.07.064

Moore, J. T., Chen, J., Han, B., Meng, Q. C., Veasey, S. C., Beck, S. G., & Kelz, M. B. (2012). Direct activation of sleep-promoting VLPO neurons by volatile anesthetics contributes to anesthetic hypnosis. *Current Biology*, *22*(21), 2008–2016. doi:10.1016/j.cub.2012.08.042

Morenilla-Palao, C., Luis, E., Fernandez-Pena, C., Quintero, E., Weaver, J. L., Bayliss, D. A., & Viana, F. (2014). Ion channel profile of TRPM8 cold receptors reveals a role of TASK-3 potassium channels in thermosensation. *Cell Reports*, *8*(5), 1571–1582. doi:10.1016/j.celrep.2014.08.003

Mulkey, D. K., Talley, E. M., Stornetta, R. L., Siegel, A. R., West, G. H., Chen, X., ... Bayliss, D. A. (2007). TASK channels determine pH sensitivity in select respiratory neurons but do not contribute to central respiratory chemosensitivity. *Journal of Neuroscience*, *27*(51), 14049–14058. doi:10.1523/JNEUROSCI.4254-07.2007

Murbartian, J., Lei, Q., Sando, J. J., & Bayliss, D. A. (2005). Sequential phosphorylation mediates receptor- and kinase-induced inhibition of TREK-1 background potassium channels. *Journal of Biological Chemistry*, *280*(34), 30175–30184. doi:10.1074/jbc.M503862200

Murtaza, G., Mermer, P., Goldenberg, A., Pfeil, U., Paddenberg, R., Weissmann, N., ... Kummer, W. (2017). TASK-1 potassium channel is not critically involved in mediating hypoxic pulmonary vasoconstriction of murine intra-pulmonary arteries. *PLoS One*, *12*(3), e0174071. doi:10.1371/journal.pone.0174071

Nicoll, R. A. (1988). The coupling of neurotransmitter receptors to ion channels in the brain. *Science*, *241*(4865), 545–551.

Nicoll, R. A., & Madison, D. V. (1982). General anesthetics hyperpolarize neurons in the vertebrate central nervous system. *Science*, *217*(4564), 1055–1057.

Nicoll, R. A., Malenka, R. C., & Kauer, J. A. (1990). Functional comparison of neurotransmitter receptor subtypes in mammalian central nervous system. *Physiological Reviews*, *70*(2), 513–565. doi:10.1152/physrev.1990.70.2.513

Niemeyer, M. I., Gonzalez-Nilo, F. D., Zuniga, L., Gonzalez, W., Cid, L. P., & Sepulveda, F. V. (2007). Neutralization of a single arginine residue gates open a two-pore domain, alkali-activated K+ channel. *Proceedings of the National Academy of Sciences of the United States of America*, *104*(2), 666–671. doi:10.1073/pnas.0606173104

Noel, J., Zimmermann, K., Busserolles, J., Deval, E., Alloui, A., Diochot, S., ... Lazdunski, M. (2009). The mechano-activated K+ channels TRAAK and TREK-1 control both warm and cold perception. *EMBO Journal*, *28*(9), 1308–1318. doi:10.1038/emboj.2009.57

O'Kelly, I., Butler, M. H., Zilberberg, N., & Goldstein, S. A. (2002). Forward transport. 14-3-3 binding overcomes retention in endoplasmic reticulum by dibasic signals. *Cell*, *111*(4), 577–588.

Olschewski, A., Li, Y., Tang, B., Hanze, J., Eul, B., Bohle, R. M., . . . Olschewski, H. (2006). Impact of TASK-1 in human pulmonary artery smooth muscle cells. *Circulation Research*, *98*(8), 1072–1080. doi:10.1161/01.RES.0000219677.12988.e9

Ortega-Saenz, P., Levitsky, K. L., Marcos-Almaraz, M. T., Bonilla-Henao, V., Pascual, A., & Lopez-Barneo, J. (2010). Carotid body chemosensory responses in mice deficient of TASK channels. *Journal of General Physiology*, *135*(4), 379–392. doi:10.1085/jgp.200910302

Pang, D. S., Robledo, C. J., Carr, D. R., Gent, T. C., Vyssotski, A. L., Caley, A., . . . Franks, N. P. (2009). An unexpected role for TASK-3 potassium channels in network oscillations with implications for sleep mechanisms and anesthetic action. *Proceedings of the National Academy of Sciences of the United States of America*, *106*(41), 17546–17551. doi:10.1073/pnas.0907228106

Patel, A., & Honore, E. (2002). The TREK two P domain K+ channels. *Journal of Physiology*, *539*(Pt 3), 647.

Patel, A. J., Honore, E., Lesage, F., Fink, M., Romey, G., & Lazdunski, M. (1999). Inhalational anesthetics activate two-pore-domain background K+ channels. *Nature Neuroscience*, *2*(5), 422–426. doi:10.1038/8084

Patel, A. J., Honore, E., Maingret, F., Lesage, F., Fink, M., Duprat, F., & Lazdunski, M. (1998). A mammalian two pore domain mechano-gated S-like K+ channel. *EMBO Journal*, *17*(15), 4283–4290. doi:10.1093/emboj/17.15.4283

Patel, A. J., Lazdunski, M., & Honore, E. (2001). Lipid and mechano-gated 2P domain K(+) channels. *Current Opinion in Cell Biology*, *13*(4), 422–428.

Patel, A. J., Maingret, F., Magnone, V., Fosset, M., Lazdunski, M., & Honore, E. (2000). TWIK-2, an inactivating 2P domain K+ channel. *Journal of Biological Chemistry*, *275*(37), 28722–28730. doi:10.1074/jbc.M003755200

Penton, D., Bandulik, S., Schweda, F., Haubs, S., Tauber, P., Reichold, M., . . . Barhanin, J. (2012). Task3 potassium channel gene invalidation causes low renin and salt-sensitive arterial hypertension. *Endocrinology*, *153*(10), 4740–4748. doi:10.1210/en.2012-1527

Pereira, V., Busserolles, J., Christin, M., Devilliers, M., Poupon, L., Legha, W., . . . Noel, J. (2014). Role of the TREK2 potassium channel in cold and warm thermosensation and in pain perception. *Pain*, *155*(12), 2534–2544. doi:10.1016/j.pain.2014.09.013

Piechotta, P. L., Rapedius, M., Stansfeld, P. J., Bollepalli, M. K., Ehrlich, G., Andres-Enguix, I., . . . Baukrowitz, T. (2011). The pore structure and gating mechanism of K2P channels. *EMBO Journal*, *30* (17), 3607–3619. doi:10.1038/emboj.2011.268

Pisani, D. F., Beranger, G. E., Corinus, A., Giroud, M., Ghandour, R. A., Altirriba, J., . . . Amri, E. Z. (2016). The K+ channel TASK1 modulates beta-adrenergic response in brown adipose tissue through the mineralocorticoid receptor pathway. *FASEB Journal*, *30*(2), 909–922. doi:10.1096/fj.15-277475

Plant, L. D., Dementieva, I. S., Kollewe, A., Olikara, S., Marks, J. D., & Goldstein, S. A. (2010). One SUMO is sufficient to silence the dimeric potassium channel K2P1. *Proceedings of the National Academy of Sciences of the United States of America*, *107*(23), 10743–10748. doi:10.1073/pnas.1004712107

Plant, L. D., Zuniga, L., Araki, D., Marks, J. D., & Goldstein, S. A. (2012). SUMOylation silences heterodimeric TASK potassium channels containing K2P1 subunits in cerebellar granule neurons. *Science Signaling*, *5*(251), ra84. doi:10.1126/scisignal.2003431

Rajan, S., Plant, L. D., Rabin, M. L., Butler, M. H., & Goldstein, S. A. (2005). Sumoylation silences the plasma membrane leak K+ channel K2P1. *Cell*, *121*(1), 37–47. doi:10.1016/j.cell.2005.01.019

Rajan, S., Preisig-Muller, R., Wischmeyer, E., Nehring, R., Hanley, P. J., Renigunta, V., ... Daut, J. (2002). Interaction with 14-3-3 proteins promotes functional expression of the potassium channels TASK-1 and TASK-3. *Journal of Physiology*, *545*(Pt 1), 13–26.

Rajan, S., Wischmeyer, E., Karschin, C., Preisig-Muller, R., Grzeschik, K. H., Daut, J., ... Derst, C. (2001). THIK-1 and THIK-2, a novel subfamily of tandem pore domain K+ channels. *Journal of Biological Chemistry*, *276*(10), 7302–7311. doi:10.1074/jbc.M008985200

Rajan, S., Wischmeyer, E., Xin Liu, G., Preisig-Muller, R., Daut, J., Karschin, A., & Derst, C. (2000). TASK-3, a novel tandem pore domain acid-sensitive K+ channel. An extracellular histidine as pH sensor. *Journal of Biological Chemistry*, *275*(22), 16650–16657. doi:10.1074/jbc.M000030200

Renigunta, V., Schlichthorl, G., & Daut, J. (2015). Much more than a leak: Structure and function of K(2)p-channels. *Pflügers Archiv*, *467*(5), 867–894. doi:10.1007/s00424-015-1703-7

Renigunta, V., Yuan, H., Zuzarte, M., Rinne, S., Koch, A., Wischmeyer, E., ... Preisig-Muller, R. (2006). The retention factor p11 confers an endoplasmic reticulum-localization signal to the potassium channel TASK-1. *Traffic*, *7*(2), 168–181. doi:10.1111/j.1600-0854.2005.00375.x

Renigunta, V., Zou, X., Kling, S., Schlichthorl, G., & Daut, J. (2014). Breaking the silence: Functional expression of the two-pore-domain potassium channel THIK-2. *Pflügers Archiv*, *466*(9), 1735–1745. doi:10.1007/s00424-013-1404-z

Reyes, R., Duprat, F., Lesage, F., Fink, M., Salinas, M., Farman, N., & Lazdunski, M. (1998). Cloning and expression of a novel pH-sensitive two pore domain K+ channel from human kidney. *Journal of Biological Chemistry*, *273*(47), 30863–30869.

Riera, C. E., Menozzi-Smarrito, C., Affolter, M., Michlig, S., Munari, C., Robert, F., ... le Coutre, J. (2009). Compounds from Sichuan and Melegueta peppers activate, covalently and non-covalently, TRPA1 and TRPV1 channels. *British Journal of Pharmacology*, *157*(8), 1398–1409. doi:10.1111/j.1476-5381.2009.00307.x

Rinne, S., Kiper, A. K., Schlichthorl, G., Dittmann, S., Netter, M. F., Limberg, S. H., ... Decher, N. (2015). TASK-1 and TASK-3 may form heterodimers in human atrial cardiomyocytes. *Journal of Molecular and Cellular Cardiology*, *81*, 71–80. doi:10.1016/j.yjmcc.2015.01.017

Rinne, S., Renigunta, V., Schlichthorl, G., Zuzarte, M., Bittner, S., Meuth, S. G., ... Preisig-Muller, R. (2014). A splice variant of the two-pore domain potassium channel TREK-1 with only one pore domain reduces the surface expression of full-length TREK-1 channels. *Pflügers Archiv*, *466*(8), 1559–1570. doi:10.1007/s00424-013-1384-z

Salinas, M., Reyes, R., Lesage, F., Fosset, M., Heurteaux, C., Romey, G., & Lazdunski, M. (1999). Cloning of a new mouse two-P domain channel subunit and a human homologue with a unique pore structure. *Journal of Biological Chemistry*, *274*(17), 11751–11760.

Sandoz, G., Douguet, D., Chatelain, F., Lazdunski, M., & Lesage, F. (2009). Extracellular acidification exerts opposite actions on TREK1 and TREK2 potassium channels via a single conserved histidine residue. *Proceedings of the National Academy of Sciences of the United States of America*, *106*(34), 14628–14633. doi:10.1073/pnas.0906267106

Sandoz, G., Tardy, M. P., Thummler, S., Feliciangeli, S., Lazdunski, M., & Lesage, F. (2008). Mtap2 is a constituent of the protein network that regulates TWIK-related K+ channel expression and trafficking. *Journal of Neuroscience*, *28*(34), 8545–8552. doi:10.1523/JNEUROSCI.1962-08.2008

Sandoz, G., Thummler, S., Duprat, F., Feliciangeli, S., Vinh, J., Escoubas, P., ... Lesage, F. (2006). AKAP150, a switch to convert mechano-, pH- and arachidonic acid-sensitive TREK K(+) channels into open leak channels. *EMBO Journal*, *25*(24), 5864–5872. doi:10.1038/sj.emboj.7601437

Sano, Y., Inamura, K., Miyake, A., Mochizuki, S., Kitada, C., Yokoi, H., . . . Furuichi, K. (2003). A novel two-pore domain K+ channel, TRESK, is localized in the spinal cord. *Journal of Biological Chemistry*, *278*(30), 27406–27412. doi:10.1074/jbc.M206810200

Schewe, M., Nematian-Ardestani, E., Sun, H., Musinszki, M., Cordeiro, S., Bucci, G., . . . Baukrowitz, T. (2016). A non-canonical voltage-sensing mechanism controls gating in K2P K(+) channels. *Cell*, *164*(5), 937–949. doi:10.1016/j.cell.2016.02.002

Sepulveda, F. V., Pablo Cid, L., Teulon, J., & Niemeyer, M. I. (2015). Molecular aspects of structure, gating, and physiology of pH-sensitive background K2P and Kir K+-transport channels. *Physiological Reviews*, *95*(1), 179–217. doi:10.1152/physrev.00016.2014

Simkin, D., Cavanaugh, E. J., & Kim, D. (2008). Control of the single channel conductance of K2P10.1 (TREK-2) by the amino-terminus: Role of alternative translation initiation. *Journal of Physiology*, *586*(23), 5651–5663. doi:10.1113/jphysiol.2008.161927

Sirois, J. E., Lei, Q., Talley, E. M., Lynch, C., 3rd, & Bayliss, D. A. (2000). The TASK-1 two-pore domain K+ channel is a molecular substrate for neuronal effects of inhalation anesthetics. *Journal of Neuroscience*, *20*(17), 6347–6354.

Sirois, J. E., Lynch, C., 3rd, & Bayliss, D. A. (2002). Convergent and reciprocal modulation of a leak K+ current and I(h) by an inhalational anaesthetic and neurotransmitters in rat brainstem motoneurones. *Journal of Physiology*, *541*(Pt 3), 717–729.

Staudacher, K., Baldea, I., Kisselbach, J., Staudacher, I., Rahm, A. K., Schweizer, P. A., . . . Thomas, D. (2011). Alternative splicing determines mRNA translation initiation and function of human K(2P)10.1 K+ channels. *Journal of Physiology*, *589*(Pt 15), 3709–3720. doi:10.1113/jphysiol.2011.210666

Steinberg, E. A., Wafford, K. A., Brickley, S. G., Franks, N. P., & Wisden, W. (2015). The role of K(2)p channels in anaesthesia and sleep. *Pflügers Archiv*, *467*(5), 907–916. doi:10.1007/s00424-014-1654-4

Talley, E. M., & Bayliss, D. A. (2002). Modulation of TASK-1 (Kcnk3) and TASK-3 (Kcnk9) potassium channels: Volatile anesthetics and neurotransmitters share a molecular site of action. *Journal of Biological Chemistry*, *277*(20), 17733–17742. doi:10.1074/jbc.M200502200

Talley, E. M., Lei, Q., Sirois, J. E., & Bayliss, D. A. (2000). TASK-1, a two-pore domain K+ channel, is modulated by multiple neurotransmitters in motoneurons. *Neuron*, *25*(2), 399–410.

Talley, E. M., Sirois, J. E., Lei, Q., & Bayliss, D. A. (2003). Two-pore-domain (KCNK) potassium channels: Dynamic roles in neuronal function. *Neuroscientist*, *9*(1), 46–56. doi:10.1177/1073858402239590

Talley, E. M., Solorzano, G., Lei, Q., Kim, D., & Bayliss, D. A. (2001). CNS distribution of members of the two-pore-domain (KCNK) potassium channel family. *Journal of Neuroscience*, *21*(19), 7491–7505.

Thomas, D., Plant, L. D., Wilkens, C. M., McCrossan, Z. A., & Goldstein, S. A. (2008). Alternative translation initiation in rat brain yields K2P2.1 potassium channels permeable to sodium. *Neuron*, *58*(6), 859–870. doi:10.1016/j.neuron.2008.04.016

Trapp, S., Aller, M. I., Wisden, W., & Gourine, A. V. (2008). A role for TASK-1 (KCNK3) channels in the chemosensory control of breathing. *Journal of Neuroscience*, *28*(35), 8844–8850. doi:10.1523/JNEUROSCI.1810-08.2008

Tsunozaki, M., Lennertz, R. C., Vilceanu, D., Katta, S., Stucky, C. L., & Bautista, D. M. (2013). A "toothache tree" alkylamide inhibits Adelta mechanonociceptors to alleviate mechanical pain. *Journal of Physiology*, *591*(13), 3325–3340. doi:10.1113/jphysiol.2013.252106

Tulleuda, A., Cokic, B., Callejo, G., Saiani, B., Serra, J., & Gasull, X. (2011). TRESK channel contribution to nociceptive sensory neurons excitability: Modulation by nerve injury. *Molecular Pain, 7*, 30. doi:10.1186/1744-8069-7-30

Turner, P. J., & Buckler, K. J. (2013). Oxygen and mitochondrial inhibitors modulate both monomeric and heteromeric TASK-1 and TASK-3 channels in mouse carotid body type-1 cells. *Journal of Physiology, 591*(23), 5977–5998. doi:10.1113/jphysiol.2013.262022

Veale, E. L., Hassan, M., Walsh, Y., Al-Moubarak, E., & Mathie, A. (2014). Recovery of current through mutated TASK3 potassium channels underlying Birk Barel syndrome. *Molecular Pharmacology, 85*(3), 397–407. doi:10.1124/mol.113.090530

Veale, E. L., Kennard, L. E., Sutton, G. L., MacKenzie, G., Sandu, C., & Mathie, A. (2007). G(alpha)q-mediated regulation of TASK3 two-pore domain potassium channels: The role of protein kinase C. *Molecular Pharmacology, 71*(6), 1666–1675. doi:10.1124/mol.106.033241

Veale, E. L., Rees, K. A., Mathie, A., & Trapp, S. (2010). Dominant negative effects of a non-conducting TREK1 splice variant expressed in brain. *Journal of Biological Chemistry, 285*(38), 29295–29304. doi:10.1074/jbc.M110.108423

Viana, F., de la Pena, E., & Belmonte, C. (2002). Specificity of cold thermotransduction is determined by differential ionic channel expression. *Nature Neuroscience, 5*(3), 254–260. doi:10.1038/nn809

Vierra, N. C., Dadi, P. K., Jeong, I., Dickerson, M., Powell, D. R., & Jacobson, D. A. (2015). Type 2 diabetes-associated K+ channel TALK-1 modulates beta-cell electrical excitability, second-phase insulin secretion, and glucose homeostasis. *Diabetes, 64*(11), 3818–3828. doi:10.2337/db15-0280

Vivier, D., Soussia, I. B., Rodrigues, N., Lolignier, S., Devilliers, M., Chatelain, F. C., ... Ducki, S. (2017). Development of the first two-pore domain potassium channel TWIK-Related K(+) channel 1-selective agonist possessing in vivo antinociceptive activity. *Journal of Medicinal Chemistry, 60*(3), 1076–1088. doi:10.1021/acs.jmedchem.6b01285

Vu, M. T., Du, G., Bayliss, D. A., & Horner, R. L. (2015). TASK channels on basal forebrain cholinergic neurons modulate electrocortical signatures of arousal by histamine. *Journal of Neuroscience, 35*(40), 13555–13567. doi:10.1523/JNEUROSCI.1445-15.2015

Wang, S., Benamer, N., Zanella, S., Kumar, N. N., Shi, Y., Bevengut, M., ... Bayliss, D. A. (2013). TASK-2 channels contribute to pH sensitivity of retrotrapezoid nucleus chemoreceptor neurons. *Journal of Neuroscience, 33*(41), 16033–16044. doi:10.1523/JNEUROSCI.2451-13.2013

Wang, W., Kiyoshi, C. M., Du, Y., Ma, B., Alford, C. C., Chen, H., & Zhou, M. (2016). mGluR3 activation recruits cytoplasmic TWIK-1 channels to membrane that enhances ammonium uptake in hippocampal astrocytes. *Molecular Neurobiology, 53*(9), 6169–6182. doi:10.1007/s12035-015-9496-4

Warth, R., Barriere, H., Meneton, P., Bloch, M., Thomas, J., Tauc, M., ... Barhanin, J. (2004). Proximal renal tubular acidosis in TASK2 K+ channel-deficient mice reveals a mechanism for stabilizing bicarbonate transport. *Proceedings of the National Academy of Sciences of the United States of America, 101*(21), 8215–8220. doi:10.1073/pnas.0400081101

Washburn, C. P., Sirois, J. E., Talley, E. M., Guyenet, P. G., & Bayliss, D. A. (2002). Serotonergic raphe neurons express TASK channel transcripts and a TASK-like pH- and halothane-sensitive K+ conductance. *Journal of Neuroscience, 22*(4), 1256–1265.

Wilke, B. U., Lindner, M., Greifenberg, L., Albus, A., Kronimus, Y., Bunemann, M., ... Oliver, D. (2014). Diacylglycerol mediates regulation of TASK potassium channels by Gq-coupled receptors. *Nature Communications, 5*, 5540. doi:10.1038/ncomms6540

Woo, D. H., Han, K. S., Shim, J. W., Yoon, B. E., Kim, E., Bae, J. Y., . . . Lee, C. J. (2012). TREK-1 and Best1 channels mediate fast and slow glutamate release in astrocytes upon GPCR activation. *Cell, 151*(1), 25–40. doi:10.1016/j.cell.2012.09.005

Yao, C., Li, Y., Shu, S., Yao, S., Lynch, C., Bayliss, D. A., & Chen, X. (2017). TASK channels contribute to neuroprotective action of inhalational anesthetics. *Scientific Reports, 7*, 44203. doi:10.1038/srep44203

Yao, J., McHedlishvili, D., McIntire, W. E., Guagliardo, N. A., Erisir, A., Coburn, C. A., . . . Barrett, P. Q. (2017). Functional TASK-3-like channels in mitochondria of aldosterone-producing zona glomerulosa cells. *Hypertension, 70*(2), 347–356. doi:10.1161/HYPERTENSIONAHA.116.08871

Zhou, J., Yang, C. X., Zhong, J. Y., & Wang, H. B. (2013). Intrathecal TRESK gene recombinant adenovirus attenuates spared nerve injury-induced neuropathic pain in rats. *Neuroreport, 24*(3), 131–136. doi:10.1097/WNR.0b013e32835d8431

Zhou, M., Xu, G., Xie, M., Zhang, X., Schools, G. P., Ma, L., . . . Chen, H. (2009). TWIK-1 and TREK-1 are potassium channels contributing significantly to astrocyte passive conductance in rat hippocampal slices. *Journal of Neuroscience, 29*(26), 8551–8564. doi:10.1523/JNEUROSCI.5784-08.2009

Zuzarte, M., Heusser, K., Renigunta, V., Schlichthorl, G., Rinne, S., Wischmeyer, E., . . . Preisig-Muller, R. (2009). Intracellular traffic of the K+ channels TASK-1 and TASK-3: Role of N- and C-terminal sorting signals and interaction with 14-3-3 proteins. *Journal of Physiology, 587*(Pt 5), 929–952. doi:10.1113/jphysiol.2008.164756

CHAPTER 17

TRPC CHANNELS—INSIGHT FROM THE *DROSOPHILA* LIGHT SENSITIVE CHANNELS

BEN KATZ, WILLIAM L. PAK, AND BARUCH MINKE

Introduction

It is now widely recognized that TRP channels were discovered in the study of *Drosophila* eye. Cosens and Manning reported on a spontaneously occurring mutant, which they called "Type A," with stimulus-intensity–dependent behavioral and electrophysiological anomalies (Cosens & Manning, 1969). With a dim light stimulus, the electroretinogram (ERG: extracellularly recorded summed responses of the eye) looked nearly normal, but under a continuous bright light stimulus, the response decayed to, or nearly to, the baseline. This finding was potentially interesting to us because the mutant phenotype might suggest a defect in phototransduction (the process by which light signals are converted to electrical signals), and we were starting to isolate potential phototransduction mutants (Pak, 2010; Pak, Grossfield, & White, 1969). However, the report did not generate widespread interest at the time.

For one thing, the cellular origin of the ERG components was not well established at the time: it was not clear at what signaling level the mutation caused the defect. It could affect the activation of photoreceptors, or be involved in the chain of events subsequent to photoreceptor activation. In fact, Cosens and Manning themselves alluded to the possibility of "a breakdown in a transmitter system," though it was not clear what this transmitter system might mean in the present-day context. They also suggested that the defect might be in the visual pigment turnover. In addition, the report was based on the study of a single allele. The phenotypes of different alleles in a given gene can be very different, and studies based on a single allele need to be interpreted with caution.

The most direct way to determine whether the abnormal ERG response originated in the photoreceptor was to record directly from the photoreceptors. The only way of doing this at the time was by intracellular recording. Fortunately, we had developed the techniques for recording intracellularly from *Drosophila* photoreceptors (Alawi & Pak, 1971; Wu & Pak, 1975). Intracellular recordings showed unequivocally that the phenotypes seen in the ERG were present in intracellular recordings as well, suggesting that the defect in the mutant originated in the photoreceptors (Minke, 1982; Minke, Wu, & Pak, 1975; Figure 17.1A).

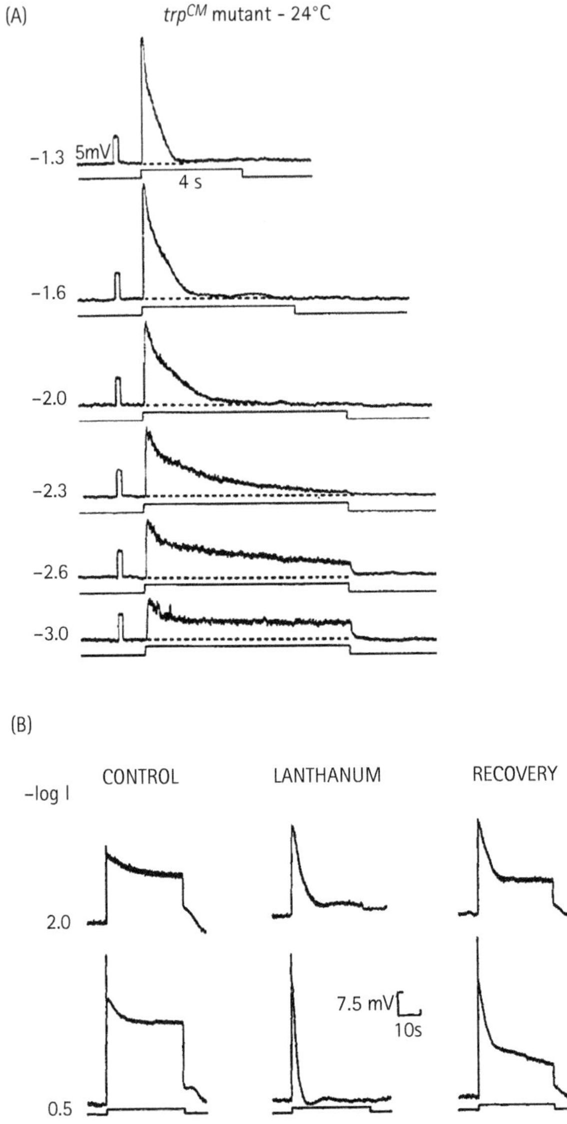

FIGURE 17.1 *The phenotype of the trp mutant is mimicked by lanthanum (La^{3+}) in wild-type fly.* A. Intracellular recordings from single photoreceptor cell of white-eyed trp^{CM} raised at 24°C showing voltage responses to increasing intensities of orange lights (in relative log scale).

From Minke, 1982; and Suss Toby et al., 1991.

These authors also measured the amount of photopigment activated by light in the mutant fly during the decline of the response to the low steady-state level. They found that, during the decay phase, only a small fraction of pigment was activated, ruling out the possibility that the mechanism for the mutant phenotype was due to a defect in photopigment turnover (Minke, 1982; Minke et al., 1975).

As for the single-allele issue, this objection could be countered by isolating additional alleles and showing that results from these alleles are consistent with those obtained with the allele discovered by Cosens and Manning. As we mentioned, a few years before the study by Cosens and Manning appeared, we began to isolate mutants defective in the ERG response using chemically mutagenizing, isogenized, Oregon R wild-type stock for studying *Drosophila* phototransduction (Pak, 2010; Pak et al., 1969). We mutagenized each major chromosome separately, the third chromosome (where *trp* is localized) being the last to be mutagenized. Nevertheless, by 1974, we began recovering mutants with similar ERG phenotypes, which did not complement the mutant discovered by Cosens/Manning. Ultimately, we isolated nine allelic mutants, including a true null mutant, trp^{P343} (Scott, Sun, Beckingham, & Zuker, 1997) and a constitutively active mutant, trp^{P365} (Yoon et al., 2000). These mutants were freely made available to other investigators. The results showed that, indeed, the phenotypes of allelic mutants of this gene do vary, but the phenotype described by Cosens and Manning represented the general features of many mutant alleles of this gene. With these results on hand, and in consultation with Dr. Cosens, we decided to name the Type A mutant *transient receptor potential* (*trp*, [Minke et al., 1975]) and this particular allele, trp^{CM}.

The *trp* gene was cloned and molecularly characterized by Montell and Rubin (1989) and shortly afterward by Wong and colleagues (1989), for a detailed summary of *trp* cloning and sequencing (see Minke, 2010). This was an important achievement, as it allowed the cloning of *trp*-related genes, ultimately leading to the identification of a new superfamily of *trp* genes. However, cloning of this gene did not lead immediately to the recognition of its function as an ion channel, or to the enzymatic cascade leading to its activation. Sequencing revealed that the *trp* gene encodes a 1275 amino-acid membrane protein with no homologies to any known protein in the database. It had eight transmembrane domains (later revised to six transmembrane [TM] and a pore domain) and displayed many topological features reminiscent of receptor-transporter-channel proteins. However, the possibility that it might encode a light-activated channel was ruled out because light responses were present in mutants that showed no protein product in Western blot analyses (Montell & Rubin, 1989; Wong et al., 1989).

The nature of TRP function began to emerge in the following several years. Hochstrate showed that application of the non-specific Ca^{2+} channel blocker, lanthanum (La^{3+}), to the extracellular space of the blowfly *Calliphora* retina caused a dramatic decline in the receptor potential during light, making a wild-type receptor potential of *Calliphora* resemble that of a *trp* mutant (Hochstrate, 1989). These results suggested to Minke and colleagues a potential clue to TRP function (Suss Toby, Selinger, & Minke, 1991). They verified the preceding observations in three species of flies and proposed that the stated effect of La^{3+} might arise from its ability to block a Ca^{2+} transporter protein, which normally allows Ca^{2+} entry into photoreceptors to cause photoreceptor excitation and to replenish the internal Ca^{2+} pool (Suss Toby et al., 1991; Figure 17.1B).

In subsequent review papers, Minke and Selinger, (Minke & Selinger, 1991, 1992) further elaborated on their "conformational coupling model," which is based on that of Berridge (Berridge, 1995). In this model, the TRP protein was seen as a Ca^{2+} transporter/channel, which oscillates between conducting and non-conducting states through interactions with IP_3 receptor protein, which also acts as a Ca^{2+} sensor of the filling state of the IP_3-sensitive Ca^{2+} stores. By this time, it had already been established that phototransduction in *Drosophila* utilizes a phosphoinositide-mediated cascade (Bloomquist et al., 1988; Devary et al., 1987; Selinger & Minke, 1988), in which Gq-activated (Blumenfeld, Erusalimsky, Heichal, Selinger, & Minke, 1985; Devary et al., 1987) phospholipase C (PLC) plays a central role. This was established in large part through biochemical analyses of light-activated *Drosophila* and *Musca* signaling proteins and the mutants isolated in the screen mentioned previously. Therefore, it seemed reasonable to assume that both IP_3 and IP_3-receptor proteins could also be involved. Subsequently, it was shown that the IP_3 receptor has no role in phototransduction in *Drosophila* (Acharya, Jalink, Hardy, Hartenstein, & Zuker, 1997; Raghu, Colley, et al., 2000; but see Kohn et al., 2015). Nevertheless, the model correctly highlighted the major role of TRP channel in light-mediated Ca^{2+} entry.

Application of the patch clamp technique to *Drosophila* photoreceptors (Hardie, 1991; Ranganathan, Harris, Stevens, & Zuker, 1991; for the method, see Katz, Gutorov, Rhodes-Mordov, Hardie, & Minke, 2017) led to the evidence that the *trp* gene encodes the major component of the light-activated channels (Hardie & Minke, 1992). Whole-cell patch clamp recordings showed that the light-activated channels of *Drosophila* are Ca^{2+} permeable (Hardie, 1991), and this finding was later extended by using Ca^{2+} selective microelectrodes and microfluorimetry by Ca^{2+} indicator fluorescence in whole-cell recordings (Peretz, Sandler, Kirschfeld, Hardie, & Minke, 1994; Peretz, Suss-Toby, et al., 1994). Ion permeability measurements led to the finding that the primary defect in *trp* mutants or in La^{3+} treated wild-type photoreceptors is a drastic reduction in the Ca^{2+} permeability of the light-sensitive channels themselves (Hardie & Minke, 1992). These authors concluded that the light response of a wild-type photoreceptor consists of two distinct conductances: one is highly Ca^{2+} permeable and is encoded by the *trp* gene, and the other, encoded by another gene, is responsible for the residual current in *trp* mutants. The nature of the second conductance was clarified by the work of Kelly and colleagues (Phillips, Bull, & Kelly, 1992). While searching for calmodulin-binding *Drosophila* proteins, they discovered a membrane protein, which they called "TRP-like" (TRPL) with homology to TRP (~40% overall identity and 74% identity within the transmembrane segments). They showed that it is smaller than TRP (900 vs. 1275 amino acids), has two calmodulin-binding domains (vs. one in TRP) in the C-terminal region, and an ankyrin repeat domain in the N-terminal region. Their analysis further revealed a transmembrane topology reminiscent of voltage-gated channels: six transmembrane segments, S1–S6, and a putative pore region between S5 and S6, except that the charged residues in S4 of voltage-gated channels are replace by non-polar ones. Moreover, within S5 and S6, they found several short stretches of amino acid identical to the mammalian voltage-gated Ca^{2+} channels. They suggested that both *trp* and *trpl* genes encode light-activated channels (Phillips et al., 1992).

The results of Hardie and Minke and those of Kelly and colleagues dovetailed each other in showing that TRP and TRPL channels together contribute to the light-activated conductance of *Drosophila* photoreceptors. A subsequent important study by Zuker and colleagues (Niemeyer, Suzuki, Scott, Jalink, & Zuker, 1996) isolated a null *trpl* allele, $trpl^{302}$, and showed that the light response is largely abolished in the double mutant, $trpl^{302};trp^{P301}$. The residual response in the double mutant arose from trp^{P301}, which is not a complete null allele. It disappeared in the presence of low concentrations of La^{3+}, which would block TRP but not TRPL channels. Subsequently, Zuker and colleagues (Scott et al., 1997) identified a true null *trp* mutant, trp^{P343}, among the *trp* mutants Pak and colleagues had previously isolated. They showed that the light response is completely abolished in the double mutant, $trpl^{302};trp^{P343}$. This series of experiments convincingly demonstrated that light-activated conductance consists solely of TRP and TRPL channels, and no other channels appear to contribute.

Thus, *Drosophila* TRP became the founding member of the TRP ion channel superfamily. The first mammalian members of the family were cloned by homology to *Drosophila* TRP (Wes et al., 1995; Zhu, Chu, Peyton, & Birnbaumer, 1995). We now know that TRP channels are widely evolutionarily conserved from yeast to mammals, are found in almost all organisms and tissues. They were classified into seven subfamilies: TRPC (classical or canonical), TRPM (Melastatin), TRPV (Vanilloid), TRPA (Ankyrin), TRPN (NOMPC), TRPML (Mucolipin), and TRPP (Polycistin). The most closely related to *Drosophila* TRP and TRPL are members of the TRPC subfamily, typified by the ones cloned by Montell and colleagues (Wes et al., 1995) and by Birnbaumer and colleagues (Zhu et al., 1995) (the TRPC1).

In retrospect, it is now possible to evaluate the contribution of the original trp^{CM} allele to the TRP field. In spite of the difficulties to the TRP field stemming from using the trp^{CM} allele, it was a crucial tool for cloning and sequencing of the *trp* gene. The mapping of the *trp* locus to the edge of the third chromosome (Levy, Ganguly, Ganguly, & Manning, 1982; Wong, Hokanson, & Chang, 1985) was conducted on trp^{CM}. The subsequent rescue of the trp^{CM} phenotype (i.e., the transient receptor potential phenotype) by P-element-mediated germline transformation (Montell, Jones, Hafen, & Rubin, 1985) led to cloning and sequencing of the *Drosophila trp* gene (Montell & Rubin, 1989; Wong et al., 1989). In spite of this important contribution to the TRP field, several specific properties of the trp^{CM} severely hampered recognizing the TRP protein as a light-activated ion channel, which probably prevented a widespread interest in TRP at that time. Specifically, electrophysiological analysis of the trp^{CM} mutant revealed that its single-photon responses (quantum bumps) were normal (Minke et al., 1975). However, Western blot analysis suggested that the TRP protein is absent in trp^{CM} (Montell & Rubin, 1989).

The main reason for this confusion arose from the fact that *trp*^{CM} turned out to be a developmental temperature-sensitive allele (Minke, 1983) that expresses ~30% functional TRP at 19°C (Figure 17.2B) (Yoon et al., 2000), while at the restrictive temperature of 24°C, it was almost null. This property led to light-induced generation of normal quantum bumps when *trp*^{CM} was raised at 19°C, the temperature used for bump analyses, and to the generation of a pronounced light-induced current (LIC; Reuss, Mojet, Chyb, & Hardie, 1997) (Figure 17.2A, bottom) and yet the receptor potential declined towards baseline during intense prolonged light.

This review does not intend to give a comprehensive outline of functional properties of TRPC channels. Recommended recent excellent reviews on TRPC channels are the reviews of Svobodova and Groschner (2016); Voolstra and Huber (2014); Dietrich, Kalwa, and Gudermann (2010); Takahashi, Kozai, and Mori (2012); and Birnbaumer (2009). We will concentrate on some features of the TRPC channel subfamily, which have been investigated in detail in *Drosophila* photoreceptors. The advantage of *Drosophila* photoreceptors is that the physiological role of TRP/TRPL channels as the light-activated channels has been unequivocally established. Therefore, the effects of deleterious mutations or post-translational modifications on these channels can guide and provide insight into similar studies on mammalian TRPC channels.

In Vivo Mutation Analyses Provide Insights into Properties of TRPC Channels

Pak and colleagues had carried out a mutagenic screen, using the alkylating agent ethyl methane sulfonate (EMS) as mutagen, for visually defective mutants by searching for defects in phototaxis or ERG response (Pak et al., 1969; Pak, 2010). The ERG screening included the use of a specific light stimulating protocol that induced the Prolonged Depolarizing Afterpotential (PDA) (Pak, Shino, & Leung, 2012; Minke, 2012). For this reason all flies had their screening pigments removed genetically to allow the development of PDA and avoid any potential complications associated with the pigments. In the PDA protocol, the flies were illuminated with intense blue light to induce large photopigment conversion (Minke, 2012; Pak, Shino, & Leung, 2012). Such stimulation protocol brings the phototransduction cascade to its upper limit of activation for an extended time, during which millions of molecules of signaling proteins are activated (Minke, 2012). Therefore, any reduction in a molecularly-dependent functional process required for the generation of maximal PDA, such as biogenesis of rhodopsin or biogenesis and proper insertion of the TRP channel into the membrane of the signaling compartment (the rhabdomere) would lead to an abnormal PDA. Hence, the PDA screen has a great power for isolating visual mutants with a clear phenotype.

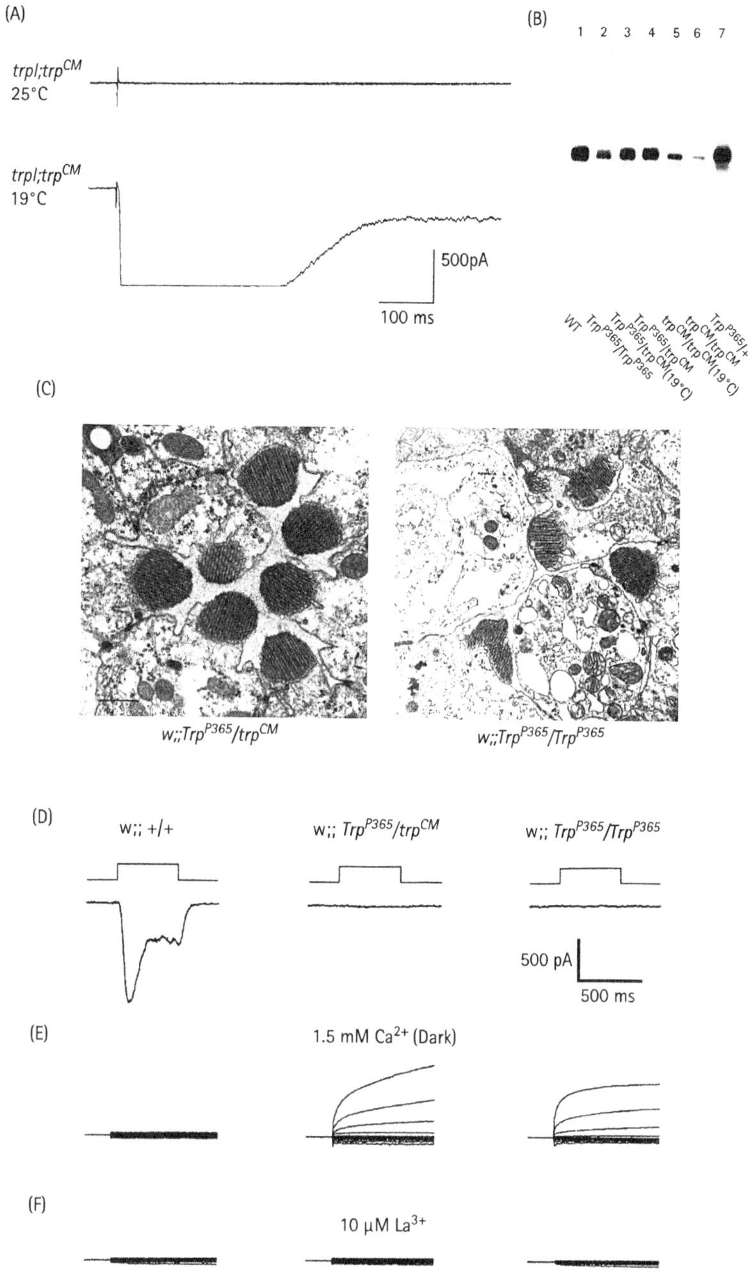

FIGURE 17.2 *Functional analysis of the developmental temperature sensitive trp*^{CM} *mutant by whole-cell recordings and Western blot analysis.*

A. Intense light stimulation elicited no responses in the *trpl*^{302};*trp*^{CM} double null mutant, when raised at 25°C *(upper trace),* while a large LIC was elicited from the same double mutant when raised at 19°C (*bottom trace*), indicating that *trp*^{CM} is a *developmental temperature sensitive trp* mutant, showing a null phenotype when raised at 25°C, but not when raised at 19°C. Note that the light induced current (*bottom trace*) is truncated.

From Reuss et al., 1997.

B. Western blot analyses showing the expression level of the TRP channel in wild-type (*lane 1*) and *trp*^{CM}/*trp*^{CM} homozygot raised at 19°C (*lanes 5)* or 25°C (*lane 6*) flies. Note, the highly reduced expression of *trp*^{CM}/*trp*^{CM} in both conditions compared to WT flies and the reduction in expression level between 19°C and 25°C. The other lanes appear again in Figure 17.4B

The Discovery of Multi-Molecular Signaling Complexes

The genetic visual screen and particularly the PDA screen, have produced a large number of visual mutants, which helped researchers discover key proteins that participate in invertebrate phototransduction with wide implications for general signaling mechanisms. One such discovery was the isolation of the *inaD* mutant, *InaD^{P215}*, which was isolated using the PDA screen (Pak, 1995) and was subsequently cloned and sequenced by Shieh and Niemeyer (Shieh & Niemeyer, 1995).

The Drosophila PDZ Scaffold Protein, Inactivation but no Afterpotential D (INAD)

The *inaD* gene encodes a 674 amino acid protein that was found to be highly enriched in the photoreceptor cells. The original sequence analysis revealed two protein interacting motifs called "PDZ (PSD95, DLG, ZO1) domains" (Shieh & Niemeyer, 1995). These domains are recognized as protein modules that bind to a variety of signaling, cell adhesion, and cytoskeletal proteins by specific binding to target sequences, typically, though not always, in the final three residues of the C'-terminus (Shieh, Zhu, Lee, Kelly, & Bahiraei, 1997). Immunoprecipitation technique showed that INAD binds two proteins, one of which was the TRP channel. *InaD^{P215}* harboring a single missense mutation, M422K, which disrupt the TRP–INAD interaction (Shieh & Zhu, 1996). The INAD-interacting domain on the TRP channel was localized at first to the last 19 residues of its C'-terminus but later was suggested to compose only the last three residues (Chevesich, Kreuz, & Montell, 1997; Shieh et al., 1997). Studies on *Calliphora* have extended these findings, showing that INAD binds not only TRP but also the No Receptor Potential A (NORPA), which encodes Phospholipase Cβ (PLCβ) and the Inactivation but no Afterpotential C (INAC), which encodes an eye specific Protein Kinase C (eyePKC) (Huber, Sander, & Paulsen, 1996). The interaction of INAD with TRP, PLCβ, and eyePKC was later confirmed in *Drosophila,* and a thorough analysis of the protein sequence revealed that it contained five PDZ domains instead of two (Tsunoda et al., 1997; Figure 17.3). Subsequent studies suggested that, in addition to PLCβ, eyePKC, and TRP, other signaling molecules such as calmodulin (CaM), the major rhodopsin (RH1), TRPL, and Neither Inactivation Nor Afterpotential C (NINAC), which encodes for a myosin III, also bind to the INAD signaling complex (Chevesich et al., 1997; Xu, Choudhury, Li, & Montell, 1998). However, such diverse binding partners without physiologically demonstrated functions must be dynamic.

The identification of INAD as a scaffold protein provides a mechanism for the co-localization of major phototransduction components in spatial proximity. However, the functional role of the INAD protein was unknown at that time. Using immunofluorescent staining, the function of INAD (Tsunoda, Sun, Suzuki, & Zuker, 2001) and TRP (Chevesich et al., 1997; Xu et al., 1998) in the localization of the major phototransduction proteins was examined. Accordingly, the localization of the signaling proteins INAD, TRP, NORPA, INAC, NINAC, and RH1 was studied using the *InaD* or *trp* null mutant

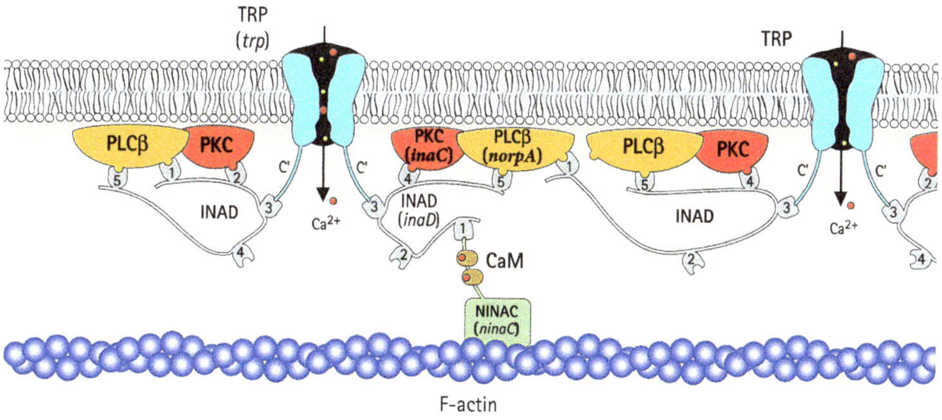

FIGURE 17.3 *Schematic representation of the INAD protein complex.*
The INAD sequence contained five consensus PDZ domains (indicated by numbers 1–5) and identified specific interactions between PKC (INAC, encoded by the *inaC* gene) and PDZ2 (or PDZ4), TRP and PDZ3, and PLC (NORPA, encoded by the *norpA* gene) with PDZ1 and PDZ5. This binding pattern is still in debate due to several contradictory reports. It was also reported that the INAD contains a Ca^{2+}-calmodulin binding site, which may be involve in its regulation. It also binds the actin cytoskeleton via myosin III (NINAC, encoded by the *ninaC* gene).

From Katz & Minke, 2009.

flies. The results showed that the INAD is correctly localized to the rhabdomeres in the *inaC* null mutants (where eyePKC is missing) and in *norpA* virtually null mutant (where PLCβ is virtually missing), but severely mislocalized in the null *trp* mutant, thus indicating that TRP, but not NORPA or INAC, is essential for localization of the signaling complex to the rhabdomere. Furthermore, in the absence of INAD, the TRP, NORPA, INAC, but not the NINAC or RH1, were mislocalized to the cell body, showing that the INAD protein is essential for the retention of the bound PLCβ and INAC signaling proteins to the rhabdomere (Tsunoda et al., 2001; Xu et al., 1998). The study of these mutants was also used to show that TRP and INAD do not depend on each other to be targeted to the rhabdomeres. Thus, INAD–TRP interaction is not required for targeting but for anchoring and retention of the signaling complex (Tsunoda et al., 2001). Additional experiments on TRP, NORPA, and INAD further showed that INAD has an important function in preventing NORPA degradation (Xu et al., 1998), which is especially important for response termination due to the GTPase activating protein (GAP) activity of NORPA (Cook et al., 2000).

A structural study of INAD has suggested that the binding of signaling proteins to INAD may be a dynamic process that constitutes an additional level of phototransduction regulation (Mishra et al., 2007). This study showed two crystal structural states of isolated INAD PDZ5 domain, differing mainly by the presence of a disulfide bond. This conformational change has light-dependent dynamics that were demonstrated by the use of transgenic *Drosophila* flies expressing an INAD having a point mutation that disrupts the formation of the disulfide bond. In this study, a model was proposed in which eyePKC phosphorylation at a still-unknown site promotes the light-dependent conformational change of PDZ5, distorting its ligand-binding groove to PLCβ and thus regulating phototransduction. Further studies showed that the redox potential of PDZ5 is allosterically regulated by its interaction with PDZ4 (Liu et al., 2011). Whereas isolated PDZ5 is stable in the oxidized state, formation of a PDZ4-5 "supramodule" locks PDZ5 in the reduced state by raising the redox potential of a disulfide bond. Acidification, potentially mediated via light-dependent PLCβ hydrolysis of Phosphatidylinositol 4,5 bis phosphate (PIP_2), disrupts the interaction between PDZ4 and PDZ5, leading to PDZ5 oxidation and dissociation from the TRP channel (Liu et al., 2011). However, demonstration of the physiological significance of these light-dependent changes in INAD is still lacking.

The Mammalian PDZ Scaffold Protein NHERF Interacts with TRPC4 and TRPC5 Channel Proteins

Studies have shown that some TRPC channels are organized in supra-molecular complexes similar to the INAD signaling complex. The scaffolding protein Na^+/H^+ -exchanger regulatory factor 1 (NHERF) was isolated as a co-factor required for inhibition of type 3 Na1/H1 exchanger by protein kinase A, and was localized to the renal brush-border (Yun et al., 1997). Later the NHERF protein was found to interact with TRPC4/5. The NHERF protein family is composed of NHERF1 (also known as ERM-Binding Protein 50 [EBP50]) and NHERF2, which shares 44% sequence homology. These proteins contain two PDZ domains and a sequence at the C'-terminus that binds several members of the ERM (ezrin-radixin-moesin) family of membrane-cytoskeletal adapters (Ardura & Friedman, 2011; Dunn & Ferguson, 2015; Murthy et al., 1998; Terawaki, Maesaki, & Hakoshima, 2006). In a biochemical study, it was shown that the first PDZ domain of NHERF binds murine TRPC4 or TRPC5 as well as PLC-β1 and PLC-β2 (Tang et al., 2000). The interaction of PLC-β1, TRPC4, and NHERF was demonstrated in the HEK293 cell line stably expressing TRPC4, and in adult mouse brain by co-immunoprecipitation experiments. Since NHERF binds also to the cytoskeleton via ERM proteins, the cytoskeleton seems to be part of this supramolecular organization (Tang et al., 2000). The binding of two partners to the same PDZ domain suggests that NHERF can form a homodimer via PDZ2 and the PDZ1 domains, bringing TRPC4 or TRPC5 in vicinity of the PLCβ1 and PLCβ2 (Suh, Hwang, Ryu, Donowitz, & Kim, 2001; Tang et al., 2000; for a review, see Constantin, 2016).

The last three C-terminal amino acids (TRL) of TRPC4 compose a PDZ-interacting domain that binds to the scaffold protein EBP50. In order to explore the role of TRPC4–EBP50 interaction on the subcellular localization of TRPC4, a truncated TRPC4 lacking the last three amino acids was examined. Accordingly, immunofluorescence microscopy analysis showed that wild-type TRPC4 channels were distributed evenly on the cell surface, while the TRPC4 mutant lacking the PDZ motif accumulated into cell outgrowths with a punctate distribution pattern. Cell surface biotinylation revealed a 2.4-fold reduction in plasma membrane expression of the truncated TRPC4 mutant compared to wild-type TRPC4. Furthermore, the consequences of the interaction between NHERF and the membrane-cytoskeletal adaptors of the ERM family on cell surface expression of TRPC4 was examined. In cells co-expressing TRPC4 and a NHERF mutant lacking the ERM-binding site, TRPC4 was not present in the plasma membrane but co-localized with the truncated NHERF in a perinuclear compartment and in vesicles associated with actin filaments. Hence, the TRPC4-NHERF-ERM complex regulates TRPC4 localization and surface expression in transfected HEK293 cells (Mery, Strauss, Dufour, Krause, & Hoth, 2002). The effects of EBP50–TRPC5 interaction on the activity and cellular distribution of TRPC5 were also investigated. In this study, rat TRPC5 (rTRPC5) and a mutant TRPC5 with a deletion of the PDZ-binding domain Val, Thr, Thr, Arg, Leu (VTTRL) were examined in the HEK293 cell expression system (Obukhov & Nowycky, 2004). Accordingly, both wild-type and mutant TRPC5 were localized to the plasma membrane, and deletion of the VTTRL motif had no detectable effect on the biophysical properties of the channel, when studied with patch-clamp technique. Co-expression of EBP50 with rTRPC5 led to a significant delay in the time-to-peak of the histamine-evoked, transient, large inward current. However, EBP50 did not modify the activation kinetics of the VTTRL-deletion mutant.

Immunohistochemical studies demonstrated expression of TRPC4 and NHERF-2 proteins in both the endothelial cells and pericytes and co-localized in some cells of the renal medullary descending *vasa recta* (DVR). Co-immunoprecipitation experiments from renal medullary lysates indicated physical interaction of TRPC4 and NHERF-2 proteins (Lee-Kwon, Wade, Zhang, Pallone, & Weinman, 2005). These results suggest that the scaffold protein NHERF-2 assembles with TRPC4 in renal medullary DVR.

Together, the interaction of EBP50 (NHERF) with TRPC4 or TRPC5 channels has different effects on the cellular distribution and activation of the channels in different tissues.

Interaction of the Mammalian PDZ-Scaffold Protein NHERF, with TRPC4 and TRPC5 Is Required for DAG Activation

In a recent study, it was demonstrated that TRPC4 and TRPC5 sensitivity to diacylglycerol (DAG) is dependent on the association of NHERF to the TRPC at the C'-terminus. Accordingly, TRPC4/5 becomes DAG sensitive when NHERF is dissociated from these TRPC channels. The interaction of NHERF and TRPC4/5 is dependent on PKC phosphorylation at the C-*terminal*. Strikingly, DAG sensitivity was achieved under several experimental paradigms, such as: (i) PKC inhibition, (ii) removal of a C-terminal PKC phosphorylation site in the PDZ-binding motif, (iii) NHERF1 and NHERF 2 down-regulation, (iv) co-expression of a NHERF1 mutant (NHERF1–E68A) incapable of interacting with the C-terminal of TRPC5, and (v) co-expression of Gq/11-coupled receptors. Importantly, C-terminal conformational rearrangements engendered by PIP_2 depletion were also required. The experiments of this study based on electrophysiology, co-immunoprecipitations, and intermolecular Fluorescence Resonance Energy Transfer (FRET) have thus suggested that a crucial step in TRPC5 activation is the dissociation of NHERF proteins from the channel C'-terminus, conferring DAG sensitivity (Storch et al., 2017). Collectively, C-terminal NHERF and PIP_2 interaction stabilize a DAG-insensitive channel conformation. During receptor activation, PIP_2 level at the plasma membrane is reduced by PLC, resulting in an active TRPC5 conformation, characterized by C-terminal rearrangements and the ensuing dissociation of NHERF1 and NHERF2, thereby conferring DAG sensitivity on TRPC4 and TRPC5 channels. The possibility that a similar mechanism may lead to DAG sensitivity of *Drosophila* TRP has not been investigated (see the section "Anoxia Activation of *Drosophila* TRP/TRPL Channels in the Dark" in this chapter).

The Mammalian Caveolin-1 Scaffold Protein Interacts with TRPC1, TRPC4, and TRPC3

Caveolae are glycosphingolipid- and cholesterol-enriched membrane microdomains found in many vertebrate cells, which are enriched with Caveolin, a transmembrane scaffolding protein. Caveolin-1 interacts with TRPC1, TRPC4, and TRPC3 channels via binding to both the N' and C' termini. A caveolin-1 conserved binding motif was identified in all TRPC members at the N-terminal part close to the first transmembrane domain TM1 (amino acids 271–349 or 322–349). Deletion of this region prevented the targeting of TRPC1 to the plasma membrane and exerted a dominant negative effect on endogenous inward current induced by intracellular Ca^{2+} store depletion, designated Store Operated Calcium Entry (SOCE). The expression of truncated caveolin-1 (Cav1Δ51-169), lacking its protein scaffolding and membrane-anchoring domains, disrupted plasma membrane targeting of TRPC1 and suppressed thapsigargin- and carbachol-stimulated Ca^{2+} entry (Brazer, Singh, Liu, Swaim, & Ambudkar, 2003).

Using the caveolin-1 (Cav1)-deficient mice (Razani & Lisanti, 2001), it was shown that in endothelial cells that the scaffolding protein governs the localization and interactions of TRPC1 and TRPC4. Furthermore, Cav-1 is associated with a dynamic protein complex consisting of TRPC4, TRPC1, and IP_3 receptors (IP_3Rs), while the loss of Cav-1 impairs the localization of TRPC4 and ACh-mediated calcium entry (Murata et al., 2007). In general, caveolae are thought to organize a multiprotein calcium signaling complex containing TRPC1, which is anchored to caveolin-1 and is associated with signaling proteins such as the IP_3R, calmodulin (CaM), plasma membrane calcium pump (PMCA), and Gαq/11 (Ambudkar & Ong, 2007; Lockwich et al., 2000).

Although Caveolin-1 appears to play a role in the interaction of mammalian TRPC with a variety of signaling proteins, and to underlie their retention in the plasma membrane regions, similar functions of Caveolin have not been investigated in *Drosophila*, and therefore are not elaborated on in this review.

Mutations Causing Constitutive Activity of TRPC Channels

Many TRP channels exhibit constitutive activity, which is mostly observed in cell-based expression systems. This constitutive activity can lead, in many cases, to cellular degeneration, which can be readily observed morphologically and by biochemical assays. In cell-based expression system, it is difficult to know if the constitutive activity is physiologically relevant. In the *Drosophila* photoreceptor cells, the TRP channels are closed in the dark and open upon illumination. Therefore, the isolation of the *Drosophila* trp^{P365} mutant showing constitutive activity of the channel in the dark has raised a widespread interest.

The Drosophila trp^{P365} *Mutant Fly Shows Constitutive Activity of the TRP Channel*

The *P365* mutant isolated in the EMS screen of Pak and colleagues was highly unusual because it showed an extremely fast retinal degeneration phenotype even at the pupa stage, and unreliability of the complementation test due to its semi-dominant nature. Later, the P365 mutation was mapped to the edge of chromosome 3R, an area that also harbors the *trp* locus. Electron micrograph (EM) studies of *P365* retinae showed light-independent retinal degeneration, while heterozygote flies showed normal morphology (Figure 17.4A), but illumination induced retinal degeneration. Initially, the *P365* mutant was not identified as a *trp* allele because *trp* alleles did not show fast retinal degeneration. Also, most *trp* mutants did not show significant TRP channel protein expression in Western blot analysis, unlike the *P365* mutant at both homozygote and heteroallelic combination (Figure 17.4B). In addition, ERG measurements showed that, although the sensitivity to light of the *P365* mutant was highly reduced in both homozygote and heterozygote mutant flies, still it did not show the typical transient receptor potential (*trp*) phenotype (Yoon et al., 2000). In order to show that the P365 mutant phenotype is caused by mutations in the TRP channel, a transgenic fly carrying the mutation of *P365* only in the *trp* locus (P[Trp^{P365}]) was constructed. The P[Trp^{P365}] on Wild Type (WT) background showed reduced light sensitivity, similar to that found in the *P365* mutant fly, supporting the notion that mutations in the TRP channel caused the observed phenotype. Whole-cell current measurements during voltage steps from photoreceptor cells revealed similarity between currents of *P365* mutant measured in darkness (Figure 17.4D) and light-induced wild-type TRP currents. Since 10 μM La^{3+} blocked these currents (Figure 17.4E), it further supported the notion that the TRP channels of the *P365* mutant are constitutively opened in the dark (Figure 17.4C–D). Sequencing the Trp^{P365} gene revealed four missense mutations: P500T, H531N, F550I, and S867F (Yoon et al., 2000). It was therefore important to determine which of the mutations cause the P365 phenotype. Using transgenic flies harboring different combinations of these mutations, it was possible to determine the F550I mutation as causing the P365

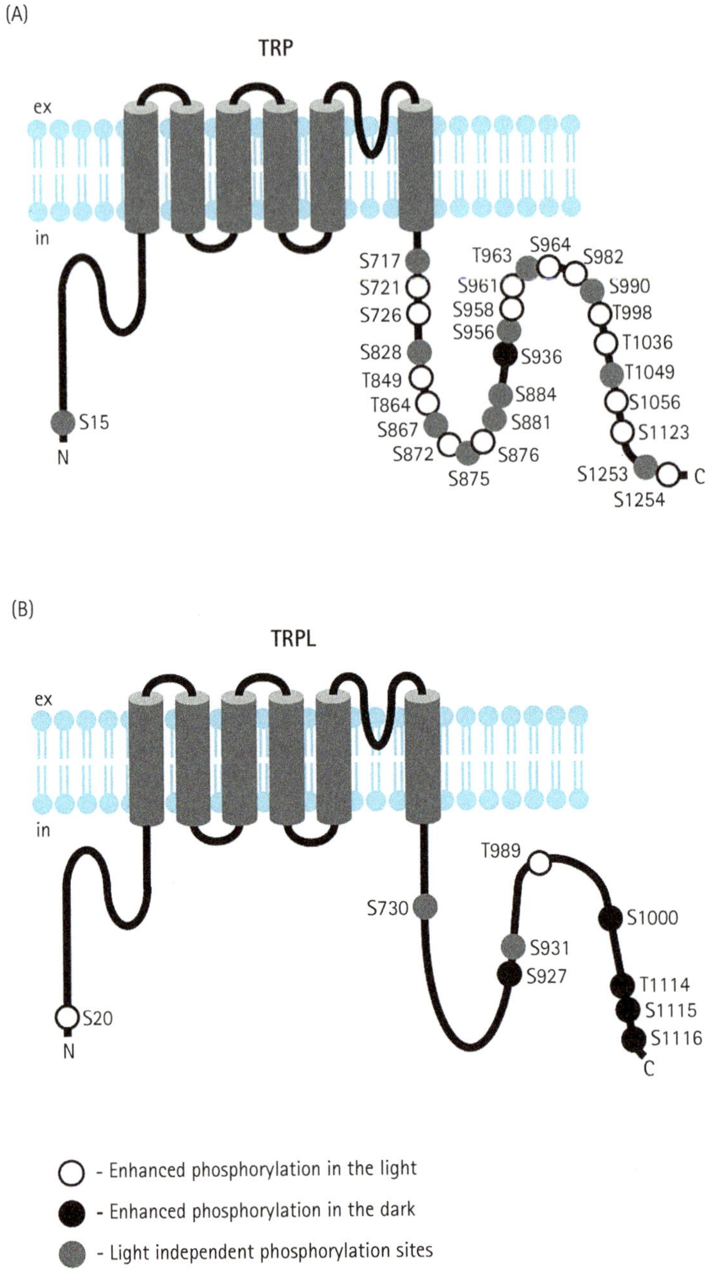

FIGURE 17.4 *Functional analysis of the* Trp^{P365} *mutant by whole-cell recordings electron microscopy and Western blot analysis.*

A. Electron micrographs (EM) of transverse sections through the ommatidial layer (at the level of R7 photoreceptor nuclei) of Trp^{P365}/trp^{CM} (*left*) and Trp^{P365}/Trp^{P365} (*right*), both raised at 19°C. All samples were obtained from newly enclosed adult flies, and all flies were on a *w (white-eyed)* background. The Trp^{P365}/Trp^{P365} mutant retina appears highly degenerated, and the degeneration was delayed in the heteroallele Trp^{P365}/trp^{CM}. Scale bar, 1 μm.

phenotypes (Hong et al., 2002). In conclusion, the *Trp*^{P365} mutant fly harbors a missense F550I mutation (Figure 17.4F) in the TRP channel, which causes constitutive activity, retinal degeneration and highly reduced sensitivity to light. It was unclear, however, whether the retinal degeneration is a consequence of massive Ca^{2+} influx through the constitutively active TRP channels, or if the mutation causes retinal degeneration, which promotes the constitutive activity of the channels. Using over-expression of the sodium–calcium exchanger, CalX, it was shown that the retinal degeneration of *Trp*^{P365}/+ could be partially rescued (Liu et al., 2007). Hence, elevated extrusion of Ca^{2+} rescues the degeneration, supporting the notion that retinal degeneration is a consequence of massive Ca^{2+} influx through the constitutively active TRP channels.

The TRP F550I mutation is located at S5 (transmembrane helix number 5) adjacent to the S4–S5 loop, and it is conserved in TRPL (F557, Figure 17.4F). This amino acid position shows only hydrophobic conservation among mammalian TRPC channels, while the two flanking Phe residues are more conserved (see sequence alignment, Figure 17.4F). Interestingly, two independent studies using a random chemically induced mutation in *Drosophila* TRP (Hong et al., 2002; Yoon et al., 2000) and a high-throughput mutagenic screen of TRPV1 (Myers, Bohlen, & Julius, 2008) both found mutant channels where this position was mutated and gave rise to constitutively active channels. In both cases, the mutation induced cell death, and in the case of *Drosophila*, the mutation resulted in retinal degeneration (Figure 17.4A). The exact mechanism of

---

B *The P365 mutant shows large channel protein expression in both homozygote and heteroallelic combination.* Western blot analyses showing the expression level of the TRP channel in heteroallelic mutant *Trp*^{P365}/*trp*^{CM} raised at 19°C (*lanes 3*) or 25°C (*lane 4*) and controls: wild-type (*lane 1*), *Trp*^{P365}/*Trp*^{P365} homozygotes (*lane 2*), *Trp*^{P365}/+ heterozygotes (*lane 7*). All raised at 25°C
C. A typical light-induced current (LIC) of a wild-type cell (*left trace*) in response to an orange stimulus, and the absence of any responses in *Trp*^{P365}/*trp*^{CM} and *Trp*^{P365}/*Trp*^{P365} (*middle and right traces, respectively*). The duration of the orange light stimulus is indicated above each trace.

D. Families of current traces elicited in the dark from photoreceptor cells by a series of voltage steps of 10 mV from -100 mV to +80 of wild-type (*left column*), the light-insensitive cells *Trp*^{P365}/*trp*^{CM} (*middle column*), and *Trp*^{P365}/*Trp*^{P365} homozygotes (*right column*). Membrane currents were recorded 30 s after establishing the whole-cell configuration with physiological concentrations (1.5 mM) of Ca^{2+} in the bath. The initial holding potential was -20 mV. Note the outward current in the dark in the *Trp*^{P365}/*trp*^{CM} and *Trp*^{P365}/*Trp*^{P365} mutants (but not in wild-type fly) due to the constitutive activity of the mutant channel.

E. Application of 10 μM La^{3+} to the bath suppressed the dark voltage elicited membrane currents.

Traces B-F are from Yoon et al., 2000.

F. Multiple sequence alignment of TRP (NP_476768.1), TRPL (NP_476895.1), TRPC1 (NP_001238774.1), TRPC3 (NP_001124170.1), TRPC4 (NP_003297.1), TRPC5 (NP_036603.1), TRPC6 (NP_004612.2), and TRPC7 (NP_065122.1) as measured by Geneious™ alignment using the Blosum62 cost matrix. Shown: the amino acid multiple sequence alignment at the S4–S5 loop and part of S5 G540, D545, K548—yellow, and F550—cyan, of TRP and the corresponding amino acids in TRPL and in their human counterparts are highlighted. The numbering represents the position of the last amino acid.

how those mutations cause constitutive channel activity is still unknown. Nevertheless, the *Trp*^{P365} mutation turned out to be extremely useful for introducing Ca^{2+} into *Drosophila* photoreceptors in the dark in a variety of *in vivo* studies. In these studies, elevated cellular Ca^{2+} triggers important cellular processes such as light- (and Ca^{2+})-dependent TRP channel dephosphorylation (Voolstra et al., 2017), and light- (and Ca^{2+})-dependent TRPL translocation (Meyer, Joel-Almagor, Frechter, Minke, & Huber, 2006; and see further in this chapter).

N-Linked Glycosylation of TRPC3 and TRPC5 Causing Constitutive Activity

N-linked glycosylation, the enzymatic reaction in which proteins are converted into glycoproteins, occurs at Asn residues within a specific sequence context (Asn-X-Ser/Thr, where X represents any amino acid except proline [Pless & Lennarz, 1977]). The effect of different N-glycosylation patterns on the function of TRPC channels was compared between TRPC3 and TRPC6. These channels, together with TRPC7, constitute a TRPC subfamily, whose members are activated by DAG in a membrane-delimited fashion (Hofmann et al., 1999). TRPC6 reveals a very low constitutive basal activity. In contrast, TRPC3 reveals pronounced constitutive basal activity in the absence of a receptor agonist. TRPC6 reveals glycosylation sites in both heterologous expression systems and in pulmonary vascular smooth muscle cells (PASMC), where TRPC6 is involved in PASMC proliferation (Weissmann et al., 2006). Two NX(S/T) motifs in TRPC6 were mutated (Asn to Gln), deleting one or both extracellular N-linked glycosylation sites. Immunoblotting analysis of wild-type and mutant TRPC6 channels expressed in HEK293 cell revealed that TRPC6 is dually glycosylated within the first (designated e1) and second (designated e2) extracellular loops as opposed to the monoglycosylated TRPC3 channel (Vannier, Zhu, Brown, & Birnbaumer, 1998). Elimination of the e2 glycosylation site, missing in the monoglycosylated TRPC3, was sufficient to convert the tightly receptor-regulated TRPC6 into a constitutively active channel, displaying functional characteristics similar to that of TRPC3. Reciprocally, construction of an additional second glycosylated site in TRPC3, to mimic the glycosylation pattern of TRPC6, markedly reduced TRPC3 basal activity. Usually N-glycosylation affects protein folding, intracellular trafficking, or membrane targeting. For the case of TRPC3 and TRPC6 channels, the glycosylation mutant channels were inserted into the plasma membrane, indicating that the effect is on channel activity. Thus, the glycosylation pattern plays a pivotal role for the tight activation of TRPC6 through phospholipase C-activating receptors (Dietrich et al., 2010).

Functional Roles of TRPC Phosphorylation

The activity of many proteins is regulated by phosphorylation and dephosphorylation reactions. Protein kinases and phosphatases that are activated during neuronal activity orchestrate cellular events that ultimately reshape the neuronal events via phosphorylation and dephosphorylation of various ion channels, including many members of the TRP channel superfamily (Por, Gomez, Akopian, & Jeske, 2013; Voolstra, Bartels, Oberegelsbacher, Pfannstiel, & Huber, 2013; Voolstra, Beck, Oberegelsbacher, Pfannstiel, & Huber, 2010) (for a comprehensive review see (Voolstra & Huber, 2014). However, the physiological roles of phosphorylation and dephosphorylation in controlling TRP channel activity are largely unclear (Cao, Cordero-Morales, Liu, Qin, & Julius, 2013).

The Roles of *Drosophila* TRP and TRPL Phosphorylation

The *Drosophila* TRP channel revealed a considerable number of phosphorylation sites. Some of them directly affect the functional properties of the channel. The eye-specific protein kinase C (eyePKC, INAC), which is part of the INAD complex (see preceding discussion) was shown to phosphorylate the *Drosophila* TRP *in vitro* (Huber et al., 1996; M. Liu, Parker, Wadzinski, & Shieh, 2000). In contrast, the TRPL channel revealed lower number of phosphorylation sites, which to date were not found to affect directly the functional properties of the channel.

TRP Phosphorylation

Using quantitative mass spectrometry, 28 TRP differential phosphorylation sites from light- and dark-adapted flies were identified by Huber, Voolstra, and colleagues (Voolstra et al., 2013; Voolstra et al., 2010).

Twenty-seven phosphorylation sites resided in the C-terminus, while a single site resided at the N-terminus. Fifteen of the C-terminal phosphorylation sites exhibited enhanced phosphorylation in the light, whereas a single site, Ser936, exhibited enhanced phosphorylation in the dark (Voolstra et al., 2013; Figure 17.5A). To investigate light-dependent TRP phosphorylation, phospho-specific antibodies were generated to specifically detect TRP phosphorylation at Thr849; Thr864, which become phosphorylated in the light; and at Ser936, which becomes dephosphorylated in the light (see following discussion). To identify the stage of the phototransduction cascade that is necessary to trigger dephosphorylation of Ser936 or phosphorylation of Thr849 and Thr864, phototransduction-defective *Drosophila* mutants and the phospho-specific antibodies were used. Strong phosphorylation of Ser936 in dark-adapted WT flies was observed, and weak phosphorylation in light-adapted WT flies was detected. Conversely, weak phosphorylation of Thr849 and Thr864 was observed in dark-adapted WT flies, and strong phosphorylation was observed in light-adapted wild-type flies. Additionally, in phototransduction-defective mutants, with highly reduced light response, strong phosphorylation of Ser936 and weak phosphorylation of Thr849 and Thr864 were observed, regardless of the light conditions. Conversely, a mutant expressing a constitutively active TRP channel (*trp*^{P365}, see preceding; Hong et al., 2002) exhibited weak phosphorylation of Ser936 and strong phosphorylation of Thr849 and Thr864, regardless of illumination.

```
             650        660
             012345678901234
Consensus    LGRMVKDIFKFXFIX
     dTRP    LGRMIIDIIKFFFIY
    dTRPL    LGRMVIDIVKFFFIY
   hTRPC1    MGQMLQDFGKFLGMF
   hTRPC3    LGRTVKDIFKFMVLF
   hTRPC4    LGRMLLDILKFLFIY
   hTRPC5    LGRMLLDILKFLFIY
   hTRPC6    LGRTVKDIFKFMVIF
   hTRPC7    LGRTVKDIFKFMVIF
```

FIGURE 17.5 *Light-dependent phosphorylation of Drosophila TRP and TRPL channels.*

(A) A model of *Drosophila* TRP and its phosphorylation sites. Amino acid residues that undergo phosphorylation are shown as circles. Sites that undergo enhanced phosphorylation in the light are shown as white circles; sites that undergo enhanced phosphorylation in the dark are shown as black circles; sites that revealed no significant difference in phosphorylation between light- or dark-adapted flies or that could not be assessed quantitatively are shown as gray circles (ex = extracellular; in = intracellular).

(B) A model of *Drosophila* TRPL and its phosphorylation sites.

From Voolstra & Huber, 2014.

These data indicate that, *in vivo*, TRP dephosphorylation at Ser936 and phosphorylation at Thr849 and Thr864 depend on the phototransduction cascade, but activation of the TRP channel and most likely Ca^{2+} elevation are sufficient to trigger this process (Voolstra et al., 2013).

To identify kinases and phosphatases of Thr849, Thr864, and Ser936, a candidate screen using available mutants of kinases and phosphatases that are expressed in *Drosophila* eye was applied. It was found that Thr849 phosphorylation was compromised in light-adapted *inaC* null mutants. Diminished phosphorylation in light-adapted *PKC*^{53e} mutants was also found; suggesting that these two PKCs synergistically phosphorylate TRP at Thr849. Using a similar method, the S936 site was found to undergo light-dependent dephosphorylation by the rhodopsin phosphatase Retinal Degeneration C (RDGC). Accordingly, light-adapted *rdgC* mutant flies showed relatively high S936-TRP phosphorylation levels but maintained a light–dark phosphorylation dynamic. These findings suggest that RDGC is one, but not the only, phosphatase involved in S936-TRP dephosphorylation (Voolstra et al., 2017).

Electroretinogram (ERG) measurements of the frequency response to oscillating lights *in vivo* was performed (Voolstra et al., 2017). Dark-reared flies expressing wild-type TRP (*trp*^{WT}) exhibited a detection limit of oscillating light at relatively low frequencies, which was shifted to higher frequencies upon light adaptation (Figure 17.6A-B). It was further found that preventing phosphorylation of the S936-TRP site by Ala substitution in transgenic *Drosophila* (*trp*^{S936A}) abolished the difference in frequency response between dark- and light-adapted flies, resulting in high-frequency response

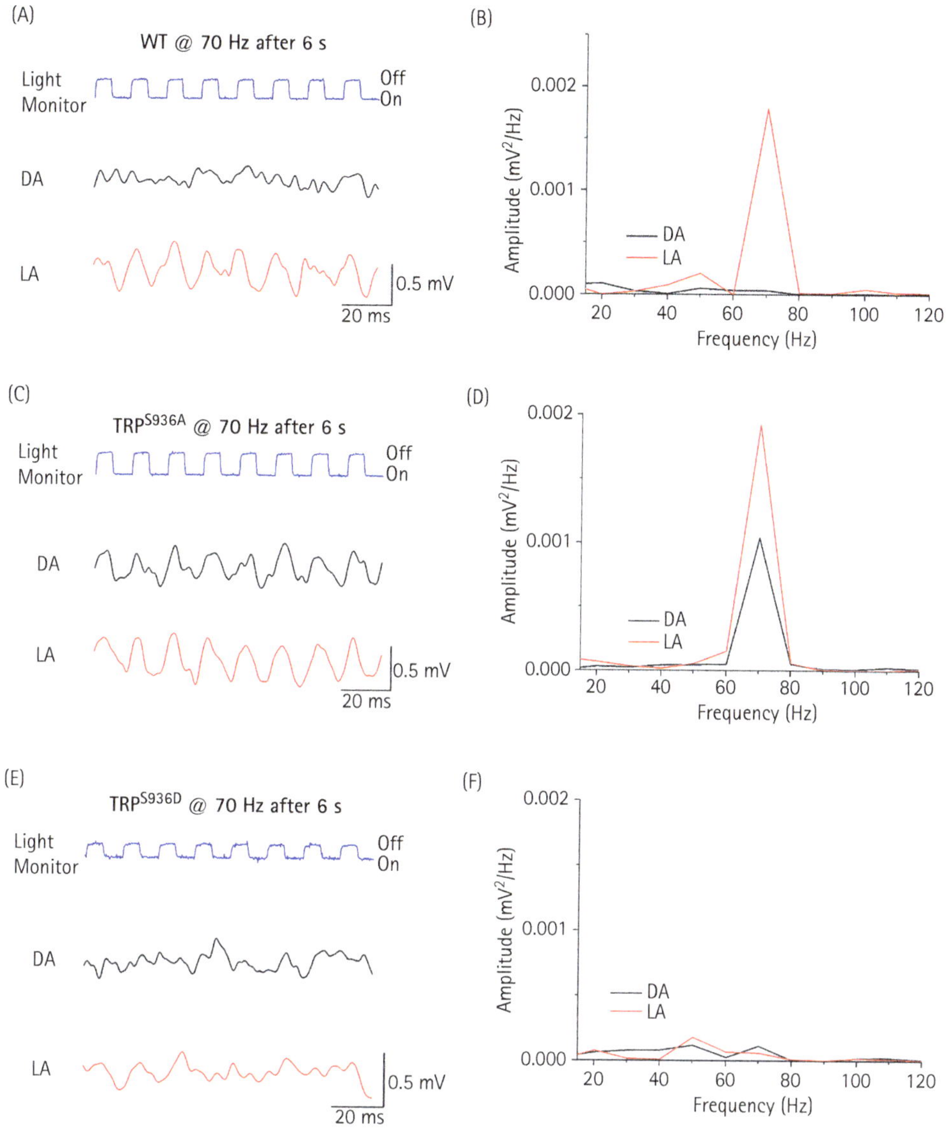

FIGURE 17.6 *Frequency response amplitude to oscillating light in dark- and light-adapted wild-type and transgenic flies in which TRP phosphorylation is prevented at specific sites.* Representative short segments of electroretinogram (ERG) responses to intense (near saturation) oscillating 527 nm light of 70 Hz, measured 6 s after light onset in dark-adapted (DA) and light-adapted (LA) fly strains: **(A)** Wild type (WT), **(C)** *trp*^{S936A}, and **(E)** *trp*^{S936D} transgenic flies in which the S936 site is replaced with Ala (preventing dephosphorylation) or Asp (mimicking constant phosphorylation), as indicated. **B, D, F)** The amplitude of the Fourier transform of the ERG responses to oscillating light of 70 Hz, measured 6 s after light onset in dark-adapted (DA) and light-adapted (LA) flies, is presented. The Fourier transform was calculated from time segments of 200 ms, 6 s after oscillating light onset. The graphs plot the Fourier transform of ERG responses as a function of frequency obtained from light- and dark-adapted fly strains, as indicated. A prominent peak at 70 Hz is observed in all light-adapted fly strains except for the *trp*^{S936D} transgenic fly. However, in dark-adapted flies, only the *trp*^{S936A} fly showed a pronounced peak at 70 Hz.

From Voolstra et al., 2017.

also in dark-adapted flies (Figure 17.6C-D). In contrast, inserting a phospho-mimetic mutation by substituting the S936-TRP site to Asp (*trp*^{S936D}) set the frequency response of light-adapted flies to low frequencies typical of dark-adapted flies (Voolstra et al., 2017; Figure 17.6E-F). In addition, measurement of the response latency, using whole-cell voltage clamp, showed that *trp*^{S936A} had a shorter latency compared with *trp*^{WT} and *trp*^{S936D} transgenic flies (Katz et al. 2017). Together, these studies indicate that TRP channel dephosphorylation is a regulatory process that affects the detection limit of oscillating light according to the light rearing condition, thus adjusting dynamic processing of visual information under varying light conditions.

Phosphorylation of TRPL

Using mass spectrometry, nine phosphorylated Ser and Thr residues were identified in TRPL channels (Cerny et al., 2013). Eight of these phosphorylation sites resided within the C-terminus, and a single site, Ser20, was located at the N-terminus (Figure 17.5B). Relative quantification revealed that Ser20 and Thr989 exhibited enhanced phosphorylation in the light, whereas Ser927, Ser1000, Ser1114, Thr1115, and Ser1116 exhibited enhanced phosphorylation in the dark. The phosphorylation state of Ser730 and Ser931 was light-independent. To further investigate the function of the eight C-terminal phosphorylation sites, the Ser and Thr residues were mutated either to Ala, eliminating phosphorylation (TRPL8x), or to Asp, mimicking phosphorylation (TRPL8xD). The mutated TRPL channels were transgenically expressed in R1-6 photoreceptor cells of flies as *trpl-eGFP* fusion constructs. The mutated channels formed multimeres with WT TRPL and produced electrophysiological responses when expressed in *trpl;trp* double null mutant background indistinguishable from responses produced by WT TRPL. These findings indicated that TRPL channels devoid of their C-terminal phosphorylation sites form fully functional channels and argue against a role of TRPL phosphorylation in channel-gating or regulation of its biophysical properties. Since TRPL undergoes light-dependent translocation (Bähner et al., 2002), subcellular localization of the phosphorylation-deficient as well as the phospho-mimetic *trpl*-eGFP fusion constructs were analyzed. Dark-adapted TRPL^{WT}-eGFP show marked eGFP signal in the rhabdomere, while in light-adapted conditions the eGFP signal translocate to the cell body. In contrast, the TRPL8x-eGFP displayed markedly different translocation from WT TRPL-eGFP. After initial dark adaptation, a faint eGFP signal was observed in the cell body, but no eGFP signal was present in the rhabdomeres. After 16 hours of light adaptation, a strong eGFP signal was observed in the cell body as in WT flies. This indicated that TRPL8x-eGFP fusion construct was newly synthesized during light adaptation. After four hours of dark adaptation, TRPL8x-eGFP was present in the rhabdomere, but 20 hours later, only faint eGFP fluorescence was observable in the cell bodies, and none in the rhabdomeres (Cerny et al., 2013).

In conclusion, the localization of light and dark phosphorylation sites in both TRP and TRPL channels is well established. However, the physiological roles of these post-translation modifications need further studies.

Regulation of Mammalian TRPC Channels by PKC Phosphorylation

Phosphorylation plays an important role in regulation of mammalian TRPC channels. The activation of TRPC3, TRPC6, and TRPC7 by the DAG analogue, 1-oleoyl-2-acetyl-sn-glycerol (OAG), is reversed by the PKC activator 12-myristate 13-acetate (PMA (Okada et al., 1999; Trebak, St J Bird, McKay, Birnbaumer, & Putney, 2003; Venkatachalam, Zheng, & Gill, 2003; Zhang & Saffen, 2001)).

PKC Phosphorylation of TRPC3, TRPC6, TRPC7, and TRPC5

Application of the PKC activator Phorbol 12-myristate 13-acetate (PMA) was shown to increase the phosphorylation of TRPC3 *in vivo* (Trebak et al., 2003), while mutating Ser712 to Ala in TRPC3, identified by comparison of conserved putative PKC phosphorylation sites between TRPC3, TRPC6, and TRPC7 channels, abolished both the PMA increase in phosphorylation and PKC-mediated inhibition of the channel (Trebak et al., 2005). Hence, the phosphorylation of Ser712 by PKC negatively regulates the activity of TRPC3.

The negative regulation of TRPC3 by PKC phosphorylation was also suggested to explain the phenotype of the *moonwalker* mouse mutant. The *moonwalker* mouse harbors a missense, gain-of-function mutation Thr635Ala in TRPC3, which causes cerebellar ataxia and abnormal development of cerebellar Purkinje neurons (Becker et al., 2009). In the cerebellum, TRPC3 channels expressed in Purkinje cell underlie the slow excitatory postsynaptic potential observed after parallel fiber stimulation (Hartmann et al., 2008; Hartmann et al., 2014). In these cells, TRPC3 channel opening requires stimulation of metabotropic glutamate receptor 1, activation of which can also lead to the induction of long-term depression (LTD) underlying cerebellar motor learning (Hartmann et al., 2008; Hartmann et al., 2014). Using *in vitro* kinase assay, it was shown that Thr635 is phosphorylated by PKCγ. This result raised the hypothesis that the gain-of-function phenotype observed in the TRPC3-Thr635A mutant was a consequence of a decrease in PKC-mediated inhibition of the mutated channel lacking the phosphorylation site (Becker et al., 2009). However, neither the phosphorylation event at Thr635 nor the inhibitory physiological consequence have been confirmed *in vivo*. Furthermore, native TRPC3-dependent currents elicited in cerebellar Purkinje cells were not inhibited by conventional PKC or PKG kinases, arguing against the hypothesis that the gain-of-function phenotype is due to lack in phosphorylation-dependent inhibition of the TRPC3 channels (Nelson & Glitsch, 2012). A recent study using computational modeling and functional characterization supports a mechanism by which hydrogen bonding of Thr635 plays a significant role in maintaining a stable, closed state of the channels (Hanson, Sansom, & Becker, 2015). However, we are still lacking the hydrogen bonding partner of Thr635 for further support of this hypothesis in the mechanism underlying the gain of function of the *moonwalking* mouse and its relationship to Thr635 Ala substitution.

Activation of the mouse TRPC5 by carbachol via the muscarinic receptors (Zhu et al., 2005) results in rapid desensitization of TRPC5, which is blocked by inhibitors of PKC. Mutation of the putative PKC phosphorylation site Thr972Ala of TRPC5 resulted in a large decrease of carbachol-mediated desensitization of the channel. TRPC6 is also inhibited by PKC-mediated phosphorylation (Zhang & Saffen, 2001), while phosphorylation by Fyn (the Src family protein kinase) increased its activity (Hisatsune et al., 2004). TRPC6 phosphorylation by Ca^{2+}-calmodulin-dependent kinase II (CaMKII) is required for channel activation, as inhibition of CaMKII prevented channel activation by carbachol (Shi et al., 2004).

Together, phosphorylation of TRPC channels by a variety of protein kinases is important for regulating channel activity. However, this regulation affects a variety of functional channel properties, which are different for *Drosophila* and mammalian TRPC channels.

Hypoxia/Anoxia-Sensing TRPC Channels

It was initially found in *Drosophila* eye that several processes that induced metabolic stress, rapidly activated the TRP and TRPL channels in the dark *in vivo*. The robust effect of metabolic stress revealed a rare property of the TRP channel proteins. Although it is unlikely that depletion of ATP is the physiological mechanism underlying TRP and TRPL activation, this striking phenomenon provides an insight into the physiological properties of TRP and TRPL channels and also, into specific mammalian TRPC channel.

Anoxia Activation of *Drosophila* TRP/TRPL Channels in the Dark

The activation of TRP and TRPL channels in the dark *in vivo* can be induced by applying continuous N_2 flow on the living intact fly, causing anoxia. Accordingly, ERG measurements from intact flies showed opening of the channels in the dark by anoxia, which was completely reversible upon termination of the anoxic condition (Agam et al., 2000; Figure 17.7). The openings of the light-sensitive channels in the dark could also be obtained in photoreceptor cells of isolated ommatidia by application of mitochondrial uncouplers or by omitting ATP from the recording pipette. Furthermore, the effects of illumination and all forms of metabolic exhaustion were additive. Effects similar to those found in wild-type flies were also found in strong hypomorph mutant alleles of rhodopsin, Gq-protein, or phospholipase C, while genetic elimination of both TRP and

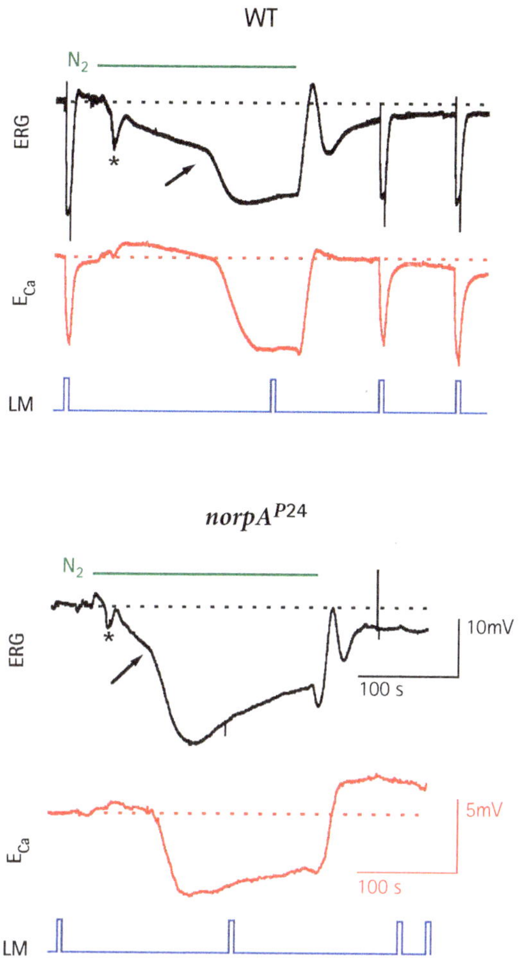

FIGURE 17.7 *Anoxia activated the TRP and TRPL channels in both WT and the PLC null mutant (norpA^{P24}), as monitored by ERG and Ca^{2+}-selective microelectrodes.*

Extracellular voltage change (ERG, *top traces in black*) and potentiometric measurements with Ca^{2+}-selective microelectrode (ECa, *bottom traces in red*) in response to orange lights (*blue pulses* (LM) *and anoxia* N_2 *green*), in WT (*upper*) *Drosophila* and *norpA*^{P24} *(lower)* mutant, are shown. Note that there is no response to light in the *norpA*^{P24} mutant, and the initial slow phase of the electrical response to anoxia due to accumulation of K^+ is missing in the Ca^{2+} signals of both the WT and the mutant. The calibrations for the ERG records are indicated in black, and the calibrations for the potentiometric measurements with the Ca^{2+}-selective microelectrode are indicated in red.

From Agam et al., 2000.

TRPL channels prevented the effects of anoxia, mitochondrial uncouplers, and ATP depletion, thus demonstrating that the TRP and TRPL channels are sensitive targets of metabolic stress (Agam et al., 2000). In the presence of Ca^{2+}, ATP depletion or inhibition of protein kinase C activated the TRP channels in the dark, while photo-release of caged ATP or application of the PKC activator PMA antagonized channel openings as measured by whole cell recordings. Furthermore, Mg^{2+}-dependent stable phosphorylation events by ATPγS or protein phosphatase inhibition by calyculin A abolished activation of the TRP and TRPL channels, either by light or by metabolic inhibition. The attenuation of the light response and channel openings by metabolic inhibition was also observed when Ca^{2+} was highly buffered by 10 mM 1,2-bis(*o*-aminophenoxy)ethane-*N,N,N′,N′*-tetraacetic acid (BAPTA). However, subsequent application of Ca^{2+} to the extracellular medium (in the presence of 7.5–10 mM BAPTA in the recording pipette) combined with ATP depletion induced a robust and fast-rising *dark* current that was reminiscent of light responses (Agam, Frechter, & Minke, 2004).

A specific mechanism for activation of the TRP and TRPL channels by metabolic inhibition in the dark has been proposed (Hardie, Martin, Chyb, & Raghu, 2003; Raghu, Usher, et al., 2000). According to this mechanism, metabolic inhibition primarily attenuates DAG kinase activity due to ATP depletion, leading to a failure of DAG inactivation via prevention of DAG phosphorylation, resulting in channel activation by DAG accumulation. A support for the hypothesis that DAG activates the TRP and TRPL channels comes from the analysis of the *rdgA* mutant fly lacking DAG kinase (Masai, Okazaki, Hosoya, & Hotta, 1993), in which the TRP and TRPL channels remain constitutively open in the dark and do not respond to light (Hardie et al., 2003; Raghu, Usher, et al., 2000). Following experiments on the *rdgA* mutant, it was suggested that DAG, or its metabolites, polyunsaturated fatty acids (PUFAs), are second messengers of excitation (Hardie, 2003). In support of this model, the LIC of mutants with minimal PLC activity and very small light response are partially rescued when combined in a double mutant with *rdgA* (Raghu, Usher, et al., 2000). Furthermore, PUFAs, the lipid product of DAG lipases, activate the native TRP/TRPL channels in *Drosophila* photoreceptors and TRPL in heterologously expressed systems (Chyb, Raghu, & Hardie, 1999). However, so far, it has proven to be virtually impossible to activate the TRP and TRPL channels in normally light-responding photoreceptor cells by exogenous application of DAG or its surrogates (but see Delgado, Muñoz, Peña-Cortés, Giavalisco, & Bacigalupo, 2014).

In summary, activation of the TRP and TRPL channels by metabolic inhibition in the dark has been a useful tool to decouple the activity of the TRP and TRPL channels from the phototransduction machinery and to study their properties in isolation in the native system. Furthermore, these studies show that a combined action of cellular Ca^{2+} and experimental conditions that presumably inhibit DAG kinase, causing DAG accumulation lead to channel activation suggesting that DAG is required for channel activation. It is not clear, however, if activation of the channels by metabolic inhibition is equivalent to the mechanism underlying physiological activation of the channels by light.

Hypoxia-Sensing TRPC6 Channels in Pulmonary Smooth Muscle Cells

Regional alveolar hypoxia causes local vasoconstriction in the lungs, shifting blood flow from hypoxic to normoxic areas, thereby matching blood perfusion to alveolar ventilation and optimizing gas exchange. This mechanism is known as "hypoxic pulmonary vasoconstriction" (HPV; Jeffery & Wanstall, 2001; Wanstall, Jeffery, Gambino, Lovren, & Triggle, 2001). In isolated pulmonary arteries and isolated perfused lungs, the HPV response is typically biphasic. The first phase is characterized by a fast-transient vasoconstrictor response that starts within seconds and reaches a maximum within minutes. The second phase is characterized by a sustained pulmonary vasoconstriction. Severe hypoxia of 1% O_2 was shown to induce cation influx and currents in pulmonary arterial smooth muscle cells (PASMC) from wild-type mice, which were absent in *Trpc6* knockout mice. The acute HPV response in isolated pulmonary arteries was missing in *Trpc6* knockout, but not the second sustained vasoconstrictor response, nor the pulmonary vasoconstriction response elicited by thromboxane mimetic using U46619. These experiments strongly suggest that TRPC6 is involved in acute HPV response in isolated pulmonary arteries. However, recombinant TRPC6 expressed heterologously cannot be activated by hypoxia (Weissmann et al., 2006), suggesting that hypoxia does not activate the TRPC6 channels directly. This same situation was also observed for the *Drosophila* TRPL channel expressed heterologously. Accordingly, anoxic conditions and mitochondria uncouplers activated the TRPL channels in intact *Drosophila* eye and isolated photoreceptors, respectively, while mitochondria uncouplers did not activate TRPL channel expressed heterologously (unpublished data).

Hypoxia-induced TRPC6 activation in smooth muscle cells is probably mediated by PLC activation and thereby DAG accumulation. This suggestion was supported by a subsequent study in which the DAG analog OAG increased normoxic vascular tone in lungs from wild-type mice in isolated perfused and ventilated mouse lungs, but not in lungs from TRPC6-deficient mice. Under conditions of repetitive hypoxic ventilation, OAG, as well as the DAG kinase inhibitor R59949, reduced the strength of acute HPV in a dose-dependent manner, whereas thromboxane mimetic-induced vasoconstriction using U46619 was not reduced. Similar to OAG, R59949 mimicked HPV, since it induced vasoconstriction during normoxic ventilation in a dose-dependent manner. In contrast, the PLC inhibitor U73122, which consequently should block DAG production, inhibited acute HPV, whereas U73343, the inactive analog of U73122, had no effect on HPV. These findings support the hypothesis that DAG induces TRPC6-dependent acute HPV. The use of TRPC6-KO mice has thus provided important insights into the role of TRPC6 in normal physiology and disease states of the pulmonary vasculature. Manipulation of TRPC6 function may offer a therapeutic strategy for the control of pulmonary hemodynamics and gas exchange (Fuchs et al., 2011; for a recent review, see Malczyk et al., 2017).

Conclusion

The founding member of the TRP family of channels, the *Drosophila* light-activated TRP channel, shares a large structural and functional similarity with the mammalian TRPC subfamily. Both the *Drosophila* TRP channel and some members of the mammalian TRPC subfamily are assembled into a multi-molecular signaling complex. The *Drosophila* INAD scaffold protein assembles the TRP, PLC, and eyePKC, while the mammalian NHERF assembles the TRPC4/5, PLC, and cytoskeleton. The mammalian Cav-1 protein is also associated dynamically in a protein complex consisting of TRPC4, TRPC1, and IP_3 receptors. The scaffold proteins in both *Drosophila* and mammalian TRPC channels co-localize the signaling proteins adjacent to the channels. In addition, they are involved in the retention and targeting of the protein complex to the signaling surface membrane: in *Drosophila,* the binding to INAD prevents degradation of TRP, PLC, and eyePKC and therefore maintains a one-to-one stoichiometry among the signaling proteins, which is crucial for normal function (Cook et al., 2000). In the case of mammalian TRPC channels, the role of Cav-1 and NHERF is also to target TRPC channels to the plasma membrane, while NHERF also regulates the sensitivity to DAG of TRPC4/5 (Storch et al., 2017).

There is a striking similarity in the activation of *Drosophila* TRP/TRPL (Agam et al., 2000) and mammalian TRPC6 (Fuchs et al., 2010; Malczyk et al., 2013) by anoxia/hypoxia. Likewise, a similar underlying mechanism has been proposed for both, by which reduction of cellular ATP levels by hypoxia attenuates DAG kinase activity, thereby resulting in DAG accumulation and channel activation (Raghu, Usher, et al., 2000; Fuchs et al., 2010; Malczyk et al., 2013). The available eye-specific DAG kinase mutant, *rdgA*, in *Drosophila* showing activation of the TRP and TRPL channels in the dark, was very useful for reaching this conclusion. The critical role of DAG in mammalian TRPC channel activation has been clearly established (Hofmann et al., 1999; Storch et al., 2017). In contrast, in *Drosophila*, except for extremely slow activation of TRP channels by DAG in isolated rhabdomeric preparations used for single-channel recordings (Delgado et al., 2014), application of DAG to isolated ommatidia has failed to activate these channels due to a still-unclear reason. Nevertheless, several experimental observations suggest that DAG may play a similar excitatory role in both mammalian and *Drosophila* TRPC channels. These observations include:

i. Constitutive activity of TRP channels (most likely) due to DAG accumulation in the *rdgA* mutant (Raghu, Usher, et al., 2000).
ii. Presumed accumulation of DAG following application of metabolic inhibitors/anoxia (Agam et al., 2000), and conversely, suppressed TRP channel activity upon ATP application (Agam et al., 2004; Delgado et al., 2014).
iii. The enhanced excitation by application of DAG lipase inhibitor in tissue culture cells expressing TRPL (Lev, Katz, Tzarfaty, & Minke, 2012).

Most studies on TRPC channels have been conducted in recent years on mammalian TRPC channels. The main reason probably stems from the diverse and important physiological functions of mammalian TRPC channels (Bandyopadhyay et al., 2005; Dietrich et al., 2005; Neuner et al., 2015; Poteser et al., 2006; Poteser et al., 2011; Quick et al., 2012). Moreover, a relatively large number of physiological disorders and diseases arises from malfunction of these channels (Fuchs et al., 2011; Kim et al., 2011; Malczyk et al., 2013; Numaga-Tomita et al., 2016; Phelan, Shwe, Abramowitz, Birnbaumer, & Zheng, 2014; Smedlund, Birnbaumer, & Vazquez, 2015). Nevertheless, as elaborated in this review, the *Drosophila* TRP and TRPL channels are useful preparations for exploring mechanisms underlying fundamental properties of TRPC channels and their mode of activation and regulation in a rigorous manner. This is due to i) the highly developed genetic toolbox available for *Drosophila,* ii) *Drosophila* eye preparation being a native system expressing TRP channels, and iii) the striking similarity between *Drosophila* TRP and mammalian TRPCs, which enables translating findings from *Drosophila* to mammals.

REFERENCES

Acharya, J. K., Jalink, K., Hardy, R. W., Hartenstein, V., & Zuker, C. S. (1997). InsP3 receptor is essential for growth and differentiation but not for vision in Drosophila. *Neuron, 18*(6), 881–887.

Agam, K., von Campenhausen, M., Levy, S., Ben-Ami, H. C., Cook, B., Kirschfeld, K., & Minke, B. (2000). Metabolic stress reversibly activates the Drosophila light-sensitive channels TRP and TRPL in vivo. *Journal of Neuroscience, 20*(15), 5748–5755.

Agam, K., Frechter, S., & Minke, B. (2004). Activation of the Drosophila TRP and TRPL channels requires both Ca 2+ and protein dephosphorylation. *Cell Calcium, 35*(2), 87–105.

Alawi, A. A., & Pak, W. L. (1971). On-transient of insect electroretinogram: Its cellular origin. *Science, 172*(3987), 1055–1057.

Ambudkar, I. S., & Ong, H. L. (2007). Organization and function of TRPC channelosomes. *Pflugers Archiv, 455*(2), 187–200. doi:10.1007/s00424-007-0252-0

Ardura, J. A., & Friedman, P. A. (2011). Regulation of G protein-coupled receptor function by Na+/H+ exchange regulatory factors. *Pharmacology Review, 63*(4), 882–900. doi:10.1124/pr.110.004176

Bähner, M., Frechter, S., Da Silva, N., Minke, B., Paulsen, R., & Huber, A. (2002). Light-regulated subcellular translocation of Drosophila TRPL channels induces long-term adaptation and modifies the light-induced current. *Neuron, 34*(1), 83–93.

Bandyopadhyay, B. C., Swaim, W. D., Liu, X., Redman, R. S., Patterson, R. L., & Ambudkar, I. S. (2005). Apical localization of a functional TRPC3/TRPC6-Ca2+-signaling complex in polarized epithelial cells. Role in apical Ca2+ influx. *Journal of Biological Chemistry, 280*(13), 12908–12916. doi:10.1074/jbc.M410013200

Becker, E. B., Oliver, P. L., Glitsch, M. D., Banks, G. T., Achilli, F., Hardy, A., . . . Davies, K. E. (2009). A point mutation in TRPC3 causes abnormal Purkinje cell development and cerebellar ataxia in moonwalker mice. *Proceedings of the National Academy of Sciences of the United States of America, 106*(16), 6706–6711. doi:10.1073/pnas.0810599106

Berridge, M. J. (1995). Capacitative calcium entry. *Biochemical Journal, 312*(Pt 1), 1–11.

Birnbaumer, L. (2009). The TRPC class of ion channels: A critical review of their roles in slow, sustained increases in intracellular Ca(2+) concentrations. *Annual Review of Pharmacology and Toxicology, 49*, 395–426. doi:10.1146/annurev.pharmtox.48.113006.094928

Bloomquist, B. T., Shortridge, R. D., Schneuwly, S., Perdew, M., Montell, C., Steller, H., ... Pak, W. L. (1988). Isolation of a putative phospholipase C gene of Drosophila, norpA, and its role in phototransduction. *Cell, 54*, 723–733.

Blumenfeld, A., Erusalimsky, J., Heichal, O., Selinger, Z., & Minke, B. (1985). Light-activated guanosinetriphosphatase in Musca eye membranes resembles the prolonged depolarizing afterpotential in photoreceptor cells. *Proceedings of the National Academy of Sciences of the United States of America, 82*(20), 7116–7120.

Brazer, S. C., Singh, B. B., Liu, X., Swaim, W., & Ambudkar, I. S. (2003). Caveolin-1 contributes to assembly of store-operated Ca2+ influx channels by regulating plasma membrane localization of TRPC1. *Journal of Biological Chemistry, 278*(29), 27208–27215. doi:10.1074/jbc.M301118200

Cao, E., Cordero-Morales, J. F., Liu, B., Qin, F., & Julius, D. (2013). TRPV1 channels are intrinsically heat sensitive and negatively regulated by phosphoinositide lipids. *Neuron, 77*(4), 667–679. doi:10.1016/j.neuron.2012.12.016

Cerny, A. C., Oberacker, T., Pfannstiel, J., Weigold, S., Will, C., & Huber, A. (2013). Mutation of light-dependent phosphorylation sites of the Drosophila transient receptor potential-like (TRPL) ion channel affects its subcellular localization and stability. *Journal of Biological Chemistry, 288*(22), 15600–15613. doi:10.1074/jbc.M112.426981

Chevesich, J., Kreuz, A. J., & Montell, C. (1997). Requirement for the PDZ domain protein, INAD, for localization of the TRP store-operated channel to a signaling complex. *Neuron, 18*(1), 95–105.

Chyb, S., Raghu, P., & Hardie, R. C. (1999). Polyunsaturated fatty acids activate the Drosophila light-sensitive channels TRP and TRPL. *Nature, 397*(6716), 255–259.

Constantin, B. (2016). Role of scaffolding proteins in the regulation of TRPC-dependent calcium entry. *Advances in Experimental Medicine and Biology, 898*, 379–403. doi:10.1007/978-3-319-26974-0_16

Cook, B., Bar, Y. M., Cohen-Ben, A. H., Goldstein, R. E., Paroush, Z., Selinger, Z., & Minke, B. (2000). Phospholipase C and termination of G-protein-mediated signalling in vivo. *Nature Cell Biology, 2*(5), 296–301.

Cosens, D. J., & Manning, A. (1969). Abnormal electroretinogram from a Drosophila mutant. *Nature, 224*, 285–287.

Delgado, R., Muñoz, Y., Peña-Cortés, H., Giavalisco, P., & Bacigalupo, J. (2014). Diacylglycerol activates the light-dependent channel TRP in the photosensitive microvilli of Drosophila melanogaster photoreceptors. *Journal of Neuroscience, 34*(19), 6679–6686. doi:10.1523/JNEUROSCI.0513-14.2014

Devary, O., Heichal, O., Blumenfeld, A., Cassel, D., Suss, E., Barash, S., . . . Selinger, Z. (1987). Coupling of photoexcited rhodopsin to inositol phospholipid hydrolysis in fly photoreceptors. *Proceedings of the National Academy of Sciences of the United States of America, 84*, 6939–6943.

Dietrich, A., Kalwa, H., & Gudermann, T. (2010). TRPC channels in vascular cell function. *Thrombosis and Haemostasis, 103*(2), 262–270. doi:10.1160/TH09-08-0517

Dietrich, A., Mederos, Y. S., Gollasch, M., Gross, V., Storch, U., Dubrovska, G., ... Birnbaumer, L. (2005). Increased vascular smooth muscle contractility in TRPC6 -/- mice. *Molecular and Cellular Biology, 25*(16), 6980–6989.

Dunn, H. A., & Ferguson, S. S. (2015). PDZ protein regulation of G protein-coupled receptor trafficking and signaling pathways. *Molecular Pharmacology, 88*(4), 624–639. doi:10.1124/mol.115.098509

Fuchs, B., Dietrich, A., Gudermann, T., Kalwa, H., Grimminger, F., & Weissmann, N. (2010). The role of classical transient receptor potential channels in the regulation of hypoxic pulmonary vasoconstriction. *Advances in Experimental Medicine and Biology, 661*, 187–200. doi:10.1007/978-1-60761-500-2_12

Fuchs, B., Rupp, M., Ghofrani, H. A., Schermuly, R. T., Seeger, W., Grimminger, F., . . . Weissmann, N. (2011). Diacylglycerol regulates acute hypoxic pulmonary vasoconstriction via TRPC6. *Respiratory Research, 12*, 20. doi:10.1186/1465-9921-12-20

Hanson, S. M., Sansom, M. S., & Becker, E. B. (2015). Modeling suggests TRPC3 hydrogen bonding and not phosphorylation contributes to the ataxia phenotype of the moonwalker mouse. *Biochemistry, 54*(26), 4033–4041. doi:10.1021/acs.biochem.5b00235

Hardie, R. C. (1991). Whole-cell recordings of the light induced current in dissociated Drosophila photoreceptors: Evidence for feedback by calcium permeating the light-sensitive channels. *Proceedings of the Royal Society of London, Series B, 245*, 203–210.

Hardie, R. C. (2003). Regulation of *trp* channels via lipid second messengers. *Annual Review of Physiology, 65*, 735–759.

Hardie, R. C., Martin, F., Chyb, S., & Raghu, P. (2003). Rescue of light responses in the Drosophila null" phospholipase C mutant, norpAP24 by diacylglycerol kinase mutant, rdgA and by metabolic inhibition. *Journal of Biological Chemistry, 278*(21), 18851–18858.

Hardie, R. C., & Minke, B. (1992). The *trp* gene is essential for a light-activated Ca^{2+} channel in Drosophila photoreceptors. *Neuron, 8*, 643–651.

Hartmann, J., Dragicevic, E., Adelsberger, H., Henning, H. A., Sumser, M., Abramowitz, J., . . . Konnerth, A. (2008). TRPC3 channels are required for synaptic transmission and motor coordination. *Neuron, 59*(3), 392–398.

Hartmann, J., Karl, R. M., Alexander, R. P., Adelsberger, H., Brill, M. S., Rühlmann, C., . . . Konnerth, A. (2014). STIM1 controls neuronal Ca(2+) signaling, mGluR1-dependent synaptic transmission, and cerebellar motor behavior. *Neuron, 82*(3), 635–644. doi:10.1016/j.neuron.2014.03.027

Hisatsune, C., Kuroda, Y., Nakamura, K., Inoue, T., Nakamura, T., Michikawa, T., . . . Mikoshiba, K. (2004). Regulation of TRPC6 channel activity by tyrosine phosphorylation. *Journal of Biological Chemistry, 279*(18), 18887–18894. doi:10.1074/jbc.M311274200

Hochstrate, P. (1989). Lanthanum mimics the *trp* photoreceptor mutant of Drosophila in the blowfly Calliphora. *Journal of Comparative Physiology A, 166*, 179–187.

Hofmann, T., Obukhov, A. G., Schaefer, M., Harteneck, C., Gudermann, T., & Schultz, G. (1999). Direct activation of human TRPC6 and TRPC3 channels by diacylglycerol. *Nature, 397*(6716), 259–263.

Hong, Y. S., Park, S., Geng, C., Baek, K., Bowman, J. D., Yoon, J., & Pak, W. L. (2002). Single amino acid change in the fifth transmembrane segment of the TRP Ca 2+ channel causes massive degeneration of photoreceptors. *Journal of Biological Chemistry, 277*(37), 33884–33889.

Huber, A., Sander, P., & Paulsen, R. (1996). Phosphorylation of the InaD gene product, a photoreceptor membrane protein required for recovery of visual excitation. *Journal of Biological Chemistry, 271*(20), 11710–11717.

Jeffery, T. K., & Wanstall, J. C. (2001). Comparison of pulmonary vascular function and structure in early and established hypoxic pulmonary hypertension in rats. *Canadian Journal of Physiology and Pharmacology, 79*(3), 227–237.

Katz, B., Gutorov, R., Rhodes-Mordov, E., Hardie, R. C., & Minke, B. (2017). Electrophysiological method for whole-cell voltage clamp recordings from Drosophila photoreceptors. *Journal of Visualized Experiments* (124). 2017 Jun 13;(124).

Katz, B., & Minke, B. (2009). Drosophila photoreceptors and signaling mechanisms. *Frontiers in Cellular Neuroscience, 3*, 2.

Katz, B., Voolstra, O., Tzadok, H., Yasin, B., Rhodes-Modrov, E., Bartels, J. P., . . . Minke, B. (2017). The latency of the light response is modulated by the phosphorylation state of Drosophila TRP at a specific site. *Channels (Austin), 11*(6), 678–685.

Kim, M. S., Lee, K. P., Yang, D., Shin, D. M., Abramowitz, J., Kiyonaka, S., . . . Muallem, S. (2011). Genetic and pharmacologic inhibition of the Ca2+ influx channel TRPC3 protects secretory epithelia from Ca2+-dependent toxicity. *Gastroenterology, 140*(7), 2107–2115. doi:10.1053/j.gastro.2011.02.052

Kohn, E., Katz, B., Yasin, B., Peters, M., Rhodes, E., Zaguri, R., . . . Minke, B. (2015). Functional cooperation between the IP3 receptor and phospholipase C secures the high sensitivity to light of Drosophila photoreceptors in vivo. *Journal of Neuroscience, 35*(6), 2530–2546. doi:10.1523/JNEUROSCI.3933-14.2015

Lee-Kwon, W., Wade, J. B., Zhang, Z., Pallone, T. L., & Weinman, E. J. (2005). Expression of TRPC4 channel protein that interacts with NHERF-2 in rat descending vasa recta. *American Journal of Physiology—Cell Physiology, 288*(4), C942–C949. doi:10.1152/ajpcell.00417.2004

Lev, S., Katz, B., Tzarfaty, V., & Minke, B. (2012). Signal-dependent hydrolysis of phosphatidylinositol 4,5-bisphosphate without activation of phospholipase C: Implications on gating of Drosophila TRPL (transient receptor potential-like) channel. *Journal of Biological Chemistry, 287*(2), 1436–1447. doi:10.1074/jbc.M111.266585

Levy, L. S., Ganguly, R., Ganguly, N., & Manning, J. E. (1982). The selection, expression, and organization of a set of head-specific genes in Drosophila. *Developmental Biology, 94*(2), 451–464.

Liu, M., Parker, L. L., Wadzinski, B. E., & Shieh, B. H. (2000). Reversible phosphorylation of the signal transduction complex in Drosophila photoreceptors. *Journal of Biological Chemistry, 275*(16), 12194–12199.

Liu, C. H., Wang, T., Postma, M., Obukhov, A. G., Montell, C., & Hardie, R. C. (2007). In vivo identification and manipulation of the Ca 2+ selectivity filter in the Drosophila transient receptor potential channel. *Journal of Neuroscience, 27*(3), 604–615.

Liu, W., Wen, W., Wei, Z., Yu, J., Ye, F., Liu, C. H., . . . Zhang, M. (2011). The INAD scaffold is a dynamic, redox-regulated modulator of signaling in the Drosophila eye. *Cell, 145*(7), 1088–1101. doi:10.1016/j.cell.2011.05.015

Lockwich, T. P., Liu, X., Singh, B. B., Jadlowiec, J., Weiland, S., & Ambudkar, I. S. (2000). Assembly of Trp1 in a signaling complex associated with caveolin-scaffolding lipid raft domains. *Journal of Biological Chemistry, 275*(16), 11934–11942.

Malczyk, M., Erb, A., Veith, C., Ghofrani, H. A., Schermuly, R. T., Gudermann, T., . . . Sydykov, A. (2017). The role of transient receptor potential channel 6 channels in the pulmonary vasculature. *Frontiers in Immunology, 8*, 707. doi:10.3389/fimmu.2017.00707

Malczyk, M., Veith, C., Fuchs, B., Hofmann, K., Storch, U., Schermuly, R. T., . . . Weissmann, N. (2013). Classical transient receptor potential channel 1 in hypoxia-induced pulmonary hypertension. *American Journal of Respiratory and Critical Care Medicine, 188*(12), 1451–1459. doi:10.1164/rccm.201307-1252OC

Masai, I., Okazaki, A., Hosoya, T., & Hotta, Y. (1993). Drosophila retinal degeneration A gene encodes an eye-specific diacylglycerol kinase with cysteine-rich zinc-finger motifs and ankyrin repeats. *Proceedings of the National Academy of Sciences of the United States of America, 90*, 11157–11161.

Mery, L., Strauss, B., Dufour, J. F., Krause, K. H., & Hoth, M. (2002). The PDZ-interacting domain of TRPC4 controls its localization and surface expression in HEK293 cells. *Journal of Cell Science, 115*(Pt 17), 3497–3508.

Meyer, N. E., Joel-Almagor, T., Frechter, S., Minke, B., & Huber, A. (2006). Subcellular translocation of the eGFP-tagged TRPL channel in Drosophila photoreceptors requires activation of the phototransduction cascade. *Journal of Cell Science, 119*(Pt 12), 2592–2603.

Minke, B. (1982). Light-induced reduction in excitation efficiency in the *trp* mutant of Drosophila. *Journal of General Physiology, 79*, 361–385.

Minke, B. (1983). The *trp* is a Drosophila mutant sensitive to developmental temperature. *Journal of Comparative Physiology A, 151*, 483–486.

Minke, B. (2010). The history of the Drosophila TRP channel: The birth of a new channel superfamily. *Journal of Neurogenetics, 24*(4), 216–233.

Minke, B. (2012). The history of the prolonged depolarizing afterpotential (PDA) and its role in genetic dissection of Drosophila phototransduction. *Journal of Neurogenetics, 26*(2), 106–117. doi:10.3109/01677063.2012.666299

Minke, B., & Selinger, Z. (1991). Inositol lipid pathway in fly photoreceptors: Excitation, calcium mobilization and retinal degeneration. In N. A. Osborne & G. J. Chader (Eds.), *Progress in retinal research* (pp. 99–124). Oxford, UK: Pergamon Press.

Minke, B., & Selinger, Z. (1992). The inositol-lipid pathway is necessary for light excitation in fly photoreceptors. In D. Corey & S. D. Roper (Eds.), *Sensory transduction* (Vol. 47, pp. 202–217). New York: The Rockefeller University Press.

Minke, B., Wu, C., & Pak, W. L. (1975). Induction of photoreceptor voltage noise in the dark in Drosophila mutant. *Nature, 258*, 84–87.

Mishra, P., Socolich, M., Wall, M. A., Graves, J., Wang, Z., & Ranganathan, R. (2007). Dynamic scaffolding in a G protein-coupled signaling system. *Cell, 131*(1), 80–92.

Montell, C., Jones, K., Hafen, E., & Rubin, G. (1985). Rescue of the Drosophila phototransduction mutation *trp* by germline transformation. *Science, 230*, 1040–1043.

Montell, C., & Rubin, G. M. (1989). Molecular characterization of the Drosophila *trp* locus: A putative integral membrane protein required for phototransduction. *Neuron, 2*, 1313–1323.

Murata, T., Lin, M. I., Stan, R. V., Bauer, P. M., Yu, J., & Sessa, W. C. (2007). Genetic evidence supporting caveolae microdomain regulation of calcium entry in endothelial cells. *Journal of Biological Chemistry, 282*(22), 16631–16643. doi:10.1074/jbc.M607948200

Murthy, A., Gonzalez-Agosti, C., Cordero, E., Pinney, D., Candia, C., Solomon, F., . . . Ramesh, V. (1998). NHE-RF, a regulatory cofactor for Na(+)-H+ exchange, is a common interactor for merlin and ERM (MERM) proteins. *Journal of Biological Chemistry, 273*(3), 1273–1276.

Myers, B. R., Bohlen, C. J., & Julius, D. (2008). A yeast genetic screen reveals a critical role for the pore helix domain in TRP channel gating. *Neuron, 58*(3), 362–373. doi:10.1016/j.neuron.2008.04.012

Nelson, C., & Glitsch, M. D. (2012). Lack of kinase regulation of canonical transient receptor potential 3 (TRPC3) channel-dependent currents in cerebellar Purkinje cells. *Journal of Biological Chemistry, 287*(9), 6326–6335. doi:10.1074/jbc.M111.246553

Neuner, S. M., Wilmott, L. A., Hope, K. A., Hoffmann, B., Chong, J. A., Abramowitz, J., . . . Kaczorowski, C. C. (2015). TRPC3 channels critically regulate hippocampal excitability and contextual fear memory. *Behavioural Brain Research, 281*, 69–77. doi:10.1016/j.bbr.2014.12.018

Niemeyer, B. A., Suzuki, E., Scott, K., Jalink, K., & Zuker, C. S. (1996). The Drosophila light-activated conductance is composed of the two channels TRP and TRPL. *Cell*, *85*(5), 651–659.

Numaga-Tomita, T., Kitajima, N., Kuroda, T., Nishimura, A., Miyano, K., Yasuda, S., . . . Nishida, M. (2016). TRPC3-GEF-H1 axis mediates pressure overload-induced cardiac fibrosis. *Scientific Reports*, *6*, 39383. doi:10.1038/srep39383

Obukhov, A. G., & Nowycky, M. C. (2004). TRPC5 activation kinetics are modulated by the scaffolding protein ezrin/radixin/moesin-binding phosphoprotein-50 (EBP50). *Journal of Cellular Physiology*, *201*(2), 227–235.

Okada, T., Inoue, R., Yamazaki, K., Maeda, A., Kurosaki, T., Yamakuni, T., . . . Mori, Y. (1999). Molecular and functional characterization of a novel mouse transient receptor potential protein homologue TRP7. Ca(2+)-permeable cation channel that is constitutively activated and enhanced by stimulation of G protein-coupled receptor. *Journal of Biological Chemistry*, *274*(39), 27359–27370.

Pak, W. L. (1995). Drosophila in vision research. The Friedenwald Lecture. *Investigative Ophthalmology and Visual Science*, *36*(12), 2340–2357.

Pak, W. L. (2010). Why Drosophila to study phototransduction? *Journal of Neurogenetics*, *24*(2), 55–66. doi:10.3109/01677061003797302

Pak, W. L., Shino, S., & Leung, H. T. (2012) . PDA (Prolonged Depolarizing Afterpotential)-defective mutants: The story of Nina's and Ina's—Pinta and Santa Maria, too. *Journal of Neurogenetics*, *26*(2), 216–237.

Pak, W. L., Grossfield, J., & White, N. V. (1969). Nonphototactic mutants in a study of vision of Drosophila. *Nature*, *222*, 351–354.

Peretz, A., Sandler, C., Kirschfeld, K., Hardie, R. C., & Minke, B. (1994). Genetic dissection of light-induced Ca 2+ influx into Drosophila photoreceptors. *Journal of General Physiology*, *104*, 1057–1077.

Peretz, A., Suss-Toby, E., Rom-Glas, A., Arnon, A., Payne, R., & Minke, B. (1994). The light response of Drosophila photoreceptors is accompanied by an increase in cellular calcium: Effects of specific mutations. *Neuron*, *12*, 1257–1267.

Phelan, K. D., Shwe, U. T., Abramowitz, J., Birnbaumer, L., & Zheng, F. (2014). Critical role of canonical transient receptor potential channel 7 in initiation of seizures. *Proceedings of the National Academy of Sciences of the United States of America*, *111*(31), 11533–11538. doi:10.1073/pnas.1411442111

Phillips, A. M., Bull, A., & Kelly, L. E. (1992). Identification of a Drosophila gene encoding a calmodulin-binding protein with homology to the *trp* phototransduction gene. *Neuron*, *8*, 631–642.

Pless, D. D., & Lennarz, W. J. (1977). Enzymatic conversion of proteins to glycoproteins. *Proceedings of the National Academy of Sciences of the United States of America*, *74*(1), 134–138.

Por, E. D., Gomez, R., Akopian, A. N., & Jeske, N. A. (2013). Phosphorylation regulates TRPV1 association with β-arrestin-2. *Biochemical Journal*, *451*(1), 101–109. doi:10.1042/BJ20121637

Poteser, M., Graziani, A., Rosker, C., Eder, P., Derler, I., Kahr, H., . . . Groschner, K. (2006). TRPC3 and TRPC4 associate to form a redox-sensitive cation channel. Evidence for expression of native TRPC3-TRPC4 heteromeric channels in endothelial cells. *Journal of Biological Chemistry*, *281*(19), 13588–13595.

Poteser, M., Schleifer, H., Lichtenegger, M., Schernthaner, M., Stockner, T., Kappe, C. O., . . . Groschner, K. (2011). PKC-dependent coupling of calcium permeation through transient receptor potential canonical 3 (TRPC3) to calcineurin signaling in HL-1 myocytes. *Proceedings of the National Academy of Sciences of the United States of America, 108*(26), 10556–10561. doi:10.1073/pnas.1106183108

Quick, K., Zhao, J., Eijkelkamp, N., Linley, J. E., Rugiero, F., Cox, J. J., . . . Wood, J. N. (2012). TRPC3 and TRPC6 are essential for normal mechanotransduction in subsets of sensory neurons and cochlear hair cells. *Open Biology, 2*(5), 120068. doi:10.1098/rsob.120068

Raghu, P., Colley, N. J., Webel, R., James, T., Hasan, G., Danin, M., . . . Hardie, R. C. (2000). Normal phototransduction in Drosophila photoreceptors lacking an InsP 3 receptor gene. *Molecular and Cellular Neuroscience, 15*(5), 429–445.

Raghu, P., Usher, K., Jonas, S., Chyb, S., Polyanovsky, A., & Hardie, R. C. (2000). Constitutive activity of the light-sensitive channels TRP and TRPL in the Drosophila diacylglycerol kinase mutant, rdgA. *Neuron, 26*(1), 169–179.

Ranganathan, R., Harris, G. L., Stevens, C. F., & Zuker, C. S. (1991). A Drosophila mutant defective in extracellular calcium-dependent photoreceptor deactivation and rapid desensitization. *Nature, 354*, 230–232.

Razani, B., Lisanti, M. P. (2001). Caveolin-deficient mice: insights into caveolar function human disease. *Journal of Clinical Investigation*, Dec;*108*(11), 1553–1561.

Reuss, H., Mojet, M. H., Chyb, S., & Hardie, R. C. (1997). In vivo analysis of the Drosophila light-sensitive channels, TRP and TRPL. *Neuron, 19*, 1249–1259.

Scott, K., Sun, Y., Beckingham, K., & Zuker, C. S. (1997). Calmodulin regulation of Drosophila light-activated channels and receptor function mediates termination of the light response in vivo. *Cell, 91*(3), 375–383.

Selinger, Z., & Minke, B. (1988). Inositol lipid cascade of vision studied in mutant flies. *Cold Spring Harbor Symposia on Quantitative Biology 53* (Pt 1), 333–341.

Shi, J., Mori, E., Mori, Y., Mori, M., Li, J., Ito, Y., & Inoue, R. (2004). Multiple regulation by calcium of murine homologues of transient receptor potential proteins TRPC6 and TRPC7 expressed in HEK293 cells. *Journal of Physiology, 561*(Pt 2), 415–432.

Shieh, B. H., & Niemeyer, B. (1995). A novel protein encoded by the InaD gene regulates recovery of visual transduction in Drosophila. *Neuron, 14*(1), 201–210.

Shieh, B. H., & Zhu, M. Y. (1996). Regulation of the TRP Ca 2+ channel by INAD in Drosophila photoreceptors. *Neuron, 16*(5), 991–998.

Shieh, B. H., Zhu, M. Y., Lee, J. K., Kelly, I. M., & Bahiraei, F. (1997). Association of INAD with NORPA is essential for controlled activation and deactivation of Drosophila phototransduction in vivo. *Proceedings of the National Academy of Sciences of the United States of America, 94*, 12682–12687.

Smedlund, K. B., Birnbaumer, L., & Vazquez, G. (2015). Increased size and cellularity of advanced atherosclerotic lesions in mice with endothelial overexpression of the human TRPC3 channel. *Proceedings of the National Academy of Sciences of the United States of America, 112*(17), E2201–E2206. doi:10.1073/pnas.1505410112

Storch, U., Forst, A. L., Pardatscher, F., Erdogmus, S., Philipp, M., Gregoritza, M., . . . Gudermann, T. (2017). Dynamic NHERF interaction with TRPC4/5 proteins is required for channel gating by diacylglycerol. *Proceedings of the National Academy of Sciences of the United States of America, 114*(1), E37–E46. doi:10.1073/pnas.1612263114

Suh, P. G., Hwang, J. I., Ryu, S. H., Donowitz, M., & Kim, J. H. (2001). The roles of PDZ-containing proteins in PLC-beta-mediated signaling. *Biochemical and Biophysical Research Communications, 288*(1), 1–7.

Suss Toby, E., Selinger, Z., & Minke, B. (1991). Lanthanum reduces the excitation efficiency in fly photoreceptors. *Journal of General Physiology, 98*, 849–868.

Svobodova, B., & Groschner, K. (2016). Reprint of "Mechanisms of lipid regulation and lipid gating in TRPC channels." *Cell Calcium*, *60*(2), 133–141. doi:10.1016/j.ceca.2016.06.010

Takahashi, N., Kozai, D., & Mori, Y. (2012). TRP channels: Sensors and transducers of gasotransmitter signals. *Frontiers in Physiology*, *3*, 324. doi:10.3389/fphys.2012.00324

Tang, Y., Tang, J., Chen, Z., Trost, C., Flockerzi, V., Li, M., . . . Zhu, M. X. (2000). Association of mammalian trp4 and phospholipase C isozymes with a PDZ domain-containing protein, NHERF. *Journal of Biological Chemistry*, *275*(48), 37559–37564.

Terawaki, S., Maesaki, R., & Hakoshima, T. (2006). Structural basis for NHERF recognition by ERM proteins. *Structure*, *14*(4), 777–789. doi:10.1016/j.str.2006.01.015

Trebak, M., Hempel, N., Wedel, B. J., Smyth, J. T., Bird, G. S., & Putney, J. W. (2005). Negative regulation of TRPC3 channels by protein kinase C–mediated phosphorylation of serine 712. *Molecular Pharmacology*, *67*(2), 558–563. doi:10.1124/mol.104.007252

Trebak, M., St J Bird, G., McKay, R. R., Birnbaumer, L., & Putney, J. W. (2003). Signaling mechanism for receptor-activated canonical transient receptor potential 3 (TRPC3) channels. *Journal of Biological Chemistry*, *278*(18), 16244–16252. doi:10.1074/jbc.M300544200

Tsunoda, S., Sierralta, J., Sun, Y., Bodner, R., Suzuki, E., Becker, A., . . . Zuker, C. S. (1997). A multivalent PDZ-domain protein assembles signalling complexes in a G-protein-coupled cascade. *Nature*, *388*(6639), 243–249.

Tsunoda, S., Sun, Y., Suzuki, E., & Zuker, C. (2001). Independent anchoring and assembly mechanisms of INAD signaling complexes in Drosophila photoreceptors. *Journal of Neuroscience*, *21*(1), 150–158.

Vannier, B., Zhu, X., Brown, D., & Birnbaumer, L. (1998). The membrane topology of human transient receptor potential 3 as inferred from glycosylation-scanning mutagenesis and epitope immunocytochemistry. *Journal of Biological Chemistry*, *273*(15), 8675–8679.

Venkatachalam, K., Zheng, F., & Gill, D. L. (2003). Regulation of canonical transient receptor potential (TRPC) channel function by diacylglycerol and protein kinase C. *Journal of Biological Chemistry*, *278*(31), 29031–29040.

Voolstra, O., Bartels, J. P., Oberegelsbacher, C., Pfannstiel, J., & Huber, A. (2013). Phosphorylation of the Drosophila transient receptor potential ion channel is regulated by the phototransduction cascade and involves several protein kinases and phosphatases. *PLoS One*, *8*(9), e73787. doi:10.1371/journal.pone.0073787

Voolstra, O., Beck, K., Oberegelsbacher, C., Pfannstiel, J., & Huber, A. (2010). Light-dependent phosphorylation of the drosophila transient receptor potential ion channel. *Journal of Biological Chemistry*, *285*(19), 14275–14284.

Voolstra, O., & Huber, A. (2014). Post-translational modifications of TRP channels. *Cells*, *3*(2), 258–287. doi:10.3390/cells3020258

Voolstra, O., Rhodes-Mordov, E., Katz, B., Bartels, J. P., Oberegelsbacher, C., Schotthöfer, S. K., . . . Minke, B. (2017). The phosphorylation state of the Drosophila TRP channel modulates the frequency response to oscillating light in vivo. *Journal of Neuroscience*, *37*(15), 4213–4224. doi:10.1523/JNEUROSCI.3670-16.2017

Wanstall, J. C., Jeffery, T. K., Gambino, A., Lovren, F., & Triggle, C. R. (2001). Vascular smooth muscle relaxation mediated by nitric oxide donors: A comparison with acetylcholine, nitric oxide and nitroxyl ion. *British Journal of Pharmacology*, *134*(3), 463–472. doi:10.1038/sj.bjp.0704269

Weissmann, N., Dietrich, A., Fuchs, B., Kalwa, H., Ay, M., Dumitrascu, R., . . . Gudermann, T. (2006). Classical transient receptor potential channel 6 (TRPC6) is essential for hypoxic pulmonary vasoconstriction and alveolar gas exchange. *Proceedings of the National*

Academy of Sciences of the United States of America, 103(50), 19093–19098. doi:10.1073/pnas.0606728103

Wes, P. D., Chevesich, J., Jeromin, A., Rosenberg, C., Stetten, G., & Montell, C. (1995). TRPC1, a human homolog of a Drosophila store-operated channel. *Proceedings of the National Academy of Sciences of the United States of America, 92*, 9652–9656.

Wong, F., Hokanson, K. M., & Chang, L. T. (1985). Molecular basis of an inherited retinal defect in Drosophila. *Investigative Ophthalmology and Visual Science, 26*(2), 243–246.

Wong, F., Schaefer, E. L., Roop, B. C., LaMendola, J. N., Johnson Seaton, D., & Shao, D. (1989). Proper function of the Drosophila *trp* gene product during pupal development is important for normal visual transduction in the adult. *Neuron, 3*, 81–94.

Wu, C. F., & Pak, W. L. (1975). Quantal basis of photoreceptor spectral sensitivity of Drosophila melanogaster. *Journal of General Physiology, 66*(2), 149–168.

Xu, X. Z., Choudhury, A., Li, X., & Montell, C. (1998). Coordination of an array of signaling proteins through homo- and heteromeric interactions between PDZ domains and target proteins. *Journal of Cellular Biology, 142*(2), 545–555.

Yoon, J., Cohen Ben-Ami, H., Hong, Y. S., Park, S., Strong, L. L. R., Bowman, J., . . . Pak, W. L. (2000). Novel mechanism of massive photoreceptor degeneration caused by mutations in the *trp* gene of Drosophila. *Journal of Neuroscience, 20*, 649–659.

Yun, C. H., Oh, S., Zizak, M., Steplock, D., Tsao, S., Tse, C. M., . . . Donowitz, M. (1997). cAMP-mediated inhibition of the epithelial brush border Na+/H+ exchanger, NHE3, requires an associated regulatory protein. *Proceedings of the National Academy of Sciences of the United States of America, 94*(7), 3010–3015.

Zhang, L., & Saffen, D. (2001). Muscarinic acetylcholine receptor regulation of TRP6 Ca2+ channel isoforms. Molecular structures and functional characterization. *Journal of Biological Chemistry, 276*(16), 13331–13339. doi:10.1074/jbc.M008914200

Zhu, M. H., Chae, M., Kim, H. J., Lee, Y. M., Kim, M. J., Jin, N. G., . . . Kim, K. W. (2005). Desensitization of canonical transient receptor potential channel 5 by protein kinase C. *American Journal of Physiology—Cell Physiology, 289*(3), C591–C600. doi:10.1152/ajpcell.00440.2004

Zhu, X., Chu, P. B., Peyton, M., & Birnbaumer, L. (1995). Molecular cloning of a widely expressed human homologue for the Drosophila *trp* gene. *FEBS Letters, 373*, 193–198.

CHAPTER 18

ACID-SENSING ION CHANNELS

STEFAN GRÜNDER

INTRODUCTION

According to Wikipedia (n.d.), "an acid is a molecule or ion capable of donating a proton (hydrogen ion H^+)." Thus, although acids vary substantially in size and structure, they have the common capability to release a proton. And that is exactly what acid-sensing ion channels (ASICs) sense: protons. Therefore, ASICs sense the genuine acid only indirectly and would be more precisely described as "proton-sensing ion channels." And they do this with an exquisite sensitivity: An only minor (approximately 3-fold) increase of the proton concentration from neutral pH of 7.4 (40 nM) to pH 6.9 (125 nM) can be sensed by the most sensitive ASICs (Babini, Paukert, Geisler, & Gründer, 2002; Sutherland, Benson, Adelman, & McCleskey, 2001; Waldmann, Bassilana, et al., 1997; Waldmann, Champigny, Bassilana, Heurteaux, & Lazdunski, 1997). Protons are unique ligands in many ways: They are the smallest ligand one can imagine, small enough to hide even in crystal structures of the highest resolution; and they are always present, requiring that the sensitivity of a proton sensor needs to be tightly tuned to the "background" proton concentration (Gründer & Pusch, 2015).

These special features of protons (diverse sources, small size, ubiquitous presence) raise the question of whether a proton is really a genuine transmitter. We can now answer this question with a clear "yes." Protons are released as a ubiquitous co-agonist whenever a synaptic vesicle fuses with a presynaptic membrane, and they have been shown to make a small contribution to postsynaptic currents via activation of specific receptors—ASICs (Du et al., 2014; Gonzalez-Inchauspe, Urbano, Di Guilmi, & Uchitel, 2017; Kreple et al., 2014). Besides these rapid and short-lasting synaptic events, the proton concentration can also slowly increase to new, longer-lasting steady-state levels. Typically, such a longer-lasting increase in the proton concentration accompanies pathophysiological situations. The detection of slow, longer-lasting events and of rapid, short-lasting events imposes different requirements on a receptor, yet ASICs are capable of sensing both types of events.

The ASIC Gene Family

ASICs belong to the degenerin/epithelial Na^+ channel (DEG/ENaC) ion channel family, which evolved in metazoans (multicellular animals) (Golubovic et al., 2007; Gründer & Pusch, 2015). However, the ability to sense protons with ASICs probably evolved only later, in deuterostome animals (Lynagh, Mikhaleva, Colding, Glover, & Pless, 2018), which include chordates. Thus, the model organisms *Drosophila* and *Caenorhabditis* do not have genuine ASICs, although they have other DEG/ENaCs (Kellenberger & Schild, 2002).

In mammals, the ASIC family is composed of four genes (*Accn1–Accn4*); a fifth gene (*Accn5*) is more distantly related (Wiemuth, Assmann, & Gründer, 2013), yet it is sometimes referred to as ASIC5 (Table 18.1). For the *Accn1* and *Accn2* genes, use of alternative promoters gives rise to two splice variants each (Bässler, Ngo-Anh, Geisler, Ruppersberg, & Gründer, 2001). Consequently, the *Accn2* gene encodes ASIC1a as well as ASIC1b, and the *Accn1* gene encodes ASIC2a as well as ASIC2b. The respective splice variants differ in the large first exon, which encodes the intracellular N terminus, the first transmembrane domain (TMD), and the proximal part of the large extracellular domain (see the Molecular structure of an ASIC). For ASIC1b, a further splice variant has been identified in rats, which lacks 114 amino acids after TMD1; it is nonfunctional, however (Ugawa et al., 2001). In humans,

Table 18.1 ASIC Subunits

ASIC	Gene	Previously Used Names	Expression	Proton-Sensitivity of Homomer
ASIC1a	*Accn2*	BNaC2, ASIC, ASICα	CNS, PNS	+
ASIC1b	*Accn2*	ASICβ	PNS	+
ASIC2a	*Accn1*	BNaC1, BNC1, MDEG	CNS, PNS	+
ASIC2b	*Accn1*	BNC1b, MDEG2	CNS, PNS	–
ASIC3	*Accn3*	DRASIC	PNS	+
ASIC4	*Accn4*	SPASIC	CNS	–
ASIC5	*Accn5*	BASIC, BLINaC/INaC	CNS	–

Note: Expression refers to rodents. BNaC = brain Na^+ channel; BNC = brain Na^+ channel; MDEG = mammalian degenerin; DRASIC = dorsal root ASIC; SPASIC = spinal cord ASIC; BASIC = bile acid–sensitive ion channel; BLINaC = brain liver intestine Na^+ channel; INaC = intestine Na^+ channel.

the *Accn2* gene gives rise to three variants as well (Hoagland, Sherwood, Lee, Walker, & Askwith, 2010). In addition to variants corresponding to rodent ASIC1a and ASIC1b, respectively, another variant has a 46-amino acid insertion just before TMD2, resulting in a loss of function (Hoagland et al., 2010). ASIC3 has also three variants in humans, a–c, which differ in their C-terminal domain; but variant a is the predominant one, and the functional significance of variants b and c is unknown (Delaunay et al., 2012).

ASIC Homomers and Heteromers

ASICs assemble as trimers (Bartoi, Augustinowski, Polleichtner, Gründer, & Ulbrich, 2014; Jasti, Furukawa, Gonzales, & Gouaux, 2007), either as homo- or as heterotrimers. The molecular details of subunit recognition and channel assembly are unknown. ASIC1a, ASIC1b, ASIC2a, and ASIC3 are sensitive to protons and can form functional homomeric ASICs in heterologous systems (Table 18.1). Among them, homomeric ASIC1a and ASIC3 have the highest proton sensitivity (see Functional Properties of ASICs); from a functional point of view they are the most important ASIC subunits. Homomeric ASIC1b has never been characterized in isolated neurons, and its role is therefore not well established. Homomeric ASIC2a is activated only at pH values <5 and is, therefore, capable of detecting only strong acids like perhaps sour tastants on the tongue (Ugawa et al., 1998).

ASIC subunits can, in principle, assemble in a variety of different heteromeric channels, and several heteromeric channels have indeed been characterized in heterologous expression systems (Babinski, Catarsi, Biagini, & Séguéla, 2000; Bassilana et al., 1997; Benson et al., 2002; X. Chen, Paukert, Kadurin, Pusch, & Gründer, 2006; Gautam & Benson, 2013; Hattori et al., 2009; Hesselager, Timmermann, & Ahring, 2004; Yagi, Wenk, Naves, & McCleskey, 2006), suggesting a relatively promiscuous assembly of subunits. Only ASIC4 and ASIC5 have not been detected in functional heteromers (Gründer, Geissler, Bässler, & Ruppersberg, 2000; Sakai, Linguegliа, Champigny, Mattei, & Lazdunski, 1999). In vivo, however, the number of functional ASIC subunit combinations is probably more limited. For example, only ASIC1a and ASIC3 are highly sensitive to protons; and therefore, probably only ASIC1a- and ASIC3-containing heteromers play a physiological role. So far, only the following heteromers have been relatively well described in isolated neurons: ASIC1a/ASIC2a, ASIC1a/ASIC2b, ASIC3/ASIC2b, and ASIC1a/ASIC2a/ASIC3 (see Physiological functions of ASICs). Other ASIC heteromers may exist, though.

ASIC1a/ASIC2a heteromers assemble in a random fashion, such that both channels with a 1a–1a–2a as well as a 1a–2a–2a stoichiometry exist (Bartoi et al., 2014). Stoichiometries of other ASIC heteromers are unknown, with the exception of ASIC1a/ASIC2a/ASIC3, for which we can assume a 1:1:1 stoichiometry.

Functional Properties of ASICs

ASICs are activated by protons in a concentration-dependent fashion. The most sensitive ASICs—ASIC1a and ASIC3—are activated by mild acidification to pH ~6.9, and their response fully saturates at pH ~6.0 (Babini et al., 2002; Sutherland et al., 2001; Waldmann, Bassilana, et al., 1997; Waldmann, Champigny, et al., 1997) (Figure 18.1A). This corresponds to a rise of the proton concentration to 125 nM and 1 mM, respectively, from a resting value of 40 nM (pH 7.4). Modest pH changes down to pH 6.5 likely mimic the stimuli encountered by an ASIC in vivo. Typically, ASICs completely desensitize within a few seconds, ASIC3 and some heteromers even within <1 s (Table 18.2) (Bässler et al., 2001; Gründer & Pusch, 2015; Sutherland et al., 2001). Thus, ASICs have at least three functional states (corresponding to three principal conformations): closed (or resting), open, and desensitized. Desensitization is responsible for another typical feature of ASICs: steady-state desensitization (SSD) (Babini et al., 2002; Sutherland et al., 2001). This means that a slight but sustained acidification, for example, to pH 7.0, will slowly lead to a new equilibrium between the three states in which all channels will eventually reside in the desensitized state without apparent opening. The channels are then no longer available for activation. Because pH 7.0 corresponds to a proton concentration of 100 nM, this means that a mere 2.5-fold increase in the concentration of protons completely desensitizes sensitive ASICs like ASIC1a and ASIC3 (Babini et al., 2002; Sutherland et al., 2001). Correspondingly, SSD curves are extremely steep, with Hill coefficients of ~10 (Babini

Table 18.2 Homo- and Heteromeric ASICs Expressed in Rodent Neurons

ASIC	Expression in CNS	Expression in PNS: Small Diameter	Expression in PNS: Large Diameter	pH50	τ_{des} (s) at pH 6.0	τ_{act} (ms)
ASIC1a	+	+	–	6.4–6.6	1.2–3.5	6–13 (pH 6.0)
ASIC1a/ASIC2a	+	–	–	4.8–5.4	0.6–0.9	20 (pH 5.0)
ASIC3	–	+	+	6.5–6.7	0.3	<5 (pH 6.0)
ASIC3/ASIC2a	–	+ (mice)	–	5.6–6.1	0.1–0.3	n.d.
ASIC1a/ASIC3/ASIC2a	–	–	+	6.0–6.4	0.1–0.2	n.d.

Note: A minus sign does not exclude expression but indicates that there is no firm evidence for it. There is also evidence for ASIC1a/ASIC2b and ASIC3/ASIC-2b heteromers, but they have similar functional properties as the ASIC1a and the ASIC3 homomers, respectively. pH50 indicates the pH at which half-maximal activation is reached. Values for ASIC1a, ASIC1a/ASIC2a, and ASIC3 are from Gründer and Pusch (2015) or MacLean and Jayaraman (2016); values for ASIC3/ASIC2a are from Hattori et al. (2009) and Benson et al. (2002). n.d. = not determined.

et al., 2002) (Figure 18.1A), suggesting a high cooperativity of proton binding during SSD. SSD is functionally highly relevant because the ambient pH can decrease slightly but persistently to pH values <7.0 under certain conditions like during ischemia and inflammation. Under these conditions, ASICs should no longer be able to open.

For ASIC3, however, activation and SSD curves slightly overlap so that in the narrow pH range between 7.3 and 6.7, ASIC3 is already partially activated and not yet completely desensitized so that it generates a so-called window current (Figure 18.1A), which allows ASIC3 to sense also a slow and persistent but modest decline in pH (Yagi et al., 2006). Several modulators which shift activation or SSD curves can widen the window, extending the pH range where ASIC3 generates sustained currents even to neutral pH 7.4 (see Modulators of ASICs). In contrast, ASIC1a does not generate window currents.

Activation and SSD curves can be described by a simple linear model, in which several closed states, one open state, and one desensitized state are linearly connected (Figure 18.1B) (Gründer & Pusch, 2015). To explain the high cooperativity of SSD, binding of several protons to closed states has to be assumed. Such a linear model has been successfully used to fit experimental data and to explain the unusual pH-dependent deactivation kinetics of ASIC1a (MacLean & Jayaraman, 2017). Moreover, this model postulates that due to the process of desensitization, ASICs' "true" affinity for protons is more faithfully reflected by the experimentally determined pH of half-maximal SSD (pH50 of SSD) than by the pH50 of activation (Gründer & Pusch, 2015),

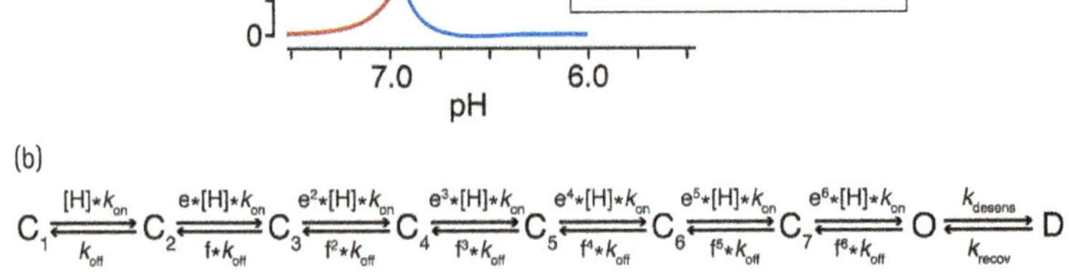

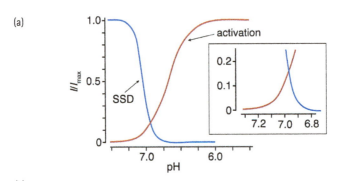

FIGURE 18.1 (A) Schematic illustration of activation and SSD curves. The inset illustrates the small region where activation and SSD curves overlap, which is typical for ASIC3. For example, at pH 7.0 ~10% of channels would already be activated and ~20% would still be available for activation so that at steady-state ~2% of all channels would be open. (B) A linear kinetic scheme as proposed by Gründer and Pusch (2015) can reproduce activation and SSD curves. Six consecutive H^+ binding steps lead to channel opening (state O), introducing substantial cooperativity in proton activation. Desensitization occurs from state O with a given rate (k_{desens}) that is independent of $[H^+]$.

revealing ASICs as exquisitely sensitive receptors for protons, in fact much more sensitive than, for example, transient receptor potential cation channel, subfamily V, member 1 (TRPV1), and a high cooperativity of ligand binding.

Extracellular divalent cations like Ca^{2+} and Mg^{2+} and polyvalent cations like spermine strongly influence ASIC gating. They reduce the apparent affinity for protons, probably by binding to the closed state of the channel, competing with protons (Babini et al., 2002; Duan et al., 2011; Immke & McCleskey, 2003). This results in a shift of activation and SSD curves to the right (lower pH).

Upon stimulation by pH 6.0, ASIC1a opens with a time constant τ_{act} of ~10 ms (Bässler et al., 2001; Sutherland et al., 2001) and ASIC3 with a τ_{act} of <5 ms (Sutherland et al., 2001). ASIC1a and ASIC1a/2a deactivate unusually fast (on a submillisecond timescale) in response to brief stimuli as they will likely encounter during synaptic transmission (MacLean & Jayaraman, 2016). This fast deactivation combined with an ~1,000-fold slower desensitization enables these ASICs to follow trains of short acidic stimuli with high frequency (MacLean & Jayaraman, 2016).

The Molecular Structure of an ASIC

Individual ASIC subunits have ~500–550 amino acids; their amino acid sequences share a 44–75% degree of identity with each other. Each subunit possesses two TMDs, intracellular N- and C-terminal domains (35–90 amino acids in length), and a large extracellular domain (ECD) connecting the two TMDs (Gründer & Chen, 2010). The overall three-dimensional structure of ASIC subunits is probably similar. ASICs have 14 conserved cysteines in the ECD that form seven disulfide bonds. The structure of homotrimeric chicken ASIC1 (cASIC) has been resolved under different conditions, providing molecular snapshots of the resting/closed (Yoder, Yoshioka, & Gouaux, 2018), the open (Baconguis, Bohlen, Goehring, Julius, & Gouaux, 2014; Baconguis & Gouaux, 2012), and the desensitized (Gonzales, Kawate, & Gouaux, 2009; Jasti et al., 2007) conformations. The closed structure has been obtained by either crystallizing a truncation mutant of cASIC1 at high pH and in the presence of Ba^{2+} or Ca^{2+} ions or by cryogenic electron microscopy (cryo-EM) of full-length cASIC1 (Yoder et al., 2018), while the open structure was stabilized by either the tarantula toxin PcTx1 (Baconguis & Gouaux, 2012; Dawson et al., 2012) or the snake toxin MitTx (Baconguis et al., 2014).

For illustration, the domain arrangement of the ECD of cASIC1 has been compared to an upright forearm and a clenched hand (Jasti et al., 2007). The two α-helical TMDs build the forearm and the large ECD the hand, protruding as much as 80 Å from the membrane surface. The clenched hand is composed of five domains: palm, thumb, finger, and knuckle domains, which together surround the central β-ball domain, composed of five β-strands (Jasti et al., 2007). The central element of the ECD is the palm domain, which spans nearly the entire height of an individual subunit and directly connects to TMD1 and TMD2 and to the thumb domain. The knuckle and the finger domains are located above the palm domain

(Jasti et al., 2007). The thumb domains sit at the edges of the three subunits, each thumb containing five of the seven disulfide bonds, providing structural rigidity. At its bottom end, close to the wrist junction, the thumb ends in a highly conserved Trp residue, which it has been proposed sits like a ball in a socket-like joint, thereby transmitting movements within the ECD to the TMDs (Jasti et al., 2007; T. Li, Yang, & Canessa, 2009). Such movements seem to be linked to opening and closing of the ion pore.

Along its central axis, the ECD contains three cavities: the extracellular vestibule, residing with its lower half in the membrane plane and providing access to the ion pore, the central vestibule in the middle of the ECD, and the upper vestibule at the top of the ECD. Ions reach the extracellular vestibule via three oval-shaped fenestrations located at the "wrist," which links the TMDs with the ECD (Gonzales et al., 2009).

One particularly remarkable feature of the ECD is the acidic pocket, a highly negatively charged cavity at the interface of two neighboring subunits (Jasti et al., 2007). The acidic pocket is far (~45 Å) from the TMDs and is formed by intrasubunit contacts between the top of the thumb, the β-ball and the finger domains together with residues from the palm domain of the adjacent subunit (Jasti et al., 2007). In the resting state structure, the thumb and finger domains are farther from the 3-fold axis than in the open and desensitized structure, resulting in an expanded conformation of the thumb and the acidic pocket (Yoder et al., 2018). In the open and desensitized conformation, however, the acidic pocket is contracted, and there are three pairs of carboxyl–carboxylate interactions between the side chains of the aspartate and glutamate residues (Jasti et al., 2007). Because the side chains are so close to each other, at least one of the acidic residues of each pair must be protonated. Thus, it appears that upon protonation the acidic pocket collapses and carboxyl–carboxylate pairs form and stabilize the interface between the thumb, finger, and palm domains (Yoder et al., 2018), bringing the thumb into a contracted conformation. Consequently, it has been proposed that the acidic pocket is involved in pH sensing (Jasti et al., 2007). The collapse of the acidic pocket is transmitted to the TMDs mainly via movements within the lower palm domain, which induce a motion of the TMDs away from the 3-fold molecular axis (Yoder et al., 2018), which ruptures the gate at the upper TMD2 and opens the ion pore (Figure 18.2). While analyses of available cASIC structures strongly suggest a pivotal role for the acidic pocket in pH sensing (Jasti et al., 2007; Yoder et al., 2018), numerous functional studies have revealed that this structural feature cannot fully account for proton sensing of ASICs and that it may rather be involved in fine-tuning it (Krauson, Rued, & Carattino, 2013; Paukert, Chen, Polleichtner, Schindelin, & Gründer, 2008; Vullo et al., 2017). It appears that many residues, probably in different domains of the ECD, contribute to pH sensing in ASICs (Krauson et al., 2013; Liechti et al., 2010; Paukert et al., 2008).

Because Ca^{2+} competes with H^+ (Babini et al., 2002; Immke & McCleskey, 2003), it has been proposed that Ca^{2+} binds to the acidic pocket in the resting conformation and is released upon binding of protons (Paukert et al., 2008). Crystallizing cASIC1 at high and low pH and in the presence of Ba^{2+} indeed revealed two Ba^{2+} binding sites in each of the acidic pockets and three symmetric binding sites in the central vestibule (Yoder & Gouaux, 2018). It appears that one of the binding sites in the acidic pocket and the

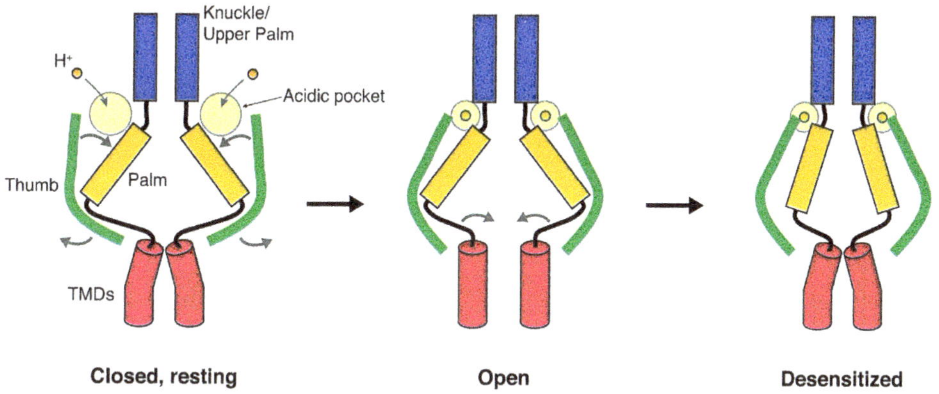

FIGURE 18.2 Schematic representation of the three principal conformations and the conformational changes that accompany gating. In the closed conformation, thumb and acidic pocket are expanded and the gate is closed. Upon binding of protons, the acidic pocket contracts and the thumb moves toward the 3-fold axis. These movements are transmitted to the TMDs and open the gate. Movement of the β1–β2 and the β11–β12 linkers, which link the palm domain with the TMDs, leads to a slight contraction of the vestibule and closes the gate during desensitization. Gray arrows highlight gating movements during opening and desensitization of the channel.

Adapted by permission from Springer Nature: *Nature*, 489(7416), 400–405, Structural plasticity and dynamic selectivity of acid-sensing ion channel–spider toxin complexes, I. Baconguis & E. Gouaux, © 2012.

binding sites in the central vestibule cannot be occupied by Ba^{2+} in the open conformation, suggesting state-dependent binding of divalent cations to these sites (Yoder & Gouaux, 2018). Evidence for the importance of these sites in modulating ASIC gating by divalent cations is so far inconclusive, however.

Comparison of open and desensitized cASIC1 structures revealed that the upper half of the ECD (upper palm and knuckle domains) adopts an almost identical conformation in the two states, providing a stable scaffold (Baconguis et al., 2014; Baconguis & Gouaux, 2012; Dawson et al., 2012). In contrast, in the lower half, there is an iris-like motion which slightly contracts the central and the extracellular vestibules during desensitization. Moreover, this movement closes the desensitization gate (see The ASIC Ion Pore). In addition, there are notable changes in the conformation of the β1–β2 and the β11–β12 linkers of the palm domain (Baconguis et al., 2014; Baconguis & Gouaux, 2012), highlighting the role of these linkers for conformational transitions during desensitization, which is supported by experimental evidence (T. Li, Yang, & Canessa, 2010; Springauf, Bresenitz, & Gründer, 2011).

The ASIC Ion Pore

ASICs are weakly selective Na^+ channels with a relative permeability $P_{Na}/P_K = 5–14$ (Gründer & Pusch, 2015). Most ASICs are impermeable for Ca^{2+}; only homomeric ASIC1a and human ASIC1b (hASIC1b) have a low Ca^{2+} permeability, with $P_{Na}/P_{Ca} >15$

for ASIC1a (Bässler et al., 2001; Gunthorpe, Smith, Davis, & Randall, 2001; Sutherland et al., 2001) and P_{Na}/P_{Ca} >2.5 for hASIC1b (Hoagland et al., 2010); the fractional Ca^{2+} current of ASIC1a is very small (Samways, Harkins, & Egan, 2009). In fact, Ca^{2+} behaves as an open channel blocker of ASICs; therefore, single-channel conductance of ASICs depends on the concentration of divalent cations (Immke & McCleskey, 2003; Paukert, Babini, Pusch, & Gründer, 2004). With 110 mM Na^{+} and 3 mM Mg^{2+} in the bath it is 7 pS for ASIC1a (Yang & Palmer, 2014), and with 130 mM Na^{+} and 1 mM Ca^{2+} it is 11 pS for ASIC3 (Gründer & Pusch, 2015; Immke & McCleskey, 2003).

It has long been known that amino acids within TMD2 make an important contribution to the selectivity of DEG/ENaC channels (reviewed in Gründer & Chen, 2010). Crystallization confirmed that TMD2 helices are closer to the 3-fold axis lining the ion pore, whereas TMD1 helices make most contacts with the lipid bilayer (Gonzales et al., 2009). Ions enter the ion pore via the lateral fenestrations and the extracellular vestibule. As it extends deeper into the membrane, the extracellular vestibule becomes progressively narrower; the narrowest part of the pore is built by the highly conserved Gly–Ala–Ser (GAS) motif in the middle of TMD2. Strikingly, the GAS motif is in an extended conformation (non-α-helical), breaking TMD2 into TMD2a and TMD2b (Figure 18.3A). Moreover, the GAS motif is perpendicular to the rest of TMD2 such that the GAS motifs of the three subunits surround the ion pore like a triangular belt (Figure 18.3B). Another consequence is a swapping of TMD2b domains between subunits (Baconguis et al., 2014; Yoder & Gouaux, 2020; Yoder et al., 2018). While structural data suggest that the Gly residue of the GAS motif coordinates monovalent cations in the ASIC pore via its carbonyl oxygen (Baconguis et al., 2014), experiments using unnatural amino acid mutagenesis failed to confirm an important contribution of this carbonyl oxygen for ion selectivity (Lynagh, Flood, et al., 2017).

A second striking structural feature of the ASIC pore is that its lower half below the GAS belt is lined by pre-TMD1 residues, which form a re-entrant loop, allowing the conserved His–Gly motif to buttress the GAS belt from below (Yoder & Gouaux, 2020) (Figure 18.3A). Although the re-entrant loop has so far only been discerned in resting and desensitized structures extracted with styrene maleic acid copolymers (Yoder & Gouaux, 2020), it is plausible that the open conformation also has a re-entrant pre-TMD1 loop. In support of a role of the pre-TMD1 residues in ion permeation, it has been reported that pre-TMD1 residues are critical for Na^{+} selectivity of ASICs (Coscoy, de Weille, Lingueglia, & Lazdunski, 1999) and Ca^{2+} permeability of ASIC1a (Bässler et al., 2001). Moreover, chemical modification of pre-TMD1 residues by methanethiosulfonate (MTS) reagents inhibit ASIC1a (Pfister et al., 2006). While the overall conformation of the ASIC pore has been revealed by recent crystallization and cryo-EM structures (Baconguis et al., 2014; Yoder & Gouaux, 2020; Yoder et al., 2018), many questions concerning the amino acids important for ion coordination remain open.

In the desensitized and resting states, a crossing of the TMD2 domains between D0′ and G3′ (see legend to figure 18.3 for numbering of amino acids) at the bottom of the extracellular vestibule keeps the ion pore shut (Gonzales et al., 2009; Yoder et al., 2018). Thus, the closing and desensitization gates are identical, as also suggested by

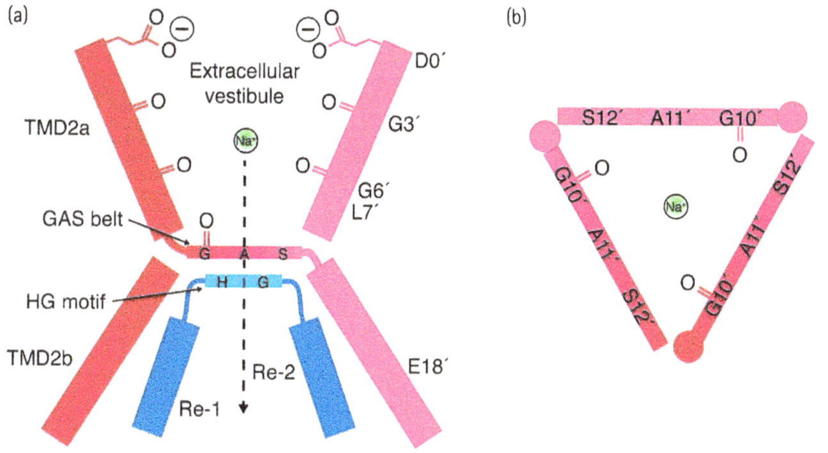

FIGURE 18.3 (A) Schematic illustration of the open ASIC ion pore. The beginning of TMD2 is marked by a conserved Asp residue (D0′) (Jasti et al., 2007). The carbonyl oxygen atoms of G3′, G6′, and G10′ could coordinate the Na^+ ions as they pass through the pore (Baconguis et al., 2014). The conserved GAS motif divides TMD2 into TMD2a and TMD2b. The cytoplasmic amino terminus forms a re-entrant loop, which consists of two short α helices (Re-1 and Re-2) connected by the highly conserved His–Gly motif. The position of the re-entrant loop in the open conformation is at present hypothetical. Numbering of amino acids is as proposed by Lynagh, Flood, and colleagues (2017). (B) The conserved GAS motif forms the narrowest part of the pore, surrounding it like a triangular belt.

state-dependent accessibility of pore residues to MTS reagents (T. Li, Yang, & Canessa, 2011). Substitutions of D0′ in lamprey ASIC1 enhance closing of the channel, providing additional support for this residue as part of the closing gate (T. Li, Yang, et al., 2011). Ca^{2+} ions bind to D0′ and may in this way stabilize the closed gate (Paukert et al., 2004), which could partially explain how Ca^{2+} competes with protons (Babini et al., 2002; Immke & McCleskey, 2003).

Pharmacology of ASICs

Different blockers and inhibitors as well as one direct activator of ASICs have been identified. They have been extremely instrumental in elucidating the role of ASICs in different in vitro and in vivo models, and drugs with a subtype preference have been used to differentiate between the ASIC subunits involved. Amiloride, a clinically approved diuretic, has even been used off-label in human volunteers to test the role of ASICs in nociception (Arun et al., 2013; Holland et al., 2012; Jones, Slater, Cadiou, McNaughton, & McMahon, 2004; McKee et al., 2019; Ugawa et al., 2002). In this section, three subtype-unselective pore blockers—amiloride, diminazene, and nafamostat—and a few subtype-selective toxins will be briefly summarized. There are several other ASIC inhibitors known, which have been described in recent reviews (Baron & Lingueglia, 2015; Rash, 2017). Moreover, a few ASIC modulators, including Zn^{2+}, which potently inhibits ASIC1a and ASIC1a/ASIC2a, will be described.

Subtype-Unselective Drugs

Several subtype-unselective ASIC inhibitors have been identified. Typically, they block the ASIC ion pore.

Amiloride

The prototypical inhibitor of ASICs is amiloride, a pyrazine with a guanidine group, which is positively charged at neutral pH (pKa = 8.7). Apparent affinity of amiloride for ASIC1a is 10 μM (Schmidt, Rossetti, Joussen, & Gründer, 2017; Waldmann, Champigny, et al., 1997), and that for ASIC3 is 60 μM (Waldmann, Bassilana, et al., 1997). Although amiloride is not specific for ASICs and ENaC has an approximately 100-fold higher affinity for it (Canessa et al., 1994), it is still useful to identify ASICs under many conditions as it is cheap and its effect readily reversible. Amiloride blocks the pore of ASICs (Adams, Snyder, & Welsh, 1999; Schmidt et al., 2017), and the voltage dependence of its block indicates that it senses 50% of the membrane electric field (Schmidt et al., 2017), suggesting that it binds deep into the ASIC pore. Amino acids G3′ and L7′ in TMD2a of ASIC1a (Figure 18.3) were predicted to form hydrogen bonds with amiloride and make an important contribution to its binding (Schmidt et al., 2017).

Amiloride does not block sustained ASIC3 currents (Waldmann, Bassilana, et al., 1997), suggesting that the binding site within the ASIC pore adopts a different conformation during sustained currents. It should be noted, however, that in native neurons, sustained acid-induced currents, which are not sensitive to amiloride, are more likely to be carried by ion channels other than ASIC3.

At modest acidification (for example, pH 7.0) and low concentrations of amiloride (100 μM), amiloride has a paradoxical potentiating effect on ASIC3 (Yagi et al., 2006). Even at pH 7.4, amiloride can induce currents with ASIC3 but not with ASIC1a, ASIC1b, or ASIC2a (W. G. Li, Yu, Huang, Cao, & Xu, 2011). Mutational analysis indicates that binding of amiloride into the central vestibule of ASIC3 is responsible for potentiation of ASIC3 but not for block (W. G. Li, Yu, et al., 2011). A similar scenario with an inhibitory site in the ion pore and a potentiating site in the ECD had previously been suggested for ASIC2a (Adams et al., 1999). The fact that amiloride does not potentiate all ASICs suggests subtle structural differences of their central vestibule.

Diminazene, Nafamostat, and A-317567

Diminazene, characterized by two aromatic rings each with an amidine group conferring a positive charge, is another subtype-unspecific blocker of ASICs (X. Chen, Qiu, et al., 2010). It senses ~30% of the membrane electric field of ASIC1a and ASIC5, suggesting that its binding site within the ASIC pore overlaps with that of amiloride (Krauson, Rooney, & Carattino, 2018; Schmidt et al., 2017; Wiemuth & Gründer, 2011). Amino acids G3′, G6′, and L7′ in TMD2a of ASIC1a (Figure 18.3) have indeed been predicted to form hydrogen bonds also with diminazene and have been shown to be important for its binding (Krauson et al., 2018; Schmidt et al., 2017). While the apparent affinity of diminazene for ASIC5 is ~2 μM (Wiemuth & Gründer, 2011), it is in the submicromolar range (0.3 μM at −70 mV) for ASIC1a (Schmidt et al., 2017), making it one of the most potent small-molecule inhibitors of ASICs. Moreover, diminazene does not block ENaC (X. Chen, Qiu, et al., 2010). It binds with a relatively slow kinetics (~0.5 s at 3 μM) into the pore of open ASIC1a (Schmidt et al., 2017). Because block by diminazene is still faster than desensitization, current decline in the presence of diminazene is faster than in its absence, leading to an apparently accelerated desensitization of ASIC1a (X. Chen, Qiu, et al., 2010; Schmidt et al., 2017).

While diminazene is being used as an antiprotozoal drug in veterinary medicine (X. Chen, Orser, & MacDonald, 2010), it is not approved for use in humans. Moreover, it likely binds into the minor groove of DNA, in particular of mitochondria (X. Chen, Orser, et al., 2010), thereby potentially limiting its usefulness also in chronic in vitro models as it might induce unspecific effects in cells. Therefore, despite its lower affinity, amiloride has major advantages over diminazene.

The protease inhibitor nafamostat, which has a structure similar to that of diminazene but with one amidine and one guanidine group, also inhibits ASICs with a potency comparable to that of amiloride (2–15 μM) (Ugawa et al., 2007; Wiemuth & Gründer, 2011). The structural similarity to diminazene suggests that nafamostat may also be a pore blocker of ASICs, as has been experimentally proven for ASIC5 (Wiemuth & Gründer, 2011).

Abbott Laboratories characterized the amidine A-317567 as an inhibitor of ASICs with a half maximal inhibitory concentration (IC50) for different ASICs ranging from 2 to 30 μM (Dube et al., 2005). In mice, it does not induce diuresis, suggesting that it does not block ENaC (Dube et al., 2005). Although its mode of inhibition has not been characterized yet, based on its structure with an amidine group one can imagine that it also acts by blocking the ASIC pore.

Subtype-Selective Drugs

Different animal peptide toxins have been isolated and characterized that inhibit ASICs with high potency, often in the nanomolar range. They bind to specific conformations of the ECD, thereby stabilizing select conformations and modifying the gating of ASICs. Toxins that differentiate between ASIC1 and ASIC3 have been isolated.

Peptide Toxins Specific for ASIC1 and ASIC1-Containing Heteromers

Peptide toxins, which are specific for ASICs containing the ASIC1a subunit, are (1) the tarantula toxin psalmotoxin 1 and related toxins, (2) the snake toxin mambalgin, and

(3) the snake toxin MitTx. While psalmotoxin 1 and mambalgin inhibits these ASICs, MitTx activates them.

Psalmotoxin

The prototypical ASIC1a inhibitor is the tarantula toxin psalmotoxin 1 (PcTx1), which has been isolated from the venom of the South American tarantula *Psalmopoeus cambridgei* (Escoubas et al., 2000). PcTx1 is a 40-amino acid peptide with three disulfide bridges that stabilize the compact inhibitor cystine knot structure (Escoubas, Bernard, Lambeau, Lazdunski, & Darbon, 2003; Escoubas et al., 2000). At pH 7.4, PcTx1 inhibits ASIC1a with an IC50 of 1 nM, but ASIC2a or ASIC3 are not inhibited up to a concentration of 100 nM (Escoubas et al., 2000). Co-crystals of PcTx with cASIC1 revealed that it binds to the acidic pocket at subunit interfaces within the ECD (Baconguis & Gouaux, 2012; Dawson et al., 2012) (Figure 18.4). Moreover, functional characterization revealed that PcTx1 increases the apparent proton affinity of ASIC1a, shifting SSD curves to the left, such that all channels are transferred to the desensitized conformation at neutral pH 7.4 (X. Chen, Kalbacher, & Gründer, 2005). Thus, inhibition by PcTx1 is strongly pH-dependent (X. Chen et al., 2005). Mechanistically, it has been proposed that PcTx1 binds most tightly to the desensitized conformation of rat ASIC1a, thereby stabilizing this conformation (X. Chen, Kalbacher, & Gründer, 2006). Thus, PcTx1 does not facilitate binding of protons but rather modulates gating of ASIC1. This mode of inhibition can also explain why inhibition by PcTx1 is slow at pH 7.4 (complete inhibition by 30 nM after 150 s) (X. Chen et al., 2005) despite its nanomolar affinity: Only the few channels occasionally entering the desensitized state will bind PcTx1 and will then be trapped in this state. It was reported that PcTx1 modifies also the gating of ASIC1b, binding most tightly to its open state, thereby promoting opening of ASIC1b much more than its desensitization (X. Chen, Kalbacher, et al., 2006). PcTx1 also opens cASIC1 at slight acidification, allowing its use for crystallizing cASIC1 in an open conformation (Baconguis & Gouaux, 2012).

At neutral pH 7.4, PcTx1 is a specific inhibitor of homomeric ASIC1a, the subtype specificity resulting from the pH dependence of inhibition. Because SSD curves of ASIC1a/ASIC2a are acid-shifted in comparison with that of ASIC1a, PcTx1 inhibits the heteromer only at more acidic pH of 6.95 (Joeres, Augustinowski, Neuhof, Assmann, & Gründer, 2016). The pH dependence of PcTx1 inhibition needs also to be considered when planning the use of PcTx1 or derivatives in humans because the SSD of human ASIC1a is acid-shifted by 0.2 pH units compared to its rodent orthologs (Sherwood & Askwith, 2008).

PcTx1 has been very useful to decipher the role of ASIC1a in several animal models of disease, such as stroke (see Pathophysiological roles of ASICs–Ischemic stroke) and is a lead compound for translational use in humans. Importantly, it can be administered intranasally to reach the brain (Pignataro, Simon, & Xiong, 2007). Moreover, a toxin has been isolated from the Togo starburst tarantula *Heteroscodra maculata*, Hm3a, that is almost identical to PcTx1 and has similar potency and mechanism as PcTx1 but higher stability (Er, Cristofori-Armstrong, Escoubas, & Rash, 2017). And finally, structural analysis of the PcTx1–cASIC co-crystals allowed the building of a pharmacophore model of the PcTx1 binding site and screening of a small-molecule library to identify a new inhibitor of ASIC1a, which has been called *compound 5b* and inhibits ASIC1a at pH 6.7 with an IC50 of about 30 nM (Buta et al., 2015).

Hi1a

The venom gland transcriptome from the Australian funnel-web spider *Hadronyche infensa* contains four peptides with marked similarity to PcTx1. They are longer (75–77 residues) and comprise two tandem PcTx1-like sequences joined by a short linker (Chassagnon et al., 2017). The two repeats of Hi1a, the most abundantly expressed of the four peptides, share 62% and 50% identity, respectively, with PcTx1. Hi1a inhibits rat and human ASIC1a with an IC50 of 0.5 nM. Like PcTx1, at pH 7.4 Hi1a does not inhibit ASIC2a or ASIC3 and potentiates ASIC1b (Chassagnon et al., 2017). But there are also marked differences from the inhibition by PcTx1: (1) The inhibition is incomplete (~50% inhibition with 1 nM Hi1a); (2) the inhibition is also observed at alkaline pH; and (3) the inhibition is only very slowly reversible (Chassagnon et al., 2017). It has been proposed that Hi1a binds tightly to the closed conformation of ASIC1a and stabilizes it (Chassagnon et al., 2017), which would be remarkable because in the closed channel the acidic pocket is in an expanded conformation (Yoder et al., 2018). It is unclear whether both tandem repeats of the toxin bind to the channel; because the linker is very short, it is highly unlikely that they bind to two PcTx1 binding sites on neighboring subunit interfaces.

Mambalgins

The two isopeptides mambalgin-1 and mambalgin-2 have been isolated from the venom of the black mamba (*Dendroaspis polylepis*) and are potent and reversible inhibitors of ASIC1a (Diochot et al., 2012). They are composed of 57 amino acids with eight cysteine residues; mamb-1 and mamb-2 differ by only one amino acid (Diochot et al., 2012). They have no sequence homology to PcTx1 and belong to the family of three-finger toxins. In contrast to PcTx1, at pH 7.4 mambalgins inhibit not only ASIC1a but also ASIC1b and different ASIC1a- and ASIC1b-containing heteromeric channels, namely ASIC1a/ASIC2a, ASIC1a/ASIC2b, and ASIC1a/ASIC1b, with IC50 values ranging from 50 to 250 nM (Diochot et al., 2012). Similar to PcTx1, mambalgins left-shift the SSD curve of ASIC1a, suggesting they are gating modifiers (Diochot et al., 2012); but like Hi1a, mambalgins inhibit ASIC1a also at alkaline pH, suggesting that they bind tightly to the closed conformation of ASIC1a and stabilize it (Salinas et al., 2014). A cryo-EM structure of the cASIC1–mambalgin-1 complex revealed that each mambalgin-1 molecule interacts via its fingers I and II almost exclusively with the thumb domain of a single cASIC1 subunit and stabilizes a closed conformation with expanded thumb and acidic pockets (D. Sun et al., 2018) (Figure 18.4). Mambalgins have been very instrumental in discovering a role of ASIC1b in peripheral nociception (see also ASICs in small-diameter sensory neurons—nociceptors and metaboreceptors) (Diochot et al., 2012).

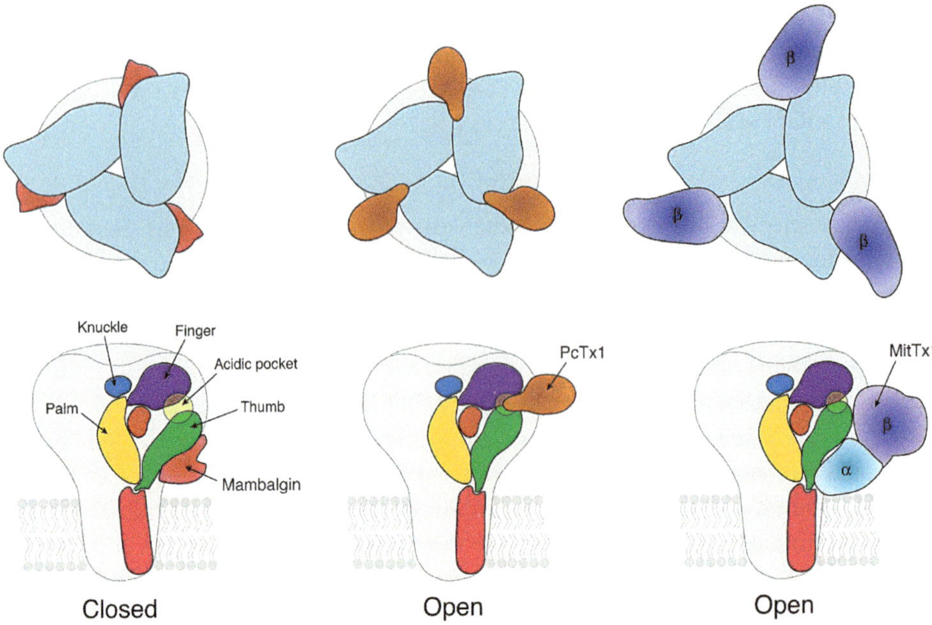

FIGURE 18.4 Schematic representations of toxins binding to different conformations of ASIC1. Top, top views; bottom, side views. Left, mambalgins bind to the thumb domain, stabilizing the closed conformation. Middle, PcTx1 binds to the acidic pocket, stabilizing the open conformation. But PcTx1 also stabilizes the desensitized conformation, which has a similar structure in the upper ECD including the acidic pocket. Right, the MitTx dimer binds to and stabilizes the open conformation.

Adapted by permission of Taylor & Francis: Toxin binding reveals two open state structures for one acid-sensing ion channel, S. Gründer & K. Augustinowski, *Channels (Austin)*, 6(6), 409–413, © 2012 Landes Bioscience.

MitTx

MitTx-α and MitTx-β have been identified from the venom of the Texas coral snake (*Micrurus tener*) in a screen for novel toxins that activate nociceptors (Bohlen et al., 2011). MitTx-α (64 amino acids) has similarity to Kunitz domains, and MitTx-β (127 amino acids) has similarity to phospholipase A2. Both proteins associate with high affinity (K_d = 12 nM) into a heteromeric complex with a 1:1 stoichiometry to form the active toxin (MitTx) (Bohlen et al., 2011). MitTx directly activates ASIC1a and ASIC1b at pH 7.4 with high affinity (half maximal effective concentration [EC50] = 10 and 25 nM, respectively), which, at least in the case of ASIC1a, is probably limited by the association constant of the α/β complex (Bohlen et al., 2011). Apparent affinity for ASIC3 is much lower (~1 μM), and MitTx does not activate ASIC2a at pH 7.4 but increases apparent proton affinity ~1,000-fold (Bohlen et al., 2011). MitTx exhibits high efficacy at ASIC1, and MitTx-induced currents do not desensitize (Bohlen et al., 2011).

Co-crystals of MitTx with cASIC1 revealed that, similar to mambalgins, the MitTx dimer almost exclusively interacts with a single cASIC subunit, protruding from the "edges" of the channel trimer (Baconguis et al., 2014). As expected for the huge toxin, the toxin–channel interactions extend from the wrist region close to the plasma membrane to the thumb and knuckle domains, as far as 60 Å from the membrane surface (Baconguis et al., 2014) (Figure 18.4). Binding of MitTx stabilizes the open conformation of cASIC1 (Baconguis et al., 2014).

Peptide Toxins Specific for ASIC3 and ASIC3-Containing Heteromers

The prototypical ASIC1a inhibitor is the toxin APETx2 from the sea anemone *Anthopleura elegantissima*. APETx is composed of 42 amino acids and is stabilized by three disulfide bridges. APETx2 inhibits the transient ASIC3 current with an IC50 of ~60 nM, but it does not inhibit the sustained current, not even at 3 μM (Diochot et al., 2004). Inhibition of ASIC3 by APETx2 is faster than inhibition of ASIC1a by PcTx1 (at pH 7.4, complete inhibition after 30 s). APETx2 does not inhibit ASIC1a, ASIC1b, or ASIC2a; but it inhibits the following ASIC3-containing heteromers: ASIC3/ASIC1a, ASIC3/ASIC1b, and ASIC3/2b with IC50 values of 0.1–2 μM (Salinas et al., 2014); surprisingly, at pH 7.4 APETx2 (3 μM) does not inhibit ASIC3/ASIC2a (Salinas et al., 2014). Compared with the affinity of PcTx1, Hi1a, and MitTx for their ASIC targets, APETx2 has a lower affinity. It also has a limited specificity because it also inhibits Na_V1.8 (IC50, 2.5 μM) (Blanchard, Rash, & Kellenberger, 2012) and HERG (IC50, 1 μM) (Jensen et al., 2014). This should be taken into account when using APETx2 as a pharmacological tool.

Nonsteroidal Anti-Inflammatory Drugs

Nonsteroidal anti-inflammatory drugs such as acetylsalicylic acid, ibuprofen, and diclofenac also inhibit ASICs with low potency (IC50 typically >100 μM) in a subtype-selective manner (Voilley, de Weille, Mamet, & Lazdunski, 2001). It has been proposed that ibuprofen, which has a low subtype selectivity, binds to a binding pocket in the ECD just above the membrane, where it allosterically inhibits proton-induced activation (Lynagh, Romero-Rojo, Lund, & Pless, 2017).

Modulators of ASICs

Various modulators of ASICs have been described. ASICs have homology to ion channels that are directly activated by neuropeptides, such as the FMRFamide-activated Na^+ channel (FaNaCh) (Lingueglia, Champigny, Lazdunski, & Barbry, 1995), the *Hydra* Na^+ channels (HyNaCs) (Assmann, Kuhn, Dürrnagel, Holstein, & Gründer, 2014; Golubovic et al., 2007), and the myoinhibitory peptide-gated ion channel (Schmidt, Bauknecht, et al., 2018). Therefore, neuropeptides represent a group of particularly interesting and well-characterized ASIC modulators. There are two main classes of neuropeptides that modulate ASICs: RFamide neuropeptides and opioid neuropeptides. Interestingly, the binding site and mode of action of both classes are completely different. Also, modulation by the divalent cation Zn^{2+} and by some lipids will be discussed briefly.

RFamide Neuropeptides and Other Modulators Binding to the Central Vestibule of ASIC3

The prototypical RFamide is FMRFamide, a neurotransmitter in the nervous system of some invertebrates. Besides FMRFamide, the mammalian neuropeptide FF and neuropeptide AF and the cono-RFamide RPRFamide from the venom of the cone snail *Conus textile* modulate ASICs (Askwith et al., 2000; Reimers et al., 2017). The main effect of RFamides is a deceleration of desensitization and an induction of a variable sustained current (Askwith et al., 2000). Although RFamides modulate different ASICs (Askwith et al., 2000), ASIC3 and ASIC3-containing heteromers are usually more strongly modulated than other ASICs (X. Chen, Paukert, et al., 2006; Deval et al., 2003; Reimers et al., 2017; Vick & Askwith, 2015). RFamides bind into the central vestibule in the lower palm domain of ASIC3 and prevent contraction of this domain during desensitization. Thus, RFamides have to unbind before desensitization can proceed, providing a mechanistic explanation for the apparently slower desensitization (Reiners et al., 2018). Binding of RFamides to the central vestibule is slow (on the order of several seconds), and they cannot bind to the central vestibule in the desensitized conformation. Therefore, they need to bind to the closed state of ASICs for efficient modulation.

The central vestibule had initially been identified as a binding pocket for synthetic compounds, such as 2-guanidine-4-methylquinazoline (GMQ) and has therefore also been named the *non-proton ligand sensing domain* (Yu et al., 2010). At pH 7.4, GMQ quickly opens ASIC3 but not ASIC1a or ASIC2. GMQ-induced currents do not desensitize at pH 7.4 or at pH values slightly lower but during stronger acidification (Yu et al., 2010). Although initially introduced as an ASIC3 activator, it was later shown that GMQ actually modulates proton gating of ASIC1a, ASIC1b, ASIC2a, and ASIC3. Only for ASIC3, however, does GMQ generate a window current at pH 7.4 (Alijevic & Kellenberger, 2012).

Besides amiloride, RFamides, and GMQ, also various natural compounds bind to the central vestibule of ASIC3, such as the arginine metabolite agmatine (W. G. Li, Yu, Zhang, Cao, & Xu, 2010; Yu et al., 2010), serotonin (X. Wang et al., 2013), and the pruritogenic peptide SLIGRLamide (Peng et al., 2015). Although the central vestibule of all ASICs can potentially bind ligands, it seems to be particularly relevant in the case of ASIC3.

Opioid Neuropeptides and ASIC1a

The dynorphin opioid peptide big dynorphin (BigDyn), a 32–amino acid peptide that is derived from the pre-prodynorphin opioid precursor peptide, potently shifts SSD curves of ASIC1a to the right, such that ASIC1a remains in its closed state and available for activation also during slight acidification (Sherwood & Askwith, 2009). This might be relevant, for example, during ischemia. While at pH 7.0, BigDyn has an EC50 of 30 nM, dynorphin A, a fully processed cleavage product of BigDyn, has a 1,000-fold higher EC50 (30 μM) (Sherwood & Askwith, 2009). Homomeric ASIC1b is also modulated by BigDyn but heteromeric ASICs much less (Sherwood & Askwith, 2009). Recently, it has been shown that BigDyn stabilizes a closed/resting conformation of ASIC1a by binding to the acidic pocket (Borg et al., 2020).

Zn^{2+}

In the brain, Zn^{2+}, an essential trace element for humans, is stored in synaptic vesicles, for example, of glutamatergic neurons, in up to millimolar concentrations and is released during neuronal activation. Therefore, it is potentially of great relevance that the two most abundant ASICs in the brain, ASIC1a and ASIC1a/ASIC2a, are potently inhibited by extracellular Zn^{2+} (IC50 ~10 nM), due to a reduced apparent affinity for protons (Chu et al., 2004). In addition, ASIC2a and heteromeric ASIC2a-containing channels, such as ASIC1a/ASIC2a, are potentiated by extracellular Zn^{2+} but with a much lower potency (EC50 ~100 μM) (Baron, Schaefer, Lingueglia, Champigny, & Lazdunski, 2001). While chelation of Zn^{2+} to relieve inhibition can increase ASIC currents in hippocampal neurons 2-fold, 300 μM Zn^{2+} potentiates ASIC1a/ASIC2a currents 3-fold (Baron et al., 2001). Thus, Zn^{2+} strongly inhibits ASIC1a and modulates ASIC1a/ASIC2a heteromers in both directions (inhibition and potentiation). Under which circumstances Zn^{2+} concentrations dynamically change to either inhibit or potentiate ASICs is not clear. High-affinity Zn^{2+} block depends on K133 in the finger domain of ASIC1a, and low-affinity Zn^{2+} potentiation depends on H162 and H339 close to the acidic pocket of ASIC2a (Baron et al., 2001; Chu et al., 2004).

Lipids

The polyunsaturated fatty acid arachidonic acid (AA) is released from phospholipids by phospholipase A2, for example, during inflammation. For ASIC3, AA increases apparent proton affinity of activation but not of SSD, resulting in an increased window current (Deval et al., 2008). In this way, AA can generate sustained ASIC3 currents also at neutral pH 7.4 (Deval et al., 2008; Marra et al., 2016), similar to GMQ. The hydrolysis of phospholipids also generates lysophosphatidylcholines (LPCs) and other lysophospholipids. Similar to AA, LPC strongly shifts activation curves of ASIC3, inducing even larger sustained ASIC3 currents at pH 7.4 than AA; addition of AA further increases this current (Marra et al., 2016). Although LPC does not induce ASIC1a, ASIC1b, or ASIC2a currents at pH 7.4, AA can potentiate ASIC1a currents induced by slight acidification (for example, pH 6.9 or 6.7) (Allen & Attwell, 2002; Smith, Cadiou, & McNaughton, 2007), likely by increasing the apparent affinity for protons (Smith et al., 2007). This is relevant because AA is also released during brain ischemia. The potentiation by AA and LPC is relatively slow (increasing during several minutes) and does not saturate with increasing concentrations of the lipids (Allen & Attwell, 2002; Marra et al., 2016). Therefore, an indirect mechanism via membrane insertion has been proposed (Allen & Attwell, 2002). It has indeed been shown that ASIC1a and ASIC3 are sensitive to membrane alterations (Schmidt, Alsop, et al., 2018). Other studies, however, found no changes in membrane shape by relevant concentrations of AA and LPC and modulation of ASICs also in isolated membrane patches, suggesting a mechanism by more direct interaction with the channel (Marra et al., 2016; Smith et al., 2007).

Bile acids, steroid molecules, can directly activate ASIC5 (Wiemuth et al., 2012), probably also by an indirect membrane-mediated mechanism (Schmidt et al., 2014). While bile acids might be a physiological ASIC5 activator in bile ducts, their potency is far too low to directly activate ASIC5 in the brain.

Physiological Functions of ASICs

With the exception of ASIC5 (Sakai et al., 1999; Schaefer, Sakai, Mattei, Lazdunski, & Lingueglia, 2000; Wiemuth et al., 2012), ASICs are almost exclusively expressed in neuronal tissue. It is difficult to find a neuron in the central nervous system (CNS) that does not express an ASIC. In the dorsal root ganglia (DRG), >50% of the sensory neurons express an ASIC (Benson, Eckert, & McCleskey, 1999). Although usually glial cells do not express ASICs (Alvarez De La Rosa et al., 2003; Lingueglia et al., 1997; Sontheimer et al., 1989), some glial cells, like oligodendrocyte-precursor cells (also known as *NG2* cells) express functional ASIC1a (Feldman et al., 2008; Y. C. Lin, Liu, Huang, & Lien, 2010; Sontheimer et al., 1989). Endogenous ASICs have been mostly characterized in neurons from rodents (mice and rats). Therefore, the following discussion mainly summarizes results from rodents; data from humans are presented in a separate paragraph at the end of this section.

ASICs in the Brain

ASIC1a and ASIC2 mRNA are co-expressed in most parts of the brain, with particularly high expression in cortex, cerebellum, hippocampus, amygdala, and olfactory bulb (García-Añoveros, Derfler, Neville-Golden, Hyman, & Corey, 1997; Lingueglia et al., 1997; Price, Snyder, & Welsh, 1996; Waldmann, Champigny, et al., 1997). Antibody staining of ASIC1 in mouse brain, which was verified by staining tissue from knockout mice, revealed strong immunoreactivity in several cortical regions, the molecular layer of the cerebellum, the dentate gyrus of the hippocampus, and the basal ganglia, in particular the striatum, the amygdala, and the glomeruli of the olfactory bulb, among others (Wemmie et al., 2002). Many of these structures receive strong excitatory corticofugal input (Wemmie et al., 2002). The expression of ASICs in these structures has been confirmed by electrophysiological characterization of ASICs in rat cerebellar Purkinje neurons (Allen & Attwell, 2002), which send their dendritic tree into the molecular layer; in rat and mouse cortical neurons (Chu et al., 2004; Herrera et al., 2008; M. Li, Kratzer, Inoue, Simon, & Xiong, 2010; Xiong et al., 2004); in rat and mouse hippocampal neurons (Baron, Waldmann, & Lazdunski, 2002; Cho & Askwith, 2008; Vukicevic & Kellenberger, 2004; W. Wang, Duan, Xu, Xu, & Xu, 2006; Wemmie et al., 2002); in rat hippocampal interneurons (Weng, Lin, & Lien, 2010); in medium spiny neurons of mouse striatum (Jiang et al., 2009); in mouse amygdala (Du et al., 2014; Wemmie et al., 2003); and in mouse olfactory bulb mitral/tufted neurons (M. H. Li et al., 2014). Moreover, ASICs have been functionally characterized in neurons of rat inferior colliculus (Zhang et al., 2008) and in rat hypothalamic neurons of the suprachiasmatic and the supraoptic nuclei (C. H. Chen, Hsu, Chen, & Huang, 2009; Ohbuchi et al., 2010). Intriguingly, functional ASICs are expressed in rat tectum at the earliest stages of neuronal development, well before voltage- and glutamate-gated ion channels (Grantyn, Perouansky, Rodriguez-Tebar, & Lux, 1989), suggesting that ASICs might have an important role during neuronal development.

In line with their broad expression pattern, homomeric ASIC1a and heteromeric ASIC1a/ASIC2a are the main ASICs in the CNS (Askwith, Wemmie, Price, Rokhlina, & Welsh, 2004; Baron, Voilley, Lazdunski, & Lingueglia, 2008; Baron et al., 2002; Chu et al., 2004; Jiang et al., 2009; M. Li, Kratzer, et al., 2010; L. J. Wu et al., 2004). Homomeric ASIC1a constitutes a substantial fraction of ASICs in the CNS (J. Wu et al., 2016), and ASIC1a is also pivotal for a high proton sensitivity of heteromeric ASICs (Askwith et al., 2004; Wemmie et al., 2002). ASIC1a protein outnumbers ASIC2, but it has been estimated that in most mouse brain regions ASIC1a/ASIC2a heteromers, predominantly with the 1a–1a–2a configuration, constitute 50–60% of all ASICs, the exceptions being cerebellum and striatum, where homomeric ASIC1a dominates (60% and 85%, respectively, of all ASICs) (J. Wu et al., 2016). The relative expression of ASIC1a/ASIC2a likely changes during development of the nervous system (M. Li, Kratzer, et al., 2010).

The case of ASIC2b is less clear. ASIC2b is not sensitive to protons on its own but forms a heteromer with ASIC1a (Sherwood, Lee, Gormley, & Askwith, 2011). One study reported that approximately 50% of isolated hippocampal neurons express ASICs resembling heteromeric ASIC1a/ASIC2b (Sherwood et al., 2011). The ASIC1a/ASIC2b heteromer has functional properties very similar to those of homomeric ASIC1a (Lingueglia et al., 1997; Sherwood et al., 2011), however, and is therefore difficult to identify unambiguously. Moreover, one study reported that ASIC2b is mainly found in intracellular compartments in the brain and therefore does not contribute substantially to functional ASICs (J. Wu et al., 2016). Thus, evidence for functional ASIC1a/ASIC2b channels is currently ambiguous.

ASIC3 mRNA is only very sparsely expressed in the CNS (C. C. Chen, England, Akopian, & Wood, 1998; C. C. Chen, Zimmer, Sun, Hall, & Brownstein, 2002; Waldmann, Bassilana, et al., 1997). mRNAs for ASIC4 and ASIC5, which are not sensitive to protons, are expressed in the CNS (Akopian, Chen, Ding, Cesare, & Wood, 2000; Gründer et al., 2000; Sakai et al., 1999); and the proteins could, in principle, associate with ASIC1a to form heteromeric channels (Table 18.1), but so far there has been no clear evidence for such functional heteromers (Akopian et al., 2000; Gründer et al., 2000). ASIC4, which is mainly expressed in subclasses of interneurons, oligodendrocyte-precursor cells, and cerebellar granule cells (S. H. Lin et al., 2015), may downregulate ASIC1a (Donier, Rugiero, Jacob, & Wood, 2008). The precise functions of ASIC4 and ASIC5, however, remain unknown.

Functions of ASICs in the Brain—Postsynaptic Receptors for Protons

Although an early study reported that ASIC1 is not enriched in postsynaptic densities (Alvarez De La Rosa et al., 2003), there is now clear evidence for a postsynaptic location of ASIC1a. In addition, many studies have recorded ASIC currents from the soma of isolated neurons, but it is possible that this reflects an artificial distribution in vitro. In vitro, a strong stimulus that fully activates ASICs can actually depolarize a neuron so strongly that it inactivates voltage-gated channels, leading to a depolarization block. While the subcellular distribution of ASIC1a in neuronal networks needs more research, it remains an attractive possibility that ASIC1a on the soma of neurons mediates tonic responses to acidic pH, while synaptically localized ASIC1 mediates phasic responses (Soto, Ortega-Ramirez, & Vega, 2018). Such a distribution on synaptic and extrasynaptic sites is well established for gamma-aminobutyric acid A ($GABA_A$) receptors.

Synaptic vesicles are acidic (pH ~5.5) (Miesenböck, De Angelis, & Rothman, 1998), and there is evidence that the fusion of vesicles with the presynaptic membrane leads to an acidification of the synaptic cleft (DeVries, 2001; Dietrich & Morad, 2010; Krishtal, Osipchuk, Shelest, & Smirnoff, 1987; Palmer, Hull, Vigh, & von Gersdorff, 2003), which could also activate ASICs on the pre- or the postsynaptic membrane. At excitatory synapses, ASIC1a and ASIC2a most likely have a postsynaptic location. In hippocampal slices, ASIC1a and ASIC2a are preferentially targeted to dendrites and dendritic spines (Wemmie et al., 2002; Zha et al., 2009; Zha, Wemmie, Green, & Welsh, 2006) and enriched in synaptosomal fractions (Wemmie et al., 2002; Zha et al., 2009). ASIC2a increases synaptic localization of ASIC1a via an association with PSD-95 (Zha et al., 2009). A low amount of ASIC2a also localizes to axons, which slightly increases localization of ASIC1a in axons (Zha et al., 2009). Thus, there is convincing evidence that homomeric ASIC1a and heteromeric ASIC1a/ASIC2a localize to the postsynapse of excitatory synapses. Moreover, ASIC1a-mediated currents have been directly recorded at excitatory glutamatergic synapses in the lateral amygdala (Du et al., 2014), the nucleus accumbens (Kreple et al., 2014), and the calyx of Held synapse in the medial nucleus of the trapezoid body (Gonzalez-Inchauspe et al., 2017), unequivocally establishing ASICs as postsynaptic receptors for protons. The contribution of the ASIC current to the total excitatory postsynaptic current at these synapses is small (5–10%). Nevertheless, ASIC1a is crucial for long-term potentiation, a form of synaptic plasticity involved in learning and memory, at Schaffer collateral—CA1 pyramidal cell synapses in the hippocampus (Wemmie et al., 2002). Supporting a role of ASIC1a in synaptic plasticity, overexpression of ASIC1a increases spine densities in hippocampal neurons, mediated by increased Ca^{2+} concentrations and activation of Ca^{2+}/calmodulin-dependent protein kinase II (Zha et al., 2006). Although ASIC currents have so far been recorded at only a few synapses and their contribution to the postsynaptic current has not always been confirmed (Cho & Askwith, 2008), it is still plausible that ASICs make a contribution to the postsynaptic current at many excitatory synapses, which, depending on the experimental conditions, may just be below detection limit. Unlike other synaptic ion channel receptors, ASIC1a and ASIC1a/ASIC2a do not desensitize during high-frequency stimulation (MacLean & Jayaraman, 2016), which could increase their contribution to synaptic transmission under high-frequency stimulation.

The density of ASIC1a-mediated currents is higher in amygdala neurons than hippocampal neurons, and ASIC1a contributes to fear conditioning (Wemmie et al., 2003). Moreover, inhaled CO_2 reduces brain pH and evokes fear-related behavior in mice in an ASIC1a-dependent manner, identifying the amygdala as an important chemosensor, detecting extracellular acidification via ASIC1a (Ziemann et al., 2009). The important role of ASIC1a in fear conditioning, which is thought to have common mechanisms with anxiety disorders and depression-related disorders, also renders it a therapeutic target in these disorders (Coryell et al., 2009).

Synaptic vesicles containing GABA are also acidic. Thus, in theory, protons could serve as a transmitter and ASICs as their receptor also at inhibitory, GABAergic synapses. Because the ASIC current will depolarize a cell, its activation would counteract the inhibitory current, however. While it is known that ASICs are also expressed in hippocampal GABAergic interneurons (Cho & Askwith, 2008; Ziemann et al., 2008), they are located on the dendrites of these cells (Weng et al., 2010), where they receive excitatory input. It has indeed been suggested that ASIC activation increases the inhibitory tone in a network (Ziemann et al., 2008), by exciting inhibitory interneurons. In contrast, at inhibitory synapses themselves, an ASIC current has not yet been described, although there is some evidence for modulation of GABAergic transmission by ASICs (X. Chen, Whissell, Orser, & MacDonald, 2011; Storozhuk, Kondratskaya, Nikolaenko, & Krishtal, 2016; Zhao et al., 2014).

In contrast to the postsynaptic membrane of excitatory synapses, there is no clear evidence for ASIC expression at presynaptic sites. One study reported increased presynaptic release probability in hippocampal neurons from ASIC1 knockout mice (Cho & Askwith, 2008), but the mechanism for this effect is unknown.

Recently, a different release mechanism for protons that can activate ASICs has been identified: release from glial cells (Nomura et al., 2019). Glial cells in the organum vasculosum lamina terminalis, which lacks a blood–brain barrier, sense an increase in brain [Na^+] and release lactate and protons via a monocarboxylate transporter. The protons activate ASIC1a on neighboring neurons that project to the paraventricular nucleus of the hypothalamus to induce secretion of antidiuretic hormone, and lactate enhances the activation of these neurons (Nomura et al., 2019). It is possible that also in other regions of the brain protons released by glial cells can activate ASICs.

ASICs in the Spinal Cord

Neurons of the dorsal horn of the spinal cord are an important relay station for nociceptive signals. Central sensitization processes in the spinal cord contribute to enhanced and to persistent pain. In mice, most dorsal spinal horn neurons express the ASIC1a/

ASIC2a heteromer, some express homomeric ASIC1a, and some both (Baron et al., 2008). In contrast, in rats it appears that dorsal spinal horn neurons mainly express homomeric ASIC1a (Duan et al., 2007; L. J. Wu et al., 2004).

In mice, spinal (intrathecal) administration of PcTx1 or antisense knockdown of ASIC1a has potent analgesic effects on acute thermal and chemical pain as well as on inflammatory and neuropathic pain (Mazzuca et al., 2007). Spinal administration of mambalgins has similar effects and uncovered a contribution by ASIC2a in transmitting pain signals in the spinal cord of mice (Diochot et al., 2012). In rats, spinal administration of PcTx1 or antisense knockdown of ASIC1a revealed that ASIC1a participates in inflammatory hyperalgesia but not in normal transmission of acute pain signals (Duan et al., 2007). Mechanistically, ASIC1a promotes central sensitization in the spinal cord (Duan et al., 2007). In line with this result, ASIC1a is also important for brain-derived neurotrophic factor–induced hypersensitivity of spinal dorsal horn neurons in mice and rats (Duan et al., 2012). Although there might be species differences regarding the role in acute pain, these results reveal an important role for ASIC1a and ASIC2a in the transmission of pain signals in the spinal cord.

ASICs in the Peripheral Nervous System

Sensory neurons of the DRG and the trigeminal ganglia (TG) have a typical pseudounipolar morphology and are functionally diverse. They innervate the skin, muscles, and internal organs. Unmyelinated, small-diameter C fibers are mainly polymodal nociceptors or thermoreceptors; myelinated, small-diameter (Aδ) fibers are nociceptors or cold sensors; and large-diameter fibers (Aβ and Aα) are low-threshold mechanoreceptors or proprioceptors. The internal organs of the upper body (down to the colon) are also innervated by sensory fibers following the vagal or the glossopharyngeal nerve. Their somata lie in the inferior ganglion of the vagus nerve (nodose ganglion) or of the glossopharyngeal nerve (petrosal ganglion).

In the peripheral nervous system (PNS), the diversity of ASICs is larger than that in the CNS. mRNAs of ASIC1a, ASIC1b, ASIC2a, and ASIC3 are all individually expressed by 40–50% of small-diameter (15–30 μm) DRG neurons, with similar proportions in peptidergic and isolectin B4-positive neurons (Voilley et al., 2001), suggesting a significant overlap of expression of individual ASIC subunits. While ASIC1a mRNA and protein seem to be predominantly expressed in small-diameter neurons (C. C. Chen et al., 1998; Olson, Riedl, Vulchanova, Ortiz-Gonzalez, & Elde, 1998; Ugawa, Ueda, Yamamura, & Shimada, 2005; Waldmann, Champigny, et al., 1997), presumably nociceptors, ASIC1b and ASIC3 are expressed in cells with small and large diameters (Chang et al., 2019; C. C. Chen et al., 1998; Fukuda et al., 2006; Ichikawa & Sugimoto, 2002; Molliver et al., 2005; Omori et al., 2008; Price et al., 2001; Ugawa et al., 2001, 2005). ASIC2 mRNA appears to be more abundant in large- than small-diameter DRG neurons, with a higher expression of ASIC2b than ASIC2a (García-Añoveros, Samad, Zuvela-Jelaska, Woolf, & Corey, 2001; Kawamata et al., 2006; Price et al., 2001). The precise expression pattern may vary for DRGs from different regions of the spine as well as between DRGs and TG. ASIC4 mRNA either is not expressed in DRGs (Gründer et al., 2000) or is expressed only at very low levels (Akopian et al., 2000); ASIC5 mRNA is not expressed in DRGs (Sakai et al., 1999).

ASIC1a and ASIC2a apparently are absent from the central terminals in the spinal cord (Duan et al., 2007; García-Añoveros et al., 2001). The presence of ASIC3 in the central terminal of DRG neurons has not been reported.

ASICs in Small-Diameter Sensory Neurons—Nociceptors and Metaboreceptors

Small-diameter sensory neurons, mainly nociceptors, are often sensitive to acid, which can induce pain (Steen, Reeh, Anton, & Handwerker, 1992; Ugawa et al., 2002). Thus, the expression of ASICs in small-diameter sensory neurons is consistent with their role in detecting extracellular acidification. A particularly well-investigated case is ASIC3. This channel is abundant in nociceptors expressing skeletal and heart muscle, and its functional properties make it an ideal sensor of small but persistent decreases of the extracellular pH, as occur during muscle ischemia.

ASIC1a—Nociceptors

It is well established that a subpopulation of small-diameter DRG neurons express functional homomeric ASIC1a (C. C. Chen et al., 1998; Escoubas et al., 2000; Poirot, Berta, Decosterd, & Kellenberger, 2006; Voilley et al., 2001; Waldmann, Champigny, et al., 1997). Most of the neurons expressing ASIC1a mRNA also express TRPV1 mRNA (Ugawa et al., 2005), a marker of nociceptors, supporting the idea that ASIC1a-expressing neurons are nociceptors. Genetic deletion or pharmacological blockade of ASIC1a does not affect responses to acute mechanical and thermal noxious stimuli, however (Duan et al., 2007; Page et al., 2004; Staniland & McMahon, 2009). On the other hand, it has been shown that injection of the ASIC1 agonist MitTx into the hindpaw of mice produces pain-related behavior and that this effect is dependent on ASIC1 and on TRPV1-expressing nociceptors, suggesting it was mediated by small-diameter fibers expressing ASIC1a (Bohlen et al., 2011). Although these results establish a role for ASIC1a in nociceptors, our knowledge about the role of ASIC1a in acute pain is still surprisingly sparse, and this role may be modulatory (Staniland & McMahon, 2009).

ASIC1b—Nociceptors

ASIC1b mRNA is expressed in small- as well as large-diameter neurons (Chang et al., 2019; C. C. Chen et al., 1998). While in small-diameter neurons from rat DRGs, ASIC1b mRNA is not co-expressed with ASIC1a or TRPV1 (Ugawa et al., 2005), most ASIC1b-expressing neurons in mouse DRGs co-express ASIC1a (Chang et al., 2019). Only recently have ASIC1b currents been described for the first time in isolated mouse DRG neurons (Chang et al., 2019). Although many details are still unclear, functional characterization of ASIC currents in ASIC1b-expressing neurons indicates the presence of ASIC1b heteromers with ASIC1a or ASIC3 (Chang et al., 2019). The ASIC1b/ASIC3 heteromer has also been characterized in heterologous systems, revealing acid-shifted SSD curves compared with ASIC3 and a prominent modulation by RFamide neuropeptides (X. Chen, Paukert, et al., 2006). In contrast, it is uncertain whether homomeric ASIC1b channels are expressed in the PNS to form low-sensitivity acid sensors (Bässler et al., 2001; C. C. Chen et al., 1998). Strikingly, peripheral application of mambalgin-1 revealed that ASIC1b, and not ASIC1a, is important for cutaneous nociception and inflammatory hyperalgesia to heat (Diochot et al., 2012), which indicates a nociceptive role for ASIC1b in sensory neurons. The precise subunit stoichiometry of the relevant channels, however, remains unknown.

ASIC3—Metaboreceptors and Nociceptors

ASIC3 is expressed in sensory neurons innervating heart and skeletal muscle and, at least in mice, to a lesser extent the skin. Its expression is particularly high in afferents innervating the heart muscle. In most (>90%) rat cardiac afferent neurons, acid evokes extraordinarily large currents (Benson et al., 1999) that match ASIC3 or ASIC3/ASIC2b currents, which are difficult to functionally differentiate, or a combination of both (Sutherland et al., 2001). These currents are enhanced by extracellular lactate, probably via chelating Ca^{2+} (Immke & McCleskey, 2001, 2003). Because lactate is produced by anaerobic metabolism during myocardial ischemia, it is likely that ASIC3 is an acid sensor that mediates chest pain (angina) (Benson et al., 1999; Immke & McCleskey, 2001; Sutherland et al., 2001). Consistent with this role, it has been shown that ASIC3 plays a protective role in a mouse model of myocardial ischemia (Cheng et al., 2011). Mechanistically, it has been shown that the ASIC3 window current, which arises from the overlap of activation and SSD curves, occurs precisely in the limited range of extracellular pH (7.3–6.7) that also occurs during cardiac ischemia (Yagi et al., 2006). This property allows ASIC3 to sense a slow and persistent but modest decline in pH. In ASIC3/ASIC2a heteromers, this window current is much larger and acid-shifted compared with ASIC3 homomers (Yagi et al., 2006). It has been reported that in mouse cardiac afferent neurons ASICs most closely resemble the ASIC3/ASIC2a heteromer (Hattori et al., 2009).

In rats, about 50% of small-diameter (<25 mm) sensory neurons innervating skeletal muscle express ASIC3 (Molliver et al., 2005). Many of these neurons innervate small arteries in muscle, and over 80% co-express the vasodilatory calcitonin gene-related peptide (CGRP) (Molliver et al., 2005). Comparable to the proposed function of ASIC3 in cardiac afferent neurons, it was suggested that these afferents are muscle metaboreceptors that sense the metabolic state of muscle and trigger vasodilation, via an axon reflex and release of CGRP, and muscle pain (myalgia) when there is insufficient oxygen supply and increased anaerobic metabolism (Molliver et al., 2005). Ischemia–reperfusion of mouse muscle leads to enhanced release of the cytokine interleukin 1β, which in turn upregulates expression of ASIC3, thereby further sensitizing the muscle afferents and contributing to ischemic myalgia (Ross et al., 2016). In mice, similarly to rats, about 50% of afferents innervating muscle respond to acid (Light et al., 2008). These neurons respond much more sensitively to a combination of protons, lactate, and adenosine triphosphate (ATP) than to individual components (Gregory, Whitley, & Sluka, 2015; Light et al., 2008). Intriguingly, it has been proposed that in muscle metaboreceptors, ASICs integrate acid and ATP signals via an assembly of ASICs with a P2X receptor, which serves as the receptor for ATP. In this way, ATP could increase the pH sensitivity of ASIC3 (Birdsong et al., 2010). As this would be an unconventional assembly of two unrelated ion channels, it still needs further experimental support, however.

The exercise pressor reflex, which increases arterial pressure during static exercise, is evoked by metabolic stimuli (in particular lactate) as well as by mechanical stimuli; in cats, ASICs mediate only the reflex response to metabolic, but not to mechanical, stimuli (Hayes, Kindig, & Kaufman, 2007; Hayes, McCord, Rainier, Liu, & Kaufman, 2008). In contrast, in rodents and humans, ASIC3 triggers the vasodilation mediated by a mechanical stimulus, low pressure, to the skin (Fromy, Lingueglia, Sigaudo-Roussel, Saumet, & Lazdunski, 2012). This vasodilatory reflex likely involves the release of CGRP from sensory nerve endings and delays the reduced blood flow, which is caused by the pressure to the skin. Over longer periods of time, obstructed blood flow can lead to pressure ulcers (bedsores), which are often observed in persons who are on chronic bed rest or regularly use a wheelchair. Another clinically relevant aspect is that muscle incision during surgery decreases pH and that ASIC3 triggers postoperative pain after muscle incision in rats (Deval et al., 2011).

In mice, only about 10% of afferent neurons innervating the skin express ASIC3 (Y. W. Lin et al., 2008) compared with ~60% in rats (Deval et al., 2008). Consequently, in mice ASIC3 does not seem to trigger cutaneous pain to acid (C. C. Chen et al., 2002; Price et al., 2001), while in rats it does (Deval et al., 2008). In rats, in 50% of unmyelinated C fibers innervating skin, a moderate acidification to pH 6.9 triggers an action potential, which can be inhibited by the ASIC3-specific toxin APETx2 (Deval et al., 2008). Moreover, hyperosmolarity and AA, which are commonly found in the interstitial fluid of inflamed tissue, strongly enhance the currents evoked by modest pH changes (Deval et al., 2008). These results suggest that ASIC3 is also a key acid sensor in cutaneous nociceptors of rats but apparently not of mice. In mice, protons do not act as a pruritogen; instead, they potentiate the itch induced by a pruritogenic peptide, and ASIC3 functions as a coincidence detector, binding both the pruritogenic peptide and protons (Peng et al., 2015).

ASICs in Large-Diameter Sensory Neurons—Mechanoreceptors

In large-diameter sensory neurons, the role of ASICs is less clear because these neurons typically do not detect acid but do detect mechanical stimuli. Upon isolation of neurons and analysis in vitro, acid can nevertheless elicit currents also in large-dimeter neurons. Analysis of acid-activated ASIC currents in larger-diameter (~30–35 μm) DRG neurons of mice with knockout of either the ASIC3, the ASIC1a, or the ASIC2a gene revealed that ASIC3-containing heteromers are the most likely ASIC candidates in these cells (Price et al., 2001; Xie, Price, Berger, & Welsh, 2002): either ASIC1a/ASIC2a/ASIC3 or, less likely, ASIC3/ASIC2a together with ASIC1a/ASIC2a (Benson et al., 2002). Skeletal muscle afferents in mice also express the ASIC1a/ASIC2a/ASIC3 heteromer (Gautam & Benson, 2013).

What could be the function of ASICs in large-diameter sensory neurons, many of which are low-threshold mechanoreceptors or proprioceptors? ASICs are homologs of mechanosensitive ion channels in *Caenorhabditis elegans* (Driscoll & Chalfie, 1991; Geffeney et al., 2011; O'Hagan, Chalfie, & Goodman, 2005), and therefore, a role of ASICs in mechanosensation has been repeatedly proposed. Ion channels of the DEG/ENaC gene family are functionally diverse (Assmann et al., 2014; Kellenberger & Schild, 2002), however; and ASICs are not the species orthologs of mechanosensitive DEG/ENaC channels from *C. elegans*. Current evidence indeed suggests that ASICs sensitize mechanoreceptors or amplify their response, rather than constituting the pore-forming unit of a mechanosensor themselves.

ASIC2

Immunoreactivity of ASIC2a co-localizes with a marker for myelinated A fibers and was found in a variety of dedicated mechanosensory terminals, including Meissner corpuscles, penicillate endings, Merkel discs, and hair follicle afferents (García-Añoveros et al., 2001). Expression of ASIC2 in hair follicles was confirmed in an independent study, where the specificity of the immunostaining was verified with tissue from ASIC2 knockout mice (Price et al., 2000). Moreover, it was reported that ASIC2 knockout mice have a specific reduction in the sensitivity of low-threshold, rapidly adapting mechanoreceptors (Price et al., 2000). Mechanoreceptor responses are not fully eliminated in ASIC2 knockout mice, however; and another study using ASIC2 knockout mice found no impairment of visceral mechanosensation, hearing, or cutaneous mechanosensation, including in rapidly adapting fibers (Roza et al., 2004). Moreover, currents activated by direct mechanical stimulation via a blunt probe are not different in cultured sensory neurons with a double knockout of ASIC2 and ASIC3 than in wild-type neurons (Drew et al., 2004). These findings call into question a role of ASIC2 as a central component of a mechanosensory complex. Because it is now established that the ion channel Piezo2 mediates mechansosensitivity of most low-threshold mechanoreceptors in DRGs (Ranade et al., 2014), it is more likely that ASIC2 increases, by an unknown mechanism, the sensitivity of some low-threshold mechanoreceptors innervating the skin. It is of interest, however, that in *Drosophila* larvae, Piezo functions in parallel pathways with the DEG/ENaC pickpocket 1 in mechanical nociception (Kim, Coste, Chadha, Cook, & Patapoutian, 2012).

In addition to its expression in DRG neurons, it has been reported that ASIC2 mRNA together with ASIC1 and ASIC3 mRNA is expressed in neurons of the nodose ganglion. Neurons from the nodose ganglion innervate, among other structures, the baroreceptors in the aortic arch. ASICs indeed localize to the baroreceptor nerve endings (Lu et al., 2009). ASIC2 knockout mice develop increased blood pressure and heart rate, a reduced gain of the baroreceptor reflex, and a reduced frequency of engagement of the baroreceptor reflex (Lu et al., 2009). As the transmission of the baroreflex is not impaired in these mice, the deficits are most likely due to an impaired peripheral sensitivity of the baroreceptor reflex (Lu et al., 2009). Although these results are compatible with ASIC2 as a pressure-sensing molecule, substantial residual baroreflex was still observed in ASIC2 knockout mice (Lu et al., 2009; McCleskey, 2009). Because it has recently been shown that mice with a knockout of Piezo1 and Piezo2 in nodose and petrosal ganglia have a failure of the baroreflex (Zeng et al., 2018), the function of ASIC2 in these neurons may be to amplify mechanosensitive currents, thereby increasing the sensitivity of baroreceptors. The details of this mechanism are unclear, however.

ASIC3

In addition to small-diameter afferent neurons, ~50% of large-diameter afferent neurons, possibly low-threshold mechanoreceptors, express ASIC3 (S. H. Lin et al., 2016; Molliver et al., 2005; Price et al., 2001). Similar to ASIC2, ASIC3 immunoreactivity was found in a variety of dedicated mechanosensory terminals, including Meissner corpuscles, lanceolate nerve endings of hairy skin, and Merkel discs (Price et al., 2001). In this study, the specificity of the immunostaining was verified with tissue from ASIC3 knockout mice (Price et al., 2001). Moreover, it was indeed found that the sensitivity of rapidly adapting mechanoreceptors was increased in ASIC3 knockout mice, while the sensitivity of Aδ mechanonociceptors was significantly reduced (Price et al., 2001), pointing to a role of ASIC3 in mechanosensation in cutaneous sensory neurons.

This idea received fresh input from a study on proprioceptors. A small group of large-diameter sensory neurons innervate muscle and mediate proprioception. Parvalbumin (Pv), a small Ca^{2+}-binding protein, is a commonly used molecular marker for muscle proprioceptors. Although, one study found no co-expression of ASIC3 with Pv in rat DRGs (Molliver et al., 2005), a more recent study using mice with a green fluorescent protein (GFP) reporter under the control of the *Accn3* reporter found that the majority of Pv^+ neurons express ASIC3 (S. H. Lin et al., 2016). Moreover, GFP localized to muscle spindles and Golgi tendon organs (S. H. Lin et al., 2016), suggesting a role in muscle proprioceptors. When Pv^+ neurons were grown on a fibronectin-coated elastic substrate, their response to stretching of the neurite via substrate indentation was blocked by amiloride and APETx1. Moreover, Pv^+ neurons with a genetic knockout of ASIC3 failed to respond to stretching of their neurites (S. H. Lin et al., 2016), strongly suggesting that ASIC3 is crucial for low-threshold mechanosensitivity by neurite stretching in Pv^+ neurons. In contrast to the dramatic effects of ASIC3 knockout on eliciting action potentials via neurite stretching in vitro, ASIC3 deletion did not abolish stretch-evoked responses of intact muscle proprioceptors (S. H. Lin et al., 2016), highlighting important differences when investigating mechanosensation in isolated neurons and in neurons in their natural environment. The dynamic responses of the proprioceptors were, however, affected, and ASIC3 knockout mice showed deficits in their walking behavior when visual cues were eliminated (S. H. Lin et al., 2016), confirming a contribution of ASIC3 to proprioception. The subtle defects in intact muscle recall the subtle defects in cutaneous mechanoreceptors with a defect in ASIC3.

In summary, it appears that ASIC3 has a specific and important role in the mechanosensitivity of some low-threshold mechanoreceptors. As for ASIC2, the precise role, however, remains elusive. Considering that Piezo2 mediates the mechanosensitivity of most low-threshold mechanoreceptors in DRGs (Ranade et al., 2014) and in the absence of more direct evidence, at present it is safer to conclude that ASIC3 is not the pore-forming unit of a mechanosensor, and the mechanistic details of its function in low-threshold mechanoreceptors still need to be clearly elaborated.

Similar to expression in baroreceptors, it has been reported that ASIC1 and ASIC3 mRNAs are expressed in rat carotid body and that responses to acid-evoked currents in glomus cells of the rat carotid body are inhibited by amiloride, potentially implicating ASIC3 (and perhaps ASIC1) together with TWIK-related acid-sensitive K^+ channels in proton sensing in chemoreceptive glomus cells (Tan et al., 2007).

Expression of ASICs in Humans

The predominant ASIC transcript in the human brain is ASIC1, being ~7-fold higher than that of ASIC3 or ASIC2 (Delaunay et al., 2012). Like in rodents, ASIC1b mRNA is not expressed in the human brain (Hoagland et al., 2010). In contrast, human ASIC3 mRNA is expressed in the brain to a similar extent as in DRGs. In fact, ASIC3 mRNA abundance in the human brain is as high as that of ASIC2a or ASIC2b (Delaunay et al., 2012). Endogenous human ASICs have been functionally characterized in cortical neurons obtained from patients undergoing craniotomy for removal of a brain tumor, revealing that ASICs in these neurons were predominantly homomeric ASIC1a, consistent with a reduced expression of ASIC2a in the human brain compared to rodents (M. Li, Inoue, et al., 2010). A study using cortical neurons freshly resected from patients undergoing surgery for intractable epilepsy, however, found a similar ratio of ASIC1a/ASIC2a protein expression as in mice but a reduced expression of ASIC2b protein. But human ASIC1a was more efficiently expressed on the surface than mouse ASIC1a (Xu et al., 2018). Apart from immortalized cell lines, like human embryonic kidney cells (Gunthorpe et al., 2001), human ASICs have also been characterized in cancer stem cell lines originating from glioblastoma multiforme (Tian et al., 2017). In these heterogenous cell lines, the presence of functional homomeric ASIC1a and, to a lesser extent, of ASIC3-containing channels was demonstrated; ASIC2 was downregulated and did not contribute to ASICs in these cell lines (Tian et al., 2017).

In contrast to rodents, ASIC2a and ASIC2b mRNA seem to be only faintly expressed in human DRGs as assessed by quantitative polymerase chain reaction (Delaunay et al., 2012). Transcriptomic analysis of human DRGs confirmed that mRNAs for ASIC1 and ASIC3 are equally expressed, that ASIC2 expression is low, and that ASIC4 and ASIC5 are virtually absent (North et al., 2019). Subpopulations of neurons in the DRG might still express ASIC2 mRNA, however. ASICs have been functionally characterized in sensory neurons of a nociceptor phenotype that were derived from human pluripotent stem cells (Young et al., 2014). The differentiated sensory cells express mRNAs of ASIC1, ASIC2, and, at low levels, ASIC3; functional characterization revealed the expression of heteromeric ASICs containing ASIC1 (Young et al., 2014).

Thus, overall it appears that there are some differences of ASIC expression in humans compared to rodents, with an even more predominant expression of ASIC1a and ASIC3 in humans. In humans, ASIC3 mRNA has also been found in non-neuronal tissue like lung, kidney, and testis (Babinski, Le, & Seguela, 1999; Ishibashi & Marumo, 1998).

In heterologous expression systems, SSDs of human ASIC1a and ASIC1b are acid-shifted by 0.2 pH units compared to their mouse orthologs (Hoagland et al., 2010; Sherwood & Askwith, 2008). In addition, human ASIC1b is permeable to Ca^{2+} ($P_{Na}/P_{Ca} = 2.4$) (Hoagland et al., 2010).

ASICs and Nociception in Humans

In psychophysical experiments, infusion of an acidic solution in the upper forearm of humans elicits pain, and this pain is potently blocked by amiloride but not by capsazepine, an inhibitor of the pain sensor TRPV1 (Ugawa et al., 2002). In agreement, delivery of protons using a noninvasive method, transcutaneous iontophoresis, also induces pricking pain in human forearm. This acid-induced pain is reduced not by desensitization of TRPV1 but by application of amiloride (Jones et al., 2004). Although many details are unknown, these studies suggest that ASICs are major acid sensors in human skin.

Pathophysiological Roles of ASICs

ASICs also have a role in different pathophysiological situations, which have in common that they are accompanied by a persistent acidosis. This acidosis can occur during ischemia, during chronic inflammation, or during pathological pain.

Ischemic Stroke

Ischemic stroke is caused by the occlusion of a blood vessel and reduced blood flow, leading to hypoxia, anaerobic metabolism, and acidosis. During ischemic stroke, tissue pH can fall to values close to 6.5 (Pignataro et al., 2007). A strongly decreased blood supply eventually leads to tissue death (infarction). Neurons surrounding the core region of infarction, the ischemic penumbra, are ischemic but can be saved from cell death. A pioneering study by Xiong and colleagues (2004) provided the first evidence, which has been reproduced in numerous studies, that acidosis induces neuronal death via ASIC1a. First, in isolated cortical neurons, 1 hr of oxygen glucose deprivation, an in vitro model of ischemia, increased ASIC current amplitude and slowed desensitization (Xiong et al., 2004). Second, acidosis (pH 6.0) induced cell death in neurons, which was significantly reduced by amiloride and PcTx1; neurons from ASIC1 knockout mice were protected from acidosis-induced cell death (Xiong et al., 2004). Transient occlusion of the middle cerebral artery (MCA) is a model for transient focal ischemia. Injection into the cerebrospinal fluid of amiloride or PcTx1 significantly reduced the infarct volume in this model; similarly, ASIC1a$^{-/-}$ mice also had a significantly reduced infarct volume (Xiong et al., 2004). These results provided compelling evidence that acidosis leads to neurodegeneration via ASIC1a. It was later shown that the therapeutic time window is as long as 5 hr after 1 hr of MCA occlusion and that PcTx1 can also be administered through the nasal cavity to provide a neuroprotective effect (Pignataro et al., 2007). The toxin Hi1a, an ASIC1a inhibitor with higher potency than PcTx1, affords neuroprotection in a rat MCA occlusion model even as long as 8 hr after stroke and not only in the penumbra but strikingly also in the core of the infarct area (Chassagnon et al., 2017). These results clearly show that ASIC1a is a very promising drug target for the treatment of stroke.

But how does activation of ASIC1a trigger cell death? The initial hypothesis was that activation of ASIC1a leads to Ca^{2+} influx, which then triggers cell death (Xiong et al., 2004). But the acidosis due to ischemia develops slowly and should completely desensitize ASIC1a. Although the biophysical properties of ASIC1a may change during acidosis, leading to enhanced currents and slower desensitization (Gao et al., 2005; Xiong et al., 2004), it has so far not been convincingly shown that ASIC1a can sustain substantial Ca^{2+} influx during ischemia. In this regard, a very attractive alternative explanation has recently been proposed. ASIC1a has been found to specifically interact via its cytoplasmic C terminus with receptor-interacting protein kinase 1 (RIPK1) (Y. Z. Wang et al., 2015). Although the regulation of RIPK1 is complex (Degterev, Ofengeim, & Yuan, 2019; Delanghe, Dondelinger, & Bertrand, 2020; Green, 2019), it can recruit RIPK3 to activate the effector protein mixed lineage kinase domain-like, which then forms pores in the membrane, leading to proinflammatory necroptotic cell death (H. Wang et al., 2014). Strikingly, the interaction of ASIC1a with RIPK1 is independent of ion fluxes and identifies a completely new function for ASIC1a: as a metabotropic receptor initiating necroptotic cell death (Y. Z. Wang et al., 2015). The ASIC1a–RIPK1 interaction would be triggered by slow intracellular conformational changes that follow a different trajectory from the conformational changes associated with gating of the ion pore (J. J. Wang et al., 2020). Although many details remain to be determined, accumulating evidence suggests the involvement of necroptotic cell death in stroke (Y. Chen et al., 2018; Degterev et al., 2005; Naito et al., 2020; Ni et al., 2018), and a possible role for ASIC1a in this pathway offers a fresh perspective to our understanding of acid-induced neuronal death during stroke.

Multiple Sclerosis

Multiple sclerosis (MS) is characterized by demyelination in the CNS, by inflammation, and by neuronal loss. In MS lesions, axons are damaged, and axonal degeneration is considered to be the major cause of neurological deficits. In an experimental autoimmune encephalomyelitis (EAE) mouse model of MS, induced by immunization with a peptide from myelin oligodendrocyte glycoprotein, pH in the spinal cord is substantially lower than in control mice (6.6 vs. 7.4), and clinical disease severity is significantly lower in ASIC1 knockout than in wild-type mice. Moreover, amiloride has strong neuroprotective effects in this model (Friese et al., 2007), suggesting that acidosis contributes to axonal degeneration via activation of ASIC1a in the EAE mouse model of MS. ASIC1a mRNA is upregulated in the inflamed tissue (Friese et al., 2007) and is ectopically expressed on axons in the EAE model (Vergo et al., 2011). Also in postmortem tissue from human MS patients, ASIC1a is ectopically expressed in axons within inflamed lesions (Arun et al., 2013; Vergo et al., 2011). Interestingly, it has recently been shown that necroptosis contributes to axonal degeneration (Arrazola et al., 2019), suggesting that necroptosis should be explored as one mechanism of ASIC1a-mediated axonal degeneration. Excitingly, a translational study indeed found evidence that amiloride has neuroprotective effects in primary progressive MS (Arun et al., 2013). However, in a phase 2 randomized controlled trial, amiloride had no protective effects in acute optic neuritis, which shares pathophysiological features with MS (McKee et al., 2019). Although discouraging, it was noted that amiloride might have been applied too late in these patients (McKee et al., 2019). Thus, it remains to be seen whether targeting ASIC1a has a clinical benefit in MS.

The *Accn1* gene, coding for ASIC2, has been genetically linked to MS in the relatively isolated population of Nuoro on Sardinia (Bernardinelli et al., 2007). However, knockout of the ASIC2 gene has only a mild benefit in the EAE model compared with knockout of ASIC1a (Fazia et al., 2019), suggesting that in the pathophysiology of MS ASIC2a may have a modulatory role and ASIC1a the more prominent role.

Pain

There are several reviews on the role of ASICs in pain (Deval & Lingueglia, 2015; Dussor, 2015; W. G. Li & Xu, 2015; Sluka & Gregory, 2015; W. H. Sun & Chen, 2016), and therefore, only a few illustrative cases will be briefly summarized here.

ASIC1 and Neuropathic Pain

Damage or dysfunction of nociceptive pathways can produce neuropathic pain, which is often severe and typically resistant to standard treatments for pain. Chronic constriction of the sciatic nerve is one model for neuropathic pain. Spinal application of PcTx1 has analgesic effects in this model, implicating ASIC1a in the generation of neuropathic pain (Mazzuca et al., 2007). In agreement, mambalgins have similar effects, and experimental evidence suggests the contribution of peripheral ASIC1b and spinal ASIC1a (Diochot et al., 2016).

ASIC1/ASIC3 and Migraine

Migraine is the second most common cause of headache. The pain associated with migraine requires activation of nociceptors of the TG innervating the cranial meninges. ASIC3 is co-expressed with CGRP in rat TGs (Ichikawa & Sugimoto, 2002). CGRP is a marker for small-diameter peptidergic neurons, and CGRP released by the peripheral nerve endings contributes to migraine. In rats, 80% of dural afferents respond to acid stimuli with huge amiloride-sensitive ASIC-like currents, which have properties of ASIC3-like currents (Yan et al., 2011; Yan, Wei, Bischoff, Edelmayer, & Dussor, 2013). Moreover, ASIC3 blockade with APETx2 reduces trigeminal nociception in rats and cutaneous allodynia in mice (Holton et al., 2020). On the other hand, mambalgin-1 is effective at reducing mechanical allodynia in a rat model of migraine (Verkest et al., 2018), clearly implicating ASIC1 in the allodynia associated with migraine. Interestingly, ASIC1a is not needed for the anti-allodynic effects of mambalgin-1 (Verkest et al., 2018), strongly suggesting that this effect is mediated by peripheral ASIC1b. It is also noteworthy that ASIC currents are potentiated by nitric oxide (Cadiou et al., 2007), which triggers migraine in humans. While this potentiation is readily reversible for other ASICs, it is persistent for ASIC1b (Cadiou et al., 2007).

ASIC3 and Chronic Muscle Pain

A paradigm in which two injections of acidic saline into one muscle, 5 days apart, induce persistent hyperalgesia in rodents was established as a model of noninflammatory chronic muscle pain (Sluka, Kalra, & Moore, 2001). The first acidic stimulus induces acute pain and primes the nociceptor, while the second pain stimulus induces long-lasting hyperalgesia (W. N. Chen et al., 2014; Sluka et al., 2001). The hyperalgesia is associated with sensitization of central neurons in the dorsal horn (secondary hyperalgesia), and the development, but not the maintenance, of the chronic hyperalgesia in this model critically depends on ASIC3 but not ASIC1a (W. N. Chen et al., 2014; Sluka et al., 2003). These results underline the importance of ASIC3 as an acid sensor in muscle and suggest that it might be a target to prevent the transition from acute to chronic muscle pain (Sluka & Gregory, 2015; W. H. Sun & Chen, 2016). Surprisingly, recent results also implicate ASIC1b in the acute pain to the first acid injection (Chang et al., 2019). It has been proposed that an ASIC1b/ASIC3 heteromer might be involved, but the details are not clear yet (Chang et al., 2019).

Conclusion

Recent years have seen a tremendous increase in our knowledge of ASICs. We now know that they are fascinating and extremely sensitive receptors of extracellular protons. While different subtypes appear to be ideally adapted to their specific functions, we still have to learn a lot about the details of subtype-specific expression patterns and functions. Whenever the extracellular proton concentration in the microenvironment of neurons rises, be it in physiological situations like synaptic transmission or slight muscle ischemia or in pathophysiological situations such as severe ischemia associated with stroke or pathological pain, ASICs are called into action—sometimes with deadly consequences for the affected cells. The numerous roles they serve in our body promise further exciting discoveries in the future.

Acknowledgments

I thank L. Leisle and D. Wiemuth for many helpful comments on the manuscript and all the present and former members of my lab for many helpful discussions on ASICs. Work in my lab was supported by the Deutsche Forschungsgemeinschaft (grants GR1771/3 and GR1771/8-1).

References

Adams, C. M., Snyder, P. M., & Welsh, M. J. (1999). Paradoxical stimulation of a DEG/ENaC channel by amiloride. *Journal of Biological Chemistry*, *274*(22), 15500–15504.

Akopian, A. N., Chen, C. C., Ding, Y., Cesare, P., & Wood, J. N. (2000). A new member of the acid-sensing ion channel family. *Neuroreport*, *11*(10), 2217–2222.

Alijevic, O., & Kellenberger, S. (2012). Subtype-specific modulation of acid-sensing ion channel (ASIC) function by 2-guanidine-4-methylquinazoline. *Journal of Biological Chemistry*, *287*(43), 36059–36070. doi:10.1074/jbc.M112.360487

Allen, N. J., & Attwell, D. (2002). Modulation of ASIC channels in rat cerebellar Purkinje neurons by ischaemia-related signals. *Journal of Physiology*, *543*(Pt. 2), 521–529.

Alvarez De La Rosa, D., Krueger, S. R., Kolar, A., Shao, D., Fitzsimonds, R. M., & Canessa, C. M. (2003). Distribution, subcellular localization and ontogeny of ASIC1 in the mammalian central nervous system. *Journal of Physiology*, *546*(Pt. 1), 77–87.

Arrazola, M. S., Saquel, C., Catalan, R. J., Barrientos, S. A., Hernandez, D. E., Martinez, N. W., . . . Court, F. A. (2019). Axonal degeneration is mediated by necroptosis activation. *Journal of Neuroscience*, *39*(20), 3832–3844. doi:10.1523/JNEUROSCI.0881-18.2019

Arun, T., Tomassini, V., Sbardella, E., de Ruiter, M. B., Matthews, L., Leite, M. I., . . . Palace, J. (2013). Targeting ASIC1 in primary progressive multiple sclerosis: Evidence of neuroprotection with amiloride. *Brain*, *136*(Pt. 1), 106–115. doi:10.1093/brain/aws325

Askwith, C. C., Cheng, C., Ikuma, M., Benson, C., Price, M. P., & Welsh, M. J. (2000). Neuropeptide FF and FMRFamide potentiate acid-evoked currents from sensory neurons and proton-gated DEG/ENaC channels. *Neuron*, *26*(1), 133–141.

Askwith, C. C., Wemmie, J. A., Price, M. P., Rokhlina, T., & Welsh, M. J. (2004). Acid-sensing ion channel 2 (ASIC2) modulates ASIC1 H^+-activated currents in hippocampal neurons. *Journal of Biological Chemistry*, *279*(18), 18296–18305.

Assmann, M., Kuhn, A., Dürrnagel, S., Holstein, T. W., & Gründer, S. (2014). The comprehensive analysis of DEG/ENaC subunits in *Hydra* reveals a large variety of peptide-gated channels, potentially involved in neuromuscular transmission. *BMC Biology*, *12*, 84. doi:10.1186/s12915-014-0084-2

Babini, E., Paukert, M., Geisler, H. S., & Gründer, S. (2002). Alternative splicing and interaction with di- and polyvalent cations control the dynamic range of acid-sensing ion channel 1 (ASIC1). *Journal of Biological Chemistry*, *277*(44), 41597–41603.

Babinski, K., Catarsi, S., Biagini, G., & Séguéla, P. (2000). Mammalian ASIC2a and ASIC3 subunits co-assemble into heteromeric proton-gated channels sensitive to $Gd3^+$. *Journal of Biological Chemistry*, *275*(37), 28519–28525.

Babinski, K., Le, K. T., & Seguela, P. (1999). Molecular cloning and regional distribution of a human proton receptor subunit with biphasic functional properties. *Journal of Neurochemistry*, *72*(1), 51–57.

Baconguis, I., Bohlen, C. J., Goehring, A., Julius, D., & Gouaux, E. (2014). X-ray structure of acid-sensing ion channel 1–snake toxin complex reveals open state of a Na^+-selective channel. *Cell, 156*(4), 717–729. doi:10.1016/j.cell.2014.01.011

Baconguis, I., & Gouaux, E. (2012). Structural plasticity and dynamic selectivity of acid-sensing ion channel–spider toxin complexes. *Nature, 489*(7416), 400–405. doi:10.1038/nature11375

Baron, A., & Lingueglia, E. (2015). Pharmacology of acid-sensing ion channels—Physiological and therapeutical perspectives. *Neuropharmacology, 94*, 19–35. doi:10.1016/j.neuropharm.2015.01.005

Baron, A., Schaefer, L., Lingueglia, E., Champigny, G., & Lazdunski, M. (2001). Zn^{2+} and H^+ are coactivators of acid-sensing ion channels. *Journal of Biological Chemistry, 276*(38), 35361–35367.

Baron, A., Voilley, N., Lazdunski, M., & Lingueglia, E. (2008). Acid sensing ion channels in dorsal spinal cord neurons. *Journal of Neuroscience, 28*(6), 1498–1508.

Baron, A., Waldmann, R., & Lazdunski, M. (2002). ASIC-like, proton-activated currents in rat hippocampal neurons. *Journal of Physiology, 539*(Pt. 2), 485–494.

Bartoi, T., Augustinowski, K., Polleichtner, G., Gründer, S., & Ulbrich, M. H. (2014). Acid-sensing ion channel (ASIC) 1a/2a heteromers have a flexible 2:1/1:2 stoichiometry. *Proceedings of the National Academy of Sciences of the United States of America, 111*, 8281–8286. doi:10.1073/pnas.1324060111

Bassilana, F., Champigny, G., Waldmann, R., de Weille, J. R., Heurteaux, C., & Lazdunski, M. (1997). The acid-sensitive ionic channel subunit ASIC and the mammalian degenerin MDEG form a heteromultimeric H^+-gated Na^+ channel with novel properties. *Journal of Biological Chemistry, 272*(46), 28819–28822.

Bässler, E. L., Ngo-Anh, T. J., Geisler, H. S., Ruppersberg, J. P., & Gründer, S. (2001). Molecular and functional characterization of acid-sensing ion channel (ASIC) 1b. *Journal of Biological Chemistry, 276*, 33782–33787.

Benson, C. J., Eckert, S. P., & McCleskey, E. W. (1999). Acid-evoked currents in cardiac sensory neurons: A possible mediator of myocardial ischemic sensation. *Circulation Research, 84*(8), 921–928.

Benson, C. J., Xie, J., Wemmie, J. A., Price, M. P., Henss, J. M., Welsh, M. J., & Snyder, P. M. (2002). Heteromultimers of DEG/ENaC subunits form H^+-gated channels in mouse sensory neurons. *Proceedings of the National Academy of Sciences of the United States of America, 99*(4), 2338–2343.

Bernardinelli, L., Murgia, S. B., Bitti, P. P., Foco, L., Ferrai, R., Musu, L., ... Berzuini, C. (2007). Association between the ACCN1 gene and multiple sclerosis in central east Sardinia. *PLoS One, 2*(5), e480. doi:10.1371/journal.pone.0000480

Birdsong, W. T., Fierro, L., Williams, F. G., Spelta, V., Naves, L. A., Knowles, M., ... McCleskey, E. W. (2010). Sensing muscle ischemia: Coincident detection of acid and ATP via interplay of two ion channels. *Neuron, 68*(4), 739–749. doi:10.1016/j.neuron.2010.09.029

Blanchard, M. G., Rash, L. D., & Kellenberger, S. (2012). Inhibition of voltage-gated Na^+ currents in sensory neurones by the sea anemone toxin APETx2. *British Journal of Pharmacology, 165*(7), 2167–2177. doi:10.1111/j.1476-5381.2011.01674.x

Bohlen, C. J., Chesler, A. T., Sharif-Naeini, R., Medzihradszky, K. F., Zhou, S., King, D., ... Julius, D. (2011). A heteromeric Texas coral snake toxin targets acid-sensing ion channels to produce pain. *Nature, 479*(7373), 410–414. doi:10.1038/nature10607

Borg, C. B., Braun, N., Heusser, S. A., Bay, Y., Weis, D., Galleano, I., ... Pless, S. A. (2020). Mechanism and site of action of big dynorphin on ASIC1a. *Proceedings of the National*

Academy of Sciences of the United States of America, *117*(13), 7447–7454. doi:10.1073/pnas.1919323117

Buta, A., Maximyuk, O., Kovalskyy, D., Sukach, V., Vovk, M., Ievglevskyi, O., ... Krishtal, O. (2015). Novel potent orthosteric antagonist of ASIC1a prevents NMDAR-dependent LTP induction. *Journal of Medicinal Chemistry*, *58*(11), 4449–4461. doi:10.1021/jm5017329

Cadiou, H., Studer, M., Jones, N. G., Smith, E. S., Ballard, A., McMahon, S. B., & McNaughton, P. A. (2007). Modulation of acid-sensing ion channel activity by nitric oxide. *Journal of Neuroscience*, *27*(48), 13251–13260. doi:10.1523/JNEUROSCI.2135-07.2007

Canessa, C. M., Schild, L., Buell, G., Thorens, B., Gautschi, I., Horisberger, J. D., & Rossier, B. C. (1994). Amiloride-sensitive epithelial Na^+ channel is made of three homologous subunits. *Nature*, *367*(6462), 463–467.

Chang, C. T., Fong, S. W., Lee, C. H., Chuang, Y. C., Lin, S. H., & Chen, C. C. (2019). Involvement of acid-sensing ion channel 1b in the development of acid-induced chronic muscle pain. *Frontiers in Neuroscience*, *13*, 1247. doi:10.3389/fnins.2019.01247

Chassagnon, I. R., McCarthy, C. A., Chin, Y. K., Pineda, S. S., Keramidas, A., Mobli, M., ... King, G. F. (2017). Potent neuroprotection after stroke afforded by a double-knot spider-venom peptide that inhibits acid-sensing ion channel 1a. *Proceedings of the National Academy of Sciences of the United States of America*, *114*(14), 3750–3755. doi:10.1073/pnas.1614728114

Chen, C. C., England, S., Akopian, A. N., & Wood, J. N. (1998). A sensory neuron-specific, proton-gated ion channel. *Proceedings of the National Academy of Sciences of the United States of America*, *95*(17), 10240–10245.

Chen, C. C., Zimmer, A., Sun, W. H., Hall, J., & Brownstein, M. J. (2002). A role for ASIC3 in the modulation of high-intensity pain stimuli. *Proceedings of the National Academy of Sciences of the United States of America*, *99*(13), 8992–8997.

Chen, C. H., Hsu, Y. T., Chen, C. C., & Huang, R. C. (2009). Acid-sensing ion channels in neurones of the rat suprachiasmatic nucleus. *Journal of Physiology*, *587*(Pt. 8), 1727–1737. doi:10.1113/jphysiol.2008.166918

Chen, W. N., Lee, C. H., Lin, S. H., Wong, C. W., Sun, W. H., Wood, J. N., & Chen, C. C. (2014). Roles of ASIC3, TRPV1, and NaV1.8 in the transition from acute to chronic pain in a mouse model of fibromyalgia. *Molecular Pain*, *10*, 40. doi:10.1186/1744-8069-10-40

Chen, X., Kalbacher, H., & Gründer, S. (2005). The tarantula toxin psalmotoxin 1 inhibits acid-sensing ion channel (ASIC) 1a by increasing its apparent H^+ affinity. *Journal of General Physiology*, *126*(1), 71–79.

Chen, X., Kalbacher, H., & Gründer, S. (2006). Interaction of acid-sensing ion channel (ASIC) 1 with the tarantula toxin psalmotoxin 1 is state dependent. *Journal of General Physiology*, *127*(3), 267–276.

Chen, X., Orser, B. A., & MacDonald, J. F. (2010). Design and screening of ASIC inhibitors based on aromatic diamidines for combating neurological disorders. *European Journal of Pharmacology*, *648*(1–3), 15–23. doi:10.1016/j.ejphar.2010.09.005

Chen, X., Paukert, M., Kadurin, I., Pusch, M., & Gründer, S. (2006). Strong modulation by RFamide neuropeptides of the ASIC1b/3 heteromer in competition with extracellular calcium. *Neuropharmacology*, *50*(8), 964–974.

Chen, X., Qiu, L., Li, M., Dürrnagel, S., Orser, B. A., Xiong, Z. G., & MacDonald, J. F. (2010). Diarylamidines: High potency inhibitors of acid-sensing ion channels. *Neuropharmacology*, *58*(7), 1045–1053.

Chen, X., Whissell, P., Orser, B. A., & MacDonald, J. F. (2011). Functional modifications of acid-sensing ion channels by ligand-gated chloride channels. *PLoS One*, *6*(7), e21970. doi:10.1371/journal.pone.0021970

Chen, Y., Zhang, L., Yu, H., Song, K., Shi, J., Chen, L., & Cheng, J. (2018). Necrostatin-1 improves long-term functional recovery through protecting oligodendrocyte precursor cells after transient focal cerebral ischemia in mice. *Neuroscience*, *371*, 229–241. doi:10.1016/j.neuroscience.2017.12.007

Cheng, C. F., Chen, I. L., Cheng, M. H., Lian, W. S., Lin, C. C., Kuo, T. B., & Chen, C. C. (2011). Acid-sensing ion channel 3, but not capsaicin receptor TRPV1, plays a protective role in isoproterenol-induced myocardial ischemia in mice. *Circulation Journal*, *75*(1), 174–178. doi:10.1253/circj.cj-10-0490

Cho, J. H., & Askwith, C. C. (2008). Presynaptic release probability is increased in hippocampal neurons from ASIC1 knockout mice. *Journal of Neurophysiology*, *99*(2), 426–441. doi:10.1152/jn.00940.2007

Chu, X. P., Wemmie, J. A., Wang, W. Z., Zhu, X. M., Saugstad, J. A., Price, M. P., . . . Xiong, Z. G. (2004). Subunit-dependent high-affinity zinc inhibition of acid-sensing ion channels. *Journal of Neuroscience*, *24*(40), 8678–8689.

Coryell, M. W., Wunsch, A. M., Haenfler, J. M., Allen, J. E., Schnizler, M., Ziemann, A. E., . . . Wemmie, J. A. (2009). Acid-sensing ion channel-1a in the amygdala, a novel therapeutic target in depression-related behavior. *Journal of Neuroscience*, *29*(17), 5381–5388.

Coscoy, S., de Weille, J. R., Lingueglia, E., & Lazdunski, M. (1999). The pre-transmembrane 1 domain of acid-sensing ion channels participates in the ion pore. *Journal of Biological Chemistry*, *274*(15), 10129–10132.

Dawson, R. J., Benz, J., Stohler, P., Tetaz, T., Joseph, C., Huber, S., . . . Ruf, A. (2012). Structure of the acid-sensing ion channel 1 in complex with the gating modifier psalmotoxin 1. *Nature Communications*, *3*, 936. doi:10.1038/ncomms1917

Degterev, A., Huang, Z., Boyce, M., Li, Y., Jagtap, P., Mizushima, N., . . . Yuan, J. (2005). Chemical inhibitor of nonapoptotic cell death with therapeutic potential for ischemic brain injury. *Nature Chemical Biology*, *1*(2), 112–119. doi:10.1038/nchembio711

Degterev, A., Ofengeim, D., & Yuan, J. (2019). Targeting RIPK1 for the treatment of human diseases. *Proceedings of the National Academy of Sciences of the United States of America*, *116*(20), 9714–9722. doi:10.1073/pnas.1901179116

Delanghe, T., Dondelinger, Y., & Bertrand, M. J. M. (2020). RIPK1 kinase-dependent death: A symphony of phosphorylation events. *Trends in Cell Biology*, *30*(3), 189–200. doi:10.1016/j.tcb.2019.12.009

Delaunay, A., Gasull, X., Salinas, M., Noel, J., Friend, V., Lingueglia, E., & Deval, E. (2012). Human ASIC3 channel dynamically adapts its activity to sense the extracellular pH in both acidic and alkaline directions. *Proceedings of the National Academy of Sciences of the United States of America*, *109*(32), 13124–13129. doi:10.1073/pnas.1120350109

Deval, E., Baron, A., Lingueglia, E., Mazarguil, H., Zajac, J. M., & Lazdunski, M. (2003). Effects of neuropeptide SF and related peptides on acid sensing ion channel 3 and sensory neuron excitability. *Neuropharmacology*, *44*(5), 662–671.

Deval, E., & Lingueglia, E. (2015). Acid-sensing ion channels and nociception in the peripheral and central nervous systems. *Neuropharmacology*, *94*, 49–57. doi:10.1016/j.neuropharm.2015.02.009

Deval, E., Noel, J., Gasull, X., Delaunay, A., Alloui, A., Friend, V., . . . Lingueglia, E. (2011). Acid-sensing ion channels in postoperative pain. *Journal of Neuroscience*, *31*(16), 6059–6066. doi:10.1523/JNEUROSCI.5266-10.2011

Deval, E., Noel, J., Lay, N., Alloui, A., Diochot, S., Friend, V., ... Lingueglia, E. (2008). ASIC3, a sensor of acidic and primary inflammatory pain. *EMBO Journal, 27*(22), 3047–3055.

DeVries, S. H. (2001). Exocytosed protons feedback to suppress the Ca^{2+} current in mammalian cone photoreceptors. *Neuron, 32*(6), 1107–1117. doi:S0896-6273(01)00535-9

Dietrich, C. J., & Morad, M. (2010). Synaptic acidification enhances GABAA signaling. *Journal of Neuroscience, 30*(47), 16044–16052. doi:10.1523/JNEUROSCI.6364-09.2010

Diochot, S., Alloui, A., Rodrigues, P., Dauvois, M., Friend, V., Aissouni, Y., ... Baron, A. (2016). Analgesic effects of mambalgin peptide inhibitors of acid-sensing ion channels in inflammatory and neuropathic pain. *Pain, 157*(3), 552–559. doi:10.1097/j.pain.0000000000000397

Diochot, S., Baron, A., Rash, L. D., Deval, E., Escoubas, P., Scarzello, S., ... Lazdunski, M. (2004). A new sea anemone peptide, APETx2, inhibits ASIC3, a major acid-sensitive channel in sensory neurons. *EMBO Journal, 23*(7), 1516–1525.

Diochot, S., Baron, A., Salinas, M., Douguet, D., Scarzello, S., Dabert-Gay, A. S., ... Lingueglia, E. (2012). Black mamba venom peptides target acid-sensing ion channels to abolish pain. *Nature, 490*(7421), 552–555. doi:10.1038/nature11494

Donier, E., Rugiero, F., Jacob, C., & Wood, J. N. (2008). Regulation of ASIC activity by ASIC4—New insights into ASIC channel function revealed by a yeast two-hybrid assay. *European Journal of Neuroscience, 28*(1), 74–86. doi:10.1111/j.1460-9568.2008.06282.x

Drew, L. J., Rohrer, D. K., Price, M. P., Blaver, K. E., Cockayne, D. A., Cesare, P., & Wood, J. N. (2004). Acid-sensing ion channels ASIC2 and ASIC3 do not contribute to mechanically activated currents in mammalian sensory neurones. *Journal of Physiology, 556*(Pt. 3), 691–710.

Driscoll, M., & Chalfie, M. (1991). The mec-4 gene is a member of a family of *Caenorhabditis elegans* genes that can mutate to induce neuronal degeneration. *Nature, 349*(6310), 588–593.

Du, J., Reznikov, L. R., Price, M. P., Zha, X. M., Lu, Y., Moninger, T. O., ... Welsh, M. J. (2014). Protons are a neurotransmitter that regulates synaptic plasticity in the lateral amygdala. *Proceedings of the National Academy of Sciences of the United States of America, 111*(24), 8961–8966. doi:10.1073/pnas.1407018111

Duan, B., Liu, D. S., Huang, Y., Zeng, W. Z., Wang, X., Yu, H., ... Xu, T. L. (2012). PI3-kinase/Akt pathway-regulated membrane insertion of acid-sensing ion channel 1a underlies BDNF-induced pain hypersensitivity. *Journal of Neuroscience, 32*(18), 6351–6363. doi:10.1523/JNEUROSCI.4479-11.2012

Duan, B., Wang, Y. Z., Yang, T., Chu, X. P., Yu, Y., Huang, Y., ... Xu, T. L. (2011). Extracellular spermine exacerbates ischemic neuronal injury through sensitization of ASIC1a channels to extracellular acidosis. *Journal of Neuroscience, 31*(6), 2101–2112. doi:10.1523/JNEUROSCI.4351-10.2011

Duan, B., Wu, L. J., Yu, Y. Q., Ding, Y., Jing, L., Xu, L., ... Xu, T. L. (2007). Upregulation of acid-sensing ion channel ASIC1a in spinal dorsal horn neurons contributes to inflammatory pain hypersensitivity. *Journal of Neuroscience, 27*(41), 11139–11148.

Dube, G. R., Lehto, S. G., Breese, N. M., Baker, S. J., Wang, X., Matulenko, M. A., ... Brioni, J. D. (2005). Electrophysiological and in vivo characterization of A-317567, a novel blocker of acid sensing ion channels. *Pain, 117*(1–2), 88–96.

Dussor, G. (2015). ASICs as therapeutic targets for migraine. *Neuropharmacology, 94*, 64–71. doi:10.1016/j.neuropharm.2014.12.015

Er, S. Y., Cristofori-Armstrong, B., Escoubas, P., & Rash, L. D. (2017). Discovery and molecular interaction studies of a highly stable, tarantula peptide modulator of acid-sensing ion channel 1. *Neuropharmacology, 127*, 185–195. doi:10.1016/j.neuropharm.2017.03.020

Escoubas, P., Bernard, C., Lambeau, G., Lazdunski, M., & Darbon, H. (2003). Recombinant production and solution structure of PcTx1, the specific peptide inhibitor of ASIC1a proton-gated cation channels. *Protein Science, 12*(7), 1332–1343.

Escoubas, P., De Weille, J. R., Lecoq, A., Diochot, S., Waldmann, R., Champigny, G., . . . Lazdunski, M. (2000). Isolation of a tarantula toxin specific for a class of proton-gated Na^+ channels. *Journal of Biological Chemistry, 275*(33), 25116–25121.

Fazia, T., Pastorino, R., Notartomaso, S., Busceti, C., Imbriglio, T., Cannella, M., . . . Bernardinelli, L. (2019). Acid sensing ion channel 2: A new potential player in the pathophysiology of multiple sclerosis. *European Journal of Neuroscience, 49*(10), 1233–1243. doi:10.1111/ejn.14302

Feldman, D. H., Horiuchi, M., Keachie, K., McCauley, E., Bannerman, P., Itoh, A., . . . Pleasure, D. (2008). Characterization of acid-sensing ion channel expression in oligodendrocyte-lineage cells. *Glia, 56*(11), 1238–1249.

Friese, M. A., Craner, M. J., Etzensperger, R., Vergo, S., Wemmie, J. A., Welsh, M. J., . . . Fugger, L. (2007). Acid-sensing ion channel-1 contributes to axonal degeneration in autoimmune inflammation of the central nervous system. *Nature Medicine, 13*(12), 1483–1489.

Fromy, B., Lingueglia, E., Sigaudo-Roussel, D., Saumet, J. L., & Lazdunski, M. (2012). Asic3 is a neuronal mechanosensor for pressure-induced vasodilation that protects against pressure ulcers. *Nature Medicine, 18*(8), 1205–1207. doi:10.1038/nm.2844

Fukuda, T., Ichikawa, H., Terayama, R., Yamaai, T., Kuboki, T., & Sugimoto, T. (2006). ASIC3-immunoreactive neurons in the rat vagal and glossopharyngeal sensory ganglia. *Brain Research, 1081*(1), 150–155. doi:10.1016/j.brainres.2006.01.039

Gao, J., Duan, B., Wang, D. G., Deng, X. H., Zhang, G. Y., Xu, L., & Xu, T. L. (2005). Coupling between NMDA receptor and acid-sensing ion channel contributes to ischemic neuronal death. *Neuron, 48*(4), 635–646. doi:10.1016/j.neuron.2005.10.011

García-Añoveros, J., Derfler, B., Neville-Golden, J., Hyman, B. T., & Corey, D. P. (1997). BNaC1 and BNaC2 constitute a new family of human neuronal sodium channels related to degenerins and epithelial sodium channels. *Proceedings of the National Academy of Sciences of the United States of America, 94*(4), 1459–1464.

García-Añoveros, J., Samad, T. A., Zuvela-Jelaska, L., Woolf, C. J., & Corey, D. P. (2001). Transport and localization of the DEG/ENaC ion channel BNaC1alpha to peripheral mechanosensory terminals of dorsal root ganglia neurons. *Journal of Neuroscience, 21*(8), 2678–2686.

Gautam, M., & Benson, C. J. (2013). Acid-sensing ion channels (ASICs) in mouse skeletal muscle afferents are heteromers composed of ASIC1a, ASIC2, and ASIC3 subunits. *FASEB Journal, 27*(2), 793–802. doi:10.1096/fj.12-220400

Geffeney, S. L., Cueva, J. G., Glauser, D. A., Doll, J. C., Lee, T. H., Montoya, M., . . . Goodman, M. B. (2011). DEG/ENaC but not TRP channels are the major mechanoelectrical transduction channels in a *C. elegans* nociceptor. *Neuron, 71*(5), 845–857. doi:10.1016/j.neuron.2011.06.038

Golubovic, A., Kuhn, A., Williamson, M., Kalbacher, H., Holstein, T. W., Grimmelikhuijzen, C. J., & Gründer, S. (2007). A peptide-gated ion channel from the freshwater polyp *Hydra*. *Journal of Biological Chemistry, 282*(48), 35098–35103.

Gonzales, E. B., Kawate, T., & Gouaux, E. (2009). Pore architecture and ion sites in acid-sensing ion channels and P2X receptors. *Nature, 460*(7255), 599–604. doi:10.1038/nature08218

Gonzalez-Inchauspe, C., Urbano, F. J., Di Guilmi, M. N., & Uchitel, O. D. (2017). Acid-sensing ion channels activated by evoked released protons modulate synaptic transmission at the mouse calyx of held synapse. *Journal of Neuroscience, 37*(10), 2589–2599. doi:10.1523/JNEUROSCI.2566-16.2017

Grantyn, R., Perouansky, M., Rodriguez-Tebar, A., & Lux, H. D. (1989). Expression of depolarizing voltage- and transmitter-activated currents in neuronal precursor cells from the rat brain is preceded by a proton-activated sodium current. *Brain Research Develomental Brain Research, 49*(1), 150–155.

Green, D. R. (2019). The coming decade of cell death research: Five riddles. *Cell, 177*(5), 1094–1107. doi:10.1016/j.cell.2019.04.024

Gregory, N. S., Whitley, P. E., & Sluka, K. A. (2015). Effect of intramuscular protons, lactate, and ATP on muscle hyperalgesia in rats. *PLoS One, 10*(9), e0138576. doi:10.1371/journal.pone.0138576

Gründer, S., & Augustinowski, K. (2012). Toxin binding reveals two open state structures for one acid-sensing ion channel. *Channels (Austin), 6*(6), 409–413. doi:10.4161/chan.22154

Gründer, S., & Chen, X. (2010). Structure, function, and pharmacology of acid-sensing ion channels (ASICs): Focus on ASIC1a. *International Journal of Physiology, Pathophysiology and Pharmacology, 2*(2), 73–94.

Gründer, S., Geissler, H. S., Bässler, E. L., & Ruppersberg, J. P. (2000). A new member of acid-sensing ion channels from pituitary gland. *Neuroreport, 11*(8), 1607–1611.

Gründer, S., & Pusch, M. (2015). Biophysical properties of acid-sensing ion channels (ASICs). *Neuropharmacology, 94*, 9–18. doi:10.1016/j.neuropharm.2014.12.016

Gunthorpe, M. J., Smith, G. D., Davis, J. B., & Randall, A. D. (2001). Characterisation of a human acid-sensing ion channel (hASIC1a) endogenously expressed in HEK293 cells. *Pflugers Archiv, 442*(5), 668–674.

Hattori, T., Chen, J., Harding, A. M., Price, M. P., Lu, Y., Abboud, F. M., & Benson, C. J. (2009). ASIC2a and ASIC3 heteromultimerize to form pH-sensitive channels in mouse cardiac dorsal root ganglia neurons. *Circulation Research, 105*(3), 279–286. doi:10.1161/CIRCRESAHA.109.202036

Hayes, S. G., Kindig, A. E., & Kaufman, M. P. (2007). Blockade of acid sensing ion channels attenuates the exercise pressor reflex in cats. *Journal of Physiology, 581*(Pt. 3), 1271–1282. doi:10.1113/jphysiol.2007.129197

Hayes, S. G., McCord, J. L., Rainier, J., Liu, Z., & Kaufman, M. P. (2008). Role played by acid-sensitive ion channels in evoking the exercise pressor reflex. *American Journal of Physiology Heart and Circulatory Physiology, 295*(4), H1720–H1725. doi:10.1152/ajpheart.00623.2008

Herrera, Y., Katnik, C., Rodriguez, J. D., Hall, A. A., Willing, A., Pennypacker, K. R., & Cuevas, J. (2008). Sigma-1 receptor modulation of acid-sensing ion channel a (ASIC1a) and ASIC1a-induced Ca^{2+} influx in rat cortical neurons. *Journal of Pharmacology and Experimental Therapeutics, 327*(2), 491–502. doi:10.1124/jpet.108.143974

Hesselager, M., Timmermann, D. B., & Ahring, P. K. (2004). pH dependency and desensitization kinetics of heterologously expressed combinations of acid-sensing ion channel subunits. *Journal of Biological Chemistry, 279*(12), 11006–11015.

Hoagland, E. N., Sherwood, T. W., Lee, K. G., Walker, C. J., & Askwith, C. C. (2010). Identification of a calcium permeable human acid-sensing ion channel 1 transcript variant. *Journal of Biological Chemistry, 285*(53), 41852–41862. doi:10.1074/jbc.M110.171330

Holland, P. R., Akerman, S., Andreou, A. P., Karsan, N., Wemmie, J. A., & Goadsby, P. J. (2012). Acid-sensing ion channel 1: A novel therapeutic target for migraine with aura. *Annals of Neurology, 72*(4), 559–563. doi:10.1002/ana.23653

Holton, C. M., Strother, L. C., Dripps, I., Pradhan, A. A., Goadsby, P. J., & Holland, P. R. (2020). Acid-sensing ion channel 3 blockade inhibits durovascular and nitric oxide-mediated trigeminal pain. *British Journal of Pharmacology*, *177*(11), 2478–2486. doi:10.1111/bph.14990

Ichikawa, H., & Sugimoto, T. (2002). The co-expression of ASIC3 with calcitonin gene-related peptide and parvalbumin in the rat trigeminal ganglion. *Brain Research*, *943*(2), 287–291.

Immke, D. C., & McCleskey, E. W. (2001). Lactate enhances the acid-sensing Na^+ channel on ischemia-sensing neurons. *Nature Neuroscience*, *4*(9), 869–870.

Immke, D. C., & McCleskey, E. W. (2003). Protons open acid-sensing ion channels by catalyzing relief of Ca^{2+} blockade. *Neuron*, *37*(1), 75–84.

Ishibashi, K., & Marumo, F. (1998). Molecular cloning of a DEG/ENaC sodium channel cDNA from human testis. *Biochemical and Biophysical Research Communications*, *245*(2), 589–593. doi:10.1006/bbrc.1998.8483

Jasti, J., Furukawa, H., Gonzales, E. B., & Gouaux, E. (2007). Structure of acid-sensing ion channel 1 at 1.9 A resolution and low pH. *Nature*, *449*(7160), 316–323.

Jensen, J. E., Cristofori-Armstrong, B., Anangi, R., Rosengren, K. J., Lau, C. H., Mobli, M., ... Rash, L. D. (2014). Understanding the molecular basis of toxin promiscuity: The analgesic sea anemone peptide APETx2 interacts with acid-sensing ion channel 3 and hERG channels via overlapping pharmacophores. *Journal of Medicinal Chemistry*, *57*(21), 9195–9203. doi:10.1021/jm501400p

Jiang, Q., Li, M. H., Papasian, C. J., Branigan, D., Xiong, Z. G., Wang, J. Q., & Chu, X. P. (2009). Characterization of acid-sensing ion channels in medium spiny neurons of mouse striatum. *Neuroscience*, *162*(1), 55–66. doi:10.1016/j.neuroscience.2009.04.029

Joeres, N., Augustinowski, K., Neuhof, A., Assmann, M., & Gründer, S. (2016). Functional and pharmacological characterization of two different ASIC1a/2a heteromers reveals their sensitivity to the spider toxin PcTx1. *Scientific Reports*, *6*, 27647. doi:10.1038/srep27647

Jones, N. G., Slater, R., Cadiou, H., McNaughton, P., & McMahon, S. B. (2004). Acid-induced pain and its modulation in humans. *Journal of Neuroscience*, *24*(48), 10974–10979.

Kawamata, T., Ninomiya, T., Toriyabe, M., Yamamoto, J., Niiyama, Y., Omote, K., & Namiki, A. (2006). Immunohistochemical analysis of acid-sensing ion channel 2 expression in rat dorsal root ganglion and effects of axotomy. *Neuroscience*, *143*(1), 175–187. doi:10.1016/j.neuroscience.2006.07.036

Kellenberger, S., & Schild, L. (2002). Epithelial sodium channel/degenerin family of ion channels: A variety of functions for a shared structure. *Physiological Reviews*, *82*(3), 735–767.

Kim, S. E., Coste, B., Chadha, A., Cook, B., & Patapoutian, A. (2012). The role of *Drosophila* Piezo in mechanical nociception. *Nature*, *483*(7388), 209–212. doi:10.1038/nature10801

Krauson, A. J., Rooney, J. G., & Carattino, M. D. (2018). Molecular basis of inhibition of acid sensing ion channel 1A by diminazene. *PLoS One*, *13*(5), e0196894. doi:10.1371/journal.pone.0196894

Krauson, A. J., Rued, A. C., & Carattino, M. D. (2013). Independent contribution of extracellular proton binding sites to ASIC1a activation. *Journal of Biological Chemistry*, *288*(48), 34375–34383. doi:10.1074/jbc.M113.504324

Kreple, C. J., Lu, Y., Taugher, R. J., Schwager-Gutman, A. L., Du, J., Stump, M., ... Wemmie, J. A. (2014). Acid-sensing ion channels contribute to synaptic transmission and inhibit cocaine-evoked plasticity. *Nature Neuroscience*, *17*(8), 1083–1091. doi:10.1038/nn.3750

Krishtal, O. A., Osipchuk, Y. V., Shelest, T. N., & Smirnoff, S. V. (1987). Rapid extracellular pH transients related to synaptic transmission in rat hippocampal slices. *Brain Research*, *436*(2), 352–356.

Li, M., Inoue, K., Branigan, D., Kratzer, E., Hansen, J. C., Chen, J. W., ... Xiong, Z. G. (2010). Acid-sensing ion channels in acidosis-induced injury of human brain neurons. *Journal of Cerebral Blood Flow & Metabolism*, *30*(6), 1247–1260. doi:10.1038/jcbfm.2010.30

Li, M., Kratzer, E., Inoue, K., Simon, R. P., & Xiong, Z. G. (2010). Developmental change in the electrophysiological and pharmacological properties of acid-sensing ion channels in CNS neurons. *Journal of Physiology*, *588*(Pt. 20), 3883–3900. doi:10.1113/jphysiol.2010.192922

Li, M. H., Liu, S. Q., Inoue, K., Lan, J., Simon, R. P., & Xiong, Z. G. (2014). Acid-sensing ion channels in mouse olfactory bulb M/T neurons. *Journal of General Physiology*, *143*(6), 719–731. doi:10.1085/jgp.201310990

Li, T., Yang, Y., & Canessa, C. M. (2009). Interaction of the aromatics Tyr-72/Trp-288 in the interface of the extracellular and transmembrane domains is essential for proton gating of acid-sensing ion channels. *Journal of Biological Chemistry*, *284*(7), 4689–4694.

Li, T., Yang, Y., & Canessa, C. M. (2010). Two residues in the extracellular domain convert a nonfunctional ASIC1 into a proton-activated channel. *American Journal of Physiology Cell Physiology*, *299*(1), C66–C73. doi:10.1152/ajpcell.00100.2010

Li, T., Yang, Y., & Canessa, C. M. (2011). Outlines of the pore in open and closed conformations describe the gating mechanism of ASIC1. *Nature Communications*, *2*, 399. doi:10.1038/ncomms1409

Li, W. G., & Xu, T. L. (2015). Acid-sensing ion channels: A novel therapeutic target for pain and anxiety. *Current Pharmaceutical Design*, *21*(7), 885–894. doi:10.2174/1381612820666141027124506

Li, W. G., Yu, Y., Huang, C., Cao, H., & Xu, T. L. (2011). Nonproton ligand sensing domain is required for paradoxical stimulation of acid-sensing ion channel 3 (ASIC3) channels by amiloride. *Journal of Biological Chemistry*, *286*(49), 42635–42646. doi:10.1074/jbc.M111.289058

Li, W. G., Yu, Y., Zhang, Z. D., Cao, H., & Xu, T. L. (2010). ASIC3 channels integrate agmatine and multiple inflammatory signals through the nonproton ligand sensing domain. *Molecular Pain*, *6*, 88. doi:10.1186/1744-8069-6-88

Liechti, L. A., Berneche, S., Bargeton, B., Iwaszkiewicz, J., Roy, S., Michielin, O., & Kellenberger, S. (2010). A combined computational and functional approach identifies new residues involved in pH-dependent gating of ASIC1a. *Journal of Biological Chemistry*, *285*(21), 16315–16329. doi:10.1074/jbc.M109.092015

Light, A. R., Hughen, R. W., Zhang, J., Rainier, J., Liu, Z., & Lee, J. (2008). Dorsal root ganglion neurons innervating skeletal muscle respond to physiological combinations of protons, ATP, and lactate mediated by ASIC, P2X, and TRPV1. *Journal of Neurophysiology*, *100*(3), 1184–1201. doi:10.1152/jn.01344.2007

Lin, S. H., Cheng, Y. R., Banks, R. W., Min, M. Y., Bewick, G. S., & Chen, C. C. (2016). Evidence for the involvement of ASIC3 in sensory mechanotransduction in proprioceptors. *Nature Communications*, *7*, 11460. doi:10.1038/ncomms11460

Lin, S. H., Chien, Y. C., Chiang, W. W., Liu, Y. Z., Lien, C. C., & Chen, C. C. (2015). Genetic mapping of ASIC4 and contrasting phenotype to ASIC1a in modulating innate fear and anxiety. *European Journal of Neuroscience*, *41*(12), 1553–1568. doi:10.1111/ejn.12905

Lin, Y. C., Liu, Y. C., Huang, Y. Y., & Lien, C. C. (2010). High-density expression of Ca^{2+}-permeable ASIC1a channels in NG2 glia of rat hippocampus. *PLoS One*, *5*(9), e12665. doi:10.1371/journal.pone.0012665

Lin, Y. W., Min, M. Y., Lin, C. C., Chen, W. N., Wu, W. L., Yu, H. M., & Chen, C. C. (2008). Identification and characterization of a subset of mouse sensory neurons that express acid-sensing ion channel 3. *Neuroscience*, *151*(2), 544–557. doi:10.1016/j.neuroscience.2007.10.020

Linguegia, E., Champigny, G., Lazdunski, M., & Barbry, P. (1995). Cloning of the amiloride-sensitive FMRFamide peptide-gated sodium channel. *Nature*, *378*(6558), 730–733.

Lingueglia, E., de Weille, J. R., Bassilana, F., Heurteaux, C., Sakai, H., Waldmann, R., & Lazdunski, M. (1997). A modulatory subunit of acid sensing ion channels in brain and dorsal root ganglion cells. *Journal of Biological Chemistry*, *272*(47), 29778–29783.

Lu, Y., Ma, X., Sabharwal, R., Snitsarev, V., Morgan, D., Rahmouni, K., . . . Abboud, F. M. (2009). The ion channel ASIC2 is required for baroreceptor and autonomic control of the circulation. *Neuron*, *64*(6), 885–897. doi:10.1016/j.neuron.2009.11.007

Lynagh, T., Flood, E., Boiteux, C., Wulf, M., Komnatnyy, V. V., Colding, J. M., . . . Pless, S. A. (2017). A selectivity filter at the intracellular end of the acid-sensing ion channel pore. *Elife*, *6*, e24630. doi:10.7554/eLife.24630

Lynagh, T., Mikhaleva, Y., Colding, J. M., Glover, J. C., & Pless, S. A. (2018). Acid-sensing ion channels emerged over 600 mya and are conserved throughout the deuterostomes. *Proceedings of the National Academy of Sciences of the United States of America*, *115*(33), 8430–8435. doi:10.1073/pnas.1806614115

Lynagh, T., Romero-Rojo, J. L., Lund, C., & Pless, S. A. (2017). Molecular basis for allosteric inhibition of acid-sensing ion channel 1a by ibuprofen. *Journal of Medicinal Chemistry*, *60*(19), 8192–8200. doi:10.1021/acs.jmedchem.7b01072

MacLean, D. M., & Jayaraman, V. (2016). Acid-sensing ion channels are tuned to follow high-frequency stimuli. *Journal of Physiology*, *594*(10), 2629–2645. doi:10.1113/JP271915

MacLean, D. M., & Jayaraman, V. (2017). Deactivation kinetics of acid-sensing ion channel 1a are strongly pH-sensitive. *Proceedings of the National Academy of Sciences of the United States of America*, *114*(12), E2504–E2513. doi:10.1073/pnas.1620508114

Marra, S., Ferru-Clement, R., Breuil, V., Delaunay, A., Christin, M., Friend, V., . . . Deval, E. (2016). Non-acidic activation of pain-related acid-sensing ion channel 3 by lipids. *EMBO Journal*, *35*(4), 414–428. doi:10.15252/embj.201592335

Mazzuca, M., Heurteaux, C., Alloui, A., Diochot, S., Baron, A., Voilley, N., . . . Lazdunski, M. (2007). A tarantula peptide against pain via ASIC1a channels and opioid mechanisms. *Nature Neuroscience*, *10*(8), 943–945.

McCleskey, E. W. (2009). A molecular sensor for the baroreceptor reflex? *Neuron*, *64*(6), 776–777. doi:10.1016/j.neuron.2009.12.020

McKee, J. B., Cottriall, C. L., Elston, J., Epps, S., Evangelou, N., Gerry, S., . . . Craner, M. (2019). Amiloride does not protect retinal nerve fibre layer thickness in optic neuritis in a phase 2 randomised controlled trial. *Multiple Sclerosis*, *25*(2), 246–255. doi:10.1177/1352458517742979

Miesenböck, G., De Angelis, D. A., & Rothman, J. E. (1998). Visualizing secretion and synaptic transmission with pH-sensitive green fluorescent proteins. *Nature*, *394*(6689), 192–195.

Molliver, D. C., Immke, D. C., Fierro, L., Pare, M., Rice, F. L., & McCleskey, E. W. (2005). ASIC3, an acid-sensing ion channel, is expressed in metaboreceptive sensory neurons. *Molecular Pain*, *1*, 35. doi:10.1186/1744-8069-1-35

Naito, M. G., Xu, D., Amin, P., Lee, J., Wang, H., Li, W., . . . Yuan, J. (2020). Sequential activation of necroptosis and apoptosis cooperates to mediate vascular and neural pathology in stroke. *Proceedings of the National Academy of Sciences of the United States of America*, *117*(9), 4959–4970. doi:10.1073/pnas.1916427117

Ni, Y., Gu, W. W., Liu, Z. H., Zhu, Y. M., Rong, J. G., Kent, T. A., . . . Zhang, H. L. (2018). RIP1K contributes to neuronal and astrocytic cell death in ischemic stroke via activating autophagic-lysosomal pathway. *Neuroscience*, *371*, 60–74. doi:10.1016/j.neuroscience.2017.10.038

Nomura, K., Hiyama, T. Y., Sakuta, H., Matsuda, T., Lin, C. H., Kobayashi, K., . . . Noda, M. (2019). [Na^+] increases in body fluids sensed by central Nax induce sympathetically mediated blood pressure elevations via H^+-dependent activation of ASIC1a. *Neuron, 101*(1), 60–75. doi:10.1016/j.neuron.2018.11.017

North, R. Y., Li, Y., Ray, P., Rhines, L. D., Tatsui, C. E., Rao, G., . . . Dougherty, P. M. (2019). Electrophysiological and transcriptomic correlates of neuropathic pain in human dorsal root ganglion neurons. *Brain, 142*(5), 1215–1226. doi:10.1093/brain/awz063

O'Hagan, R., Chalfie, M., & Goodman, M. B. (2005). The MEC-4 DEG/ENaC channel of *Caenorhabditis elegans* touch receptor neurons transduces mechanical signals. *Nature Neuroscience, 8*(1), 43–50.

Ohbuchi, T., Sato, K., Suzuki, H., Okada, Y., Dayanithi, G., Murphy, D., & Ueta, Y. (2010). Acid-sensing ion channels in rat hypothalamic vasopressin neurons of the supraoptic nucleus. *Journal of Physiology, 588*(Pt. 12), 2147–2162. doi:10.1113/jphysiol.2010.187625

Olson, T. H., Riedl, M. S., Vulchanova, L., Ortiz-Gonzalez, X. R., & Elde, R. (1998). An acid sensing ion channel (ASIC) localizes to small primary afferent neurons in rats. *Neuroreport, 9*(6), 1109–1113.

Omori, M., Yokoyama, M., Matsuoka, Y., Kobayashi, H., Mizobuchi, S., Itano, Y., . . . Ichikawa, H. (2008). Effects of selective spinal nerve ligation on acetic acid–induced nociceptive responses and ASIC3 immunoreactivity in the rat dorsal root ganglion. *Brain Research, 1219*, 26–31. doi:10.1016/j.brainres.2008.03.040

Page, A. J., Brierley, S. M., Martin, C. M., Martinez-Salgado, C., Wemmie, J. A., Brennan, T. J., . . . Blackshaw, L. A. (2004). The ion channel ASIC1 contributes to visceral but not cutaneous mechanoreceptor function. *Gastroenterology, 127*(6), 1739–1747. doi:10.1053/j.gastro.2004.08.061

Palmer, M. J., Hull, C., Vigh, J., & von Gersdorff, H. (2003). Synaptic cleft acidification and modulation of short-term depression by exocytosed protons in retinal bipolar cells. *Journal of Neuroscience, 23*(36), 11332–11341. doi:23/36/11332

Paukert, M., Babini, E., Pusch, M., & Gründer, S. (2004). Identification of the Ca^{2+} blocking site of acid-sensing ion channel (ASIC) 1: Implications for channel gating. *Journal of General Physiology, 124*(4), 383–394.

Paukert, M., Chen, X., Polleichtner, G., Schindelin, H., & Gründer, S. (2008). Candidate amino acids involved in H^+ gating of acid-sensing ion channel 1a. *Journal of Biological Chemistry, 283*(1), 572–581.

Peng, Z., Li, W. G., Huang, C., Jiang, Y. M., Wang, X., Zhu, M. X., . . . Xu, T. L. (2015). ASIC3 mediates itch sensation in response to coincident stimulation by acid and nonproton ligand. *Cell Reports, 13*(2), 387–398. doi:10.1016/j.celrep.2015.09.002

Pfister, Y., Gautschi, I., Takeda, A. N., van Bemmelen, M., Kellenberger, S., & Schild, L. (2006). A gating mutation in the internal pore of ASIC1a. *Journal of Biological Chemistry, 281*(17), 11787–11791.

Pignataro, G., Simon, R. P., & Xiong, Z. G. (2007). Prolonged activation of ASIC1a and the time window for neuroprotection in cerebral ischaemia. *Brain, 130*(Pt. 1), 151–158. doi:10.1093/brain/awl325

Poirot, O., Berta, T., Decosterd, I., & Kellenberger, S. (2006). Distinct ASIC currents are expressed in rat putative nociceptors and are modulated by nerve injury. *Journal of Physiology, 576*(Pt. 1), 215–234. doi:10.1113/jphysiol.2006.113035

Price, M. P., Lewin, G. R., McIlwrath, S. L., Cheng, C., Xie, J., Heppenstall, P. A., . . . Welsh, M. J. (2000). The mammalian sodium channel BNC1 is required for normal touch sensation. *Nature, 407*(6807), 1007–1011.

Price, M. P., McIlwrath, S. L., Xie, J., Cheng, C., Qiao, J., Tarr, D. E., ... Welsh, M. J. (2001). The DRASIC cation channel contributes to the detection of cutaneous touch and acid stimuli in mice. *Neuron*, *32*(6), 1071–1083.

Price, M. P., Snyder, P. M., & Welsh, M. J. (1996). Cloning and expression of a novel human brain Na^+ channel. *Journal of Biological Chemistry*, *271*(14), 7879–7882.

Ranade, S. S., Woo, S. H., Dubin, A. E., Moshourab, R. A., Wetzel, C., Petrus, M., ... Patapoutian, A. (2014). Piezo2 is the major transducer of mechanical forces for touch sensation in mice. *Nature*, *516*(7529), 121–125. doi:10.1038/nature13980

Rash, L. D. (2017). Acid-sensing ion channel pharmacology, past, present, and future. *Advances in Pharmacology*, *79*, 35–66. doi:10.1016/bs.apha.2017.02.001

Reimers, C., Lee, C. H., Kalbacher, H., Tian, Y., Hung, C. H., Schmidt, A., ... Gründer, S. (2017). Identification of a cono-RFamide from the venom of *Conus textile* that targets ASIC3 and enhances muscle pain. *Proceedings of the National Academy of Sciences of the United States of America*, *114*(17), E3507–E3515. doi:10.1073/pnas.1616232114

Reiners, M., Margreiter, M. A., Oslender-Bujotzek, A., Rossetti, G., Gründer, S., & Schmidt, A. (2018). The conorfamide RPRFa stabilizes the open conformation of acid-sensing ion channel 3 via the nonproton ligand-sensing domain. *Molecular Pharmacology*, *94*(4), 1114–1124. doi:10.1124/mol.118.112375

Ross, J. L., Queme, L. F., Cohen, E. R., Green, K. J., Lu, P., Shank, A. T., ... Jankowski, M. P. (2016). Muscle IL1β drives ischemic myalgia via ASIC3-mediated sensory neuron sensitization. *Journal of Neuroscience*, *36*(26), 6857–6871. doi:10.1523/JNEUROSCI.4582-15.2016

Roza, C., Puel, J. L., Kress, M., Baron, A., Diochot, S., Lazdunski, M., & Waldmann, R. (2004). Knockout of the ASIC2 channel in mice does not impair cutaneous mechanosensation, visceral mechanonociception and hearing. *Journal of Physiology*, *558*(Pt. 2), 659–669.

Sakai, H., Lingueglia, E., Champigny, G., Mattei, M. G., & Lazdunski, M. (1999). Cloning and functional expression of a novel degenerin-like Na^+ channel gene in mammals. *Journal of Physiology*, *519*(Pt. 2), 323–333.

Salinas, M., Besson, T., Delettre, Q., Diochot, S., Boulakirba, S., Douguet, D., & Lingueglia, E. (2014). Binding site and inhibitory mechanism of the mambalgin-2 pain-relieving peptide on acid-sensing ion channel 1a. *Journal of Biological Chemistry*, *289*(19), 13363–13373. doi:10.1074/jbc.M114.561076

Samways, D. S., Harkins, A. B., & Egan, T. M. (2009). Native and recombinant ASIC1a receptors conduct negligible Ca^{2+} entry. *Cell Calcium*, *45*(4), 319–325.

Schaefer, L., Sakai, H., Mattei, M., Lazdunski, M., & Lingueglia, E. (2000). Molecular cloning, functional expression and chromosomal localization of an amiloride-sensitive Na^+ channel from human small intestine. *FEBS Letters*, *471*(2–3), 205–210.

Schmidt, A., Alsop, R. J., Rimal, R., Lenzig, P., Joussen, S., Gervasi, N. N., ... Wiemuth, D. (2018). Modulation of DEG/ENaCs by amphiphiles suggests sensitivity to membrane alterations. *Biophysical Journal*, *114*(6), 1321–1335. doi:10.1016/j.bpj.2018.01.028

Schmidt, A., Bauknecht, P., Williams, E. A., Augustinowski, K., Gründer, S., & Jekely, G. (2018). Dual signaling of Wamide myoinhibitory peptides through a peptide-gated channel and a GPCR in *Platynereis*. *FASEB Journal*, *32*(10), 5338–5349. doi:10.1096/fj.201800274R

Schmidt, A., Lenzig, P., Oslender-Bujotzek, A., Kusch, J., Lucas, S. D., Gründer, S., & Wiemuth, D. (2014). The bile acid-sensitive ion channel (BASIC) is activated by alterations of its membrane environment. *PLoS One*, *9*(10), e111549. doi:10.1371/journal.pone.0111549

Schmidt, A., Rossetti, G., Joussen, S., & Gründer, S. (2017). Diminazene is a slow pore blocker of acid-sensing ion channel 1a (ASIC1a). *Molecular Pharmacology, 92*(6), 665–675. doi:10.1124/mol.117.110064

Sherwood, T. W., & Askwith, C. C. (2008). Endogenous arginine-phenylalanine-amide-related peptides alter steady-state desensitization of ASIC1a. *Journal of Biological Chemistry, 283*(4), 1818–1830.

Sherwood, T. W., & Askwith, C. C. (2009). Dynorphin opioid peptides enhance acid-sensing ion channel 1a activity and acidosis-induced neuronal death. *Journal of Neuroscience, 29*(45), 14371–14380.

Sherwood, T. W., Lee, K. G., Gormley, M. G., & Askwith, C. C. (2011). Heteromeric acid-sensing ion channels (ASICs) composed of ASIC2b and ASIC1a display novel channel properties and contribute to acidosis-induced neuronal death. *Journal of Neuroscience, 31*(26), 9723–9734. doi:10.1523/JNEUROSCI.1665-11.2011

Sluka, K. A., & Gregory, N. S. (2015). The dichotomized role for acid sensing ion channels in musculoskeletal pain and inflammation. *Neuropharmacology, 94*, 58–63. doi:10.1016/j.neuropharm.2014.12.013

Sluka, K. A., Kalra, A., & Moore, S. A. (2001). Unilateral intramuscular injections of acidic saline produce a bilateral, long-lasting hyperalgesia. *Muscle Nerve, 24*(1), 37–46.

Sluka, K. A., Price, M. P., Breese, N. M., Stucky, C. L., Wemmie, J. A., & Welsh, M. J. (2003). Chronic hyperalgesia induced by repeated acid injections in muscle is abolished by the loss of ASIC3, but not ASIC1. *Pain, 106*(3), 229–239.

Smith, E. S., Cadiou, H., & McNaughton, P. A. (2007). Arachidonic acid potentiates acid-sensing ion channels in rat sensory neurons by a direct action. *Neuroscience, 145*(2), 686–698. doi:10.1016/j.neuroscience.2006.12.024

Sontheimer, H., Perouansky, M., Hoppe, D., Lux, H. D., Grantyn, R., & Kettenmann, H. (1989). Glial cells of the oligodendrocyte lineage express proton-activated Na^+ channels. *Journal of Neuroscience Research, 24*(4), 496–500.

Soto, E., Ortega-Ramirez, A., & Vega, R. (2018). Protons as messengers of intercellular communication in the nervous system. *Frontiers in Cellular Neuroscience, 12*, 342. doi:10.3389/fncel.2018.00342

Springauf, A., Bresenitz, P., & Gründer, S. (2011). The interaction between two extracellular linker regions controls sustained opening of acid-sensing ion channel 1. *Journal of Biological Chemistry, 286*(27), 24374–24384. doi:10.1074/jbc.M111.230797

Staniland, A. A., & McMahon, S. B. (2009). Mice lacking acid-sensing ion channels (ASIC) 1 or 2, but not ASIC3, show increased pain behaviour in the formalin test. *European Journal of Pain, 13*(6), 554–563. doi:10.1016/j.ejpain.2008.07.001

Steen, K. H., Reeh, P. W., Anton, F., & Handwerker, H. O. (1992). Protons selectively induce lasting excitation and sensitization to mechanical stimulation of nociceptors in rat skin, in vitro. *Journal of Neuroscience, 12*(1), 86–95.

Storozhuk, M., Kondratskaya, E., Nikolaenko, L., & Krishtal, O. (2016). A modulatory role of ASICs on GABAergic synapses in rat hippocampal cell cultures. *Molecular Brain, 9*(1), 90. doi:10.1186/s13041-016-0269-4

Sun, D., Yu, Y., Xue, X., Pan, M., Wen, M., Li, S., . . . Tian, C. (2018). Cryo-EM structure of the ASIC1a–mambalgin-1 complex reveals that the peptide toxin mambalgin-1 inhibits acid-sensing ion channels through an unusual allosteric effect. *Cell Discovery, 4*, 27. doi:10.1038/s41421-018-0026-1

Sun, W. H., & Chen, C. C. (2016). Roles of proton-sensing receptors in the transition from acute to chronic pain. *Journal of Dental Research*, *95*(2), 135–142. doi:10.1177/0022034515618382

Sutherland, S. P., Benson, C. J., Adelman, J. P., & McCleskey, E. W. (2001). Acid-sensing ion channel 3 matches the acid-gated current in cardiac ischemia-sensing neurons. *Proceedings of the National Academy of Sciences of the United States of America*, *98*(2), 711–716.

Tan, Z. Y., Lu, Y., Whiteis, C. A., Benson, C. J., Chapleau, M. W., & Abboud, F. M. (2007). Acid-sensing ion channels contribute to transduction of extracellular acidosis in rat carotid body glomus cells. *Circulation Research*, *101*(10), 1009–1019. doi:10.1161/CIRCRESAHA.107.154377

Tian, Y., Bresenitz, P., Reska, A., El Moussaoui, L., Beier, C. P., & Gründer, S. (2017). Glioblastoma cancer stem cell lines express functional acid sensing ion channels ASIC1a and ASIC3. *Scientific Reports*, *7*(1), 13674. doi:10.1038/s41598-017-13666-9

Ugawa, S., Ishida, Y., Ueda, T., Inoue, K., Nagao, M., & Shimada, S. (2007). Nafamostat mesilate reversibly blocks acid-sensing ion channel currents. *Biochemical and Biophysical Research Communications*, *363*(1), 203–208.

Ugawa, S., Minami, Y., Guo, W., Saishin, Y., Takatsuji, K., Yamamoto, T., ... Shimada, S. (1998). Receptor that leaves a sour taste in the mouth. *Nature*, *395*(6702), 555–556.

Ugawa, S., Ueda, T., Ishida, Y., Nishigaki, M., Shibata, Y., & Shimada, S. (2002). Amiloride-blockable acid-sensing ion channels are leading acid sensors expressed in human nociceptors. *Journal of Clinical Investigation*, *110*(8), 1185–1190.

Ugawa, S., Ueda, T., Takahashi, E., Hirabayashi, Y., Yoneda, T., Komai, S., & Shimada, S. (2001). Cloning and functional expression of ASIC-beta2, a splice variant of ASIC-beta. *Neuroreport*, *12*(13), 2865–2869.

Ugawa, S., Ueda, T., Yamamura, H., & Shimada, S. (2005). In situ hybridization evidence for the coexistence of ASIC and TRPV1 within rat single sensory neurons. *Brain Research Molecular Brain Research*, *136*(1–2), 125–133.

Vergo, S., Craner, M. J., Etzensperger, R., Attfield, K., Friese, M. A., Newcombe, J., ... Fugger, L. (2011). Acid-sensing ion channel 1 is involved in both axonal injury and demyelination in multiple sclerosis and its animal model. *Brain*, *134*(Pt. 2), 571–584. doi:10.1093/brain/awq337

Verkest, C., Piquet, E., Diochot, S., Dauvois, M., Lanteri-Minet, M., Lingueglia, E., & Baron, A. (2018). Effects of systemic inhibitors of acid-sensing ion channels 1 (ASIC1) against acute and chronic mechanical allodynia in a rodent model of migraine. *British Journal of Pharmacology*, *175*(21), 4154–4166. doi:10.1111/bph.14462

Vick, J. S., & Askwith, C. C. (2015). ASICs and neuropeptides. *Neuropharmacology*, *94*, 36–41. doi:10.1016/j.neuropharm.2014.12.012

Voilley, N., de Weille, J., Mamet, J., & Lazdunski, M. (2001). Nonsteroid anti-inflammatory drugs inhibit both the activity and the inflammation-induced expression of acid-sensing ion channels in nociceptors. *Journal of Neuroscience*, *21*(20), 8026–8033.

Vukicevic, M., & Kellenberger, S. (2004). Modulatory effects of acid-sensing ion channels on action potential generation in hippocampal neurons. *American Journal of Physiology Cell Physiology*, *287*(3), C682–C690.

Vullo, S., Bonifacio, G., Roy, S., Johner, N., Berneche, S., & Kellenberger, S. (2017). Conformational dynamics and role of the acidic pocket in ASIC pH-dependent gating. *Proceedings of the National Academy of Sciences of the United States of America*, *114*(14), 3768–3773. doi:10.1073/pnas.1620560114

Waldmann, R., Bassilana, F., de Weille, J., Champigny, G., Heurteaux, C., & Lazdunski, M. (1997). Molecular cloning of a non-inactivating proton-gated Na^+ channel specific for sensory neurons. *Journal of Biological Chemistry, 272*(34), 20975–20978.

Waldmann, R., Champigny, G., Bassilana, F., Heurteaux, C., & Lazdunski, M. (1997). A proton-gated cation channel involved in acid-sensing. *Nature, 386*(6621), 173–177.

Wang, H., Sun, L., Su, L., Rizo, J., Liu, L., Wang, L. F., ... Wang, X. (2014). Mixed lineage kinase domain-like protein MLKL causes necrotic membrane disruption upon phosphorylation by RIP3. *Molecular Cell, 54*(1), 133–146. doi:10.1016/j.molcel.2014.03.003

Wang, J. J., Liu, F., Yang, F., Wang, Y. Z., Qi, X., Li, Y., ... Xu, T. L. (2020). Disruption of auto-inhibition underlies conformational signaling of ASIC1a to induce neuronal necroptosis. *Nature Communications, 11*(1), 475. doi:10.1038/s41467-019-13873-0

Wang, W., Duan, B., Xu, H., Xu, L., & Xu, T. L. (2006). Calcium-permeable acid-sensing ion channel is a molecular target of the neurotoxic metal ion lead. *Journal of Biological Chemistry, 281*(5), 2497–2505.

Wang, X., Li, W. G., Yu, Y., Xiao, X., Cheng, J., Zeng, W. Z., ... Xu, T. L. (2013). Serotonin facilitates peripheral pain sensitivity in a manner that depends on the nonproton ligand sensing domain of ASIC3 channel. *Journal of Neuroscience, 33*(10), 4265–4279. doi:10.1523/JNEUROSCI.3376-12.2013

Wang, Y. Z., Wang, J. J., Huang, Y., Liu, F., Zeng, W. Z., Li, Y., ... Xu, T. L. (2015). Tissue acidosis induces neuronal necroptosis via ASIC1a channel independent of its ionic conduction. *Elife, 4*, e05682. doi:10.7554/eLife.05682

Wemmie, J. A., Askwith, C. C., Lamani, E., Cassell, M. D., Freeman, J. H., Jr., & Welsh, M. J. (2003). Acid-sensing ion channel 1 is localized in brain regions with high synaptic density and contributes to fear conditioning. *Journal of Neuroscience, 23*(13), 5496–5502.

Wemmie, J. A., Chen, J., Askwith, C. C., Hruska-Hageman, A. M., Price, M. P., Nolan, B. C., ... Welsh, M. J. (2002). The acid-activated ion channel ASIC contributes to synaptic plasticity, learning, and memory. *Neuron, 34*(3), 463–477.

Weng, J. Y., Lin, Y. C., & Lien, C. C. (2010). Cell type–specific expression of acid-sensing ion channels in hippocampal interneurons. *Journal of Neuroscience, 30*(19), 6548–6558. doi:10.1523/JNEUROSCI.0582-10.2010

Wiemuth, D., Assmann, M., & Gründer, S. (2013). The bile acid–sensitive ion channel (BASIC), the ignored cousin of ASICs and ENaC. *Channels (Austin), 8*(1), 29–43. doi:10.4161/chan.27493

Wiemuth, D., & Gründer, S. (2011). The pharmacological profile of brain liver intestine Na^+ channel: Inhibition by diarylamidines and activation by fenamates. *Molecular Pharmacology, 80*(5), 911–919. doi:10.1124/mol.111.073726

Wiemuth, D., Sahin, H., Falkenburger, B. H., Lefevre, C. M., Wasmuth, H. E., & Gründer, S. (2012). BASIC—A bile acid–sensitive ion channel highly expressed in bile ducts. *FASEB Journal, 26*(10), 4122–4130. doi:10.1096/fj.12-207043

Wikipedia. (n.d.). Acid. Retrieved from https://en.wikipedia.org/wiki/Acid

Wu, J., Xu, Y., Jiang, Y. Q., Xu, J., Hu, Y., & Zha, X. M. (2016). ASIC subunit ratio and differential surface trafficking in the brain. *Molecular Brain, 9*(1), 4. doi:10.1186/s13041-016-0185-7

Wu, L. J., Duan, B., Mei, Y. D., Gao, J., Chen, J. G., Zhuo, M., ... Xu, T. L. (2004). Characterization of acid-sensing ion channels in dorsal horn neurons of rat spinal cord. *Journal of Biological Chemistry, 279*(42), 43716–43724.

Xie, J., Price, M. P., Berger, A. L., & Welsh, M. J. (2002). DRASIC contributes to pH-gated currents in large dorsal root ganglion sensory neurons by forming heteromultimeric channels. *Journal of Neurophysiology, 87*(6), 2835–2843.

Xiong, Z. G., Zhu, X. M., Chu, X. P., Minami, M., Hey, J., Wei, W. L., ... Simon, R. P. (2004). Neuroprotection in ischemia: Blocking calcium-permeable acid-sensing ion channels. *Cell, 118*(6), 687–698.

Xu, Y., Jiang, Y. Q., Li, C., He, M., Rusyniak, W. G., Annamdevula, N., ... Zha, X. M. (2018). Human ASIC1a mediates stronger acid-induced responses as compared with mouse ASIC1a. *FASEB Journal, 32*(7), 3832–3843. doi:10.1096/fj.201701367R

Yagi, J., Wenk, H. N., Naves, L. A., & McCleskey, E. W. (2006). Sustained currents through ASIC3 ion channels at the modest pH changes that occur during myocardial ischemia. *Circulation Research, 99*(5), 501–509.

Yan, J., Edelmayer, R. M., Wei, X., De Felice, M., Porreca, F., & Dussor, G. (2011). Dural afferents express acid-sensing ion channels: A role for decreased meningeal pH in migraine headache. *Pain, 152*(1), 106–113. doi:10.1016/j.pain.2010.09.036

Yan, J., Wei, X., Bischoff, C., Edelmayer, R. M., & Dussor, G. (2013). pH-evoked dural afferent signaling is mediated by ASIC3 and is sensitized by mast cell mediators. *Headache, 53*(8), 1250–1261. doi:10.1111/head.12152

Yang, L., & Palmer, L. G. (2014). Ion conduction and selectivity in acid-sensing ion channel 1. *Journal of General Physiology, 144*(3), 245–255. doi:10.1085/jgp.201411220

Yoder, N., & Gouaux, E. (2018). Divalent cation and chloride ion sites of chicken acid sensing ion channel 1a elucidated by X-ray crystallography. *PLoS One, 13*(8), e0202134. doi:10.1371/journal.pone.0202134

Yoder, N., & Gouaux, E. (2020). Conserved His-Gly motif of acid-sensing ion channels resides in a reentrant "loop" implicated in gating and ion selectivity. *Elife, 9*, e56527. doi:10.1101/2020.03.02.974154

Yoder, N., Yoshioka, C., & Gouaux, E. (2018). Gating mechanisms of acid-sensing ion channels. *Nature, 555*(7696), 397–401. doi:10.1038/nature25782

Young, G. T., Gutteridge, A., Fox, H., Wilbrey, A. L., Cao, L., Cho, L. T., ... Stevens, E. B. (2014). Characterizing human stem cell–derived sensory neurons at the single-cell level reveals their ion channel expression and utility in pain research. *Molecular Therapy, 22*(8), 1530–1543. doi:10.1038/mt.2014.86

Yu, Y., Chen, Z., Li, W. G., Cao, H., Feng, E. G., Yu, F., ... Xu, T. L. (2010). A nonproton ligand sensor in the acid-sensing ion channel. *Neuron, 68*(1), 61–72. doi:10.1016/j.neuron.2010.09.001

Zeng, W. Z., Marshall, K. L., Min, S., Daou, I., Chapleau, M. W., Abboud, F. M., ... Patapoutian, A. (2018). PIEZOs mediate neuronal sensing of blood pressure and the baroreceptor reflex. *Science, 362*(6413), 464–467. doi:10.1126/science.aau6324

Zha, X. M., Costa, V., Harding, A. M., Reznikov, L., Benson, C. J., & Welsh, M. J. (2009). ASIC2 subunits target acid-sensing ion channels to the synapse via an association with PSD-95. *Journal of Neuroscience, 29*(26), 8438–8446.

Zha, X. M., Wemmie, J. A., Green, S. H., & Welsh, M. J. (2006). Acid-sensing ion channel 1a is a postsynaptic proton receptor that affects the density of dendritic spines. *Proceedings of the National Academy of Sciences of the United States of America, 103*(44), 16556–16561.

Zhang, M., Gong, N., Lu, Y. G., Jia, N. L., Xu, T. L., & Chen, L. (2008). Functional characterization of acid-sensing ion channels in cultured neurons of rat inferior colliculus. *Neuroscience, 154*(2), 461–472. doi:10.1016/j.neuroscience.2008.03.040

Zhao, D., Ning, N., Lei, Z., Sun, H., Wei, C., Chen, D., & Li, J. (2014). Identification of a novel protein complex containing ASIC1a and GABAA receptors and their interregulation. *PLoS One, 9*(6), e99735. doi:10.1371/journal.pone.0099735

Ziemann, A. E., Allen, J. E., Dahdaleh, N. S., Drebot, II, Coryell, M. W., Wunsch, A. M., . . . Wemmie, J. A. (2009). The amygdala is a chemosensor that detects carbon dioxide and acidosis to elicit fear behavior. *Cell, 139*(5), 1012–1021.

Ziemann, A. E., Schnizler, M. K., Albert, G. W., Severson, M. A., Howard, M. A., 3rd, Welsh, M. J., & Wemmie, J. A. (2008). Seizure termination by acidosis depends on ASIC1a. *Nature Neuroscience, 11*(7), 816–822.

CHAPTER 19

TMEM16 CA^{2+} ACTIVATED CL^{-} CHANNELS AND CLC CHLORIDE CHANNELS AND TRANSPORTERS

ANNA BOCCACCIO AND MICHAEL PUSCH

Overview of Non-Receptor Type Chloride Channels in the Brain

SEVERAL general aspects have to be considered regarding the role of chloride transport in neurons and in glia. The physiological function of chloride channels is strictly related to the equilibrium potential for chloride, E_{Cl}, which can differ among cells or cellular compartments. For neurons, the intracellular chloride concentration ($[Cl]_i$) is essential to determine the generally inhibitory response of GABA and glycine receptor channels (Hübner & Holthoff, 2013), and secondary active cation chloride transporters play essential roles in setting $[Cl]_i$. However, in immature neurons, GABA is excitatory as a result of a high $[Cl]_I$ (Ben-Ari, 2002), and in some neuronal types, high $[Cl]_i$ is maintained, as in dorsal root ganglia neurons (Deschenes, Feltz, & Lamour, 1976) and olfactory sensory neurons (Kaneko, 2004; Reuter, Zierold, Schröder, & Frings, 1998). The resting membrane chloride permeability is an important additional factor in determining $[Cl]_i$ as a function of the membrane potential. In addition, a background chloride conductance will lower the neuronal input resistance and thus neuronal excitability. A second important role of chloride ions in neurons (as in all cells) regards their role in osmo/volume regulation.

In glia, the role of chloride channels is still very little understood. In general, they appear to be important for osmo/volume regulation as well as for spatial K^+ buffering, also called "K^+ siphoning," a mechanism for the removal of K^+ ions released from axons during neuronal activity via the pan glial syncytium (van der Knaap, Boor, & Estevez, 2012).

Finally, endo/lysosomal chloride/proton antiporters appear to have important roles, particularly in neurons, as shown primarily by mouse knock-out (KO) models (Jentsch & Pusch, 2018).

Members of three gene families encoding anion channels different from GABA and glycine receptors are expressed in the brain: the CLC family of chloride channels and chloride/proton antiporters (Jentsch & Pusch, 2018); the TMEM16 family of Ca^{2+} -activated chloride channels and lipid scramblases (Pedemonte & Galietta, 2014), and the LRRC8 family of volume-regulated anion channels (Jentsch, 2016). The current chapter focuses on the role of TMEM16A and TMEM16B channels and CLC channels/transporters.

Overview of the TMEM16 Family

The family of integral membrane proteins named TMEM16 or anoctamins (the latter name refers to the anion-selectivity [AN] and the presumed eight [OCT] transmembrane domains of the first proposed topology for this protein family (Huang, Wong, & Jan, 2012; Pedemonte & Galietta, 2014; Yang et al., 2008), includes ten members in mammals (labeled with letters from A to K, excluding I) that are functionally split into two categories: TMEM16A/Ano1 and TMEM16B/Ano2 function as Ca^{2+}-activated chloride channels, while others, among them TMEM16F/Ano6, are Ca^{2+}-dependent phospholipid scramblases mediating the transfer of phospholipids between the leaflets of the membrane bilayer (Gyobu, Ishihara, Suzuki, Segawa, & Nagata, 2017; Gyobu et al., 2015; Suzuki, Umeda, Sims, & Nagata, 2010).

In the following discussion, we will refer exclusively to the TMEM16 nomenclature.

TMEM16A and TMEM16B form Ca^{2+}-activated Cl^- channels that are involved in a variety of physiological functions, such as transepithelial ion transport, olfaction, phototransduction, smooth muscle contraction, nociception, cell proliferation, and control of neuronal excitability (Pedemonte & Galietta, 2014).

The search of the molecular counterpart of Ca^{2+}-activated Cl^- currents lasted more than 25 years after they were first observed in the early 1980s in the salamander retina and in *Xenopus laevis* oocytes (Bader, Bertrand, & Schwartz, 1982), and only in 2008 were TMEM16A (Caputo et al., 2008; Schroeder, Cheng, Jan, & Jan, 2008; Yang et al., 2008) and TMEM16B (Schroeder et al., 2008) cloned and shown to mediate this type of anion current. TMEM16B was confirmed as a Ca^{2+}-activated Cl^- channel in several studies (Pifferi, Dibattista, & Menini, 2009; Sagheddu et al., 2010; Stephan et al., 2009; Stohr et al., 2009).

According to the X-ray crystal structure of the fungal homologue nhTMEM16 (Brunner, Lim, Schenck, Duerst, & Dutzler, 2014), which functions as a lipid scramblase, and the cryo-electron microscopy data from mouse TMEM16A (Paulino et al., 2017), which functions as a Ca^{2+} activated Cl^- channel, TMEM16 proteins share the same general architecture, with 10 transmembrane-spanning helical domains and intracellular N and C terminals (Figure 19.1A and B).

The protein is a homodimer (Brunner et al., 2014; Fallah et al., 2011; Paulino et al., 2017; Sheridan et al., 2011) (Figure 19.1A and B), and each subunit has a highly conserved Ca^{2+}-binding site that coordinates two Ca^{2+} ions (Brunner et al., 2014), formed by five acidic and one polar residue in transmembrane domains 6, 7, and 8 (Brunner et al., 2014). Mutating these acidic residues in TMEM16A greatly altered the apparent Ca^{2+}

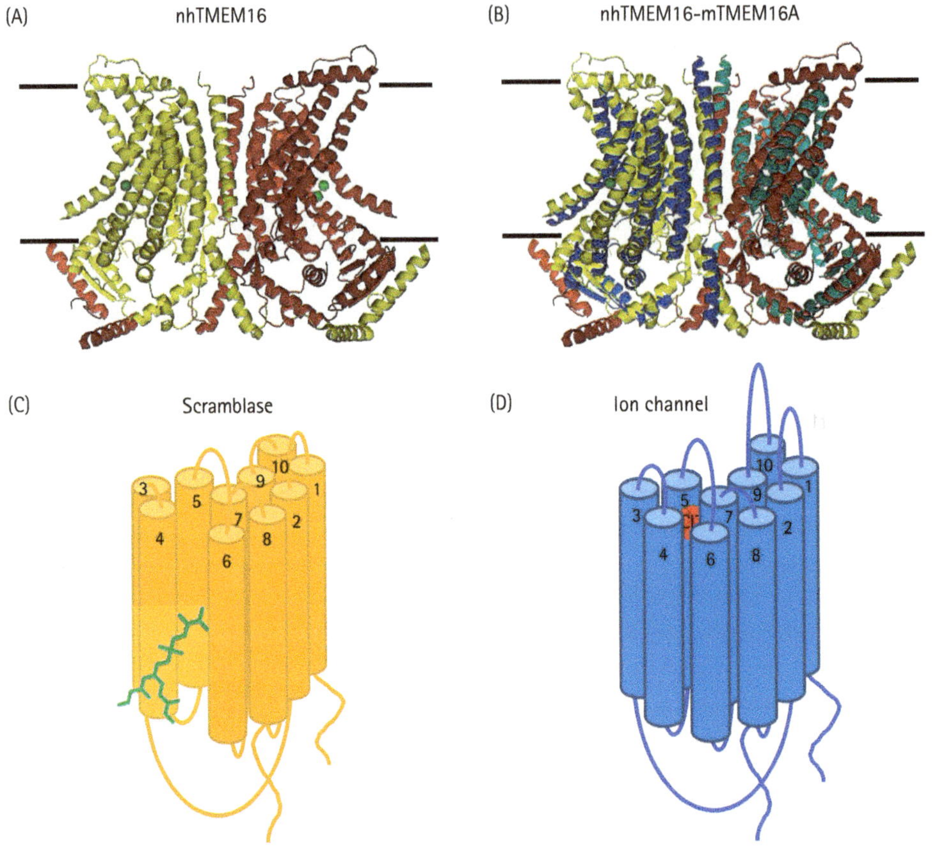

FIGURE 19.1 TMEM16 structure: (**A**): Ribbon representation of nhTMEM16 dimer (Brunner et al 2014). Bound Ca^{2+} are shown as green spheres. The view is from within the membrane with the extracellular side at the top. The membrane boundary is indicated by a straight line. (**B**): Superposition of the nhTMEM16 dimer (yellow and red, colors as in [**A**]) with the mTMEM16A dimer (blue and light blue; Paulino et al 2017). (**C**) and (**D**): Cartoon showing the location of transmembrane helices of a single subunit in the scramblase nhTMEM16 and in the ion channel TMEM16A. In nhTMEM16, helix 4 is quite close to helix 3, shaping, together with helices 5 and 6, a furrow through which lipids can be transported (phosphatidylserine is shown in green in [**C**]). In TMEM16A, helix 4 is moved away from helix 3 and is closer to helix 6, forming, together with helices 3, 5, 6, and 7, a pore that can be permeated by chloride ions (red sphere) (**D**).

affinity of the channel (Yu, Duran, Qu, Cui, & Hartzell, 2012). The voltage dependence of the Ca^{2+}-sensitivity observed in TMEM16A and TMEM16B (Pifferi et al., 2009; Stephan et al., 2009; Tien et al., 2014; Yang et al., 2012) can be therefore explained by the localization of the Ca^{2+}-binding site in the transmembrane part of the protein.

Each subunit has at the periphery of the dimer a narrow crevice that spans the membrane, named the *subunit cavity*, which differs significantly in the scramblase nhTMEM16 (Brunner et al., 2014) (Figure 19.1C) and in the channel mTMEM16A (Paulino et al., 2017) (Figure 19.1D). The open furrow formed by helices 4–6 faces the hydrophobic core of the membrane in nhTMEM16, while the helices are rearranged in TMEM16A to form a hydrophilic conduction pathway suitable for ion permeation (Fisher & Hartzell, 2017; Paulino et al., 2017) (Figure 19.1C and D). Indeed, mutagenesis of the residues facing this crevice affects ion selectivity (Peters et al., 2015; Pifferi, 2017). Each TMEM16A dimer has therefore two separate ion conduction pores that are independently activated by calcium (Jeng, Aggarwal, Yu, & Chen, 2016; Lim, Lam, & Dutzler, 2016) similarly to the double-barreled CLC channels.

Although scramblases of the TMEM16 family do mediate ionic currents (Lee, Menon, & Accardi, 2016; Yang et al., 2012), these might not be physiologically relevant. We will concentrate on the well-characterized TMEM16A and TMEM16B channels as components of Ca^{2+}-activated chloride channels in the nervous system.

When expressed in heterologous systems, TMEM16A and TMEM16B recapitulate the characteristics of classical Ca^{2+}-activated Cl^- currents (Frings, Reuter, & Kleene, 2000; Hartzell, Putzier, & Arreola, 2005). They generate Cl^- currents activated by both membrane depolarization and sub-micromolar/micromolar concentrations of cytosolic Ca^{2+}; the apparent Ca^{2+}-sensitivity depends on membrane voltage; at low Ca^{2+} concentrations, currents show time-dependent kinetics and are strongly outwardly rectifying, whereas they become "leak-like" with a linear current–voltage relationship at higher intracellular Ca^{2+} concentrations; currents show a relatively poor selectivity among halide anions, with a preference for larger anions compared to chloride.

However, TMEM16A and TMEM16B have significantly different properties (Figure 19.2): TMEM16B currents show considerably faster activation and deactivation kinetics than TMEM16A, and TMEM16B has a lower Ca^{2+}-sensitivity, requiring micromolar cytoplasmic Ca^{2+}-concentrations (Pifferi et al., 2009; Stephan et al., 2009), while TMEM16A can be activated by a few hundreds of nanomolar Ca^{2+} (Caputo et al., 2008).

Alternative splicing is an important mechanism to regulate TMEM16A and TMEM16B current activity and increases the molecular diversity of TMEM16 in different tissues. Human TMEM16A cDNA contains four alternatively spliced segments, all of which influence the sensitivity to Ca^{2+} and voltage (Ferrera et al., 2009). In olfactory receptor neurons, TMEM16B exists in two isoforms, each occurring in two distinct splice variants (Ponissery Saidu, Stephan, Talaga, Zhao, & Reisert, 2013). Differences in Ca^{2+} sensitivity were discovered both between the isoforms and between the splice variants, and co-expression of different isoforms is necessary to recapitulate the native olfactory channel. It is worth noticing that the olfactory TMEM16B isoform, and not

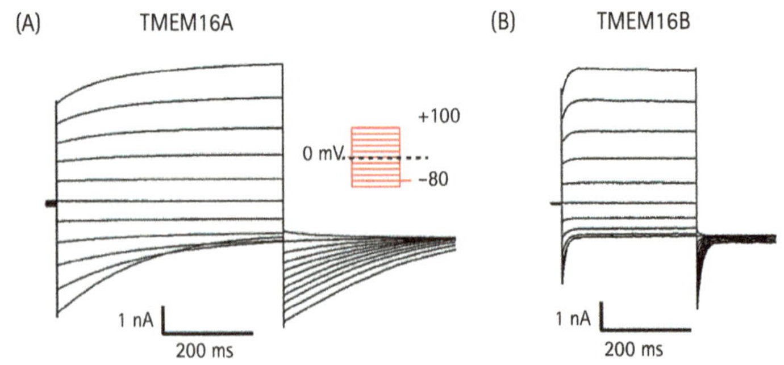

FIGURE 19.2 **TMEM16A and TMEM16B current recordings:** Representative currents recorded with the patch clamp technique in the whole-cell configuration from HEK293 cells transfected with mouse TMEM16A (isoform ac) (**A**), and mouse TMEM16B (retinal isoform) (**B**). Stimulation protocol is shown in the *inset* in panel **A**. Pipette solution contained 0.5 (**A**) and 9 μM (**B**) calculated free Ca^{2+} concentration, buffered respectively with EGTA (N-((4-methoxy)-2-naphthyl)-5-nitroanthranilic acid) and HEDTA (N-(2-Hydroxyethyl)ethylenediamine-N,N′,N′-triacetic acid). Note the faster activation and deactivation kinetics of TMEM16B compared to TMEM16A.

the retinal one, is the prevalent form in brain regions such as cortex, cerebellum, hippocampus, and brainstem (Vocke et al., 2013).

Some reports have suggested involvement of calmodulin in the regulation of Ca^{2+}-activated chloride channels. However, how calmodulin regulates TMEM16 proteins remains controversial.

Different groups found evidence for a direct interaction of calmodulin with TMEM16A (Jung et al., 2013; Tien et al., 2014; Vocke et al., 2013) and TMEM16B (Vocke et al., 2013); however, this is in conflict with data from other laboratories (Terashima, Picollo, & Accardi, 2013; Yu et al., 2012). More recently, it has been reported that Ca^{2+} free calmodulin, apo-calmodulin, is pre-associated with TMEM16A and TMEM16B channel complexes at resting Ca^{2+} concentrations (Yang & Colecraft, 2016; Yang, Hendrickson, & Colecraft, 2014). Interestingly, the interaction would depend on the specific splicing variant, since the individual segments affect apoCaM binding and regulation of TMEM16A.

Pharmacology is a powerful tool for dissecting the contribution of specific ion conductances to the overall electrical cell activity. The lack of specific blockers for Ca^{2+}-activated chloride currents contributed to limiting the studies of their role in the nervous system. Several compounds described as blocking Ca^{2+}-activated Cl^- currents in diverse tissues indeed affect TMEM16A- and TMEM16B-induced currents, including the fenamates niflumic acid (NFA) and flufenamic acid, NPPB (5-nitro-2-[3-phenylpropyl-amino] benzoic acid), the stilbene derivative DIDS (4,4-diisothiocyanato-stilbene-2,2-disulfonic acid), and A9C (anthracene-9-carboxylic acid) (Cherian, Menini, & Boccaccio, 2015; Liu et al., 2015; Ta, Adomaviciene, Rorsman, Garnett, & Tammaro, 2016). Unfortunately, these compounds are poorly selective, have a low affinity, and in some cases have complex effects, both inhibiting and activating the channel (Bradley et al., 2014; Cherian et al., 2015; Liu et al., 2015; Ta et al., 2016). Other blockers have been identified more recently, such as eugenol (Yao et al., 2012), digallic and tannic acids (Namkung, Thiagarajah, Phuan, & Verkman, 2010), benzbromarone (Huang et al., 2009), CaCC inh-A01 (De La Fuente, Namkung, Mills, & Verkman, 2008), T16A inh-A01 (Namkung, Phuan, & Verkman, 2011), MONNA (N-((4-methoxy)-2-naphthyl)-5-nitroanthranilic acid; Oh et al., 2013), Ani9 (2-(4-chloro-2-methylphenoxy)-N-[(2-methoxyphenyl)methylideneamino]-acetamide; Seo et al., 2016). However, the investigation of their specificity and blocking mode (Boedtkjer, Kim, Jensen, Matchkov, & Andersson, 2015; Bradley et al., 2014; Liu et al., 2015) is still in progress, and care should be taken before drawing conclusions exclusively based on pharmacological approach.

In the following section, the current knowledge of the physiological role of TMEM16A and TMEM16B in the central and peripheral nervous systems will be discussed.

TMEM16A and TMEM16B in Neurons

Although Ca^{2+}-activated Cl^{-} channels play a major physiological role in various types of cells and tissues (Hartzell et al., 2005), data regarding their expression, function, and cellular and sub-cellular localization in the CNS are still limited, mainly due to delay in their molecular identification, the poor pharmacology, and difficulties in the identification of ionic currents mediated by these channels, since they might be masked by the most known and ubiquitous Ca^{2+}-activated K^{+} currents.

Only a very few reports described the occurrence of Ca^{2+}-activated Cl^{-} currents in central neurons (exhaustively reviewed in Frings et al., 2000) before the cloning of TMEM16A and TMEM16B. After the first description in retina (Bader et al., 1982; Miledi, 1982), a Ca^{2+}-activated Cl^{-} current ha been described in cerebellar Purkinje cells (Llano, Leresche, & Marty, 1991), in a cell line derived from the anterior pituitary gland (Korn, Bolden, & Horn, 1991), in amygdala neurons (Sugita, Tanaka, & North, 1993), and in cingulate cortical neurons (Higashi, Tanaka, Inokuchi, & Nishi, 1993). However, the physiological importance of Ca^{2+}-activated Cl^{-} currents in neurons has been overlooked.

TMEM16A has been detected in cerebellum (Cho, Jeon, Chun, Yeo, & Kim, 2014; W. Zhang, Schmelzeisen, Parthier, Frings, & Mohrlen, 2015), and it is strongly expressed in two brainstem nuclei that are key auditory relay nuclei for sound source localization.

TMEM16B expression in the CNS is found in rod photoreceptor terminals (Jeon, Paik, Chun, Oh, & Kim, 2013; Stohr et al., 2009), in olfactory bulb (Billig, Pal, Fidzinski, & Jentsch, 2011), in cerebellum (Vocke et al., 2013; W. Zhang et al., 2015), in hippocampal pyramidal neurons (W. C. Huang et al., 2012), in thalamocortical neurons (Ha et al., 2016), and in the striatum (Song, Beatty, & Wilson, 2016).

Up to now, no reports associate functional recordings of Ca^{2+}-activated Cl^{-} currents in central neurons to TMEM16A. Conversely, recent data on thalamocortical and CA1 hippocampal neurons show a contribution of TMEM16B to spike-frequency adaptation in the CNS (Ha et al., 2016; W. C. Huang et al., 2012).

Outside the CNS, Ca^{2+}-activated Cl^- channels received more attention. They are present in many sensory neurons and in particular in olfactory sensory neurons (Kleene & Gesteland, 1991; Lowe & Gold, 1993), and nociceptive neurons (Mayer, 1985), where they have been investigated in great detail.

TMEM16B has been shown to be expressed in olfactory sensory neurons, and Ca^{2+}-activated Cl^-currents are abolished in TMEM16B KO mice (Billig et al., 2011), while TMEM16A is important for detection of nociceptive stimuli in dorsal root ganglion (DRG) neurons (although a possible involvement of TMEM16B has not been excluded).

Most of these papers rely on global TMEM16A and TMEM16B KO animal models and on pharmacological analysis. However mice with a global TMEM16A KO survive only few weeks after birth (Rock, Futtner, & Harfe, 2008), due to respiration problems, limiting its use for functional investigation.

Central Nervous System

After years of neglect, following the identification of TMEM16A and TMEM16B as Ca^{2+}-activated Cl^- channels, several groups finally started to investigate their presence in the CNS.

Retina

Despite the fact that Ca^{2+}-activated Cl^- currents have been described in 1982 in rod inner segments isolated from the retina of tiger salamander (Bader et al., 1982), and early experiments showed TMEM16A (Yang et al., 2008) and TMEM16B (Stohr et al., 2009) expression in the mammalian retina, there is little established information on their cellular localization and function in retina.

TMEM16A and TMEM16B are expressed in the outer plexiform layer, at the photoreceptor presynaptic terminals (in tiger salamander [Mercer et al., 2011] and mouse [Caputo et al., 2015; Jeon et al., 2013; Stohr et al., 2009]). TMEM16A is expressed both in rod and cone terminals (in tiger salamander [Mercer et al., 2011], mouse [Jeon et al., 2013], but apparently not in rat [Dauner, Mobus, Frings, & Mohrlen, 2013]), in close proximity with voltage gated Ca^{2+} channels (Caputo et al., 2015), while TMEM16B is expressed specifically in rod terminals (Dauner et al., 2013).

Ca^{2+}-activated Cl^- currents have been described in photoreceptors of the vertebrate retina (Barnes & Hille, 1989; MacLeish & Nurse, 2007) and they are thought to regulate synaptic transmission at synaptic terminals. However, their function, which is strictly linked to E_{Cl} (Frings, 2009), has still to be clarified, and the specific contribution of either TMEM16A or TMEM16B has not been investigated yet.

Additionally, TMEM16A is expressed in the inner plexiform layer (Jeon et al., 2013; Mercer et al., 2011), in particular in bipolar cell axon terminals and in amacrine cell processes (Jeon et al., 2013). Consistently, a Ca^{2+}-activated Cl^- tail current has been recorded in isolated rod bipolar cells, while it is absent in cells missing the axon terminal (Jeon et al., 2013). In rod bipolar cells, E_{Cl} is more negative than the resting potential and Ca^{2+}-activated Cl^- currents activation in response to depolarization-induced Ca^{2+} -entry can contribute to shortening membrane depolarization or facilitating cell membrane repolarization (Jeon et al., 2013).

Hippocampus

Before the identification of TMEM16A and TMEM16B as Ca^{2+}-activated Cl^- channels, the presence of Ca^{2+}-activated Cl^- currents in hippocampal neurons had never been addressed. W. C. Huang et al. (2012) reported that Ca^{2+}-activated Cl^- currents are present in pyramidal hippocampal neurons and are activated by Ca^{2+}-influx through voltage-gated Ca^{2+} channels or N-methyl-D-aspartate (NMDA) receptors. Neuronal depolarization resulted in activation of Ca^{2+} currents followed by tail currents whose amplitude depended on the amount of Ca^{2+} influx, which reversed at the equilibrium potential for Cl^- and were sensitive to chloride channel blockers such as NFA(W. C. Huang et al., 2012). Ca^{2+}-activated Cl^- currents' activation had an inhibitory function (in cultured neurons and in slices) since they reduced amplitude and temporal summation of excitatory postsynaptic potentials (EPSP), decreased action potential duration, and EPSP–spike coupling elevating the threshold for spike generation by EPSP (W. C. Huang et al., 2012).

Several pieces of evidence, including reverse transcriptase-polymerase chain reaction, in situ hybridization, western blots (W. C. Huang et al., 2012), and immunohistochemistry (Song et al., 2016), suggest that hippocampal Ca^{2+}-activated Cl^- currents are carried by TMEM16B and not TMEM16A. Additionally, there is no significant difference in Ca^{2+}-activated Cl^- currents between wild type and TMEM16A KO mice, while it is strongly reduced in neurons infected with small hairpin RNAs targeting TMEM16B (W. C. Huang et al., 2012).

Ca^{2+}-activated Cl^- currents, mediated by TMEM16B, account for spike-frequency adaptation in pyramidal neurons. Indeed, CA1 hippocampal pyramidal neurons infected with small hairpin RNAs targeting TMEM16B showed a decrease in spike-frequency adaptation pattern, similar to what is observed in thalamocortical neurons (see further in this chapter) (Ha et al., 2016). Further studies on hippocampal neurons will establish the contribution of Ca^{2+}-activated Cl^- channels to afterhyperpolarization (AHP).

Cerebellum

TMEM16A and TMEM16B have been reported to be expressed in cerebellum at the mRNA and protein levels (Cho et al., 2014; Neureither, Ziegler, Pitzer, Frings, & Mohrlen, 2017; Vocke et al., 2013; W. Zhang et al., 2015).

TMEM16A and TMEM16B showed a differential expression pattern in cerebellum in immunohistochemistry experiments. While TMEM16A is expressed in GABAergic interneurons in the molecular and granule-cell layer of cerebellar cortex and in soma of Purkinje cells (W. Zhang et al., 2015), TMEM16B is localized in somata and in the dendritic tree of Purkinje cells (Neureither, Ziegler, et al., 2017; W. Zhang et al., 2015).

However, it is worth noting that other groups failed in detecting TMEM16B in cerebellum (Billig et al., 2011; Stohr et al., 2009) and in particular in Purkinje cells (Y. Zhang et al., 2017).

To our knowledge, there are only two reports describing Ca^{2+}-activated Cl^- currents in cerebellum, probably overlooked in favor of Ca^{2+}-activated K-currents, mediated by BK and SK channels. A Ca^{2+}-activated Cl^- current has been first described in rat Purkinje cells by Llano et al. (Llano et al., 1991) and is responsible for tail currents following depolarizing voltage pulses. The amplitude of the tail current increases with repetition of the stimulation, due to increased Ca^{2+}-entry. It was hypothesized that Ca^{2+}-activated Cl^- currents could play a role in the repolarization of dendritic membrane following Ca-spikes under physiological conditions (Llano et al., 1991). More recently, it has been reported that in rat Purkinje cells, Ca^{2+} -activated Cl^- currents modulate the efficacy of synaptic input from inhibitory interneurons, a plasticity form named *depolarization-induced depression* (Satoh, Qu, Suzuki, & Saitow, 2013).

Recent data (W. Zhang et al., 2015) suggest that the Ca^{2+}-activated Cl^- current in mouse Purkinje cells is mediated by TMEM16B: depression of GABAergic inhibitory postsynaptic currents (IPSCs) recorded after electrical stimulation of climbing fiber was absent in TMEM16B KO neurons (W. Zhang et al., 2015). Interestingly, behavioral tests revealed a cerebellar dysfunction in TMEM16B KO mice (Neureither, Ziegler, et al., 2017). Indeed, animals were affected by a mild ataxia during spontaneous voluntary walking and displayed motor coordination and motor learning deficits in voluntary and enforced locomotion tasks. The presence of a cerebellar dysfunction in TMEM16B KO animals is in agreement with the belief that the inhibitory regulation of Purkinje cells by molecular layer interneurons is important for motor ability.

However, a contribution to the TMEM16B KO phenotype from other brain regions is not excluded, since TMEM16B has been reported to be expressed in striatum (Song et al., 2016) and in inferior olivary neurons in the brainstem that send climbing fibers to innervate cerebellar Purkinje cells (Y. Zhang et al., 2017). Indeed, TMEM16B-deficient mice have a reduced motor learning capability (Y. Zhang et al., 2017), but Zhang et al. causally attributed this effect to a modified firing activity of inferior olivary neurons in the brainstem, which in turn affects Purkinje cells' firing activity through climbing fibers (Y. Zhang et al., 2017).

The use of a tissue-specific TMEM16 KO mouse model would help to clarify these inconsistencies.

Brainstem

Both TMEM16A and TMEM16B have been detected at low levels in the brainstem (Billig et al., 2011; Cho et al., 2014; Y. Zhang et al., 2017). Recently, Y. Zhang et al. (2017) described a new Ca^{2+}-sensitive conductance distinct from Ca^{2+}-activated BK and SK K^+ channels in inferior olivary neurons. This conductance is activated by Ca^{2+}- influx through the dendritic high-threshold voltage-gated Ca^{2+} channels and is insensitive to blockers of BK and SK channels, while it is blocked by NFA and its reversal potential is affected by the extracellular Cl^- concentration. Cl^- influx during afterdepolarization (ADP) through the newly identified Ca^{2+}-activated Cl^- channels will cause hyperpolarization and provide negative feedback to the high-threshold Ca^{2+} channels, contributing to ADP termination. Immunofluorescence experiments showed that TMEM16B, but not TMEM16A, is expressed in inferior olivary neurons (Y. Zhang et al., 2017). In accordance with expression data, in neurons from TMEM16B KO mice, Ca^{2+}-activated Cl^- currents are abolished, ADP duration is prolonged, and firing activity is reduced, confirming that the Ca^{2+}-activated Cl^- current serves as a major repolarizing current that controls inferior olivary neurons' excitability.

TMEM16A is strongly expressed at the mRNA and protein levels in two brainstem nuclei that are important auditory relay nuclei for sound source localization: the medial nucleus of the trapezoid body and the anterior ventral cochlear nucleus (Cho et al., 2014). TMEM16A is localized in presynaptic endings within these two auditory brainstem nuclei (Cho et al., 2014), suggesting a possible role for TMEM16A in presynaptic modulation of synaptic transmission in auditory processing. However, evidence for the existence of Ca^{2+}-activated Cl^- channels in auditory processing is still missing.

Striatum

Recently, a Ca^{2+}-activated Cl^- current has been recorded in striatal low-threshold spiking interneurons (Song et al., 2016). These interneurons spontaneously transition between a spiking and a persistently depolarized state, characterized by oscillations and resonance. The membrane potential oscillations depend on voltage-gated Ca^{2+} channels that in turn activate a Ca^{2+}-activated Cl^- current, which contributes to membrane hyperpolarization after Ca^{2+} channel activation, therefore providing the restorative current for oscillation. Immunohistochemistry data showed that TMEM16B, and not TMEM16A, is expressed in these interneurons (Song et al., 2016).

Thalamus

In thalamocortical neurons in the ventrobasal nuclei, a Ca^{2+}-activated tail current primarily mediated by Cl^- ions with a minor contribution of apamin-sensitive SK channels has been described (Ha et al., 2016). This hyperpolarizing Ca^{2+}-activated Cl^- current affects spike frequency adaptation (Ha et al., 2016), a Ca^{2+}-dependent gradual prolongation of interspike intervals in response to a long, depolarizing current input.

Several pieces of evidence demonstrate that this Ca^{2+}-activated Cl^- current is mediated by TMEM16B. First TMEM16B, and not TMEM16A, is expressed at the mRNA and protein levels in the thalamic region, and is localized at the membrane near the soma of thalamocortical neurons (Ha et al., 2016). Secondly, thalamocortical neurons from mice infected with adenovirus containing TMEM16B small-interference RNA have a lower TMEM16B expression, and it is noteworthy that they show a reduced NFA-sensitive current, a reduction in slow-type AHP following a depolarizing step, higher tonic firing rates, and less spike accommodation (Ha et al., 2016).

Spike adaptation is present in several neuronal types and is ascribed to slow type AHP, which is frequently mediated by Ca^{2+}-activated K^+ channels. To our knowledge, this is the first case in which a Ca^{2+}-activated Cl^- current has been shown to be involved in AHP.

Behavioral experiments showed also in vivo that TMEM16B function is important for transmission from the thalamus to the cortex of sensory information received from nociceptive nerves. Indeed, mice with thalamic restricted knockdown of TMEM16B exhibited enhanced pain response in persistent types of pain, such as visceral pain, but not clearly in acute pain (Ha et al., 2016).

Olfactory Bulb

Several groups reported the expression of TMEM16B in the olfactory bulb (Billig et al., 2011; Y. Zhang et al., 2017). It is expressed in the axons of olfactory sensory neurons entering the olfactory glomeruli, where they coalesce. Expression in other cell types has not been reported.

Neurons Outside the Central Nervous System

Outside the CNS, Ca^{2+}-activated Cl^- currents are present in many mammalian sensory neurons—in particular, olfactory sensory neurons and DRG neurons—and in these cell types, the currents have been investigated in detail.

Dorsal Root Ganglion and Trigeminal Neurons

In 1985, a Ca^{2+}-activated Cl^- current was described in a subset of DRG sensory neurons, in which Ca^{2+} entry due to depolarizing stimuli caused Ca^{2+}-activated Cl^- current activation, leading to depolarizing after-potentials (Mayer, 1985). Since the intracellular Cl^- concentration is high in DRG neurons (Alvarez-Leefmans, Gamino, Giraldez, & Nogueron, 1988; Kenyon, 2000; Rocha-Gonzalez, Mao, & Alvarez-Leefmans, 2008), E_{Cl} is more positive than the resting-membrane potential, therefore the Ca^{2+} -induced Cl conductance is excitatory, contributing to membrane depolarization.

In particular, small-diameter DRG neurons, which respond specifically to noxious stimuli, have been shown to express a TMEM16A-mediated Ca^{2+}-activated Cl^- current (Liu et al., 2010). In these nociceptive DRG neurons, TMEM16A is located in specific signaling microdomains tethered to the endoplasmic reticulum (ER) and are consequently activated after Ca^{2+}-release from ER stores, induced by exposure to bradykinin noxious stimuli (Jin et al., 2013).

TMEM16A is co-expressed in these neurons with transient receptor potential vanilloid 1 (TRPV1) channels (Cho et al., 2012), cationic channels activated by heat, acid, and capsaicin. The interaction of TRPV1 and TMEM16A constitutes a significant pain-enhancing mechanism, since Ca^{2+} influx through TRPV1 channels, activated by noxious stimulus, activates anionic efflux through TMEM16A, which further depolarizes the neuron (Takayama, Uta, Furue, & Tominaga, 2015).

Additionally, TMEM16A responds also to heat, being activated by temperatures above 44°C (Cho et al., 2012) and its activation by heat is sufficient to depolarize $Trpv1^{-/-}$ DRG neurons (Cho et al., 2012). In accordance, RNAi knockdown or genetic deletion of TMEM16A in DRG neurons leads to reduced neuropathic pain-related behaviors induced by heat (Cho et al., 2012) or chemical noxious stimuli, like bradykinin (Lee et al., 2014).

On one side, these results make TMEM16A an interesting drug target, since specific inhibitors could be used to induce analgesia. On the other side, activators of TMEM16A that could be useful in other situations, such as in treatment of cystic fibrosis, could have unpleasant side effects.

Although the expression of not only TMEM16A, but also TMEM16B, has been reported in mouse DRG neurons (Boudes et al., 2009), its contribution to Ca^{2+}-activated Cl^- current in DRG neurons has not been demonstrated yet.

Trigeminal ganglia also contain a subset of neurons expressing Ca^{2+}-activated Cl^- channels (Bader, Bertrand, & Schlichter, 1987; Schlichter, Bader, Bertrand, Dubois-Dauphin, & Bernheim, 1989) and expressing TMEM16A (Schöbel et al., 2012). Small and medium-sized trigeminal neurons innervating the tongue express TMEM16A, in most cases together with TRPV1 (Kanazawa & Matsumoto, 2014), suggesting a cooperation between the two channels in signaling pathways involved in nociception in trigeminal, as in DRG neurons.

Olfactory Sensory Neurons

Ca^{2+}-activated Cl^- currents in vertebrate olfactory sensory neurons were first described in the early 1990s (Kleene, 1993; Kleene & Gesteland, 1991; Kurahashi & Yau, 1993; Lowe

& Gold, 1993) and are the first neuronal Ca^{2+}-activated Cl^- channels extensively studied (reviewed in Dibattista, Pifferi, Boccaccio, Menini, & Reisert, 2017). The binding of odorants to olfactory receptors in the cilia of olfactory sensory neurons produces an increase in cyclic adenosine monophosphate (cAMP) concentration: Ca^{2+} enters in the cilia through cyclic nucleotide gated (CNG) channels and activates an excitatory anionic current, due to the high intraciliar Cl^- concentration (55 to 69 mM; Kaneko, 2004; Reuter et al., 1998). The Ca^{2+}-activated Cl^- current carries most of the odorant-induced receptor current (Li, Ben-Chaim, Yau, & Lin, 2016); up to 90% in rodents (Boccaccio & Menini, 2007; Lowe & Gold, 1993).

TMEM16B was found to be abundantly expressed in the cilia of olfactory sensory neurons (Stephan et al., 2009), to have biophysical properties like those of the native olfactory channel (Pifferi, Cenedese, & Menini, 2012; Pifferi et al., 2009; Sagheddu et al., 2010; Stephan et al., 2009); and finally, the Ca^{2+}-activated Cl^- current is completely abolished in neurons from TMEM16B KO mice (Billig et al., 2011; Pietra, Dibattista, Menini, Reisert, & Boccaccio, 2016).

The transduction current elicited by odorant recognition produces a depolarization leading to generation of action potentials; additionally, cAMP fluctuations, due to basal activity of the olfactory receptors, are sufficient to cause CNG and TMEM16B channel activation with generation of spontaneous firing activity.

In TMEM16B KO mice, olfactory sensory neurons expressing the same olfactory receptor showed (1) a prolonged firing activity in response to odor stimulus, (2) a reduced spontaneous firing activity, and (3) altered wiring between olfactory epithelium and olfactory bulb (Pietra et al., 2016), which may affect the spatial representation of odors in the bulb.

Although earlier behavioral experiments suggested a limited role for olfactory physiology and behavior (Billig et al., 2011), more recent data suggest that TMEM16B is important to locating unfamiliar (Pietra et al., 2016) or weak odor sources (Neureither, Stowasser, Frings, & Mohrlen, 2017), and their memorization (Neureither, Stowasser, et al., 2017).

However, at this stage, a contribution to the behavioral deficit of other brain areas where TMEM16B is expressed cannot be excluded, although it is unlikely, since TMEM16B KO mice did perform normally well with familiar odors and did not show deficits in olfactory learning (Billig et al., 2011; Pietra et al., 2016).

Vomeronasal Sensory Neurons

Vomeronasal sensory neurons serve as pheromone detectors in the vomeronasal sensory organ (VNO). Ca^{2+}-activated Cl^- currents have been described in mouse vomeronasal sensory neurons (Dibattista et al., 2012; Kim, Ma, & Yu, 2011; Yang & Delay, 2010), and contribute to the response to urine (Kim et al., 2011; Yang & Delay, 2010). TMEM16A and TMEM16B are both expressed at the apical surface of the vomeronasal neuroepithelium (Dauner, Lissmann, Jeridi, Frings, & Mohrlen, 2012; Dibattista et al., 2012); in particular, in the microvilli of vomeronasal sensory neurons (Dibattista et al., 2012). Although early experiments suggested that Ca^{2+}-activated Cl^- currents in vomeronasal sensory neurons were mainly mediated by TMEM16B (Billig et al., 2011), more recent data obtained with conditional KO mice for TMEM16A show that TMEM16A is on the contrary a necessary component of Ca^{2+}-activated Cl^- channels (Amjad et al., 2015). Since both TMEM16A and TMEM16B are expressed in single vomeronasal sensory neurons, these data could be explained by the presence of heteromeric channels composed of TMEM16A and TMEM16B VNO-specific isoforms, both required to form functional channels. An alternative explanation is that the two studies, the two studies investigated different populations of vomeronasal neurons.

Cochlea

Both TMEM16A and TMEM16B are expressed in spiral ganglion neurons from mouse cochlea sections from pups and young animals (X. D. Zhang et al., 2015). These neurons face remarkable changes in firing activity during development, correlated with changes in intracellular Cl^- concentration, from 90 mM in the pre-hearing stage to 20 mM in post-hearing mature neurons (X. D. Zhang et al., 2015). Pre-hearing neurons can show a rhythmic and burst-patterned spontaneous action potential activity (Jones & Jones, 2000; Jones, Jones, & Paggett, 2001), which occurs less frequently in mature neurons (X. D. Zhang et al., 2015). Ca^{2+}-activated Cl^- currents depolarize the resting membrane potential favoring spontaneous activity that is indeed attenuated in neurons from TMEM16A KO-animals (X. D. Zhang et al., 2015). Further studies are also needed to clarify a possible involvement of TMEM16B. Additionally, TMEM16A was found in the organ of Corti, where it is localized at medial olivocochlear efferent nerve endings, which form synapses on the outer hair cell somata (Jeon et al., 2011).

Overview of the CLC Family of Chloride Channels and Transporters

CLC proteins are found in all phyla. They are homodimeric membrane proteins with separate anion transport pathways in each subunit. Such a "double-barreled" structure (Figure 19.3) had been inferred from functional analysis of the prototype ClC-0 channel from the electric organ of *Torpedo* (Ludewig, Pusch, & Jentsch, 1996; Middleton, Pheasant, & Miller, 1996; Miller, 1982), and fully confirmed by all available structures of CLC proteins from various organisms (Dutzler, Campbell, Cadene, Chait, & MacKinnon, 2002; Feng, Campbell, Hsiung, & Mackinnon, 2010; Jayaram, Robertson, Wu, Williams, & Miller, 2011; Lim, Stockbridge, & Miller, 2013; Park, Campbell, & MacKinnon, 2017) (Figure 19.4). Based on the fact that the

Torpedo ClC-0 is a Cl^- channel of sizable conductance (around 8 pS in physiological solutions; Bauer, Steinmeyer, Schwarz, & Jentsch, 1991; Ludewig et al., 1996; Miller, 1982; Figure 19.3), and that also the first cloned mammalian CLCs, including ClC-1 (Pusch, Steinmeyer, & Jentsch, 1994), ClC-2 (Weinreich & Jentsch, 2001), and ClC-Ka/Kb (Fischer, Janssen, & Fahlke, 2010; L'Hoste et al., 2013), are clearly passive Cl^- channels with conductances ranging from ~1.5 pS to 40 pS, it was assumed that all CLC proteins are anion channels. This included the *E. coli* EcClC-1 protein, whose crystal structure was determined in 2002 (Dutzler et al., 2002). The structure fully confirmed the homodimeric double-barreled architecture and revealed important molecular insights into CLC function. However, it was a big surprise that EcClC-1 is actually not a passive anion channel but a strictly coupled, secondary, active Cl^-/H^+ antiporter with a 2 Cl^- : 1 H^+ stoichiometry (Accardi & Miller, 2004). Turnover of EcClC-1 is relatively high for a typical transporters (around 4000 sec^{-1}; Walden et al., 2007), but about two to three orders of magnitude "slower" compared to the previously mentioned CLC channels. Later it was found that five mammalian CLCs—i.e., ClC-3–ClC-7—are actually Cl^-/H^+ antiporters (Guzman, Grieschat, Fahlke, & Alekov, 2013; Leisle, Ludwig, Wagner, Jentsch, & Stauber, 2011; Neagoe, Stauber, Fidzinski, Bergsdorf, & Jentsch, 2010; Picollo & Pusch, 2005; Scheel, Zdebik, Lourdel, & Jentsch, 2005). For some of these, the stoichiometry was determined and found to be 2 Cl^- : 1 H^+, identical to the bacterial homologue (Leisle et al., 2011; Zifarelli & Pusch, 2009a). All these mammalian CLC antiporters are localized to the endo/lysosomal system; i.e., they are predominantly expressed in intracellular membranes, where they are involved in pH and/or intraluminal [Cl^-] regulation (Jentsch & Pusch, 2018).

Interestingly, despite the different thermodynamic properties of CLC channels and transporters, they share several structural features. In particular, a highly conserved glutamate residue in the center of the pore serves as a gating element in CLC channels and is strictly essential for coupled Cl^-/H^+ exchange in CLC transporters (Accardi & Miller, 2004; Dutzler, Campbell, & MacKinnon, 2003; Traverso, Elia, & Pusch, 2003). This residue is also called the "gating glutamate" (Zdebik et al., 2008) (see Figure 19.4).

CLC channels and transporters all depend on voltage, pH, and extra- and intracellular chloride concentrations in a manner that is highly interrelated with the pore occupancy of anions and the protonation state of the gating glutamate (Pusch, Ludewig, Rehfeldt, & Jentsch, 1995; Zifarelli & Pusch, 2007, 2010). To add to this complexity, gating of CLC channels comes in two flavors: a fast, or protopore, gate acts on single pore of the double-barreled channel, while a slow, or common, gate depends on both subunits (Figure 19.3). The protopore gate mostly depends on the gating glutamate, whereas the common gate depends on many protein structures, including the pore, the dimer interface, and the cytoplasmic C-terminus (Zifarelli & Pusch, 2007). The C-terminus of all eukaryotic CLC proteins contains two so-called cystathionine-ß-synthase (CBS) domains (Ponting, 1997). The CBS domains are organized in pairs that dimerize to form a stable globular domain. As in other proteins, CBS1 and CBS2 form

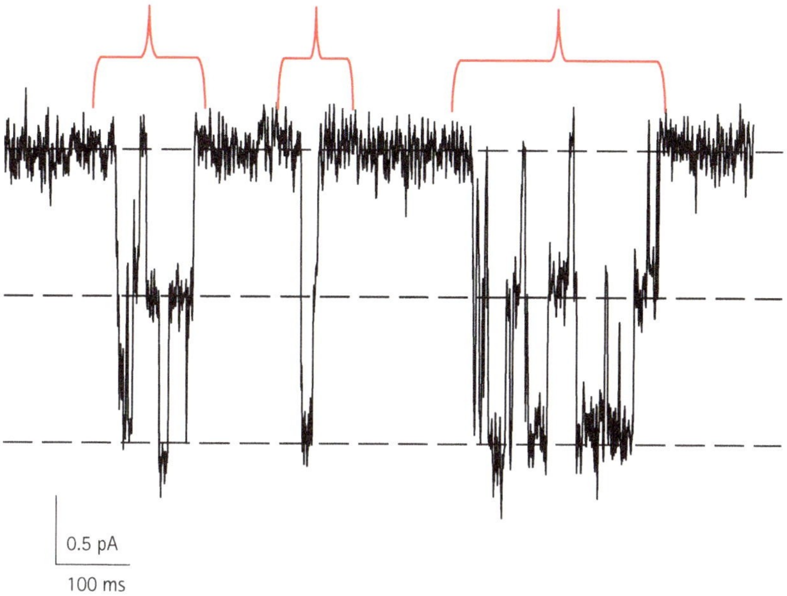

FIGURE 19.3 **Double barreled structure of CLC-channels.** Single ClC-0 channel recorded at −100 mV in an inside-out patch from a cRNA injected oocyte. Bursts of opening are indicated by red brackets.

a sandwich structure via their beta strands (Figure 19.4), with an adenine nucleotide bound between the two domains in C-terminal fragments of ClC-5 (Meyer, Savaresi, Forster, & Dutzler, 2007). Interestingly, CBS1 is relatively far away from the membrane domain, whereas CBS2 is in close contact with the membrane part (Feng et al., 2010; Park et al., 2017) (Figure 19.4). CBS domain pairs from a variety of proteins, including AMP-activated protein kinase, inosine monophosphate dehydrogenase-2, ClC-2, ClC-5, and CBS itself bind adenosine monophosphate, adenosine triphosphate (ATP), and other adenine nucleotides, suggesting that CBS domains serve as sensors of the cellular energy status (Scott et al., 2004; Wellhauser et al., 2006). No nucleotides were found in X-ray structures of C-terminal fragments from ClC-0 and ClC-Ka or the full-length structures of CmClC and ClC-K (Feng et al., 2010; Markovic & Dutzler, 2007; Meyer & Dutzler, 2006; Park et al., 2017), but adenosine triphosphate or adenosine diphosphate molecules were tightly bound between the two CBS domains in crystal structures of the isolated C-terminal fragment from ClC-5 (Meyer et al., 2007). Functional studies confirmed a regulatory role of nucleotide binding for ClC-1 and ClC-5 (Bennetts et al., 2005; Zifarelli & Pusch, 2009b). The physiological role of nucleotide regulation remains unclear, however, since all adenine nucleotides are almost equally effective, and their summated concentration is generally in the millimolar range.

In the following section, the knowledge on the physiological role of CLC proteins in the brain will be discussed.

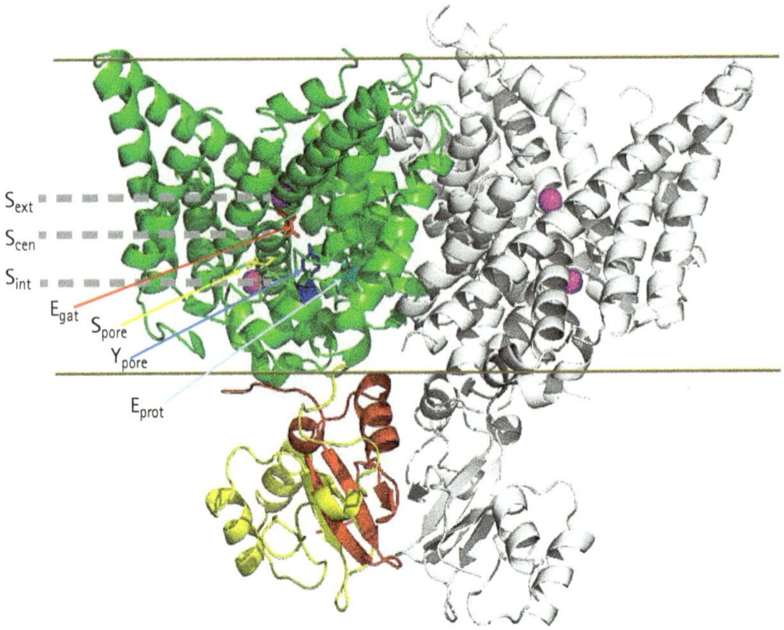

FIGURE 19.4 **Structure of a CLC transporter.** The crystal structure of CmClC (entry in Brookhaven pdb database: 3ORG) is shown in a lateral view in cartoon representation with the approximate position of the membrane indicated by horizontal lines. One monomer of the dimeric transporter is shown in light gray, while the membrane embedded part of the other subunit is shown in green. The intracellular CBS domains of the same subunit are colored in yellow (CBS1) and red (CBS2), respectively. Bound chloride ions are shown as pink spheres. Positions of chloride binding sites are indicated by dashed lines. Functionally important amino acids are shown colored in stick representation and include in red E_{gat} (the gating glutamate, E210 in CmClC), in yellow S_{pore}, the pore serine (S165 in CmClC), in dark blue Y_{pore}, the pore tyrosine (Y515 E210 in CmClC), and in light blue E_{prot}, the proton glutamate (whose equivalent is threonine 269 in CmClC).

ClC-2: The Only CLC Channel That Is Significantly Expressed in the Brain

Among the CLC channels—i.e., ClC-1, ClC-2, ClC-Ka, and ClC-Kb—only ClC-2 is significantly expressed in the CNS (Steinmeyer, Ortland, & Jentsch, 1991; Thiemann, Gründer, Pusch, & Jentsch, 1992) (apart from CLC-K channels, which are important for fluid transport in inner-ear epithelia; Estévez et al., 2001; Jentsch, 2005; Jentsch, Poët, Fuhrmann, & Zdebik, 2005; Sage & Marcus, 2001)). In heterologous expression systems, ClC-2 is characterized by an inwardly rectifying phenotype; i.e., the channel is practically closed at voltages greater than = −30 mV, and currents activate slowly (time scale in the hundreds of milliseconds to seconds range) at negative voltages (Gründer, Thiemann, Pusch, & Jentsch, 1992; Thiemann et al., 1992). ClC-2 activity is also strongly dependent on extracellular pH in a biphasic manner with inhibition at alkaline pH and inhibition by acidification to pH < 6.5 (Arreola, Begenisich, & Melvin, 2002; Jordt & Jentsch, 1997). Inhibition by low pH is mediated by an extracellularly accessible histidine at the N-terminus of helix Q (Niemeyer, Cid, Yusef, Briones, & Sepúlveda, 2009), whereas the activating effect is mediated by protonation of the "gating glutamate" (Niemeyer et al., 2009). Furthermore, activation of ClC-2 is also strongly dependent on the intracellular Cl^- concentration (Niemeyer, Cid, Zúñiga, Catalán, & Sepúlveda, 2003; Pusch, Jordt, Stein, & Jentsch, 1999). The voltage-dependence of ClC-2 is most likely indirectly caused by a voltage-dependent displacement of the gating glutamate by intracellular Cl^- ion and a possible contribution from the protonation of the gating glutamate (De Jesus-Perez et al., 2016; Niemeyer et al., 2009; Sanchez-Rodriguez, De Santiago-Castillo, & Arreola, 2010; Sanchez-Rodriguez et al., 2012). Raising the intracellular Cl^- concentration shifts the voltage-dependence of ClC-2 to more positive potentials and thereby opens the channel (De Jesus-Perez et al., 2016; Niemeyer et al., 2003; Pusch et al., 1999; Sanchez-Rodriguez et al., 2010; Sanchez-Rodriguez et al., 2012). The dependence on the intracellular Cl^- concentration is of likely physiological relevance. A peculiar and unique feature of the gating of ClC-2 is that it strongly depends on the N-terminal region, deletions in which cause it to be constitutively open (Gründer et al., 1992). The mechanisms underlying this regulation are unknown, but they may involve a region encompassing the intracellular loop between helices J and K (Jordt & Jentsch, 1997).

ClC-2 KO mice show testicular and retinal degeneration (Bösl et al., 2001) and a leukoencephalopathy (Blanz et al., 2007). The channel has been proposed to play a role in ion homeostasis in tissues with small extracellular spaces.

ClC-2 in Neurons

The ClC-2 protein is expressed in neurons (Sik, Smith, & Freund, 2000), and ClC-2 like, hyperpolarization activated Cl^- currents have been recorded in several neuronal cell types (Chesnoy-Marchais, 1983; Clark, Jordt, Jentsch, & Mathie, 1998; Clayton, Staley, Wilcox, Owens, & Smith, 1998; Enz, Ross, & Cutting, 1999; Földy, Lee, Morgan, & Soltesz, 2010; Madison, Malenka, & Nicoll, 1986; Petheo, Molnar, Roka, Makara, & Spat, 2001; Rinke, Artmann, & Stein, 2010; Smith, Clayton, Wilcox, Escudero, & Staley, 1995; Staley, 1994). Proof that the channel underlying the current in CA1 neurons is indeed ClC-2 was provided by Rinke et al., who showed that the current is absent in ClC-2 KO mice (Rinke et al., 2010).

In neurons, ClC-2 has been proposed to be important for the maintenance of a low intracellular Cl^- concentration ($[Cl^-]_{int}$), which is important to assure the inhibitory effect of $GABA_A$ receptor activation. Indeed, overexpression of ClC-2 stabilizes a low $[Cl^-]_I$ in DRG cells (Staley, Smith, Schaack, Wilcox, & Jentsch, 1996), and Rinke et al. suggested that ClC-2 serves a dual role in neurons. These authors recorded rather large, typical ClC-2 currents from CA1 pyramidal cells and found that ClC-2 contributes about 40% to their resting conductance (Rinke et al., 2010). In agreement with such a large conductance, CA1 neurons from ClC-2 KO mice were intrinsically hyperexcitable. Surprisingly, however, in field recordings this did not result in increased basal synaptic transmission or to decreased input/output relations, results that were hypothesized to be caused by an increased feed-forward inhibition by hyperexcitable interneurons (Rinke et al., 2010). Rinke et al. concluded that ClC-2 provides a background Cl^- conductance that dampens excitability, and that the channel provides an efflux pathway for Cl^- ions, thus helping in keeping a low $[Cl^-]_{int}$ (Rinke et al., 2010). Also, Földy et al., comparing results on WT and ClC-2 KO mice, concluded that ClC-2 helps in keeping a low $[Cl^-]_i$ in parvalbumin-expressing, fast-spiking basket cells (Földy et al., 2010). The direction of Cl^- flow through any Cl^- channel depends on the electrochemical driving force; i.e., Cl^- will flow out of the cell if the membrane potential is more negative than the Nernst potential for Cl^-. Indeed, using computer modeling, Ratté and Prescott challenged the hypothesis that ClC-2 serves as an efflux pathway (Ratté & Prescott, 2011). However, their model appeared to lack two important ingredients. First, the activity of the Na-K ATPase was not considered. This pump will increase intracellular K^+, leading to a more negative membrane potential via the activity of background K^+ channels. In addition, this pump directly hyperpolarizes the cell due to its electrogenicity. Furthermore, the model of Ratté and Prescott did not incorporate the pronounced dependence of the open probability of ClC-2 on $[Cl^-]_i$ (Niemeyer et al., 2003; Pusch et al., 1999). Thus, the predictions of the model remain difficult to interpret.

In any case, the ClC-2-mediated Cl^- conductance does not seem to be fundamental for neuronal function, because ClC-2 KO mice do not show symptoms that appear to be directly correlated with neuronal dysfunction, but are rather caused by its defective function in glia (Blanz et al., 2007; Hoegg-Beiler et al., 2014) (see further in this chapter). A suspected role of ClC-2 in epilepsy could not be confirmed (Niemeyer et al., 2010).

ClC-2 in Glia

In the brain, ClC-2 is of utmost importance in glia. The protein had been localized in astrocytes, especially in astrocytic end feet (Sik et al., 2000). ClC-2 like currents are seen in astrocytes, which are particularly large when astrocytes cultured for several days in the presence of membrane permeable cAMP analogs (Ferroni, Marchini, Nobile, & Rapisarda, 1997; Nobile, Pusch, Rapisarda, & Ferroni, 2000). Smaller ClC-2-like currents were seen in "complex" astrocytes in acute slices or acutely isolated astrocytes (Makara et al., 2003). These currents were absent in astrocytes from ClC-2 KO mice (Makara et al., 2003).

The physiological importance of ClC-2 in glia became evident from the brain phenotype of ClC-2 KO mice that showed fluid-filled "vacuoles" between myelin sheaths of the central but not the peripheral nervous system (Blanz et al., 2007). This phenotype was similar to what is seen in patients with the rare genetic disease called megalencephalic leukoencephalopathy with subcortical cysts (MLC). However, no mutations were found in the *CLCN2* gene in MLC patients (Scheper et al., 2010). Most MLC patients carry mutations in *MLC1*, encoding a membrane protein of unknown function (Leegwater et al., 2001). Later, it was found that the glia-specific cell-adhesion molecule GlialCAM binds to MLC1, and that most MLC patients without mutations in *MLC1* had mutations in the *GLIALCAM* gene (Lopez-Hernandez et al., 2011). GlialCAM, previously called hepaCAM, is a type 1 transmembrane protein with two extracellular Ig domains (Moh, Zhang, Luo, Lee, & Shen, 2005) that is specifically expressed in glia (Favre-Kontula et al., 2008). Surprisingly, GlialCAM also binds to ClC-2 (Jeworutzki et al., 2012). In heterologous expression systems, GlialCAM strongly clusters at cell–cell contacts and also drags MLC1 and ClC-2 at the cell contacts when co-expressed (Jeworutzki et al., 2012). I n vivo, co-expression and clustering of GlialCAM and ClC-2 are seen in astrocyte–astrocyte and astrocyte–oligodendrocyte contacts (Hoegg-Beiler et al., 2014; Jeworutzki et al., 2012). In heterologous expression systems, and in oligodendrocytes, but not in Bergmann glia, GlialCAM also increases ClC-2 currents at positive voltages, partially abolishing the inward rectification by activating the common gate of the channel (Barrallo-Gimeno, Gradogna, Zanardi, Pusch, & Estévez, 2015; Hoegg-Beiler et al., 2014; Jeworutzki et al., 2014; Jeworutzki et al., 2012). Many MLC causing *GLIALCAM* mutations abolish the clustering of GlialCAM at cell contacts but retain the ability to increase ClC-2 currents (Arnedo, Aiello, et al., 2014; Arnedo, Lopez-Hernandez, et al., 2014; Jeworutzki et al., 2014). Biochemical interaction of GlialCAM with ClC-2 and MLC1 involves the extracellular Ig domains but also the cytoplasmic C-terminus, whereas the initial part of the transmembrane domain is mostly responsible for the functional alterations of ClC-2 currents (Capdevila-Nortes et al., 2015). The physiological importance of the interaction of GlialCAM with ClC-2 and MLC1 was confirmed by several KO mouse models (Bugiani et al., 2017; Dubey et al., 2015; Hoegg-Beiler et al., 2014; Sirisi et al., 2014). A detailed investigation of the physiological roles and interactions of ClC-2, GlialCAM, and MLC1 was performed by Hoegg-Beiler et al. (2014), who studied KO mice of all three genes in various combinations (see Figure 19.5). GlialCAM was found to be important for the targeting of MLC1 and of ClC-2 to specific glial domains. Surprisingly, ClC-2 currents were inwardly rectifying in astrocytes, but almost linear in oligodendrocytes, even though GlialCAM is expressed in both cell types. Interestingly, even though MLC1 and ClC-2 do not appear to directly interact, the KO of MLC1 led to a mis-localization of GlialCAM and of ClC-2 (Figure 19.5) and abolished the activation of ClC-2 currents at positive voltages in oligodendrocytes (Hoegg-Beiler et al., 2014). In vitro, treatment of cultured astrocytes with depolarizing high K^+ solutions induced the appearance of a ternary complex of GlialCAM, ClC-2 and MLC1 (Sirisi et al., 2017). However, the precise role of the localization of ClC-2 channels at cell junctions is unclear, as is the role of MLC1. In humans, a small number of patients with homozygous or compound heterozygous loss-of-function mutations in *CLCN2* have been described that exhibit leukodystrophy (among other symptoms), confirming the role of ClC-2 in glia (Depienne et al., 2013; Di Bella et al., 2014; Giorgio et al., 2017; Hanagasi et al., 2015; Zeydan et al., 2017).

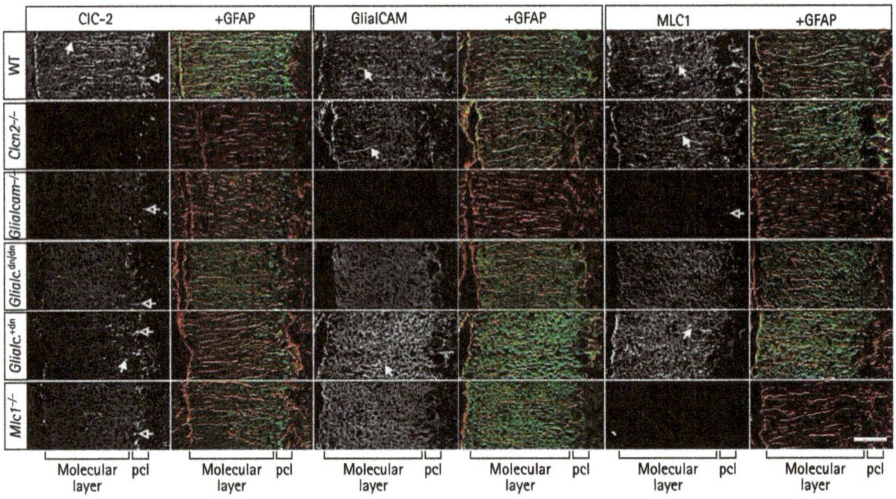

FIGURE 19.5 **Localization of ClC-2, GlialCAM, and MLC1 in Bergmann glia and effects of their knockout.** Immunohistochemical staining shows that KO of GlialCAM reduces ClC-2 as well as MLC1 expression, KO of MLC1 reduces ClC-2 expression and leads to GlialCAM mislocalization, whereas ClC-2 KO has no appreciable effect on GlialCAM or MLC1 expression. Co-staining for the astrocytic cytoskeletal protein GFAP (red) visualizes Bergmann glia processes. For details, see Hoegg-Beiler et al (2014).

ClC-3–ClC-7: Overview

It is now well established that ClC-3 to ClC-7 are all endolysosomal Cl^-/H^+ antiporters (Jentsch & Pusch, 2018). Other functional properties have been ascribed especially to ClC-3. These include, e.g., being volume-regulated anion channels, Ca^{2+}-activated Cl^- channels, or acid-activated Cl^- channels. However, all these alternative hypotheses can been safely ruled out (see, e.g., Jentsch, 2005; Stauber & Jentsch, 2013; Zifarelli & Pusch, 2007; Jentsch & Pusch, 2018). Despite their normally intracellular localization, ClC-4 and ClC-5 reach the plasma membrane in heterologous expression systems and can be functionally studied (Friedrich, Breiderhoff, & Jentsch, 1999; Steinmeyer, Schwappach, Bens, Vandewalle, & Jentsch, 1995). The same was found for a specific splice variant of ClC-3 (Guzman, Miranda-Laferte, Franzen, & Fahlke, 2015). ClC-3, ClC-6, and ClC-7 can be redirected to the plasma membrane by interfering with targeting signals (Guzman et al., 2015; Leisle et al., 2011; Neagoe et al., 2010; Stauber & Jentsch, 2010). Among the intracellular CLC transporters, ClC-3, ClC-4, and ClC-5 share rather high sequence similarity and exhibit similar functional properties in heterologous expression systems (Friedrich et al., 1999; Guzman et al., 2013; X. Li, Shimada, Showalter, & Weinman, 2000; Picollo & Pusch, 2005; Steinmeyer et al., 1995; Zdebik et al., 2008), but only ClC-3 and ClC-4 are significantly expressed in neurons

(Jentsch, Günther, Pusch, & Schwappach, 1995; Stobrawa et al., 2001; van Slegtenhorst et al., 1994). All intracellular CLC transporters exhibit a striking "outward rectification," corresponding to an inward movement of protons into the endo-/lysosomal lumen and a corresponding movement of Cl^- ions into the cytosol (Friedrich et al., 1999; Guzman et al., 2015; Leisle et al., 2011; Li et al., 2000; Neagoe et al., 2010; Steinmeyer et al., 1995). This outward rectification is puzzling because transmembrane voltages in endo/lysosomal membranes are not expected to reach values necessary for the activation of these transporters. The transport stoichiometry has been quantitatively determined as 2 Cl^-/1 H^+ for ClC-5 (De Stefano, Pusch, & Zifarelli, 2013; Zifarelli & Pusch, 2009a), and for ClC-7 (Leisle et al., 2011), exactly like the bacterial EcClC-1 (Accardi & Miller, 2004). Most likely, ClC-3, ClC-4, and ClC-6 exhibit the same stoichiometry. ClC-7 is unique among the CLC transporters in that it is obligatorily associated with a small beta subunit called Ostm1 (osteopetrosis-associated transmembrane protein 1) (Barrallo-Gimeno et al., 2015; Lange, Wartosch, Jentsch, & Fuhrmann, 2006). Ostm1 is a single-span transmembrane protein that is heavily glycosylated on its extracellular N-terminus (Lange et al., 2006). Ostm1 strictly co-localizes with ClC-7, and it is needed for the protein stability of ClC-7 (Lange et al., 2006). Lysosomal targeting of the ClC-7/Ostm1 complex depends mostly on ClC-7 (Lange et al., 2006).

Two major hypotheses have been put forward for the role of intracellular CLC transporters. First, they could help in an efficient acidification, providing a charge compensation for the electrogenic V-type H^+ ATPase (Jentsch, 2008). Even though *a priori* counterintuitive, under certain assumptions, a 2:1 Cl^-/H^+ antiporter can allow in principle a more acidic luminal pH compared to the activity of a simple shunting Cl^- channel (Marcoline, Ishida, Mindell, Nayak, & Grabe, 2016; Weinert et al., 2010). In line with this hypothesis, defects in acidification rates and in steady-state pH (in osteoclasts) have been observed in several KO models (Günther, Piwon, & Jentsch, 2003; Hara-Chikuma, Wang, Guggino, Guggino, & Verkman, 2005; Kornak et al., 2001; Stobrawa et al., 2001). In disagreement with an essential role for vesicle acidification, normal lysosomal pH was found in ClC-7 KO brain cells (Kasper et al., 2005; Weinert et al., 2010). In fact, an alternative role of a coupled 2:1 Cl^-/H^+ antiporter could be to set the luminal Cl^- concentration according to the pH gradient (Steinberg et al., 2010; Weinert et al., 2010; Jentsch & Pusch, 2018). A high luminal Cl^- concentration could fuel (unknown) secondary active solute transporters, or it could be important to directly activate luminal enzymes like, for example, cathepsin C (Cigic & Pain, 1999). Evidence in favor of this hypothesis was recently provided by Chakraborty et al., who found that lysosomes have high luminal Cl concentrations and that a reduction of luminal Cl impaired lysosomal function (Chakraborty, Leung, & Krishnan, 2017).

ClC-3 and ClC-4: Endosomal Cl^-/H^+ Antiporters

ClC-3 and ClC-4 are both rather ubiquitously expressed in practically all tissues, including brain (Borsani, Rugarli, Taglialatela, Wong, & Ballabio, 1995; Kawasaki et al., 1999; Kawasaki et al., 1994; Mohammad-Panah et al., 2002; Stobrawa et al., 2001; van Slegtenhorst et al.,

1994). A prominent role of ClC-3 in the CNS was deduced from the severe brain phenotype of ClC-3 seen in ClC-3 KO mice (Stobrawa et al., 2001). The mice develop initially rather normally, but severe neurodegeneration leads to a complete loss of the hippocampi by three months of age (Stobrawa et al., 2001). In addition, the mice show retinal degeneration due to a loss of photoreceptors (Stobrawa et al., 2001), and a generalized neurodegeneration was seen also in two other KO models (Dickerson et al., 2002; Yoshikawa et al., 2002). ClC-3 KO mice show behavioral deficits, but only a slightly reduced lifespan (Dickerson et al., 2002; Stobrawa et al., 2001; Yoshikawa et al., 2002). In agreement with a neurodegenerative phenotype, KO mice exhibited impaired motor skills but were able to improve their performance despite the almost complete absence of the hippocampus (Stobrawa et al., 2001). The precise reason for the neurodegenerative phenotype in ClC-3 KO mice is unknown. It can be speculated that the transporter is involved in endosomal trafficking, which is critically important in the neuronal cell population.

ClC-3 has been shown to be localized to synaptic vesicles (Maritzen, Keating, Neagoe, Zdebik, & Jentsch, 2008; Salazar et al., 2004; Stobrawa et al., 2001) and to synaptic microvesicles in neuroendocrine cells (Maritzen et al., 2008; Salazar et al., 2004), and it has been proposed to be important for glutamatergic (Guzman, Alekov, Filippov, Hegermann, & Fahlke, 2014; Stobrawa et al., 2001) and GABAergic (Riazanski et al., 2011) neurotransmission. Since glutamate uptake into synaptic vesicles is Cl^--dependent, a lack of ClC-3 might lead to reduced neurotransmitter contents. However, for unknown reasons, expression of the vesicular glutamate transporter vGlut1 is reduced in ClC-3 KO mice (Stobrawa et al., 2001). In addition, vGlut1 has been shown to mediate a Cl^- conductance itself that is probably sufficient to ensure glutamate uptake (Schenck, Wojcik, Brose, & Takamori, 2009); and the localization of ClC-3 to synaptic vesicles has been questioned (Schenck et al., 2009). Thus, it cannot be excluded that the observed effects on neurotransmission are indirectly caused by a general dysfunction of vesicular trafficking.

Despite its prominent expression in the brain, mice lacking ClC-4 did not show any overt phenotype (Rickheit et al., 2010). Recently, however, in humans, mutations in the *CLCN4* gene have been linked to hereditary forms of intellectual disability and epilepsy (Hu et al., 2016; Palmer et al., 2016), indicating that ClC-4 plays an important role in the brain. Because in humans the gene encoding ClC-4 is located on the X-chromosome, these syndromes mostly affect males, but certain disease symptoms are also observed in heterozygous females (Palmer et al., 2016). Most mutations associated with these syndromes lead to a loss of function in heterologous expression systems (Hu et al., 2016; Palmer et al., 2016; Veeramah et al., 2013). The mechanisms underlying these diseases remain largely unclear. Due to their high similarity, it is also conceivable that ClC-3 and ClC-4 form heterodimers, possibly in a cell type–specific manner.

ClC-6: A Neuronal Late Endosomal Cl^-/H^+ Antiporter

Initially, ClC-6 was described as being rather ubiquitously expressed (Brandt & Jentsch, 1995; Kida, Uchida, Miyazaki, Sasaki, & Marumo, 2001), but the protein is predominantly observed in the CNS in the late endosomal cellular compartment (Poët et al., 2006). ClC-6 KO mice are fertile and have a rather normal life span (Poët et al., 2006). However, they develop lysosomal storage disease associated with lipofuscin accumulation, which was particularly prominent in dorsal root ganglia (Poët et al., 2006), a feature that might be related to a reduced pain sensitivity in tail flick assays. Like for ClC-3, it can be speculated that ClC-6 is critically important for neuronal endolysosomal trafficking.

ClC-7: A Ubiquitous Lysosomal Cl^{-}/H^{+} Antiporter

Mutations in the genes encoding ClC-7 and Ostm1 underlie osteopetrosis (Chalhoub et al., 2003; Kornak et al., 2001; Sobacchi, Schulz, Coxon, Villa, & Helfrich, 2013), caused by a defective osteoclast-mediated bone resorption. As in ClC-7 and Ostm1 KO mice (Chalhoub et al., 2003; Kasper et al., 2005; Kornak et al., 2001; Wartosch, Fuhrmann, Schweizer, Stauber, & Jentsch, 2009), the human disease is also associated with a neurological phenotype that is in general more severe for *OSTM1* mutations (Barrallo-Gimeno et al., 2015). The neurodegeneration in ClC-7 KO mice initiates in the CA3 region of the hippocampus and is associated with a severe lysosomal storage disease, indicative of neuronal ceroid lipofuscinosis (Kasper et al., 2005; Wartosch et al., 2009). In addition to its function in neurons, ClC-7 is critical also in microglia (Kasper et al., 2005). In fact, ClC-7 is expressed in all neuronal types studied and in glia (Kasper et al., 2005). However, using area specific neuron-only KO models, Wartosch et al. could show that the neurodegeneration caused by loss of ClC-7 is cell autonomous, and that the astrogliosis and activation of microglia represent a secondary response to the primary neurodegeneration (Wartosch et al., 2009).

The retinal degeneration is a primary event caused by photoreceptor death independent of compression of the optic nerve by the osteopetrotic phenotype. The phenotype of Ostm1 is largely overlapping with that of ClC-7 KO, in agreement with the idea that, most likely, the only function of Ostm1 is to serve as a beta subunit of ClC-7, which is necessary to guarantee protein stability in the acidic environment of lysosomes, probably aided by the heavy glycosylation of Ostm1 (Lange et al., 2006; Wartosch et al., 2009).

ClC-7 can be efficiently targeted to the plasma membrane by mutating N-terminal sorting motifs (Stauber & Jentsch, 2010). Currents mediated by plasma-membrane localized ClC-7/Ostm1 transporters are strongly voltage- and time-dependent: they activate very slowly (seconds time scale) at positive voltages, and tail currents decay relatively slowly at negative voltages (Leisle et al., 2011). This behavior clearly shows that the transporter possesses a voltage-dependent gating process, which has been shown to be dependent on both subunits of the homodimeric protein (Ludwig, Ullrich, Leisle, Stauber, & Jentsch, 2013), analogous to the "common" gate of CLC channels. A qualitatively similar, but kinetically much faster, gating mechanism was also found for ClC-5 (De Stefano et al., 2013). Surprisingly, several osteopetrosis causing *CLCN7* mutations induce faster gating kinetics in ClC-7, which is *a priori* a "gain-of-function" effect (Leisle et al., 2011). The physiological role of the gating process is not understood.

Somewhat surprisingly, lysosomes from cultured cells form ClC-7 and Ostm1 KO mice show a normal luminal pH (Kasper et al., 2005; Lange et al., 2006; Weinert et al., 2010), suggesting that the 2 Cl^-/1 H^+ antiporter is not critical for lysosomal acidification. Based on the transport properties of lysosomes from KO animals and knock-in mice carrying an "uncoupled," passively Cl^- transporting ClC-7 mutation, Weinert et al. proposed that the most relevant function of ClC-7/Ostm1 is to increase the lysosomal Cl^- concentration above its equilibrium value, exploiting the pH gradient (Weinert et al., 2010).

It is fair to conclude that, while several CLC antiporters are critically important for neuronal function, the precise role of these endolysosomal proteins is far from being understood.

References

Accardi, A., & Miller, C. (2004). Secondary active transport mediated by a prokaryotic homologue of ClC Cl^- channels. *Nature, 427*(6977), 803–807.

Alvarez-Leefmans, F. J., Gamino, S. M., Giraldez, F., & Nogueron, I. (1988). Intracellular chloride regulation in amphibian dorsal root ganglion neurones studied with ion-selective microelectrodes. *Journal of Physiology, 406*, 225–246.

Amjad, A., Hernandez-Clavijo, A., Pifferi, S., Maurya, D. K., Boccaccio, A., Franzot, J., . . . Menini, A. (2015). Conditional knockout of TMEM16A/anoctamin1 abolishes the calcium-activated chloride current in mouse vomeronasal sensory neurons. *Journal of General Physiology, 145*, 285–301. doi:10.1085/jgp.201411348

Arnedo, T., Aiello, C., Jeworutzki, E., Dentici, M. L., Uziel, G., Simonati, A., . . . Estevez, R. (2014). Expanding the spectrum of megalencephalic leukoencephalopathy with subcortical cysts in two patients with GLIALCAM mutations. *Neurogenetics, 15*(1), 41–48. doi:10.1007/s10048-013-0381-x

Arnedo, T., Lopez-Hernandez, T., Jeworutzki, E., Capdevila-Nortes, X., Sirisi, S., Pusch, M., & Estevez, R. (2014). Functional analyses of mutations in HEPACAM causing megalencephalic leukoencephalopathy. *Human Mutation, 35*(10), 1175–1178. doi:10.1002/humu.22622

Arreola, J., Begenisich, T., & Melvin, J. E. (2002). Conformation-dependent regulation of inward rectifier chloride channel gating by extracellular protons. *Journal of Physiology, 541*(Pt 1), 103–112.

Bader, C. R., Bertrand, D., & Schlichter, R. (1987). Calcium-activated chloride current in cultured sensory and parasympathetic quail neurones. *Journal of Physiology, 394*, 125–148.

Bader, C. R., Bertrand, D., & Schwartz, E. A. (1982). Voltage-activated and calcium-activated currents studied in solitary rod inner segments from the salamander retina. *Journal of Physiology, 331*, 253–284.

Barnes, S., & Hille, B. (1989). Ionic channels of the inner segment of tiger salamander cone photoreceptors. *Journal of General Physiology, 94*(4), 719–743.

Barrallo-Gimeno, A., Gradogna, A., Zanardi, I., Pusch, M., & Estévez, R. (2015). Regulatory-auxiliary subunits of CLC chloride channel-transport proteins. *Journal of Physiology, 593*(18), 4111–4127. doi:10.1113/JP270057

Bauer, C. K., Steinmeyer, K., Schwarz, J. R., & Jentsch, T. J. (1991). Completely functional double-barreled chloride channel expressed from a single *Torpedo* cDNA. *Proceedings of the National Academy of Sciences of the United States of America, 88*(24), 11052–11056.

Ben-Ari, Y. (2002). Excitatory actions of GABA during development: The nature of the nurture. *Nature Reviews Neuroscience, 3*(9), 728–739. doi:10.1038/nrn920

Bennetts, B., Rychkov, G. Y., Ng, H.-L., Morton, C. J., Stapleton, D., Parker, M. W., & Cromer, B. A. (2005). Cytoplasmic ATP-sensing domains regulate gating of skeletal muscle ClC-1 chloride channels. *Journal of Biological Chemistry, 280*(37), 32452–32458.

Billig, G. M., Pal, B., Fidzinski, P., & Jentsch, T. J. (2011). $Ca2^{+}$-activated Cl^{-} currents are dispensable for olfaction. *Nature Neuroscience, 14*(6), 763–769.

Blanz, J., Schweizer, M., Auberson, M., Maier, H., Muenscher, A., Hubner, C. A., & Jentsch, T. J. (2007). Leukoencephalopathy upon disruption of the chloride channel ClC-2. *Journal of Neuroscience, 27*(24), 6581–6589.

Boccaccio, A., & Menini, A. (2007). Temporal development of cyclic nucleotide-gated and Ca2+-activated Cl- currents in isolated mouse olfactory sensory neurons. *Journal of Neurophysiology, 98*, 153–160. doi:10.1152/jn.00270.2007

Boedtkjer, D. M. B., Kim, S., Jensen, a. B., Matchkov, V. M., & Andersson, K. E. (2015). New selective inhibitors of calcium-activated chloride channels-T16A inh -A01, CaCC inh -A01, and MONNA—what do they inhibit? *British Journal of Pharmacology, 172*(16), 4158–4172. doi:10.1111/bph.13201

Borsani, G., Rugarli, E. I., Taglialatela, M., Wong, C., & Ballabio, A. (1995). Characterization of a human and murine gene (CLCN3) sharing similarities to voltage-gated chloride channels and to a yeast integral membrane protein. *Genomics, 27*(1), 131–141. doi:10.1006/geno.1995.1015

Bösl, M. R., Stein, V., Hübner, C., Zdebik, A. A., Jordt, S. E., Mukhopadhyay, A. K., . . . Jentsch, T. J. (2001). Male germ cells and photoreceptors, both dependent on close cell-cell interactions, degenerate upon ClC-2 Cl(-) channel disruption. *EMBO Journal, 20*(6), 1289–1299.

Boudes, M., Sar, C., Menigoz, A., Hilaire, C., Pequignot, M. O., Kozlenkov, A., . . . Scamps, F. (2009). Best1 is a gene regulated by nerve injury and required for Ca2+-activated Cl- current expression in axotomized sensory neurons. *Journal of Neuroscience, 29*(32), 10063–10071. doi:10.1523/JNEUROSCI.1312-09.2009

Bradley, E., Fedigan, S., Webb, T., Hollywood, M. A., Thornbury, K. D., McHale, N. G., & Sergeant, G. P. (2014). Pharmacological characterization of TMEM16A currents. *Channels (Austin), 8*(4), 308–320.

Brandt, S., & Jentsch, T. J. (1995). ClC-6 and ClC-7 are two novel broadly expressed members of the CLC chloride channel family. *FEBS Letters, 377*(1), 15–20.

Brunner, J. D., Lim, N. K., Schenck, S., Duerst, A., & Dutzler, R. (2014). X-ray structure of a calcium-activated TMEM16 lipid scramblase. *Nature, 516*, 207–212. doi:10.1038/nature13984

Bugiani, M., Dubey, M., Breur, M., Postma, N. L., Dekker, M. P., Ter Braak, T., . . . van der Knaap, M. S. (2017). Megalencephalic leukoencephalopathy with cysts: The Glialcam-null mouse model. *Annals of Clinical and Translational Neurology, 4*(7), 450–465. doi:10.1002/acn3.405

Capdevila-Nortes, X., Jeworutzki, E., Elorza-Vidal, X., Barrallo-Gimeno, A., Pusch, M., & Estevez, R. (2015). Structural determinants of interaction, trafficking and function in the ClC-2/MLC1 subunit GlialCAM involved in leukodystrophy. *Journal of Physiology, 593*(18), 4165–4180. doi:10.1113/JP270467

Caputo, A., Caci, E., Ferrera, L., Pedemonte, N., Barsanti, C., Sondo, E., ... Galietta, L. J. (2008). TMEM16A, a membrane protein associated with calcium-dependent chloride channel activity. *Science, 322*(5901), 590–594.

Caputo, A., Piano, I., Demontis, G. C., Bacchi, N., Casarosa, S., Della Santina, L., & Gargini, C. (2015). TMEM16A is associated with voltage-gated calcium channels in mouse retina and its function is disrupted upon mutation of the auxiliary alpha2delta4 subunit. *Frontiers in Cellular Neuroscience*, 9, 422. doi:10.3389/fncel.2015.00422

Chakraborty, K., Leung, K., & Krishnan, Y. (2017). High lumenal chloride in the lysosome is critical for lysosome function. *Elife*, 6, pii: e28862. doi:10.7554/eLife.28862

Chalhoub, N., Benachenhou, N., Rajapurohitam, V., Pata, M., Ferron, M., Frattini, A., ... Vacher, J. (2003). Grey-lethal mutation induces severe malignant autosomal recessive osteopetrosis in mouse and human. *Nature Medicine*, 9(4), 399–406.

Cherian, O. L., Menini, A., & Boccaccio, A. (2015). Multiple effects of anthracene-9-carboxylic acid on the TMEM16B/anoctamin2 calcium-activated chloride channel. *Biochimica et Biophysica Acta, 1848*, 1005–1013. doi:10.1016/j.bbamem.2015.01.009

Chesnoy-Marchais, D. (1983). Characterization of a chloride conductance activated by hyperpolarization in Aplysia neurones. *Journal of Physiology, 342*, 277–308.

Cho, H., Yang, Y. D., Lee, J., Lee, B., Kim, T., Jang, Y., ... Oh, U. (2012). The calcium-activated chloride channel anoctamin 1 acts as a heat sensor in nociceptive neurons. *Nature Neuroscience, 15*(7), 1015–1021. doi:10.1038/nn.3111

Cho, S. J., Jeon, J. H., Chun, D. I., Yeo, S. W., & Kim, I. B. (2014). Anoctamin 1 expression in the mouse auditory brainstem. *Cell Tissue Research, 357*(3), 563–569. doi:10.1007/s00441-014-1897-6

Cigic, B., & Pain, R. H. (1999). Location of the binding site for chloride ion activation of cathepsin C. *European Journal of Biochemistry, 264*(3), 944–951.

Clark, S., Jordt, S. E., Jentsch, T. J., & Mathie, A. (1998). Characterization of the hyperpolarization-activated chloride current in dissociated rat sympathetic neurons. *Journal of Physiology, 506*(Pt 3), 665–678.

Clayton, G. H., Staley, K. J., Wilcox, C. L., Owens, G. C., & Smith, R. L. (1998). Developmental expression of C1C-2 in the rat nervous system. *Brain Research. Developmental Brain Research, 108*(1–2), 307–318.

Dauner, K., Lissmann, J., Jeridi, S., Frings, S., & Mohrlen, F. (2012). Expression patterns of anoctamin 1 and anoctamin 2 chloride channels in the mammalian nose. *Cell Tissue Research, 347*(2), 327–341. doi:10.1007/s00441-012-1324-9

Dauner, K., Mobus, C., Frings, S., & Mohrlen, F. (2013). Targeted expression of anoctamin calcium-activated chloride channels in rod photoreceptor terminals of the rodent retina. *Investigative Ophthalmology & Visual Science, 54*(5), 3126–3136. doi:10.1167/iovs.13-11711

De Jesus-Perez, J. J., Castro-Chong, A., Shieh, R. C., Hernandez-Carballo, C. Y., De Santiago-Castillo, J. A., & Arreola, J. (2016). Gating the glutamate gate of CLC-2 chloride channel by pore occupancy. *Journal of General Physiology, 147*(1), 25–37. doi:10.1085/jgp.201511424

De La Fuente, R., Namkung, W., Mills, A., & Verkman, A. S. (2008). Small-molecule screen identifies inhibitors of a human intestinal calcium-activated chloride channel. *Molecular Pharmacology, 73*, 758–768. doi:10.1124/mol.107.043208

De Stefano, S., Pusch, M., & Zifarelli, G. (2013). A single point mutation reveals gating of the human ClC-5 Cl^-/H^+ antiporter. *Journal of Physiology, 591*(Pt 23), 5879–5893. doi:10.1113/jphysiol.2013.260240

Depienne, C., Bugiani, M., Dupuits, C., Galanaud, D., Touitou, V., Postma, N., . . . van der Knaap, M. S. (2013). Brain white matter oedema due to ClC-2 chloride channel deficiency: An observational analytical study. *Lancet Neurology, 12*(7), 659–668. doi:10.1016/S1474-4422(13)70053-X

Deschenes, M., Feltz, P., & Lamour, Y. (1976). A model for an estimate in vivo of the ionic basis of presynaptic inhibition: An intracellular analysis of the GABA-induced depolarization in rat dorsal root ganglia. *Brain Research, 118*(3), 486–493.

Di Bella, D., Pareyson, D., Savoiardo, M., Farina, L., Ciano, C., Caldarazzo, S., . . . Salsano, E. (2014). Subclinical leukodystrophy and infertility in a man with a novel homozygous CLCN2 mutation. *Neurology, 83*(13), 1217–1218. doi:10.1212/WNL.0000000000000812

Dibattista, M., Amjad, A., Maurya, D. K., Sagheddu, C., Montani, G., Tirindelli, R., & Menini, A. (2012). Calcium-activated chloride channels in the apical region of mouse vomeronasal sensory neurons. *The Journal of General Physiology, 140*, 3–15. doi:10.1085/jgp.201210780

Dibattista, M., Pifferi, S., Boccaccio, A., Menini, A., & Reisert, J. (2017). The long tale of the calcium activated Cl- channels in olfactory transduction. *Channels (Austin, Texas)*, 1–16. doi:10.1080/19336950.2017.1307489

Dickerson, L. W., Bonthius, D. J., Schutte, B. C., Yang, B., Barna, T. J., Bailey, M. C., . . . Lamb, F. S. (2002). Altered GABAergic function accompanies hippocampal degeneration in mice lacking ClC-3 voltage-gated chloride channels. *Brain Research, 958*(2), 227–250.

Dubey, M., Bugiani, M., Ridder, M. C., Postma, N. L., Brouwers, E., Polder, E., . . . van der Knaap, M. S. (2015). Mice with megalencephalic leukoencephalopathy with cysts: A developmental angle. *Annals of Neurology, 77*(1), 114–131. doi:10.1002/ana.24307

Dutzler, R., Campbell, E. B., Cadene, M., Chait, B. T., & MacKinnon, R. (2002). X-ray structure of a ClC chloride channel at 3.0 Å reveals the molecular basis of anion selectivity. *Nature, 415*(6869), 287–294.

Dutzler, R., Campbell, E. B., & MacKinnon, R. (2003). Gating the selectivity filter in ClC chloride channels. *Science, 300*(5616), 108–112.

Enz, R., Ross, B. J., & Cutting, G. R. (1999). Expression of the voltage-gated chloride channel ClC-2 in rod bipolar cells of the rat retina. *Journal of Neuroscience, 19*(22), 9841–9847.

Estévez, R., Boettger, T., Stein, V., Birkenhäger, R., Otto, E., Hildebrandt, F., & Jentsch, T. J. (2001). Barttin is a Cl^- channel beta-subunit crucial for renal Cl^- reabsorption and inner ear K^+ secretion. *Nature, 414*(6863), 558–561.

Fallah, G., Romer, T., Detro-Dassen, S., Braam, U., Markwardt, F., & Schmalzing, G. (2011). TMEM16A(a)/anoctamin-1 shares a homodimeric architecture with CLC chloride channels. *Molecular & Cellular Proteomics, 10*(2), M110 004697. doi:10.1074/mcp.M110.004697

Favre-Kontula, L., Rolland, A., Bernasconi, L., Karmirantzou, M., Power, C., Antonsson, B., & Boschert, U. (2008). GlialCAM, an immunoglobulin-like cell adhesion molecule is expressed in glial cells of the central nervous system. *Glia, 56*(6), 633–645.

Feng, L., Campbell, E. B., Hsiung, Y., & Mackinnon, R. (2010). Structure of a eukaryotic CLC transporter defines an intermediate state in the transport cycle. *Science, 330*, 635–641.

Ferrera, L., Caputo, A., Ubby, I., Bussani, E., Zegarra-Moran, O., Ravazzolo, R., . . . Galietta, L. J. (2009). Regulation of TMEM16A chloride channel properties by alternative splicing. *Journal of Biological Chemistry, 284*(48), 33360–33368. doi:10.1074/jbc.M109.046607

Ferroni, S., Marchini, C., Nobile, M., & Rapisarda, C. (1997). Characterization of an inwardly rectifying chloride conductance expressed by cultured rat cortical astrocytes. *Glia, 21*(2), 217–227.

Fischer, M., Janssen, A. G., & Fahlke, C. (2010). Barttin activates ClC-K channel function by modulating gating. *Journal of the American Society of Nephrology, 21*, 1281–1289.
Fisher, S. I., & Hartzell, H. C. (2017). Poring over furrows. *eLife, 6*, pii: e27933. doi:10.7554/eLife.27933
Földy, C., Lee, S. H., Morgan, R. J., & Soltesz, I. (2010). Regulation of fast-spiking basket cell synapses by the chloride channel ClC-2. *Nature Neuroscience, 13*(9), 1047–1049.
Friedrich, T., Breiderhoff, T., & Jentsch, T. J. (1999). Mutational analysis demonstrates that ClC-4 and ClC-5 directly mediate plasma membrane currents. *Journal of Biological Chemistry, 274*(2), 896–902.
Frings, S. (2009). Chloride-based signal amplification in olfactory sensory neurons. In F. J. Alvarez-Leefmans & E. Delpire (Eds.), *Physiology and pathology of chloride transporters and channels in the nervous system* (pp. 413–424). Amsterdam: Academic Press Elsevier.
Frings, S., Reuter, D., & Kleene, S. J. (2000). Neuronal $Ca2^{+}$ -activated Cl^{-} channels—homing in on an elusive channel species. *Progress in Neurobiology, 60*(3), 247–289.
Giorgio, E., Vaula, G., Benna, P., Lo Buono, N., Eandi, C. M., Dino, D., . . . Brusco, A. (2017). A novel homozygous change of CLCN2 (p.His590Pro) is associated with a subclinical form of leukoencephalopathy with ataxia (LKPAT). *Journal of Neurology, Neurosurgery, & Psychiatry, 88*(10), 894–896. doi:10.1136/jnnp-2016-315525
Gründer, S., Thiemann, A., Pusch, M., & Jentsch, T. J. (1992). Regions involved in the opening of CIC-2 chloride channel by voltage and cell volume. *Nature, 360*(6406), 759–762.
Günther, W., Piwon, N., & Jentsch, T. J. (2003). The ClC-5 chloride channel knock-out mouse—an animal model for Dent's disease. *Pflügers Archiv, 445*(4), 456–462.
Guzman, R. E., Alekov, A. K., Filippov, M., Hegermann, J., & Fahlke, C. (2014). Involvement of ClC-3 chloride/proton exchangers in controlling glutamatergic synaptic strength in cultured hippocampal neurons. *Frontiers in Cellular Neuroscience, 8*, 143. doi:10.3389/fncel.2014.00143
Guzman, R. E., Grieschat, M., Fahlke, C., & Alekov, A. K. (2013). ClC-3 is an intracellular chloride/proton exchanger with large voltage-dependent nonlinear capacitance. *ACS Chemical Neuroscience, 4*(6), 994–1003. doi:10.1021/cn400032z
Guzman, R. E., Miranda-Laferte, E., Franzen, A., & Fahlke, C. (2015). Neuronal ClC-3 splice variants differ in subcellular localizations, but mediate identical transport functions. *Journal of Biological Chemistry, 290*(43), 25851–25862. doi:10.1074/jbc.M115.668186
Gyobu, S., Ishihara, K., Suzuki, J., Segawa, K., & Nagata, S. (2017). Characterization of the scrambling domain of the TMEM16 family. *Proceedings of the National Academy of Sciences of the United States of America, 114*(24), 6274–6279. doi:10.1073/pnas.1703391114
Gyobu, S., Miyata, H., Ikawa, M., Yamazaki, D., Takeshima, H., Suzuki, J., & Nagata, S. (2015). A role of TMEM16E carrying a scrambling domain in sperm motility. *Molecular and Cellular Biology, 36*(4), 645–659. doi:10.1128/MCB.00919-15
Ha, G. E., Lee, J., Kwak, H., Song, K., Kwon, J., Jung, S. Y., . . . Cheong, E. (2016). The $Ca2^{+}$-activated chloride channel anoctamin-2 mediates spike-frequency adaptation and regulates sensory transmission in thalamocortical neurons. *Nature Communications, 7*, 13791. doi:10.1038/ncomms13791
Hanagasi, H. A., Bilgic, B., Abbink, T. E., Hanagasi, F., Tufekcioglu, Z., Gurvit, H., . . . Emre, M. (2015). Secondary paroxysmal kinesigenic dyskinesia associated with CLCN2 gene mutation. *Parkinsonism & Related Disorders, 21*(5), 544–546. doi:10.1016/j.parkreldis.2015.02.013
Hara-Chikuma, M., Wang, Y., Guggino, S. E., Guggino, W. B., & Verkman, A. S. (2005). Impaired acidification in early endosomes of ClC-5 deficient proximal tubule. *Biochemical and Biophysical Research Communications, 329*(3), 941–946.

Hartzell, C., Putzier, I., & Arreola, J. (2005). Calcium-activated chloride channels. *Annual Review of Physiology*, *67*(1), 719–758.

Higashi, H., Tanaka, E., Inokuchi, H., & Nishi, S. (1993). Ionic mechanisms underlying the depolarizing and hyperpolarizing afterpotentials of single spike in guinea-pig cingulate cortical neurons. *Neuroscience*, *55*(1), 129–138.

Hoegg-Beiler, M. B., Sirisi, S., Orozco, I. J., Ferrer, I., Hohensee, S., Auberson, M., . . . Jentsch, T. J. (2014). Disrupting MLC1 and GlialCAM and ClC-2 interactions in leukodystrophy entails glial chloride channel dysfunction. *Nature Communications*, *5*, 3475. doi:10.1038/ncomms4475

Hu, H., Haas, S. A., Chelly, J., Van Esch, H., Raynaud, M., de Brouwer, A. P., . . . Kalscheuer, V. M. (2016). X-exome sequencing of 405 unresolved families identifies seven novel intellectual disability genes. *Molecular Psychiatry*, *21*(1), 133–148. doi:10.1038/mp.2014.193

Huang, F., Rock, J. R., Harfe, B. D., Cheng, T., Huang, X., Jan, Y. N., & Jan, L. Y. (2009). Studies on expression and function of the TMEM16A calcium-activated chloride channel. *Proceedings of the National Academy of Sciences*, *106*, 21413–21418. doi:10.1073/pnas.0911935106

Huang, F., Wong, X., & Jan, L. Y. (2012). International Union of Basic and Clinical Pharmacology. LXXXV: Calcium-activated chloride channels. *Pharmacological Reviews*, *64*, 1–15. doi:10.1124/pr.111.005009

Huang, W. C., Xiao, S., Huang, F., Harfe, B. D., Jan, Y. N., & Jan, L. Y. (2012). Calcium-activated chloride channels (CaCCs) regulate action potential and synaptic response in hippocampal neurons. *Neuron*, *74*, 179–192. doi:10.1016/j.neuron.2012.01.033

Hübner, C. A., & Holthoff, K. (2013). Anion transport and GABA signaling. *Frontiers in Cellular Neuroscience*, *7*, 177. doi:10.3389/fncel.2013.00177

Jayaram, H., Robertson, J. L., Wu, F., Williams, C., & Miller, C. (2011). Structure of a slow CLC Cl^{-}/H^{+} antiporter from a Cyanobacterium. *Biochemistry*, *50*(5), 788–794.

Jeng, G., Aggarwal, M., Yu, W. P., & Chen, T. Y. (2016). Independent activation of distinct pores in dimeric TMEM16A channels. *Journal of General Physiology*, *148*(5), 393–404. doi:10.1085/jgp.201611651

Jentsch, T. J. (2005). Chloride transport in the kidney: Lessons from human disease and knockout mice. *Journal of the American Society of Nephrology*, *16*(6), 1549–1561.

Jentsch, T. J. (2008). CLC chloride channels and transporters: From genes to protein structure, pathology and physiology. *Critical Reviews in Biochemistry and Molecular Biology*, *43*(1), 3–36.

Jentsch, T. J. (2016). VRACs and other ion channels and transporters in the regulation of cell volume and beyond. *Nat Rev Molecular and Cellular Biology*, *17*(5), 293–307. doi:10.1038/nrm.2016.29

Jentsch, T. J., Günther, W., Pusch, M., & Schwappach, B. (1995). Properties of voltage-gated chloride channels of the ClC gene family. *Journal of Physiology*, *482*, 19S–25S.

Jentsch, T. J., Poët, M., Fuhrmann, J. C., & Zdebik, A. A. (2005). Physiological functions of CLC Cl channels gleaned from human genetic disease and mouse models. *Annual Review of Physiology*, *67*(1), 779–807.

Jentsch, T. J. & Pusch, M. (2018). CLC chloride channels and transporters: structure, function, physiology, and disease. *Physiological Reviews*, *98*(3), 1493–1590.

Jeon, J. H., Paik, S. S., Chun, M.-H., Oh, U., & Kim, I.-B. (2013). Presynaptic localization and possible function of calcium-activated chloride channel anoctamin 1 in the mammalian retina. *PLoS One*, *8*, e67989–e67989. doi:10.1371/journal.pone.0067989

Jeon, J. H., Park, J. W., Lee, J. W., Jeong, S. W., Yeo, S. W., & Kim, I. B. (2011). Expression and immunohistochemical localization of TMEM16A/anoctamin 1, a calcium-activated chloride channel in the mouse cochlea. *Cell Tissue Research, 345*(2), 223–230. doi:10.1007/s00441-011-1206-6

Jeworutzki, E., Lagostena, L., Elorza-Vidal, X., Lopez-Hernandez, T., Estevez, R., & Pusch, M. (2014). GlialCAM, a CLC-2 Cl(-) channel subunit, activates the slow gate of CLC chloride channels. *Biophysical Journal, 107*(5), 1105–1116. doi:10.1016/j.bpj.2014.07.040

Jeworutzki, E., López-Hernández, T., Capdevila-Nortes, X., Sirisi, S., Bengtsson, L., Montolio, M., ... Estévez, R. (2012). GlialCAM, a protein defective in a leukodystrophy, serves as a ClC-2 Cl^- channel auxiliary subunit. *Neuron, 73*(5), 951–961.

Jin, X., Shah, S., Liu, Y., Zhang, H., Lees, M., Fu, Z., ... Gamper, N. (2013). Activation of the Cl^- channel ANO1 by localized calcium signals in nociceptive sensory neurons requires coupling with the IP3 receptor. *Science Signaling, 6*(290), ra73. doi:10.1126/scisignal.2004184

Jones, T. A., & Jones, S. M. (2000). Spontaneous activity in the statoacoustic ganglion of the chicken embryo. *Journal of Neurophysiology, 83*(3), 1452–1468.

Jones, T. A., Jones, S. M., & Paggett, K. C. (2001). Primordial rhythmic bursting in embryonic cochlear ganglion cells. *Journal of Neuroscience, 21*(20), 8129–8135.

Jordt, S. E., & Jentsch, T. J. (1997). Molecular dissection of gating in the ClC-2 chloride channel. *EMBO Journal, 16*(7), 1582–1592.

Jung, J., Nam, J. H., Park, H. W., Oh, U., Yoon, J. H., & Lee, M. G. (2013). Dynamic modulation of ANO1/TMEM16A HCO3(-) permeability by $Ca2^+$/calmodulin. *Proceedings of the National Academy of Sciences of the United States of America, 110*(1), 360–365. doi:10.1073/pnas.1211594110

Kanazawa, T., & Matsumoto, S. (2014). Expression of transient receptor potential vanilloid 1 and anoctamin 1 in rat trigeminal ganglion neurons innervating the tongue. *Brain Research Bulletin, 106*, 17–20. doi:10.1016/j.brainresbull.2014.04.015

Kaneko, H. (2004). Chloride accumulation in mammalian olfactory sensory neurons. *Journal of Neuroscience, 24*, 7931–7938. doi:10.1523/JNEUROSCI.2115-04.2004

Kasper, D., Planells-Cases, R., Fuhrmann, J. C., Scheel, O., Zeitz, O., Ruether, K., ... Jentsch, T. J. (2005). Loss of the chloride channel ClC-7 leads to lysosomal storage disease and neurodegeneration. *EMBO Journal, 24*(5), 1079–1091.

Kawasaki, M., Fukuma, T., Yamauchi, K., Sakamoto, H., Marumo, F., & Sasaki, S. (1999). Identification of an acid-activated Cl(-) channel from human skeletal muscles. *American Journal of Physiology, 277*(5 Pt 1), C948–C954.

Kawasaki, M., Uchida, S., Monkawa, T., Miyawaki, A., Mikoshiba, K., Marumo, F., & Sasaki, S. (1994). Cloning and expression of a protein kinase C-regulated chloride channel abundantly expressed in rat brain neuronal cells. *Neuron, 12*(3), 597–604.

Kenyon, J. L. (2000). The reversal potential of Ca(2+)-activated Cl(-) currents indicates that chick sensory neurons accumulate intracellular Cl(-). *Neuroscience Letters, 296*(1), 9–12.

Kida, Y., Uchida, S., Miyazaki, H., Sasaki, S., & Marumo, F. (2001). Localization of mouse CLC-6 and CLC-7 mRNA and their functional complementation of yeast CLC gene mutant. *Histochemistry and Cell Biology, 115*(3), 189–194.

Kim, S., Ma, L., & Yu, C. R. (2011). Requirement of calcium-activated chloride channels in the activation of mouse vomeronasal neurons. *Nature Communications, 2*, 365–365. doi:10.1038/ncomms1368

Kleene, S. J. (1993). Origin of the chloride current in olfactory transduction. *Neuron, 11*, 123–132. doi:10.1016/0896-6273(93)90276-W

Kleene, S. J., & Gesteland, R. C. (1991). Calcium-activated chloride conductance in frog olfactory cilia. *Journal of Neuroscience, 11*, 3624–3629.

Korn, S. J., Bolden, A., & Horn, R. (1991). Control of action potentials and Ca2+ influx by the Ca(2+)-dependent chloride current in mouse pituitary cells. *Journal of Physiology*, *439*, 423–437.

Kornak, U., Kasper, D., Bösl, M. R., Kaiser, E., Schweizer, M., Schulz, A., . . . Jentsch, T. J. (2001). Loss of the ClC-7 chloride channel leads to osteopetrosis in mice and man. *Cell*, *104*(2), 205–215.

Kurahashi, T., & Yau, K.-W. (1993). Co-existence of cationic and chloride components in odorant-induced current of vertebrate olfactory receptor cells. *Nature*, *363*, 71–74. doi:10.1038/363071a0

L'Hoste, S., Diakov, A., Andrini, O., Genete, M., Pinelli, L., Grand, T., . . . Lourdel, S. (2013). Characterization of the mouse ClC-K1/Barttin chloride channel. *Biochimica et Biophysica Acta*, *1828*, 2399–2409. doi:10.1016/j.bbamem.2013.06.012

Lange, P. F., Wartosch, L., Jentsch, T. J., & Fuhrmann, J. C. (2006). ClC-7 requires Ostm1 as a beta-subunit to support bone resorption and lysosomal function. *Nature*, *440*(7081), 220–223.

Lee, B.-C., Menon, A. K., & Accardi, A. (2016). The nhTMEM16 scramblase is also a nonselective ion channel. *Biophysical Journal*, *111*, 1919–1924. doi:10.1016/j.bpj.2016.09.032

Lee, B., Cho, H., Jung, J., Yang, Y. D., Yang, D. J., & Oh, U. (2014). Anoctamin 1 contributes to inflammatory and nerve-injury induced hypersensitivity. *Molecular Pain*, *10*, 5. doi:10.1186/1744-8069-10-5

Leegwater, P. A., Yuan, B. Q., van der Steen, J., Mulders, J., Konst, A. A., Boor, P. K., . . . van der Knaap, M. S. (2001). Mutations of MLC1 (KIAA0027), encoding a putative membrane protein, cause megalencephalic leukoencephalopathy with subcortical cysts. *American Journal of Human Genetics*, *68*(4), 831–838.

Leisle, L., Ludwig, C. F., Wagner, F. A., Jentsch, T. J., & Stauber, T. (2011). ClC-7 is a slowly voltage-gated 2Cl(-)/1H(+)-exchanger and requires Ostm1 for transport activity. *EMBO Journal*, *30*(11), 2140–2152.

Li, R. C., Ben-Chaim, Y., Yau, K. W., & Lin, C. C. (2016). Cyclic-nucleotide-gated cation current and Ca2+-activated Cl current elicited by odorant in vertebrate olfactory receptor neurons. *Proceedings of the National Academy of Sciences of the United States of America*, *113*(40), 11078–11087. doi:10.1073/pnas.1613891113

Li, X., Shimada, K., Showalter, L. A., & Weinman, S. A. (2000). Biophysical properties of ClC-3 differentiate it from swelling-activated chloride channels in Chinese hamster ovary-K1 cells. *Journal of Biological Chemistry*, *275*(46), 35994–35998.

Lim, H. H., Stockbridge, R. B., & Miller, C. (2013). Fluoride-dependent interruption of the transport cycle of a CLC Cl-/H+ antiporter. *Nature Chemical Biology*, *9*(11), 721–725. doi:10.1038/nchembio.1336

Lim, N. K., Lam, A. K., & Dutzler, R. (2016). Independent activation of ion conduction pores in the double-barreled calcium-activated chloride channel TMEM16A. *Journal of General Physiology*, *148*(5), 375–392. doi:10.1085/jgp.201611650

Liu, B., Linley, J. E., Du, X., Zhang, X., Ooi, L., Zhang, H., & Gamper, N. (2010). The acute nociceptive signals induced by bradykinin in rat sensory neurons are mediated by inhibition of M-type K+ channels and activation of Ca2+-activated Cl- channels. *Journal of Clinical Investigation*, *120*(4), 1240–1252. doi:10.1172/JCI41084

Liu, Y., Zhang, H., Huang, D., Qi, J., Xu, J., Gao, H., . . . Gamper, N. (2015). Characterization of the effects of Cl(-) channel modulators on TMEM16A and bestrophin-1 Ca(2)(+) activated Cl(-) channels. *Pflügers Archiv*, *467*(7), 1417–1430. doi:10.1007/s00424-014-1572-5

Llano, I., Leresche, N., & Marty, A. (1991). Calcium entry increases the sensitivity of cerebellar Purkinje cells to applied GABA and decreases inhibitory synaptic currents. *Neuron, 6*(4), 565–574.

Lopez-Hernandez, T., Ridder, M. C., Montolio, M., Capdevila-Nortes, X., Polder, E., Sirisi, S., . . . van der Knaap, M. S. (2011). Mutant GlialCAM causes megalencephalic leukoencephalopathy with subcortical cysts, benign familial macrocephaly, and macrocephaly with retardation and autism. *American Journal of Human Genetics, 88*(4), 422–432.

Lowe, G., & Gold, G. H. (1993). Nonlinear amplification by calcium-dependent chloride channels in olfactory receptor cells. *Nature, 366*(6452), 283–286. doi:10.1038/366283a0

Ludewig, U., Pusch, M., & Jentsch, T. J. (1996). Two physically distinct pores in the dimeric ClC-0 chloride channel. *Nature, 383*(6598), 340–343.

Ludwig, C. F., Ullrich, F., Leisle, L., Stauber, T., & Jentsch, T. J. (2013). Common gating of both CLC transporter subunits underlies voltage-dependent activation of the $2Cl^-/1H^+$ exchanger ClC-7/Ostm1. *Journal of Biological Chemistry, 288*(40), 28611–28619. doi:10.1074/jbc.M113.509364

MacLeish, P. R., & Nurse, C. A. (2007). Ion channel compartments in photoreceptors: Evidence from salamander rods with intact and ablated terminals. *Journal of Neurophysiology, 98*(1), 86–95. doi:10.1152/jn.00775.2006

Madison, D. V., Malenka, R. C., & Nicoll, R. A. (1986). Phorbol esters block a voltage-sensitive chloride current in hippocampal pyramidal cells. *Nature, 321*(6071), 695–697. doi:10.1038/321695a0

Makara, J. K., Rappert, A., Matthias, K., Steinhauser, C., Spat, A., & Kettenmann, H. (2003). Astrocytes from mouse brain slices express ClC-2-mediated Cl- currents regulated during development and after injury. *Molecular & Cellular Neuroscience, 23*(4), 521–530.

Marcoline, F. V., Ishida, Y., Mindell, J. A., Nayak, S., & Grabe, M. (2016). A mathematical model of osteoclast acidification during bone resorption. *Bone, 93*, 167–180. doi:10.1016/j.bone.2016.09.007

Maritzen, T., Keating, D. J., Neagoe, I., Zdebik, A. A., & Jentsch, T. J. (2008). Role of the vesicular chloride transporter ClC-3 in neuroendocrine tissue. *Journal of Neuroscience, 28*(42), 10587–10598. doi:10.1523/JNEUROSCI.3750-08.2008

Markovic, S., & Dutzler, R. (2007). The structure of the cytoplasmic domain of the chloride channel ClC-Ka reveals a conserved interaction interface. *Structure, 15*(6), 715–725. doi:S0969-2126(07)00180-3 [pii] 10.1016/j.str.2007.04.013

Mayer, M. L. (1985). A calcium-activated chloride current generates the after-depolarization of rat sensory neurones in culture. *Journal of Physiology, 364*, 217–239.

Mercer, A. J., Rabl, K., Riccardi, G. E., Brecha, N. C., Stella, S. L., Jr., & Thoreson, W. B. (2011). Location of release sites and calcium-activated chloride channels relative to calcium channels at the photoreceptor ribbon synapse. *Journal of Neurophysiology, 105*(1), 321–335. doi:10.1152/jn.00332.2010

Meyer, S., & Dutzler, R. (2006). Crystal structure of the cytoplasmic domain of the chloride channel ClC-0. *Structure, 14*(2), 299–307. doi:S0969-2126(06)00042-6 [pii] 10.1016/j.str.2005.10.008

Meyer, S., Savaresi, S., Forster, I. C., & Dutzler, R. (2007). Nucleotide recognition by the cytoplasmic domain of the human chloride transporter ClC-5. *Nature Structural & Molecular Biology, 14*(1), 60–67. doi:nsmb1188 [pii] 10.1038/nsmb1188

Middleton, R. E., Pheasant, D. J., & Miller, C. (1996). Homodimeric architecture of a ClC-type chloride ion channel. *Nature, 383*(6598), 337–340.

Miledi, R. (1982). A calcium-dependent transient outward current in *Xenopus laevis* oocytes. *Proceedings of the Royal Society of London Series B, Containing Papers of a Biological Character. Royal Society (Great Britain)*, *215*, 491–497.

Miller, C. (1982). Open-state substructure of single chloride channels from *Torpedo* electroplax. *Philosophical Transactions of the Royal Society of London. Series B, Biological Sciences*, *299*(1097), 401–411.

Moh, M. C., Zhang, C., Luo, C., Lee, L. H., & Shen, S. (2005). Structural and functional analyses of a novel ig-like cell adhesion molecule, hepaCAM, in the human breast carcinoma MCF7 cells. *Journal of Biological Chemistry*, *280*(29), 27366–27374. doi:10.1074/jbc.M500852200

Mohammad-Panah, R., Ackerley, C., Rommens, J., Choudhury, M., Wang, Y., & Bear, C. E. (2002). The chloride channel ClC-4 co-localizes with cystic fibrosis transmembrane conductance regulator and may mediate chloride flux across the apical membrane of intestinal epithelia. *Journal of Biological Chemistry*, *277*(1), 566–574.

Namkung, W., Phuan, P.-W., & Verkman, A. S. (2011). TMEM16A inhibitors reveal TMEM16A as a minor component of calcium-activated chloride channel conductance in airway and intestinal epithelial cells. *Journal of Biological Chemistry*, *286*, 2365–2374. doi:10.1074/jbc.M110.175109

Namkung, W., Thiagarajah, J. R., Phuan, P.-W., & Verkman, A. S. (2010). Inhibition of Ca2+-activated Cl- channels by gallotannins as a possible molecular basis for health benefits of red wine and green tea. *FASEB Journal 24*, 4178–4186. doi:10.1096/fj.10-160648

Neagoe, I., Stauber, T., Fidzinski, P., Bergsdorf, E. Y., & Jentsch, T. J. (2010). The late endosomal ClC-6 mediates proton/chloride countertransport in heterologous plasma membrane expression. *Journal of Biological Chemistry*, *285*(28), 21689–21697.

Neureither, F., Stowasser, N., Frings, S., & Mohrlen, F. (2017). Tracking of unfamiliar odors is facilitated by signal amplification through anoctamin 2 chloride channels in mouse olfactory receptor neurons. *Physiology Report*, *5*(15), pii: e13373. doi:10.14814/phy2.13373

Neureither, F., Ziegler, K., Pitzer, C., Frings, S., & Mohrlen, F. (2017). Impaired motor coordination and learning in mice lacking anoctamin 2 calcium-gated chloride channels. *Cerebellum*, *16*(5–6), 929–937. doi:10.1007/s12311-017-0867-4

Niemeyer, M. I., Cid, L. P., Sepúlveda, F. V., Blanz, J., Auberson, M., & Jentsch, T. J. (2010). No evidence for a role of CLCN2 variants in idiopathic generalized epilepsy. *Nature Genetics*, *42*(1), 3.

Niemeyer, M. I., Cid, L. P., Yusef, Y. R., Briones, R., & Sepúlveda, F. V. (2009). Voltage-dependent and -independent titration of specific residues accounts for complex gating of a ClC chloride channel by extracellular protons. *Journal of Physiology*, *587*(Pt 7), 1387–1400.

Niemeyer, M. I., Cid, L. P., Zúñiga, L., Catalán, M., & Sepúlveda, F. V. (2003). A conserved pore-lining glutamate as a voltage- and chloride-dependent gate in the ClC-2 chloride channel. *Journal of Physiology*, *553*(Pt 3), 873–879.

Nobile, M., Pusch, M., Rapisarda, C., & Ferroni, S. (2000). Single-channel analysis of a ClC-2-like chloride conductance in cultured rat cortical astrocytes. *FEBS Letters*, *479*(1–2), 10–14.

Oh, S.-J., Hwang, S. J., Jung, J., Yu, K., Kim, J., Choi, J. Y., ... Lee, C. J. (2013). MONNA, a potent and selective blocker for transmembrane protein with unknown function 16/anoctamin-1. *Molecular Pharmacology*, *84*, 726–735. doi:10.1124/mol.113.087502

Palmer, E. E., Stuhlmann, T., Weinert, S., Haan, E., Van Esch, H., Holvoet, M., ... Kalscheuer, V. M. (2016). De novo and inherited mutations in the X-linked gene CLCN4 are associated with syndromic intellectual disability and behavior and seizure disorders in males and females. *Molecular Psychiatry*. doi:10.1038/mp.2016.135

Park, E., Campbell, E. B., & MacKinnon, R. (2017). Structure of a CLC chloride ion channel by cryo-electron microscopy. *Nature*, *541*(7638), 500–505. doi:10.1038/nature20812

Paulino, C., Neldner, Y., Lam, A. K., Kalienkova, V., Brunner, J. D., Schenck, S., & Dutzler, R. (2017). Structural basis for anion conduction in the calcium-activated chloride channel TMEM16A. *eLife*, *6*, pii: e26232. doi:10.7554/eLife.26232

Pedemonte, N., & Galietta, L. J. V. (2014). Structure and function of TMEM16 proteins (anoctamins). *Physiological Reviews*, *94*, 419–459. doi:10.1152/physrev.00039.2011

Peters, C. J., Yu, H., Tien, J., Jan, Y. N., Li, M., & Jan, L. Y. (2015). Four basic residues critical for the ion selectivity and pore blocker sensitivity of TMEM16A calcium-activated chloride channels. *Proceedings of the National Academy of Sciences of the United States of America*, *112*(11), 3547–3552. doi:10.1073/pnas.1502291112

Petheo, G. L., Molnar, Z., Roka, A., Makara, J. K., & Spat, A. (2001). A pH-sensitive chloride current in the chemoreceptor cell of rat carotid body. *Journal of Physiology*, *535*(Pt 1), 95–106.

Picollo, A., & Pusch, M. (2005). Chloride/proton antiporter activity of mammalian CLC proteins ClC-4 and ClC-5. *Nature*, *436*(7049), 420–423.

Pietra, G., Dibattista, M., Menini, A., Reisert, J., & Boccaccio, A. (2016). The Ca2$^+$-activated Cl$^-$ channel TMEM16B regulates action potential firing and axonal targeting in olfactory sensory neurons. *Journal of General Physiology*, *148*(4), 293–311. doi:10.1085/jgp.201611622

Pifferi, S. (2017). Permeation mechanisms in the TMEM16B calcium-activated chloride channels. *PLoS One*, *12*, e0169572. doi:10.1371/journal.pone.0169572

Pifferi, S., Cenedese, V., & Menini, A. (2012). Anoctamin 2/TMEM16B: A calcium-activated chloride channel in olfactory transduction. *Experimental Physiology*, *97*, 193–199. doi:10.1113/expphysiol.2011.058230

Pifferi, S., Dibattista, M., & Menini, A. (2009). TMEM16B induces chloride currents activated by calcium in mammalian cells. *Pflügers Archiv: European Journal of Physiology*, *458*, 1023–1038. doi:10.1007/s00424-009-0684-9

Poët, M., Kornak, U., Schweizer, M., Zdebik, A. A., Scheel, O., Hoelter, S., . . . Jentsch, T. J. (2006). Lysosomal storage disease upon disruption of the neuronal chloride transport protein ClC-6. *Proceedings of the National Academy of Sciences of the United States of America*, *103*(37), 13854–13859.

Ponissery Saidu, S., Stephan, A. B., Talaga, A. K., Zhao, H., & Reisert, J. (2013). Channel properties of the splicing isoforms of the olfactory calcium-activated chloride channel Anoctamin 2. *Journal of General Physiology*, *141*, 691–703. doi:10.1085/jgp.201210937

Ponting, C. P. (1997). CBS domains in CIC chloride channels implicated in myotonia and nephrolithiasis (kidney stones). *Journal of Molecular Medicine*, *75*(3), 160–163.

Pusch, M., Jordt, S. E., Stein, V., & Jentsch, T. J. (1999). Chloride dependence of hyperpolarization-activated chloride channel gates. *Journal of Physiology*, *515*(Pt 2), 341–353.

Pusch, M., Ludewig, U., Rehfeldt, A., & Jentsch, T. J. (1995). Gating of the voltage-dependent chloride channel CIC-0 by the permeant anion. *Nature*, *373*(6514), 527–531.

Pusch, M., Steinmeyer, K., & Jentsch, T. J. (1994). Low single channel conductance of the major skeletal muscle chloride channel, ClC-1. *Biophysical Journal*, *66*(1), 149–152.

Ratté, S., & Prescott, S. A. (2011). ClC-2 channels regulate neuronal excitability, not intracellular chloride levels. *Journal of Neuroscience*, *31*(44), 15838–15843. doi:10.1523/JNEUROSCI.2748-11.2011

Reuter, D., Zierold, K., Schröder, W. H., & Frings, S. (1998). A depolarizing chloride current contributes to chemoelectrical transduction in olfactory sensory neurons in situ. *Journal of Neuroscience*, *18*, 6623–6630.

Riazanski, V., Deriy, L. V., Shevchenko, P. D., Le, B., Gomez, E. A., & Nelson, D. J. (2011). Presynaptic CLC-3 determines quantal size of inhibitory transmission in the hippocampus. *Nature Neuroscience, 14*(4), 487–494. doi:10.1038/nn.2775

Rickheit, G., Wartosch, L., Schaffer, S., Stobrawa, S. M., Novarino, G., Weinert, S., & Jentsch, T. J. (2010). Role of ClC-5 in renal endocytosis is unique among ClC exchangers and does not require PY-motif-dependent ubiquitylation. *Journal of Biological Chemistry, 285*(23), 17595–17603. doi:10.1074/jbc.M110.115600

Rinke, I., Artmann, J., & Stein, V. (2010). ClC-2 voltage-gated channels constitute part of the background conductance and assist chloride extrusion. *Journal of Neuroscience, 30*(13), 4776–4786.

Rocha-Gonzalez, H. I., Mao, S., & Alvarez-Leefmans, F. J. (2008). Na+,K+,2Cl- cotransport and intracellular chloride regulation in rat primary sensory neurons: Thermodynamic and kinetic aspects. *Journal of Neurophysiology, 100*(1), 169–184. doi:10.1152/jn.01007.2007

Rock, J. R., Futtner, C. R., & Harfe, B. D. (2008). The transmembrane protein TMEM16A is required for normal development of the murine trachea. *Developmental Biology, 321*(1), 141–149. doi:10.1016/j.ydbio.2008.06.009

Sage, C. L., & Marcus, D. C. (2001). Immunolocalization of ClC-K chloride channel in strial marginal cells and vestibular dark cells. *Hearing Research, 160*(1–2), 1–9.

Sagheddu, C., Boccaccio, A., Dibattista, M., Montani, G., Tirindelli, R., & Menini, A. (2010). Calcium concentration jumps reveal dynamic ion selectivity of calcium-activated chloride currents in mouse olfactory sensory neurons and TMEM16b-transfected HEK 293T cells. *The Journal of Physiology, 588*, 4189–4204. doi:10.1113/jphysiol.2010.194407

Salazar, G., Love, R., Styers, M. L., Werner, E., Peden, A., Rodriguez, S., . . . Faundez, V. (2004). AP-3-dependent mechanisms control the targeting of a chloride channel (ClC-3) in neuronal and non-neuronal cells. *Journal of Biological Chemistry, 279*(24), 25430–25439.

Sanchez-Rodriguez, J. E., De Santiago-Castillo, J. A., & Arreola, J. (2010). Permeant anions contribute to voltage dependence of ClC-2 chloride channel by interacting with the protopore gate. *Journal of Physiology, 588*(Pt 14), 2545–2556.

Sanchez-Rodriguez, J. E., De Santiago-Castillo, J. A., Contreras-Vite, J. A., Nieto-Delgado, P. G., Castro-Chong, A., & Arreola, J. (2012). Sequential interaction of chloride and proton ions with the fast gate steer the voltage-dependent gating in ClC-2 chloride channels. *Journal of Physiology, 590*(Pt 17), 4239–4253.

Satoh, H., Qu, L., Suzuki, H., & Saitow, F. (2013). Depolarization-induced depression of inhibitory transmission in cerebellar Purkinje cells. *Physiology Report, 1*(3), e00061. doi:10.1002/phy2.61

Scheel, O., Zdebik, A. A., Lourdel, S., & Jentsch, T. J. (2005). Voltage-dependent electrogenic chloride/proton exchange by endosomal CLC proteins. *Nature, 436*(7049), 424–427.

Schenck, S., Wojcik, S. M., Brose, N., & Takamori, S. (2009). A chloride conductance in VGLUT1 underlies maximal glutamate loading into synaptic vesicles. *Nature Neuroscience, 12*(2), 156–162. doi:10.1038/nn.2248

Scheper, G. C., van Berkel, C. G., Leisle, L., de Groot, K. E., Errami, A., Jentsch, T. J., & Van der Knaap, M. S. (2010). Analysis of CLCN2 as candidate gene for megalencephalic leukoencephalopathy with subcortical cysts. *Genetic Testing and Molecular Biomarkers, 14*(2), 255–257.

Schlichter, R., Bader, C. R., Bertrand, D., Dubois-Dauphin, M., & Bernheim, L. (1989). Expression of substance P and of a Ca2+-activated Cl- current in quail sensory trigeminal neurons. *Neuroscience, 30*(3), 585–594.

Schöbel, N., Radtke, D., Lubbert, M., Gisselmann, G., Lehmann, R., Cichy, A., . . . Wetzel, C. H. (2012). Trigeminal ganglion neurons of mice show intracellular chloride accumulation and chloride-dependent amplification of capsaicin-induced responses. *PLoS One*, *7*(11), e48005. doi:10.1371/journal.pone.0048005

Schroeder, B. C., Cheng, T., Jan, Y. N., & Jan, L. Y. (2008). Expression cloning of TMEM16A as a calcium-activated chloride channel subunit. *Cell*, *134*(6), 1019–1029.

Scott, J. W., Hawley, S. A., Green, K. A., Anis, M., Stewart, G., Scullion, G. A., . . . Hardie, D. G. (2004). CBS domains form energy-sensing modules whose binding of adenosine ligands is disrupted by disease mutations. *Journal of Clinical Investigation*, *113*(2), 274–284.

Seo, Y., Lee, H. K., Park, J., Jeon, D. K., Jo, S., Jo, M. & Namkung W. (2016). Ani9, A Novel Potent Small-Molecule ANO1 Inhibitor with Negligible Effect on ANO2. *PloS one*, *11*(5), e0155771. doi:10.1371/journal.pone.0155771

Sheridan, J. T., Worthington, E. N., Yu, K., Gabriel, S. E., Hartzell, H. C., & Tarran, R. (2011). Characterization of the oligomeric structure of the Ca(2+)-activated Cl- channel Ano1/TMEM16A. *Journal of Biological Chemistry*, *286*(2), 1381–1388. doi:10.1074/jbc.M110.174847

Sik, A., Smith, R. L., & Freund, T. F. (2000). Distribution of chloride channel-2-immunoreactive neuronal and astrocytic processes in the hippocampus. *Neuroscience*, *101*(1), 51–65.

Sirisi, S., Elorza-Vidal, X., Arnedo, T., Armand-Ugon, M., Callejo, G., Capdevila-Nortes, X., . . . Estevez, R. (2017). Depolarization causes the formation of a ternary complex between GlialCAM, MLC1 and ClC-2 in astrocytes: Implications in megalencephalic leukoencephalopathy. *Human Molecular Genetics*, *26*(13), 2436–2450. doi:10.1093/hmg/ddx134

Sirisi, S., Folgueira, M., Lopez-Hernandez, T., Minieri, L., Perez-Rius, C., Gaitan-Penas, H., . . . Barrallo-Gimeno, A. (2014). Megalencephalic leukoencephalopathy with subcortical cysts protein 1 regulates glial surface localization of *GLIALCAM* from fish to humans. *Human Molecular Genetics*, *23*(19), 5069–5086. doi:10.1093/hmg/ddu231

Smith, R. L., Clayton, G. H., Wilcox, C. L., Escudero, K. W., & Staley, K. J. (1995). Differential expression of an inwardly rectifying chloride conductance in rat brain neurons: A potential mechanism for cell-specific modulation of postsynaptic inhibition. *Journal of Neuroscience*, *15*(5 Pt 2), 4057–4067.

Sobacchi, C., Schulz, A., Coxon, F. P., Villa, A., & Helfrich, M. H. (2013). Osteopetrosis: Genetics, treatment and new insights into osteoclast function. *Nature Reviews Endocrinology*, *9*(9), 522–536. doi:10.1038/nrendo.2013.137

Song, S. C., Beatty, J. A., & Wilson, C. J. (2016). The ionic mechanism of membrane potential oscillations and membrane resonance in striatal LTS interneurons. *Journal of Neurophysiology*, *116*(4), 1752–1764. doi:10.1152/jn.00511.2016

Staley, K. (1994). The role of an inwardly rectifying chloride conductance in postsynaptic inhibition. *Journal of Neurophysiology*, *72*(1), 273–284.

Staley, K., Smith, R., Schaack, J., Wilcox, C., & Jentsch, T. J. (1996). Alteration of $GABA_A$ receptor function following gene transfer of the CLC-2 chloride channel. *Neuron*, *17*(3), 543–551.

Stauber, T., & Jentsch, T. J. (2010). Sorting motifs of the endosomal/lysosomal CLC chloride transporters. *Journal of Biological Chemistry*, *285*(45), 34537–34548.

Stauber, T., & Jentsch, T. J. (2013). Chloride in vesicular trafficking and function. *Annual Review of Physiology*, *75*, 453–477. doi:10.1146/annurev-physiol-030212-183702

Steinberg, B. E., Huynh, K. K., Brodovitch, A., Jabs, S., Stauber, T., Jentsch, T. J., & Grinstein, S. (2010). A cation counterflux supports lysosomal acidification. *Journal of Cellular Biology*, *189*(7), 1171–1186.

Steinmeyer, K., Ortland, C., & Jentsch, T. J. (1991). Primary structure and functional expression of a developmentally regulated skeletal muscle chloride channel. *Nature*, *354*(6351), 301–304.
Steinmeyer, K., Schwappach, B., Bens, M., Vandewalle, A., & Jentsch, T. J. (1995). Cloning and functional expression of rat CLC-5, a chloride channel related to kidney disease. *Journal of Biological Chemistry*, *270*(52), 31172–31177.
Stephan, A. B., Shum, E. Y., Hirsh, S., Cygnar, K. D., Reisert, J., & Zhao, H. (2009). ANO2 is the cilial calcium-activated chloride channel that may mediate olfactory amplification. *Proceedings of the National Academy of Sciences of the United States of America*, *106*, 11776–11781. doi:10.1073/pnas.0903304106
Stobrawa, S. M., Breiderhoff, T., Takamori, S., Engel, D., Schweizer, M., Zdebik, A. A., . . . Jentsch, T. J. (2001). Disruption of ClC-3, a chloride channel expressed on synaptic vesicles, leads to a loss of the hippocampus. *Neuron*, *29*(1), 185–196.
Stohr, H., Heisig, J. B., Benz, P. M., Schoberl, S., Milenkovic, V. M., Strauss, O., . . . Schulz, H. L. (2009). TMEM16B, a novel protein with calcium-dependent chloride channel activity, associates with a presynaptic protein complex in photoreceptor terminals. *The Journal of Neuroscience*, *29*, 6809–6818. doi:10.1523/JNEUROSCI.5546-08.2009
Sugita, S., Tanaka, E., & North, R. A. (1993). Membrane properties and synaptic potentials of three types of neurone in rat lateral amygdala. *Journal of Physiology*, *460*, 705–718.
Suzuki, J., Umeda, M., Sims, P. J., & Nagata, S. (2010). Calcium-dependent phospholipid scrambling by TMEM16F. *Nature*, *468*, 834–838. doi:10.1038/nature09583
Ta, C. M., Adomaviciene, A., Rorsman, N. J. G., Garnett, H., & Tammaro, P. (2016). Mechanism of allosteric activation of TMEM16A/ANO1 channels by a commonly used chloride channel blocker. *British Journal of Pharmacology*, *173*(3), 511–528. doi:10.1111/bph.13381
Takayama, Y., Uta, D., Furue, H., & Tominaga, M. (2015). Pain-enhancing mechanism through interaction between TRPV1 and anoctamin 1 in sensory neurons. *Proceedings of the National Academy of Sciences of the United States of America*, *112*(16), 5213–5218. doi:10.1073/pnas.1421507112
Terashima, H., Picollo, A., & Accardi, A. (2013). Purified TMEM16A is sufficient to form $Ca2^+$ -activated Cl^- channels. *Proceedings of the National Academy of Sciences of the United States of America*, *110*(48), 19354–19359. doi:10.1073/pnas.1312014110
Thiemann, A., Gründer, S., Pusch, M., & Jentsch, T. J. (1992). A chloride channel widely expressed in epithelial and non-epithelial cells. *Nature*, *356*(6364), 57–60.
Tien, J., Peters, C. J., Wong, X. M., Cheng, T., Jan, Y. N., Jan, L. Y., & Yang, H. (2014). A comprehensive search for calcium binding sites critical for TMEM16A calcium-activated chloride channel activity. *eLife*, *3*, e02772. doi:10.7554/eLife.02772
Traverso, S., Elia, L., & Pusch, M. (2003). Gating competence of constitutively open CLC-0 mutants revealed by the interaction with a small organic Inhibitor. *Journal of General Physiology*, *122*(3), 295–306.
van der Knaap, M. S., Boor, I., & Estevez, R. (2012). Megalencephalic leukoencephalopathy with subcortical cysts: Chronic white matter oedema due to a defect in brain ion and water homoeostasis. *Lancet Neurology*, *11*(11), 973–985. doi:10.1016/S1474-4422(12)70192-8
van Slegtenhorst, M. A., Bassi, M. T., Borsani, G., Wapenaar, M. C., Ferrero, G. B., de Conciliis, L., . . . et al. (1994). A gene from the Xp22.3 region shares homology with voltage-gated chloride channels. *Human Molecular Genetics*, *3*(4), 547–552.
Veeramah, K. R., Johnstone, L., Karafet, T. M., Wolf, D., Sprissler, R., Salogiannis, J., . . . Hammer, M. F. (2013). Exome sequencing reveals new causal mutations in children with epileptic encephalopathies. *Epilepsia*, *54*(7), 1270–1281. doi:10.1111/epi.12201

Vocke, K., Dauner, K., Hahn, A., Ulbrich, A., Broecker, J., Keller, S., . . . Mohrlen, F. (2013). Calmodulin-dependent activation and inactivation of anoctamin calcium-gated chloride channels. *Journal of General Physiology, 142*, 381–404. doi:10.1085/jgp.201311015

Walden, M., Accardi, A., Wu, F., Xu, C., Williams, C., & Miller, C. (2007). Uncoupling and turnover in a Cl^-/H^+ exchange transporter. *Journal of General Physiology, 129*(4), 317–329.

Wartosch, L., Fuhrmann, J. C., Schweizer, M., Stauber, T., & Jentsch, T. J. (2009). Lysosomal degradation of endocytosed proteins depends on the chloride transport protein ClC-7. *FASEB Journal, 23*(12), 4056–4068. doi:10.1096/fj.09-130880

Weinert, S., Jabs, S., Supanchart, C., Schweizer, M., Gimber, N., Richter, M., . . . Jentsch, T. J. (2010). Lysosomal pathology and osteopetrosis upon loss of H^+-driven lysosomal Cl^- accumulation. *Science, 328*, 1401–1403. doi:10.1126/science.1188072

Weinreich, F., & Jentsch, T. J. (2001). Pores formed by single subunits in mixed dimers of different CLC chloride channels. *Journal of Biological Chemistry, 276*(4), 2347–2353.

Wellhauser, L., Kuo, H. H., Stratford, F. L., Ramjeesingh, M., Huan, L. J., Luong, W., . . . Bear, C. E. (2006). Nucleotides bind to the C-terminus of ClC-5. *Biochemical Journal, 398*(2), 289–294.

Yang, C., & Delay, R. J. (2010). Calcium-activated chloride current amplifies the response to urine in mouse vomeronasal sensory neurons. *Journal of General Physiology, 135*, 3–13. doi:10.1085/jgp.200910265

Yang, H., Kim, A., David, T., Palmer, D., Jin, T., Tien, J., . . . Jan, L. Y. (2012). TMEM16F forms a Ca2+-activated cation channel required for lipid scrambling in platelets during blood coagulation. *Cell, 151*, 111–122. doi:10.1016/j.cell.2012.07.036

Yang, T., & Colecraft, H. M. (2016). Calmodulin regulation of TMEM16A and 16B Ca(2+)-activated chloride channels. *Channels (Austin, Texas), 10*, 38–44. doi:10.1080/19336950.2015.1058455

Yang, T., Hendrickson, W. A., & Colecraft, H. M. (2014). Preassociated apocalmodulin mediates $Ca2^+$-dependent sensitization of activation and inactivation of TMEM16A/16B $Ca2^+$-gated Cl^- channels. *Proceedings of the National Academy of Sciences of the United States of America, 111*, 18213–18218. doi:10.1073/pnas.1420984111

Yang, Y. D., Cho, H., Koo, J. Y., Tak, M. H., Cho, Y., Shim, W. S., . . . Oh, U. (2008). TMEM16A confers receptor-activated calcium-dependent chloride conductance. *Nature, 455*(7217), 1210–1215.

Yao, Z., Namkung, W., Ko, E., Park, J., Tradtrantip, L., & Verkman, A. S. (2012). Fractionation of a herbal antidiarrheal medicine reveals eugenol as an inhibitor of $Ca2^+$-activated Cl^- channel TMEM16A. *PLoS One, 7*, e38030. doi:10.1371/journal.pone.0038030

Yoshikawa, M., Uchida, S., Ezaki, J., Rai, T., Hayama, A., Kobayashi, K., . . . Sasaki, S. (2002). CLC-3 deficiency leads to phenotypes similar to human neuronal ceroid lipofuscinosis. *Genes to Cells, 7*(6), 597–605.

Yu, K., Duran, C., Qu, Z., Cui, Y. Y., & Hartzell, H. C. (2012). Explaining calcium-dependent gating of anoctamin-1 chloride channels requires a revised topology. *Circulation Research, 110*(7), 990–999. doi:10.1161/CIRCRESAHA.112.264440

Zdebik, A. A., Zifarelli, G., Bergsdorf, E. Y., Soliani, P., Scheel, O., Jentsch, T. J., & Pusch, M. (2008). Determinants of anion-proton coupling in mammalian endosomal CLC proteins. *Journal of Biological Chemistry, 283*(7), 4219–4227.

Zeydan, B., Uygunoglu, U., Altintas, A., Saip, S., Siva, A., Abbink, T. E. M., . . . Yalcinkaya, C. (2017). Identification of 3 novel patients with CLCN2-related leukoencephalopathy due to CLCN2 mutations. *European Neurology, 78*(3–4), 125–127. doi:10.1159/000478089

Zhang, W., Schmelzeisen, S., Parthier, D., Frings, S., & Mohrlen, F. (2015). Anoctamin calcium-activated chloride channels may modulate inhibitory transmission in the cerebellar cortex. *PLoS One*, *10*, e0142160. doi:10.1371/journal.pone.0142160

Zhang, X. D., Lee, J. H., Lv, P., Chen, W. C., Kim, H. J., Wei, D., . . . Yamoah, E. N. (2015). Etiology of distinct membrane excitability in pre- and posthearing auditory neurons relies on activity of Cl- channel TMEM16A. *Proceedings of the National Academy of Sciences of the United States of America*, *112*(8), 2575–2580. doi:10.1073/pnas.1414741112

Zhang, Y., Zhang, Z., Xiao, S., Tien, J., Le, S., Le, T., . . . Yang, H. (2017). Inferior olivary TMEM16B mediates cerebellar motor learning. *Neuron*, *95*(5), 1103–1111 e1104. doi:10.1016/j.neuron.2017.08.010

Zifarelli, G., & Pusch, M. (2007). CLC chloride channels and transporters: A biophysical and physiological perspective. *Reviews of Physiology, Biochemistry and Pharmacology*, *158*, 23–76.

Zifarelli, G., & Pusch, M. (2009a). Conversion of the 2 Cl(-)/1 H(+) antiporter ClC-5 in a NO(3)(-)/H(+) antiporter by a single point mutation. *EMBO Journal*, *28*, 175–182.

Zifarelli, G., & Pusch, M. (2009b). Intracellular regulation of human ClC-5 by adenine nucleotides. *EMBO Reports*, *10*(10), 1111–1116.

Zifarelli, G., & Pusch, M. (2010). The role of protons in fast and slow gating of the Torpedo chloride channel ClC-0. *European Biophysical Journal*, *39*(6), 869–875.

INDEX

Tables and figures are indicated by *t* and *f* following the page number

1K channel, 16*t*, 17
2 arachidonylglycerol (2-AG), 431–432
2-Bn-Tet-AMPA, 302
2-guanidine-4-methylquinazoline (GMQ), 662, 663
5-HT_2 receptors, on prefrontal cortex pyramidal cell sodium current, 6, 7*f*
5-HT_3 receptors, 390

A

A-317567, 657
Abdelnour, E., 177
Abramson, J., 79
Accn1, 647, 647*t*
Accn2, 647–648, 647*t*
Accn3, 647, 647*t*
Accn4, 647, 647*t*
Accn5, 647, 647*t*
acetazolamide, 152–153, 172
acetylcholine (ACh), 374. *See also* nicotinic acetylcholine receptors (nAChRs)
  hydrolysis, 375
  kinetic model binding, 381, 381*f*
  nicotinic acetylcholine receptor binding, 374–375
  receptor types, 375
  structure, 376*f*
  studies, pioneering, 374–375
acetylcholine binding protein (AChBP), 392
acetylcholine esterase (AChE), 375
acid, defined, 646
acidification
  acid-sensing ion channel activation, 649, 656, 658, 662, 663, 666, 667, 669, 671
  potassium channels, tandem pore, 580, 582, 585
  sodium channels, voltage-dependent, 207
  TMEM16Ca^{2+} activated Cl^- channels, 713, 717, 720
  TRPC channels, 620
acid-sensing ion channels (ASICs), 646–678
  functional properties, 649–651, 649*t*, 650*f*
  gene family, 647–648, 647*f*
  homomers and heteromers, 647*t*, 648
  modulation
    arachidonic acid, 663
    bile acids and steroid molecules, 662
    lipids, 663–664
    lysophosphatidylcholines, 663
    RFamide neuropeptides and $ASIC_3$ central vestibule binding, 662
    Zn^{2+}, 663
  molecular structure, 651–655
    acidic pockets, 652, 653*f*
    conformations, principal, 651–653, 653*f*
    extracellular domain, 651–652
    ion pore, 652, 653–655, 655*f*
    subunits and domain arrangement, 651–652
    transmembrane domains, 651–652
  protons, 646
    activation, 649
    concentration, 646–647
    sensing, 646
  steady-state desensitization, 649–651, 650*f*, 658, 659, 662, 663, 670, 675
acid-sensing ion channels (ASICs), pathophysiological roles, 675–678
  ischemic stroke, 658, 675–676
  multiple sclerosis, 676–677
  pain, 677–678
    chronic muscle, $ASIC_3$, 678
    migraine, $ASIC_1$/$ASIC_3$, 677–678
    neuropathic, $ASIC_1$, 677

acid-sensing ion channels (ASICs), pharmacology, 655–664
modulators, 661–664
lipids, 663–664
opioid neuropeptides and $ASIC_{1a}$, 662–663
RFamide neuropeptides and $ASIC_3$ central vestibule binding, 662
Zn^{2+}, 663
subtype-selective drugs, 657–661
$ASIC_3$ and $ASIC_3$-containing heteromers, 661
Hi_{1a}, 659
mambalgins, 659, 660*f*
MitTx, 660–661, 660*f*
NSAIDs, 661
PcTx1, 659, 660*f*
peptide toxins specific for $ASIC_1$ and $ASIC_1$-containing heteromers, 657–658
psalmotoxin, 658–659
subtype-unselective drugs, 656–657
A-317567, 657
amiloride, 656–657, 673, 674, 675, 676, 677, 678
diminazene, 656–657
nafamostat, 657
acid-sensing ion channels (ASICs), physiological functions, 664–675
brain
general, 664–665
postsynaptic receptors, proton, 666–667
humans
expression, 674–675
nociception, 674–675
peripheral nervous system, 668–674
peripheral nervous system, large-diameter sensory neurons, mechanoreceptors, 671–674
$ASIC_2$, 672–673
$ASIC_3$, 673–674
peripheral nervous system, small-diameter sensory neurons, nociceptors and metaboreceptors, 669–671
$ASIC_{1a}$ nociceptors, 669
$ASIC_{1b}$ nociceptors, 669–670
$ASIC_3$ metaboreceptors and nociceptors, 670–671
spinal cord, 667–668
action potentials. *See also specific types*
modulation, neurotransmitter, 6
varieties, 4, 4*f*
ADAM22, 112
addictions
ethanol, BK channel, 504*t*, 507–508
hyperpolarization-activated cyclic nucleotide-gated channel, 558–559
nicotine, nicotinic acetylcholine receptors, 385, 388, 400, 401
adenosine triphosphate (ATP), 458. *See also* P2X receptors
P2X receptors, activation, 462–464
SL_{O2} channel regulation, 513
adult-onset neuronal ceroid lipofuscinosis (ANCL), 504*t*, 506
SL_{O1} channelopathies, 504*t*, 506
Aeropyrum pernix (K_VAP), 68, 69*f*
agglutinins, 309
Ahern, C., 201
Ahmad, S., 228
Akk, G., 429
alcohol
hyperexcitability from, $GABA_A$ receptor plasticity, 439–440
target, tonic current-carrying $GABA_A$ receptors, 434–436
alkaline pH
acid-sensing ions, 659
hyperpolarization-activated cyclic nucleotide-gated channels, 549
potassium channels, tandem pore domain, 578, 594
TALK family, 588–590
TMEM16Ca^{2+} activated Cl^- channels, 713
all-atom molecular dynamics (MD) methodology, 36, 37*f*
allopregnanolone, 428
allosteric coupling, SL_{O1} channels
CTD-PD, 493
Horrigan and Aldrich allosteric model, 491
VSD-CTD, 493
VSD-PD, 492–493
$(\alpha + \beta_1)$BK channels, 498
$(\alpha + \beta_2)$BK channels, 498

expression patterns, role in neurons and, 499
nervous system, 496–498, 497*t*, 498*f*
($\alpha + \beta_4$)BK channels, 500–501
expression patterns, role in neuronal soma and, 500–501
nerve terminals, role, 501
alphaxalone, 428
α5 receptor measurement, GABA concentration on tonic currents mediated by, 437–438
α7 receptor, homomeric, drugs targeting, 402
α-Amino-3-Hydroxy-5-Methyl-4-Isoxazolepropionic Acid (AMPA), 291
α-thujone, 428
alternative translation initiation (ATI), 93, 579
Althoff, K., 79
Alzheimer's disease
dihydromyricetin on, 431
nicotinic acetylcholine receptors, 400, 401–402, 403
SL_{O1} channelopathies, 504*t*, 506
TRP channel mutations, 71
Aman, T. K., 81
Ambrosino, P., 178
amidine A-317567, 657
AMIGO-1, 113, 114*f*
amiloride, 656–657, 673, 674, 675, 676, 677, 678
amino-terminal domain (ATD), IGluR, 293–294, 294*f*
dimers/dimerization, 308–309, 314–315, 314*f*
pore opening, 316, 317
structure and function, 308–310
protein interactions mediated by, 309–310
receptor assembly, 308–309
AMPA and kainate receptors (KARs), 291–323
auxiliary subunits, 317–323
AMPAR functional modulation, by TARPs, 319
AMPAR–TARP assembly and stoichiometry, 319–321, 320*f*
cornichon homologs 2 and 3, 321
cystine-knot AMPAR-modulating proteins, 322
germ cell-specific gene1-like, 321–322
KAR auxiliary proteins, 322–323
other AMPAR, 321–322
TARP as AMPAR chaperones, 317–318, 318*f*
channel pore regulation, 311–313
fatty acids modulation, 313
polyamine block, 311–313
channel pore selectivity filter, 311
channel pore structure, 311
heteromerization on channel gating, 310–311
iGluR amino-terminal domain, structure and function, 308–310
protein interactions mediated by, 309–310
receptor assembly, 308–309
iGluR ligand-binding domain, structure and function, 302–307, 304*f*
agonist efficacy and desensitization, molecular mechanisms, 302–305, 304*f*
AMPARs, allosteroid modulators, 305
KAR channel gating, external ion requirement, 306–307
LBD dimer interface and receptor desensitization, 305–306
ionotropic glutamate receptor family, 291–297
fundamentals, 291–293, 292*f*
stoichiometry, 294*f*, 297
topology, 293–296, 294*f*, 295*f*
ionotropic glutamate receptor family, gating and pharmacology, 297–302
channel gating, iGluR subunits, 298–299
deactivation, 297
desensitization, 297–298
gating, 297
gating behavior, native AMPARs and KARs, 299–300
gating behavior, recombinant AMPARs and KARs, 300–301, 300*t*
permeation, 297
pharmacology, AMPARs and KARs, 301–302, 303*t*
pre-desensitization, 298
terminology, 297
tetrameric complex organization, intact AMPA and kainate structures, 313–317
discovery and elucidation, 313–315, 314*f*

AMPA and kainate receptors (KARs) (*cont.*)
GluA2/GluA3 apo state structures, 314–315, 314*f*
LBD layer rotation, during sensitization, 316–317
pore opening mechanism, 316
AMPA receptors (AMPARs), 291, 292, 292*t*. *See also* AMPA and kainate receptors
functional modulation, TARP receptor, 319
synaptic, brain auxiliary/regulatory proteins, 297
AMPAR–TARP assembly and stoichiometry, 319–321, 320*f*
amphotericin-induced gene and ORF 1 (AMIGO-1), 113–114, 114*f*
ANAp incorporation, 80–81
anesthetics
GABA$_A$ receptor, 429–430
glycine and glycine gating, 429
transmembrane domain, 429, 430
volatile, GABA$_A$ receptors, 414*f*, 422–423
anesthetics, general
GABA$_A$ receptor, 429–430
inhalational, on tandem-pore potassium channels, 593–594
TASK family, 582, 584*f*–585*f*, 585, 587, 588
THIK family, 590
TREK family, 581
TRESK family, 592
anesthetics, local, 212, 231–232, 380, 397
aniracetam, 305
ankyrin-G, 118, 122*f*, 206, 240, 265
anomalous mole fraction effect, 50
anoxia activation of *Drosophila* TRP/TRPL channels in dark, 632–634, 633*f*
antagonists, noncompetitive, 298, 299–300, 394, 397, 397*f*
Anthopleura elegantissima, 661
antiepileptic drugs (AEDs), 144, 152–153
antiporters
Cl^-/H^+, 717–720
endo/lysosomal chloride/proton, Cl^- channels, 696
APETx2, 661
arachidonic acid, acid-sensing ion channel modulation by, 663
2 arachidonylglycerol (2-AG), 431–432
architecture, ion channel, 65–70
historical view, 65–68
ion-conducting pore, 65, 66*f*
K^+ channel pore and gate, structural evidence, 68–70, 69*f*
sensor, 65, 66*f*
X-ray crystallographic and cryo-EM data, 35, 65
Armstrong, C. M., 67
ASIC$_{1a}$, opioid neuropeptides, 662–663
ASIC$_{1a}$ nociceptors, 669
ASIC$_{1b}$ nociceptors, 669–670
ASIC$_1$ peptide toxins specific for, 657–658
ASIC$_2$, 672–673
ASIC$_3$, 661, 662, 673–674
ASIC$_3$-containing heteromers, 661
ASIC$_3$ metaboreceptors and nociceptors, 670–671
ast-inactivationf, 225, 226*f*
astrocytes, voltage-dependent sodium channel, 207
ataxia
episodic ataxia type 1, 113, 151–152, 152*f*
myoclonus epilepsy and ataxia due to potassium channel mutation, 158–159
SeSAME (seizures, sensorineural deafness, ataxia, mental retardation, and electrolyte imbalance) syndrome, 180
SL$_{O1}$ channelopathies, 504*t*, 505–506
spinocerebellar ataxia type 13, 118
atomic-resolution structures, 35
atropine, nAChR blocking by, 384
Australian funnel-web spider, 659
autism spectrum disorders
KCNB1-related encephalopathy epilepsy with, 156
sodium channelopathies, 266*f*, 274–275
autoimmune diseases, 676. *See also specific types*
main immunogenic region, nACH receptors, 395, 396*f*
P2X7R blockade/deletion on, 474
P2X7 receptors, 474
autosomal-dominant nocturnal frontal lobe epilepsy (ADNFLE), 174–175, 400
SL$_{O2}$ channelopathies, 517, 518*t*
axonal degeneration, 6, 676

axon initial segment (AIS)
hyperpolarization-activated cyclic nucleotide-gated channel, 552
potassium channels, voltage-dependent, 103*t*, 104*t*, 108*f*, 109, 111*f*, 112, 114*f*, 115, 122*f*, 123, 130
sodium channelopathies, 270, 274
autism spectrum disorder, 274, 275
epilepsy syndromes, monogenic, 273
epileptic encephalopathy, 270
sodium channels
CNS, 264–265
$Na_V1.6$, 240
voltage-dependent, 206

B

Bailey, C. S., 172
Ballivet, Marc, 385
BAPTAs, 494, 634
barbiturates, 402, 429–430
bark scorpions, 234, 235*f*
baroreflex, 672–673
Bauer, C. K., 173, 185
Baukrowitz, T., 81
Bearden, D., 176
Beers, W. H., 375
benign familial infantile epilepsy (BFIE), 164
benign familial neonatal infantile seizures (BFNIS; *SCN2A*), 273–274
benign familiar neonatal epilepsy (BFNE, KCNQ2-BFNE), 123–124, 162–165, 166*f*, 186
benzodiazepines, 402, 425
BZ binding site, 422, 425–427, 425*f*
diazepam, 422, 425, 434
$GABA_A$ receptor binding site, 425–427
zolpidem, 426, 427
Berridge, M. J., 614
beta-carbolines, 426
βγ subunits, 18
β-selective compounds, 432–433
β subunits
potassium channels, large conductance nervous system, 495–496, 497*f*
sodium channels, 6
Bezanilla, F., 67, 73, 78
bicuculline, 427
big-conductance (BK) channels, 12, 13*f*
big dynorphin (BigDyn), $ASIC_{1a}$ and, 662–663
Big K (BK) channel. *See* Sl_{O1} (BK) channels
bi-ionic potential, 468
bile acids
$ASIC_5$ activation, 664
steroid molecules, acid-sensing ion channel modulation by, 662
"binding jaw" tightening, agonist action, P2X receptor, 460*f*, 461–462
BK (big potassium) channels. *See* Sl_{O1} (BK) channels
black mamba, 659
BMA (bis-maleimido azobenzene), 463
Bockenhauer, D., 180
Bosmans, F., 201
Boulter, Jim, 385
brain, acid-sensing ion channels
general, 664–665
postsynaptic receptors for protons, 666–667
brainstem, TME16 Ca^{2+} activated Cl^- channels, 705
Brownian dynamics (BD) simulations, 36, 39, 46, 51–52
Brugada syndrome, 107, 107*f*, 162, 205
bundle crossing, 75–76
potassium channels, 68, 70
large conductance, 490, 510–511
tandem pore domain, 575, 593
burst analysis, 380

C

Ca^{2+}
fractional current, 467
potassium channels, large conductance nervous system, 494–495
Ca^{2+} channels. *See* calcium (Ca^{2+}) channels
calcitonin gene-related peptide (CGRP), 515, 670, 671, 677–678
calcium channels, voltage-gated (Ca_V), 66*f*, 70
calcium (Ca^{2+}) channels, 8–11, 9*t*, 10*f*. *See also specific topics and types*
action potential modulation, 6
Aplysia, 10*f*, 11
electrostatic interactions, 52
genes, 3
L-type, 8–9

calcium (Ca^{2+}) channels (*cont.*)
  N-type, 8–9
  permeation and selectivity, 49–53, 52*f*
  P/Q-type, 8–9
  R-type, 8–9
  second messenger pathways, 9, 10*f*
  SNARE protein interactions, 9–11
  synprint site, 9
  T-type, 9
  unitary, types, 3
  voltage-activated
    high-voltage, 8
    low-voltage, 9
  voltage-dependent
    genes, 3, 8, 9*t*
    opening and action potential, 8
    T-type, 9
  voltage-gated mechanisms, 33
calmodulin (CaM), 9
  calcium channel, 9
  chloride channels, TMEM16A Ca^{2+}-activated, 700
  epilepsy, 163*f*, 173
  K_{V7} channel, 122–123, 122*f*
  NMDA receptors, 357
  potassium channel, voltage-dependent, 14*t*
    K_{V7}, 122, 122*f*
  sodium channel, voltage-dependent, 206
  sodium channelopathies, 265
    autism spectrum disorder, 274
  TRPC channels, 618, 619*f*, 622, 632
Cameron, J. M., 158
cannabidiol (CBD), 431–432
cannabinoids, 431–432
cardiac cells, voltage-dependent sodium channels, 208–209
cardiac channel phagosomes/endosomes, sodium channel, 206–207
Carr, D. B., 7*f*
Casida, J. E., 422, 428
cation channels, nonselective, 21–26, 22*t*
  cyclic nucleotide–regulated, 21–24, 22*t*, 23*f*
  transient receptor potential, 21, 22*t*, 24–26, 25*f*
$Ca_V1.2$, 8, 9*t*
$Ca_V1.3$, 8, 9*t*
$Ca_V1.4$, 8, 9*t*
$Ca_V2.1$, 8, 9*t*
$Ca_V2.2$, 8, 9*t*
$Ca_V2.3$, 8, 9*t*
$Ca_V3.1$, 8, 9*t*
$Ca_V3.2$, 8, 9*t*
$Ca_V3.3$, 8, 9*t*
CCAT (calcium channel-associated transcriptional regulator), 11
central nervous system
  BK channel expression pattern, 492–493
  sodium channelopathies, 265–277, 266*f* (*see also* sodium channelopathies, central nervous system)
    autism spectrum disorders, 266*f*, 274–275
    benign familial neonatal infantile seizures, 273–274
    cryptogenic focal epilepsy, 273–274
    epilepsy, 266–267, 266*f*, 267*f*
    epileptic encephalopathy, 268–272, 269*f*
    familial hemiplegic migraine, 266*f*, 275–277
    genetic epilepsy with febrile seizures plus, 272–273
  sodium channels, 263–265
central pathway, P2X receptors, 465
ceratotoxin-1 (CcoTx1), 213
cerebellar ataxia, 33
cerebellar atrophy, developmental delay, and seizures, 171–172
cerebellum, TMEM16Ca^{2+} activated Cl^- channels, 703–704
Cha, A., 78
Chanda, B., 78, 79, 201
Chandy, K. G., 102
channel gating, 64. *see also* gating, ion channel
channelopathies, 33, 503–509, 504*t*–505*t*. *See also specific types*
  epilepsy, 503–505, 504*t*
  permeation and selectivity, 33
  sodium, CNS, 265–277, 266*f* (*see also* sodium channelopathies, central nervous system)
channel pore blocker, $GABA_A$ receptor, 427–428
channel pores, AMPA and kainate receptors
  regulation, 311–313
    fatty acids modulation, 313
    polyamine block, 311–313
  structure, 311

channel pores, K^+ channel, 68–70, 69*f*
chaperones, 398–399
chemical sensors, 71–72
chloride channels, 5
antiporters, endo/lysosomal chloride/ proton, 696
equilibrium potential, 696
resting membrane Cl^- permeability, 696
chloride transport. *See also* TMEM16Ca^{2+} activated Cl^- channels
neurons and glia, 696–697
cholinergic drugs, 375–376, 376*f*
Clarke, P. B. S., 384–385
CLC chloride channels and transporters, 709–720
ClC_{-2}
brain expressed, 712–713
glia, 714–715, 716*f*
neurons, 713–714
ClC_{-3}-ClC_{-7}, 716–720
overview, 716–720
CLC family, overview
CLC transporter structure, 710, 712*f*
double barreled structure, 709–710, 711*f*
Cl^-/H^+ antiporters, 717–720
ClC_{-3} and ClC_{-4}: endosomal, 717–718
ClC_{-6}: neuronal late endosomal, 718–719
ClC_{-7}: ubiquitous lysosomal, 719–720
overview, 716–717
closed-state fast inactivation, 227–228, 231, 233, 236, 240, 241
CNG channels, 22–23, 22*t*, 23*f*
cobra toxin, 383–384
cochlea neurons, TMEM16Ca^{2+} activated Cl^- channels, 709
Colquhoun, David, 379
compound 5b, 659
concanavalin-A, 309
concentration-clamp systems, 299
cone-dystrophy with supernormal rod electroretinogram, 126
cone snail, 662
conformational coupling model and phospholipase C, 614
congenitally insensitive to pain (CIP), sodium channels, 209–210, 212, 228–229, 230*f*, 232
cono-RFamide, 662
continuum-based theories, 52–53
Conus textile, 662
coral snake, Texas, 660
Corbett, M. A., 153
cornichon homologs 2 and 3 (CNIH-2/-3), 321
Cosens, D. J., 611, 613
Coulomb knock-on mechanism, direct, 40, 41*f*
counter-adaptive evolution, 234
Cowgill, J., 79
Cox, J. J., 228
CREB (cyclic adenosine monophosphate [AMP] response element binding protein), 11
Cross, J. H., 182
cryo-electron microscopy, 35, 65
cryptogenic focal epilepsy *(SCN3A),* 273–274
C-terminal domain (CTD)
allosteric coupling
with PD, 493
with VSD, 493
AMPA and kainate receptors, 293, 294*f*, 296, 320
ionotropic glutamate receptors, 293, 294*f*, 296, 320
NMDA receptors, 346, 347*f*, 349, 350, 357
potassium channels, large conductance, 487–489, 489*f*, 514, 518
allosteric coupling with PD, 493
allosteric coupling with VSD, 493
sodium channelopathies, CNS, 262
C-terminal (Na^+ sensor) domain, Sl_{O2} channels, 512
Cui, J., 85
curare, 376, 382, 384, 399
current-voltage relationship, reversal potential, 35
Curtis, D. R., 344
cyclic adenosine monophosphate (cAMP)
Ca^{2+}-activated Cl^- currents
glial, 713
olfactory, 708
on hyperpolarization-activated cyclic nucleotide-gated channel, 548–549, 550*f*, 552, 553, 555, 558
cyclic nucleotide binding domain (CNBD), HCN channels, 547–549

cyclic nucleotide-binding homology domain (CNBHD), 126, 128*f*
cyclic nucleotide-gated channels, 22–23, 22*t*, 23*f*
  hyperpolarization-activated, 208, 545–559 (*see also* hyperpolarization-activated cyclic nucleotide-gated [HCN] channel)
cyclic nucleotide–regulated channels, 21–24, 22*t*, 23*f*
cyclothiazide, 295*f*, 296, 299, 300*t*, 304*f*, 305–306, 316, 319
cys-loop pentameric ligand-gated ion channel, 420
cys-loop superfamily, ligand-gated ion channels, 390–391
cystathionine-ß-synthase (CBS) domains, 710–711
cysteine-scanning mutagenesis, 74–76, 75*f*
cystine-knot AMPAR-modulating proteins (CKAMPs), 322
cytoplasmic cap, open state structure stabilization, P2X receptor, 460*f*, 464–465

D
d'Adamo, M. C., 152
Dai, G., 81
Dale, Sir Henry, 375
deactivation, 297
deafness, 33
degenerin/epithelial Na^+ channel (DEG/ENaC) ion channel family, 647, 654, 672
dehydration penalty, 42
DEKA ring, 45, 48–49, 200*f*, 202*f*
de Kovel, C. G. F., 156, 157
delayed rectifier, 3
delta-specific 1/2 (DS1/DS2), 437
dementia, nicotinic acetylcholine receptor, 400, 401–402
Dendroaspis polylepis, 659
DEND syndrome (developmental delay, epilepsy and neonatal diabetes), 183
denervation sensitivity, 385
depolarization, 33. *See also specific channels*
  paradoxical, 578
depolarization-activated channels. *See also specific types*
  classic, 547
  neuronal hyperpolarization-activated cyclic nucleotide-gated channels, 547–548
depression, cholinergic hyperfunction, 401–402
desensitization. *See also specific channels*
  acid-sensing ion channels, 649–651, 650*f*, 653, 653*f*, 654, 657, 658, 662, 675–676
  AMPA and kainate receptors, 296–301, 300*t*, 304*f*, 315, 318–319, 321–322
    LBD layer rotation, 316–317, 318*f*
  $GABA_A$ receptors, 420, 433
  ionotropic glutamate receptors, 297–298
  nACH receptors, 376, 378, 380*f*, 381, 399, 402
  NMDA receptors, 353*f*, 354
  P2X receptors, 459, 465, 470, 472
  potassium channels, large conductance nervous system, 507–508
  steady-state, 649–651, 650*f*, 658, 659, 662, 663, 670, 675
  TRPC channels, 632
developmental and epileptic encephalopathy (DEE), 144, 187
  acetazolamide on seizure frequency, 153
  vs. ADNFLE, 176
  EIMFS, 175
  gain-of-function mutations, 163
  $KCNA_1$, mutation, 152*f*
  KCNT2, 178
  $K_V1.2$ mutation, 153
  $K_V3.1$ mutation, 159*f*
  $K_V4.5$ mutation, 161
  $K_V7.5$ mutation, 167
$DFNA_2$, 124
diabetes mellitus, permanent neonatal, 182–183
diazepam, 422, 425, 434
dielectric material, structureless continuum, 36, 37*f*
dihydromyricetin (DHM), 431
[^{3}H]dihydropicrotoxin (DHP), 422
DiMaio, F., 81
diminazene, 656–657
dipeptidyl peptidase-like (DPPL) subunits, 119*f*, 120–121
direct Coulomb knock-on mechanism, 40, 41*f*
distance-encoding photoinduced electron transfer, 79

DKEA ring, 48
DMBXA, 403
domain swapping, 72, 82, 84*f*, 489*f*, 574
domoate, 301, 302, 303*t*
dopamine, 432
dopaminergic neurons
drug abuse, 559
$GABA_A$, 432
nicotinic acetylcholine receptors, 387*f*, 389, 397, 400–402
Parkinson's disease, 557
potassium channels
hyperpolarization-activated cyclic nucleotide-gated, 549, 557
voltage-dependent, 119*f*
dorsal root ganglia
sensory neurones, voltage-dependent sodium channels, 200*f*, 201*f*, 209–210, 211*f*
sensory nociceptors, 224
TMEM16Ca^{2+} activated Cl^- channels, 706–707
Dravet syndrome, 241, 243, 267, 268–270, 269*f*, 275, 432
Drosophila light sensitive channels, 611–637. *See also* TRPC (*Drosophila* light sensitive) channels
Drosophila melanogaster Shaker mutant, potassium channel cloning, 102
Drosophila PDZ scaffold protein, Inactivation but No Afterpotential D (INAD), 618–620, 619*f*, 627, 636
dynamic ion selectivity, 578, 593
dynamic simulation, ion permeation events, 36

E

E1211K childhood epilepsy, 270
EAG channel, 82
eag (K_V10) K^+ channel, 126
eag-like (elk, K_{V12}) K^+ channel, 126
eag-related gene (erg) (K_{V11}) K^+ channel, 126
early infantile epileptic encephalopathy-32 (EIEE32), 113
EAST syndrome, 180–182, 181*f*
EEEE locus, 51
EEEE ring, 45–48, 51
Eisenman, G., 42–43
electric signaling, 64
electron-density maps, 35
electron paramagnetic resonance (EPR), 64, 69–70
Electrophorus electricus, 67
Na_V channel, 260–262, 261*f*
electrostatic interactions
calcium channels, 52
potassium channels, 40, 42
ELIC (*Erwinia chrysanthemi* pentameric ligand-gated ion channel), 391, 423
encephalic epilepsy, pediatric, 116
endocannabinoids, 431–432
endo/lysosomal chloride/proton antiporters, Cl^- channels, 696
epilepsy, 33, 144–145
benign familial neonatal infantile seizures, 273–274
characteristics, 144
cryptogenic focal epilepsy, 273–274
developmental and epileptic encephalopathy, 144
diagnosis and etiology, 266
E1211K, childhood epilepsy, 270
early infantile epileptic encephalopathy type 13, 271
$GABA_A$ receptor plasticity, 439–440
genes, 144–145
antiepileptic drugs and, 145–146
monogenic causes, 145
potassium subunit encoding, 148, 149*t*–150*t*
genetic epilepsy with febrile seizures plus, 272–273
hyperpolarization-activated cyclic nucleotide-gated channel, 556–557
KCNQ2-related seizures, 145
molecular etiology, 145
P2X7R blockade/deletion on, 474
potassium channel structures, 146–148, 147*f*
prevalence, 266
Sl_{O1} channelopathies, 503–505, 504*t*
Sl_{O2} channels, 517–518, 518*t*
sodium channelopathies, 266–267, 266*f*, 267*f*
sudden unexplained death in epilepsy, 144, 145
treatment, drug, 144, 145–146

epilepsy, potassium channel mutations (channelopathies), 148–188
conclusions and future directions, 185–188
episodic ataxia type 1, 151–152, 152*f*
four transmembrane domain family, 184–185
KCNK$_4$ encoding K_{2P}4.1, 184–185, 184*f*
genotype-phenotype correlations, 186
six transmembrane domain family: Ca- and Na-activated potassium channels, 150*t*, 170–179
KCNMA$_1$ encoding K_{Ca}1.1, 170–173, 171*f*
KCNN3 Encoding K_{Ca}2.2 , 173, 174*f*
KCNT$_1$ encoding K_{Na}1.1, 174–178, 175*f*
KCNT2 Encoding K_{Na}1.2, 178–179, 179*f*
six transmembrane domain family: voltage-gated potassium channels, 149*t*, 151–170
KCNA$_1$ encoding K_V1.1, 151–153, 152*f*
KCNA$_2$ encoding K_V1.1, 153–154, 154*f*
KCNA$_4$ encoding K_V1.4, 154–155, 155*f*
KCNB$_1$ encoding K_V2.1, 155–157, 157*f*
KCNC$_1$ encoding K_V3.1, 157–159, 159*f* (*see also KCNC*$_1$ encoding K_V3.1)
KCNC$_3$ encoding K_V3.3, 159
KCND$_2$ encoding K_V4.2, 159–160, 160*f*
KCND$_3$ encoding K_V4.3, 160–162
KCNH$_1$ encoding K_V10.1, 168–169, 169*f*
KCNH$_2$ encoding K_V10.2, 169–170
KCNQ$_2$ encoding K_V7.2, 162–164, 163*f*
KCNQ$_3$ encoding K_V7.3, 164–166, 166*f*
KCNQ$_5$ encoding K_V7.5, 166–168, 167*f*
targeted therapies, 146, 163–164, 172, 176, 186, 187
two transmembrane domain family, including inwardly rectifying potassium channels, 150*t*, 179–184
KCNJ$_{10}$ encoding K_{ir}4.1, 179–182, 181*f*
KCNJ$_{11}$ encoding K_{ir}6.2, 182–184
epilepsy of infancy with migrating focal seizures *(EIMFS)*, 174, 175
epileptic encephalopathy
early onset, 271
sodium channelopathies, 268–272, 269*f*
episodic ataxia type 1 (EA1), 1, 113, 151–152, 152*f*
estradiol, Sl_{O2} channel regulation, 514–515
etazolate, 422
ethanol addiction, Sl_{O1} channelopathies, 504*t*, 507–508
ethyl alcohol, $GABA_A$ receptor, 429
etomidate, 429–430, 432
eukaryotic Na_V sodium channels, archetypal, 257–263
evolutionary origins and structures, 257–258
$Na_V\beta$ subunit complex, 262–263
structural elements, 260–262, 261*f*
structure–function relationship, 258–260, 259*f*
Evans, R. H., 344
excitability, intrinsic, 4, 5
HCN channels, 551–553, 552*f*
nonselective cation channels, 21
potassium channels, voltage-dependent, 116
excitable membrane properties, neurons, 3–30
calcium channels, 8–11, 9*t*, 10*f* (*see also* calcium channels)
firing patterns, 3–4, 4*f*
genes, 3
large number, rationales, 3–5
nonselective cation channels, 21–26, 22*t* (*see also* nonselective cation channels)
cyclic nucleotide–regulated, 21–24, 22*t*, 23*f*
transient receptor potential, 21, 22*t*, 24–26, 25*f*
potassium channels, 11–21, 13*f* (*see also* potassium channels)
calcium-activated, 16–17, 16*t*
inward rectifier, 17–19, 18*t*
sodium-activated, 17, 17*t*
two-pore domain K^+ channel subunits, 19–21, 20*t*
voltage-dependent, 12–16, 14*t*, 15*f*
sodium channels, 5–6, 7*f*, 8*t* (*see also* sodium channels)
excitatory postsynaptic current
AMPARs, 305, 321
NMDA receptor
kinetic properties, 351–352, 353*f*
postsynaptic signal, neuronal, 358–359
slow, 25*f*

exercise pressor reflex, 671
experimental autoimmune encephalomyelitis (EAE) mouse model, 676
expression pattern, CNS BK channel, 493–494
extracellular cap, tandem pore domain potassium channels, 184, 573*f*, 574, 592
extrasynaptic δ $GABA_A$ receptors, 434
Eyring rate theory, 379, 380*f*
ezogabine, 164

F

familial hemiplegic migraine, sodium channelopathies, 266*f*, 275–277
fast inactivation
  closed-state, 227–228, 231, 233, 236, 240, 241
  potassium channels, 118, 499
  sodium channels, 204, 206
    central nervous system, 264
    eukaryotic, 260, 262, 264
    $Na_V1.1$, 242
    $Na_V1.6$, 240, 241
    $Na_V1.7$, 227–228, 231
    $Na_V1.8$, 233
    $Na_V1.9$, 236
    peripheral pain-sensing neurones, 210
    voltage gated, 226, 226*f*
Fatt, P., 376–377
fatty acid modulation, on AMPA and kainate receptors, 313
Feldberg, W., 382
Fessard, A., 382
FHEIG (facial dysmorphism, hypertrichosis, epilepsy, intellectual disability/ developmental delay, and gingival overgrowth), 185
fipronil, 427
firing patterns, 3–4, 4*f*
5-hydroxytryptamine 2 ($5\text{-}HT_2$ receptors), on prefrontal cortex pyramidal cell sodium current, 6, 7*f*
5-hydroxytryptamine 3 ($5\text{-}HT_3$ receptors), 390
flavan, 431
flavonoids, $GABA_A$ receptors, 430–431
flip/flop cassette, 295–296, 295*f*, 305, 310, 315, 316, 323
flumazenil, 427
fluorescence resonance energy transfer (FRET), 79, 80–81
flux ratio, 35, 40
flux ratio exponent, 35
FMFRFamide, 662
FMR1, 174
FMRP (fragile X mental retardation protein), 17
Földy, C., 714
four transmembrane (4THM) domain family, K^+ channels, 147–148, 147*f*, 184–185, 184*f*
fractional Ca^{2+} current, 467
fragile X mental retardation protein (FMRP), 17, 174, 176, 504*t*, 507, 517, 518*t*
fragile X syndrome (FXS), 428, 507, 517, 518*t*
fragrant dioxane derivatives (FDDs), 433
Fritsch, Gustav Theodor, 343
funny current, 208, 545

G

$GABA_A$ receptors, 419–423
  abbreviations, 419
  barbiturates and CNS depressants on, 422
  clinical allosteric modulators, 421
  cys-loop pentameric ligand-gated ion channel family and subtypes, 420–421
  DHP and "picrotoxinin receptor," 422
  early work, Olsen lab, 421–422
  general anesthetics, 422
  identification and binding sites, 421–422
  mechanisms of action, 420
  neurosteroids on, 422
  neurotransmission, 420
  overview, 420
  plasticity, in epilepsy and alcohol-induced hyperexcitability, 439–440
  positive allosteric modulations, 438–439
  positive allosteric modulators, 426
  selective inverse agonists, 427
  volatile anesthetics and long-chain alcohols, 414*f*, 422–423

GABA$_A$ receptors, pharmacology, 423–440
anesthetics, 429–430
benzodiazepines binding site and subtype-selective modulators, 422, 425–427, 425*f*
β-selective compounds, 432–433
binding sites, 423–425, 424*f*
cannabinoids, 431–432
channel pore blocker, 427–428
extrasynaptic δ GABA$_A$ receptors, 434
flavonoids, 430–431
GABA analogs THIP and muscimol, δ-GABA$_A$ receptor selective ligands, 436
GABA binding, 425, 425*f*
neurotransmitter, amino acids, 432
plasticity, epilepsy and alcohol-induced hyperexcitability, 439–440
steroids, neuroactive, 428–429
synaptic, extrasynaptic & intermediate inhibition, functional selectivivity to GABAergic modulators, 433–434
tonic current-carrying GABA$_A$ receptors, as low-dose alcohol targets, 434–436
tonic currents mediated by α5 subunit-containing receptor measurement, GABA concentration on, 437–438
tonic *vs.* phasic currents, functional distinction, shifted towards tonic inhibition by GABA$_A$ receptor PAMs, 438–439
δ-GABA$_A$, recombinantly expressed, low efficacy and tonic neuronal GABA currents, 436–437
GABAergic inhibition
intermediate, 433–434
synaptic, 433
tonic, 433
gaboxadol. *See* THIP [4, 5, 6,7-tetrahydroisoxazolo(5,4-c) pyridin-3-ol]
galectins, 309
Galvani, Luigi, 374
γ-aminobutyric acid (GABA), 420
epilepsy genes, 152
GABA$_A$ receptor binding, 425, 425*f*
inhibitory response, intracellular Cl$^-$ concentration, 696
neurotransmission, 420
structure, 425*f*
γ-hydroxybutyric acid (GHB), 433
γ subunit family, 496, 497*f*
gating, ion channel, 64–85. *See also specific types*
architecture, 65–70
historical view, 65–68
ion-conducting pore, 65, 66*f*
K$^+$ channel pore and gate, 68–70, 69*f*
membrane topology, 65, 66*f*
sensor, 65, 66*f*
X-ray crystallographic and cryo-EM data, 35, 65
definition, 64, 297
electric signaling, mechanisms, 64
ligand-gated, 297–298
NMDA receptor, 353, 353*f*
nonclassical
selectivity filter as activating gate, 81–82
VSD-to-PD coupling, 82–85, 84*f*
sensors, gating domain/pore region, 71–73
physical and chemical sensors, 71–72
voltage sensing, 73
structure-function models, 64–65
structure/function study, biophysical and chemical tools, 73–76
cysteine-scanning mutagenesis, 74–76, 75*f*
patch-clamp fluorometry, 79–81, 80*f*
voltage-clamp fluorometry, 76–79, 77*f*
gating charges, 258
gating glutamate, 710
gating kinetics
AMPA and kainate receptors, 293, 295*f*
ClC-7, 720
definition, 548
neuronal hyperpolarization-activated cyclic nucleotide-gated channels, 548, 550*f*, 551, 555
NMDA receptors, 352, 356, 359
potassium channels, large conductance: (α+β2)BK, 498
generalized epilepsy, *KCNMA$_1$* encoding K$_{Ca}$1.1, 171
genes. *See also specific genes and channels*
ion channels, 3–5

genetic epilepsy with febrile seizures plus (GEFS+; *SCN1A*), 272–273
germ cell-specific gene1-like (GSG1L), 321–322
Gilbert, Wally, 421
glia
  chloride transport, 696–697
  NMDA receptors, 361–362
GlialCAM, 715, 716*f*
GLIC (*Gloebacter violaceous* pentameric ligand-gated channel), 391, 423
GluA, 346
GluD, 345, 346, 347*f*
GluK, 346, 348
GluN, 346, 347*f*, 348, 350, 357
GluN1 subunit, NMDA receptor, 315, 346, 347*f*, 350, 356, 357
  disease-associated mutations, 355, 355*f*
GluN2 subunit, NMDA receptor, 308, 315, 346, 347*f*, 357
GluN3 subunit, NMDA receptor, 346, 347*f*, 350
glutamate, 344
glutamate-gated excitatory channels, 344
glutamatergic synapse, 291, 292*f*
glutamine-binding protein (QBP), 262, 347*f*, 354
glycine and glycine gating
  anesthetics, 429
  BK channels, pain disorders, 508
  $GABA_A$ receptors, 420, 429, 431
  nicotinic acetylcholine receptors, 390, 394
  NMDA receptors, 343, 346, 351, 353, 357
  P2X receptors, 464
glycine receptors
  BK channels, pain disorders, 508
  brain, 696, 697
  cys-loop pentameric ligand-gated ion channel, 420
Goldman-Hodgkin-Katz (GHK), 467, 575
G-protein-coupled metabotropic P2Y receptors. *See* P2X receptors
G protein-coupled receptors (GPCRs), 18, 21
  $GABA_B$ receptors, 420
  potassium channels, tandem pore domain, 581
G proteins, Sl_{O_2} channel regulation, 513–514
grid cells, 553
GTS-21 (DMBXA), 403
2-guanidine-4-methylquinazoline (GMQ), 662, 663
GYKI compounds, 299–300

H
Hadronyche infensa, 659
Hardie, R. C., 614–615
HCN channels. *See* hyperpolarization-activated cyclic nucleotide-gated (HCN) channels
hearing, sensory receptors, large conductance potassium channels, 501–502
HEK293/HEK293 cells
  $GABA_A$ receptor, 431, 432
  potassium channels
    large conductance, nervous system, 513, 514
    tandem pore domain, 573*f*, 584*f*–585*f*, 591
    voltage-dependent, 114*f*, 125*f*
  sodium channels, pain transmission, 228, 229, 230*f*, 236
  $TMEM16CA^{2+}$, 700*f*
  TRPC channels, 621, 626
helical screw model, 203
hERG, 78, 82
heterodimers/heterodimerization. *See also specific types*
  amino-terminal domain, 309
  ClC-3 and ClC-4, 719
  family, 592–593
  GluN1/GluN2B receptors, 355
  iGluRs, 315
  K2P subunits, 574
  TASK, 583, 586
  THIK, 590, 591
  TREK, 579
  TWIK, 577, 578
  two-pore domain K^+ channels, 19
heteromerization
  AMPA and kainate receptors, 310–311
  K_{V_2} channel, 113, 114*f*
  P2X receptors, 470
heteromers, 12. *See also specific types*
  acid-sensing ion channels, 648
  potassium channels, inward rectifier, 18
Heteroscodra maculata, 658
hetero-tetramerization, 121–122, 127

heterotrimers, 471, 648. *See also specific types*
acid-sensing ion channels, 648
P2X receptors, 471
Hi_{1a}, 659
high-pass filtering, 551, 555
Hille, B., 67
hippocampus, TMEM16Ca^{2+} activated Cl^- channels, 703, 718, 719
hispidulin, 431
histamine, 432
Hitzig, Eduard, 343
hNE9. *See* $Na_V1.7$
Hodgkin, A. L., 35, 38–39, 40, 65, 101, 377, 379
Hodgkin–Huxley current, 3, 12
Hoegg-Beiler, M. B., 715, 716*f*
Hold, K. M., 428
homomeric α7 receptor, drug targeting, 402
homomers, 12. *See also specific types*
acid-sensing ion channels, 648
P2X receptors, 471
potassium channels, inward rectifier, 18
homotrimers, 458, 648, 651
Horrigan and Aldrich (H-A) allosteric model, 491
Hosie, A. M., 428
Huang, W. C., 703
Huang, Z., 556
human ether-a-go-go-related gene (hERG), 78, 82
huwentoxin-IV (HwTx-IV), 212–213
Huxley, A. F., 65, 101, 377, 379
hybrid QM/MM techniques, 36
6-hydroxyflavones, 431
5-hydroxytryptamine 2 (5-HT_2) neurotransmitter receptors, on prefrontal cortex pyramidal cell sodium current, 6, 7*f*
5-hydroxytryptamine 3 (5-HT_3), 390
hyperexcitability, alcohol-induced, $GABA_A$ receptor plasticity in, 439–440
hyperpolarization-activated cyclic nucleotide-gated (HCN) channels, 22–24, 22*t*, 23*f*, 545–559
basic facts, 546–549
expression pattern, cellular and subcellular, 546
expression pattern, central nervous system, 546
expression pattern, peripheral nervous system, 546–547
functional modulation, multiple heterogeneous mechanisms, 548–549
neuronal, biophysical features, 547–548
CNS pathologic states, 556–559
addictions, 558–559
epilepsy, 556–557
neuropathy, 558
Parkinson's disease, 557
retinal pathology, 558
funny current, 208, 545
neuronal, types, 545
properties and function, nervous system, 549–556, 550*f*
neurotransmitter release, 555–556
resonance properties, 555
resting membrane potential regulation and intrinsic excitability, 551–553, 552*f*
rhythmogenesis, 553–554
synaptic excitability and plasticity, 554–555
unique biophysical properties, neurophysiological implications, 550–551
structure and properties, 545, 550*f*
hypoxia-sensing TRPC6 channels, pulmonary smooth muscle cells, 635–636

I
I381T, 239
I1461T, 231
IFMT (isoleucine-phenylalanine-methionine-threonine) motif, 200*f*, 204, 210, 225, 230, 233
Ikeda, T., 428
IKM channel, activating mechanisms, 83
inactivation
eukaryotic Na_V channels, 260
fast (*see* fast inactivation)
Inactivation but No Afterpotential C (INAC), 618–619, 619*f*, 627
Inactivation but No Afterpotential D (INAD), 618–620, 619*f*, 627, 636
inactivation gate, 78, 204
hyperpolarization-activated cyclic nucleotide-gated channels, 547

sodium channel, voltage-dependent, 200, 204, 206, 210
inaD gene, 618
inflammasome, 465, 474
inflammatory bowel disease, P2X7R blockade/deletion on, 474
inhalational anesthetics, on tandem-pore potassium channels, 593–594
TASK family, 582, 584*f*–585*f*, 585, 587, 588
THIK family, 590
TREK family, 581
TRESK family, 592
inherited erythromelalgia (IEM) mutation/syndrome, sodium channels, 210, 229–230, 230*f*, 231
inhibitory postsynaptic currents (IPSCs), 433–434
intellectual disability without epilepsy, 158
intermediate-conductance (IK) channel, 12, 13*f*
intermediate GABAergic inhibition, 433–434
intrinsic excitability, 4, 5
HCN channels, 551–553, 552*f*
nonselective cation channels, 21
potassium channels, voltage-dependent, K_{V2}, 116
inwardly rectifying bacterial K^+ channel (KirBac), 68
inward rectification
AMPAR and KAR channel pore, 312
GlialCAM on, 715
potassium channels, 17–19, 18*t*, 147
potassium channels, tandem pore domain, 576, 582, 584*f*–585*f*, 589
inward rectifier potassium channels (K_{ir}), 17–19, 18*t*, 147
ion channels. *See also specific types*
definition, 64
gating, theory, 297–302
membrane topology, 65, 66*f*
permeation and selectivity, 33–54 (*see also* permeation and selectivity, ion channel)
ion-conducting pore, 65, 66*f*
ionotropic glutamate receptors (iGluRs), 291–297, 343
gating, 297–301
AMPARs and KARs, native, 299–300
AMPARs and KARs, recombinant, 300–301, 300*t*
deactivation, 297
desensitization, 297–298
iGluR subunits, 298–299
permeation, 297
pre-desensitization, 298
terminology, 297
genes, 346, 347*f*
glutamatergic synapse, 291, 292*f*
mechanisms, 291
nomenclature, 292
pharmacology, AMPARs and KARs, 301–302, 303*t*
receptor antagonism, 293
stoichiometry, 294*f*, 297
subunits, 291–293, 292*t*, 346, 347*f*
AMPARs, 291, 292*t*
KARs, 292–293, 292*t*
NMDA receptors, 291
orphan/delta subunits, 291–292
topology and subunit assembly, 293–296, 294*f*, 295*f*
amino-terminal domain, 293–294, 294*f*
C-terminal tail, 293, 294*f*, 296
flip/flop cassette, 295–296, 295*f*
ligand-binding domain, 293, 294–296, 294*f*
TM domain, 294*f*, 296
ionotropic neurotransmitter receptors, 5
ion pore. *See also specific types*
acid-sensing ion channels, 652, 653–655, 655*f*
inhibitors blocking, 656
calcium channels, 50
multi-ion pore, 35, 50
nicotinic acetylcholine receptors, 392
single-ion pore, 50
voltage-gated cation channels, 66*f*
ischemia
Sl_{O1} channelopathies, 505*t*, 509
Sl_{O2} channels, 518*t*, 519
ischemic stroke, 71
acid-sensing ion channels, 658, 675–676
benzodiazepines for, 426
cerebral, BK channels, 509
potassium channel mutations, epilepsy, 185
TRP channel mutations, 71

isoform-selective inhibitors, 212
isoleucine-phenylalanine-methionine (IFM) motif, 262
ivabradine, 553, 558

J
Jacob, F., 421
Jayakar, S. S., 427–428
Jervell and Lange-Nielsen syndrome, 124

K
K_2P channels, 19–21, 20*t*
  selectivity filter as activating gate, 81–82
kainic acid (KA), 291
Kang, S. K., 156, 157
KA receptors (KARs), 291, 292–293, 292*t*. *See also* AMPA and kainate receptors (KARs)
  auxiliary proteins, 322–323
  synaptic, brain auxiliary/regulatory proteins, 297
Katz, B., 376–377
$K_{Ca}1.1$, 16, 16*t*
$K_{Ca}2.1$, 16–17, 16*t*
$K_{Ca}2.2$, 16–17, 16*t*
$K_{Ca}2.3$, 16–17, 16*t*
$K_{Ca}3.1$, 16–17, 16*t*
$K_{Ca}5.1$, 486
K^+ channel. *See* potassium channel
KCN3 gene, 118
$KCNA_1$ encoding $K_V1.1$, 151–153, 152*f*
KCNA1 gene, 113
$KCNA_1$ mutations, 113
$KCNA_2$ encoding $K_V1.1$, 153–154, 154*f*
$KCNA_2$ mutations, 113
$KCNA_4$ encoding $K_V1.4$, 154–155, 155*f*
$KCNB_1$ encoding $K_V2.1$, 155–157, 157*f*
KCNB1 gene, 116
KCNB1 mutations, 116
$KCNC_1$ encoding $K_V3.1$, 157–159, 159*f*
  developmental and epileptic encephalopathy, 158
  intellectual disability without epilepsy, 158
  myoclonus epilepsy and ataxia due to potassium channel mutation, 158–159
  nonprogressive myoclonus without epilepsy or intellectual disability, 158
$KCNC_3$ mutations, 118
  encoding $K_V3.3$, 159
$KCND_2$ encoding $K_V4.2$, 159–160, 160*f*
$KCND_3$ encoding $K_V4.3$, 160–162
KCNH1 channel, 82
KCNH1-encephalopathy, 168
$KCNH_1$ encoding $K_V10.1$, 168–169, 169*f*
$KCNH_1$ mutations, 129
$KCNH_2$ encoding $K_V10.2$, 169–170
KCNH2 gene, 129
$KCNH_2$ mutations, 129
$KCNJ_{10}$ encoding $K_{ir}4.1$, 179–182, 181*f*
  EAST syndrome, 180–182, 181*f*
  SeSAME syndrome, 180, 181*f*
$KCNJ_{11}$ encoding $K_{ir}6.2$, 182–184
  DEND syndrome, 180–182
$KCNK_4$ encoding $K_{2P}4.1$, 184–185, 184*f*
KCNK gene family, 572, 573*f*
$KCNMA_1$, 486, 487
  Alzheimer's disease, 506
  ethanol dependence, 507
  mental retardation, 506–507
$KCNMA_1$ encoding $K_{Ca}1.1$, 170–173, 171*f*
  cerebellar atrophy, developmental delay, and seizures, 171–172
  generalized epilepsy, 171
  Liang-Wang syndrome, 171
  paroxysmal nonkinesigneic dyskinesia, with or without generalized epilepsy, 171
$KCNMB_1$, ethanol dependence, 507
$KCNMB_3$, *epilepsy, 503–505*
$KCNMB_4$, *epilepsy, 505*
KCNN3 Encoding $K_{Ca}2.2$, 173, 174*f*
$KCNQ_2$
  encoding $K_V7.2$, 162–164, 163*f*
    KCNQ2-BFNE, 162–164
    KCNQ2-NEE, 162–164
  seizures, 145
KCNQ2-BFNE, 162–164
KCNQ2-NEE, 162–164
$KCNQ_3$
  developmental disability, 164, 165, 166*f*
  encoding $K_V7.3$, 164–166, 166*f*
    developmental disability, 164, 165, 166*f*
    $KCNQ_3$-BFIE, 165
    $KCNQ_3$-BFNE, 165

KCNQ$_5$ encoding K_V7.5, 166–168, 167*f*
KCNQ gene, 121
KCNT$_1$, 486
KCNT$_1$ encoding K_{Na}1.1, 174–178, 175*f*
  autosomal dominant nocturnal frontal lobe epilepsy, 174–175
  epilepsy of infancy with migrating focal seizures, 174, 175
KCNT$_1$ gain-of-function variants, 146
KCNT$_2$, 486
KCNT2 Encoding K_{Na}1.2, 178–179, 179*f*
KCNU$_1$, 486
KCNV2 gene, 126
KCNV$_2$ mutations, 126
KcsA channel, selectivity filter as activating gate, 81–82
Keynes, R. D., 35, 38–39, 40
K_{ir} channels, 17–19, 18*t*
K_{Na}1.1, 17, 17*t*, 486
K_{Na}1.2, 17, 17*t*, 486
knock-off mechanism, 51, 52*f*
knock-on mechanism
  potassium channels
    direct Coulomb, 40, 41*f*
    traditional, 38
  sodium channels, 49
    loosely coupled, 46, 47*f*
    strong, 46, 47*f*
Kortüm, F., 168
K^+ siphoning, 696
K_{V1} channel, 103*t*, 109–113
  accessory proteins, 110–112, 111*f*
  expression and activation kinetics, 109–110, 111*f*
  function, neuronal, 112–113
  members, names, and locations, 109
  structure, 109, 111*f*
  subcellular localization, 111*f*, 112
K_{V2} channel, 103*t*, 113–116
  accessory proteins, 113–114, 114*f*
  expression and activation, 113, 114*f*
  function, neuronal, 116
  structure, heteromerization, and subunit assembly, 113, 114*f*
  subcellular localization, 114*f*, 115
K_{V3} channel, 103*t*–104*t*, 116–118
  accessory proteins, 116–117, 117*f*
  currents, activation threshold, and inactivation, 116, 117*f*
  functions, neuronal, 118
  structure, 116, 117*f*
  subcellular localization, 117–118, 117*f*
K_{V4} channel, 104*t*, 118–121
  accessory proteins, 119–120, 119*f*
  function, neuronal, 121
  structure, 118, 119*f*
  subcellular localization, 119*f*, 120–121
K_V7.1 channel, activating gating mechanisms, 84–85
K_V7.2 channel, activating gating mechanisms, 83–85, 84*f*
K_{V7} channel, 104*t*, 121–124, 122*f*
  accessory proteins, 122–123, 122*f*
  activating gating mechanisms, 83–85, 84*f*
  function, neuronal, 122*f*, 123–124
  structure and tetramerization, 121–122, 122*f*
  subcellular localization, 122*f*, 123
K_{V10-12} channels, 105*t*, 126–128, 128*f*
  accessory proteins, 127, 128*f*
  expression, activation, and inactivation, 126–127, 128*f*
  function, neuronal, 128–129
  members and structure, 126, 128*f*
  regulation, 127
  subcellular localization, 127, 128*f*
K_V channel-interacting proteins (KChIPs), 119–121, 119*f*
K_V channels, 12–15, 13*f*, 14*t*, 15*f*. *See also* *specific types*
  selectivity filter, 106–107, 107*f*
  voltage-sensing domain, 107, 107*f*
Kyte-Doolittle hydropathy plots, 391–392, 393*f*

L

L811P, 238–239
L1320F, 238
Lange-Nielsen syndrome, 124
Langley, John, 375
large conductance. *See also* potassium (K^+) channels, large conductance, nervous system
  molecular determinants, 490–491, 492*f*
lateral pathway, P2X receptors, 465
Latorre, R., 491

LBD-TMD linkers, 314, 314*f*, 316, 355
"leak" K^+ conductance, potassium channels, 19, 81
  epilepsy, 148
  tandem pore domain, 571–572, 575–576, 578, 579, 581, 582, 583, 588, 589, 591, 593
learning defects, Sl_{O2} channels, 518–519, 518*t*
Lehman, A., 166
Lennox-Gastaut syndrome, 172, 178
leucine-isoleucine-valine-binding protein (LIVBP), 347*f*, 349, 354
leucine-methionine-phenylalanine (LMF) motif, 262
leucine-rich glioma-inactivated 1 (LGI1) protein, 111–112
L-glutamate, 291. *See also* ionotropic glutamate receptors (iGluRs)
Li, W., 469
Liang, L., 171
Liang-Wang syndrome, 171
ligand-binding domain (LBD), ionotropic glutamate receptors, 293, 294–296, 294*f*
  layer rotation, during sensitization, 316–317
  structure and function, 302–307, 304*f*
    agonist efficacy and desensitization, molecular mechanisms, 302–305, 304*f*
    AMPARs, allosteroid modulators, 305
    KAR channel gating, external ion requirement, 306–307
    LBD dimer interface and receptor desensitization, 305–306
ligand-gated ion channels. *See also specific types*
  cys-loop superfamily, 390–391
  gating, 297–298
  iGluRs, mechanisms, 297–298
light-activated channel, recognizing TRPC as, 615–616, 617*f*. *See also* TRPC (*Drosophila* light sensitive) channels
lindane, 427
Lindia, J. A., 242
lipids, acid-sensing ion channel modulation by, 663–664
liquid-like model, 43
local anesthetics, 212, 231–232, 380, 397
Loewi, Otto, 374–375
long QT syndrome, 33, 124
  "eag" potassium channel, 162
loreclazole, 432
loss-of-function variants, 186
lymphocyte antigen 6 (Ly6) proteins, 399
lysophosphatidylcholines (LPCs)
  acid-sensing ion channels, 663
  potassium channels, tandem pore domain, 580, 593
lysophospholipids
  acid-sensing ion channels, 663
  potassium channels, tandem pore domain, 580, 582
lysosomal storage disease, 719

M

M136V, 241
M1627K, 231
main immunogenic region (MIR), 395*f*, 399
MAM (Maleimide Azobenzene Maleimide), 463–464
mambalgins, 659, 660*f*
Manning, A., 611, 613
Mao, X., 179
Markov models, 379, 380*f*
massively parallel sequencing (MPS), 145
maxi K channel. *See* Sl_{O1} (BK) channels
McTague, A., 176
M current, 14*t*, 123, 162–163, 164, 166, 167
MEA-TMA gating, 463–464
mechanoreceptors. *See specific types*
megalencephalic leukoencephalopathy with subcortical cysts (MLC), 714–715, 716*f*
membrane potential
  resting (*see* resting membrane potential)
  sag, 24
membrane-re-entrant pore (P) loop, 68, 69*f*
  resting, HCN channel, 551–553, 552*f*
memory
  acid-sensing ion channels, 666
  GABAergic inhibition, 427
  glutamatergic synaptic transmission, 291
  HCN channels, 552, 554–555
  nACh receptors, 403
  NMDA receptors, 351
  potassium channels
    calcium-activated, 17

epilepsy, 161
large conductance, 518
reference, 518
sodium channels, 6
spatial, 427
working, 6, 518, 555
mental retardation
potassium channels
epilepsy mutations, 174, 180
sodium-activated, 17
tandem pore domain, 588
Sl_{O1} channelopathies, 504*t*, 506–507
methanethiosulfonate (MTS), 68, 74, 463–464, 654–655
Methanobacterium thermoautotrophicum (MthK), 68, 69*f*, 491
microtubule affinity-regulating kinases (MARKs), 592
Micrurus tener, 660
migraine
$ASIC_1/ASIC_3$, 677–678
familial hemiplegic
$Na_V1.1$, 243
sodium channelopathies, 266*f*, 275–277
pain, $ASIC_1/ASIC_3$, 677–678
potassium channels
tandem pore domain, TRESK, 592–593
two pore domain, 20*t*
Mihic, S. J., 429
Mihic residues, 428–429
Miledi, Ricardo, 384
miniature endplate potentials (mEPPs), 377
Minke, B., 613–615
Mir, A., 182
MitTx, 660–661, 660*f*
Mohapatra, D. P., 15*f*
molecular dynamics (MD)
all-atom methodology, 36, 37*f*
simulations, 39–40, 42, 45–46, 48–49, 51
molecular mechanics (MM) techniques, 36
molecular modeling. *See also specific molecules*
ion-channel architecture, 65
monoclonal antibodies, voltage-gated sodium channel blocking, 213–214
Monod, J., 421
monogenic epilepsy syndromes
etiology, 266
sodium channelopathies, 272–274
MTSET, 74, 75*f*, 83
Mullen, S. A., 177
multi-molecular signaling complexes, TRPC, 618–622
Drosophila PDZ scaffold protein, Inactivation but No Afterpotential D, 618–620, 619*f*, 627, 636
Inactivation but no Afterpotential C, 618–619, 619*f*, 627
inaD gene and PDZ domains, 618
interaction of mammalian PDZ scaffold protein NHERF with $TRPC_4$ and $TRPC_5$, for DAG activation, 621–622
mammalian caveolin-1 scaffold protein interacts with $TRPC_1$, $TRPC_4$ and $TRPC_3$, 622
mammalian PDZ scaffold protein NHERF interacts with $TRPC_4$ and $TRPC_5$ channel proteins, 620–621
Neither Inactivation Nor Afterpotential C, 618, 619, 619*f*
No Receptor Potential A, 618–619, 619*f*
rhodopsin, 618, 619
multiple sclerosis, acid-sensing ion channels, 676–677
muscarine, 375, 376*f*
muscarinic acetylcholine receptors (mAChR), 375
muscimol, 423, 427
structure, 423, 425*f*
δ-$GABA_A$ receptor selective ligand, 436
muscle pain, chronic, $ASIC_3$, 670, 678
myasthenia gravis, 395*f*, 399
myoclonus epilepsy and ataxia due to potassium channel mutation (MEAK), 158–159

N

Na^+-Ca^{2+} bursts, 499
Na^+ channel. *See* sodium (Na^+) channels
nafamostat, 657
NaK2CNG channel, 43
nano-tweezers, 463, 469
N-arachidonylglycine, 431–432
Narahashi, T., 428
Nassar, M. A., 242

Na_V. *See* sodium (Na^+) channels, voltage-dependent (Na_V)
Na_V1.1, 5–6, 8*t*, 243
Na_V1.2, 5–6, 8*t*
Na_V1.3, 5–6, 8*t*, 241–242
Na_V1.6, 5–6, 8*t*, 240–241
Na_V1.7, 5–6, 8*t*, 209–210, 211*f*, 212, 227–232, 230*f*
Na_V1.8, 5–6, 8*t*, 209, 210, 212, 232–236, 235*f*
Na_V1.9, 5–6, 8*t*, 209, 210, 226, 236–240, 237*f*
Na_x, 243–244
Na_X1, 6
NBQX, 299
N-cadherin, 310
ND7/23, 229, 230*f*, 236
negative allosteric modulators (NAM), $GABA_A$ receptors, 421
negative resting membrane potential, maintaining, K^+ channels, 571
Neher, Erwin, 378
Neither Inactivation Nor Afterpotential C (NINAC), 618, 619, 619*f*
neurogliaform cells (NGFCs), 433–434
neuromuscular junction, acetylcholine/nicotinic acetylcholine receptor, 375–378, 382, 385, 389, 390, 398
neuropathic pain
  $ASIC_1$, 677
  $ASIC_{1a}$, 668
  K_{V7} channels, 83
  Na_V1.3, 242
  Na_V1.6, 241
  Na_V1.7, 244
  Na_V1.8, 234, 235
  Na_V1.9, 237
  P2X4 receptors, 472
  P2X7 receptors, 474
  potassium channels
    large conductance, 508, 518*t*
    tandem pore domain, 592, 593
  TMEM16A, 707
neuropathy, hyperpolarization-activated cyclic nucleotide-gated channel, 558
neurosteroids, 428–429
neurotransmission, 343. *See also specific types and topics*
nicotinamide adenine dinucleotide (NAD^+), Sl_{O2} channel regulation, 513
nicotine, 376, 376*f*
  addiction, nicotinic acetylcholine receptors, 385, 388, 400, 401
nicotinic acetylcholine receptors (nAChRs), 374–404. *See also* acetylcholine (ACh)
  acetylcholine, 374–375, 376*f*
  cholinergic drugs, 375–376, 376*f*
  curare blocking, 376
  cys-loop superfamily, ligand-gated ion channels, 390–391
  derivatives, modern, 399
  discovery and initial characterization, 374–376, 376*f*
  electrophysiology, 376–382, 380*f*, 381*f*, 383*f*
  neuromuscular junction, 375–378, 382, 385, 389, 390, 398
  pharmacology and therapeutic indications, 399–404
    Alzheimer's disease and dementias, 400, 401–402
    autosomal-dominant nocturnal frontal lobe epilepsy, 400
    curare derivatives, 399
    depression and cholinergic hyperfunction, 401–402
    GTS-21, 403
    homomeric α7 receptor, drug targeting, 402–403
    NS6740, 403
    succinylcholine, 399
    tobacco use/nicotine addiction, 400–401
  proteins
    chaperones, 398–399
    lymphocyte antigen 6, 399
    membrane-bound/intracellular partners, 399
    rapsyn, 398
  proteins, biochemical isolation and molecular cloning, 382–390, 383*f*, 386*f*, 387*f*
    α2, 385–386, 387*f*
    α4, 385
    α4β2, 389
    α6, 387*f*, 388–389
    α9, 387*f*, 389
    α10, 387*f*, 389
    α-β combinations, 388

β, 385–388, 387*f*
β2, 385
β3, 388
β4, 387*f*, 388, 389
brain, 385
nicotinic gene family, 384–385
protein subunits, 383–385
structure, 391–398, 393*f*, 395*f*, 397*f*
acetylcholine binding protein, 392
C-loop, 394, 396
crystallization, $\alpha_4\beta_2$-related pentamers, 387*f*, 393*f*, 394, 395*f*
crystal structures, 392
Kyte-Doolittle hydropathy plots, 391–392, 393*f*
TM2 domains, 396–397, 397*f*
nicotinic gene family, 384–385
nicotinic pharmacophore, 375*f*, 400
Nilius, B., 72
N-linked glycosylation, 198, 309
of $TRPC_3$ and $TRPC_5$, constitutive activity from, 626
N-methyl-D-aspartate (NMDA) receptors
AMPARs, 291–292, 292*f*, 293, 297–299, 303*t*, 308, 311, 343–363
calmodulin, 357
cloning, 344, 345
evolutionary origin, 346–350, 347*f*
expression: life span, cell types, and specialized membrane segments, 350–351
functional properties, unique, 351–358
excitatory postsynaptic current, kinetic properties, 351–352, 353*f*
permeation properties, 356
reaction mechanism, 353–355, 353*f*, 355*f*
signaling hubs, nonionic, 357–358
functions, critical, 344
GluN1 subunit, 315, 346, 350, 356, 357
disease-associated mutations, 355, 355*f*
GluN2 subunit, 308, 315, 346, 357
GluN3 subunit, 346, 350
glycine and glycine gating, 343, 346, 351, 353, 353*f*, 357
homology groups, 346, 347*f*
introduction and historical perspective, 343–344
molecular diversity and modular structure, 346
physiological roles, 358–363
erythrocytes, 362
glial cells, 361–362
neuroepithelial cells, 362
non-neuronal cell types, 362–363
postsynaptic signal, neuronal, 358–360
presynaptic signal, neuronal, 360–361
polyamine block, 312
receptor distribution, 345
research, early, 345
subunit identification, 345
N-methyl-D-aspartic acid (NMDA), 291
nociception, human. *See also specific types*
acid-sensing ion channels, 674–675
nociceptors, 224. *See also* specific types
sensory, 224
noise analysis, 378
noncompetitive antagonists, 298, 299–300, 394, 397, 397*f*
nonconductive functions, 5
nonionic signaling hubs, 357–358
nonprogressive myoclonus without epilepsy or intellectual disability, 158
non-proton ligand sensing domain, ASICs, 662
nonselective cation channels, 21–26, 22*t*
cyclic nucleotide–regulated, 21–24, 22*t*, 23*f*
transient receptor potential, 21, 22*t*, 24–26, 25*f*
No Receptor Potential A (NORPA), 618–619, 619*f*
NS6740, 403
NSAIDs, 661

O

Ohtahara syndrome, 267, 270
Olcese, R., 79
olfactory bulb, TMEM16Ca^{2+} activated Cl^- channels, 706
olfactory sensory neurons, TMEM16Ca^{2+} activated Cl^- channels, 707–708
Olsen, R. W., 421–422
opioid neuropeptides, $ASIC_{1a}$, 662–663
optogating, 463–464
orphan/delta subunits (GluD1-2), 291–292

Ostm1 (osteopetrosis-associated transmembrane protein 1), 717, 719–720
outward rectification, 576, 584*f*–585*f*, 717
oxaliplatin, 239, 241

P

P2X receptors, 458–475
  adenosine triphosphate, 458
  channel gating, 459–465
    "binding jaw" tightening, agonist action, 460*f*, 461–462
    cytoplasmic cap, open state structure stabilization, 460*f*, 464–465
    definition, 459
    desensitization, 459
    functional state, cycling, 459
    gating cycle, 459–461
    molecular mechanisms, 459–461
    other gating motions, TM2 helices, 464
    pore expansion, ATP activation, 462–464
  gene products, 458
  ion conduction, 465–469
    conduction pathway, 460*f*, 465–466
    feedback modulation, 465
    ion permeability, 465
    mechanisms, 465
    reversal potentials, time-dependent changes, 468–469
    selectivity filters, 466–468
  MEA-TMA gating, 463–464
  physiological function, 469–474
    general, 469–470
    P2X1R, 470–471
    P2X2R, 471
    P2X3Rs, 471–472
    P2X4Rs, 472–473
    P2X5R, 473
    P2X6Rs, 473
    P2X7Rs, 474
  structure, overview, 458–459, 460*f*
  subunits, 458
P2Y receptors, G-protein-coupled metabotropic. *See* P2X receptors
pain
  acid-sensing ion channels, 677–678
    chronic muscle, $ASIC_3$, 678
    migraine, $ASIC_1/ASIC_3$, 677–678
    neuropathic, $ASIC_1$, 677
    spinal cord, 667–671, 675, 677–678
  as adaptive, 234
  chronic, 187, 210, 230, 426
    muscle, $ASIC_3$, 678
    TRP channel mutations, 71
  as counter-adaptive, 234
  definition and function, 224
  disorders, large conductance potassium channels, 505*t*, 508–509, 518*t*, 519
    Sl_{O2} channels, 518*t*, 519
  dorsal root ganglia, 224
  hyperpolarization-activated cyclic nucleotide gated channels, 556, 558
  K2P18, 592
  K_{V7} channels, 83
  Na_V channels, 198, 212, 213, 214
    peripheral pain-sensing neurones, 209–211, 211*f*
  nociceptors, 224
  P2X receptors, 470–472, 474
  P2X3 receptors, 471
  sodium channel transmission, 224–245 (*see also* sodium channels, pain transmission)
  TMEM16A Ca^{2+}-activated Cl^- channels, 706, 707, 719
  TREK family, 582
  TRESK family, 592
  TRP channels, 26, 71
pain-sensing neurones, peripheral, voltage-dependent sodium channels in, 200*f*, 201*f*, 209–210, 211*f*
Pantazis, A., 79
paradoxical depolarization, 578
Park, J., 158
Parkinson's disease
  hyperpolarization-activated cyclic nucleotide-gated channel, 557
  nicotinic acetylcholine receptors, 400, 402
  TRP channel mutations, 71
paroxysmal extreme pain disorder (PEPD), 210, 229–231, 230*f*
paroxysmal nonkinesigenic dyskinesia, 506
paroxysmal nonkinesigneic dyskinesia
  with or without generalized epilepsy, 171

parvalbumin (Pv), 118, 264, 268, 275, 673, 714
patch clamp studies
  discovery, 378
  fluorometry, 79–81, 80*f*
  single-channel recordings, 34
Payandeh, J., 201
PcTx1, 658–659, 660*f*
PDZ domains, 618
PDZ scaffold protein
  Drosophila, Inactivation but No Afterpotential D, 618–620, 619*f*, 627, 636
  mammalian NHERF
    interaction with $TRPC_4$ and $TRPC_5$, for DAG activation, 621–622
    $TRPC_4$ and $TRPC_5$ channel proteins interaction, 620–621
pentameric ligand-gated ion channels (pLGIC), 390–391
pentraxin, 310
peripheral nervous system, acid-sensing ion channels, 668–674
  large-diameter sensory neurons, mechanoreceptors, 671–674
    $ASIC_2$, 672–673
    $ASIC_3$, 673–674
  small-diameter sensory neurons, nociceptors and metaboreceptors, 669–671
    $ASIC_{1a}$ nociceptors, 669
    $ASIC_{1b}$ nociceptors, 669–670
    $ASIC_3$ metaboreceptors and nociceptors, 670–671
peripheral pain-sensing neurones, voltage-dependent sodium channels, 200*f*, 201*f*, 209–210, 211*f*
Periplaneta americana, 260–261
periplasmic bacterial proteins (PBPs), 349
permanent neonatal diabetes mellitus (PNDM), 182–183
permeability ratio, 34
permeation, defined, 297
permeation and selectivity, ion channel, 33–54
  calcium channels, 49–53, 52*f*
  channelopathies, 33
  gating, 33
  ion transport mechanisms and steps, 33
  permeation, defined, 34
  permeation control, 34
  potassium channels, 38–44, 39*f*, 41*f*
  selectivity, defined, 34
  sodium channels, 44–49, 47*f*
  study approaches, 34–38
    experimental, 34–35
    theoretical, 35–38, 37*f*
Perozo, E., 69
Pf%, 467
Phactr-1 (phosphatase and actin regulator protein-1), 17
pharmacophore, 375
2-phenylchromane, 431
phosphatidylinositol 4,5-bisphosphate (PIP_2)
  hyperpolarization-activated cyclic nucleotide-gated channel, 549, 553, 558
  Sl_{O2} channel regulation, 514
  TRPC channels, 25*f*, 620–622
phosphatidylinositol 4,5-bisphosphate (PIP_2), potassium channels
  large conductance, 514
  tandem pore domain, 580–581, 583
  voltage-dependent
    epilepsy, 163
    K_{V7}, 121
phosphenes, 558
phospholipase C (PLC)
  TRP, 614, 619, 619*f*, 622, 633*f*, 634, 635, 636
  TRPC channels, 614, 619, 619*f*, 622, 633*f*, 634, 635, 636
phospholipid phosphatidylinositol-4,5-bisphosphate (PIP_2), 121
phosphorylation, 4
  calcium channels, 9–11, 10*f*
  potassium channels, 14, 16, 21
  sodium channels, 6
  TRPC channels, functional roles, 626–632, 628*f*, 629*f*
    Drosophila TRPL phosphorylation, 630
    Drosophila TRP phosphorylation, 627–630, 628*f*, 629*f*
    mammalian TRPC channel regulation, 631–632
    PKC phosphorylation of $TRPC_3$, $TRPC_6$, $TRPC_7$, and $TRPC_5$, 631–632
photoswitchable tweezers, 464

physical sensors, gating, 71–72
physiological function
  P2X1R, 470–471
  P2X2R, 471
  P2X3Rs, 471–472
  P2X4Rs, 472–473
  P2X5R, 473
  P2X6Rs, 473
  P2X7Rs, 474
picrotoxinin, 422, 427
PLC, 614, 619, 619*f*, 622, 633*f*, 634, 635, 636
PLC-β1, 619*f*, 620
PLC-β2, 619*f*, 620
PN1. *See* $Na_V1.7$
PN4. *See* $Na_V1.6$
Poirier, K., 158
Poisson-Nernst-Planck (PNP) simulations, 36
polyamines, 311–313
polyunsaturated fatty acids (PUFAs)
  potassium channels, tandem pore domain, 580, 582
  TRPC channels, 634
Porcupine, 322
pore domain (PD), 66*f*
  ion channel, 66*f*, 72, 84*f*
  potassium channels, 38
    epilepsy, 155
    tandem, 571–594, 573*f*, 584*f*–585*f*, 591
    tandem, *Xenopus laevis* oocyte studies, 576
    two-pore subunits, 12, 19–21, 20*t*
  potassium channels, voltage-gated, 106–107, 107*f*
  K_{V7}, 121
  Sl_{O1} channels, 487, 489*f*, 490, 492*f*
  Sl_{O2} channels, 510–511
  sodium channelopathies, CNS, 160
  sodium channels, 44
  TRPC channels, 613
pore expansion, ATP activation, P2X receptor, 462–464
pore-forming (P) loop, potassium channel, 51
pore module (PM), voltage-dependent sodium channel, 199, 202–204
positive allosteric modulators (PAMs)
  $GABA_A$ cys-loop receptors, 402
  $GABA_A$ receptor, 421, 426
  α7 receptor, 402–403
postsynaptic signal, neuronal, NMDA receptors, 358–360
posttranslational modification, 4
potassium (K^+) channelopathies, epilepsy, 148–188
  conclusions and future directions, 185–188
  episodic ataxia type 1, 151–152, 152*f*
  four transmembrane domain family, 184–185
    $KCNK_4$ encoding $K_{2P}4.1$, 184–185, 184*f*
  genotype-phenotype correlations, 186
  six transmembrane domain family: Ca- and Na-activated potassium channels, 150*t*, 170–179
  six transmembrane domain family: voltage-gated potassium channels, 149*t*, 151–170
  targeted therapies, 146, 163–164, 172, 176, 186, 187
  two transmembrane domain family, including inwardly rectifying potassium channels, 150*t*, 179–184
potassium (K^+) channelopathies, large conductance, 503–509, 504*t*–505*t*
  Sl_{01} channels, 503–509, 504*t*–505*t*
    adult-onset neuronal ceroid lipofuscinosis, 504*t*, 506
    Alzheimer's disease, 504*t*, 506
    ataxia, 504*t*, 505–506
    epilepsy, 503–505, 504*t*
    ethanol addiction, 504*t*, 507–508
    ischemia, 505*t*, 509
    mental retardation, 504*t*, 506–507
    pain disorders, 505*t*, 508–509
potassium (K^+) channels, 11–21, 13*f*. *See also specific topics and types*
  action potential modulation, 6
  atomic structures, 38–39, 39*f*
  calcium-activated, 16–17, 16*t*
  electrostatic interactions, 40, 42
  epilepsy mutations, 144–188 (*see also* epilepsy, potassium channel mutations [channelopathies])
  fast inactivation, 118, 499
  four transmembrane family, 147–148, 147*f*
  genes, pore-forming subunits, 3
  inward rectifier, 17–19, 18*t*
  knock-on mechanism
    direct Coulomb, 40, 41*f*

traditional, 38
mechanisms, 11–12
permeation and selectivity, 38–44, 39*f*, 41*f*
pore and gate architecture, structural evidence, 68–70, 69*f*
selectivity filter, 38–43, 39*f*, 41*f*
Shaker, 67–68, 70, 73, 74, 76, 491
six transmembrane family, 146–147, 147*f*
sodium-activated, 17, 17*t*
structure, epilepsy, 146–148, 147*f*
two-pore domain K^+ channel subunits, 19–21, 20*t*
two transmembrane family, 148
TXGYG sequence, 38, 41*f*
potassium (K^+) channels, large conductance, nervous system, 486–520
future directions, 520
locations, mediators, and activation, 520
Sl_{O_1} channels, 487–509, 488*t* (*see also* Sl_{O_1} (BK) channels)
Sl_{O_2} channels, 488*t*, 509–519 (*see also* Sl_{O_2} channels)
Sl_{O_3} channels, 486, 488*t*
types and names, 486
potassium (K^+) channels, tandem pore domain, 571–594
general features, 572–576
functional characteristics, unique, 575–576
KCNK gene family, 572, 573*f*
structural considerations, unique, 573–575, 573*f*
"leak" K^+ conductance, 571–572, 575–576, 578, 579, 581, 582, 583, 588, 589, 591, 593
negative resting membrane potential, maintaining, 571
subgroup properties, 588–593
TALK family, 588–590
TASK family, 582–588, 584*f*–585*f*
THIK family, 590–591
TREK family, 578–582
TRESK family, 591–593
TWIK family, 576–578
potassium (K^+) channels, voltage-dependent (K_V), 12–16, 14*t*, 15*f*, 101–131
axon initial segment, 109, 111*f*, 112, 114*f*, 115, 122*f*, 123, 130
cloning, 102–106
cryo-EM study, 83–84, 84*f*
diversity, 108–109, 108*f*
future directions, 130–131
genes, 14*t*, 101
history, 101
Hodgkin & Huxley experiments, 101
K_V, 12–15, 13*f*, 14*t*, 15*f*
K_{V1}, 103*t*, 109–113
accessory proteins, 110–112, 111*f*
expression and activation kinetics, 109–110, 111*f*
function, neuronal, 112–113
members, names, and locations, 109
structure, 109, 111*f*
subcellular localization, 111*f*, 112
K_{V2}, 103*t*, 113–116
accessory proteins, 113–114, 114*f*
expression and activation, 113, 114*f*
function, neuronal, 116
structure, heteromerization, and subunit assembly, 113, 114*f*
subcellular localization, 114*f*, 115
K_{V3}, 103*t*–104*t*, 116–118
accessory proteins, 116–117, 117*f*
currents, activation threshold, and inactivation, 116, 117*f*
functions, neuronal, 118
structure, 116, 117*f*
subcellular localization, 117–118, 117*f*
K_{V4}, 104*t*, 118–121
accessory proteins, 119–120, 119*f*
function, neuronal, 121
structure, 118, 119*f*
subcellular localization, 119*f*, 120–121
K_{V7}, 104*t*, 121–124, 122*f*
accessory proteins, 122–123, 122*f*
function, neuronal, 122*f*, 123–124
structure and tetramerization, 121–122, 122*f*
subcellular localization, 122*f*, 123
$K_{V10\text{-}12}$, 105*t*, 126–128, 128*f*
accessory proteins, 127, 128*f*
expression, activation, and inactivation, 126–127, 128*f*
function, neuronal, 128–129
members and structure, 126, 128*f*
regulation, 127
subcellular localization, 127, 128*f*
K_V channel complexes, diverse, 101
K_V channels, distinct, 101

potassium (K^+) channels, voltage-dependent (K_V) (*cont.*)
K_VS (K_{V5}, K_{V6}, K_{V8}, and K_{V9}), 104*t*–105*t*, 124–126, 125*f*
accessory proteins, 125, 125*f*
activation, inactivation, deactivation, 124, 125*f*
function, neuronal, 126
structure, expression, and tetramerization, 124, 125*f*
subcellular localization, 125–126, 125*f*
mechanisms, 33
modulation, dynamic, 101–102
naming, subcellular localization, current features, and disease associations, 103*t*–105*t*
nomenclature
HUGO gene, 102
IUPHAR, 102
permeation control, 34
regulation, on neuronal firing patterns, 14–16, 15*f*
Shaker
cloning, 102
electrical resistance, 491
gating mechanisms, 7, 67–68, 70, 73, 74, 76, 80, 85
structure, 106
subfamilies, K_{V1}, 109
structure, 106–108, 107*f*
subunits and groupings, 12, 13*f*
potassium ion (K^+), "leak" K^+ conductance, 571–572, 575–576, 578, 579, 581, 582, 583, 588, 589, 591, 593
potential-energy profile, 37, 37*f*
pre-densensitization, 298
Prescott, S. A., 714
presynaptic signal, neuronal, NMDA receptors, 360–361
prokaryotic Na_V sodium channels, archetypal
assembly, 258
evolutionary origins, 257
structure–function relationship, 258–260, 259*f*
prolonged depolarizing afterpotential (PDA) screen, 616
pro-loop ligand-gated ion channels ligand-gated ion channels (pLGIC), 390–391, 392
propofol, 429–430, 432
protein kinase A (PKA), Sl_{O2} channel regulation, 514
protein kinase C (PKC), Sl_{O2} channel regulation, 513–514
protons. *See also specific types*
acid-sensing ion channels, 646–647
proton sensing, 646
protoxin-II (ProTx-II), 212–213
proximal restriction and clustering (PRC) motif, 115
Psalmopoeus cambridgei, 658
psalmotoxin, 658–659

Q

Q/R site, 312
quantum mechanical (QM) techniques, 36
quaternary ammonium (QA) derivatives, 67
quinidine, for epilepsy, 146, 176, 177, 179
quinolones, 426
quisqualate, 301, 302, 303*t*, 304*f*, 316
Quraishi, I. H., 176

R

Rana pipiens, 67
rapsyn, 398
rate-theory models, 51
Ratté, S., 714
receptor-interacting protein kinase 1 ($RIPK_1$), 676
reference memory, 518
Reich, E., 375
repolarization, 33
resonance properties
Ca^{2+}-activated Cl^- current, 705
HCN channels, 555
resting membrane potential
HCN channels, 551–553, 552*f*
hyperpolarization-activated cyclic nucleotide-gated channel, 549, 551–553, 552*f*
negative, maintaining, 571
potassium channels
epilepsy, 148, 185
K_{V7}, 83, 123
large conductance, Sl_{O2}, 516
tandem pore, 571, 575
two-pore domain subunits, 19, 21
sodium channels
$Na_V1.6$, 240

Na$_V$1.7, 228
Na$_V$1.9, 236–239
TMEM16Ca^{2+} activated Cl$^-$ channels, 707, 709
resurgent current, 5
retina
pathology, hyperpolarization-activated cyclic nucleotide-gated channel, 558
TMEM16Ca^{2+} activated Cl$^-$ channels, 702–703
retrotrapezoid nucleus (RTN), 589
reversal potentials
current-voltage relationship, 35
time-dependent changes, P2X receptors, 468–469
RFamide neuropeptides, ASIC$_3$ central vestibule binding and, 662
rheumatoid arthritis, P2X7R blockade/deletion on, 474
rhodopsin (RH$_1$), 618, 619
rhythmogenesis, HCN channels, 553–554
Rinke, I., 714
Rosenmund, C., 298
Rosti, G., 167

S

sag, 24
Sakmann, Bert, 378, 379, 381
salicylidene salicylhydrazide, 433
Samad, O. A., 242
SAN pacemaker, 208
Scholl, U. I., 180
Schultz, P. G., 79
SCN1A, 8*t*, 267
autism spectrum disorders, 274–275
Dravet syndrome, 241, 243, 267, 268–270, 269*f*, 275, 432
encephalopathy, 267
epilepsy, 146, 266, 267
epilepsy syndromes, monogenic, 272–273
eukaryotic Na$_V$ channels, 260
familial hemiplegic migraine, 275–276
Na$_V$1.1, 243
Na$_V$ channels, eukaryotic, 260
sodium channels, central nervous system, 264
SCN1A-10A, 198
SCN1B, 6, 7*f*, 261
SCN2A, 8*t*
autism spectrum disorder, 274–275
benign familial neonatal infantile seizures, 273–274
epilepsy, 267
epileptic encephalopathy, 270–271
monogenic epilepsy syndromes, 272, 273
Ohtahara syndrome, 270
SCN2B, 6, 7*f*, 262
SCN3A, 6, 8*t*, 241, 271, 272–274
autism spectrum disorders, 274
cryptogenic focal epilepsy, 273–274
early onset epileptic encephalopathy, 271
SCN3B, 6, 7*f*, 262
SCN4A, 6, 8*t*
SCN4B, 6, 7*f*, 262
SCN5A, 6, 8*t*, 209, 236
SCN6A, 6, 8*t*
SCN7A, 6, 8*t*
SCN8A, 6, 8*t*, 146, 267
early infantile epileptic encephalopathy type 13, 271
epilepsy, 267
epileptic encephalopathy, 271
Na$_V$1.6, 240
SCN9A, 6, 8*t*, 227, 228
SCN10A, 6, 8*t*, 232, 236
SCN11A, 6, 8*t*, 236, 260
eukaryotic Na$_V$ channels, 260
sea anemone, 661
selectivity, ion channel, 33–54. *See also* permeation and selectivity, ion channel
selectivity filters, 65, 67–68, 69*f*, 72, 81
as activating gate, 81–82
AMPA and kainate receptors, 296, 311, 313, 316
calcium channels, 49–54, 52*f*
electron-density mapping, 35
full-quantum mechanical treatments, 36
hyperpolarization-activated cyclic nucleotide-gated channel, 547
P2X receptors, 466–468
sodium channelopathies, CNS, 259–260, 261*f*, 262
sodium channels, 44–49, 47*f*
voltage-dependent, 199, 200*f*, 202–204, 202*f*
theoretical approaches, 36
VSD-to-PD coupling, 84*f*

selectivity filters, potassium channels, 38–43, 39*f*, 41*f*
large conductance, 490–491, 492*f*
Sl_{O_2}, 510–511
tandem pore, 574–575, 593
TASK family, 583
TREK family, 580
TWIK family, 578
voltage-dependent, 106–107, 107*f*
voltage-gated, 106–107, 107*f*
epilepsy, 155–156, 184, 186
selectivity measurements, 34–35
Selinger, Z., 613–614
sensors, gating, 64, 65, 66*f*. *See also* *specific types*
architecture, 65, 66*f*
gating domain/pore region, 71–73
physical and chemical sensors, 71–72
voltage sensing, 73
sensory neurons, PNS
large-diameter mechanoreceptors, acid-sensing ion channels
$ASIC_2$, 672–673
$ASIC_3$, 673–674
small-diameter, acid-sensing ion channels, nociceptors and metaboreceptors, 669–671
$ASIC_{1a}$ nociceptors, 669
$ASIC_{1b}$ nociceptors, 669–670
$ASIC_3$ metaboreceptors and nociceptors, 670–671
sensory neuron-specific (SNS) sodium channel, 232
sensory receptors, large conductance potassium channels, 501–503
hearing, 501–502
vision, 502–503
SeSAME (seizures, sensorineural deafness, ataxia, mental retardation, and electrolyte imbalance) syndrome, 180
Shaker K^+ channel, 67–68, 70, 73, 74, 76, 491
Shaker subfamily, 109. *See also* K_{V_1} channel
Sicca, F., 181
signaling hubs, nonionic, 357–358
Sigworth, Fred, 379, 381
Simons, C., 168, 169
single-channel patch-clamp recordings, 34
Sirisoma, N. S., 428
6-hydroxyflavones, 431
six transmembrane (6TM) family, K^+ channels, 146–147, 147*f*
SK channels (surrounding Ca^{2+}-activated K^+ channels), 16, 17, 173, 359, 494, 704, 705
skeletal muscle, calcium channel excitation–contraction coupling, 11
Slack, 509–510. *See also* potassium channels, large conductance, nervous system
Slack, 486
SLACK (sequence like a calcium-activated K^+) channel, 174
Slick, 509–510. *See also* potassium channels, large conductance, nervous system
Slick, 486
sliding helix model, 203
Slo_1, 486, 487
Sl_{O_1} (BK) channels, 16, 16*t*, 70, 487–509, 488*t*
accessory subunit modulation, diversity from, 495
acetazolamide on, 172
activation, 170, 487
allosteric coupling
CTD-PD, 493
VSD-CTD, 493
VSD-PD, 492–493
allosteric nature
Ca^{2+} and voltage-dependent BK channel activation, 491
Horrigan and Aldrich model, 491
$(\alpha + \beta_1)$BK channels, 498
$(\alpha + \beta_2)$BK channels, 498
expression patterns and role in neurons, 499
nervous system, 496–498, 497*t*, 498*f*
$(\alpha + \beta_4)$BK channels, 500–501
β subunit family, 495–496, 497*f*
Ca^{2+} sources, 494–495
channelopathies, 503–509, 504*t*–505*t*
adult-onset neuronal ceroid lipofuscinosis, 504*t*, 506
Alzheimer's disease, 504*t*, 506
ataxia, 504*t*, 505–506
epilepsy, 503–505, 504*t*

ethanol addiction, 504*t*, 507–508
ischemia, 505*t*, 509
mental retardation, 504*t*, 506–507
pain disorders, 505*t*, 508–509
C-terminal domain, 487, 488–489, 489*f*
expression pattern, CNS BK channel, 493–494
γ subunit family, 496, 497*f*
genes and encoding, 487
large conductance, molecular determinants, 490–491, 492*f*
pore domain and gate, 487, 489*f*, 490
sensory receptors, 501–503
hearing, 501–502
vision, 502–503
structural features, 487–488, 489*f*
voltage sensor domain, 487, 488, 489*f*, 490*f*
Slo$_{2.1}$, 486
Slo$_{2.2}$, 486
Sl_{O2} channels, 488*t*, 509–519
C-terminal domain, 512
differential regulation, intracellular factor, 513–515
adenosine triphosphate, 513
cell volume, 515
estradiol, 514–515
nicotinamide adenine dinucleotide, 513
phosphatidylinositol 4,5-bisphosphate, 514
protein kinase A, 514
protein kinase C and G proteins, 513–514
expression patterns, 515–516
gene encoding, 509
Na^+ entry pathways, 512–513
neuronal disease, 517–519, 518*t*
autosomal dominant nocturnal frontal lobe epilepsy, 517, 518*t*
epilepsy, 517–518, 518*t*
fragile X syndrome, 517, 518*t*
ischemia, 518*t*, 519
learning defects, 518–519, 518*t*
pain disorders, 518*t*, 519
pore domain and gate, 510–511
role, nervous system, 516–517
Slick and Slack, overview, 509–510
structural features, common, 510
types and activation, 509–510
voltage sensory domain, canonical, absence of, 511–512
Sl_{O3} channels, 486, 488*t*
Slowpoke, 486
small-conductance (SK), 12, 13*f*
small-conductance ($K_{Ca}2$ and $K_{Ca}3$) channels, 12, 13*f*
small ubiquitin-like modulator (SUMO), 21, 577
Smets, K., 161
smoking cessation, nicotinic acetylcholine receptor drugs, 400–401
SNARE (*N*-ethylmaleimide-sensitive factor activating protein receptor) proteins, 9–11
SNS_2/NaN. *See* $Na_V1.9$
snug-fit model, 42–43, 45
sodium (Na^+) channelopathies, 265–277, 266*f*
axon initial segment, 270, 274
central nervous system, 265–277, 266*f*
autism spectrum disorders, 266*f*, 274–275
benign familial neonatal infantile seizures, 273–274
cryptogenic focal epilepsy, 273–274
epilepsy, 266–267, 266*f*, 267*f*
epilepsy syndromes, monogenic, 272–274
epileptic encephalopathy, 268–272, 269*f*
familial hemiplegic migraine, 266*f*, 275–277
genetic epilepsy with febrile seizures plus, 272–273
mutations, biophysical properties, 266, 267*f*
voltage-gated sodium channel structure, 258–263
archetypal prokaryotic Na_V channel assembly, 258
eukaryotic Na_V channel, structural elements, 260–262, 261*f*
eukaryotic $Na_V\beta$ subunit complex, 262–263
evolutionary origins, 257
structure–function relationship, 258–260, 259*f*
sodium channel protein VIII. *See* $Na_V1.6$

sodium (Na^+) channels, 5–6, 7*f*, 8*t*. *See also specific topics and types*
action potential modulation, 6
central nervous system, 263–265
mutations, biophysical properties, 266, 267*f*
eukaryotic Na_V, archetypal, 257–263
evolutionary origins and structures, 257–258
$Na_V\beta$ subunit complex, 262–263
structural elements, 260–262, 261*f*
structure–function relationship, 258–260, 259*f*
fast inactivation
eukaryotic, channelopathies, 260, 262, 264
$Na_V1.1$, 243
$Na_V1.6$, 240, 241
$Na_V1.7$, 227–228, 231
$Na_V1.8$, 233
$Na_V1.9$, 236
genes, 5–6, 8*t*
knock-on mechanism, 49
loosely coupled, 46, 47*f*
strong, 46, 47*f*
permeation and selectivity, 44–49, 47*f*
pore-forming subunits, 5, 7*f*
resurgent current, 5
selectivity filter, 44–49, 47*f*
sensory neuron-specific, 232
structure and mechanisms, 226
molecular basis, 204
$Na_V1.8$, 210
paroxysmal extreme pain disorder, 210
β1 and β3 subunits, 206
β subunits, 6
sodium (Na^+) channels, pain transmission, 224–245
congenitally insensitive to pain, 209–210, 212, 228–229, 230*f*, 232
fast inactivation, 227–228, 231, 233, 236, 240, 241
future directions, 244–245
isoforms, 226–227, 226*f*
$Na_V1.1$, 243
$Na_V1.3$, 241–242
$Na_V1.6$, 240–241
$Na_V1.7$, 227–232, 230*f*
$Na_V1.8$, 232–236, 235*f*
$Na_V1.9$, 236–240, 237*f*
Nax, 243–244
paroxysmal extreme pain disorder, 210, 229–231, 230*f*
peripheral pain-sensing neurones, 200*f*, 201*f*, 209–210, 211*f*
SUMOylation, 244
voltage-gated, overview, 225–227, 226*f*
sodium (Na^+) channels, voltage-dependent (Na_V), 66*f*, 70, 198–214
action potential modulation, 6
activation mechanism, 198
ast-inactivationf, 225, 226*f*
astrocytes, 207
auxiliary subunits, 204–206
cardiac channel phagosomes/endosomes, 206–207
drug development, 198
earliest, 257
eukaryotic, slow and fast inactivation, 260, 262
evolutionary origins, 257–258
genes and modifications, 6, 8*t*, 198–199, 200*f*
isoform combinations, physiological, 207–211
cardiac cells, 208–209
peripheral pain-sensing neurones, 200*f*, 201*f*, 209–210, 211*f*
mechanisms, 33
mutations
biophysical properties, 266, 267*f*
diseases, and syndromes, 198
"non-classical" roles, 206–207
overview, 225–227, 226*f*
structure, 198–200, 200*f*, 201*f*, 225, 226*f*, 258–263
archetypal prokaryotic Na_V channel assembly, 258
eukaryotic Na_V channel, structural elements, 260–262, 261*f*
eukaryotic $Na_V\beta$ subunit complex, 262–263
evolutionary origins, 257
gating mechanism, 202–204, 202*f*, 203*f*

ion-selectivity filter, 199, 202–204, 202*f*
pore module, 199, 202–204
structure–function relationship, 258–260, 259*f*
voltage-sensing module, 199, 200*f*, 202–204, 205, 210, 212, 213
voltage sensor, 200*f*, 203, 203*f*, 212, 213, 225
targeting, toxins and antibodies, 212–214
tetrodotoxin on, 6, 8*t*, 209, 226–227
tetrodotoxin-sensitive isoforms, 227
sodium (Na^+) current, 5-HT_2 neurotransmitter receptor inactivation, 6, 7*f*
SOL-1, 2, 322
South American tarantula, 658
spinal cord, acid-sensing ion channels, 667–668
spinocerebellar ataxia type 13 (SCA13), 118
squid giant axon, 3, 65, 67, 82, 101, 129, 377
stargazin, 317
steady-state desensitization (SSD), 649–651, 650*f*, 658, 659, 662, 663, 670, 675
Stefani, E., 73
steroids
molecules, $ASIC_5$ activation, 664
neuroactive, 428–429
"sticky pore" hypothesis, 50
Streptomyces A (KcsA), 68, 69*f*, 70
striatum, TMEM16Ca^{2+} activated Cl^- channels, 705
stroke, ischemic, 71
acid-sensing ion channels, 658, 675–676
benzodiazepines for, 426
cerebral, BK channels, 509
potassium channel mutations, epilepsy, 185
TRP channel mutations, 71
structureless continuum dielectric material, 36, 37*f*
substantia nigra pars compacta, 557
sudden unexplained death in epilepsy (SUDEP), 144, 145
SUMO (small ubiquitin-like modifier), 21
SUMOylation, 20*t*, 244, 577, 578, 593
SUR1-selective sulfonylureas, 183–184
SUR1/SUR2A/SUR2B, 19
Suss Toby, E., 613–614
SVmAb1, 213
SYM2081, 301
synaptic excitability, HCN channels, 549, 551, 554–555, 556, 557
synaptic plasticity, 17, 323
acid-sensing ion channels, 666
HCN channels, 554–555
NMDA receptors, 344, 356, 357, 358–361
P2X3 receptors, 471
P2X4 receptors, 473
synaptic receptors, $GABA_A$, 439
SynDIG1, 322
synprint site, 9

T

Takai, T., 384
TALK family (K2Pd, K2P16, K2P17), 19, 20*t*, 588–590
tandem pore domain potassium channels, 571–594. *See also* potassium (K^+) channels, tandem pore domain
tarantula ceratotoxin-1 (CcoTx1), 213
TARPs. *See* transmembrane AMPAR regulatory proteins (TARPs)
TASK family, 19, 20*t*, 582–588, 584*f*–585*f*
taurine, 432
Temple-Baraitser syndrome (TBS), 129, 168
tetrahydro-deoxycorticosterone, 428–429
tetramerization, 309
AMPAR/KAR complex organization/IGluRs, 291, 297, 298–299, 306, 308–310, 313–317, 314*f*
K_{V1}, 110
K_{V2}, 113
K_{V3}, 116
K_{V4}, 119, 119*f*
K_{V7}, 121–122
K_{V10-12}, 127
K_VS, 124
potassium channels, voltage-sensitive, epilepsy, 147*f*, 148, 151, 155, 157, 170, 173, 174, 177, 179
sodium channelopathies, voltage-dependent, 258, 260
sodium channels, voltage-dependent, 199
tetrodotoxin (TTX), 6
sodium channel blocker, 6, 8*t*, 209, 226–227

tetrodotoxin (TTX)-sensitive voltage-gated sodium channels, $Na_V1.7$, 227–228
Texas coral snake, 660
thalamus, TMEM16Ca^{2+} activated Cl^- channels, 705–706
thermo-TRP channels, 72
THIK family (K2P12, K2P13), 19, 20*t*, 590–591
THIP [4, 5, 6, 7-tetrahydroisoxazolo(5,4-c) pyridin-3-ol], 423, 427
  structure, 425*f*
  δ-$GABA_A$ receptor selective ligand, 436
TLESWS residues, 44, 46, 47*f*, 48, 51
TLSSWE, 45, 48
TM1/TM2, 622
  BK channels, 498
  calcium channels, 49
  caveolin-1 conserved binding motif, 622
  GABA receptor, 419, 424*f*
    anesthetics, 429
  GluK2, 311
  iGluR, 293, 296
  P2X2Rs, 459
  potassium channels, 38, 39*f*
    pore and gate, 68, 69*f*
  sodium channels, 44
  TM domain, 392, 393*f*
  α + β2 BK channels, 497*f*, 498
TMEM16A, neuronal, 701–702
TMEM16B, neuronal, 701–702
TMEM16-C (transmembrane protein 16-C), 17
TMEM16Ca^{2+} activated Cl^- channels, 696–709
TMEM16Ca^{2+}activated Cl^- channels
  central nervous system, 702–706
    brainstem, 705
    cerebellum, 703–704
    hippocampus, 703
    olfactory bulb, 706
    retina, 702–703
    striatum, 705
    thalamus, 705–706
  gene families, 697
  neurons, outside CNS, 706–709
    cochlea, 709
    dorsal root ganglion, 706–707
    olfactory sensory, 707–708
    trigeminal, 707
    vomeronasal sensory, 708–709
  non-receptor type brain chloride channels, 696–697
  TMEM16A and TMEM16B, neuronal, 701–702
  TMEM16 family, 697–701, 698*f*, 700*f*
    discovery and elucidation, 697
    mechanisms of action, 699
    members, 697
    nomenclature, 697
    pharmacology studies, 700–701
    properties, regulation, and interactions, 699–700, 700*f*
    structure and subunits, 697–699, 698*f*
TMEM16 family, 697–701
  discovery and elucidation, 697
  mechanisms of action, 699
  members, 697
  nomenclature, 697
  pharmacology studies, 700–701
  properties, regulation, and interactions, 699–700, 700*f*
  structure and subunits, 697–699, 698*f*
tobacco use
  addiction, nicotinic acetylcholine receptors, 385, 388, 400, 401
  nicotinic acetylcholine receptors, 385, 388, 400, 401
Togo starburst tarantula, 658
tonic current-carrying $GABA_A$ receptors, as low-dose alcohol targets, 434–436
tonic inhibition, GABAergic, 433
Torkamani, A., 155
Torpedo
  chloride families/transporter, CLC, 709–710
  $GABA_A$ receptor, 387*f*, 391–396, 398, 423, 428
  nicotinic acetylcholine receptors, 382–384, 385
TRAAK (TWIK-related arachidonic acid-stimulated K^+ channel), 184
Tracey, Kevin, 399
transient receptor potential (TRP) channels, 21, 22*t*, 24–26, 25*f*, 71–72
transition metal ion fluorescence resonance energy transfer (TmFRET), 80–81

transmembrane AMPAR regulatory proteins (TARPs), 297
as AMPAR chaperones, 317–318, 318*f*
AMPAR functional modulation by, 319
AMPAR–TARP assembly and stoichiometry, 319–321, 320*f*
transmembrane (TM) domain (TMD)
anesthetics, 429, 430
ASICs, 647
BK channels, 489*f*
cornichon homologs, 321
fatty acids on, 313
$GABA_AR$ ligand binding sites, 423, 424*f*
iGluR, 294*f*, 296, 309
ionotropic glutamate receptors, 294*f*, 296
KCNK, 573*f*
NMDA receptor, 346
steroids, neuroactive, 428
TARP enhancement of AMPAR, 320
TM regions, 311
TREK-1 channels, 19–21, 20*t*
TREK family (K2P2, K2P4, K2P10), 19–21, 20*t*, 578–582
TRESK family (K2P18), 19, 20*t*, 591–593
triazolopyridazines and quinolones, 426
trigeminal neuron, TMEM16Ca^{2+} activated Cl^- channels, 707
2-(trimethylammonium)ethyl methanethiosulfonate (MTSET), 74, 75*f*, 83
TRIP8b, 549, 556
TRPA1, 71
$TRPC_3$, 71
TRPC6, 71
TRPC (*Drosophila* light sensitive) channels, 611–637
conformational coupling model and phospholipase C, 614
discovery and elucidation, 611–613, 612*f*
hypoxia/anoxia-sensing TRPC channels, 632–635
anoxia activation of *Drosophila* TRP/TRPL channels in dark, 632–634, 633*f*
hypoxia-sensing TRPC6 channels, pulmonary smooth muscle cells, 635–636
inaD gene, 618
as light-activated channel, recognizing, 615–616, 617*f*
No Receptor Potential A, 618–619, 619*f*
PDZ domains, 618
phospholipase C, 614, 619, 619*f*, 622, 633*f*, 634, 635, 636
phosphorylation, functional roles, 626–632, 628*f*, 629*f*
Drosophila TRPL phosphorylation, 630
Drosophila TRP phosphorylation, 627–630, 628*f*, 629*f*
mammalian TRPC channel regulation, 631–632
PKC phosphorylation of $TRPC_3$, $TRPC_6$, $TRPC_7$, and $TRPC_5$, 631–632
PLC-β1, 619*f*, 620
PLC-β2, 619*f*, 620
properties, in vivo mutation analyses on, 616–626
Drosophila trp^{P365} mutant fly shows constitutive activity of TRP channel, 623–625, 624*f*–625*f*
multi-molecular signaling complexes, discovery, 618–622 (*see also* multi-molecular signaling complexes, TRPC)
mutations causing constitutive activity, TRPC channel, 623–626
N-linked glycosylation of TRP_3 and $TRPC_5$ causing constitutive activity, 626
prolonged depolarizing afterpotential screen, 616
subfamilies, 615
TRP function, elucidation, 613–614
trp genes
discovery and cloning, 613
encoding by, 613–614
TRP-like membrane protein, 614–615
TRPC (*Drosophila* light sensitive) channels
Trp channels, 21, 22*t*, 24–26, 25*f*, 72
trp genes, 613–614
TRP-like (TRPL), 614
TRP-like membrane protein, 614–615
TRPM2, 71
TRPM8, 71
TRPV1-4, 71
Truty, R., 146

TSC1/2, 145
TVGY(F), 106
TVGYG, 41*f*, 68, 107*f*
TVGYGD, 490, 492*f*
TWIK family (K2P1, K2P6, K2P7), 19–21, 20*t*, 576–578
2 arachidonylglycerol (2-AG), 431–432
2-guanidine-4-methylquinazoline (GMQ), 662, 663
two-electrode voltage-clamp (TEVC), 75*f*, 78
two P domain potassium channels, 184–185
  KCNK$_4$ encoding $K_{2P}4.1$, 184–185, 184*f*
two-pore channels (TPC). *See also* *specific types*
  early voltage-gated sodium channels, 257
two transmembrane (2TM) domain family, including inwardly rectifying K^+ channels, 150*t*, 179–184
  KCNJ$_{10}$ encoding $K_{ir}4.1$, 179–182, 181*f*
  KCNJ$_{11}$ encoding $K_{ir}6.2$, 182–184
two transmembrane (2TM) domain family, K^+ channels, 148
TXGYG, 38
TXGYG sequence, 38

U
umbrella sampling, 37–38, 45, 46, 48
unnatural amino acids (UAAs) mutagenesis, 79–80, 80*f*

V
V1184A, 239
ventral tegmental area (VTA), 558–559
vesicle-associated membrane-associated proteins isoform A and B (VAPA/VAPB), 115
vision sensory receptors, large conductance potassium channels, 502–503
volatile anesthetics, $GABA_A$ receptor, 429
voltage-clamp fluorometry, 76–79, 77*f*
voltage clamping, nicotinic acetylcholine receptors, 377
voltage-gated human proton (Hv) channels, 66*f*, 74, 77
voltage-gated ion channels. *See also* *specific types*
  depolarization, 33
  functions, 33
  phylogenetic superfamily, 199
  repolarization, 33
  studies, pioneering, 297
  superfamily, 199
voltage-gated pore-loop channels family, 545. *See also* hyperpolarization-activated cyclic nucleotide-gated (HCN) channel
voltage-sensing module (VSM), voltage-dependent sodium channel, 199, 200*f*, 202–204, 205, 210, 212, 213
  activated states, 259, 259*f*
  archetypal prokaryotic Na_V channels, 258–259
  ion selectivity mechanism, origins, 259–260
voltage-sensing particles, 82
voltage-sensing phosphatase channel, 82
voltage sensing/sensor domain (VSD), 39*f*, 66*f*, 70, 72, 81
  allosteric coupling
    with CTD, 493
    with PD, 492–493
  potassium channels
    epilepsy mutations, 155, 158, 161
    large conductance, nervous system, 487, 488, 489*f*, 490*f*
    large conductance, nervous system, canonical VSD absent, 511–512
    voltage-dependent, 107, 107*f*, 120, 121
  sodium channelopathies, CNS, 262
  sodium channels, 44
    voltage-dependent, 212
  VSD-to-PD coupling, 82–85, 84*f*
voltage sensors, 225
  gating, 73
  sodium channel, voltage-dependent, 200*f*, 203, 203*f*, 212, 213
vomeronasal sensory neuron, TMEM16Ca^{2+} activated Cl^- channels, 708–709

W
Watkins, J. C., 344
Westerberg, R., 79
West syndrome, 178

willardiine agonist series, 302, 304
Williams, D. K., 379
window current, 572, 650
Wong, F., 613

X

Xenopus laevis oocyte studies
GABA$_A$ receptor, 431, 432
ion channel gating, 77, 78, 80*f*
K^+ channel, voltage-dependent, 106, 111*f*, 119*f*
K_{V1}-K_{V12}, 128*f*
K_{V7}, 122*f*
nicotinic acetylcholine receptors, 384, 385, 388–389, 398
potassium channels
large conductance, 513, 516, 518*t*, 519
mutations, epilepsy, 153, 155, 158, 160, 178, 180
tandem pore domain, 576
TMEM16, 697
Xiong, Z. G., 675
X-ray crystallography, 35, 65. *See also specific applications*

Y

Yellen, G., 67–68, 70, 74–76

Z

Zagotta, W. N., 81
zaleplon, 426
Zhang, Y., 705
Zimmermann-Laband syndrome (ZLS), 129, 168, 169*t*, 173
Zn^{2+}, acid-sensing ion channel modulation by, 663
zolpidem, 426, 427
zopiclone, 426